Joemon M. Jose · Emine Yilmaz ·
João Magalhães · Pablo Castells ·
Nicola Ferro · Mário J. Silva
Flávio Martins (Eds.)

Advances in Information Retrieval

42nd European Conference on IR Research, ECIR 2020
Lisbon, Portugal, April 14–17, 2020
Proceedings, Part I

Editors
Joemon M. Jose
University of Glasgow
Glasgow, UK

João Magalhães
Universidade NOVA de Lisboa
Lisbon, Portugal

Nicola Ferro
University of Padua
Padua, Italy

Flávio Martins
Universidade NOVA de Lisboa
Lisbon, Portugal

Emine Yilmaz
University College London
London, UK

Pablo Castells
Universidad Autónoma de Madrid
Madrid, Spain

Mário J. Silva
Universidade de Lisboa
Lisbon, Portugal

ISSN 0302-9743 ISSN 1611-3349 (electronic)
Lecture Notes in Computer Science
ISBN 978-3-030-45438-8 ISBN 978-3-030-45439-5 (eBook)
https://doi.org/10.1007/978-3-030-45439-5

LNCS Sublibrary: SL3 – Information Systems and Applications, incl. Internet/Web, and HCI

This Springer imprint is published by the registered company Springer Nature Switzerland AG
The registered company address is: Gewerbestrasse 11, 6330 Cham, Switzerland

Lecture Notes in Computer Science **12035**

More information about this series at http://www.springer.com/series/7409

Preface

The 42nd European Conference on Information Retrieval (ECIR) was held on April 14–17, 2020, and brought together hundreds of researchers from all over the world. ECIR 2020 was to be held in Lisbon, Portugal, but due to the COVID-19 lockdown and travel restrictions enforced worldwide, the conference was held online. The conference was organized by Universidade NOVA de Lisboa, Portugal, and Universidad Autónoma de Madrid, Spain, in cooperation with the British Computer Society's Information Retrieval Specialist Group (BCS-IRSG). It was supported by the ACM Special Interest Group on Information Retrieval (ACM SIGIR), Bloomberg, Amazon, Salesforce, TextKernel, NTENT, Google, and Levi Strauss.

These proceedings contain the papers presented and the summaries of workshops and tutorials given during the conference. This year the ECIR 2020 program boasted a variety of novel works from contributors located all around the world and also provided a platform for information retrieval-related (IR) activities from the CLEF Initiative. In addition, a new collaboration was instated between BCS-IRSG and the *Information Retrieval Journal*, whereby selected papers from the journal were presented at the conference, and a selection of ECIR 2020 papers were invited to submit an extended version for publication in a special issue of the journal.

In total, 457 submissions were fielded across all tracks from 57 different countries – adding the papers submitted to workshops, ECIR 2020 broke the 500 submissions barrier. The final program included 55 full papers (26% acceptance rate), 46 short papers (28% acceptance rate), 10 demonstration papers (30% acceptance rate), 8 reproducibility papers (38% acceptance rate), and 12 invited CLEF papers. All submissions were peer reviewed by at least three international Program Committee members to ensure that only submissions of the highest quality were included in the final program. The acceptance decisions were further informed by discussions among the reviewers for each submitted paper, led by a senior Program Committee member. A call for reviewers was set forth aiming to strengthen and update the Program Committee, integrating and catching up with both new and accomplished reviewing workforce in the field.

The accepted papers cover the state of the art in IR: deep learning based information retrieval techniques, use of entities and knowledge graphs, recommender systems, retrieval methods, information extraction, question answering, topic and prediction models, multimedia retrieval, etc. As with tradition, the ECIR 2020 program has seen a high proportion of papers with students as first authors, as well as papers from a variety of universities, research institutes, and commercial organizations.

In addition to the papers, the program also included three keynotes, three tutorials, four workshops, a doctoral consortium, and an industry day. The first keynote was presented by this year's BCS IRSG Karen Sparck Jones Award winner, Chirag Shah, the second keynote was presented by Jamie Callan, and the third keynote by Joana Gonçalves de Sá. The tutorials covered a range of topics including entity repositories, similar-question retrieval, and geographic IR, while the workshops brought together participants around such areas as narrative extraction, bibliometric IR, algorithmic bias, and health IR. ECIR 2020 also featured a CLEF session to enable CLEF organizers to report on and promote their upcoming tracks. The program introduced a new activity where recently published papers from the *Information Retrieval Journal* were presented at the conference – a selection of nine papers was included in this track. Such links between related forums added to the success and diversity of ECIR and helped build bridges between communities. The Industry Day was held on the last conference day, bringing together academic researchers and industry, offering a mix of talks by industry leaders (including Farfetch, Doctrine, Microsoft, TigerGraph, eBay) and presentations of novel and innovative ideas from industry research.

The success of ECIR 2020 would not have been possible without all the help from the team of volunteers and reviewers. We wish to thank all our track chairs for coordinating the different tracks, along with the teams of meta-reviewers and reviewers who helped ensure the high quality of the program. Thanks are due to the demo chairs: Nuno Correia and Ana Freire; reproducibility track chairs: Edleno Moura and Pável Calado; doctoral consortium chairs: Stefan Rueger and Suzan Verberne; workshop chairs: Suzane Little and Sérgio Nunes; tutorial chairs: Laura Dietz and Allan Hanbury; industry day chairs: Vanessa Murdock and Bruno Martins; publicity chair: Carla Teixeira Lopes and Ricardo Campos; and sponsorship chair: Dyaa Albakour. We would like to thank our webmaster, Flávio Martins, and our local chairs, Rui Nóbrega and Filipa Peleja, along with all the student volunteers who helped to create an excellent online and offline experience for participants and attendees.

ECIR 2020 was sponsored by: Bloomberg, Amazon, Salesforce, Google, Textkernel, Ntent, Levi Strauss, Signal AI, SIGIR, and Springer. We thank them all for their support and contributions to the conference. Finally, we wish to thank all the authors, contributors, and participants in the conference.

April 2020

Joemon M. Jose
Emine Yilmaz
João Magalhães
Pablo Castells
Nicola Ferro
Mário J. Silva
Flávio Martins

Organization

General Chairs

João Magalhães	Universidade NOVA de Lisboa, Portugal
Christina Lioma	University of Copenhagen, Denmark
Pablo Castells	Universidad Autónonoma de Madrid, Spain

Program Chairs

Joemon M. Jose	University of Glasgow, UK
Emine Yilmaz	University College London, UK

Short Paper Chairs

Nicola Ferro	University of Padua, Italy
Mário J. Silva	Universidade de Lisboa, Portugal

Workshop Chairs

Suzanne Little	Dublin City University, Ireland
Sérgio Nunes	Universidade do Porto, Portugal

Tutorial Chairs

Laura Dietz	University of New Hampshire, USA
Allan Hanbury	Technische Universität Wien, Austria

Demo Chairs

Nuno Correia	Universidade NOVA de Lisboa, Portugal
Ana Freire	Universitat Pompeu Fabra, Spain

Industry Day Chairs

Vanessa Murdock	Amazon, USA
Bruno Martins	Universidade de Lisboa, Portugal

Proceedings Chair

Flávio Martins	Universidade NOVA de Lisboa, Portugal

Reproducibility Track Chairs

Edleno Moura | Universidade Federal de Minas Gerais, Brazil
Pável Calado | Universidade de Lisboa, Portugal

Doctoral Consortium Chairs

Stefan Rueger | KMI, The Open University, UK
Suzan Verberne | Leiden University, Netherlands

Test of Time Award Chair

Maristella Agosti | University of Padua, Italy

Best Paper Award Chair

Ben Carterette | Spotify, USA

Publicity Chairs

Carla Teixeira Lopes | Universidade do Porto, Portugal
Ricardo Campos | INESC-TEC and Instituto Politécnico Tomar, Portugal

Sponsorship Chairs

Dyaa Albakour | Signal AI, UK
João Magalhães | Universidade NOVA de Lisboa, Portugal

Local Organization

Rui Nóbrega (Local Chair) | Universidade NOVA de Lisboa, Portugal
Filipa Peleja (Local Chair) | Levi Strauss & Co. Europe, Belgium
Flávio Martins (Web Chair) | Universidade NOVA de Lisboa, Portugal
João Magalhães (Support) | Universidade NOVA de Lisboa, Portugal
Nuno Correia (Support) | Universidade NOVA de Lisboa, Portugal
David Semedo (Support) | Universidade NOVA de Lisboa, Portugal
Pablo Castells (Support) | Universidad Autónonoma de Madrid, Spain

Program Committee

Full-Paper Meta-Reviewers

Ioannis Arapakis | Telefonica Research, Spain
Krisztian Balog | University of Stavanger, Norway
Ben Carterette | Spotify, USA
Fabio Crestani | University of Lugano (USI), Switzerland
Bruce Croft | University of Massachusetts Amherst, USA

Nicola Ferro	University of Padua, Italy
Norbert Fuhr	University of Duisburg-Essen, Germany
Lorraine Goeuriot	Grenoble Alpes University, France
Julio Gonzalo	UNED, Spain
Cathal Gurrin	Dublin City University, Ireland
Morgan Harvey	Northumbria University, UK
Claudia Hauff	Delft University of Technology, The Netherlands
Grace Huiyang	Georgetown University, USA
Gareth Jones	Dublin City University, Ireland
Diane Kelly	University of Tennessee, USA
Liadh Kelly	Maynooth University, Ireland
Udo Kruschwitz	University of Regensburg, Germany
Oren Kurland	Technion – Israel Institute of Technology, Israel
David Losada	University of Santiago de Compostela, Spain
Massimo Melucci	University of Padua, Italy
Boughanem Mohand	IRIT - Université Paul Sabatier Toulouse 3, France
Yashar Moshfeghi	University of Strathclyde, UK
Henning Müller	HES-SO, Switzerland
Iadh Ounis	University of Glasgow, UK
Gabriella Pasi	Università degli Studi di Milano-Bicocca, Italy
Benjamin Piwowarski	CNRS and University Pierre et Marie Curie, France

Full-Paper Program Committee

Mohamed Abdel Maksoud	Codoma.tech Advanced Technologies, Egypt
Ahmed Abdelali	Research Administration, Qatar
Karam Abdulahhad	GESIS, Germany
Dirk Ahlers	NTNU, Norway
Qingyao Ai	University of Utah, USA
Mohammad Akbari	University College London, UK
Ahmet Aker	University of Duisburg Essen, Germany
Navot Akiva	Bar Ilan University, Israel
Mehwish Alam	FIZ Karlsruhe, AIFB Institute, KIT, Germany
M-Dyaa Albakour	Signal AI, UK
Mohammad Aliannejadi	University of Amsterdam, The Netherlands
Pegah Alizadeh	University of Caen Normandy, France
Giambattista Amati	Fondazione Ugo Bordoni, Italy
Linda Andersson	Vienna University of Technology, Austria
Hassina Aouidad	CERIST, Algiers
Avi Arampatzis	Democritus University of Thrace, Greece
Ioannis Arapakis	Telefonica Research, Spain
Jaime Arguello	University of North Carolina at Chapel Hill, USA
Mozhdeh Ariannezhad	University of Amsterdam, The Netherlands
Nihal Yağmur Aydın	GREYC, France
Ebrahim Bagheri	Ryerson University, Canada
Seyed-Ali Bahrainian	Idsia, Swiss AI Lab, Switzerland

Krisztian Balog	University of Stavanger, Norway
Alvaro Barreiro	University of A Coruña, Spain
Alberto Barrn-Cede	University of Bologna, Italy
Alejandro Bellogin	Universidad Autónoma de Madrid, Spain
Patrice Bellot	Aix-Marseille Université, CNRS, LSIS, France
Anis Benammar	Regim, Tunisia
Klaus Berberich	Saarbruecken University of Applied Sciences, Germany
Pablo Bermejo	Universidad de Castilla-La Mancha, Spain
Catherine Berrut	LIG, Université Joseph Fourier Grenoble I, France
Sumit Bhatia	IBM, India
Pierre Bonnet	CIRAD, France
Gloria Bordogna	National Research Council of Italy (CNR), Italy
Larbi Boubchir	University of Paris 8, France
Pavel Braslavski	Ural Federal University, Russia
David Brazier	Northumbria University, UK
Paul Buitelaar	National University of Ireland Galway, Ireland
Guillaume Cabanac	IRIT, Université Paul Sabatier Toulouse 3, France
Luis Adrián Cabrera-Diego	Edge Hill University, UK
Fidel Cacheda	Universidade da Coruña, Spain
Sylvie Calabretto	LIRIS, CNRS, France
Pável Calado	INESC-ID, Universidade de Lisboa, Portugal
Rodrigo Calumby	University of Feira de Santana, Brazil
Ricardo Campos	INESC-TEC and Instituto Politécnico Tomar, Portugal
Fazli Can	Bilkent University, Turkey
Ivàn Cantador	Universidad Autónoma de Madrid, Spain
Annalina Caputo	University College Dublin, Ireland
Cornelia Caragea	University of Illinois at Chicago, USA
Ben Carterette	Spotify, USA
Pablo Castells	Universidad Autónoma de Madrid, Spain
Long Chen	University of Glasgow, UK
Max Chevalier	IRIT, France
Adrian-Gabriel Chifu	Aix-Marseille Université, Université de Toulon, France
Manoj Chinnakotla	Microsoft, India
Malcolm Clark	University of the Highlands and Islands, UK
Vincent Claveau	IRISA, CNRS, France
Jeremie Clos	University of Nottingham, UK
Fabio Crestani	University of Lugano (USI), Switzerland
Bruce Croft	University of Massachusetts Amherst, USA
Alfredo Cuzzocrea	ICAR-CNR and University of Calabria, Italy
Arthur Câmara	Delft University of Technology, The Netherlands
Zhuyun Dai	Carnegie Mellon University, USA
Jeffery Dalton	University of Glasgow, UK
Tirthankar Dasgupta	Tata Consultancy Services, India
Hélène De Ribaupierre	Cardiff University, UK
Martine DeCock	University of Washington, USA

Yashar Deldjoo	Polytechnic University of Bari, Italy
Kostantinos Demertzis	Democritus University of Thrace, Greece
José Devezas	University of Porto, Portugal
Kuntal Dey	IBM Research Lab, India
Emanuele DiBuccio	University of Padua, Italy
Giorgio DiNunzio	University of Padua, Italy
Laura Dietz	University of New Hampshire, USA
Inês Domingues	IPO Porto and Universidade de Coimbra, Portugal
Pan Du	University of Montreal, Canada
Mateusz Dubiel	University of Strathclyde, UK
Carsten Eickhoff	Brown University, USA
Mehdi Elahi	Free University of Bozen - Bolzano, Italy
Tamer Elsayed	Qatar University, Qatar
Liana Ermakova	Université de Bretagne Occidentale, France
Jose Esquivel	Signal AI, UK
Michael Faerber	University of Freiburg, Germany
Hui Fang	University of Delaware, USA
Hossein Fani	University of New Brunswick, Canada
Paulo Fernandes	Roberts College, USA
Nicola Ferro	University of Padua, Italy
Mustansar Fiaz	Kyungpook National University, South Korea
Sebastien Fournier	LSIS, France
Christoph M. Friedrich	University of Applied Science and Arts Dortmund, Germany
Ingo Frommholz	University of Bedfordshire, UK
Norbert Fuhr	University of Duisburg-Essen, Germany
Luke Gallagher	RMIT University, Australia
Patrick Gallinari	LIP6, University of Paris 6, France
Shreyansh Gandhi	WalmartLabs, USA
Debasis Ganguly	IBM Research Lab, Ireland
Wei Gao	Victoria University of Wellington, New Zealand
Dario Garigliotti	University of Stavanger, Norway
Anastasia Giachanou	Universitat Politècnica de València, Spain
Giorgos Giannopoulos	IMSI Institute, Athena Research Center, Greece
Alessandro Giuliani	University of Cagliari, Italy
Lorraine Goeuriot	Grenoble Alpes University, France
Julio Gonzalo	UNED, Spain
Pawan Goyal	IIT Kharagpur, India
Michael Granitzer	University of Passau, Germany
Guillaume Gravier	CNRS, IRISA, France
Adrien Guille	ERIC Lyon 2, EA 3083, Université de Lyon, France
Shashank Gupta	IIIT, India
Rajeev Gupta	Microsoft, India
Cathal Gurrin	Dublin City University, Ireland
Matthias Hagen	Martin-Luther-Universität Halle-Wittenberg, Germany
Lei Han	The University of Queensland, Australia

Shuguang Han	Google, USA
Preben Hansen	Stockholm University, Sweden
Donna Harman	NIST, USA
Morgan Harvey	Northumbria University, UK
Helia Hashemi	University of Massachusetts Amherst, USA
Claudia Hauff	Delft University of Technology, The Netherlands
Jer Hayes	Accenture, Ireland
Ben He	University of Chinese Academy of Sciences, China
Nathalie Hernandez	IRIT, France
Andreas Hotho	University of Würzburg, Germany
Gilles Hubert	IRIT, France
Grace Hui Yang	Georgetown University, USA
Ali Hürriyetoğlu	Koc University, Turkey
Adrian Iftene	Alexandru Ioan Cuza University of Iaşi, Romania
Dmitry Ignatov	National Research University Higher School of Economics, Russia
Bogdan Ionescu	University Politehnica of Bucharest, Romania
Radutudor Ionescu	University of Bucharest, Romania
Amir H. Jadidinejad	University of Glasgow, UK
Shoaib Jameel	Kent University, UK
Adam Jatowt	Kyoto University, Japan
Shen Jialie	Queen's University, Belfast, UK
Jiepu Jiang	Virginia Tech, USA
Alexis Joly	Inria, France
Gareth Jones	Dublin City University, Ireland
Jaap Kamps	University of Amsterdam, The Netherlands
Nattiya Kanhabua	Kasikorn Business - Technology Group, Thailand
Jaana Kekäläinen	University of Tampere, Finland
Diane Kelly	University of Tennessee, USA
Liadh Kelly	Maynooth University, Ireland
Roman Kern	Graz University of Technology, Austria
Daniel Kershaw	Elsevier, UK
Julia Kiseleva	Microsoft Research AI, USA
Prasanna Lakshmi Kompalli	GRIET, India
Yiannis Kompatsiaris	ITI-CERTH, Greece
Panos Kostakos	University of Oulu, Finland
Ralf Krestel	HPI, University of Potsdam, Germany
Kriste Krstovski	University of Massachusetts Amherst, USA
Udo Kruschwitz	University of Regensburg, Germany
Vaibhav Kumar	Carnegie Mellon University, USA
Oren Kurland	Technion – Israel Institute of Technology, Israel
Saar Kuzi	University of Illinois at Urbana-Champaign, USA
Wang-Chien Lee	The Pennsylvania State University, USA
Teerapong Leelanupab	King Mongkut's Institute of Technology Ladkrabang, Thailand
Mark Levene	Birkbeck, University of London, UK

Nut Limsopatham	Amazon, USA
Chunbin Lin	Amazon, USA
Aldo Lipani	University College London, UK
Nedim Lipka	Adobe Research, USA
Fernando Loizides	Cardiff University, UK
David Losada	University of Santiago de Compostela, Spain
Natalia Loukachevitch	Moscow State University, Russia
Bernd Ludwig	University Regensburg, Germany
Mihai Lupu	Research Studios Austria, Austria
Sean Macavaney	Georgetown University, USA
Craig Macdonald	University of Glasgow, UK
Andrew Macfarlane	City University of London, UK
João Magalhães	Universidade NOVA de Lisboa, Portugal
Walid Magdy	The University of Edinburgh, UK
Marco Maggini	University of Siena, Italy
Shikha Maheshwari	Chitkara University, India
Maria Maistro	University of Copenhagen, Denmark
Antonio Mallia	University of Pisa, Italy
Thomas Mandl	University of Hildesheim, UK
Behrooz Mansouri	Rochester Institute of Technology, USA
Jiaxin Mao	Tsinghua University, China
Stefania Marrara	Consorzio C2T, Italy
Miguel Martinez-Alvarez	Signal AI, UK
Bruno Martins	INESC-ID, Universidade de Lisboa, Portugal
Flávio Martins	Universidade NOVA de Lisboa, Portugal
Fernando Martínez-Santiago	Universidad de Jaén, Spain
Yosi Mass	IBM Haifa Research Lab, Israel
Sérgio Matos	IEETA, Universidade de Aveiro, Portugal
Philipp Mayr	GESIS, Germany
Parth Mehta	IRSI, India
Edgar Meij	Bloomberg L.P., UK
Massimo Melucci	University of Padua, Italy
Marcelo Mendoza	Universidad Técnica Federico Santa María, Chile
Zaiqiao Meng	University of Glasgow, UK
Alessandro Micarelli	Roma Tre University, Italy
Dmitrijs Milajevs	Queen Mary University of London, UK
Malik Muhammad Saad Missen	Research Lab L3I, Université de la Rochelle, France
Boughanem Mohand	IRIT, Université Paul Sabatier Toulouse 3, France
Ludovic Moncla	LIRIS, CNRS, France
Felipe Moraes	Delft University of Technology, The Netherlands
Ajinkya More	Netflix, USA
Jose Moreno	IRIT, UPS, France
Yashar Moshfeghi	University of Strathclyde, UK
Josiane Mothe	IRIT, Université de Toulouse, France
André Mourão	Universidade NOVA de Lisboa, Portugal

Henning Müller	HES-SO, Switzerland
Franco Maria Nardini	ISTI-CNR, Italy
Rekabsaz Navid	Johannes Kepler University Linz, Austria
Wolfgang Nejdl	L3S and University of Hannover, Germany
Massimo Nicosia	Google, Switzerland
Jian-Yun Nie	University of Montreal, Canada
Qiang Ning	University of Illinois at Urbana-Champaign, USA
Andreas Nuernberger	Otto von Guericke University Magdeburg, Germany
Neil O'Hare	Yahoo Research, USA
Anais Ollagnier	University of Exeter, UK
Teresa Onorati	Universidad Carlos III de Madrid, Spain
Salvatore Orlando	Università Ca' Foscari Venezia, Italy
Iadh Ounis	University of Glasgow, UK
Mourad Oussalah	University of Oulu, Finland
Deepak P.	Queen's University Belfast, UK
Jiaul Paik	IIT Kharagpur, India
Joao Palotti	Qatar Computing Research Institute, Qatar
Girish Palshikar	Tata Research Development and Design Centre, India
Panagiotis Papadakos	FORTH-ICS, Greece
Javier Parapar	University of A Coruña, Spain
Gabriella Pasi	Università degli Studi di Milano-Bicocca, Italy
Arian Pasquali	University of Porto, Portugal
Bidyut Kr. Patra	National Institute of Technology, India
Virgil Pavlu	Northeastern University, USA
Pavel Pecina	Charles University in Prague, Czech Republic
Gustavo Penha	Delft University of Technology, The Netherlands
Avar Pentel	Tallinn University, Estonia
Raffaele Perego	ISTI-CNR, Italy
Benjamin Piwowarski	CNRS, Pierre et Marie Curie University, France
Animesh Prasad	Amazon Alexa, UK
Chen Qu	University of Massachusetts Amherst, USA
Hossein A. Rahmani	University of Zanjan, Iran
Pengjie Ren	University of Amsterdam, The Netherlands
Kamal Sarkar	Jadavpur University, India
Ramit Sawhney	Netaji Subhas Institute of Technology, India
Philipp Schaer	TH Köln – University of Applied Sciences, Germany
Fabrizio Sebastiani	Italian National Council of Research, Italy
Florence Sedes	IRIT, Université Paul Sabatier Toulouse 3, France
Giovanni Semeraro	University of Bari, Italy
Procheta Sen	Indian Statistical Institute, India
Armin Seyeditabari	UNC Charlotte, USA
Gautam Kishore Shahi	University of Duisburg-Essen, Germany
Mahsa Shahshahani	University of Amsterdam, The Netherlands
Azadeh Shakery	University of Tehran, Iran
Ritvik Shrivastava	Columbia University, USA
Manish Shrivastava	IIIT Hyderabad, India

Gianmaria Silvello	University of Padua, Italy
Laure Soulier	Sorbonne Université, UPMC-LIP6, France
Marc Spaniol	Université de Caen Normandie, France
Günther Specht	University of Innsbruck, Austria
Rene Spijker	Cochran Netherlands, The Netherlands
Efstathios Stamatatos	University of the Aegean, Greece
L. Venkata Subramaniam	IBM Research Lab, India
Hanna Suominen	The ANU, Australia
Pascale Sébillot	IRISA, France
Lynda Tamine	IRIT, France
Thibaut Thonet	Naver Labs Europe, France
Antonela Tommasel	ISISTAN Research Institute, CONICET-UNCPBA, Argentina
Nicola Tonellotto	University of Pisa, Italy
Alina Trifan	University of Aveiro, Portugal
Theodora Tsikrika	ITI-CERTH, Greece
Ferhan Ture	Comcast Labs, USA
Yannis Tzitzikas	University of Crete and FORTH-ICS, Greece
Md Zia Ullah	CNRS, France
Julián Urbano	Delft University of Technology, The Netherlands
Daniel Valcarce	Google, Switzerland
Sumithra Velupillai	Institute of Psychiatry, Psychology and Neuroscience, UK
Nadimpalli Venkata Ganapathi Raju	GRIET, India
Suzan Verberne	Leiden University, The Netherlands
Manisha Verma	Verizon Media, USA
Vishwa Vinay	Adobe Research, India
Marco Viviani	Università degli Studi di Milano-Bicocca, Italy
Duc-Thuan Vo	Ryerson University, Canada
Stefanos Vrochidis	Information Technologies Institute, Greece
Shuohang Wang	Singapore Management University, Singapore
Christa Womser-Hacker	University of Hildesheim, UK
Peilin Yang	Twitter Inc., USA
Tao Yang	University of California at Santa Barbara, USA
Andrew Yates	Max Planck Institute for Informatics, Germany
Hai-Tao Yu	University of Tsukuba, Japan
Fattane Zarrinkalam	Ferdowsi University, Iran
Guido Zuccon	The University of Queensland, Australia
Arjen de Vries	Radboud University, The Netherlands

Short-Paper Program Committee

Dirk Ahlers	NTNU, Norway
Mehwish Alam	FIZ Karlsruhe, AIFB Institute, KIT, Germany
Mohammad Aliannejadi	University of Amsterdam, The Netherlands

Giambattista Amati	Fondazione Ugo Bordoni, Italy
Giuseppe Amato	ISTI-CNR, Italy
Maurizio Atzori	University of Cagliari, Italy
Ebrahim Bagheri	Ryerson University, Canada
Alvaro Barreiro	University of A Coruña, Spain
Alberto Barrón-Cedeño	University of Bologna, Italy
Alejandro Bellogin	Universidad Autónoma de Madrid, Spain
Patrice Bellot	Aix-Marseille Université, CNRS, LSIS, France
Anis Benammar	Regim, Tunisia
Klaus Berberich	Saarbruecken University of Applied Sciences, Germany
Catherine Berrut	LIG, Université Joseph Fourier Grenoble I, France
Pierre Bonnet	CIRAD, France
Gloria Bordogna	National Research Council of Italy (CNR), Italy
Luis Adrián Cabrera-Diego	Edge Hill University, UK
Fidel Cacheda	Universidade da Coruña, Spain
Pável Calado	INESC-ID, Universidade de Lisboa, Portugal
Iván Cantador	Universidad Autónoma de Madrid, Spain
Michelangelo Ceci	Università degli Studi di Bari, Italy
Adrian-Gabriel Chifu	Aix-Marseille Université, Université de Toulon, France
Malcolm Clark	University of the Highlands and Islands, UK
Arthur Câmara	Delft University of Technology, The Netherlands
Zhuyun Dai	Carnegie Mellon University, USA
Emanuele Di Buccio	University of Padua, Italy
Giorgio Maria Di Nunzio	University of Padua, Italy
Dennis Dosso	University of Padua, Italy
Tamer Elsayed	Qatar University, Qatar
Liana Ermakova	Université de Bretagne Occidentale, France
Andrea Esuli	Istituto di Scienza e Tecnologie dell'Informazione, Italy
Fabrizio Falchi	ISTI-CNR, Italy
Norbert Fuhr	University of Duisburg-Essen, Germany
Debasis Ganguly	IBM Research Lab, Ireland
Dario Garigliotti	University of Stavanger, Norway
Anastasia Giachanou	Universitat Politècnica de València, Spain
Giorgos Giannopoulos	IMSI Institute, Athena Research Center, Greece
Lorraine Goeuriot	Grenoble Alpes University, France
Adrien Guille	ERIC Lyon 2, Université de Lyon, France
Matthias Hagen	Martin-Luther-Universität Halle-Wittenberg, Germany
Donna Harman	NIST, USA
Helia Hashemi	University of Massachusetts Amherst, USA
Adrian Iftene	Alexandru Ioan Cuza University of Iaşi, Romania
Bogdan Ionescu	University Politehnica of Bucharest, Romania
Amir Jadidinejad	University of Glasgow, UK
Adam Jatowt	Kyoto University, Japan
Jaap Kamps	University of Amsterdam, The Netherlands

Mat Kelly	Drexel University, USA
Diane Kelly	University of Tennessee, USA
Daniel Kershaw	Elsevier, UK
Ioannis Kompatsiaris	ITI-CERTH, Greece
Kriste Krstovski	University of Massachusetts Amherst, USA
Vaibhav Kumar	Carnegie Mellon University, USA
Aldo Lipani	University College London, UK
Natalia Loukachevitch	Moscow State University, Russia
Claudio Lucchese	Ca' Foscari University of Venice, Italy
Mihai Lupu	Research Studios Austria, Austria
Andrew Macfarlane	City, University of London, UK
Marco Maggini	University of Siena, Italy
Maria Maistro	University of Copenhagen, Denmark
Antonio Mallia	University of Pisa, Italy
Jiaxin Mao	Tsinghua University, China
Stefano Marchesin	University of Padua, Italy
Bruno Martins	INESC-ID, Universidade de Lisboa, Portugal
Philipp Mayr	GESIS, Germany
Graham Mcdonald	University of Glasgow, UK
Edgar Meij	Bloomberg L.P., UK
Zaiqiao Meng	University of Glasgow, UK
Franco Maria Nardini	ISTI-CNR, Italy
Jian-Yun Nie	University of Montreal, Canada
Andreas Nuernberger	Otto von Guericke University Magdeburg, Germany
Sérgio Nunes	University of Porto, Portugal
Neil O'Hare	Yahoo Research, USA
Anais Ollagnier	University of Exeter, UK
Joao Palotti	Qatar Computing Research Institute, Qatar
Javier Parapar	University of A Coruña, Spain
Pavel Pecina	Charles University in Prague, Czech Republic
Raffaele Perego	ISTI-CNR, Italy
Martin Potthast	Leipzig University, Germany
Pengjie Ren	University of Amsterdam, The Netherlands
Paolo Rosso	Universitat Politècnica de València, Spain
Pasquale Savino	ISTI-CNR, Italy
Philipp Schaer	TH Köln University of Applied Sciences, Germany
Azadeh Shakery	University of Tehran, Iran
Gianmaria Silvello	University of Padua, Italy
Laure Soulier	Sorbonne Université, UPMC-LIP6, France
Marc Spaniol	Université de Caen Normandie, France
Hanna Suominen	The ANU, Australia
Lynda Tamine	IRIT, France
Thibaut Thonet	Naver Labs Europe, France
Nicola Tonellotto	University of Pisa, Italy
Theodora Tsikrika	ITI-CERTH, Greece
Yannis Tzitzikas	University of Crete and FORTH-ICS, Greece

Md Zia Ullah	CNRS, France
Julián Urbano	Delft University of Technology, The Netherlands
Gaurav Verma	Adobe Research, India
Marco Viviani	Università degli Studi di Milano-Bicocca, Italy
Christa Womser-Hacker	University of Hildesheim, UK
Yikun Xian	Rutgers University, USA
Andrew Yates	Max Planck Institute for Informatics, Germany
Justin Zobel	The University of Melbourne, Canada
Guido Zuccon	The University of Queensland, Australia
Arjen de Vries	Radboud University, The Netherlands

Reproducibility Track Reviewers

Jalal Alowibdi	University of Jeddah, Saudi Arabia
Leandro Balby Marinho	Federal University of Campina Grande, Brazil
José Borbinha	INESC-ID, Universidade de Lisboa, Portugal
Andre Carvalho	Universidade Federal do Amazonas, Brazil
André Carvalho	Universidade de Lisboa, Portugal
Edgar Chaves	CICESE, Mexico
Thierson Couto Rosa	Universidade Federal de Goiás, Brazil
Antonio Fariña	University of A Coruña, Spain
Juan M. Fernández-Luna	University of Granada, Spain
Marcos Goncalves	Federal University of Minas Gerais, Brazil
Xiaodong Liu	Microsoft, USA
Maria Maistro	University of Copenhagen, Denmark
David Matos	INESC-ID, Universidade de Lisboa, Portugal
Sérgio Matos	IEETA, Universidade de Aveiro, Portugal
Ajinkya More	Netflix, USA
Viviane P. Moreira	Universidade Federal do Rio Grande Do Sul, Brazil
Wolfgang Nejdl	L3S and University of Hannover, Germany
Özlem Özgöbek	NTNU, Norway
Altigran S. Da Silva	Universidade Federal do Amazonas, Brazil
Mahsa Shahshahani	University of Amsterdam, The Netherlands
Fei Sun	Alibaba Group, China
Ricardo Torres	NTNU, Norway
Alina Trifan	University of Aveiro, Portugal
Guido Zuccon	The University of Queensland, Australia

Demonstration Reviewers

Ahmed Abdelali	Research Administration, Qatar
Qingyao Ai	University of Utah, USA
M-Dyaa Albakour	Signal AI, UK
Diogo Cabral	ITI, LARSyS, Universidade de Lisboa, Portugal
Sylvie Calabretto	LIRIS, CNRS, France
Long Chen	University of Glasgow, UK

Manoj Chinnakotla	Microsoft, India
Alfredo Cuzzocrea	ICAR-CNR and University of Calabria, Italy
Yashar Deldjoo	Polytechnic University of Bari, Italy
Kuntal Dey	IBM India Research Lab, India
Jose Alberto Esquivel	Signal AI, UK
Hossein Fani	University of New Brunswick, Canada
Manuel Fonseca	Universidade de Lisboa, Portugal
Ingo Frommholz	University of Bedfordshire, UK
Michael Färber	University of Freiburg, Germany
Rui Nóbrega	Universidade NOVA de Lisboa, Portugal
时 冯	Northeastern University, China

Doctoral Consortium Reviewers

Carsten Eickhoff	Brown University, USA
Norbert Fuhr	University of Duisburg-Essen, Germany
Claudia Hauff	Delft University of Technology, The Netherlands
Gareth Jones	Dublin City University, Ireland
Udo Kruschwitz	University of Regensburg, Germany
Haiming Liu	University of Bedfordshire, UK
Philipp Mayr	GESIS, Germany
Josiane Mothe	IRIT, Université de Toulouse, France
Henning Müller	HES-SO, Switzerland

CLEF Track Reviewers

Giorgio Maria Di Nunzio	University of Padua, Italy
Nicola Ferro	University of Padua, Italy
Norbert Fuhr	University of Duisburg-Essen, Germany
Lorraine Goeuriot	University Grenoble Alpes, France
Donna Harman	NIST, USA
Bogdan Ionescu	University Politehnica of Bucharest, Romania
Mihai Lupu	Research Studios Austria, Austria
Maria Maistro	University of Copenhagen, Denmark
Jian-Yun Nie	University of Montreal, Canada
Aurélie Névéol	LIMSI, CNRS, Université Paris-Saclay, France
Raffaele Perego	ISTI-CNR, Italy
Martin Potthast	Leipzig University, Germany
Andreas Rauber	Vienna University of Technology, Austria
Paolo Rosso	Universitat Politècnica de València, Spain
Fabrizio Sebastiani	Italian National Council of Research, Italy
Laure Soulier	Sorbonne Université, UPMC-LIP6, France
Ellen Voorhees	NIST, USA

Additional Reviewers

Aditya Chandrasekar
Afraa Ahmad Alyosef
Alakananda Vempala
Alberto Purpura
Alessandra T. Cignarella
Alfonso Landin
Amir Jadidinejad
Amit Kumar Jaiswal
Ana Sabina Uban
Anastasia Moumtzidou
Andrea Iovine
Angelo Impedovo
Anna Nguyen
Behrooz Omidvar-Tehrani
Benjamin Murauer
Bilal Ghanem
Bishal Santra
Boteanu Bogdan Andrei
Cagri Toraman
Cataldo Musto
Charles Jochim
Christophe Rodrigues
Claudio Biancalana
Claudio Vairo
Daniel Campbell
Daniel Zoller
Dario Del Fante
David Otero
Debanjan Mahata
Despoina Chatzakou
Diana Nurbakova
Dilek Küçük
Dimitrios Effrosynidis
Disen Wang
Elisabeth Fischer
Emanuele Pio Barracchia
Eugene Yang
Fabian Hoppe
Fabio Carrara
Fang He
Fatima Haouari
Fedelucio Narducci
Felice Antonio Merra
Genet Asefa Gesese
Ghazal Fazelnia
Giulio Ermanno Pibiri
Giuseppe Sansonetti
Graziella De Martino
Gretel Sarracén
Guglielmo Faggioli
Himanshu Sharma
Hui-Ju Hung
Ilias Gialampoukidis
Janek Bevendorff
Jean-Michel Renders
Jinjin Shao
Johannes Jurgovsky
Johannes Kiesel
Johannes Schwerdt
Jun Ho Shin
Jussi Karlgren
Kalyani Roy
Konstantin Kobs
Kristian Noullet
Kuang Lu
Leopoldo Melo
Liviu-Daniel Ştefan
Lucia Vadicamo
Mahdi Dehghan
Maik Fröbe
Malte Bonart
Mandy Neumann
Manoj Kilaru
Maram Hasanain
Marco Polignano
Marco Ponza
Marcus Thiel
Matti Wiegmann
Michael Kotzyba
Mihai Dogariu
Mihai Gabriel Constantin
Mucahid Kutlu
Nirmal Roy
Paolo Mignone
Polina Panicheva
Reem Suwaileh
Reynier Ortega Bueno
Richard Mccreadie
Rob Koeling
Roberto Trani
Russa Biswas
Satarupa Guha
Sayantan Polley
Sevil Çalışkan
Shahbaz Syed
Shahrzad Naseri
Shikib Mehri
Shiyu Ji
Silvia Corbara
Silvio Moreira
Siwei Liu
Stefano Souza
Suresh Kumar Kaswan
Symeon Papadopoulos
Symeon Symeonidis
Tao-Yang Fu
Tarek Saier
Thanassis Mavropoulos
Thierson Couto-Rosa
Timo Breuer
Ting Su
Vaibhav Kasturia
Vikas Raunak
Wei-Fan Chen
Xiaoqi Ren
Yagmur Gizem Cinar
Yash Kumar Lal
Zuohui Fu

Test of Time Award Committee

Catherine Berrut	LIG, Université Joseph Fourier Grenoble I, France
Paul Clough	University of Sheffield, UK
Fabio Crestani	University of Lugano (USI), Switzerland
Gareth Jones	Dublin City University, Ireland
Josiane Mothe	IRIT, Université de Toulouse, France
Stefan Rueger	KMI, The Open University, UK

Best Paper Award Committee

Nicola Ferro	University of Padua, Italy
Udo Kruschwitz	University of Regensburg, Germany
Gabriella Pasi	Università degli Studi di Milano-Bicocca, Italy
Grace Hui Yang	Georgetown University, USA

Diamond Sponsors

Bloomberg

Engineering

Gold Sponsors

Silver Sponsors

Bronze Sponsor

ntent

textkernel
Machine Intelligence for People and Jobs

Industry Impact Award Sponsor

SIGNAL

With Generous Support From

Abstracts of Keynotes

Task-Based Intelligent Retrieval and Recommendation

Chirag Shah

University of Washington, Seattle, USA
chirags@uw.edu

Abstract. While the act of looking for information happens within a context of a task from the user side, most search and recommendation systems focus on user actions ('what'), ignoring the nature of the task that covers the process ('how') and user intent ('why'). For long, scholars have argued that IR systems should help users accomplish their tasks and not just fulfill a search request. But just as keywords have been good enough approximators for information need, satisfying a set of search requests has been deemed to be good enough to address the task. However, with changing user behaviors and search modalities, specifically found in conversational interfaces, the challenge and opportunity to focus on task have become critically important and central to IR. In this talk, I will discuss some of the key ideas and recent works – both theoretical and empirical – to study and support aspects of task. I will show how we could derive user's search path or strategy and intentions, and how they could be instrumental in not only creating more personalized search and recommendation solutions, but also solving problems not possible otherwise. Finally, I will extend this to the realm of intelligent assistants with our recent work in a new area called Information Fostering, where our knowledge of the user and the task can help us address another classical problem in IR – people don't know what they don't know.

Keywords: Task-based IR · Recommendation systems · Information Fostering

Better Representations for Search Tasks

Jamie Callan

Language Technologies Institute, Carnegie Mellon University,
Pittsburgh, USA
callan@cs.cmu.edu

Abstract. Neural models are having a major impact on Information Retrieval (IR), as much as they have recently had on other language technologies. Neural language models and continuous term representations provide new and more effective paths to overcoming vocabulary mismatch, probability estimation, and other core problems in IR. Some classic Natural Language Processing (NLP) tasks are now treated as text similarity problems, and techniques developed for NLP are being applied to classic IR problems, which reduces some of the past differences between IR and NLP. Everything uses machine learning. This technology shift is a good time to think about what is unique and distinct about IR as a field compared to neighboring fields.

From its earliest days, IR has studied document collections, information seekers, and information seeking tasks. These topics are embedded deeply in our experimental methodology and how we think about research problems. Neighboring fields focus more attention and computational effort on understanding individual documents, and less on how individual documents should be understood in the context of specific people, tasks, and collections.

This talk describes several recent research activities at CMU's Language Technologies Institute. Although each has a different focus, the unifying theme is using knowledge of the search task, context, or corpus to develop more effective representations and models. We find that neural techniques offer new tools for understanding and modeling these core elements of search, in some cases reinvigorating research in stable areas and challenging old assumptions, but do not reduce their importance.

Keywords: Information retrieval · Text understanding · Neural IR

Much of the work discussed in this talk was done with Zhuyun Dai.

Focusing the Macroscope: How We Can Use Data to Understand Behavior

Joana Gonçalves-Sá

NOVA School of Business and Economics, Carcavelos, Portugal
joana.sa@novasbe.pt

Abstract. Individual decisions can have a large impact on society as a whole. This is obvious for political decisions, but still true for small, daily decisions made by common citizens. Individuals decide how to vote, whether to stay at home when they feel sick, to drive or to take the bus. In isolation, these individual decisions have a negligible social outcome, but collectively they determine the results of an election and the start of an epidemic. For many years, studying these processes was limited to observing the outcomes or to analyzing small samples. New data sources and data analysis tools have created a "macroscope" and made it possible to start studying the behavior of large numbers of individuals, enabling the emergence of large-scale quantitative social research. At the Data Science and Policy (DS&P) research group we are interested in understanding these decision-making events, expecting that this deeper knowledge will lead to a better understanding of human nature, and to improved public decisions. During the talk I will offer some examples of how can use this macroscope to study psychology and human behavior. At the end, and recognizing that these tools might also have a very negative impact on society, I will present new ideas in distributed computing and how it can help us in privacy protection.

Contents – Part I

Deep Learning I

Entities

Evaluation

Recommendation

Information Extraction

Deep Learning II

Retrieval

Multimedia

Deep Learning III

Queries

IR - General

Question Answering, Prediction, and Bias

Deep Learning IV

Abstracts of the IR Journal Papers

Contents – Part II

Reproducibility Papers

Short Papers

Demonstration Papers

CLEF Organizers Lab Track

Doctoral Consortium Papers

Workshops

Tutorials

Deep Learning I

Seed-Guided Deep Document Clustering

Mazar Moradi Fard[1(✉)], Thibaut Thonet[2], and Eric Gaussier[1]

[1] Univ. Grenoble Alpes, CNRS - LIG, Grenoble, France
{maziar.moradi-fard,eric.gaussier}@univ-grenoble-alpes.fr
[2] NAVER LABS Europe, Meylan, France
thibaut.thonet@naverlabs.com

Abstract. Different users may be interested in different clustering views underlying a given collection (e.g., topic and writing style in documents). Enabling them to provide constraints reflecting their needs can then help obtain tailored clustering results. For document clustering, constraints can be provided in the form of seed words, each cluster being characterized by a small set of words. This seed-guided constrained document clustering problem was recently addressed through topic modeling approaches. In this paper, we jointly learn deep representations and bias the clustering results through the seed words, leading to a Seed-guided Deep Document Clustering approach. Its effectiveness is demonstrated on five public datasets.

Keywords: Document clustering · Representation learning · Dataless text classification

1 Introduction

Clustering traditionally consists in partitioning data into subsets of similar instances with no prior knowledge on the clusters to be obtained. However, clustering is an ill-defined problem in the sense that the data partitions output by clustering algorithms have no guarantee to satisfy end users' needs. Indeed, different users may be interested in different views underlying the data [25]. For example, considering either the topics or the writing style in a collection of documents leads to different clustering results. In this study, we consider a setting where clustering is guided through user-defined constraints, which is known as *constrained clustering* [2]. Enabling users to provide clustering constraints in the context of an exploratory task can help obtain results better tailored to their needs. Typically, must-link and cannot-link constraints are considered (e.g., see [27,29]), which state whether two data instances should be (respectively, should not be) in the same cluster. However, important manual annotation efforts may still be required to provide such constraints in sufficient number. In the specific case of document clustering, constraints can otherwise be provided in the form of *seed words*: each cluster that the user wishes to obtain is described by a small set of words (e.g., 3 words) which characterize the cluster. For example,

J. M. Jose et al. (Eds.): ECIR 2020, LNCS 12035, pp. 3–16, 2020.
https://doi.org/10.1007/978-3-030-45439-5_1

a user who wants to explore a collection of news articles might provide the set of seed words {'sport', 'competition', 'champion'}, {'finance', 'market', 'stock'}, {'technology', 'innovation', 'science'} to guide the discovery of three clusters on sport, finance, and technology, respectively. Recent studies which include seed word constraints for document clustering are mostly focused on topic modeling approaches [8,16,17,19], inspired by the Latent Dirichlet Allocation model [3]. Concurrently, important advances on clustering were recently enabled through its combination with deep representation learning (e.g., see [12,23,30,31]), which is now known as *deep clustering*. A common approach to deep clustering is to jointly train an autoencoder and perform clustering on the learned representations [23,30,31]. One advantage of deep clustering approaches lies in their ability to leverage semantic representations based on word embeddings, enabling related documents to be close in the embedding space even when they use different (but related) words.

The main contributions of this study can be summarized as follows: (a) We introduce the **S**eed-guided **D**eep **D**ocument **C**lustering (SD2C) framework,[1] the first attempt, to the best of our knowledge, to constrain clustering with seed words based on a deep clustering approach; and (b) we validate this framework through experiments based on automatically selected seed words on five publicly available text datasets with various sizes and characteristics.

The remainder of the paper is organized as follows. In Sect. 2, we describe existing works on seed-guided constrained document clustering, also known as dataless text classification. Section 3 then introduces the seed-guided deep document clustering framework, which is then evaluated in Sect. 4. Section 5 concludes the paper and provides some perspectives on SD2C.

2 Related Work

One way to address the above problem is to try and identify multiple clustering views from the data in a purely unsupervised fashion [7,24,25]. While such an approach provides users with several possible clustering results to choose from, there is still no guarantee that the obtained clusters are those the users are interested in.

The constrained clustering problem we are addressing in fact bears strong similarity with the one of seed-guided dataless text classification, which consist in categorizing documents based on a small set of seed words describing the classes/clusters. For a more general survey on constrained clustering, we invite the reader to refer to [2]. The task of dataless text classification was introduced independently by Liu *et al.* [20] and Ko *et al.* [14]. In [20], the seed words are provided by a user and exploited to automatically label a part of the unlabeled documents. On the other hand, in [14], seed words initially correspond to labels/titles for the classes of interest and are extended based on co-occurrence patterns. In both cases, a Naive Bayes classifier is applied to estimate the documents' class assignments. In the wake of these seminal works, several studies

[1] The code is available at https://github.com/MaziarMF/SD2C.

further investigated the exploitation of seed words for text classification [6,9,10]. Chang *et al.* [6] introduced both an 'on-the-fly' approach and a bootstrapping approach by projecting seed words and documents in the same space. The former approach simply consists in assigning each document to the nearest class in the space, whereas the latter learns a bootstrapping Naive Bayes classifier with the class-informed seed words as initial training set. Another bootstrapping approach is studied in [10], where two different methods are considered to build the initial training set from the seed words: Latent Semantic Indexing and Gaussian Mixture Models. The maximum entropy classifier proposed in [9] instead directly uses seed words' class information by assuming that documents containing seed words from a class are more likely to belong to this class.

More recently, the dataless text classification problem was addressed through topic modeling approaches [8,16,17,19], extending the Latent Dirichlet Allocation model [3]. The topic model devised by Chen *et al.* [8] integrates the seed words as pseudo-documents, where each pseudo-document contains all the seed words given for a single class. The co-occurrence mechanism underlying topic models along with the known class membership of pseudo-documents help guide the actual documents to be classified towards their correct class. In [17], the Seed-guided Topic Model (STM) distinguishes between two types of topics: category topics and general topics. The former describe the class information and are associated with an informed prior based on the seed words, whereas the latter correspond to the general topics underlying the whole collection. The category topics assigned to a document are then used to estimate its class assignment. STM was extended in [16] to simultaneously perform classification and document filtering – which consists in identifying the documents related to a given set of categories while discarding irrelevant documents – by further dividing category topics into relevant and non-relevant topics. Similarly to STM, the Laplacian Seed Word Topic Model (LapSWTM) introduced by Li *et al.* [19] considers both category topics and general topics. It however differs from previous models in that it enforces a document manifold regularization to overcome the issue of documents containing no seed words. If these models outperform previously proposed models, they suffer from a lack of flexibility on the input representations they rely on. Indeed, topic models require documents to be organized as sets of discrete units – the word tokens. This prohibits the use of representation learning techniques such as word embeddings (e.g., word2vec [22] and GloVe [26]).

To the best of our knowledge, only one deep learning-based approach was proposed to address a problem similar to dataless text classification [18]. In this recent work, Li *et al.* devised a deep relevance model for zero-shot document filtering – which consists at test time in predicting the relevance of documents with respect to a category unseen in the training set, where each category is characterized by a set of seed words. This problem is nonetheless different from dataless text classification as it focuses on estimating documents' relevance (or lack thereof) instead of class membership.

3 Seed-Guided Deep Document Clustering

Deep clustering consists in jointly performing clustering and deep representation learning in an unsupervised fashion (e.g., with an auto-encoder). All deep clustering approaches aim at obtaining representations that are both faithful to the original documents and are more suited to document clustering purposes than the original document representation. To do so, they trade off between a reconstruction loss, denoted $\mathcal{L}_{\text{rec}}$, and a clustering loss, denoted $\mathcal{L}_{\text{clust}}$, through a joint optimization problem of the form: $\mathcal{L}_{\text{rec}} + \lambda_0 \mathcal{L}_{\text{clust}}$, where λ_0 is an hyperparameter balancing the contribution of the reconstruction and clustering losses.

In the remainder, $\mathcal{X}$ will denote the set of documents to cluster. Each document $x \in \mathcal{X}$ is associated with a representation $\mathbf{x}$ in $\mathbb{R}^d$ – thereafter, the *input space* – defined as the average of the (precomputed) embeddings of the words in x, where d is the dimension of the word embedding space. Each word w is thus represented as a d-dimensional vector $\mathbf{w}$ corresponding to its embedding (Sect. 4 further discusses the different word embeddings considered). Let $f_\theta : \mathbb{R}^d \rightarrow \mathbb{R}^p$ and $g_\eta : \mathbb{R}^p \rightarrow \mathbb{R}^d$ be an encoder and a decoder with parameters θ and η, respectively; $g_\eta \circ f_\theta$ then defines an auto-encoder (AE). $\mathbb{R}^p$ denotes the space in which we wish to embed the learned document representations – thereafter, the *embedding space*. Lastly, we denote by $\mathcal{R}$ the parameters of the clustering algorithm. With a slight abuse of notations in which $f_\theta(\mathcal{X})$ corresponds to the application of the function f_θ to each element of the set $\mathcal{X}$, the overall deep clustering (DC) optimization problem takes the form:

$$\underset{\theta,\eta,\mathcal{R}}{\operatorname{argmin}} \underbrace{\mathcal{L}_{\text{rec}}(\mathcal{X}, g_\eta \circ f_\theta(\mathcal{X})) + \lambda_0 \mathcal{L}_{\text{clust}}(f_\theta(\mathcal{X}), \mathcal{R})}_{\mathcal{L}_{\text{dc}}(\mathcal{X},\theta,\eta,\mathcal{R})}. \tag{1}$$

We propose to integrate constraints on seed words in this framework by biasing the embedding representations, which guarantees that the information pertaining to seed words will be used in the clustering process. This can be done by enforcing that seed words have more influence either on the learned document embeddings, a solution we refer to as **SD2C-Doc**, or on the cluster representatives, a solution we refer to as **SD2C-Rep**. Note that the second solution can only be used when the clustering process is based on cluster representatives (i.e., $\mathcal{R} = \{r_k\}_{k=1}^K$ with K the number of clusters), which is indeed the case for most current deep clustering methods [1].

In addition to the notations introduced previously, we will denote by s_k the subset of seed words corresponding to cluster k, and by $\mathcal{S} = \{s_k\}_{k=1}^K$ the complete set of seed words defining the prior knowledge on the K clusters to recover. We further define $\overline{\mathcal{S}} = \bigcup_{k=1}^K s_k$, the set of seed words from all clusters.

SD2C-Doc. One way to bias the document representations according to the seed words is to reduce the gap in the embedding space between the representation of the documents and the representation of the seed words occurring in these documents. For that purpose, we first define, for each document, a masked version of it that is based on seed words. This can be done *aggressively*, by retaining, in the masked version, only the words that correspond to seed words and by

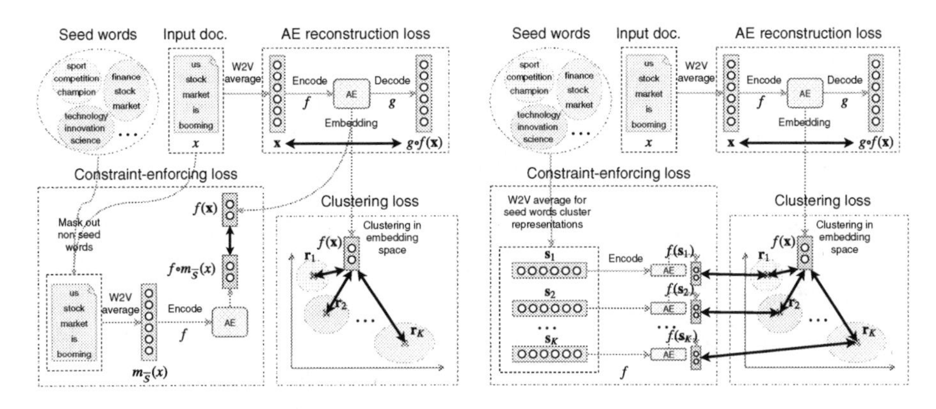

Fig. 1. Illustration of SD2C-Doc (left) and SD2C-Rep (right). Thick double arrows indicate the computation of a distance between two vectors.

computing an average of their word embeddings, or *smoothly* by reweighing all words in the original document according to their proximity with seed words. A weighted average of their embeddings then defines the smooth, masked version of the documents. The equation below formalizes these two approaches:

$$m_{\overline{\mathcal{S}}}(x) = \begin{cases} \frac{1}{\sum_{w \in \overline{\mathcal{S}}} \mathrm{tf}_x(w)} \sum_{w \in \overline{\mathcal{S}}} \mathrm{tf}_x(w) \cdot \mathbf{w} \\ \frac{1}{|\overline{\mathcal{S}}| \cdot |x|} \sum_{w' \in x} \sum_{w \in \overline{\mathcal{S}}} \frac{1 + \cos(\mathbf{w}, \mathbf{w}')}{2} \cdot \mathbf{w}' \end{cases} \tag{2}$$

where cos denotes the cosine similarity. If a document x does not contain any seed word, $m_{\overline{\mathcal{S}}}(x)$ is ill-defined when using the first version of Eq. 2 as $\sum_{w \in \overline{\mathcal{S}}}$ is null in that case. To address this issue, one can simply discard the documents without seed words. In practice, the two masked versions of Eq. 2 yielded the same results in our experiments. Because of its simplicity, we rely on the first one in the remainder of the paper, which is illustrated in Fig. 1 (left). One can then force the embedding representation of documents to be close to the embedding of their masked version by minimizing the dissimilarity in the embedding space, denoted by δ^E, between $f_\theta(\mathbf{x})$ and $f_\theta \circ m_{\overline{\mathcal{S}}}(x)$, leading to:

$$\underset{\theta, \eta, \mathcal{R}}{\operatorname{argmin}} \, \mathcal{L}_{\mathrm{dc}}(\mathcal{X}, \theta, \eta, \mathcal{R}) + \lambda_1 \sum_{x \in \mathcal{X}} \delta^E(f_\theta(\mathbf{x}), f_\theta \circ m_{\overline{\mathcal{S}}}(x)), \tag{3}$$

where λ_1 is an hyperparameter controlling the importance of the deep clustering loss $\mathcal{L}_{dc}$ and the loss associated to seed words.

SD2C-Rep. The other bias one can consider in the embedding space is the one related to cluster representatives. Here, one can naturally push cluster representatives towards the representation of seed words, in order to ensure that the discovered clusters will account for the prior knowledge provided by them.

For that purpose, we first build a representation for each subset of seed words by averaging the word embeddings of the seed words it contains:

$$\mathbf{s}_k = \frac{1}{|s_k|} \sum_{w \in s_k} \mathbf{w}.$$

$\mathbf{s}_k$ thus corresponds to the seed word-based representation of cluster k in $\mathbb{R}^d$. The optimization problem solved by SD2C-Rep, depicted in Fig. 1 (right), then takes the form:

$$\operatorname*{argmin}_{\theta,\eta,\mathcal{R}} \mathcal{L}_{\text{dc}}(\mathcal{X}, \theta, \eta, \mathcal{R}) + \lambda_1 \sum_{k=1}^{K} \delta^E(\mathbf{r}_k, f_\theta(\mathbf{s}_k)), \tag{4}$$

As before, δ^E denotes a dissimilarity in the embedding space. The last term in Eq. 4 forces cluster representatives to be close to subsets of seed words, the alignment between the two being defined by the initialization of the cluster representatives performed after pretraining (see Sect. 3.1 below).

3.1 Training

In practice, we use fully differentiable formulations of Problems 3 and 4. In the context of the k-Means algorithm, a popular clustering method, such differentiable formulations can be directly developed on top of the algorithms provided in [31] (called DCN) and [23] (called DKM), the latter proposing a truly joint formulation of the deep clustering problem. Other state-of-the-art deep clustering approaches, as IDEC [12], also based on cluster representatives, could naturally be adopted as well. The comparison between these approaches performed in [23] nevertheless suggests that DKM outperforms the other approaches. This difference was confirmed on the text collections retained in this study. We thus focus here on the DKM algorithm introduced in [23] with:

$$\mathcal{L}_{\text{rec}}(\mathcal{X}, g_\eta \circ f_\theta(\mathcal{X})) = \sum_{x \in \mathcal{X}} \delta^I(\mathbf{x}, g_\eta \circ f_\theta(\mathbf{x})), \tag{5}$$

where δ^I denotes a dissimilarity in the input space, and:

$$\mathcal{L}_{\text{clust}}(f_\theta(\mathcal{X}), \mathcal{R}) = \sum_{x \in \mathcal{X}} \sum_{k=1}^{K} \delta^E(f_\theta(\mathbf{x}), \mathbf{r}_k)\, G_k(f_\theta(\mathbf{x}), \alpha; \mathcal{R}) \tag{6}$$

where α is an inverse temperature parameter and $G_k(f_\theta(\mathbf{x}), \alpha; \mathcal{R})$ is a softmax function parameterized by α defined as follows:

$$G_k(f_\theta(\mathbf{x}), \alpha; \mathcal{R}) = \frac{\exp\big(-\alpha \cdot \delta^E(f_\theta(\mathbf{x}), \mathbf{r}_k)\big)}{\sum_{k'=1}^{K} \exp\big(-\alpha \cdot \delta^E(f_\theta(\mathbf{x}), \mathbf{r}_{k'})\big)}. \tag{7}$$

The k-means solution is recovered when α tends to $+\infty$.

Following prior deep clustering works [12,23,30,31], we initialize the auto-encoder parameters through pretraining by first only optimizing the reconstruction loss of the auto-encoder. In the pretraining of SD2C-Doc, we also include the constraint-enforcing term (second term in Problem 3) so that learned representations are impacted by seed words early in the training. At the end of pretraining, the cluster centers are initialized by the seed words cluster embeddings $\{\mathbf{s}_k\}_{k=1}^{K}$.[2] Then, in the fine-tuning phase, the whole loss – including the clustering loss and the constraint-enforcing loss (for SD2C-Doc and SD2C-Rep) – is optimized.

4 Experiments

The experiments we performed to evaluate the proposed SD2C framework are based on five publicly available datasets with various sizes and characteristics that have been extensively used in the context of text classification and clustering: The 20 Newsgroups[3] dataset, referred to as *20NEWS*; the Reuters-21578[4] dataset, referred to as *REUTERS*, from which, similarly to [8,16,17,19], we use only the 10 largest (and highly imbalanced) categories; the Yahoo! Answers dataset [32], referred to as *YAHOO*, from which we use only the test set comprising 60,000 documents evenly split into 10 classes; the DBPedia dataset [32], referred to as *DBPEDIA*, from which we also only use the test set made of 70,000 documents uniformly distributed in 14 classes; and the AG News dataset, introduced as well in [32] and referred to as *AGNEWS*, from which we use the training set, composed of 120,000 documents evenly split into 4 classes. After preprocessing, which includes removing stop words and words made of less than 2 characters, Porter stemming and discarding the empty documents, the number of documents in 20NEWS, REUTERS, YAHOO, DBPEDIA, AGNEWS are respectively 18,846, 7,964, 59,978, 70,000, 120,000. 20NEWS and REUTERS contain the documents with the greatest and most varied length whereas DBPEDIA, YAHOO, and AGNEWS are made of rather short documents.

Baselines and SD2C Variants. For both SD2C-Doc and SD2C-Rep, different dissimilarities can be adopted for δ^I and δ^E. As the cosine distance performed consistently better for δ^E than the Euclidean distance in our preliminary experiments, it is adopted here. We nevertheless did not observe such a clear trend for δ^I, and we indicate here the results obtained both for the cosine distance and Euclidean distance. This yields two versions for each method, which we denote as SD2C-Doc-e/SD2C-Rep-e and SD2C-Doc-c/SD2C-Rep-c, depending on whether the Euclidean (*-e) or the cosine (*-c) distance is used for δ^I, respectively.

To compare against SD2C, we considered the following baseline methods:

[2] This is especially important for SD2C-Rep which is based on the assumption that the clusters defined by the seed words and those defined by the cluster representatives are aligned.

[3] http://qwone.com/~jason/20Newsgroups/.

[4] http://www.daviddlewis.com/resources/testcollections/reuters21578/.

- *KM*, *AE-KM* and *DKM*: KM corresponds to k-Means [21] applied on the same input for documents as the one used for SD2C (average of documents' word embeddings); AE-KM first trains an auto-encoder on the collection and then applies k-Means to the document embeddings learned by the auto-encoder; DKM is the deep k-Means algorithm[5] presented in [23] which we also study under the two variants DKM-e and DKM-c.[6]
- *NN*: This method is similar to the 'on-the-fly' nearest neighbor-like classification described in [6]. Each document, represented by its word embeddings average, is assigned to the nearest class, in terms of the cosine distance, which outperformed the Euclidean distance, represented by the class' average seed word embeddings (denoted as $\{\mathbf{s}_k\}_{k=1}^K$ in Sect. 3).
- *STM*: In our experiments, we ran the Java implementation of the Seed-guided Topic Model [17] provided by the authors[7] and used the standard hyperparameters indicated in the paper. Given that this approach was not scalable when the whole vocabulary is used, we only kept the 2000 most frequent words (after preprocessing) for each dataset[8].

Seed Word Selection. Recent works on dataless text classification [8,16,17,19] only considered the 20NEWS and REUTERS datasets in their experiments, relying respectively on the seed words induced by the class labels and on the manually curated seed words from [8]. To perform an evaluation on all the collections retained here, we devised a simple heuristics based on tf-idf to propose seed words. For a given collection and for each class k of the collection, all words w in the vocabulary are scored according to:

$$\text{score}(w,k) = \left(\text{tf}_k(w) - \frac{1}{K-1} \sum_{\substack{k'=1 \\ k' \neq k}}^{K} \text{tf}_{k'}(w) \right) \times \text{idf}(w), \tag{8}$$

where $\text{idf}(w)$ is the inverse document frequency computed on the documents of the whole collection and $\text{tf}_k(w)$ is the term frequency for class k, which we define as the sum of $\text{tf}_x(w)$ for all documents x in class k. The rationale for this score is that one wishes to select words that are frequent in class k and unfrequent in other classes, hence the penalization term inside the brackets.

[5] https://github.com/MaziarMF/deep-k-means.
[6] Seed words are not utilized in these approaches.
[7] https://github.com/ly233/Seed-Guided-Topic-Model.
[8] Very recently, another topic modeling approach, the Laplacian Seed Word Topic Model (LapSWTM), was proposed in [19]. However, firstly, LapSWTM counts 8 hyperparameters that were empirically optimized in the original paper, and it is not straightforward how these hyperparameters should be tuned on the additional datasets used here. Secondly, LapSWTM shares a lot with the STM model in its construction and performance. Thirdly, the code for LapSWTM is, as far as we are aware, not publicly available. For these different reasons, we simply chose STM to represent the state of the art in topic modeling-based dataless text classification.

Table 1. Macro-average results in terms of accuracy (ACC) and adjusted rand index (ARI). The double vertical line separates approaches which leverage seed words (right) from approaches which do not (left). Bold values correspond to the best results.

Metric	KM	AE-KM	DKM-e	DKM-c	NN	STM	SD2C-Doc-e	SD2C-Doc-c	SD2C-Rep-e	SD2C-Rep-c
ACC	61.4	64.0	**65.5**	65.0	72.6	73.3	73.6	**75.9**	75.6	74.1
ARI	45.4	48.8	**50.4**	48.9	50.9	53.6	56.2	**57.1**	55.7	53.5

Based on this score, one can then select the top words for each class as seed words. We emphasize that such heuristics is only adopted for the purpose of simulating seed words during the evaluation: it is not destined to be used to identify seed words in a real-world application, where ground truth is unknown.

Architecture and Hyperparameters. The auto-encoder used in our experiments on all datasets is similar to the ones adopted in prior deep clustering works [12,23,30,31]. The encoder and decoder are mirrored fully-connected neural networks with dimensions d-500-500-2000-50 and 50-2000-500-500-d, respectively – d is the input space dimension and 50 corresponds to the dimension p of the auto-encoder embedding space. Neural networks' weights are initialized based on the Xavier scheme [11]. The SD2C, DKM, and AE-KM models are trained with the Adam optimizer [13] with standard hyperparameters ($\eta = 0.001, \beta_1 = 0.9$, and $\beta_2 = 0.999$) and minibatches of 256 documents. The number of epochs for the auto-encoder pretraining and model finetuning are fixed to 50 and 200, respectively, as in [23]. We also use the inverse temperature $\alpha = 1000$ from [23] for the parameterized softmax-based differentiable reformulations of SD2C models. The balancing hyperparameters λ_0 and λ_1 of SD2C-Doc and SD2C-Rep were both set to 10^{-5}. We experimented with different word embedding techniques including word2vec [22], doc2vec [15], and FastText [4] trained either on an external large corpus (e.g., Google News) or individually on the datasets used in the experiments. We found that training the word embedding models on the experiments' collections consistently improved in terms of clustering performance on external corpus-based training. Among the word embedding techniques we tested, word2vec and FastText performed evenly and significantly better than doc2vec. Since word2vec is faster to train than FastText, which operates at the character level, we chose the former technique (in practice, Gensim[9] word2vec Python implementation) trained on each of our experiments' datasets to compute the word embeddings. The word embedding size was fixed to 100. The Skip-Gram model was trained with a window size of 50 words on 20NEWS and 10 words on other datasets. Note that a word2vec model is trained once for each dataset so that all approaches rely on the same word embeddings.

[9] https://radimrehurek.com/gensim/.

Table 2. Seed-guided constrained clustering results with 3 seed words per cluster. Bold results denote the best, as well as not significantly different from the best, results. Underlined SD2C results indicate a significant improvement over STM.

Model	20NEWS		REUTERS		YAHOO		DBPEDIA		AGNEWS	
	ACC	ARI	ACC	ARI	ACC	ARI	ACC	ARI	ACC	ARI
NN	72.3±0.0	53.2±0.0	79.0±0.0	58.3±0.0	54.7±0.0	26.5±0.0	79.9±0.0	64.7±0.0	77.3±0.0	51.9±0.0
STM	65.7±0.9	47.5±1.0	**83.0±0.7**	**66.3±1.2**	57.1±0.1	29.3±0.2	**80.9±0.4**	**72.7±0.4**	79.7±0.2	55.8±0.3
SD2C-Doc-e	**80.5±0.6**	**66.4±0.7**	66.3±3.7	53.0±3.2	60.4±0.3	**34.3±0.3**	76.1±0.2	63.0±0.2	**84.8±0.2**	**64.4±0.5**
SD2C-Doc-c	77.0±1.5	61.7±2.2	78.1±1.8	60.2±1.1	**61.1±0.8**	**34.4±1.3**	79.4±1.9	66.3±2.1	**84.1±1.1**	**63.3±2.3**
SD2C-Rep-e	*76.1±0.3*	60.1±0.5	80.2±0.8	59.9±0.8	60.2±0.3	**33.5±0.4**	80.3±0.5	67.0±0.6	81.1±0.3	58.1±0.5
SD2C-Rep-c	72.1±1.5	55.7±1.7	81.4±0.7	61.1±0.9	57.8±1.7	29.7±2.7	79.8±1.4	66.1±1.8	79.4±2.0	55.0±3.6

4.1 Results

We measure the clustering performance in terms of clustering accuracy (ACC) and adjusted rand index (ARI), which are standard clustering metrics [5]. Table 1 first provides the macro-average (over the 5 datasets) of these measures for all methods, using the top 3 automatically selected seed words per cluster. As one can note, the use of seed words is beneficial to the clustering. Indeed, the approaches which use seed words (NN, STM, SD2C) have markedly higher ACC and higher ARI than those which do not (KM, AE-KM, DKM). Among these latter methods, DKM is the best ones (as a comparison, DCN and IDEC, mentioned in Sect. 3, respectively obtain 64.8 and 64.1 for ACC, and 49.3 and 47 for ARI). Among the methods exploiting seed words, SD2C methods are the best ones, outperforming the baseline NN and the STM method by up to 2.6 points for ACC and 3.5 points for ARI.

We further provide in Table 2 a detailed account of the performance of the methods based on seed words. The results have been averaged over 10 runs and are reported with their standard deviation. We furthermore performed an unpaired Student t-test with a significance level of 0.01 to study whether differences are significant or not (all results in bold are not not statistically different from the best result). As one can note, the proposed SD2C models compare favorably against STM, the strongest baseline. Indeed, all SD2C approaches significantly outperform STM on 20NEWS, and SD2C-Doc-e/c as well as SD2C-Rep-e also significantly outperform STM on YAHOO and AGNEWS. On the other hand, STM obtained significantly better results in terms of both ACC and ARI on REUTERS and DBPEDIA, the difference on these collections (and especially on DBPEDIA) being nevertheless small. Among the SD2C methods, SD2C-Doc-c yields the best performance overall (as shown in Table 1).

Runtime. We further compared, in Table 3, the efficiency of STM and the SD2C methods on a machine with eight i7-7700HQ CPUs at 2.80 GHz, 16 GB RAM, and an NVIDIA GeForce GTX 1070 (only used for the deep learning approaches). The runtime of the SD2C approaches is lower than that of STM on most datasets. STM was only faster on AGNEWS (between 2 and 3 times), yet far slower on 20NEWS (about 10 times). This discrepancy can be explained by the fact that STM's complexity is dominated by the total number of tokens

Table 3. Execution time per run (in seconds) for each model on 20NEWS, REUTERS, YAHOO, DBPEDIA, and AGNEWS.

Model	20NEWS	REUTERS	YAHOO	DBPEDIA	AGNEWS
NN	15	3	24	37	22
STM	2008	177	762	686	338
SD2C-Doc	221	72	549	736	1048
SD2C-Rep	186	60	459	590	808

Table 4. Seed-guided constrained clustering results with manual seed words.

Model	20NEWS		REUTERS	
	ACC	ARI	ACC	ARI
STM	66.4±0.2	48.4±0.3	83.2±0.4	68.4±0.8
SD2C-Doc-e	**78.5±0.8**	**64.9±0.9**	67.8±0.4	56.8±2.3
SD2C-Doc-c	75.6±1.0	60.6±1.8	75.5±2.2	61.2±1.7
SD2C-Rep-e	75.2±0.5	60.0±0.4	82.9±0.7	67.7±1.5
SD2C-Rep-c	73.7±0.6	57.5±0.9	**84.2±0.8**	**71.2±1.6**

(large for a small number of documents in 20NEWS), whereas SD2C models only depend on the number of documents (large with few tokens per document in AGNEWS). As to the SD2C models, SD2C-Rep runs faster than SD2C-Doc. This is due to the complexity of the constraint-enforcing loss term being lower for the former than for the latter.

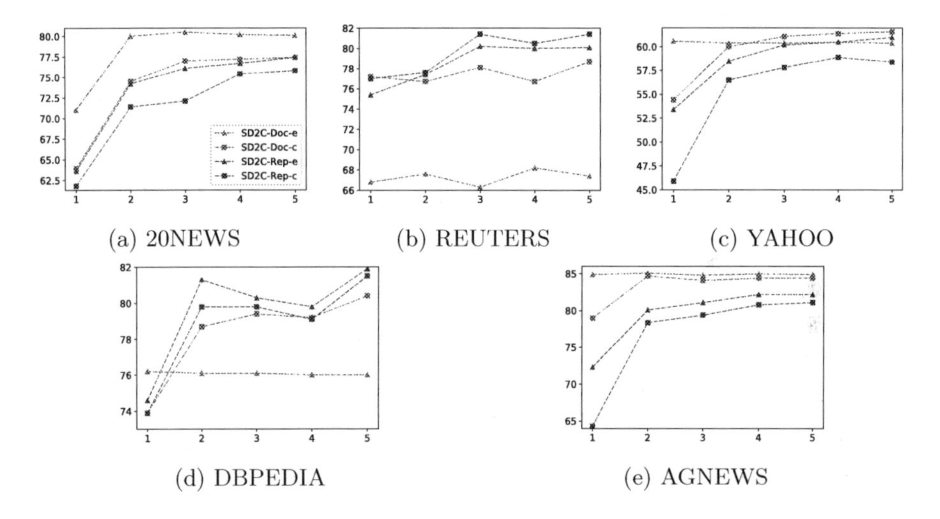

Fig. 2. Clustering results in terms of ACC for SD2C-Doc-e, SD2C-Doc-c, SD2C-Rep-e and SD2C-Rep-c with 1 to 5 seed words.

Impact of the Number of Seed Words. In our general setting used to report the previous results, the number of seed words per class was arbitrarily set to 3. For comprehensiveness, we study the clustering results of the SD2C models when the number of (automatically selected) seed words per cluster is varied from 1 to 5. The evolution of the performance for the SD2C models in terms of accuracy is illustrated in Fig. 2. We observe that using more seed words leads to notable improvements in most cases – with the exception of SD2C-Doc-e, which seems to be less influenced by the number of seed words. This trend is particularly apparent when the number of seed words is increased from 1 to 2. Although slight performance gain is observed between 2 and 5 seed words, the results exhibit greater stability. This suggests that providing as few as 2 seed words per cluster – which constitutes a modest annotation effort for humans – can prove highly beneficial for the clustering results obtained by our SD2C approaches.

Comparing Automatic and Manual Seed Words. In order to check that the method we retained to automatically extract seed words is appropriate, we also computed the results obtained by STM and the SD2C methods using the manual seed words available for 20NEWS and REUTERS and presented in, *e.g.*, [8,17,28] (denoted as $\mathbb{S}^D$ in the latter). The corresponding list of seed words contain in average 5.1 words per category for 20NEWS and 6.8 words per category for REUTERS. The procedure to constitute these lists of descriptive seed words is detailed in [8]. Table 4 summarizes the results obtained with such seed words. These results first show that the scores obtained by the different methods using the manual seed words are close to the ones obtained with the automatically selected ones. For example, the difference in ACC for STM amounts to only 0.7 points on 20NEWS and 0.2 points on REUTERS. This shows that the automatic selected seed words are a reasonable substitute to manual seed words for evaluation purposes. In addition, SD2C methods still significantly outperform STM on 20NEWS. SD2C-Rep-c is here significantly better, even though the difference is not important, than STM on REUTERS – this is in line with our comment on Table 2 on the small differences between STM and SD2C on REUTERS.

5 Conclusion

We have introduced in this paper the SD2C framework, the first attempt, to the best of our knowledge, to constrain document clustering with seed words using a deep clustering approach. To do so, we have integrated constraints associated to seed words in the Deep k-Means optimization problem [23], modifying either the document embeddings, the cluster representatives or the input representations to make them closer to the seed words retained. The new methods thus derived have been evaluated on five text collections widely used for text classification purposes. For this evaluation, we have proposed a simple method to automatically select seed words that behaves comparably to manual seed words for evaluation purposes.

Several perspectives for this work can be envisaged. First of all, it is possible to extend the current framework with a 'garbage' cluster to collect documents that do not fit well within the clusters defined by the seed words. This can be useful in particular for document filtering [16]. Other types of autoencoders and other attention mechanisms can also be designed to try and improve the results of the SD2C methods. Combinations of the different approaches can also be studied so as to benefit from their respective strengths. Lastly, if the SD2C-Doc-c method overall outperforms the other approaches in terms of accuracy and adjusted rand index, we want to better understand when it is beneficial to bias the document representations and when to bias the cluster representative ones.

Acknowledgment. This research was partly funded by the ANR project LOCUST and the AURA project AISUA.

References

1. Aljalbout, E., Golkov, V., Siddiqui, Y., Cremers, D.: Clustering with deep learning: taxonomy and new methods. arXiv:1801.07648 (2018)
2. Basu, S., Davidson, I., Wagstaff, K.: Constrained Clustering: Advances in Algorithms, Theory, and Applications, 1st edn. Chapman & Hall/CRC, Boca Raton (2008)
3. Blei, D.M., Ng, A.Y., Jordan, M.I.: Latent dirichlet allocation. J. Mach. Learn. Res. **3**, 993–1022 (2003)
4. Bojanowski, P., Grave, E., Joulin, A., Mikolov, T.: Enriching word vectors with subword information. Trans. Assoc. Comput. Linguist. **5**, 135–146 (2017)
5. Cai, D., He, X., Han, J.: Locally consistent concept factorization for document clustering. IEEE Trans. Knowl. Data Eng. **23**(6), 902–913 (2011)
6. Chang, M.W., Ratinov, L., Roth, D., Srikumar, V.: Importance of semantic representation: dataless classification. In: Proceedings of AAAI, pp. 830–835 (2008)
7. Chang, Y., Chen, J., Cho, M.H., Castaldi, P.J., Silverman, E.K., Dy, J.G.: Multiple clustering views from multiple uncertain experts. In: Proceedings of ICML, pp. 674–683 (2017)
8. Chen, X., Xia, Y., Jin, P., Carroll, J.: Dataless text classification with descriptive LDA. In: Proceedings of AAAI, pp. 2224–2231 (2015)
9. Druck, G., Mann, G., Mccallum, A.: Learning from labeled features using generalized expectation criteria. In: Proceedings of SIGIR, pp. 595–602 (2008)
10. Gliozzo, A., Strapparava, C., Dagan, I.: Improving text categorization bootstrapping via unsupervised learning. ACM Trans. Speech Lang. Process. **6**(1), 1–24 (2009)
11. Glorot, X., Bengio, Y.: Understanding the difficulty of training deep feedforward neural networks. In: Proceedings of AISTATS, pp. 249–256 (2010)
12. Guo, X., Gao, L., Liu, X., Yin, J.: Improved deep embedded clustering with local structure preservation. In: Proceedings of IJCAI, pp. 1753–1759 (2017)
13. Kingma, D.P., Ba, J.L.: Adam: a method for stochastic optimization. In: Proceedings of ICLR (2015)
14. Ko, Y., Seo, J.: Learning with unlabeled data for text categorization using bootstrapping and feature projection techniques. In: Proceedings of ACL, pp. 255–262 (2004)
15. Le, Q., Mikolov, T.: Distributed representations of sentences and documents. In: Proceedings of ICML, pp. 1188–1196 (2014)
16. Li, C., Chen, S., Xing, J., Sun, A., Ma, Z.: Seed-guided topic model for document filtering and classification. ACM Trans. Inf. Syst. **37**(1) (2018)
17. Li, C., Xing, J., Sun, A., Ma, Z.: Effective document labeling with very few seed words: a topic model approach. In: Proceedings of CIKM, pp. 85–94 (2016)
18. Li, C., Zhou, W., Ji, F., Duan, Y., Chen, H.: A deep relevance model for zero-shot document filtering. In: Proceedings of ACL, pp. 2300–2310 (2018)
19. Li, X., Li, C., Chi, J., Ouyang, J., Li, C.: Dataless text classification: a topic modeling approach with document manifold. In: Proceedings of CIKM, pp. 973–982 (2018)
20. Liu, B., Li, X., Lee, W.S., Yu, P.S.: Text classification by labeling words. In: Proceedings of AAAI/IAAI, pp. 425–430 (2004)

21. MacQueen, J.: Some methods for classification and analysis of multivariate observations. In: Proceedings of BSMSP, pp. 281–297 (1967)
22. Mikolov, T., Sutskever, I., Chen, K., Corrado, G., Dean, J.: Distributed representations of words and phrases and their compositionality. In: Proceedings of NeurIPS, pp. 3111–3119 (2013)
23. Fard, M.M., Thonet, T., Gaussier, E.: Deep k-means: jointly clustering with k-means and learning representations. arXiv:1806.10069 (2018)
24. Niu, D., Dy, J.G., Ghahramani, Z.: A nonparametric Bayesian model for multiple clustering with overlapping feature views. In: Proceedings of AISTATS, pp. 814–822 (2012)
25. Niu, D., Dy, J.G., Jordan, M.I.: Multiple non-redundant spectral clustering views. In: Proceedings of ICML, pp. 831–838 (2010)
26. Pennington, J., Socher, R., Manning, C.D.: GloVe: global vectors for word representation. In: Proceedings of EMNLP, pp. 1532–1543 (2014)
27. Shental, N., Bar-Hillel, A., Hertz, T., Weinshall, D.: Computing Gaussian mixture models with EM using equivalence constraints. In: Proceedings of NeurIPS, pp. 465–472 (2003)
28. Song, Y., Roth, D.: On dataless hierarchical text classification. In: Proceedings of the 28th AAAI Conference on Artificial Intelligence, AAAI 2014, pp. 1579–1585 (2014)
29. Wagstaff, K., Cardie, C., Rogers, S., Schrödl, S.: Constrained K-means clustering with background knowledge. In: Proceedings of ICML, pp. 577–584 (2001)
30. Xie, J., Girshick, R., Farhadi, A.: Unsupervised deep embedding for clustering analysis. In: Proceedings of ICML, pp. 478–487 (2016)
31. Yang, B., Fu, X., Sidiropoulos, N.D., Hong, M.: Towards K-means-friendly spaces: simultaneous deep learning and clustering. In: Proceedings of ICML, pp. 3861–3870 (2017)
32. Zhang, X., Zhao, J., LeCun, Y.: Character-level convolutional networks for text classification. In: Proceedings of NeurIPS, pp. 649–657 (2015)

Improving Knowledge Graph Embedding Using Locally and Globally Attentive Relation Paths

Ningning Jia, Xiang Cheng(✉), and Sen Su

State Key Laboratory of Networking and Switching Technology,
Beijing University of Posts and Telecommunications,
Beijing, People's Republic of China
chengxiang@bupt.edu.cn

Abstract. Knowledge graphs' incompleteness has motivated many researchers to propose methods to automatically infer missing facts in knowledge graphs. Knowledge graph embedding has been an active research area for knowledge graph completion, with great improvement from the early TransE to the current state-of-the-art ConvKB. ConvKB considers a knowledge graph as a set of triples, and employs a convolutional neural network to capture global relationships and transitional characteristics between entities and relations in the knowledge graph. However, it only utilizes the triple information, and ignores the rich information contained in relation paths. In fact, a path of one relation describes the relation from some aspect in a fine-grained way. Therefore, it is beneficial to take relation paths into consideration for knowledge graph embedding. In this paper, we present a novel convolutional neural network-based embedding model PConvKB, which improves knowledge graph embedding by incorporating relation paths locally and globally. Specifically, we introduce attention mechanism to measure the local importance of relation paths. Moreover, we propose a simple yet effective measure DIPF to compute the global importance of relation paths. Experimental results show that our model achieves substantial improvements against state-of-the-art methods.

Keywords: Knowledge graph embedding · Link prediction · Triple classification · Convolutional neural network · Attention mechanism

1 Introduction

Large-scale knowledge graphs such as Freebase [3], DBpedia [1], and Wikidata [38] store real-world facts in the form of triples (*head, relation, tail*), abbreviated as (h, r, t), where *head* and *tail* are entities and *relation* represents the relationship between *head* and *tail*. They are important resources for many intelligence applications like question answering and web search. Although current knowledge graphs consist of billions of triples, they are still far from complete and missing crucial

J. M. Jose et al. (Eds.): ECIR 2020, LNCS 12035, pp. 17–32, 2020.
https://doi.org/10.1007/978-3-030-45439-5_2

facts, e.g., 75% of the person entities in Freebase have no known nationality [8], which hampers their usefulness in the aforementioned applications.

Various methods are proposed to address this problem, and the knowledge graph embedding methods have attracted increasing attention in recent years. The main idea of knowledge graph embedding is to embed entities and relations of a knowledge graph into a continuous vector space and predict missing facts by manipulating the entity and relation embeddings involved. Among knowledge graph embedding methods, the translation-based models are simple and efficient, also perform well. For example, given a triple (h, r, t), the most well-known translation-based model TransE [5] models the relation r as a translation vector $\mathbf{r}$ connecting the embeddings $\mathbf{h}$ and $\mathbf{t}$ of the two entities, i.e., $\mathbf{h} + \mathbf{r} \approx \mathbf{t}$. It performs well on simple relations, i.e., 1-to-1 relations. but poorly on complicated relations, i.e., 1-to-N, N-to-1 and N-to-N relations. To address this issue, TransH [41], TransR [20] and TransD [14] are proposed. Unfortunately, these models are less simplicity and efficiency than TransE. Nickel et al. [26] present HolE, which uses circular correlation to combine the expressive power of the tensor product with the simplicity and efficiency of TransE.

Recently, several convolutional neural network (CNN)-based models [7,22,23] have been proposed to learn the embeddings of entities and relations in knowledge graphs, in which [22] reserves the transitional characteristic in translation-based models and is comparably simple and efficient, achieves state-of-the-art performance. However, it only focuses on knowledge triples, ignoring the rich knowledge contained in relation paths. In fact, a path of one entity pair describes the relation connecting the entity pair from some aspect in a fine-grained way, and the importance of each path is different. For example, in Fig. 1, the two paths *place of birth – country* and *friend – nationality* of entity pair *(Tom Cruise, America)* describes the relation *nationality* from the location and social way, respectively. Since the path *place of birth – country* is more essential than *friend – nationality* to express the relation *nationality*, thus it is more important from the local view. Moreover, from the global view the path *friend – nationality* also occurs in entity pair *(Tom Cruise, England)*, which is connecting by the relation *travel*, thus it is less important than the path *place of birth – country* to express the relation *nationality*.

In this paper, we present a path-augmented CNN-based model, which incorporates relation paths for knowledge graph embedding. Specifically, we first introduce the attention mechanism to automatically measure the local importance of each path for the given entity pair, then inspired by inverse document frequency, we propose degree-guided inverse path frequency to compute the global importance of each path. Finally, we improve knowledge graph embedding by incorporating locally and globally attentive relation paths.

Our contributions in this paper are summarized as follows:

- We present a path-augmented CNN-based knowledge graph embedding model, which improves embedding model by incorporating relation paths locally and globally.

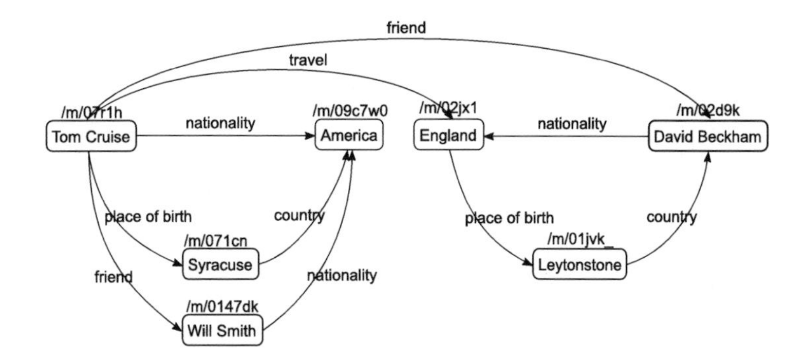

Fig. 1. An illustration that a path of one relation describes the relation from some aspect in a fine-grained way, and the importance of each path is different.

- We introduce attention mechanism to model the local importances of relation paths for knowledge graph embedding.
- We propose a simple yet effective measure, degree-guided inverse path frequency, to compute the global importances of relation paths for knowledge graph embedding.
- In addition, we apply three pooling operations to aggregate convolutional feature maps, which reduces the number of parameters greatly.
- The experimental results on four benchmark datasets show that our model achieves state-of-the-art performance.

2 Preliminaries

2.1 Problem Definition

Given a knowledge graph $\mathcal{G}$, which is a collection of valid factual triples (h, r, t), where $h, t \in \mathcal{E}$ and $r \in \mathcal{R}$. $\mathcal{E}$ is the entity set and $\mathcal{R}$ is the relation set. In knowledge graph completion, embedding methods aim to define a score function f that gives an implausibility score for each triple (h, r, t) such that valid triples receive lower scores than invalid triples.

2.2 ConvKB

In this section, we briefly describe the state-of-the-art CNN-based model ConvKB, and choose it as the base of our model.

For each triple (h, r, t), ConvKB denotes the dimensionality of embeddings by k, such that each embedding triple $(\boldsymbol{v}_h, \boldsymbol{v}_r, \boldsymbol{v}_t)$ can be viewed as a matrix $\mathbf{A} = [\boldsymbol{v}_h, \boldsymbol{v}_r, \boldsymbol{v}_t] \in \mathbb{R}^{k \times 3}$. A filter $\boldsymbol{\omega} \in \mathbb{R}^{1 \times 3}$ is repeatedly operated over every row of $\mathbf{A}$ to generate a feature map $\boldsymbol{v} = [v_1, v_2, \ldots, v_k] \in \mathbb{R}^k$, in which $v_i = g(\boldsymbol{\omega} \cdot \mathbf{A}_{i,:} + b)$, where $\cdot$ denotes a dot product, $\mathbf{A}_{i,:}$ is the i-th row of $\mathbf{A}$, b is a bias term, and g is the non-linear activation function ReLU. In particular, if $\boldsymbol{\omega} = [1, 1, -1]$, $b = 0$, and $g(x) = |x|$ or $g(x) = x^2$, ConvKB reduces to the plain

TransE. Hence, in some point of view, ConvKB is an extension of TransE, which models triple more globally and comprehensively. The overview of ConvKB is shown in Fig. 2.

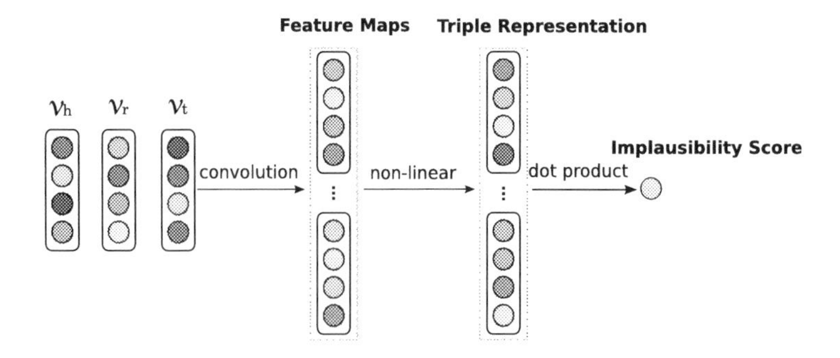

Fig. 2. The architecture of ConvKB.

Let $\boldsymbol{\Omega}$ and n denote the set of filters and the number of filters, respectively. ConvKB uses n filters to generate n feature maps. These feature maps are concatenated into a single vector, which is then calculated using the dot product with a weight vector $\mathbf{w} \in \mathbb{R}^{nk \times 1}$ to give an implausibility score for the triple (h, r, t). Formally, the score function of ConvKB is defined as follows:

$$f_{ConvKB}(h, r, t) = \text{concat}(g([\boldsymbol{v}_h, \boldsymbol{v}_r, \boldsymbol{v}_t] * \boldsymbol{\Omega})) \cdot \mathbf{w} \tag{1}$$

where $\boldsymbol{\Omega}$ and $\mathbf{w}$ are shared parameters, independent of h, r and t, $*$ denotes the convolution operator, and concat denotes the concatenation operator.

It is obvious that ConvKB only learns from triples, ignoring the rich knowledge contained in relation paths, which can lead to poor performance.

3 Our Proposed Model

3.1 PConvKB

In this section, we present our model PConvKB, which learns the embeddings by taking relation paths into consideration. Moreover, we also take into account the local and global importances of the relation paths. The architecture of our model is shown in Fig. 3.

We denote relation paths between the head entity h and the tail entity t as $P(h, t) = \{p_1, p_2, \ldots, p_N\}$, where relation path $p = (r_1, \ldots, r_m)$ is a series of interconnected relations between the entities, i.e., $h \xrightarrow{r_1} \ldots \xrightarrow{r_m} t$. Similar to ConvKB, for each triple (h, r, t), the score function of our model PConvKB is defined as follows:

$$f_{PConvKB}(h, r, t) = \sigma(\psi([\boldsymbol{v}_h, \sum_{i=1}^{N} \Phi_{G_i} \times \Phi_{L_i} \times \boldsymbol{p}_i + \boldsymbol{v}_r, \boldsymbol{v}_t] * \boldsymbol{\Omega})) \cdot \mathbf{w} \tag{2}$$

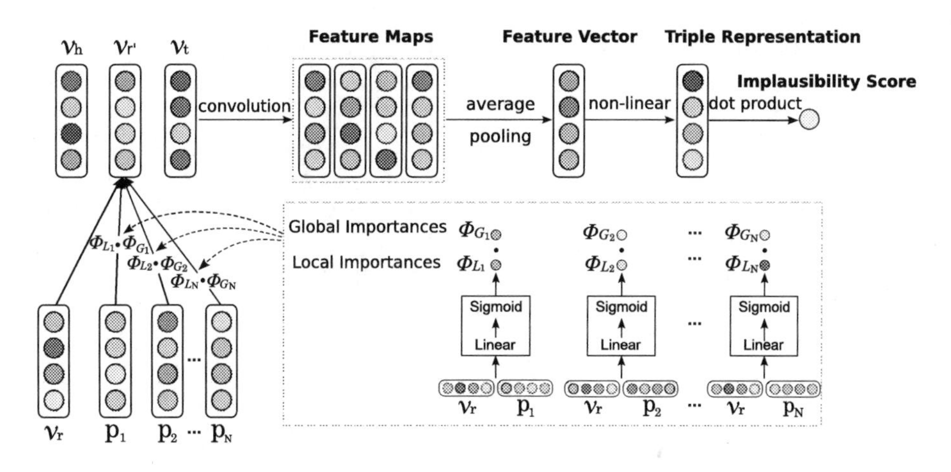

Fig. 3. The architecture of our model PConvKB.

where σ denotes the non-linear function, i.e., sigmoid, ψ denotes the average pooling operation, Φ_{G_i} denotes the global importance of the i-th path, Φ_{L_i} denotes the local importance of the i-th path, $\boldsymbol{p}_i$ is the embedding of the i-th path, which is computed as $\sum_{i=1}^{m} \boldsymbol{v}_{r_i}$, $\boldsymbol{\Omega}$ and $\mathbf{w}$ are shared parameters.

The computation of local and global importances is detailed in Sects. 3.2 and 3.3, respectively.

3.2 Measuring Local Importances of Relation Paths by Attention Mechanism

Attention mechanism [2] is designed to improve the performance of encoder-decoder model on machine translation, which assigns different weights to different data to allow the model focusing on important data. In recent years, attention mechanism has been widely used in several research topics, such as question answering [18] and image captioning [40]. In this paper, we apply attention mechanism to measure the local importances of relation paths for knowledge graph embedding. Given a triple (h, r, t) and its set of relation paths $P(h, t) = \{p_1, p_2, \ldots, p_N\}$, we compute the local importance of each path as:

$$\Phi_{L_i} = sigmoid(\boldsymbol{v}_r W_L \boldsymbol{p}_i) \tag{3}$$

where $W_L \in \mathbb{R}^{k \times k}$ is the parameter matrix. Similar to [19], we set the maximum length of each path to 3.

3.3 Measuring Global Importances of Relation Paths by Degree-Guided Inverse Path Frequency

Since the attention mechanism only focuses on the set of relation paths $P(h, t)$ of the given entity pair (h, t) that connects by the relation r. It does not consider that the path in the set of relation paths may also occur in other entity pairs that connects by other relations. Typically, the more set of relation paths a path occurs in, the less importance the path is. Therefore, inspired by inverse document frequency [10,16], which is a weighting function that has been widely used for measuring how informative each word is in a set of documents. We propose the **D**egree-guided **I**nverse **P**ath **F**requency (**DIPF**) to model the global importance of each path in the set of relation paths.

For each relation $r \in \mathcal{R}$ in the knowledge graph $\mathcal{G}$, we first find its corresponding entity pairs $(h^r, t^r)_i, i = 1, 2, \ldots, n^r$, where n^r is the number of entity pairs connecting by the relation r. i.e.,

$$(h^r, t^r)_i \in \mathcal{G} \quad and \quad (h^r, r, t^r)_i \in \mathcal{G}. \tag{4}$$

Then, we choose the entity pair $(h^r, t^r)_b$, which has the biggest node degree and is computed as:

$$\text{NodeDegree}((h^r, t^r)_b) = \max[\text{NodeDegree}((h^r, t^r)_i)], i = 1, 2, \ldots, n^r \tag{5}$$

in which,

$$\text{NodeDegree}((h^r, t^r)_i) = deg(h_i^r) + deg(t_i^r) \tag{6}$$

where NodeDegree$(\cdot)$ is the function to compute the node degree of an entity pair, max$[\cdot]$ is the maximum function, $deg(\cdot)$ is the node degree of an entity, which is computed as the number of the edges connected with the entity.

Next, we count the set of relation paths for entity pair $(h^r, t^r)_b$, and is denoted as $P((h^r, t^r)_b)$:

$$P((h^r, t^r)_b) = \{p_1^r, p_2^r, \ldots, p_{m^r}^r\} \tag{7}$$

where m^r is the number of paths of entity pair $(h^r, t^r)_b$. Similar to local importance computation, we set the maximum length of each path to 3.

Finally, the global importance of each path in the set of relation paths $P(h, t) = \{p_1, p_2, \ldots, p_N\}$ of the given triple (h, r, t) is computed as:

$$\Phi_{G_i} = log \frac{|\mathcal{R}|}{pt_i} \tag{8}$$

where $|\mathcal{R}|$ is the cardinality of $\mathcal{R}$ (i.e., total number of relations in $\mathcal{R}$), pt_i is the number of times the path p_i occurs in the set of $\{P((h^r, t^r)_b), r \in \mathcal{R}\}$.

3.4 Aggregating Feature Maps Using Pooling Operation

As mentioned in Sect. 2.1, ConvKB uses concatenate operation to aggregate feature maps. However, previous works [30,35] demonstrate that pooling operation can better aggregate feature maps than simply concatenate operation, and reduce the number of parameters greatly. In this paper, we adopt the following three pooling operations to replace the concatenate operation, respectively:

$$\psi_{sum} = \sum_{i=1}^{n} \boldsymbol{v}_i \tag{9}$$

$$\psi_{ave} = \frac{1}{n}\sum_{i=1}^{n} \boldsymbol{v}_i \tag{10}$$

$$\psi_{max} = \max([\boldsymbol{v}_1, \ldots, \boldsymbol{v}_n]) \tag{11}$$

The average pooling operation is finally chosen due to its superior performance in the experiments.

3.5 Model Training

The objective is to ensure that a triple in the golden set $\mathcal{G}$ should have a lower implausibility score than a triple in the corrupted triple set $\mathcal{G}'$. Similar to [22], we adopt Adam optimizer [17] to train PConvKB, and minimize the loss function with L_2 regularization on the weight vector $\mathbf{w}$ as follows:

$$\mathcal{L} = \sum_{(h,r,t)\in\mathcal{G}\bigcup\mathcal{G}'} log(1 + \exp(l_{(h,r,t)} \cdot f_{PConvKB}(h, r, t))) + \frac{\lambda}{2}\|\mathbf{w}\|_2^2 \tag{12}$$

in which, $l_{(h,r,t)} = \begin{cases} 1, & for\ (h, r, t) \in \mathcal{G} \\ -1, & for\ (h, r, t) \in \mathcal{G}'. \end{cases}$

3.6 Complexity Analysis

We compare the parameter size and computational complexity of our model PConvKB with ConvKB. Let N_e denote the number of entities, N_r the number of relations, K the embedding dimension, S the number of triples for learning, P the expected number of relation paths connecting two entities, and L the expected length of relation paths. The parameter size of PConvKB is equal to the parameter size of ConvKB, i.e., $(N_e + N_r)K$. For each iteration in optimization, the computational complexity of PConvKB is $O(SKPL)$, and the computational complexity of ConvKB is $O(SK)$.

4 Experiments

For a fair comparison, we evaluate our model on two tasks: link prediction [5], and triples classification [33]. Both of them evaluate the accuracy of predicting unseen triples from different viewpoints.

4.1 Datasets

We evaluate our model on four benchmark datasets WN18 [5], FB15k [5], WN18RR [7] and FB15k-237 [36]. WN18 is extracted from WordNet [21], which contains word concepts and lexical relations between the concepts. FB15k is a subset of Freebase constructed by Bordes et al. [5]. As noted by Toutanova and Chen [36], WN18 and FB15k have problematic reversible triples causing abnormally high results. This is the reason that the refined version of WN18 and FB15k, i.e., WN18RR and FB15k-237, are widely used in state-of-the-art methods. Table 1 shows the statistics of the datasets used in our experiments.

Table 1. Statistics of the experimental datasets

Dataset	#Entity	#Relation	#Train	#Valid	#Test
WN18	40,943	18	141,442	5,000	5,000
FB15k	14,951	1,345	483,142	50,000	59,071
WN18RR	40,943	11	86,835	3,034	3,134
FB15k-237	14,541	237	272,115	17,535	20,466

4.2 Comparison Methods

To demonstrate the effectiveness of our model, we compare PConvKB against a variety of knowledge graph embedding methods developed in recent years.

- **TransE** [5] is one of the most widely used knowledge graph embedding methods.
- **TransH** [41] associates each relation with a relation-specific hyperplane to alleviate the complex relations problem.
- **TransD** [14] not only considers the complex relations, but also the diversity of entities, by embedding entities and relations into separate entity space and relation-specific spaces.
- **HolE** [26] uses circular correlation, a novel compositional operator, to capture rich interactions of embeddings.
- **ConvE** [7] is the first CNN-based model for knowledge graph embedding.
- **ConvKB** [22] improves ConvE by taking the transitional characteristic (i.e., one of the most useful intuitions for knowledge graph completion) into consideration.
- **CapsE** [23] combines convolutional neural network with capsule network [29] for knowledge graph embedding.

4.3 Link Prediction

Link prediction task is to complete a triple (h, r, t) with h or t missing, i.e., to predict the missing h given (r, t) or the missing t given (h, r).

Evaluation Protocol. To evaluate the performance in link prediction, we follow the standard protocol used in [5]. For each test triple (h, r, t), we replace either h or t by each of other entities in $\mathcal{E}$ to create a set of corrupted triples, and calculate implausibility scores on the corrupted triples. Ranking these scores in ascending order, we can get the rank of the test triple. Notice that a corrupted triple may exist in train, validation or test set, we use the Filtered setting protocol [5] to eliminate its misleading effect, i.e., not taking any corrupted triples that appear in the knowledge graph into accounts. We employ two common evaluation metrics: mean rank (MR) and Hits@10. MR is the mean of the test triples' ranks. Hits@10 is the percentage of test triples that are ranked within top 10.

Implementation Details. Following the previous work [41], we use the common Bernoulli trick to generate the head or tail entities when sampling invalid triples. Like in ConvKB [22], we also use entity and relation embeddings produced by TransE to initialize entity and relation embeddings in PConvKB. We use the pre-trained 100-dimensional glove word embeddings [28] to train TransE model, and employ the TransE implementation provided by [25]. We select the learning rate in $\{5e^{-6}, 1e^{-5}, 5e^{-5}, 1e^{-4}\}$, the number of filters in $\{50, 100, 200, 400\}$. We fix the batch size at 128 and set the L_2-regularizer λ at 0.001 in our objective function. We run PConvKB up to 150 epochs and monitor the Hits@10 score after every 10 training epochs to choose optimal hyper-parameters. We obtain the highest Hits@10 scores on the validation set when learning rate at $5e^{-5}$, the number of filters at 400 on WN18; and learning rate at $1e^{-5}$, the number of filters at 50 on FB15k; and the learning rate at $5e^{-6}$, the number of filters at 400 on WN18RR; and the learning rate at $1e^{-5}$, the number of filters at 200 on FB15k-237. For comparison methods, we use the codes released by [11], [7] and [22].

Table 2. Experiments results on link prediction. Hits@10 is reported in %. The best score is in **bold**, while the second best score is in <u>underline</u>. For comparison methods, the values in black color are the results listed in the original publication, except ConvKB uses the [23] implemented version, which has been reported significantly better performance than the original one. The values in blue color are obtained by implementations from the OpenKE repository.

Model	WN18		FB15k		WN18RR		FB15k-237	
	MR	Hits@10	MR	Hits@10	MR	Hits@10	MR	Hits@10
TransE	–	–	125	47.1	–	–	–	–
	733	60.3	168	60.6	5766	32.4	349	42.2
TransH	388	82.3	87	64.4	–	–	–	–
	602	87.7	137	63.1	7019	36.3	358	40.0
TransD	212	92.2	91	77.3	–	–	–	–
	617	88.5	151	62.9	7050	36.8	418	40.2
HolE	–	94.9	–	73.9	–	–	–	–
	589	79.6	76	76.2	5992	38.5	369	43.3
ConvE	504	<u>95.5</u>	64	87.3	5277	48.0	<u>246</u>	49.1
	549	83.7	70	82.4	5385	46.9	286	48.4
CapsE	–	–	–	–	719	56.0	303	59.3
	244	93.9	72	88.3	<u>711</u>	56.2	301	<u>59.4</u>
ConvKB	–	–	–	–	763	56.7	254	53.2
	<u>204</u>	94.7	66	88.7	759	56.8	275	54.7
PConvKB (local)	212	95.3	<u>58</u>	<u>89.6</u>	733	<u>57.0</u>	267	57.5
PConvKB (global)	249	93.8	63	89.1	749	56.8	283	56.2
PConvKB	**196**	**96.3**	**54**	**91.4**	**691**	**57.4**	**245**	**59.8**

Results. Table 2 shows the link prediction results of our model and the comparison methods on the four benchmark datasets. From the results, we can observe that:

1. PConvKB obtains the best MR and highest Hits@10 scores on the four benchmark datasets, demonstrating the effectiveness of incorporating relation paths for knowledge graph embedding.
2. Among PConvKB, PConvKB (local) and PConvKB (global), PConvKB obtains the best performance, which indicates that considering relation paths locally and globally is beneficial for knowledge graph embedding.
3. PConvKB does better than the closely related model ConvKB on all experimental datasets, especially on FB15k where PConvKB gains significant improvements of $275 - 247 = 28$ in MR (which is about 10.1% relative improvement) and $59.8\% - 54.7\% = 5.1\%$ absolute improvement in Hits@10.

4.4 Triple Classification

Triple classification task is to determine whether a given triple (h, r, t) is correct or not, i.e., binary classification on a triple.

Evaluation Protocol. We follow the same protocol in [33]. For each triple in test set and validation set, we construct one negative triple by switching entities from test triples and validation triples, respectively. The triple classification decision rule is: for a triple (h, r, t), if its implausibility score is below the relation-specific threshold σ_r, predict positive, otherwise negative. The relation-specific threshold σ_r is determined by maximizing classification accuracy on the validation set. The triple classification accuracy is the percentage of triples in the test set that are classified correctly.

Implementation Details. We use TransE to initialize entity and relation embeddings in PConvKB, select the learning rate in $\{5e^{-6}, 1e^{-5}, 5e^{-5}, 1e^{-4}\}$, the number of filters in $\{50, 100, 200, 400\}$. We set the batch size at 128 and set the L_2-regularizer λ at 0.001 in our objective function. We run PConvKB up to 150 epochs and monitor the accuracy after every 10 training epochs to choose optimal hyper-parameters. We obtain the highest accuracy on the validation set when learning rate at $5e^{-5}$, the number of filters at 400 on WN18; and learning rate at $1e^{-5}$, the number of filters at 50 on FB15k; and the learning rate at $5e^{-6}$, the number of filters at 400 on WN18RR; and the learning rate at $1e^{-5}$, the number of filters at 200 on FB15k-237. For comparison methods, we implement them by the codes released by [11], [7] and [22].

Results. Table 3 shows the triple classification results of our model and the comparison methods on the four benchmark datasets. From the results, we can observe that:

Table 3. Experiments results on triple classification (%). The best score is in **bold**, while the second best score is in underline.

Model	WN18	FB15k	WN18RR	FB15k-237
TransE	87.6	82.9	74.0	75.6
TransH	96.5	85.7	77.0	77.0
TransD	96.4	86.1	76.3	77.0
HolE	88.1	82.6	71.4	70.3
ConvE	95.4	87.3	78.3	78.2
CapsE	96.5	88.4	79.6	79.5
ConvKB	96.4	87.9	79.1	80.1
PConvKB (local)	97.5	88.1	79.7	80.6
PConvKB (global)	96.9	87.6	79.4	80.9
PConvKB	**97.6**	**89.5**	**80.3**	**82.1**

1. On the whole, PConvKB yields the best performance on the four benchmark datasets, which is consistent with the results of link prediction, and further illustrates taking the relation paths into consideration is beneficial for knowledge graph embedding.
2. More specifically, on FB15k-237, the accuracy of triple classification improves from 80.6% of PConvKB(locally) to 82.1% PConvKB, and 80.9% of PConvKB (global) to 82.1% PConvKB. It demonstrates that considering the importances of relation paths locally and globally can better improve the knowledge graph embedding.

5 Related Work

Various methods have been proposed for knowledge graph embedding, such as general linear-based models [6], bilinear-based models [13,27,34], translation-based models [5,9,14,15,20,41,43], and neural network-based models [4,7,22, 23,31–33]. We refer to [24,39] for a recent survey. In this section, we focus on the most relevant neural network-based models, and briefly review the other related methods.

Socher et al. [33] introduce neural tensor networks for knowledge graph embedding, which allows mediated interaction of entity embeddings via a tensor. Schlichtkrull et al. [31] present relational graph convolutional networks for knowledge graph completion. Shi and Weninger [32] present a shared variable neural network model called ProjE, which fills-in missing facts in a knowledge graph by learning joint embeddings of entities and relations. Dettmers et al. [7] present a multi-layer convolutional network model, namely ConvE, which uses 2D convolutions over embeddings to predict missing links in knowledge graphs. Nguyen et al. [22] present a CNN-based embedding model, i.e., ConvKB. It applies CNN to explore the global relationships among same dimensional entries in each embedding triple, which generalizes the transitional characteristics in the transition-based embedding models. Nguyen et al. [23] present CapsE, which combines CNN with capsule networks [29] for knowledge graph embedding. All these models treat a knowledge graph as a collection of triples, and disregard the rich information exist in relation paths.

There are several translation-based models [12,19,37,42,44] incorporating relation paths to improve the embeddings of entities and relations. However, they fully rely on hand-designed features to measure the importance of each path, which is not differentiable and cannot adjust during training. Moreover, they all based on translation-based models, which are not suitable for CNN-based model. To the best of our knowledge, our model PConvKB is the first attempt which incorporates relation paths in CNN-based embedding model.

6 Conclusion

In this paper, we present a novel CNN-based embedding model PConvKB, which improves knowledge graph embedding by incorporating relation paths locally and globally. In particular, we introduce attention mechanism to measure the local importance of relation paths. Moreover, we propose a simple yet effective measure DIPF to compute the global importance of relation paths. We evaluate our model on link prediction and triple classification. Experimental results show that our model achieves substantial improvements against state-of-the-art methods.

Acknowledgments. We acknowledge anonymous reviewers for their valuable comments. This work was supported by the National Natural Science Foundation of China (Grant No. 61872045), the Foundation for Innovative Research Groups of the National Natural Science Foundation of China (Grant No. 61921003).

References

1. Auer, S., Bizer, C., Kobilarov, G., Lehmann, J., Cyganiak, R., Ives, Z.: DBpedia: a nucleus for a web of open data. In: Aberer, K., et al. (eds.) ASWC/ISWC -2007. LNCS, vol. 4825, pp. 722–735. Springer, Heidelberg (2007). https://doi.org/10.1007/978-3-540-76298-0_52
2. Bahdanau, D., Cho, K., Bengio, Y.: Neural machine translation by jointly learning to align and translate. In: 3rd International Conference on Learning Representations, ICLR 2015, Conference Track Proceedings, San Diego, CA, USA, 7–9 May 2015 (2015). http://arxiv.org/abs/1409.0473
3. Bollacker, K., Evans, C., Paritosh, P., Sturge, T., Taylor, J.: Freebase: a collaboratively created graph database for structuring human knowledge. In: Proceedings of the 2008 ACM SIGMOD International Conference on Management of Data, pp. 1247–1250. ACM (2008)
4. Bordes, A., Glorot, X., Weston, J., Bengio, Y.: A semantic matching energy function for learning with multi-relational data - application to word-sense disambiguation. Mach. Learn. **94**(2), 233–259 (2014). https://doi.org/10.1007/s10994-013-5363-6

5. Bordes, A., Usunier, N., Garcia-Duran, A., Weston, J., Yakhnenko, O.: Translating embeddings for modeling multi-relational data. In: Advances in Neural Information Processing Systems, pp. 2787–2795 (2013)
6. Bordes, A., Weston, J., Collobert, R., Bengio, Y.: Learning structured embeddings of knowledge bases. In: Proceedings of the Twenty-Fifth AAAI Conference on Artificial Intelligence, AAAI 2011, San Francisco, California, USA, 7–11 August 2011 (2011). http://www.aaai.org/ocs/index.php/AAAI/AAAI11/paper/view/3659
7. Dettmers, T., Minervini, P., Stenetorp, P., Riedel, S.: Convolutional 2D knowledge graph embeddings. In: Thirty-Second AAAI Conference on Artificial Intelligence (2018)
8. Dong, X., et al.: Knowledge vault: a web-scale approach to probabilistic knowledge fusion. In: Proceedings of the 20th ACM SIGKDD International Conference on Knowledge Discovery and Data Mining, pp. 601–610. ACM (2014)
9. Ebisu, T., Ichise, R.: TorusE: knowledge graph embedding on a lie group. In: Proceedings of the Thirty-Second AAAI Conference on Artificial Intelligence (AAAI-2018), the 30th Innovative Applications of Artificial Intelligence (IAAI-2018), and the 8th AAAI Symposium on Educational Advances in Artificial Intelligence (EAAI-2018), New Orleans, Louisiana, USA, 2–7 February 2018, pp. 1819–1826 (2018). https://www.aaai.org/ocs/index.php/AAAI/AAAI18/paper/view/16227
10. Ghosh, S., Desarkar, M.S.: Class specific TF-IDF boosting for short-text classification: application to short-texts generated during disasters. In: Companion of the The Web Conference 2018 on The Web Conference 2018, WWW 2018, Lyon, France, 23–27 April 2018, pp. 1629–1637 (2018). https://doi.org/10.1145/3184558.3191621
11. Han, X., et al.: OpenKE: an open toolkit for knowledge embedding. In: Proceedings of the 2018 Conference on Empirical Methods in Natural Language Processing, EMNLP 2018: System Demonstrations, Brussels, Belgium, 31 October–4 November 2018, pp. 139–144 (2018). https://aclanthology.info/papers/D18-2024/d18-2024
12. Huang, W., Li, G., Jin, Z.: Improved knowledge base completion by the path-augmented TransR model. In: Li, G., Ge, Y., Zhang, Z., Jin, Z., Blumenstein, M. (eds.) KSEM 2017. LNCS (LNAI), vol. 10412, pp. 149–159. Springer, Cham (2017). https://doi.org/10.1007/978-3-319-63558-3_13
13. Jenatton, R., Roux, N.L., Bordes, A., Obozinski, G.: A latent factor model for highly multi-relational data. In: Advances in Neural Information Processing Systems 25: 26th Annual Conference on Neural Information Processing Systems 2012. Proceedings of a Meeting Held at 3–6 December 2012, Lake Tahoe, Nevada, United States, pp. 3176–3184 (2012). http://papers.nips.cc/paper/4744-a-latent-factor-model-for-highly-multi-relational-data
14. Ji, G., He, S., Xu, L., Liu, K., Zhao, J.: Knowledge graph embedding via dynamic mapping matrix. In: Proceedings of the 53rd Annual Meeting of the Association for Computational Linguistics and the 7th International Joint Conference on Natural Language Processing (Volume 1: Long Papers), vol. 1, pp. 687–696 (2015)
15. Ji, G., Liu, K., He, S., Zhao, J.: Knowledge graph completion with adaptive sparse transfer matrix. In: Proceedings of the Thirtieth AAAI Conference on Artificial Intelligence, Phoenix, Arizona, USA, 12–17 February 2016, pp. 985–991 (2016). http://www.aaai.org/ocs/index.php/AAAI/AAAI16/paper/view/11982
16. Kim, D., Seo, D., Cho, S., Kang, P.: Multi-co-training for document classification using various document representations: TF-IDF, LDA, and Doc2Vec. Inf. Sci. **477**, 15–29 (2019). https://doi.org/10.1016/j.ins.2018.10.006

17. Kingma, D.P., Ba, J.: Adam: a method for stochastic optimization. In: 3rd International Conference on Learning Representations, ICLR 2015, Conference Track Proceedings, San Diego, CA, USA, 7–9 May 2015 (2015). http://arxiv.org/abs/1412.6980
18. Li, X., et al.: Beyond RNNs: positional self-attention with co-attention for video question answering. In: The Thirty-Third AAAI Conference on Artificial Intelligence, AAAI 2019, The Thirty-First Innovative Applications of Artificial Intelligence Conference, IAAI 2019, The Ninth AAAI Symposium on Educational Advances in Artificial Intelligence, EAAI 2019, Honolulu, Hawaii, USA, 27 January–1 February 2019, pp. 8658–8665 (2019). https://aaai.org/ojs/index.php/AAAI/article/view/4887
19. Lin, Y., Liu, Z., Luan, H., Sun, M., Rao, S., Liu, S.: Modeling relation paths for representation learning of knowledge bases. In: Proceedings of the 2015 Conference on Empirical Methods in Natural Language Processing, EMNLP 2015, Lisbon, Portugal, 17–21 September 2015, pp. 705–714 (2015). http://aclweb.org/anthology/D/D15/D15-1082.pdf
20. Lin, Y., Liu, Z., Sun, M., Liu, Y., Zhu, X.: Learning entity and relation embeddings for knowledge graph completion. In: Twenty-Ninth AAAI Conference on Artificial Intelligence (2015)
21. Miller, G.A.: WordNet: a lexical database for English. Commun. ACM **38**(11), 39–41 (1995)
22. Nguyen, D.Q., Nguyen, T.D., Nguyen, D.Q., Phung, D.: A novel embedding model for knowledge base completion based on convolutional neural network. In: Proceedings of the 2018 Conference of the North American Chapter of the Association for Computational Linguistics: Human Language Technologies, Volume 2 (Short Papers), vol. 2, pp. 327–333 (2018)
23. Nguyen, D.Q., Vu, T., Nguyen, T.D., Nguyen, D.Q., Phung, D.: A capsule network-based embedding model for knowledge graph completion and search personalization. arXiv preprint arXiv:1808.04122 (2018)
24. Nguyen, D.Q.: An overview of embedding models of entities and relationships for knowledge base completion. CoRR abs/1703.08098 (2017). http://arxiv.org/abs/1703.08098
25. Nguyen, D.Q., Sirts, K., Qu, L., Johnson, M.: STransE: a novel embedding model of entities and relationships in knowledge bases. In: NAACL HLT 2016: The 2016 Conference of the North American Chapter of the Association for Computational Linguistics: Human Language Technologies, San Diego California, USA, 12–17 June 2016, pp. 460–466 (2016). https://www.aclweb.org/anthology/N16-1054/
26. Nickel, M., Rosasco, L., Poggio, T.: Holographic embeddings of knowledge graphs. In: Thirtieth AAAI Conference on Artificial Intelligence (2016)
27. Nickel, M., Tresp, V., Kriegel, H.: A three-way model for collective learning on multi-relational data. In: Proceedings of the 28th International Conference on Machine Learning, ICML 2011, Bellevue, Washington, USA, 28 June–2 July 2011, pp. 809–816 (2011). https://icml.cc/2011/papers/438_icmlpaper.pdf
28. Pennington, J., Socher, R., Manning, C.D.: GloVe: global vectors for word representation. In: Proceedings of the 2014 Conference on Empirical Methods in Natural Language Processing, EMNLP 2014, A Meeting of SIGDAT, a Special Interest Group of the ACL, Doha, Qatar, 25–29 October 2014, pp. 1532–1543 (2014). https://www.aclweb.org/anthology/D14-1162/

29. Sabour, S., Frosst, N., Hinton, G.E.: Dynamic routing between capsules. In: Advances in Neural Information Processing Systems 30: Annual Conference on Neural Information Processing Systems 2017, Long Beach, CA, USA, 4–9 December 2017, pp. 3856–3866 (2017). http://papers.nips.cc/paper/6975-dynamic-routing-between-capsules
30. Saeedan, F., Weber, N., Goesele, M., Roth, S.: Detail-preserving pooling in deep networks. In: 2018 IEEE Conference on Computer Vision and Pattern Recognition, CVPR 2018, Salt Lake City, UT, USA, 18–22 June 2018, pp. 9108–9116 (2018). http://openaccess.thecvf.com/content_cvpr_2018/html/Saeedan_Detail-Preserving_Pooling_in_CVPR_2018_paper.html
31. Schlichtkrull, M., Kipf, T.N., Bloem, P., van den Berg, R., Titov, I., Welling, M.: Modeling relational data with graph convolutional networks. In: Gangemi, A., et al. (eds.) ESWC 2018. LNCS, vol. 10843, pp. 593–607. Springer, Cham (2018). https://doi.org/10.1007/978-3-319-93417-4_38
32. Shi, B., Weninger, T.: ProjE: embedding projection for knowledge graph completion. In: Proceedings of the Thirty-First AAAI Conference on Artificial Intelligence, San Francisco, California, USA, 4–9 February 2017, pp. 1236–1242 (2017). http://aaai.org/ocs/index.php/AAAI/AAAI17/paper/view/14279
33. Socher, R., Chen, D., Manning, C.D., Ng, A.Y.: Reasoning with neural tensor networks for knowledge base completion. In: Advances in Neural Information Processing Systems 26: 27th Annual Conference on Neural Information Processing Systems 2013. Proceedings of a Meeting Held at 5–8 December 2013, Lake Tahoe, Nevada, United States, pp. 926–934 (2013). http://papers.nips.cc/paper/5028-reasoning-with-neural-tensor-networks-for-knowledge-base-completion
34. Sutskever, I., Salakhutdinov, R., Tenenbaum, J.B.: Modelling relational data using Bayesian clustered tensor factorization. In: Advances in Neural Information Processing Systems 22: 23rd Annual Conference on Neural Information Processing Systems 2009. Proceedings of a Meeting Held at 7–10 December 2009, Vancouver, British Columbia, Canada, pp. 1821–1828 (2009). http://papers.nips.cc/paper/3863-modelling-relational-data-using-bayesian-clustered-tensor-factorization
35. Tong, Z., Tanaka, G.: Hybrid pooling for enhancement of generalization ability in deep convolutional neural networks. Neurocomputing **333**, 76–85 (2019). https://doi.org/10.1016/j.neucom.2018.12.036
36. Toutanova, K., Chen, D.: Observed versus latent features for knowledge base and text inference. In: Proceedings of the 3rd Workshop on Continuous Vector Space Models and their Compositionality, pp. 57–66 (2015)
37. Toutanova, K., Lin, V., Yih, W., Poon, H., Quirk, C.: Compositional learning of embeddings for relation paths in knowledge base and text. In: Proceedings of the 54th Annual Meeting of the Association for Computational Linguistics, ACL 2016, Volume 1: Long Papers, Berlin, Germany, 7–12 August 2016 (2016). http://aclweb.org/anthology/P/P16/P16-1136.pdf
38. Vrandečić, D., Krötzsch, M.: Wikidata: a free collaborative knowledge base. Commun. ACM **57**, 78–85 (2014)
39. Wang, Q., Mao, Z., Wang, B., Guo, L.: Knowledge graph embedding: a survey of approaches and applications. IEEE Trans. Knowl. Data Eng. **29**(12), 2724–2743 (2017). https://doi.org/10.1109/TKDE.2017.2754499

40. Wang, W., Chen, Z., Hu, H.: Hierarchical attention network for image captioning. In: The Thirty-Third AAAI Conference on Artificial Intelligence, AAAI 2019, The Thirty-First Innovative Applications of Artificial Intelligence Conference, IAAI 2019, The Ninth AAAI Symposium on Educational Advances in Artificial Intelligence, EAAI 2019, Honolulu, Hawaii, USA, 27 January–1 February 2019, pp. 8957–8964 (2019). https://aaai.org/ojs/index.php/AAAI/article/view/4924
41. Wang, Z., Zhang, J., Feng, J., Chen, Z.: Knowledge graph embedding by translating on hyperplanes. In: Twenty-Eighth AAAI Conference on Artificial Intelligence (2014)
42. Xiong, S., Huang, W., Duan, P.: Knowledge graph embedding via relation paths and dynamic mapping matrix. In: Woo, C., Lu, J., Li, Z., Ling, T.W., Li, G., Lee, M.L. (eds.) ER 2018. LNCS, vol. 11158, pp. 106–118. Springer, Cham (2018). https://doi.org/10.1007/978-3-030-01391-2_18
43. Yuan, J., Gao, N., Xiang, J.: TransGate: knowledge graph embedding with shared gate structure. In: The Thirty-Third AAAI Conference on Artificial Intelligence, AAAI 2019, The Thirty-First Innovative Applications of Artificial Intelligence Conference, IAAI 2019, The Ninth AAAI Symposium on Educational Advances in Artificial Intelligence, EAAI 2019, Honolulu, Hawaii, USA, 27 January–1 February 2019, pp. 3100–3107 (2019). https://aaai.org/ojs/index.php/AAAI/article/view/4169
44. Zhang, M., Wang, Q., Xu, W., Li, W., Sun, S.: Discriminative path-based knowledge graph embedding for precise link prediction. In: Pasi, G., Piwowarski, B., Azzopardi, L., Hanbury, A. (eds.) ECIR 2018. LNCS, vol. 10772, pp. 276–288. Springer, Cham (2018). https://doi.org/10.1007/978-3-319-76941-7_21

ReadNet: A Hierarchical Transformer Framework for Web Article Readability Analysis

Changping Meng[1(✉)], Muhao Chen[2], Jie Mao[3], and Jennifer Neville[1]

[1] Department of Computer Science, Purdue University, West Lafayette, USA
{meng40,neville}@purdue.edu
[2] Department of Computer Science, University of California, Los Angeles, USA
muhaochen@ucla.edu
[3] Google Inc., Mountain View, USA
mjmjmtl@gmail.com

Abstract. Analyzing the *readability* of articles has been an important sociolinguistic task. Addressing this task is necessary to the automatic recommendation of appropriate articles to readers with different comprehension abilities, and it further benefits education systems, web information systems, and digital libraries. Current methods for assessing readability employ empirical measures or statistical learning techniques that are limited by their ability to characterize complex patterns such as article structures and semantic meanings of sentences. In this paper, we propose a new and comprehensive framework which uses a hierarchical self-attention model to analyze document readability. In this model, measurements of sentence-level difficulty are captured along with the semantic meanings of each sentence. Additionally, the sentence-level features are incorporated to characterize the overall readability of an article with consideration of article structures. We evaluate our proposed approach on three widely-used benchmark datasets against several strong baseline approaches. Experimental results show that our proposed method achieves the state-of-the-art performance on estimating the readability for various web articles and literature.

1 Introduction

Readability is an important linguistic measurement that indicates how easily readers can comprehend a particular document. Due to the explosion of web and digital information, there are often hundreds of articles describing the same topic, but vary in levels of readability. This can make it challenging for users to find the articles online that better suit their comprehension abilities. Therefore, an automated approach to assessing readability is a critical component

C. Meng—This work was done during the summer internships of CM and MC at Google, Mountain View. We thank the anonymous reviewers for their insightful comments.

J. M. Jose et al. (Eds.): ECIR 2020, LNCS 12035, pp. 33–49, 2020.
https://doi.org/10.1007/978-3-030-45439-5_3

for the development of recommendation strategies for web information systems, including digital libraries and web encyclopedias.

Text readability is defined as the overall effect of language usage and composition on readers' ability to easily and quickly comprehend the document [14]. In this work, we focus on evaluating document difficulty based on the composition of words and sentences. Consider the following two descriptions of the concept *rainbow* as an example.

1. **A more rigid scientific definition from *English Wikipedia*:** A rainbow is a meteorological phenomenon that is caused by reflection, refraction and dispersion of light in water droplets resulting in a spectrum of light appearing in the sky.
2. **A more generic description from the *Simple English Wikipedia*:** A rainbow is an arc of color in the sky that can be seen when the sun shines through falling rain. The pattern of colors starts with red on the outside and changes through orange, yellow, green, blue, to violet on the inside.

Clearly, the first description provides more rigidly expressed contents, but is more sophisticated due to complicated sentence structures and the use of professional words. In contrast, the second description is simpler, with respect to both grammatical and document structures. From the reader's perspective, the first definition is more appropriate for technically sophisticated audiences, while the second one is suitable for general audiences, such as parents who want to explain rainbows to their young children.

The goal of *Readability Analysis* is to provide a rating regarding the difficulty of an article for average readers. As the above example illustrates that, many approaches for automatically judging the difficulty of the articles are rooted in two factors: the difficulty of the words or phrases, and the complexity of syntax [11]. To characterize these factors, existing works [3,29] mainly rely on some explicit features such as *Average Syllables Per Word*, *Average Words Per Sentence*, etc. For example, the Flesch-Kincaid index is a representative empirical measure defined as a linear combination of these factors [4]. Some later approaches mainly focus on proposing new features with the latest CohMetrix 3.0 [36] providing 108 features, and they combine and use the features using either linear functions or statistical models such as Support Vector Machines or multilayer perceptron [12,40,41,43,51]. While these approaches have shown some merits, they also lead to several drawbacks. Specifically (1) they do not consider sequential and structural information, and (2) they do not capture sentences-level or document-level semantics that are latent but essential to the task [11].

To address these issues, we propose ReadNet, a comprehensive readability classification framework that uses a hierarchical transformer network. The self-attention portion of the transformer encoder is better able to model long-range and global dependencies among words. The hierarchical structure can capture how words form sentences, and how sentences form documents, meanwhile reduce the model complexity exponentially. Moreover, explicit features indicating the readability of different granularities of text can be leveraged and aggregated from multiple levels of the model. We compare our proposed model to a number of widely-adopted document encoding techniques, as well as traditional readability analysis approaches based on explicit features. Experimental results on three benchmark datasets show that our work properly identifies the document representation techniques, and achieves the state-of-the-art performance by significantly outperform previous approaches.

2 Related Work

Existing computational methods for readability analysis [3,11,29,40,53] mainly use empirical measures on the symbolic aspects of the text, while ignoring the sequence of words and the structure of the article. The Flesch-Kincaid index [28] and related variations use a linear combination of explicit features.

Although models based on these traditional features are helpful to the quantification of readability for small and domain-specific groups of articles, they are far from generally applicable for a larger body of web articles [10,17,45]. Because those features or formulas generated from a small number of training text specifically selected by domain experts, they are far from generally representing the readability of large collections of corpora. Recent machine learning methods on readability evaluation are generally in the primitive stage. [18] proposes to combine language models and logistic regression. The existing way to integrate features is through a statistical learning method such as SVM [12,20,40,41,43,51]. These approaches ignore the sequential or structural information on how sentences construct articles. Efforts have also been made to select optimal features from current hundreds of features [15]. Some computational linguistic methods have been developed to extract higher-level language features. The widely-adopted Coh-Metrix [22,37] provides multiple features based on cohesion such as referential cohesion and deep cohesion.

Plenty of works have been conducted on utilizing neural models for sentimental or topical document classification or ranking, while few have paid attention to the readability analysis task. The convolutional neural network (CNN) [27] is often adopted in sentence-level classification which leverages local semantic features of sentence composition that are provided by word representation approaches. In another line of approaches, a recursive neural network [46] is adopted, which focuses on modeling the sequence of words or sentences. Hierarchical structures of such encoding techniques are proposed to capture structural information of articles, and have been widely used in tasks of document classification [7,32,48], and sequence generation [30] and sub-article matching [6]. Hierarchical attention network [52] is the current state-of-the-art method for document classification, which employs attention mechanisms on both word and sentence levels to capture the uneven contribution of different words and sentences to the overall meaning of the document. The Transformer model [50] uses multi-head self-attention to perform sequence-to-sequence translation. Self-attention is also adopted in text summarization, entailment and representation [31,38]. Unlike topic and sentiment-related document classification tasks that focus on leveraging portions of lexemes that are significant to the overall meanings and sentiment of the document, readability analysis requires the aggregation of difficulty through all sentence components. Besides, precisely capturing the readability of documents requires the model to incorporate comprehensive readability-aware features, including difficulty, sequence and structure information, to the corresponding learning framework.

3 Preliminary

In this section, we present the problem definition, as well as some representative explicit features that are empirically adopted for the readability analysis task.

3.1 Problem Definition

The readability analysis problem is defined as an ordinal regression problem for articles. Given an article with up to n sentences and each sentence with up to m words, an article can be represented as a matrix $\boldsymbol{A}$ whose i-th row $\boldsymbol{A}_{i,:}$ corresponds to the i-th sentences, and $A_{i,j}$ denotes the j-th word of the i-th sentence. Given an article $\boldsymbol{A}$, a label will be provided to indicate the readability of this article.

We consider the examples introduced in Sect. 1, where two articles describe the same term "*rainbow*". The first rigorous scientific article can be classified as "difficult", and the second general description article can be classified as "easy".

Instead of classifying articles into binary labels like "easy" or "difficult", more fine-grained labels can help people better understand the levels of readability. For instance, we can map the articles in standardization systems of English tests such as 5-level Cambridge English Exam (CEE), where articles from professional level English exam (CPE) are regarded than those from introductory English exam (KET).

3.2 Explicit Features

Previous works [11,21,22,24,25,28,34] have proposed empirical features to evaluate readability. Correspondingly, we divide these features into sentence-level features and document-level features. Sentence-level features seek to evaluate the difficulty of sentences. For instance, the sentence-level feature "number of words" for sentences can be averaged into "number of words per sentence" to evaluate the difficulty of documents. Document-level features include the traditional readability indices and cohesion's proposed by Coh-Metrix [22]. These features are listed in Table 1.

Current approaches [12,41,43] average the sentence-level features of each sentence to construct document level features. Furthermore, these features are concatenated with document-level features, and use an SVM to learn on these features. The limitation lies in failing to capture the structure information of sentences and documents. For instance, in order to get the sentence level features for the document, it averages all these features of each sentence. It ignores

Table 1. Explicit features

Name	Description
Sentence-level features	
#characters_per_word	The average number of characters per word, which provides a character-level measure for the difficulty of words
#syllabi_per_word	The average number of syllabi per word, which measures the difficulty of words from the syllabus level
#words	The number of words that measures the verbosity of the sentence
#long_words	The number of words longer than 6 characters in a sentence
#difficult_words	The number of difficult word in a sentence. Difficult word is a word not listed in the 3000 words for fourth-grade American students
#pronoun	The number of pronoun in a sentence
Document-level features	
Flesch Reading Ease [28]	The United States Military Standard of readability scoring for technical manuals, which is calculated as $206.835 - 1.015 \times \frac{\#\text{words}}{\#\text{sentences}} - 84.6 \times \frac{\#\text{syllables}}{\#\text{words}}$
Flesch—Kincaid grade level [28]	An empirical readability metric which maps to a U.S. school grade level, calculated as $0.39 \times \frac{\#\text{words}}{\#\text{sentences}} + 11.8 \times \frac{\#\text{syllables}}{\#\text{words}} - 15.59$
Automated Readability Index [44]	A metric that also produces an approximate representation of the US grade level needed to comprehend the text, calculated as $4.71 \times \frac{\#\text{characters}}{\#\text{words}} + 0.5 \times \frac{\#\text{words}}{\#\text{sentences}} - 21.43$. Instead of considering syllables, this metric more generally characterizes on the character level
Coleman-Liau Index [9]	An index used to gauge the understandability of a text from the character-level: $0.0588 \times \frac{\#\text{letters}}{\#\text{words} \times 100} + 0.296 \times \frac{\#\text{sentences}}{\#\text{words}} \times 100$
Gunning Fog Index [23]	$0.4 \times (\frac{\#\text{words}}{\#\text{sentences}} + 100 \times \frac{\#\text{complex\_words}}{\#\text{words}})$; It estimates the years of formal education a person needs to understand the text on the first reading
LIX [2]	A measure indicating the difficulty of reading a text based on the proportions of long words and verbosity of sentences: $\frac{\text{word longer than 6 letters }\#}{\#\text{words}} + \frac{\#\text{words}}{\#\text{sentences}}$
RIX [1]	A metric based on the proportion of long words in text, $\frac{\#\text{ long words}}{\#\text{sentences}}$
SMOG Index [35]	A measure of readability that seeks to estimate the years of education needed to understand a piece of writing: $1.0430 \times \sqrt{\#\text{ of polysyllables} \times \frac{30}{\#\text{sentences}}} + 3.1291$
Dale Chall Index [19]	$0.1579 \times \frac{\#\text{difficult\_words}}{\#\text{words}} \times 100 + 0.0496 \times \frac{\#\text{words}}{\#\text{sentences}}$. Difficult word is a word not listed in the 3000 words for fourth-grade American students.
Incidence of connectives [33]	5 numerical features indicate additive, logic, temporal, causal and negative connectives
Logic operator connectivity [13]	Logical connectives between logical particles such as "and", "if" proposed by Coh-Metrix
Lexical diversity	The character-level density of the lexicon: $\frac{\#\text{unique\_words}}{\#\text{words}}$
Content diversity	$\frac{\#\text{content\_words}}{\#\text{words}}$. It measures the diversity of content. Content words are adjectives, nouns, verbs and adverbs
Incidence of part-of-speech elements	Incidence of word categories (adjectives, nouns, verbs, adverbs, pronouns) per 1000 words in the text

how these sentences construct an article and which parts of the document more significantly decides the readability of the document. While cohesion features provided by Coh-Metrix tries to captures relationships between sentences, these features mainly depend on the repeat of words across multiple sentences. They did not directly model how these sentences construct a document in perspectives of structure and sequence.

Briefly speaking, existing works are mainly contributing more features as shown in Table 1. But the current models used to aggregate these features are based on SVM and linear models. In this work, we target to propose a more advanced model to better combine these features with document information.

4 Hierarchical Transformer for Readability Analysis

In order to address the limitations of traditional approaches, we propose ReadNet: the Hierarchical Transformer model for readability analysis as shown in Fig. 1.

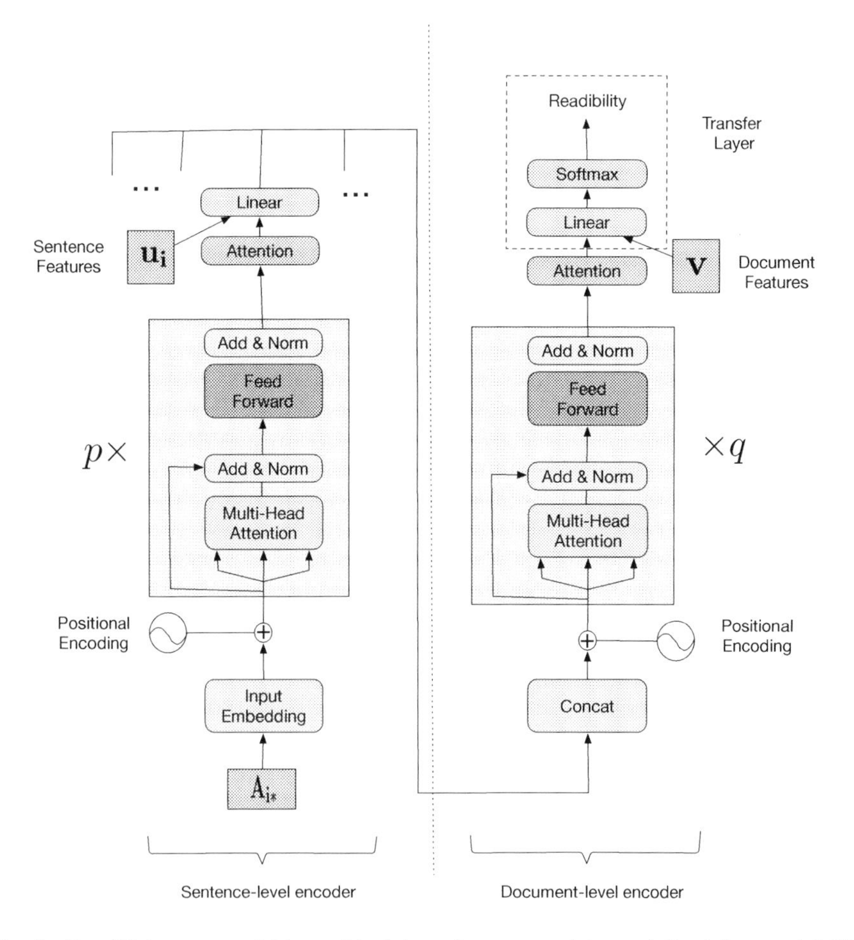

Fig. 1. ReadNet: proposed hierarchical transformer model specialized for readability analysis

The proposed model incorporates the explicit features with a hierarchical document encoder that encodes the sequence and structural information of an article. The first level of the hierarchical learning architecture models the formation of sentences from words. The second level models the formation of the article from sentences. The self-attention encoder (to be described in Subsect. 4.1) is adapted from the vanilla Transformer encoder [50]. The hierarchical structure, attention aggregation layer, combination with explicit features and transfer layer are specially designed for this readability analysis task.

4.1 From Words to Sentences

In this subsection, we introduce the encoding process of sentences in hierarchical mutli-head self-attention. The encoding process has three steps: *(1)* the self-attention encoder transforms the input sequence into a series of latent vectors; *(2)* the attention layer aggregates the encoded sequential information based on the induced significance of input units; *(3)* The encoded information is combined with the explicit features.

Transformer Self-attention Encoder. This encoder is adapted from the vanilla Transformer encoder [50]. The input for this encoder is $\boldsymbol{A}_{i,:}$, which represents the i-th sentence.

The Embedding layer encodes each word $A_{i,j}$ into a d-dimensional vector based on word embedding. The output is a $m \times d$-dimensional matrix $\boldsymbol{B}$ where d is the embedding dimension and m is the number of words.

The position encoding layer indicates the relative position of each word $A_{i,j}$. The elements of positional embedding matrix $\boldsymbol{P}$ where values in the i-th row j-th column is defined as follows.

$$P_{i,j} = \begin{cases} \sin(i/10^{4j/d}) & j \text{ is even} \\ \cos(i/10^{4(j-1)/d}) & j \text{ is odd} \end{cases} \tag{1}$$

The embedded matrix $\boldsymbol{B}$ and positional embedding matrix $\boldsymbol{P}$ are added into the initial hidden state matrix $\boldsymbol{H}^{(0)} = \boldsymbol{B} + \boldsymbol{P}$. $\boldsymbol{H}^{(0)}$ will go through a stack of p identical layers. Each layer contains two parts: (i) the Multi-Head Attention donated as function f_{MHA} defined in Eq. 2, and (ii) the Position-wise Feed-Forward f_{FFN} defined in Eq. 4. Layer normalization is used to avoid gradient vanishing or explosion.

Multi-head Self-Attention function (f_{MHA}) [50] encodes the relationship among query matrix $\boldsymbol{Q}$, key matrix $\boldsymbol{K}$ and value matrix $\boldsymbol{V}$ from different representation subspaces at different positions. $d_k = d/h$. $\boldsymbol{W}$ is a $d \times d$ weight matrix. $\oplus$ denotes concatenation. $\boldsymbol{W}_{Ki}, \boldsymbol{W}_{Vi}, \boldsymbol{W}_{Qi}$ are $d \times d_k$ weight matrix for head function g_i.

$$f_{MHA}(\boldsymbol{Q}, \boldsymbol{K}, \boldsymbol{V}) = (g_1(\boldsymbol{Q}, \boldsymbol{K}, \boldsymbol{V})) \oplus \ldots \oplus g_h(\boldsymbol{Q}, \boldsymbol{K}, \boldsymbol{V}))\boldsymbol{W} \tag{2}$$

$$g_i(\boldsymbol{Q}, \boldsymbol{K}, \boldsymbol{V}) = \text{softmax}(\frac{\boldsymbol{Q}\boldsymbol{W}_{Qi}(\boldsymbol{K}\boldsymbol{W}_{Ki})^T}{\sqrt{d_k}})(\boldsymbol{V}\boldsymbol{W}_{Vi}) \tag{3}$$

Position-wise Feed-Forward Function f_{FFN} [50] adopts two 1-Dimensional convolution layers with kernel size 1 to encode input matrix $\boldsymbol{X}$.

$$f_{FFN}(\boldsymbol{X}) = \text{Conv1D}(\text{ReLU}(\text{Conv1D}(\boldsymbol{X}))) \tag{4}$$

For the l-th encoder layer, $\boldsymbol{H}^{(l)}$ is encoded into $\boldsymbol{H}^{(l+1)}$ according to Eq. 5

$$\boldsymbol{H}^{(l+1)} = f_{FFN}(f_{MHA}(\boldsymbol{H}^{(l)}, \boldsymbol{H}^{(l)}, \boldsymbol{H}^{(l)})) \tag{5}$$

Attention Aggregation Layer. After p transformer encoder layers, each sentence $\boldsymbol{A}_{i,:}$ is encoded into a $m \times d$-dimensional matrix $\boldsymbol{H}^{(p)}$.

We first pass $\boldsymbol{H}^{(p)}$ through a feed forward layer with $d \times d$ dimensional weights $\boldsymbol{W}_1$ and bias term b_1 to obtain a hidden representation as $\boldsymbol{U}$:

$$\boldsymbol{U} = \tanh(\boldsymbol{H}^{(p)}\boldsymbol{W}_1 + b_1),$$

then compute the similarity between $\boldsymbol{U}$ and the trainable $d \times 1$ dimensional context matrix $\boldsymbol{C}$ via

$$\boldsymbol{w} = \text{softmax}(\boldsymbol{U}\boldsymbol{C}),$$

which we use as importance weights to obtain the final embedding of the sentence $\boldsymbol{A}_{i,:}$:

$$\boldsymbol{h}_i = \sum_{byRow} \boldsymbol{H}^{(p)} \cdot \boldsymbol{w} \tag{6}$$

Combination of Explicit Features. The sentence level features $\boldsymbol{u}_i$ introduced in Sect. 3.2 Table 1 for i-th sentence are concatenated by $\boldsymbol{h}_i^* = \boldsymbol{h}_i \oplus \boldsymbol{u}_i$.

4.2 From Sentences to Articles

The second level of the hierarchical learning architecture is on top of the first layer. n encoded vector $\boldsymbol{h}_i^* (1 \leq i \leq n)$ are concatenated as the input for this layer. The structure of second level is the same as the first level. The output of this level is a vector $\boldsymbol{y}$ as the overall embedding of this article.

4.3 Transfer Layer

The goal of the transfer layer is to improve prediction quality on a target task where training data are scarce, while a large amount of other training data are available for a set of related tasks.

The readability analysis problem suffers from the lack of labeled data. Traditional benchmark datasets labeled by domain experts typically contain a small number of articles. For instance, CEE contains 800 articles and Weebit contains around 8 thousand articles. Such quantities of articles are far smaller than those for sentiment or topic-related document classification tasks which typically involve over ten thousand articles even for binary classification [7,27]. On the other hand, with the emerging of online encyclopedia applications such as Wikipedia, it provides a huge amount of training dataset. For instance, English Wikipedia and Simple-English Wikipedia contain more than 100 thousand articles which can be used to train a deep learning model.

One fully connected layer combines the article embedding vector $\boldsymbol{y}$ and document-level features $\boldsymbol{v}$ from Table 1 to output the readability label vector $\boldsymbol{r}$ after a Softmax function. $\boldsymbol{W}_t$ is the weight of the fully connected layer. For dataset with m categories of readability ratings, each document is embedded into $\boldsymbol{r}$ with $m-1$ dimensions.

$$\boldsymbol{r} = \text{softmax}(\boldsymbol{W}_t(\boldsymbol{y} \oplus \boldsymbol{v}))$$

If transfer learning is needed, instead of random initialization, this network is initialized with a pre-trained network based on a larger corpus. During the training process, update the transfer layer while keeping all other layers frozen. If transfer learning is not needed, all layers are updated during the training process.

4.4 Learning Objective

Given dataset with m categories of readability ratings, the goal is to minimize ordinal regression loss [42] defined as Eq. 7. $\boldsymbol{r}_k$ represents the k-th dimension of the $\boldsymbol{r}$ vector. y is the true label. The threshold parameter $\theta_1, \theta_2, \dots \theta_{m-1}$ are also learned automatically from the data.

$$L(\boldsymbol{r}; y) = -\sum_{k=1}^{m-1} f(s(k;y)(\theta_k - \boldsymbol{r}_k)), \quad where \quad s(k;y) = \begin{cases} -1 & k < y \\ +1 & k \geq y \end{cases} \tag{7}$$

Here, the objective of learning the readability analysis model is essentially different from that of a regular document classification model, since the classes here do form a partial-order. However, the case of two classes degenerates the learning to the same as that of a binary classifier.

4.5 Why Hierarchical Self-attention

For self-attention, the path length in the computation graph between long-range dependencies in the network is $O(1)$ instead of $O(n)$ for recurrent models such as LSTM. Shorter path length in the computation graph makes it easier to learn the interactions between any elements in the sequence. For readability analysis, modeling the overall interaction between words is more important than modeling the consequent words. For semantic understanding, the consequence of two words such as "very good" and "not good" make distinct semantic meanings. While for readability analysis, it does not make difference in difficulty to understand it. The overall evaluation of the words difficulties in the sentences matters.

The hierarchical learning structure benefits in two ways. First, it mimics human reading behaviors, since the sentence is a reasonable unit for people to read, process and understand. People rarely check the interactions between arbitrary words across different sentences in order to understand the article. Second, the hierarchical structure can reduce parameter complexity. For a document with n sentences, m words per sentence, d dimension per word, the parameter complexity of the model is $O((nm)^2 d)$ for single level structure. While for the hierarchical structure, the parameter complexity is $O(m^2 d + n^2 d)$.

5 Experiments

In this section, we present the experimental evaluation of the proposed approach. We first introduce the datasets used for the experiments, followed by the comparison of the proposed approach and baselines based on held-out evaluation, as well as detailed ablation analysis of different techniques enabled by our approach.

5.1 Datasets

We use the following three datasets in our experiment. Table 2 reports the statistics of the three datasets including the average number of sentences per article n_{sent} and the average number of words per sentence n_{word}.

Wiki dataset [26] contains *English Wikipedia* and *Simple English Wikipedia*. Simple English Wikipedia thereof is a simplified version of English Wikipedia which only uses simple English words and grammars. This dataset contains 59,775 English Wikipedia articles and 59,775 corresponding Simple English Wikipedia articles.

Cambridge English Exam (CEE) [51] categorizes articles based on the criteria of five Cambridge English Exam level (KET, PET, FCE, CAE, CPE). The five ratings are sequentially from the easiest KET to the hardest CPE. In total, it contains 110 KET articles, 107 PET articles, 153 FCE articles, 263 CAE articles and 155 CPE articles. Even though this dataset designed for non-native speakers may differ from materials for native English speakers, the difficulty between five levels is still comparable. We test our model on this dataset in order to check whether our model can effectively evaluate the difficulty of English articles according to an existing standard.

Weebit [49] is one of the largest dataset for readability analysis. It contains 7,676 articles targeted at different age group readers from Weekly Reader magazine and BBC-Bitesize website. Weekly Reader magazine categorizes articles according to the ages of targeted readers in 7–8, 8–9 and 9–10 years old. BBC-Bitesize has two levels for age 11–14 and 15–16. The targeted age is used to evaluate readability levels.

Table 2. Statistics of datasets Wiki, Cambridge English Exam and Weebit

Datasets	Wiki		Cambridge English Exam					WeeBit				
	En	Simple En	KET	PET	FCE	CAE	CPE	WR 2	WR 3	WR 4	KS3	GCSE
n_{sent}	37.46	7.74	6.30	8.80	16.47	10.63	16.69	23.41	23.28	28.12	22.71	27.85
n_{word}	17.03	14.41	9.40	16.63	17.96	16.39	23.47	12.56	13.48	16.29	20.04	18.62

Table 3. Cross-validation classification accuracy and standard deviation (in parentheses) on Wikipedia (Wiki), Cambridge English Exam (CEE) and Weebit dataset. We report accuracy on three groups of models: (1) Statistical classification algorithms including multi-class logistic regression, Linear SVM and Multilayer Perceptron (MLP); (2) Three types of document classifier CNN, hierarchical GRNN using LSTM cells (LSTM), Hierarchical Attention Network (HATT); (3) Hierarchical Attention Network combined with explicit features (HATT+), and our proposed approach which combines explicit features and semantics with Hierarchical Self-Attention (ReadNet). Transfer learning is not used, and all parameters in the model are initialized randomly (transfer learning is evaluated separately in Table 5).

Accuracy	Explicit features			Semantic features			Explicit+semantic	
	Logistic	SVM	MLP	CNN	LSTM	HATT	HATT+	ReadNet
Wiki	0.822 (±0.006)	0.848 (±0.008)	0.819 (±0.007)	0.583 (±0.035)	0.849 (±0.007)	0.877 (±0.007)	0.898 (±0.007)	**0.912 (±0.006)**
CEE	0.462 (±0.027)	0.492 (±0.041)	0.475 (±0.044)	0.277 (±0.031)	0.473 (±0.047)	0.512 (±0.043)	0.513 (±0.041)	0.528 (±0.045)
Weebit	0.724 (±0.007)	0.846 (±0.006)	0.845 (±0.006)	0.635 (±0.043)	0.886 (±0.005)	0.884 (±0.007)	0.902 (±0.006)	**0.917 (±0.006)**

5.2 Evaluation

In this subsection, we provide a detailed evaluation of the proposed approach.

Baseline Approaches. We compare our proposed approach (denoted ReadNet) against the following baseline methods.

- Statistical classification algorithms based on explicit features: this category of baselines including the statistical classification algorithms that are widely adopted in a line of previous works [12,20,40,41,43,51], such as multi-class Logistic Regression, the Linear SVM, and the Multilayer Perceptron (MLP) [49]. Explicit features on which these models are trained have been introduced in Sect. 3.2. Since this work targets at proposing a more advanced model to utilize features instead of proposing new features, all these features from Table 1 are used.
- Neural document classifiers: this category of baselines represents the other line of previous works that adopt variants of neural document models for sentence or document classification. Corresponding approaches including the Convolutional Neural Networks (CNN) [27], the Hierarchical Gated Neural Network with Long Short-term Memory (LSTM) [48], and the Hierarchical Attention Network (HATT) [52].

Table 4. Average readability scores of 10 randomly selected articles in Cambridge English Test predicted by our model trained using Wikipedia. PET, KET, FCE, CPE and CAE have increasing difficulty levels according to Cambridge English. The scores are the confidence scores of classified as regular English Wikipedia instead of simple English Wikipedia.

	KET	PET	FCE	CAE	CPE
Scores	0.381 ± 0.078	0.544 ± 0.092	0.620 ± 0.054	0.671 ± 0.085	0.837 ± 0.071

- The Hierarchical Attention Network combined with explicit features (HATT+), for which we use the same mechanism as our proposed approach to incorporate the explicit features into the representation of each sentence by the attentive RNN.

Model Configurations. For article encoding, we limit the number of sentences of each article to up to 50, zero-pad short ones and truncate over-length ones. According to the data statistics in Table 2, 50 sentences are enough to capture the majority of information of articles in the datasets. For each sentence, we also normalize the number of words to be fed into the model as 50, also via zero-padding and truncating. We fix the batch size to 32, and use Adam [16] as the optimizer with a learning rate 0.001. The epochs of training for the neural models are limited to 300. We set the number of encoder layers p and q to 6. The embedding dimension $d = 100$. Number of heads h in f_{MHA} is 3. CNN adopts the same configuration as [27]. Other statistical classification algorithms are trained until converge. Source code will be available in the final version.

Evaluation Protocol. We formalize the task as a classification task following previous works on the three benchmark datasets. In order to provide a valid quantitative evaluation, we have to follow the existing evaluation method to show the advantage of our proposed model compared with the baselines. We adopt 5-fold cross-validation to evaluate the proposed model and baselines. We report the classification accuracy that is aggregated on all folds of validation.

Results. The results are reported in Table 3. Traditional explicit features can provide satisfying results. Since the multi-class logistic regression, SVM and MLP models can combine the features *number of words per sentence* and *number of syllabi per word* which are included in Flesch-Kincaid score, they provide the reasonable result. CNN is only slightly better than random guess. We assume that this is because CNN does not capture the sequential and structural information of documents. The HATT approach provides the best among models without explicit features. The reasons root in the structure of the model which is able to capture length and structural information of the article. Since it also adopted a hierarchical structure, the conciseness of each sentence and that of the overall article structure is captured, which appears to be significant to the task. The explicit features further improve the results of HATT as shown by HATT+. Even without explicit features, our proposed approach is better than

HATT+. HATT has appeared to be successful at highlighting some lexemes and sentence components that are significant to the overall meanings or sentiment of a document. However, unlike topic and sentiment-related document classification tasks, readability does not rely on several consecutive lexemes, but the aggregation of all sentence components. The path length in the computation graph between arbitrary components dependencies in ReadNet is $O(1)$ instead of $O(n)$ for HATT. Shorter path length in the computation graph makes it easier to learn the interactions between any arbitrary words in sentence level, or sentences in document-level.

Compared with traditional approaches, the main advantage of the proposed approach is that it uses the document encoder to learn how words are connected into sentences and how sentences are connected into documents. Baseline approaches only use the averaged explicit features of all the sentences. For these datasets, several extremely difficult and complicated sentences usually determine the readability of a document. This useful information is averaged and weakened by the total number of sentences in baselines.

5.3 Analysis on Transfer Learning

As shown in Table 3, the standard deviation of the CEE task is large compared with those in Wiki and Weebit tasks since the quantity of CEE articles is not enough to train a complex deep learning model. Transfer layer in ReadNet is utilized in three steps. First is to train and save the model from larger datasets such as Wiki or Weebit. Then, we initialize the model for CEE task and load the parameter weights from the saved model except for the transfer layer. Eventually on the target task, the transfer layer is trained while keeping all other layers fixed. As shown in Table 5, loading a pre-trained model based on Weebit or Wiki can increase the accuracy and decrease standard deviation on the CEE task. It is shown that a more accurate and stable model can be achieved by utilizing the transfer layer and well-trained models from related tasks.

Table 5. Accuracy for CEE classification using the transfer layer. Original is the model not using transfer learning, and without loading trained weights from other dataset. *Load Weebit* is to load the parameters weights trained in Weebit except the transfer layer. *Load Wiki* is to load the parameters weights trained in Wiki except the transfer layer.

	Original	Load Weebit	Load Wiki
Accuracy	0.528 (0.045)	0.568 (0.012)	0.561 (0.014)

Besides directly training and evaluating the same dataset, we also tried the model trained using Wikipedia dataset and evaluate on Cambridge English dataset. 10 articles are randomly selected from each level of Cambridge English Test. The probability of being classified as regular English Wikipedia instead of simple English Wikipedia is treated as the difficulty score. The average difficulty scores predicted by the model are shown in Table 4, which shows that our produced readability score implies correctly the difficulty of English documents for different levels of exams. A larger score indicates higher difficulty. These scores correctly indicate the difficulty levels of these exams.

6 Conclusion and Future Work

We have proposed a model to evaluate the readability of articles which can make great contributions to a variety of applications. Our proposed Hierarchical Self-Attention framework outperforms existing approaches by combining hierarchical document encoders with the explicit features proposed by linguistics. For future works, we are interested in providing the personalized recommendation of articles based on the combination of article readability and the understanding ability of the user. Currently, readability of articles only evaluate the texts of articles, other modalities such as images [39] and taxonomies [8] considered to improve readers' understanding. More comprehensive document encoders such as RCNN [5] and tree LSTM [47] may also be considered.

References

1. Anderson, J.: Lix and Rix: variations on a little-known readability index. J. Read. **26**(6), 490–496 (1983)
2. Brown, J., Eskenazi, M.: Student, text and curriculum modeling for reader-specific document retrieval. In: Proceedings of the IASTED International Conference on Human-Computer Interaction, Phoenix, AZ (2005)
3. Chall, J.S.: Readability: an appraisal of research and application, no. 34 (1958)
4. Chall, J.S., Dale, E.: Readability Revisited: The New Dale-Chall Readability Formula. Brookline Books (1995)
5. Chen, M., et al.: Multifaceted protein-protein interaction prediction based on Siamese residual RCNN. Bioinformatics **35**(14), i305–i314 (2019)
6. Chen, M., Meng, C., Huang, G., Zaniolo, C.: Neural article pair modeling for Wikipedia sub-article matching. In: Brefeld, U., et al. (eds.) ECML PKDD 2018. LNCS (LNAI), vol. 11053, pp. 3–19. Springer, Cham (2019). https://doi.org/10.1007/978-3-030-10997-4_1
7. Chen, M., Meng, C., Huang, G., Zaniolo, C.: Learning to differentiate between main-articles and sub-articles in Wikipedia. In: Proceedings of the IEEE International Conference on Big Data (2019)
8. Chen, M., Tian, Y., Chen, X., Xue, Z., Zaniolo, C.: On2Vec: embedding-based relation prediction for ontology population. In: Proceedings of the 2018 SIAM International Conference on Data Mining, pp. 315–323. SIAM (2018)
9. Coleman, M., Liau, T.L.: A computer readability formula designed for machine scoring. J. Appl. Psychol. **60**(2), 283 (1975)
10. Collins-Thompson, K., Callan, J.: A language modeling approach to predicting reading difficulty. In: Proceedings of the Human Language Technology Conference of the North American Chapter of the Association for Computational Linguistics, HLT-NAACL 2004 (2004)

11. Collins-Thompson, K.: Computational assessment of text readability: a survey of current and future research. ITL-Int. J. Appl. Linguist. **165**(2), 97–135 (2014)
12. Collins-Thompson, K., Callan, J.: Predicting reading difficulty with statistical language models. J. Am. Soc. Inform. Sci. Technol. **56**(13), 1448–1462 (2005)
13. Coxhead, A.: A new academic word list. TESOL Q. **34**(2), 213–238 (2000)
14. Dale, E., Chall, J.S.: The concept of readability. Elem. Engl. **26**(1), 19–26 (1949)
15. De Clercq, O., Hoste, V.: All mixed up? Finding the optimal feature set for general readability prediction and its application to English and Dutch. Comput. Linguist. **42**(3), 457–490 (2016)
16. Duchi, J., Hazan, E., Singer, Y.: Adaptive subgradient methods for online learning and stochastic optimization. J. Mach. Learn. Res. **12**(Jul), 2121–2159 (2011)
17. Feng, L., Elhadad, N., Huenerfauth, M.: Cognitively motivated features for readability assessment. In: Proceedings of the 12th Conference of the European Chapter of the Association for Computational Linguistics, pp. 229–237. Association for Computational Linguistics (2009)
18. François, T.L.: Combining a statistical language model with logistic regression to predict the lexical and syntactic difficulty of texts for FFL. In: Proceedings of the 12th Conference of the European Chapter of the Association for Computational Linguistics: Student Research Workshop, pp. 19–27. Association for Computational Linguistics (2009)
19. Fry, E.: A readability formula that saves time. J. Read. **11**(7), 513–578 (1968)
20. Fry, E.B.: The varied uses of readability measurement today. J. Read. **30**(4), 338–343 (1987)
21. Gibson, E.: Linguistic complexity: locality of syntactic dependencies. Cognition **68**(1), 1–76 (1998)
22. Graesser, A.C., McNamara, D.S., Louwerse, M.M., Cai, Z.: Coh-Metrix: analysis of text on cohesion and language. Behav. Res. Methods Instrum. Comput. **36**(2), 193–202 (2004). https://doi.org/10.3758/BF03195564
23. Gunning, R.: The fog index after twenty years. J. Bus. Commun. **6**(2), 3–13 (1969)
24. Heilman, M., Collins-Thompson, K., Eskenazi, M.: An analysis of statistical models and features for reading difficulty prediction. In: 3rd Workshop on Innovative Use of NLP for Building Educational Applications (2008)
25. Heilman, M., et al.: Combining lexical and grammatical features to improve readability measures for first and second language texts. In: Human Language Technologies (2007)
26. Kauchak, D.: Improving text simplification language modeling using un simplified text data. In: Proceedings of the 51st Annual Meeting of the Association for Computational Linguistics (Volume 1: Long Papers), vol. 1, pp. 1537–1546 (2013)
27. Kim, Y.: Convolutional neural networks for sentence classification. In: Empirical Methods in Natural Language Processing (2014)
28. Kincaid, J.P., Fishburne Jr, R.P., Rogers, R.L., Chissom, B.S.: Derivation of new readability formulas for navy enlisted personnel (1975)
29. Klare, G.R.: The measurement of readability: useful information for communicators. ACM J. Comput. Doc. (JCD) **24**(3), 107–121 (2000)
30. Li, J., Luong, M.T., Jurafsky, D.: A hierarchical neural autoencoder for paragraphs and documents. arXiv preprint arXiv:1506.01057 (2015)
31. Li, Z., Wei, Y., Zhang, Y., Yang, Q.: Hierarchical attention transfer network for cross-domain sentiment classification. In: Thirty-Second AAAI Conference on Artificial Intelligence (2018)

32. Lin, R., Liu, S., Yang, M., Li, M., Zhou, M., Li, S.: Hierarchical recurrent neural network for document modeling. In: Proceedings of the 2015 Conference on Empirical Methods in Natural Language Processing, pp. 899–907 (2015)
33. Louwerse, M.: An analytic and cognitive parametrization of coherence relations. Cogn. Linguist. **12**(3), 291–316 (2001)
34. Malvern, D., Richards, B.: Measures of lexical richness. In: The Encyclopedia of Applied Linguistics (2012)
35. Mc Laughlin, G.H.: SMOG grading-a new readability formula. J. Read. **12**(8), 639–646 (1969)
36. McNamara, D.S., Graesser, A.C., McCarthy, P.M., Cai, Z.: Automated Evaluation of Text and Discourse with Coh-Metrix. Cambridge University Press, Cambridge (2014)
37. McNamara, D.S., Louwerse, M.M., McCarthy, P.M., Graesser, A.C.: Coh-Metrix: capturing linguistic features of cohesion. Discourse Process. **47**(4), 292–330 (2010)
38. Parikh, A.P., Täckström, O., Das, D., Uszkoreit, J.: A decomposable attention model for natural language inference. arXiv preprint arXiv:1606.01933 (2016)
39. Pezeshkpour, P., Chen, L., Singh, S.: Embedding multimodal relational data for knowledge base completion. In: Proceedings of the 2018 Conference on Empirical Methods in Natural Language Processing, pp. 3208–3218 (2018)
40. Pilán, I., Volodina, E., Zesch, T.: Predicting proficiency levels in learner writings by transferring a linguistic complexity model from expert-written coursebooks. In: Proceedings of COLING 2016, the 26th International Conference on Computational Linguistics: Technical Papers, pp. 2101–2111 (2016)
41. Pitler, E., Nenkova, A.: Revisiting readability: a unified framework for predicting text quality. In: Proceedings of the Conference on Empirical Methods in Natural Language Processing, pp. 186–195. Association for Computational Linguistics (2008)
42. Rennie, J.D., Srebro, N.: Loss functions for preference levels: regression with discrete ordered labels. In: Proceedings of the IJCAI Multidisciplinary Workshop on Advances in Preference Handling, pp. 180–186. Kluwer Norwell (2005)
43. Schwarm, S.E., Ostendorf, M.: Reading level assessment using support vector machines and statistical language models. In: Proceedings of the 43rd Annual Meeting on Association for Computational Linguistics, pp. 523–530. Association for Computational Linguistics (2005)
44. Senter, R., Smith, E.A.: Automated readability index. Technical report, Cincinnati University, OH (1967)
45. Si, L., Callan, J.: A statistical model for scientific readability. In: CIKM, vol. 1, pp. 574–576 (2001)
46. Socher, R., et al.: Recursive deep models for semantic compositionality over a sentiment treebank. In: Proceedings of the 2013 Conference on Empirical Methods in Natural Language Processing, pp. 1631–1642 (2013)
47. Tai, K.S., Socher, R., Manning, C.D.: Improved semantic representations from tree-structured long short-term memory networks. In: Proceedings of the 53rd Annual Meeting of the Association for Computational Linguistics and the 7th International Joint Conference on Natural Language Processing (Volume 1: Long Papers), pp. 1556–1566 (2015)
48. Tang, D., Qin, B., Liu, T.: Document modeling with gated recurrent neural network for sentiment classification. In: Proceedings of the 2015 Conference on Empirical Methods in Natural Language Processing, pp. 1422–1432 (2015)

49. Vajjala, S., Meurers, D.: On improving the accuracy of readability classification using insights from second language acquisition. In: Proceedings of the Seventh Workshop on Building Educational Applications Using NLP, pp. 163–173. Association for Computational Linguistics (2012)
50. Vaswani, A., et al.: Attention is all you need. In: Advances in Neural Information Processing Systems, pp. 5998–6008 (2017)
51. Xia, M., Kochmar, E., Briscoe, T.: Text readability assessment for second language learners. In: Proceedings of the 11th Workshop on Innovative Use of NLP for Building Educational Applications, pp. 12–22 (2016)
52. Yang, Z., Yang, D., Dyer, C., He, X., Smola, A., Hovy, E.: Hierarchical attention networks for document classification. In: Proceedings of the 2016 Conference of the North American Chapter of the Association for Computational Linguistics: Human Language Technologies, pp. 1480–1489 (2016)
53. Zakaluk, B.L., Samuels, S.J.: Readability: Its Past, Present, and Future. ERIC (1988)

Variational Recurrent Sequence-to-Sequence Retrieval for Stepwise Illustration

Vishwash Batra[1(✉)], Aparajita Haldar[1], Yulan He[1], Hakan Ferhatosmanoglu[1], George Vogiatzis[2], and Tanaya Guha[1]

[1] University of Warwick, Coventry CV4 7AL, UK
{v.batra,aparajita.haldar,yulan.he}@warwick.ac.uk
[2] Aston University, Birmingham B4 7ET, UK

Abstract. We address and formalise the task of *sequence-to-sequence (seq2seq) cross-modal retrieval*. Given a sequence of text passages as query, the goal is to retrieve a sequence of images that best describes and aligns with the query. This new task extends the traditional cross-modal retrieval, where each image-text pair is treated independently ignoring broader context. We propose a novel *variational recurrent seq2seq (VRSS) retrieval model* for this seq2seq task. Unlike most cross-modal methods, we generate an image vector corresponding to the latent topic obtained from combining the text semantics and context. This synthetic image embedding point associated with every text embedding point can then be employed for either image generation or image retrieval as desired. We evaluate the model for the application of *stepwise illustration* of recipes, where a sequence of relevant images are retrieved to best match the steps described in the text. To this end, we build and release a new *Stepwise Recipe* dataset for research purposes, containing 10K recipes (sequences of image-text pairs) having a total of 67K image-text pairs. To our knowledge, it is the first publicly available dataset to offer rich semantic descriptions in a focused category such as food or recipes. Our model is shown to outperform several competitive and relevant baselines in the experiments. We also provide qualitative analysis of how semantically meaningful the results produced by our model are through human evaluation and comparison with relevant existing methods.

Keywords: Semantics · Multimodal datasets · Sequence retrieval

1 Introduction

There is growing interest in cross-modal analytics and search in multimodal data repositories. A fundamental problem is to associate images with some corresponding descriptive text. Such associations often rely on semantic understanding, beyond traditional similarity search or image labelling, to provide human-like visual understanding of the text and reflect abstract ideas in the image.

J. M. Jose et al. (Eds.): ECIR 2020, LNCS 12035, pp. 50–64, 2020.
https://doi.org/10.1007/978-3-030-45439-5_4

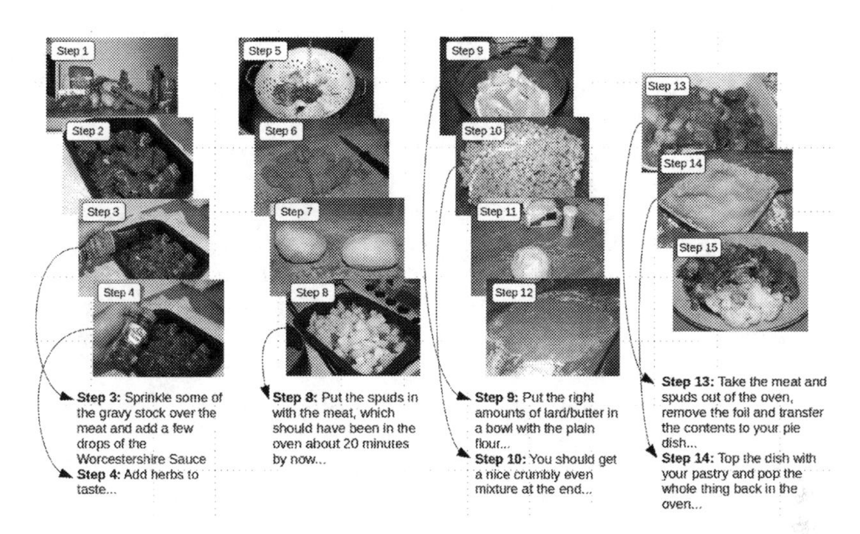

Fig. 1. Stepwise Recipe illustration example showing a few text recipe instruction steps alongside one full sequence of recipe images. Note that retrieval of an accurate illustration of Step 4, for example, depends on previously acquired context information.

Cross-modal retrieval systems must return outputs of one modality from a data repository, while a different modality is used as the input query. The multi-modal repository usually consists of paired objects from two modalities, but may be labelled or unlabelled. Classical approaches to compare data across modalities include canonical correlation analysis [12], partial least squares regression [28], and their numerous variants. More recently, various deep learning models have been developed to learn shared embedding spaces from paired image-text data, either unsupervised, or supervised using image class labels. The deep models popularly used include deep belief networks [23], correspondence autoencoders [9], deep metric learning [13], and convolutional neural networks (CNNs) [33]. With all these models it is expected that by learning from pairwise aligned data, the common representation space will capture semantic similarities across modalities.

Most such systems, however, do not consider sequences of related data in the query or result. In traditional image retrieval using text queries, for example, each image-text pair is considered in isolation ignoring any broader 'context'. A context-aware image-from-text retrieval model must look at pairwise associations and also consider sequential relationships. Such *sequence-to-sequence (seq2seq) cross-modal retrieval* is possible when contextual information and semantic meaning are both encoded and used to inform the retrieval step.

For *stepwise recipe illustration*, an effective retrieval system must identify and align a set of relevant images corresponding to each step of a given text sequence of recipe instructions. More generally, for the task of automatic *story picturing*, a series of suitable images must be chosen to illustrate the events and abstract concepts found in a sequential text taken from a story. An example of the instruction steps and illustrations of a recipe taken from our new Stepwise Recipe dataset is shown in Fig. 1.

In this paper, we present a variational recurrent learning model to enable seq2seq retrieval, called Variational Recurrent Sequence-to-Sequence (VRSS) model. VRSS produces a joint representation of the image-text repository, where the semantic associations are grounded in context by making use of the sequential nature of the data. Stepwise query results are then obtained by searching this representation space. More concretely, we incorporate the global context information encoded in the entire text sequence (through the attention mechanism) into a variational autoencoder (VAE) at each time step, which converts the input text into an image representation in the image embedding space. To capture the semantics of the images retrieved so far (in a story/recipe), we assume the prior of the distribution of the topic given the text input follows the distribution conditional on the latent topic from the previous time step. By doing so, our model can naturally capture sequential semantic structure.

Our main contributions can be summarised below:

- We formalise the task of *sequence-to-sequence (seq2seq) retrieval* for stepwise illustration of text.
- We propose a new *variational recurrent seq2seq (VRSS) retrieval model* for seq2seq retrieval, which employs temporally-dependent latent variables to capture the sequential semantic structure of text-image sequences.
- We release a new *Stepwise Recipe* dataset ($10K$ recipes, $67K$ total image-text pairs) for research purposes, and show that VRSS outperforms several cross-modal retrieval alternatives on this dataset, using various performance metrics.

2 Related Work

Our work is related to: cross-modal retrieval, story picturing, variational recurrent neural networks, and cooking recipe datasets.

Cross-Modal Retrieval. A number of pairwise-based methods over the years have attempted to address the cross-modal retrieval problem in different ways, such as metric learning [26] and deep neural networks [32]. For instance, an alignment model [16] was devised that learns inter-modal correspondences using MS-COCO [19] and Flickr-30k [25] datasets. Other work [18] proposed unifying joint image-text embedding models with multimodal neural language models, using an encoder-decoder pipeline. A later method [8] used hard negatives to improve their ranking loss function, which yielded significant gains in retrieval performance. Such systems focus only on isolated image retrieval when given a text query, and do not address the seq2seq retrieval problem that we study here.

In a slight variation [2], the goal was to retrieve an image-text multimodal unit when given a text query. For this, they proposed a gated neural architecture to create an embedding space from the query texts and query images along with the multimodal units that form the retrieval results set, and then performed semantic matching in this space. The training minimized structured hinge loss, and there was no sequential nature to the data used.

Story Picturing. An early story picturing system [15] retrieved landscape and art images to illustrate ten short stories based on key terms in the stories and image descriptions as well as a similarity linking of images. The idea was pursued further with a system [11] for helping people with limited literacy to read, which split a sentence into three categories and then retrieved a set of explanatory pictorial icons for each category.

To our knowledge, an application [17] that ranks and retrieves image sequences based on longer text paragraphs as queries was the first to extend the pairwise image-text relationship to matching image sequences with longer paragraphs. They employed a structural ranking support vector machine with latent variables and used a custom-built Disneyland dataset, consisting of blog posts with associated images as the parallel corpus from which to learn joint embeddings. We follow a similar approach, creating our parallel corpus from sequential stepwise cooking recipes rather than unstructured blog posts, and design an entirely new seq2seq model to learn our embeddings.

The Visual Storytelling Dataset (VIST) [14] was built with a motivation similar to our own, but for generating text descriptions of image sequences rather than the other way around. Relying on human annotators to generate captions, VIST contains sequential image-text pairs with a focus on abstract visual concepts, temporal event relations, and storytelling. In our work, we produce a similar sequenced dataset in a simple, automated manner.

A recent joint sequence-to-sequence model [20] learned a common image-text semantic space and generated paragraphs to describe photo streams. This bidirectional attention recurrent neural network was evaluated on both the above datasets. Despite being unsuitable for our inverse problem, VIST has also been used for retrieving images when given text, in work related to ours. In an approach called Coherent Neural Story Illustration (CNSI), an encoder-decoder network [27] was built to first encode sentences using a hierarchical two-level sentence-story gated recurrent unit (GRU), and then sequentially decode into a corresponding sequence of illustrative images. A previously proposed coherence model [24] was used to explicitly model co-references between sentences.

Variational Recurrent Neural Networks. Our model is partly inspired by the variational recurrent neural network (VRNN) [6], which introduces latent random variables into the hidden state of an RNN by combining it with a variational autoencoder (VAE). They showed that using high level latent random variables, VRNN can model the variability observed in structured sequential data such as natural speech and handwriting. VRNN has recently been applied to other sequential modelling tasks such as machine translation [31].

Our proposed VRSS model introduces temporally-dependent latent variables to capture the sequential semantic structure of text/image sequences. Different from existing approaches, we take into account the global context information encoded in the entire query sequence. We use VAE for cross-modal generation by converting the text into a representation in the image embedding space instead of using it to reconstruct the text input. Finally, we use the max-margin hinge loss to enforce similarity between text and paired image representations.

Cooking Recipe Datasets. The first attempt at automatic classification of food images was the Food-101 dataset [3] having $101K$ images across 101 categories. Since then, the new Recipe1M dataset [29] gained wide attention, which paired each recipe with several images to build a collection of $13M$ food images for $1M$ recipes. Recent work [4] proposed a cross-modal retrieval model that aligns Recipe1M images and recipes in a shared representation space. As this dataset does not offer any sequential data for stepwise illustration, this association is between images of the final dish and the corresponding entire recipe text. Our Stepwise Recipe dataset, by comparison, provides an image for each instruction step, resulting in a sequence of image-text pairs for each recipe.

In [5] they release a dataset of sequenced image-text pairs in the cooking domain, with focus on text generation conditioned on images. RecipeQA [34] is another popular dataset, used for multimodal comprehension and reasoning, with 36K questions about the 20K recipes and illustrative images for each step of the recipes. Recent work [1] used it to analyse image-text coherence relations, thereby producing a human-annotated corpus with coherence labels to characterise different inferential relationships. The RecipeQA dataset reveals associations between image-text pairs much like our Stepwise Recipe dataset, and we therefore utilise it to augment our own dataset.

3 Stepwise Recipe Dataset Construction

We construct the *Stepwise Recipe* dataset, composed of illustrated, step-by-step recipes from three websites[1]. Recipes were automatically web-scraped and cleaned of HTML tags. The information about data and scripts will be made available on GitHub[2]. The construction of such an image-text parallel corpus has several challenges as highlighted in previous work [17]. The text is often unstructured, without information about the canonical association between image-text pairs. Each image is semantically associated with some portion of the text in the same recipe, and we assume that the images chosen by the author to augment the text are semantically meaningful. We thus perform text segmentation to divide the recipe text and associate segments with a single image each.

We perform text-based filtering [30] to ensure text quality: (1) descriptions should have a high unique word ratio covering various part-of-speech tags, therefore descriptions with high noun ratio are discarded; (2) descriptions with high repetition of tokens are discarded; and (3) some predefined boiler-plate prefix-suffix sequences are removed. Our constructed dataset consists of about 2K recipes with 44K associated images.

[1] simplyrecipes.com, visualrecipes.com, olgasflavorfactory.com.
[2] https://github.com/vishwerine/StepRecipe.

Furthermore, we augment our parallel corpus using similarly filtered RecipeQA data [34], which contains images for each step of the recipes in addition to visual question answering data. The final dataset contains over 10K recipes in total and 67K images.

4 Variational Recurrent Seq2seq (VRSS) Retrieval Model

The seq2seq retrieval task is formalised as follows: given a sequence of text passages, $\boldsymbol{x} = \{x_1, x_2, ..., x_T\}$, retrieve a sequence of images $\boldsymbol{i} = \{i_1, i_2, ..., i_T\}$ (from a data repository) which best describes the semantic meanings of the text passages, i.e., $p(\boldsymbol{i}|\boldsymbol{x}) = \prod_{t=1}^{T} p(i_t|\boldsymbol{x}, i_{<t})$. The training set (e.g., recipes or stories) is $S = \{S^1, S^2, \cdots S^N\}$, where each S^n consists of a sequence of images and their associated text. Each such sequence $S^n = \{(x_1^n, i_1^n), (x_2^n, i_2^n), \cdots, (x_{|S^n|}^n, i_{|S^n|}^n)\}$ is paired element-wise where each text sequence $\boldsymbol{x^n} = \{x_1^n, x_2^n, ..., x_T^n\}$ and each image sequence $\boldsymbol{i^n} = \{i_1^n, i_2^n, ..., i_T^n\}$.

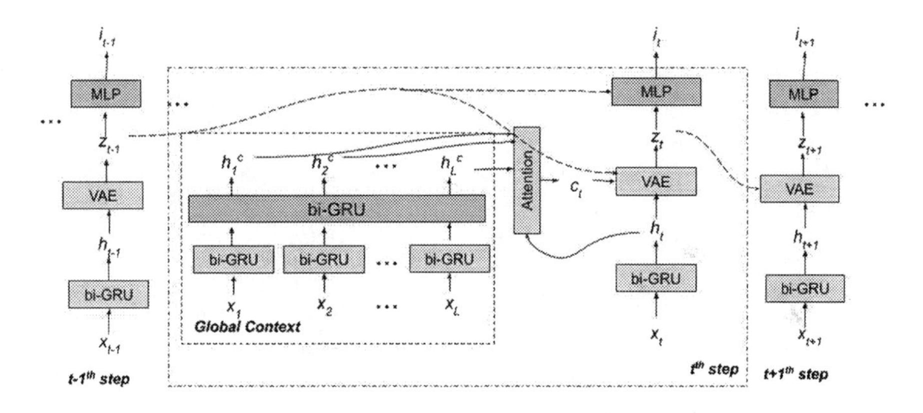

Fig. 2. Variational Recurrent Sequence-to-Sequence (VRSS) model architecture.

We address the seq2seq retrieval problem by considering three aspects: (1) encoding the contextual information of text passages; (2) capturing the semantics of the images retrieved (in a story/recipe); and (3) learning the relatedness between each text passage and its corresponding image.

It is natural to use RNNs to encode a sequence of text passages. Here, we encode a text sequence using a bi-directional GRU (bi-GRU). Given a text passage, we use the attention mechanism to capture the contextual information of the whole recipe. We map the text embedding into a latent topic z_t by using a VAE. In order to capture the semantics of the images retrieved so far (in a story/recipe), we assume the prior of the distribution of the topic given the text input follows a distribution conditional on the latent topic z_{t-1} from the previous step. We decode the corresponding image vector i_t conditional on the latent topic, to learn the relatedness between text and image with a multi-layer perceptron and obtain a synthetic image embedding point generated from its associated text embedding point. Our proposed *Variational Recurrent Seq2seq (VRSS) model* is illustrated in Fig. 2.

Below, we describe each of the main components of the VRSS model.

Text Encoder. We use a bi-GRU to learn the hidden representations of the text passage (e.g. one recipe instruction) in the forward and backward directions. The two learned hidden states are then concatenated to form the text segment representation $\{x_t = [\overrightarrow{h_T}, \overleftarrow{h_T}]\}$. To encode a sequence of such text passages (e.g. one recipe), a hierarchical bi-GRU is used which first encodes each text segment and subsequently combines them.

Image Encoder. To generate the vector representation of an image, we use the pre-trained modified ResNet50 CNN [22]. In experiments, this model produced a well distributed feature space when trained on the limited domain, namely food related images. This was verified using t-SNE visualisations [21], which showed less clustering in the generated embedding space as compared to embeddings obtained from models pre-trained on ImageNet [7].

Incorporating Context. To capture global context, we feed the bi-GRU encodings into a top level bi-GRU. Assuming the hidden state output of each text passage x_l in the global context is h_l^c, we use an attention mechanism to capture its similarity with the hidden state output of the tth text passage h_t as $\alpha_l = \text{softmax}(h_t^T W h_l^c)$. The context vector is encoded as the combination of L text passages weighted by the attentions as $c_t = \sum_{l=1}^{L} \alpha_l h_l^c$. This ensures that any given text passage is influenced more by others that are semantically similar.

Latent Topic Modeling. At the tth step text x_t of the text sequence, the bi-GRU output h_t is combined with the context c_t and fed into a VAE to generate the latent topic z_t. Two prior networks f_{μ_θ} and f_{Σ_θ} define the prior distribution of z_t conditional on the previous z_{t-1}. We also define two inference networks f_{μ_ϕ} and f_{Σ_ϕ} which are functions of h_t, c_t, and z_{t-1}:

$$p_\theta(z_t|\boldsymbol{z}_{<t}, \boldsymbol{x}_{<t}) = \mathcal{N}(z_t|f_{\mu_\theta}(z_{t-1}), f_{\Sigma_\theta}(z_{t-1})) \tag{1}$$

$$q_\phi(z_t|\boldsymbol{z}_{<t}, \boldsymbol{x}_{\leq t}) = \mathcal{N}(z_t|f_{\mu_\phi}(z_{t-1}, h_t, c_t), f_{\Sigma_\phi}(z_{t-1}, h_t, c_t)) \tag{2}$$

Unlike the typical VAE setup where the text input x_t is reconstructed by generation networks, here we generate the corresponding image vector i_t. To generate the image vector conditional on z_t, the generation networks are defined which are also conditional on z_{t-1}:

$$p_\varphi(i_t|\boldsymbol{z}_{\leq t}, \boldsymbol{x}_{\leq t}) = \mathcal{N}(i_t|f_{\mu_\varphi}(z_{t-1}, z_t), f_{\Sigma_\varphi}(z_{t-1}, z_t)) \tag{3}$$

The generation loss for image i_t is then:

$$\begin{aligned}\mathcal{L}_{recons.}(i_t) = \mathbb{E}_{q(\boldsymbol{z}_{\leq T}|\boldsymbol{x}_{\leq T})} \log p(i_t|\boldsymbol{z}_{\leq t}, \boldsymbol{x}_{<t}) \\ - KL(q(z_t|\boldsymbol{x}_{\leq t}, \boldsymbol{z}_{<t}) \| p(z_t|\boldsymbol{x}_{<t}, \boldsymbol{z}_{<t}))\end{aligned} \tag{4}$$

Image Retrieval. We enable the search process by a timestep-wise hinge loss to model $p(i_t|\boldsymbol{x}, z_t, i_{<t})$. The latent semantic variable z_t is used to predict the image at the given timestep t, with a hinge loss max-margin objective:

$$\mathcal{L}_{HL}(i_t) = \sum_j \max(0, \alpha - s(i_t, \hat{i}_t) + s(i_j, \hat{i}_t)) \tag{5}$$

where α is the margin parameter, i_t is the image vector generated by the model, $\hat{i}_t$ is the vector representation of the gold-standard image at time step t, i_j is the negative images, and $s(\cdot)$ denotes the similarity measurement function. In our experiments, we use the cosine distance function.

Overall Objective Function. The overall objective function is the total of the image reconstruction loss and the image retrieval hinge loss summing over all the time steps for the whole image sequence, with β as the weighting factor:

$$\mathcal{L}_{overall} = \sum_{t=1}^{T} \mathcal{L}_{recons.}(i_t) + \beta \mathcal{L}_{HL}(i_t) \tag{6}$$

Parameter Configuration. As the initial parameter setting of the VRSS architecture, we use bi-GRU with the hidden dimension of 500 and set the dimension of latent topics to 500. We also introduce a dropout layer in the RNNs with probability of 0.3. Each word in the text is represented in the 500 dimensional embedding space. The image encoder projects images to a $2,048$ dimensional feature space. For training the objective function, we use AdaDelta optimisation function, with a learning rate of 1.0. The values of hyperparameters α and β were set to be 0.2 and 1.7 respectively.

5 Experimental Setup

We create a train-test split of $60k/6k$ image-text pairs and $9k/1k$ recipes in the Stepwise Recipe dataset. The split is done author-wise to ensure style consistency, but having overlapping authors in train and test splits.

5.1 Models for Comparison

- **LDA.** We re-implement the topic modelling based approach [10] to jointly generate words in text and visual words in image assuming each image-text pair share the same set of topics.
- **Visual Semantic Embeddings (VSE++).** Following [8], we implement a deep neural network approach which maps the text representations and image vectors into the same semantic embedding space.
- **Coherence Neural Story Illustration (CNSI).** We use the encoder-decoder CNSI model proposed in [27], with coherence capturing the co-reference relations among sentences, to retrieve a sequence of images illustrating a passage of text.

- **VRSS-VAE.** This follows the same encoder-decoder architecture of our VRSS model, using two bi-GRU architectures as encoders and decoders with the same learning objective, but without latent variables. Therefore, it is treated as an ablation study of our VRSS model without the VAE module.
- **VRSS-globalCon.** This is a variant of our VRSS model without the incorporation of the global context.

In all the neural models evaluated here, the image representation are extracted using the ResNet50 model [22] pre-trained on food-related images.

5.2 Evaluation Methods

Recall@k indicates that the retrieved image was among the top k best matches out of the set of candidate images. We also define *Story Recall@k*, which considers the retrieved image as correct if it is from the same data sequence. Further, we provide *Visual Saliency Recall@k* values. We implement *Visual Saliency Recall* following [27] and train a VGG-19 network to classify the images of the story test set, with visual features from [22] for initialization. We also report *Visual Feature Similarity* using the average cosine similarity between gold-standard image and retrieved image, considering image features generated by [22].

Previous work [27] highlights that existing quantitative retrieval metrics may be too harsh for a task of this description. Therefore, it is imperative that we use human evaluators to judge how appropriate and coherent the retrieved illustration sequences are. For our human evaluation, we pick a random sample (164 recipes, 1564 image-text pairs) from the test set ($1K$ recipes, $6K$ image-text pairs). We present each evaluator with a sequence of recipe instruction steps that make up one complete recipe. Alongside each text segment, they are given three possible illustrations that depict that step, which are randomly shuffled images of the gold-standard, the non-context model, and the proposed VRSS model. The evaluator is asked to select all image options that may be appropriate illustrations for the corresponding text segment. A total of $5.1K$ ratings are obtained from 12 evaluators, ensuring that every sample receives at least 2 ratings.

6 Results and Discussion

6.1 Automatic Evaluation

Table 1 reports the retrieval performance of different methods using *Recall@k* and *Story Recall@k* metrics. LDA gives the worst results, which shows that using a generative model for capturing the semantic topics from both text and image does not work well in the seq2seq retrieval task. By mapping both text and image into the same embedding space, VSE++ outperforms LDA. Our VRSS model without the VAE component (VRSS-VAE) gives similar performance compared

Table 1. Text illustration performance using *Recall@k (R@k)* and *Story Recall@k (StR@k)* and *Visual Saliency Recall@k (VSR@k)* on the Stepwise Recipe dataset. The best result in each column is highlighted in **bold**.

	Recall@k			Story Recall@k	Visual Saliency Recall@k		
Models	R@1	R@5	R@10	StR@1	VSR@1	VSR@5	VSR@10
Non-context models							
LDA	1.4	3.4	8.9	4.1	3.2	6.7	12.5
VSE++	7.7	18.6	24.6	21.3	8.1	23.1	26.6
Context models							
CNSI	3.6	8.9	13.7	18.4	16.6	31.8	39.8
VRSS-VAE	6.4	19.7	23.1	18.1	11.3	29.2	33.2
VRSS-GlobalCon	5.2	19.9	26.5	21.1	15.1	28.9	32.7
VRSS	**8.2**	**21.3**	**29.8**	**24.4**	**18.4**	**33.4**	**45.1**

to the non-context model VSE++ despite considering the contextual information. VRSS without the incorporation of global context (VRSS-GlobalCon.) performs similarly as VRSS-VAE. CNSI gives worse results compared to both VRSS variants in *Recall@k* and *Story Recall@1*. Our new VRSS model, which maps each hidden state of the RNN into a latent topic and also further incorporates global context information, gives the best results across all metrics. This indicates the importance of representing semantics encoded in both text and images in a more abstract manner and the benefit of incorporating global context.

Recall@k and *Story Recall@k* metrics only measure the degree of exact matches of the retrieved images with regards to the gold-standard images. This might not be appropriate for our text illustration task since a given text segment could be illustrated by multiple images expressing similar semantics. Example image retrieval results are shown in Fig. 3 where both the gold-standard and the VRSS retrieved images are displayed for some recipe instructions. It can be observed that although VRSS failed to retrieve the gold-standard images in these examples, its output images are still appropriate illustrations of the corresponding texts. For this reason, we also report the evaluation results using more semantics-based and feature-based metric, *Visual Saliency Recall@k*.

It can be observed from Table 1 that VRSS performs significantly better than baselines on *Visual Saliency Recall@k*. These recall scores indicate that VRSS is able to retrieve images that are described by text segments that are semantically related to the query text, even if the images themselves do not match the gold-standard image. We also calculate the *Visual Feature Similarity* which measures the average cosine similarity between the gold-standard image and the retrieved image in the feature space. For VRSS, this is 0.51 and for VSE++ it is 0.37, and for CNSI it is 0.45 This confirms that VRSS retrieves illustrations that are visually similar to the gold-standard image.

6.2 Human Evaluation

For the human evaluation, we count the number of votes received for the gold-standard images, the VRSS model output images, and the VSE++ (non-context based) model output images. We only count a vote if there is majority consensus among the evaluators. Hence, in Table 2, the '# Votes' column indicates the number that constitutes a majority among voters.

Table 2. Human Evaluation results. The cell values indicate the number of images output by the corresponding model(s) that receive x number of votes ($x \in \{2, 3, 4, 5\}$) as majority.

# Votes	2	3	4	5
Gold-standard only	0	442	171	47
Gold-standard and VRSS	255	41	0	0
Gold-standard and VSE++	88	9	0	0
Gold-standard, VRSS and VSE++	75	0	0	0

In Table 2, we see the preference results obtained from human evaluation of the retrieved recipe illustrations. Considering majority agreement as 2 votes, gold-standard was never preferred in isolation. Rather, in 61% of the cases, both the gold-standard image and the image retrieved using VRSS were deemed to be appropriate illustrations for the given text query. In 18% of the cases, gold-standard as well as the retrieved images from both models were considered appropriate. In the remaining 21% of the cases, the VRSS output was not judged as being appropriate. Taking 3 votes as the majority, gold-standard alone was picked in 88% of the cases, and picked in combination with the VRSS output in 8% of the remaining cases, with a negligible number of cases for the other combinations. Where the majority consensus is above 4 votes, evaluators chose gold-standard alone in every case. Therefore, VRSS outperforms other models particularly in ambiguous cases where the text is likely to contain an indirect description of the image. The VRSS output is about 3 times more likely to be selected compared to the VSE++ output. Over 60% of the time, at least 2 human evaluators believe that the VRSS output is as appropriate as the gold-standard image. These results indicate that the context based VRSS model significantly outperforms the non-context based model.

Figure 3 shows examples where the VRSS output was preferred by human evaluators. It also highlights cases where metrics other than recall are beneficial such as semantically related entities and paired images having *Visual Feature Similarity*. The last text segment implicitly refers to the previous, with this retrieved image counted favourably when using the context-aware *Story Recall* metric. We also perform a qualitative error analysis, and find that the attention mechanism sometimes misdirects the image retrieval (Figs. 4 and 5).

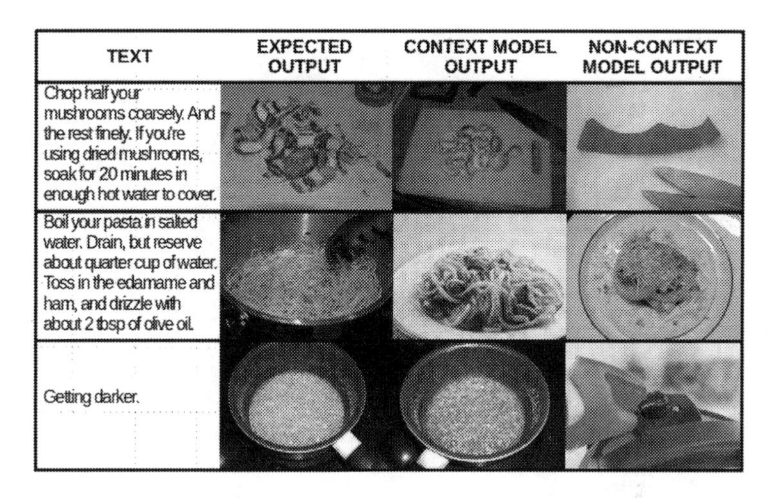

Fig. 3. Illustrative comparison of non-context (VSE++) and context models (VRSS) - VRSS result preferred by human evaluators.

Fig. 4. Illustrative comparison of non-context (VSE++) and context models (VRSS) - VSE++(R) result preferred by human evaluators.

Fig. 5. Illustrative comparison of non-context (VSE++) and context models (VRSS) - Neither VRSS nor VSE++(R) result preferred by human evaluators.

7 Conclusion

We presented VRSS model that given a sequence of text passages, identifies a sequence of images best describing the semantic content of text. We introduced the Stepwise Recipe dataset to facilitate further research on this problem. Our results on the Stepwise Recipe dataset show that VRSS significantly outperforms competitive baselines in terms of both automatic and human evaluations.

References

1. Alikhani, M., Chowdhury, S.N., de Melo, G., Stone, M.: CITE: a corpus of image-text discourse relations. arXiv preprint arXiv:1904.06286 (2019)
2. Balaneshin-kordan, S., Kotov, A.: Deep neural architecture for multi-modal retrieval based on joint embedding space for text and images. In: Proceedings of the Eleventh ACM International Conference on Web Search and Data Mining, pp. 28–36. ACM (2018)
3. Bossard, L., Guillaumin, M., Van Gool, L.: Food-101 – mining discriminative components with random forests. In: Fleet, D., Pajdla, T., Schiele, B., Tuytelaars, T. (eds.) ECCV 2014. LNCS, vol. 8694, pp. 446–461. Springer, Cham (2014). https://doi.org/10.1007/978-3-319-10599-4_29
4. Carvalho, M., Cadène, R., Picard, D., Soulier, L., Thome, N., Cord, M.: Cross-modal retrieval in the cooking context: learning semantic text-image embeddings. In: The 41st International ACM SIGIR Conference on Research & Development in Information Retrieval, pp. 35–44. ACM (2018)
5. Chandu, K., Nyberg, E., Black, A.W.: Storyboarding of recipes: grounded contextual generation. In: Proceedings of the 57th Annual Meeting of the Association for Computational Linguistics, pp. 6040–6046. Association for Computational Linguistics, Florence, Italy, July 2019. https://doi.org/10.18653/v1/P19-1606. https://www.aclweb.org/anthology/P19-1606
6. Chung, J., Kastner, K., Dinh, L., Goel, K., Courville, A.C., Bengio, Y.: A recurrent latent variable model for sequential data. In: Advances in Neural Information Processing Systems, pp. 2980–2988 (2015)
7. Deng, J., Dong, W., Socher, R., Li, L.J., Li, K., Fei-Fei, L.: ImageNet: a large-scale hierarchical image database. In: 2009 IEEE Conference on Computer Vision and Pattern Recognition, pp. 248–255. IEEE (2009)
8. Faghri, F., Fleet, D.J., Kiros, J.R., Fidler, S.: VSE++: improving visual-semantic embeddings with hard negatives. arXiv preprint arXiv:1707.05612 (2017)
9. Feng, F., Wang, X., Li, R.: Cross-modal retrieval with correspondence autoencoder. In: Proceedings of the 22nd ACM International Conference on Multimedia, pp. 7–16. ACM (2014)
10. Feng, Y., Lapata, M.: Topic models for image annotation and text illustration. In: Human Language Technologies: The 2010 Annual Conference of the North American Chapter of the Association for Computational Linguistics (HLT 2010), pp. 831–839. Association for Computational Linguistics, Stroudsburg, PA, USA (2010). http://dl.acm.org/citation.cfm?id=1857999.1858124
11. Goldberg, A.B., Zhu, X., Dyer, C.R., Eldawy, M., Heng, L.: Easy as ABC?: facilitating pictorial communication via semantically enhanced layout. In: Proceedings of the Twelfth Conference on Computational Natural Language Learning (CoNLL 2008), pp. 119–126. Association for Computational Linguistics, Stroudsburg, PA, USA (2008). http://dl.acm.org/citation.cfm?id=1596324.1596345

12. Hardoon, D.R., Szedmak, S., Shawe-Taylor, J.: Canonical correlation analysis: an overview with application to learning methods. Neural Comput. **16**(12), 2639–2664 (2004)
13. He, Y., Xiang, S., Kang, C., Wang, J., Pan, C.: Cross-modal retrieval via deep and bidirectional representation learning. IEEE Trans. Multimed. **18**(7), 1363–1377 (2016)
14. Huang, T.H.K., et al.: Visual storytelling. In: Proceedings of the 2016 Conference of the North American Chapter of the Association for Computational Linguistics: Human Language Technologies, pp. 1233–1239 (2016)
15. Joshi, D., Wang, J.Z., Li, J.: The story picturing engine–a system for automatic text illustration. ACM Trans. Multimed. Comput. Commun. Appl. **2**(1), 68–89 (2006). https://doi.org/10.1145/1126004.1126008
16. Karpathy, A., Fei-Fei, L.: Deep visual-semantic alignments for generating image descriptions. IEEE Trans. Pattern Anal. Mach. Intell. **39**(4), 664–676 (2017). https://doi.org/10.1109/TPAMI.2016.2598339
17. Kim, G., Moon, S., Sigal, L.: Ranking and retrieval of image sequences from multiple paragraph queries. In: Proceedings of the IEEE Conference on Computer Vision and Pattern Recognition, pp. 1993–2001 (2015)
18. Kiros, R., Salakhutdinov, R., Zemel, R.S.: Unifying visual-semantic embeddings with multimodal neural language models. CoRR abs/1411.2539 (2014). http://arxiv.org/abs/1411.2539
19. Lin, T., et al.: Microsoft COCO: common objects in context. CoRR abs/1405.0312 (2014). http://arxiv.org/abs/1405.0312
20. Liu, Y., Fu, J., Mei, T., Chen, C.W.: Let your photos talk: generating narrative paragraph for photo stream via bidirectional attention recurrent neural networks. In: Thirty-First AAAI Conference on Artificial Intelligence (2017)
21. van der Maaten, L., Hinton, G.: Visualizing data using t-SNE. J. Mach. Learn. Res. **9**(Nov), 2579–2605 (2008)
22. Marin, J., et al.: Recipe1M+: a dataset for learning cross-modal embeddings for cooking recipes and food images. arXiv preprint arXiv:1810.06553 (2018)
23. Ngiam, J., Khosla, A., Kim, M., Nam, J., Lee, H., Ng, A.Y.: Multimodal deep learning. In: Proceedings of the 28th International Conference on Machine Learning (ICML 2011), pp. 689–696 (2011)
24. Park, C.C., Kim, G.: Expressing an image stream with a sequence of natural sentences. In: Advances in Neural Information Processing Systems, pp. 73–81 (2015)
25. Plummer, B.A., Wang, L., Cervantes, C.M., Caicedo, J.C., Hockenmaier, J., Lazebnik, S.: Flickr30k entities: collecting region-to-phrase correspondences for richer image-to-sentence models. In: 2015 IEEE International Conference on Computer Vision (ICCV), pp. 2641–2649 (2015)
26. Quadrianto, N., Lampert, C.: Learning multi-view neighborhood preserving projections. In: Proceedings of the 28th International Conference on Machine Learning, Washington, USA, 28 June–2 July 2011, pp. 425–432. Association for Computing Machinery (2011)
27. Ravi, H., Wang, L., Muniz, C., Sigal, L., Metaxas, D., Kapadia, M.: Show me a story: towards coherent neural story illustration. In: Proceedings of the IEEE Conference on Computer Vision and Pattern Recognition, pp. 7613–7621 (2018)
28. Rosipal, R., Krämer, N.: Overview and recent advances in partial least squares. In: Saunders, C., Grobelnik, M., Gunn, S., Shawe-Taylor, J. (eds.) SLSFS 2005. LNCS, vol. 3940, pp. 34–51. Springer, Heidelberg (2006). https://doi.org/10.1007/11752790_2

29. Salvador, A., et al.: Learning cross-modal embeddings for cooking recipes and food images. In: Proceedings of the IEEE Conference on Computer Vision and Pattern Recognition, pp. 3020–3028 (2017)
30. Sharma, P., Ding, N., Goodman, S., Soricut, R.: Conceptual captions: a cleaned, hypernymed, image alt-text dataset for automatic image captioning. In: Proceedings of the 56th Annual Meeting of the Association for Computational Linguistics (Volume 1: Long Papers), pp. 2556–2565. Association for Computational Linguistics (2018). http://aclweb.org/anthology/P18-1238
31. Su, J., Wu, S., Xiong, D., Lu, Y., Han, X., Zhang, B.: Variational recurrent neural machine translation. In: Thirty-Second AAAI Conference on Artificial Intelligence (2018)
32. Wang, J., He, Y., Kang, C., Xiang, S., Pan, C.: Image-text cross-modal retrieval via modality-specific feature learning. In: Proceedings of the 5th ACM on International Conference on Multimedia Retrieval, pp. 347–354. ACM (2015)
33. Wang, W., Yang, X., Ooi, B.C., Zhang, D., Zhuang, Y.: Effective deep learning-based multi-modal retrieval. VLDB J. **25**(1), 79–101 (2015). https://doi.org/10.1007/s00778-015-0391-4
34. Yagcioglu, S., Erdem, A., Erdem, E., Ikizler-Cinbis, N.: RecipeQA: a challenge dataset for multimodal comprehension of cooking recipes. arXiv preprint arXiv:1809.00812 (2018)

A Hierarchical Model for Data-to-Text Generation

Clément Rebuffel[1,2](✉), Laure Soulier[1], Geoffrey Scoutheeten[2], and Patrick Gallinari[1,3]

[1] LIP6, Sorbonne Université, Paris, France
{clement.rebuffel,laure.soulier,patrick.gallinari}@lip6.fr
[2] BNP Paribas, Paris, France
{clement.rebuffel,geoffrey.scoutheeten}@bnpparibas.com
[3] Criteo AI Lab, Paris, France

Abstract. Transcribing structured data into natural language descriptions has emerged as a challenging task, referred to as "data-to-text". These structures generally regroup multiple elements, as well as their attributes. Most attempts rely on translation encoder-decoder methods which linearize elements into a sequence. This however loses most of the structure contained in the data. In this work, we propose to overpass this limitation with a hierarchical model that encodes the data-structure at the element-level and the structure level. Evaluations on RotoWire show the effectiveness of our model w.r.t. qualitative and quantitative metrics.

Keywords: Data-to-text · Hierarchical encoding · Language generation

1 Introduction

Knowledge and/or data is often modeled in a structure, such as indexes, tables, key-value pairs, or triplets. These data, by their nature (e.g., raw data or long time-series data), are not easily usable by humans; outlining their crucial need to be synthesized. Recently, numerous works have focused on leveraging structured data in various applications, such as question answering [24,34] or table retrieval [7,32]. One emerging research field consists in transcribing data-structures into natural language in order to ease their understandablity and their usablity. This field is referred to as "data-to-text" [8] and has its place in several application domains (such as journalism [22] or medical diagnosis [25]) or wide-audience applications (such as financial [26] and weather reports [30], or sport broadcasting [4,39]). As an example, Fig. 1 shows a data-structure containing statistics on NBA basketball games, paired with its corresponding journalistic description.

Designing data-to-text models gives rise to two main challenges: (1) understanding structured data and (2) generating associated descriptions. Recent data-to-text models [18,28,29,39] mostly rely on an encoder-decoder architecture [2] in which the data-structure is first encoded sequentially into a fixed-size

J. M. Jose et al. (Eds.): ECIR 2020, LNCS 12035, pp. 65–80, 2020.
https://doi.org/10.1007/978-3-030-45439-5_5

TEAM	H/V	WINS	LOSSES	PTS	REB	AST	...
Hawks	H	46	12	95	42	27	...
Magic	V	19	41	88	40	22	...

PLAYER	PTS	REB	AST	STL	BLK	CITY	...
Al Horford	17	13	4	2	0	Atlanta	...
Kyle Korver	8	3	2	1	2	Atlanta	...
Jeff Teague	17	0	7	2	0	Atlanta	...
N. Vucevic	21	15	3	1	1	Orlando	...
Tobias Harris	15	4	1	2	1	Orlando	...
...	...	...	...	...	...	...	...

H/V: home or visiting; PTS: points; REB: rebounds; AST: assists; STL: steals; BLK: blocks

The **Atlanta Hawks (46-12)** beat the **Orlando Magic (19-41) 95-88** on Friday. **Al Horford** had a good all-around game, putting up **17 points, 13 rebounds, four assists and two steals** in a tough matchup against **Nikola Vucevic**. **Kyle Korver** was the lone Atlanta starter not to reach double figures in points. **Jeff Teague** bounced back from an illness, he scored **17 points** to go along with **seven assists and two steals**. After a rough start to the month, the **Hawks** have won three straight and sit atop the Eastern Conference with a nine game lead on the second place Toronto Raptors. The **Magic** lost in devastating fashion to the Miami Heat in overtime Wednesday. They blew a seven point lead with 43 seconds remaining and they might have carried that with them into Friday's contest against the **Hawks**. **Vucevic** led the **Magic** with **21 points and 15 rebounds**. **Aaron Gordon** (ankle) and **Evan Fournier** (hip) were unable to play due to injury. The **Magic** have four teams between them and the eighth and final playoff spot in the Eastern Conference. The **Magic** will host the Charlotte Hornets on Sunday, and the **Hawks** with take on the Heat in Miami on Saturday.

Fig. 1. Example of structured data from the RotoWire dataset. Rows are entities (either a team or a player) and each cell a record, its key being the column label and its value the cell content. Factual mentions from the table are boldfaced in the description.

vectorial representation by an encoder. Then, a decoder generates words conditioned on this representation. With the introduction of the attention mechanism [19] on one hand, which computes a context focused on important elements from the input at each decoding step and, on the other hand, the copy mechanism [11,33] to deal with unknown or rare words, these systems produce fluent and domain comprehensive texts. For instance, Roberti et al. [31] train a character-wise encoder-decoder to generate descriptions of restaurants based on their attributes, while Puduppully et al. [28] design a more complex two-step decoder: they first generate a plan of elements to be mentioned, and then condition text generation on this plan. Although previous work yield overall good results, we identify two important caveats, that hinder precision (*i.e.* factual mentions) in the descriptions:

1. *Linearization of the data-structure.* In practice, most works focus on introducing innovating decoding modules, and still represent data as a unique sequence of elements to be encoded. For example, the table from Fig. 1 would be linearized to [(Hawks, H/V, H), ..., (Magic, H/V, V), ...], effectively leading to losing distinction between rows, and therefore entities. To the best of our knowledge, only Liu et al. [17,18] propose encoders constrained by the structure but these approaches are designed for single-entity structures.
2. *Arbitrary ordering of unordered collections in recurrent networks (RNN).* Most data-to-text systems use RNNs as encoders (such as GRUs or LSTMs), these architectures have however some limitations. Indeed, they require in practice their input to be fed sequentially. This way of encoding unordered sequences (*i.e.* collections of entities) implicitly assumes an arbitrary order within the collection which, as demonstrated by Vinyals et al. [37], significantly impacts the learning performance.

To address these shortcomings, we propose a new structured-data encoder assuming that structures should be hierarchically captured. Our contribution focuses on the encoding of the data-structure, thus the decoder is chosen to be a classical module as used in [28,39]. Our contribution is threefold:

- We model the general structure of the data using a two-level architecture, first encoding all entities on the basis of their elements, then encoding the data structure on the basis of its entities;
- We introduce the Transformer encoder [36] in data-to-text models to ensure robust encoding of each element/entities in comparison to all others, no matter their initial positioning;
- We integrate a hierarchical attention mechanism to compute the hierarchical context fed into the decoder.

We report experiments on the RotoWire benchmark [39] which contains around $5K$ statistical tables of NBA basketball games paired with human-written descriptions. Our model is compared to several state-of-the-art models. Results show that the proposed architecture outperforms previous models on BLEU score and is generally better on qualitative metrics.

In the following, we first present a state-of-the art of data-to-text literature (Sect. 2), and then describe our proposed hierarchical data encoder (Sect. 3). The evaluation protocol is presented in Sect. 4, followed by the results (Sect. 5). Section 6 concludes the paper and presents perspectives.

2 Related Work

Until recently, efforts to bring out semantics from structured-data relied heavily on expert knowledge [6,30]. For example, in order to better transcribe numerical time series of weather data to a textual forecast, Reiter et al. [30] devise complex template schemes in collaboration with weather experts to build a consistent set of data-to-word rules.

Modern approaches to the wide range of tasks based on structured-data (*e.g.* table retrieval [7,41], table classification [9], question answering [12]) now propose to leverage progress in deep learning to represent these data into a semantic vector space (also called embedding space). In parallel, an emerging task, called "data-to-text", aims at describing structured data into a natural language description. This task stems from the neural machine translation (NMT) domain, and early work [1,15,39] represent the data records as a single sequence of facts to be entirely translated into natural language. Wiseman et al. [39] show the limits of traditional NMT systems on larger structured-data, where NMT systems fail to accurately extract salient elements.

To improve these models, a number of work [16,28,40] proposed innovating decoding modules based on planning and templates, to ensure factual and coherent mentions of records in generated descriptions. For example, Puduppully et al. [28] propose a two-step decoder which first targets specific records and then use them as a plan for the actual text generation. Similarly, Li et al. [16] proposed a delayed copy mechanism where their decoder also acts in two steps: (1) using a classical LSTM decoder to generate delexicalized text and (2) using a pointer network [38] to replace placeholders by records from the input data.

Closer to our work, very recent work [17,18,29] have proposed to take into account the data structure. More particularly, Puduppully et al. [29] follow entity-centric theories [10,20] and propose a model based on dynamic entity representation at decoding time. It consists in conditioning the decoder on entity representations that are updated during inference at each decoding step. On the other hand, Liu et al. [17,18] rather focus on introducing structure into the encoder. For instance, they propose a dual encoder [17] which encodes separately the sequence of element names and the sequence of element values. These approaches are however designed for single-entity data structures and do not account for delimitation between entities.

Our contribution differs from previous work in several aspects. First, instead of flatly concatenating elements from the data-structure and encoding them as a sequence [18,28,39], we constrain the encoding to the underlying structure of the input data, so that the delimitation between entities remains clear throughout the process. Second, unlike all works in the domain, we exploit the Transformer architecture [36] and leverage its particularity to directly compare elements with each others in order to avoid arbitrary assumptions on their ordering. Finally, in contrast to [5,29] that use a complex updating mechanism to obtain a dynamic representation of the input data and its entities, we argue that explicit hierarchical encoding naturally guides the decoding process via hierarchical attention.

3 Hierarchical Encoder Model for Data-to-Text

In this section we introduce our proposed hierarchical model taking into account the data structure. We outline that the decoding component aiming to generate descriptions is considered as a black-box module so that our contribution is focused on the encoding module. We first describe the model overview, before detailing the hierarchical encoder and the associated hierarchical attention.

3.1 Notation and General Overview

Let's consider the following notations:

• An *entity* e_i is a set of J_i unordered records $\{r_{i,1}, ..., r_{i,j}, ..., r_{i,J_i}\}$; where record $r_{i,j}$ is defined as a pair of *key* $k_{i,j}$ and *value* $v_{i,j}$. We outline that J_i might differ between entities.

• A *data-structure* s is an unordered set of I entities e_i. We thus denote $s := \{e_1, ..., e_i, ..., e_I\}$.

• For each data-structure, a textual *description* y is associated. We refer to the first t words of a description y as $y_{1:t}$. Thus, the full sequence of words can be noted as $y = y_{1:T}$.

• The *dataset* $\mathcal{D}$ is a collection of N aligned (data-structure, description) pairs (s, y).

For instance, Fig. 1 illustrates a data-structure associated with a description. The data-structure includes a set of entities (*Hawks, Magic, Al Horford, Jeff Teague, ...*). The entity Jeff Teague is modeled as a set of records {(PTS, 17), (REB, 0), (AST, 7) ...} in which, e.g., the record (PTS, 17) is characterized by a *key* (PTS) and a *value* (17).

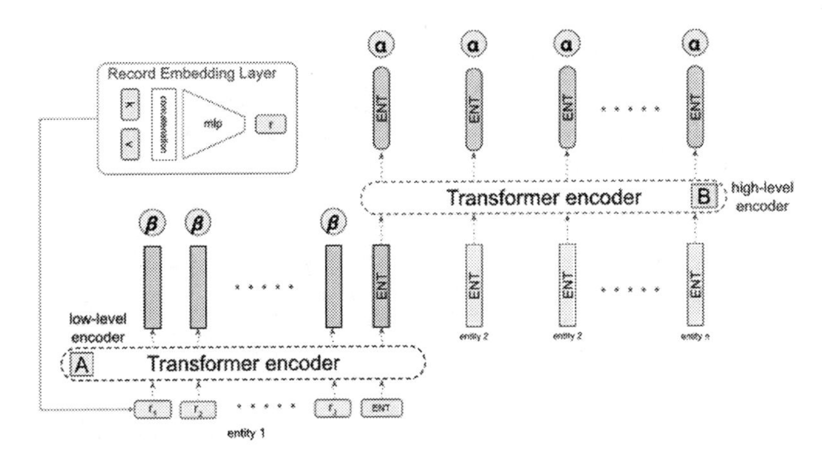

Fig. 2. Diagram of the proposed hierarchical encoder. Once the records are embedded, the low-level encoder works on each entity independently (A); then the high-level encoder encodes the collection of entities (B). In circles, we represent the hierarchical attention scores: the α scores at the entity level and the β scores at the record level.

For each data-structure s in $\mathcal{D}$, the objective function aims to generate a description $\hat{y}$ as close as possible to the ground truth y. This objective function optimizes the following log-likelihood over the whole dataset $\mathcal{D}$:

$$\arg\max_{\theta} \mathcal{L}(\theta) = \arg\max_{\theta} \sum_{(s,y)\in\mathcal{D}} \log P(\hat{y} = y \mid s; \theta) \tag{1}$$

where θ stands for the model parameters and $P(\hat{y} = y \mid s; \theta)$ the probability of the model to generate the adequate description y for table s.

During inference, we generate the sequence $\hat{y}^*$ with the maximum a posteriori probability conditioned on table s. Using the chain rule, we get:

$$\hat{y}^*_{1:T} = \arg\max_{\hat{y}_{1:T}} \prod_{t=1}^{T} P(\hat{y}_t | \hat{y}_{1:t-1}; s; \theta) \tag{2}$$

This equation is intractable in practice, we approximate a solution using beam search, as in [17,18,28,29,39].

Our model follows the encoder-decoder architecture [2]. Because our contribution focuses on the encoding process, we chose the decoding module used in [28,39]: a two-layers LSTM network with a copy mechanism. In order to supervise this mechanism, we assume that each record value that also appears in the target is copied from the data-structure and we train the model to switch between freely generating words from the vocabulary and copying words from the input. We now describe the hierarchical encoder and the hierarchical attention.

3.2 Hierarchical Encoding Model

As outlined in Sect. 2, most previous work [16,28,29,39,40] make use of flat encoders that do not exploit the data structure. To keep the semantics of each element from the data-structure, we propose a hierarchical encoder which relies on two modules. The first one (module A in Fig. 2) is called *low-level encoder* and encodes entities on the basis of their records; the second one (module B), called *high-level encoder*, encodes the data-structure on the basis of its underlying entities. In the low-level encoder, the traditional embedding layer is replaced by a record embedding layer as in [18,28,39]. We present in what follows the record embedding layer and introduce our two hierarchical modules.

Record Embedding Layer. The first layer of the network consists in learning two embedding matrices to embed the record keys and values. Keys $k_{i,j}$ are embedded to $\mathbf{k}_{i,j} \in \mathbb{R}^d$ and values $v_{i,j}$ to $\mathbf{v}_{i,j} \in \mathbb{R}^d$, with d the size of the embedding. As in previous work [18,28,39], each record embedding $\mathbf{r}_{i,j}$ is computed by a linear projection on the concatenation $[\mathbf{k}_{i,j};\ \mathbf{v}_{i,j}]$ followed by a non linearity:

$$\mathbf{r}_{i,j} = \mathrm{ReLU}(\mathbf{W}_r[\mathbf{k}_{i,j};\ \mathbf{v}_{i,j}] + \mathbf{b}_r) \tag{3}$$

where $\mathbf{W}_r \in \mathbb{R}^{2d \times d}$ and $\mathbf{b}_r \in \mathbb{R}^d$ are learnt parameters.

The low-level encoder aims at encoding a collection of records belonging to the same entity while the high-level encoder encodes the whole set of entities. Both the low-level and high-level encoders consider their input elements as unordered. We use the Transformer architecture from [36]. For each encoder, we have the following peculiarities:

- the **Low-level encoder** encodes each entity e_i on the basis of its record embeddings $\mathbf{r}_{i,j}$. Each record embedding $\mathbf{r}_{i,j}$ is compared to other record embeddings to learn its final hidden representation $\mathbf{h}_{i,j}$. Furthermore, we add a special record [ENT] for each entity, illustrated in Fig. 2 as the last record. Since entities might have a variable number of records, this token allows to aggregate final hidden record representations $\{\mathbf{h}_{i,j}\}_{j=1}^{J_i}$ in a fixed-sized representation vector $\mathbf{h}_i$.
- the **High-level encoder** encodes the data-structure on the basis of its entity representation $\mathbf{h}_i$. Similarly to the **Low-level encoder**, the final hidden state $\mathbf{e_i}$ of an entity is computed by comparing entity representation $\mathbf{h}_i$ with each others. The data-structure representation $\mathbf{z}$ is computed as the mean of these entity representations, and is used for the decoder initialization.

3.3 Hierarchical Attention

To fully leverage the hierarchical structure of our encoder, we propose two variants of hierarchical attention mechanism to compute the context fed to the decoder module.

- *Traditional Hierarchical Attention.* As in [29], we hypothesize that a dynamic context should be computed in two steps: first attending to entities, then to records corresponding to these entities. To implement this hierarchical attention, at each decoding step t, the model learns a first set of attention scores $\alpha_{i,t}$ over entities e_i and a second set of attention scores $\beta_{i,j,t}$ over records $r_{i,j}$ belonging to entity e_i. The $\alpha_{i,t}$ scores are normalized to form a distribution over all entities e_i, and $\beta_{i,j,t}$ scores are normalized to form a distribution over records $r_{i,j}$ of entity e_i. Each entity is then represented as a weighted sum of its record embeddings, and the entire data structure is represented as a weighted sum of the entity representations. The dynamic context is computed as:

$$\mathbf{c_t} = \sum_{i=1}^{I}(\alpha_{i,t}(\sum_{j}\beta_{i,j,t}\mathbf{r}_{i,j})) \quad (4)$$

$$where \quad \alpha_{i,t} \propto exp(\mathbf{d}_t \mathbf{W}_\alpha \mathbf{e}_i) \quad and \quad \beta_{i,j,t} \propto exp(\mathbf{d}_t \mathbf{W}_\beta \mathbf{h}_{i,j}) \quad (5)$$

where $\mathbf{d_t}$ is the decoder hidden state at time step t, $\mathbf{W}_\alpha \in \mathbb{R}^{d \times d}$ and $\mathbf{W}_\beta \in \mathbb{R}^{d \times d}$ are learnt parameters, $\sum_i \alpha_{i,t} = 1$, and for all $i \in \{1, ..., I\}$ $\sum_j \beta_{i,j,t} = 1$.

• *Key-guided Hierarchical Attention.* This variant follows the intuition that once an entity is chosen for mention (thanks to $\alpha_{i,t}$), only the type of records is important to determine the content of the description. For example, when deciding to mention a player, all experts automatically report his score without consideration of its specific value. To test this intuition, we model the attention scores by computing the $\beta_{i,j,t}$ scores from Eq. (5) solely on the embedding of the *key* rather than on the full record representation $\mathbf{h}_{i,j}$:

$$\hat{\beta}_{i,j,t} \propto exp(\mathbf{d}_t \mathbf{W}_{\alpha_2} \mathbf{k}_{i,j}) \quad (6)$$

Please note that the different embeddings and the model parameters presented in the model components are learnt using Eq. 1.

4 Experimental Setup

4.1 The Rotowire Dataset

To evaluate the effectiveness of our model, and demonstrate its flexibility at handling heavy data-structure made of several types of entities, we used the RotoWire dataset [39]. It includes basketball games statistical tables paired with journalistic descriptions of the games, as can be seen in the example of Fig. 1. The descriptions are professionally written and average 337 words with a vocabulary size of 11.3K. There are 39 different record keys, and the average number of records (resp. entities) in a single data-structure is 628 (resp. 28). Entities are of two types, either team or player, and player descriptions depend on their involvement in the game. We followed the data partitions introduced with the dataset and used a train/validation/test sets of respectively 3, 398/727/728 (data-structure, description) pairs.

4.2 Evaluation Metrics

We evaluate our model through two types of metrics. The BLEU score [23] aims at measuring to what extent the generated descriptions are literally closed to the ground truth. The second category designed by [39] is more qualitative.

BLEU Score. The **BLEU score** [23] is commonly used as an evaluation metric in text generation tasks. It estimates the correspondence between a machine output and that of a human by computing the number of co-occurrences for ngrams ($n \in 1, 2, 3, 4$) between the generated candidate and the ground truth. We use the implementation code released by [27].

Information Extraction-Oriented Metrics. These metrics estimate the ability of our model to integrate elements from the table in its descriptions. Particularly, they compare the gold and generated descriptions and measure to what extent the extracted relations are aligned or differ. To do so, we follow the protocol presented in [39]. First, we apply an information extraction (IE) system trained on labeled relations from the gold descriptions of the RotoWire train dataset. Entity-value pairs are extracted from the descriptions. For example, in the sentence *Isaiah Thomas led the team in scoring, totaling 23 points [...].*, an IE tool will extract the pair (Isaiah Thomas, 23, PTS). Second, we compute three metrics on the extracted information:

- **Relation Generation (RG)** estimates how well the system is able to generate text containing factual (i.e., correct) records. We measure the precision and absolute number (denoted respectively RG-P% and RG-#) of unique relations r extracted from $\hat{y}_{1:T}$ that also appear in s.
- **Content Selection (CS)** measures how well the generated document matches the gold document in terms of mentioned records. We measure the precision and recall (denoted respectively CS-P% and CS-R%) of unique relations r extracted from $\hat{y}_{1:T}$ that are also extracted from $y_{1:T}$.
- **Content Ordering (CO)** analyzes how well the system orders the records discussed in the description. We measure the normalized Damerau-Levenshtein distance [3] between the sequences of records extracted from $\hat{y}_{1:T}$ that are also extracted from $y_{1:T}$.

CS primarily targets the "what to say" aspect of evaluation, CO targets the "how to say it" aspect, and RG targets both. Note that for CS, CO, RG-% and BLEU metrics, higher is better; which is not true for RG-#. The IE system used in the experiments is able to extract an average of 17 factual records from gold descriptions. In order to mimic a human expert, a generative system should approach this number and not overload generation with brute facts.

4.3 Baselines

We compare our hierarchical model against three systems. For each of them, we report the results of the best performing models presented in each paper.

- *Wiseman* [39] is a standard encoder-decoder system with copy mechanism.
- *Li* [16] is a standard encoder-decoder with a delayed copy mechanism: text is first generated with placeholders, which are replaced by salient records extracted from the table by a pointer network.
- *Puduppully-plan* [28] acts in two steps: a first standard encoder-decoder generates a plan, *i.e.* a list of salient records from the table; a second standard encoder-decoder generates text from this plan.
- *Puduppully-updt* [29]. It consists in a standard encoder-decoder, with an added module aimed at updating record representations during the generation process. At each decoding step, a gated recurrent network computes which records should be updated and what should be their new representation.

Model Scenarios. We test the importance of the input structure by training different variants of the proposed architecture:

- *Flat*, where we feed the input sequentially to the encoder, losing all notion of hierarchy. As a consequence, the model uses standard attention. This variant is closest to *Wiseman*, with the exception that we use a Transformer to encode the input sequence instead of an RNN.
- *Hierarchical-kv* is our full hierarchical model, with traditional hierarchical attention, *i.e.* where attention over records is computed on the full record encoding, as in Eq. (5).
- *Hierarchical-k* is our full hierarchical model, with key-guided hierarchical attention, *i.e.* where attention over records is computed only on the record key representations, as in Eq. (6).

Table 1. Evaluation on the RotoWire testset using relation generation (RG) count (#) and precision (P%), content selection (CS) precision (P%) and recall (R%), content ordering (CO), and BLEU. –: number of parameters unavailable.

	BLEU	RG		CS			CO	Nb
		P%	#	P%	R%	F1		Params
Gold descriptions	100	96.11	17.31	100	100	100	100	
Wiseman	14.5	75.62	**16.83**	32.80	39.93	36.2	15.62	45M
Li	16.19	84.86	19.31	30.81	38.79	34.34	16.34	–
Pudupully-plan	16.5	87.47	34.28	34.18	51.22	41	18.58	35M
Puduppully-updt	16.2	**92.69**	30.11	38.64	48.51	43.01	**20.17**	23M
Flat	$16.7_{.2}$	76.62_{1}	$18.54_{.6}$	$31.67_{.7}$	42.9_{1}	$36.42_{.4}$	$14.64_{.3}$	14M
Hierarchical-kv	$17_{.3}$	89.04_{1}	$21.46_{.9}$	$38.57_{1.2}$	$51.50_{.9}$	$44.19_{.7}$	$18.70_{.7}$	14M
Hierarchical-k	$\mathbf{17.5}_{.3}$	$89.46_{1.4}$	$21.17_{1.4}$	$\mathbf{39.47}_{1.4}$	$\mathbf{51.64}_{1}$	$\mathbf{44.7}_{.6}$	$18.90_{.7}$	14M

4.4 Implementation Details

The decoder is the one used in [28,29,39] with the same hyper-parameters. For the encoder module, both the low-level and high-level encoders use a two-layers multi-head self-attention with two heads. To fit with the small number of record keys in our dataset (39), their embedding size is fixed to 20. The size of the record value embeddings and hidden layers of the Transformer encoders are both set to 300. We use dropout at rate 0.5. The models are trained with a batch size of 64. We follow the training procedure in [36] and train the model for a fixed number of 25K updates, and average the weights of the last 5 checkpoints (at every 1K updates) to ensure more stability across runs. All models were trained with the Adam optimizer [13]; the initial learning rate is 0.001, and is reduced by half every 10K steps. We used beam search with beam size of 5 during inference. All the models are implemented in OpenNMT-py [14]. All code is available at https://github.com/KaijuML/data-to-text-hierarchical.

5 Results

Our results on the RotoWire testset are summarized in Table 1. For each proposed variant of our architecture, we report the mean score over ten runs, as well as the standard deviation in subscript. Results are compared to baselines [28,29,39] and variants of our models. We also report the result of the oracle (metrics on the gold descriptions). Please note that gold descriptions trivially obtain 100% on all metrics expect RG, as they are all based on comparison with themselves. RG scores are different, as the IE system is imperfect and fails to extract accurate entities 4% of the time. RG-# is an absolute count.

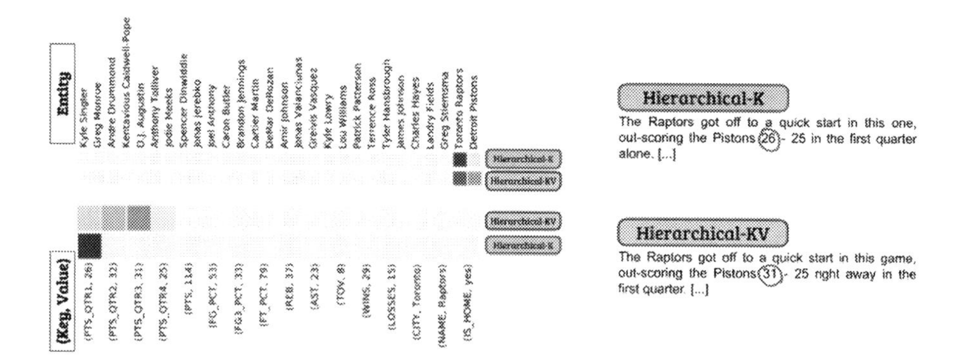

Fig. 3. Right: Comparison of a generated sentence from *Hierarchical-k* and *Hierarchical-kv*. Left: Attention scores over entities (top) and over records inside the selected entity (bottom) for both variants, during the decoding of respectively 26 or 31 (circled in red). (Color figure online)

Ablation Studies. To evaluate the impact of our model components, we first compare scenarios *Flat*, *Hierarchical-k*, and *Hierarchical-kv*. As shown in Table 1, we can see the lower results obtained by the *Flat* scenario compared to the other scenarios (*e.g.* BLEU 16.7 vs. 17.5 for resp. *Flat* and *Hierarchical-k*), suggesting the effectiveness of encoding the data-structure using a hierarchy. This is expected, as losing explicit delimitation between entities makes it harder a) for the encoder to encode semantics of the objects contained in the table and b) for the attention mechanism to extract salient entities/records.

Second, the comparison between scenario *Hierarchical-kv* and *Hierarchical-k* shows that omitting entirely the influence of the record values in the attention mechanism is more effective: this last variant performs slightly better in all metrics excepted CS-R%, reinforcing our intuition that focusing on the structure modeling is an important part of data encoding as well as confirming the intuition explained in Sect. 3.3: once an entity is selected, facts about this entity are relevant based on their key, not value which might add noise. To illustrate this intuition, we depict in Fig. 3 attention scores (recall $\alpha_{i,t}$ and $\beta_{i,j,t}$ from Eqs. (5) and (6)) for both variants *Hierarchical-kv* and *Hierarchical-k*. We particularly focus on the timestamp where the models should mention the number of points scored during the first quarter of the game. Scores of *Hierarchical-k* are sharp, with all of the weight on the correct record (PTS_QTR1, 26) whereas scores of *Hierarchical-kv* are more distributed over all PTS_QTR records, ultimately failing to retrieve the correct one.

Comparison w.r.t. Baselines. From a general point of view, we can see from Table 1 that our scenarios obtain significantly higher results in terms of BLEU over all models; our best model *Hierarchical-k* reaching 17.5 vs. 16.5 against the best baseline. This means that our models learns to generate fluent sequences of words, close to the gold descriptions, adequately picking up on domain lingo. Qualitative metrics are either better or on par with baselines. We show in Fig. 4 a text generated by our best model, which can be directly compared to the gold

The **Atlanta Hawks** (46 - 12) defeated the **Orlando Magic** (19 - 41) 95 - 88 on Monday at Philips Arena in Atlanta. The **Hawks** got out to a quick start in this one, out - scoring the **Magic** 28 - 16 in the first quarter alone. Along with the quick start, the **Hawks** were able to hold off the **Magic** late in the fourth quarter, out - scoring the **Magic** 19 - 21. The **Hawks** were led by **Nikola Vucevic**, who went 10 - for - 16 from the field and 0 - for - 0 from the three-point line to score a team - high of 21 points, while also adding 15 rebounds in 37 minutes. It was his second double - double in a row, a stretch where he's averaging 22 points and 17 rebounds. Notching a double - double of his own, **Al Horford** recorded 17 points (7 - 9 FG , 0 - 0 3Pt , 3 - 4 FT), 13 rebounds and four steals. He's now averaging 15 points and 6 rebounds on the year. **Paul Millsap** had a strong showing , posting 20 points (8 - 17 FG , 4 - 7 3Pt , 0 - 2 FT), four rebounds and three blocked shots. He's been a pleasant surprise for the **Magic** in the second half, as he's averaged 14 points and 5 rebounds over his last three games. **DeMarre Carroll** was the other starter in double figures, finishing with 15 points (6 - 12 FG , 3 - 6 3Pt), eight rebounds and three steals. He's had a nice stretch of three games , averaging 24 points, 3 rebounds and 2 assists over that span. **Tobias Harris** was the only other **Magic** player to reach double figures, scoring 15 points (5 - 9 FG , 2 - 4 3Pt , 3 - 4 FT). The **Magic** 's next game will be at home against the Miami Heat on Wednesday, while the **Magic** will travel to Charlotte to play the Hornets on Wednesday.

Fig. 4. Text generated by our best model. Entites are boldfaced, factual mentions are in green, erroneous mentions in red and hallucinations are in blue. (Color figure online)

description in Fig. 1. Generation is fluent and contains domain-specific expressions. As reflected in Table 1, the number of correct mentions (in green) outweights the number of incorrect mentions (in red). Please note that, as in previous work [16,28,29,39], generated texts still contain a number of incorrect facts, as well hallucinations (in blue): sentences that have no basis in the input data (e.g. *"[...] he's now averaging 22 points [...]."*). While not the direct focus of our work, this highlights that any operation meant to enrich the semantics of structured data can also enrich the data with incorrect facts.

Specifically, regarding all baselines, we can outline the following statements.

• Our hierarchical models achieve significantly better scores on all metrics when compared to the flat architecture *Wiseman*, reinforcing the crucial role of structure in data semantics and saliency. The analysis of RG metrics shows that *Wiseman* seems to be the more naturalistic in terms of number of factual mentions (RG#) since it is the closest scenario to the gold value (16.83 vs. 17.31 for resp. *Wiseman* and *Hierarchical-k*). However, *Wiseman* achieves only 75.62% of precision, effectively mentioning on average a total of 22.25 records (wrong or accurate), where our model *Hierarchical-k* scores a precision of 89.46%, leading to 23.66 total mentions, just slightly above *Wiseman*.

• The comparison between the *Flat* scenario and *Wiseman* is particularly interesting. Indeed, these two models share the same intuition to flatten the data-structure. The only difference stands on the encoder mechanism: bi-LSTM vs. Transformer, for *Wiseman* and *Flat* respectively. Results shows that our *Flat* scenario obtains a significant higher BLEU score (16.7 vs. 14.5) and generates fluent descriptions with accurate mentions (RG-P%) that are also included in the gold descriptions (CS-R%). This suggests that introducing the Transformer architecture is promising way to implicitly account for data structure.

• Our hierarchical models outperform the two-step decoders of *Li* and *Puduppully-plan* on both BLEU and all qualitative metrics, showing that capturing structure in the encoding process is more effective that predicting a structure in the decoder (i.e., planning or templating). While our models sensibly outperform in precision at factual mentions, the baseline *Puduppully-plan* reaches 34.28 mentions on average, showing that incorporating modules dedicated to entity extraction leads to over-focusing on entities; contrasting with our models that learn to generate more balanced descriptions.

• The comparison with *Puduppully-updt* shows that dynamically updating the encoding across the generation process can lead to better Content Ordering (CO) and RG-P%. However, this does not help with Content Selection (CS) since our best model *Hierarchical-k* obtains slightly better scores. Indeed, *Puduppully-updt* updates representations after each mention allowing to keep track of the mention history. This guides the ordering of mentions (CO metric), each step limiting more the number of candidate mentions (increasing RG-P%). In contrast, our model encodes saliency among records/entities more effectively (CS metric). We note that while our model encodes the data-structure once and for all, *Puduppully-updt* recomputes, via the updates, the encoding at each step and therefore significantly increases computation complexity. Combined with their

RG-# score of 30.11, we argue that our model is simpler, and obtains fluent description with accurate mentions in a more human-like fashion.

We would also like to draw attention to the number of parameters used by those architectures. We note that our scenarios relies on a lower number of parameters (14 millions) compared to all baselines (ranging from 23 to 45 millions). This outlines the effectiveness in the design of our model relying on a structure encoding, in contrast to other approach that try to learn the structure of data/descriptions from a linearized encoding.

6 Conclusion and Future Work

In this work we have proposed a hierarchical encoder for structured data, which (1) leverages the structure to form efficient representation of its input; (2) has strong synergy with the hierarchical attention of its associated decoder. This results in an effective and more light-weight model. Experimental evaluation on the RotoWire benchmark shows that our model outperforms competitive baselines in terms of BLEU score and is generally better on qualitative metrics. This way of representing structured databases may lead to automatic inference and enrichment, e.g., by comparing entities. This direction could be driven by very recent operation-guided networks [21,35]. In addition, we note that our approach can still lead to erroneous facts or even hallucinations. An interesting perspective might be to further constrain the model on the data structure in order to prevent inaccurate of even contradictory descriptions.

Acknowledgements. We would like to thank the H2020 project AI4EU (825619) which partially supports Laure Soulier and Patrick Gallinari.

References

1. Agarwal, S., Dymetman, M.: A surprisingly effective out-of-the-box char2char model on the E2E NLG challenge dataset. In: Proceedings of the 18th Annual SIGdial Meeting on Discourse and Dialogue, Saarbrücken, Germany, 15–17 August 2017, pp. 158–163 (2017). https://www.aclweb.org/anthology/W17-5519/
2. Bahdanau, D., Cho, K., Bengio, Y.: Neural machine translation by jointly learning to align and translate (2014). http://arxiv.org/abs/1409.0473, cite arxiv:1409.0473Comment. Accepted at ICLR 2015 as oral presentation
3. Brill, E., Moore, R.C.: An improved error model for noisy channel spelling correction. In: Proceedings of the 38th Annual Meeting on Association for Computational Linguistics (ACL 2000), pp. 286–293. Association for Computational Linguistics, Stroudsburg, PA, USA (2000). https://doi.org/10.3115/1075218.1075255
4. Chen, D.L., Mooney, R.J.: Learning to sportscast: a test of grounded language acquisition. In: Proceedings of the 25th International Conference on Machine Learning (ICML 2008), pp. 128–135. ACM, New York (2008). https://doi.org/10.1145/1390156.1390173

5. Clark, E., Ji, Y., Smith, N.A.: Neural text generation in stories using entity representations as context. In: Proceedings of the 2018 Conference of the North American Chapter of the Association for Computational Linguistics: Human Language Technologies, Volume 1 (Long Papers), pp. 2250–2260. Association for Computational Linguistics, New Orleans, Louisiana, June 2018. https://doi.org/10.18653/v1/N18-1204. https://www.aclweb.org/anthology/N18-1204
6. Deng, D., Jiang, Y., Li, G., Li, J., Yu, C.: Scalable column concept determination for web tables using large knowledge bases. In: Proceedings of the VLDB Endowment, vol. 6, no. 13, pp. 1606–1617, August 2013. https://doi.org/10.14778/2536258.2536271. http://dl.acm.org/citation.cfm?doid=2536258.2536271
7. Deng, L., Zhang, S., Balog, K.: Table2Vec: neural word and entity embeddings for table population and retrieval. In: Proceedings of the 42nd International ACM SIGIR Conference on Research and Development in Information Retrieval (SIGIR 2019), pp. 1029–1032. ACM Press, Paris (2019). https://doi.org/10.1145/3331184.3331333. http://dl.acm.org/citation.cfm?doid=3331184.3331333
8. Gatt, A., Krahmer, E.: Survey of the state of the art in natural language generation: core tasks, applications and evaluation. J. Artif. Int. Res. **61**(1), 65–170 (2018). http://dl.acm.org/citation.cfm?id=3241691.3241693
9. Ghasemi-Gol, M., Szekely, P.A.: TabVec: table vectors for classification of web tables. CoRR abs/1802.06290 (2018). http://arxiv.org/abs/1802.06290
10. Grosz, B., Joshi, A., Weinstein, S.: Centering: a framework for modelling the coherence of discourse. Technical Reports (CIS), January 1995
11. Gulcehre, C., Ahn, S., Nallapati, R., Zhou, B., Bengio, Y.: Pointing the unknown words. In: Proceedings of the 54th Annual Meeting of the Association for Computational Linguistics (Volume 1: Long Papers), pp. 140–149. Association for Computational Linguistics, Berlin, Germany, August 2016. https://doi.org/10.18653/v1/P16-1014
12. Haug, T., Ganea, O.-E., Grnarova, P.: Neural multi-step reasoning for question answering on semi-structured tables. In: Pasi, G., Piwowarski, B., Azzopardi, L., Hanbury, A. (eds.) ECIR 2018. LNCS, vol. 10772, pp. 611–617. Springer, Cham (2018). https://doi.org/10.1007/978-3-319-76941-7_52
13. Kingma, D.P., Ba, J.: Adam: a method for stochastic optimization (2014). http://arxiv.org/abs/1412.6980, cite arxiv:1412.6980Comment. Published as a conference paper at the 3rd International Conference for Learning Representations, San Diego (2015)
14. Klein, G., Kim, Y., Deng, Y., Senellart, J., Rush, A.M.: OpenNMT: open-source toolkit for neural machine translation. In: Proceedings of the ACL (2017). https://doi.org/10.18653/v1/P17-4012.
15. Lebret, R., Grangier, D., Auli, M.: Neural text generation from structured data with application to the biography domain. In: Proceedings of the 2016 Conference on Empirical Methods in Natural Language Processing, pp. 1203–1213. Association for Computational Linguistics, Austin, Texas, November 2016. https://doi.org/10.18653/v1/D16-1128. https://www.aclweb.org/anthology/D16-1128
16. Li, L., Wan, X.: Point precisely: towards ensuring the precision of data in generated texts using delayed copy mechanism. In: Proceedings of the 27th International Conference on Computational Linguistics, pp. 1044–1055. Association for Computational Linguistics, Santa Fe, New Mexico, USA, August 2018
17. Liu, T., Luo, F., Xia, Q., Ma, S., Chang, B., Sui, Z.: Hierarchical encoder with auxiliary supervision for neural table-to-text generation: learning better representation for tables. In: Proceedings of the AAAI Conference on Artificial Intelligence, vol. 33, pp. 6786–6793, July 2019. https://doi.org/10.1609/aaai.v33i01.33016786

18. Liu, T., Wang, K., Sha, L., Chang, B., Sui, Z.: Table-to-text generation by structure-aware Seq2seq learning. In: AAAI (2018)
19. Luong, T., Pham, H., Manning, C.D.: Effective approaches to attention-based neural machine translation. In: Proceedings of the 2015 Conference on Empirical Methods in Natural Language Processing, pp. 1412–1421. Association for Computational Linguistics, Lisbon, Portugal, September 2015. https://doi.org/10.18653/v1/D15-1166. https://www.aclweb.org/anthology/D15-1166
20. Mann, W.C., Thompson, S.A.: Rhetorical structure theory: toward a functional theory of text organization. Text - Interdisc. J. Study Discourse **8**, 243–281 (1988)
21. Nie, F., Wang, J., Yao, J., Pan, R., Lin, C.: Operation-guided neural networks for high fidelity data-to-text generation. In: Proceedings of the 2018 Conference on Empirical Methods in Natural Language Processing, Brussels, Belgium, 31 October–4 November 2018, pp. 3879–3889 (2018). https://www.aclweb.org/anthology/D18-1422/
22. Oremus, W.: The First News Report on the L.A. Earthquake Was Written by a Robot (2014). https://slate.com/technology/2014/03/quakebot-los-angeles-times-robot-journalist-writes-article-on-la-earthquake.html
23. Papineni, K., Roukos, S., Ward, T., Zhu, W.J.: BLEU: a method for automatic evaluation of machine translation. In: Proceedings of the 40th Annual Meeting on Association for Computational Linguistics (ACL 2002), pp. 311–318. Association for Computational Linguistics, Stroudsburg, PA, USA (2002). https://doi.org/10.3115/1073083.1073135
24. Pasupat, P., Liang, P.: Compositional semantic parsing on semi-structured tables. In: Proceedings of the 53rd Annual Meeting of the Association for Computational Linguistics and the 7th International Joint Conference on Natural Language Processing (Volume 1: Long Papers), pp. 1470–1480. Association for Computational Linguistics, Beijing, China, July 2015. https://doi.org/10.3115/v1/P15-1142. https://www.aclweb.org/anthology/P15-1142
25. Pauws, S., Gatt, A., Krahmer, E., Reiter, E.: Making effective use of healthcare data using data-to-text technology: methodologies and applications. Data Science for Healthcare, pp. 119–145. Springer, Cham (2019). https://doi.org/10.1007/978-3-030-05249-2_4
26. Plachouras, V., et al.: Interacting with financial data using natural language. In: Proceedings of the 39th International ACM SIGIR Conference on Research and Development in Information Retrieval (SIGIR 2016), pp. 1121–1124. ACM, New York (2016). https://doi.org/10.1145/2911451.2911457
27. Post, M.: A call for clarity in reporting BLEU scores. In: Proceedings of the Third Conference on Machine Translation: Research Papers, pp. 186–191. Association for Computational Linguistics, Belgium, Brussels, October 2018. https://doi.org/10.18653/v1/W18-6319. https://www.aclweb.org/anthology/W18-6319
28. Puduppully, R., Dong, L., Lapata, M.: Data-to-text generation with content selection and planning. In: AAAI (2018)
29. Puduppully, R., Dong, L., Lapata, M.: Data-to-text generation with entity modeling. In: Proceedings of the 57th Conference of the Association for Computational Linguistics (ACL 2019), Florence, Italy, 28 July–2 August 2019, Volume 1: Long Papers, pp. 2023–2035 (2019). https://www.aclweb.org/anthology/P19-1195/
30. Reiter, E., Sripada, S., Hunter, J., Yu, J., Davy, I.: Choosing words in computer-generated weather forecasts. Artif. Intell. **167**(1–2), 137–169 (2005). https://doi.org/10.1016/j.artint.2005.06.006

31. Roberti, M., Bonetta, G., Cancelliere, R., Gallinari, P.: Copy mechanism and tailored training for character-based data-to-text generation. CoRR abs/1904.11838 (2019). http://arxiv.org/abs/1904.11838
32. Sarma, A.D., et al.: Finding related tables. In: SIGMOD (2012). http://i.stanford.edu/~anishds/publications/sigmod12/modi255i-dassarma.pdf
33. See, A., Liu, P.J., Manning, C.D.: Get to the point: summarization with pointer-generator networks. In: Proceedings of the 55th Annual Meeting of the Association for Computational Linguistics (Volume 1: Long Papers), pp. 1073–1083. Association for Computational Linguistics, Vancouver, Canada, July 2017. https://doi.org/10.18653/v1/P17-1099
34. Sun, H., Ma, H., He, X., Yih, W.T., Su, Y., Yan, X.: Table cell search for question answering. In: Proceedings of the 25th International Conference on World Wide Web (WWW 2016), pp. 771–782. ACM Press (2016)
35. Trask, A., Hill, F., Reed, S.E., Rae, J.W., Dyer, C., Blunsom, P.: Neural arithmetic logic units. CoRR abs/1808.00508 (2018). http://dblp.uni-trier.de/db/journals/corr/corr1808.html#abs-1808-00508
36. Vaswani, A., et al.: Attention is all you need. In: Proceedings of the 31st International Conference on Neural Information Processing Systems (NIPS 2017), pp. 6000–6010. Curran Associates Inc., USA (2017). http://dl.acm.org/citation.cfm?id=3295222.3295349
37. Vinyals, O., Bengio, S., Kudlur, M.: Order matters: Sequence to sequence for sets. In: International Conference on Learning Representations (ICLR) (2016). http://arxiv.org/abs/1511.06391
38. Vinyals, O., Fortunato, M., Jaitly, N.: Pointer networks. In: Cortes, C., Lawrence, N.D., Lee, D.D., Sugiyama, M., Garnett, R. (eds.) Advances in Neural Information Processing Systems 28, pp. 2692–2700. Curran Associates, Inc. (2015). http://papers.nips.cc/paper/5866-pointer-networks.pdf
39. Wiseman, S., Shieber, S., Rush, A.: Challenges in data-to-document generation. In: Proceedings of the 2017 Conference on Empirical Methods in Natural Language Processing, pp. 2253–2263. Association for Computational Linguistics, Copenhagen, Denmark, September 2017. https://doi.org/10.18653/v1/D17-1239. https://www.aclweb.org/anthology/D17-1239
40. Wiseman, S., Shieber, S., Rush, A.: Learning neural templates for text generation. In: Proceedings of the 2018 Conference on Empirical Methods in Natural Language Processing, pp. 3174–3187. Association for Computational Linguistics, Brussels, Belgium, October–November 2018. https://doi.org/10.18653/v1/D18-1356. https://www.aclweb.org/anthology/D18-1356
41. Zhang, S., Balog, K.: Web table extraction, retrieval and augmentation. In: Proceedings of the 42nd International ACM SIGIR Conference on Research and Development in Information Retrieval (SIGIR 2019), pp. 1409–1410. ACM Press (2019)

Entities

Context-Guided Learning to Rank Entities

Makoto P. Kato[1(✉)], Wiradee Imrattanatrai[2], Takehiro Yamamoto[3], Hiroaki Ohshima[3], and Katsumi Tanaka[2]

[1] University of Tsukuba/JST, PRESTO, Tsukuba, Japan
mpkato@acm.org
[2] Kyoto University, Kyoto, Japan
wiradee@db.soc.i.kyoto-u.ac.jp, tanaka.katsumi.85e@st.kyoto-u.ac.jp
[3] University of Hyogo, Kobe, Japan
t.yamamoto@sis.u-hyogo.ac.jp, ohshima@ai.u-hyogo.ac.jp

Abstract. We propose a method for learning entity orders, for example, safety, popularity, and livability orders of countries. We train linear functions by using samples of ordered entities as training data, and attributes of entities as features. An example of such functions is $f(\text{Entity}) = +0.5$ (Police budget) -0.8 (Crime rate), for ordering countries in terms of safety. As the size of training data is typically small in this task, we propose a machine learning method referred to as *context-guided learning* (CGL) to overcome the over-fitting problem. Exploiting a large amount of contexts regarding relations between the labeling criteria (*e.g.* safety) and attributes, CGL guides learning in the correct direction by estimating a roughly appropriate weight for each attribute by the contexts. This idea was implemented by a regularization approach similar to support vector machines. Experiments were conducted with 158 kinds of orders in three datasets. The experimental results showed high effectiveness of the contextual guidance over existing ranking methods.

1 Introduction

Entity search is one of the emerging trends in major search engines [19,32], and has been powered by large-scale knowledge bases such as DBpedia, Wikidata, and YAGO. A wide variety of entity attributes are stored in knowledge bases and have enabled search engines to support entity search queries such as "european countries" and "movies starring emma watson".

On the other hand, the current entity search systems have not supported various kinds of rankings yet, which can be found on the Web, for example, the most *livable* countries, *innovative* companies, and *high-performance* cameras. If such diverse rankings were integrated into entity search and explained objectively with some evidences, users could be more efficient for accomplishing complex tasks such as decision making, comparison, and planning. For example, a user is planning to visit several European countries and inputs a query "european countries safety" to know how safe each country is. If an entity search engine

J. M. Jose et al. (Eds.): ECIR 2020, LNCS 12035, pp. 83–96, 2020.
https://doi.org/10.1007/978-3-030-45439-5_6

provided a list of countries ranked by public safety and factors used to determine the ranking (*e.g.* crime rate and police budget), they would be helpful for the user to make his/her travel plan.

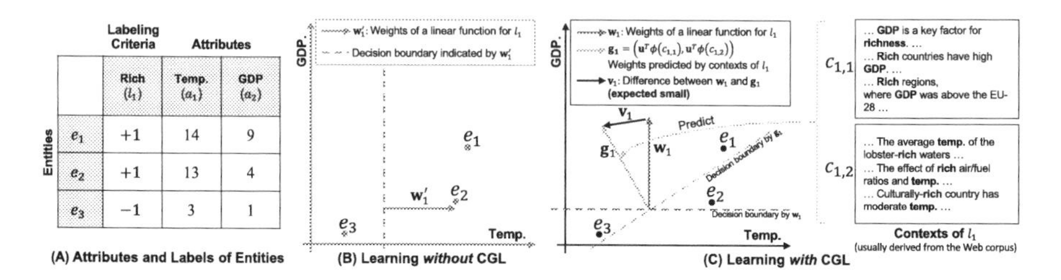

Fig. 1. (A) Entities e_1 and e_2 are rich countries, and e_3 is not a rich country. They have only two attributes a_1 (temperature) and a_2 (GDP). (B) Every entity can be expressed as a point in a two dimensional space by their attribute values in this example. Our goal is to learn a linear function for the labeling criterion l_1, which is defined as $f_1(\mathbf{x}) = \mathbf{w}_1^T\mathbf{x}$. One of the possible weights that perfectly classify the training examples is $\mathbf{w}_1' = (1, 0)$, but not necessarily effective for the other examples. (C) Contexts are used to produce a "rough" prediction $\mathbf{g}_1$ of the ideal weights. CGL determines the weights $\mathbf{w}_1$ such that $\mathbf{v}_1$, the difference between $\mathbf{w}_1$ and $\mathbf{g}_1$, is small and training examples are separated well. The weights $\mathbf{w}_1$ are expected to be effective for the other cases, since a strong correlation between richness and GDP is suggested by their contexts.

In this paper, we propose a method for learning orders of entities using samples of ordered entities as training data and attributes of entities as features. Entity orders are expressed in several forms on the Web: comparative sentences (*e.g.* "DiCaprio is taller than Pitt"), scores (*e.g.* "[Camera A] portrait: 9.2, landscape: 7.5, and sports: 8.5"), and rankings (*e.g.* "1st: Iceland, 2nd: Denmark, and 3rd: Austria"). These expressions can be interpreted with a uniform model, *i.e.* a subset of entity pairs that defines an entity order, and be used as training data to learn entity orders. The learned models can be used not only to rank entities but also to explain rankings by correlated attributes. We assume that entity orders can be represented as a linear function of attributes (denoted by f), primarily because of the high explanatory capacity for users. For example, given a list of entities ordered by labeling criterion "safety", (Iceland, Denmark, Austria), and their attributes such as "GDP", "Crime rate", and "Police budget", we learn function $f(\text{Entity}) = +0.5$ (Police budget) -0.8 (Crime rate).

A major challenge for this problem is the lack of training data. Many Web sites do not present all the ordered entities (see Table 1). Moreover, the size of training data might not be sufficiently large for some entity classes, even if all the ordered entities are described (*e.g.* only 50 states in the United States). As the number of attributes should be large enough to explain diverse orders, and can be increased easily with existing techniques [11,28], the problem of learning to rank entities can suffer from serious over-fitting problems.

To cope with this essential problem, we propose a learning method referred to as *context-guided learning* (*CGL*). This method uses not only ordered entities but also *contexts of labeling criteria and attributes* to learn the function f. A labeling criterion refers to a textual representation to determine labels (or an order in a ranking problem). The context can provide the models with additional information, and guide learning in the correct direction by preventing over-fitting. Figure 1 illustrates how CGL is applied to a classification problem. (As can be seen later, CGL is first explained for a classification problem and later extended to a ranking problem). Our goal in this example is to learn a linear function for the labeling criterion l_1 (richness), which is defined as $f_1(\mathbf{x}) = \mathbf{w}_1^T\mathbf{x}$ (an intercept is omitted for simplicity). When we simply apply an ordinary learning algorithm, learned weights can be $\mathbf{w}_1' = (1, 0)$ in (B) of Fig. 1, indicating that the attribute a_1 is useful for this classification. Although these weights seem reasonable as their decision boundary perfectly separates positive (e_1 and e_2) and negative (e_3) examples, it is easy to anticipate that the attribute a_1 can be useless for the other cases if we know the meaning of the labeling criterion (*i.e.* richness) and attribute a_1 (*i.e.* temperature). CGL, on the other hand, incorporates contexts of the labeling criterion and attributes for making a "rough" prediction of the ideal weights, and expects the weights $\mathbf{w}_1$ to be close to the "rough" prediction (denoted as $\mathbf{g}_1$ in (C) of Fig. 1). Although the prediction based on contexts cannot be always accurate (indeed, the decision boundary of $\mathbf{g}_1$ fails to classify examples well), $\mathbf{g}_1$ suggests that the attribute a_1 is not strongly related to the labeling criterion, and guides the learning of the weights $\mathbf{w}_1$. Thus, the learning can be successful even if sufficient training data are not available. CGL does not require any annotations for the contexts. Alternatively, CGL learns multiple functions at the same time for learning the relationship between contexts and weights in the function f.

To the best of our knowledge, CGL is the first attempt to leverage contexts of labeling criteria and features directly in machine learning (ML) problems. CGL is a general ML method and can be applied not only to ranking problems but also to classification and regression problems as long as relations between labeling criteria and features are described in a particular corpus.

Our contributions in this paper can be summarized as follows: (1) we introduced the problem of learning to rank entities by using attributes as features, in order to rank entities by various criteria and precisely understand labeling criteria; (2) we proposed CGL, a general ML method using contexts of labeling criteria and features for preventing over-fitting; and (3) we conducted experiments with a wide variety of orders, and demonstrated the effectiveness of CGL in the task of learning to rank entities.

2 Related Work

We review related work on entity ranking and discuss the difference between CGL and existing ML methods, in particular, multi-task learning methods.

2.1 Entity Ranking

Entity ranking has been addressed in some tracks in INEX and TREC. The INEX Entity Ranking track held two tasks: entity ranking and entity list completion tasks [12–14]. The entity ranking task expected systems to return relevant entities in response to a given query, while the entity list completion task expected systems to return entities related to given example entities. The TREC Entity track offered related entity finding tasks, in which systems were expected to find entities related to a given entity, with the type of the target entity and nature of their relation [2–4]. Those tasks only expect that retrieved entities are ordered by the relatedness to given example entities, and do not expect different kinds of orders within related entities.

Apart from the evaluation campaigns, there are some work that addresses learning to rank entities. Kang *et al.* used a ranking algorithm based on a boosted tree model for finding entities related to a given query [24]. Tran *et al.* proposed a method of ranking entities based on salience and informativeness for timeline summarization of events [30]. Zhou *et al.* addressed a problem of finding entities that have a specified relation with an input entity [34]. They trained a ranker for each relation based on training queries and labeled entities by using features derived from search snippets regarding pairs of entities. Although this work and ours use contexts (or search snippets) for learning to rank entities, our rankers are built primarily on attributes of entities and does not use contexts of entity pairs. Jameel *et al.* proposed an entity embedding method for entity retrieval [22]. Their method is mainly based on the co-occurrence between entities and words, and does not directly model entity attributes.

Some NLP tasks are also related to our task. Iwanari *et al.* tackled a problem of ordering entities in terms of a given adjective by using some evidences extracted from texts [20]. Their task is similar to ours as both address entity ranking in terms of a particular labeling criterion. While their method uses contexts of labeling criteria and entities, our method uses contexts of labeling criteria and *attributes* of entities.

2.2 Multi-task Learning

The important characteristics of CGL are summarized as follows: (1) weights in the function f are learned based on labels as well as contexts regarding labeling criteria and features, and (2) multiple functions are learned at the same time to learn the relationship between the contexts and weights in the function f. Below, we review several ML methods and discuss their relationship to CGL.

Multi-task learning is an approach to improving learning in each task by learning multiple tasks simultaneously [9]. CGL is considered as an instance of multi-task learning. Regularized multi-task learning, which was proposed by Evgeniou and Pontil, assumes that weights of multiple tasks are similar [15]. As explained later, their model is a special case of our model when contexts are all the same. Other models assume that weights are sampled from a common prior [10,27,33]. Argyriou *et al.* used an assumption that weights are represented in a low subspace common to multiple tasks [1]. In contrast to these methods using an assumption that all the tasks are related, some work selectively decides which tasks are related and are expected to share similar weights [21,25]. Similarly, CGL uses contexts to measure the similarity between tasks implicitly, and tends to estimate similar weights for similar tasks. An interesting difference between CGL and the other multi-task learning methods is that *CGL still works even if any pairs of tasks are not similar.* CGL only requires that some contexts are similar among multiple tasks. Thus, the applicable scope of CGL is not limited to problems targeted by existing multi-task learning methods.

3 Methodology

In this section, we first explain the problem of learning to rank entities from samples of ordered entities with attributes. We then introduce CGL, apply it to our problem, and explain some approaches to modeling contexts for CGL.

3.1 Problem Definition

Letting E be a set of entities of a particular class, we define an entity order as a total order on E, denoted by $\preceq_k$. Each order has a labeling criterion (or an ordering criterion in this case) denoted by l_k. For example, labeling criteria could include "livability", "innovativeness", "beauty", and "performance". A set of all $(e_i, e_j) \in E \times E$ for which $e_i \preceq_k e_j$ holds is called a graph[1] of an entity order, denoted by $G_{\preceq_k}$. Orders are usually expressed on the Web as subsets of their graphs. Thus, we can observe and use only $G'_{\preceq_k} \subseteq G_{\preceq_k}$ for learning entity orders. For example, a ranking of safe countries "1st: Iceland, 2nd: Denmark, and 3rd: Austria" implies $G'_{\preceq_k} = \{$("Denmark", "Iceland"), ("Austria", "Iceland"), ("Austria", "Denmark")$\}$ and $l_k =$ "safety".

Our principal purpose is to learn a linear function $f_k(\mathbf{e}_i) = \mathbf{w}_k^T \mathbf{e}_i$ based on a subset of a graph $G'_{\preceq_k}$ for each entity order $\preceq_k$, where $\mathbf{e}_i$ is an M-dimensional vector representing attributes of entity $e_i \in E$, and the d-th value of the vector represents a value of attribute a_d. We expect that the function f_k *preserves* the entity order $\preceq_k$: $e_i \preceq_k e_j \Rightarrow f_k(\mathbf{e}_i) \leq f_k(\mathbf{e}_j)$ for any $e_i, e_j \in E$, so that entities can be ranked by entity order $\preceq_k$ with learned function f_k. Moreover, attributes whose weights are non-zero are expected to explain the entity order well.

As we explained earlier, the key challenge of this problem is lack of training data: $|G'_{\preceq_k}|$ is typically small compared with the number of attributes M. For example, $M = 83$ for countries and $M = 137$ for cities in our experiments. Ranked lists of ten or fewer entities can provide only at most 45 entity pairs as training data, which are not considered as sufficiently large for learning. Moreover, M must be as large as possible for modeling a wide range of orders. Thus, some approaches are necessary for preventing the over-fitting problem caused by lack of training data.

---

[1] *Graph of a function*, a subset of the Cartesian product of two sets defining an order.

The key idea in our work is to use data other than $G'_{\preceq_k}$ for learning $\mathbf{w}_k$ effectively. One of the unique characteristics or assumptions in our problem is that textual representations for labeling criteria and attributes are available. Therefore, given a labeling criterion, it is possible to estimate a roughly appropriate weight for each attribute by leveraging the contexts regarding relations between the labeling criterion and attribute. This idea is instantiated as CGL, which is explained in the next subsection.

3.2 Context-Guided Learning

We introduce CGL, our proposed learning method that leverages contexts of labeling criteria and features. We begin with CGL for classification problems and then extend it to be used for ranking problems.

The input for a classification problem is $\mathcal{D} = \{D_k\}_{k=1}^{K}$, where $D_k = \{(\mathbf{x}_{k,i}, y_{k,i})\}_{i=1}^{N_k}$, $\mathbf{x}_{k,i} \in \mathbb{R}^M$, $y_{k,i} \in \{-1, +1\}$, K is the number of labeling criteria, and N_k is the number of examples for the k-th labeling criteria. Labeling criterion l_k is a textual representation to determine values for $y_{k,i}$. For example, if $\mathbf{x}_{k,i}$ represents a feature of a city and $y_{k,i} = +1$ if the city is a metropolitan city, the labeling criterion l_k could be "metropolitan city". Another example can be found in Fig. 1. The d-th value of a vector should correspond to a particular feature and have a name denoted by a_d. Example names include "population" and "GDP".

The requirements for CGL are summarized as follows: (1) A labeling criterion l_k is expressed in language, (2) Features $A = \{a_d\}_{d=1}^{M}$ are expressed in language, and (3) There is a corpus including contexts regarding relations between labeling criteria and feature names. It is not necessary that all the labeling criteria and feature are expressed in language. In contrast to multi-task learning, *CGL does not require that tasks (or labeling criteria in CGL) are similar.*

A classification problem can be formalized as learning function f_k for each labeling criterion $k = 1, \ldots, K$ such that $f_k(\mathbf{x}_{k,i}) \simeq y_{k,i}$. To solve this problem, we use a linear function $f_k(\mathbf{x}_{k,i}) = \mathbf{w}_k^T \mathbf{x}_{k,i}$. Letting $c_{k,d}$ represent contexts for labeling criterion l_k and feature $a_d \in A$, we can use the contexts for estimating $\mathbf{w}_k$ as follows:

$$w_{k,d} = \mathbf{u}^T \phi(c_{k,d}) + v_{k,d}, \tag{1}$$

where $w_{k,d}$ is the d-th value of $\mathbf{w}_k$, ϕ is a feature map function that transforms a context to a vector, and $\mathbf{u}$ is a weight vector that does not depend on labeling criteria. The equation above indicates that the weight for the labeling criterion l_k and feature a_d is estimated by their context $c_{k,d}$ and an intercept $v_{k,d}$. Equation 1 is generalization of $w_{k,d} = z_d + v_{k,d}$ in the regularized multi-task learning [15], where z_d is a weight common to multiple tasks. Equation 1 is reduced to their model if all the contexts are the same. If contexts for two labeling criteria are similar, or equivalently, labeling criteria are similar, $w_{k,d}$ tends to be similar for these labeling criteria. This property is similar to some multi-task learning methods [21,25].

Based on Eq. 1, $\mathbf{w}_k$ can be expressed as follows:

$$\mathbf{w}_k = \mathbf{g}_k + \mathbf{v}_k, \tag{2}$$

where $\mathbf{v}_k = (v_{k,1}, \ldots, v_{k,M})$, $\mathbf{g}_k = \Phi_k^T \mathbf{u}$, and $\Phi_k = (\phi(c_{k,1}), \ldots, \phi(c_{k,M}))$. This equation is illustrated in (C) of Fig. 1. We expect that the "rough" prediction $\mathbf{g}_k$ can be given by contexts of labeling criterion l_k, and ideal weights are close to $\mathbf{g}_k$; in other words, $\mathbf{v}_k$ is *not large*.

We propose to learn the linear function using a regularization approach similar to support vector machines (SVMs) and the regularized multi-task learning [15]. The optimization problem is shown below:

Problem 1

$$\min_{\mathbf{u}, \mathbf{v}_k, \xi_{k,i}} \|\mathbf{u}\|^2 + \frac{c}{K} \sum_{k=1}^{K} \|\mathbf{v}_k\|^2 + C \sum_{k=1}^{K} \sum_{i=1}^{N_k} \xi_{k,i}, \tag{3}$$

subject, for $k = 1, \ldots, K$ *and* $i = 1, \ldots, N_k$*, to the constraints that* $y_{k,i} f_k(\mathbf{x}_{k,i}) \geq 1 - \xi_{k,i}$, $\xi_{k,i} \geq 0$, *where* c *and* C *are hyper parameters.*

Slack variables $\xi_{k,i}$ measure the error of the linear functions on the training data, while the other terms are regularization terms for the weights $\mathbf{u}$ and $\mathbf{v}_k$. Hyper parameters c and C can control the effect of the contexts on the model and the sensitivity for the error on the training data: a large value for c increases the effect of the contexts, while a large value for C tends to inhibit misclassification of the training data. We learn multiple functions f_k for $k = 1, \ldots, K$ with the single objective function so that we can learn the weight $\mathbf{u}$ based on the whole training data.

We show that Problem 1 can be solved in the same manner as would be used with the standard SVM. To this end, we first define a single function to be learned that summarizes functions f_k for $k = 1, \ldots, K$ as $F(\mathbf{x}, k) = f_k(\mathbf{x})$. This function, $F : \mathbb{R}^M \times \{1, \ldots, K\} \to \mathbb{R}$, can be written as a linear function:

$$F(\mathbf{x}, k) = \mathbf{w}^T \psi(\mathbf{x}, k), \tag{4}$$

by using the following settings:

$$\mathbf{w} = (\mathbf{u}^T, \sqrt{\frac{c}{K}} \mathbf{v}^T)^T, \; \psi(\mathbf{x}, k) = ((\Phi_k \mathbf{x})^T, \underbrace{\mathbf{0}^T, \ldots, \mathbf{0}^T}_{k-1}, \sqrt{\frac{K}{c}} \mathbf{x}^T, \underbrace{\mathbf{0}^T, \ldots, \mathbf{0}^T}_{K-k})^T, \tag{5}$$

where ψ is a feature map function, and $\mathbf{0}$ is an M-dimensional vector whose values are all zeros.

Reassigning $\mathbf{x}_i$ to $\mathbf{x}_{k,i'}$, y_i to $y_{k,i'}$, and ξ_i to $\xi_{k,i'}$ ($i = \sum_{k'=1}^{k-1} N_{k'} + i'$), we can reduce Problem 1 to the standard SVM problem, as follows.

Theorem 1. *The optimization of Problem 1 is equivalent to solving the following problem:*

Problem 2. *Given* $D = \{((\mathbf{x}_i, k_i), y_i)\}_{i=1}^N$ *where* $N = \sum_{k=1}^K N_k$ *such that* $D = \bigcup_{k=1}^K \{((x_{k,i}, k), y_{k,i}) | (x_{k,i}, y_{k,i}) \in D_k\}$,

$$\min_{\mathbf{w}, \xi_i} \frac{1}{2} \|\mathbf{w}\|^2 + C' \sum_{i=1}^N \xi_i, \tag{6}$$

subject, for $i = 1, \ldots, N$, *to the constraints that* $y_i F(\mathbf{x}_i, k_i) \geq 1 - \xi_i,\ \xi_i \geq 0$, *where* $C' = C/2$ *and* ξ_i *is a slack variable for* $((x_i, k_i), y_i) \in D$.

Proof. The norm of $\mathbf{w}$ is $\|\mathbf{w}\|^2 = \|\mathbf{u}\|^2 + \frac{c}{K}\|\mathbf{v}\|^2$. Therefore, the objective function of Problem 2 is rewritten as:

$$\frac{1}{2} \left\{ \|\mathbf{u}\|^2 + \frac{c}{K} \sum_{k=1}^K \|\mathbf{v}_k\|^2 + C \sum_{i=1}^N \xi_i \right\}, \tag{7}$$

which is equivalent to the objective function of Problem 1.

Since Problem 2 is the standard SVM problem, we can use the standard SVM dual problem for solving Problem 1. Furthermore, we can use an important characteristic of SVMs: *i.e.* non-linear functions can be used by means of kernels. While the linear function for classification (*i.e.* f_k) cannot be a non-linear function owing to the form of the model, we can use a non-linear function for estimating the weights based on contexts (see Eq. 1). The kernel method for CGL provides us with a wide range of choices for the representation of contexts. They can be represented as vectors, sets of vectors, trees, *etc.* as long as the kernel function is appropriately designed for two contexts.

3.3 Context-Guided Learning for Ranking

We extend CGL to the ranking problem and explain how it can be applied to the problem of ranking entities.

The input for the ranking problem is $\mathcal{D} = \{D_k\}_{k=1}^K$, where $D_k \subseteq \mathbb{R}^M \times \mathbb{R}^M$, and K is the number of labeling criteria. Labeling criterion l_k is a textual representation to determine the order for D_k: *i.e.* $(\mathbf{x}_{k,i}, \mathbf{x}_{k,j})$ in D_k indicates that $\mathbf{x}_{k,j}$ is higher than $\mathbf{x}_{k,i}$ in terms of the labeling criterion l_k. The d-th value of vectors in D_k must correspond to a particular feature and have a name denoted by a_d. The requirements are the same as those explained in regard to CGL for classification. A ranking problem can be formalized as a learning function f_k for each labeling criterion $k = 1, \ldots, K$ such that $f_k(\mathbf{x}_{k,j}) - f_k(\mathbf{x}_{k,i}) \simeq 1$ for $(\mathbf{x}_{k,i}, \mathbf{x}_{k,j})$ in D_k. As assumed in the classification problem, we use a linear function $f_k(\mathbf{x}_{k,i}) = \mathbf{w}_k^T \mathbf{x}_{k,i}$.

It is clear that the ranking problem can be reduced to the classification problem if we redefine D_k as follows: $D'_k = \{(\mathbf{x}_{k,j} - \mathbf{x}_{k,i}, 1) | (\mathbf{x}_{k,i}, \mathbf{x}_{k,j}) \in D_k\}$, since $f_k(\mathbf{x}_{k,j} - \mathbf{x}_{k,i}) = f_k(\mathbf{x}_{k,j}) - f_k(\mathbf{x}_{k,i})$.

We can apply CGL for ranking to the problem in *Problem Definition* section by using vectors of entity pairs in $G'_{\preceq_k}$ as the training data, *i.e.* $D_k = \{(\mathbf{e}_i, \mathbf{e}_j) | (e_i, e_j) \in G'_{\preceq_k}\}$.

3.4 Context Models

Having described the learning method for the problem of ranking entities, we explain the context models used in the learning. Contexts can be a set of sentences or a set of documents regarding a labeling criterion and a feature. In this work, we describe methods of modeling contexts by using sentences retrieved from Web search results.

Given labeling criterion l_k and feature a_d, we create a query combining l_k and a_d with an AND operator, and use the query to retrieve the top $N^{(c)}$ search results using a particular Web search engine ($N^{(c)} = 500$ in our experiments). We then split snippets of the search results into sentences and find sentences including both the labeling criterion l_k and the feature a_d.

We use two basic methods for modeling sentences. One is a vector representation based on the TF-IDF weighting, and the other is a distributional representation of sentences [26]. The vector representation based on the TF-IDF weighting is sparse, and not sensitive to the order of words, but it can represent exact words appearing in the context. In contrast, the distributed representation of sentences is dense, and sensitive to the word order, but it might not retain the exact words appearing in the context.

4 Experiments

This section explains data used in the experiment, describes experimental settings, and shows the experimental results.

4.1 Data

Since there is no publicly available dataset for our task, we first explain our development of a dataset and its statistics.

Various kinds of entity orders in three datasets were mined from the Web and from magazines both automatically and manually. The three datasets include *City* (more specifically, Japanese prefectures), *Country*, and *Camera* entities, respectively. These classes were selected primarily for the following reasons: (1) availability of a wide range of entity orders, (2) availability of attributes, and (3) diversity of statistics. The language scope of our dataset was Japanese, as we used a Japanese crowd-sourcing service in the evaluation. Entity names and attribute names were Japanese and translated into English for this paper.

Entity orders were mined from Web pages for *City* and *Country* datasets, and from ten Japanese camera magazines for *Camera* dataset. The retrieved ranked lists were converted into a set of pairs for each entity orders. We excluded entity sets including less than five entities.

Attributes for *City* and *Country* datasets were mined from tables in Web documents. We chose *Web tables* as a resource for obtaining attributes because (1) the extraction method can be accurate and language-independent, and (2) standardization of numerical values was not necessary as units of numerical

Table 1. Statistics of the datasets and examples of entities, orders, and attributes.

	City	Country	Camera
# Entities	47	138	149
# Orders	64	40	54
# Entities/Order	13.3	17.7	14.4
# Attributes	137	83	16
Entity examples	Tokyo	Denmark	EOS 5DS
	Kyoto	Iceland	Nikon D3300
Attribute examples	Population	# Tourists	Resolution
	Crime rate	# Suicides	Weight
Order examples	Attractive	Livable	Portable
	Rich	Happy	Tough

Table 2. Accuracy in the three datasets (±SEM).

	Accuracy			
	City	Country	Camera	Total
RankNet [5]	0.482	0.478	0.530	0.497
	(0.023)	(0.025)	(0.030)	(0.015)
RankBoost [16]	0.513	0.636	0.552	0.557
	(0.028)	(0.024)	(0.036)	(0.018)
LinearFeature [29]	0.566	0.670	0.614	0.609
	(0.019)	(0.024)	(0.034)	(0.015)
LambdaMART [31]	0.614	0.659	0.697	0.654
	(0.021)	(0.019)	(0.024)	(0.013)
ListNet [7]	0.559	0.518	0.504	0.530
	(0.020)	(0.022)	(0.031)	(0.014)
CGL (TF-IDF, Linear)	**0.661**	0.716	**0.823**	**0.730**
	(0.017)	(0.022)	(0.019)	(0.012)
CGL (TF-IDF, RBF)	**0.661**	0.725	0.799	0.724
	(0.019)	(0.021)	(0.019)	(0.012)
CGL (Distributed, Linear)	0.646	0.701	0.798	0.712
	(0.020)	(0.023)	(0.021)	(0.013)
CGL (Distributed, RBF)	**0.661**	**0.731**	0.804	0.728
	(0.018)	(0.022)	(0.021)	(0.013)

values are usually consistent within a table. Attributes for *Camera* dataset were scraped from Web pages of a Japanese Web site, Kakaku.com[2], which provides prices and specifications of products. All the numerical values for each attribute were normalized into $[0, 1]$.

Table 1 shows statistics and examples of entities, orders, and attributes. There are 158 entity orders in total. For most of the orders, we could not find all of the entities in a class in a ranking on the Web. There were many Web pages presenting the top three or ten entities for an order. Thus, the average number of entities per order is much less than the total number of entities.

4.2 Experimental Settings

We selected as baseline methods for this experiment some existing ranking methods that do not use contexts: (1) **RankNet** [5]: a pairwise ranking method that uses a neural network model and optimizes the cross entropy loss, (2) **RankBoost** [16]: application of AdaBoost [17] to pairwise preferences, (3) **LinearFeature** [29]: a linear feature-based model optimized by coordinate ascent, (4) **LambdaMART** [31]: a combination of the ranking model, LambdaRank [6], and the boosted tree model, MART [18], and (5) **ListNet** [7]: a listwise ranking method using a neural network model. We used these methods implemented in RankLib[3]. We used normalized discounted cumulative gain (nDCG@10) [23] as an evaluation metric to be optimized for some methods.

[2] http://kakaku.com/.

[3] https://www.lemurproject.org/ranklib.php.

We conducted experiments using the developed dataset in the following settings. For each set of ordered entities $G_{\preceq_k}$, we split entities in the set E into 50:50, E_{train} and E_{test}, and obtained training data $G_{\text{train}} = \{(e_i, e_j) | (e_i, e_j) \in G_{\preceq_k} \wedge e_i \in E_{\text{train}} \wedge e_j \in E_{\text{train}}\}$ and test data $G_{\text{test}} = G_{\preceq_k} - G_{\text{train}}$. Our task in this experiment is to learn a model based on G_{train}, and to predict the pairwise preference of e_i and e_j for $(e_i, e_j) \in G_{\text{test}}$. We measured the accuracy defined as the fraction of correctly predicted pairwise preferences. We used five-fold cross validation on entity orders *within* E_{train} *of the same dataset* to determine the best parameters for each method.

We configured CGL with the following settings. Two context models were used: **TF-IDF** and **Distributed** (distributed representation with 400 dimensional vectors). Parameters c and C were determined using the cross validation explained above. A linear kernel (**Linear**) and an RBF kernel (**RBF**) were used for the kernel in CGL.

4.3 Experimental Results

Table 2 shows the accuracy in the three datasets with the standard error of the mean (SEM). CGL in any settings were better than any of the baseline methods. Among the CGL-based methods, the best method was CGL (TF-IDF, Linear), followed by CGL (Distributed, RBF). The total improvement over the best baseline method, LambdaMart, was 11.6%. According to a randomized Tukey HSD test [8][4] ($\alpha = 0.01$), the differences between CGL (TF-IDF, Linear) and all the baseline methods were found to be statistically significant, while there was no statistically significant difference across methods based on CGL.

CGL (TF-IDF, Linear) achieved 8%, 11%, and 18% improvements over LambdaMART for *City*, *Country*, and *Camera*, respectively. We hypothesize that the quality and amount of contexts are the main factors that determine the effectiveness of CGL, based on the observation that the number of sentences used for modeling contexts per attribute was 36.0, 45.7, and 137 for *City*, *Country*, and *Camera*, respectively.

Table 3. Examples of linear functions learned by CGL, in which three attributes for the highest absolute weights are shown.

Class	Learned linear model		
City	Attractiveness = +0.035 Women's life expectancy	−0.032 # Accident fatalities	−0.031 Population/family
City	Avg. savings = −0.174 Highest temperature	+0.160 Healthy life-span	+0.148 # Country inns
Country	Reputation = +0.058 Happiness	−0.057 # Applicants for asylum	−0.045 # Suicides
Country	Peace = +0.170 Grain harvest	+0.166 GDP growth rate	−0.126 # Suicides
Camera	Operability = −0.240 Weight	−0.213 Height	+0.133 Max. shutter speed

We also conducted evaluation of the attributes used in the learned functions. Five attributes with the highest absolute weights for each entity order

[4] http://www.f.waseda.jp/tetsuya/tools.html.

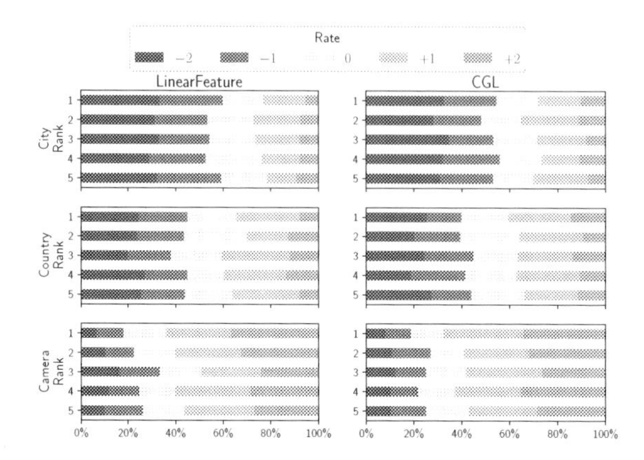

Fig. 2. Distribution of rates for five attributes with the highest absolute weights.

were pooled, and then presented to users in a Japanese crowd-sourcing service, Lancers[5]. In this evaluation, we aimed to understand to what extent the learned attributes could explain the orders. The instruction was as follows: "If you agree that there is a correlation between <labeling criterion> and <attribute>, please assign a score +2. If you disagree, please assign a score −2. If you cannot agree or disagree, please assign a score 0." Users could choose a rate from −2, −1, 0, +1, and +2. We assigned five users for each pair of a labeling criterion and an attribute. The best CGL method, CGL (TF-IDF, Linear), was selected for this evaluation. LinearFeature was used as a baseline method, since only this method used a linear function among the baseline methods.

Figure 2 shows the distribution of rates for five attributes with the highest absolute weights. The average rates of CGL were −0.455, −0.166, and +0.581, while those of LinearFeature were −0.560, −0.204, +0.516 for *City*, *Country*, and *Camera* datasets, respectively. These average rates show a high correlation with the accuracy of the models. Even though CGL could find more reasonable attributes in all of the classes than LinearFeature, their differences were small for those datasets. The average rates for *City* and *Country* datasets were negative indicating low explainability of the attributes. This is partially because some attributes only correlate to a particular labeling criterion, but were not considered as causes for increasing the criterion. Although CGL could learn a more accurate model than the baseline methods, it is still challenging to find *highly explanatory* attributes for a given label criterion.

Finally, we show some examples of linear functions learned by CGL in Table 3. Most of the attributes seem explainable and can possibly affect the entity order. While the others do not seem explanatory for the labeling criteria (*e.g.* "population/family" for "attractiveness" and "highest temperature" for "avg. savings"), they correlate well to the labeling criteria in our dataset, and are examples of attributes that were considered unreasonable in the subjective evaluation, but highly contributed to the prediction.

[5] http://www.lancers.jp/.

5 Conclusions

In this paper, we addressed the problem of learning orders of entities, by using partially observed orders as training data and attributes of entities as features. We proposed a learning method called context-guided learning (CGL) to avoid the over-fitting problem caused by lack of training data, and demonstrated the effectiveness of CGL for 158 orders in three datasets. Our future work includes theoretical analysis of CGL, application of CGL to the other problems (*e.g.* a fact verification task), exploration of better context models, and improvement of the efficiency of CGL for a large amount of data.

Acknowledgments. This work was supported by JSPS KAKENHI Grant Numbers JP16H02906, JP17H00762, JP18H03243, and JP18H03244, and JST PRESTO Grant Number JPMJPR1853, Japan.

References

1. Argyriou, A., Evgeniou, T., Pontil, M.: Convex multi-task feature learning. Mach. Learn. **73**(3), 243–272 (2008)
2. Balog, K., Serdyukov, P., De Vries, A.P.: Overview of the TREC 2010 entity track. In: TREC (2010)
3. Balog, K., Serdyukov, P., De Vries, A.P.: Overview of the TREC 2011 entity track. In: TREC (2010)
4. Balog, K., De Vries, A.P., Serdyukov, P., Thomas, P., Westerveld, T.: Overview of the TREC 2009 entity track. In: TREC (2009)
5. Burges, C., et al.: Learning to rank using gradient descent. In: ICML, pp. 89–96 (2005)
6. Burges, C.J., Ragno, R., Le, Q.V.: Learning to rank with nonsmooth cost functions. In: NIPS, pp. 193–200 (2006)
7. Cao, Z., Qin, T., Liu, T.Y., Tsai, M.F., Li, H.: Learning to rank: from pairwise approach to listwise approach. In: ICML, pp. 129–136 (2007)
8. Carterette, B.A.: Multiple testing in statistical analysis of systems-based information retrieval experiments. ACM TOIS **30**(1), 4 (2012)
9. Caruana, R.: Multitask learning. Mach. Learn. **28**(1), 41–75 (1997)
10. Daumé III, H.: Bayesian multitask learning with latent hierarchies. In: UAI, pp. 135–142 (2009)
11. Davidov, D., Rappoport, A.: Extraction and approximation of numerical attributes from the web. In: ACL, pp. 1308–1317 (2010)
12. de Vries, A.P., Vercoustre, A.-M., Thom, J.A., Craswell, N., Lalmas, M.: Overview of the INEX 2007 entity ranking track. In: Fuhr, N., Kamps, J., Lalmas, M., Trotman, A. (eds.) INEX 2007. LNCS, vol. 4862, pp. 245–251. Springer, Heidelberg (2008). https://doi.org/10.1007/978-3-540-85902-4_22
13. Demartini, G., Iofciu, T., de Vries, A.P.: Overview of the INEX 2009 entity ranking track. In: Geva, S., Kamps, J., Trotman, A. (eds.) INEX 2009. LNCS, vol. 6203, pp. 254–264. Springer, Heidelberg (2010). https://doi.org/10.1007/978-3-642-14556-8_26
14. Demartini, G., de Vries, A.P., Iofciu, T., Zhu, J.: Overview of the INEX 2008 entity ranking track. In: Geva, S., Kamps, J., Trotman, A. (eds.) INEX 2008. LNCS, vol. 5631, pp. 243–252. Springer, Heidelberg (2009). https://doi.org/10.1007/978-3-642-03761-0_25

15. Evgeniou, T., Pontil, M.: Regularized multi-task learning. In: KDD, pp. 109–117 (2004)
16. Freund, Y., Iyer, R., Schapire, R.E., Singer, Y.: An efficient boosting algorithm for combining preferences. J. Mach. Learn. Res. **4**, 933–969 (2003)
17. Freund, Y., Schapire, R.E.: A decision-theoretic generalization of on-line learning and an application to boosting. J. Comput. Syst. Sci. **1**(55), 119–139 (1997)
18. Friedman, J.H.: Greedy function approximation: a gradient boosting machine. Ann. Stat. **29**(5), 1189–1232 (2001)
19. Guo, J., Xu, G., Cheng, X., Li, H.: Named entity recognition in query. In: SIGIR, pp. 267–274 (2009)
20. Iwanari, T., Yoshinaga, N., Kaji, N., Nishina, T., Toyoda, M., Kitsuregawa, M.: Ordering concepts based on common attribute intensity. In: IJCAI, pp. 3747–3753 (2016)
21. Jacob, L., Vert, J.p., Bach, F.R.: Clustered multi-task learning: a convex formulation. In: NIPS, pp. 745–752 (2009)
22. Jameel, S., Bouraoui, Z., Schockaert, S.: Member: Max-margin based embeddings for entity retrieval. In: SIGIR, pp. 783–792 (2017)
23. Järvelin, K., Kekäläinen, J.: Cumulated gain-based evaluation of ir techniques. ACM TOIS **20**(4), 422–446 (2002)
24. Kang, C., Yin, D., Zhang, R., Torzec, N., He, J., Chang, Y.: Learning to rank related entities in web search. Neurocomputing **166**, 309–318 (2015)
25. Kumar, A., Daumé III, H.: Learning task grouping and overlap in multi-task learning. In: ICML, pp. 1383–1390 (2012)
26. Le, Q.V., Mikolov, T.: Distributed representations of sentences and documents. In: ICML, pp. 1188–1196 (2014)
27. Lee, S.I., Chatalbashev, V., Vickrey, D., Koller, D.: Learning a meta-level prior for feature relevance from multiple related tasks. In: ICML, pp. 489–496 (2007)
28. Madaan, A., Mittal, A., Mausam, G.R., Ramakrishnan, G., Sarawagi, S.: Numerical relation extraction with minimal supervision. In: AAAI, pp. 2764–2771 (2016)
29. Metzler, D., Croft, W.B.: Linear feature-based models for information retrieval. Inf. Retrieval **10**(3), 257–274 (2007)
30. Tran, T.A., Niederée, C., Kanhabua, N., Gadiraju, U., Anand, A.: Balancing novelty and salience: Adaptive learning to rank entities for timeline summarization of high-impact events. In: CIKM, pp. 1201–1210 (2015)
31. Wu, Q., Burges, C.J., Svore, K.M., Gao, J.: Adapting boosting for information retrieval measures. Inf. Retrieval **13**(3), 254–270 (2010)
32. Yin, X., Shah, S.: Building taxonomy of Web search intents for name entity queries. In: WWW, pp. 1001–1010 (2010)
33. Yu, K., Tresp, V., Schwaighofer, A.: Learning gaussian processes from multiple tasks. In: ICML, pp. 1012–1019 (2005)
34. Zhou, M., Wang, H., Change, K.C.C.: Learning to rank from distant supervision: exploiting noisy redundancy for relational entity search. In: ICDE, pp. 829–840 (2013)

Graph-Embedding Empowered Entity Retrieval

Emma J. Gerritse(✉), Faegheh Hasibi, and Arjen P. de Vries

Institute for Computing and Information Sciences, Radboud University, Nijmegen, The Netherlands
emma.gerritse@ru.nl, {f.hasibi,a.devries}@cs.ru.nl

Abstract. In this research, we improve upon the current state of the art in entity retrieval by re-ranking the result list using graph embeddings. The paper shows that graph embeddings are useful for entity-oriented search tasks. We demonstrate empirically that encoding information from the knowledge graph into (graph) embeddings contributes to a higher increase in effectiveness of entity retrieval results than using plain word embeddings. We analyze the impact of the accuracy of the entity linker on the overall retrieval effectiveness. Our analysis further deploys the cluster hypothesis to explain the observed advantages of graph embeddings over the more widely used word embeddings, for user tasks involving ranking entities.

Keywords: Entity retrieval · Graph embeddings · Word embeddings

1 Introduction

Many information needs are entity-oriented, and with the rise of knowledge graphs in Web and enterprise search [20], the role of entities has gained importance, both in the UI/UX where so-called entity cards are shown in response to entity-oriented queries, and in the ranking, where presence and absence of entity mentions is weighted differently from traditional term occurrences.

Recently, word embeddings have been shown to be helpful for a number of information retrieval problems. In the case of entity retrieval, a natural representation would however not just represent words in context of their textual neighborhood, but in context of the knowledge graph instead. Here, we would want to apply graph embeddings instead of word embeddings, where the semantic space constructed by graph embeddings does not only encode the textual context of an entity mention, but also the context as defined through the knowledge graph. Considering Wikipedia as the knowledge graph to define the entities of interest, for example, creating a graph embedding representation does not just take the entity's page itself as context, but also its anchor text, presence in lists and/or tables, *etc.* It is therefore likely that graph embeddings capture more of the entity's semantic roles and as a result may distinguish better between ambiguous entities than a plain word embedding based representation.

J. M. Jose et al. (Eds.): ECIR 2020, LNCS 12035, pp. 97–110, 2020.
https://doi.org/10.1007/978-3-030-45439-5_7

Exploring the use of graph embeddings in entity retrieval, we have studied a two-stage entity retrieval approach where the second stage employs graph embeddings for re-ranking the retrieval results of state-of-the-art entity ranking methods. We investigate the following research questions:

RQ1: Does adding graph embeddings improve entity retrieval methods?
RQ2: Which queries are helped the most?

To our knowledge, we are the first to investigate how the structural information captured in graph embeddings can contribute to improved retrieval effectiveness in entity-oriented search. The contributions of this paper are as follows: We have build graph embeddings from Wikipedia as a knowledge graph[1] and evaluated the contribution of these embeddings as a representation of entities in the ranking algorithm, using the DBpedia-Entity V2 collection [12]. For every query, we re-rank the results of state-of-the-art entity retrieval methods using the similarity between the entity embeddings of the candidate entities retrieved in stage one with the entity embeddings of the entities identified in the query (using an off-the-shelf entity linker). We show that re-ranking using graph embeddings improves retrieval effectiveness, and investigate how to explain this result by comparing the structure of the two types of embeddings. We also analyze why some queries are helped by this method while others are not.

2 Related Work

2.1 Word and Graph Embeddings

Distributional representations of language have been object of study for many years in natural language processing (NLP), because of their promise to represent words not in isolation, but 'semantically', with their immediate context. Algorithms like Word2Vec [19] and Glove [21] construct a vector space of word domains where similar words are mapped together (based on their linguistic context). Word2Vec uses neural networks to predict words based on the context (continuous bag of words) or context based on a word (skip gram). These word embedding representations have turned out to be highly effective in a wide variety of NLP tasks.

Word embeddings have been shown to help effectiveness in document retrieval [6,7]. In [7], locally trained word embeddings are used for query expansion. Here queries are expanded with terms highly similar to the query, and it is shown that this method beats several other neural methods. In [6], embeddings are used for weak supervision of documents. This paper uses query embeddings and document embeddings to predict relevance between queries and documents, when given BM25 scores as labels. It is able to improve on BM25.

Word embeddings consider the immediate linguistic context of the word occurrences. Going beyond just the text itself, researchers have proposed to

[1] Downloadable at https://github.com/informagi/GEEER.

develop so-called *graph embeddings* to encode not just words in text, but words in context of semi-structured documents represented as graphs - for example, to distinguish the occurrence of a word in the title of a document from its occurrences in a paragraph, or in a document's anchor text.

Different methods to produce graph embeddings have been proposed. Methods like Deepwalk [22] expect non-labeled edges and can be considered extensions of the word embedding approaches discussed before. Other approaches include the well-known method Trans-E [4], where edges in the graph are denoted as triples *(head, label, tail)*, where *label* is the value of the edge. Adding graph embedding vectors of the *head* and the *label* should result in the vector of the *tail*. The embeddings here are learned by gradient descent.

Wikipedia2Vec [26] applies graph embeddings to Wikipedia, creating embeddings that jointly capture link structure and text. The Wikipedia knowledge graph is indeed a natural resource for using graph embeddings, because it represents entities in a graph of interlinked Wikipedia pages and their text. The method proposed in [26] embeds words and entities in the same vector space by using word context and graph context. The word-word context is modeled using the Word2Vec approach, entity-entity context considers neighboring entities in the link graph, and word-entity context takes the words in the context of the anchor that links to an entity. The authors of Wikipedia2Vec demonstrate performance improvements on a variety of NLP tasks, although they did not consider entity retrieval in their work.

2.2 Entity Retrieval

An entity is an object or concept in the real world that can be distinctly identified [2]. Knowledge graphs like Wikipedia enrich the representation of entities by modeling the relations between them. Methods for document retrieval such as BM25 have been applied successfully to entity retrieval. However, since knowledge bases are semi-structured resources, this structural information may be used as well, for example by viewing entities as fielded documents extracted from the knowledge graph. A well-known example of this approach applies the fielded probabilistic model (BM25F [23]), where term frequencies between different fields in documents are normalized to the length of each field. Another effective model for entity retrieval uses the fielded sequential dependence model (FSDM [27]), which estimates the probability of relevance using information from single terms and bigrams, normalized per field.

2.3 Using Entity Linking for Entity Retrieval

Linking entities mentioned in the query to the knowledge graph [3,9] enables the use of relationships encoded in the knowledge graph, helping improve the estimation of relevance of candidate entities. Previous work has shown empirically that entity linking can increase effectiveness of entity retrieval. In [10], for example, entity retrieval has been combined with entity linking to improve retrieval effectiveness over state-of-the-art methods like FSDM.

Our research uses the TAGME entity linker [8] because it is especially suited to annotate short and poorly composed text like the queries we need to link to. TAGME adds Wikipedia hyperlinks to parts of the text, together with a confidence score.

2.4 Using Embeddings for Entity Retrieval

Very recent work has applied Trans-E graph embeddings to the problem of entity retrieval, and shown consistent but small improvements [15]. However, Trans-E graph embeddings are not a good choice if the graph has 1-to-many, transitive or symmetric relations, which is the case in knowledge graphs [1]. In our research, we also look into improving entity retrieval using graph embeddings, but use the Wikipedia2Vec representation to address these shortcomings.

3 Embedding Based Entity Retrieval

3.1 Graph Embeddings

We base the training of our entity embeddings on Wikipedia2Vec [25,26]. Taking a knowledge graph as the input, Wikipedia2Vec extends the skip-gram variant of Word2Vec [18,19] and learns word and entity embeddings jointly. The objective function of this model is composed of three components. The first component infers optimal embeddings for words W in the corpus. Given a sequence of words $w_1 w_2 ... w_T$ and a context window of size c, the word-based objective function is:

$$\mathcal{L}_w = \sum_{t=1}^{T} \sum_{-c \leq j \leq c, j \neq 0} \log \frac{\exp(\mathbf{V}_{w_t}^T \mathbf{U}_{w_{t+j}})}{\sum_{w \in W} \exp(\mathbf{V}_{w_t}^T \mathbf{U}_w)}, \tag{1}$$

where matrices $\mathbf{U}$ and $\mathbf{V}$ represent the input and output vector representations, deriving the final embeddings from matrix $\mathbf{V}$.

The two other components of the objective function take the knowledge graph into account. One addition considers a link-based measure estimated from the knowledge graph (i.e., Wikipedia). This measure captures the relatedness between entities in the knowledge base, based on the similarity between their incoming links:

$$\mathcal{L}_e = \sum_{e_i \in \mathcal{E}} \sum_{e_o \in C_{e_i}, e_i \neq e_i} \log \frac{\exp(\mathbf{V}_{e_i}^T \mathbf{U}_{e_o})}{\sum_{e \in \mathcal{E}} \exp(\mathbf{V}_{e_i}^T \mathbf{U}_e)}. \tag{2}$$

Here, C_e denotes entities linked to an entity e, and $\mathcal{E}$ represents all entities in the knowledge graph.

The last addition to the objective function places similar entities and words near each other by considering the context of the anchor text. The intuition is the same as in classic Word2Vec, but here, words in the vicinity of the anchor text have to predict the entity mention. Considering a knowledge graph with anchors A and an entity e the goal is to predict context words of the entity:

$$\mathcal{L}_a = \sum_{e_i \in A} \sum_{w_o \in a(e_i)} \log \frac{\exp(\mathbf{V}_{e_i}^T \mathbf{U}_{w_o})}{\sum_{w \in W} \exp(\mathbf{V}_{e_i}^T \mathbf{U}_w)}, \quad (3)$$

where $a(e)$ gives the previous and next c words of the referent entity e.

These three components (word context, link structure, and anchor context) are then combined linearly into the following objective function:

$$\mathcal{L} = \mathcal{L}_w + \mathcal{L}_e + \mathcal{L}_a. \quad (4)$$

3.2 Re-ranking Entities

Training the Wikipedia2Vec model on a Wikipedia knowledge graph results in a single graph embedding vector for every Wikipedia entity. The next question to answer is how to use these graph embeddings in the setting of entity retrieval.

We propose a two-stage ranking model, where we first produce a ranking of candidate entities using state-of-the-art entity retrieval models (see Sect. 2.2), and then use the graph embeddings to reorder these entities based on their similarity to the query entities, as measured in the derived graph embedding space.

Following the related work discussed in Sect. 2.3, we use the TAGME entity linker to identify the entities mentioned in the query. Given input query Q, we obtain a set of linked entities $E(Q)$ and a confidence score $s(e)$ for each entity, which represents the strength of the relationship between the query and the linked entity. We then compute an embedding-based score for every query Q and entity E:

$$F(E, Q) = \sum_{e \in E(Q)} s(e) \cdot cos(\overrightarrow{E}, \overrightarrow{e}), \quad (5)$$

where $\overrightarrow{E}, \overrightarrow{e}$ denote the embeddings vectors for entities E and e.

The rationale for this approach is the hypothesis that relevant entities for a given query are situated close (in graph embedding space) to the query entities identified by the entity linker.

Consider for example the query *"Who is the daughter of Bill Clinton married to."* TAGME links the query to entities BILL CLINTON with a confidence of 0.66, DAUGHTER with a confidence of 0.13, and SAME-SEX MARRIAGE with a confidence score of 0.21. Highly ranked entities then have a large similarity to these entities, where similarity to BILL CLINTON adds more to the score than similarity to DAUGHTER or SAME-SEX MARRIAGE (as the confidence score of BILL CLINTON is higher than the other two). The relevant entities for this query (according to the DBpedia-Entity V2 test collection [12]) are CHELSEA CLINTON, who is Bill Clinton's daughter, and CLINTON FAMILY. We can reasonably expect these entities to have similarity to the linked entities, confirming our intuition.

To produce our final score, we interpolate the embedding-based score computed using Eq. (5) with the score of the state-of-the-art entity retrieval model used to produce the candidate entities in stage one:

$$score_{total}(E, Q) = (1 - \lambda) \cdot score_{other}(E, Q) + \lambda \cdot F(E, Q) \quad \lambda \in [0, 1]. \quad (6)$$

4 Experimental Setup

4.1 Test Collection

In our experiments, we used the DBpedia-Entity V2 test collection [12]. The collection consists of 467 queries and relevance assessments for 49280 query-entity pairs, where the entities are drawn from the DBpedia 2015-10 dump. The relevance assessments are graded values of 2, 1, and 0 for highly relevant, relevant, and not relevant entities, respectively. The queries are categorized into 4 different groups: **SemSearch ES** consisting of short and ambiguous keyword queries (e.g., *"Nokia E73"*), **INEX-LD** containing IR-Style keyword queries (e.g., *"guitar chord minor"*), **ListSearch** consisting of queries seeking for a list of entities (e.g., *"States that border Oklahoma"*), and **QALD-2** containing entity-bearing natural language queries (e.g., *"Which country does the creator of Miffy come from"*). Following the baseline runs curated with the DBpedia-Entity V2 collection, we used the stopped version of queries, where stop patterns like "which" and "who" are removed from the queries.

4.2 Embedding Training

Wikipedia2Vec provides pre-trained embeddings. These embeddings, however, are not available for all entities in Wikipedia; e.g., 25% of the assessed entities in DBpedia-Entity V2 collection have no pre-trained embedding. The reasons for these missing embeddings are two-fold: (i) "rare" entities were excluded from the training data, and, (ii) entity identifiers evolve over time, resulting in entity mismatches with those in the DBpedia-Entity collection.

For training new graph embeddings, we used Wikipedia 2019-07 dump. This was the newest version at the time of training. We address the entity mismatch problem by identifying the entities that have been renamed in the new Wikipedia dump. Some of these entities were obtained using the redirect API of Wikipedia.[2] Others were found by matching the Wikipedia page IDs of the two Wikipedia dumps. The page IDs of Wikipedia 2019-07 were available on the Wikipedia website. For the dump where DBpedia-Entity is based on, however, these IDs are not available anymore; we obtained them from the Nordlys package [11].

To avoid excluding rare entities and generate embeddings for a wide range of entities, we changed several Wikipedia2Vec settings. The two settings that resulted in the highest coverage of entities are: (i) minimum number of times an entity appears as a link in Wikipedia, (ii) whether to include or exclude disambiguation pages. Table 1 shows the effect of these settings on the number of missing entities; specifically the number of entities that are assessed in the DBpedia-Entity collection, but have missing embeddings. We categorize these missing entities into two groups:

- *No-page*: Entities without any pages. These entities neither were found by the Wikipedia redirect API nor could be matched by their page IDs.

[2] https://wikipedia.readthedocs.io/en/latest/.

Table 1. Missing entities with different settings

Settings	No-emb	No-page	Total
min-entity-count = 5, disambiguation = False	9640	608	10248
min-entity-count = 1, disambiguation = False	1220	398	1618
min-entity-count = 1, disambiguation = True	1220	377	1597
min-entity-count = 0, disambiguation = False	**724**	380	1104
min-entity-count = 0, disambiguation = True	**724**	**333**	**1057**

- *No-emb*: Entities that could be found by their identifiers, but were not included in the Wikipedia2Vec embeddings.

The first line in Table 1 corresponds to the default setting of Wikipedia2Vec, which covers only 75% of assessed entities in the DBpedia-Entity collection. When considering all entities in the knowledge graph, this setting discards an even larger number of entities, which is not an ideal setup for entity ranking. By choosing the right settings (the last line of Table 1), we increased the coverage of entities to 97.6%.

We trained two versions of embeddings: with and without link graph; i.e., using Eq. (4) with and without the $\mathcal{L}_e$ component.

4.3 Parameter Setting

Our entity re-ranking approach involves free parameter λ that needs to be estimated (see Eq. (6)). To set this parameter, we employed the Coordinate Ascent algorithm [17] with random restart of 3, optimized for NDCG@100. All experiments were performed using 5-fold cross-validation, where the folds were obtained from the collection (DBpedia-Entity V2). This makes our results comparable to the DBpedia-Entity V2 baseline runs, as the same folds are used for all the methods. Entity re-ranking was performed on top 1000 entities ranked by two state-of-the-art term-based entity retrieval models: FSDM and BM25F-CA [12]. For all experiments, we used the embedding vectors of 100 dimensions, which were trained using the settings described in Sect. 4.2.

5 Results and Analysis

5.1 Overall Performance

To answer our first research question, whether embeddings improve the score of entity retrieval, we compare our entity re-ranking approach with a number of baseline entity retrieval models. Table 2 shows the results for different models with respect to NDCG@10 and NDCG@100, the default evaluation measures for DBpedia-entity V2. In this table, the embedding-based similarity component (Eq. (5)) is denoted by *ESim*, where *c* and *cg* subscripts refer to the two versions of our entity embeddings: without and with link graph.

Table 2. Results of embedding-based entity re-ranking approach on different query subsets of DBpedia-Entity V2 collection. Significance of results is explained in running text.

Model	SemSearch		INEX-LD		ListSearch		QALD-2		Total	
NDCG	@10	@100	@10	@100	@10	@100	@10	@100	@10	@100
Reranking the FSDM top 1000 entities										
ESim_c	0.365	0.412	0.194	0.252	0.210	0.288	0.192	0.255	0.239	0.300
ESim_{cg}	0.397	0.462	0.216	0.282	0.211	0.311	0.213	0.286	0.258	0.334
FSDM	0.652	0.722	0.421	0.504	0.420	0.495	0.340	0.436	0.452	0.534
+ELR	0.656	0.726	0.435	0.513	0.422	0.496	0.347	0.446	0.459	0.541
$+\text{ESim}_c$	0.659	0.725	0.433	0.513	0.432	0.509	0.353	0.447	0.463	0.543
$+\text{ESim}_{cg}$	**0.672**	0.733	0.440	0.528	0.424	0.507	0.349	0.451	0.465	0.549
Reranking the BM25F-CA top 1000 entities										
ESim_c	0.381	0.424	0.194	0.253	0.211	0.283	0.192	0.252	0.243	0.301
ESim_{cg}	0.417	0.478	0.217	0.286	0.211	0.302	0.212	0.282	0.262	0.335
BM25F-CA	0.628	0.720	0.439	0.530	0.425	0.511	0.369	0.461	0.461	0.551
$+\text{ESim}_c$	0.658	0.730	0.462	0.545	0.448	0.529	0.380	0.469	0.481	0.563
$+\text{ESim}_{cg}$	0.660	**0.736**	**0.466**	**0.552**	**0.452**	**0.535**	**0.390**	**0.483**	**0.487**	**0.572**

The results of our method are presented for components ESim_c and ESim_{cg} by themselves (i.e., $\lambda = 1$ in Eq. (6)), and also in combination with FSDM and BM25F-CA. The mean and standard deviation of λ found by the Coordinate Ascent algorithm over all folds are: 0.34 ± 0.02 for FSDM+ESim_c, 0.61 ± 0.01 for FSDM+ESim_{cg}, 0.81 ± 0.03 for BM25F-CA+ESim_c, and 0.88 ± 0.00 for BM25F-CA+ESim_{cg}. The results show that the embedding-based scores alone do not perform very well, however, when combining them with other scores, the performance improves by a large margin. We determine the statistical significance of the difference in effectiveness for both the NDCG@10 and the NDCG@100 values, using the two-tailed paired t-test with $\alpha < 0.05$. The results show that both versions of FSDM+ESim and BM25-CA+ESim models yield significant improvements over FSDM and BM25-CA models (with respect to all metrics), respectively. Also, FSDM+ESim_{cg} improves significantly over FSDM+ELR with respect to NDCG@100, showing that our embedding based method captures entity similarities better than the strong entity ID matching approach used in the ELR method.

When considering the query subsets, we observe that FSDM+ESim_{cg} significantly outperforms FSDM for SemSearch and QALD queries with respect to NDCG@10, and for INEX-LD queries with respect to NDCG@100. Improvements over BM25F-CA were more substantial: BM25F-CA+ESim_{cg} brings significant improvements for all categories (with respect to all metrics) except for SemSearch queries for NDCG@100.

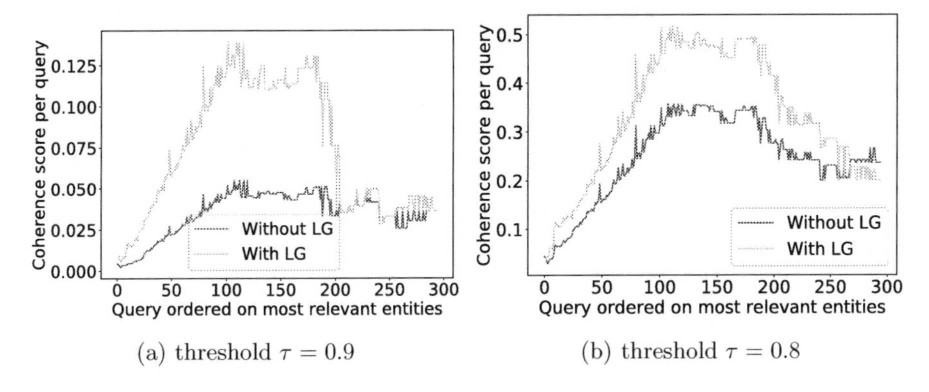

(a) threshold $\tau = 0.9$ (b) threshold $\tau = 0.8$

Fig. 1. Coherence score of all relevant entities per query, computed for the versions of entity embeddings (without and with link graph). The queries are ordered by the number of their relevant entities in x-axis.

5.2 Entity Embeddings Analysis

The results of Table 2 suggest that graph-based entity embeddings yield better performance compared to context only entity embeddings. To analyze why graph-based entity embeddings are beneficial for entity retrieval models, we conduct a set of experiments and investigate properties of embeddings with and without the graph structure.

According to the cluster hypothesis [14], documents relevant to the same query should cluster together. We consider the embeddings as data-points to be clustered and compare the resulting clusters in several ways. First, we compute the Davies Bouldin index [5] and the Silhouette index [24], which are: 3.16 and 0.08 for the embeddings with link graph, and 3.98 and −0.05 for the embeddings without link graph, respectively. Both measures indicate that better clusters arise for the embeddings that capture graph structure.

To get an indication of how coherent the clusters are, we compute for each query the coherence score defined in [13]. This score measures the similarity between item pairs of a cluster and returns the percentage of items with similarity score higher than a threshold, thereby assigning high scores to the clusters that are coherent. Formally, given a document set D, the coherence score is computed as:

$$Co(D) = \frac{\sum_{i \neq j \in 1,\ldots,M} \delta(d_i, d_j)}{\frac{1}{2}M(M-1)}, \tag{7}$$

where M is total number of documents and the δ function for each document pair d_i and d_j is defined as:

$$\delta(d_i, d_j) = \begin{cases} 1, & \text{if } sim(d_i, d_j) \geq \tau \\ 0, & \text{otherwise.} \end{cases} \tag{8}$$

We compute the coherence score with thresholds 0.8, 0.9, using *cosine* for similarity function $sim(d_i, d_j)$, where d_i and d_j correspond to entities. Figure 1

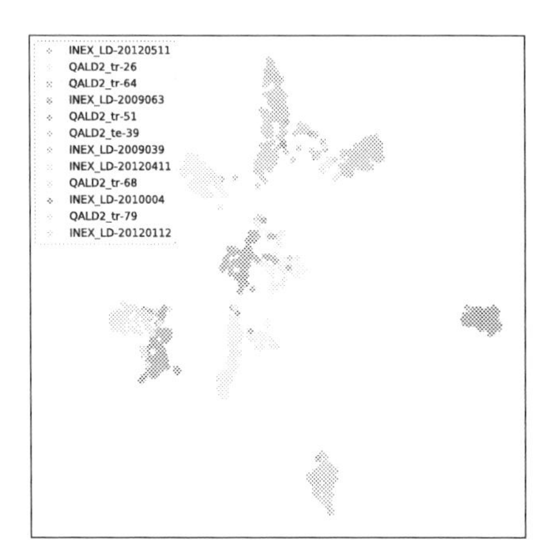

(a) Embeddings with link graph

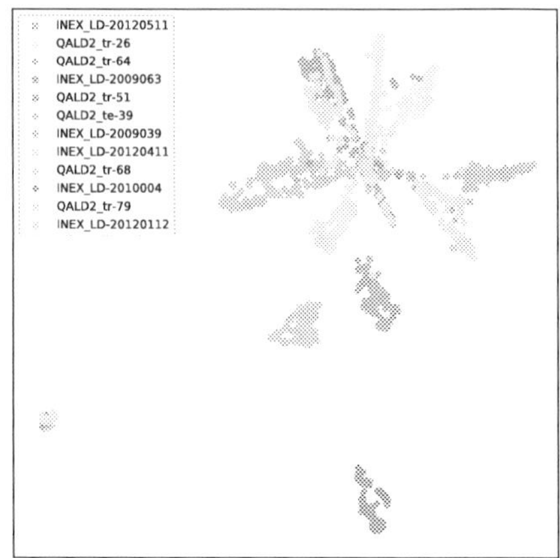

(b) Embeddings without link graph

Fig. 2. UMAP visualization of entity embeddings for a subset of queries. Color-codes correspond to the relevant entities per query. Queries per code are listed in Table 5 of the Appendix. Default settings of UMAP in python were used. (Color figure online)

shows the results of coherence score for all queries in our collection. Each point represents the coherence score of all relevant entities (according to the qrels) for a query. We considered only queries with more than 10 relevant entities, for clusters large enough to compute a meaningful score. Queries are sorted on the x-axis by the number of relevant entities. The plots clearly show that the coherence score for graph-based entity embeddings is higher than for context only ones. Based on these performance improvements we conclude that adding the graph structure results in embeddings that are more suitable for entity-oriented tasks.

Figures 2 helps to visually understand how clusters of entities differ for the two methods (a subset of all entities is shown for clarity). The data points correspond to the entities with a relevance grade higher than 0, for 12 queries with 100–200 relevant entities in the ground truth data. We use Uniform Manifold Approximation and Projection (UMAP) [16] to reduce the embeddings dimensions from 100 to two and plot the projected entities for each query. In Fig. 2b most of the clusters are overlapping in a star-like shape, while in Fig. 2a the clusters are more separated and the ones with similar search intents are close to each other; e.g., queries `QALD2_te-39` and `QALD2_tr-64` (which are both about companies), or `INEX_LD-20120112` and `INEX_LD-2009063` (which are both about war) are situated next to each other. To observe how false positive entities are placed in the embedding space, we added the 10 highest ranked false positives to the data and created new UMAP plots. In the obtained plots, false positive entities that are semantically similar to the true positive entities are close to each other. For example, two false positive entities for the query *"South Korean girl groups"* are: SHINee (a South Korean boy band) and Hyuna (a South Korean female singer). Both of these entities are semantically similar to the relevant

Table 3. Top queries with the highest gains and losses in NDCG at cut-offs 10 and 100, BM25F + ESim_{cg} vs. BM25F.

Query	Gain in NDCG	
	@10	@100
st paul saints	0.716	0.482
continents in the world	0.319	0.362
What did Bruce Carver die from?	0.307	0.307
spring shoes canada	−0.286	−0.286
vietnam war movie	−0.470	−0.240
mr rourke fantasy island	−0.300	−0.307

Table 4. Top queries with the highest gains and losses in NDCG at cut-offs 10 and 100, BM25F + ESim_{cg} vs. BM25F + ESim_{c}.

Query	Gain in NDCG	
	@10	@100
What did Bruce Carver die from?	0.307	0.307
Which other weapons did the designer of the Uzi develop?	0.236	0.248
Which instruments did John Lennon play?	0.154	0.200
Companies that John Hennessey serves on the board of	−0.173	−0.173
Which European countries have a constitutional monarchy?	−0.101	−0.197
vietnam war movie	−0.276	−0.222

entities of the query and are also placed in the vicinity of them, although they do not address the information needs of the query. This is consistent with the plots of Fig. 2 and in line with our conclusion on the effect of graph embeddings for entity-oriented search.

5.3 Query Analysis

Next, we investigate our second research question and analyse queries that are helped and hurt the most by our embedding-based method. Table 3 shows six queries that are affected the most by BM25F-CA+ESim_{cg} compared to BM25F-CA (on NDCG@100). Each of the three queries with highest gains are linked to at least one relevant entity (according to the assessments). The losses can be attributed to various sources of errors. For the query *"spring shoe canada"*, the only relevant entity belongs to the 2.4% of entities that have no embedding (cf. §4.2). Query *"vietnam war movie"* is linked to entities VIETNAM WAR and

War film, with confidence scores of 0.7 and 0.2, respectively. This emphasizes Vietnam war facts instead of its movies, and could be resolved by improving the accuracy of the entity linker and/or employing a re-ranking approach that is more robust to linking errors. The query *"mr rourke fantasy island"* is linked to a wrong entity due to a spelling mistake. To conclude, errors in entity linking form one of the main reasons of performance loss in our approach.

To further understand the difference between the two versions of the embeddings at the query-level, we selected the queries with the highest and lowest gain in NDCG@100 (i.e., comparing BM25F+ESim$_{cg}$ and BM25F+ESim$_{c}$). For the query *"Which instruments did John Lennon play?"*, the two linked entities (with the highest confidence score) are John Lennon and Musical Instruments. Their closest entity in graph embedding space is John Lennon's musical instruments, relevant to the query. This entity, however, is not among the most similar entities when we consider the context-only case.

For the other queries in Table 4, the effect is similar but less large than in the BM25F and BM25F + ESim$_{cg}$ case, probably due to the lower value of λ.

6 Conclusion

We investigated the use of entity embeddings for entity retrieval. We trained entity embeddings with Wikipedia2Vec, combined these with state-of-the-art entity ranking models, and find empirically that using graph embeddings leads to increased effectiveness of query results on DBpedia-Entity V2.

The empirical findings can be interpreted as evidence for the cluster hypothesis. Including a representation of the graph structure in the entity embeddings leads to better clusters and higher effectiveness of retrieval results. We further see that queries which get linked to relevant entities or pages neighboring to relevant entities get helped the most, while queries with wrongly linked entities are helped the least.

We conclude that enriching entity retrieval methods with entity embeddings leads to improved effectiveness, but acknowledge the following limitations of this study. Not all query categories lead to improvements on NDCG. While the state-of-the-art in entity-linking has made significant progress in recent years, we applied Tagme to identify the entities in queries. As we observed that lower performance of queries can often be attributed to erroneously linked entities, we expect better results by replacing this component for a state-of-the-art approach. Finally, we have only experimented using the embeddings constructed by Wikipedia2Vec, and plan to continue our experiments using alternative entity embedding methods like TransE.

A Queries

Table 5. Queries mentioned by their query ID.

Query ID	Query text
INEX_LD-20120511	female rock singers
QALD2_tr-26	Which bridges are of the same type as the Manhattan Bridge?
QALD2_tr-64	Which software has been developed by organizations founded in California?
INEX_LD-2009063	D-Day normandy invasion
QALD2_tr-51	Give me all school types
QALD2_te-39	Give me all companies in Munich
INEX_LD-2009039	roman architecture
INEX_LD-20120411	bicycle sport races
QALD2_tr-68	Which actors were born in Germany?
INEX_LD-2010004	Indian food
QALD2_tr-79	Which airports are located in California, USA?
INEX_LD-20120112	vietnam war facts

References

1. Paulheim, H.: Machine learning & embeddings for large knowledge graphs, July 2019. https://www.slideshare.net/heikopaulheim/machine-learning-embeddings-for-large-knowledge-graphs
2. Balog, K.: Entity-Oriented Search. Springer, Cham (2018). https://doi.org/10.1007/978-3-319-93935-3
3. Blanco, R., Ottaviano, G., Meij, E.: Fast and space-efficient entity linking in queries. In: Proceedings of the Eighth ACM International Conference on Web Search and Data Mining, pp. 179–188 (2015)
4. Bordes, A., Usunier, N., Garcia-Duran, A., Weston, J., Yakhnenko, O.: Translating embeddings for modeling multi-relational data. In: Proceedings of the 26th International Conference on Neural Information Processing Systems, pp. 2787–2795. ACM (2013)
5. Davies, D.L., Bouldin, D.W.: A cluster separation measure. IEEE Trans. Pattern Anal. Mach. Intell. **1**(2), 224–227 (1979)
6. Dehghani, M., Zamani, H., Severyn, A., Kamps, J., Croft, W.B.: Neural ranking models with weak supervision. In: Proceedings of the 40th International ACM SIGIR Conference on Research and Development in Information Retrieval, pp. 65–74 (2017)
7. Diaz, F., Mitra, B., Craswell, N.: Query expansion with locally-trained word embeddings. In: Proceedings of the 54th Annual Meeting of the Association for Computational Linguistics, pp. 367–377 (2016)
8. Ferragina, P., Scaiella, U.: TAGME: on-the-fly annotation of short text fragments (by wikipedia entities). In: Proceedings of the 19th ACM International IW3C2 on Information and Knowledge Management, pp. 1625–1628. ACM (2010)

9. Hasibi, F., Balog, K., Bratsberg, S.E.: Entity linking in queries: tasks and evaluation. In: Proceedings of the 2015 International Conference on The Theory of Information Retrieval, p. 171–180 (2015)
10. Hasibi, F., Balog, K., Bratsberg, S.E.: Exploiting entity linking in queries for entity retrieval. In: Proceedings of the 2016 ACM International Conference on the Theory of Information Retrieval, pp. 209–218. ACM (2016)
11. Hasibi, F., Balog, K., Garigliotti, D., Zhang, S.: Nordlys: a toolkit for entity-oriented and semantic search. In: Proceedings of the 40th International ACM SIGIR Conference on Research and Development in Information Retrieval, pp. 1289–1292. ACM (2017)
12. Hasibi, F., et al.: DBpedia-Entity V2: a test collection for entity search. In: Proceedings of the 40th International ACM SIGIR Conference on Research and Development in Information Retrieval, pp. 1265–1268. ACM (2017)
13. He, J., et al.: Exploring topic structure: coherence, diversity and relatedness. SIKS (2011)
14. Jardine, N., van Rijsbergen, C.J.: The use of hierarchic clustering in information retrieval. Inf. Storage Retrieval **7**(5), 217–240 (1971)
15. Liu, Z., Xiong, C., Sun, M., Liu, Z.: Explore entity embedding effectiveness in entity retrieval. arXiv preprint arXiv:1908.10554 (2019)
16. McInnes, L., Healy, J., Saul, N., Grossberger, L.: UMAP: uniform manifold approximation and projection. J. Open Source Softw. **3**(29), 861 (2018)
17. Metzler, D., Bruce Croft, W.: Linear feature-based models for information retrieval. Inf. Retrieval **10**(3), 257–274 (2007)
18. Mikolov, T., Chen, K., Corrado, G., Dean, J.: Efficient estimation of word representations in vector space. In: 1st International Conference on Learning Representations, ICLR, pp. 1–12 (2013)
19. Mikolov, T., Sutskever, I., Chen, K., Corrado, G.S., Dean, J.: Distributed representations of words and phrases and their compositionality. In: Advances in Neural Information Processing Systems (NIPS), pp. 3111–3119 (2013)
20. Noy, N., Gao, Y., Jain, A., Narayanan, A., Patterson, A., Taylor, J.: Industry-scale knowledge graphs: lessons and challenges. Commun. ACM **62**(8), 36–43 (2019)
21. Pennington, J., Socher, R., Manning, C.: Glove: global vectors for word representation. In: Proceedings of the 2014 Conference on Empirical Methods in Natural Language Processing, pp. 1532–1543. ACL (2014)
22. Perozzi, B., Al-Rfou, R., Skiena, S.: DeepWalk: online learning of social representations. In: Proceedings of the 20th ACM SIGKDD International Conference on Knowledge Discovery and Data Mining, pp. 701–710. ACM (2014)
23. Robertson, S., Zaragoza, H., et al.: The probabilistic relevance framework: BM25 and beyond. Found. Trends® Inf. Retrieval **3**(4), 333–389 (2009)
24. Rousseeuw, P.J.: Silhouettes: a graphical aid to the interpretation and validation of cluster analysis. J. Comput. Appl. Math. **20**, 53–65 (1987)
25. Yamada, I., Asai, A., Shindo, H., Takeda, H., Takefuji, Y.: Wikipedia2Vec: an optimized tool for learning embeddings of words and entities from Wikipedia. arXiv preprint 1812.06280 (2018)
26. Yamada, I., Shindo, H., Takeda, H., Takefuji, Y.: Joint learning of the embedding of words and entities for named entity disambiguation. In: The SIGNLL Conference on Computational Natural Language Learning (2016)
27. Zhiltsov, N., Kotov, A., Nikolaev, F.: Fielded sequential dependence model for ad-hoc entity retrieval in the web of data. In: Proceedings of the 38th International ACM SIGIR Conference on Research and Development in Information Retrieval, pp. 253–262. ACM (2015)

Learning Advanced Similarities and Training Features for Toponym Interlinking

Giorgos Giannopoulos[1(✉)], Vassilis Kaffes[1,2], and Georgios Kostoulas[1]

[1] IMSI/Athena Research Center, Marousi, Greece
giann@athenarc.gr
[2] University of the Peloponnese, Tripoli, Greece

Abstract. Interlinking of spatio-textual entities is an open and quite challenging research problem, with application in several commercial fields, including geomarketing, navigation and social networks. It comprises the process of identifying, between different data sources, entity descriptions that refer to the same real-world entity. In this work, we focus on toponym interlinking, that is we handle spatio-textual entities that are exclusively represented by their name; additional properties, such as categories, coordinates, etc. are considered as either absent or of too low quality to be exploited in this setting. Toponyms are inherently heterogeneous entities; quite often several alternative names exist for the same toponym, with varying degrees of similarity between these names. State of the art approaches adopt mostly generic, domain-agnostic similarity functions and use them as is, or incorporate them as training features within classifiers for performing toponym interlinking. We claim that capturing the specificities of toponyms and exploiting them into elaborate meta-similarity functions and derived training features can significantly increase the effectiveness of interlinking methods. To this end, we propose the *LGM-Sim* meta-similarity function and a series of novel, similarity-based and statistical training features that can be utilized in similarity-based and classification-based interlinking settings respectively. We demonstrate that the proposed methods achieve large increases in accuracy, in both settings, compared to several methods from the literature in the widely used Geonames toponym dataset.

Keywords: Interlinking · Machine learning · Toponym · Spatio-textual entities · Feature extraction · String similarity

1 Introduction

Interlinking (alt. deduplication, entity matching/linking, record linkage), in its most common form, is the task of identifying, from two entity sources, pairs of entity descriptions that correspond to the same real world entities. Interlinking is a crucial task in several domains, since it is quite often the case that real world entities are modeled, represented and gathered by different stakeholders, following different schemas, procedures and quality standards. As a result, multiple

J. M. Jose et al. (Eds.): ECIR 2020, LNCS 12035, pp. 111–125, 2020.
https://doi.org/10.1007/978-3-030-45439-5_8

databases might exist for representing the same groups of real world entities, in heterogeneous ways. Examples include: person names, product names and spatio-textual entities (toponyms, POIs, addresses) in different data providers.

In this paper, we examine the problem of interlinking spatio-textual entities, based solely on their name, i.e. we handle the problem of *toponym interlinking*. A toponym might refer to a broad range of spatio-textual entities, from small places, to countries. In our problem setting, the name of a spatio-textual entity/toponym, is its only *reliable* attribute that can be used for identifying same entities; the other attributes, such as spatial coordinates, categories, extended textual descriptions, are either non-existent or of too low quality/accuracy to be used for interlinking. Consider the scenario of a toponym data provider that maintains a proprietary toponym database, and periodically enriches/extends it with toponym entities extracted from user check-ins in social media. It is quite often the case that a check-in regarding a specific place is performed in locations that are considerably distant from the actual place. In this case, the coordinates extracted for the specific place are inaccurate and might even hurt the interlinking process, e.g. by leading to the rejection of a link between two toponyms that are actually the same, but appear to have distant locations. The authors of [4] elegantly describe the problem of dealing with *extremely noisy location coordinates* in the Facebook database. In another scenario, the recognition and extraction of toponyms might be performed on documents (e.g. travel guides), where the coordinates of toponyms are non-existent. Further, in either scenario, no other properties of the toponyms, such as categories or extended textual descriptions, can generally be retrieved for the majority of the extracted toponyms.

Competitive approaches from the literature utilize generic string similarity measures to solve the problem and limit their contributions to tuning their parameters or using them as training features in machine learning (ML) algorithms for classification. We take a different approach, claiming that domain knowledge is a critical factor for toponym interlinking that needs to be captured and incorporated within the interlinking process. To this end, we analyse a large toponym dataset, Geonames[1], which contains, among other toponym metadata, alternative names for millions of toponyms. Based on the insights we gain, we build an elaborate meta-similarity function, `LGM-Sim`, that takes into account and incorporates within its processing steps the specificities of toponym names. Additionally, we derive training features from `LGM-Sim`, that can be used for interlinking via classification. We demonstrate the superiority of the proposed models in two settings: similarity-based and classification-based toponym interlinking.

The rest of the paper is organized as follows. Section 2 discusses related work on interlinking, emphasizing on spatio-textual data. Section 3 defines the toponym interlinking problem and presents the two generic methodologies for solving it and, further, briefly discusses our findings and insights on the domain specificities of toponyms. Section 4 presents our proposed methods that incorporate the aforementioned insights, including domain specific similarity functions and training features, for toponym interlinking. Section 5 evaluates the effectiveness of

[1] https://www.geonames.org/.

the proposed methods in two different interlinking settings: similarity-based and classification-based. Finally, Sect. 6 concludes the paper.

2 Related Work

Considering the more general problem of name matching, various methods are proposed in the literature, with several previous studies [2,3] performing thorough evaluations of the most prominent ones in several datasets and demonstrating that there are no distinctively better methods that surpass all others in all settings/datasets. A generic framework for named entity interlinking is presented in [10], aiming to properly handle a variety of generic tasks where accuracy is not the most important criterion. In this frame, an enhancement of the Soft-TFIDF measure, combined with the Levenshtein similarity, is proposed.

Metrics specifically designed for toponym interlinking that mostly correspond to variations of the procedures used for generic name matching are proposed in [5,8]. The DAS similarity measure [8] comprises a hybrid, three-stages method that combines features from token-based and edit-based approaches. The meta-similarity proposed in [5] takes into account accentuation and other language-specific aspects of toponym names, in a four-stages process. The set of algorithms evaluated in [2] were assessed in the toponym interlinking problem by [12]. The authors experimented on place names listed in the GEOnet Names Server, that contains romanized toponyms from 11 different countries. Based on their study, no similarity measure achieves the highest accuracy in all datasets. In [4], the problem of business places deduplication is studied, taking a different approach. In the proposed solution, certain words (*core terms*) are identified, that are of higher significance in the name deduplication process. Based on these terms, a name model is constructed and properly combined with a spatial context model, using unsupervised learning algorithms.

Several works apply supervised machine learning approaches by extracting training features on name, coordinates, category, as well as textual, topological and semantic similarity, and utilizing them within classifiers [1,9,15,16]. A framework for improving duplicate detection using learnable text distance functions is presented in [1], however, in this work too, generic string similarity measures are used for the feature scores computation. The authors of [15] extract features for location name, coordinates, location type and demographic information. Then, machine learning algorithms are used to weight all features to solve the spatial entity matching problem. The work in [16] proposes a machine learning based approach to detect duplicate location entities. For this, a proposed metric is calculated for each key feature consisting of name, address and category similarity that describe entity pairs. Then, these extracted features are fed in a classification model to decide whether two entities are duplicates or not. Suport Vector Machine (SVM) and an alternating Decision Tree classifiers are used to combine different similarity features in [9]. The authors consider a variety of features corresponding to place name similarity, geospatial footprint similarity, place type similarity, similarity measures corresponding to semantic relations and temporal similarity. Contrary to these works, our study bases the matching process exclusively on toponym names. Utilizing richer spatio-textual profiles that enable the construction of features based on location-based, temporal or categorical similarity is beyond the scope of this work.

Most closely comparable to ours is the work presented in [14], where a thorough overview of the literature and an extended comparison of 13 different string similarity functions on toponym interlinking is performed. Additionally, these similarity functions are also assessed as training features within state of the art supervised machine learning algorithms for classification. Similarly, most works reviewed in [14] do not take into account the specificities of toponyms, however, using these methods as baselines for comparison allows as to compare with a large part of the literature. Our preliminary work on the problem [6], presents a first version of the `LGM-Sim`, a meta-similarity function aiming to capture the specificities of toponyms. In [6] we demonstrate that applying `LGM-Sim` on top of several baseline similarity functions improves their interlinking accuracy by a large extent. In outr current work, we fine-tune the `LGM-Sim` function and we exploit it into deriving training features to be used within classification algorithms for toponym interlinking, demonstrating further increases in accuracy.

Deep Learning methods for toponym interlinking are also being proposed in the literature. [13] present such a method, were Siamese RNNs are applied, yielding better accuracy results than traditional classifiers on similarity-based training features. Incorporating Deep Learning methods eliminates the need for feature extraction and engineering, however, it requires large amounts of data to train proper models, as well as engineering proper DNN architectures. The goal of the currently presented work is to demonstrate the potential and the gains of incorporating domain knowledge into generic similarity measures and classifiers for toponymm interlinking. Comparing traditional classification methods with Deep Learning methods, as well as devising approaches for exploiting both worlds comprises part of our ongoing work. Thus, in the current manuscript, the approach of [13] is considered orthogonal but potentially complementary.

A different research strand focus on improving the efficiency of the interlinking process [11], since, in its naive version, it is an $O(n^2)$ problem, and thus prohibitive for large datasets. Such methods consider as a given that the similarity functions they apply will be sufficiently accurate on identifying same entities, and focus on developing indexes, structures and schemes for optimizing the performance of the interlinking process, in terms of time-efficiency and scalability. Following a different rationale, the Magellan system, presented in [7], aims to provide the tools to end users to exploit a wide range of techniques related to the entity matching process under a unified framework. However, the implemented similarity functions are generic, widely adopted similarities that are not specialized or properly tuned for the setting of comparing geospatial entities. Further, an integral concept in Magellan is user interaction, prescribing pipelines where automated interlinking tasks and user feedback are iteratively combined. These categories of works on interlinking solve orthogonal/complementary problems and are not directly comparable to our proposed methods.

3 Background and Domain Knowledge

3.1 Problem Formulation

Given a set of candidate toponym pairs, toponym interlinking can be formalized as the problem of learning a decision function that decides whether a candidate toponym pair contains two toponyms that correspond to the same real world spatio-textual entity[2]. One can identify two generic methodologies for solving the above problem.

Similarity comparison methods apply, on every candidate pair of toponyms, a string similarity function, in conjunction with a similarity threshold to compare the names of the two toponyms. If the similarity score, produced by the similarity function, surpasses or equals the defined threshold, then the pair of toponyms is marked as `True`, otherwise, it is marked as `False`. Depending on the similarity function, several parameters, such as thresholds and weights, need to be tuned either by learning them on a dedicated training set or by being empirically selected by domain experts via tuning/trial-and-error procedures.

Classification methods train a binary classification algorithm that takes as input a candidate pair of toponyms and classifies it in either of the two available classes: {`True`, `False`}. In this case, a feature extraction process needs to be performed before the classification algorithm is deployed, in order to represent candidate pairs into an appropriate feature space, capturing meaningful toponym properties and relations with respect to the task at hand. After the above representations are defined, the classifier is trained on historical data, i.e. candidate toponym pairs for which we already know their class. The trained classifier is then able to decide whether a new pair of toponyms is `True` or `False`.

The literature on toponym interlinking includes methods adopting both presented methodologies. A detailed comparison of most of the proposed methods is performed in [14], as well as in our evaluation, which follows a similar experimental setting. The goal of this paper is to establish the value of learning/elaborate domain specific similarities for toponym interlinking. Given that most of the aforementioned methodologies are either based or can benefit from similarity measures and/or respective training features, the contribution of our methods touches and can potentially improve a wide range of interlinking methods.

3.2 Concepts and Intuitions

For our analysis, it is useful to first discuss the concept of *core name* and *core term*, also similarly defined, but differently handled in [4]. Core name consists in the subset of terms from the toponym name that are the most important in distinguishing the toponym from other ones, and, respectively, identifying same toponyms. A core term is a single term contained in the core name. The concept of core name cannot be strictly/formally defined, since it largely depends on human understanding on toponyms and domain knowledge. However, we believe

[2] Obtaining the set of candidate toponym pairs is an orthogonal problem, with several efficient solutions in the literature, like *blocking* [11]; in what follows, we consider this set available and focus on the problem of toponym interlinking, as defined above.

that approximating the identification of core terms and handling them differently within a meta-similarity function may yield increased interlinking accuracy.

One of the major specificities of toponyms, compared to namings of other entities, is the fact that different terms within their name might largely vary with respect to their significance in deciding whether two toponyms are the same. This variability might also take several forms, such as: (i) The existence of a frequent term that provides categorical information of the toponym, like "community" or "square". Such terms do not comprise a part of the core name of the toponym, and need to be handled differently; (ii) Terms that comprise part of the core name of a toponym might be of different significance and, thus, some of them might be omitted from its name in one data source, while maintained in the name in another source. An indicative example would be the toponym: "St. Paul's German Lutheran Church". In this case, "Church" could be considered a non-core term, however, it is expected to be a frequent term. Also, "St. Paul's" potentially has higher significance than "German Lutheran" in distinguishing/interlinking the toponym from/with another one.

Additionally, toponyms are inherently characterized by large variability in the terms that actually comprise their core name. That is, the same term might be spelled in different ways, or be expressed in abbreviated, or generally altered, forms. This fact complicates the identification of proper matchings between name terms within a toponym. The aforementioned variability extends to punctuation and accentuation too. A representative example is the following matching pair "Solovejcev Kljuch - Soloveytsev Klyuch".

Another issue lies in the order of the toponym terms. Two variations of the same toponym might contain the same terms in different order; term ordering becomes even more cumbersome in case some of these terms are missing in one of the names. In this case, it is uncertain whether sorting the terms of each toponym aphanumerically, before comparing them, will facilitate or hinder their similarity comparison process. A representative example is the pair: "Lake Thompson - Thompson Lake Reservoir". Not only the two toponyms contain core terms in different order but also only one of them contains the term "Reservoir".

We note that all the presented examples are drawn from Geonames, a large toponym dataset on which we perform our analysis and which contains hundreds of thousands of such cases and specificities. The `LGM-Sim` meta-similarity function, proposed next, comprises several string processing steps that take into account and handle the aforementioned toponym specificities.

4 Models for Toponym Interlinking

In this section, we present `LGM-Sim`, a meta-similarity function for toponym interlinking, that incorporates domain knowledge on toponyms within its processing. Consequently, we discuss how `LGM-Sim` can be transformed into training features and utilized, along with previously proposed similarity-based features and additional statistical feratures, within classifiers for toponym interlinking.

4.1 LGM-Sim Meta-Similarity for Toponym Interlinking

A high level description of the `LGM-Sim` meta-similarity is provided in Algorithm 1. `LGM-Sim` takes as input two toponym strings, while it has a set of parameters regarding comparison thresholds and individual score weighting. Additionally, it considers a set of frequent terms that can be automatically gathered by the corpus of toponyms that are to be interlinked. All the parameters of `LGM-Sim` can be automatically learned by evaluating the effectiveness of different parameterizations on a small training dataset, and selecting the parameters that yield the highest accuracy. We show in our experiments that it is sufficient to train `LGM-Sim` in a much smaller train dataset than the deployment dataset.

`LGM-Sim` aims at properly splitting the compared toponym strings into discrete lists of terms, with each list containing terms of different semantics. First, `LGM-Sim` initializes a list of punctuation marks and two initially empty lists of terms for the two toponyms, $\mathcal{S}_1, \mathcal{S}_2$ respectively (lines 6–7). Next, an initial preprocessing step on the two strings is performed by `TransformNames`, including lowercasing, transliteration and punctuation/accentuation alignment (line 8).

Algorithm 1: LGM-Sim

Input: name of toponym A, s_1; name of toponym B, s_2
1 set of frequent terms in the corpus of names $\mathcal{S}_{FT}$;
2 total similarity comparison threshold θ_{sim};
3 threshold used in sorting terms θ_{sort};
4 threshold used in splitting names θ_{split};
5 weights for each name part $w_{base}, w_{mis}, w_{freq}$;
Output: ($True$ || $False$) on whether toponyms A and B are the same

6 *punct* ← *[‘’“”‴!?;/‘,.-]*
7 $\mathcal{S}_1, \mathcal{S}_2 \leftarrow []$
8 $(s_1, s_2) \leftarrow$ `TransformNames`$(s_1, s_2, punct)$
9 $(\mathcal{S}_1, \mathcal{S}_2) \leftarrow$ `SortTerms`(s_1, s_2)
10 $(\mathcal{S}_1^{base}, \mathcal{S}_2^{base}, \mathcal{S}_1^{freq}, \mathcal{S}_2^{freq}) \leftarrow$ `ExtractFrequentTerms`$(\mathcal{S}_1, \mathcal{S}_2, \mathcal{S}_{FT})$
11 $(\mathcal{S}_1^{base}, \mathcal{S}_2^{base}, \mathcal{S}_1^{mis}, \mathcal{S}_2^{mis}) \leftarrow$ `CompareCoreTerms`$(\mathcal{S}_1^{base}, \mathcal{S}_2^{base}, \theta_{split})$
12 $matchScore \leftarrow$ `WeightedSim`$(\mathcal{S}_1^{base}, \mathcal{S}_2^{base}, w_{base}, \mathcal{S}_1^{mis}, \mathcal{S}_2^{mis}, w_{mis},$
$\mathcal{S}_1^{freq}, \mathcal{S}_2^{freq}, w_{freq})$
13 **return** $matchScore \geq \theta_{sim}$

Then, the terms within the two toponym strings are sorted alphanumerically and stored to the initialized lists (line 9). The first step of `SortTerms` is to concatenate all the (unsorted) terms in the initial strings and then compare the two concatenated strings with a loose threshold θ_{sort}. If the two concatenated strings are similar enough, then the function returns the two lists of terms *unsorted*. The rationale is that, if the initial, unsorted strings are similar enough, then sorting their terms might reduce their similarity, e.g. by re-ordering small terms that are not common in both strings or that start with different alphabet letter.

On the other hand, if the two initial strings are not similar enough, the function returns two alphanumerically sorted lists of their terms, since, in this case, it is quite probable that sorting will increase their similarity.

Then, the first step of splitting the two toponym strings into separate lists is performed (line 10). `ExtractFrequentTerms` identifies frequent terms within $\mathcal{S}_1, \mathcal{S}_2$, removes them and adds them in two new lists, $\mathcal{S}_1^{freq}, \mathcal{S}_2^{freq}$. The remaining lists are now called $\mathcal{S}_1^{base}, \mathcal{S}_2^{base}$ and contain terms more probable to be *core* ones.

Thereafter, the second step of splitting the toponyms into separate lists is performed (line 11). Specifically, `CompareCoreTerms` further splits each of the *base* toponym lists into the new lists, containing matching and non-matching terms, as follows. First, four empty lists are initialized, so as to be filled with the matching and non-matching terms of each toponym. Then, the two input lists of *base* terms of the two toponyms are parsed simultaneously, term by term. At each step, if the two considered terms from the two lists (loosely) match, then they are permanently stored as *base* terms. If the two terms do not match, the parsing proceeds only in the list with the alphanumerically lowest term, while the specific term is stored in the respective *mismatch* list. Finally, the remaining, mismatched terms are added to the respective *mismatch* lists, and the function returns the four new lists $\mathcal{S}_1^{base}, \mathcal{S}_2^{base}, \mathcal{S}_1^{mis}, \mathcal{S}_2^{mis}$, containing matching base terms and mismatching terms from the two toponym strings.

At this stage, the two initial toponym strings have been split in three lists of terms each: (i) Two *base* lists containing potential *core terms* of the toponyms that are identified to match between the two toponyms; (ii) Two *mismatch* lists containing potential *core terms*, that, however, have not been matched between the two toponyms; (iii) Two *frequent terms* lists that contain terms from the toponym that are frequently found in the corpus, and thus might not belong to the *core names* of the toponyms, functioning auxiliary to them.

Next, `CompareCoreTerms` calculates three similarity scores, comparing individually the three different types of lists, and properly weights the three individual similarity scores in order to produce the final similarity score for the toponyms (line 12). For this, three individual weights are utilized, w_b, w_m and w_f, which represent the significance of the similarity scores calculated on the three term lists that are individually compared for each pair of toponyms. Each comparison process first examines whether the input lists are empty, so as to re-adjust the corresponding weights for the individual similarity scores. For example, if for a pair of toponyms no frequent terms have been identified, then the score weight w_f is set to null and the rest weights w_b and w_m are proportionately increased according to the lengths of the respective term lists. Next, the significance weights w_b, w_m and w_f are re-calculated, taking into account the lengths of the corresponding lists they refer to. This process is performed in order to compensate for large discrepancies between the lengths of lists (measured in total number of characters) of different types. Finally, the individual scores are summed and the final similarity score is returned by the function (line 13).

In the last step (line 13), `LGM-Sim` compares the final score calculated with a similarity comparison threshold θ_{sim}. Depending on the result of the comparison, the value `True` or `False` is returned denoting whether the two toponyms are the same or not. LGM-Sim is a meta-similarity function, thus it can be applied on top of any generic similarity measure. Following the evaluation paradigm of [14], we consider a large set of similarity measures, presented in Table 1. Presenting the specifics of these similarity measures is out of the scope of this paper. However, these measures are well studied in the literature and the reader can refer to [14] for a short presentation of each of them.

Table 1. Considered similarity measures

Damerau-Levenshtein	Jaro
Jaro–Winkler	Jaro–Winkler Reversed
Sorted Jaro–Winkler	Cosine N-Grams
Jaccard N-Grams	Dice Bi-Grams
Jaccard Skip-grams	Monge–Elkan
Soft–Jaccard	Davis and De Salles
Tuned Jaro-Winkler	Tuned Jaro-Winkler Reversed

We note that the last line of Table 1 presents two similarity measures that are not presented in [14] or [6], rather than comprise a variation of the respective Jaro-Winkler measures, that we propose in the current work, taking into account the characteristics of toponyms that we have studied in Geonames. In particular, it adds the notion of skip-grams into the Jaro-Winkler metric and, especially, the Winkler part, that gives higher scores to strings that match from the beginning up to a given prefix length. Thus, we allow for a gap of one character in the matching of the prefixes between two strings. The same applies to its reverse variation by considering the endings of the strings.

4.2 LGM-Sim Based Classifiers for Toponym Interlinking

The generic process of training classifiers for toponym interlinking is described in Sect. 3.1. Here, we discuss the training features we introduce, for better capturing and exploiting the domain knowledge of toponyms. One of the major merits of training a classifier is the combinatorial exploitation of several features within a model. While in the generic similarity comparison based setting, only one similarity function at a time can be examined/deployed, in the classification based setting, several similarity functions can be encoded as training features of the model. Past approaches, as well as the study presented in [14], have used combinations of the similarity measures presented in Table 1 as training features.

In this work, we adopt the aforementioned set of training features, however, we enrich it with a corresponding set of features generated by applying the LGM-Sim on all the similarity measures presented in Table 1. Further, we consider an "intermediate" set of training features, corresponding again to the similarity measures of Table 1, however, having performed only the sorting function of

LGM-Sim, before comparing the toponym strings. Additionally, we consider the three individual similarity scores calculated on the three individual lists that LGM-Sim splits the two toponym names. That is, we consider $score_b$, $score_m$ and $score_f$, derived by individually comparing $(\mathcal{S}_1^{base}, \mathcal{S}_2^{base})$, $(\mathcal{S}_1^{mis}, \mathcal{S}_2^{mis})$ and $(\mathcal{S}_1^{freq}, \mathcal{S}_2^{freq})$, as separate features, allowing the model to learn the significance of the similarity of each of these individual parts of the toponym names, along with the significance of their total similarity.

Table 2. Considered training features

Feature Type	Number of Features
Basic similarity measures	14
Sorted similarity measures	13
LGM-Sim based similarity measures	13
Individual matching scores from LGM-Sim based on Damerau	3
Statistical features	44

Further, we define a set of non-similarity based features, that concern statistical aspects of the compared toponyms. Specifically, for each pair of toponyms, the number of terms contained in each forms two integer features; the existence of a frequent term in each forms two boolean features, while the existence of one or more of the 20 more frequent terms in the whole dataset in each toponym forms 40 additional boolean features. Eventually, we consider five groups of training features (Table 2): (i) the *basic* similarity features as presented in [14]; (ii) the *sorted* similarity features; (iii) the *LGM-Sim* similarity features; (iv) the individual scores on the split toponyms produced by applying *LGM-Sim* based Damerau-Levenshtein similarity; and (v) the statistical features on toponyms.

5 Evaluation

This section presents the evaluation of the proposed methods for toponym interlinking, with respect to the two different settings presented in Sect. 3.1: *similarity-based* and *classification-based*. In the former setting, we compare the interlinking effectiveness of our proposed LGM-Sim meta-similarity against the traditional similarity measures and functions that are used in several works of the literature (Table 1). In the latter setting, we compare the interlinking effectiveness of our proposed method, that uses additional, novel, LGM-Sim-derived and statistical features within classifiers, against the approach presented in [14], that only uses traditional similarity measures as features[3].

[3] We note that the baselines we compare with cover a large part of the presented literature, presented in the following papers: [2,3,5,12,14].

The evaluation dataset is drawn from Geonames, a database that contains more than 11 Million toponyms from around 250 countries. For each toponym in the dataset, there exists its main name and a list of alternate names. By following the exact procedure of [14], we construct a balanced dataset of 5M `True` and `False` toponym pairs. The False toponym pairs are created by selecting a name and an alternate name from different toponym records of the initial dataset. The True toponym pairs are created by selecting the name and the alternate name from the same toponym record, but ensuring, to some extent, that some of the created pairs vary in their name. To measure the effectiveness of the evaluated methods, four standard IR measures are adopted: *Accuracy*, *Precision*, *Recall*, *F1-Score*. To better evaluate the generalization capacity of the compared methods, we slightly modify the setting used in [14], introducing a separate training set, where compared similarity methods are trained[4]. We keep the 5M toponym pairs test set the same as in [14], for evaluating the trained models. The training set contains 100 K toponym pairs, equally balanced between `True` and `False`. The code and the respective datasets are available on GitHub[5].

5.1 Evaluation Results

We denote as `Basic` all baseline models as presented in [14], while as `LGM/LGM-Sim` all our proposed models. Moreover, we mark with bold the best reported value, per evaluation measure, and per compared approach (`Basic` vs. `LGM-Sim`). We note that we exclude from our experiments the *Permuted Jaro-Winkler* similarity measure, since it is reported in [14] that it is orders of magnitude slower than the rest similarity measures, without any substantial gains in interlinking effectiveness. Also, we do not report values for the `LGM-Sim` version of the *Jaro-Winkler Sorted* similarity, since it would conflict with the fact that `LGM-Sim` incorporates its own mechanism for deciding whether to sort or not two toponym strings.

Examining the effectiveness of the compared methods in the test set[6], Table 3 demonstrates that `LGM-Sim` improves the Accuracy of all baseline models by 8–15%. Specifically, in the similarity-based setting, the best `LGM-Sim` model increases the Accuracy of the best `Basic` model by 14.9%; in the classification setting, the respective increase is 8.1%. Similar observations stand for the rest of the evaluation measures where both Precision and Recall of the models increase, noting that `LGM-Sim` meta-similarity seems to give a large boost to Recall, whereas Precision

[4] To learn the hyper-parameters for each classification model, we perform 5-fold cross-validation on the train set, averaging the Accuracy score for each examined hyper-parameterization; the one with the higher Accuracy is selected for the classifier. The weights and thresholds used within the similarity measures are also handled as parameters of the models and learned on the training set. Reporting the optimal values for these is omitted due to lack of space, however, they can be reproduced by executing the referenced GitHub code.

[5] https://github.com/LinkGeoML/ToponymInterlinking.

[6] Similar numbers and differences are also reported in the training set but omitted due to lack of space.

is the one significantly boosted on the LGM-Sim derived features within classifiers. Another observation is that the LGM-Sim meta-similarity methods close the gap between similarity comparison and classification based methods, making the former an acceptable solution in scenarios where the more heavy-weight classification models are not an option. A third observation is that the introduced *Tuned Jaro-Winkler Reversed* similarity marginally increases the effectiveness of the *Jaro-Winkler Reverse* in the similarity based setting (79.8% vs. 79.6%), comprising a seemingly insignificant boost in Accuracy.

Table 3. Evaluation results on test dataset (5M)

	Accuracy		Precision		Recall		F1-score	
	Basic	LGM	Basic	LGM	Basic	LGM	Basic	LGM
Damerau-Levenshtein	0.645	0.780	0.791	**0.830**	0.393	0.704	0.526	0.762
Jaro	0.634	0.771	0.776	0.826	0.377	0.686	0.508	0.750
Jaro-Winkler	0.632	0.768	0.722	0.778	**0.431**	0.748	**0.540**	0.763
Jaro-Winkler Reverse	0.646	0.796	0.782	0.830	0.405	0.744	0.533	0.784
Jaro-Winkler Sorted	0.615	—	0.719	—	0.377	—	0.495	—
Cosine n-grams	0.614	0.718	0.710	0.741	0.386	0.669	0.500	0.703
Jaccard n-grams	0.609	0.709	0.753	0.785	0.326	0.575	0.455	0.664
Dice bi-grams	0.621	0.731	0.761	0.758	0.352	0.678	0.481	0.716
Jaccard skip-grams	0.625	0.738	0.741	0.753	0.385	0.710	0.507	0.730
Monge-Elkan	0.595	0.764	0.664	0.775	0.385	0.745	0.488	0.759
Soft-Jaccard	0.594	0.762	0.705	0.767	0.322	0.751	0.442	0.759
Davis/De Salles	0.617	0.771	0.716	0.787	0.389	0.742	0.504	0.764
Tuned Jaro-Winkler	0.630	0.770	0.728	0.796	0.413	0.725	0.527	0.759
Tuned Jaro-Winkler Reverse	**0.649**	**0.798**	**0.808**	0.822	0.390	**0.761**	0.526	**0.791**
Gradient Boosted Trees	**0.773**	**0.854**	0.764	0.876	0.790	**0.824**	**0.777**	**0.849**
SVM	0.719	0.824	0.688	0.864	**0.802**	0.768	0.741	0.813
Random Forests	0.770	0.849	0.769	0.877	0.772	0.811	0.770	0.843
Extr. Rand. Trees	0.766	0.844	**0.769**	**0.878**	0.760	0.799	0.765	0.837
Decision Tree	0.718	0.778	0.708	0.779	0.741	0.775	0.724	0.777

Finally, we examine the most informative training features utilized by the best classifier on our problem, in terms of higher Accuracy, i.e. the *Gradient Boosted Trees*. Table 4 presents the top-10 features from higher to lower order of importance. A first observation is that the majority (7/10) of the top-10 training features are introduced in this work. This is consistent with the findings from Table 3. Another observation is that variations of our proposed *Tuned Jaro-Winkler* measure comprise 3/10 most informative features, demonstrating that the proposed modification, despite its marginal effect in the similarity-based setting, it is rather useful in the classification based setting. Further, it becomes evident that appropriate sorting of toponym strings, following the processing of

`SortTerms`, is of high importance, since 5/10 features incorporate this processing (either plain *Sorted* or *LGM-Sim* features). Finally, although the introduced statistical features seem promising, it is evident that they require more elaboration, since they occupy only the last two positions in the list.

Table 4. Evaluation of top-10 most important training features for the best classifier

Ranking	Gradient Boosted Trees
1	Damerau-Levenshtein Sorted
2	Tuned Jaro-Winkler
3	Sorted Jaro-Winkler Reverse
4	LGM-Sim-Damerau-Levenshtein
5	Tuned Jaro-Winkler Reverse
6	LGM-Sim-Tuned Jaro-Winkler Reverse
7	LGM-Sim-Jaro-Winkler Reverse
8	Jaccard n-grams
9	Top-20 Freq. Term exists in ($\mathcal{S}_1$)
10	Number of terms in string ($\mathcal{S}_1$)

Regarding computation times, we present some indicative runtimes to showcase that the `LGM-Sim` methods do not introduce prohibitive overheads. Indicatively, for 5M toponym pairs, *JW Reversed* similarity runs in 10 s, while its `LGM-Sim` version in 114 s; both times are marginal considering the magnitude of the dataset. Respectively, *Gradient Boosted Trees* with the baseline features runs in 113 min., while its `LGM-Sim` version in 135 min., introducing less that 20% overhead, which is negligible considering the Accuracy gains presented above.

6 Conclusion

In this paper, we presented domain specific models that can be applied for toponym interlinking. We demonstrated that the proposed meta-similarity, `LGM-Sim` and the training features derived from it consistently, and to a large extent, improve the interlinking effectiveness of widely used baseline models. As future work, further examining and refining non-similarity based training features (e.g. structural, statistical), as well as combining traditional features and methods with Deep Learning/embedding-based methods comprise promising directions.

Acknowledgments. This research has been co-financed by the European Regional Development Fund of the European Union and Greek national funds through the Operational Program Competitiveness, Entrepreneurship and Innovation, under the call RESEARCH – CREATE – INNOVATE (project codeT1EDK-04568).

References

1. Bilenko, M., Mooney, R.J.: Adaptive duplicate detection using learnable string similarity measures. In: Proceedings of the Ninth ACM SIGKDD International Conference on Knowledge Discovery and Data Mining, KDD 2003, pp. 39–48. ACM, New York (2003). https://doi.org/10.1145/956750.956759, http://doi.acm.org/10.1145/956750.956759
2. Christen, P.: A comparison of personal name matching: techniques and practical issues. In: Sixth IEEE International Conference on Data Mining - Workshops (ICDMW 2006), pp. 290–294, December 2006. https://doi.org/10.1109/ICDMW.2006.2
3. Cohen, W.W., Ravikumar, P., Fienberg, S.E.: A comparison of string distance metrics for name-matching tasks. In: Proceedings of the 2003 International Conference on Information Integration on the Web, IIWEB 2003, pp. 73–78. AAAI Press (2003). http://dl.acm.org/citation.cfm?id=3104278.3104293
4. Dalvi, N., Olteanu, M., Raghavan, M., Bohannon, P.: Deduplicating a places database. In: Proceedings of the 23rd International Conference on World Wide Web, WWW 2014, pp. 409–418. ACM, New York (2014). https://doi.org/10.1145/2566486.2568034, http://doi.acm.org/10.1145/2566486.2568034
5. Davis, C.A., de Salles, E.: Approximate string matching for geographic names and personal names. In: GeoInfo, pp. 49–60, January 2007
6. Kaffes, V., Giannopoulos, G., Karagiannakis, N., Tsakonas, N.: Learning domain specific models for toponym interlinking. In: Proceedings of the 27th ACM SIGSPATIAL International Conference on Advances in Geographic Information Systems, SIGSPATIAL 2019, Chicago, IL, USA, 5–8 November 2019, pp. 504–507 (2019). https://doi.org/10.1145/3347146.3359339
7. Konda, P., et al.: Magellan: toward building entity matching management systems. PVLDB **9**, 1197–1208 (2016)
8. KilinçS, D.: An accurate toponym-matching measure based on approximate string matching. J. Inf. Sci. **42**(2), 138–149 (2016). https://doi.org/10.1177/0165551515590097
9. Martins, B.: A supervised machine learning approach for duplicate detection over gazetteer records. In: Claramunt, C., Levashkin, S., Bertolotto, M. (eds.) GeoS 2011. LNCS, vol. 6631, pp. 34–51. Springer, Heidelberg (2011). https://doi.org/10.1007/978-3-642-20630-6_3. http://dl.acm.org/citation.cfm?id=2008664.2008669
10. Moreau, E., Yvon, F., Cappé, O.: Robust similarity measures for named entities matching. In: Proceedings of the 22nd International Conference on Computational Linguistics - Volume 1, COLING 2008, pp. 593–600. Association for Computational Linguistics, Stroudsburg (2008). http://dl.acm.org/citation.cfm?id=1599081.1599156
11. Papadakis, G., Alexiou, G., Papastefanatos, G., Koutrika, G.: Schema-agnostic vs schema-based configurations for blocking methods on homogeneous data. PVLDB **9**(4), 312–323 (2015). https://doi.org/10.14778/2856318.2856326. http://www.vldb.org/pvldb/vol9/p312-papadakis.pdf
12. Recchia, G., Louwerse, M.: A comparison of string similarity measures for toponym matching. In: COMP 2013 - ACM SIGSPATIAL International Workshop on Computational Models of Place, pp. 54–61, November 2013. https://doi.org/10.1145/2534848.2534850
13. Santos, R., Murrieta-Flores, P., Calado, P., Martins, B.: Toponym matching through deep neural networks. Int. J. Geogr. Inf. Sci. **32**, 1–25 (2017). https://doi.org/10.1080/13658816.2017.1390119

14. Santos, R., Murrieta-Flores, P., Martins, B.: Learning to combine multiple string similarity metrics for effective toponym matching. Int. J. Digit. Earth **11**, 1–26 (2017). https://doi.org/10.1080/17538947.2017.1371253
15. Sehgal, V., Getoor, L., Viechnicki, P.D.: Entity resolution in geospatial data integration. In: Proceedings of the 14th Annual ACM International Symposium on Advances in Geographic Information Systems, GIS 2006, pp. 83–90. ACM, New York (2006). https://doi.org/10.1145/1183471.1183486, http://doi.acm.org/10.1145/1183471.1183486
16. Zheng, Y., Fen, X., Xie, X., Peng, S., Fu, J.: Detecting nearly duplicated records in location datasets. In: Proceedings of the 18th SIGSPATIAL International Conference on Advances in Geographic Information Systems, GIS 2010, pp. 137–143. ACM, New York (2010). https://doi.org/10.1145/1869790.1869812, http://doi.acm.org/10.1145/1869790.1869812

Patch-Based Identification of Lexical Semantic Relations

Nesrine Bannour[1], Gaël Dias[1(✉)], Youssef Chahir[1], and Houssam Akhmouch[1,2]

[1] Normandie Univ, UNICAEN, ENSICAEN, CNRS, GREYC, 14000 Caen, France
gael.dias@unicaen.fr

[2] Crédit Agricole Brie Picardie, 77700 Serris, France

Abstract. The identification of lexical semantic relations is of the utmost importance to enhance reasoning capacities of Natural Language Processing and Information Retrieval systems. Within this context, successful results have been achieved based on the distributional hypothesis and/or the paradigmatic assumption. However, both strategies solely rely on the input words to predict the lexical semantic relation. In this paper, we make the hypothesis that the decision process should not only rely on the input words but also on their K closest neighbors in some semantic space. For that purpose, we present different binary and multi-task classification strategies that include two distinct attention mechanisms based on PageRank. Evaluation results over four gold-standard datasets show that average improvements of 10.6% for binary and 8% for multi-task classification can be achieved over baseline approaches in terms of F_1. The code and the datasets are available upon demand.

Keywords: Patches · PageRank · Attention mechanism · Multi-task learning

1 Introduction

Recognizing the exact nature of the semantic relation holding between a pair of words is crucial for many applications such as taxonomy induction [10], question answering [6,22], query expansion [15] or text summarization [8]. The most studied lexical semantic relations are synonymy, co-hyponymy, hypernymy, or meronymy, but more exist [37]. Numerous approaches have been proposed to identify one particular semantic relation of interest following either the paradigmatic approach [28,29,33,39], the distributional model [9,31,36,37], or their combination [25,32].

In all these studies, the decision process relies on finding representation regularities between two input words. In this paper, we make the assumption that finding the lexical semantic relation that holds between two words does not solely rely on the pair itself, but also on the semantically related neighboring words. Our hypothesis relies on two different ideas. First, studies about the mental lexicon [13,23] theorize that words are highly interconnected within a mental semantic network, such that conceptual information is encoded in one's mind rather

J. M. Jose et al. (Eds.): ECIR 2020, LNCS 12035, pp. 126–140, 2020.
https://doi.org/10.1007/978-3-030-45439-5_9

than single words alone. Second, studies in image segmentation show that dividing up an image into a patch work of regions, each of which being homogeneous, leads to successful results [18,20]. Analogously, we propose to define a word patch as a source word augmented by its semantically close related neighbors, and expect that performance gains can be achieved by grounding the decision process on finding representation regularities between word patches.

However, as the information contained in patches may be voluminous, noisy information may be embedded possibly due to semantic shifts, so that concept centrality may be lost. Consequently, we propose to define two attention mechanisms (one inside patches and one between patches) based on the PageRank algorithm [26] to account for the valuable information present in large patch-based input representation vectors.

In order to test our hypotheses, we follow the distributional approach, although we acknowledge that its combination with the pattern-based approach could lead to improved results as stated in [25,32]. However, integrating continuous pattern representations within the patch paradigm is not straightforward and requires specific further analysis, which is out of the scope of this paper. But, to overcome the main drawback of the distributional hypothesis that conflates different semantic relations between words [9], we design different multi-task neural learning strategies, as recently introduced by [1], together with their binary classification counterparts.

Results over four gold-standard datasets i.e. RUMEN [1], ROOT9 [30], WEEDS [39] and BLESS [3] show that the patch representation leads to significant improvements, in particular when attention mechanisms are applied - 10.6% for binary and 8% for multi-task classification over baselines in terms of F_1 score. Moreover, multi-task learning strategies evidence slightly improved performance results as well as more coherent behaviors when compared to binary configurations.

2 Related Work

Two main approaches have been intensively studied to classify word pairs into the lexical semantic relation they share, or to categorize them as unrelated (or random). On the one hand, pattern-based methods rely on lexico-syntactic patterns, which connect a pair of words [12,17,25,29,32,33]. On the other hand, the distributional approach consists in characterizing the semantic relation between two words based on their distributional representations, thus following Harris distributional hypothesis [11]. In this case, a word pair can be represented by the concatenation of the context vectors of the individual words [2,28,32,39] or by their difference [7,37,39]. The main drawback of the distributional hypothesis is that it conflates different semantic relations between words. Therefore, different solutions have been proposed to overcome this issue. First, specialized similarity measures can be defined to distinguish different relations [29]. Another solution is to specialize word embeddings for particular relations using external knowledge [9,36]. However, these methods are one-relation specific and cannot differentiate between multiple semantic relations at a time. So, multi-task learning strategies have been proposed [1], which concurrently learn different semantic relations with the assumption that the learning process of a given semantic relation may be improved by jointly learning another semantic relation.

In this paper, we propose to look at the problem from a different point of view, the underlying idea being that if each word is augmented by its set of K nearest neighbors (i.e. a patch), improved performance results may be attained. A similar idea is proposed by [14], who presented a set cardinality-based method, which exploits WordNet [24] to compute related neighboring words. In particular, they show that the features extracted from set cardinalities produce competitive results compared to word embedding approaches for a large set of word similarity tasks. However, their methodology relies on the pre-existence of a knowledge base, which is not available for a vast majority of languages. Moreover, their hypothesis builds on discrete representations of words, which cannot account for word continuous similarities and thus highly relies on representative[1] set intersections. Also, only word similarity tasks are tested and it is difficult to access to what extent their methodology can adapt to lexical semantic relation identification. Furthermore, in their proposal, all neighboring words receive the same importance for the decision process, although by extending the semantic scope of each individual word, semantic shifts may occur as well as concept centrality may be lost.

To overcome all these situations, we propose binary and multi-task classification strategies grounded on continuous input representations, that combine two attention mechanisms to evidence word centrality within and between patches based on the PageRank algorithm. In particular, word pairs are represented by the concatenation of their word embeddings as suggested by [32], augmented by their cosine similarity, which is an important feature for lexical semantic relation identification [31].

3 Patch-Based Classification

In this section, we present the overall learning architecture, which consists in the definition of (1) the new input representations based on patches, (2) two attention mechanisms grounded on the PageRank algorithm, and (3) the binary and multi-task neural parallel and sequential classification configurations.

3.1 Definition of Patch and Similarity Between Patches

Definition of Patch. A patch consists of the K most similar words w_j to a source word w_0 in terms of cosine similarity in some latent semantic space (embedding). As such, the patch $P^K_{w_0}$ corresponding to the source word w_0 is a set of $K+1$ words defined as in Eq. 1, where $cos(\overrightarrow{w_0}, \overrightarrow{w_j})$ stands for the cosine similarity between the two representation vectors of w_0 and w_j in some semantic space.

$$P^K_{w_0} = \{w_0\} \cup \{w_j | argmax^K_{cos(\overrightarrow{w_0}, \overrightarrow{w_j})}\} \tag{1}$$

[1] In terms of occurrence and variety.

The next step consists in transforming a patch into a learning input. For that purpose, we follow a simple strategy that consists in concatenating the distributional representations of all words within a patch in their order of similarity measure with the source word[2]. Such input is defined in Eq. 2, where $\overrightarrow{w_i}$ stands for the distributional representation of the i_{th} word in the patch $P^K_{w_0}$.

$$\bigoplus_{i=0}^{K} \overrightarrow{w_i} \tag{2}$$

Similarity Between Patches. Many similarity measures exist to account for the lexical semantic relation that links two words [31]. Within this scope, the cosine similarity measure has evidenced successful results for a variety of semantic relations [35]. As a consequence, we propose to extend the cosine similarity to patches in a straightforward manner. The similarity between two patches is the set of one-to-one cosine similarity measures between all words in their respective patches. It is formally defined in Eq. 3, where $P^K_{w_0}$ and $P^K_{w'_0}$ are two patches, and the $(K+1) \times (K+1)$ matrix noted $SP(P^K_{w_0}, P^K_{w'_0})$ summarizes all values.

$$SP(P^K_{w_0}, P^K_{w'_0}) = \begin{bmatrix} cos(\overrightarrow{w_0}, \overrightarrow{w'_0}) & cos(\overrightarrow{w_0}, \overrightarrow{w'_1}) & \dots & cos(\overrightarrow{w_0}, \overrightarrow{w'_K}) \\ cos(\overrightarrow{w_1}, \overrightarrow{w'_0}) & cos(\overrightarrow{w_1}, \overrightarrow{w'_1}) & \dots & cos(\overrightarrow{w_1}, \overrightarrow{w'_K}) \\ \vdots & \vdots & \ddots & \vdots \\ cos(\overrightarrow{w_K}, \overrightarrow{w'_0}) & cos(\overrightarrow{w_K}, \overrightarrow{w'_1}) & \dots & cos(\overrightarrow{w_K}, \overrightarrow{w'_K}) \end{bmatrix} \tag{3}$$

Similarly to the transformation of a patch into a learning input, we concatenate all $(K+1) \times (K+1)$ values of the $SP(P^K_{w_0}, P^K_{w'_0})$ matrix to be fed to the decision process. Such input is defined in Eq. 4.

$$\bigoplus_{i=0}^{K} \bigoplus_{j=0}^{K} cos(\overrightarrow{w_i}, \overrightarrow{w'_j}) \tag{4}$$

3.2 Attention Mechanisms

Attention Mechanism Within a Patch. A patch should ideally represent a semantic concept centered around its source word. However, this may not be the case as semantic shifts may occur when augmenting the source word with K likely related neighbors, such that centrality may be lost. In order to measure centrality within a patch, we propose to run the PageRank algorithm [26] over the patch defined as a weighted complete graph. Thus, a patch is defined as $G_{P^K_{w_0}} = (V_{P^K_{w_0}}, E_{P^K_{w_0}})$, where $V_{P^K_{w_0}}$ is the set of $K+1$ vertices (words) within $P^K_{w_0}$, and $E_{P^K_{w_0}}$ is the complete set of edges that link all vertices together, weighted by their corresponding cosine similarity. The result of the PageRank algorithm over

[2] Other representations have been tested, but this solution proved to lead to better results.

$G_{P^K_{w_0}}$ is a vector of $(K+1)$ dimensions, where each vertex (word) within the patch receives a centrality score in $\mathbb{R}$, and it is noted $\overrightarrow{\alpha^K_{w_0}} = \langle \alpha_{w_0}, \alpha_{w_1}, \alpha_{w_2}, \ldots, \alpha_{w_K} \rangle$.

The output of the PageRank algorithm $\overrightarrow{\alpha^K_{w_0}}$ stands for the attention mechanism within a patch. Indeed, if we represent a patch (as a learning input) by the concatenation of its words embeddings, all input words are given the same importance for the decision process, letting the classification algorithm decide upon which input dimensions should be discriminant. To guide the learning process, attention mechanisms have shown successful results [34]. As a consequence, we propose to weight each input embedding by its PageRank value so that word centrality scores are included in the decision process. Such an attention-based input is defined in Eq. 5.

$$\bigoplus_{i=0}^{K} \alpha_{w_i}.\overrightarrow{w_i} \tag{5}$$

Attention Mechanism Between Patches. While the first attention mechanism focuses on word centrality within a given patch, the second attention mechanism spotlights on word centrality between patches. Indeed, it is important to acknowledge, which words are central in a set of two patches in order to verify if two words are in a lexical semantic relation. As such, if two patches share a set of close semantically related words that are central to both concepts, the decision process may be more reliable. In order to measure word centrality between two patches $(P^K_{w_0}, P^K_{w'_0})$, we propose to run the PageRank algorithm over the graph defined by the $SP(P^K_{w_0}, P^K_{w'_0})$ matrix, which results in a vector of $2 \times (K+1)$ dimensions, where each word of both patches receives a centrality score in $\mathbb{R}$, and it is noted $\overrightarrow{\beta^K_{w_0,w'_0}} = \langle \beta_{w_0}, \beta_{w_1}, \ldots, \beta_{w_K}, \beta_{w'_0}, \beta_{w'_1}, \ldots, \beta_{w'_K} \rangle$.
The output of the PageRank algorithm $\overrightarrow{\beta^K_{w_0,w'_0}}$ stands for the attention mechanism between patches. So, similarly to the previous attention mechanism, we propose to weight each input embedding of a pair of patches by its PageRank value so that inter-patch word centrality scores are fed to the decision process. Such a second attention-based input is defined in Eq. 6.

$$\bigoplus_{i=0}^{K} \beta_{w_i}.\overrightarrow{w_i} \bigoplus_{i=0}^{K} \beta_{w'_i}.\overrightarrow{w'_i} \tag{6}$$

Combined Attention Mechanisms. Based on the previous definitions of attention mechanisms, we can acknowledge that all input word embeddings may receive two centrality scores: one within patches (first attention) and one between patches (second attention). As a consequence, both attention mechanisms can be combined in a unique learning representation to be fed to the decision process, and it is defined in Eq. 7.

$$\bigoplus_{i=0}^{K} \alpha_{w_i}.\beta_{w_i}.\overrightarrow{w_i} \bigoplus_{i=0}^{K} \alpha_{w'_i}.\beta_{w'_i}.\overrightarrow{w'_i} \tag{7}$$

3.3 Learning Framework

In order to perform binary and multi-task classification, we define two distinct learning input representations, X_p and X_s, which combine patches representations, attention mechanisms and similarity between patches. The first learning input configuration X_p builds on the individual attention mechanisms and it is defined in Eq. 8. In particular, it is composed of the concatenated embeddings of $P^K_{w_0}$ and $P^K_{w'_0}$ weighted by their inside patch attention, plus the concatenated embeddings of $P^K_{w_0}$ and $P^K_{w'_0}$ weighted by their in between patch attention, plus the concatenated values of the cosine similarity between both patches $P^K_{w_0}$ and $P^K_{w'_0}$.

$$X_p = (\bigoplus_{i=0}^{K} \alpha_{w_i}.\overrightarrow{w_i}, \bigoplus_{j=0}^{K} \alpha_{w'_j}.\overrightarrow{w'_j}, \bigoplus_{i=0}^{K} \beta_{w_i}.\overrightarrow{w_i} \bigoplus_{i=0}^{K} \beta_{w'_i}.\overrightarrow{w'_i}, \bigoplus_{i=0}^{K}\bigoplus_{j=0}^{K} cos(\overrightarrow{w_i}, \overrightarrow{w'_j})) \tag{8}$$

The second learning input X_s is grounded on the combined attention mechanism, which allows a more compact representation. It is defined in Eq. 9 and it consists in the concatenated embeddings of $P^K_{w_0}$ and $P^K_{w'_0}$ weighted by their combined inside and in between patch attentions, plus the concatenated values of the cosine similarity between both patches $P^K_{w_0}$ and $P^K_{w'_0}$.

$$X_s = (\bigoplus_{i=0}^{K} \alpha_{w_i}.\beta_{w_i}.\overrightarrow{w_i} \bigoplus_{i=0}^{K} \alpha_{w'_i}.\beta_{w'_i}.\overrightarrow{w'_i}, \bigoplus_{i=0}^{K}\bigoplus_{j=0}^{K} cos(\overrightarrow{w_i}, \overrightarrow{w'_j})) \tag{9}$$

With respect to binary and multi-task classification algorithms, we adapted the (hard parameter sharing architecture) feed-forward neural networks proposed in [1], as the code is freely available[3], and as a consequence allows direct comparison and reproducibility. In particular, Adam [16] is used as the optimizer with default parameters of Keras [5]. For multi-task classification, the network is trained with batches of 64 examples and the number of iterations is optimized to maximize the F_1 score on the validation set. Word embeddings are initialized with the 300-dimensional representations of GloVe [27]. The overall architectures are illustrated in Fig. 1 both for X_p (parallel architecture) and X_s (sequential architecture).

4 Evaluation Results

In this section, we present overall classification results for four gold-standard datasets, namely RUMEN [1], ROOT9 [32], WEEDS [39] and BLESS [3].

[3] https://bit.ly/2Qitasd.

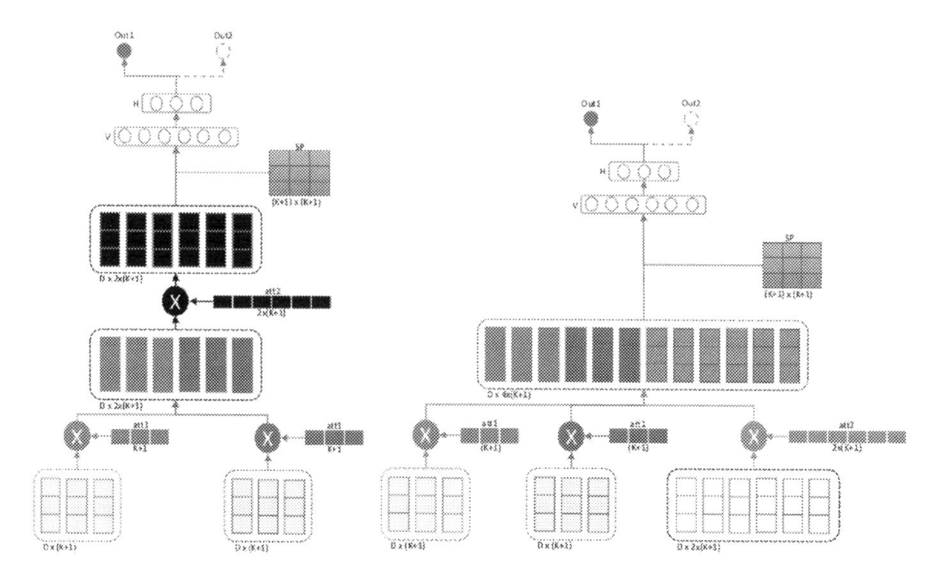

Fig. 1. On the **left**, the sequential architecture. On the **right**, the parallel architecture. Note that the dotted red line between the last fully connected layer of the neural network and the output layer stands for the multi-task architecture only. (Color figure online)

In particular, we test 8 different configurations for both binary and multi-task classification strategies: (1) *Concat*, where the learning input is the concatenation of the word pair embeddings, (2) *Concat + cos* stands for the same configuration as *Concat* plus the cosine similarity measure between the two word embeddings, (3) *Patches* consists of the concatenation of the embeddings of the word pair plus all its respective neighbors, (4) *Concat + SP* is similar to *Concat + cos*, but the cosine similarity is replaced by the concatenated SP matrix, (5) *Patches + SP* represents the exact counterpart of the *Concat + cos* input for patches, and combines the concatenation of all embeddings within patches plus the concatenated SP matrix, (6) *Patches + SP +* att_1 is similar to *Patches + SP*, where each individual patch is weighted by the attention mechanism within patch, (7) *Patches + SP +* att_{12} represents the sequential architecture illustrated in Fig. 1 (left), and (8) *Patches + SP +* att_{1+2} stands for the parallel architecture illustrated in Fig. 1 (right). Classification performance is evaluated through Accuracy, F_1 score, Precision and Recall. Note that lexical split is applied to avoid vocabulary intersection between training, validation and test datasets to avoid lexical memorization [19].

Binary Classification: Results for binary classification are given in Table 1. They clearly evidence the superiority of the sequential architecture, which produces best results in 6 learning situations out of 8 in terms of F_1 score[4]. In particular,

[4] F_1 is not sensitive to unbalanced datasets.

Table 1. Accuracy, F_1, Precision and Recall scores on all datasets for binary classification.

	Inputs		Synonym vs Random					Hypernym vs Random			
			Acc	F_1	Prec	Rec		Acc	F_1	Prec	Rec
RUMEN	Concat	–	0.756	0.760	0.744	0.776	–	0.776	0.768	0.741	0.714
	Concat + cos	–	0.817	0.819	0.806	0.833	–	0.790	0.766	0.741	0.792
	Patches	K = 1	0.713	0.718	0.701	0.736	K = 3	0.765	0.728	0.728	0.729
	Concat + SP	K = 6	0.829	0.830	0.819	**0.841**	K = 9	0.801	0.778	0.753	**0.805**
	Patches + SP	K = 1	0.792	0.793	0.786	0.799	K = 7	0.779	0.747	0.740	0.754
	Patches + SP + att_1	K = 9	0.833	0.831	0.835	0.828	K = 2	0.818	**0.789**	0.792	0.786
	Patches + SP + att_{1+2}	K = 9	0.830	0.827	0.839	0.814	K = 2	0.811	0.780	0.782	0.779
	Patches + SP + att_{12}	K = 9	**0.855**	**0.851**	**0.873**	0.829	K = 4	**0.819**	0.783	**0.813**	0.756
			Co-hynonym vs Random					Hypernym vs Random			
	Inputs		Acc	F_1	Prec	Rec		Acc	F_1	Prec	Rec.
ROOT9	Concat	–	0.907	0.938	0.949	0.888	–	0.852	0.895	0.901	0.927
	Concat + cos	–	0.911	0.941	0.951	0.931	–	0.835	0.883	0.891	0.874
	Patches	K = 10	0.911	0.941	0.948	0.935	K = 2	0.876	0.913	0.908	0.918
	Concat + SP	K = 10	0.926	0.951	0.955	0.947	K = 9	0.893	0.924	0.927	0.942
	Patches + SP	K = 10	0.920	0.946	0.957	0.936	K = 6	0.903	0.931	0.938	0.923
	Patches + SP + att_1	K = 9	0.945	0.964	0.961	0.966	K = 3	0.919	0.943	0.938	0.949
	Patches + SP + att_{1+2}	K = 9	**0.950**	**0.967**	0.965	**0.970**	K = 6	0.908	0.934	**0.943**	0.925
	Patches + SP + att_{12}	K = 3	0.946	0.964	**0.968**	0.961	K = 2	**0.931**	**0.952**	0.941	**0.963**
			Co-hyponym vs Random					Hypernym vs Random			
	Inputs		Acc	F_1	Prec	Rec		Acc	F_1	Prec	Rec.
WEEDS	Concat	–	0.718	0.429	0.414	0.444	–	0.810	0.422	0.433	0.412
	Concat + cos	–	0.789	0.587	0.548	0.632	–	0.842	0.527	0.530	0.524
	Patches	K = 8	0.711	0.406	0.397	0.416	K = 1	0.789	0.376	0.374	0.377
	Concat + SP	K = 9	0.841	0.671	0.660	**0.682**	K = 9	0.901	0.653	0.796	0.554
	Patches + SP	K = 9	0.794	0.592	0.560	0,628	K = 9	0.871	0.602	0.623	0.583
	Patches + SP + att_1	K = 8	0.859	0.680	**0.789**	0.629	K = 7	0.904	0.672	0.793	0.583
	Patches + SP + att_{1+2}	K = 10	0.861	0.685	0.743	0.635	K = 9	0.903	0.661	0.799	0.564
	Patches + SP + att_{12}	K = 4	**0.870**	**0.713**	0.751	0.679	K = 9	**0.926**	**0.753**	**0.856**	**0.670**
			Meronym vs Random					Hypernym vs Random			
	Inputs		Acc	F_1	Prec	Rec		Acc	F_1	Prec	Rec.
BLESS	Concat	–	0.844	0.757	0.804	0.716	–	0.868	0.468	0.515	0.429
	Concat + cos	–	0.862	0.785	0.833	0.743	–	0.881	0.521	0.568	0.481
	Patches	K = 3	0.848	0.764	0.808	0.725	K = 2	0.922	0.692	0.742	0.647
	Concat + SP	K = 8	0.887	0.829	0.857	0.803	K = 4	0.933	0.741	0.780	0.705
	Patches + SP	K = 7	0.881	0.818	0.855	0.785	K = 2	0.939	0.759	0.806	0.718
	Patches + SP + att_1	K = 9	0.896	0.835	0.905	0.776	K = 3	0.949	0.794	0.870	0.731
	Patches + SP + att_{1+2}	K = 3	0.889	0.835	0.850	**0.820**	K = 2	0.948	0.798	0.838	0.763
	Patches + SP + att_{12}	K = 6	**0.898**	**0.837**	**0.915**	0.772	K = 2	**0.955**	**0.822**	**0.882**	**0.769**

for RUMEN, it outperforms the second best strategy by 2% for synonymy. For ROOT9, improvements of 0.9% are obtained for hypernymy. For WEEDS, increases of 2.8% for co-hyponymy and 8.1% for hypernymy are obtained. For BLESS, enhancements of 0.2% and 2.4% are respectively achieved for meronymy and hypernymy over the second best approach. Interestingly, the parallel architecture is not capable of taking advantage of the second attention mechanism (i.e. centrality between patches) as it is the case for the sequential model. The inability of the parallel architecture to combine both attention mechanisms can be explained by two factors. First, in the parallel architecture, individual word embeddings can receive different PageRank weights depending on the attention mechanism. But, as these weights are not explicitly combined, the neural network may not be able to compute regularities between different weights for the same input. The other reason can be explained by the tree structure of human knowledge [38], that theorizes that the cognitive process of human acquisition is agglomerative. Results also evidence clear superiority of attention-based strategies compared to attention-unaware patch-based models i.e., *Patches*, *Concat + SP* and *Patches + SP*. On average overall all datasets and all classification tasks, an increase of 3.5% is achieved by the best attention-based model compared to the best attention-unaware configuration containing neighbor information. Moreover, if we compare the sequential architecture to the current non-patch baseline i.e. *Concat + cos*, the difference in performance is much more important. In particular, an average improvement of 10.6% can be attained, with a minimum increase of 1.7% for RUMEN (hypernymy) and a maximum gain of 30.1% for BLESS (hypernymy). Results also show that the simple introduction of neighbors, i.e. without attention mechanisms, cannot account systematically for increase in performance. Indeed, if we compare *Concat + cos* and its direct counterpart *Patches + SP* that includes the neighbor information both in terms of semantic content and similarity, best results are obtained for the second strategy in 6 cases out of 8. This means that in two situations, the introduction of more information does not lead to improvements. By looking in more details and comparing *Concat* to *Patches*, we can see that better results can be obtained in only half of the cases by the introduction of neighbor embeddings. In fact, going further in the analysis and looking at the results of *Concat + SP*, we clearly understand that the improvement in results comes from the SP matrix and not from the concatenation of the embeddings. Indeed, the *Concat + SP* strategy shows improved results in 6 cases out of 8 over the more complete *Patches + SP* configuration, with an average increase of 3.6%. This situation can be explained by the inability of the neural network to focus on the meaningful word embeddings. Indeed, by just concatenating all neighbor embeddings, all words become equal for the decision process, although this should not be the case. As such, attention mechanisms allow to overcome this situation.

Finally, we can observe that different values of K are obtained for all tested situations. First, if we compare the best attention-based model with the best attention-unaware configuration[5] for all the classification tasks, it is clear that smaller values of K are needed for models with attention. On average, attention-aware models find optimal results for $K = 5.4$, while configurations without attention attain maximum performance for $K = 7.4$. This situation can easily be explained as the best attention-unaware model is *Concat + SP*, which does not include the embeddings of the neighbors. As such, the only information from the neighbors is given by the SP matrix, which must give trace of all the information alone between patches. As a consequence, only large matrices can account for the extra information included in the patches. Results also show that different values of K can be obtained for the same semantic

[5] Which may not include any information about neighbors.

relation depending on the dataset, although some regularities seem to exist. For instance, for hypernymy, best results are obtained for $K=9$ for WEEDS, but the best results for RUMEN, BLESS and ROOT9 are obtained for $K=2$. Furthermore, by comparing symmetric (synonymy, co-hyponymy) and asymmetric (hypernymy, meronymy) relations, results show that best results are obtained for $K=7.4$ on average in the first case, and for $K=6$ on average in the second case, thus suggesting the need for less extra-knowledge for asymmetric semantic relations. This can be explained by the semantic shift phenomenon [4]. When dealing with asymmetric relations, it is likely that one of the words within the pair is a general word. As a consequence, when expanding the general word with its neighbors, it is likely that a semantic shift occurs. As a consequence, it is conceivable that smaller values of K are preferred for asymmetric relations.

Multi-task Classification: Results for multi-task classification are given in Table 2. They clearly evidence the increase in performance from the sequential architecture compared to all other strategies. Indeed, best results are achieved by this configuration in almost all cases. In particular, average improvements of 15.4% and 8.1% are attained when compared to the recent work of [1] i.e. *FS Concat* and its extended version including the cosine similarity measure, i.e. *FS Concat + cos*. Results also show that similar behaviors to binary classification can be observed. First, the introduction of the cosine similarity measure greatly boosts performance. Second, most of the information gain obtained by strategies without attention mechanisms comes from the SP matrix and not from the concatenation of the embeddings of the neighbors present in a given patch. Note that in this case, that *Concat + SP* is the best attention-unaware strategy, which achieves better results in terms of F_1 than *Patches + SP +* att_1 in 2 cases out of 8. Third, if we compare the sequential architecture to the current multi-task non-patch baseline, i.e. *FS Concat + cos*, similar differences in performance are obtained compared to the binary situation. In particular, an average improvement of 8% can be attained, with a minimum increase of 1.7% for RUMEN (hypernymy) and a maximum gain of 21.1% for BLESS (hypernymy). Finally, likewise the binary situation, the parallel architecture can not compete with the sequential counterpart and does not outperform the configuration with only within patch attention, i.e. *Patches + SP +* att_1. When compared to the sequential binary classification model, the multi-task sequential counterpart based on the fully-shared architecture evidences best results in 5 cases out of 8. Nevertheless, overall improvements are small with an average increase in performance of 0.7%. Similarly, in the 3 other cases where the binary strategy offers best results, average gains of 0.4% are obtained. This can easily be explained by the small difference in architectures as already evidenced in [1]. Interestingly, the multi-task model evidences steady improvements in terms of Recall, with best values than the binary counterpart in 7 cases out of 8. In parallel, Precision shows worst results in 6 cases out of 8. This can also be explained by the fully-shared architecture that produces its decision based on a single shared layer, i.e. features that could be specific to one single task maybe lost. As a consequence, Precision may be affected while Recall boosted.

Table 2. Accuracy, F_1, Precision and Recall scores on all datasets for multi-task classification.

	Inputs	K	Synonym vs Random				Hypernym vs Random			
			Acc	F_1	Prec	Rec	Acc	F_1	Prec	Rec
RUMEN	FS Concat [1]	–	0.740↓	0.744↓	0.727↓	0.764↓	0.790↑	0.763↓	0.746↑	0.780↑
	FS Concat [1] + cos	–	0.824↑	0.827↑	0.810↑	0.844↑	0.794↑	0.769↑	0.748↑	0.792=
	FS Patches	1	0.717↑	0.722↑	0.704↑	0.742↑	0.742↓	0.705↓	0.696↓	0.715↓
	FS Concat + SP	9	0.835↑	0.834↑	0.834↑	0.835↓	0.807↑	0.780↑	0.767↑	0.792↓
	FS Patches + SP	1	0.800↑	0.801↑	0.790↑	0.812↑	0.775↓	0.743↓	0.733↓	0.753↓
	FS Patches + SP + att_1	4	0.828↓	0.823↓	0.840↑	0.807↓	0.815↓	0.782↓	0.795↑	0.769↓
	FS Patches + SP + att_{1+2}	4	0.820↓	0.816↓	0.829↓	0.802↓	0.815↑	**0.790**↑	0.775↓	**0.806** ↑
	FS Patches + SP + att_{12}	8	**0.852**↓	**0.850**↓	**0.861**↓	**0.839**↑	**0.821**↑	0.786↑	**0.810**↓	0.764↑
			Co-hyponym vs Random				Hypernym vs Random			
	Inputs		Acc	F_1	Prec	Rec	Acc	F_1	Prec	Rec.
ROOT9	FS Concat [1]	–	0.896↓	0.932↓	0.936↓	0.928↑	0.807↓	0.863↓	0.868↓	0.858↓
	FS Concat [1] + cos	–	0.901↓	0.935↓	0.938↓	0.933↑	0.835=	0.882↓	0.893↑	0.872↓
	FS Patches	1	0.903↓	0.937↓	0.932↓	0.942↑	0.850↓	0.895↓	0.888↓	0.902↓
	FS Concat + SP	5	0.922↓	0.949↓	0.952↓	0.945↓	0.883↓	0.917↓	0.918↓	0.916↓
	FS Patches + SP	2	0.917↓	0.946=	0.941↓	0.951↓	0.898↓	0.929↓	0.918↓	0.940↑
	FS Patches + SP + att_1	3	0.926↓	0.951↓	0.954↑	0.949↓	0.909↓	0.937↓	0.925↓	0.949=
	FS Patches + SP + att_{1+2}	5	0.919↓	0.947↓	0.944↓	0.950↓	0.914↑	0.939↑	**0.941**↓	0.937↑
	FS Patches + SP + att_{12}	3	**0.941**↓	**0.961**↓	**0.960**↓	**0.963**↑	**0.921**↓	**0.945**↓	0.938↓	**0.951**↓
			Co-hyponym vs Random				Hypernym vs Random			
	Inputs		Acc	F_1	Prec	Rec	Acc	F_1	Prec	Rec.
WEEDS	FS Concat [1]	–	0.732↑	0.446↑	0.439↑	0.454↑	0.829↑	0.455↑	0.489↑	0.426↑
	FS Concat [1] + cos	–	0.810↑	0.627↑	0.615↑	0.670↑	0.871↑	0.616↑	0.615↑	0.618↑
	FS Patches	1	0.701↓	0.391↓	0.379↓	0.403↓	0.816↑	0.415↑	0.446↑	0.387↑
	FS Concat + SP	8	0.841=	0.672↑	0.659↓	0.686↑	0.897↓	0.668↑	0.728↓	0.618↑
	FS Patches + SP	9	0.800↑	0.604↑	0.571↑	0.641↑	0.827↓	0.563↓	0.489↓	0.662↑
	FS Patches + SP + att_1	3	0.832↓	0.657↓	0.640↓	0.676↑	0.907↑	0.685↑	0.794↑	0.603↑
	FS Patches + SP + att_{1+2}	5	0.853↓	0.679↓	0.709↓	0.670↑	0.877↓	0.639↓	0.631↓	0.647↑
	FS Patches + SP + att_{12}	6	**0.876**↑	**0.731**↑	**0.756**↑	**0.708**↑	**0.927**↑	**0.759**↑	**0.848**↓	**0.686**↑
			Meronym vs Random				Hypernym vs Random			
	Inputs		Acc	F_1	Prec	Rec	Acc	F_1	Prec	Rec.
BLESS	FS Concat [1]	–	0.835↓	0.740↓	0.800↓	0.689↓	0.891↑	0.526↑	0.636↑	0.448↑
	FS Concat [1] + cos	–	0.862=	0.783↓	0.845↑	0.729↓	0.908↑	0.616↑	0.699↑	0.551↑
	FS Patches	2	0.842↓	0.750↓	0.817↑	0.692↓	0.927↑	0.695↑	0.789↑	0.622↓
	FS Concat + SP	8	0.882↓	0.816↓	0.872↑	0.766↓	0.931↓	0.722↓	0.788↑	0.667↓
	FS Patches + SP	2	0.853↓	0.769↓	0.828↓	0.718↓	0.942↑	0.774↑	0.816↑	0.737↑
	FS Patches + SP + att_1	3	0.872↓	0.803↓	0.844↓	0.766↑	0.955↑	0.818↑	0.900↑	**0.750**↑
	FS Patches + SP + att_{1+2}	3	0.867↓	0.794↓	0.843↓	0.750↓	0.951↑	0.801↑	0.878↑	0.737↓
	FS Patches + SP + att_{12}	3	**0.895**↓	**0.839**↑	**0.882**↓	**0.789**↑	**0.958**↑	**0.827**↑	**0.921**↑	**0.750**↓

In order to better understand the impact of K on the results, we show performance results for the binary and multi-task sequential architectures for $K = 1..10$ in Fig. 2. Note that results in Table 2 are given for the best K on average and cannot account for differences in K between two semantic relations. Overall, we can notice different behaviors depending on the dataset. For RUMEN, performance values vary minimally with respect to K, independently of the semantic relation. For ROOT9 and BLESS, similar situations can be observed. Indeed, for both datasets, performance highly degrades with higher values of K for hypernymy,

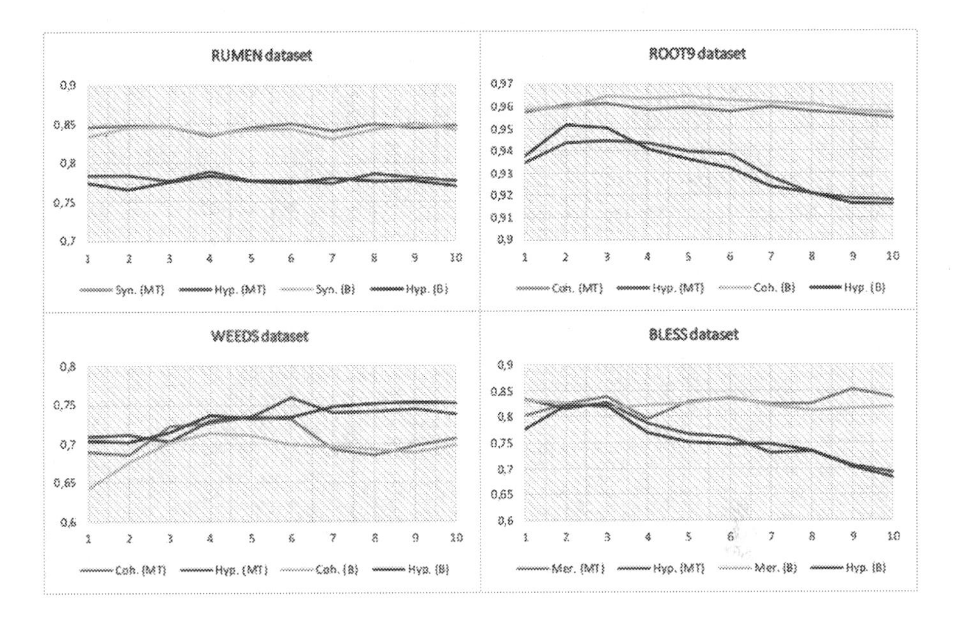

Fig. 2. F_1 score distribution by K ($K = 1..10$) for all datasets for the sequential architecture, both for binary (B) and multi-task (MT) classification.

while the impact of K on co-hyponymy is more limited. For WEEDS, the situation is the opposite for hypernymy and co-hyponymy, as higher performance values are obtained for higher values of K. As a consequence, no clear conclusion can be drawn as results drastically change depending on the dataset. It is clear that further efforts are needed to propose better evaluation standards. Nevertheless, some regularities emerge. Indeed, over all values of K, the multi-task architecture evidences steady improvements over the binary counterpart for all datasets, except for ROOT9 where best results are obtained by the binary classification for both semantic relations. Also, in 2 cases out of 8, the best value of K is the same for the binary and the multi-task situations. In 3 cases, the difference in K value is one, and in the remaining 3 cases, the difference of K equals to 3, with 2 cases in which the multi-task architecture needs less neighbors than the binary version. This confirms the fact that small differences in terms of K values are evidenced on average between both strategies, although there seems to be a slight tendency to use less neighbors in the multi-task strategy.

5 Conclusions

In this paper, we presented a patch-based classification strategy to tackle lexical semantic relation identification. In particular, we showed that attention mechanisms (if correctly combined) drastically boost results compared to attention-unaware configurations. Indeed, average improvements can reach 10.6% for binary and 8% for multi-task classification over non-patch baseline approaches in terms of F_1 for the sequential architecture, when tested over four gold-standard datasets, namely RUMEN, ROOT9, WEEDS and BLESS. Moreover, results witness that small but steady improvements in classification performance can be attained by multi-task architectures. As such, immediate future work should include (1) the design of new multi-task architectures following the ideas of [21], (2) the combination of the distributional approach with the paradigmatic model as suggested in [25], and (3) the evaluation comparison to the very recent work proposed by [14], which follows similar ideas with discrete word representations.

References

1. Balikas, G., Dias, G., Moraliyski, R., Akhmouch, H., Amini, M.-R.: Learning lexical-semantic relations using intuitive cognitive links. In: Azzopardi, L., Stein, B., Fuhr, N., Mayr, P., Hauff, C., Hiemstra, D. (eds.) ECIR 2019. LNCS, vol. 11437, pp. 3–18. Springer, Cham (2019). https://doi.org/10.1007/978-3-030-15712-8_1
2. Baroni, M., Bernardi, R., Do, N.Q., Shan, C.C.: Entailment above the word level in distributional semantics. In: 13th Conference of the European Chapter of the Association for Computational Linguistics (EACL), pp. 23–32 (2012)
3. Baroni, M., Lenci, A.: How we Blessed distributional semantic evaluation. In: Workshop on Geometrical Models of Natural Language Semantics (GEMS) associated to Conference on Empirical Methods on Natural Language Processing (EMNLP), pp. 1–10 (2011)
4. Blank, A.: Why do new meanings occur? A cognitive typology of the motivations for lexical semantic change. Cogn. Linguist. Res. **13**, 61–90 (1999)
5. Chollet, F.: Keras. https://keras.io (2015)
6. Dong, L., Mallinson, J., Reddy, S., Lapata, M.: Learning to paraphrase for question answering. In: Conference on Empirical Methods in Natural Language Processing (EMNLP), pp. 875–886 (2017)
7. Fu, R., Guo, J., Zhao, Y., Che, W., Wang, H., Liu, T.: Learning semantic hierarchies: a continuous vector space approach. IEEE/ACM Trans. Audio Speech Lang. Process. **23**, 461–471 (2015)
8. Gambhir, M., Gupta, V.: Recent automatic text summarization techniques: a survey. Artif. Intell. Rev. **47**(1), 1–66 (2016). https://doi.org/10.1007/s10462-016-9475-9
9. Glavas, G., Vulic, I.: Generalized tuning of distributional word vectors for monolingual and cross-lingual lexical entailment. In: 57th Conference of the Association for Computational Linguistics (ACL), pp. 4824–4830 (2019)
10. Gupta, A., Lebret, R., Harkous, H., Aberer, K.: Taxonomy induction using hypernym subsequences. In: Conference on Information and Knowledge Management (CIKM), pp. 1329–1338 (2017)
11. Harris, Z.S.: Distributional structure. Word **10**(2–3), 146–162 (1954)
12. Hearst, M.: Automatic acquisition of hyponyms from large text corpora. In: 14th Conference on Computational Linguistics (COLING), pp. 539–545 (1992)
13. Jackendoff, R.: Foundations of Language: Brain, Meaning, Grammar, and Evolution. Oxford University Press, Oxford (2002)
14. Jiménez, S., González, F.A., Gelbukh, A.F., Dueñas, G.: Word2set: WordNet-based word representation rivaling neural word embedding for lexical similarity and sentiment analysis. IEEE Comput. Intell. Mag. **14**, 41–53 (2019)

15. Kathuria, N., Mittal, K.: A comprehensive survey on query expansion techniques, their issues and challenges. Int. J. Comput. Appl. **168**, 17–20 (2017)
16. Kingma, D., Ba, J.: Adam: a method for stochastic optimization. In: 3rd International Conference on Learning Representations (ICLR) (2015)
17. Kozareva, Z., Hovy, E.: A semi-supervised method to learn and construct taxonomies using the web. In: Conference on Empirical Methods in Natural Language Processing (EMNLP), pp. 1110–1118 (2010)
18. Ledig, C., Shi, W., Bai, W., Rueckert, D.: Patch-based evaluation of image segmentation. In: Conference on Computer Vision and Pattern Recognition (CVPR), pp. 3065–3072 (2014)
19. Levy, O., Remus, S., Biemann, C., Dagan, I.: Do supervised distributional methods really learn lexical inference relations? In: Conference of the North American Chapter of the Association for Computational Linguistics: Human Language Technologies (HLT-NAACL), pp. 970–976 (2015)
20. Lézoray, O.: Patch-Based mathematical morphology for image processing, segmentation and classification. In: Battiato, S., Blanc-Talon, J., Gallo, G., Philips, W., Popescu, D., Scheunders, P. (eds.) ACIVS 2015. LNCS, vol. 9386, pp. 46–57. Springer, Cham (2015). https://doi.org/10.1007/978-3-319-25903-1_5
21. Liu, P., Qiu, X., Huang, X.: Adversarial multi-task learning for text classification. In: 55th Annual Meeting of the Association for Computational Linguistics (ACL) (2017)
22. Lu, P., Ji, L., Zhang, W., Duan, N., Zhou, M., Wang, J.: R-VQA: learning visual relation facts with semantic attention for visual question answering. In: 24th International Conference on Knowledge Discovery and Data Mining (KDD), pp. 1880–1889 (2018)
23. MIkołajczak-Matyja, N.: The associative structure of the mental lexicon: hierarchical semantic relations in the minds of blind and sighted language users. Psychol. Lang. Commun. **19**, 1–18 (2015)
24. Miller, G.A., Beckwith, R., Fellbaum, C., Gross, D., Miller, K.J.: Introduction to WordNet: an on-line lexical database. Int. J. Lexicogr. **3**(4), 235–244 (1990)
25. Nguyen, K.A., Schulte im Walde, S., Vu, N.T.: Distinguishing antonyms and synonyms in a pattern-based neural network. In: 15th Conference of the European Chapter of the Association for Computational Linguistics (EACL), pp. 76–85 (2017)
26. Page, L., Brin, S., Motwani, R., Winograd, T.: The PageRank citation ranking: bringing order to the web. Technical report 1999-66, Stanford InfoLab, November 1999
27. Pennington, J., Socher, R., Manning, C.D.: GloVe: global vectors for word representation. In: Conference on Empirical Methods on Natural Language Processing (EMNLP), pp. 1532–1543 (2014)
28. Roller, S., Erk, K., Boleda, G.: Inclusive yet selective: supervised distributional hypernymy detection. In: 25th International Conference on Computational Linguistics (COLING), pp. 1025–1036 (2014)
29. Roller, S., Kiela, D., Nickel, M.: Hearst patterns revisited: automatic hypernym detection from large text corpora. In: 56th Annual Meeting of the Association for Computational Linguistics (ACL), pp. 358–363 (2018)
30. Santus, E., Lenci, A., Chiu, T., Lu, Q., Huang, C.: Nine features in a random forest to learn taxonomical semantic relations. In: 10th International Conference on Language Resources and Evaluation (LREC), pp. 4557–4564 (2016)

31. Santus, E., Shwartz, V., Schlechtweg, D.: Hypernyms under siege: linguistically-motivated artillery for hypernymy detection. In: 15th Conference of the European Chapter of the Association for Computational Linguistics, pp. 65–75 (2017)
32. Shwartz, V., Goldberg, Y., Dagan, I.: Improving hypernymy detection with an integrated path-based and distributional method. In: 54th Annual Meeting of the Association for Computational Linguistics (ACL), pp. 2389–2398 (2016)
33. Snow, R., Jurafsky, D., Ng, A.Y.: Learning syntactic patterns for automatic hypernym discovery. In: 17th International Conference on Neural Information Processing Systems (NIPS), pp. 1297–1304 (2004)
34. Vaswani, A., et al.: Attention is all you need. In: Advances in Neural Information Processing Systems (2017)
35. Vulic, I., Mrksic, N.: Specialising word vectors for lexical entailment. In: Conference of the North American Chapter of the Association for Computational Linguistics: Human Language Technologies (NAACL-HLT), pp. 1134–1145 (2018)
36. Vulic, I., Mrksic, N., Reichart, R., Séaghdha, D.Ó., Young, S.J., Korhonen, A.: Morph-fitting: fine-tuning word vector spaces with simple language-specific rules. In: 55th Annual Meeting of the Association for Computational Linguistics (ACL), pp. 56–68 (2017)
37. Vylomova, E., Rimell, L., Cohn, T., Baldwin, T.: Take and took, gaggle and goose, book and read: evaluating the utility of vector differences for lexical relation learning. In: 54th Annual Meeting of the Association for Computational Linguistics, pp. 1671–1682 (2016)
38. Wang, Y.: On cognitive foundations of creativity and the cognitive process of creation. Int. J. Cogn. Inform. Nat. Intell. **3**, 1–18 (2009)
39. Weeds, J., Clarke, D., Reffin, J., Weir, D.J., Keller, B.: Learning to distinguish hypernyms and co-hyponyms. In: 5th International Conference on Computational Linguistics (COLING), pp. 2249–2259 (2014)

Joint Word and Entity Embeddings for Entity Retrieval from a Knowledge Graph

Fedor Nikolaev[1,2] and Alexander Kotov[1](✉)

[1] Wayne State University, Detroit, MI 48202, USA
{fedor,kotov}@wayne.edu
[2] Kazan Federal University, Kazan, Russia

Abstract. Recent years have witnessed the emergence of novel models for ad-hoc entity search in knowledge graphs of varying complexity. Since these models are based on direct term matching, their accuracy can suffer from a mismatch between vocabularies used in queries and entity descriptions. Although successful applications of word embeddings and knowledge graph entity embeddings to address the issues of vocabulary mismatch in ad-hoc document retrieval and knowledge graph noisiness and incompleteness, respectively, have been reported in recent literature, the utility of joint word and entity embeddings for entity search in knowledge graphs has been relatively unexplored. In this paper, we propose Knowledge graph Entity and Word Embedding for Retrieval (KEWER), a novel method to embed entities and words into the same low-dimensional vector space, which takes into account a knowledge graph's local structure and structural components, such as entities, attributes, and categories, and is designed specifically for entity search. KEWER is based on random walks over the knowledge graph and can be considered as a hybrid of word and network embedding methods. Similar to word embedding methods, KEWER utilizes contextual co-occurrences as training data, however, it treats words and entities as different objects. Similar to network embedding methods, KEWER takes into account knowledge graph's local structure, however, it also differentiates between structural components. Experiments on publicly available entity search benchmarks and state-of-the-art word and joint word and entity embedding methods indicate that a combination of KEWER and BM25F results in a consistent improvement in retrieval accuracy over BM25F alone.

1 Introduction

Entity search is an information retrieval (IR) task aimed at addressing information needs focused on abstract or material objects, such as people, organizations, products and book characters. Such information needs include finding a particular entity (e.g. *"Einstein relativity theory"*), an attribute or a property of an entity (e.g. *"Who founded Intel?"*), an entity by its property (e.g. *"England football player highest paid"*) or a list of entities matching a description (e.g.

J. M. Jose et al. (Eds.): ECIR 2020, LNCS 12035, pp. 141–155, 2020.
https://doi.org/10.1007/978-3-030-45439-5_10

"Formula 1 drivers that won the Monaco Grand Prix") and can be formulated as short or "telegraphic" keyword queries or natural language questions [2,18]. Target entity or a list of entities for these information needs can be retrieved from a knowledge graph, such as Wikipedia, DBpedia, Freebase or Wikidata.

In prior research, the problem of entity retrieval from a knowledge graph was cast into a special case of structured document retrieval [21,22,25,45], database search [18,32] or a combination of the two [5,35]. Such methods take into account only the directly adjacent structural components of a knowledge graph (entities, predicates, categories, and literals) when constructing a document for an entity from a knowledge graph. As a result, a significant amount of potentially useful information can be lost by not taking into account the local structure of knowledge graphs or structural components that are separated by more than one edge in a knowledge graph. For example, the entities *Michelangelo* and *Sistine Chapel* are separated by 3 edges in DBpedia. Another drawback of existing methods is that they only consider entities as points (i.e. entity documents) in a high-dimensional space, with the number of dimensions equal to vocabulary size. This can lead to a well-known problem of vocabulary "gap" between queries and documents for relevant entities. For example, searchers looking for the entities that are related to the Musée National d'Art Moderne can pose the queries containing the terms "Beaubourg" or "MNAM", some or all of which may not be in the documents corresponding to the relevant entities. In ad-hoc document retrieval, vocabulary mismatch has been successfully addressed by employing the methods that create a low-dimensional representation of words and documents, such as Latent Semantic Indexing [8], Latent Dirichlet Allocation [38], word2vec [19], and GloVe [26]. Many recently proposed approaches [10,15,36,40,41] have successfully utilized word embeddings to address vocabulary mismatch in ad-hoc document IR.

At the same time, to address the issues of graph incompleteness and noisiness, a number of methods have been proposed for knowledge graph embedding, such as RESCAL [24], TransE [3], and NTN [33]. These methods represent knowledge graph entities and relations as vectors in the same embedding space with geometrical constraints that encode the local structure of knowledge graphs. Although these methods have been shown to be effective for the task of knowledge graph link prediction, the embedding spaces constructed by these methods as well as network embedding methods, such as DeepWalk [28], LINE [34], and node2vec [11] do not consider words and, therefore, *cannot be utilized in the tasks that involve both words and entities, such as entity search.* **However, the methods to construct joint embedding spaces for both words and entities that are effective for entity search in knowledge graph have not yet been explored.**

To address this issue, we propose **K**nowledge graph **E**ntity and **W**ord **E**mbeddings for **R**etrieval (KEWER), a novel method to create a low-dimensional representation of entities and words in the same embedding space that *takes into account both local structure and structural components of knowledge graphs.* KEWER samples random walks over a given knowledge graph and thus can be considered as a hybrid between word and network embedding methods. Similar to word embedding methods, KEWER utilizes contextual co-occurrences as training data but differentiates between words and entities. Similar to network embedding methods, KEWER explicitly models the local structure of a knowledge graph, but unlike these methods, it takes into account various structural components of a knowledge graph, which allows us to jointly model both entities from the graph, such as DBpedia, and keyword queries.

We perform a series of experiments with KEWER to answer the following research questions: **RQ1:** How to learn joint word and entity embeddings that are effective for entity retrieval from a knowledge graph? **RQ2:** Which structural components of a knowledge graph are the most effective when learning joint entity and word embeddings for entity retrieval? **RQ3:** How does joint word and entity embeddings affect the retrieval accuracy of standard term matching based retrieval models for different types of entity search queries when they are utilized along with these models? **RQ4:** How does retrieval accuracy of the methods using joint word and entity embeddings compare with that of the methods using only word embeddings?

2 Related Work

Entity Search. Entity search approaches can be categorized into the ones that utilize structured information from knowledge graphs and the ones that do not. While earlier studies [5,32,35] heavily utilized knowledge graph's structure during retrieval, more recent studies [21,22,25,45] only use it to construct fielded entity representations, effectively casting entity search into an instance of structured document retrieval. Entity similarity information obtained from entity embeddings was successfully utilized for re-ranking the results of term-based retrieval models in [14,17,44] using a learning-to-rank approach. A publicly available benchmark for entity search based on DBpedia [16] and its more recent version [13], which provides graded relevance judgments obtained using crowdsourcing and subsequent conflict resolution by experts, are standard test collections for evaluating entity search methods.

Word Embeddings in IR. Significant research efforts in the IR community were devoted to assessing the utility of word embeddings for different IR tasks. While the initial and some of the recent works in this area directly utilize word embeddings obtained using the methods such as word2vec, several word embedding models specifically targeting IR [9,30,42] have been recently proposed. The Dual Embedding Space Model [20] utilizes embedding matrices, which correspond to the two layers of the CBOW or Skip-gram architectures, to re-rank retrieval results. Experiments with this model indicate that utilizing IN-OUT instead of IN-IN similarity between embeddings of a query and document words allows for better modeling of *aboutness* of a document with respect to a query.

Network Embeddings. Network embedding models aim to embed network nodes into a low-dimensional vector space. A common idea underlying these methods is the adoption of the embedding methods from language modeling to sequences obtained using random walks on a given network. DeepWalk [28] is the first method that is based on this idea. DeepWalk trains the Skip-Gram architecture on sequences of vertices generated by random walks of specified length starting from each node in the network. The resulting embeddings can be used for various classification tasks, such as group labeling in social networks. Other notable network embedding methods are LINE [34], node2vec [11], and struc2vec [31].

Knowledge Graph Embeddings. Knowledge graph embeddings are a popular way to obtain low-dimensional dense representations for entities and predicates. A widely known TransE model [3] was proposed as a way to greatly reduce the number of parameters required to train the Structured Embeddings model [4] by using vector algebra. MEmbER [14] is an extension of GloVe [26] to learn conceptual spaces consisting of word and entity embeddings, in which the salient words in a given domain are associated with separating hyperplanes. Several studies [37,39,46] proposed a hybrid between entity and word embeddings by employing a loss function, which includes both a TransE-based component to model relations between entities and a word2vec-based component to model semantic relations between the words along with the third component, whose purpose is to align entity and word embeddings obtained by the first two components. In [23] authors take a different approach by learning word and entity embeddings without utilizing relations between entities from a knowledge graph and instead relying only on an unannotated corpus of text. None of the previously proposed approaches for learning joint word and entity embedding spaces were proposed specifically for entity search in a knowledge graph, and thus ignore important information, such as knowledge graph structural components.

3 Method

The primary goal of the proposed method is to learn joint word and entity embeddings that are effective for entity retrieval from a knowledge graph. The proposed method is based on the idea that a knowledge graph consists of key structural components. Structural components are loosely related to the fields of entity documents used extensively in knowledge graph entity search [21,27,45], but are defined in a more general way as a set of components of a knowledge graph that are directly or indirectly related to entities.

A given knowledge graph is formally defined as $G = \{E, R, A, C, S\}$, where $E = \{e_1, \ldots, e_{|E|}\}$ is a set of entities; R is a set of subject-predicate-object triples (s, p, o) where $s, o \in E$ are entities and p is a predicate; A is a set of triples (e, p, a) where $e \in E$ is an entity, p is a predicate, and a is a textual attribute that contains words $\{w_1, \ldots, w_k\}$ from vocabulary V; C is a set of entity-category pairs $(e, c), c \in K$; S is a combined set of entity-surface form (e, s), category-surface form (c, s), and predicate-surface form (p, s) pairs, where $s = \{n_1, \ldots, n_k\}$ is a set of word tokens in a surface form. The most commonly available surface form for an entity or category is its name or label (with $k \approx 3$).

Another example of a surface form for an entity is its anchor text. We do not use long surface forms, such as entity descriptions, in this study. The vocabulary of all distinct surface form tokens is denoted as N. We also define the following three **structural components** of a knowledge graph: **categories** (C), **literals** (A), and **predicates** ($P = \{p : (s, p, o) \in R \text{ or } (e, p, a) \in A\}$).

3.1 Knowledge Graph Entity and Word Embedding for Retrieval

To address **RQ1**, we propose KEWER, a method to jointly embed knowledge graph entities and words for entity search that takes into account the local structure of a knowledge graph, as well as its structural components. KEWER is based on a neural architecture that utilizes as input a set of sequences of word tokens and entity URIs produced by the following two-step procedure: (1) perform random walks over a knowledge graph to generate sequences consisting of structural components of a given knowledge graph (entities, predicates, attributes, and categories) of specified length t (2) randomly with probability r replace URIs in sequences resulting from random walks with their respective surface forms obtained from the same knowledge graph.

3.2 Proposed Method

In our approach, sequences generated from random walks can be viewed as short descriptions of entities that are accessed by them. For example, a random walk over DBpedia $\textit{Pierre\_Curie} \xrightarrow{spouse} \textit{Marie\_Curie} \xrightarrow{knownFor} \textit{Radioactivity}$ can be seen as a short description of Marie Curie, who was the wife of Pierre Curie and is known for discovering radioactivity. The objective that is used during training when given a current element from a sequence is to predict its surrounding context. In our example sequence, if *Marie_Curie* is the current element, the model will try to minimize the distance between embeddings of the context elements *Pierre_Curie*, *spouse*, *knownFor*, *Radioactivity* and an embedding of the current entity, *Marie_Curie*. The resemblance of this objective to the entity search task, when we need to predict target entity *Marie_Curie* from the user query such as "Who is known for her research on radioactivity and was the wife of Pierre Curie?" is a primary motivation for using random walks over a knowledge graph in the proposed method.

Random Walks Generation. Formally, the random walks are generated in the following way: starting from each entity e we generate γ random walks of length $\leq t$. Each random walk is independently generated by repeatedly following directed edges $(s, p, o) \in R$, such that the same node is not visited more than once during each random walk. If the **predicates** component is used, we add predicate-object pair (p, o) to the walk sequence, otherwise, we only add an object o. The walk procedure terminates when it either already contains t nodes or all the nodes adjacent to the current entity have already been visited. In the end, the randomly chosen attribute a of the last visited entity e, such that

$(e, p, a) \in A$ from the **literals** structural component is added to the sequence. If **categories** are considered, then the pairs $(e, c) \in C$ are treated as undirected edges during the walking procedure so that it does not need to be terminated when category nodes, which typically do not have outgoing edges, are reached.

Mixing with Surface Forms. To fulfill the need to work with user queries constructed in natural language in the typical ad-hoc entity retrieval scenario, the model should have the ability to properly embed the words that can be found in entity and category names. For that, after walks are generated, entity and category URIs are randomly replaced with probability r by their respective surface forms consisting of word tokens. If an entity or category has more than one surface form, the surface form for URI replacement is chosen uniformly at random from the set of available surface forms. In theory, this might create an issue with not utilizing all available surface forms for an entity, but in practice, this doesn't happen, since the number of generated random walks γ is typically much larger than the number of available surface forms for any given entity. If the **predicates** component is used, then the same procedure is also performed for the predicate URIs in the sequences.

Training Objective. Finally, to obtain the embeddings for words, entities, and, optionally, categories and predicates (if the corresponding knowledge graph structural components were used for sequence generation), the Skip-Gram-based model with Negative sampling [19] is trained on the resulting set of $|E| * \gamma$ random walks consisting of elements $\xi_1, \ldots, \xi_T$, where $\xi_{1 \ldots T}$ are either URIs or words, and $T \geq t$ is the length of a random walk after replacement of the URIs with their surface forms. The model maximizes the probability of observing elements ξ_O from the context of the current element ξ_I by using the following objective:

$$\frac{1}{T} \sum_{i=1}^{T} \sum_{-c \leq j \leq c, j \neq 0} \log p(\xi_{i+j} | \xi_i), \ \xi_{1 \ldots T} \in \Xi, \\ \Xi = E \cup N \begin{cases} \cup K, \text{ if } \textbf{categories} \text{ are used} \\ \cup V, \text{ if } \textbf{literals} \text{ are used} \\ \cup P, \text{ if } \textbf{predicates} \text{ are used.} \end{cases} \tag{1}$$

where c is the size of the training context and the probability of observing context element $p(\xi_{i+j} | \xi_i)$ is defined using softmax as: $p(\xi_O | \xi_I) = \frac{\exp(\mathbf{v}'^{\top}_{\xi_O} \mathbf{v}_{\xi_I})}{\sum_{k=1}^{|\Xi|} \exp(\mathbf{v}'^{\top}_{\xi_k} \mathbf{v}_{\xi_I})}$. Note that each element ξ has two different IN and OUT [20] embeddings: $\mathbf{v}_\xi$ and $\mathbf{v}'_\xi$. In practice, calculating the softmax denominator of $p(\xi_O | \xi_I)$ is infeasible, and it is approximated using negative sampling.

The objective from Eq. (1) is maximized using stochastic gradient descent to learn IN and OUT embeddings of size d with derivatives estimated using back-propagation. To better utilize cross-dependencies between IN and OUT spaces [20], we use a concatenation of IN and OUT embeddings for words and OUT and IN embeddings for entities. Thus, our final embeddings are vectors of size $d * 2$. Note that the proposed method can scale to large knowledge graphs, since all three steps of it are easily parallelizable.

3.3 Embedding-Based Entity Search

The obtained embeddings can be used to score entities with respect to a given user query in the following way. For a query Q consisting of the query terms $q_1, \ldots, q_k$, we compute embedding of the entire query $\mathbf{q}$ by calculating the weighted sum of embeddings of individual query words:

$$\mathbf{q} = \sum_{i=1}^{k} \frac{a}{p(q_i) + a} \mathbf{v}_{q_i}, \tag{2}$$

where $p(q_i)$ is a unigram probability of the query term q_i in the corpus of knowledge graph literals, and a is a free parameter [1]. The ranking score $KEWER(Q, e)$ of an entity e is then calculated as cosine similarity between entity embedding $\mathbf{v}_e$ and query embedding $\mathbf{q}$:

$$KEWER(Q, e) = \cos(\mathbf{q}, \mathbf{v}_e) \tag{3}$$

This score can be used directly to score all entities in a given knowledge graph, or used in a re-ranking scenario by combining it with traditional retrieval models such as BM25F-CA with the score $BM25F(Q, e)$ that uses term counts in the fields of a textual description of entity e. To parameterize the degree of influence of KEWER on the final ranking, its score can be multiplied by the importance weight β:

$$MM(Q, e) = \beta KEWER(Q, e) + (1 - \beta) BM25F(Q, e), \; 0 \leq \beta \leq 1 \tag{4}$$

Utilizing Entity Linking in Queries. Besides considering only words from queries, we can perform entity linking in queries to find the URIs of entities mentioned in them. For DBpedia, this can be done by using DBpedia Spotlight [7], SMAPH [6], or Nordlys toolkits [12]. After that, the embeddings of linked entities $e_1, \ldots, e_m$ are used in conjunction with the embeddings of query words to calculate the embedding of the entire query as follows:

$$\mathbf{q}_{el} = \sum_{i=1}^{k} \frac{a}{p(q_i) + a} \mathbf{v}_{q_i} + \sum_{i=1}^{m} s(e_i) \mathbf{v}_{e_i}, \tag{5}$$

where $s(e_i)$ is the entity linker's annotation score for the entity e_i. For linked entities' embeddings, we use a concatenation of IN and OUT embedding vectors.

We refer to the method that uses Eq. (5) to obtain query embedding as KEWER$_{el-tool}$, where *tool* is either *Sp* for Spotlight, *SM* for SMAPH, or *N* for Nordlys LTR method, depending on which toolkit was used for entity linking.

4 Experiments

We performed a series of experiments to answer the research questions stated in the introduction and to find the best configuration of our model by implementing KEWER and evaluating it on the DBpedia-Entity v2 dataset [13]. We implemented random walk generation ourselves and used gensim [29] for the Skip-Gram-based optimization step. The source code of the proposed method and all the baselines used in the experiments, detailed dataset construction instructions, as well as the resulting embeddings and runs are available at https://github.com/teanalab/kewer.

4.1 Dataset

DBpedia-Entity v2 collection [13] was used in all the experiments reported in this paper. Following the creators of that dataset, we used the English subset of DBpedia 2015-10 and only considered entities that have both *rdfs:label* and *rdfs:comment* predicates as our entity set E. Detailed statistics of this collection are provided in Table 1.

Table 1. Collection statistics.

Statistic	Value
#Entities $\|E\|$	4,612,277
#Categories $\|K\|$	981,499
#Predicates $\|P\|$	40,750
Avg. # of connected entities for entity	6.62
Avg. # of categories for entity	4.61
Avg. # of surface forms for entity	3.80
Avg. # of tokens in entity surface form	2.72
Avg. # of literals for entity	7.53
Avg. # of tokens in literal	2.45

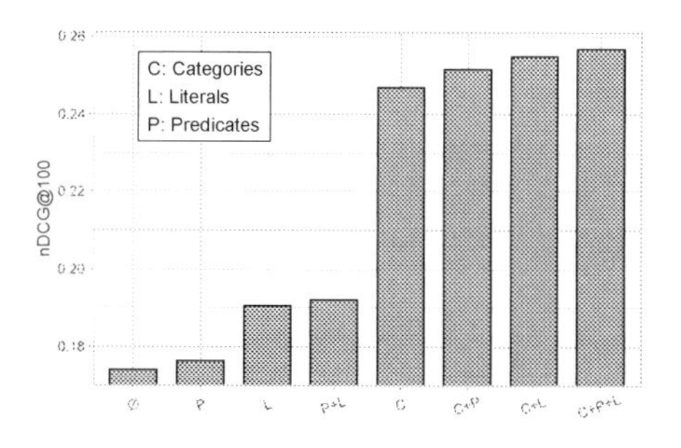

Fig. 1. $nDCG_{100}$ for different combinations of source information used to train embeddings.

Entity search experiments were conducted using four query sets from [13]: **SemSearch ES** contains 113 named entity queries; **INEX-LD** contains 99 keyword-style IR queries; **ListSearch** contains 115 list search queries; **QALD2** contains 140 more complex question answering queries. Following DBpedia-Entity v2 creators, we mainly focus on $nDCG_{100}$, $nDCG_{10}$, and MAP evaluation metrics. The cutoff of 1000 is used for calculating MAP.

4.2 Parameter Sensitivity

To find the optimal values for the length of the random walk t and the replacement probability r, we performed a parameter sweep over the values of $t \in \{2, 3, \ldots, 10\}$ and $r \in \{0.1, 0.2, \ldots, 0.9\}$ to find out a setting that results in the highest nDCG on the query set. We found that the model always performs better with higher values of t, and the performance saturates around $t = 10$, which we use as the parameter value in the experiments. For replacement probability, the model also performs better with higher values of r reaching top nDCG_{10} with $r = 0.9$, the value we use in the experiments. Note that we can't use $r = 1$ since there won't be any URIs for training entity embeddings ($\mathbf{v}_e$ in Eq. (3)) in this case, since they all will be replaced with surface forms. Similarly, we perform a sweep for the context size $c \in \{1, 3, \ldots, 15\}$ to find out the optimal value ($c = 5$) and term weighting parameter $a \in \{10^{-i}, 3 \times 10^{-i} : 1 \leq i \leq 5\}$ to find the optimal value ($a = 3 \times 10^{-4}$). In all subsequent experiments, we generate 100 random walks for each entity and use 5 negative samples during training.

4.3 Usefulness of Structural Components

Since it is unclear which structural components will result in embeddings that are the most useful for entity search, in the first experiment, we attempt to answer **RQ2** by trying all possible combinations of using **categories**, **literals**, and **predicates** structural components for training word and entity embeddings by KEWER. Figure 1 illustrates nDCG_{100} of KEWER averaged over all queries from four query sets when different combinations of structural components are used. In this figure, Ø corresponds to the configuration when only R is used to generate random walks.

From Fig. 1 it can be concluded that using all three structural components is helpful for entity search, with **categories** providing the most benefit and **predicates** providing only a slight increase in retrieval accuracy. Regarding the specific query sets, we observed that using **predicates** on SemSearch ES decreased performance, which can be explained by their ineffectiveness for named entity queries that only contain entities' surface forms. On INEX-LD, we observed that not using **literals** resulted in better performance, which can be explained by the lack of attribute mentions in this query set's keyword queries. In the following experiments we use the word and entity embeddings that are trained using all three knowledge graph's structural components (**categories**, **literals** and **predicates**) and are made publicly available.

4.4 Jointly Embedding Model

As a baseline for learning embeddings, we used our implementation of the Jointly (desp) [46]. $\mathcal{L}_J$, the loss function for Jointly consists of the knowledge and the text component losses ($\mathcal{L}_K$ and $\mathcal{L}_T$, respectively) and the alignment loss $\mathcal{L}_A$:

$$\mathcal{L}_J = \mathcal{L}_K + \mathcal{L}_T + \mathcal{L}_A$$

The knowledge component is formulated similar to TransE [3] with a single embedding space for entities and relations R. Both text and alignment components use textual descriptions of entities obtained from the short abstracts of entities using the *rdfs:comment* property. The text component in our implementation is formulated as a CBOW model with a single embedding space for words. The alignment component predicts the entity embedding given the sum of embeddings of words in entity description. As an alternative to using entity descriptions, we also implemented Jointly (sf) model, where alignment and text models are trained using all available surface forms for entities from S. As in the Sect. 3.3, we define three entity linking extensions of Jointly (Jointly$_{el-Sp}$, Jointly$_{el-SM}$, and Jointly$_{el-N}$) using three different entity linking tools.

4.5 Entity Linking

We annotated all 467 queries using public DBpedia Spotlight API with confidence = 0.5, SMAPH, and Nordlys LTR and report the results for all entity linking models in Table 2. We don't weight linked entities by their scores in Jointly, since we have found that entity weighting is not beneficial to this model. Results indicate that using the SMAPH entity linker results in the best performance for both KEWER and Jointly. For Jointly, using entity descriptions results in better performance than using surface forms.

Table 2. Retrieval performance with entity linking. The best result is in bold.

Model	$nDCG_{10}$	$nDCG_{100}$	MAP
KEWER	0.2102	0.2569	0.1449
KEWER$_{el-Sp}$	0.2417	0.2803	0.1579
KEWER$_{el-SM}$	**0.2704**	**0.3098**	**0.1780**
KEWER$_{el-N}$	0.2660	0.3083	0.1775
Jointly (desp)	0.0486	0.0547	0.0211
Jointly$_{el-Sp}$ (desp)	0.1603	0.1587	0.0838
Jointly$_{el-SM}$ (desp)	**0.1981**	**0.1924**	**0.1014**
Jointly$_{el-N}$ (desp)	0.1870	0.1814	0.0981
Jointly (sf)	0.0291	0.0393	0.0137
Jointly$_{el-Sp}$ (sf)	0.1365	0.1357	0.0684
Jointly$_{el-SM}$ (sf)	0.1685	0.1627	0.0795
Jointly$_{el-N}$ (sf)	0.1624	0.1598	0.0836

Table 2 shows that, even without entity linking, KEWER outperforms both Jointly and Jointly with entity linking based on all metrics. A significant increase in performance of Jointly after performing entity linking suggests that word embeddings learned by Jointly are not useful for entity search, and most of its performance comes from the TransE-based component. This situation is particularly dangerous for queries that do not have entity mentions, such as "Who produced the most films?" or "What is the highest mountain?".

4.6 Mixture Model

It is clear from the above results that KEWER can be a weak ranker by itself. To achieve state-of-the-art results for ad-hoc entity search and to answer **RQ3**, KEWER can be combined with the BM25F-CA model [43], which showed good results in [13]. We implemented BM25F by indexing entities with Galago using 5 fields (*names*, *categories*, *similar entity names*, *attributes*, and *related entity names*) for entity descriptions, as was proposed in [45]. Parameters of the model were separately optimized with a coordinate ascent on each query set using $nDCG_{10}$ as the target metric and 5 cross-validation folds from DBpedia-Entity v2. For each query, we scored the top 1000 results obtained with BM25F using $MM(Q, e)$ score from Eq. (4). The parameter β was optimized using cross-fold validation by sweeping between zero and one with 0.025 increments and choosing the setting that results in the highest nDCG$_{100}$ on each fold's training set. BM25F results were not significantly improved by re-ranking them using Jointly, and we don't report these results. However, in our attempt to answer **RQ4**, we were able to obtain good results by applying word embeddings trained with word2vec's Skip-Gram with the hyperparameter values from Sect. 4.2, trained on the corpus of entity descriptions, where 5 aforementioned fields were combined into one textual description of an entity. The best results with word2vec for entity ranking were obtained when entity embeddings were obtained by summing up without weighting the OUT embeddings of words from their name (*rdfs:label* property), and IN embeddings were used for query terms with weighting. Results on each query set for BM25F, BM25F+word2vec, BM25F+KEWER, BM25F+KEWER$_{el-SM}$ are presented in Table 3.

Table 3. Re-ranking results per query set for KEWER with and without entity linking, and word2vec. Statistically significant improvements (determined by a randomized test with $\alpha = 0.05$) over BM25F and BM25F+word2vec are indicated by "$\star$" and "$\dagger$", respectively. The best result in each column is boldfaced.

SemSearch ES

Model	nDCG$_{10}$	nDCG$_{100}$	MAP
BM25F	0.6606	0.7391	0.5693
BM25F+word2vec	**0.6798***	**0.7445**	**0.5712**
BM25F+KEWER	0.6606	0.7333	0.5627
BM25F+KEWER$_{el\text{-}SM}$	0.6619	0.7409	0.5690

INEX-LD

Model	nDCG$_{10}$	nDCG$_{100}$	MAP
BM25F	0.4456	0.5127	0.3271
BM25F+word2vec	0.4591	0.5227	0.3406*
BM25F+KEWER	**0.4676***	**0.5298***	**0.3417***
BM25F+KEWER$_{el\text{-}SM}$	0.4577*	0.5215*	0.3363*

ListSearch

Model	nDCG$_{10}$	nDCG$_{100}$	MAP
BM25F	0.4287	0.4989	0.3506
BM25F+word2vec	0.4235	0.5055*	0.3551
BM25F+KEWER	0.4402†	0.5210*†	0.3752*†
BM25F+KEWER$_{el\text{-}SM}$	**0.4451***†	**0.5251***†	**0.3777***†

QALD-2

Model	nDCG$_{10}$	nDCG$_{100}$	MAP
BM25F	0.3442	0.4375	0.2861
BM25F+word2vec	0.3567*	0.4504*	0.2986*
BM25F+KEWER	**0.3859***†	**0.4743***†	**0.3154***†
BM25F+KEWER$_{el\text{-}SM}$	0.3800*†	0.4700*†	0.3081*†

All queries

Model	nDCG$_{10}$	nDCG$_{100}$	MAP
BM25F	0.4631	0.5416	0.3792
BM25F+word2vec	0.4730*	0.5504*	0.3874*
BM25F+KEWER	**0.4831***†	**0.5602***†	**0.3955***†
BM25F+KEWER$_{el\text{-}SM}$	0.4807*†	0.5601*†	0.3944*†

The results demonstrate that re-ranking by KEWER is particularly useful for complex question answering queries from QALD-2, list queries from ListSearch, and keyword queries from INEX-LD, while being less useful for simple named entity queries from SemSearch ES, where word2vec thrives. For queries from ListSearch, KEWER is particularly useful when used in combination with entity linker, while for QALD-2 and INEX-LD using entity linking provides lower performance gain. This can be explained by the lack of useful entity mentions in QALD-2 and INEX-LD queries. In QALD-2 queries, mentioned entities are often of a different category than the entity of user's interest and have a complex relationship with it. Using the embeddings of linked entities, in this case, would skew results in the wrong direction. Instead, using plain KEWER helps to clarify the query's intent directly from its keywords.

4.7 Success/Failure Analysis

To illustrate the positive effect of using KEWER on retrieval accuracy, we analyze a sample query SemSearch_LS-50 "wonders of the ancient world" where employing KEWER embeddings resulted in a performance boost. The top results for BM25F and KEWER (without interpolation with BM25F) are presented in Table 4. From these results, it is evident that BM25F failed to capture the conceptual focus of the query by using term matching and most of its top results are only marginally relevant to the query's main focus. On the other hand, KEWER correctly identified the query's main focus on the ancient world, providing five highly relevant results in the ranking.

Table 4. Top 10 ranked entities for the query "wonders of the ancient world" for different models. Relevant results are *italicized* and highly relevant results are **boldfaced**.

BM25F	KEWER
Seven Wonders of the Ancient World	**Colossus of Rhodes**
7 Wonders of the Ancient World (video game)	**Statue of Zeus at Olympia**
Wonders of the World	**Temple of Artemis**
Seven Ancient Wonders	List of archaeoastronomical sites by country
The Seven Fabulous Wonders	**Hanging Gardens of Babylon**
The Seven Wonders of the World (album)	Antikythera mechanism
Times of India's list of seven wonders of India	Timeline of ancient history
Lighthouse of Alexandria	*Wonders of the World*
7 Wonders (board game)	*Lighthouse of Alexandria*
Colossus of Rhodes	**Great Pyramid of Giza**

An example of a query where KEWER was unable to identify query focus is "goodwill of michigan", where it returns entities that are related to Goodwill Games instead of Goodwill Industries. This is caused by the fact that there exist a lot of entities with words "Goodwill Games" in their surface forms, which makes the model believe that token "goodwill" has a strong association with games.

5 Conclusion

This paper proposed KEWER, a method to learn joint word and entity embeddings that was experimentally shown to be effective for entity search, which addresses **RQ1**.

To answer **RQ2**, we compared the effectiveness of embeddings trained on various combinations of knowledge graph structural components and found out that using a combination of **categories**, **literals**, and **predicates** results in the highest retrieval accuracy on DBpedia-Entity v2.

To answer **RQ3** and **RQ4**, we performed an evaluation of KEWER in the re-ranking scenario where it was used in combination with the BM25F retrieval model. Experimental results indicate that KEWER is particularly suitable for improving the ranking of results for complex entity search queries, such as question answering, list search, and keyword queries.

References

1. Arora, S., Liang, Y., Ma, T.: A simple but tough-to-beat baseline for sentence embeddings. In: ICLR 2017 (2017)
2. Blanco, R., et al.: Entity search evaluation over structured web data. In: SIGIR 2011 (2011)
3. Bordes, A., Usunier, N., García-Durán, A., Weston, J., Yakhnenko, O.: Translating embeddings for modeling multi-relational data. In: NIPS 2013. pp. 2787–2795 (2013)
4. Bordes, A., Weston, J., Collobert, R., Bengio, Y.: Learning structured embeddings of knowledge bases. In: AAAI 2011, pp. 301–306 (2011)
5. Ciglan, M., Nørvåg, K., Hluchý, L.: The semsets model for ad-hoc semantic list search. In: WWW 2012, pp. 131–140 (2012)
6. Cornolti, M., Ferragina, P., Ciaramita, M., Rüd, S., Schütze, H.: A piggyback system for joint entity mention detection and linking in web queries. In: WWW 2016, pp. 567–578 (2016). https://doi.org/10.1145/2872427.2883061
7. Daiber, J., Jakob, M., Hokamp, C., Mendes, P.N.: Improving efficiency and accuracy in multilingual entity extraction. In: Proceedings of the 9th International Conference on Semantic Systems, I-SEMANTICS 2013, pp. 121–124 (2013). https://doi.org/10.1145/2506182.2506198
8. Deerwester, S.C., Dumais, S.T., Landauer, T.K., Furnas, G.W., Harshman, R.A.: Indexing by latent semantic analysis. JASIS **41**(6), 391–407 (1990)
9. Diaz, F., Mitra, B., Craswell, N.: Query expansion with locally-trained word embeddings. In: ACL 2016 (2016)
10. Embedding-based Query Expansion for Weighted Sequential Dependence Retrieval Model. https://doi.org/10.1145/3077136.3080764
11. Grover, A., Leskovec, J.: Node2vec: Scalable feature learning for networks. In: KDD 2016, pp. 855–864 (2016). https://doi.org/10.1145/2939672.2939754
12. Hasibi, F., Balog, K., Garigliotti, D., Zhang, S.: Nordlys: a toolkit for entity-oriented and semantic search. In: SIGIR 2017, pp. 1289–1292 (2017). https://doi.org/10.1145/3077136.3084149
13. Hasibi, F., et al.: DBpedia-entity v2: a test collection for entity search. In: SIGIR 2017, pp. 1265–1268 (2017). https://doi.org/10.1145/3077136.3080751
14. Jameel, S., Bouraoui, Z., Schockaert, S.: MEmbER: max-margin based embeddings for entity retrieval. In: SIGIR 2017, pp. 783–792 (2017). https://doi.org/10.1145/3077136.3080803
15. Kuzi, S., Shtok, A., Kurland, O.: Query expansion using word embeddings. In: CIKM 2016, pp. 1929–1932 (2016). https://doi.org/10.1145/2983323.2983876

16. Lehmann, J., et al.: DBpedia - a large-scale, multilingual knowledge base extracted from wikipedia. Semant. Web **6**(2), 167–195 (2015). https://doi.org/10.3233/SW-140134
17. Liu, Z., Xiong, C., Sun, M., Liu, Z.: Explore entity embedding effectiveness in entity retrieval. In: Sun, M., Huang, X., Ji, H., Liu, Z., Liu, Y. (eds.) CCL 2019. LNCS (LNAI), vol. 11856, pp. 105–116. Springer, Cham (2019). https://doi.org/10.1007/978-3-030-32381-3_9
18. Lopez, V., Unger, C., Cimiano, P., Motta, E.: Evaluating question answering over linked data. Web Semant. **21**, 3–13 (2013). https://doi.org/10.1016/j.websem.2013.05.006
19. Mikolov, T., Sutskever, I., Chen, K., Corrado, G., Dean, J.: Distributed representations of words and phrases and their compositionality. In: NIPS 2013, pp. 3111–3119 (2013)
20. Mitra, B., Nalisnick, E.T., Craswell, N., Caruana, R.: A dual embedding space model for document ranking. CoRR abs/1602.01137 (2016). http://arxiv.org/abs/1602.01137
21. Neumayer, R., Balog, K., Nørvåg, K.: On the modeling of entities for ad-hoc entity search in the web of data. In: Baeza-Yates, R., et al. (eds.) ECIR 2012. LNCS, vol. 7224, pp. 133–145. Springer, Heidelberg (2012). https://doi.org/10.1007/978-3-642-28997-2_12
22. Neumayer, R., Balog, K., Nørvåg, K.: When Simple is (more than) good enough: effective semantic search with (almost) no semantics. In: Baeza-Yates, R., de Vries, A.P., Zaragoza, H., Cambazoglu, B.B., Murdock, V., Lempel, R., Silvestri, F. (eds.) ECIR 2012. LNCS, vol. 7224, pp. 540–543. Springer, Heidelberg (2012). https://doi.org/10.1007/978-3-642-28997-2_59
23. Newman-Griffis, D., Lai, A.M., Fosler-Lussier, E.: Jointly embedding entities and text with distant supervision. In: Rep4NLP@ACL 2018, pp. 195–206 (2018)
24. Nickel, M., Tresp, V., Kriegel, H.P.: Factorizing YAGO: scalable machine learning for linked data. In: WWW 2012, pp. 271–280 (2012). https://doi.org/10.1145/2187836.2187874
25. Nikolaev, F., Kotov, A., Zhiltsov, N.: Parameterized fielded term dependence models for ad-hoc entity retrieval from knowledge graph. In: SIGIR 2016, pp. 435–444 (2016). https://doi.org/10.1145/2911451.2911545
26. Pennington, J., Socher, R., Manning, C.D.: GloVe: global vectors for word representation. In: EMNLP 2014, pp. 1532–1543 (2014)
27. Pérez-Agüera, J.R., Arroyo, J., Greenberg, J., Iglesias, J.P., Fresno, V.: Using BM25F for semantic search. In: SEMSEARCH 2010, pp. 2:1–2:8 (2010). https://doi.org/10.1145/1863879.1863881
28. Perozzi, B., Al-Rfou, R., Skiena, S.: DeepWalk: online learning of social representations. In: KDD 2014, pp. 701–710 (2014). https://doi.org/10.1145/2623330.2623732
29. Řehůřek, R., Sojka, P.: Software framework for topic modelling with large corpora. In: NLP Frameworks at LREC 2010, pp. 45–50. May 2010
30. Rekabsaz, N., Mitra, B., Lupu, M., Hanbury, A.: Toward incorporation of relevant documents in word2vec. CoRR abs/1707.06598 (2017). http://arxiv.org/abs/1707.06598
31. Ribeiro, L.F., Saverese, P.H., Figueiredo, D.R.: Struc2vec: learning node representations from structural identity. In: KDD 2017, pp. 385–394 (2017). https://doi.org/10.1145/3097983.3098061

32. Shekarpour, S., Ngonga Ngomo, A.C., Auer, S.: Question answering on interlinked data. In: WWW 2013, pp. 1145–1156 (2013). https://doi.org/10.1145/2488388.2488488
33. Socher, R., Chen, D., Manning, C.D., Ng, A.: Reasoning with neural tensor networks for knowledge base completion. In: NIPS 2013, pp. 926–934 (2013)
34. Tang, J., Qu, M., Wang, M., Zhang, M., Yan, J., Mei, Q.: Line: large-scale information network embedding. In: WWW 2015, pp. 1067–1077 (2015). https://doi.org/10.1145/2736277.2741093
35. Tonon, A., Demartini, G., Cudré-Mauroux, P.: Combining inverted indices and structured search for ad-hoc object retrieval. In: SIGIR 2012, pp. 125–134 (2012). https://doi.org/10.1145/2348283.2348304
36. Vulić, I., Moens, M.F.: Monolingual and cross-lingual information retrieval models based on (bilingual) word embeddings. In: SIGIR 2015, pp. 363–372 (2015)
37. Wang, Z., Zhang, J., Feng, J., Chen, Z.: Knowledge graph and text jointly embedding. In: EMNLP 2014, pp. 1591–1601 (2014)
38. Wei, X., Croft, W.B.: LDA-based document models for ad-hoc retrieval. In: SIGIR 2006, pp. 178–185 (2006). https://doi.org/10.1145/1148170.1148204
39. Xie, R., Liu, Z., Jia, J., Luan, H., Sun, M.: Representation learning of knowledge graphs with entity descriptions. In: AAAI 2016, pp. 2659–2665 (2016)
40. Zamani, H., Croft, W.B.: Embedding-based query language models. In: ICTIR 2016, pp. 147–156 (2016). https://doi.org/10.1145/2970398.2970405
41. Zamani, H., Croft, W.B.: Estimating embedding vectors for queries. In: ICTIR 2016, pp. 123–132 (2016). https://doi.org/10.1145/2970398.2970403
42. Zamani, H., Croft, W.B.: Relevance-based word embedding. In: SIGIR 2017, pp. 505–514 (2017). https://doi.org/10.1145/3077136.3080831
43. Zaragoza, H., Craswell, N., Taylor, M.J., Saria, S., Robertson, S.E.: Microsoft Cambridge at TREC 13: Web and hard tracks. In: TREC 2004 (2004)
44. Zhiltsov, N., Agichtein, E.: Improving entity search over linked data by modeling latent semantics. In: CIKM 2013, pp. 1253–1256 (2013). https://doi.org/10.1145/2505515.2507868
45. Zhiltsov, N., Kotov, A., Nikolaev, F.: Fielded sequential dependence model for ad-hoc entity retrieval in the web of data. In: SIGIR 2015, pp. 253–262 (2015). https://doi.org/10.1145/2766462.2767756
46. Zhong, H., Zhang, J., Wang, Z., Wan, H., Chen, Z.: Aligning knowledge and text embeddings by entity descriptions. In: EMNLP 2015, pp. 267–272 (2015)

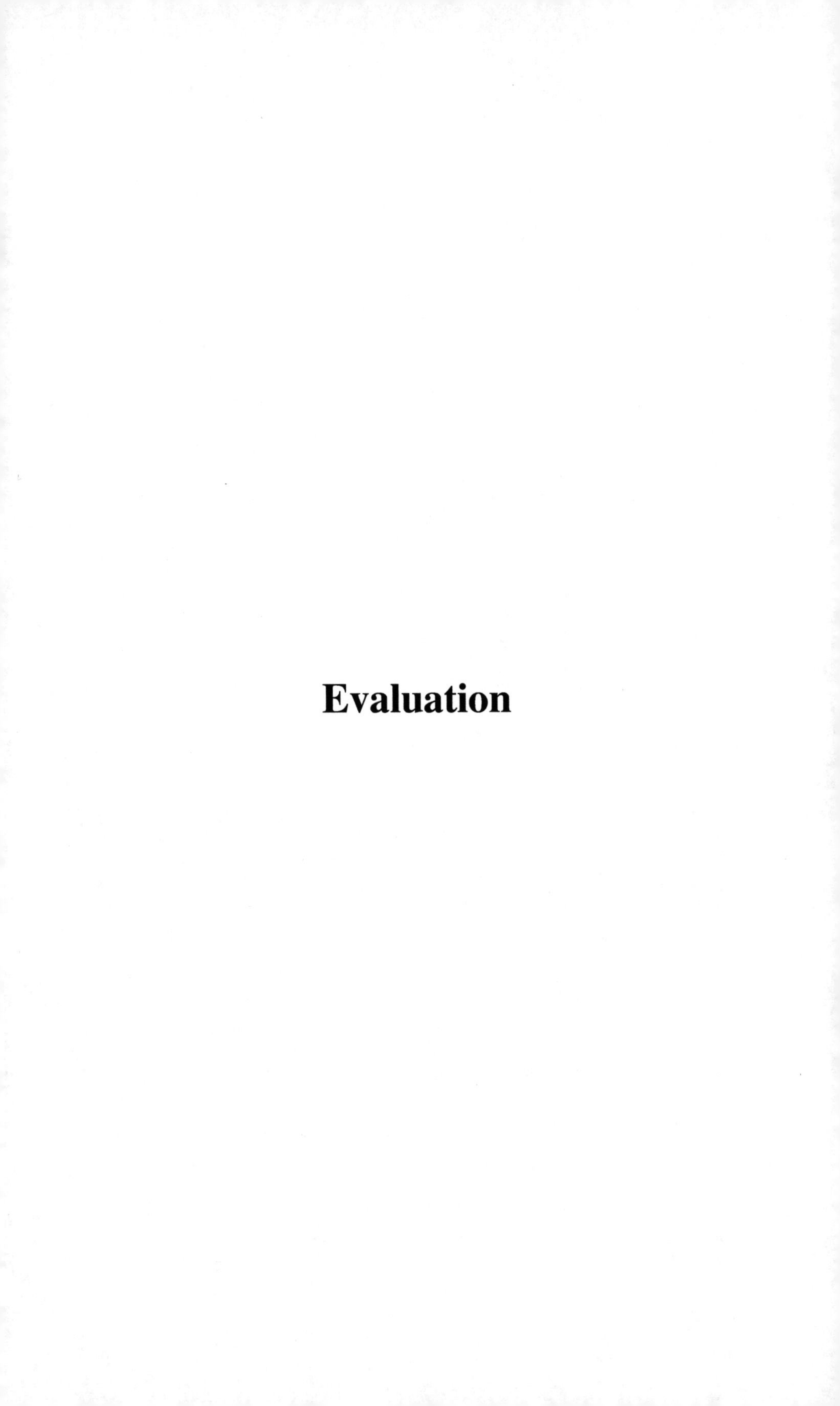

Evaluation

Evaluating the Effectiveness of the Standard Insights Extraction Pipeline for Bantu Languages

Mathibele Nchabeleng and Joan Byamugisha(✉)

IBM Research Africa, Johannesburg, South Africa
{mathibele.nchabeleng,joan.byamugisha}@ibm.com

Abstract. Extracting insights from data obtained from the web in order to identify people's views and opinions on various topics is a growing practice. The standard insights extraction pipeline is typically an unsupervised machine learning task composed of processes that preprocess the text, visualize it, cluster and identify the topics and sentiment in each cluster, and then graph the network. Given the increasing amount of data being generated on the internet in Africa today, and the multilingual state of African countries, we evaluated how well the standard pipeline works when applied to text wholly or partially written in indigenous African languages, specifically Bantu languages. We carried out an exploratory investigation using Twitter data and compared the outputs from each step of the pipeline for an English dataset and a mixed Bantu language dataset. We found that for Bantu languages, due to their complex grammatical structure, extra preprocessing steps such as part-of-speech tagging and morphological analysis are required during data cleaning, threshold values should be adjusted during topic modeling, and semantic analysis should be performed before completing text preprocessing.

Keywords: Insights extraction · Bantu languages · Twitter data

1 Introduction

The growing penetration of mobile telephony and internet services in Africa has led to an increased presence of African user-generated content, especially on social media platforms (such as Facebook, Twitter, and WhatsApp). According to Internet World Stats [8], by the end of 2018, over 460 million out of the continent's 1.3 billion people used the internet, and there were over 200 million Facebook subscribers at the end of 2017. This represents a 35.2% internet penetration rate and a 15.5% Facebook penetration rate [8]. The user-generated content has been leveraged to obtain insights about elections [23], design marketing strategies [1], and monitor the aftermath of epidemics [19]. However, only the content that is written in languages with high-quality linguistic resources

J. M. Jose et al. (Eds.): ECIR 2020, LNCS 12035, pp. 159–172, 2020.
https://doi.org/10.1007/978-3-030-45439-5_11

such as English, French, Portuguese, and Arabic are used for such analyses and content generated in indigenous African languages is largely excluded.

It has been found that even though the amount of content generated in indigenous African languages is significantly lower than non-indigenous language content, it nonetheless contains valuable insights, especially relevant to the local context [12]. Hence, it is extremely important that we develop resources and tools that can be used to parse out useful information from free-text written in any language. In this paper, we investigated whether the standard insights extraction pipeline is sufficient when applied to a single language family indigenous to Africa, Bantu languages, using the following questions: (1) how well does the standard insights extraction pipeline apply to Bantu languages; and (2) if found to be inadequate, why, and how can the pipeline be modified so as to be applicable to Bantu languages?

Two datasets of 20,000 tweets each were included in the study: one was comprised solely of English text and the other a mixed batch of six Bantu languages and English text. Both datasets were analysed using a seven-step pipeline: (1) text preprocessing and normalization, (2) dimensionality reduction, (3) visualization, (4) clustering, (5) topic modeling, (6) sentiment analysis, and (7) network graphing; and the differences in outcomes were measured.

We found that: (1) there is a need to differentiate between conjunctively and disjunctively written languages; (2) sentiment analysis should be performed before verb stemming during text preprocessing, before any present negation morpheme is removed; (3) during text preprocessing and normalization, stemming verbs and adjectives is crucial to avoiding very high levels of sparsity in the representation matrix; (4) stemming nouns must be avoided so as to prevent the loss of important semantic information; and (5) during topic modeling, some threshold values must be adjusted to account for agglutination. This evaluation has, to the best of our knowledge, never been done for Bantu languages.

The rest of the paper is arranged as follows: in Sect. 2, a brief background on Bantu languages and their grammatical structure is presented; Sect. 3 presents related work on extracting insights using the standard pipeline; and the methods, investigation, and results of the evaluation are presented in Sect. 4. The implications of our findings are discussed in Sect. 5, and we conclude in Sect. 6.

2 Brief Background on Bantu Languages

Bantu languages are indigenous to Africa, geographically extending from the south, below Nigeria, to most of central, east, and southern Africa, they are found in 27 of the continent's 54 countries, and range in number from 300 to 680 [21]. Bantu languages have an agglutinating morphology, where words consist of several morphemes, and each affix agglutinated with the root word carries meaning such as tense and aspect [21]. The writing system of Bantu languages is either conjunctive or disjunctive [25]. In the former case, several orthographic words, 'I love them', are written as a single word, for example, *Mbakunda* in Runyankore (a language indigenous to Uganda). The latter case writes different orthographic words as separate words. For example, the same translation for 'I love them' is *Kea ba rata* in Sepedi (a language indigenous to South Africa).

The hallmark of Bantu nominal morphology is the noun class (NC), where all nouns are assigned to a class; and there are over 20 NCs, although some have fallen into disuse in most languages [17,21]. A simple noun comprises a prefix and a stem [11]; for example, *omuntu*, 'person' in Runyankore, can be analyzed as the prefix *o-mu-* and stem *-ntu*. However, not all Bantu languages have the initial vowel on the prefix [11,17]; for example, 'person' in Sepedi is *motho*, with prefix *mo-* and stem *-tho*. Noun classes are also at the heart of an extensive system of concordial agreement that governs grammatical agreement in verbs, adjectives, possessives, subject, object, etc. [11,25]; this is a pivotal constituent of the whole Bantu sentence structure [25].

The morphological and phonological structure of Bantu verbs is very regular in most languages [20,24], with a typical verbal form consisting of: one or more bound morphemes, a verb-root, and one or more extensions [24]. The morphemes preceding the verb-root specify the person, noun class, tense, aspect, time, negation, etc., while the extensions specify valency-changing categories,—the arguments controlled by a verb—which can be as many as eleven [24]. Additionally, Bantu languages typically have a large number of tenses, with up to four observed past tenses, and up to three observed discrete future tenses [20].

This complex grammatical structure is partly what has led to Bantu languages being largely computationally under-resourced, despite still primarily being a first language throughout the continent. In the next section, we present related work on extracting insights from collections of data.

3 Standard Insights Extraction Pipeline

Social media data mining has become a common tool used to extract opinions from a large population in order to monitor, understand, and predict people's reactions to an event, and to measure the diffusion of ideas within the social network [15]. In this section, related work on extracting insights from collections of documents is presented. The scope here is limited to social media data, specifically textual data, and more so, Twitter data, because the vast amount of content generated and shared through social media contains rich knowledge and covers a wide spectrum of social dynamics [33]. In their socio-semantic analysis of Twitter data, Lipizzi et al. [15] stated that the following processes are necessary to extract complete and valuable insights from data: (1) preprocessing the text, (2) identifying and classifying opinions in the network, (3) analyzing the sentiment of individual or groups of text, (4) visualization of the large amounts of data; and (5) extracting conversational maps from social streams. We subdivided these processes into seven steps, including dimensionality reduction and clustering, shown in Fig. 1.

Text preprocessing is necessary because of the strong heterogeneity and noisiness characteristic of social media texts [26]. It involves dealing with incorrect spelling, contractions and abbreviations, stop words, inflectional variants, user

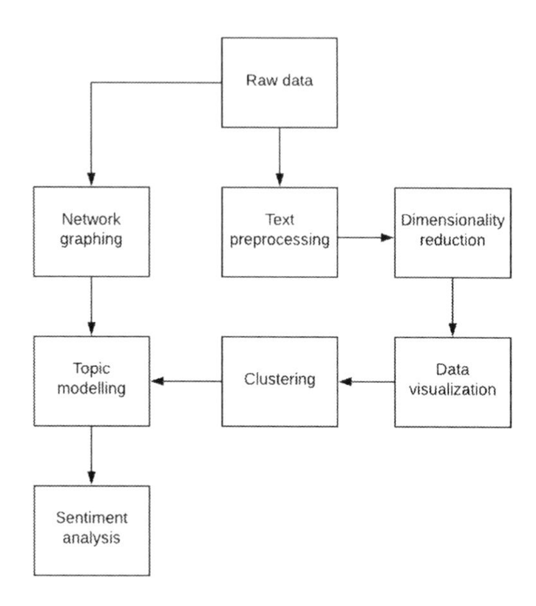

Fig. 1. The seven steps for extracting insights from social media data

tags, hyperlinks, numbers, and email addresses [26,32]. However, the steps taken during preprocessing depend on the quality, quantity, and style of the underlying text [32].

Data visualization requires that the large amount of data be compacted in an expressive fashion [15]. Because text documents are high dimensional objects, effectively visualizing such data requires it to be projected to a lower-dimensional space [18]. Thus, dimensionality reduction, which involves transforming high-dimensional data into a meaningful representation of reduced dimensionality, is an essential part of text mining [6,31]. However, for the lower dimensional representation to be meaningful, it must be a good approximation of the original document set given in its full space [6]. The commonly used techniques for dimensionality reduction are: Principal Component Analysis (PCA), which both minimizes information loss and increases interpretability [6,10]; Singular Value Decomposition (SVD), a stable and robust technique [9]; and T-Distributed Stochastic Neighbor Embedding (T-SNE), which is particularly well suited for the visualization of high dimensional datasets [30].

Document clustering aims to efficiently organize, navigate, summarize, and retrieve documents [3]. It can either be done using partitioning algorithms, where the number of clusters is specified before clustering takes place (for example, K-Means [2,4], Locally Adaptive Clustering [3]), and Non-negative Matrix Factorization [13,14,29]; or hierarchical algorithms, which start by either considering each document as a cluster (agglomerative clustering) or all documents as belonging to a single cluster (divisive clustering) [2,35]. In the former case, documents are continually assigned to the nearest cluster until no further improvement is achieved, while the latter either decreases or increases the number of clusters until a stopping condition is met [2,35].

Topic modeling is the application of probabilistic models to uncover the underlying semantic structure of a collection of documents, where each topic is defined as a distribution over a set of words [2,34]. There are several topic modeling algorithms, but the most commonly applied are Latent Dirichlet Allocation (LDA) [2,4,26,33,34] and Non-negative Matrix Factorization (NMF) [7,13,14,22,29].

Sentiment analysis involves the computational study of people's opinions, appraisals, attitudes, and emotions about events, entities, individuals, and topics [16,27]. Features found to be important during sentiment analysis include: terms and their frequency, adjectives, negation, and opinion words and phrases [16]. Sentiment analysis can be formulated either as a supervised learning problem that can be solved using well-known classification algorithms such as Naive Bayes or Support Vector Machines [16,27], or as an unsupervised problem where opinion words and phrases are used as the dominating indicators of sentiment [16].

Network graphing is used to provide structure to the information exchanged in a social network, and has mostly been used to identify influential users on a topic for marketing or advertising services [4]. Here, each user in a social network is considered as a node in a graph, and the relationships between users (follow, retweet, like, etc.) as directed edges between nodes in the graph [4].

4 Evaluation of Suitability of Pipeline to Bantu Languages

The above processes have been found to be sufficient to extract insights from text in other languages beyond English, such as French [26], Chinese [34], and Arabic [2]. However, to the best of our knowledge, no work has been done completely to apply the described pipeline to Bantu languages. Here, the methodology and results of evaluating the suitability of the standard pipeline for use with Bantu languages are presented.

4.1 Materials and Methods

We used two datasets in this evaluation, each comprising 20,000 tweets; the first, an English dataset composed of customer reviews[1]; the second, composed of tweets in English, mixed code, and six Bantu languages, was archived directly from live South African and Ugandan tweets covering the period February 2019 to May 2019. The live tweets were archived based on the trending hashtags during the period of data collection. The six Bantu languages targeted were IsiZulu, Luganda, Runyankore, Sepedi, Sesotho, and Setswana. These languages were

[1] The English dataset is available from https://www.kaggle.com/thoughtvector/customer-support-on-twitter.

selected because they cover both conjunctive and disjunctive writing styles, and they are understood by the authors. However, due to the use of the mixed code writing style, we found tweets that contained terms in other Bantu languages beyond the six considered.

Our investigation was limited to Twitter data due to the inherent difficulty of performing opinion mining on it, resulting from the informal writing style used and limited tweet length. We hypothesize that the findings based on Twitter data are generalizable to other social media platforms. We further limited the size of each dataset to 20,000 tweets, as the results based on a limited dataset are also generalizable to a larger dataset. Both datasets were run through the seven processes in the standard pipeline, and analyzed for any significant differences. For text preprocessing, we used the same techniques as described in [26,32]. However, no stemming/lemmatizing was performed on either dataset because, to the best of our knowledge, two of the Bantu languages (Luganda and Runyankore) do not have tools for this[2]. We used multiple approaches for dimensionality reduction (PCA, and a combination of SVD followed by T-SNE), clustering (K-Means and NMF) and topic modeling (LDA and NMF), in order to consider the approach which gives the better result. Gephi[3] was used to graph the network.

At each step in the pipeline, the results between the two datasets were compared, with emphasis placed on any observed differences, significant or otherwise. Where a process in the pipeline was found to be insufficient to process the Bantu language dataset, we then investigated if and how the complex grammatical structure of these languages causes the observed limitations. We further investigated what needs to be done in order to adapt that process to fulfill the same task for Bantu languages. On the other hand, where a process in the pipeline was found to adequately apply to the Bantu language dataset, we noted this finding and proceeded to the next step.

4.2 Results

At the end of the evaluation, the processes of text preprocessing, topic modeling, and sentiment analysis were found to require some modification in order to sufficiently extract meaningful insights from textual data in Bantu languages. The processes of dimensionality reduction, data visualization, and clustering, though being language independent, were also found to be affected by the term-document matrix, which is itself language dependent. Only network graphing was found to be completely language independent. The following subsections provide details on the limitations found during text preprocessing, topic modeling, and sentiment analysis, and explain the findings based on the grammatical structure of Bantu languages.

[2] This fact is true for the majority of Bantu languages, and is yet another example of their under-resourced state.

[3] Gephi is available from https://gephi.org/users/download/.

Text Preprocessing. During text preprocessing, after converting the data to lower case, it underwent the removal of HTML tags, URLs, numbers, email addresses, Twitter handles, and hashtags; then the expansion of contractions (such as *can't* and *we're*) and abbreviations (such as *lol*, *dm*, and *tbh*); and finally, the elimination of non-alphanumeric characters and stop words. The text in both datasets was not stemmed or lemmatized due to the lack of such resources for some of the Bantu languages considered in this investigation.

For the English-only dataset, the preprocessing performed was found to be sufficient. However, we found that several additional processes are necessary to fully preprocess the mixed Bantu language dataset. These processes are: distinguishing conjunctively versus disjunctively written languages, part-of-speech tagging, and stemming/lemmatizing only verbs and adjectives.

Distinguishing Between Conjunctively and Disjunctively Written Languages. The mixed Bantu language dataset comprised three conjunctively written languages (isiZulu, Luganda, and Runyankore) and three disjunctively written languages (Sepedi, Sesotho, and Setswana). As explained in Sect. 2, Bantu languages are written either conjunctively or disjunctively, and therefore, there is a need to differentiate between them in order to perform the appropriate preprocessing. Taljard and Bosch [25] identified that a word-class tagger is sufficient for disjunctively written languages, while a morphological analyzer is required for the conjunctively written languages. This is because the disjunctive system of writing requires bound morphemes to be written as orthographically distinct units (*Kea ba rata* 'I love them' in Sepedi), thus making morphological information explicit in the orthography [25]. On the other hand, the conjunctive writing style requires a morphological analyzer to make the different morphemes in the orthography explicit [25], for example from *Mbakunda* to *m-ba-kunda* 'I love them' in Runyankore. The authors concluded that the differences in writing systems necessitate the use of different architectures specifically for part-of-speech tagging. The need for part-of-speech tagging was identified as crucial during text preprocessing and therefore, the type of writing style first needs to be identified before this can be performed.

Part-Of-Speech Tagging. Though neither stemming nor lemmatization were performed on both datasets during preprocessing, we nonetheless recognize the need to stem/lemmatize the verbs and adjectives because of their numerous grammatical forms. Nouns, on the other hand, should not be stemmed as this would result in the loss of their core semantics. As explained in Sect. 2, a noun is composed of a prefix and a stem. However, the stem of a noun is not unique, but rather gets its full semantics from the prefix. Table 1 shows examples of tweets from the dataset where stemming the noun will result in a meaningless stem[4] (the nouns of interest are in bold font, with the prefix underlined).

Table 1. The noun stems *-pedi*, *-tswana*, and *-ntu* are not unique and are meaningless on their own

Language	Tweet	Stem
Sepedi	*o* ***mopedi*** *empa o palela kego ngwala* ***sepedi***	*-pedi*
Setswana	*south african* ***setswana*** *eseng* ***setswana*** *sa* ***botswana***	*-tswana*
isiZulu	***umuntu*** *ng****umuntu*** *ng****abantu***	*-ntu*

The examples shown in Table 1 highlight the problem that can result if nouns are stemmed during text preprocessing. For Sepedi, *sepedi* (a language) would be indistinguishable from *mopedi* (a member of the Bapedi tribe); for Setswana, *setswana* (a language) would have the same stem as *Botswana* (a country); for isiZulu, *umuntu* (person) would be reduced to the same stem as *abantu* (people). Additionally, for isiZulu, a conjunctively written language, the example also shows the need for morphological analysis, to separate the copulative *ng* from the noun.

With the semantics of the noun removed through stemming, the resultant stems *-pedi*, *-tswana*, and *-ntu* are meaningless without a prefix. This in turn would affect topic modeling downstream. Part-of-speech tagging is therefore required to differentiate between nouns that should not be stemmed and other parts-of-speech that should.

Stemming Verbs and Adjectives. A typical Bantu language verbal form consists of one or more bound morphemes, a verb-root, and one or more extensions [24]. The bound morphemes include the subject and object, which are determined by the noun class, as is the full adjectival form [11,25]. Therefore, for a language like Runyankore with 20 noun classes, there are 400 different ways of conjugating a single verb stem for subject and object. Additionally, the number of extensions can be as many as nine, as shown in Table 2, where a single verb stem *reeb-* in Runyankore and *bon-* in Sepedi for 'see' is extended.

In addition to the increasing number of verb forms owing to the extensions shown in Table 2 and the noun class system, Bantu languages typically have a very large number of tenses [20]. For example, Runyankore has 14 tenses [28] and these too are part of the verb form. This complex grammatical structure results in a single verb root having thousands of possible verb forms. Therefore, verb stemming/lemmatizing is a crucial step during preprocessing, which, if not performed results in very high levels of sparsity in the resultant matrix.

[4] The translations to the text in Table 1 are:
Sepedi: O mopedi empa o palela kego ngwala sepedi, 'Your native tongue is Sepedi but you can't even write the language'
Setswana: South African Setswana eseng Setswana sa Botswana, 'South African Setswana not the Setswana from Botswana'
isiZulu: Umuntu ngumuntu ngabantu, 'A person is a person through/because of (other) people'.

Table 2. Different verb extensions for the verb stems *reeb-* in Runyankore and *bon-* in Sepedi (the dashes between the letters represent the separation between the verb root and the extensions)

Runyankore	Sepedi
Reeb-a (See)	*Bon-a* (See)
Reeb-er-a (See for)	*Bon-a-ng* (See, both or all of you)
Reeb-erer-a (Look after)	*Bon-a-ne* (See each other, must)
Reeb-w-a (Seen by)	*Bon-a-le* (Be visible, must)
Reeb-an-a (Look at each other)	*Bon-a-na* (See each other)
Reeb-ek-a (Materialize)	*Bon-a-la* (Be visible)
Reeb-uur-a (Observe)	*Bon-a-la-ng* (Who/which shows)
Reeb-agur-a (Stare)	*Bon-a-gala* (Become visible)
Reeb-a-reeb-a (Look around)	*Bon-a-gala-go* (Who/which become(s) visible)
Reeb-es-a (See with)	*Bon-a-la-go* (Who/which are visible)

Adjectives also require stemming because the full form of an adjective depends on the noun class of the noun being described. Therefore, the number of forms that a single adjective can take depend on the number of noun classes in that language. Runyankore, for example, has 20 different forms for each adjectival stem because it has 20 noun classes. Table 3 shows some examples of the forms that the adjective 'beautiful' in Runyankore *-rungi* and Sepedi *-botse* (the adjective prefix is underlined).

Table 3. The different adjectival forms for the stems *-rungi* in Runyankore and *-botse* in Sepedi

English	Runyankore	Sepedi
Beautiful woman	*Omukazi murungi*	*Mosadi yo mobotse*
Beautiful children	*Abaana barungi*	*Bana ba ba botse*
Beautiful guava	*Eipeera rirungi*	*Kwaba ye botse*
Beautiful eyes	*Amaisho marungi*	*Mahlo a mabotse*
Beautiful building	*Ekizimbe kirungi*	*Moago o mobotse*
Beautiful leg	*Okuguru kurungi*	*Leoto le lebotse*

Topic Modeling. Topic modeling was performed using LDA and NMF. In the case of the mixed Bantu language dataset, we used all tweets and tokens during the modeling. The average tweet length in the corpus by Dela Rosa et al. [5] was 15.22, so we considered a very low threshold of at least five tokens per tweet. From these datasets, it was found that 21.26% of tweets in the mixed Bantu language dataset were below this threshold, compared to 16.22% in the English dataset. While this difference is not significant, we emphasize the agglutinative structure of Bantu languages presented in Sect. 2, where a word consists of several morphemes, and each affix agglutinated with the root word carries meaning such as subject, object, tense, aspect, negation, etc. [21]. As a result, for the conjunctive writing style, an entire sentence can be represented as a single word. Consider the following Runyankore example from [28]: *Titukakimureeterahoganu*, meaning 'We have never ever brought it to him', and comprises the morphemes *ti-tu-ka-ki-mu-reet-er-a-ho-ga-nu*. For this reason, all tweets, despite their length, were included in the topic modeling of the mixed Bantu language dataset.

We also included all tokens in the mixed Bantu language dataset during topic modeling. Although, this is contrary to the recommended minimum token count of three, it was done because, as explained in Sect. 4.2, a single verb stem can be inflected into thousands of verb forms, and it should therefore be expected that, without performing verb stemming, such tokens will be extremely rare in the dataset. From measuring the number of tokens below the recommended threshold count of three in both datasets, we found that 72.04% of tokens in the mixed Bantu language dataset were below this threshold, compared to 0.00% in the English dataset. This is a significant result, again pointing to the importance of verb stemming during preprocessing. Conversely, the English dataset, which was not stemmed either, does not show such an adverse need for it.

Table 4. Negation in Runyankore and Sepedi (the negation morphemes are underlined)

Conjunctive (Runyankore)	Disjunctive (Sepedi)
*oru runyankore nanye tinaru**kyeng**a*	*ka sepedi ga re **berekis**e c q z le x*
*tiha**ri**ho border erikwatanitsa uganda na rwanda*	*ga re **buw**e sesotho mo limpopo*
*konkashi eki otarikuki**reeb**a noha*	*ka sepedi bare tshwene ga e **ipon**e makopo*

Sentiment Analysis. There are currently no publicly available sentiment analysis implementations for any of the Bantu languages used during this investigation. However, we assessed the currently available tools to evaluate whether sentiment analysis could be done following the standard pipeline. In Sect. 3, four features were identified as important for sentiment analysis; three of these (terms and their frequency, adjectives, and opinion words and phrases) are also applicable to a Bantu language dataset. However, if verb stemming is performed during text preprocessing (as we recommend in Sect. 4.2), negation will present a differentiating factor for Bantu languages. This is because, for conjunctively written languages, the negation morpheme(s) is agglutinated to the verb stem, while for disjunctively written languages, the negation morpheme is not necessarily only used in the context of negation. Consider the excerpts from the dataset shown in Table 4[5] writing styles (the negation morphemes are underlined and the verb roots are in bold font).

In the standard pipeline, sentiment analysis is performed after text preprocessing, visualization, clustering, and topic modeling, in order to assess the sentiment associated within a specific cluster or topic. However, for Bantu languages, once verb stemming is performed during text preprocessing, then the verbs in Table 4 are reduced to their roots (shown in bold font); thus losing the negation morphemes *ti*, *ta*, *ga*, and *e*.

Further complexities during sentiment analysis arise from:(1) multiple rules regarding negation, and (2) the negation morpheme being applicable to other parts of speech other than negation. For the former case, consider the example of Runyankore, where *ti* is the primary negative and *ta* the secondary negative; Sepedi, in addition to the negation morpheme *ga*, encodes negation in the change of the final vowel from *a* to *e*. Losing such morphemes would in turn skew the results on sentiment analysis further down the pipeline.

5 Discussion

From the findings presented in Sect. 4, we have shown that Bantu languages require a different architecture from the 'standard'. We therefore propose an alternative architecture shown in Fig. 2.

The following are the areas where differences arise (note that the other processes maintain their original placement in Fig. 1):

(1) During text preprocessing, identifying the writing style of a language is done first, to determine whether to perform part-of-speech tagging for Disjunctively written languages or morphological analysis for conjunctively written languages.
(2) Next, part-of-speech tagging and morphological analysis are performed to prevent nouns from being stemmed, thus avoiding the loss of their semantics encoded in the noun prefix, and ensure that verbs and adjectives are stemmed in order to avoid noise in the data and high levels of sparsity in the resultant matrix.
(3) Sentiment analysis is performed during text preprocessing, before any negation morphemes are lost during verb stemming. Further, it is performed after part-of-speech tagging and morphological analysis, in order to avoid the ambiguity of the negation morpheme identified for some disjunctively

[5] The translations to the text in Table 4 are:
Oru Runyankore nanye tinarukyenga, 'I have also not understood this Runyankore'
Tihariho border erikwatanitsa Uganda na Rwanda, 'There is no border that joins Uganda and Rwanda'
Konkashi eki otarikukireeba noha, 'But honestly, who does not see this'
Ka sepedi ga re berekise c q z le x, 'In sepedi, we don't use the letters c, q, z, and x'
Ga re buwe sesotho mo limpopo, 'We don't speak Sesotho in Limpopo'
Ka sepedi bare tshwene ga e ipone makopo, 'In Sepedi, they say, "A monkey does not see its own forehead".'.

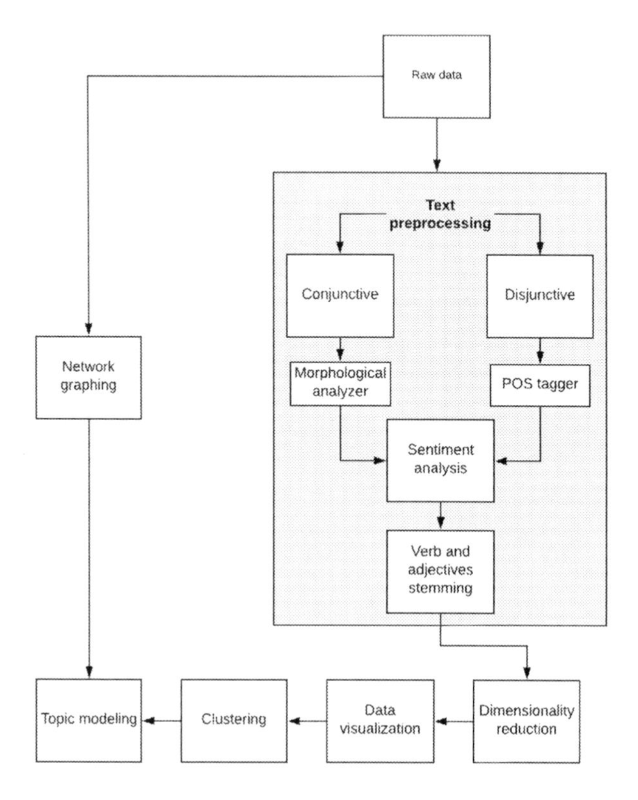

Fig. 2. The revised insights extraction pipeline for Bantu languages

written languages, while also making the negation morpheme explicit for conjunctively written languages.

(4) Finally, during topic modeling, without stemming verbs and adjectives, the threshold counts should not be applied because a significant amount of the dataset will be excluded.

6 Conclusion

In this paper, the standard insights extraction pipeline was evaluated for how well it applies to a grammatically complex and under-resourced family of languages, Bantu languages. Seven processes were identified as belonging to the standard pipeline (text preprocessing, dimensionality reduction, visualization, clustering, topic modeling, sentiment analysis, and network graphing) and tested for their effectiveness on two datasets of 20,000 tweets each, one composed of English and the other a mixture of English and six Bantu languages. Results showed that: conjunctively written languages should be distinguished from disjunctively written languages, because they require different preprocessing steps; verbs and adjectives, but not nouns, should be stemmed; threshold counts should be revised during topic modeling; and sentiment analysis should be done before verb stemming, in order to prevent the loss of the negation morpheme. Future work will include implementing these recommendations and assessing their effectiveness.

References

1. Afolabi, A.: Social Media Marketing: The Case of Africa. Master's thesis, Carleton University, Ontario, Canada (2016)
2. Alhawarat, M., Hegazi, M.: Revisiting k-means and topic modeling: a comparison study to cluster arabic documents. IEEE Access **6**, 42740–42749 (2018)
3. AlSumait, L., Domeniconi, C.: Text clustering with local semantic kernel. In: Berry, M.W., Castellanos, M. (eds.) Survey of Text Mining, Clustering, Classification and Retrieval, 2nd edn, pp. 87–105. Springer, London (2007). https://doi.org/10.1007/978-1-84800-046-9_5
4. Cha, Y., Cho, J.: Social network analysis using topic models. In: 35th Annual SIGIR Conference (SIGIR 2012), pp. 565–574. ACM, Portland (2012)
5. Dela Rosa, K., Shah, R., Lin, B., Gershman, A., Frederking, R.: Topical clustering of tweets. In: Proceedings of the ACM SIGIR: SWSM, vol. 63 (2011)
6. Howland, P., Park, H.: Cluster preserving dimension reduction methods for document classification. In: Berry, M.W., Castellanos, M. (eds.) Survey of Text Mining, Clustering, Classification and Retrieval, 2nd edn, pp. 3–23. Springer, London (2007). https://doi.org/10.1007/978-1-84800-046-9_1
7. Hoyer, P.O.: Non-negative matrix factorization with sparsity constraints. J. Mach. Learn. Res. **5**, 1457–1469 (2004)
8. Internet World Stats: Africa Internet user stats in 2019 population by country (2019). https://internetworldstats.com/africa.htm. Accessed 11 Apr 2019
9. Jackson, J.E.: A Users Guide to Principal Components Analysis. Wiley, New York (1997)
10. Jolliffe, I.T., Cadima, J.: Principal component analysis: a review and recent developments. Philos. Trans. R. Soc. A. Math. Phys. Eng. Sci. **374**(2065), 20150202 (2016)
11. Katamba, F.: Bantu nominal morphology. In: The Bantu Languages: Routledge Language Family Series, vol. 4, chap. 7, pp. 103–120. Taylor and Francis/Routledge, London (2003)
12. Kende, M., Quast, B.: Promoting content in Africa (2016). https://www.internetsociety.org/wp-content/uploads/2017/08/Promoting20Content20In20Africa.pdf. Accessed 11 Apr 2019
13. Kim, J., Park, H.: Sparse nonnegative matrix factorization for clustering. Technical report, Georgia Institute of Technology (2008)
14. Kuang, D., Choo, J., Park, H.: Nonnegative matrix factorization for interactive topic modeling and document clustering. In: Celebi, M.E. (ed.) Partitional Clustering Algorithms, pp. 215–243. Springer, Cham (2015). https://doi.org/10.1007/978-3-319-09259-1_7
15. Lipizzi, C., Iandoli, L., Ramirez-Marquez, J.E.: Extracting and evaluating conversational patterns in social media: a socio-semantic analysis of customers' reactions to the launch of new products using Twitter streams. Int. J. Inf. Manag. **35**, 490–503 (2015)
16. Liu, B., Zhang, L.: A survey of opinion mining and sentiment analysis. In: Aggarwal, C., Zhai, C. (eds.) Mining text data, pp. 415–463. Springer, Boston (2012). https://doi.org/10.1007/978-1-4614-3223-4_13

17. Maho, J.: A comparative study of bantu noun classes. Ph.D. thesis, Goteborg University, Goteborg, Sweden (1999)
18. Mao, Y., Balasubramanian, K., Lebanon, G.: Dimensionality reduction for text using domain knowledge. In: 23rd International Conference on Computational Linguistics (COLING 2010), pp. 801–809. Association for Computational Linguistics (2010)
19. Morin, C., Most, I., Mercier, A., Dozon, J.P., Atlani-Duault, L.: Information circulation in times of Ebola: Twitter and the sexual transmission of Ebolaby survivors. PLoS Currents **10** (2018)
20. Nurse, D.: Aspect and tense in bantu languages. In: The Bantu Languages: Routledge Language Family Series, vol. 4, chap. 6, pp. 90–102. Taylor and Francis/Routledge, London (2003)
21. Nurse, D., Philippson, G.: Introduction. In: The Bantu Languages: Routledge Language Family Series, vol. 4, chap. 1, pp. 1–9. Taylor and Francis/Routledge, London (2003)
22. Peharz, R., Stark, M., Purnkopf, F.: Sparse non-negative matrix factorization using l0-constraints. In: IEEE International Workshop on Machine Learning for Signal Processing. IEEE, Kittila (2010)
23. Portland Africa: How Africa tweets 2018 (2018). https://portland-communications.com/publications/how-africa-tweets-2018/. Accessed 08 Feb 2019
24. Schadeberg, C.T.: Derivation. In: The Bantu Languages: Routledge Language Family Series, vol. 4, chap. 5, pp. 71–89. Taylor and Francis/Routledge, London (2003)
25. Taljard, E., Bosch, S.: A comparison of approaches to word class tagging: conjunctively versus disjunctively written Bantu languages. Nord. J. Afr. Stud. **15**, 428–442 (2006)
26. Tapi-Nzali, M.D., Bringay, S., Lavergne, C., Mollevi, C., Opitz, T.: What patients can tell us: topic analysis for social media on breast cancer. J. Med. Internet Res. (JMIR) **5**(3), e23 (2017)
27. Thakkar, H., Patel, D.: Approaches for sentiment analysis on Twitter: a state-of-art study. Computer Science, Social and Information Networks. arXiv:1512.01043 (2015)
28. Turamyomwe, J.: Tense and Aspect in Runyankore-Rukiga: Linguistic Resources and Analysis. Master's thesis, Norwegian University of Science and Technology, Norway (2011)
29. Túrkmen, A.C.: A review of non-negative matrix factorization methods for clustering. Stat.ML. arXiv:1507.03194v2 (2015)
30. van der Maaten, L., Hinton, G.E.: High-dimensional data using t-SNE. J. Mach. Learn. Res. **9**, 2579–2605 (2008)
31. van der Maaten, L., Postma, E., van der Herik, H.: Dimensionality reduction: a comparative review. Tiburg Centre for Creative Computing, Tilburg University, Technical report (2009)
32. Wesslen, R.: Computer assisted text analysis for social science: topic models and beyond. Computation, and Language, Computer Science (2018)
33. Wu, Y., Cao, N., Gotz, D., Tan, Y.P., Kim, D.A.: A survey on visual analytics of social media data. IEEE Trans. Multimedia **99**(1) (2016)
34. Wu, Y., Ding, Y., Wang, X., Xu, J.: A comparative study of topic models for topic clustering of Chinese web news. In: 3rd IEEE International Conference on Computer Science and Information Technology (ICCSIT), vol. 5 (2010)
35. Xu, R., Wunsch, D.: Survey of clustering algorithms. IEEE Trans. Neural Netw. **16**(3), 645–678 (2005)

Recommendation

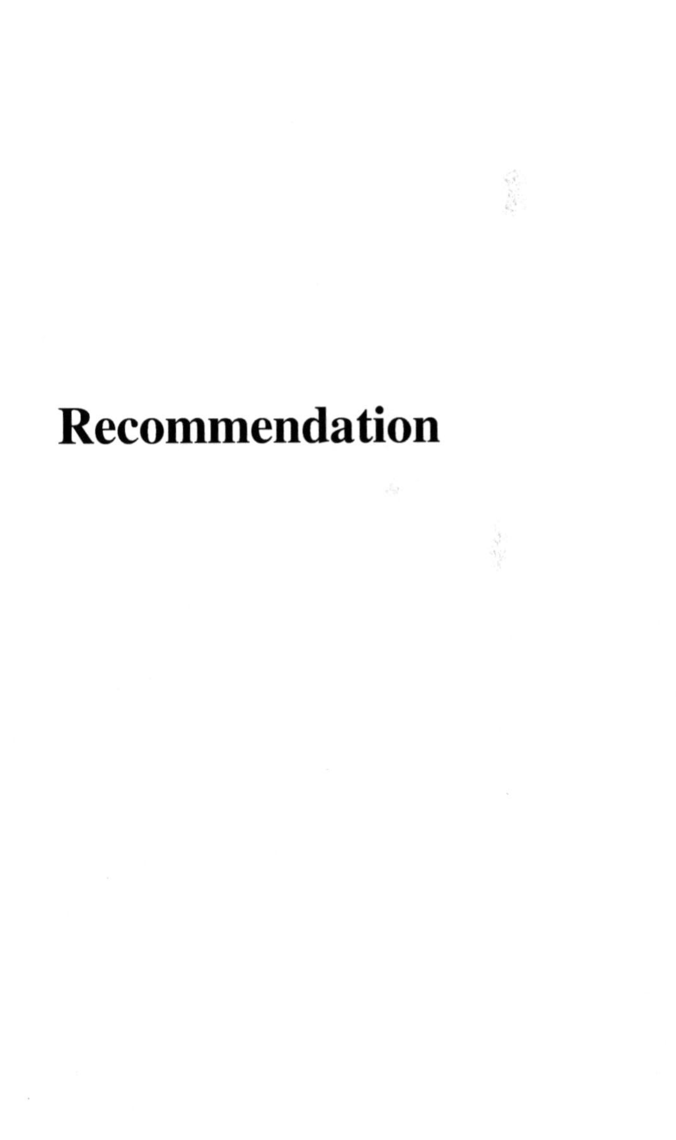

Axiomatic Analysis of Contact Recommendation Methods in Social Networks: An IR Perspective

Javier Sanz-Cruzado[1](✉), Craig Macdonald[2], Iadh Ounis[2], and Pablo Castells[1]

[1] Universidad Autónoma de Madrid, Madrid, Spain
{javier.sanz-cruzado,pablo.castells}@uam.es
[2] University of Glasgow, Glasgow, UK
{craig.macdonald,iadh.ounis}@glasgow.ac.uk

Abstract. Contact recommendation is an important functionality in many social network scenarios including Twitter and Facebook, since they can help grow the social networks of users by suggesting, to a given user, people they might wish to follow. Recently, it has been shown that classical information retrieval (IR) weighting models – such as BM25 – can be adapted to effectively recommend new social contacts to a given user. However, the exact properties that make such adapted contact recommendation models effective at the task are as yet unknown. In this paper, inspired by new advances in the axiomatic theory of IR, we study the existing IR axioms for the contact recommendation task. Our theoretical analysis and empirical findings show that while the classical axioms related to term frequencies and term discrimination seem to have a positive impact on the recommendation effectiveness, those related to length normalization tend to be not desirable for the task.

1 Introduction

With the large-scale growth of social network platforms such as Twitter or Facebook, recommender systems technology that targets explicit social scenarios has seen a surge of interest [32,37]. As part of this trend, the adaptation of Information Retrieval (IR) approaches to recommend people to connect to in the network have been particularly studied [17,34]. This specific class of recommender systems has the interesting property that users play a dual role: they are the users to whom we want to provide recommendations, but they are also the items we want to recommend [32]. Recently, it has been shown that classical IR weighting models – such as BM25 – can not only be used, but are also effective and efficient for the contact recommendation task [34].

In fact, recommender systems have always had strong connections with textual information retrieval (IR), since both tasks can be considered as particular cases of information filtering [9]. These ties have been materialized in the design and development of recommendation approaches based on IR models [2,10,39].

J. M. Jose et al. (Eds.): ECIR 2020, LNCS 12035, pp. 175–190, 2020.
https://doi.org/10.1007/978-3-030-45439-5_12

Content-based recommender systems [2] have been the most direct realization of such ties. However, we also note the collaborative filtering methods of [10,39], which employed the vector space model or query likelihood to their advantage.

In this paper, we analyze the reasons behind the effectiveness of IR approaches for the task of recommending contacts in social networks, through an exploratory analysis of the importance and validity of the fundamental IR axioms [13]. We start our analysis by examining contact recommendation methods that directly adapt IR models [34], as they provide a bridge between existing work on axiomatic analysis in IR models, and this new task. In particular, we empirically analyze whether satisfying the IR axioms leads to an increase in the performances of the algorithms. Interestingly, we find that while this is generally true, the axioms related to length normalization negatively impact the contact recommendation performance, since they interfere with a key evolutionary principle in social networks, namely preferential attachment [8].

2 Related Work

By identifying the set of properties that an IR model must (at least) follow to provide effective results, axiomatic thinking as developed by Fang et al. [12] has permitted to guide the development of both sound and effective IR approaches by explaining, diagnosing and improving them. In their seminal work, Fang et al. [12] proposed several heuristics (known as axioms) addressing different properties of the models such as the frequency of the query terms in the retrieved documents, the relative discrimination between query terms, or how a model deals with long documents. They also analyzed the effect such properties had on the effectiveness of state-of-the-art models such as BM25 [29] or query likelihood [27], and found that, with minor modifications to adhere to the different proposed axioms, the modified IR models achieved an improved retrieval performance.

Since the seminal work of Fang et al., the original axioms have been refined and expanded [13,35], and other additional properties of effective IR models have been studied, such as the semantic relations between queries and documents [14] or term proximity [38]. Recently, axiomatic analysis has been applied on neural IR models: Rennings et al. [28] proposed a method for empirically checking if the learned neural models fulfil the different IR axioms, while Rosset et al. [30] used the axioms as constraints for guiding the training of neural models. Beyond IR, axiomatic analysis has also expanded to other areas such as recommender systems, where Valcarce et al. [39,40] explored the benefits of penalizing users who rate lots of items when selecting neighbors in user-based kNN approaches.

In this paper, using the IR-based contact recommendation framework proposed by Sanz-Cruzado and Castells [34] as a basis, we map the IR axioms of Fang et al. [13] into the task of recommending people in social networks, and empirically analyze how valid and meaningful each axiom is for this task.

3 Preliminaries

We first introduce the notations we use during the rest of the paper. Given a social network, we represent its structure as a graph $\mathcal{G} = \langle \mathcal{U}, E \rangle$, where $\mathcal{U}$ denotes the set of people in the network and E is the set of relationships between users. For each user $u \in \mathcal{U}$, we denote by $\Gamma(u)$ the set of users with whom u has established relationships (the neighborhood of user u). In directed networks, three different neighborhoods can be considered depending on the link *orientation*: users who have a link towards u, $\Gamma_{in}(u)$; users towards whom u has a link, $\Gamma_{out}(u)$; and the union of both, $\Gamma_{und}(u)$. We define $\Gamma_{inv}(u)$ as the inverse neighborhood of u, i.e. the neighborhood u would have if the orientation of the links is reversed. Weighted networks additionally include a function $w : \mathcal{U}^2 \to \mathbb{R}$, where $w(u, v) > 0 \Leftrightarrow (u, v) \in E$. Unweighted networks can be seen as a particular case where $w : \mathcal{U}^2 \to \{0, 1\}$. Then, given a target user u, the contact recommendation task consists of suggesting a subset of users $\hat{\Gamma}_{out}(u) \subset \mathcal{U} \setminus \Gamma_{out}(u)$ towards whom u has no links but who might be of interest for u. We define the recommendation task as a ranking problem, in which the result set $\hat{\Gamma}_{out}(u)$ is obtained and sorted by a ranking function $f_u : \mathcal{U} \setminus \Gamma_{out}(u) \to \mathbb{R}$.

Relation Between IR and Contact Recommendation. Since we explore the importance of IR axioms for contact recommendation, we need to establish connections between both tasks. We take for this purpose the mapping proposed in [34]: we fold the three spaces in the IR task (documents, queries and terms) into a single space for people to people recommendation, namely the users in the network. We map queries and documents to the target and candidate users, respectively. We also use the neighbors of both target and candidate users as equivalent to the terms contained in the queries and documents. As proposed by Sanz-Cruzado and Castells [34], we might use different neighborhoods to represent the target and candidate users (we could take either Γ_{in}, Γ_{out} or Γ_{und} for each of them). We denote by $\Gamma^q(u)$ the neighborhood representing the target user, and by $\Gamma^d(v)$ the one for the candidate user. The frequency of a term t in a document is represented as an edge weight $w^d(v, t)$ in our mapping:

$$\text{freq}(t, v) = w^d(v, t) = w(v, t) \cdot \mathbb{1}_{[\Gamma^d \neq \Gamma_{in}]} + w(t, v) \cdot \mathbb{1}_{[\Gamma^d \neq \Gamma_{out}]} \tag{1}$$

where $\mathbb{1}_x$ is equal to one when the condition x is true, or 0 otherwise.

In textual IR, the frequency is the basis to establish a measure of how important a term is for a document, and it is always positive. Therefore, we assume that $w^d \geq 0$, and $w^d(v, t) = 0$ if and only if $t \notin \Gamma^d(v)$. The higher the importance of the link (v, t), the higher the weight $w^d(v, t)$ should be. In our experiments (described in Sect. 6), we use the number of interactions (i.e. retweets, mentions) between users as an example definition of $w^d(v, t)$. In those network datasets where this type of information is not available, we simply use binary weights.

Finally, the document length is mapped to the sum of the weights of the neighborhood of the target user: $\text{len}(v) = \sum_{t \in \Gamma^l(v)} w^l(v, t)$, which can be seen as a generalized notion of vertex degree in the social graph. For some methods (such as BM25 [29]), we may consider a different neighborhood orientation when computing the user "size"; This explains the different symbols Γ^l, w^l (not necessarily equal to Γ^d, w^d) in the definition of $\text{len}(v)$. In this framework, as the IR models rely on common neighbors between the target and the candidate user, they can only recommend people at distance 2[1]. Table 1 summarizes the relation between the IR and contact recommendation tasks. Further details about the mapping are described in [34].

Table 1. Relation between the IR and contact recommendation tasks.

Information retrieval	Contact recommendation
Document collection, D	Set of users, $\mathcal{U}$
Query, q	Target user's neighborhood, $\Gamma^q(u)$
Document, d	Candidate user's neighborhood, $\Gamma^d(u)$
Term $t \in q/d$	Neighbor user $t \in \Gamma^q(u)/\Gamma^d(v)$
Documents containing a term, D_t	User's inverse neighborhood, $\Gamma^d_{inv}(t)$
Frequency of a term, $\text{freq}(t,d)$	Weight of a link, $w^d(v,t)$
Document length, $\lvert d' \rvert$	Length of the user, $\text{len}(v)$

4 IR Axioms in Contact Recommendation

Before analyzing the importance of the IR axioms in the recommendation task, we first recall the IR axioms, and reformulate them using the mapping from IR to contact recommendation. In the remainder of this section, we take the seven axioms proposed by Fang et al. [13], divided into four categories, and analyze them.

4.1 Term Frequency Constraints (TFC)

The first family of axioms analyzes the role of the frequency of the query terms in the retrieved documents. Since term frequencies are represented as edge weights in our framework, we rename them as "edge weight constraints" (EWC) in our reformulation. The first constraint, TFC1, establishes that if the only difference between two documents is the frequency of a query term, then, the document with the higher term frequency should be ranked atop of the other. The intuition behind this axiom is naturally translated to contact recommendation by considering the "common friends" principle in social bonding: all things being equal, you are more likely to connect to people who have stronger bonds to common friends. This principle can be expressed as follows:

---

[1] Distance is the minimum number of links you need to traverse from the target user to the candidate user, regardless of the *orientation* (direction) of the link.

EWC1: If the target user u has a single neighbor $\Gamma^q(u) = \{t\}$, and we have two different candidate users v_1, v_2 such that $\text{len}(v_1) = \text{len}(v_2)$, and $w^d(v_1, t) > w^d(v_2, t)$, then we should have $f_u(v_1) > f_u(v_2)$.

The second term frequency constraint (TFC2) establishes that the ranking score increment produced by increasing term frequency should decrease with the frequency (i.e. ranking scores should have a dampened growth on term frequency, as in a diminishing returns pattern). This also has a direct meaning in the contact recommendation space: the difference in scores between two candidate contacts should decrease with the weights of their common friends with the target user. Formally, this constraint is expressed as:

EWC2: For a target user u with a single neighbor $\Gamma^q(u) = \{t\}$, and three candidate users v_1, v_2, v_3 such that $\text{len}(v_1) = \text{len}(v_2) = \text{len}(v_3)$, and $w^d(v_3, t) = w^d(v_2, t) + 1$ and $w^d(v_2, t) = w^d(v_1, t) + 1$, then $f_u(v_2) - f_u(v_1) > f_u(v_3) - f_u(v_2)$.

Finally, the third axiom reflects the following property: occurrence frequencies and discriminative power being equal, the document that covers more distinct query terms should attain a higher score. In people recommendation, this translates to the triadic closure principle [25,26]: all other things being equal, the more common friends a candidate contact has with the target user, the higher the chance that a new link between them exists. Formally:

EWC3: Let $\{t_1, t_2\} \subset \Gamma^q(u)$ be two neighbors of target user u, with $\text{td}(t_1) = \text{td}(t_2)$. Given two candidate users v_1, v_2 with $\text{len}(v_1) = \text{len}(v_2)$, if $w^d(v_1, t_1) = w^d(v_2, t_1) + w^d(v_2, t_2)$, $t_2 \notin \Gamma^d(v_1)$, and $\{t_1, t_2\} \subset \Gamma^d(v_2)$, then $f_u(v_1) < f_u(v_2)$.

where $\text{td}(t)$ is a measure of the informativeness of the common neighbors of the target and candidate users, as can be obtained from an IDF measure.

These three axioms are interdependent: if we take $\Gamma^q(u) = \{t\}$ and we fix the values for $\text{td}(t)$ and $\text{len}(v)$, we could rewrite $f_u(v)$ as a function of the document weight, $f_u(w^d(v,t))$. If $f_u(w^d(v,t))$ is positive, it is easy to see that EWC1 $\Leftrightarrow$ $f_u(w^d(v,t))$ is an increasing function, EWC2 $\Leftrightarrow$ $f_u(w^d(v,t))$ is strictly concave, and EWC3 $\Leftrightarrow$ $f_u(w^d(v,t))$ is strictly subadditive. Given a function g, g positive and concave $\Rightarrow$ g is increasing and subadditive. Therefore, for such functions (as is the case for most of the classic IR functions), EWC2 $\Rightarrow$ EWC1 $\wedge$ EWC3. However, if EWC2 is not satisfied, either EWC1 or EWC3 could still be satisfied.

4.2 Term Discrimination Constraint (TDC)

The term discrimination constraint is an axiom that formalizes the intuition that penalizing popular words in the collection (such as stopwords) and assigning higher weights to more discriminative query terms should produce better search results. This principle makes sense in contact recommendation: sharing a very popular and highly connected friend (e.g. two people following Katy Perry on Twitter) may be a rather weak signal to infer that these two people would relate to each other. A less social common friend, however, may suggest the two people may indeed have more interests in common. This idea is in fact reflected in some contact recommendation algorithms such as Adamic-Adar [1,22].

Hence, we rename the axiom as "neighbor discrimination constraint" (NDC), and we adapt the version of the axiom proposed by Shi et al. [35], which simplifies the translation to our domain, as follows:

NDC: Let u be the target user, with $\Gamma^q(u) = \{t_1, t_2\}$. Given two candidate users v_1, v_2 where where $\text{len}(v_1) = \text{len}(v_2)$, and $w^d(v_1, t_1) = w^d(v_2, t_2)$ and $w^d(v_1, t_2) = w^d(v_2, t_1)$, if $w^d(v_1, t_1) > w^d(v_1, t_2)$ and $\text{td}(t_1) > \text{td}(t_2)$, then $f_u(v_1) > f_u(v_2)$.

4.3 Length Normalization Constraints (LNC)

The third family of IR axioms studies how algorithms should deal with the length of the documents. As defined in Sect. 3, in our mapping, the length of the document is translated to the sum of the edge weights between the candidate user and its neighbors: $\text{len}(v)$. As we only study the length of the candidate user, we will rename this family of constraints as "candidate length normalization constraints" (CLNC). Fang et al. [13] proposed two different LNCs.

The first axiom states that for two documents with the same query term occurrence frequency, we should choose the shorter one, since it contains the least amount of query-unrelated information. In contact recommendation, this means penalizing popular, highly connected candidate users with many neighbors not shared with the target user. We hence reformulate this axiom as:

CLNC1: Given a target user u and two candidate users v_1, v_2, if $w^d(v_2, t) > w^d(v_1, t)$ for some user $t \notin \Gamma^q(u)$, but $w^d(v_1, x) = w^d(v_2, x)$ for any other user $x \neq t$, then $f_u(v_1) > f_u(v_2)$.

The second constraint aims to avoid over-penalizing long documents: it states that if a document is concatenated to itself multiple times, the resulting document should not get a lower score than the original. In contact recommendation, this means that, if we multiply all the edge weights of a candidate user by a positive number, the score for the candidate user should not decrease. Formally:

CLNC2: If two candidate users v_1, v_2 are such that $w^d(v_1, x) = k \cdot w^d(v_2, x)$ for all users x and some constant $k > 1$, and $w^d(v_1, t) > 0$ for some neighbor $t \in \Gamma^q(u)$ of the target user u, then we have $f_u(v_1) \geq f_u(v_2)$.

4.4 Term Frequency – Length Normalization Constraint (TF-LNC)

The last heuristic aims to provide a balance between query term frequency in documents and length normalization. The axiom states that if we add more occurrences of a query term to a document, its retrieval score should increase. For contact recommendation, the intuition is similar: if the link weight between two users v and t increases, then v's score as a candidate for target users having t in their neighborhood should increase. This axiom is then expressed as follows:

EW-CLNC: Given a target user u with a single neighbor $\Gamma^q(u) = \{t\}$, if two candidates v_1 and v_2 are such that $w^d(v_1, t) > w^d(v_2, t)$ and $\text{len}(v_1) = \text{len}(v_2) + w^d(v_1, t) - w^d(v_2, t)$, then $f_u(v_1) > f_u(v_2)$.

5 Theoretical Analysis

The first step to undertake an analysis of the IR axioms in contact recommendation is to determine the set of algorithms for which the different axioms are applicable, and, for those, to identify which constraints they satisfy and under which conditions. In this section, we provide an overview of different contact recommendation methods and their relation with the axioms.

We divide the approaches into two groups: friends of friends approaches, which only recommend people at network distance 2 from the target user, and methods which might recommend more distant users. The first group includes all IR models, as well as other approaches such as the most common neighbors (MCN) and Adamic-Adar's approach [22], whereas the second group includes matrix factorization [18,21], random walk-based methods [16,41] and kNN [2].

The proposed set of constraints is not applicable to the algorithms in the second group, since the constraints are based on the idea that the weighting functions depend on the common users between the target and the candidate users. Therefore, in the rest of the article, we focus on the algorithms in the first family. As future work, we envisage the formulation of new constraints tailored for algorithms that recommend users at distance greater than 2, possibly as a generalization of the set of constraints we study in this paper (see e.g. the formal analysis of pseudo-relevance feedback by Clinchant and Gaussier [11], which in our mapping would correspond to distance greater than 2).

We start analyzing the friends of friends methods by studying the IR models. In the adaptation of these models by Sanz-Cruzado and Castells [34], the components of the ranking functions (frequency/weight, discriminative power functions, document/user length) maintain the basic properties on which the formal analysis by Fang et al. [12,13] has relied. Therefore, the adapted methods satisfy the same constraints in the social network as those satisfied in the text IR space, and, if they are only satisfied under certain conditions, we can find the new conditions just by adapting them for the contact recommendation task. Then, models like PL2 [3,7], the pivoted normalization vector space model (VSM) [36] query likelihood with Dirichlet (QLD) [42] or Jelinek-Mercer smoothing (QLJM) [27] keep their original properties in this new space.

We find however one point of difference related to a possibility considered by Sanz-Cruzado and Castells in the definition of the candidate user length; namely, that we can define the length of the candidate users by selecting a different neighborhood $\Gamma^l(v)$ than the one used for defining the candidate user, $\Gamma^d(v)$, as explained in Sect. 3. As the only difference between the original and the version of BM25 defined by Sanz-Cruzado and Castells is just the definition of the candidate length, it is straightforward to prove that all edge weight constraints and NDC are satisfied in the same way as they are for textual IR: NDC is unconditionally true, whereas all EWC axioms depend just on the condition:

$$C_1 : |\Gamma^d_{inv}(t)| < |\mathcal{U}|/2 \tag{2}$$

which, in contact recommendation, is likely to be true – indeed, as of 2019, Twitter has >300 M users, and, the most followed user has just 107 M followers.

On the other hand, differences arise when we study the constraints involving length normalization: CLNCs and EW-CLNC. If we keep the same orientation for the user length and neighborhood selection for the candidate user, the mapping maintains the same components as the original ranking function, and, consequently, the condition for satisfying the three axioms is the same as the original: satisfying condition C_1. However, if the orientation for the length is changed, it is easy to show that, for CLNC1, BM25 satisfies the axiom if both conditions C_1 and C_2 are true, or both are false, where:

$$C_2 : \left(\Gamma^l := \Gamma^d\right) \vee (\text{len}(v_2) > \text{len}(v_1)) \tag{3}$$

and, for the EW-CLNC, the constraint is kept if conditions C_1 and C_3 are met, or none of them are, where:

$$C_3 : \left(\Gamma^l := \Gamma^d\right) \vee \left(\Gamma^l := \Gamma_{und}\right) \vee \left(\frac{1-b}{b} \operatorname*{avg}_{v'} \text{len}(v') > w^d(t, v_2) - \text{len}(v_2)\right) \tag{4}$$

The only length normalization-related constraint that is satisfied under the same conditions as the original BM25 model is the CLNC2 constraint, since it does not really depend on the definition of user length. Table 3 shows the differences between the original version and this adaptation of the BM25 model for contact recommendation. Hence, we introduce a new IR-based approach, namely the Extreme BM25 (EBM25) method, a variant of BM25 where we make the k parameter tend to infinity. In comparison with BM25, all constraints are satisfied under the conditions specified for BM25, except EWC2 and EWC3, which are not satisfied at all for EBM25. In the BM25 model, under the conditions of EWC2, the k parameter establishes how $f_u(v)$ grows as a function of the weight of the only common neighbor between the target and candidate users. The greater the value of k, the more the growth function approximates a linear function. When $k \to \infty$, the growth becomes linear, and as a consequence, the model does not meet the EWC2 constraint. A similar issue occurs with EWC3.

Beyond the IR models, other approaches such as Adamic-Adar or MCN do operate at distance 2. In the particular case of these methods, they consider neither weights nor any means of normalization; only EWC3 and CLNC2 are applicable here: under the conditions of EWC3, both methods just measure the number of common neighbors, satisfying the constraint. For CLNC2, if we multiply all the weights of the link for a candidate by any number $k \neq 0$, the score of the functions would not vary (and, consequently, they meet the axiom).

We summarize this analysis in Table 2, where we identify whether a method satisfies (fully or conditionally) or not the different axioms. In the case of the models not described in this section (pivoted normalization VSM, PL2, QLD), we refer to the article by Fang et al. [13] for further information on the conditions to satisfy the axioms. Next, we empirically analyze whether satisfying the axioms leads to an improvement of the performance of such algorithms.

Table 2. Constraint satisfaction for different contact recommendation algorithms.

Algorithm	EWC1	EWC2	EWC3	NDC	CLNC1	CLNC2	EW-CLNC
BM25	Cond.	Cond.	Cond.	Yes	Cond.	Cond.	Cond.
EBM25	Cond.	No	No	Yes	Cond.	Cond.	Cond.
Pivoted	Yes	Yes	Yes	Yes	Yes	Cond.	Cond.
PL2	Cond.	Cond.	Cond.	Cond.	Cond.	Cond.	Cond.
QLD	Yes	Yes	Yes	Yes	Yes	Cond.	Yes
QLJM	Yes	Yes	Yes	Yes	Yes	Yes	Yes
MCN	No	No	Yes	No	No	Yes	No
Adamic-Adar	No	No	Yes	No	No	Yes	No

Table 3. Constraint analysis results for BM25. By the equivalence notation e.g. $C_1 \equiv C_2$ we mean that C_1 and C_2 can only be either both true or both false.

	TFC/EWC			TDC/NDC	LNC/CULNC		TF-LNC/EW-CULNC
	1	2	3		1	2	
Text IR	C_1	C_1	C_1	Yes	C_1	C_1	C_1
Contact rec.	C_1	C_1	C_1	Yes	$C_1 \equiv C_2$	C_1	$C_1 \equiv C_3$

6 Empirical Analysis

Prior work on axiomatic thinking [12,13] has analyzed to which extent the satisfaction of a suitable set of constraints correlates with effectiveness. This is also a mechanism to validate such constraints, showing that it is useful to predict, explain or diagnose why an IR system is working well or badly. Taking up this perspective, we undertake next such an empirical analysis of constraints in the contact recommendation setting, using a set of friends-of-friends algorithms.

6.1 Experimental Setup

Data: We use different network samples from Twitter and Facebook: the ego-Facebook network released in the Stanford Large Network Dataset collection [24], and two Twitter data downloads described in [34] as 1-month and 200-tweets. The Twitter downloads include each two different sets of edges for the same set of users: the follow network (where $(u, v) \in E$ if u follows v), and the interaction network (where $(u, v) \in E$ if u retweeted or mentioned v). The datasets are described in more detail in [32–34].

For evaluation purposes, we partition each network into a training graph that is supplied as input to the recommendation algorithms, and a test graph that is held out for evaluation. Using the test graph, IR metrics such as precision, recall or nDCG can be computed, as well as other accuracy metrics such as AUC [15], by considering test edges as binary relevance judgements: a user v is relevant to

Table 4. Dataset statistics

	Twitter 1-month		Twitter 200-tweets		Facebook
	Interactions	Follows	Interactions	Follows	
Directed	Yes	Yes	Yes	Yes	No
Users with links	9,528	9,770	9,985	9,964	4,039
Training edges	170,425	645,022	137,850	475,730	70,566
Validation edges	33,867	46,628	29,131	46,760	14,100
Test edges	54,335	81,110	21,598	98,519	17,643

a user u if – and only if – the edge (u, v) appears in the test graph. We further divide the training graph into a smaller training graph and a validation graph for parameter tuning. Table 4 shows the size of the different resulting subgraphs.

For all Twitter networks, temporal splits are applied: the training data includes edges created before a given time, and the test set includes links created afterwards. Edges appearing in both sides of the split are removed from the test network. For the interaction network, two different temporal points are selected to generate the split: July 5th and July 12th in the 1-month dataset, and July 24th and July 29th in 200-tweets. Weights for the training graphs were computed by counting the number of interactions before the splits.

For the follow networks, the edges between the users of the interaction network were downloaded three times: the first download is used as training graph for parameter tuning; the new links in the second snapshot (not present in the initial one), downloaded four months later, are used as the validation set; the complete second snapshot is given as input to the recommendation algorithms under evaluation; finally, the new edges in the third download (not present in the second), obtained two years afterwards, are used as the test data for evaluation.

For the Facebook data, since temporal information is not available, we apply a simple random split: 80% of links are sampled as training and 20% as test; within the training data, we use 25% of the edges as the validation subset.

Algorithms: We focus on contact recommendation approaches that recommend users at distance 2. From that set, as representative IR models, we include adaptations for the pivoted normalization vector space model [36]; BIR and BM25 [29] as probabilistic models based on the probability ranking principle; query likelihood [27] with Jelinek-Mercer [20], Dirichlet [23] and Laplace [39] smoothing as language models; and PL2 [3,7], DFRee, DFReeKLIM [6], DPH [4] and DLH [5] as divergence from randomness approaches. In addition, we include adaptations of a number of link prediction methods [22] (following [34]): Adamic-Adar [1], Jaccard [19], most common neighbors [22] and cosine similarity [31].[2]

[2] Code and additional details about the experimental configuration are available at https://github.com/ir-uam/contact-rec-axioms.

6.2 Experiments and Results

Edge Weight Constraints (EWCs): We start by analyzing the edge weight constraints. Since weights are binary in the Twitter follow graphs and Facebook, we focus here on interaction graphs, where the interaction frequency provides a natural basis for edge weighting.

Table 5. Average AUC values for the most common neighbors algorithm for the different datasets, using $\Gamma^q := \Gamma_{und}$ and $\Gamma^d := \Gamma_{in}$ in the directed networks.

Twitter 1-month		Twitter 200-tweets		Facebook
Interactions	Follows	Interactions	Follows	
0.7545	0.8327	0.7064	0.7951	0.9218

A first natural question that arises when we study these axioms is whether the weights are useful or not for providing good recommendations. This is equivalent to test the importance of the first axiom for the contact recommendation task. To answer that question, we compare the two options (binarized vs. not binarized weights) in all algorithms which make use of weights: cosine similarity between users and all the IR models except BIR. We show the results in Fig. 1(a), where each dot represents a different approach. In the x axis, we show the nDCG@10 value for the unweighted approaches, whereas the y axis shows nDCG@10 for the weighted ones. We can see that using weights results in an inferior performance in all algorithms except for BM25 and the simple cosine similarity. These observations suggest that EWC1 does not appear to be a reliable heuristic for contact recommendation in networks.

However, once the weight is important for a model (and, therefore, EWC1 is important) does satisfying the rest of the edge weight constraints provide more accurate recommendations? To check that, similarly to Fang et al. [12,13], we compare an algorithm that satisfies all three EWCs (and benefits from weights) with another one that does not satisfy EWC2 and EWC3: we compare BM25 vs. EBM25. Fixing the k parameter for the BM25 model (using the optimal configuration from our experiments), we compare different parameter configurations for BM25 and EBM25. Results are shown in Fig. 1(b), where every dot in the plot corresponds to a different model configuration, the x axis represents the nDCG@10 values for BM25, and the y axis those of the EBM25 model. As it can be observed, EBM25 does not improve over BM25 for almost every configuration (dots are all below the $y = x$ plane), thus showing that, as long as EWC1 is important for the model, both EWC2 and EWC3 are relevant.

As explained in Sect. 4, EWC3 can also be satisfied independently of EWC1 and EWC2, so we finally check its importance. For that purpose, we address the following question: for any friends-of-friends algorithm, such as Adamic-Adar [1] or the IR models, is it beneficial to reward the number of common users between the target and the candidate users? To analyze this, we compare the MCN approach (which satisfies the constraint) with a binarized version of MCN which returns all people at distance 2 regardless of the common neighbor count. Restricting the test set to people at distance 2, Table 5 shows the resulting AUC [15] of the MCN algorithm, averaged over users on each network. Under these conditions, the binarized version would have an AUC value of 0.5. Hence, our results show that the number of common neighbors seem to be a strong signal for providing accurate recommendations (and, therefore, EWC3 seems to be important on its own for the contact recommendation task).

Neighbor Discrimination Constraint (NDC): As previously explained, this constraint suggests penalizing highly popular common neighbors. In IR approaches, this constraint is satisfied or not depending on the presence or absence of a term discrimination element (such as the Robertson-Spärck-Jones in BM25/EBM25 or the $p_c(t)$ term in query likelihood approaches). Therefore, to check the effectiveness benefit of this axiom, we compare – in terms of nDCG@10 – the BM25, EBM25, QLD, QLJM and the pivoted normalization VSM models with variants of them that lack term discrimination.

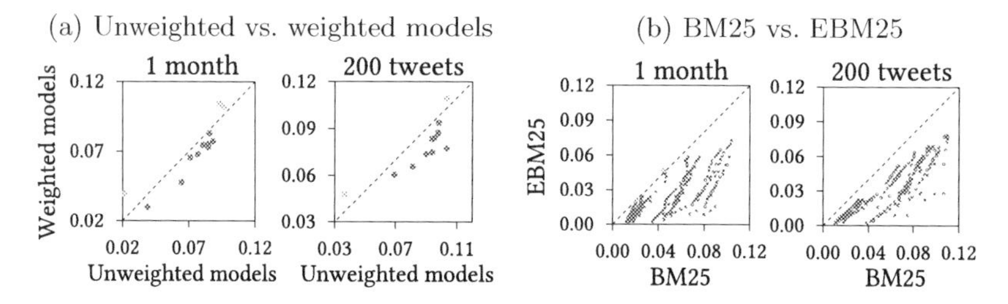

Fig. 1. For the Twitter interaction datasets: (a) nDCG@10 comparison between the weighted (y axis) and unweighted (x axis) versions of different contact recommendation algorithms. (b) nDCG@10 comparison between weighted versions of BM25 (x axis) and EBM25 (y axis). In both graphs, red dots represent those elements such that the value of nDCG@10 is greater for the y axis than for the x axis. (Color figure online)

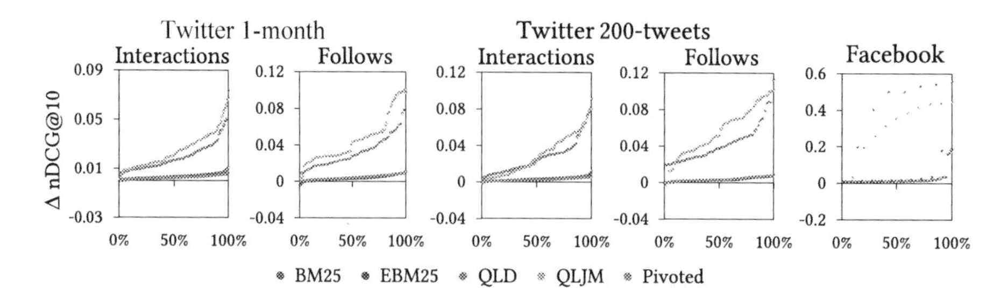

Fig. 2. Difference in nDCG with and without term discrimination for different configurations of IR-based algorithms, sorted by difference value. Each dot represents a different configuration of the corresponding algorithm. A positive value indicates that the variant with term discrimination is more effective.

Figure 2 shows the difference between different variants of each model. In the figure, a positive value indicates that the original version (with term discrimination) performs better. We observe that in an overwhelming majority of points the original versions achieve a better accuracy, hence NDC appears to be key to providing good contact recommendations. This confirms the hypothesis in many recommendation approaches that using high-degree users to discriminate which users are recommended does not seem to be a good idea [1,43].

Length Normalization Constraints (CLNCs and EW-CLNC): Finally, we study the effect of normalizing by candidate user length. For that purpose, similarly to the previous section, we compare the BM25, EBM25, QLJM, QLD and the pivoted normalization VSM models with versions of the models lacking the normalization by the candidate user length (which do not satisfy CLNC1 and EW-CLNC) using nDCG@10. We show a graph showing the differences in accuracy between different variants of the algorithms in Fig. 3(a). Since there are few differences between datasets, we only show results for the interactions network of the Twitter 1-month dataset. In the figure, we observe an opposite trend to what was expected: instead of performing worse, the algorithms without normalization do improve the results. Therefore, it seems that the different length normalization constraints are not useful for contact recommendation.

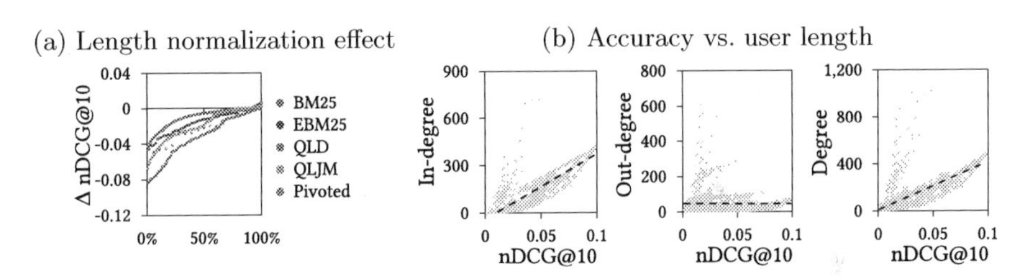

Fig. 3. For the Twitter 1-month interaction network: (a) Difference in nDCG with and without length normalization for different configurations of IR-based algorithms, sorted by difference value. A positive value indicates that the variant with length normalization is more effective. (b) Comparison between nDCG@10 and the average in-degree and out-degree of the recommended users.

These observations are consistent with the preferential attachment phenomenon in social networks [8], whereby high-degree users are more likely to receive new links than long-tail degree users. As an example, we check this in Fig. 3(b), where we compare the performances of the recommendation approaches listed in Section 6.1 with the average in-degree, out-degree and (undirected) degree of the recommended people. We observe that, in general, in-degree and degree are clearly correlated with the performances of the methods, as the principle indicates. With out-degree this is not so clear though. This explains the few configurations in Fig. 3(a) that do not improve when we remove the normalization: all of them normalize by the sum of the weights of the outgoing links of the candidate users. Similar trends are observed in other networks.

7 Conclusions

We have theoretically and empirically analyzed the importance of the fundamental IR axioms for the contact recommendation task in social networks. Theoretically, we have translated the different axioms proposed in [13] to the contact recommendation task, and we have checked whether the mapping introduced in [34] is sound and complete. We have found that, in general, the properties of the IR models are held in the recommendation task when we apply this mapping, unless we use a different definition for the document length from the usual. Empirically, we have conducted several experiments over various Twitter and Facebook networks to check if those axioms have any positive effect on the accuracy of the recommenders. We showed that satisfying the constraints related to term frequencies and term discrimination have a positive impact on the accuracy. However, those related to length normalization tend to have the opposite effect, as they interfere with a basic evolutionary principle of social networks, namely preferential attachment [8].

Acknowledgements. J. Sanz-Cruzado and P. Castells were partially supported by the Spanish Government (TIN2016-80630-P). C. Macdonald and I. Ounis were partially supported by the European Community's Horizon 2020 programme, under grant agreement nº 779747 entitled BigDataStack.

References

1. Adamic, L.A., Adar, E.: Friends and neighbors on the Web. Soc. Netw. **25**(3), 211–230 (2003)
2. Adomavicius, G., Tuzhilin, A.: Toward the next generation of recommender systems: a survey of the state-of-the-art and possible extensions. IEEE Trans. Knowl. Data Eng. **17**(6), 734–749 (2005)
3. Amati, G.: Probability information models for retrieval based on divergence from randomness. Ph.D. thesis, University of Glasgow (2003)
4. Amati, G.: Frequentist and Bayesian approach to information retrieval. In: Lalmas, M., et al. (eds.) ECIR 2006. LNCS, vol. 3936, pp. 13–24. Springer, Heidelberg (2006). https://doi.org/10.1007/11735106_3
5. Amati, G., Ambrosi, E., Bianchi, M., Gaibisso, C., Gambosi, G.: FUB, IASI-CNR and University of Tor Vergata at TREC 2007 blog track. In: Proceedings of the 16th Text REtrieval Conference (TREC 2007). NIST (2007)
6. Amati, G., et al.: FUB, IASI-CNR, UNIVAQ at TREC 2011 microblog track. In: Proceedings of the 20th Text REtrieval Conference (TREC 2011). NIST (2011)
7. Amati, G., Van Rijsbergen, C.J.: Probabilistic models of information retrieval based on measuring the divergence from randomness. ACM Trans. Inf. Syst. **20**(4), 357–389 (2002)
8. Barabàsi, A.L., Albert, R.: Emergence of scaling in random networks. Science **286**(5439), 509–512 (1999)
9. Belkin, N.J., Croft, W.B.: Information filtering and information retrieval: two sides of the same coin? Commun. ACM **35**(12), 29–38 (1992)
10. Bellogín, A., Wang, J., Castells, P.: Bridging memory-based collaborative filtering and text retrieval. Inf. Retrieval **16**(6), 697–724 (2013)

11. Clinchant, S., Gaussier, E.: A theoretical analysis of pseudo-relevance feedback models. In: Proceedings of the 2013 Conference on the Theory of Information Retrieval (ICTIR 2013), pp. 6–13. ACM (2013)
12. Fang, H., Tao, T., Zhai, C.: A formal study of information retrieval heuristics. In: Proceedings of the 27th annual International ACM SIGIR Conference on Research and Development in Information Retrieval (SIGIR 2004), pp. 49–56. ACM (2004)
13. Fang, H., Tao, T., Zhai, C.: Diagnostic evaluation of information retrieval models. ACM Trans. Inf. Syst. **29**(2), 1–42 (2011)
14. Fang, H., Zhai, C.: Semantic term matching in axiomatic approaches to information retrieval. In: Proceedings of the 29th Annual International ACM SIGIR Conference on Research and Development in Information Retrieval (SIGIR 2006), pp. 115–122. ACM (2006)
15. Fawcett, T.: An introduction to ROC analysis. Pattern Recogn. Lett. **27**(8), 861–874 (2006)
16. Goel, A., Gupta, P., Sirois, J., Wang, D., Sharma, A., Gurumurthy, S.: The who-to-follow system at Twitter: strategy, algorithms, and revenue impact. Interfaces **45**(1), 98–107 (2015)
17. Hannon, J., Bennett, M., Smyth, B.: Recommending Twitter users to follow using content and collaborative filtering approaches. In: Proceedings of the 4th ACM Conference on Recommender Systems (RecSys 2010), pp. 199–206. ACM (2010)
18. Hu, Y., Koren, Y., Volinsky, C.: Collaborative filtering for implicit feedback datasets. In: Proceedings of the 8th IEEE International Conference on Data Mining (ICDM 2008), pp. 263–272. IEEE (2008)
19. Jaccard, P.: Étude comparative de la distribution florale dans une portion des Alpes et des Jura. Bulletin de la Société Vaudoise des Sciences Naturelles **37**(142), 547–579 (1901)
20. Jelinek, F., Mercer, R.: Interpolated estimation of Markov source parameters from sparse data. In: Gelsema, E.S., Kanal, L.N. (eds.) Pattern Recognition in Practice, pp. 381–402. North-Holland (1980)
21. Koren, Y., Bell, R., Volinsky, C.: Matrix factorization techniques for recommender systems. Computer **42**(8), 30–37 (2009)
22. Liben-Nowell, D., Kleinberg, J.: The link-prediction problem for social networks. J. Am. Soc. Inf. Sci. Technol. **58**(7), 1019–1031 (2007)
23. MacKay, D.J.C., Peto, L.C.B.: A hierarchical Dirichlet language model. Nat. Lang. Eng. **1**(3), 289–307 (1995)
24. McAuley, J., Leskovec, J.: Learning to discover social circles in ego networks. In: Proceedings of the 25th International Conference on Neural Information Processing Systems (NIPS 2012), pp. 539–547. Curran Associates Inc. (2012)
25. Newman, M.E.J.: Clustering and preferential attachment in growing networks. Phys. Rev. E **64**, 025102 (2001)
26. Newman, M.E.J.: Networks: An Introduction, 1st edn. Oxford University Press, Oxford (2010)
27. Ponte, J.M., Croft, W.B.: A language modeling approach to information retrieval. In: Proceedings of the 21st Annual International ACM SIGIR Conference on Research and Development in Information Retrieval (SIGIR 1998), pp. 275–281. ACM (1998)
28. Rennings, D., Moraes, F., Hauff, C.: An axiomatic approach to diagnosing neural IR models. In: Azzopardi, L., et al. (eds.) ECIR 2019. LNCS, vol. 11437, pp. 489–503. Springer, Cham (2019). https://doi.org/10.1007/978-3-030-15712-8_32
29. Robertson, S.E., Zaragoza, H.: The probabilistic relevance framework: BM25 and beyond. Found. Trends Inf. Retrieval **3**(4), 333–389 (2009)

30. Rosset, C., Mitra, B., Xiong, C., Craswell, N., Song, X., Tiwary, S.: An axiomatic approach to regularizing neural ranking models. In: Proceedings of the 42nd International ACM SIGIR Conference on Research and Development in Information Retrieval (SIGIR 2019), pp. 981–984. ACM (2019)
31. Salton, G., Wong, A., Yang, C.: A vector space model for automatic indexing. Commun. ACM **18**(11), 613–620 (1975)
32. Sanz-Cruzado, J., Castells, P.: Contact recommendations in social networks. In: Berkovsky, S., Cantador, I., Tikk, D. (eds.) Collaborative Recommendations: Algorithms, Practical Challenges and Applications, pp. 519–569. World Scientific Publishing (2018)
33. Sanz-Cruzado, J., Castells, P.: Enhancing structural diversity in social networks by recommending weak ties. In: Proceedings of the 12th ACM Conference on Recommender Systems (RecSys 2018), pp. 233–241. ACM (2018)
34. Sanz-Cruzado, J., Castells, P.: Information retrieval models for contact recommendation in social networks. In: Azzopardi, L., et al. (eds.) ECIR 2019. LNCS, vol. 11437, pp. 148–163. Springer, Cham (2019). https://doi.org/10.1007/978-3-030-15712-8_10
35. Shi, S., Wen, J.R., Yu, Q., Song, R., Ma, W.Y.: Gravitation-based model for information retrieval. In: Proceedings of the 28th Annual International ACM SIGIR Conference on Research and Development in Information Retrieval (SIGIR 2005), pp. 488–495. ACM (2005)
36. Singhal, A., Choi, J., Hindle, D., Lewis, D.D., Pereira, F.C.N.: AT&T at TREC-7. In: Proceedings of the 7th Text REtrieval Conference (TREC 1998), pp. 186–198. NIST (1998)
37. Tang, J., Hu, X., Liu, H.: Social recommendation: a review. Soc. Netw. Anal. Min. **3**(4), 1113–1133 (2013). https://doi.org/10.1007/s13278-013-0141-9
38. Tao, T., Zhai, C.: An exploration of proximity measures in information retrieval. In: Proceedings of the 30th Annual International ACM SIGIR Conference on Research and Development in Information Retrieval (SIGIR 2007), pp. 295–302. ACM (2007)
39. Valcarce, D., Parapar, J., Barreiro, Á.: Axiomatic analysis of language modelling of recommender systems. Int. J. Uncertainty Fuzziness Knowl.-Based Syst. **25**(Suppl. 2), 113–127 (2017)
40. Valcarce, D., Parapar, J., Barreiro, Á.: Finding and analysing good neighbourhoods to improve collaborative filtering. Knowl.-Based Syst. **159**, 193–202 (2018)
41. White, S., Smyth, P.: Algorithms for estimating relative importance in networks. In: Proceedings of the 9th ACM SIGKDD International Conference on Knowledge Discovery and Data Mining (KDD 2003), pp. 266–275. ACM (2003)
42. Zhai, C., Lafferty, J.: A study of smoothing methods for language models applied to information retrieval. ACM Trans. Inf. Syst. **22**(2), 179–214 (2004)
43. Zhou, T., Lü, L., Zhang, Y.C.: Predicting missing links via local information. Eur. Phys. J. B **71**(4), 623–630 (2009)

Recommending Music Curators: A Neural Style-Aware Approach

Jianling Wang(✉) and James Caverlee

Texas A&M University, College Station, TX, USA
{jlwang,caverlee}@tamu.edu

Abstract. We propose a framework for personalized music curator recommendation to connect users with curators who have matching *curation style*. Three unique features of the proposed framework are: (i) models of curation style to capture the coverage of music and curator's individual style in assigning tracks to playlists; (ii) a curation-based embedding approach to capture inter-track agreement, beyond the audio features, resulting in models of music tracks that pair well together; and (iii) a novel neural pairwise ranking model for personalized music curator recommendation that naturally incorporates both curator style models and track embeddings. Experiments over a Spotify dataset show significant improvements in precision, recall, and F1 versus state-of-the-art.

1 Introduction

Music streaming platforms provide access to a diverse, incredibly large, and ever growing collection of music tracks. To make sense of the millions of available tracks (e.g., 40 million on Apple Music and 30 million on Spotify), playlists have become an essential feature of many music streaming platforms for organizing music, mediating how users experience the service. Across platforms, most playlists are manually curated and managed by a group of *music curators*, which consists of both "regular" users and expert tastemakers. To benefit from the power of human curation [15], many platforms enable users to follow these music curators to receive updates of their listening activities, e.g., to discover new tracks, albums, or playlists (as illustrated in Fig. 1(a)).

While recommendation systems have been widely deployed in many music streaming platforms for tasks like recommending individual music tracks [3,19,23] or playlists [2,14], they are not well-suited for real-world scenarios like (i) discovering new tracks with little or no feedback; (ii) finding relevant playlists that are frequently updated (and hence, out-of-sync with respect to a learned recommendation model); and (iii) recommending playlist creators themselves who can provide direct access to new tracks, albums, or playlists. As a step toward supporting these scenarios, we focus on the task of *curator recommendation* to create a personalization layer to help users discover vast amounts of new tracks, fresh playlists, and interesting curators.

While some services highlighting highly-rated or popular curators [7,16,22] (e.g., Spotify's recommendation of "featured" curators with high popularity),

J. M. Jose et al. (Eds.): ECIR 2020, LNCS 12035, pp. 191–204, 2020.
https://doi.org/10.1007/978-3-030-45439-5_13

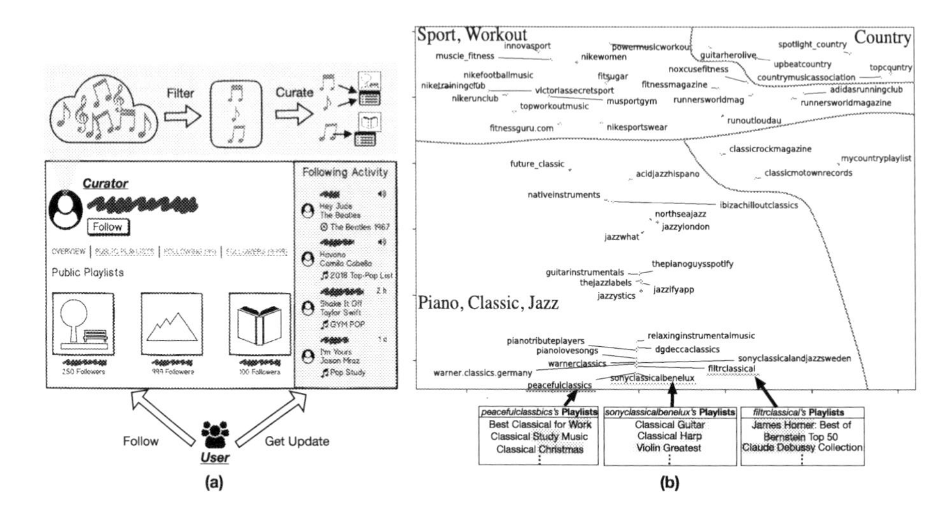

Fig. 1. (a) Users follow *music curators* with matching curation style to receive updates. (b) We randomly sample curators with IDs containing the keywords ("workout", "sport", "run", "fit", "gym", "country", "piano", "instrument", "classic", "jazz"). We show the 2D visualization (with t-SNE [11]) of selected curators based on average audio features of music tracks they curate.

identifying personally-relevant music curators is a daunting task due to the following challenges. First, music curators themselves are complex amalgamations of the playlists they create, the tracks they select, and their unique *style*. For example, some curators may focus on specific emotions (like happy or excited), eras (like the 80s or 90s), or situations (like workouts, parties, or road trips), while others cross boundaries (like happy 80s music, or 90s road trip). Hence the first challenge is: How can we build models that capture these stylistic differences across music curators taking both their curating coverage and individual style into consideration? To illustrate, we conduct an initial exploration of Spotify curators whose areas of interest can be inferred from their user IDs. We represent each of the music curators using the average of audio features (see Sect. 3.2 for details) for all the tracks in playlists they curate and plot the 2D t-SNE distribution in Fig. 1(b). We see that there are clear patterns of curator *coverage*: country music curators cluster together in the top-right, curators focusing on active music for sports and workouts cluster in the top-left, while classical and instrumental music curators dominate the lower portion. We see that *curators have preferences in the coverage of music* they would curate. However, coverage alone is insufficient to distinguish between curators. We must also consider individual *style*. Considering the playlists curated by three classical music curators (in the bottom of the figure), while all drawing from the same musical coverage area, each of them displays a unique style – one groups tracks for activities like work or study, one collects tracks featuring the same instrument, while the third groups by artists. We see that *curators who curate similar types of music can have different style in deciding how tracks go with each other.*

Second, user preferences for music curators may be driven by many factors, including preferences for curator coverage and style. These preferences may only be revealed through extremely sparse user feedback. For example, in a sample of Spotify playlists (see Sect. 4.1) we find that only 0.20% of all user-curator pairs have a following relationship. And for pairs without a following relationship, it may mean the user dislikes the curator or just has not known her yet. Furthermore, because anyone can be a curator on these platforms, there are many long-tail candidate curators who may be invisible to most users. Hence the second challenge is: How can we uncover the hidden taste preferences that connect users to the music curators they may prefer?

Toward answering these questions, we propose a novel personalized **M**usic **C**urator **R**anking (MCR) framework to recommend music curators to users based on the style of each curator and on each user's taste profile. There are three unique features of the proposed MCR framework: (i) We propose to model *curation style* through a novel neural pairwise framework that considers how each curator assigns tracks to different playlists, toward uncovering each curator's latent style; (ii) Based on how crowds of curators compose their playlists, we propose an embedding model for tracks to capture inter-track agreement to uncover hidden connections among tracks, which can assist in characterizing the coverage of each curator; (iii) We propose a novel neural ranking model for personalized music curator recommendation that naturally incorporates both curator style models and track embeddings to identify personally relevant curators. Through experiments over a Spotify dataset, we find the proposed framework results in a 20.5% and 5.7% improvement in top-k F1 score compared to Bayesian Personalized Ranking (BPR) and Neural Personalized Ranking (NPR), and in a 24.9% and 21.4% improvement in cold start scenarios.

2 Related Work

Music Recommendation and Playlist Generation. Complementary to our focus on recommending curators, many researchers have explored music continuation and automated playlist generation. For example, [12] predicts the next track based on a listener's preferences and most-recently played tracks. DJ-MC [14] aims to recommend track sequences based on reinforcement learning. Groove Radio [1] generates personalized playlists based on seed artists. EFM [2] recommends both tracks and playlists through a new embedding approach. There are also efforts on training an embedding model on users' historical music playing sequences to estimate the similarity between songs for recommendation [5,24]. In contrast, we propose a *style-based* recommender that links users directly to curators rather than specific tracks or playlists.

User Profiling and Expert Recommendation. Somewhat similar to our notion of curator is research on finding expertise to improve search and recommendation [9,27]. By uncovering the latent preferences of users or building up profiles for them, these approaches aim to identify related experts. For music, one effort has aimed to identify curators [13] based on a Linked Data graph capturing listening history and other factors. However, it cannot scale to larger datasets demonstrating sparsity as in our case.

Neural-Based Recommendation with Implicit Feedback. To make recommendations in such sparse, implicit feedback scenarios, methods like Bayesian Personalized Recommendation (BPR) [21] and a recently introduced variant called Neural Personalized Ranking (NPR) [18] have shown good success. These and other neural approaches have demonstrated their power in recommenders, including [4,6,25]. Other approaches include the autoencoder-based CDAE [26] and Neural Collaborative Filtering [10] that adds nonlinearities to traditional Matrix Factorization (MF). In this work, we propose to take advantage of the benefits of neural architectures for recommenders, while carefully incorporating special properties of music curation, including curator style, coverage, and curation-based embeddings.

3 MCR: Music Curator Ranking

In this section, we start from problem setting and then present the design of our MCR framework, organized around three guiding research questions.

Problem Setting: Let $U = \{u_1, u_2,..., u_N\}$ be a set of N users and $C = \{c_1, c_2,..., c_M\}$ be a set of M curators. A curator c can create a playlist L_q^c composed of tracks drawn from a collection of possible tracks $T = \{t_1, t_2, ..., t_{|T|}\}$. Further, curator c can create multiple playlists $P_c = \{L_1^c, L_2^c, ..., L_Q^c\}$. Users may express their interest in a curator through an action such as a "like" or "follow". Our goal is to recommend a personalized ranked list of music curators to each user.

Research Question: RQ1: How to model the hidden *curation style* that guides how playlist curators select tracks for their playlists? **RQ2:** How can we use these curation decisions to model individual tracks, to capture how curators view tracks beyond their particular audio characteristics? **RQ3:** How to model users preferences on both curation style and coverage for improved recommendation?

3.1 RQ1: Model of Curation Style

As shown in the previous section, curators have their own style in deciding how tracks go together. For example, some classical music lovers may create playlists based on the time period (e.g., 1800 s) or for different artists. Some may focus on the particular instrument featured, while others may curate based on feelings, activities (e.g., for studying or focusing), or locations (e.g., for "coffee shop"). Modeling these styles is important for accurately connecting users to their preferred curators. A purely content-based approach to model style (e.g., based on composer, time period, or audio features of the track) may face challenges in determining these subtle stylistic choices that motivate a curator.

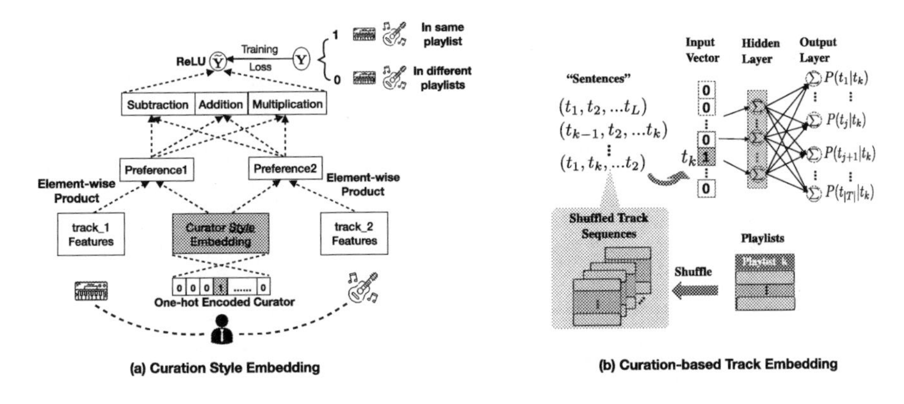

Fig. 2. Curation style modeling and curation-based track embedding.

Hence, we propose to model *curation style* through a pairwise framework intended to discern why a curator chooses one track over another. Given tracks t_1, t_2, and t_3, curator c may put t_1 and t_2 in the same list while putting t_3 in a different list. We propose a neural network (see Fig. 2(a)) to simulate how curators choose what tracks go in what playlists. By iterating among and across playlists of different curators, we can generate input tuples to represent their behaviors in assigning tracks to playlists. Each input tuple has three components: the index of the curator, the features representing t_1, and the features representing t_2. If curator c puts t_1 and t_2 into the same playlist (a positive pair), then the ground truth of prediction is set to be 1. However, if t_1 and t_2 appear in different playlists of curator c (a negative pair), then the ground truth is set to be 0. We can then train a binary classification model to predict how curators will relate two tracks during curation. The intermediate embedding layer of this model – $\mathbf{e}_c^{style}$ – can then characterize curator c's curation style.

Concretely, such a model requires a vector representation as input. Let $\mathbf{i}_c$ denote the one-hot encoding of curator c. Since the space of tracks is large, instead of a one-hot approach, we represent each track t_i by a dense vector $\mathbf{F}_{t_i}$ (more details in Sect. 3.2). After feeding the input tuple $< c, \mathbf{F}_{t_1}, \mathbf{F}_{t_2} >$ to the neural network, to learn the representation of c's curation style, the vector $\mathbf{i}_c$ is passed to the embedding layer: $\mathbf{e}_c^{style} = \mathbf{W}_{style}\mathbf{i}_c$, where $\mathbf{W}_{style}$ is the embedding matrix and the resulting embedding $\mathbf{e}_c^{style}$ will be used to characterize c. Then the element-wise multiplication between embedding of the curator and music track is analogous to matrix factorization with $\mathbf{H}_c^{t_1} = \mathbf{e}_c^{style} \circ \mathbf{F}_{t_1}$ and $\mathbf{H}_c^{t_2} = \mathbf{e}_c^{style} \circ \mathbf{F}_{t_2}$. Thus the resulting vectors $\mathbf{H}_c^{t_1}$ and $\mathbf{H}_c^{t_2}$ represent c's preferences in curating. To decide whether the curator will put t_1 and t_2 into the same or different playlists, we calculate the element-wise absolute difference between them $|\mathbf{H}_c^{t_1} - \mathbf{H}_c^{t_2}|$ for comparison. We also calculate the element-wise multiplication $\mathbf{H}_c^{t_1} \circ \mathbf{H}_c^{t_2}$ to represent the cosine distance between them. Then the concatenated vector will be fully connected to a dense layer activated with ReLU function. Finally a one-dimension output is generated as the prediction of likelihood that t_1 and t_2 will be in the same playlist by curator c. Binary cross-entropy is used to calculate the training loss. After reaching an accurate prediction model, the embedding vector $\mathbf{e}_c^{style}$ can be used to represent c's curation style.

Curator Coverage. Complementary to curation style, we can also characterize a curator's coverage as a simple aggregate of the tracks associated with a curator. Let $T_c = \{t_i | t_i \in L_x^c, \forall L_x^c \in P_c\}$ represent the set of tracks that curator c adds to at least one playlist, where P_c denotes the set of playlists curated by c and L_x^c is each of those playlists. Recall that each track t_i can be represented by a dense vector $\mathbf{F}_{t_i}$. Concretely, we can use the weighted sum of T_c to model c's curating coverage, in which the weight of each track is the frequency with which the track appears across all of curator c's playlists. Then we can describe the range of musical style curator c covers using the weighted average of set T_c:

$$\mathbf{cov}_c = \frac{1}{\sum_{t_i \in T_c} count_c(t_i)} \sum_{t_i \in T_c} count_c(t_i) \mathbf{F}_{t_i}, \tag{1}$$

where $count_c(\cdot)$ counts the frequency of appearing across all the playlists of c.

3.2 RQ2: Curation-Based Track Embeddings

A natural question raised in the last section is how to model individual tracks in the first place? One alternative is to use the *audio features* of music tracks. For example, there are 13 audio features provided by Spotify, including *danceability, energy, key, loudness, mode, speechless, acousticness, instrumentalness, liveness, valence, tempo, duration in ms*, and *time signature*. However, these features are not always available due to the cost of analyzing audio signals.

Instead, we seek to characterize tracks by how groups of curators compose their playlists. The intuition is that when users create or follow playlists, they provide implicit linkages between tracks in the same playlist. Hence, a key hypothesis is that tracks in a particular playlist are in high coherence, regardless of their underlying audio-based signature. Inspired by word2vec [17], we propose to learn a vector representation for each track from the curation-based perspective of how tracks cohere with other tracks. We can treat tracks as "words" and find neighboring tracks within a window. In many music services, users can shuffle the playlists, meaning that tracks arrive in a random order. Hence, tracks within a playlist will have high coherence, regardless of their immediate order in the playlist. To simulate this shuffle activity, we randomly reorder the music tracks in each playlist to generate the shuffled track sequences (see Fig. 2(b)). Then we treat each shuffled track sequence as a "sentence" and each music track as a "word". For instance, given a music sequence (similar to a "sentence") $\boldsymbol{m} = (t_1, t_2, ...t_{|\boldsymbol{m}|})$, the log probability $l_{\boldsymbol{m}}$ is calculated as $l_{\boldsymbol{m}} = \frac{1}{|\boldsymbol{m}|} \sum_{0 \leq k < |\boldsymbol{m}|} \sum_{-w \leq j \leq w, j \neq 0} \log P(t_{k+j}|t_k)$, where $P(t_{k+j}|t_k)$ represents the probability that track t_{k+j} is the neighbor of t_k, given that track t_k is listened. Here, w is the window size in observing neighboring tracks. This skip-gram model aims to maximize the log probability across the set of all generated track "sentences". The hidden layer (in Fig. 2(b)) that is learned will be used as the low-dimensional vector representations for tracks. Tracks with larger pairwise-similarity are more likely to be listened to together from the perspective of how curators create their playlists. After generating the embedding representation $\mathbf{E}_t$ for each track t, we can use it as track features for *curation style embedding* in Fig. 2(a) and *coverage* in Eq. 1.

3.3 RQ3: Neural Personalized Curator Ranking

Given these models of curation style and embedding of individual tracks, we now turn to the challenge of connecting users with the right music curators. The main insight of the proposed MCR approach is that when deciding between two curators, a user will consider the style and coverage of each curator.

Users leave only implicit feedback on curators in the form of "following". That is, we can assume that if user u follows curator c, then u is interested in c. However, if u does not follow curator c, we cannot conclude that u is not interested in c because it is also possible that u is unaware of c. Hence, to overcome this implicit feedback challenge, we propose a neural pairwise ranking model inspired by Bayesian Personalized Recommendation (BPR) [21] and Neural Personalized Ranking (NPR) [18]. Following BPR and NPR, the key assumption of this proposed approach is that users prefer the observed positive items (following a curator) to the unobserved items.

The proposed MCR model consists of two symmetric branches as shown in Fig. 3. Suppose that user u has already followed curator c and hasn't followed curator q yet. We denote this relationship as $c >_u q$. Given user u and a pair of curators (c, q), the left branch is designed to estimate a user's overall preference for curator c, while the right branch (the transparent part) aims to estimate a user's overall preference for curator q.

There are 7 inputs to the symmetric structure, denoted as $(u, c, q, \mathbf{cov}_c, \mathbf{cov}_q, \mathbf{e}_c^{style}, \mathbf{e}_q^{style})$. Besides the index of the user u, the curator c and another curator q, we also feed in the coverage $\mathbf{cov}_c$, $\mathbf{cov}_q$ and style $\mathbf{e}_c^{style}$, $\mathbf{e}_q^{style}$ of curator c and q. While constructing the input tuple for model training, we select a curator followed by u and another curator for whom u does not leave feedback. Given the tuple (u, c, q), its ground truth label $y(u, c, q)$ is equal to 1 if $c >_u q$, while $y(u, c, q) = -1$ if $q >_u c$. Then the personalized ranking problem is transformed into a binary classification problem.

Model Details. Since MCR is symmetric, we focus on one of the branches in detail. To fit into the neural structure, the index of u and c are one-hot encoded as $\mathbf{i}_u$ and $\mathbf{i}_c$ directly after input. First, we model the direct preference relationships between users and curators. $\mathbf{i}_u$ and $\mathbf{i}_c$ are connected to the corresponding embedding layers to learn the compact and vectorized representations. The resulting embeddings $\mathbf{e}_u$ and $\mathbf{e}_c$ act as the latent factors of u and c. To simulate the traditional matrix factorization, we calculate the element-wise product $\mathbf{e}_u \circ \mathbf{e}_c$, which captures the interaction between u and c.

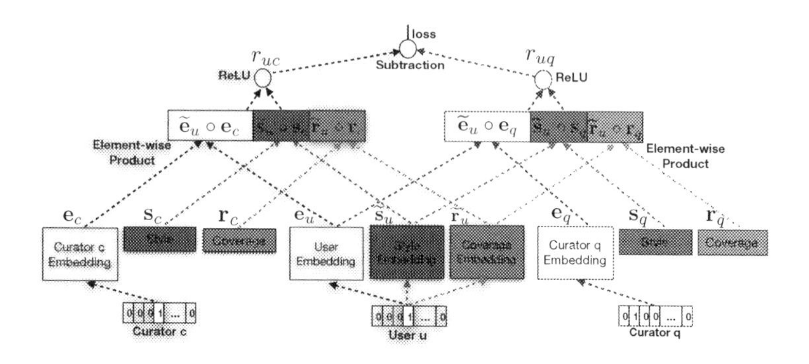

Fig. 3. Neural Music Curator Ranking (MCR) structure.

Then we need to determine how a user's preferences are aligned with each curator's coverage and style. We use ReLU function to activate curator c's auxiliary features. The activated coverage and style of curator c are denoted as $\mathbf{r}_c$ and $\mathbf{s}_c$. Given the one-hot encoding of user u as $\mathbf{i}_u$, u's preferences on coverage and style are learned by the corresponding embedding layers $\widetilde{\mathbf{r}}_u = \mathbf{W}_r \mathbf{i}_u$ and $\widetilde{\mathbf{s}}_u = \mathbf{W}_s \mathbf{i}_u$, where $\mathbf{W}_r$ and $\mathbf{W}_s$ are the embedding weight matrices for coverage and style. The resulting embeddings share the same size with the coverage and style features of curators. We use the element-wise multiplications to capture user u's preferences on curator c. We will concatenate all the element-wise products together to represent the preference of u on c and get $\mathbf{p}_{uc} = [\mathbf{e}_u \circ \mathbf{e}_c \quad \widetilde{\mathbf{r}}_u \circ \mathbf{r}_c \quad \widetilde{\mathbf{s}}_u \circ \mathbf{s}_c]^T$. Then this concatenation is passed to a fully connected layer. The one-dimension output from this layer can be represented as $r_{uc} = f(\mathbf{W}\mathbf{p}_{uc}+b)$. Here $f(\cdot)$ denotes the ReLU function. $\mathbf{W}$ is the weight matrix and b is the bias term for the single perceptron. The output r_{uc} represents the preference score of u on c. Because the model is symmetric, r_{uq}, the preference score of u on curator q, is calculated in the same process with the other branch (the transparent part in Fig. 3). Thus we apply a subtraction layer to estimate the difference between u's preferences on curator c and q: $d(u, c, q) = r_{uc} - r_{uq}$. Since our objective is to make $d(u, c, q)$ share the same sign as $y(u, c, q)$, we train the entire model to minimize the loss: $L = \frac{1}{|\boldsymbol{V}|} \sum_{(u,c,q) \in \boldsymbol{V}} - \ln(\delta(d(u, c, q) \cdot y(u, c, q)))$, in which $\boldsymbol{V}$ is the set of training tuples and δ is the sigmoid function. If the product $d(u, c, q) \cdot y(u, c, q)$ is larger, it means that u's preference for the curator pair (c, q) is more likely to be predicted correctly by the model. We adopt the L2-norm for regularization.

During prediction, we can input the tuple (u, c, c, $\mathbf{cov}_c$, $\mathbf{cov}_c$, $\mathbf{e}_c^{style}$, $\mathbf{e}_c^{style}$) to calculate user u's preference on curator c. We denote the preference scores generated from the symmetric components as r_{uc}^{left} and r_{uc}^{right}. Then the overall preference score can be predicted as $r_{uc} = \frac{1}{2}(r_{uc}^{left} + r_{uc}^{right})$. Thus for prediction, we only need to replace the last subtraction layer with an addition layer. With the preference score of a user over each curator, we will be able to generate a personalized ranked list of curators for this user.

4 Experiments

In this section, we conduct a series of experiments to evaluate the proposed framework by answering these questions: (i) How does the proposed MCR perform in curator recommendation compared with state-of-the-art? (ii) How does each component of MCR contribute to the quality of curator recommendation? and (iii) Are the proposed *style* features able to alleviate the cold start issue?

4.1 Setup

Data. Our experiments are based on a dataset sampled from Spotify. Initially, we create a seed list of playlists by issuing 200 keyword queries representing popular topics on Spotify (e.g., pop, coffee, trip) and then randomly selecting 0.2 million returned playlists. We then identify the creators of these playlists and crawl their followees, arriving at a list of 5 million valid user IDs. We then identify active users who have followed more than 5 music curators. We end up with a dataset with 19,760 users and 6,821 music curators who have curated 5,413,478 music tracks in total. Further, the resulting dataset is extremely sparse, in which only 0.20% of the user-curator pairs have a following relationship. We use 60% of user-curator following data for training, 10% for validation and the remaining 30% for testing.

Metrics and Baselines. We adopt Precision@k (Prec@k), Recall@k (Rec@k) and F-1 score@k (F1@k) as metrics to evaluate the personalized recommendation. Prec@k represents the percentage of correctly predicted curators among the top-k recommendations, and Rec@k represents the fraction of relevant curators which are discovered by the top-k recommendations. F1@k is a weighted combination of Prec@k and Rec@k, that is $F1@k = \frac{2 \cdot Prec@k \cdot Rec@k}{Prec@k + Rec@k}$.

We first consider 3 classic recommendation models and a graph-based model for curator recommendation as baselines:

- *Popular (MP)* recommends the curators with highest popularity.
- *User-based Collaborative Filtering (UCF)* estimates user u's preference on curator c with a weighted aggregation of his/her neighboring users' feedback.
- *Bayesian Personalized Ranking (BPR)* [21] is a basic pairwise ranking model with Matrix Factorization.
- *node2vec* [8] explores diverse neighborhoods on a graph through a biased random walk. We use *node2vec* to find embeddings for users and curators with the user-user (curators) following graph. Based on the cosine similarities between each user and all the curators, we can recommend nearby curators.

We also want to compare MCR with its simplified variants:

- *Neural Personalized Ranking (NPR)* [18] ignores both *style* and *coverage*. Recommendation is purely based on user-curator following relationships.
- *Coverage-based MCR (C-MCR)* relies only on the *coverage* features based on track embeddings and ignores *style*.

- *Audio feature-based MCR (A-MCR).* Let $\mathbf{A}_{t_i}$ denote the 13-dimension audio feature (details in Sect. 3.2) vector of music track t_i. To compare our curation-based track embeddings with the traditional audio-based features, we replace $\mathbf{F}_{t_i}$ in Eq. 1 with track embedding $\mathbf{A}_{t_i}$. This model ignores the curator *style* features.
- *Style-based MCR (S-MCR)* integrates the *style* embedding $\mathbf{e}^{style}$ as a contextual feature for MCR, while ignoring the simpler *coverage* feature.

Parameter Settings. For the curator *style* embedding, we randomly pick 10 tracks from each playlist and iterate among all playlists to get the positive tuples. Then for each curator, we iterate across playlists he/she creates to generate negative tuples. We keep the number of positive tuples and the number of negative tuples to be the same for each curator by random sampling. With this method, we generate a training set and a validation set with the same size. We train the model for 20 epochs, with the loss generally converging within 10 epochs. We use the intermediate embedding layer in the well-trained model to represent a curator's *style*. For the curation-based track embeddings, we set the window and feature vector size following the default settings of Gensim [20].

We adopt Adam optimizer and mini-batch approach of gradient descent, in which the batch size is 3072. We grid search the regularization parameters over $\{1, 10^{-1}, ...10^{-6}\}$ in the validation set. We set the dimension of latent factors for "following" feedback to be 100 for all the models. The dimensions of *coverage* based on track embeddings and curation style embedding $\mathbf{e}^{style}$ are also set to be 100. We keep them to be the same for fair comparison. To train the pairwise ranking model, we select 5 negative curators (unfollow/unobserved) for each user to construct the training tuples; we find that increasing these negative samples leads to longer training time but little improvement on recommendation quality.

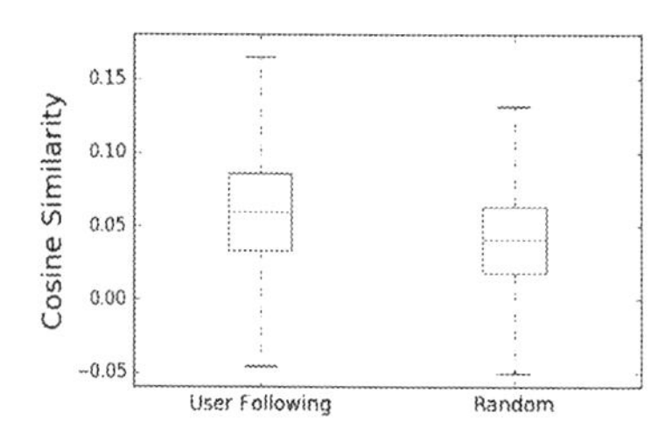

Fig. 4. Similarity of users and curators on curation style.

Exploring Curator Style. Firstly, we want to explore what the *style* embeddings discover about curators. Although users may not create playlists themselves, they can reveal their "curation style" by following different playlists. For a user u, we calculate the cosine similarity of the curation style embedding between u and two different sets of curators: (i) we select the set of K curators u is *following*; and (ii) we *randomly* select K curators. We summarize the results for all the users in Fig. 4. We observe that *users have a higher similarity with curators they follow rather than random curators.* Thus, we see that users do have clear preferences based on style, and there exists clear pattern of users sharing style with curators.

4.2 Evaluating Curator Recommenders

In our first experiment, we compare MCR versus the baselines. We report the precision, recall and F1 over the testing set in Fig. 5. We see that the proposed MCR approach results in the highest precision, recall and F1 for k = 5, 10, 15.

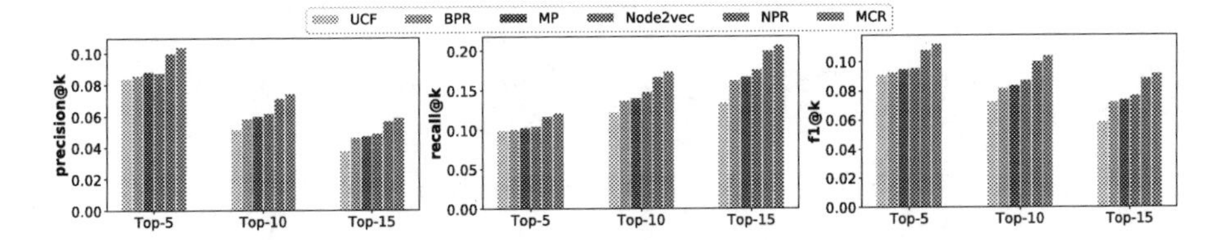

Fig. 5. Comparing models: top-k precision, recall and F1.

Beginning on the left-side of each figure, we see that user-based collaborative filtering (UCF) performs the worst. Since this curation network is so sparse, users have provided insufficient implicit feedback for this recommender to figure out their similarities purely based on the curators they followed. Applying matrix factorization to find latent factors for both users and curators, BPR can outperform UCF by 11.9% on average for the top-k recall and precision, though it still lags the neural models. Relative to traditional collaborative filtering, we see that BPR's relaxed assumption of users' unobserved implicit feedback and pairwise ranking model is more effective for this task. Perhaps surprisingly, the Most Popular (MP) recommender performs slightly better than both UCF and BPR. Currently, Spotify suggests users with popular verified curators and friends on Facebook. Thus, although MP can result in better precision and recall than BPR and UCF, it necessarily ignores the long-tail challenge of uncovering curators who are not widely known.

NPR adds nonlinearity and improves over BPR by 20.3% without any auxiliary information. NPR also outperforms MP by 17.4% on average. The neural architectures are beneficial for this recommendation scenario. In addition, by aggregating the proposed curation style and coverage features, the final MCR model can further improve NPR by 5.7%. From the experiment, we observe the effectiveness of adopting a neural pairwise ranking model for the music curator ranking problem and incorporating curator *style* and *coverage*.

We compare with node2vec, which has demonstrated good performance in link prediction by efficiently exploring diverse neighborhoods on the graph. We find node2vec performs worse than MCR; indeed, it under-performs NPR by more than 10%. Both NPR and the node2vec method rely only on user-user (curators) following information, so the comparison can be fair. These results indicate that graph structure alone may be insufficient in this domain; in contrast, our MCR method can capture both graph structure (via the underlying matrix factorization which can uncover links between user-user and user-item pairs) and user preferences for curation style/coverage.

4.3 Comparison of Different Features

Here, we compare MCR with its simplified variants and the baseline NPR to examine how the curation style and coverage features perform in generating recommendations (see Table 1 *Normal Setting*). In the t-tests between NPR and MCR or its simplified variants on F1, we can get $p < 0.01$, which indicates a statistically significant difference.

Table 1. Comparison of different features of MCR under normal setting and cold start with **F1@K** scores.

Method	Normal setting				Cold start			
	@5	@10	@15	Ave Δ	@5	@10	@15	Ave Δ
NPR	.1082	.1003	.0889	–	.0615	.0651	.0585	–
A-MCR	.1117	.1028	.0907	+2.0%	.0744	.0753	.0626	+14.5%
C-MCR	.1125	.1034	.0913	+3.3%	.0818	.0696	.0709	+17.2%
S-MCR	.1127	.1037	.0913	+3.4%	.0781	.0753	.0692	+20.3%
MCR	.1124	.1043	.0920	+5.7%	.0818	.0753	.0676	+21.4%

Firstly, we observe that curation-based track embeddings (C-MCR) perform better than the audio-based features (A-MCR) in characterizing a curator's *coverage*. The track embeddings are learned from curation choices with skip-gram embedding model to extract the collective wisdom of curators, while the audio-based track features relying on audio analysis of individual tracks. The superior results from our track embeddings indicate the importance of curation decisions.

Additionally, the curation style features $\mathbf{e}^{style}$ (S-MCR) increase F1 score by 3.4% while the *coverage* with track embeddings (C-MCR) increasing it by 3.3%. Combining both the coverage and style, the proposed MCR framework is able to improve NPR by 5.7% in top-k F1 score. This nearly additive improvement suggests that coverage and style are complementary perspectives on curators and so both need to be properly modeled for curator recommendation.

4.4 Cold Start

In practice, it can be challenging to infer preferences of new users with very little feedback. We want to investigate whether the coverage and style contextual features can also help in the cold-start setting. For this experiment, we select users who follow fewer than 8 curators and examine how MCR and its simplified variants work for those users (in Table 1 *Cold-Start*). We find that the curating coverage with track embeddings (*C-MCR*) and the curation style (*S-MCR*) can improve NPR by 17.2% and 20.3% separately. Combining both of these features leads to an average improvement of 21.4% in F1 score. For each of the features and their combination, we see larger improvements compared with the original setting. This suggests that these features are critical in cold start scenarios, which are typical in real-world curation settings.

5 Conclusion

In this work, we tackle the problem of personalized music curator recommendation through a style-aware framework. We introduce the curation style and coverage features to capture a curator's individual approach for curating music. We also propose a curation-based embedding approach to capture inter-track agreement for music that pair well together. Through experiments, we observe that MCR results in the best precision, recall and F1 versus state-of-the-art. Also, the proposed style and coverage features can alleviate the challenges posed by cold start scenarios. In the future, we are interested in extending the models to support other curation platforms like Pinterest and Flipboard.

References

1. Ben-Elazar, S., et al.: Groove radio: a Bayesian hierarchical model for personalized playlist generation. In: Proceedings of the Tenth ACM International Conference on Web Search and Data Mining. ACM (2017)
2. Cao, D., Nie, L., He, X., Wei, X., Zhu, S., Chua, T.S.: Embedding factorization models for jointly recommending items and user generated lists. In: Proceedings of the 40th International ACM SIGIR Conference on Research and Development in Information Retrieval, pp. 585–594. ACM (2017)
3. Celma, O.: Music recommendation. In: Music Recommendation and Discovery, pp. 43–85. Springer, Heidelberg (2010). https://doi.org/10.1007/978-3-642-13287-2_3
4. Cheng, H.T., et al.: Wide & deep learning for recommender systems. In: Proceedings of the 1st Workshop on Deep Learning for Recommender Systems, pp. 7–10. ACM (2016)
5. Cheng, Z., Shen, J., Zhu, L., Kankanhalli, M., Nie, L.: Exploiting music play sequence for music recommendation. In: IJCAI: International Joint Conferences on Artificial Intelligence (2017)
6. Ding, D., Zhang, M., Li, S.Y., Tang, J., Chen, X., Zhou, Z.H.: BayDNN: friend recommendation with Bayesian personalized ranking deep neural network. In: Proceedings of the 2017 ACM on Conference on Information and Knowledge Management, pp. 1479–1488. ACM (2017)
7. Fazel-Zarandi, M., Devlin, H.J., Huang, Y., Contractor, N.: Expert recommendation based on social drivers, social network analysis, and semantic data representation. In: Proceedings of the 2nd International Workshop on Information Heterogeneity and Fusion in Recommender Systems, pp. 41–48. ACM (2011)
8. Grover, A., Leskovec, J.: node2vec: scalable feature learning for networks. In: Proceedings of the 22nd ACM SIGKDD International Conference on Knowledge Discovery and Data Mining, pp. 855–864. ACM (2016)
9. Hannon, J., Bennett, M., Smyth, B.: Recommending Twitter users to follow using content and collaborative filtering approaches. In: Proceedings of the Fourth ACM Conference on Recommender Systems, pp. 199–206. ACM (2010)
10. He, X., Liao, L., Zhang, H., Nie, L., Hu, X., Chua, T.S.: Neural collaborative filtering. In: WWW. International World Wide Web Conferences Steering Committee (2017)
11. Hinton, G.E.: Visualizing high-dimensional data using t-SNE. Vigiliae Christianae **9**(2), 2579–2605 (2008)

12. Jannach, D., Lerche, L., Kamehkhosh, I.: Beyond hitting the hits: generating coherent music playlist continuations with the right tracks. In: Proceedings of the 9th ACM Conference on Recommender Systems, pp. 187–194. ACM (2015)
13. Kitaya, K., Huang, H.H., Kawagoe, K.: Music curator recommendations using linked data. In: INTECH. IEEE (2012)
14. Liebman, E., Saar-Tsechansky, M., Stone, P.: DJ-MC: a reinforcement-learning agent for music playlist recommendation. In: Proceedings of the 2015 International Conference on Autonomous Agents and Multiagent Systems, pp. 591–599. International Foundation for Autonomous Agents and Multiagent Systems (2015)
15. Liu, Y., Chechik, D., Cho, J.: Power of human curation in recommendation system. In: Proceedings of the 25th International Conference Companion on World Wide Web, pp. 79–80. International World Wide Web Conferences Steering Committee (2016)
16. McDonald, D.W., Ackerman, M.S.: Expertise recommender: a flexible recommendation system and architecture. In: Proceedings of the 2000 ACM Conference on Computer Supported Cooperative Work, pp. 231–240. ACM (2000)
17. Mikolov, T., Sutskever, I., Chen, K., Corrado, G.S., Dean, J.: Distributed representations of words and phrases and their compositionality. In: Advances in Neural Information Processing Systems, pp. 3111–3119 (2013)
18. Niu, W., Caverlee, J., Lu, H.: Neural personalized ranking for image recommendation. In: Proceedings of the Eleventh ACM International Conference on Web Search and Data Mining. ACM (2018)
19. Van den Oord, A., Dieleman, S., Schrauwen, B.: Deep content-based music recommendation. In: Advances in Neural Information Processing Systems, pp. 2643–2651 (2013)
20. Řehůřek, R., Sojka, P.: Software framework for topic modelling with large corpora. In: Proceedings of the LREC 2010 Workshop on New Challenges for NLP Frameworks. ELRA (2010)
21. Rendle, S., Freudenthaler, C., Gantner, Z., Schmidt-Thieme, L.: BPR: Bayesian personalized ranking from implicit feedback. In: Proceedings of the Twenty-Fifth Conference on Uncertainty in Artificial Intelligence, pp. 452–461. AUAI Press (2009)
22. Sacheti, A., et al.: Recommending a content curator, US Patent App. 14/839,385, 2 March 2017
23. Schedl, M., Zamani, H., Chen, C.-W., Deldjoo, Y., Elahi, M.: Current challenges and visions in music recommender systems research. Int. J. Multimedia Inf. Retrieval **7**(2), 95–116 (2018). https://doi.org/10.1007/s13735-018-0154-2
24. Wang, D., Deng, S., Zhang, X., Xu, G.: Learning music embedding with metadata for context aware recommendation. In: Proceedings of the 2016 ACM on International Conference on Multimedia Retrieval, pp. 249–253. ACM (2016)
25. Wang, X., Wang, Y.: Improving content-based and hybrid music recommendation using deep learning. In: Proceedings of the 22nd ACM International Conference on Multimedia, pp. 627–636. ACM (2014)
26. Wu, Y., DuBois, C., Zheng, A.X., Ester, M.: Collaborative denoising auto-encoders for top-N recommender systems. In: Proceedings of the Ninth ACM International Conference on Web Search and Data Mining. ACM (2016)
27. Zhao, Z., Cheng, Z., Hong, L., Chi, E.H.: Improving user topic interest profiles by behavior factorization. In: Proceedings of the 24th International Conference on World Wide Web, pp. 1406–1416. International World Wide Web Conferences Steering Committee (2015)

Joint Geographical and Temporal Modeling Based on Matrix Factorization for Point-of-Interest Recommendation

Hossein A. Rahmani[1(✉)], Mohammad Aliannejadi[2], Mitra Baratchi[3], and Fabio Crestani[1]

[1] Università della Svizzera Italiana, Lugano, Switzerland
srahmani@znu.ac.ir, fabio.crestani@usi.ch
[2] Univiersity of Amsterdam, Amsterdam, The Netherlands
m.aliannejadi@uva.nl
[3] Leiden University, Leiden, The Netherlands
m.baratchi@liacs.leidenuniv.nl

Abstract. With the popularity of Location-based Social Networks, Point-of-Interest (POI) recommendation has become an important task, which learns the users' preferences and mobility patterns to recommend POIs. Previous studies show that incorporating contextual information such as geographical and temporal influences is necessary to improve POI recommendation by addressing the data sparsity problem. However, existing methods model the geographical influence based on the physical distance between POIs and users, while ignoring the temporal characteristics of such geographical influences. In this paper, we perform a study on the user mobility patterns where we find out that users' check-ins happen around several centers depending on their current temporal state. Next, we propose a spatio-temporal activity-centers algorithm to model users' behavior more accurately. Finally, we demonstrate the effectiveness of our proposed contextual model by incorporating it into the matrix factorization model under two different settings: (i) static and (ii) temporal. To show the effectiveness of our proposed method, which we refer to as STACP, we conduct experiments on two well-known real-world datasets acquired from Gowalla and Foursquare LBSNs. Experimental results show that the STACP model achieves a statistically significant performance improvement, compared to the state-of-the-art techniques. Also, we demonstrate the effectiveness of capturing geographical and temporal information for modeling users' activity centers and the importance of modeling them jointly.

Keywords: Contextual information · Point-of-Interest recommendation · Recommender system

Work done while Mohammad Aliannejadi was affiliated with Università della Svizzera italiana (USI).

J. M. Jose et al. (Eds.): ECIR 2020, LNCS 12035, pp. 205–219, 2020.
https://doi.org/10.1007/978-3-030-45439-5_14

1 Introduction

With the availability of Location-based Social Networks (LBSNs) such as Yelp and Foursquare users can share their locations, experiences, and content associated with the Point-of-Interests (POIs) via check-ins. Employing the successes in the area of Recommender Systems (RSs), POI recommendation helps to improve the user experiences on LBSNs, suggesting POIs according to users' past check-in history. POI recommendation helps users explore new interesting POIs while helping businesses to increase their revenues by providing context-aware advertisements. As such, POI recommendation has attracted much attention from both research and industry [4,6,22].

One of the most important challenges that limit the accuracy of POI recommendation is the data sparsity problem [1,25,26]. Numerous users are active on LBSNs with millions of POIs already being listed on these platforms. However, in practice, users are able to only visit a very limited number of POIs. Hence, the user-POI matrix used in different Collaborative Filtering (CF) approaches becomes sparse, limiting the attainable recommendation accuracy [3,24]. To address this problem, several studies have incorporated contextual information such as geographical and temporal influences separately into their model [7,21,30,31]. For example, relevant studies have tried to incorporate geographical [9,21,28] and temporal influences [16,20,27] in their proposed model. Moreover, as argued in [9–11], users commonly check in to POIs around several geographical centers. While modeling these centers it is assumed that they are static and do not change according to temporal information. This assumption may not be correct and taking into account both geographical and temporal influences might help to model the users' behavior with a higher accuracy. For instance, if we consider working time and leisure time, this suggests that users tend to explore POIs around their activity centers in leisure time, while they prefer to visit the same locations more often while they are at work. For example, a person would go to the same restaurant every day to have lunch while working during the weekdays. However, the same user might decide to visit a more diverse set of POIs and visit new places while on holidays (i.e., leisure time). Therefore, the users' check-in behavior and activity centers are dependent on their temporal states (e.g., working time vs. leisure time).

To elaborate more, in Fig. 1, we have depicted a randomly selected user's check-ins from the Gowalla dataset [22] during working and leisure time. As seen, this user follows a temporal center-based check-in pattern; that is, the activity centers are different at different temporal states. Also, while we compare Fig. 1a with Figs. 1b and c, we see that the activity centers are different from each temporal state, compared to all the check-ins. Based on these observations, we conclude that joint modeling of geographical and temporal information is an effective approach for defining users' activity centers. In this paper, we take a step for joint modeling of geographical and temporal information. Our contributions can be summarized as follows:

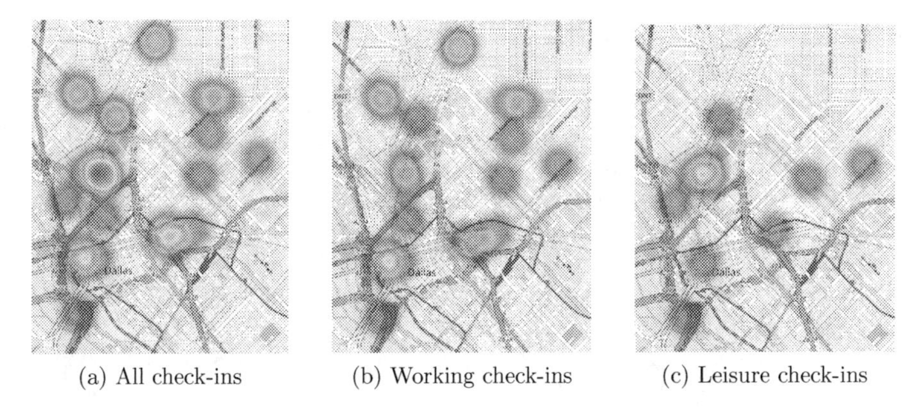

(a) All check-ins (b) Working check-ins (c) Leisure check-ins

Fig. 1. A typical user's spatio-temporal activity centers from the Gowalla dataset. As we see, (a) shows all check-ins of the user whereas (b) and (c) show the check-ins for working and leisure time are focused in different centers. (best viewed in color) (Color figure online)

- We propose a novel contextual model that jointly considers both geographical and temporal information.
- We propose a spatio-temporal activity-centers model that consider users' center-based behavior in different temporal states.
- We propose static and temporal MF models to study the users' preference and behavior both in static and temporal manners. In the static MF, we train the model on the whole user-POI matrix, whereas in the temporal MF, we train the model using different user-POI matrices for every time slot.
- We address the data sparsity problem by incorporating the proposed contextual model into the traditional MF model and propose a novel MF framework.

We conduct several experiments on two well-known real-world datasets, namely, Gowalla and Foursquare, demonstrating the improvement of the proposed method in the accuracy of POI recommendation compared to a number state-of-the-art approaches. Our experiments show that joint modeling of the geographical and temporal influence improves the performance of POI recommendation. Finally, to enable reproducibility of the results, we have made our codes open source.[1]

2 Related Work

POI recommendation plays an essential role in improving LBSNs user experience. Much work has been carried out in this area based on the core idea behind recommendation systems, assuming that *users with similar behavioral histories tend to act similarly* [28]. Collaborative Filtering (CF-based) recommendation approaches aim to base recommendations on the similarity between users

[1] https://github.com/rahmanidashti/STACP.

and items [14,16]. POI recommendation considers a large number of available POIs (items) while a single user can only visit a few of them. Hence, CF-based approaches applied to POI recommendation often suffer from the data sparsity problem. This leads to poor performance in POI recommendation. Many studies have tried to address the data sparsity problem of CF approaches incorporating additional information into the model [5,21,22,28]. As the users' check-in behavior follows a spatio-temporal mobility pattern, much work has incorporated this critical information. Considering users' movement trajectories between POIs, many of the previous studies have shown that geographical influence is one of the most important factors in POI recommendation [22,24].

More specifically, Ye et al. [28] argued that a user's check-in behavior is affected by the geographical influence of POIs, following the power-law distribution and proposed a unified POI recommender system incorporating spatial and social influences to address the data sparsity problem. Ference et al. [14] took into consideration several factors such as user preference, geographical proximity, and social influences for out-of-town POI recommendation. This work, however, did not take into account users' temporal information and in-town users' behavior. Cheng et al. [9,10] modeled users' check-ins via center-based Gaussian distribution to capture users' movement patterns based on the assumption that users' movements consist of several centers. Li et al. [21], in another work, modeled the POI recommendation task as a pairwise ranking loss, where they exploited the geographical information using an extra factor matrix. Zhang et al. [31] proposed a method that considered the geographical influence on each user separately. To this end, they proposed a model based on kernel density estimation of the distance distributions between POI check-ins per user. Aliannejadi et al. [2] proposed a ranking model and predicted the appropriateness of a POI given a user's context into the ranking process. Yuan et al. [29] addressed the data sparsity problem based on the idea that users tend to rank higher those POIs that are geographically closer to their visited POIs. Guo et al. [17] proposed a location neighborhood-aware weighted matrix factorization model to exploit the location perspective that incorporates the geographical relationships among POIs. More recently, Aliannejadi et al. [6] proposed a two-phase collaborative ranking algorithm for POI recommendation that takes into account the geographical influence of POIs in the same neighborhood.

Another line of research studies the temporal influence on users' preferences [13,30,32]. Temporal information has been shown to improve POI recommendation accuracy and alleviate the problem of data sparsity [22]. Griesner et al. [16] proposed an approach to integrate temporal influences into matrix factorization. Gao et al. [15] computed the similarity between users by dividing users' check-ins into different hourly time slots and finding the same POIs at the same time slots in their check-in history to train a user-based CF model. Yao et al. [27] matched the temporal regularity of users with the popularity of POIs to improve a factorization-based algorithm. Le et al. [20] proposed a time-aware personalized model adopting a fourth-order tensor factorization-based ranking, which enables to capture short-term and long-term preferences. Yuan et al. [30] preserved the similarity of personal preference in consecutive time slots by considering different latent variables at each time slot per user. Zhao et al. [33] proposed a latent ranking method to model the temporal interactions among users and POIs explicitly. In particular, the proposed model builds upon a ranking-based pairwise tensor factorization framework.

These previous approaches mainly explored the geographical and temporal information separately. Differently from these studies, our work addresses the data sparsity problem by jointly modeling the geographical and temporal contextual information. Moreover, the previous research modeled users' center-based behavior based on geographical influence. In contrast, we consider the formation of spatio-temporal activity centers for each user. Therefore, we model the users' center-based behavior based on different temporal states.

3 Proposed Approach

In this section, we propose a **S**patio-**T**emporal **A**ctivity **C**enter **P**OI recommendation model called STACP, which models users' preference and users' context together. In the users' preference model, we design two preference functions for each user to consider both static and temporal users' preferences in the model. Moreover, in the users' context model, we incorporate the influence of geographical and temporal information jointly. In what follows, we first describe an overview of our STACP model and further explain how each part is implemented and which challenges are addressed at each part.

Formally, let $\mathcal{U} = \{u_1, u_2, u_3, ..., u_m\}$ be the set of users and $\mathcal{L} = \{l_1, l_2, l_3, ..., l_n\}$ be the set of POIs. Further, let m and n be the number of users and POIs, respectively. Then, the users visit-frequency can be encoded in $R_{m \times n}$, where entries $r_{u,l} \in R$ can represent the previous POI check-ins of user $u \in \mathcal{U}$ to POI $l \in \mathcal{L}$. Also, $\mathcal{L}_u$ shows all POIs checked-in by user u. To address the data sparsity problem and explore the contextual influence we need to fuse the users' context with the users' preference model in a fusion framework. We fuse users' static and temporal preferences on a POI and the score of whether a user will visit that place based on our contextual influence model. STACP is proposed to estimate the recommendation score that a user u visits a POI l as follows:

$$STACP_{u,l} = U_u^T L_l \times P(u, l|C_{u,t}) \times \hat{R_{u,l}} \quad (1)$$

where $U_u^T L_l$ and $\hat{R_{u,l}}$ respectively denote static and temporal users' preference model, and $P(u, l|C_{u,t})$ denotes the users' context model.

In the following, we first introduce the user context model, where we show users' behavior in a joint model of geographical and temporal influences. Moreover, we propose our temporal center allocation method. Finally, we describe our users' static and temporal preference model.

Spatio-Temporal Activity Centers. As shown in Fig. 1, users' behaviors are center-based and these centers are different based on the periodicity of temporal information (see Fig. 1b and c). This phenomenon points to the shortcoming of the previous geographical and temporal models that considered geographical and temporal influences separately. As shown previously, the second characteristic of users' behavior is that users tend to visit POIs that are near their current centers. We apply these two characteristics jointly to model users' check-in behavior and propose the spatio-temporal activity-centers model. That is, the score of a user u, visiting a POI l, given the temporal multi-center set $C_{u,t}$ of user u in time t and temporal state T, is defined as follows:

$$P(u, l|C_{u,t}; T) = \sum_{c_{u,t}}^{|C_{u,t}|} \frac{1}{dist(l, c_{u,t})} \frac{freq_{c_{u,t}}}{\sum_{i \in C_{u,t}} freq_i} \tag{2}$$

where l denotes a POI and $C_{u,t}$ is the set of centers for the user u in time t, given the temporal state T. For each center, calculating (2) consists of the multiplication of two terms. The first term determines the score of the POI l belonging to the center $c_{u,t}$, which is related to the distance between the POI l and the center $c_{u,t}$. The second term denotes the effect of check-in frequency $freq_{c_{u,t}}$, on the center $c_{u,t}$.

Further, we define the multi-center activity function $P(u, l|C_{u,t})$ as a linear interpolation under two temporal states, as follows:

$$P(u, l|C_{u,t}) = \lambda \times P(u, l|C_{u,t}; WT) + (1 - \lambda) \times P(u, l|C_{u,t}; LT) \tag{3}$$

where we consider it for working time $P(u, l|C_{u,t}; WT)$ and leisure time $P(u, l|C_{u,t}; LT)$ where λ shows the impact of each temporal state. The model can be generalized to define other temporal states. For example, we could apply it for weekday/weekend, monthly, or daily patterns.

Activity Center Allocation. As argued earlier, the users' activities follow a center-based pattern. Furthermore, these centers are different depending on the temporal state. To model the users' behavior in a spatio-temporal manner, we propose a temporal multi-center clustering algorithm among each user's check-ins based on the Pareto principle [18], as the most visited POIs account for a few users. First, for each user u and temporal state t, we rank all POIs L_u according to the check-in frequency. Next, we select the most visited POI and combine all other visited POIs within d kilometers from the selected POI, to create a region. Let N_u be the user u's total check-in numbers, r be the current region and $N_{r,u}$ be the total check-in number of current region of user u. To decide if a center should be added to the user's profile, we consider a threshold of α. A new center is considered if $\frac{N_{r,u}}{N_u} > \alpha$. We repeat this procedure until we cover all of the user u's checked-in POIs.

Static and Temporal Users' Preferences. To model the user's preference based on check-in data, we apply Matrix Factorization (MF) in two ways: a static model of user's preference (SMP) and a temporal model of user's preference (TMP). In SMP, we consider the traditional matrix factorization method to model the static behavior of users. The goal of MF is to find two low-rank matrices $U \in \mathbb{R}^{K \times |\mathcal{U}|}$ and $L \in \mathbb{R}^{K \times |\mathcal{L}|}$ based on the frequency matrix R such that $R \approx U^T L$. The predicted recommendation score of a user u, like a POI l, is determined by:

$$P_{u,l} \propto U_u^T L_l \tag{4}$$

via solving the following optimization problem which places Beta distributions as priors on the latent matrices U and L, while defining a Poisson distribution on the frequency:

$$\begin{aligned} min_{\{U,L|R\}} = &\sum_{i=1}^{|\mathcal{U}|}\sum_{k=1}^{K}((\sigma_k - 1)\ln(U_{ik}/\rho_k) - U_{ik}/\rho_k) \\ &+ \sum_{j=1}^{|\mathcal{L}|}\sum_{k=1}^{K}((\sigma_k - 1)\ln(L_{jk}/\rho_k) - L_{jk}/\rho_k) \\ &+ \sum_{i=1}^{|\mathcal{U}|}\sum_{j=1}^{|\mathcal{L}|}((R_{ij}\ln(U^T L)_{ij} - (U^T L)_{ij}) + c \end{aligned} \tag{5}$$

where $\sigma = \{\sigma_1, ..., \sigma_K\}$ and $\rho = \{\rho_1, ..., \rho_K\}$ are parameters for Beta distributions, and c is a constant term. In TMP, to model the temporal behavior of users, inspired by [15], we divide the original user-POI frequency matrix R into t sub-matrices according to the different temporal states T. Then each sub-matrix only containing check-in actions that happened at the corresponding temporal state. For example, we can consider $t = 2$ for working time and leisure time in our case. Then, we apply MF on each R_t to compute user u's preference on POI l at time t. Finally, we sum them into $\hat{R}$, representing the user check-in preferences of each POI. It should be mentioned that a more advanced method of automated periodic pattern extraction from spatio-temporal data as proposed in [8] can also be used for a more data-informed decision to be made for the parameter (t).

4 Experiments

In this section, several experiments are conducted to compare the performance of STACP with the other state-of-the-art POI recommendation methods.

4.1 Experimental Setup

Datasets. We use two real-world check-in datasets from Gowalla and Foursquare provided by [22][2]. The Gowalla dataset consists of 620,683 number of world-wide check-ins made by 5,628 number of users on 31,803 POIs with 99.78% sparsity in the period of Feb. 2009 to Oct. 2010. The Foursquare dataset, on the other hand, includes 512,523 check-ins made by 7,642 users on 28,483 POIs

[2] http://spatialkeyword.sce.ntu.edu.sg/eval-vldb17/.

with 99.87% sparsity in the United States from Apr. 2012 to Sep. 2013. Every check-in contains a user-id, POI-id, time, and geographical coordinates.

Evaluation Metrics. To evaluate the performance of the recommendation methods, we used three evaluation metrics: Precision@N, Recall@N, and nDCG@N with $N \in \{10, 20\}$. We partition each dataset into training data, validation data, and test data. For each user, we use the earliest 70% check-ins as training data, the most recent 20% check-ins as test data, and the remaining 10% as validation data. We determine the statistically significant differences in the results using the two-tailed paired t-test at a 95% confidence interval ($p < 0.05$).

Table 1. Performance comparison with baselines in terms of Precision@N, Recall@N, and nDCG@N for $N \in \{10, 20\}$ on Gowalla and Foursquare. The superscripts † and ‡ denote significant improvements compared to baselines and model variations, respectively ($p < 0.05$).

		Precision		Recall		nDCG	
		@10	@20	@10	@20	@10	@20
Gowalla	TopPopular	0.0192	0.0146	0.0176	0.0270	0.0088	0.0079
	PFM	0.0181	0.0143	0.0161	0.0252	0.0077	0.0068
	PFMMGM	0.0240	0.0207	0.0258	0.0442	0.0140	0.0144
	LRT	0.0249	0.0182	0.0220	0.0321	0.0105	0.0093
	PFMPD	0.0217	0.0184	0.0223	0.0373	0.0099	0.0101
	LMFT	0.0315	0.0269	0.0303	0.0515	0.0157	0.0150
	iGLSR	0.0297	0.0242	0.0283	0.0441	0.0153	0.0145
	MLP	0.0243	0.0215	0.0237	0.0396	0.0982	0.0127
	Rank-GeoFM	0.0352	0.0297	0.0379	0.0602	0.0187	0.0179
	L-WMF	0.0341	0.0296	0.0351	0.0582	0.0183	0.0178
	STACP-NoCTX	0.0219	0.0167	0.1920	0.0293	0.0092	0.0081
	STACP-NoTC	0.0282	0.0236	0.0281	0.0457	0.0147	0.0151
	STACP	**0.0383**†‡	**0.0318**‡	**0.0404**†‡	**0.0651**†‡	**0.0212**†‡	**0.0211**†‡
Foursquare	TopPopular	0.0200	0.0155	0.0272	0.0429	0.0114	0.0121
	PFM	0.0213	0.0154	0.0290	0.0424	0.0125	0.0129
	PFMMGM	0.0170	0.0150	0.0283	0.0505	0.0109	0.0126
	LRT	0.0199	0.0155	0.0265	0.0425	0.0117	0.0124
	PFMPD	0.0214	0.0155	0.0290	0.0426	0.0124	0.0128
	LMFT	0.0241	0.0194	0.0359	0.0568	0.0150	0.0161
	MLP	0.0248	0.0204	0.0309	0.0373	0.0135	0.0152
	Rank-GeoFM	0.0263	0.0241	0.0399	0.0625	0.0183	0.0197
	L-WMF	0.0248	0.0197	0.0387	0.0591	0.0162	0.0174
	STACP-NoCTX	0.0207	0.0184	0.0196	0.0285	0.0094	0.0085
	STACP-NoTC	0.0242	0.0217	0.0331	0.0552	0.0142	0.0157
	STACP	**0.0312**†‡	**0.0285**†‡	**0.0453**†‡	**0.0671**†‡	**0.0203**‡	**0.0227**†‡

Comparison Methods. We compared the proposed STACP model with the state-of-the-art POI recommendation approaches that consider geographical or temporal influences in the recommendation process. The details of the compared methods are listed below:

- **TopPopular** [12]**:** A simple and non-personalized method that recommends the most popular POIs to users. Popularity is measured by the number of check-ins.
- **PFM** [23]**:** A MF method, which can model the frequency data directly. PFM places Beta distributions as priors on the latent matrices U and V, while defining a Poisson distribution on the frequency.
- **PFMMGM** [9]**:** A method based on the observation that a user's check-ins follow a Gaussian distribution that combines geographical and social influence with MF.
- **LRT** [15]**:** A method that incorporates temporal information in a latent ranking model and learns the user's preferences based on temporal influence.
- **PFMPD:** A geographical method using the Power-law distribution [28] that models people's tendency to visit nearby POIs. We integrate this geographical model with the Probabilistic Factor Model (PFM).
- **LMFT** [26]**:** A method that applies temporal information on the user's recent activities and multiple visits to a POI.
- **iGLSR**[3] [31]: A method that personalizes social and geographical influences on POI recommendation using a Kernel Density Estimation (KDE) approach.
- **Rank-GeoFM** [21]**:** A ranking-based MF model that includes the geographical influence of neighboring POIs while learning user preference rankings for POIs.
- **MLP** [19]**:** A component of the NeuMF framework that models the user-POI interaction using the concatenation of latent factors via Multi-layers Neural Network.
- **L-WMF** [17]**:** A location neighborhood-aware weighted probabilistic matrix factorization model. L-WMF incorporates the geographical relationships among POIs as regularization to exploit the geographical characteristics from a location perspective.
- **STMCP-NoCTX:** A variation of our model which excludes the contextual model. We include this model as a baseline to demonstrate the effectiveness of our contextual model.
- **STMCP-NoTC:** A variation of our model in which we remove the temporal states and consider geographical centers without temporal differences. We include this model to show the effectiveness of temporal centers in our model.

4.2 Results

Performance Evaluation Against Compared Methods. Table 1 shows the results of experiments based on the Gowalla and Foursquare datasets. As seen,

[3] We evaluate iGLSR only on Gowalla as we do not have access to the social data of the Foursquare dataset.

STACP obtains the best performance compared to the other POI recommendation methods in terms of all evaluation metrics on both of the datasets. Dacrema et al. in [12] show that even some state-of-the-art deep-learning-based methods are not able to outperform a non-personalized method such as TopPopular. This is the reason why we have also selected this method as a baseline to compare with STACP. It seems that in comparison to two non-personalized baseline methods, TopPopular and PFM, our method achieves significantly better performance. Comparing with other geographical-based methods, PFMPD, PFMMGM, iGLSR, and L-WMF, it is seen that STACP followed by Rank-GeoFM perform best. One reason for the performance of Rank-GeoFM is that it considers the geographical neighborhoods as a major element in the factorization method. Also, Rank-GeoFM takes a ranking approach to modeling the interactions between users and POIs. This means that instead of considering a point-wise loss function, it applies a pairwise loss function in geographical factorization.

Results show that STACP beats all geographical-based methods in terms of all metrics for all different values of N in both datasets. This is expected as the previous models only consider the basic idea of the geographical influence that users tend to visit nearby POIs. Also, our proposed model outperforms the LRT and LMFT that modeled temporal information to improve the accuracy of POI recommendation. The reason is that these methods do not consider the geographical information. Compared to the neural baseline, MLP, the improvements of STACP in terms of Recall@20 and nDCG@20 on Gowalla dataset are 65%

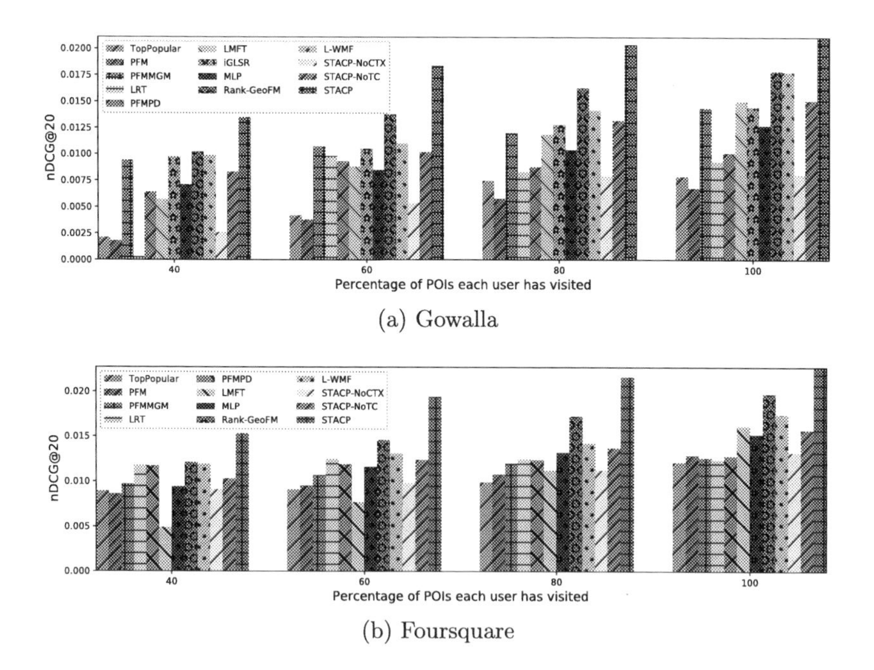

(a) Gowalla

(b) Foursquare

Fig. 2. Effect on nDCG@20 by varying the percentage of POIs that each user has visited for (a) Gowalla and (b) Foursquare.

and 66%, respectively. This shows the effectiveness of our users' preference and users' context models, which considers both geographical and temporal information jointly to model users' activity centers.

Effect of Activity Centers. In this experiment, we compare the performance of STACP with its variation where we only consider geographical information in allocating activity centers (i.e., STACP-NoTC). Therefore, we remove temporal states t from Eq. (2). The goal is to demonstrate the effect of the spatio-temporal activity centers on the performance of STACP. As seen in Table 1, STACP exhibits a significant improvement over STACP-NoTC in terms of all evaluation metrics for both datasets. We see that STACP improves STACP-NoTC by 42% in terms of Recall@20. This indicates that users follow a spatio-temporally centered behavior. This objectively validates our analysis in Sect. 1, in which users' centers are different based on the different temporal states.

Effect of Contextual Model. Next, we study the effect of the contextual model. To this end, we compare the performance of STACP with its variation where no contextual information is used while training the model (i.e., STACP-NoCTX). In other words, in this experiment Eq. (2) is excluded from Eq. (1). As seen in Table 1, a statistically significant improvement of STACP over STACP-NoCTX is observed in terms of all evaluation metrics on both datasets. This observation suggests that using contextual information enables STACP to model the users' behavior more accurately. Moreover, it indicates that by incorporating the contextual information, we can address the data sparsity problem.

Effect of Number of Visited POIs. In this experiment, our goal is to study the effect of data size on the performance of our model. As such, we train STACP, as well as all the baseline methods with different data sizes. To do so, for each user, we only consider a certain percentage of visited POIs in the training set randomly, ranging from 40% to 100%. We see in Figs. 2a and b the performance of STACP and all baseline models in terms of nDCG@20 for different training data sizes. The results show that STACP is more effective in comparison with the baselines as the size of the training data varies, indicating that it addresses the data sparsity problem more effectively. As we see in Fig. 2a, when we change the data size from 100% to 40% on Gowalla, the performance of STACP decreases by about 35%, while for the competitor baseline method Rank-GeoFM the value of decrease is 45%. This shows that STACP is more robust when we do not have access to enough data from users. More interesting, the performance of LMFT, a temporal-information-based competitor baseline, decreases by 65% when the size of data changes to 40%. This indicates the unsuitability of the methods that only consider the temporal information. Also, it is worth noting that we observe a more robust behavior of STACP compared to the STACP-NoCN. Thus, the proposed context model enables STACP to deal with noise and data sparsity more effectively. This is clearer when STACP outperforms the best competitor baseline (i.e., Rank-GeoFM) with a larger margin, 33%, in terms of nDCG@20 on Gowalla.

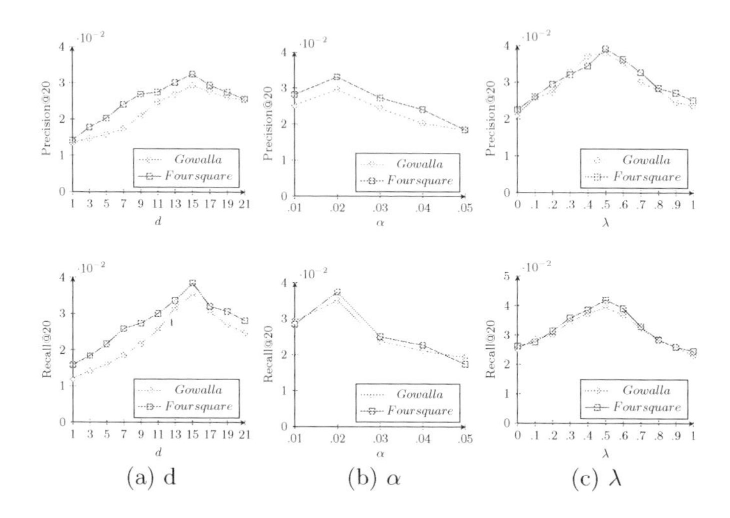

Fig. 3. Effect of different model parameters on the performance of STACP

Effect of Model Parameters. Figure 3 shows the performance of STACP for different values of d, α and λ. We report in Fig. 3a the effect of different values of d on the performance of STACP in terms of Precision@20 and Recall@20 metrics, respectively. It can be seen that the optimal value of d for both datasets is 15. These results show that users tend to visit nearby POIs to their centers, which are formed in regions. Figure 3b, on the other hand, shows the effect of different α values on the performance of STACP. We can see that the optimal value is achieved at $\alpha = 0.02$ for both datasets. More importantly, as seen in Fig. 3c, the optimal value for λ that shows the impact of different temporal states is 0.5. This confirms our assumption and shows that users follow a spatio-temporal activity centered behavior. In fact, when we set this parameter to 1 (i.e., just working time) or 0 (i.e., just leisure time), the performance decreases.

5 Discussion and Conclusion

In this paper, we study the problem of POI recommendation. We have investigated in detail the characteristics of the user mobility behavior on a large-scale check-in dataset. Based on the extracted properties, we propose a novel spatio-temporal activity-centers model to jointly model the geographical and temporal influence of the user's check-in behavior. We then consider the user's temporal information and the user's preferences on POIs. Finally, we propose the STACP model as a uniform framework for combining these three components to recommend POIs. Experimental results on the datasets show that our model significantly outperforms the state-of-the-art models. In our future work, we may consider more information such as the user's comments and social relations to improve the performance.

Acknowledgment. This work was partially supported by a Swiss State Secretariat for Education, Research and Innovation (SERI) mobility grant between Switzerland and Iran.

References

1. Adomavicius, G., Tuzhilin, A.: Toward the next generation of recommender systems: a survey of the state-of-the-art and possible extensions. IEEE Trans. Knowl. Data Eng. **6**, 734–749 (2005)
2. Aliannejadi, M., Crestani, F.: Venue appropriateness prediction for personalized context-aware venue suggestion. In: Proceedings of the 40th International ACM SIGIR Conference on Research and Development in Information Retrieval, pp. 1177–1180. ACM (2017)
3. Aliannejadi, M., Crestani, F.: Personalized context-aware point of interest recommendation. ACM Trans. Inf. Syst. (TOIS) **36**(4), 45 (2018)
4. Aliannejadi, M., Mele, I., Crestani, F.: A cross-platform collection for contextual suggestion. In: Proceedings of the 40th International ACM SIGIR Conference on Research and Development in Information Retrieval, pp. 1269–1272. ACM (2017)
5. Aliannejadi, M., Rafailidis, D., Crestani, F.: Personalized keyword boosting for venue suggestion based on multiple LBSNs. In: Jose, J.M., et al. (eds.) ECIR 2017. LNCS, vol. 10193, pp. 291–303. Springer, Cham (2017). https://doi.org/10.1007/978-3-319-56608-5_23
6. Aliannejadi, M., Rafailidis, D., Crestani, F.: A joint two-phase time-sensitive regularized collaborative ranking model for point of interest recommendation. IEEE Trans. Knowl. Data Eng. (2019)
7. Baral, R., Wang, D., Li, T., Chen, S.C.: GeoTeCS: exploiting geographical, temporal, categorical and social aspects for personalized POI recommendation. In: 2016 IEEE 17th International Conference on Information Reuse and Integration (IRI), pp. 94–101. IEEE (2016)
8. Baratchi, M., Meratnia, N., Havinga, P.J.M.: Recognition of periodic behavioral patterns from streaming mobility data. In: Stojmenovic, I., Cheng, Z., Guo, S. (eds.) MindCare 2014. LNICST, vol. 131, pp. 102–115. Springer, Cham (2014). https://doi.org/10.1007/978-3-319-11569-6_9
9. Cheng, C., Yang, H., King, I., Lyu, M.R.: Fused matrix factorization with geographical and social influence in location-based social networks. In: Twenty-Sixth AAAI Conference on Artificial Intelligence (2012)
10. Cheng, C., Yang, H., King, I., Lyu, M.R.: A unified point-of-interest recommendation framework in location-based social networks. ACM Trans. Intell. Syst. Technol. (TIST) **8**(1), 10 (2016)
11. Cho, E., Myers, S.A., Leskovec, J.: Friendship and mobility: user movement in location-based social networks. In: Proceedings of the 17th ACM SIGKDD International Conference on Knowledge Discovery and Data Mining, pp. 1082–1090. ACM (2011)
12. Dacrema, M.F., Cremonesi, P., Jannach, D.: Are we really making much progress? A worrying analysis of recent neural recommendation approaches. In: Proceedings of the 13th ACM Conference on Recommender Systems, pp. 101–109. ACM (2019)
13. Ding, Y., Li, X.: Time weight collaborative filtering. In: Proceedings of the 14th ACM International Conference on Information and Knowledge Management, pp. 485–492. ACM (2005)

14. Ference, G., Ye, M., Lee, W.C.: Location recommendation for out-of-town users in location-based social networks. In: Proceedings of the 22nd ACM International Conference on Information & Knowledge Management, pp. 721–726. ACM (2013)
15. Gao, H., Tang, J., Hu, X., Liu, H.: Exploring temporal effects for location recommendation on location-based social networks. In: Proceedings of the 7th ACM Conference on Recommender Systems, pp. 93–100. ACM (2013)
16. Griesner, J.B., Abdessalem, T., Naacke, H.: POI recommendation: towards fused matrix factorization with geographical and temporal influences. In: Proceedings of the 9th ACM Conference on Recommender Systems, pp. 301–304. ACM (2015)
17. Guo, L., Wen, Y., Liu, F.: Location perspective-based neighborhood-aware POI recommendation in location-based social networks. Soft Comput. **23**(22), 11935–11945 (2019). https://doi.org/10.1007/s00500-018-03748-9
18. Hafner, A.W.: Pareto's principle: the 80-20 rule (2001). Accessed 26 December 2001
19. He, X., Liao, L., Zhang, H., Nie, L., Hu, X., Chua, T.S.: Neural collaborative filtering. In: Proceedings of the 26th International Conference on World Wide Web, pp. 173–182. International World Wide Web Conferences Steering Committee (2017)
20. Li, X., Jiang, M., Hong, H., Liao, L.: A time-aware personalized point-of-interest recommendation via high-order tensor factorization. ACM Trans. Inf. Syst. (TOIS) **35**(4), 31 (2017)
21. Li, X., Cong, G., Li, X.L., Pham, T.A.N., Krishnaswamy, S.: Rank-GeoFM: a ranking based geographical factorization method for point of interest recommendation. In: Proceedings of the 38th International ACM SIGIR Conference on Research and Development in Information Retrieval, pp. 433–442. ACM (2015)
22. Liu, Y., Pham, T.A.N., Cong, G., Yuan, Q.: An experimental evaluation of point-of-interest recommendation in location-based social networks. Proc. VLDB Endow. **10**(10), 1010–1021 (2017)
23. Ma, H., Liu, C., King, I., Lyu, M.R.: Probabilistic factor models for web site recommendation. In: Proceedings of the 34th International ACM SIGIR Conference on Research and Development in Information Retrieval, pp. 265–274. ACM (2011)
24. Rahmani, H.A., Aliannejadi, M., Ahmadian, S., Baratchi, M., Afsharchi, M., Crestani, F.: LGLMF: local geographical based logistic matrix factorization model for POI recommendation. In: Wang, F.L., et al. (eds.) AIRS 2019. LNCS, vol. 12004, pp. 66–78. Springer, Cham (2020). https://doi.org/10.1007/978-3-030-42835-8_7
25. Rahmani, H.A., Aliannejadi, M., Mirzaei Zadeh, R., Baratchi, M., Afsharchi, M., Crestani, F.: Category-aware location embedding for point-of-interest recommendation. In: Proceedings of the 2019 ACM SIGIR International Conference on Theory of Information Retrieval, pp. 173–176. ACM (2019)
26. Stepan, T., Morawski, J.M., Dick, S., Miller, J.: Incorporating spatial, temporal, and social context in recommendations for location-based social networks. IEEE Trans. Comput. Soc. Syst. **3**(4), 164–175 (2016)
27. Yao, Z., Fu, Y., Liu, B., Liu, Y., Xiong, H.: POI recommendation: a temporal matching between POI popularity and user regularity. In: 2016 IEEE 16th International Conference on Data Mining (ICDM), pp. 549–558. IEEE (2016)
28. Ye, M., Yin, P., Lee, W.C., Lee, D.L.: Exploiting geographical influence for collaborative point-of-interest recommendation. In: Proceedings of the 34th International ACM SIGIR Conference on Research and Development in Information Retrieval, pp. 325–334. ACM (2011)

29. Yuan, F., Jose, J.M., Guo, G., Chen, L., Yu, H., Alkhawaldeh, R.S.: Joint geo-spatial preference and pairwise ranking for point-of-interest recommendation. In: 2016 IEEE 28th International Conference on Tools with Artificial Intelligence (ICTAI), pp. 46–53. IEEE (2016)
30. Yuan, Q., Cong, G., Ma, Z., Sun, A., Thalmann, N.M.: Time-aware point-of-interest recommendation. In: Proceedings of the 36th International ACM SIGIR Conference on Research and Development in Information Retrieval, pp. 363–372. ACM (2013)
31. Zhang, J.D., Chow, C.Y.: iGSLR: personalized geo-social location recommendation: a kernel density estimation approach. In: Proceedings of the 21st ACM SIGSPATIAL International Conference on Advances in Geographic Information Systems, pp. 334–343. ACM (2013)
32. Zhao, S., Zhao, T., King, I., Lyu, M.R.: Geo-Teaser: geo-temporal sequential embedding rank for point-of-interest recommendation. In: Proceedings of the 26th International Conference on World Wide Web Companion, pp. 153–162. International World Wide Web Conferences Steering Committee (2017)
33. Zhao, S., Zhao, T., Yang, H., Lyu, M.R., King, I.: STELLAR: spatial-temporal latent ranking for successive point-of-interest recommendation. In: Thirtieth AAAI Conference on Artificial Intelligence (2016)

Semantic Modelling of Citation Contexts for Context-Aware Citation Recommendation

Tarek Saier(✉) and Michael Färber

Institute AIFB, Karlsruhe Institute of Technology (KIT), Karlsruhe, Germany
{tarek.saier,michael.faerber}@kit.edu

Abstract. New research is being published at a rate, at which it is infeasible for many scholars to read and assess everything possibly relevant to their work. In pursuit of a remedy, efforts towards automated processing of publications, like semantic modelling of papers to facilitate their digital handling, and the development of information filtering systems, are an active area of research. In this paper, we investigate the benefits of semantically modelling citation contexts for the purpose of citation recommendation. For this, we develop semantic models of citation contexts based on entities and claim structures. To assess the effectiveness and conceptual soundness of our models, we perform a large offline evaluation on several data sets and furthermore conduct a user study. Our findings show that the models can outperform a non-semantic baseline model and do, indeed, capture the kind of information they're conceptualized for.

Keywords: Recommender systems · Semantics · Digital libraries

1 Introduction

Citations are a central building block of scholarly discourse. They are the means by which scholars relate their research to existing work—be it by backing up claims, criticising, naming examples, or engaging in any other form. Citing in a meaningful way requires an author to be aware of publications relevant to their work. Here, the ever increasing rate of new research being published poses a serious challenge. With the goal of supporting researchers in their choice of what to read and cite, approaches to paper recommendation and citation recommendation have been an active area of research for some time now [2].

In this paper, we focus on the task of context-aware citation recommendation (see e.g. [7,10,11,14]). That is, recommending publications for the use of citation within a specific, confined context (e.g. one sentence)—as opposed to global citation recommendation and paper recommendation, where publications are recommended with respect to whole documents or user profiles. Within context-aware citation recommendation, we specifically investigate the *explicit* semantic modelling of citation contexts. While *implicit* semantic information (such as what is

J. M. Jose et al. (Eds.): ECIR 2020, LNCS 12035, pp. 220–233, 2020.
https://doi.org/10.1007/978-3-030-45439-5_15

captured by word embeddings) greatly benefits scenarios like keyword search, we argue that the specificity of information needs in academia—e.g. finding publications that use a certain data set or address a specific problem—require a more rigidly modelled knowledge representations, such as those proposed in [21] or [8]. Regarding quality, such knowledge representations (e.g. machine readable annotations of scientific publications) would ideally be created manually by the researchers themselves (see [13]). However, neither have such ideals become the norm in academic writing so far, nor are large scale data sets with manually created annotations available. Thus, we create semantic models of citation contexts using NLP techniques to automatically derive such knowledge representations. Using our models we investigate if and when such novel representation formats are beneficial for context-aware citation recommendation.

Overall, we make the following contributions:

1. We propose novel ways of deriving semantically-structured representations from citation contexts, based on entities and claims, intended for context-aware citation recommendation. To the best of our knowledge, this is the first approach of its kind, as previous uses of semantically-structured representations for citation recommendation were only ever applied to whole papers (i.e. in a setting where richer information including authors, list of references, venue, etc. is available).
2. We perform a large-scale offline evaluation using four data sets, in which we test the effectiveness of our models.
3. We also perform a user study to further evaluate the performance of our models and assess their conceptual soundness.
4. We make the code for our models and details of our evaluation publicly available.[1]

The rest of the paper is structured as follows: In Sect. 2 we outline existing works on citation recommendation. We then describe in Sect. 3 the novel semantic approaches to citation recommendation. Sect. 4 is dedicated for the evaluation of our approaches. We conclude in Sect. 5.

2 Related Work

Citation recommendation can be classified into *global* citation recommendation and *context-aware* (sometimes also referred to as *"local")* citation recommendation [10]. Various approaches have been published in both areas, but there is, to the best of our knowledge, not one that both (a) is

Table 1. Overview of related work.

Semantic	Citation recommendation type	
	Context-aware	Global
Yes	This paper	[19,28,29]
No	[5–7,10–12,14,16]	(Not considered)

[1] See https://github.com/IllDepence/ecir2020.

Table 2. Semantic approaches to global citation recommendation.

Paper	Recommendation approach	Semantic paper model
[19]	Content-based filtering and collaborative filtering	Topic ontology (used to classify papers)
[28]	Hybrid recommender system	Enrich metadata using LOD sources
[29]	Content-based filtering	Semantic distance measure based on relational features between papers

Table 3. Non-semantic approaches to context-aware citation recommendation.

Paper	Recommendation approach	Citation contex model
[10]	Content-based filtering	TF-IDF weighted VSM vectors
[12]	Translational model	"Source language" of translational model
[11]	Neural probabilistic model	Distributed word representation
[5]	Content-based filtering	TF-IDF weighted VSM vectors
[6]	Content-based filtering	TF-IDF weighted VSM vectors with weights for rhetorical functions
[7]	Neural citation network	Word embeddings plus author embeddings
[16]	Content-based filtering	Word embeddings along three discourse facets **+1 given citation**
[14]	Graph Convolutional Network + BERT	Word embeddings

context-aware and (b) uses explicit semantic representations of citation contexts. We illustrate this in Table 1. In the following, we therfore outline the most related works on (semantic) global citation recommendation (upper right cell in Table 1), and (non-semantic) context-aware citation recommendation (lower left cell in Table 1).

Global Citation Recommendation. Global citation recommendation is characterized as a task for which the input of the recommendation engine is not a specific citation context but a whole paper. Various approaches have been published for global citation recommendation, some of which can also be used for paper recommendation, i.e., for recommending papers for the purpose of reading [1]. A few *semantic* approaches to global citation recommendation exist (see Table 2). They are based on a semantically-structured representation of papers' metadata (e.g., authors, title, abstract) [28,29] and/or papers' contents [19]. Note that the approaches proposed in this paper are not using any of the papers' metadata or full text, as our goal is to provide fine-grained, semantically suitable recommendations for specific citation contexts.

Context-Aware Citation Recommendation. Context-aware citation recommendation approaches recommend publications for a specific citation context and are thus also called "local" citation recommendation approaches. Existing context-aware citation recommendation approaches soly rely on lexical and syntactic features (n-grams, part-of-speech tags, word embeddings etc.) but do not attempt to model citation contexts in an explicit semantic fashion. Table 3 gives an overview of context-aware citation recommendation approaches. We can mention SemCir [29] as the only approach we are aware of that *could* be regarded as a semantic approach to context-aware citation recommendation. The explicit semantic representations are, however, not generated from citation contexts (not context-aware), but from papers (global), that are textually (not necessarily semantically) similar to the citation contexts. We therefore categorize it as a semantic global approach.

3 Approach

To ensure wide coverage and applicability of our citation context models, we base our selection of structures to model on a typology of citation functions from the field of citation context analysis ([22], built upon [26]). Note that, because citation context analysis is primarily concerned with the *intent* of the author rather than the *content* of the citation context, we cannot use the functions as a basis for our models directly. Instead we inspect sample contexts of each function type and thereby identify *named entity* (NE) and *claim* as semantic structures of interest, as illustrated in Table 4. Note that the example contexts listed to have no structure ("-" in the *Structure* column) may contain named entities and claims as well (e.g. "DBLP" or "Lamers et al. base their definition of the author's name"), but these are (in the case of NEs) not representative of the cited work or (in the case of claims) just statements *about* a publication rather than statements being backed by the cited work.

The following sections will describe our entity-based and claim-based models for context-aware citation recommendation.

3.1 Entity-Based Recommendation

The intuition behind an entity-based approach is that there exists a reference publication for a named entity mentioned in the citation context. For instance, this can be a data set ("CiteSeer^{x} [37]"), a tool ("Neural ParsCit [23]"), or a (scientific) concept ("Semantic Web [37]"). In a more loose sense this can also include publications being referred to as examples ("approaches to context-aware citation recommendation [5–7,10–12,14,16]"). Because names of methods, data sets, tools, etc. in academia often are neologisms and only the most widely used ones of them are reflected in resources like DBpedia, we use a set of noun phrases found in academic publications as surrogates for named entities (instead of performing entity linking). For this, we extract noun phrases from the arXiv

Table 4. Semantic structures identified in citation contexts from a range of citation functions used in the field of citation context analysis (NE = named entity).

Function [22]	Structure	Examples (semantic structure *highlighted*)
Attribution	Claim	"Berners-Lee et al. [37] argue that *structured collections of information and sets of inference rules are prerequisites for the semantic web to function*"
	NE	"A variation of this task is '*context-based co-citation recommendation*' [16]"
	-	"In [5] Duma et al. test the effectiveness of using a variety of document internal and external text inputs with a TFIDF model"
Exemplification	NE	"We looked into approaches to *context-aware citation recommendation* such as [5–7, 10–12, 14, 16] for our investigation"
Further reference	-	"See [37] for a comprehensive overview"
Statement of use	NE	"We use *CiteSeer^x* [37] for our evaluation"
Application	NE	"Using this mechanism we perform '*context-based co-citation recommendation*' [16]"
Evaluation	-	"The use of DBLP in [37] restricts their data set to the field of computer science"
Establishing links between sources	Claim	"A common motivation brought forward for research on citation recommendation is that *finding proper citations is a time consuming task* [7,9,10,16]"
	-	"Lamers et al. [37] base their definition on the author's name whereas Thompson [26] focuses on the grammatical role of the citation marker"
Comparison of own work with sources	Claim	"Like [37] we find that, albeit written in a structured language, *parsing LaTeX sources is a non trivial task*"

publications provided by [24] and filter out items that appear only once. In doing so we end up with a set of 2,835,929 noun phrases[2] (NPs) that we use.

In the following, we define two NP-based representations of citation contexts, R_{NP} and R^{2+}_{NPmrk}. For this, $\mathcal{P}$ shall denote our set of NPs and c shall denote a citation context.

***R*NP.** We define $R_{\text{NP}}(c)$ as the set of maximally long NPs contained in c. Formally, $R_{\text{NP}}(c) = \{t | t \text{ appears in } c \wedge t \in \mathcal{P} \wedge t^{+pre} \notin \mathcal{P} \wedge t^{+suc} \notin \mathcal{P}\}$ where t^{+pre} and t^{+suc} denote an extension of t using its preceding or succeeding word respectively. A context *"This has been done for language model training [37]"*, for

[2] See https://github.com/IllDepence/ecir2020 for a full list.

Listing 1.1. PredPatt example output.

```
?a shows ?b
    ?a: The paper
    ?b: SOMETHING := context-based methods can outperform
        global approaches
?a can outperform ?b
    ?a: context-based methods
    ?b: global approaches
```

example, would therefore have "language model training" in its representation, but not "language model".

$\boldsymbol{R^{2+}_{\mathbf{NPmrk}}}$. We define $R^{2+}_{\text{NPmrk}}(c)$ as a subset of $R_{\text{NP}}(c)$ containing, if present, the NP of minimum word length 2 directly preceding the citation marker which a recommendation is to be made for. Formally, $R^{2+}_{\text{NPmrk}}(c) = \{t | t \in R_{\text{NP}}(c) \wedge len(t) \geq 2 \wedge t \text{ directly precedes } m\}$ where m is the citation marker in c that a prediction is to be made for.

Recommendation. As is typical in context-aware citation recommendation [5,6,10] we aggregate citation contexts referencing a publication to describe it as a recommendation candidate. To that end, we define frequency vector representations for single citation contexts and documents as follows. A citation context vector is $V(R(c)) = (t_1, t_2, ..., t_{|\mathcal{P}|})$, where t_i denotes how often the ith term in $\mathcal{P}$ appears in $R(c)$. A document vector then is a sum of citation context vectors $\sum_{c \in \varrho(d)} V(R(c))$, where $\varrho(d)$ denotes the set of citation contexts referencing d. Similarities can then be calculated as the cosine of context and document vectors.

3.2 Claim-Based Recommendation

Our claim-based approach is motivated by the fact that citations are used to back up claims (see Table 4). These can, for example, be findings presented in a specific publication ("It has been shown, that ... [37].") or more general themes found across multiple works ("... is still an unsolved task [37–39].").

For the extraction of claims, we considered a total of four state of the art [30] information extraction tools (PredPatt [27], Open IE 5.0 [17], ClausIE [4] and Ollie [18]) and found PredPatt to give the best quality results[3]. For the simple sentence *"The paper shows that context-based methods can outperform global approaches."*, Listing 1.1 shows the user interface output of PredPatt and Fig. 1 its internal representation using Universal Dependencies (UD) [20].

[3] See https://github.com/IllDepence/ecir2020 for details on the evaluation.

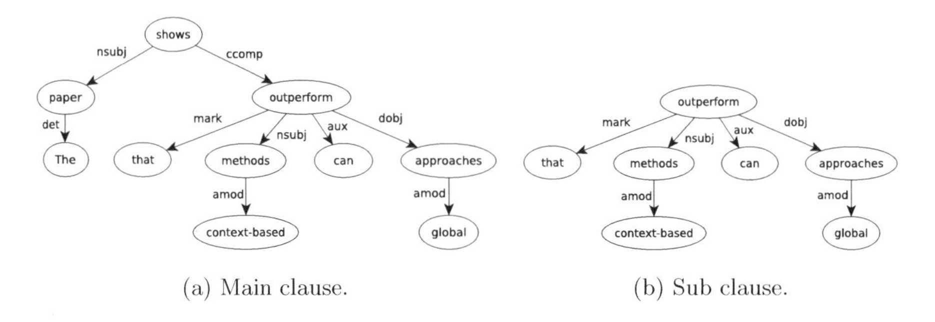

(a) Main clause. (b) Sub clause.

Fig. 1. UD trees as generated by PredPatt.

Algorithm 1. Construction of $R_{\text{claim}}(c)$

```
c ← strip_quotation_marks(c)                    ▷ remove quotation marks
c ← merge_citation_markers(c)                   ▷ e.g. "[x], [y]"⇒"[xy]"
pp_trees_c ← predpatt(c)                        ▷ get PredPatt output
output ← []
resolve_rels ← ['name', 'goeswith', 'mwe',      ▷ (b1) UD relations
                'compound', 'conj', 'amod',
                'advmod']
foreach t ∈ pp_trees_c do                       ▷ for all claims identified
    pred ← identify_predicate(t)                ▷ (a) resolve copula if present
    pred ← lemmatize(pred)
    foreach n ∈ traverse(t) do
        if pos_tag(n) == 'NOUN' then
            arg ← resolve_all(n, resolve_rels)  ▷ (b2) resolve compounds etc.
            output.append(pred + ':' + arg)     ▷ build pred:arg tuple
        end if
    end for
end for
return output
```

Because the predicates and especially arguments in the PredPatt user interface output can get very long—e.g. *"can outperform"* (including the auxiliary verb *"can"*) and *"context-based methods can outperform global approaches"* (unlikely to appear in another citation context with the exact same wording)—we build our claim-based representation R_{claim} from UD trees, as explained in the following section.

R**claim.** For each claim that PredPatt detects, it internally builds one UD tree. To construct our claim-based representation R_{claim}, we traverse each tree, identify the predicate and its arguments (subject and object) and save these in tuples. The exact procedure for this is given in Algorithm 1. If a sentence uses a copula (*be*, *am*, *is*, *are*, *was*), the actual predicate is a child node of the root with the relation type *"cop"*. This is resolved at marker **(a)**. For the identification of useful arguments (markers **(b$_1$)** and **(b$_2$)** in Algorithm 1), we look at all nouns

Listing 1.2. $R_{\text{claim}}(c)$ for $c =$ *"The paper shows that context-based methods can outperform global approaches."*.

```
[
  'show:paper',
  'show:context based methods',
  'show:global approaches',
  'outperform:context based methods',
  'outperform:global approaches',
]
```

within the UD tree and resolve compounds (*"compound"*, *"mwe"*, *"name"* relations), phrases split by formatting (*"goeswith"*), conjunctions (*"conj"*) as well as adjectival and adverbial modifiers (*"amod"*, *"advmod"*). To give an example for this, the noun *"methods"* in both trees in Fig. 1 has the adjectival modifier *"context-based"*. In such a case our model would not choose *"methods"* as an argument to *"outperform"* but *"context-based methods"*. Listing 1.2 shows the complete representation generated for the example sentence.

Recommendation. For a set of predicate-argument tuples $\mathcal{T}$, we define frequency vector representations of citation contexts and documents as follows. A citation context vector is $V(R(c)) = (t_1, t_2, ..., t_{|\mathcal{T}|})$, where t_i denotes how often the ith tuple in $\mathcal{T}$ appears in $R(c)$. A document vector, again, is a sum of citation context vectors $\sum_{c \in \varrho(d)} V(R(c))$, where $\varrho(d)$ is the set of citation contexts referencing d. Similarities are then calculated as the cosine of TF-IDF weighted context and document vectors.

R**claim+BoW.** In addition to R_{claim}, we define a combined model $R_{\text{claim+BoW}}$ as a linear combination of similarity values given by R_{claim} and an bag-of-words model (BoW). Similarities in the combined model are calculated as $sim(A, B) = \sum_{m \in \mathcal{M}} \alpha_m sim_m(A, B)$ of the models $\mathcal{M} = \{R_{\text{claim}}, \text{BoW}\}$ with the coefficients $\alpha_{R_{\text{claim}}} = 1$ and $\alpha_{\text{BoW}} = 2$.

4 Evaluation

We evaluate our models in a large offline evaluation as well as a user study. In total, we compare four models—R_{NP}, R^{2+}_{NPmrk}, R_{claim} and $R_{\text{claim+BoW}}$—against a bag-of-words baseline (BoW). Our choice of a simple BoW model for a baseline is motivated as follows. Because our entity-based and claim-based models are, in their current form, string based, they can be seen as a semantically informed selection of words from the citation context. In this sense, they work akin to what is done in reference scope identification [15]. To evaluate the validity of the selections of words that our models lay focus on, we use the complete set

Table 5. Citation context sources and filter criteria.

Data source	Citing doc	Cited doc
arXiv [24]	Computer science	≥5 citing docs
MAG [25]	Computer science, English, abstract not NULL	≥50 citing docs
RefSeer [11]	Title, venue, venuetype, abstract, and year in DB not NULL	Title, venue, venuetype, abstract, and year in DB not NULL
ACL-ARC [3]	-	Has a DBLP ID

Table 6. Key properties of data used for evaluation.

Data set	Train/test split	#Candidate docs	#Test set items	Mean CC/RC (SD)
arXiv	≤2016/≥2017	63,239	490,018	21.7 (51.2)
MAG	≤2017/≥2018	81,320	141,631	104.1 (198.6)
RefSeer	≤2011/≥2012	184,539	53,401	18.2 (47.0)
ACL-ARC	≤2005/=2006	2,431	3,881	6.8 (9.5)

of words contained in the context (BoW) to compare against. Comparing our models against deep learning based approaches (e.g. based on embeddings) would not provide a comparison in this selection behavior, and are therefore was not considered.

4.1 Offline Evaluation

Our offline evaluation is performed in a citation re-prediction setting. That is, we take existing citation contexts from scientific publications and split them into training and test subsets. The training contexts are used to learn the representations of the cited documents. The test contexts are stripped of their citations, used as input to our recommender systems and the resulting recommendations checked against the original citations.

Table 5 shows the four data sources we use as well as applied filter criteria. RefSeer and ACL-ARC are often used in related work (e.g. [5,7]), we therefore use *both* of them and *two additional* large data sets to ensure a thorough evaluation. Table 6 gives an overview of key properties of the training and test data for the evaluation. We split our data according to the citing paper's publication date and report *#Candidate docs*: the number of candidate documents to rank for a recommendation; *#Test set items*: the number of test set items (unit: citation contexts); *Mean CC/RC*: the mean number of citation contexts per recommendation candidate in the training set (i.e., a measure for how well the recommendation candidates are described, giving insight into how difficult the recommendation task for each of the data sets is).

Figure 2 shows the results of our evaluation. We measure NDCG, MAP, MRR and Recall at cut-offs from 1 to 10. Note that the evaluation using the arXiv data

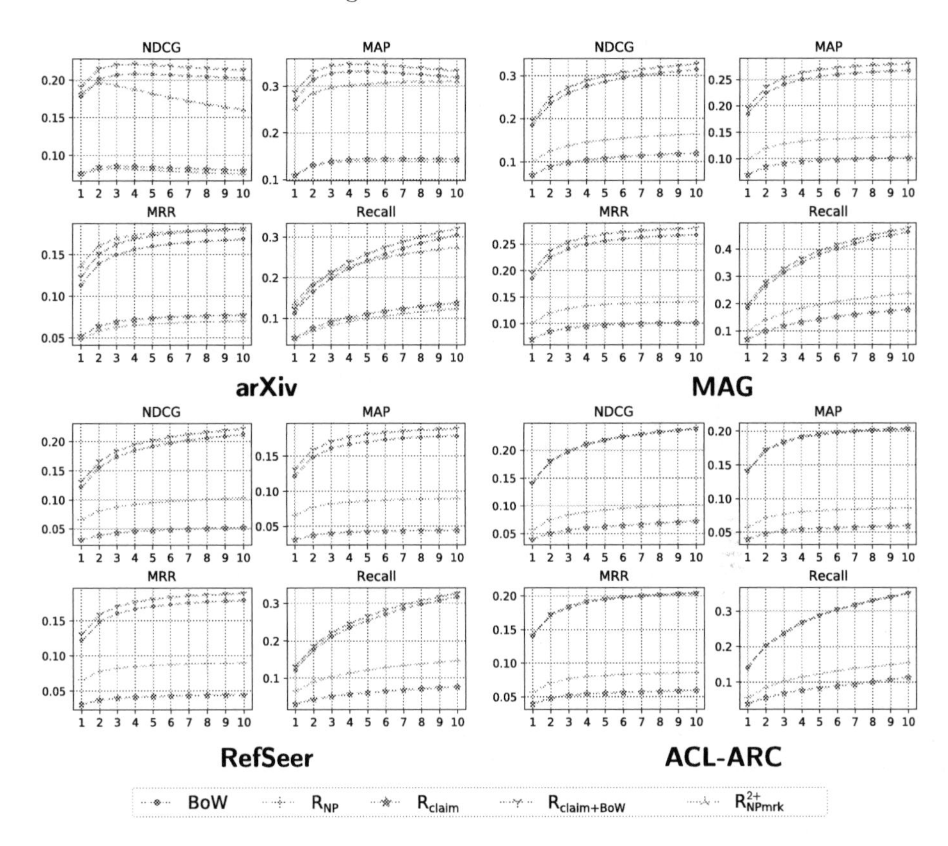

Fig. 2. Evaluation using arXiv, MAG, RefSeer and ACL-ARC. Showing NDCG, MAP, MRR and Recall scores at cut of values from 1 to 10.

differs from the other cases in two aspects. First, it is the only case where we can apply R^{2+}_{NPmrk}, because citation marker positions are given. Second, because for citation contexts with several citations (cf. Table 4, "Exemplification") the data set lists several cited documents (instead of just a single one), we are able to treat more than a single re-predicted citation as valid. We do this by counting re-predicted "co-citations" as relevant when calculating MAP scores and give them a relevance of 0.5 in the NDCG calculation. This also means that, looking at higher cut-offs, NDCG and MAP values can decrease because ideal recommendations require relevant re-predictions on all ranks above the cut-off.

As for the performance of our models shown in Fig. 2, we see that for each of the data sets $R_{\text{claim+BoW}}$ outperforms the BoW baseline in each metric and for all cut-off values.[4] R_{claim} and R_{NP} do not compare in performance with the two aforementioned. This suggests that the claim structures we model with

[4] To validate our findings, we further analyze the NDCG@5 results and note a statistically significant improvement for the arXiv, MAG and RefSeer data but no significant difference for the ACL-ARC data set.

R_{claim} are not enough for well performing recommendations on their own, but do capture important information that non-semantic models (BoW) miss. R^{2+}_{NPmrk}, only present in the arXiv evaluation, gives particularly good results for lower cut-offs and performs especially well in the MRR metric. It performs the worst at high cut-offs measured by NDCG. Note that R^{2+}_{NPmrk} is only evaluated for test set items, where the model was applicable (i.e. where a noun phrase of minimum length 2 is directly preceding the citation marker; cf. Sect. 3.1). For our evaluation this was the case for 100,308 out of the 490,018 test set items (20.5%). The evaluation results for the citation marker-aware model R^{2+}_{NPmrk} indicate that it is comparatively well suited to recommend citations where there is one particularly fitting publication (e.g. a reference paper) and less suited for exemplifications (cf. Table 4).

4.2 User Study

To obtain more insights into the nature of our evaluation data, as well as a better understanding of our models, we perform a user study in which two human raters (the two authors) judge input-output pairs of our offline evaluation (i.e. citation contexts and the recommendations given for them). For this, we randomly choose 100 citation contexts from the arXiv evaluation, so that we can include R^{2+}_{NPmrk}. For each input context, we show raters the top 5 recommendations of the 3 best performing models of the offline evaluation, i.e., BoW, $R_{\text{claim+BoW}}$ and R^{2+}_{NPmrk} models (resulting in $100 \times 5 \times 3 = 1500$ items). Judgments are performed by looking at each citation context and the respective recommended paper. In addition, we let the raters judge the type of citation (Claim, NE, Exemplification, Other; cf. Table 4).

Table 7. User study evaluation scores at cut-off 5.

Model	Recall@5	MRR@5	MAP@5	NDCG@5
All contexts (138)				
Claim+BoW	**0.53**	0.44	0.41	0.46
BoW	0.51	**0.46**	**0.44**	**0.48**
NPmarker	0.35	0.35	0.33	0.34
Only contexts of type "claim" (38)				
Claim+BoW	**0.63**	0.46	0.42	0.49
BoW	0.58	**0.48**	**0.46**	**0.51**
NPmarker	0.20	0.13	0.13	0.15
Only contexts of type "NE" (45)				
Claim+BoW	0.46	0.44	0.41	0.44
BoW	0.47	0.45	0.41	0.35
NPmarker	**0.52**	**0.53**	**0.48**	**0.51**
Only contexts of type "exemplification" (38)				
Claim+BoW	**0.56**	0.52	0.47	0.52
BoW	0.54	**0.53**	**0.49**	**0.54**
NPmarker	0.21	0.24	0.24	0.24
Only contexts of type "other" (17)				
Claim+BoW	0.44	0.29	0.29	0.33
BoW	0.41	0.33	0.33	0.36
NPmarker	**0.50**	**0.50**	**0.44**	**0.47**

Table 7 shows the results based on the raters' relevance judgments. We present measurements for all contexts, as well as each of the citation classes on its own. We note that $R_{\text{claim+BoW}}$ and BoW are close, but in contrast to the offline evaluation, $R_{\text{claim+BoW}}$ only outperforms BoW in the Recall metric. In the case of NE type citations, the R^{2+}_{NPmrk} model performs better than the other two models in all metrics. Furthermore, we can see that both $R_{\text{claim+BoW}}$ and R^{2+}_{NPmrk} achieve their best results for the type of citation they're designed for—Claim and NE respectively. This indicates that both models actually capture the kind of information they're conceptualized for. Compared to the offline evaluation, we measure higher numbers overall. While the user study is of considerably smaller scale and a direct comparison therefore not necessarily possible, the notably higher numbers indicate, that a re-prediction setting involves a non-negligible number of false negatives (actually relevant recommendations counted as not relevant).

4.3 Main Findings

The entity-based model R^{2+}_{NPmrk}, which captures noun phrases preceding the citation marker, performs best at low cut-offs and in the MRR metric. Low cut-offs and measuring the MRR can be interpreted as emulating citations for reference publications. This interpretation is also backed by the results of the user study, where R^{2+}_{NPmrk} outperformed all other models when recommending for citation contexts that referenced a named entity or concept. We therefore conclude that R^{2+}_{NPmrk} is well suited for recommending such types of citations. Our claim-based model R_{claim} does not compare in performance to a BoW baseline, but $R_{\text{claim+BoW}}$ outperforms aforementioned. We take this as an indication that the claim representation encodes important information which the non-semantic BoW model is not able to capture. In the user study $R_{\text{claim+BoW}}$ performs best for citation contexts, in which a claim is backed by the target citation. This suggests that the model indeed captures information related to claim structures.

5 Conclusion

In the field of context-aware citation recommendation, the explicit semantic modeling of citation contexts is not well explored yet. In order to investigate the merit of such approaches, we developed semantic models of citation contexts based on entities as well as claim structures. We then evaluated our models on several data sets in a citation re-prediction setting and furthermore conducted a user study. In doing so, we could demonstrate their applicability and conceptual soundness. The next step from hereon is to move from semantically informed text-based models to explicit knowledge representations. Our research also shows, that differentiating between different semantic representations of citation contexts due to varying ways of citing information is reasonable. Developing different citation recommendation approaches, depending on the semantic citation types, might therefore be a promising next step in our research.

References

1. Beel, J., Dinesh, S.: Real-world recommender systems for academia: the pain and gain in building, operating, and researching them. In: Proceedings of the Fifth Workshop on Bibliometric-enhanced Information Retrieval (BIR) co-located with the 39th European Conference on Information Retrieval (ECIR 2017), pp. 6–17 (2017)
2. Beel, J., Gipp, B., Langer, S., Breitinger, C.: Research-paper recommender systems: a literature survey. Int. J. Digit. Libr. **17**(4), 305–338 (2015). https://doi.org/10.1007/s00799-015-0156-0
3. Bird, S., et al.: The ACL anthology reference corpus: a reference dataset for bibliographic research in computational linguistics. In: Proceedings of the 6th International Conference on Language Resources and Evaluation, LREC 2008 (2008)
4. Corro, L.D., Gemulla, R.: ClausIE: clause-based open information extraction. In: Proceedings of the 22nd International World Wide Web Conference, WWW 2013, pp. 355–366 (2013)
5. Duma, D., Klein, E.: Citation resolution: a method for evaluating context-based citation recommendation systems. In: Proceedings of the 52nd Annual Meeting of the Association for Computational Linguistics, ACL 2014, pp. 358–363 (2014)
6. Duma, D., Klein, E., Liakata, M., Ravenscroft, J., Clare, A.: Rhetorical classification of anchor text for citation recommendation. D-Lib Mag. **22** (2016). https://doi.org/10.1045/september2016-duma. http://www.dlib.org/dlib/september16/09contents.html
7. Ebesu, T., Fang, Y.: Neural citation network for context-aware citation recommendation. In: Proceedings of the 40th International ACM SIGIR Conference on Research and Development in Information Retrieval, pp. 1093–1096 (2017)
8. Gábor, K., Buscaldi, D., Schumann, A., QasemiZadeh, B., Zargayouna, H., Charnois, T.: SemEval-2018 task 7: semantic relation extraction and classification in scientific papers. In: Proceedings of the 12th International Workshop on Semantic Evaluation, SemEval@NAACL-HLT 2018, pp. 679–688 (2018)
9. He, Q., Kifer, D., Pei, J., Mitra, P., Giles, C.L.: Citation recommendation without author supervision. In: Proceedings of the Forth International Conference on Web Search and Web Data Mining, WSDM 2011, pp. 755–764 (2011)
10. He, Q., Pei, J., Kifer, D., Mitra, P., Giles, L.: Context-aware citation recommendation. In: Proceedings of the 19th International Conference on World Wide Web, WWW 2010, pp. 421–430. ACM, New York (2010)
11. Huang, W., Wu, Z., Liang, C., Mitra, P., Giles, C.L.: A neural probabilistic model for context based citation recommendation. In: Proceedings of the Twenty-Ninth AAAI Conference on Artificial Intelligence, AAAI 2015, pp. 2404–2410 (2015)
12. Huang, W., Wu, Z., Mitra, P., Giles, C.L.: Refseer: a citation recommendation system. In: Proceedings of the IEEE/ACM Joint Conference on Digital Libraries, JCDL 2014, pp. 371–374 (2014)
13. Jaradeh, M.Y., et al.: Open research knowledge graph: next generation infrastructure for semantic scholarly knowledge. In: Proceedings of the 10th International Conference on Knowledge Capture, K-CAP 2019, pp. 243–246 (2019)
14. Jeong, C., Jang, S., Shin, H., Park, E., Choi, S.: A Context-Aware Citation Recommendation Model with BERT and Graph Convolutional Networks (2019). http://arxiv.org/abs/1903.06464
15. Jha, R., Jbara, A.A., Qazvinian, V., Radev, D.R.: NLP-driven citation analysis for scientometrics. Nat. Lang. Eng. **23**(1), 93–130 (2017)

16. Kobayashi, Y., Shimbo, M., Matsumoto, Y.: Citation recommendation using distributed representation of discourse facets in scientific articles. In: Proceedings of the 18th ACM/IEEE on Joint Conference on Digital Libraries, JCDL 2018, pp. 243–251 (2018)
17. Mausam, M.: Open information extraction systems and downstream applications. In: Proceedings of the Twenty-Fifth International Joint Conference on Artificial Intelligence, IJCAI 2016, pp. 4074–4077 (2016)
18. Schmitz, M., Bart, R., Soderland, S., Etzioni, O.: Open language learning for information extraction. In: Proceedings of the 2012 Joint Conference on Empirical Methods in Natural Language Processing and Computational Natural Language Learning, EMNLP-CoNLL 2012, Stroudsburg, USA, pp. 523–534 (2012)
19. Middleton, S.E., Roure, D.D., Shadbolt, N.: Capturing knowledge of user preferences: ontologies in recommender systems. In: Proceedings of the First International Conference on Knowledge Capture, K-CAP 2001, pp. 100–107 (2001)
20. Nivre, J., et al.: Universal dependencies v1: a multilingual treebank collection. In: Proceedings of the Tenth International Conference on Language Resources and Evaluation, LREC 2016, pp. 1659–1666 (2016)
21. Peroni, S., Shotton, D.M.: FaBiO and CiTO: ontologies for describing bibliographic resources and citations. J. Web Semant. **17**, 33–43 (2012)
22. Petrić, B.: Rhetorical functions of citations in high- and low-rated master's theses. J. Engl. Acad. Purp. **6**(3), 238–253 (2007)
23. Prasad, A., Kaur, M., Kan, M.Y.: Neural ParsCit: a deep learning based reference string parser. Int. J. Digit. Libr. **19**, 323–337 (2018). https://doi.org/10.1007/s00799-018-0242-1
24. Saier, T., Färber, M.: Bibliometric-enhanced arXiv: a data set for paper-based and citation-based tasks. In: Proceedings of the 8th International Workshop on Bibliometric-enhanced Information Retrieval, BIR 2019, pp. 14–26 (2019)
25. Sinha, A., et al.: An overview of Microsoft academic service (MAS) and applications. In: Proceedings of the 24th International Conference on World Wide Web, WWW 2015, pp. 243–246 (2015)
26. Thompson, P.: A pedagogically-motivated corpus-based examination of PhD theses: macrostructure, citation practices and uses of modal verbs. Ph.D. thesis, University of Reading (2001)
27. White, A.S., et al.: Universal decompositional semantics on universal dependencies. In: Proceedings of the 2016 Conference on Empirical Methods in Natural Language Processing, EMNLP 2016, pp. 1713–1723 (2016)
28. Zarrinkalam, F., Kahani, M.: A multi-criteria hybrid citation recommendation system based on linked data. In: Proceedings of the 2nd International eConference on Computer and Knowledge Engineering, ICCKE 2012, pp. 283–288 (2012)
29. Zarrinkalam, F., Kahani, M.: SemCiR: a citation recommendation system based on a novel semantic distance measure. Program: Electron. Libr. Inf. Syst. **47**, 92–112 (2013)
30. Zhang, S., Rudinger, R., Durme, B.V.: An evaluation of PredPatt and open IE via stage 1 semantic role labeling. In: Proceedings of the 12th International Conference on Computational Semantics, IWCS 2017 (2017)

TransRev: Modeling Reviews as Translations from Users to Items

Alberto García-Durán[1(✉)], Roberto González[2], Daniel Oñoro-Rubio[2], Mathias Niepert[2], and Hui Li[3]

[1] EPFL, Lausanne, Switzerland
alberto.duran@epfl.ch
[2] NEC Labs Europe, Heidelberg, Germany
{roberto.gonzalez,daniel.onoro,mathias.niepert}@neclab.eu
[3] Xiamen University, Xiamen, China
huili.xmu@gmail.com

Abstract. The text of a review expresses the sentiment a customer has towards a particular product. This is exploited in sentiment analysis where machine learning models are used to predict the review score from the text of the review. Furthermore, the products costumers have purchased in the past are indicative of the products they will purchase in the future. This is what recommender systems exploit by learning models from purchase information to predict the items a customer might be interested in. The underlying structure of this problem setting is a bipartite graph, wherein customer nodes are connected to product nodes via 'review' links. This is reminiscent of knowledge bases, with 'review' links replacing relation types. We propose TransRev, an approach to the product recommendation problem that integrates ideas from recommender systems, sentiment analysis, and multi-relational learning into a joint learning objective.

TransRev learns vector representations for users, items, and reviews. The embedding of a review is learned such that (a) it performs well as input feature of a regression model for sentiment prediction; and (b) it always translates the reviewer embedding to the embedding of the reviewed item. This is reminiscent of TransE [5], a popular embedding method for link prediction in knowledge bases. This allows TransRev to approximate a review embedding at test time as the difference of the embedding of each item and the user embedding. The approximated review embedding is then used with the regression model to predict the review score for each item. TransRev outperforms state of the art recommender systems on a large number of benchmark data sets. Moreover, it is able to retrieve, for each user and item, the review text from the training set whose embedding is most similar to the approximated review embedding.

Keywords: Recommender systems · Knowledge graphs · Sentiment analysis

J. M. Jose et al. (Eds.): ECIR 2020, LNCS 12035, pp. 234–248, 2020.
https://doi.org/10.1007/978-3-030-45439-5_16

1 Introduction

Online retail is a growing market with sales accounting for $394.9 billion or 11.7% of total US retail sales in 2016 [35]. In the same year, e-commerce sales accounted for 41.6% of all retail sales growth [15]. For some entertainment products such as movies, books, and music, online retailers have long outperformed traditional in-store retailers. One of the driving forces of this success is the ability of online retailers to collect purchase histories of customers, online shopping behavior, and reviews of products for a very large number of users. This data is driving several machine learning applications in online retail, of which personalized recommendation is the most important one. With recommender systems online retailers can provide personalized product recommendations and anticipate purchasing behavior.

Fig. 1. (Left) A typical product summary with review score and 'Pros'. Image taken from www.bestbuy.com. (Right) A small bipartite graph modeling customers (users), products (items), product reviews, and review scores.

In addition, the availability of product reviews allows users to make more informed purchasing choices and companies to analyze costumer sentiment towards their products. The latter was coined sentiment analysis and is concerned with machine learning approaches that map written text to scores. Nevertheless, even the best sentiment analysis methods cannot help in determining which *new* products a costumer might be interested in. The obvious reason is that costumer reviews are not available for products they have not purchased yet.

In recent years the availability of large corpora of product reviews has driven text-based research in the recommender system community (e.g. [3,19,21]). Some of these novel methods extend latent factor models to leverage review text by employing an explicit mapping from text to either user or item factors. At prediction time, these models predict product ratings based on some operation (typically the dot product) applied to the user and product representations. Sentiment analysis, however, is usually applied to some representation (e.g. bag-of-words) of review text but in a recommender system scenario the review is not available at prediction time.

With this paper we propose TRANSREV, a method that combines a personalized recommendation learning objective with a sentiment analysis objective into a joint learning objective. TRANSREV learns vector representations for

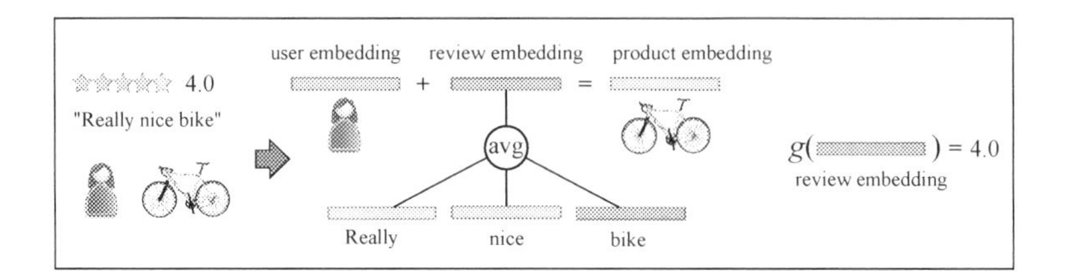

Fig. 2. At training time, a function's parameters are learned to compute the review embedding from the word token embeddings such that the embedding of the user translated by the review embedding is similar to the product embedding. At the same time, a regression model g is trained to perform well on predicting ratings.

users, items, and reviews jointly. The crucial advantage of TRANSREV is that the review embedding is learned such that it corresponds to a translation that moves the embedding of the reviewing user to the embedding of the item the review is about. This allows TRANSREV to approximate a review embedding at test time as the difference of the item and user embedding despite the absence of a review from the user for that item. The approximated review embedding is then used in the sentiment analysis model to predict the review score. Moreover, the approximated review embedding can be used to retrieve reviews in the training set deemed most similar by a distance measure in the embedding space. These retrieved reviews could be used for several purposes. For instance, such reviews could be provided to users as a starting point for a review, lowering the barrier to writing reviews.

2 TransRev: Modeling Reviews as Translations in Vector Space

We address the problem of learning prediction models for the product recommendation problem. A small example of the input data typical to such a machine learning system is depicted in Fig. 1. This reminds of knowledge bases, with 'reviews' replacing relation types. Two nodes in a knowledge base may be joined by a number of links, each representing one relation type from a small vocabulary. Here, if two nodes are connected they are linked by one single edge type, in which case it is represented by a number of words from a (very) large vocabulary.

There are a set of users $\mathbf{U}$, a set of items $\mathbf{I}$, and a set of reviews $\mathbf{R}$. Each $\mathtt{rev}_{(\mathtt{u},\mathtt{i})} \in \mathbf{R}$ represents a review written by user $\mathtt{u}$ for item $\mathtt{i}$. Hence, $\mathtt{rev}_{(\mathtt{u},\mathtt{i})} = [\mathtt{t}_1, \cdots, \mathtt{t}_n]$, that is, each review is a sequence of n tokens. In the following we refer to $(\mathtt{u}, \mathtt{rev}_{(\mathtt{u},\mathtt{i})}, \mathtt{i})$ as a *triple*. Each such triple is associated with the review score $\mathtt{r}_{(\mathtt{u},\mathtt{i})}$ given by the user $\mathtt{u}$ to item $\mathtt{i}$.

TRANSREV embeds all users, items and reviews into a latent space where the embedding of a user plus the embedding of the review is learned to be close to the embedding of the reviewed item. It simultaneously learns a regression model to predict the rating given a review text. This is illustrated in Fig. 2.

At prediction time, reviews are not available, but the modeling assumption of TRANSREV allows to predict the review embedding by taking the difference of the embedding of the item and user. Then this approximation is used as input feature of the regression model to perform rating prediction—see Fig. 3.

TRANSREV embeds all nodes and reviews into a latent space $\mathbb{R}^k$ (k is a model hyperparameter). The review embeddings are computed by applying a learnable function f to the token sequence of the review

$$\mathbf{h}_{\mathtt{rev}_{(\mathtt{u},\mathtt{i})}} = f(\mathtt{rev}_{(\mathtt{u},\mathtt{i})}).$$

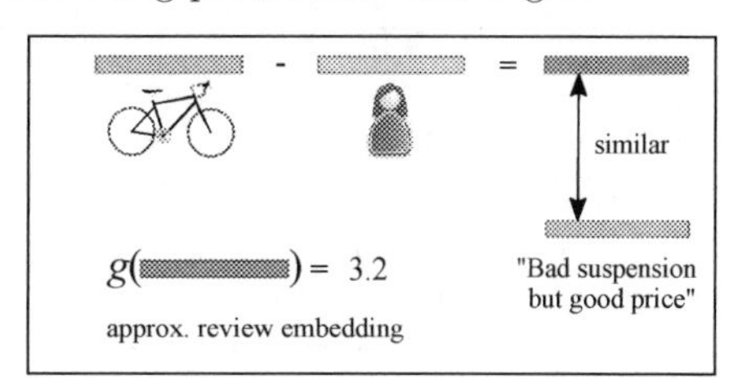

Fig. 3. At test time, the review embedding is approximated as the difference between the product and user embeddings. The approximated review embedding is used to predict the rating and to retrieve similar reviews.

The function f can be parameterized (typically with a neural network such as a recursive or convolutional neural network) but it can also be a simple parameter-free aggregation function that computes, for instance, the element-wise average or maximum of the token embeddings.

We propose and evaluate a simple instance of f where the review embedding $\mathbf{h}_{\mathtt{rev}_{(\mathtt{u},\mathtt{i})}}$ is the average of the embeddings of the tokens occurring in the review. More formally,

$$\mathbf{h}_{\mathtt{rev}_{(\mathtt{u},\mathtt{i})}} = f(\mathtt{rev}_{(\mathtt{u},\mathtt{i})}) = \frac{1}{|\mathtt{rev}_{(\mathtt{u},\mathtt{i})}|} \sum_{\mathtt{t} \in \mathtt{rev}_{(\mathtt{u},\mathtt{i})}} \mathbf{v}_{\mathtt{t}} + \mathbf{h}_0, \tag{1}$$

where $\mathbf{v}_{\mathtt{t}}$ is the embedding associated with token $\mathtt{t}$ and $\mathbf{h}_0$ is a review bias which is common to all reviews and takes values in $\mathbb{R}^k$. The review bias is of importance since there are some reviews all of whose tokens are not in the training vocabulary. In these cases we have $\mathbf{h}_{\mathtt{rev}_{(\mathtt{u},\mathtt{i})}} = \mathbf{h}_0$.

The learning of the item, review, and user embeddings is determined by two learning objectives. The first objective guides the joint learning of the parameters of the regression model and the review embeddings such that the regression model performs well at review score prediction

$$\min \mathcal{L}_1 = \min \sum_{((\mathtt{u},\mathtt{rev}_{(\mathtt{u},\mathtt{i})},\mathtt{i}),\mathtt{r}_{(\mathtt{u},\mathtt{i})}) \in S} (g(\mathbf{h}_{\mathtt{rev}_{(\mathtt{u},\mathtt{i})}}) - \mathtt{r}_{(\mathtt{u},\mathtt{i})})^2, \tag{2}$$

where S is the set of training triples and their associated ratings, and g is a learnable regression function $\mathbb{R}^k \to \mathbb{R}$ that is applied to the representation of the review $\mathbf{h}_{\mathtt{rev}_{(\mathtt{u},\mathtt{i})}}$.

While g can be an arbitrary complex function, the instance of g used in this work is as follows

$$g(\mathbf{h}_{\mathtt{rev}_{(\mathtt{u},\mathtt{i})}}) = \sigma(\mathbf{h}_{\mathtt{rev}_{(\mathtt{u},\mathtt{i})}})\mathbf{w}^T + \mathtt{b}_{(\mathtt{u},\mathtt{i})}, \tag{3}$$

where $\mathbf{w}$ are the learnable weights of the linear regressor, σ is the sigmoid function $\sigma(x) = \frac{1}{1+e^{-x}}$, and $\mathtt{b}_{(\mathtt{u},\mathtt{i})}$ is the shortcut we use to refer to the sum of the bias terms, namely the user, item and overall bias: $\mathtt{b}_{(\mathtt{u},\mathtt{i})} = \mathtt{b}_{\mathtt{u}} + \mathtt{b}_{\mathtt{i}} + \mathtt{b}_0$. Later we motivate the application of the sigmoid function to the review embedding.

Of course, in a real-world scenario a recommender system makes rating predictions on items that users have *not rated yet* and, consequently, reviews are not available for those items. The application of the regression model of Eq. (3) to new examples, therefore, is not possible at test time. Our second learning procedure aims at overcoming this limitation by leveraging ideas from embedding-based knowledge base completion methods. We want to be able to approximate a review embedding at test time such that this review embedding can be used in conjunction with the learned regression model. Hence, in addition to the learning objective (2), we introduce a second objective that forces the embedding of a review to be close to the difference between the item and user embeddings. This translation-based modeling assumption is followed in TransE [5] and several other knowledge base completion methods [10,13]. We include a second term in the objective that drives the distance between (a) the user embedding translated by the review embedding and (b) the embedding of the item to be small

$$\min \mathcal{L}_2 = \min \sum_{((\mathtt{u},\mathtt{rev}_{(\mathtt{u},\mathtt{i})},\mathtt{i}),\mathtt{r}_{(\mathtt{u},\mathtt{i})}) \in S} ||\mathbf{e}_{\mathtt{u}} + \mathbf{h}_{\mathtt{rev}_{(\mathtt{u},\mathtt{i})}} - \mathbf{e}_{\mathtt{i}}||_2, \tag{4}$$

where $\mathbf{e}_{\mathtt{u}}$ and $\mathbf{e}_{\mathtt{i}}$ are the embeddings of the user and item, respectively. In the knowledge base embedding literature (cf. [5]) it is common the representations are learned via a margin-based loss, where the embeddings are updated if the score (the negative distance) of a positive triple (e.g. (`Berlin`, `located_in`, `Germany`)) is not larger than the score of a negative triple (e.g. (`Berlin`, `located_in`, `Portugal`)) plus a margin. Note that this type of learning is required to avoid trivial solutions. The minimization problem of Eq. (4) can easily be solved by setting $\mathbf{e}_{\mathtt{u}} = \mathbf{h}_{\mathtt{rev}_{(\mathtt{u},\mathtt{i})}} = \mathbf{e}_{\mathtt{i}} = \mathbf{0}\ \forall \mathtt{u},\mathtt{i}$. However, this kind of trivial solutions is avoided by jointly optimizing Eqs. (2) and (4), since a degenerate solution like the aforementioned one would lead to a high error with respect to the regression objective (Eq. (2)). The overall objective can now be written as

$$\min_{\Theta} \mathcal{L} = \min_{\Theta}(\mathcal{L}_1 + \lambda \mathcal{L}_2 + \mu ||\Theta||_2), \tag{5}$$

where λ is a term that weights the approximation loss due to the modeling assumption formalized in Eq. (4). In our model, Θ corresponds to the parameters $\mathbf{w}$, $\mathbf{e}$, $\mathbf{v}$, $\mathbf{h}_0 \in \mathbb{R}^k$ and the bias terms $\mathtt{b}$.

At test time, we can now approximate review embeddings of $(\mathtt{u},\mathtt{i})$ pairs *not seen* during training by computing

$$\boxed{\hat{\mathbf{h}}_{\mathtt{rev}_{(\mathtt{u},\mathtt{i})}} = \mathbf{e}_{\mathtt{i}} - \mathbf{e}_{\mathtt{u}}.} \tag{6}$$

With the trained regression model g we can make rating predictions $\hat{\mathtt{r}}_{(\mathtt{u},\mathtt{i})}$ for *unseen* $(\mathtt{u},\mathtt{i})$ pairs by computing

$$\hat{\mathtt{r}}_{(\mathtt{u},\mathtt{i})} = g(\hat{\mathbf{h}}_{\mathtt{rev}_{\mathtt{u},\mathtt{i}}}). \quad (7)$$

Contrary to training, now the regression model g is applied to $\hat{\mathbf{h}}_{\mathtt{rev}_{\mathtt{u},\mathtt{i}}}$, instead of $\mathbf{h}_{\mathtt{rev}_{\mathtt{u},\mathtt{i}}}$, which is not available at test time. The sigmoid function of the regression function g adds a non-linear interaction between the user and item representation. Without such activation function, the model would consist of a linear combination of bias terms and the (ranking of) served recommendations would be identical to all users.

All parameters of the parts of the objective are jointly learned with stochastic gradient descent. More details regarding the parameter learning are contained in the experimental section.

2.1 On the Choice of TRANSE as Modeling Assumption

The choice of TRANSE as underlying modeling assumption to this recommendation problem is not arbitrary. Given the user and item embeddings, and without further constraints, it allows to distinctively compute the approximate review embedding via Eq. (6). Another popular knowledge graph embedding method is DISTMULT [16]. In applying such modeling assumption to this problem one would obtain the approximate review embedding by solving the following optimization problem: $\hat{\mathbf{h}}_{\mathtt{rev}_{(\mathtt{u},\mathtt{i})}} = \max_{\mathbf{h}}(\mathbf{e}_{\mathtt{i}} \circ \mathbf{e}_{\mathtt{u}})\mathbf{h}$, where $\circ$ is the element-wise multiplication. The solution to that problem would be any vector with infinite norm. Therefore, one should impose constraints in the norm of the embeddings to obtain a non-trivial solution. However, previous work [11] shows that such constraint harms performance. Similarly, most of the knowledge graph embedding methods would require to impose constraints in the norm of the embeddings. The translation modeling assumption of TRANSE facilitates the approximation of the review embedding without additional constraints, while its performance is on par with, if not better, than most of all other translation-based knowledge graph embedding methods [11].

3 Related Work

There are three lines of research related to our work: knowledge graph completion, recommender systems and sentiment analysis.

The first research theme related to TRANSREV is knowledge graph completion. In the last years, many embedding-based methods have been proposed to infer missing relations in knowledge graphs based on a function that computes a likelihood score based on the embeddings of entities and relation types. Due to its simplicity and good performance, there is a large body of work on translation-based scoring functions [5,13]. [14] propose an approach to large-scale sequential sales prediction that embeds items into a transition space where user embeddings are modeled as translation vectors operating on item sequences. The associated optimization problem is formulated as a sequential Bayesian ranking problem [28]. To the best of our knowledge, [14] is the first work in leveraging ideas from knowledge graph completion methods for recommender system. Whereas TransRev addresses the problem of rating prediction by incorporating review text, [14] addresses the different problem of sequential recommendation. Therefore the experimental comparison to that work is not possible. In TransRev the review embedding translates the user embedding to the product embedding. In [14], the user embedding translates a product embedding to the embedding of the next purchased product. Moreover, TransRev gets rid of the margin-based loss (and consequently of the negative sampling) due to the joint optimization of Eqs. (2) and (4), whereas [14] is formalized as a ranking problem in a similar way to [5]. Subsequently, there has been additional work on translation-based models in recommender systems [25,33]. However, these works cannot incorporate users' feedback other than ratings into the learning, which has been shown to boost performance [21].

There is an extensive body of work on recommender systems [1,6,29]. Singular Value Decomposition (SVD) [17] computes the review score prediction as the dot product between the item embeddings and the user embeddings plus some learnable bias terms. Due to its simplicity and performance on numerous data sets—including winning solution to the Netflix prize—it is still one of the most used methods for product recommendations. Most of the previous research that explored the utility of review text for rating prediction can be classified into two categories.

Semi-supervised Approaches. HFT [21] was one of the first methods combining a supervised learning objective to predict ratings with an unsupervised learning objective (e.g. latent Dirichlet allocation) for text content to regularize the parameters of the supervised model. The idea of combining two learning objectives has been explored in several additional approaches [3,9,19]. The methods differ in the unsupervised objectives, some of which are tailored to a specific domain. For example, JMARS [9] outperforms HFT on a movie recommendation data set but it is outperformed by HFT on data sets similar to those used in our work [36].

Supervised Approaches. Methods that fall into this category such as [31,32] learn latent representations of users and items from the text content so as to perform well at rating prediction. The learning of the latent representations is done via a deep architecture. The approaches differences lie mainly in the neural architectures they employ.

There is one crucial difference between the aforementioned methods and TransRev. TransRev predicts the review score based on an approximation of the review embedding computed at test time. Moreover, since TransRev is able to approximate a review embedding, we can use this embedding to retrieve reviews in the training set deemed most similar by a distance metric in the embedding space.

Similar to sentiment analysis methods, TransRev trains a regression model that predicts the review rating from the review text. Contrary to the typical setting in which sentiment analysis methods operate, however, review text is not available at prediction time in the recommender system setting. Consequently, the application of sentiment analysis to recommender systems is not directly possible. In the simplest case, a sentiment analysis method is a linear regressor applied to a text embedding (Eq. (3)).

4 Experimental Setup

We conduct several experiments to empirically compare TransRev to state of the art methods for product recommendation. Moreover, we provide some qualitative results on retrieving training reviews most similar to the approximated reviews at test time.

4.1 Data Sets

We evaluate the various methods on data sets from the Amazon Product Data[1], which has been extensively used in previous works [21–23]. The data set consists of reviews and product metadata from Amazon from May 1996 to July 2014. We focus on the 5-core versions (which contain at least 5 reviews for each user and item) of those data sets. There are 24 product categories from which we have randomly picked 18. As all previously mentioned works, we treat each of these resulting 18 data sets independently in our experiments. Ratings in all benchmark data sets are integer values between 1 and 5. As in previous work, we randomly sample 80% of the reviews as training, 10% as validation, and 10% as test data. We remove reviews from the validation and test splits if they involve either a product or a user that is not part of the training data.

4.2 Review Text Preprocessing

We follow the same preprocessing steps for each data set. First, we lowercase the review texts and apply the regular expression "$\backslash w+$" to tokenize the text data, discarding those words that appear in less than 0.1% of the reviews of the data set under consideration. For all the Amazon data sets, both full reviews and short summaries (rarely having more than 30 words) are available. Since classifying short documents into their sentiment is less challenging than doing the same for longer text [4], we have used the reviews summaries for our work. We truncate these reviews to the first 200 words. For lack of space we cannot include statistics of the preprocessed data sets.

[1] http://jmcauley.ucsd.edu/data/amazon.

4.3 Baselines

We compare to the following methods: a SVD matrix factorization; HFT, which has not often been benchmarked in previous works; and DeepCoNN [38], which learns user and item representations from reviews via convolutional neural networks. We also include MPCN [34] (which stands for multi-pointer co-attention networks) in the comparison, however, as indicated in previous work [8] MPCN is a non-reproducible work[2]. Therefore, we simply copy numbers from [34], since they used the same data sets as the ones used in this work. Additionally, we also include performance for TransNets (T-Nets) [7], whose numbers are also copied from [34]. T-Nets is similar to TransRev in that it also infers review latent representations from user and item representations. Different to TransRev, it does not have any underlying graph-based modeling assumption among users, items and reviews.

Table 1. Performance comparison (MSE) on 18 datasets. The asterisk $*$ indicates the macro MSE across all the Amazon data sets.

	HFT	SVD	DeepCoNN	T-Nets	MPCN	TransRev
Amazon Instant Video	0.888	0.904	0.943	1.007	0.997	**0.884**
Automotive	0.862	0.857	**0.853**	0.946	0.861	0.855
Baby	1.104	1.108	1.154	1.338	1.304	**1.100**
Cds and Vinyl	**0.854**	0.863	0.888	1.010	1.005	**0.854**
Grocery and Gourmet Food	0.961	0.964	0.973	1.129	1.125	**0.957**
Health and personal care	1.014	1.016	1.081	1.249	1.238	**1.011**
Kindle Store	**0.593**	0.607	0.648	0.797	0.775	0.599
Musical Instruments	0.692	0.694	0.723	1.100	0.923	**0.690**
Office Products	0.727	0.727	0.738	0.840	0.779	**0.724**
Patio, Lawn and Garden	0.956	0.950	1.070	1.123	1.011	**0.941**
Pet Supplies	1.194	1.198	1.281	1.346	1.328	**1.191**
Tools and Home Improvement	0.884	0.884	0.946	1.122	1.096	**0.879**
Toys and Games	**0.784**	0.788	0.851	0.974	0.973	**0.784**
Beauty	1.165	1.168	1.184	1.404	1.387	**1.158**
Digital Music	0.793	0.797	0.835	1.004	0.970	**0.782**
Video Games	1.086	1.093	1.133	1.276	1.257	**1.082**
Sports and Outdoors	0.824	0.828	0.882	0.994	0.980	**0.823**
Cell Phones and Accessories	1.285	1.290	1.365	1.431	1.413	**1.279**
	0.926*	0.930*	0.969*	1.116*	1.079*	**0.921***

[2] A work is considered to be reproducible if a working version of the source code is available, and at least one dataset used in the original paper is available.

4.4 Parameter Setting

We set the dimension k of the embedding space to 16 for all methods. We evaluated the robustness of TransRev to changes in Sect. 4.6. Alternatively, one could use off-the-shelf word embeddings (e.g. word2vec [24] or ELMO [26]), but this would require to assume the existence of a large collection of text for effectively learning good word representations in an unsupervised manner. However, such a corpus may not be available for some low-resource languages or domain-specific use cases. For HFT we used the original implementation of the authors[3] and validated the trade-off term from the values $[0.001, 0.01, 0.1, 1, 10, 50]$. For TransRev we validated λ among the values $[0.05, 0.1, 0.25, 0.5, 1]$ and the learning rate of the optimizer and regularization term (μ in our model) from the values $[0.001, 0.005, 0.01, 0.05, 0.1]$ and $[0.00001, 0.00005, 0.0001, 0.0005, 0.001]$, respectively. TransRev's parameters were randomly initialized [12] and learned with vanilla stochastic gradient descent. A single learning iteration performs SGD with all review triples in the training data and their associated ratings. For TransRev we used a batch size of 64. We ran TransRev for a maximum of 500 epochs and validated every 10 epochs. For SVD we used the Python package Surprise[4], and chose the learning rate and regularization term from the same range of values. Parameters for HFT were learned with L-BFGS, which was run for 2,500 learning iterations and validated every 50 iterations. For DeepCoNN the original authors' code is not available and we used a third-party implementation[5]. We applied the default hyperparameters values for dropout and L2 regularization and used the same embedding dimension as for all other methods. All methods are validated according to the Mean Squared Error (MSE).

4.5 Results

The experimental results are listed in Table 1 where the best performance is in bold font. TransRev achieves the best performance on all data sets with the exception of the Kindle Store and Automotive categories. Surprisingly, HFT is more competitive than more recent approaches that also take advantage of review text. Most of these recent approaches do not include HFT in their baselines. TransRev is competitive with and often outperforms HFT on the benchmark data sets under consideration. To quantify that the rating predictions made by HFT and TransRev are significantly different we have computed the dependent t-test for paired samples and for all data sets where TransRev outperforms HFT. The p-value is always smaller than 0.01.

It is remarkable the low performance of DeepCoNN, MPCN and T-Nets in almost all datasets. This is in line with the findings reported in very recent work [8], where authors' analysis reveals that deep recommender models are systematically outperformed by simple heuristic recommender methods. These results only confirm the existing problem reported in [8].

[3] http://cseweb.ucsd.edu/jmcauley/code/code_RecSys13.tar.gz.
[4] http://surpriselib.com/.
[5] https://github.com/chenchongthu/DeepCoNN.

4.6 Hyperparameters

We randomly selected the 4 data sets *Baby, Digital Music, Office* and *Tools&Home Improvement* from the Amazon data and evaluated different values of k for user, item and word embedding sizes. We increase k from 4 to 64 and always validate all hyperparameters, including the regularization term. Table 2 list the MSE scores. We only observe small differences in the corresponding model's performances. This observation is in line with [21].

Table 2. Sensitivity to latent dimension.

k	Baby	Digital Music	Office products	Tools& Home Improv.
4	1.100	0.782	0.724	0.880
8	1.100	0.782	0.723	0.878
16	1.100	0.782	0.724	0.879
32	1.102	0.785	0.722	0.888
64	1.099	0.787	0.726	0.888

For most of the data sets the validated weighting term λ takes the value of either 0.1 or 0.25. This seems to indicate that the regression objective is more important than the modeling assumption in our task, as it directly relates to the goal of the task. The regularization term is of crucial importance to obtain good performance and largely varies across data sets, as their statistics also largely differ across data sets.

4.7 Visualization of the Word Embeddings

Review embeddings, which are learned from word embeddings, are learned to be good predictors of user ratings. As a consequence the learned word embeddings are correlated with the ratings. To visualize the correlation between words and ratings we proceed as follows. First, we assign a score to each word that is computed by taking the average rating of the reviews that contain the word. Second, we compute a 2-dimensional representation of the words by applying t-SNE [20] to the 16-dimensional word embeddings learned by TransRev. Figure 4 depicts these 2-dimensional word embedding vectors learned for the Amazon *Beauty* data set. The corresponding rating scores are indicated by the color.

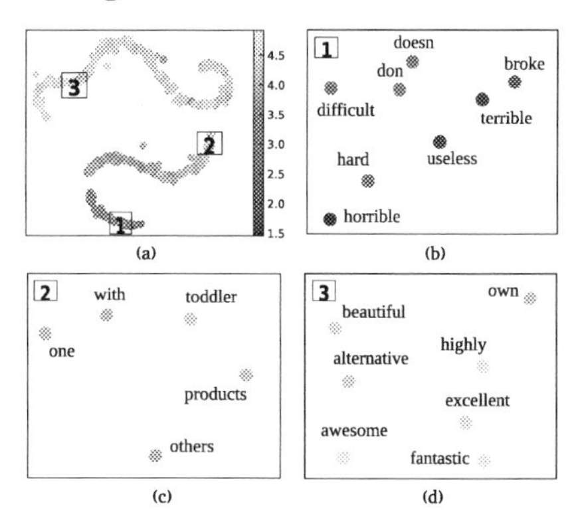

Fig. 4. (a) Two-dimensional t-SNE representations of the word embeddings learned by TransRev for the *Beauty* data set. The color bar represents the average rating of the reviews where each word appears. (b), (c) and (d) depict regions of the embedding space where negative, neutral and positive words are clustered, respectively. (Color figure online)

The clusters we discovered in Fig. 4 are interpretable. They are meaningful with respect to the score, observing that the upper right cluster is mostly made up of words with negative connotations (e.g. horrible, useless...), the lower left one contains neutral words (e.g. with, products...) and the lower right one contains words with positive connotations (e.g. awesome, excellent...).

4.8 Suggesting Reviews to Users

One of the characteristics of TRANSREV is its ability to approximate the review representation at prediction time. This approximation is used to make a rating prediction, but it can also be used to propose a tentative review on which the user can elaborate on. This is related to a number of approaches [18,27,37] on explainable recommendations. We compute the Euclidean distance between the approximated review embedding $\hat{\mathbf{h}}_{\texttt{rev}_{(u,i)}}$ and all review embeddings $\mathbf{h}_{\texttt{rev}_{(u,i)}}$ from the training set. We then retrieve the review text with the most similar review embedding. We investigate the quality of the tentative reviews that TRANSREV retrieves for the *Beauty* and *Digital Music* data sets. The example reviews listed in Table 3 show that while the overall sentiment is correct in most cases, we can also observe the following shortcomings: (a) The function f chosen in our work is invariant to word ordering and, therefore, cannot learn that bigrams such as "not good" have a negative meaning. (b) Despite matching the overall sentiment, the actual and retrieved review can refer to different aspects of the product (for example, "it clumps" and "gives me headaches"). Related work [37] extracts aspects from reviews by applying a number of grammatical and morphological analysis tools. These aspects are used later on to explain why the model suspects that a user might be interested in a certain product. We think this type of explanation is complementary to ours, and might inspire future work. (c) Reviews can be specific to a single product. A straightforward improvement could consist of retrieving only existing reviews for the specific product under consideration.

Table 3. Reviews retrieved from the *Beauty* (upper) and *Digital Music* (lower) data sets. In parenthesis the ratings associated to the reviews.

Actual test review	Closest training review in embedding space
skin improved (5)	makes your face feel refreshed (5)
love it (5)	you'll notice the difference (5)
best soap ever (5)	I'll never change it (5)
it clumps (2)	gives me headaches (1)
smells like bug repellent (3)	pantene give it up (2)
fake fake fake do not buy (1)	seems to be harsh on my skin (2)
saved my skin (5)	not good quality (2)
another great release from saliva (5)	can't say enough good things about this cd (5)
a great collection (5)	definitive collection (5)
sound nice (3)	not his best nor his worst (4)
a complete massacre of an album (2)	some great songs but overall a disappointment (3)
the very worst best of ever (1)	overall a pretty big disappointment (2)
what a boring moment (1)	overrated but still alright (3)
great cd (5)	a brilliant van halen debut album (5)

We believe that more sophisticated sentence and paragraph representations might lead to better results in the review retrieval task. As discussed, a promising line of research has to do with learning representations for reviews that are aspect-specific (*e.g.* "ease of use" or "price").

5 Conclusion

TransRev is a novel approach for product recommendation combining ideas from knowledge graph embedding methods, recommender systems and sentiment analysis. TransRev achieves state of the art performance on the data sets under consideration while having fewer (hyper)parameters than more recent works.

Most importantly, one main characteristic of TransRev is its ability to approximate the review representation during inference. This approximated representation can be used to retrieve reviews in the training set that are similar with respect to the overall sentiment towards the product. Such reviews can be dispatched to users as a starting point for a review, and thus lowering the barrier to writing new reviews. Given the known influence of product reviews in the purchasing choices of the users [2,30], we think that recommender systems will benefit from such mechanism.

Acknowledgements. The research leading to these results has received funding from the European Union's Horizon 2020 innovation action programme under grant agreement No 786741 – SMOOTH project. This publication reflects only the author's views and the European Community is not liable for any use that may be made of the information contained herein.

References

1. Allen, R.B.: User models: theory, method, and practice. Int. J. Man Mach. Stud. **32**(5), 511–543 (1990)
2. Amazon. https://www.amzinsight.com/amazon-product-review-importance/
3. Bao, Y., Fang, H., Zhang, J.: TopicMF: simultaneously exploiting ratings and reviews for recommendation. In: AAAI, pp. 2–8 (2014)
4. Bermingham, A., Smeaton, A.F.: Classifying sentiment in microblogs: is brevity an advantage? In: CIKM, pp. 1833–1836 (2010)
5. Bordes, A., Usunier, N., García-Durán, A., Weston, J., Yakhnenko, O.: Translating embeddings for modeling multi-relational data. In: NIPS, pp. 2787–2795 (2013)
6. Breese, J.S., Heckerman, D., Kadie, C.M.: Empirical analysis of predictive algorithms for collaborative filtering. In: UAI, pp. 43–52 (1998)
7. Catherine, R., Cohen, W.W.: TransNets: learning to transform for recommendation. In: RecSys, pp. 288–296 (2017)
8. Dacrema, M.F., Cremonesi, P., Jannach, D.: Are we really making much progress? A worrying analysis of recent neural recommendation approaches. In: RecSys (2019)
9. Diao, Q., Qiu, M., Wu, C., Smola, A.J., Jiang, J., Wang, C.: Jointly modeling aspects, ratings and sentiments for movie recommendation (JMARS). In: KDD, pp. 193–202 (2014)
10. García-Durán, A., Bordes, A., Usunier, N.: Composing relationships with translations. In: EMNLP, pp. 286–290. The Association for Computational Linguistics (2015)
11. García-Durán, A., Bordes, A., Usunier, N., Grandvalet, Y.: Combining two and three-way embedding models for link prediction in knowledge bases. J. Artif. Intell. Res. **55**, 715–742 (2016)

12. Glorot, X., Bengio, Y.: Understanding the difficulty of training deep feedforward neural networks. In: AISTATS. JMLR Proceedings, vol. 9, pp. 249–256 (2010)
13. Guu, K., Miller, J., Liang, P.: Traversing knowledge graphs in vector space. In: EMNLP, pp. 318–327. The Association for Computational Linguistics (2015)
14. He, R., Kang, W., McAuley, J.: Translation-based recommendation. In: RecSys, pp. 161–169 (2017)
15. Joe, M.: https://www.matrixmarketinggroup.com/product-taxonomy-ecommerce-sales/
16. Kadlec, R., Bajgar, O., Kleindienst, J.: Knowledge base completion: baselines strike back. arXiv preprint arXiv:1705.10744 (2017)
17. Koren, Y., Bell, R.M., Volinsky, C.: Matrix factorization techniques for recommender systems. IEEE Comput. **42**(8), 30–37 (2009)
18. Lawlor, A., Muhammad, K., Rafter, R., Smyth, B.: Opinionated explanations for recommendation systems. In: Bramer, M., Petridis, M. (eds.) Research and Development in Intelligent Systems XXXII, pp. 331–344. Springer, Cham (2015). https://doi.org/10.1007/978-3-319-25032-8_25
19. Ling, G., Lyu, M.R., King, I.: Ratings meet reviews, a combined approach to recommend. In: RecSys, pp. 105–112 (2014)
20. Maaten, L., Hinton, G.: Visualizing data using t-SNE. J. Mach. Learn. Res. **9**, 2579–2605 (2008)
21. McAuley, J.J., Leskovec, J.: Hidden factors and hidden topics: understanding rating dimensions with review text. In: RecSys, pp. 165–172 (2013)
22. McAuley, J.J., Pandey, R., Leskovec, J.: Inferring networks of substitutable and complementary products. In: KDD, pp. 785–794 (2015)
23. McAuley, J.J., Targett, C., Shi, Q., van den Hengel, A.: Image-based recommendations on styles and substitutes. In: SIGIR, pp. 43–52 (2015)
24. Mikolov, T., Sutskever, I., Chen, K., Corrado, G.S., Dean, J.: Distributed representations of words and phrases and their compositionality. In: Advances in Neural Information Processing Systems, pp. 3111–3119 (2013)
25. Palumbo, E., Rizzo, G., Troncy, R., Baralis, E., Osella, M., Ferro, E.: An empirical comparison of knowledge graph embeddings for item recommendation. In: DL4KGS@ ESWC, pp. 14–20 (2018)
26. Peters, M.E., et al.: Deep contextualized word representations. arXiv preprint arXiv:1802.05365 (2018)
27. Qureshi, M.A., Greene, D.: *Lit@EVE*: explainable recommendation based on wikipedia concept vectors. In: Altun, Y., et al. (eds.) ECML PKDD 2017. LNCS (LNAI), vol. 10536, pp. 409–413. Springer, Cham (2017). https://doi.org/10.1007/978-3-319-71273-4_41
28. Rendle, S., Freudenthaler, C., Schmidt-Thieme, L.: Factorizing personalized Markov chains for next-basket recommendation. In: WWW, pp. 811–820 (2010)
29. Rennie, J.D.M., Srebro, N.: Fast maximum margin matrix factorization for collaborative prediction. In: ICML. ACM International Conference Proceeding Series, vol. 119, pp. 713–719 (2005)
30. Saleh, K.: https://www.invespcro.com/blog/the-importance-of-online-customer-reviews-infographic/
31. Seo, S., Huang, J., Yang, H., Liu, Y.: Interpretable convolutional neural networks with dual local and global attention for review rating prediction. In: RecSys, pp. 297–305 (2017)

32. Seo, S., Huang, J., Yang, H., Liu, Y.: Representation learning of users and items for review rating prediction using attention-based convolutional neural network. In: 3rd International Workshop on Machine Learning Methods for Recommender Systems (MLRec) (2017)
33. Tay, Y., Anh Tuan, L., Hui, S.C.: Latent relational metric learning via memory-based attention for collaborative ranking. In: Proceedings of the 2018 World Wide Web Conference on World Wide Web, pp. 729–739. International World Wide Web Conferences Steering Committee (2018)
34. Tay, Y., Tuan, L.A., Hui, S.C.: Multi-pointer co-attention networks for recommendation. In: KDD (2018)
35. Wallace, T.: https://www.bigcommerce.com/blog/ecommerce-sales-funnel/
36. Wu, C., Beutel, A., Ahmed, A., Smola, A.J.: Explaining reviews and ratings with PACO: poisson additive co-clustering. In: WWW (Companion Volume), pp. 127–128 (2016)
37. Zhang, Y., Lai, G., Zhang, M., Zhang, Y., Liu, Y., Ma, S.: Explicit factor models for explainable recommendation based on phrase-level sentiment analysis. In: SIGIR, pp. 83–92 (2014)
38. Zheng, L., Noroozi, V., Yu, P.S.: Joint deep modeling of users and items using reviews for recommendation. In: WSDM, pp. 425–434 (2017)

Information Extraction

Domain-Independent Extraction of Scientific Concepts from Research Articles

Arthur Brack[1(✉)], Jennifer D'Souza[1(✉)], Anett Hoppe[1(✉)], Sören Auer[1(✉)], and Ralph Ewerth[1,2(✉)]

[1] TIB – Leibniz Information Centre for Science and Technology, Hannover, Germany
{arthur.brack,jennifer.dsouza,anett.hoppe,soeren.auer ralph.ewerth}@tib.eu

[2] L3S Research Center, Leibniz University, Hannover, Germany

Abstract. We examine the novel task of *domain-independent scientific concept extraction from abstracts of scholarly articles* and present two contributions. First, we suggest a set of generic scientific concepts that have been identified in a systematic annotation process. This set of concepts is utilised to annotate a corpus of scientific abstracts from 10 domains of Science, Technology and Medicine at the phrasal level in a joint effort with domain experts. The resulting dataset is used in a set of benchmark experiments to (a) provide baseline performance for this task, (b) examine the transferability of concepts between domains. Second, we present a state-of-the-art deep learning baseline. Further, we propose the active learning strategy for an optimal selection of instances from among the various domains in our data. The experimental results show that (1) a substantial agreement is achievable by non-experts after consultation with domain experts, (2) the baseline system achieves a fairly high F1 score, (3) active learning enables us to nearly halve the amount of required training data.

Keywords: Sequence labelling · Information extraction · Scientific articles · Active learning · Scholarly communication · Research knowledge graph

1 Introduction

Scholarly communication as of today is a document-centric process. Research results are usually conveyed in written articles, as a PDF file with text, tables and figures. Automatic indexing of these texts is limited and generally does not access their semantic content. There are thus severe limitations how current research infrastructures can support scientists in their work: finding relevant research works, comparing them, and compiling summaries is still a tedious and error-prone manual work. The heightened increase in the number of published research papers aggravates this situation [7].

Knowledge graphs are recognised as an effective approach to facilitate semantic search [3]. For academic search engines, Xiong et al. [47] have shown that

J. M. Jose et al. (Eds.): ECIR 2020, LNCS 12035, pp. 251–266, 2020.
https://doi.org/10.1007/978-3-030-45439-5_17

exploiting knowledge bases like Freebase can improve search results. However, the introduction of new scientific concepts occurs at a faster pace than knowledge base curation, resulting in a large gap in knowledge base coverage of scientific entities [1], e.g. the task *geolocation estimation of photos* from the Computer Vision field is neither present in Wikipedia nor in more specialised knowledge bases like Computer Science Ontology (CSO) [39] or "Papers with code" [36]. Information extraction from text helps to identify emerging entities and to populate knowledge graphs [3]. It then is a first vital step towards a fine-grained research knowledge graph in which research articles are described and interconnected through entities like tasks, materials, and methods. Our work is motivated by the idea of the automatic construction of a research knowledge graph.

Information extraction from scientific texts, obviously, differs from its general domain counterpart: Understanding a research paper and determining its most important statements demands certain expertise in the article's domain. Every domain is characterised by its specific terminology and phrasing which is hard to grasp for a non-expert reader. In consequence, extraction of scientific concepts from text would entail the involvement of domain experts and a specific design of an extraction methodology for each scientific discipline – both requirements are rather time-consuming and costly.

At present, a systematic study of these assumptions is missing. We thus present the task of *domain-independent scientific concept extraction*. We examine the intuition that most research papers share certain core concepts such as the mentions of research tasks or methods. If so, these would allow a domain-independent information extraction system to support populating a research knowledge graph, which does not reach all semantic depths of the analysed article, but still provides some science-specific structure.

In this paper, we introduce a set of common scientific concepts that we find are relevant over a set of 10 examined domains from Science, Technology, and Medicine (STM). These generic concepts have been identified in a systematic, joint effort of domain experts and non-domain experts. The inter-coder agreement is measured to ensure the adequacy and quality of concepts. A set of research abstracts has been annotated using these concepts and the results are discussed with experts from the corresponding fields. The resulting dataset serves as a basis to train two baseline deep learning classifiers. In particular, we present an active learning approach to reduce the number of required training data. The systems are evaluated in different experimental setups.

Our main contributions can be summarised as follows: (1) We introduce the novel task *domain-independent scientific concept extraction*, which aims at automatically extracting scientific entities in a domain-independent manner. (2) We release a new corpus that comprises 110 abstracts of 10 STM domains annotated at the phrasal level. (3) We present and evaluate a state-of-the-art deep learning approach for this task. Additionally, we employ active learning for an optimal selection of instances, which to our knowledge, is demonstrated for the first time on scholarly text. We find that strategic instance selection gives us the same performance with only about half of the training data. (4) We release a silver-labelled corpus with 62 K automatically annotated abstracts of Elsevier with CCBY license and 1.2 Mio. extracted unique concepts comprising 24 domains. (5) We make our corpora and source code publicly available to facilitate further research.

2 Related Work

This section gives a brief overview of existing annotated datasets for scientific information extraction, followed by related work on some exemplary applications for domain-independent information extraction from scientific papers.

2.1 Scientific Corpora

Sentence Level Annotation. Early approaches for semantic structuring of research papers focused on sentences as the basic unit of analysis. This enables, for instance, automatic highlighting of relevant paper passages to enable efficient assessment regarding quality and relevance. Several ontologies have been created that focus on the rhetorical [11,19], argumentative [31,46] or activity-based [37] structure of research papers.

Annotated datasets exist for several domains, e.g. PubMed200k [12] from biomedical randomized controlled trials, NICTA-PIBOSO [26] from evidence-based medicine, Dr. Inventor [15] from Computer Graphics, Core Scientific Concepts (CoreSC) [31] from Chemistry and Biochemistry, and Argumentative Zoning (AZ) [46] from Chemistry and Computational Linguistics, Sentence Corpus [8] from Biology, Machine Learning and Psychology. Most datasets cover only a single domain, while few other datasets cover three domains. Several machine learning methods have been proposed for scientific sentence classification [12,15,24,30].

Phrase Level Annotation. More recent corpora have been annotated at phrasal level (e.g. noun phrases). SciCite [9] and ACL ARC [25] are datasets for citation intent classification from Computer Science, Medicine, and Computational Linguistics. ACL RD-TEC [20] from Computational Linguistics aims at extracting scientific technology and non-technology terms. ScienceIE-17 [2] from Computer Science, Material Sciences, and Physics contains three concepts PROCESS, TASK and MATERIAL. SciERC [32] from the machine learning domain contains six concepts TASK, METHOD, METRIC, MATERIAL, OTHER-SCIENTIFIC-TERM and GENERIC. Each corpus covers at most three domains.

Experts vs. Non-experts. The aforementioned datasets were usually annotated by domain experts [2,12,20,26,31,32]. In contrast, Teufel et al. [46] explicitly use non-experts in their annotation tasks, arguing that text understanding systems can use general, rhetorical and logical aspects also when qualifying scientific text. According to this line of thought, more researchers used (presumably cheaper) non-expert annotation as an alternative [8,15].

Snow et al. [43] provide a study on expert versus non-expert performance for general, non-scientific annotation tasks. They state that about four non-experts (Mechanical Turk workers, in their case) were needed to rival the experts' annotation quality. However, systems trained on data generated by non-experts showed to benefit from annotation diversity and to suffer less from annotator bias. A recent study [38] examines the agreement between experts and non-experts for visual concept classification and person recognition in historical video data. For the task of face recognition, training with expert annotations lead to an increase of only 1.5% in classification accuracy.

Active Learning in Natural Language Processing (NLP). To the best of our knowledge, active learning has not been utilised in classification approaches for scientific text yet. Recent publications demonstrate the effectiveness of active learning for NLP tasks such as *Named Entity Recognition* (NER) [41] and sentence classification [49]. Siddhant and Lipton [42] and Shen et. al. [41] compare several sampling strategies on NLP tasks and show that *Maximum Normalized Log-Probability* (MNLP) based on uncertainty sampling performs well in NER.

2.2 Applications for Domain-Independent Scientific Information Extraction

Academic Search Engines. Academic search engines such as Google Scholar [18], Microsoft Academic [34] and Semantic Scholar [40] specialise in search of scholarly literature. They exploit graph structures such as the Microsoft Academic Knowledge Graph [35], SciGraph [45], or the Semantic Scholar Corpus [1]. These graphs interlink the papers through meta-data such as citations, authors, venues, and keywords, but not through deep semantic representation of the articles' content.

However, first attempts towards a more semantic representation of article content exist: Ammar et al. [1] interlink the Semantic Scholar Corpus with DBpedia [29] and Unified Medical Language System (UMLS) [6] using entity linking techniques. Yaman et al. [48] connect SciGraph with DBpedia person entities. Xiong et al. [47] demonstrate that academic search engines can greatly benefit from exploiting general-purpose knowledge bases. However, the coverage of science-specific concepts is rather low [1].

Research Paper Recommendation Systems. Beel et al. [4] provide a comprehensive survey about research paper recommendation systems. Such systems usually employ different strategies (e.g. content-based and collaborative filtering) and several data sources (e.g. text in the documents, ratings, feedback, stereotyping). Graph-based systems, in particular, exploit citation graphs and genes mentioned in the papers [27]. Beel et al. conclude that it is not possible to determine the most effective recommendation approach at the moment. However, we believe that a fine-grained research knowledge graph can improve such systems. Although "Papers with code" [36] is not a typical recommendation system, it allows researchers to browse easily for papers from the field of machine learning that address a certain task.

3 Corpus for Domain-Independent Scientific Concept Extraction

In this section, we introduce the novel task of *domain-independent extraction of scientific concepts* and present an annotated corpus. As the discussion of related work reveals, the annotation of scientific resources is not a novel task. However, most researchers focus on at most three scientific disciplines and on expert-level annotations. In this work, we explore the domain-independent annotation of lexical phrasal units indicating scientific knowledge, i.e. scientific concepts, in abstracts from ten different science domains. Since other studies have also shown that non-expert annotations are feasible for the scientific domain, we go for a cost-efficient middle course: annotations by non-experts with scientific proficiency, and consultation with domain-experts. Finally, we explore how well a state-of-the-art deep learning model performs on this novel information extraction task and whether active learning can help to reduce the amount of required training data. Our novel corpus and the annotation process are described below.

3.1 OA-STM Corpus

The OA-STM corpus [14] is a set of open access (OA) articles from various domains in Science, Technology and Medicine (STM). It was published in 2017 as a platform for benchmarking methods in scholarly article processing, amongst other scientific information extraction. The dataset contains a selection of 110 articles from 10 domains, namely Agriculture (*Agr*), Astronomy (*Ast*), Biology (*Bio*), Chemistry (*Che*), Computer Science (*CS*), Earth Science (*ES*), Engineering (*Eng*), Materials Science (*MS*), Mathematics (*Mat*), and Medicine (*Med*). This first annotation cycle focuses on the articles' abstracts as they contain a condensed summary of the article.

3.2 Annotation Process

The OA-STM Corpus is used as a base for (a) the identification of potential domain-independent concepts; (b) a first annotated corpus for baseline classification experiments. The annotation task was mainly performed by two post-doctoral researchers with a background in Computer Science (acting as non-expert annotators); their basic annotation assumptions were checked by domain experts.

Pre-annotation. A literature review of annotation schemes [2,11,30,31] provided a seed set of potential candidate concepts. Both non-experts independently annotated a subset of the STM abstracts with these concepts (non-overlapping) and discussed the outcome. In a three-step process, the concept set was pruned to only contain those which seemed suitably transferable between domains. Our set of *generic* scientific concepts consists of PROCESS, METHOD, MATERIAL, and DATA (see Table 1 for their definitions). We also identified TASK [2], OBJECT [30],

Table 1. The four core scientific concepts that were derived in this study

Process	Natural phenomenon or activities, e.g. growing (*Bio*), reduction (*Mat*), flooding (*ES*)
Method	A commonly used procedure that acts on entities, e.g. powder X-ray (*Che*), the PRAM analysis (*CS*), magnetoencephalography (*Med*)
Material	A physical or abstract entity used in scientific experiments or proofs, e.g. soil (*Agr*), the moon (*Ast*), the carbonator (*Che*)
Data	The data themselves, measurements, or quantitative or qualitative characteristics of entities, e.g. rotational energy (*Eng*), tensile strength (*MS*), 3D time-lapse seismic data (*ES*)

and Results [11], however, in this study we do not consider nested span concepts, hence we leave them out since they were almost always nested with the other scientific entities (e.g. a Result may be nested with Data).

Phase I. Five abstracts per domain (i.e. 50 abstracts) were annotated by both annotators and the inter-annotator agreement was computed using Cohen's κ [10] at exact annotated spans. Results showed a moderate inter-annotator agreement of 0.52 κ.

Phase II. The annotations were then presented to subject specialists who each reviewed (a) the choice of concepts and (b) annotation decisions on the respective domain corpus. The interviews mostly confirmed the concept candidates as generally applicable. The experts' feedback on the annotation was even more valuable: The comments allowed for a more precise reformulation of the annotation guidelines, including illustrating examples from the corpus.

Consolidation. Finally, the 50 abstracts from phase I were reannotated by the non-experts. Based on the revised annotation guidelines, a substantial agreement of 0.76 κ could be reached (see Table 2). Similar annotation tasks for scientific entities, i.e. SciERC [32] considering one domain and ScienceIE-17 [2] considering three domains achieved agreements of 0.76 κ and 0.6 κ, respectively. Subsequently, the remaining 60 abstracts (six per domain) were annotated by one annotator. This phase also involved reconciliation of the previously annotated 50 abstracts to obtain a gold standard corpus.

Table 2. Per-domain and overall inter-annotator agreement (Cohen's Kappa κ) for Process, Method, Material, and Method scientific concept annotation

	Med	*MS*	*CS*	*ES*	*Eng*	*Che*	*Bio*	*Agr*	*Mat*	*Ast*	*Overall*
κ	0.94	0.90	0.85	0.81	0.79	0.77	0.75	0.60	0.58	0.57	0.76

3.3 Corpus Characteristics

Table 3 shows some characteristics of the resulting corpus. The corpus has a total of 6,127 scientific entities, including 2,112 Process, 258 Method, 2,099 Material, and 1,658 Data concept entities. The number of entities per abstract in our corpus directly correlates with the length of the abstracts (Pearson's R 0.97). Among the concepts, Process and Material directly correlate with abstract length (R 0.8 and 0.83, respectively), while Data has only a slight correlation (R 0.35) and Method has no correlation (R 0.02). The domains *Bio*, *CS*, *Ast*, and *Eng* contain the most of Process, Method, Material, and Data concepts, respectively.

Table 3. The annotated corpus characteristics containing 11 abstracts per domain in terms of size and the number of scientific concept phrases

	Ast	*Agr*	*Eng*	*ES*	*Bio*	*Med*	*MS*	*CS*	*Che*	*Mat*
Avg. # Tokens/Abstract	382	333	303	321	273	274	282	253	217	140
# Process	241	252	248	243	281	244	178	220	149	56
# Method	19	28	27	9	15	33	27	66	27	7
# Material	296	292	208	249	291	191	231	102	188	51
# Data	235	169	258	197	62	132	138	165	119	183
# Gold scientific concept phrases	791	741	741	698	649	600	574	553	483	297
# Unique gold scientific concept phrases	663	631	618	633	511	518	493	482	444	287

4 Automatic Domain-Independent Scientific Concept Extraction

The current state-of-the-art for scientific entity extraction is Beltagy et al.'s deep learning system with SciBERT word embeddings [5], which were pre-trained on scientific texts using the BERT [13] architecture. It consists of three components: (a) a token embedding layer comprising a per-sentence sequence of tokens, where each token is represented as a concatenation of SciBERT word embedding and CNN-based character embeddings [33], (b) a token-level encoder with two stacked bidirectional LSTMs [21], and (c) a Conditional Random Field (CRF) based tag decoder [33] with BILOU (beginning, inside, last, outside, unit) tagging scheme. This deep learning architecture is implemented in AllenNLP [17] and uses spaCy [44] for text preprocessing, i.e. for tokenisation and sentence-splitting.

4.1 Supervised Learning with Full Training Dataset

Using the above mentioned architecture, we train one model with data from all domains combined. We refer to this model as the *domain-independent* classifier.

Similarly, we train 10 models for each domain in our corpus – the *domain-specific* classifier.

To obtain a robust evaluation of models, we perform five-fold cross-validation experiments. In each fold experiment, we train a model on 8 abstracts per domain (i.e. 80 abstracts), tune hyperparameters on 1 abstract per domain (i.e. 10 abstracts), and test on the remaining 2 abstracts per domain (i.e. 20 abstracts) ensuring that the data splits are not identical between the folds. All results reported in the paper are averaged over the five folds. We still obtain reliably trained domain-specific classifiers since on average they are trained on 400 concepts.

4.2 Active Learning with Training Data Subset

In this setting, we employ an active learning strategy [42,49] to train a new *domain-independent* classifier. Active learning is usually applied to determine the optimal set of sufficiently distinct instances to minimise annotation costs. With our application of active learning we find which proportion of our annotations suffice for training a robust classifier. We decide to use the MNLP [41] sampling strategy. We prefer it over its contemporary, Bayesian Active Learning by Disagreement (BALD) [22], since it has less computational requirements. The MNLP objective involves greedy sampling of sentences preferring those with the least logarithmic likelihood of the predicted tag sequence output by the CRF tag decoder, normalised by the number of tokens to avoid preferring longer sentences. In our experiments, we found that adding 4% of the data to be the most discriminative selection of classifier performance. Therefore, we run 25 iterations of active learning in each stage adding 4% training data. We perform five-fold cross validation as before and the per-fold models are retrained after data resampling.

5 Experimental Results and Discussion

In this section, we discuss the results obtained with our trained classifiers and the correlation analysis between inter-annotator agreement and performance of the classifiers.

5.1 Domain-Independent and Domain-Specific Classifiers: Full Training Dataset

Table 4 shows an overview of the *domain-independent* classifier results. The system achieves an overall $F1$ of 65.5% and has low standard deviation 1.26 across the five folds. For this classifier, MATERIAL was the easiest concept with an $F1$ of 71% (±1.88), whereas METHOD was the hardest concept with an $F1$ of 43% (±6.30). METHOD is also the most underrepresented in our corpus, which partly explains the poor extraction performance. Best reported results for similar datasets, ScienceIE17 [2] and SciERC [32] (both have 500 abstracts), have

Table 4. The *domain-independent* classifier results in terms of Precision (P), Recall (R), and F1-score on scientific concepts, respectively, and *Overall*

	PROCESS	METHOD	MATERIAL	DATA	*Overall*
P	65.5 (±4.22)	45.8 (±13.50)	69.2 (±3.55)	60.3 (±4.14)	64.3 (±1.73)
R	68.3 (±1.93)	44.1 (±8.73)	73.2 (±4.27)	60.0 (±4.84)	66.7 (±0.92)
$F1$	66.8 (±2.07)	43.0 (±6.30)	71.0 (±1.88)	59.8 (±1.75)	65.5 (±1.26)

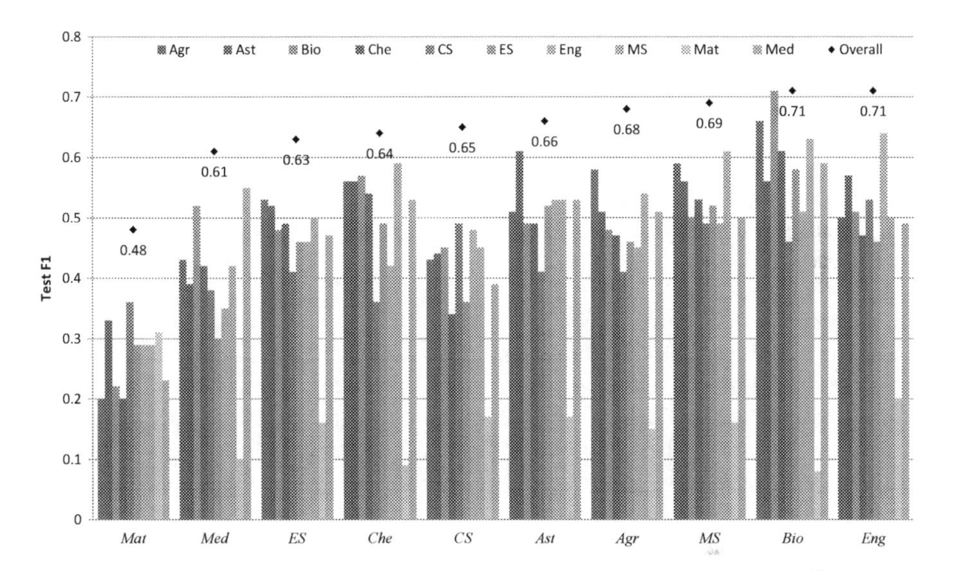

Fig. 1. $F1$ per domain of the 10 *domain-specific* classifiers (as bar plots) and of the *domain-independent* classifier (as scatter plots) for scientific concept extraction; the x-axis represents the 10 test domains

an $F1$ score of 65.6% [5] and 44.7% [32], respectively, indicating that the size of our dataset with only 110 abstracts is sufficient.

Next, we compare and contrast the 10 *domain-specific* classifiers (see Fig. 1) by their capability to extract the concepts from their own domains and in other domains.

Most Robust Domain. *Bio* (third bar in each domain in Fig. 1) extracts scientific concepts from its own domain at the same performance as the *domain-independent* classifier with an $F1$ score of 71% (±9.0) demonstrating a robust domain. It comprises only 11% of the overall data, yet the *domain-independent* classifier trained on all data does not outperform it.

Most Generic Domain. *MS* (the third last bar in each domain in Fig. 1) exhibits a high degree of domain independence since it is among the top 3 classifiers for seven of the 10 domains (viz. *ES*, *Che*, *CS*, *Ast*, *Agr*, *MS*, and *Bio*).

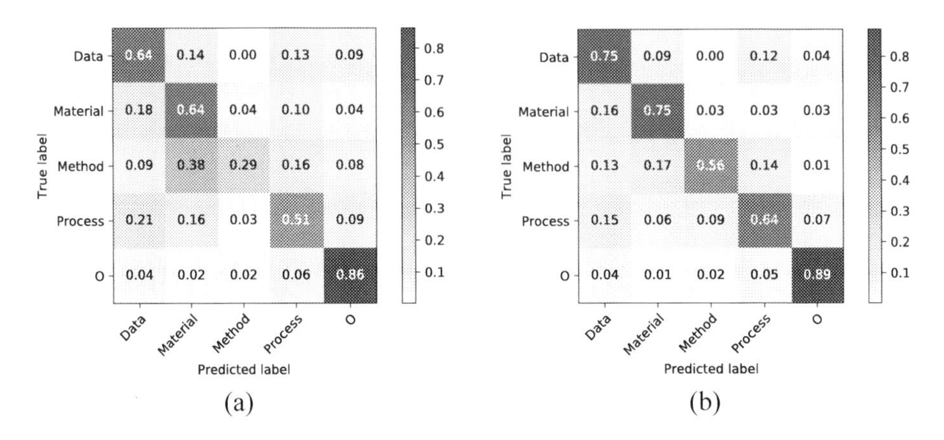

Fig. 2. Confusion matrix for (a) the *CS* classifier and (b) *domain-independent classifier* on *CS* domain predicting concept-type of tokens

Most Specialised Domain. *Mat* (the second last bar in each domain in Fig. 1) shows the lowest performance in extracting scientific concepts from all domains except itself. Hence it shows to be the most specialised domain in our corpus. Notably, a characteristic feature of this domain is that it has short abstracts (nearly a third of the size of the longest abstracts), so it is also the most underrepresented in our corpus. Also, distinct from the other domains, *Mat* has triple the number of DATA entities compared to each of its other concepts, where in the other domains PROCESS and MATERIAL are consistently predominant.

Medical and Life Science Domains. The *Med*, *Agr*, and *Bio* domains show strong domain relatedness. Their respective *domain-specific* classifiers show top five system performances among the three domains, when applied to another domain. For instance, the *Med* domain shows the strongest domain relatedness and is classified best by *Med* (last bar), followed by *Bio* (third bar) and *Agr* (first bar).

Domain-Independent vs. Domain-Specific Classifier. Except for *Bio* the *domain-independent* classifier clearly outperforms the *domain-specific* one in extracting concepts from their respective domains. We attribute this, in part, to the improved span-detection performance. Span-detection merely relies on syntactic regularity, thus the *domain-independent* classifier can benefit from more training data of other domains. E.g., the CS classifier shows a relative improvement of 49.5% *domain-specific* $F1$ score to 65.9% in the *domain-independent* setting, which is supported by the enhanced span-detection performance from 73.4% to 82.0% in $F1$. Accuracy on token-level also improves from 67.7% to 77.5% $F1$ for CS, that is correct labelling of the tokens also benefits from other domains. This is also supported by the results in the confusion matrix depicted in Fig. 2 for the *CS* and the *domain-independent* classifier on token-level.

Scientific Concept Extraction. Figure 3 depicts the 10 *domain-specific* classifier results for extracting each of the four scientific concepts. It can be observed that *Agr*, *Med*, *Bio*, and *Ast* classifiers are the best in extracting their respective PROCESS, METHOD, MATERIAL, and DATA concepts.

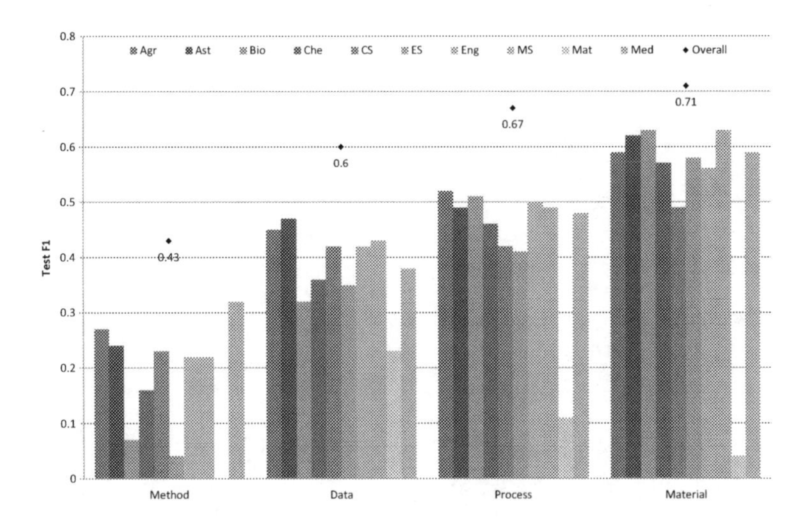

Fig. 3. $F1$ scores of the 10 *domain-specific* classifiers (bar plots) and the *domain-independent* classifiers (scatter plots) for extracting each scientific concept; the x-axis represents the evaluated concepts

5.2 Domain-Independent Classifier with Active Learning

The results of the active learning experiment over the full dataset plotted over the 25 iterations are depicted in Fig. 4, showing that MNLP clearly outperforms the random baseline. While using only 52% of the training data, the best result of the *domain-independent* classifier trained with all training data is surpassed with an $F1$ score of 65.5% (±1.0). The random baseline achieves an $F1$ score of only 62.5% (±2.6) with the same proportion of training data. When 76% of the data are sampled by MNLP, the best active learning performance across all steps is achieved with an $F1$ score of 69.0% on the validation set, having the best $F1$ of 66.4% (±2.0) on the test set. Thus, 76% of our annotated sentences suffice to train an optimal performing model.

Analysing the distribution of sentences in the training data sampled by MNLP, shows (*Math*, *CS*) as the most preferred domains and (*Eng*, *MS*) the least preferred ones. Nonetheless, all domains are represented, that is a non-uniformly mix of sentences sampled by MNLP yields the most generic model with less training data. In contrast, the random sampling strategy uniformly samples sentences from all domains.

Further, we show in Table 5 the proportion of training data for MNLP when the performance using the entire training dataset is achieved for related SciERC [32] and ScienceIE-17 [2] datasets. The results indicate, that also for related datasets on scientific texts MNLP can significantly reduce the amount of labelled training data.

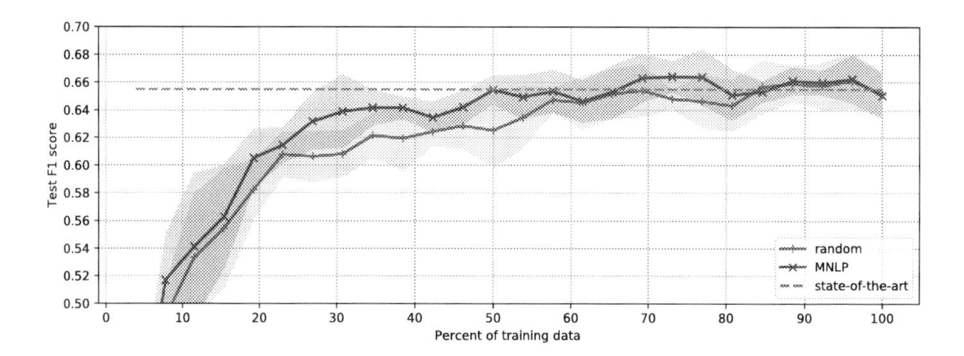

Fig. 4. Progress of active learning with MNLP and random sampling strategy; the areas represent the standard deviation (std) of the F1 score across 5 folds for MNLP and random sampling strategy, respectively

Table 5. Performance of active learning with MNLP and random sampling strategy for the fraction of training data when the performance with entire training dataset is achieved; for SciERC and ScienceIE-17 results are reported across 5 random restarts

	Training data	$F1$ (MNLP)	$F1$ (random)	$F1$ (full data)
STM (our corpus)	52%	65.5 (±1.0)	62.5 (±2.6)	65.5 (±1.3)
SciERC [32]	62%	65.3 (±1.5)	62.3 (±1.5)	65.6 (±1.0)
ScienceIE17 [2]	38%	43.9 (±1.2)	42.2 (±1.8)	43.8 (±1.0)

5.3 Correlations Between Inter-annotator Agreement and Performance

In this section, we analyse the correlations (Pearson's R) of inter-coder agreement κ and the number of annotated concepts per domain (#) on (1) the performance $F1$ and (2) variance resp. standard deviation (std) of the classifiers across five-fold cross validation.

Table 6 summarises the results of our correlation analysis. The active learning classifier (AL-trained) has been trained with 52% training data sampled by MNLP since it is the point at which the performance of the full data trained model is surpassed (see Table 5). For the domain-specific, domain-independent and AL-trained classifier we observe a strong correlation between F1 and number of concepts per domain (R 0.70, 0.76, 0.68) and a weak correlation between κ and F1 (R 0.20, 0.28, 0.23). Thus, we surmise that the number of annotated concepts in a particular domain has more influence on the performance than the inter-annotator agreement.

Table 6. Inter-annotator agreement (κ) and the number of concept phrases (#) per domain; F1 and std of domain-specific classifiers on their domains; F1 and std of domain-independent and AL-trained classifier on each domain; the right side depicts correlation coefficients (R) of each row with κ and the number of concept phrases

	Agr	*Ast*	*Bio*	*Che*	*CS*	*ES*	*Eng*	*MS*	*Mat*	*Med*	R κ	R #
Inter-annotator agreement (κ)	0.6	0.57	0.75	0.77	0.85	0.81	0.79	0.9	0.58	0.94	1.00	−0.02
# concept phrases (#)	741	791	649	483	553	698	741	574	297	600	−0.02	1.00
Domain-specific (F1)	0.58	0.61	0.71	0.54	0.49	0.46	0.64	0.61	0.31	0.55	0.20	0.70
Domain-independent (F1)	0.68	0.66	0.71	0.64	0.65	0.63	0.71	0.69	0.48	0.61	0.28	0.76
AL-trained (F1)	0.65	0.67	0.74	0.65	0.62	0.63	0.72	0.69	0.50	0.60	0.23	0.68
Domain-specific (std)	0.06	0.06	0.09	0.08	0.05	0.06	0.04	0.11	0.06	0.07	0.29	0.28
Domain-independent (std)	0.04	0.04	0.11	0.08	0.07	0.05	0.03	0.04	0.06	0.03	−0.11	−0.05
AL-trained (std)	0.04	0.04	0.09	0.08	0.07	0.04	0.07	0.05	0.15	0.02	−0.41	−0.72

The correlation values for the variance are different between the classifier types. For the domain-specific classifier the correlation between κ and std, and the number of concepts per domain and std are slightly positive (R 0.29, 0.28), i.e. the higher the agreement and the size of the domain, the higher the variance of the domain-specific classifier. For the domain-independent classifier, there is no correlation (R 0.11, −0.05) and for the AL-trained classifier, the correlations become negative (R −0.41, −0.72), i.e. higher agreement and more annotated concepts per domain lead to less variance for the AL-trained classifier. In summary, we hypothesise that more diverse training data from several domains lead to better performance and lower variance by introducing an inductive bias.

6 Conclusions

In this paper, we have introduced the novel task of *domain-independent concept extraction* from scientific texts. During a systematic annotation procedure involving domain experts, we have identified four general core concepts that are relevant across the domains of Science, Technology and Medicine. To enable and foster research on these topics, we have annotated a corpus for the domains. We have verified the adequacy of the concepts by evaluating the human annotator agreement for our broad STM domain corpus. The results indicate that the identification of the *generic* concepts in a corpus covering 10 different scholarly domains is feasible by non-experts with moderate agreement and after consultation of domain experts with substantial agreement (0.76 κ).

We evaluated a state-of-the-art system on our annotated corpus which achieved a fairly high F1 score (65.5% overall). The domain-independent system noticeably outperforms the domain-specific systems, which indicates that the model can generalise well across domains. We also observed a strong correlation between the number of annotated concepts per domain and classifier performance, and only a weak correlation between inter-annotator agreement per domain and the performance. It is assumed that more annotated data positively influence the performance in the respective domain.

Furthermore, we have suggested active learning for our novel task. We have shown that only approx. 5 annotated abstracts per domain serving as training data are sufficient to build a performant model. Our active learning results for SciERC [32] and ScienceIE17 [2] datasets were similar. The promising results suggest that we do not need a large annotated dataset for scientific information extraction. Active learning can significantly save annotation costs and enable fast adaptation to new domains.

We make our annotated corpus, a silver-labelled corpus with 62K abstracts comprising 24 domains, and source code publicly available.[1] Thereby, we hope to facilitate research on the task of scientific information extraction and its several applications, e.g. academic search engines or research paper recommendation systems.

In the future, we plan to extend and refine the concepts for certain domains. We also intend to apply and evaluate our automatic scientific concept extraction system to expand an open research knowledge graph [23]. For this purpose, we plan to extend the corpus with additional relevant annotation layers such as with coreference links [28] and relations [16,32].

References

1. Ammar, W., et al.: Construction of the literature graph in semantic scholar. In: NAACL-HLT (2018)
2. Augenstein, I., Das, M., Riedel, S., Vikraman, L., McCallum, A.: Semeval 2017 task 10: Scienceie - extracting keyphrases and relations from scientific publications. In: SemEval@ACL (2017)
3. Balog, K.: Entity-oriented search. The Information Retrieval Series. Springer, Heidelberg (2018). https://doi.org/10.1007/978-3-319-93935-3
4. Beel, J., Gipp, B., Langer, S., Breitinger, C.: Research-paper recommender systems: a literature survey. Int. J. Digit. Libr. **17**(4), 305–338 (2015). https://doi.org/10.1007/s00799-015-0156-0
5. Beltagy, I., Lo, K., Cohan, A.: SciBERT: pretrained language model for scientific text. In: EMNLP (2019)
6. Bodenreider, O.: The unified medical language system (UMLS): integrating biomedical terminology. Nucleic Acids Res. **32**(Database issue), D267-70 (2004)
7. Bornmann, L., Mutz, R.: Growth rates of modern science: a bibliometric analysis based on the number of publications and cited references. J. Assoc. Inf. Sci. Technol. **66**(11), 2215–2222 (2015)

[1] https://gitlab.com/TIBHannover/orkg/orkg-nlp/tree/master/STM-corpus.

8. Chambers, A.: Statistical models for text classification and clustering: applications and analysis. Ph.D. thesis, University of California, Irvine (2013)
9. Cohan, A., Ammar, W., van Zuylen, M., Cady, F.: Structural scaffolds for citation intent classification in scientific publications. In: NAACL-HLT (2019)
10. Cohen, J.: A coefficient of agreement for nominal scales. Educ. Psychol. Measur. **20**(1), 37–46 (1960)
11. Constantin, A., Peroni, S., Pettifer, S., Shotton, D.M., Vitali, F.: The document components ontology (DoCO). Semant. Web **7**, 167–181 (2016)
12. Dernoncourt, F., Lee, J.Y.: Pubmed 200k RCT: a dataset for sequential sentence classification in medical abstracts. In: IJCNLP (2017)
13. Devlin, J., Chang, M.W., Lee, K., Toutanova, K.: BERT: pre-training of deep bidirectional transformers for language understanding. CoRR abs/1810.04805 (2018)
14. Elsevier OA STM Corpus. https://github.com/elsevierlabs/OA-STM-Corpus. Accessed 12 Apr 2019
15. Fisas, B., Saggion, H., Ronzano, F.: On the discoursive structure of computer graphics research papers. In: LAW@NAACL-HLT (2015)
16. Gábor, K., Buscaldi, D., Schumann, A.K., QasemiZadeh, B., Zargayouna, H., Charnois, T.: Semeval-2018 task 7: semantic relation extraction and classification in scientific papers. In: Proceedings of The 12th International Workshop on Semantic Evaluation, pp. 679–688 (2018)
17. Gardner, M., et al.: AllenNLP: a deep semantic natural language processing platform. arXiv preprint arXiv:1803.07640 (2018)
18. Google scholar. https://scholar.google.com/. Accessed 12 Sept 2019
19. Groza, T., Kim, H., Handschuh, S.: Salt: semantically annotated latex. In: SAAW@ISWC (2006)
20. Handschuh, S., Zadeh, B.Q.: The ACL RD-TEC: a dataset for benchmarking terminology extraction and classification in computational linguistics. In: COLING 2014: 4th International Workshop on Computational Terminology (2014)
21. Hochreiter, S., Schmidhuber, J.: Long short-term memory. Neural Comput. **9**, 1735–1780 (1997)
22. Houlsby, N., Huszar, F., Ghahramani, Z., Lengyel, M.: Bayesian active learning for classification and preference learning. CoRR abs/1112.5745 (2011)
23. Jaradeh, M.Y., et al.: Open research knowledge graph: next generation infrastructure for semantic scholarly knowledge. In: K-CAP 2019 (2019)
24. Jin, D., Szolovits, P.: Hierarchical neural networks for sequential sentence classification in medical scientific abstracts. In: EMNLP (2018)
25. Jurgens, D., Kumar, S., Hoover, R., McFarland, D.A., Jurafsky, D.: Measuring the evolution of a scientific field through citation frames. Trans. Assoc. Comput. Linguist. **6**, 391–406 (2018)
26. Kim, S., Martínez, D., Cavedon, L., Yencken, L.: Automatic classification of sentences to support evidence based medicine. In: BMC Bioinformatics (2011)
27. Lao, N., Cohen, W.W.: Relational retrieval using a combination of path-constrained random walks. Mach. Learn. **81**, 53–67 (2010)
28. Lee, K., He, L., Lewis, M., Zettlemoyer, L.S.: End-to-end neural coreference resolution. In: EMNLP (2017)
29. Lehmann, J., et al.: DBpedia - a large-scale, multilingual knowledge base extracted from Wikipedia. Semant. Web **6**, 167–195 (2015)
30. Liakata, M., Saha, S., Dobnik, S., Batchelor, C., Rebholz-Schuhmann, D.: Automatic recognition of conceptualization zones in scientific articles and two life science applications. Bioinformatics **28**(7), 991–1000 (2012)

31. Liakata, M., Teufel, S., Siddharthan, A., Batchelor, C.R.: Corpora for the conceptualisation and zoning of scientific papers. In: LREC (2010)
32. Luan, Y., He, L., Ostendorf, M., Hajishirzi, H.: Multi-task identification of entities, relations, and coreference for scientific knowledge graph construction. In: EMNLP (2018)
33. Ma, X., Hovy, E.H.: End-to-end sequence labeling via bi-directional LSTM-CNNS-CRF. CoRR abs/1603.01354 (2016)
34. Microsoft Academic. https://academic.microsoft.com/home. Accessed 12 Sept 2019
35. Microsoft Academic Knowledge Graph. http://ma-graph.org/. Accessed 12 Sept 2019
36. Papers with code. https://paperswithcode.com/. Accessed 12 Sept 2019
37. Pertsas, V., Constantopoulos, P.: Scholarly ontology: modelling scholarly practices. Int. J. Digit. Libr. **18**(3), 173–190 (2017)
38. Pustu-Iren, K., et al.: Investigating correlations of inter-coder agreement and machine annotation performance for historical video data. In: TPDL (2019)
39. Salatino, A.A., Thanapalasingam, T., Mannocci, A., Osborne, F., Motta, E.: The computer science ontology: a large-scale taxonomy of research areas. In: International Semantic Web Conference (2018)
40. Semantic scholar. https://www.semanticscholar.org/. Accessed 12 Sept 2019
41. Shen, Y., Yun, H., Lipton, Z.C., Kronrod, Y., Anandkumar, A.: Deep active learning for named entity recognition. In: ICLR (2017)
42. Siddhant, A., Lipton, Z.C.: Deep Bayesian active learning for natural language processing: results of a large-scale empirical study. In: EMNLP (2018)
43. Snow, R., O'Connor, B.T., Jurafsky, D., Ng, A.Y.: Cheap and fast - but is it good? Evaluating non-expert annotations for natural language tasks. In: EMNLP (2008)
44. spaCy: Industrial-strength natural language processing. http://www.spacy.io. Accessed 02 Sep 2019
45. Springer Nature SciGraph. https://www.springernature.com/gp/researchers/scigraph. Accessed 12 Sept 2019
46. Teufel, S., Siddharthan, A., Batchelor, C.: Towards discipline-independent argumentative zoning: evidence from chemistry and computational linguistics. In: Proceedings of the 2009 Conference on Empirical Methods in Natural Language Processing: Volume 3, vol. 3, pp. 1493–1502. Association for Computational Linguistics (2009)
47. Xiong, C., Power, R., Callan, J.P.: Explicit semantic ranking for academic search via knowledge graph embedding. In: WWW (2017)
48. Yaman, B., Pasin, M., Freudenberg, M.: Interlinking SciGraph and DBpedia datasets using link discovery and named entity recognition techniques. In: LDK (2019)
49. Zhang, Y., Lease, M., Wallace, B.C.: Active discriminative text representation learning. In: AAAI (2016)

Leveraging Schema Labels to Enhance Dataset Search

Zhiyu Chen(✉), Haiyan Jia, Jeff Heflin, and Brian D. Davison

Lehigh University, Bethlehem, PA, USA
{zhc415,haiyan.jia}@lehigh.edu, {heflin,davison}@cse.lehigh.edu

Abstract. A search engine's ability to retrieve desirable datasets is important for data sharing and reuse. Existing dataset search engines typically rely on matching queries to dataset descriptions. However, a user may not have enough prior knowledge to write a query using terms that match with description text. We propose a novel schema label generation model which generates possible schema labels based on dataset table content. We incorporate the generated schema labels into a mixed ranking model which not only considers the relevance between the query and dataset metadata but also the similarity between the query and generated schema labels. To evaluate our method on real-world datasets, we create a new benchmark specifically for the dataset retrieval task. Experiments show that our approach can effectively improve the precision and NDCG scores of the dataset retrieval task compared with baseline methods. We also test on a collection of Wikipedia tables to show that the features generated from schema labels can improve the unsupervised and supervised web table retrieval task as well.

Keywords: Dataset search · Table retrieval · Text normalization · Data fusion

1 Introduction

Dataset retrieval is receiving more attention as people from different fields and domains start to rely on datasets for their work. There are many data portals with the purpose of effective and efficient data management and data sharing, such as data.gov[1], datahub[2] and data.world[3]. Most of those data portals use CKAN[4] as their backend. However, there are two problems of dataset search engines using such infrastructure: First, ranking performance relies on the quality of metadata of datasets, while many datasets lack high quality metadata; second, the information in the metadata may not satisfy the user's information need or help them solve their task [3]. A user may not know the organization of a

[1] https://www.data.gov/.
[2] http://datahub.io/.
[3] https://data.world/.
[4] https://docs.ckan.org/.

J. M. Jose et al. (Eds.): ECIR 2020, LNCS 12035, pp. 267–280, 2020.
https://doi.org/10.1007/978-3-030-45439-5_18

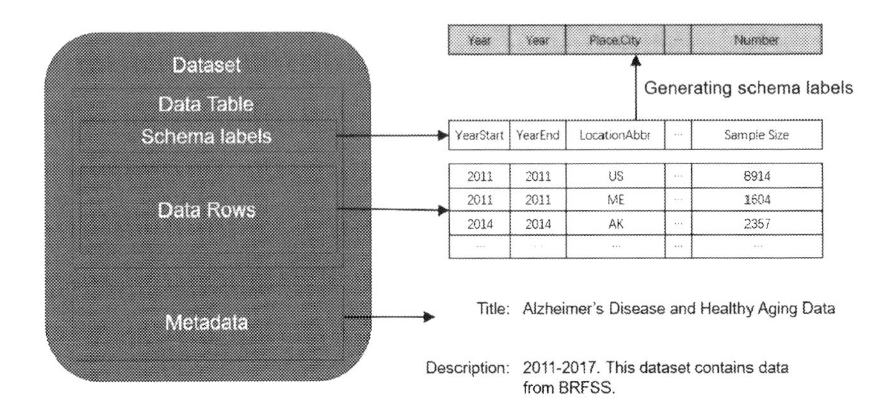

Fig. 1. The structure of a dataset. Metadata includes the title and any description. A trained schema label generator is used to generate additional schema labels (green part) from similar data tables. (Color figure online)

potentially relevant dataset, or the tags data publishers provide with a dataset. Such information can hardly be used for dataset ranking.

In this paper, we focus on the problem of dataset retrieval where dataset content is in tabular form, since tabular data is widely-used and easy to read and write. As illustrated in Fig. 1, a dataset consists of a data table (dataset content) and metadata. A data table usually has one header row, followed by one or more data rows. The header row consists of a list of **schema labels** (attribute names) whose actual values are stored in data rows. Metadata usually includes title and description of the dataset.

Schema labels, which represent high-level concepts, are underutilized if we directly score them with a user query. Consider the example in Fig. 1; the vocabulary of schema labels could be very different from other fields and user queries. "LocationAbbr", standing for "Location Abbreviation", is unlikely to appear in a user query so this dataset is less likely to be recalled. However, we can enhance this dataset by generating schema labels such as "place" and "city" appearing in other, similar datasets, which could provide a better soft-matching signal with respect to a user query, and therefore increase the chance that it can be recalled.

In this work, we first propose a new method for schema label generation. We learn latent feature representations of schema labels automatically by jointly decomposing the dataset-schema label interaction matrix and schema label-schema label interaction matrix. Then we propose a framework for enhancing dataset retrieval by schema label generation to address the problem that schema labels are not effectively used by existing dataset search engines. We create a new public benchmark[5] based on federal (U.S.) datasets and use it to demonstrate the effectiveness of our proposed framework for dataset retrieval. We additionally consider a web table retrieval task and demonstrate that the features generated from schema labels can be effective for supervised ranking.

[5] Available via https://github.com/Zhiyu-Chen/ECIR2020-dataset-search.

2 Related Work

Dataset search has become a new research field with new challenges. Chapman et al. [3] classify dataset search into *basic* and *constructive* dataset search. Basic dataset search returns a list of existing datasets based on a user's query, while constructive dataset search [5] generates datasets on-the-fly based on a user's needs and query. Google recently released a dataset search service[6]. Like many other data portals, their service relies on metadata of datasets, annotated on web pages using a standard defined by schema.org.

Other work on applications of Web tables is also related to our work. Cafarella et al. [2] proposed WebTables system which extract Web tables from top ranked pages by keyword search. Sekhavat et al. [13] proposed a probabilistic method that augments an existing knowledge base with facts from Web tables. Zhang et al. [16] developed generative probabilistic models to equip spreadsheets with smart assistance capabilities. Specifically, given a table, they recommend additional rows and column headings by leveraging the information from the Web tables. They also developed semantic matching features for table retrieval [17].

The techniques designed for Web table analysis could potentially be applied to dataset search. In our work, each dataset is associated with data in tabular form. Extracting useful information from tables such as entities and attribute names could help with the retrieval task. Trabelsi et al. [14] recently proposed custom embeddings for column headers based on multiple contexts for table retrieval, and found representing numerical cell values to be useful. Zhang et al. [16] proposed to use semantic concepts to represent queries and tables for ranking entity-focused tables. However, dataset search could be inherently more difficult since datasets do not need to be entity-focused.

3 Schema Label Enhanced Ranking

In this section, we introduce the framework of schema label enhanced dataset retrieval. As illustrated in Fig. 2, our framework has two stages: in the first stage, we first train a schema label generator with the method proposed in Sect. 3.1 and use it to generate additional schema labels for all the datasets; in the second stage, we use a mixed ranking model to combine the scores of schema labels and other fields for dataset ranking. In the following subsections, we present a detailed illustration of the two stages.

3.1 Schema Label Generation

We propose to improve dataset search by making use of generated schema labels, since these can be complementary to the original schema labels and especially valuable when they are otherwise absent from a dataset.

[6] https://toolbox.google.com/datasetsearch.

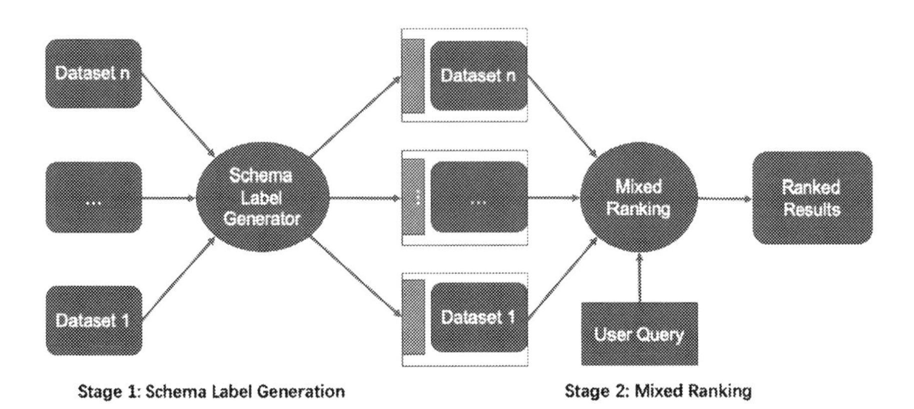

Fig. 2. The proposed schema label enhanced dataset retrieval framework. The green blocks indicate generated schema labels for different datasets. (Color figure online)

We treat schema label generation as a multi-label classification problem. Let $L = \{l_1, l_2, ..., l_k\}$ denote the labels appearing in all datasets and $D = \{(\mathbf{x}^i, \mathbf{y}^i) | 1 \leq i \leq n\}$ denote the training set. Here, for each training sample $(\mathbf{x}^i, \mathbf{y}^i)$, $\mathbf{x}^i$ is a d-dimensional feature vector of column i which can be calculated from data rows [4] or learned from matrix factorization proposed later in this section. $\mathbf{y}^i$ is k-dimensional vector $[y_1^i, y_2^i, ..., y_k^i]$ and $y_j^i = 1$ only if x_i is relevant to label l_j, otherwise $y_j^i = 0$. Our objective is to learn a function that models $P(l|x_i)$, $(l \in L)$. To generate m schema labels for column i, we can select the top m labels L_m by:

$$L_m = \underset{l \in L_m \subseteq L}{\arg\max}\, P(l|x_i)$$

We could also generate schema labels by selecting a probability threshold θ:

$$L_m = \{l \in L | P(l|x_i) \geq \theta\}$$

In practice, we could first generate the top m schema labels and filter out those results with a probability lower than the threshold.

Chen et al. [4] proposed to predict schema labels based on curated features of data values. Instead of designing curated features for schema labels, we consider learning their representations in an automated manner. Inspired by collaborative filtering methods in recommender systems, we model each dataset as a user and each schema label as an item. Then a dataset with a schema label can be considered as positive feedback between a user and an item. By exploiting the user-item co-occurrences and item-item co-occurrences, we can learn the latent representations of schema labels. In the following, we show how to construct a preference matrix in the context of schema label generation and how to learn the schema label features.

Preference Matrix Construction. With m data tables and n unique schema labels, we can construct a dataset-column preference matrix $M^{m \times n}$, where M_{up} is 1 if dataset u contains schema label p.

Matrix Factorization. MF [7] decomposes M into the product of $U^{m \times k}$ and $P^{k \times n}$ where $k < min(m, n)$. U^T can be denoted as $(\alpha_1, ..., \alpha_u ..., \alpha_m)$ where $\alpha_u \in R^k$ represents the latent factor vector of dataset u. Similarly, P^T can be denoted as $(\beta_1, ..., \beta_p ..., \beta_n)$ where $\beta_p \in R^k$ represents the latent factor vector of schema label p. Since the preference matrix actually models the implicit feedback, MF optimizes the following objective function:

$$\mathcal{L}_{mf} = \sum_{u,p} c_{up}(M_{up} - \alpha_u^T \beta_p)^2 + \lambda_\alpha \sum_u \|\alpha_u\|^2 + \lambda_\beta \sum_p \|\beta_p\|^2 \tag{1}$$

where c_{up} is a hyperparameter tuned to balance the non-zero and zero values since M is a sparse matrix. λ_α and λ_β are regularization parameters that adjust the importance of regularization terms $\sum_u \|\alpha_u\|^2$ and $\sum_p \|\beta_p\|^2$.

Label Embedding. Recently, word embedding techniques (e.g., word2vec [11]) have been valuable in natural language processing tasks. Given a sequence of words, a low-dimensional continuous representation called word embedding can be learned for each word. Word2vec's skip-gram model with negative sampling (SGNS) is equivalent to implicitly factorizing a word-context matrix, whose cells are the pointwise mutual information (PMI) of the respective word and context pairs, shifted by a global constant [9]. The PMI between word i and its context word j is defined as:

$$PMI(i, j) = log \frac{P(i, j)}{P(i) \times P(j)} = log \frac{\#(i, j) \times |D|}{\sum_j \#(i, j) \times \sum_i \#(i, j)}$$

where $\#(i, j)$ is the number of times word j appears in the context window of word i and $|D|$ is the total number of word-context pairs. Then, a shifted positive PMI (SPPMI) of word i and word j is calculated as:

$$SSPMI(i, j) = max\{PMI(i, j) - log(k), 0\} \tag{2}$$

where k is the number of negative samples of SGNS. Given a corpus, matrix M^{SPPMI} can be constructed based on Eq. (2) and factorizing it is equivalent to performing SGNS.

A schema label exists in the context of other schema labels. Therefore, we perform word embedding techniques to learn the latent representations of schema labels. However, we do not consider the order of schema labels. Therefore, given a schema label, all other schema labels which come from the same data table are considered as its context. With the constructed SSPMI matrix of co-occurring schema labels, we are able to decompose it to learn the latent representations of schema labels.

Joint Learning of Schema Label Representations. Schema label representations learned from MF capture the interactive information between datasets and schema labels, while the word2vec style representations explain the co-occurrence relationships of schema labels. We use the CoFactor model [10] to jointly learn schema label representations from both dataset-label interaction and label-label interaction:

$$\begin{aligned}\mathcal{L} = & \overbrace{\sum_{u,p} c_{up}(M_{up} - \alpha_u^T \beta_p)^2}^{MF} \\ & + \overbrace{\sum_{M_{pi}^{SPPMI} \neq 0} (M_{pi}^{SPPMI} - \beta_p^T \gamma_i - b_p - c_i)^2}^{schema\ label\ embedding} \\ & + \lambda_\alpha \sum_u \|\alpha_u\|^2 + \lambda_\beta \sum_p \|\beta_p\|^2 + \lambda_\gamma \sum_i \|\gamma_i\|^2\end{aligned} \tag{3}$$

From the objective function we can see the schema label representation β_p is shared between MF and schema label embedding. γ_i is the latent representation of context embedding. b_p and c_i are the schema label embedding bias and context embedding bias, respectively. The last line of Eq. 3 incorporates regularization terms with different λ controlling their effects. We use the vector-wise ALS algorithm [15] to optimize the parameters.

Schema Label Generation. After obtaining the jointly learned representations of schema labels, we can use them as features for schema label generation. In this paper, we use the concatenation of schema label representations introduced here and the curated features proposed by Chen et al. [4] to construct each x^i. Any multi-label classification models can be used to train the schema label generator and in this paper we choose Random Forest.

3.2 The Mixture Ranking Model

Based on the schema label generation method proposed above, we index the generated schema labels for each dataset. Now, each dataset has the following fields: metadata, data rows, schema labels and generated schema labels. A straightforward way to rank datasets is to use traditional ranking methods for documents.

Zhang and Balog [17] represent tables as single field documents or multifield documents for table retrieval task. For *single field document representation*, a dataset is treated as a single document by concatenating the text from all the fields. Then traditional methods such as BM25 can be used to score the dataset. For *multifield document representation*, each field is scored independently against the query and a weighted sum is used for ranking.

In our **Schema Label Mixed Ranking (SLMR)** model, we score schema labels differently from other fields. The focus of our work is to learn how schema labels, data rows and other metadata may differently influence dataset retrieval performance. Note that, for simplicity, we consider the other metadata (title and description) as a single text field, since title and description are homogeneous compared with schema labels and data rows. Therefore, we have the following scoring function for a dataset D:

$$score(q, D) = \sum_{i \in \{text, data\}} w_i \times score_{text}(q, F_i) + w_l \times score_l(q, F_l) \quad (4)$$

where F_{text} denotes the concatenation of title and description, F_{data} denotes the data table, and F_l denotes the generated schema labels. Each field has a corresponding weights. F_{text} and F_{data} have the same scoring function $score_{text}$ while F_l has a different scoring function $score_l$. For F_{text} and F_{data}, we can use a standard scoring function for normal documents. In the experiments, we use BM25 as $score_{text}$.

Due to the existence of a large number of non-dictionary words in schema labels [4] that would otherwise be outside of the vocabulary of a word-based embedding, we represent schema labels and query terms using fastText [1] in $score_l$, since such word embeddings are calculated from character n-grams instead of terms. To score the schema labels with respect to a query, we use the negative Word Mover's Distance (WMD) [8]. WMD measures the dissimilarity between two text documents as the minimum amount of distance that the word embeddings of one document need to "travel" to reach the word embeddings of another document. So $score_l(q, F_l) = -wmd(fasttext(q), fasttext(F_l))$ reflects the semantic similarity between a query and schema labels.

4 Data Collection

Here we describe how we construct the new benchmark for dataset retrieval in detail. We collected 2417 resources published by the U.S. federal government from Data.gov which cover a variety of topics. Each resource includes one or more CSV format data tables and corresponding metadata. Each CSV table is treated as a single dataset and we use the resource-level metadata to annotate each dataset.

4.1 Task Creation and Query Collection

We created six tasks in which each describes a separate information need to find one or more datasets. For each, we have a statement about the information need which describes what datasets are considered as relevant. We additionally verified for each task the existence of at least one relevant dataset. The dataset is public available[7].

We used Amazon Mechanical Turk[8] to obtain diverse queries for these tasks from real users. Every annotator was presented with the task descriptions and asked to provide a query for each created task. To avoid the impact of task order on the quality of annotations, we randomly shuffled the order of tasks for each annotator. We paid one dollar for each completed annotation job and 20 queries were collected for each task. Every collected query was manually examined and obviously unrelated queries were excluded from the collection.

[7] Available from https://github.com/Zhiyu-Chen/ECIR2020-dataset-search.
[8] https://www.mturk.com/.

Table 1. For each task, the number of pairs assigned to each relevance label.

Task #	Off topic	Poor	Good	Excellent
1	1006	34	37	64
2	164	248	585	308
3	300	270	456	153
4	246	324	660	289
5	162	246	355	198
6	181	303	614	367

4.2 Relevance Assessments

For each task and each suggested query, we used traditional ranking functions to score single field representations of each dataset and collect the top 100 results. The following ranking models were used: BM25, TF-IDF, Language model based on Jelinek-Mercer smoothing, and Language Model with Dirichlet smoothing. We also used each model with two different representations: the concatenation of all fields of the dataset and the concatenation of title and description. This leads to eight baselines for the pooled results.

Then, the collected task-dataset pairs were annotated for relevance using the crowdsourcing service provided by Figure Eight[9]. We did not annotate the *query*-dataset pairs because the goal of dataset retrieval is to find relevant datasets with respect to a *task* which represents the real information need.

Annotators were presented with the task title, description and link to the data table. Each task-dataset pair was judged on a four point scale: 0 (off topic), 1 (poor), 2 (good), and 3 (excellent).[10] Every annotator was paid 10 cents per task-dataset judgement.

Every single task-dataset pair was judged by three annotators and we take the majority vote as the relevance label. If no majority agreement is achieved, we take the average of the scores as the final label. The statistics of annotation results is shown in Table 1.

---

[9] https://www.figure-eight.com/.

[10] The following labeling guidance was provided to annotators: *a dataset is off topic if the information does not satisfy the information need, and should not be listed in the search results from a search engine; a dataset is poor if a search engine were to include this in the search results, but it should not be listed at the top; a dataset is good if you would expect this dataset to be included in the search results from a search engine; a dataset is excellent if you would expect this dataset ranked near the top of the search results from a search engine.*

Table 2. NDCG@k and Precision@k of different models on dataset retrieval. The superscript + shows statistically significant improvements for our SLMR model over other single and multifield document ranking models. T means title, D means description, DT means data table, G means generated schema labels.

Method	Used fields	NDCG@5	@10	@20	@50	P@5	@10	@20	@50
SDR	T+D	0.8920	0.8490	0.8222	0.8121	0.4122	0.3652	0.3452	0.3585
SDR	DT	0.7378	0.7036	0.6964	0.7107	0.2856	0.2974	0.2931	0.3122
SDR	T+D+DT	0.8435	0.7954	0.7763	0.7785	0.2574	0.2870	0.3170	0.3357
MDR	T+D+DT	0.9285	0.8874	0.8683	0.8631	0.4086	0.3612	**0.4026**	0.3767
SLMR	T+D+G	**0.9293**$^{+}$	**0.8898**	**0.8722**$^{+}$	**0.8662**	**0.5000**$^{+}$	**0.4388**$^{+}$	0.4000	0.3761
SLMR	T+D+DT+G	0.9169	0.8808	0.8680	0.8555	**0.5000**$^{+}$	0.4345^{+}	0.4013	**0.3783**

5 Evaluation

5.1 Evaluation Metrics

We evaluate dataset retrieval performance over a range of metrics: Precision at k and Normalized Discounted Cumulative Gain (NDCG) at k [6]. To test the significance of differences between model performances, we use paired t-tests with significance at the $p = 0.01$ level.

5.2 Baselines

We first present the baseline retrieval methods.

Single-Field Document Ranking (SDR). A dataset is considered as a single document. We use BM25 to score the concatenation of title and description, the text of the data table and the concatenation of all of them. By comparing the three results, we can learn about field level importance for dataset retrieval. Parameters are chosen by grid search.

Multifield Document Ranking (MDR). By setting $w_l = 0$, Eq. (4) degenerates to the Mixture of Language Models [12]. BM25 is also used here as $score_{text}()$ in order to have a fair comparison with other methods. To optimize field weights, we use coordinate ascent. Finally, smoothing parameters are optimized in the same manner as single-field document ranking.

5.3 Experimental Results

In this section, we examine the following research questions:

Q1 Does data table content help in dataset retrieval?
Q2 Do generated schema labels help in dataset retrieval?
Q3 Which fields are most important for the dataset retrieval task?

We first obtain features of schema labels as described in Sect. 3.1 and the number of latent factors is set to 40. Then we train a Random Forest with the learned schema label features. The scikit-learn implementation of Random Forest[11] is used with default parameters except the number of trees is set to 25. In practice, we could choose any multi-label classifier. For each column, we select the top 10 generated schema labels and filter those with probability lower than 0.5. For each dataset, we index the generated schema labels as an additional field. Table 2 summarizes the NDCG at k and Precision at k of different models. Note that, for Schema Label Mixed Ranking (SLMR), we trained three different models and the weights of used fields were forced to be non-zero in order to study the proposed research questions. The weights of used fields for multifield document representation are also set non-zero when optimizing the parameters.

From the results of single-field document ranking, we can see that only utilizing the data table for ranking leads to the worst performance. Scoring on the concatenation of title and description achieved the best results, which indicates that title and description are more important than the data table for ranking a dataset (**Q3**). Treating all fields of a dataset as a single-field document provides performance between the previous two models. This result is expected since the length of data tables are usually much larger than titles and descriptions, and therefore dominate the table representation.

By comparing the results of single-field and multifield document ranking, we observe that the combination of the scores of data table, title and description could improve NDCG@k. Though NDCG@k decreases when k increases, the relative improvement against single-field document ranking are more significant. In contrast, for Precision@5, Precision@10, single-field document ranking performs better than multifield document ranking, though the differences are small. So for **Q1**, under the setting of multifield document ranking, the content of the data table could help NDCG, but not help Precision of dataset retrieval results.

Without scoring data tables, our proposed schema label mixed ranking approach achieves the highest NDCG on all the rank cut-offs, which indicates that the generated schema labels can be useful to improve the NDCG of dataset retrieval results (**Q2**). Though Precision@20 of multifield document ranking are higher than our proposed model, the difference is no more than 0.4% ($p\_value > 0.9$). Significantly, our model outperforms by 21.3% for Precision@5 ($\frac{0.5-0.4122}{0.4122}$) and by 20.1% for Precision@10 ($\frac{0.4388-0.3652}{0.3652}$) than the best baseline methods ($p\_value < 0.01$). Whether data tables are scored or not, Precision@k is not significantly different for schema label mixed ranking. Therefore, under the setting of schema label mixed ranking, data tables make little contribution in this scenario (**Q1**). One possible reason could be that data tables collected from data.gov contain large quantities of numerical values and will rarely be used to match user queries.

If a schema label mixed ranking model scores only on titles and descriptions ($w_l = 0$), it is equivalent to single-field ranking model scoring on titles and

[11] http://scikit-learn.org/stable/modules/generated/sklearn.ensemble.RandomForest Classifer.html.

Table 3. Supervised ranking results on table retrieval.

Method	NDCG@5	@10	@15	@20
STR [16]	0.6366	0.6571	0.663	0.6632
Schema Label Features	0.4489	0.5201	0.534	0.5347
STR + Schema Label Feat	**0.6530**	**0.6728**	**0.6789**	**0.6761**

descriptions. Therefore, we can compare the results in first and fifth rows in Table 2. With generated schema labels, the ranking model can have a higher performance on dataset retrieval task (**Q2**).

5.4 Schema Label Generation Enhanced Search for Web Tables

The task of dataset search is similar to Web table search since both tasks use table structure to represent data. The difference is that a large amount of Web tables are entity focused and contain many named entities that can be linked to a knowledge base. However, our datasets collected from the data.gov data portal contain few useful entities in the table. Therefore, a lot of methods designed for Web table ranking cannot be applied to dataset search. The semantic table retrieval (STR) method proposed by Zhang and Balog [16] relies on features from knowledge bases (bag of entities) which are not generally available for the scenario of dataset search. However, the schema label generation based method can be applied to table search. Thus, we performed additional experiments to show the performance of our method for the table search scenario.

We first generate schema labels for the table corpus shared by Zhang and Balog [16] using the method proposed in Sect. 3.1. Then we append five additional features to their proposed features[12] based on schema labels. Each feature is one type of semantic similarity between query and schema labels. Four features are calculated using the measurement proposed by Zhang and Balog (one early fusion feature, three late fusion features) and the last feature is the negative of Word Mover's Distance. Finally, like Zhang and Balog, we use Random Forest to perform pointwise regression and the final reported results are averaged over five runs of 5-fold cross-validation and shown in Table 3.

We can see that schema label features along cannot outperform STR. But combining them results in improvement. However, by calculating the normalized feature importance measured in terms of Gini score, we find that for STR with schema label features, WMD based measurement contributes the most among all the semantic features. Thus it demonstrates that the schema labels can be valuable for the table retrieval task as well.

Notably, in this table corpus, many tables lack much table content but contain rich text descriptions, which could be unfair for schema label generation-based methods. While for dataset search, each table has values but may lack high quality dataset descriptions. We believe that our schema label generation method

[12] https://github.com/iai-group/www2018-table/tree/master/feature.

can outperform STR in the scenario where text descriptions provide less useful information than the table itself.

Table 4. Unsupervised ranking results on table retrieval.

Used fields	NDCG@5	@10	@15	@20
text	0.3724	0.3891	0.4009	0.4178
text + data table	0.3901	0.4042	0.4422	0.4686
text + data table + generated labels	**0.4006**	**0.4118**	**0.4495**	**0.4766**
text + data table + original labels	0.3930	0.4055	0.4457	0.4709
text + original labels	0.3785	0.3934	0.4110	0.4283
text + generated labels	0.3808	0.3955	0.4064	0.4197

We also show unsupervised ranking results with Eq. 4 in Table 4. Unlike Zhang and Balog [16], we consider page title, section title and caption as a single text field, in order to reduce the number of hyperparameters (field weights). The results show that generated labels are more effective than original labels for table ranking. It is unsurprising because generated labels often include not only original labels but also additional labels that can benefit the ranking model. We also notice that including the data table field achieves better results than not scoring it, which is contrary to the results of dataset ranking. It is also expected since WikiTables are entity-focused and include a lot of text information while data tables from data.gov include more numeric values.

6 Conclusion

In this paper, we have proposed a schema label enhanced ranking framework for dataset retrieval. The framework has two stages: in the first stage, a schema label generator is trained to generate additional schema labels for each dataset column; in the second stage, given a user query, datasets are ranked by their original fields together with generated schema labels. Schema label generation is treated as a multi-label classification task in which each column of a dataset is associated with multiple schema labels. Instead of using hand-curated features, we learn the latent feature representations of schema labels by a CoFactor model in which the dataset-schema label interactions and schema label-schema label interactions are captured. With the schema label mixed ranking model, the traditional ranking scores for text fields (title, description, data rows) and word embedding-based scores for generated schema labels can be used to rank the datasets.

We created a new benchmark to evaluate the performance of dataset retrieval. The experimental results demonstrate our proposed framework can effectively improve the performance on the dataset retrieval task. It achieved the highest NDCG on all the rank cut-offs compared with all baseline methods. We also apply our method to the web table retrieval task which is similar to dataset search and find that the features generated from schema labels can help in supervised ranking as well.

Acknowledgment. This material is based upon work supported by the National Science Foundation under Grant No. IIS-1816325.

References

1. Bojanowski, P., Grave, E., Joulin, A., Mikolov, T.: Enriching word vectors with subword information. Trans. Assoc. Comput. Linguist. **5**, 135–146 (2017)
2. Cafarella, M.J., Halevy, A., Wang, D.Z., Wu, E., Zhang, Y.: Webtables: exploring the power of tables on the web. Proc. VLDB Endow. **1**(1), 538–549 (2008)
3. Chapman, A., et al.: Dataset search: a survey. arXiv preprint arXiv:1901.00735 (2019)
4. Chen, Z., Jia, H., Heflin, J., Davison, B.D.: Generating schema labels through dataset content analysis. In: Companion of the The Web Conference 2018, pp. 1515–1522. International World Wide Web Conferences Steering Committee (2018)
5. Gentile, A.L., Kirstein, S., Paulheim, H., Bizer, C.: Extending RapidMiner with data search and integration capabilities. In: Sack, H., Rizzo, G., Steinmetz, N., Mladenić, D., Auer, S., Lange, C. (eds.) ESWC 2016. LNCS, vol. 9989, pp. 167–171. Springer, Cham (2016). https://doi.org/10.1007/978-3-319-47602-5_33
6. Järvelin, K., Kekäläinen, J.: Cumulated gain-based evaluation of IR techniques. ACM Trans. Inf. Syst. (TOIS) **20**(4), 422–446 (2002)
7. Koren, Y., Bell, R., Volinsky, C.: Matrix factorization techniques for recommender systems. Computer **42**(8), 30–37 (2009)
8. Kusner, M., Sun, Y., Kolkin, N., Weinberger, K.: From word embeddings to document distances. In: International Conference on Machine Learning, pp. 957–966 (2015)
9. Levy, O., Goldberg, Y.: Neural word embedding as implicit matrix factorization. In: Advances in Neural Information Processing Systems, pp. 2177–2185 (2014)
10. Liang, D., Altosaar, J., Charlin, L., Blei, D.M.: Factorization meets the item embedding: Regularizing matrix factorization with item co-occurrence. In: Proceedings of the 10th ACM Conference on Recommender Systems, pp. 59–66. ACM (2016)
11. Mikolov, T., Sutskever, I., Chen, K., Corrado, G.S., Dean, J.: Distributed representations of words and phrases and their compositionality. In: Advances in Neural Information Processing Systems, pp. 3111–3119 (2013)
12. Ogilvie, P., Callan, J.: Combining document representations for known-item search. In: Proceedings of the 26th Annual International ACM SIGIR Conference on Research and Development in Informaion Retrieval, pp. 143–150. ACM (2003)
13. Sekhavat, Y.A., Di Paolo, F., Barbosa, D., Merialdo, P.: Knowledge base augmentation using tabular data. In: LDOW (2014)
14. Trabelsi, M., Davison, B., Jeff, H.: Improved table retrieval using multiple context embeddings for attributes. In: Proceedings of IEEE Big Data 2019. IEEE (2019)
15. Yu, H.-F., Hsieh, C.-J., Si, S., Dhillon, I.S.: Parallel matrix factorization for recommender systems. Knowl. Inf. Syst. **41**(3), 793–819 (2013). https://doi.org/10.1007/s10115-013-0682-2

16. Zhang, S., Balog, K.: Entitables: smart assistance for entity-focused tables. In: Proceedings of the 40th International ACM SIGIR Conference on Research and Development in Information Retrieval, SIGIR 2017, pp. 255–264, ACM, New York (2017). https://doi.org/10.1145/3077136.3080796
17. Zhang, S., Balog, K.: Ad hoc table retrieval using semantic similarity. In: Proceedings of the 2018 World Wide Web Conference, WWW 2018, pp. 1553–1562, Republic and Canton of Geneva, Switzerland (2018). https://doi.org/10.1145/3178876.3186067

Moving from Formal Towards Coherent Concept Analysis: Why, When and How

Pavlo Kovalchuk[1,2(✉)], Diogo Proença[2], José Borbinha[1,2], and Rui Henriques[1,2]

[1] Instituto Superior Técnico, Universidade de Lisboa, Lisbon, Portugal
{pavlo.kovalchuk,jlb,rmch}@tecnico.ulisboa.pt
[2] INESC-ID, Lisbon, Portugal
diogo.proenca@tecnico.ulisboa.pt

Abstract. Formal concept analysis has been largely applied to explore taxonomic relationships and derive ontologies from text collections. Despite its recognized relevance, it generally misses relevant concept associations and suffers from the need to learn from Boolean space models. Biclustering, the discovery of coherent concept associations (subsets of documents correlated on subsets of terms and topics), is here suggested to address the aforementioned problems. This work proposes a structured view on why, when and how to apply biclustering for concept analysis, a subject remaining largely unexplored up to date. Gathered results from a large text collection confirm the relevance of biclustering to find less-trivial, yet actionable and statistically significant concept associations.

Keywords: Concept analysis · Biclustering · Topic modeling · Unsupervised knowledge discovery · Large digital libraries

1 Introduction

Concept analysis is up to date the most referred unsupervised option for content categorization in large text collections [32]. A concept is an association between attributes (terms or topics) that is coherently verified in a subset of objects (documents). Concept analysis has been largely pursued to explore taxonomic relationships within a corpus, addressing the typical limitations that peer unsupervised approaches face in high-dimensional and sparse spaces [19]. Formal concept analysis (FCA) aims at finding, in Boolean data spaces, concepts as subsets of topics that co-occur in a subset of documents. FCA is the paradigmatic approach to concept analysis [11]. Despite its well-recognized relevance to derive ontologies for content categorization, FCA is hampered by major drawbacks. First, it imposes the selection of binarization thresholds to decide whether a topic is represented in a given document, making it vulnerable to subjective choices and to the item-boundaries problem [13]. As a result, FCA is unable to retain concepts sensitive to the varying predominance of topics in a given document, neglecting the rich nature of vector space models. Also, by focusing on

J. M. Jose et al. (Eds.): ECIR 2020, LNCS 12035, pp. 281–295, 2020.
https://doi.org/10.1007/978-3-030-45439-5_19

dense regions, FCA neglects potentially relevant concepts, such as where specific topics have a preserved order of importance in a subset of documents [24].

Biclustering aims at finding coherent subspaces (subsets of attributes correlated in a subset of objects), which has been previously suggested for concept analysis in real-valued data spaces derived from text collections [5,8]. The use of biclustering for concept analysis is here termed *coherent concept analysis* (in contrast with formal concept analysis) since concepts are associations that satisfy specific homogeneity criteria of interest, therefore going beyond the strict Boolean formal view. Coherent concepts are sensitive to the predominance of each topic in a given document. In spite of its potentialities, existing research on biclustering text collections pursue specific forms of homogeneity [2,5], not offering a discussion on how different homogeneity and quality criteria affect concept analysis. In addition, existing research leaves aside current breakthroughs in the biclustering domain [12,16]. Finally, a fully structured view on why, when and how to apply biclustering in large text collections remains largely unexplored.

This work offers the first comprehensive view on the use of biclustering to explore large text collections in a fully automated and unsupervised manner, and further discusses its role for content categorization, retrieval, and navigation. The motivation is the need to support search and navigation in the official online publication of a national journal state, a digital library comprising all national laws, regulations and legal acts.

This document is organized as follows. Section 2 provides essential background on concept analysis. Section 3 surveys relevant work on the topic. Section 4 discusses why, when and how to apply biclustering. Section 5 gathers results demonstrating the role of 5 biclustering in large text collections. Finally, concluding remarks and future directions are presented.

2 Background

The process of knowledge discovery in text collections (KDT) aims at finding relevant relations in a collection of documents $D = \{d_1, .., d_n\}$, a necessary basis for content categorization, search and navigation. To this end, KDT combines principles from information retrieval, topic modeling, and concept analysis.

To preserve a sound terminology ground, *topic* denotes a semantically related set of *terms*, and *concept* is a (putative) association between terms or topics.

Representing unstructured documents as sets of terms allows subsequent queries on those terms. The *vector space model* represents documents as weighted vectors, $d_i = (w_{i1}, w_{i2}, w_{i3}, ..., w_{im})$ where w_{ij} is the frequency of term t_j in document d_i, $w_{ij} \in \mathbb{R}$ and $w_{ij} \geq 0$. Weights can be alternatively set using the classic term frequency-inverse document frequency (Tf-idf) metric [29]. Document similarity can be then computed using a loss function such as cosine distance.

Given the common high-dimensionality of vector space models, they can be reduced using principles from **topic modeling** to facilitate subsequent mining:

- *principal component analysis* (PCA) uses algebraic operations to project data into a new data space along axes (eigenvectors α_k) where data mostly vary [20], $w'_{ij} = \sum_k^m \alpha_k w_{ik}$. Semantic relations between terms are lost;

- *latent semantic analysis* (LSA) preserves semantic relations without relying on dictionaries or semantic networks. Terms in a given text document are seen as conceptually independent and linked to each other by underlying, unobserved topics. LSA algorithm identifies those topics considering both their local and global relevance [23];
- *latent Dirichlet allocation* (LDA) sees documents as probability distributions over latent topics, which in turn are described by probability distributions over terms. To this end, it places multinomial and Dirichlet assumptions to estimate the likelihood of a document to be described by a given topic;
- *hierarchical Dirichlet processes* (HDP) provides a non-parametric alternative to LDA, enabling the discovery of a non-fixed number of topics from text.

Formal Concept Analysis. The theory of FCA, first introduced by Wille [33], is currently a popular method for knowledge representation [19].

A *formal context* is a triplet (D, T, I), where D is the set of documents, T is the set of terms and/or topics, and $I \subseteq D \times T$ relates D and T (incidence relation). A *formal concept* is a pair (A, O) of a formal context (D, T, I), where A objects (extent) is the set of documents that share O attributes (intent).

A *concept lattice*, $\mathfrak{B}_{(D,T,I)}$, is the set of all concepts in a formal context. Concept lattices (also called Galois lattices) related all concepts hierarchically based on the shared elements, from less specific (concepts grouping many objects sharing few attributes) to most specific (fewer objects and more attributes).

Biclustering. Given a vector space model A defined by a set of objects (documents) $D = \{d_1, .., d_n\}$, attributes (terms and topics) $Y = \{t_1, .., t_m\}$, and elements $w_{ij} \in \mathbb{R}$ observed in d_i and t_j:

- a bicluster B = (I, J) is a $n \times m$ submatrix of A, where $I = (i_1, .., i_n) \subseteq D$ is a subset of documents and $J = (j_1, .., j_m) \subseteq Y$ is a subset of attributes;
- the biclustering task aims at identifying a set of biclusters $B = (B_1, .., B_s)$ such that each bicluster $B_k = (I_k, J_k)$ is a coherent concept that satisfies specific *homogeneity*, *dissimilarity* and *statistical significance* criteria.

Homogeneity criteria are commonly guaranteed through the use of a merit function, such as the variance of the values in a bicluster [24]. Merit functions are typically applied to guide the formation of biclusters in greedy and exhaustive searches. In stochastic approaches, a set of parameters that describe the biclustering solution are learned by optimizing a merit (likelihood) function.

The pursued homogeneity determines the coherence, quality and structure of a biclustering solution [13]. The coherence of a bicluster is determined by the observed form of correlation among its elements (coherence assumption) and by the allowed value deviations from perfect correlation (coherence strength). The quality of a bicluster is defined by the type and amount of accommodated noise. The structure of a biclustering solution is defined by the number, size, shape and positioning of biclusters. A flexible structure is characterized by an arbitrary number of (possibly overlapping) biclusters. Definitions 1 and 2 formalize these concepts, and Fig. 1 illustrates them, contrasting coherent and formal concepts.

Definition 1. *Given a vector space model A, elements in a bicluster $w_{ij} \in (I, J)$ have coherence across documents (attributes) if $w_{ij} = c_j + \gamma_i + \eta_{ij}$ ($w_{ij} = c_i + \gamma_j + \eta_{ij}$), where c_j (or c_i) is the value of attribute t_j (or document d_i), γ_i (or γ_j) is the adjustment for document d_i (or attribute y_j), and η_{ij} is the noise factor of w_{ij}.*

A bicluster has constant coherence *when $\gamma_i = 0$ (or $\gamma_j = 0$), and* additive coherence *otherwise, $\gamma_i \neq 0$ (or $\gamma_j \neq 0$).*

Let $\bar{A}$ be the amplitude of values in A, coherence strength *is a value $\delta \in [0, \bar{A}]$ such that $w_{ij} = c_j + \gamma_i + \eta_{ij}$ where $\eta_{ij} \in [-\delta/2, \delta/2]$.*

Definition 2. *Given a numeric dataset A, a bicluster (I, J) satisfies the* order-preserving coherence *assumption iff the values for each object in I (attribute in J) induce the same ordering π along the subset of attributes J (documents I).*

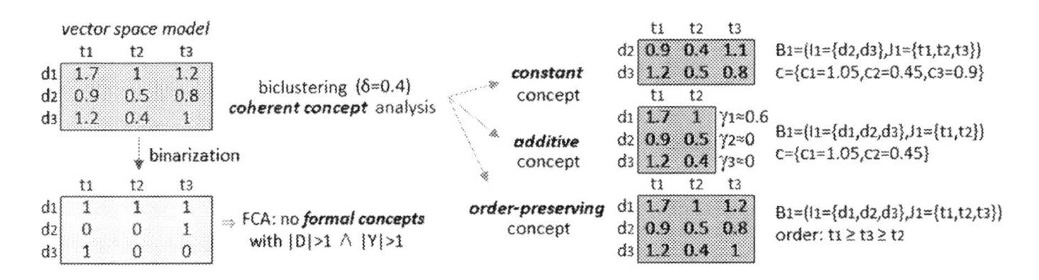

Fig. 1. Formal versus coherent concepts: biclustering with varying homogeneity criteria. Three coherent concepts were found under a constant, additive and order-preserving assumption (Definitions 1 and 2), corresponding to a set of terms with coherent importance (in value, difference and order) on a set documents. Illustrating, $t_1 \geq t_2 \geq t_3$ permutation of terms' relevance is preserved along documents $\{d_1, d_2, d_3\}$. In contrast, no formal concepts were found on the given vector space.

Statistical significance criteria, in addition to homogeneity criteria, guarantees that the probability of a bicluster's occurrence (against a null data model) deviates from expectations [17].

Dissimilarity criteria can be further placed to comprehensively cover the vector space with non-redundant biclusters [14].

3 Related Work

FCA in Digital Collections. FCA has been largely applied in Boolean space models given either by terms or (previously extracted) topics. In [4], a method is proposed, guided by both internal clustering quality metrics (Davies-Bouldin Index [7], Dunn Index [9], Silhouette coefficient [31] and The Calinski-Harabasz Index [21]) and external metrics (Reliability, Sensitivity and F-measure [1]). The experimental analysis used a collection of 2200 manually labeled tweets from 61 entities. The binary attributes are given by terms, named entities, references and

URLs. A concept lattice is inferred using the Next Neighbours [3] algorithm. Each formal concept is here seen as a topic. Still, a large number of non-relevant topics is generated. The authors thus propose the Stability metric [22] to extract the most promising formal concepts, concluding that, if considering the external evaluation, FCA show a more homogeneous performance than the LDA and Hierarchic Agglomerative Clustering (HAC), with better overall results. Ignatov in [19] and Poelmans et al. in [28] present a survey on different contributions for FCA regarding several applications. Myat and Hla [25] developed a method for web document organization based on FCA. Cimiano et al. [6] presented an approach for the automatic extraction of concept hierarchies from text data. The authors modeled the context of a certain term as a vector representing syntactic dependencies that are automatically acquired from the text corpus with a linguistic parser, producing with the FCA a lattice of partial order that constitutes the concept hierarchy.

Biclustering Digital Collections. Following the taxonomy of Madeira and Oliveira [24], biclustering algorithms can be categorized according to the pursued homogeneity and type of search. Hundreds of biclustering algorithms were proposed in the last decade, as shown by recent surveys [10,26]. In recent years, a clearer understanding of the synergies between biclustering and pattern mining paved the rise for a new class of algorithms, referred to as pattern-based biclustering algorithms [13]. Pattern-based biclustering algorithms are inherently prepared to efficiently find exhaustive solutions of biclusters and offer the unprecedented possibility to affect their structure, coherency and quality [12,14]. This behavior explains why this class of biclustering algorithms are receiving an increasing attention in recent years [13,18]. BicPAMS [14] consistently combines such state-of-the-art contributions on pattern-based biclustering.

Castro et al. [5] developed BIC-aiNet, an immune-inspired biclustering approach for document categorization that was applied over Brazilian newspapers. Despite its relevance, it is limited to Boolean spaces (presence or absence of topics per document), sharing similar limitations to FCA. Dhillon [8] proposed the use of coclustering (a restrictive variant of the biclustering task that imposes a checkboard structure of biclusters [24]) to explore text collections. Coclustering was applied to vector space models with entries given by $w_{ij} \times log(\frac{n}{n_j})$, where n is the number of documents and n_j the number of statements containing term t_j in document d_i. The author was able to identify subsets of words and documents with strong correlation along the Cranfield (1400 aeronautical documents), Medline (1033 medical documents) and Cisi (1460 information retrieval documents) collections. Despite its relevance, coclustering requires all elements to belong to a concept (exhaustive condition) and to a single concept only (exclusive condition), largely limiting the inherent flexibility of the biclustering task.

4 On Why, When and How to Apply Biclustering

As surveyed, pattern-based biclustering approaches provide the unprecedented possibility to comprehensively find concepts in vector space models with parameterizable homogeneity and guarantees of statistical significance [14]. Despite their relevance, their use to explore digital collections remains largely unassessed. This section provides a structured view on why, when and how to bicluster text data.

4.1 On *WHY*

As motivated, coherent concept analysis should be considered to:

- avoid the drawbacks of formal concept analysis related with the need to specify thresholds and the item-boundaries problems [11];
- discover concepts in real-valued data spaces sensitive to the representativity of terms and topics per document;
- pursue concepts with desirable properties by parameterizing pattern-based biclustering searches [14] with the aimed coherence, quality, dissimilarity and statistical significance criteria.

Depending on the goal, one or more coherence assumptions (Definitions 1 and 2) can be pursued [13,18]. The classic **constant coherence** can be placed to find groups of documents and topics, where each document has a similar probability to be described by a specific topic. Illustrating, documents d_1 and d_2 with $p(t_2, t_3, t_7|d_1) = \{0.32,\ 0.90,\ 0.49\}$ and $p(t_2, t_3, t_7|d_2) = \{0.29,\ 0.88,\ 0.55\}$ are coherently related under a coherence strength $\delta = 0.1$ (allowed deviations from expectations). The notion of constant association is already a generalization over the traditional Boolean formal concept. Still, it can be further generalized to allow more flexible correlations. One paradigmatic example is the **order-preserving coherence** where a subset of topics have preserved orders of predominance on a subset of documents (Fig. 1). Illustrating, documents d_1 and d_2 with $p(t_2, t_3, t_7|d_1) = \{0.32,\ 0.50,\ 0.47\}$ and $p(t_2, t_3, t_7|d_2) = \{0.29,\ 0.97,\ 0.55\}$ are coherently related since they preserve the permutation $w_{i2} \leq w_{i3} \leq w_{i7}$.

Pattern-based biclustering [14] allows the discovery of these less-trivial yet coherent, meaningful and potentially relevant concepts.

4.2 On *WHEN*

Coherent concept analysis should be applied when:

- topic representativity matters. Recovering the introduced example, in contrast with coherent concept analysis, FCA under a binarization threshold $\theta = 0.1$ is unable to differentiate $p(t_3|d_1) = w_{1,3} = 0.12$ from $p(t_3|d_5) = w_{5,3} = 0.95$;
- pursuing less-trivial forms of knowledge (including the introduced constant or order-preserving concepts);
- discretization drawbacks must be avoided;
- pursuing comprehensive solutions of concepts with diverse homogeneity and quality (noise-tolerance) criteria.

In contrast, coherent concept analysis should **not** be applied when:

- text collections are optimally represented as Boolean space models;
- extracting formal ontology structures [11]. Although pattern-based biclustering searches can also explore hierarchical relationships between biclusters, the resulting taxonomies are harder to interpret;
- the desirable binarization thresholds are known in advance and noise-tolerant FCA searches [27] can be applied to handle the noise associated with values near the boundaries of discretization.

4.3 On *HOW*

Pattern-based biclustering offers principles to find all potentially relevant concepts as they pursue multiple homogeneity criteria (including multiple coherence assumptions, coherence strength thresholds, and noise tolerance levels), and exhaustively yet efficiently explore different regions of the search space, preventing that regions with large concepts jeopardize the search [14]. As a result, less-trivial (yet coherent) topic associations are not neglected.

The possibility to allow deviations from value expectations (under limits defined by the placed coherence strength) tackles the item-boundaries problem.

Pattern-based biclustering does not require the input of support thresholds as it explores the search space at different supports [12], i.e. we do need to place expectations on the minimum number of documents per concept. Still, the minimum number of (dissimilar) concepts and topics per concept can be optionally inputted to guide the search. Dissimilarity criteria and condensed representations can be placed [14] to prevent redundant concepts.

Statistical Significance. A sound statistical testing of concepts is key to guarantee the absence of spurious relations, and ensure concept relevance when categorizing contents and making other decisions. To this end, the statistical tests proposed in BSig [17] are suggested to minimize false positives (outputted concepts yet not statistically significant) without incurring on false negatives. This is done by approximating a null model of the target vector space and appropriately testing each bicluster in accordance with its underlying coherence.

On Robustness to Noise and Missing Values. Similarly to some FCA extensions, pattern-based biclustering can pursue biclusters with a parameterizable tolerance to noise [12]. This possibility ensures robustness to the algorithm-specific fluctuations on topic likelihood per document. Also, and similarly to general FCA approaches, pattern-based biclustering is robust to missing data as it allows the discovery of biclusters with an upper bound on the allowed amount of missing values [16]. This is particularly relevant to handle topic uncertainties.

Other Opportunities. Additional benefits of pattern-based biclustering that can be carried towards concepts analysis include: (1) the possibility to remove uninformative elements in data to guarantee a focus, for instance, on coherent concepts with non-residual topic probabilities [16]; (2) incorporation of domain knowledge to guide the task in the presence of background metadata [15]; and

(3) support classification and regression task in the presence of document annotations by guaranteeing the discriminative power of biclusters [13].

5 Results

To illustrate the enumerated potentialities of coherent concept analysis, results are gathered in four major steps. First, we introduce the pursued methodology and analyze the target corpus. Second, we empirically delineate general differences of FCA and biclustering. Third, we provide evidence for the relevance of finding non-trivial (yet meaningful) concepts with constant and order-preserving forms of coherence. Finally, we show that biclustering guarantees the statistical significance of concepts, providing a trustworthy means for concept analysis.

Methodology. The target forms of concept analysis should be preceded by the preprocessing of text collections to find a proper structured data representation of relevant topics, and succeeded by the statistical and domain-driven assessment of the found concepts, which then serve as basis to support categorization and navigation by linking documents with shared concepts.

Dataset. Over 35000 legal documents issued by state bodies in the domain of agriculture were extracted from the *Diário da República Eletrónico* (DRE), the official on-line publication journal of the Portuguese state. This collection has a total of 24018518 tokens (213868 unique tokens).

Preprocessing. Each document was pre-processed to remove stop words, punctuation, numbers, links, emails and dates. Next, the Part-Of-Speech (POS) for each word is extracted, and all words that are not nouns or proper nouns are removed. Finally, words with high frequency and low TF-IDF scores are also removed. Figure 2 depicts the word distribution of the documents before (green histogram) and after (blue histogram) preprocessing.

Topic Modeling. We further used *Phrase*[1] to extract the combined words (phrasing) per document. From the obtained feature matrix, topics were extracted using LSA, LDA and HDP methods. Figure 3 shows for LDA and LSA how the quality of the approaches vary with the number of topics (HDP is non-parametric). The coherence score establishes the quality of the obtained topics by computing the probability of pairs of words in a given topic appearing together on the documents associated with a given topic. In accordance, LDA was selected. A document is then seen as a vector of probabilistic values that corresponds to the likelihood (predominance) of a given topic appear in the document.

[1] Automatic keyphrase extraction tool from Gensim: https://radimrehurek.com/gensim/.

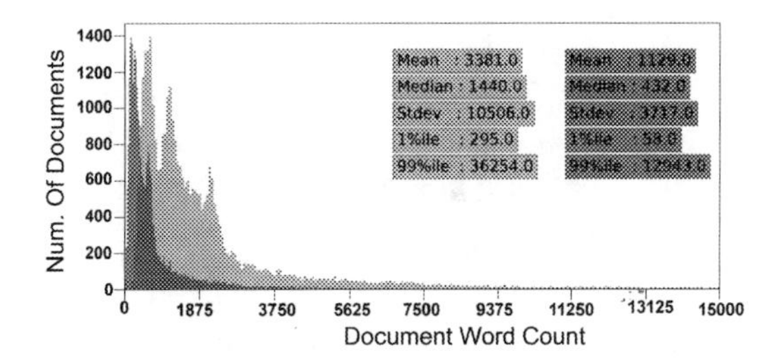

Fig. 2. Word count distribution over documents before and after preprocessing.

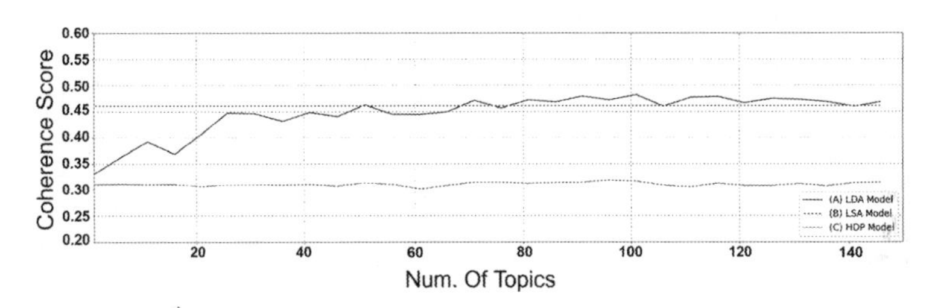

Fig. 3. Comparing topic modeling methods (LSA, LDA, HDP) w.r.t. coherence score.

Formal Concept Analysis. Figure 4 applies FCA [11] to the preprocessed dataset – a vector space model with 35000 documents and 120 topics – under a variable binarization threshold θ. θ parameterization is a highly sensible choice as evidenced by its impact on the number of formal concepts (from 230 k concepts when $\theta = 0.05$ to 48 k when $\theta = 0.1$ and 122 when $\theta = 0.5$), average number of

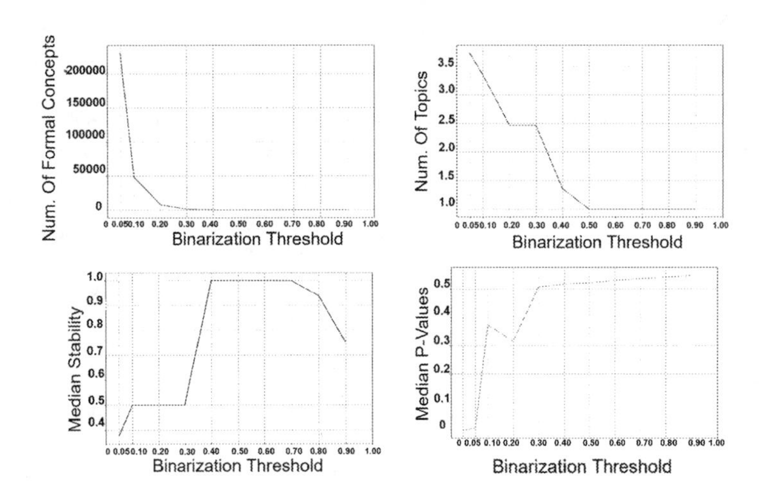

Fig. 4. FCA: binarization threshold impact on the: (a) number of concepts, (b) average number of topics per concept, (c) solution stability, and (d) median stat. significance.

topics per concept, and the stability criterion [30]. Elements in the vector space model close to θ are excluded from the concepts. By seeing topics as Bernoulli variables in a Boolean data space, binomial tail statistics [17] reveal that only a small fraction of the returned concepts are statistically significant.

Coherent Concept Analysis. BicPAMS [14] is applied as it combines state-of-the-art principles on pattern-based biclustering. BicPAMS is below used with default parameters: varying coherence strength ($\delta = \bar{A}/|\mathcal{L}|$ where $|\mathcal{L}| \in \{2, 3, 4, 5\}$), decreasing support until 100 dissimilar biclusters are found, up to 30% noisy elements, 0.05 significance level, and constant and order-preserving coherence assumptions. Two search iterations were considered by masking the biclusters discovered after the first iteration to ensure a more comprehensive exploration of the data space and a focus on less-trivial concepts. Topic-based frequency distributions were approximated, and the statistical tests proposed in [17] applied to compute the statistical significance of each concept.

Table 1 synthesizes the results produced by BicPAMS [14] on the preprocessed dataset. BicPAMS is able to efficiently find homogeneous, dissimilar and statistically significant concepts (subsets of topics with coherent predominance on a subset of documents). Illustrating, a total of 327 statistically significant concepts (p-value < 1) with constant coherence ($|\mathcal{L}| = 3$) and an average of 112 supporting documents were found. These initial results show the impact of placing coherence assumptions and coherence strength criteria on concept analysis.

Constant Concepts. Table 2 provides the details of four constant biclusters (their respective pattern, topics, coherence strength and statistical significance) using BicPAMS. Each bicluster shows a unique pattern of topic predominance. Figure 5 visually depicts these concepts using line charts and heatmaps. Each line in the chart (and row in the heatmap) represents a document and the values (colors) show the representivity of its topics. These results motivate the relevance of finding constant concepts to group topics in accordance with their representivity in a document, a possibility neglected by FCA.

A closer analysis of the found biclusters further shows their robustness to the item-boundaries problem: topics with slightly deviating likelihoods from pattern expectations are not excluded. This allows the analysis of vector space models without the drawbacks of discrete views placed by FCA approaches.

Table 1. Biclustering solutions found in DRE dataset using BicPAMS with varying homogeneity criteria.

Homogeneity	$\|\mathcal{L}\|$	#biclusters	Average #rows	Median p-value	% most freq. pattern
Constant	2	121	647.62	0.00	$I = [0, 0, 0]$(100%)
Constant	3	327	112.07	2.34e−152	$I = [0, 0, 0]$(23%)
Constant	4	165	77.72	6.18e−122	$I = [1, 0, 0]$(24%)
Constant	5	161	44.78	1.97e−74	$I = [0, 0, 0]$(30%)
Order preserving	NA	163	201.66	0.99	$I = [7, 13, 5]$(4%)

Table 2. Coherence concepts: zoom-in on 4 constant and 4 order-preserving concepts. For simplicity sake, the values of the concepts are presented in a discrete manner: $|\mathcal{L}|$ for constant coherence and 0 to 20 for order-preserving coherence. Illustrating, consider the constant concept B_1 with elements {2, 0.5, 1} for document $x_{3662117}$ in topics $\{t_{14}, t_{43}, t_{47}\}$: 0.5, 1 and 2 values correspond to topics with respectively residual, low and high probability to occur in $x_{3662117}$ document.

Bicluster properties	Pattern			
B_1 (with $\|\mathcal{L}\|$=4)	3662117	2.0	0.5	1.0
topics = [14,43,47]	3384398	2.0	0.5	0.5
#documents = 147	...			
p-value = 6.79e-170	979773	2.0	0.0	1.5
	1438820	2.0	0.0	1.0
	3459557	1.5	0.0	0.5
	75740163	2.0	0.0	1.5
B_2 (with $\|\mathcal{L}\|$=3)	235558	1.5	1.0	1.0
topics = [76,103,118]	1762073	2.0	1.5	1.0
#documents = 337	...			
p-value = 0	553876	2.0	1.0	1.0
	632429	1.5	1.0	1.5
	196216	2.0	1.5	1.5
	250617	2.0	1.0	1.5
B_3 (with $\|\mathcal{L}\|$=3)	221325	1.5	2.0	1.0
topics = [37,76,118]	547844	1.5	1.5	1.0
#documents = 363	...			
p-value = 0	572890	2.0	1.5	1.0
	3189813	1.5	2.0	1.5
	156660	2.0	1.5	1.0
	553876	2.0	2.0	1.0
B_4 (with $\|\mathcal{L}\|$=4)	361504	1.5	1.0	2.5
topics = [37, 46, 76]	221325	1.5	1.0	2.5
#documents = 183	...			
p-value = 1.81e-252	168871	2.0	1.5	1.5
	512991	1.5	1.0	2.5
	324968	1.0	1.0	2.5
	148432	1.5	1.5	3.0

Bicluster properties	Pattern			
B_1	2177091	4.5	8.5	1.5
topics = [37,76,93]	293285	2.5	15.0	1.5
#documents = 299	...			
p-value = 0.08	1181178	2.5	15.5	1.5
	74661197	15.0	17.0	5.5
Order of difficulty:	3189813	8.5	13.5	1.5
$t_{76} \geq t_{37} \geq t_{93}$	385434	12.5	15.0	1.5
B_2	453556	9.5	2.5	8.5
topics = [14,43,47]	2806956	7.5	4.5	5.5
#documents = 290	...			
p-value = 0.21	494218	13.5	1.5	2.5
	75740163	15.5	1.5	12
Order of difficulty:	279258	8.5	1.5	1.5
$t_{14} \geq t_{47} \geq t_{43}$	3551103	16.5	1.5	1.5
B_3	421452	9.5	2.5	9.5
topics = [76,93,118]	547844	8.5	1.5	5.5
#documents = 283	...			
p-value = 0.36	3189813	13.5	1.5	8.5
	553876	13.5	1.5	1.5
Order of difficulty:	385434	15.0	1.5	2.5
$t_{76} \geq t_{118} \geq t_{93}$	196216	17.5	2.5	7.5
B_4	3595682	15.0	4.5	8.5
topics = [14,19,109]	2806956	7.5	1.5	5.5
#documents = 186	...			
p-value = 0.99	2645902	14.5	2.5	10.5
	67412614	17.5	2.5	10.5
Order of difficulty:	341633	17.0	1.5	1.5
$t_{14} \geq t_{109} \geq t_{19}$	3551103	16.5	2.5	14.5

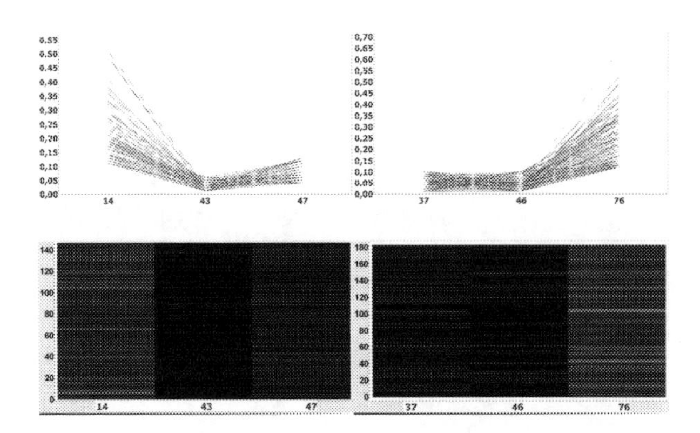

Fig. 5. Visuals of constant concepts **B1** and **B4** (Table 2): chart and heatmap views.

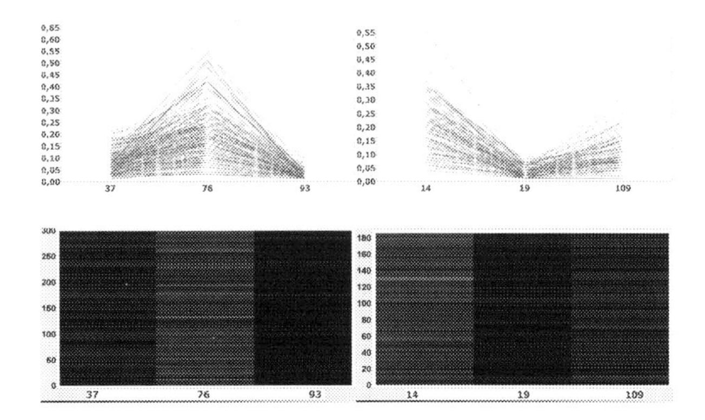

Fig. 6. Visuals of order-preserving concepts **B1** and **B4** (Table 2): chart-heatmap views.

Order-Preserving Concepts. Non-constant patterns are suggested if the focus is not on determining levels of performance but to assess the relative representativity among topics. BicPAMS [14] was applied to find such less-trivial yet relevant concepts. Table 2 details 4 order-preserving biclusters. Figure 6 visually depicts 2 of these concepts. Understandable, FCA is unable to recover such concepts given their flexible (yet meaningful) homogeneity criteria.

Robustness. Tolerance to noise can be customized to find concepts with desirable bounds on quality. In addition to noise tolerance, η_{ij}, coherence strength, $\delta = \bar{A}/|\mathcal{L}|$, can be further explored to comprehensively model associations with slight-to-moderate deviations from expectations. Figure 7 shows the impact of quality on the number of biclusters, average number of documents per bicluster and median p-values when BicPAMS is applied with constant coherence.

Statistical Significance. Table 1 shows the biclustering ability to find statistically significant concepts. A bicluster is statistically significant if the number of documents with a given pattern or permutation of topics is unexpectedly low [17]. Figure 8 provides a scatter plot of the statistical significance and area ($|I| \times |J|$) of constant ($|\mathcal{L}| = 3$) and order-preserving biclusters. This analysis

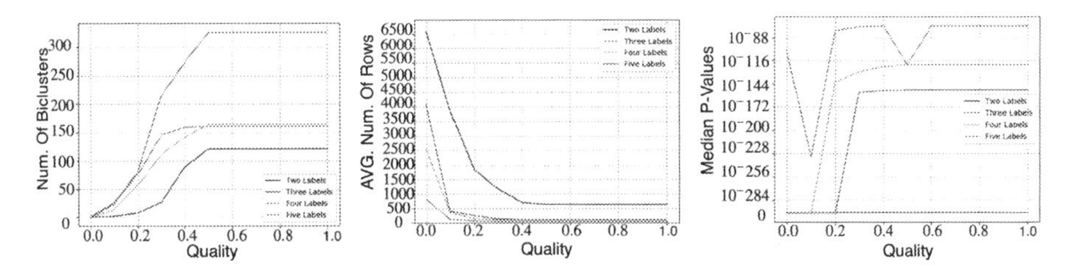

Fig. 7. Impact of the allowed noise tolerance in coherent concept analysis (BicPAMS under constant coherence and $\mathcal{L} \in \{2, 3, 4, 5\}$): number of concepts, average number documents per concept, and median p-value.

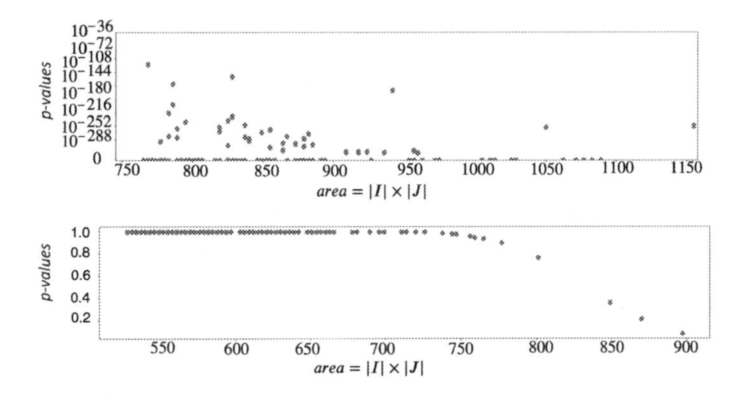

Fig. 8. Statistical significance *versus* size of constant (up) ($\mathcal{L} = \{\{0, 0.10\}, \{0.045, 1\}\}$) and order preserving (down) biclusters (using statistical tests proposed in [17]).

suggests the presence of a soft correlation between size and statistical significance. A few order-preserving concepts have low statistical significance (upper dots) and should therefore be discarded for not incorrectly bias decisions.

6 Concluding Remarks

This work proposes comprehensive principles on how to apply biclustering for content categorization in large and heterogeneous text collections. Biclustering, a form of coherent concept analysis, is suggested to tackle the limitations of FCA since it explores all potentially relevant information available in vector spaces by focusing the searches on less-trivial, yet meaningful and statistically significant concepts. Pattern-based biclustering searches are suggested since they hold unique properties of interest: efficient exploration; optimality guarantees; discovery of concepts with parameterizable coherence; tolerance to noise and missing data; incorporation of domain knowledge; complete biclustering structures without positioning restrictions; and sound statistical testing.

Results from a real corpus confirm the unique role of biclustering in finding relevant associations between topics and documents. Results further evidence the ability to unveil interpretable concepts with guarantees of statistical significance and robustness, thus providing a trustworthy context with enough feedback for content categorization in large text collections.

Acknowledgement. This work was supported by Imprensa Nacional Casa da Moeda (INCM) and national funds through Fundação para a Ciência e a Tecnologia (FCT) with references DSAIPA/DS/0111/2018 and UID/CEC/50021/2019.

References

1. Amigó, E., Gonzalo, J., Verdejo, F.: A general evaluation measure for document organization tasks. In: Proceedings of the 36th International ACM SIGIR Conference on Research and Development in Information Retrieval, pp. 643–652. ACM (2013)
2. Banerjee, A., Dhillon, I., Ghosh, J., Merugu, S., Modha, D.S.: A generalized maximum entropy approach to Bregman co-clustering and matrix approximation. In: ACM SIGKDD International Conference on Knowledge Discovery and Data Mining, pp. 509–514. ACM (2004)
3. Carpineto, C., Romano, G.: Concept Data Analysis: Theory and Applications. Wiley, Hoboken (2004)
4. Castellanos, A., Cigarrán, J., García-Serrano, A.: Formal concept analysis for topic detection: a clustering quality experimental analysis. Inf. Syst. **66**, 24–42 (2017)
5. de Castro, P.A.D., de França, F.O., Ferreira, H.M., Von Zuben, F.J.: Applying biclustering to text mining: an immune-inspired approach. In: de Castro, L.N., Von Zuben, F.J., Knidel, H. (eds.) ICARIS 2007. LNCS, vol. 4628, pp. 83–94. Springer, Heidelberg (2007). https://doi.org/10.1007/978-3-540-73922-7_8
6. Cimiano, P., Hotho, A., Staab, S.: Learning concept hierarchies from text corpora using formal concept analysis. J. Artif. Intell. Res. **24**, 305–339 (2005)
7. Davies, D.L., Bouldin, D.W.: A cluster separation measure. IEEE Trans. Pattern Anal. Mach. Intell. PAMI **1**(2), 224–227 (1979)
8. Dhillon, I.S.: Co-clustering documents and words using bipartite spectral graph partitioning. In: Proceedings of the Seventh ACM SIGKDD International Conference on Knowledge Discovery and Data Mining, pp. 269–274. ACM (2001)
9. Dunn, J.C.: Well-separated clusters and optimal fuzzy partitions. J. Cybern. **4**(1), 95–104 (1974)
10. Eren, K., Deveci, M., Küçüktunç, O., Çatalyürek, Ü.V.: A comparative analysis of biclustering algorithms for gene expression data. Briefings Bioinform. **14**(3), 279–292 (2013)
11. Ganter, B., Wille, R.: Formal Concept Analysis: Mathematical Foundations. Springer, Heidelberg (2012)
12. Henriques, R., Madeira, S.: BicPAM: pattern-based biclustering for biomedical data analysis. Algorithms Mol. Biol. **9**(1), 27 (2014)
13. Henriques, R., Antunes, C., Madeira, S.C.: A structured view on pattern mining-based biclustering. Pattern Recogn. **4**(12), 3941–3958 (2015)
14. Henriques, R., Ferreira, F.L., Madeira, S.C.: BicPAMS: software for biological data analysis with pattern-based biclustering. BMC Bioinf. **18**(1), 82 (2017)
15. Henriques, R., Madeira, S.C.: BIC2PAM: constraint-guided biclustering for biological data analysis with domain knowledge. Algorithms Mol. Biol. **11**(1), 23 (2016)
16. Henriques, R., Madeira, S.C.: BicNET: flexible module discovery in large-scale biological networks using biclustering. Algorithms Mol. Biol. **11**(1), 1–30 (2016)
17. Henriques, R., Madeira, S.C.: BSig: evaluating the statistical significance of biclustering solutions. Data Min. Knowl. Disc. **32**(1), 124–161 (2018)
18. Henriques, R., Madeira, S.C.: Triclustering algorithms for three-dimensional data analysis: a comprehensive survey. ACM Comput. Surv. **51**(5), 95:1–95:43 (2018)
19. Ignatov, D.I.: Introduction to formal concept analysis and its applications in information retrieval and related fields. In: Braslavski, P., Karpov, N., Worring, M., Volkovich, Y., Ignatov, D.I. (eds.) RuSSIR 2014. CCIS, vol. 505, pp. 42–141. Springer, Cham (2015). https://doi.org/10.1007/978-3-319-25485-2_3

20. Kalman, D.: A singularly valuable decomposition: the SVD of a matrix. Coll. Math. J. **27**(1), 2–23 (1996)
21. Kozak, M.: "A dendrite method for cluster analysis" by caliński and harabasz: a classical work that is far too often incorrectly cited. Commun. Stat. Theor. Methods **41**(12), 2279–2280 (2012)
22. Kuznetsov, S.: Stability as an estimate of the degree of substantiation of hypotheses derived on the basis of operational, similarity. Autom. Documentation Math. Linguist. **24** (1990)
23. Landauer, T.K., Foltz, P.W., Laham, D.: An introduction to latent semantic analysis. Discourse Process. **25**(2–3), 259–284 (1998)
24. Madeira, S.C., Oliveira, A.L.: Biclustering algorithms for biological data analysis: a survey. IEEE/ACM Trans. Comput. Biol. Bioinf. **1**(1), 24–45 (2004)
25. Myat, N.N., Hla, K.H.S.: Organizing web documents resulting from an information retrieval system using formal concept analysis. In: Asia-Pacific Symposium on Information and Telecommunication Technologies, pp. 198–203. IEEE (2005)
26. Oghabian, A., Kilpinen, S., Hautaniemi, S., Czeizler, E.: Biclustering methods: biological relevance and application in gene expression analysis. PLoS ONE **9**(3), e90801 (2014)
27. Pensa, R.G., Boulicaut, J.-F.: Towards fault-tolerant formal concept analysis. In: Bandini, S., Manzoni, S. (eds.) AI*IA 2005. LNCS (LNAI), vol. 3673, pp. 212–223. Springer, Heidelberg (2005). https://doi.org/10.1007/11558590_22
28. Poelmans, J., Kuznetsov, S.O., Ignatov, D.I., Dedene, G.: Formal concept analysis in knowledge processing: a survey on models and techniques. Expert Syst. Appl. **40**(16), 6601–6623 (2013)
29. Rajaraman, A., Ullman, J.D.: Data Mining, pp. 1–17. Cambridge University Press, Cambridge (2011)
30. Roth, C., Obiedkov, S., Kourie, D.: Towards concise representation for taxonomies of epistemic communities. In: Yahia, S.B., Nguifo, E.M., Belohlavek, R. (eds.) CLA 2006. LNCS (LNAI), vol. 4923, pp. 240–255. Springer, Heidelberg (2008). https://doi.org/10.1007/978-3-540-78921-5_17
31. Rousseeuw, P.J.: Silhouettes: a graphical aid to the interpretation and validation of cluster analysis. J. Comput. Appl. Math. **20**, 53–65 (1987)
32. Tan, P.N.: Introduction to data mining. Pearson Education India, New York (2018)
33. Wille, R.: Restructuring lattice theory: an approach based on hierarchies of concepts. In: Rival, I. (ed.) Ordered Sets, pp. 445–470. Springer, Dordrecht (1982). https://doi.org/10.1007/978-94-009-7798-3_15

Beyond Modelling: Understanding Mental Disorders in Online Social Media

Esteban Andrés Ríssola[1(✉)], Mohammad Aliannejadi[2], and Fabio Crestani[1]

[1] Università della Svizzera italiana, Lugano, Switzerland
{esteban.andres.rissola,fabio.crestani}@usi.ch
[2] Univiersity of Amsterdam, Amsterdam, The Netherlands
m.aliannejadi@uva.nl

Abstract. Mental disorders are a major concern in societies all over the world, and in spite of the improved diagnosis rates of such disorders in recent years, many cases still go undetected. Nowadays, many people are increasingly utilising online social media platforms to share their feelings and moods. Despite the collective efforts in the community to develop models for identifying potential cases of mental disorders, not much work has been done to provide insights that could be used by a predictive system or a health practitioner in the elaboration of a diagnosis.

In this paper, we present our research towards better visualising and understanding the factors that characterise and differentiate social media users who are affected by mental disorders from those who are not. Furthermore, we study to which extent various mental disorders, such as depression and anorexia, differ in terms of language use. We conduct different experiments considering various dimensions of language such as vocabulary, psychometric attributes and emotional indicators. Our findings reveal that positive instances of mental disorders show significant differences from control individuals in the way they write and express emotions in social media. However, there are not quantifiable differences that could be used to distinguish one mental disorder from each other.

1 Introduction

During the last decade, there has been an increasing research interest in the identification of mental state alterations through the exploitation of online digital traces. One of the main reasons is that the capabilities of public health systems to cope with the plethora of cases that emerge on a daily basis are certainly limited. However, the proliferation of online social media platforms is changing the dynamics in which mental state assessment is performed [7,23]. Individuals are using these platforms on a daily basis to share their thoughts as well as to disclose their feelings and moods [8].

Research on language and psychology has shown that various useful cues about an individuals' mental state (as well as personality, social and emotional

Work done while Mohammad Aliannejadi was affiliated with Università della Svizzera italiana (USI).

J. M. Jose et al. (Eds.): ECIR 2020, LNCS 12035, pp. 296–310, 2020.
https://doi.org/10.1007/978-3-030-45439-5_20

conditions) can be discovered by examining the patterns of their language use [6]. As a matter of fact, language attributes could act as indicators of the current mental state [22,25], personality [19,26] and even personal values [2,4]. The main reason, as argued by Pennebaker et al. [21], is because such latent mental-related variables are encoded in the words that individuals use to communicate.

The constraints dictated in reality, such as cost and time, make the efficient process of personal diagnosis unfeasible. Initiatives such as the Strategic Workshop on Information Retrieval in Lorne [11] (SWIRL) are already proposing the application of principles of core Information Retrieval for the development of decision-making systems applied to fields that years back were not easy to conceive or imagine. In particular, they highlight the potential for cross-disciplinary collaboration and impact with a number of scientific fields, including psychology. In this respect, the Early Risk Prediction on the Internet (eRisk) [14,15], as well as the Computational Linguistics and Clinical Psychology (CLPsych) [9] workshops were the first to propose benchmarks to bring together many researchers to address the automatic detection of mental disorders in online social media.

These initial efforts to address the automatic identification of potential cases of mental disorders in social media have mainly modelled the problem as classification [16]. Researchers participating in these workshops have examined a wide variety of methods to identify positive cases [27,30]; however, not much insight has been given as to why a system succeeds or fails. Moreover, the models and features used in those studies could be analysed and motivated more deeply. Therefore, we argue that even though achieving an effective performance is important, being able to track and visualise the development of the mental disorder also is. This means that an accurate system can be more useful if it provides a way of understanding the explanatory factors that lead to a certain decision.

For this reason, it is necessary to carry out experiments providing insights on how the use of language is distinctive among social media users suffering from mental disorders as well as between different disorders. Moreover, it is useful to find ways to better visualise such development. Thus, systems oriented at visualisation for risk-assessment and decision-making could be complemented with preliminary step-by-step directions for practitioners to identify high-risk individuals based on statistical and visual analyses.

In this work, we conduct a thorough study of various dimensions of language to characterise users affected by mental disorders. Also, we provide several methods for visualising the data in order to provide useful insights to psychologists. To this end, we first compare users affected by a particular disorder against control individuals. Secondly, we are interested to know whether different mental disorders share the same characteristics or they are clearly different in terms of the dimensions analysed. Our main research questions, therefore, are:

- **RQ1:** How different is the language of users with mental disorder compared to control individuals in online social media?
- **RQ2:** To what extent is the language of depression, anorexia and self-harm cases different in online social media?

- **RQ3:** How can language-specific and emotional information be visualised to be utilised by psychologists during the diagnosis process?

To the best of our knowledge, this is the first study where the expression of mental disorders in social media is analysed at this depth. Our main findings reveal that positive instances of mental disorder significantly differ from control individuals[1]. More interestingly, we discover that considering the dimensions of language analysed it is not possible to establish a difference between depression, anorexia and self-harm.

The remainder of the paper is organised as follows. Section 2 summarises the related work; Sect. 3 details the approach followed to answer the research questions; Sect. 4 describes the data used in this work; Sect. 5 presents the corresponding results and analyses; Sect. 6 concludes the work.

2 Related Work

The majority of the works in the area have been mostly focused on the automatic identification of mental disorders in social media. Here, we outline those where some effort have been devoted to better understanding the relationship between language and mental disorders in social media and are relevant to our work.

Park et al. [20] provided a preliminary study towards verifying whether online social media data were truly reflective of users' clinical depressive symptoms. To this end, they analysed the expression of depression among the general Twitter[2] population. Over a period of two months, they collected tweets which contained the word "depression". A subsequent analysis showed that depression was most frequently mentioned to describe one's depressed status and, to a lesser extent, to share general information about depression.

De Choudhury et al. [5] presented an early work on automatic depression detection by using crowd-sourcing to collect assessments from several Twitter users who reported being diagnosed with depression. They a built a depression lexicon containing words that are associated with depression and its symptoms.

Coppersmith et al. [9] used Twitter data to carry out an exploratory analysis to determine language features that could be useful to distinguish users experiencing various mental disorders from healthy individuals. Despite they were able to determine a set of useful features, they observed that language differences in communicating about the different mental health problems remain an open question.

Gkotsis et al. [12] analysed various mental disorder communities on Reddit[3] (better known as *subreddits*[4]) to discover discriminating language features

[1] In this work, the term *positive* refers to subjects who have been diagnosed with depression, anorexia or self-harm. While, *control* refers to individuals not affected by any of the aforementioned mental disorders.

[2] See: https://twitter.com.

[3] See: https://www.reddit.com.

[4] Titled forums on Reddit are denominated *subreddits*.

between the users in the different communities. They found that, overall, the subreddits that were topically unrelated had condition-specific vocabularies as well as discriminating lexical and syntactic characteristics. Such study of Reddit communities might not result in accurate discrimination between users affected by mental disorders and healthy individuals. The main reason is that many of the participants of such specific forums are individuals concerned about the disorder because they had a close relative or friend suffering from it. We are interested in studying user's language features regardless of the topic discussed.

Overall, the presented works are concerned about the ability to predict whether users in online media platforms are positive instances of a mental disorder. Little effort has been devoted to understanding and providing insight and measuring the attributes which differentiates users affected by mental disorders from healthy individuals as well as between diverse mental disorders.

3 Objectives and Method

In this section we describe how we design the experiments to study social media posts in order to answer the research questions posed in the introduction. We outline what can be learned from each experiment, focusing on the language of diagnosed subjects and how their differences can be quantified.

3.1 Open Vocabulary

Vocabulary Uniqueness: One variable we analyse to answer **RQ1** is the similarity and diversity of the unique sets of words which compose the vocabulary of positive and control classes. Analysing such dimension tell us up to which extent classes have a common vocabulary and which words, if any, could be specifically used by users belonging to a certain class.

Considering each vocabulary as a set, we inspect the relative size of the their intersection. To this end, we use Jaccard's index to measure the similarity between finite sample sets. Formally, let P be the unique set of words obtained from positive users, *e.g.* self-harm, and C be the unique set of words obtained from control users. We compute Jaccard's index as follows:

$$J(P, C) = |P \cap C| / |P \cup C|.$$

As we see, the index gives us the ratio of the size of the intersection of P and C to the size of their union. The index ranges from 0 to 1, where an index of 1 indicates that the sets completely intersect, and thus, have the same elements. As the value approaches to 0 the sets are a more diverse among themselves.

Word Usage: An important aspect when studying the language of different groups, in addition to vocabulary similarities and differences, is to understand the patterns of word usage. Here, we attempt to answer **RQ1**, **RQ2**, **RQ3** by computing and comparing the language models for each class. The goal of this analysis is to quantify the differences that might emerge between the classes in terms of the probability of using certain words more than others.

Language models are processes that capture the regularities of language across large amounts of data [10]. In its simplest form, known as a unigram language model, it is a probability distribution over the terms in the corpus. In other words, it associates a probability distribution of occurrence with every term in the vocabulary for a given collection. In order to estimate the probability for a word w_i in a document D in a collection of documents S we use

$$P(w_i|D) = (1 - \alpha_D)P(w_i|D) + \alpha_D P(w_i|S),$$

where α is a smoothing coefficient used to control the probability assigned to out-of-vocabulary words. In particular, we use the linear interpolation method[5] where $\alpha_D = \lambda$, *i.e.*, a constant. To estimate the probability for word w_i in the collection we use $s_{w_i}/|S|$, where s_{w_i} is the number of times a word occurs in the collection, and $|S|$ is the total number of words occurrences in the collection. In this work, D identifies all the documents in a specific class, *i.e.*, we concatenate all the documents of a particular class such as self-harm. While S is the union of all the documents of two classes in a corpus, *i.e.*, positive and control.

Once we computed the language models for each class, we plot the probability distributions obtained and analyse to which extent the distributions differ. Furthermore, we support our observations by computing the Kullback-Leibler divergence (KL), a well-known measure from probability theory and information theory used to quantify how much two probability distributions differ. In essence, a KL-divergence of 0 denotes that the two distributions in question are identical. The KL-divergence is always positive and is larger for distributions that are more different. Given the *true* probability distribution P and control distribution C, the KL-divergence is defined as:

$$KL(P||C) = \sum_x P(x) log \frac{P(x)}{C(x)}.$$

3.2 Psychometric Attributes and Linguistic Style

A common method for linking language with psychological variables involves counting words belonging to manually-created categories of language [5,6,9]. Conversely to the experiment described in Sect. 3.1, such method is known as "closed vocabulary" analysis [28]. In essence, we address **RQ1**, **RQ2** and **RQ3** by studying "function words[6]", and topic-specific vocabulary. On the one hand, the goal of conducting such analysis is to quantify specific stylistic patterns that could differentiate positive instances of a mental disorder from control individuals. For example, individuals suffering from depression exhibit a higher tendency

[5] Also referred to as Jelinek-Mercer smoothing.

[6] A *function word* is a word whose purpose is to contribute to the syntax rather than to the meaning of the sentence.

to focus on themselves [3], and thus, it is expected that the use of personal pronouns such as "I" would be higher. On the other hand, certain positives classes might exhibit a higher use of specific topically-related words. As we show later for the case of anorexia when compared to depression and self-harm.

It should be noted that we decide to keep the stop-words since many words such as pronouns, articles and prepositions reveal part of people's emotional state, personality, thinking style and connection with others individuals [6]. As a matter of fact such words, called *function words*, account for less than one-tenth of a percent of an individual's vocabulary but constitute almost 60 percent of the words a person uses [6].

The *Linguistic Inquiry and Word Count* [29] (LIWC)[7], provides mental health practitioners with a tool for gathering quantitative data regarding the mental state of patients from the their writing style. In essence, LIWC is equipped with a set of dictionaries manually constructed by psychologist which covers various psychologically meaningful categories and is useful to analyse the linguistic style patterns of an individual's way of writing. In our study, we measure the proportion of documents from each user that scores positively on various LIWC categories (*i.e.*, have at least one word from that category). In particular, we choose a subset of the psychometric categories included in LIWC where we found significant differences between positive and control users. Subsequently, we plot the distributions obtained using box-plots and compare them.

3.3 Emotional Expression

Individuals usually convey emotions, feelings, and attitudes through the words they use. For instance, gloomy and cry denote sadness, whereas delightful and yummy evoke the emotion of joy. Here, we address **RQ1**, **RQ2**, and **RQ3** by studying how individuals, suffering from mental disorders, emotionally express themselves in their social media posts. Furthermore, we investigate how such emotional expression could differentiate between affected and non-affected users.

We utilise the emotion lexicons built by Mohammad et al. [17,18] where each word is associated with the emotions it evokes to capture word-emotion connotations. In addition to common English terms, the lexicons include words that are more prominent in social media platforms. Moreover, they include some words that might not predominantly convey a certain emotion and still tend to co-occur with words that do. For instance, the words *failure* and *death* describe concepts that are usually accompanied by sadness and, thus, they denote some amount of sadness.

4 Data

Here, we study various collections released at different editions of the eRisk workshop [14,15]. The main goal of the workshop is to provide a common evaluation framework for researchers to address the early identification of depression,

[7] See: http://liwc.wpengine.com/.

anorexia and self-harm. The collections consist of a set of documents posted by users of Reddit, consisting of two groups of users. Positive cases of a particular mental disorder, such as anorexia, as well as control individuals. We choose to conduct our study using these collections since they have been developed and used through the various editions of eRisk and, therefore, have been extensively curated and validated. Furthermore, they are publicly available for research.

Following the methodology proposed by Coppersmith et al. [7], users of the positive class (*i.e.*, depression, anorexia or self-harm) were gathered by retrieving self-expressions of diagnoses (*e.g.*, the sentence "I was diagnosed with depression") and manually verifying if they contained a genuine statement of diagnosis. Control users were collected by randomly sampling from a large set of users available in the platform. The maximum number of posts per user is 2,000. A summary of the eRisk's collections studied in this work is shown in Table 1.

Table 1. Summary of eRisk's collections. The activity period represents the number of days passed from the first to last the document collected for each user. On average, a user's corpus spans over a period of roughly one year and half. The oldest documents in the collections date from the middle of 2006, while the latest ones are from 2017.

	Depression		Anorexia		Self-harm	
	Positive	Control	Positive	Control	Positive	Control
# of subjects	214	1,493	61	411	41	299
# of documents	89,999	982,747	24,776	227,219	7,141	161,886
Avg. # of documents/subject	420.5	658.2	406.16	552.84	174.17	541.42
Avg. # words/document	45.0	35.3	64.6	31.4	39.3	28.9
Avg. activity period (days)	≈658	≈661	≈799	≈654	≈504	≈785

5 Results and Analyses

In this section we present the results obtained from the experiments outlined in Sect. 3 on the various collections described in Sect. 4. Moreover, we analyse the corresponding outcomes towards answering the proposed research questions.

Our analyses mainly focus on language and its attributes. Nonetheless, we would like to highlight certain differences observed in terms of social engagement behaviour between the different groups.

Basic Statistics. Table 1 shows various statistics of different eRisk's collections. Interestingly, we note that among all the collections, an average positive user generates less documents than an average control user. However, as we see the average length of the documents is longer for the positive cases. In particular, it is interesting to highlight the case of users who suffer from anorexia. They even write longer documents than users affected by depression and self-harm.

Finally, it is worth noting that we spotted no meaningful differences among the various control groups. This observation is repeated for various experiments and is expected since each control group is a random sample of Reddit posts at different temporal periods. Therefore, each control group is a representative sample of users on Reddit.

5.1 Open Vocabulary

Vocabulary Uniqueness: Table 2 compares the vocabulary of the different groups of positive users (*i.e.* depression, anorexia and self-harm) against that of control users. We analyse their union, intersection and difference. We observe that positive cases of depression and anorexia exhibit more similarity to their respective control groups with a Jaccard index of 59% and 65%, respectively. Self-harm positive cases, on the other hand, use a more diverse set of words in their posts when compared to the control group (44%).

Moreover, the vocabulary size of the various positive groups gives us an idea about the words that have never been used by control users, but used by the positive groups. Among the terms that are unique to the positive groups, we find the following ones interesting: selfharm, trazodone[8] (Depression); anorexics, depersonalization[9], emetrol, pepto[10] (Anorexia).

Table 2. Vocabularies comparison between positive and control users. KL-divergence computed across the language models obtained for the documents of positive and control users. As a reference, the KL-Divergence is also calculated between the different control groups. For instance, we observe an average divergence of 0.08 between the control group of the depression dataset and the other two control groups (*i.e.*, self-harm and anorexia).

	Depression	Anorexia	Self-harm
# of unique words positive	41,986	21,448	11,324
# of unique words control	70,229	31,980	25,091
Jaccard's index (positive vs. control)	0.59	0.65	0.44
Difference size (positive vs. control)	218	229	49
Difference size (control vs. positive)	28,461	10,761	13,816
KL (postive||control)	0.18	0.18	0.18
KL (control||positive)	0.21	0.31	0.20
KL (control||control)	0.08	0.07	0.10

[8] *Trazodone* is an antidepressant medication.

[9] Depersonalization is a mental disorder in which subjects feel disconnected or detached from their bodies and thoughts.

[10] *Emetrol*, and *Peptol* are medications used to treat discomfort of the stomach.

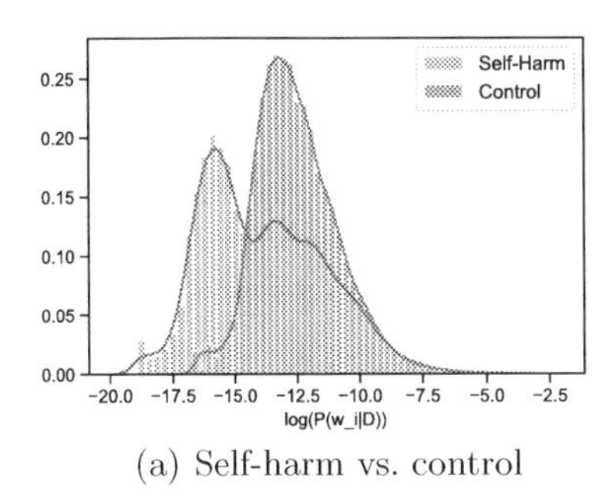

(a) Self-harm vs. control

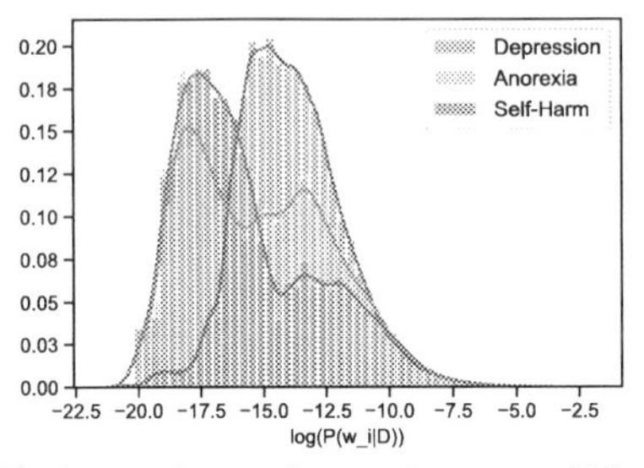

(b) Anorexia vs. depression vs. self-harm

Fig. 1. Language models probability distribution comparison (best viewed in colour). (Color figure online)

Word Usage: Figure 1 compares the language models obtained for the different classes. Note that the smoothing is necessary since, as shown before, there are terms which are present only in the positive class vocabulary but not in the control one and vice-versa. Figure 1(a) contrasts the language model of self-harm users against control individuals. We note that there are clear differences in terms of language use.

This observation is supported by the computation of KL-divergence. Table 2 shows the value of KL-divergence computed across the language models obtained for the documents of positive and control users. We note that the KL-divergence confirms the difference between the positive and control language models observed in the plots. We note similar patterns for depression and anorexia. In fact, as we compare the distribution of different control groups, we observe smaller KL-divergence values (0), indicating that these distributions are very similar.

Finally, comparing the language models of the three positive classes as in Fig. 1(b) is harder to identify noticeable differences between the distributions. Depression and anorexia language models are rather similar. While the largest noticeable difference is observed for self-harm when compared with either depression or anorexia. This reinforces the idea that the word probability distribution between the different positive classes is very similar, and thus, the way they use words. In this way, a system could compare the language model of a patient with both control and positive groups to provide the psychologists assistance in determining whether they are positive or not. The psychologist can further examine the patient to determine which disorder they are diagnosed with.

5.2 Psychometric Attributes and Linguistic Style

Using selected set of categories[11] from LIWC, we demonstrate that language use of Reddit users, as measured by LIWC, is statistically significantly different

[11] For a comprehensive list of LIWC categories see: http://hdl.handle.net/2152/31333.

between positive and control individuals. Figure 2 shows the proportion of documents from each user that scores positively on various LIWC categories (*i.e.*, have at least one word from that category). Selected categories includes function words (like pronouns and conjunctions), time orientation (like past focus and present focus) and emotionality. Bars are coloured according to the positive and control classes they represent.

The most remarkable case is the difference found in the use of the pronoun "I" between positive and control users, which in the case of depression replicates previous findings for other social media platforms [5,9]. Moreover, the proportion of messages using words related with positive emotions (*posemo*) is larger than negative (*negemo*) ones, even for the positive classes. Such circumstance could be related to the fact that English words, as they appear in natural language, are biased towards positivity [13]. Except for categories *we* and *she/he* differences reach statistical significance using Welch two sample t-test (p-value < 0.001) from each corresponding control group in Fig. 2(a). No significant differences between depression, anorexia and self-harm were found for any of the LIWC categories analysed in Fig. 2(a).

Figure 2(b) depicts categories related to *biological process*. In essence, this category includes words directly associated with the body and its main functions. We note that individuals within the anorexia group show a certainly different behaviour compared with depression and self-harm groups. Intuitively, this result is expected, given that anorexia is characterised by an intense fear of gaining weight and a distorted perception of weight. Overall, people with anorexia place a high value on controlling their weight and shape, using extreme efforts that tend to significantly interfere with their lives. Therefore, it is reasonable that such individuals talk more about themes related with their body and its function. Statistical significance (p-value < 0.001) between individuals suffering from anorexia and those affected by depression is achieved for categories *body*, *health* and *ingest*. While only for the latter category the differences are statistically significant when comparing self-harm and anorexia users.

In addition to the default categories that LIWC includes, we study other domain-specific lexicons. The first of them is the well-known depression lexicon[12] released by Choudhury et al. [5]. It consists of words closely associated with texts written by individuals discussing depression or its symptoms in online settings. The use of such words is not significantly different with respect to anorexia and self-harm groups. This suggest that those words are also frequently used by users affected by anorexia or self-harm. The same situation was observed when considering the set of absolutist terms[13] derived from the work of Al-Mosaiwi et al. [1], who concluded that the elevated use of absolutist words is a marker specific to anxiety, depression, and suicidal ideation.

[12] Examples of words included in the lexicon: *insomnia*, *grief*, *suicidal*, *delusions*.
[13] Examples of absolutist terms: *absolutely*, *constantly*, *definitely*, *never*.

5.3 Emotional Expression

Figure 3 depicts for each class the average number of documents that contain at least one word associated with a particular emotion, including the polarity (positive or negative). We note considerable differences on the expression of emotions between positive and their respective controls. One way to interpret such results is that on average positive users tend to share emotions more regularly than control individuals.

Moreover, we also analyse the frequency correlation of the different emotions for each class. Due to space constraints, we only include self-harm. However, similar observations arise from the remaining positive classes. We note that certain emotions show different correlations depending on the class under observation. For instance, in the control class *surprise* reveals a larger positive correlation with *trust*, *joy* and *positive* and *negative* orientation when compared with self-harm class. Conversely, with *surprise* and *disgust* as well as with *fear* and *disgust*. Interesting to note is that in the case of depression, *sadness* exhibits a negative correlation with *joy* and *positive* orientations. Such correlation does not hold for the corresponding controls. This kind of study allows to better understand how different emotional patterns emerge from the use of emotions (Fig. 4).

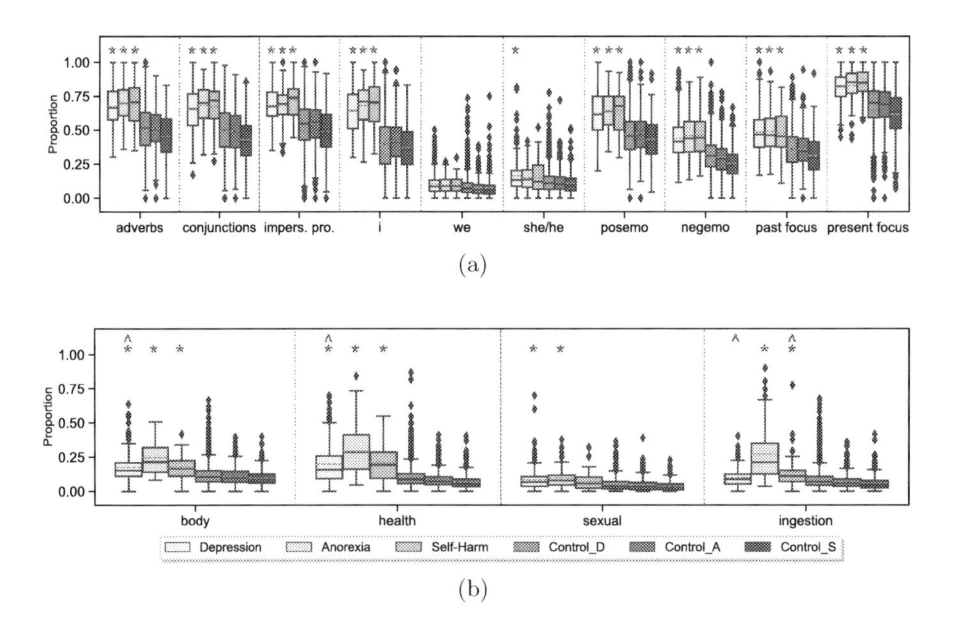

Fig. 2. Box and whiskers plot of the proportion of documents each user has (y-axis) matching various LIWC categories. Statistically significant differences between each positive and their respective control groups are denoted by * (p-value < 0.001). Also, statistically significant differences between Depression/Self-Harm and Anorexia are denoted by ˆ (p-value < 0.001).

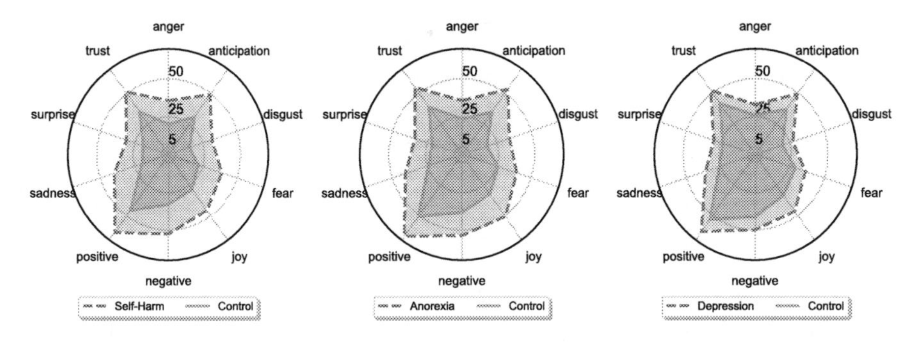

Fig. 3. Radar plot representing the average number of documents that contain at least one word associated with a particular emotion, including the polarity (positive or negative) for each positive group and its respective control.

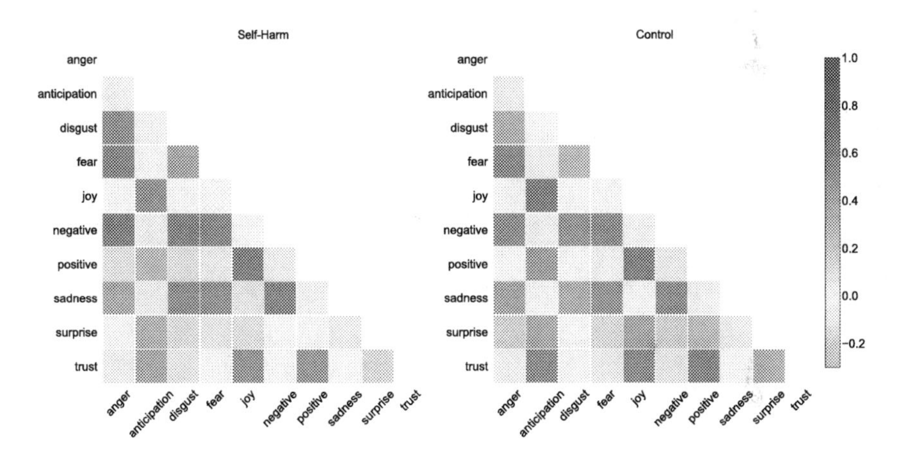

Fig. 4. Heatmap depicting the frequency correlation of the different emotions for self-harm (left) and control (right) groups (best viewed in colour). (Color figure online)

6 Conclusions

The wealth of information encoded in continually-generated social media is eager for analysis. In particular, social media data naturally occurs in a non-reactive way becoming a valuable complement for more conventional methods (such as questionnaires) used to determine the potential presence of mental disorders.

In this work, we reported results from a thorough analysis to show how users affected by mental disorders differ significantly from control individuals. We investigated the writing style, as well as how people express their emotions on social media via visualising certain probabilistic attributes.

To this aim, we analysed and visualised the activity, vocabulary, psychometric attributes, and emotional indicators in people's posts. Studying and visualising such dimensions, we discovered several interesting differences that could help a predictive system and a health practitioner to determine whether someone is affected by a mental disorder. Across different mental disorders, however, we could not find any significant indicators. Therefore, we can conclude that analysing social media posts could help a system identify people that are more likely to be diagnosed a mental disorder. On the other hand, determining the exact disorder is a much more difficult task, requiring expert judgement. Also, we found that psychometric attributes and emotional expression provides a quantifiable way to differentiate between individuals affected by mental disorders from healthy ones. The study we presented in this work has high practical impact since research should be steered towards building new metrics that can correlate with a disease before traditional symptoms arise and which clinicians can use as leading indicators of traditional later-onset symptoms.

For the future, we are interested in investigating whether the findings also hold in other social media, such as Twitter, where users are restricted to other types of constraints, such as space limitations. This could prove that the language use by individuals affected by mental disorders is independent from the social media platform they participate. Also, it would be interesting to uncover the differences in people's behaviour on different social media platforms. Based on the restrictions, objectives, and features people tend to behave differently on different platforms. Therefore, it is of high importance to see if our findings can be generalised to other social media platforms or not. Also, studying other modalities of data such as video and image can be very effective in detecting people with mental disorders [24].

Acknowledgements. We thank the anonymous reviewers for the constructive suggestions. This work was supported in part by the Swiss Government Excellence Scholarships and Hasler Foundation.

References

1. Al-Mosaiwi, M., Johnstone, T.: In an absolute state: elevated use of absolutist words is a marker specific to anxiety, depression, and suicidal ideation. Clin. Psychol. Sci. **6**(4), 529–542 (2018)
2. Aliannejadi, M., Crestani, F.: Venue suggestion using social-centric scores. In: Proceedings of ECIR Workshop on Social Aspects in Personalization and Search (2018)
3. Association, A.P.: Diagnostic and Statistical Manual of Mental Disorders, 5th edn. American Psychiatric Publishing, Washington (2013)
4. Boyd, R.L., Wilson, S.R., Pennebaker, J.W., Kosinski, M., Stillwell, D.J., Mihalcea, R.: Values in words: using language to evaluate and understand personal values. In: Proceedings of the Ninth International Conference on Web and Social Media, ICWSM 2015, Oxford, UK, pp. 31–40 (2015)
5. Choudhury, M.D., Gamon, M., Counts, S., Horvitz, E.: Predicting depression via social media. In: Proceedings of the Seventh International Conference on Weblogs and Social Media, ICWSM 2013, Cambridge, USA (2013)
6. Chung, C., Pennebaker, J.: The psychological functions of function words. In: Fiedler, K. (ed.) Social Communication. Frontiers of Social Psychology. Psychology Press, New York (2007)

7. Coppersmith, G., Dredze, M., Harman, C.: Quantifying mental health signals in Twitter. In: Proceedings of the Workshop on Computational Linguistics and Clinical Psychology: From Linguistic Signal to Clinical Reality, Baltimore, USA (2014)
8. Coppersmith, G., Dredze, M., Harman, C., Hollingshead, K.: From ADHD to SAD: analyzing the language of mental health on Twitter through self-reported diagnoses. In: Proceedings of the 2nd Workshop on Computational Linguistics and Clinical Psychology: From Linguistic Signal to Clinical Reality. Association for Computational Linguistics (2015)
9. Coppersmith, G., Dredze, M., Harman, C., Hollingshead, K., Mitchell, M.: CLPsych 2015 shared task: depression and PTSD on Twitter. In: Proceedings of the 2nd Workshop on Computational Linguistics and Clinical Psychology: From Linguistic Signal to Clinical Reality, Denver, USA (2015)
10. Croft, B., Metzler, D., Strohman, T.: Search Engines: Information Retrieval in Practice, 1st edn. Addison-Wesley Publishing Company, Boston (2009)
11. Culpepper, J.S., Diaz, F., Smucker, M.D.: Research frontiers in information retrieval: report from the third strategic workshop on information retrieval in Lorne (SWIRL 2018). SIGIR Forum **52**(1), 34–90 (2018)
12. Gkotsis, G., et al.: The language of mental health problems in social media. In: Proceedings of the 3rd Workshop on Computational Linguistics and Clinical Psychology: From Linguistic Signal to Clinical Reality (2016)
13. Kloumann, I.M., Danforth, C.M., Harris, K.D., Bliss, C.A., Dodds, P.S.: Positivity of the English language. PLoS ONE **7**(1), 1–7 (2012)
14. Losada, D.E., Crestani, F., Parapar, J.: Overview of eRisk: early risk prediction on the internet. In: Conference and Labs of the Evaluation Forum. CEUR-WS.org (2018)
15. Losada, D.E., Crestani, F., Parapar, J.: Overview of eRisk 2019 early risk prediction on the internet. In: Crestani, F., Braschler, M., Savoy, J., Rauber, A., Müller, H., Losada, D.E., Heinatz Bürki, G., Cappellato, L., Ferro, N. (eds.) CLEF 2019. LNCS, vol. 11696, pp. 340–357. Springer, Cham (2019). https://doi.org/10.1007/978-3-030-28577-7_27
16. Masood, R.: Adapting models for the case of early risk prediction on the internet. In: Azzopardi, L., Stein, B., Fuhr, N., Mayr, P., Hauff, C., Hiemstra, D. (eds.) ECIR 2019. LNCS, vol. 11438, pp. 353–358. Springer, Cham (2019). https://doi.org/10.1007/978-3-030-15719-7_48
17. Mohammad, S.: Word affect intensities. In: Proceedings of the Eleventh International Conference on Language Resources and Evaluation, LREC 2018, Miyazaki, Japan (2018)
18. Mohammad, S., Turney, P.D.: Crowdsourcing a word-emotion association lexicon. Comput. Intell. **29**(3), 436–465 (2013)
19. Neuman, Y.: Computational Personality Analysis. Introduction, Practical Applications and Novel Directions. Springer, Cham (2016). https://doi.org/10.1007/978-3-319-42460-6
20. Park, M., Cha, C., Cha, M.: Depressive moods of users portrayed in Twitter. In: Proceedings of the ACM SIGKDD Workshop on Healthcare Informatics (2012)
21. Pennebaker, J.W., Mehl, M.R., Niederhoffer, K.G.: Psychological aspects of natural language use: our words, our selves. Annu. Rev. Psychol. **54**(1), 547–577 (2003)
22. Preoţiuc-Pietro, D., et al.: The role of personality, age and gender in tweeting about mental illnesses. In: Proceedings of the 2nd Workshop on Computational Linguistics and Clinical Psychology: From Linguistic Signal to Clinical Reality (2015)

23. Prieto, V.M., Matos, S., Alvarez, M., Cacheda, F., Oliveira, J.L.: Twitter: a good place to detect health conditions. PLoS ONE **9**(1), 1–11 (2014)
24. Reece, A.G., Danforth, C.M.: Instagram photos reveal predictive markers of depression. EPJ Data Sci. **6**(1), 15 (2017)
25. Ríssola, E.A., Bahrainian, S.A., Crestani, F.: Anticipating depression based on online social media behaviour. In: Cuzzocrea, A., Greco, S., Larsen, H.L., Saccà, D., Andreasen, T., Christiansen, H. (eds.) FQAS 2019. LNCS (LNAI), vol. 11529, pp. 278–290. Springer, Cham (2019). https://doi.org/10.1007/978-3-030-27629-4_26
26. Ríssola, E.A., Bahrainian, S.A., Crestani, F.: Personality recognition in conversations using capsule neural networks. In: 2019 IEEE/WIC/ACM International Conference on Web Intelligence, WI 2019, Thessaloniki, Greece, 14–17 October 2019, pp. 180–187 (2019)
27. Sadeque, F., Xu, D., Bethard, S.: Measuring the latency of depression detection in social media. In: Proceedings of the Eleventh ACM International Conference on Web Search and Data Mining, WSDM 2018 (2018)
28. Schwartz, H.A., et al.: Personality, gender, and age in the language of social media: the open-vocabulary approach. PLoS ONE **8**(9), e73791 (2013)
29. Tausczik, Y.R., Pennebaker, J.W.: The psychological meaning of words: LIWC and computerized text analysis methods. J. Lang. Soc. Psychol. **29**(1), 24–54 (2009)
30. Trotzek, M., Koitka, S., Friedrich, C.M.: Word embeddings and linguistic metadata at the CLEF 2018 tasks for early detection of depression and anorexia. In: Working Notes of CLEF 2018 - Conference and Labs of the Evaluation Forum, Avignon, France, 10–14 September 2018 (2018)

Deep Learning II

Learning Based Methods for Code Runtime Complexity Prediction

Jagriti Sikka[1], Kushal Satya[1(✉)], Yaman Kumar[1], Shagun Uppal[2], Rajiv Ratn Shah[2], and Roger Zimmermann[3]

[1] Adobe, Noida, India
{jsikka,satya,ykumar}@adobe.com
[2] Midas Lab, IIIT Delhi, Delhi, India
{shagun16088,rajivratn}@iiitd.ac.in
[3] School of Computing, National University of Singapore, Singapore, Singapore
rogerz@comp.nus.edu.sg

Abstract. Predicting the runtime complexity of a programming code is an arduous task. In fact, even for humans, it requires a subtle analysis and comprehensive knowledge of algorithms to predict time complexity with high fidelity, given any code. As per Turing's Halting problem proof, estimating code complexity is mathematically impossible. Nevertheless, an approximate solution to such a task can help developers to get real-time feedback for the efficiency of their code. In this work, we model this problem as a machine learning task and check its feasibility with thorough analysis. Due to the lack of any open source dataset for this task, we propose our own annotated dataset, (The complete dataset is available for use at https://github.com/midas-research/corcod-dataset/blob/master/README.md) *CoRCoD: Code Runtime Complexity Dataset*, extracted from online coding platforms. We establish baselines using two different approaches: feature engineering and code embeddings, to achieve state of the art results and compare their performances. Such solutions can be highly useful in potential applications like automatically grading coding assignments, IDE-integrated tools for static code analysis, and others.

Keywords: Time complexity · Code embeddings · Code analysis

1 Introduction

Time Complexity computation is a crucial aspect in the study and design of well-structured and computationally efficient algorithms. It is a measure of the performance of a solution for a given problem. As a popular mistaken consideration, it is not the execution time of a code. Execution time depends upon a number of factors such as the operating system, hardware, processors etc. Since execution time is machine dependent, it is not used as a standard measure to analyze the efficiency of algorithms. Formally, *Time Complexity* quantifies the amount of time taken by an algorithm to process as a function of the input. For

J. M. Jose et al. (Eds.): ECIR 2020, LNCS 12035, pp. 313–325, 2020.
https://doi.org/10.1007/978-3-030-45439-5_21

a given algorithm, we consider its worst case complexity, which reflects the maximum time required to process it, given an input. Time complexity is represented in **Big O** notation, i.e., $O(n)$ denotes the asymptotic linear upper bound of an algorithm as a function of the input size n. Typically, the complexity classes in Computer Science refer to P and NP classes of decision problems, however, for the entire length of this paper, complexity class refers to a category of time complexity. The commonly considered categories in computer science as well in our work are $O(1)$, $O(logn)$, $O(n)$, $O(nlogn)$ and $O(n^2)$.

In this work, we try to predict the time complexity of a solution, given the code. This can have widespread applications, especially in the field of education. It can be used in automatic evaluation of code submissions on different online judges. It can also aid in static analyses, informing developers how optimized their code is, enabling more efficient development of industry level solutions.

Historically, there are a number of ways of predicting time complexity. For instance, master theorem [7] is effective to calculate run-time complexity of divide and conquer problems; but it is limited to only one type of problems and have several constraints on the permissible value of program's parameters.

Mathematically speaking, it is impossible to find a universal function to compute the time complexity of all programs. Rice's theorem and other works in this area [1,6] have established that it is impossible to formulate a single mathematical function that can calculate the complexity of all codes with polynomial order complexity.

Therefore, we need a Machine Learning based solution which can learn the internal structure of the code effectively. Recent research in the areas of machine learning and deep learning for programming codes provide several potential approaches which can be extended to solve this problem [5,13]. Also, several *"Big Code"* datasets have been made available publicly. The Public Git Archive is a dataset of a large collection of Github repositories [12,16] and [15] are datasets of Question-code pairs mined from Stack Overflow. However, to the best of our knowledge, at the time of writing this paper, there is no existing public dataset that, given the source code, gives runtime complexity of the source code. In our work, we have tried to address this problem by creating a Code Runtime Complexity Dataset *(CoRCoD)* consisting of 932 code files belonging to 5 different classes of complexities, namely $O(1)$, $O(logn)$, $O(n)$, $O(nlogn)$ and $O(n^2)$ (see Table 1).

We aim to substantially explore and solve the problem of code runtime complexity prediction using machine learning with the following contributions:

- Releasing a novel annotated dataset of program codes with their runtime complexities.
- Proposing baselines of ML models with hand-engineered features and study of how these features affect the computational efficiency of the codes.
- Proposing another baseline, the generation of code embeddings from Abstract Syntax Tree of source codes to perform classification.

Furthermore, we find that code embeddings have a comparable performance to hand-engineered features for classification using Support Vector Machines

(SVMs). To the best of our knowledge, CoRCoD is the first public dataset for code runtime complexity, and this is the first work that uses Machine Learning for runtime complexity prediction.

The rest of this paper is structured as follows. In Sect. 3, we talk about dataset curation and its key characteristics. We experiment using two different baselines on the dataset: classification using hand engineered features extracted from code and using graph based methods to extract the code embeddings via Abstract Syntax Tree of code. Section 4 explains the details and key findings of these two approaches. In Sect. 5, we enumerate the results of our model and data ablation experiments performed on these two baselines.

2 Related Work

In recent years, there has been extensive research in the deep learning community on programming codes. Hutter et al. [9] proposed supervised learning methods for algorithm runtime prediction. However, as explained before, execution time is not a standard measure to analyse efficiency of algorithms. Therefore, in our work, we do not consider algorithms' execution times. Most of the research in deep learning has been focused on two buckets, either on predicting some structure/attribute in the program or generating code snippets that are syntactically and/or semantically correct.

Variable/Method name prediction is a widely attempted problem, wherein Allamanis et al. [3] used a convolutional neural network with attention technique to predict method names, Alon et al. [4] suggested the use of AST paths to be used as context for generating *code embeddings* and training classifiers on top of them. Yonai et al. [17] used call graphs to compute method embeddings and recommend names of existing methods with function similar to target function.

Another popular prediction problem is that of defect prediction, given a piece of code. Li et al. [11] used Abstract Syntax Trees of programs in their CNN for feature generation which were then used for defect prediction. A major goal in all these approaches is to come up with a representation of the source program, which effectively captures the syntactic and semantic features of the program. Chen and Monperrus [8] performed a survey on word embedding techniques used on source codes. However, so far, there has been no such work for predicting time complexity of programs using code embeddings. We have established the same as one of our baselines using graph2vec [13].

Srikant and Aggarwal [14] extract hand-engineered features from Control Flow and Data Dependency graphs of programs such as number of nested loops, number of instances of *if* statements in a *loop* etc. for automatic grading of programs. They then used the grading criteria, that correct test programs would have similar programming constructs/features as those in the correct hand-graded programs. We use the same idea of identifying key features as the other baseline, which are constructs that a human evaluator would look at, to compute complexity and use them to train the classification models. Though, unlike [14], our features are problem independent. Moreover, the solution in [14] is commercially deployed, and thus, their dataset is not publicly available.

3 Dataset

To construct our dataset, we collected source codes of different problems from Codeforces[1]. Codeforces is a platform that regularly hosts programming contests. The large availability of contests having a wide variety of problems both in terms of data structures and algorithms as well as runtime complexity, made Codeforces a viable choice for our dataset.

Table 1. Classwise data distribution

Complexity class	Number of samples
$O(n)$	385
$O(n^2)$	200
$O(nlogn)$	150
$O(1)$	143
$O(logn)$	55

Table 2. Sample extracted features

Features from code samples	
Number of methods	Number of breaks
Number of switches	Number of loops
Conditional-Loop frequency	Loop-conditional frequency
Loop-Loop frequency	Conditional-conditional frequency
Nested loop depth	Recursion present
Number of variables	Number of ifs
Number of statements	Number of jumps

For the purpose of construction of our dataset, we collected Java source codes from Codeforces. We used the Codeforces API to retrieve problem and contest information, and further used web scraping to download the solution source codes. Sampling of source codes is done on the basis of data structure/algorithm tags associated with the problem, e.g., binary search, sorting etc. to ensure that the dataset contains source codes belonging to different complexity classes.

In order to ensure correctness of evaluated runtime complexity, the source codes selected should be devoid of issues such as compilation errors and segmentation faults. To meet this criterion, we filtered the source codes on the basis of their verdict and only selected the codes having verdicts *Accepted* or *Time limit exceeded* (TLE). For codes having TLE verdict, we ensured accuracy of solutions by only selecting codes that successfully passed at least four Test Cases. This criterion also allowed us to include multiple solutions for a single problem, different solutions having different runtime complexities. These codes were then manually annotated by a group of five experts, hailing from programming background each with a bachelor's degree in Computer Science. Each code was analyzed and annotated by two experts, in order to minimize the potential for error. Since calculating time complexity of a program comprises well-defined steps, inter-annotator agreement in our case was 100% (Cohen's kappa coefficient was 1). Only the order of complexity was recorded, for example, a solution having two variable inputs, n and m, and having a runtime complexity of $O(n*m)$ is labeled as *n_square* ($O(n^2)$).

[1] https://codeforces.com.

Certain agreed upon rules were followed for the annotation process. The rationale lies in the underlying implementations of these data structures in Java. Following points list down the rules followed for annotation and the corresponding rationale:

- Sorting algorithm's implementation in Java collections has worst case complexity $O(nlogn)$.
- Insertion/retrieval in HashSet and HashMap is annotated to be $O(1)$, given n elements.
- TreeSet and TreeMap are implemented as Red-Black trees and thus have $O(logn)$ complexity for insertion/retrieval.

```
class noOfNestedLoops extends ASTVisitor {
    int current = 0;
    int max_depth = 0;
    @Override
    bool visit(WhileStatement node) {
        current += 1;
        max_depth = max(current, max_depth);
        return true;
    }
    @Override
    void endVisit(WhileStatement node){
        current -= 1;
    }
}
```

Listing 1: Extracting nested loop depth using ASTVisitor

We removed few classes with insufficient data points, and ended up with 932 source codes, 5 complexity classes, corresponding annotation and extracted features. We selected nearly 400 problems from 170 contests, picking an average of 3 problems per contest. For 120 of these problems, we collected 4–5 different solutions, with different complexities.

In order to increase the size of the dataset for future work, we have created an online portal with an easy-to-use interface where contributors can upload source code and its complexity. Developers can also check the time complexity of a program predicted by our models.[2]

4 Solution Approach

The classification model is trained using two approaches: one, extracting hand-engineered features from code using static analysis and two, learning a generic representation of codes in the form of code embeddings.

[2] The portal is available for use at http://midas.center/corcod/.

4.1 Feature Engineering

Feature Extraction. We identified key coding constructs and extracted 28 features, some of them are listed in Table 2. Our feature set is inspired from [14]. We used two types of features for our feature set, basic features were obtained by counting occurrences of keywords represeting fundamental programming constructs, and sequence features captured key sequences generally present in the program, e.g. *Loop-Conditional frequency* captured number of *If* statements present inside loops in the program. We extracted these features from the Abstract Syntax Tree (AST) of source codes. AST is a tree representation of syntax rules of a programming language. ASTs are used by compilers to check codes for accuracy. We used Eclipse JDT for feature extraction. A generic representation of AST as parsed by ASTParser in JDT is shown in Fig. 1.

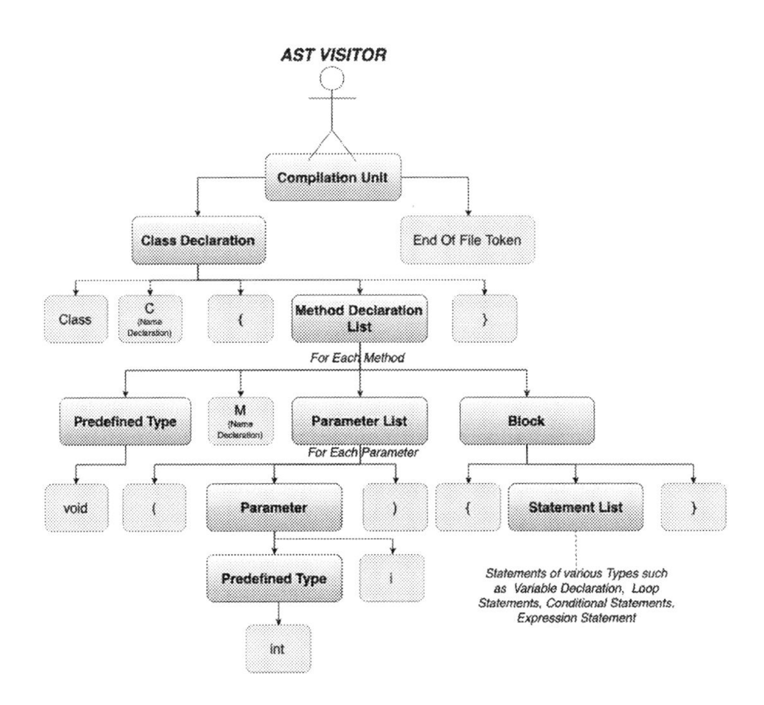

Fig. 1. Code Representation as an AST; being traversed by AST Parser

An ASTParser object creates the AST, and the ASTVisitor object *"visits"* the nodes of the tree via *visit* and *endVisit* methods using Depth First Search. One of the features chosen was the maximum depth of nested loops. Code snippet (Listing 1) depicts how the value of depth of nested loops was calculated using ASTVisitor provided by JDT. Other features were calculated in a similar manner.

We observed that our code samples often had unused code like methods or class implementations never invoked from the main function. Removing such unused code manually from each code sample is tedious. Instead, we used JDT plugins to identify the methods reachable from main function and used those methods for extracting the listed features. The same technique was also used while creating the AST for the next baseline.

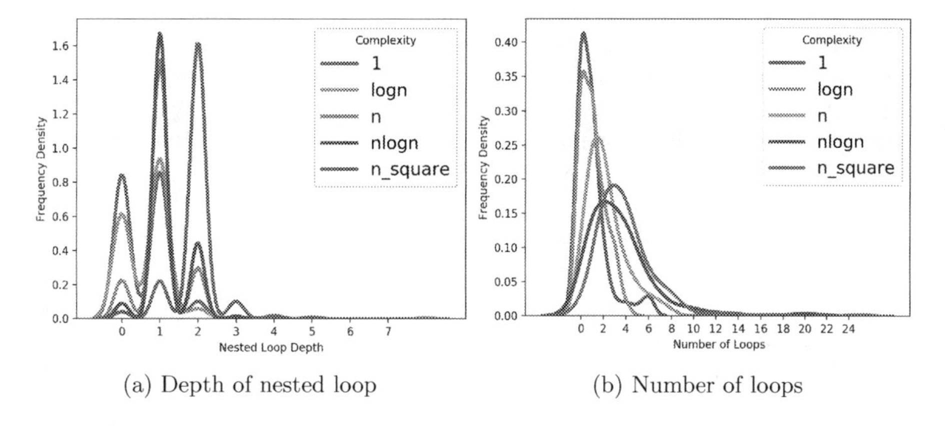

(a) Depth of nested loop (b) Number of loops

Fig. 2. Density plot for the different features

Figure 2 represents the density distribution of features across different classes. For nested loops, *n_square* has peak at depth 2 as expected; similarly n and *nlogn* have peak at depth 1 loop depth (see Fig. 2(a)). For number of loops (see Fig. 2(b)), we find that the mean value of the number of loops in code increases with the increase in complexity. On qualitative analysis, we find out that in case of $O(n)$ complexity, one loop is being used in code for processing the inputs and the other loop is being used for computing the solution to the problem. As we move towards $O(n\_square)$ codes, there is often one nested loop in the code and one loop is being used for input processing. Hence, it has a peak centered at a frequency of 3. This confirms our intuition that number of loops and nested loops are important parameters in complexity computation.

4.2 Code Embeddings

The Abstract Syntax Tree of a program captures comprehensive information regarding a program's structure, syntactic and semantic relationships between variables and methods. An effective method to incorporate this information is to compute code embeddings from the program's AST. An AST is infact a graph and thus using graph based methods for computing code embeddings was the right approach. We used graph2vec, a neural embedding framework [13], which can be used to compute embeddings for any generic graph. Graph2vec automatically generates task agnostic embeddings, and does not require a large corpus of data, making it apt for our problem. We used the graph2vec implementation from [2] to compute code embeddings.

Graph2vec is analogous to doc2vec [10] which predicts a document embedding given the sequence of words in it. The goal of graph2vec is, given a set of graphs $\mathbb{G} = \{G_1, G_2, ...G_n\}$, learn a δ-dimensional embedding vector for each graph. Here, each graph G is represented as (N, E, λ) where N are the nodes of the graph, E the edges and λ represents a function $n \rightarrow l$ which assigns a unique label from alphabet l to every node $n \in N$. To achieve the same, graph2vec extracts nonlinear substructures, more specifically, rooted subgraphs from each graph which are analogical to words in doc2vec. It uses skipgram model for learning graph embeddings which correspond to code embeddings in our scenario. The model works by considering a subgraph $s_j \in c(g_i)$ to be occurring in the context of graph g_i and tries to maximize the log likelihood in Eq. 1:

$$\sum_{j=1}^{D} log\ Pr(s_j|g_i) \tag{1}$$

where $c(g_i)$ gives all subgraphs of a graph g_i and D is the total number of subgraphs in the entire graph corpus.

We extracted AST from all codes using the JDT plugins. Each node in AST has two attributes: a Node Type and an optional Node Value. For e.g., a Method-Declaration Type node will have the declared function name as the node value. Graph2vec expects each node to have a single label. To get a single label, we followed two different representations:

1. Concatenating Node Type and Node Value.
2. Choosing selectively for each type of node whether to include node type or node value. For instance, every identifier node has a SimpleName node as its child. For all such nodes, only node value i.e. identifier name was considered as the label.

For both the AST representations, we used graph2vec to generate 1024-dimensional code embeddings. These embeddings were further used to train SVM based classification model and several experiments were performed as discussed in the next section.

5 Experiments and Results

5.1 Feature Engineering

Deep Learning (*DL*) algorithms tend to improve their performance with the amount of data available unlike classical machine learning algorithms. With lesser amount of data and correctly hand engineered features, Machine Learning (*ML*) methods outperform many *DL* models. Moreover, the former are computationally less expensive as compared to the latter. Therefore, we choose traditional *ML* classification algorithms to verify the impact of various features present in programming codes on their runtime complexities. We also perform a similar analysis on a simple Multi level Perceptron *(MLP)* classifier and compare

Table 3. Accuracy Score, Precision and Recall values for different classification algorithms

Algorithm	Accuracy %	Precision %	Recall %	F1 score
K-means	50.76	52.34	50.76	0.52
Random forest	**71.84**	78.92	71.84	0.68
Naive Bayes	67.97	68.08	67.97	0.67
k-Nearest	65.21	68.09	65.21	0.64
Logistic Regression	69.06	69.23	69.06	0.68
Decision Tree	70.75	68.88	70.75	0.69
MLP Classifier	53.37	50.69	53.37	0.47
SVM	**60.83**	**67.62**	**67.00**	**0.65**

Table 4. Per feature accuracy score, averaged over different classification algorithms.

Feature	Mean accuracy
No. of ifs	44.35
No. of switches	44.38
No. of loops	51.33
No. of breaks	43.85
Recursion present	42.38
Nested loop depth	**62.31**
No. of Variables	42.78
No. of methods	42.19
No. of jumps	43.65
No. of statements	44.18

against others. Table 3 depicts the accuracy score, weighted precision, recall and F1-score values for this classification task using 8 different algorithms, with the best accuracy score achieved using the ensemble approach of random forests.

Further, as per Table 4 showing per-feature-analysis, we distinctly make out that for the collected dataset, the most prominent feature which solely gives maximum accuracy is nested loop depth, followed by loops. Tables 5 and 6 demarcate the difference between accuracy scores considering data samples from classes $O(1)$, $O(n)$, $O(n^2)$ as compared to classes $O(1)$, $O(logn)$, $O(nlogn)$. A clear increment in accuracy scores is noticed amongst all the algorithms considered for the classification task for both sets of 3 classes as compared to the set of 5 classes except MLP classifier.

5.2 Code Embeddings

We extracted ASTs from source codes, computed 1024-dimensional code embeddings from ASTs using graph2vec and trained an SVM classifier on these embeddings. Results are tabulated in Table 7. We note that the average accuracy obtained for SVM on code embeddings is greater than that of SVM on hand-engineered features. Also, average precision and recall is higher for code embedding model. We performed statistical significance tests on results of 100 different runs of the two algorithms on the dataset. We observed that the data distribution was non-Gaussian and thus we used the Kolmogorov-Smirnov test. The p-value of the test for 100 different experimental precision scores for each algorithm was found to be $1.02e{-}13$ while for recall, it was $4.52e{-}17$. Thus, we established that the difference in precision and recall results from the two experiments is statistically significant and the code embeddings baseline has better precision and recall scores for both representations of AST.

Table 5. Accuracy, Precision and Recall values for different classification algorithms considering samples from complexity classes $O(1)$, $O(n)$ and $O(n^2)$

Algorithm	Accuracy	Precision	Recall
K-means	64.38	63.76	64.38
Random forest	**83.57**	84.19	83.57
Naive Bayes	67.82	67.69	67.82
k-Nearest	65.61	68.09	65.61
Logistic regression	80.42	80.71	80.42
Decision tree	81.08	81.85	81.08
MLP classifier	69.33	65.70	69.33
SVM	76.43	72.14	74.35

Table 6. Accuracy, Precision and Recall values for different classification algorithms considering samples from complexity classes $O(1)$, $O(logn)$ and $O(nlogn)$

Algorithm	Accuracy	Precision	Recall
K-means	52.31	53.23	52.31
Random forest	86.62	86.85	86.62
Naive Bayes	84.52	85.10	84.52
k-Nearest	76.74	80.66	76.74
Logistic regression	86.30	87.04	86.30
Decision tree	83.21	84.60	83.21
MLP classifier	47.11	22.19	47.11
SVM	69.64	70.76	67.24

5.3 Data Ablation Experiments

To get further insight into the learning framework, we performed following data ablation tests:

Label Shuffling. Training models with shuffled class labels can indicate whether the model is learning useful features pertaining to the task at hand. If the performance does not significantly decrease upon shuffling, it can imply that the model is hanging on to statistical cues that do not contain meaningful information w.r.t. the problem.

Method/Variable Name Alteration. Graph2vec uses node labels along with edge information to generate graph embeddings. Out of randomly selected 50 codes having correct prediction, if the predicted class labels before and after data ablation are different for a significant number of test samples, it would imply that the model relies on method/variable name tokens whereas it should only rely on the relationships between variables/methods.

Replacing Input Variables with Constant Literals. Program complexity is a function of input variables. Thus, to test the robustness of models, we replace the input variables with constant values making resultant complexity $O(1)$ for 50 randomly chosen codes, which earlier had non-constant complexity. A good model should have a higher percentage of codes with predicted complexity as $O(1)$.

Removing Graph Substructures. We randomly remove program elements such as *for*, *if* blocks with a probability of 0.1. The expectation is that the correctly predicted class labels should not change heavily as the complexity most likely does not change and hence a good model should have a higher percentage of codes with same correct label before and after removing graph substructures. This would imply that the model is robust to changes in code that do not change the resultant complexity.

Table 7. Accuracy, Precision, Recall values for classification of graph2vec embeddings, with and without node type & node value concatenation in node label.

AST representation	Accuracy	Precision	Recall	F1 score
Node Labels with concatenation	73.86	74	73	0.73
Node Labels without concatenation	70.45	71	70	0.70

Following are our observations regarding data ablation results in Table 8:

Label Shuffling. The drop in test performance is higher in graph2vec than that in the basic model indicating that graph2vec learns better features compared to simple statistical models.

Method/Variable Name Alteration. Table 8 shows that SVM correctly classifies most of the test samples' embeddings upon altering method and variable names, implying that the embeddings generated do not rely heavily on the actual method/variable name tokens.

Replacing Input Variables with Constant Literals. We see a significant and unexpected dip in accuracy, highlighting one of the limitations of our model.

Removing Graph Substructures. Higher accuracy for code embeddings as compared to feature engineering implies that the model must be learning the types of nodes and their effect on complexity to at least some extent, as removing substructures does not change the predicted complexity class of a program significantly.

Table 8. Data Ablation Tests Accuracy of feature engineering and code embeddings *(for two different AST representations)* baselines

Ablation technique	Accuracy		
	Feature engineering	Graph2vec: with concatenation	Graph2vec: without concatenation
Label shuffling	48.29	36.78	31.03
Method/variable name alteration	NA	84.21	89.18
Replacing input variables with constant literals	NA	16.66	13.33
Removing graph substructures	66.92	87.56	88.96

6 Limitations

The most pertinent limitation of our dataset is its size which is fairly small compared to what is considered standard today. Another limitation of our work is moderate accuracy of the models. An important point to note is that although we established that using code embeddings is a better approach, still their accuracy does not beat feature engineering significantly. One possible solution is to increase dataset size so that generated code embeddings can better model the characteristics of programs that differentiate them into multiple complexity classes, when trained on larger number of codes. However, generating a larger dataset is a challenging task since annotation process is tedious and needs people with a sound knowledge of algorithms. In order to increase the size of our dataset, we have created an online portal to crowd source the data. Lastly, we observe that replacing variables with constant literals does not change the prediction to $O(1)$ which highlights the inability of graph2vec to identify the variable on which complexity depends.

7 Usefulness of the Dataset

Computational complexity is a quantification of computational efficiency. Computationally efficient programs better utilize resources and improve software performance. With rapid advancements, there is a growing demand for resources; at the same time, there is greater need for optimizing existing solutions. Thus, writing computationally efficient programs is an asset for both students and professionals. With this dataset, we aim to analyze attributes and capture relationships that best define the computational complexity of codes. We do so, not just by heuristically picking up evident features, but by investigating their role in the quality, structure and dynamics of the problem using *ML* paradigm. We also capture relationships between various programming constructs by generating code embeddings from Abstract Syntax Trees. This dataset can not only help automate the process of predicting complexities, but we plan on using the dataset to develop a feedback based recommendation system which can help learners decide apt features for well-structured and efficient codes. It can also be used to train models that can be further integrated with IDEs and assist professional developers in writing computationally efficient programs for fast performance software development.

8 Conclusion

The dataset presented and the baseline models established should serve as guidelines for the future work in this area. The dataset presented is balanced and well-curated. Though both the baselines; Code Embeddings and Handcrafted features have comparable accuracy, we have established through data ablation tests that code embeddings learned from Abstract Syntax Tree of the code better capture relationships between different code constructs that are essential for predicting runtime complexity. Work can be done in future to increase the size of the dataset to verify our hypothesis that code embeddings will perform significantly better than hand crafted features. Moreover, we hope that the approaches discussed in this work, their usage becomes explicit for programmers and learners to bring into practice efficient and optimized codes.

References

1. Are runtime bounds in p decidable? (answer: no). https://cstheory.stackexchange.com/questions/5004/are-runtime-bounds-in-p-decidable-answer-no
2. Graph2vec implementation. https://github.com/MLDroid/graph2vec_tf
3. Allamanis, M., Peng, H., Sutton, C.: A convolutional attention network for extreme summarization of source code. In: Balcan, M.F., Weinberger, K.Q. (eds.) Proceedings of The 33rd International Conference on Machine Learning. Proceedings of Machine Learning Research, PMLR, New York, New York, USA, 20–22 June 2016, vol. 48, pp. 2091–2100. http://proceedings.mlr.press/v48/allamanis16.html
4. Alon, U., Zilberstein, M., Levy, O., Yahav, E.: A general path-based representation for predicting program properties. CoRR abs/1803.09544 (2018). http://arxiv.org/abs/1803.09544
5. Alon, U., Zilberstein, M., Levy, O., Yahav, E.: Code2vec: learning distributed representations of code. Proc. ACM Program. Lang. **3**(POPL), 40:1–40:29 (2019). https://doi.org/10.1145/3290353
6. Asperti, A.: The intensional content of Rice's theorem. In: Proceedings of the 35th Annual ACM SIGPLAN-SIGACT Symposium on Principles of Programming Languages. POPL 2008, pp. 113–119. ACM, New York (2008). https://doi.org/10.1145/1328438.1328455
7. Bentley, J.L., Haken, D., Saxe, J.B.: A general method for solving divide-and-conquer recurrences. SIGACT News **12**(3), 36–44 (1980). https://doi.org/10.1145/1008861.1008865
8. Chen, Z., Monperrus, M.: A literature study of embeddings on source code. CoRR abs/1904.03061 (2019). http://arxiv.org/abs/1904.03061
9. Hutter, F., Xu, L., Hoos, H.H., Leyton-Brown, K.: Algorithm runtime prediction: the state of the art. CoRR abs/1211.0906 (2012). http://arxiv.org/abs/1211.0906
10. Le, Q.V., Mikolov, T.: Distributed representations of sentences and documents (2014)
11. Li, J., He, P., Zhu, J., Lyu, M.R.: Software defect prediction via convolutional neural network. In: 2017 IEEE International Conference on Software Quality, Reliability and Security (QRS), pp. 318–328 (2017)
12. Markovtsev, V., Long, W.: Public git archive: a big code dataset for all. CoRR abs/1803.10144 (2018). http://arxiv.org/abs/1803.10144
13. Narayanan, A., Chandramohan, M., Venkatesan, R., Chen, L., Liu, Y., Jaiswal, S.: graph2vec: learning distributed representations of graphs. CoRR abs/1707.05005 (2017). http://arxiv.org/abs/1707.05005
14. Srikant, S., Aggarwal, V.: A system to grade computer programming skills using machine learning. In: Proceedings of the 20th ACM SIGKDD International Conference on Knowledge Discovery and Data Mining, KDD 2014, pp. 1887–1896. ACM, New York (2014). https://doi.org/10.1145/2623330.2623377
15. Yao, Z., Weld, D.S., Chen, W., Sun, H.: StaQC: a systematically mined question-code dataset from stack overflow. CoRR abs/1803.09371 (2018). http://arxiv.org/abs/1803.09371
16. Yin, P., Deng, B., Chen, E., Vasilescu, B., Neubig, G.: Learning to mine aligned code and natural language pairs from stack overflow. In: International Conference on Mining Software Repositories, MSR, pp. 476–486. ACM (2018). https://doi.org/10.1145/3196398.3196408
17. Yonai, H., Hayase, Y., Kitagawa, H.: Mercem: method name recommendation based on call graph embedding. CoRR abs/1907.05690 (2019). http://arxiv.org/abs/1907.05690

Inductive Document Network Embedding with Topic-Word Attention

Robin Brochier[1,2(✉)], Adrien Guille[1], and Julien Velcin[1]

[1] Université de Lyon, Lyon 2 ERIC EA3083, Lyon, France
{robin.brochier,adrien.guille,julien.velcin}@univ-lyon2.fr
[2] Digital Scientific Research Technology, Lyon, France

Abstract. Document network embedding aims at learning representations for a structured text corpus *i.e.* when documents are linked to each other. Recent algorithms extend network embedding approaches by incorporating the text content associated with the nodes in their formulations. In most cases, it is hard to interpret the learned representations. Moreover, little importance is given to the generalization to new documents that are not observed within the network. In this paper, we propose an interpretable and inductive document network embedding method. We introduce a novel mechanism, the Topic-Word Attention (TWA), that generates document representations based on the interplay between word and topic representations. We train these word and topic vectors through our general model, Inductive Document Network Embedding (IDNE), by leveraging the connections in the document network. Quantitative evaluations show that our approach achieves state-of-the-art performance on various networks and we qualitatively show that our model produces meaningful and interpretable representations of the words, topics and documents.

Keywords: Document network embedding · Interpretability · Attention mechanism

1 Introduction

Document networks, *e.g.* social media, question-and-answer websites, the scientific literature, are ubiquitous. Because these networks keep growing larger and larger, navigating efficiently through them becomes increasingly difficult. Modern information retrieval systems rely on machine learning algorithms to support users. The performance of these systems heavily depends on the quality of the document representations. Learning good features for documents is still challenging, in particular when they are structured in a network.

Recent methods learn the representations in an unsupervised manner by combining structural and textual information. Text-Associated DeepWalk (TADW) [28] incorporates text features into the low-rank factorization of a matrix describing the network. Graph2Gauss [2] learns a deep encoder, guided by the network,

J. M. Jose et al. (Eds.): ECIR 2020, LNCS 12035, pp. 326–340, 2020.
https://doi.org/10.1007/978-3-030-45439-5_22

that maps the nodes' attributes to embeddings. GVNR-t [3] factorizes a random walk based matrix of node co-occurrences and integrates word vectors of the documents in its formulation. CANE [25] introduces a mutual attention mechanism that builds representations of a document contextually to each of its direct neighbors in the network.

Apart from Graph2gauss, these methods are not intended to generate representations for documents with no connection to other documents and thus cannot induce *a posteriori* representations for new documents. Moreover, they provide little to no possibility to interpret the learned representations. CANE is a notable exception since its attention mechanism produces interpretable weights that highlight the words explaining the links between documents. Nevertheless, it lacks the ability to explain the representations for each document independently.

In this paper, we describe and evaluate an inductive and interpretable method that learns word, topic and document representations in a single vector space, based on a new attention mechanism. Our contributions are the following:

- we present a novel attention mechanism, Topic-Word Attention (TWA), that produces representations of a text where latent topic vectors attend to the word vectors of a document;
- we explain how to train the parameters of TWA by leveraging the links of the network. Our method, Inductive Document Network Embedding (IDNE), is able to produce representations for previously unseen documents, without network information;
- we quantitatively assess the performance of IDNE on several networks and show that our method performs better than recent methods in various settings, including when new documents, not part of the network, are inductively represented by the algorithms. To our knowledge, we are the first to evaluate this kind of inductive setting in the context of document network embedding;
- we qualitatively show that our model learns meaningful word and topic vectors and produces interpretable document representations.

The rest of the paper is organized as follows. In Sect. 2 we survey related works. We present in details our attention mechanism and show how to train it on networks of documents in Sect. 3. Next, in Sect. 4, we present a thorough experimental study, where we assess the performance of our model following the usual evaluation protocol on node classification and further evaluating its capacity of inducting representations for text documents with no connection to the network. In Sect. 5, we study the ability of our method to provide interpretable representations. Lastly, we conclude this paper and provide future directions in Sect. 6. The code for our model, the datasets and the evaluation procedure are made publicly available[1].

2 Related Work

Network embedding (NE) provides an efficient approach to represent nodes in a low dimensional vector space, suitable for solving various machine learning

[1] https://github.com/brochier/idne.

tasks. Recent techniques extend NE for document networks, showing that text and graph information can be combined to improve the resolution of classification and prediction tasks. In this section, we first cover important works in document NE and then relate recent advances in attention mechanisms.

2.1 Document Network Embedding

DeepWalk [22] and node2vec [9] are the most well-known NE algorithms. They train dense embedding vectors by predicting nodes co-occurrences through random walks by adapting the Skip-Gram model initially designed for word embedding [19]. VERSE [24] propose an efficient algorithm that can handle any type of similarity over the nodes.

Text-Associated DeepWalk (TADW) [28] extends DeepWalk to deal with textual attributes. Yang *et al.* prove, following the work in [17], that Skip-Gram with hierarchical softmax can be equivalently formulated as a matrix factorization problem. TADW then consists in constraining the factorization problem with a pre-computed representation of the documents T by using Latent Semantic Analysis (LSA) [6]. The task is to optimize the objective:

$$\text{argmin}_{W,H}||M - W^{\intercal}HT||_F^2. \tag{1}$$

where $M = (A + A^2)/2$ is a normalized second-order adjacency matrix of the network, W is a matrix of one-hot node embeddings and H a feature transformation matrix. Final document embeddings are the concatenation of W and HT. Graph2Gauss (G2G) [2] is an approach that embeds each node as a Gaussian distribution instead of a vector. The algorithm is trained by passing node attributes through a non-linear transformation via a deep neural network (encoder). GVNR-t [3] is a matrix factorization approach for document network embedding, inspired by GloVe [21], that simultaneously learns word, node and document representations. In practice, the following least-square objective is optimized:

$$\underset{U,W}{\text{argmin}} \sum_{i=1}^{n_d}\sum_{j=1}^{n_d}\Big(u_i \cdot \frac{\delta_j\ W}{|\delta_j|_1} - \log(1 + x_{ij})\Big)^2. \tag{2}$$

where x_{ij} is the number of co-occurrences of nodes i and j, u_i is a one-hot encoding of node i and $\frac{\delta_j\ W}{|\delta_j|_1}$ is the average of the word embeddings of document j. Context-Aware Network Embedding (CANE) [25] consists in a mutual attention mechanism trained on a document network. It learns several embeddings for a document according to its different contextual documents, represented by its neighbors in the network. The attention mechanism selects meaningful features from text information in pairs of documents that explain their relatedness in the graph. A similar approach is presented in [4] where the links between pairs of documents are predicted by computing the mutual contribution of their word embeddings.

In this work, we aim at constructing representations of documents that reflect their connections in a network. A key motivation behind our approach is to be able to predict a document's neighborhood given only its textual content. This allows our model to inductively produce embeddings for new documents for which no existing link is known. To that extend, Graph2Gauss is a similar approach. On the contrary, TADW and GVNR-t are not primarily designed for this purpose as they both learn one-hot embeddings for each node in the document network. Note that if some methods like GraphSage [10], SDNE [27] and GAE [13] also enable induction on new nodes, they cannot deal with nodes that have no known connection. Also, our approach differs from CANE since this latter needs the neighbors of a document to generate its representation. IDNE learns to produce a single interpretable vector for each document in the network. In the next section, we review recent works in attention mechanisms for natural language processing (NLP) that inspired the conception of our method.

2.2 Attention Mechanism

An attention mechanism uses a contextual representation to highlight or hide some parts of input data. Attention is an essential element of state-of-the-art neural machine translation (NMT) algorithms [18] by providing a powerful way to capture dependencies between words.

The Transformer [26] introduces a formalism of attention mechanisms for NMT. Given a query vector q, a set of key vectors K and a set of value vectors V, an attention vector is produced with the following formula:

$$v_a = \omega(qK^T)V. \tag{3}$$

qK^T measures the similarity between the query and each key k of K. ω is a normalization function such that all attention weights are positive and sum to 1. v_a is then the weighted sum of the values V according to the attention weights. Multiple attention vectors can be generated by using a set of queries Q.

In CANE, as for various NLP tasks [7], an attention mechanism generates attention weights that represent the strengths of relation between pairs of input words. However, in this paper, we do not seek to learn dependencies between pairs of words, but rather between words and some global topics. In this direction, the Set Transformer [16] constitutes a computationally efficient attention mechanism where the queries are replaced with a fixed-size set of learnable global inducing points. This model is originally not intended for NLP tasks, therefore we will explore the capacity of such inducing points to play the role of topic representations when applied to textual data.

Even if we introduce the concept of topic vectors, the aim of this work is not to propose another topic model [5,23]. We hypothesize that the introduction of global topic vectors in an attention mechanism can (1) lead to useful representations of documents for different tasks and (2) bring an interpretable sight on the patterns learned by the model. Interpretability can help both machine learning practitioners to better refine their models and end users to understand automated recommendations.

3 Method

We are interested in finding low dimensional vector space representations of a set of n_d documents organized in a network, described by a document-term matrix $X \in \mathbb{N}^{n_d \times n_w}$ and an adjacency matrix $A \in \mathbb{N}^{n_d \times n_d}$, where n_w stands for the number of words in our vocabulary. The method we propose, Inductive Document Network Embedding (IDNE), learns to represent the words and topics underlying the corpus in a single vector space. The document representations are computed by combining words and topics through an attention mechanism.

In the following, we first describe how to derive the document vectors from known word and topic vectors through a novel attention mechanism, the Topic-Word Attention (TWA). Next, we show how to estimate the word and topic vectors, guided by the links connecting the documents of the network.

3.1 Representing Documents with Topic-Aware Attention

We assume a p-dimensional vector space in which both words and topics are represented. We note $W \in \mathbb{R}^{n_w \times p}$ the matrix that contain the n_w word embedding vectors and $T \in \mathbb{R}^{n_t \times p}$ the matrix of n_t topic vectors. Figure 1 shows the matrix computation of the attention weights.

Topic-Word Attention. Given a document i and its bag-of-word encoding $X_i \in \mathbb{N}^{+n_w}$, we measure the attention weights between topics and words, $Z^i \in \mathbb{R}^{n_t \times n_w}$, as follows:

$$Z^i = \mathrm{g}\big(TW^\intercal \mathrm{diag}(X_i)\big). \tag{4}$$

The activation function g must satisfy two requirements: (1) all the weights are non-negative and (2) columns of Z^i sum to one. The intuition behind the first requirement is that enforcing non-negativity should lead to sparse and interpretable topics. The second requirement transforms the raw weights into word-wise relative attention weights, which can be read as probabilities similarly to what is done in neural topic models [23]. An obvious choice would be column-wise softmax, however, we empirically find that ReLU followed by a column-wise normalization performs best.

Document Representation. Given Z^i, we are able to calculate topic-specific representations of the document i. From the perspective of topic k, the p-dimensional representation of document i is:

$$D^i_k = \frac{Z^i_k \mathrm{diag}(X_i) W}{|X_i|_1}. \tag{5}$$

Similarly to Eq. 3, each topic vector, akin to a query, attends to the word vectors that play the role of keys to generate Z^i. The topic-specific representations are then the weighted sum of the values, also played by the word vectors. The final document vector is obtained by simple summation of all the topic-specific representations, which leads to $d^i = \sum_k D^i_k$. Scaling by $\frac{1}{|X_i|_1}$ in Eq. 5 ensures that the document vectors have the same order of magnitude as the word vectors.

3.2 Learning from the Network

Since the corpus is organized in a network, we propose to estimate the parameters, W and T, by leveraging the links between the documents. We posit that the representations of documents connected by a short path in the network should be more similar in the vector space than those that are far apart. Thus, we learn W and T in a supervised manner, through the training of a discriminative model.

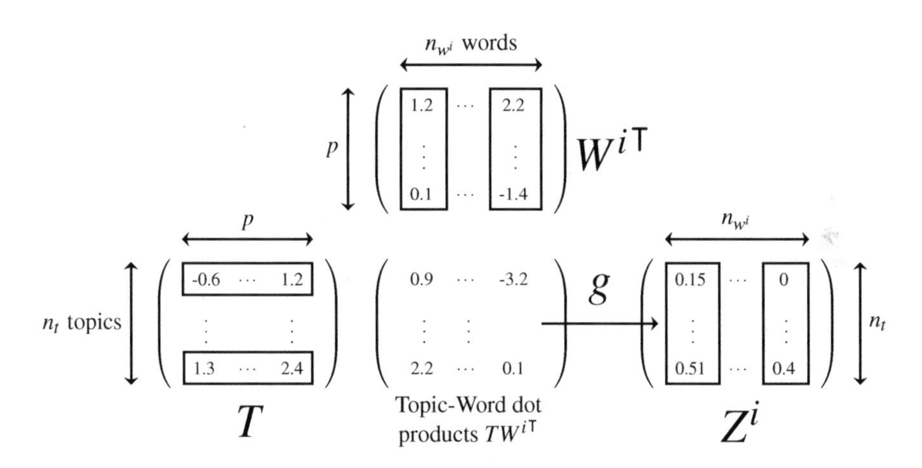

Fig. 1. Matrix computation of the attention weights. Here W^i is the compact view of $\text{diag}(X_i)W$ where zero-columns are removed since they do not impact on the result. n_{w^i} denotes the number of distinct words in document i. Each element z_{jk} of Z^i is the column-normalized rectified scalar product between the topic vector $\boldsymbol{t}_j$ and the word embedding $\boldsymbol{w^i}_k$ and represents the strength of association between the topic j and the word k in document i. The final document representation is then the sum of the topic-specific representations $D^i = \frac{Z^i W^i}{|X_i|_1}$.

Let $\Delta \in \{0,1\}^{n_d \times n_d}$ be a binary matrix, so that $\delta_{ij} = 1$ if document j is reachable from document i and $\delta_{ij} = 0$ otherwise. We model the probability of a pair of documents to be connected, given their representations, in terms of the sigmoid of the dot-product of d_i and d_j:

$$P(Y = 1|d_i, d_j; W, T) = \sigma(d_i \cdot d_j). \tag{6}$$

Assuming the document representations are i.i.d, we can express the log-likelihood of Δ given W and T:

$$\ell(W,T) = \sum_{i=1}^{n_d}\sum_{j=1}^{n_d} \log P(Y = \delta_{ij}|d_i, d_j; W, T)$$
$$= \sum_{i=1}^{n_d}\sum_{j=1}^{n_d} \delta_{ij} \log \sigma(d_i \cdot d_j) + (1 - \delta_{ij}) \log \sigma(-d_i \cdot d_j). \tag{7}$$

Through the maximization of this log-likelihood via a first-order optimization technique, we back-propagate the gradient and thus learn the word and topic vectors that lead to the document representations that best reconstruct Δ.

4 Quantitative Evaluation

Common tasks in document network embedding are classification and link prediction. We assess the quality of the representations learned with IDNE for these tasks in two different settings: (1) a traditional setting where all links and documents are observed and (2) an inductive setting where only a fraction of the links and documents is observed during training.

The first setting corresponds to a scenario where the goal is to propagate labels associated with a small portion of the documents. The second represents a scenario where we want to predict labels and links for new documents that have no network information, once the algorithm is already trained. This is common setting in real world applications. As an example, when a new user asks a new question on a Q&A website, we would like to suggest tags for its question and to recommend potential similar questions. In this case, the only information available to the algorithm is the textual content of the question.

4.1 Experimental Setup

We detail here the setup we use to train IDNE.

Computing the Δ *Matrix.* We consider paths of length up to 2 and compute the Δ matrix in the following manner:

$$\delta_{ij} = \begin{cases} 1 & \text{if}(A + A^2)_{ij} > 0, \\ 0 & \text{otherwise}. \end{cases} \tag{8}$$

This means that two documents are considered close in the network if they are direct neighbors or share at least one neighbor. Note that this matrix is the binarized version of the matrix TADW factorizes.

Optimizing the Log-Likelihood. We perform mini-batch SGD with the ADAM [12] update rule. Because most document networks are sparse, rather than uniformly sampling entries of Δ, we sample 5000 balanced mini-batches in order to favor convergence. We sample 16 positive examples ($\delta_{ij} = 1$) and 16 negative ones ($\delta_{ij} = 0$) per mini-batch. Positive pairs of documents are drawn according to the number of paths of length 1 or 2 linking them. Negative samples are uniformly drawn. The impact of the number of steps is detailed in Sect. 4.6.

4.2 Networks

We consider 4 networks of documents of various nature:

- A well-known scientific citation network extracted from Cora[2]. Each document is an article labelled with a conference.
- New York Times (NYT) titles of articles from January 2007. Articles are linked according to common tags (*e.g.* business, arts, technology) and are labeled with the section they appear in (*e.g.* opinion, news). This network is particularly dense and documents have a short length.
- Two networks of the Q&A website Stack Exchange (SE)[3] from June 2019, namely gaming.stackexchange.com and travel.stackexchange.com. We only keep questions with at least 10 user votes and that have at least one answer with 10 user votes or more. We build the network by linking questions with their answers and by linking questions and answers of the same user. The labels are the tags associated with each question (Table 1).

Table 1. General properties of the studied networks.

	# docs	# links	# labels	Vocab size	# words per doc	Density	Multi-label
Cora	2,211	4,771	7	4,333	67 ± 32	0.20%	No
NYT	5,135	3,050,513	4	5,748	24 ± 17	23.14%	No
Gaming	22,872	400,664	40	15,760	53 ± 74	0.15%	Yes
Travel	15,087	465,696	60	14,539	70 ± 73	0.41%	Yes

4.3 Tasks and Evaluation Metrics

For each network, we consider a traditional classification tasks, an inductive classification task and an inductive link prediction task.

- the traditional task refers to a setting where the model is trained on the entire network and the learned representations are used as features for a one-vs-all linear classifier with a training set of labelled documents ranging from 2% to 10% for multi-class networks and from 10% to 50% for multi-label networks.
- the inductive tasks refer to a setting where 10% of the documents are removed from the network and the model is trained on the resulting sub-network. For the classification task, a linear classifier is trained with the representations and the labels of the observed documents. Representations for hidden documents are then generated in an inductive manner, using their textual content only. Classifications and link predictions are then performed on these induced representations.

[2] https://linqs.soe.ucsc.edu/data.
[3] https://archive.org/details/stackexchange.

To classify the learned representations, we use the LIBLINEAR [8] logistic regression [14] algorithm and we cross validate the regularization parameter for each dataset and each model. Every experiment is repeated 10 times and we report the micro average of the area under the ROC curve (AUC). The AUC uses the probabilities of the logistic regression for all classes and evaluates the quality of the resulting ranking given the true labels. This metric is thus suitable for information retrieval tasks where we want to penalize wrong predictions depending on their ranks. For link prediction, we rank pairs of documents according to the cosine similarity between their representations.

4.4 Compared Representations

For all document networks, we process the documents by tokenizing text into words, discarding punctuation, stop words and words that appear less than 5 times or in more than 25% of the documents. We create document-term matrices that are used as input for 6 algorithms. Our baselines are representative of the different approaches for document NE. TADW and GVNR-t are based on matrix factorization whereas CANE and G2G are deep learning models. For each of them, we used the implementations of the authors:

- LSA: we use a 256-dimensional SVD decomposition of the tf-idf vectors as a text-only baseline;
- TADW: we follow the guidelines of the original paper by using 20 iterations and a penalty term $\lambda = 0.2$. For induction, we generate a document vector by computing the textual component HT in Eq. 1;
- Graph2gauss (G2G): we make sure the loss function converges before the maximum number of iterations;
- GVNR-t: we use $\gamma = 10$ random walks of length $t = 40$, a sliding window of size $l = 5$ and a threshold $x_{\min} = 5$ with 1 iteration. For induction, we compute $\frac{\delta_j W}{|\delta_j|_1}$ in Eq. 2;
- CANE: we use the same parameters as in the original paper;
- IDNE: we run all experiments with $n_t = 32$ topic vectors. The effect of n_t is discussed in Sect. 4.6.

4.5 Results Analysis

Tables 2 and 3 detail the AUC scores on the traditional classification task. We report the results for CANE only for Cora since the algorithm did not terminate within 10 h for the other networks. In comparison, our method takes about 5 min to run on each network on a regular laptop. The classifier performs well on the representations we learned, achieving similar or better results than the baseline algorithms on Cora, Gaming and Travel Stack Exchange. However, regarding the New York Times network, GVNR-t and TADW have a slight advantage. Because of its high density, the links in this network are little informative which may explain the relative good scores of the LSA representations. We hypothesize that (1) TADW benefits from its input LSA features and that (2) GVNR-t benefits both from its random walk based matrix of node co-occurrences [20], which captures more precisely the proximities of the nodes in such dense network, and from the short length of the documents making the word embedding averaging efficient [1,15].

Table 4 shows the AUC scores in the inductive settings. For link prediction IDNE performs best on three networks, showing its capacity to learn meaningful word and topic representations according to the network structure. For classification, LSA and GVNR-t achieve the best results while IDNE reaches similar but slightly lower scores on all datasets. On the contrary, TADW and Graph2gauss show weaknesses on NYT and Gaming SE.

In summary, IDNE shows constant performances across all settings where other methods lack of robustness against the type of network or the type of task. A surprising result is the good scores of GVNR-t for inductive classification which we didn't expect given that its textual component only is used for this setting. However, for the traditional classification, GVNR-t has difficulties to handle networks with longer documents. IDNE does not suffer the same problem because TWA carefully select discriminative words before averaging them. In Sect. 5, we further show that IDNE learns meaningful representations of words and topics and builds interpretable document representations.

4.6 Impact of the Number of Topics and Convergence Speed

Figure 2 shows the impact of the number of topic vectors n_t and of the number of steps (mini-batches) on the AUC scores obtained in traditional classification with Cora. Note that we observe a similar behavior on the other networks. We see that the scores improve from 1 to 16 topics and tend to stagnate for upper values. In a similar manner, performances improve up to 5000 iterations after which no increase is observed.

Table 2. Micro AUC scores on Cora and NYT

	Cora					NYT				
	2%	4%	6%	8%	10%	2%	4%	6%	8%	10%
LSA	67.54	81.76	88.63	89.68	91.43	79.90	82.06	81.18	83.99	86.06
TADW	65.17	74.11	80.27	83.04	86.56	85.28	**88.91**	87.49	89.39	88.72
G2G	91.12	92.38	91.98	93.79	94.09	79.74	81.41	80.91	82.37	81.42
CANE	**94.40**	**95.86**	95.90	96.37	95.88	NA	NA	NA	NA	NA
GVNR-t	87.13	92.54	94.37	95.21	95.83	**85.83**	87.67	**88.76**	**90.39**	**89.90**
IDNE	93.34	94.93	**95.98**	**96.77**	**96.68**	82.40	84.60	86.16	86.72	87.98

Table 3. Micro AUC scores on Stack Exchange networks

	Gaming					Travel				
	10%	20%	30%	40%	50%	10%	20%	30%	40%	50%
LSA	86.73	88.51	89.51	90.25	90.18	80.18	83.77	83.40	84.12	84.60
TADW	88.05	90.34	91.64	93.18	93.29	78.69	84.33	85.05	83.60	84.62
G2G	82.12	84.42	85.14	86.10	87.84	66.04	67.48	69.67	70.94	71.58
GVNR-t	89.09	92.60	94.14	**94.79**	95.24	79.47	83.47	85.06	85.85	86.58
IDNE	**92.75**	**93.53**	**94.72**	94.61	**95.57**	**86.83**	**88.86**	**89.24**	**89.31**	**89.26**

5 Qualitative Evaluation

We first show in Sect. 5.1 that IDNE is capable of learning meaningful word and topic vectors. Then, we provide visualizations of documents that highlight the ability of the topic-word attention to reveal topics of interest. For all experiments, we set the number of topics to $n_t = 6$.

Table 4. Micro AUC scores for inductive classification and inductive link prediction

	Inductive classification				Inductive Link Prediction			
	Cora	NYT	Gaming	Travel	Cora	NYT	Gaming	Travel
LSA	97.02	**89.45**	90.70	85.88	88.10	60.71	58.99	58.97
TADW	96.23	86.06	93.16	91.35	84.82	69.10	57.00	57.91
G2G	94.04	85.44	89.81	80.71	81.58	74.22	58.18	**59.50**
GVNR-t	**97.60**	88.47	**96.09**	**91.54**	82.27	71.15	59.71	58.39
IDNE	96.58	88.21	95.22	90.78	**91.66**	**77.90**	**62.82**	58.43

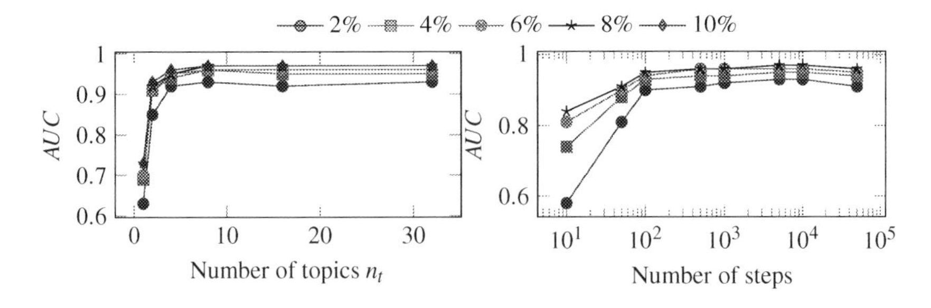

Fig. 2. Impact of the number of topics and of the number of steps on the traditional classification task on Cora with IDNE.

5.1 Word and Topic Vectors

Table 5 shows the closest words to each topic, computed as the dot product between their respective vectors, learned on Cora. Word and topic vectors are trained to predict the proximity of the nodes in a network, meaningless words are thus always dissimilar to the topic vectors, since they do not help to predict a link. This can be verified by observing the words that have the largest and the smallest norms, also reported in Table 5. Even though the topics are learned in an unsupervised manner, we notice that, when we set the number of topics close to the number of classes, each topic seems to capture the semantics of one particular class.

Table 5. Topics with their closest words produced by IDNE on Cora and words whose vector L_2 norms are the largest (resp. the smallest) reported in parenthesis. The labels in this dataset are: Case Based, Genetic Algorithms, Neural Networks, Probabilistic Methods, Reinforcement Learning, Rule Learning and Theory.

Topic 1	Casebased, reasoning, reinforcement, knowledge, system, learning, decision
Topic 2	Chain, belief, probabilistic, length, inference, distributions, markov
Topic 3	Search, ilp, problem, optimal, algorithms, heuristic, decision
Topic 4	Genetic, algorithm, fitness, evolutionary, population, algorithms, trees
Topic 5	Bayesian, statistical, error, data, linear, accuracy, distribution
Topic 6	Accuracy, induction, classification, features, feature, domains, inductive
Largest	Genetic (8.80), network (8.07), neural (7.43), networks (6.94), reasoning (6.16)
Smallest	Calculus (0.34), instability (0.34), acquiring (0.34), tested (0.34), le (0.34)

5.2 Topic Attention Weights Visualization

To further highlight the ability of our model to bring interpretability, we show in Fig. 3 the topics that most likely generated the words of a document according to TWA. The document is the abstract of this paper whose weights are inductively calculated with IDNE previously trained on Cora. We compute its attention weights Z^i and associate each word k to the maximum value of its column Z^i_k. We then colorize and underline each word associated to the two most represented topics in the document, if its weight is higher than $\frac{1}{2}$. We see that the major topic (green and single underline), that accounts for 32% of the weights, deals with the type of data, here document networks. The second topic (blue and double underline), which represents 18% of the weights, relates to text modeling, with words like "interpretable" and "topics".

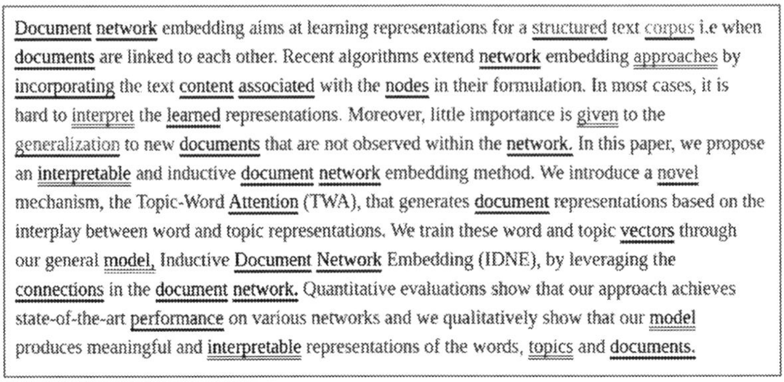

Document network embedding aims at learning representations for a structured text corpus i.e when documents are linked to each other. Recent algorithms extend network embedding approaches by incorporating the text content associated with the nodes in their formulation. In most cases, it is hard to interpret the learned representations. Moreover, little importance is given to the generalization to new documents that are not observed within the network. In this paper, we propose an interpretable and inductive document network embedding method. We introduce a novel mechanism, the Topic-Word Attention (TWA), that generates document representations based on the interplay between word and topic representations. We train these word and topic vectors through our general model, Inductive Document Network Embedding (IDNE), by leveraging the connections in the document network. Quantitative evaluations show that our approach achieves state-of-the-art performance on various networks and we qualitatively show that our model produces meaningful and interpretable representations of the words, topics and documents.

Fig. 3. Topics provided by IDNE in the abstract of this very paper trained on Cora.

6 Discussion and Future Work

In this paper, we presented IDNE, an inductive document network embedding algorithm that learns word and latent topic representations via TWA, a topic-word attention mechanism able to produce interpretable document representations. We showed that IDNE performs state-of-the-art results on various network in different settings. Moreover, we showed that our attention mechanism provides an efficient way of interpreting the learned representations. In future work, we would like to study the effect of the sampling of the documents on the learned topics. In particular, the matrix Δ could capture other types of similarities between documents such as SimRank [11] which measures structural relatedness between nodes instead of proximities. This could reveal complementary topics underlying a document network and could provide interpretable explanations of the roles played by documents in networks.

References

1. Arora, S., Liang, Y., Ma, T.: A simple but tough-to-beat baseline for sentence embeddings. In: International Conference on Learning Representations (2017)
2. Bojchevski, A., Günnemann, S.: Deep Gaussian embedding of graphs: unsupervised inductive learning via ranking. In: International Conference on Learning Representations (2018). https://openreview.net/forum?id=r1ZdKJ-0W
3. Brochier, R., Guille, A., Velcin, J.: Global vectors for node representations. In: The World Wide Web Conference, pp. 2587–2593. ACM (2019)
4. Brochier, R., Guille, A., Velcin, J.: Link prediction with mutual attention for text-attributed networks. In: Companion Proceedings of the 2019 World Wide Web Conference, pp. 283–284. ACM (2019)
5. Chang, J., Blei, D.: Relational topic models for document networks. In: Artificial Intelligence and Statistics, pp. 81–88 (2009)
6. Deerwester, S., Dumais, S.T., Furnas, G.W., Landauer, T.K., Harshman, R.: Indexing by latent semantic analysis. J. Am. Soc. Inf. Sci. **41**(6), 391–407 (1990)

7. Devlin, J., Chang, M.W., Lee, K., Toutanova, K.: BERT: pre-training of deep bidirectional transformers for language understanding. arXiv preprint arXiv:1810.04805 (2018)
8. Fan, R.E., Chang, K.W., Hsieh, C.J., Wang, X.R., Lin, C.J.: LIBLINEAR: a library for large linear classification. J. Mach. Learn. Res. **9**, 1871–1874 (2008)
9. Grover, A., Leskovec, J.: node2vec: scalable feature learning for networks. In: Proceedings of the 22nd ACM SIGKDD International Conference on Knowledge Discovery and Data Mining, pp. 855–864. ACM (2016)
10. Hamilton, W., Ying, Z., Leskovec, J.: Inductive representation learning on large graphs. In: Advances in Neural Information Processing Systems, pp. 1024–1034 (2017)
11. Jeh, G., Widom, J.: SimRank: a measure of structural-context similarity. In: Proceedings of the Eighth ACM SIGKDD International Conference on Knowledge Discovery and Data Mining, pp. 538–543. ACM (2002)
12. Kingma, D., Ba, J.: Adam: a method for stochastic optimization. In: International Conference on Learning Representations (2014)
13. Kipf, T.N., Welling, M.: Variational graph auto-encoders. In: NIPS Workshop on Bayesian Deep Learning (2016)
14. Kleinbaum, D.G., Klein, M.: Logistic Regression. Statistics for Biology and Health. Springer, New York (2002). https://doi.org/10.1007/b97379
15. Le, Q., Mikolov, T.: Distributed representations of sentences and documents. In: International Conference on Machine Learning, pp. 1188–1196 (2014)
16. Lee, J., Lee, Y., Kim, J., Kosiorek, A.R., Choi, S., Teh, Y.W.: Set transformer (2019)
17. Levy, O., Goldberg, Y.: Neural word embedding as implicit matrix factorization. In: Advances in Neural Information Processing Systems, pp. 2177–2185 (2014)
18. Luong, T., Pham, H., Manning, C.D.: Effective approaches to attention-based neural machine translation. In: Proceedings of the 2015 Conference on Empirical Methods in Natural Language Processing, EMNLP 2015, Lisbon, Portugal, 17–21 September 2015, pp. 1412–1421 (2015). https://www.aclweb.org/anthology/D15-1166/
19. Mikolov, T., Sutskever, I., Chen, K., Corrado, G.S., Dean, J.: Distributed representations of words and phrases and their compositionality. In: Advances in Neural Information Processing Systems, pp. 3111–3119 (2013)
20. Page, L., Brin, S., Motwani, R., Winograd, T.: The pagerank citation ranking: bringing order to the web. Technical report, Stanford InfoLab (1999)
21. Pennington, J., Socher, R., Manning, C.: GloVe: global vectors for word representation. In: Proceedings of the 2014 Conference on Empirical Methods in Natural Language Processing (EMNLP), pp. 1532–1543 (2014)
22. Perozzi, B., Al-Rfou, R., Skiena, S.: Deepwalk: online learning of social representations. In: Proceedings of the 20th ACM SIGKDD International Conference on Knowledge Discovery and Data Mining, pp. 701–710. ACM (2014)
23. Srivastava, A., Sutton, C.: Autoencoding variational inference for topic models. In: Proceedings of International Conference on Learning Representations (ICLR) (2017)
24. Tsitsulin, A., Mottin, D., Karras, P., Müller, E.: VERSE: versatile graph embeddings from similarity measures. In: Proceedings of the 2018 World Wide Web Conference, pp. 539–548. International World Wide Web Conferences Steering Committee (2018)

25. Tu, C., Liu, H., Liu, Z., Sun, M.: CANE: context-aware network embedding for relation modeling. In: Proceedings of the 55th Annual Meeting of the Association for Computational Linguistics (Volume 1: Long Papers), pp. 1722–1731 (2017)
26. Vaswani, A., et al.: Attention is all you need. In: Advances in Neural Information Processing Systems, pp. 5998–6008 (2017)
27. Wang, D., Cui, P., Zhu, W.: Structural deep network embedding. In: Proceedings of the 22nd ACM SIGKDD International Conference on Knowledge Discovery and Data Mining, pp. 1225–1234. ACM (2016)
28. Yang, C., Liu, Z., Zhao, D., Sun, M., Chang, E.: Network representation learning with rich text information. In: Twenty-Fourth International Joint Conference on Artificial Intelligence (2015)

Multi-components System for Automatic Arabic Diacritization

Hamza Abbad and Shengwu Xiong(✉)

Wuhan University of Technology, Wuhan, Hubei, China
{hamza.abbad,xiongsw}@whut.edu.cn

Abstract. In this paper, we propose an approach to tackle the problem of the automatic restoration of Arabic diacritics that includes three components stacked in a pipeline: a deep learning model which is a multi-layer recurrent neural network with LSTM and Dense layers, a character-level rule-based corrector which applies deterministic operations to prevent some errors, and a word-level statistical corrector which uses the context and the distance information to fix some diacritization issues. This approach is novel in a way that combines methods of different types and adds edit distance based corrections.

We used a large public dataset containing raw diacritized Arabic text (Tashkeela) for training and testing our system after cleaning and normalizing it. On a newly-released benchmark test set, our system outperformed all the tested systems by achieving DER of 3.39% and WER of 9.94% when taking all Arabic letters into account, DER of 2.61% and WER of 5.83% when ignoring the diacritization of the last letter of every word.

Keywords: Arabic · Diacritization · Diacritics restoration · Deep learning · Rule-based · Statistical methods · Natural Language Processing

1 Introduction

Arabic is the largest Semitic language today, used by more than 422 millions persons around the world, as a first or second language, making it the fifth most spoken language in the world.

The Arabic language uses a writing system consisted of 28 letters but represented by 36 characters due to 2 letters which have more than one form[1]. Unlike Latin, Arabic is always written in a cursive style where most of the letters are joined together with no upper case letters, from right to left (RTL).

The writing system is composed of letters and other marks representing phonetic information, known as **diacritics**, which are small marks that should be placed above or below most of the letters. They are represented as additional

[1] The letter ت has another form represented as ة, and the letter ا has the following forms: أ, إ, آ, ء, ؤ, ئ, ى, depending on its pronunciation and position in the word.

J. M. Jose et al. (Eds.): ECIR 2020, LNCS 12035, pp. 341–355, 2020.
https://doi.org/10.1007/978-3-030-45439-5_23

Arabic characters in UTF-8 encoding. There are eight diacritics in the *Modern Standard Arabic* (MSA), arranged into three main groups:

Short vowels. Three marks: *Fatha*, *Damma*, *Kasra*.
Doubled case endings (Tanween). Three marks: *Tanween Fath* (*Fathatan*), *Tanween Damm* (*Dammatan*), *Tanween Kasr* (*Kasratan*).
Syllabification marks. Two marks: *Sukoon* and *Shadda* [46].

Shadda is a secondary diacritic indicating that the specified consonant is doubled, rather than making a primitive sound. The *Tanween* diacritics can appear only at the end of the word, and *Sukoon* cannot appear in the first letter. Besides, short vowels can be placed in any position. Furthermore, some characters cannot accept any diacritics at all (ex: ى), and some others cannot do that in specified grammatical contexts (ex: the definitive ال at the beginning of the word). The diacritics are essential to indicate the correct pronunciation and the meaning of the word. They are all presented on the letter د in Table 1.

Table 1. The diacritics of the Modern Standard Arabic

Diacritic	Arabic name	Transliteration	Diacritic	Arabic name	Transliteration
دَ	فَتْحَة	Fatha	دُ	ضَمَّة	Damma
دِ	كَسْرَة	Kasra	دْ	سُكُون	Sukoon
دً	تَنْوِينُ فَتْحٍ	Tanween Fath	دٌ	تَنْوِينُ ضَمٍّ	Tanween Damm
دٍ	تَنْوِينُ كَسْرٍ	Tanween Kasr	دّ	شَدَّة	Shadda

These marks are dropped from almost all the written text today, except the documents intolerant to pronunciation errors, such as religious texts and Arabic teaching materials. The native speakers can generally infer the correct diacritization from their knowledge and the context of every word. However, this is still not a trivial task for a beginner learner or NLP applications [12].

The automatic diacritization problem is an essential topic due to the high ambiguity of the undiacritized text and the free word order nature of the grammar. Table 2 illustrates the differences made by the possible diacritizations of the

Table 2. The diacritizations of علم and their meanings

Diacritized form of علم	Meaning	Diacritized form of علم	Meaning
عَلِمَ	He knew	عَلَمٌ	A flag (nominative)
عَلَّمَ	He taught	عَلَمٍ	A flag (genitive)
عُلِمَ	It was known	عِلْمٌ	A science (nominative)
عُلِّمَ	It was taught	عِلْمٍ	A science (genitive)

word علم. As one might see, the diacritization defines many linguistic features, such as the part-of-speech (POS), the active/passive voice, and the grammatical case.

The full diacritization problem includes two sub-problems: *morphological diacritization* and *syntactic diacritization.* The first indicates the meaning of the word, and the second shows the grammatical case.

Two metrics are defined to calculate the quantitative performance of an automated diacritics restoration system: **Diacritization Error Rate** (DER) and **Word Error Rate** (WER). The first one measures the ratio of the number of incorrectly diacritized characters to the number of all characters. The second metric applies the same principle considering the whole word as a unit, where a word is considered incorrect if any of its characters has a wrong diacritic. Both metrics have two variants: One includes the diacritics of all characters (DER1 and WER1), and another excludes the diacritics of the last character of every word (DER2 and WER2).

We propose a new approach to restore the diacritics of a raw Arabic text using a combination of deep learning, rule-based, and statistical methods.

2 Related Works

Many works were done in the automatic restoration of the Arabic diacritics using different techniques. They can be classified into three groups.

Rule-based approaches. The used methods include cascading *Weighted Finite-State Transducers*[33], lexicon retrieval and rule-based morphological analysis [7]. One other particular work [9] used diacritized text borrowing from other sources to diacritize a highly cited text.

Statistical approaches. This type of approaches includes using *Hidden Markov Models* both on word level and on character level [8,18,21], *N-grams* models on word level and on character level as well [10], *Dynamic Programming* methods [24–26], classical Machine learning models such as *Maximum-entropy* classifier [46], and *Deep Learning* methods like the *Deep Neural Networks*, both the classical *Multi-Layer Perceptron* and the advanced *Recurrent Neural Networks*[6,14,32,36].

Hybrid approaches. They are a combination of rule-based methods and statistical methods in the same system. They include hybridization of rules and dictionary retrievals with morphological analysis, N-grams, Hidden Markov Models, Dynamic Programming and Machine Learning methods [5,15,17,20,23,31,35,37–39,42]. Some Deep Learning models improved by rules [2,3] have been developed as well.

Despite a large number of works done on this topic, the number of available tools for Arabic diacritization is still limited because most researchers do not release their source code or provide any practical application. Therefore, we will compare the performance of our system to these available ones:

Farasa [4] is a text processing toolkit which includes an automatic diacritics restoration module, in addition to other tools. It is based on the segmentation of the words based on separating the prefixes and suffixes using SVM-ranking and performing dictionary lookups.

MADAMIRA [34] is a complete morphological analyser that generates possible analyses for every word with their diacritization and uses an SVM and n-gram language models to select the most probable one.

Mishkal [44] is an application which diacritize a text by generating the possible diacritized word forms through the detection of affixes and the use of a dictionary, then limiting them using semantic relations, and finally choosing the most likely diacritization.

Tashkeela-Model [11] uses a basic N-gram language model on character level trained on the Tashkeela corpus [45].

Shakkala [13] is a character-level deep learning system made of an embedding, three bidirectional LSTM, and dense layers. It was trained on Tashkeela corpus as well. To the best of our knowledge, this is the system that achieves state-of-the-art results.

3 Dataset

In this work, the Tashkeela corpus [45] was mainly used for training and testing our model. This dataset is made of 97 religious books written in the *Classical Arabic* style, with a small part of web crawled text written in the *Modern Standard Arabic* style. The original dataset has over 75.6 million words, where over 67.2 million are diacritized Arabic words.

The structure of the data in this dataset is not consistent since its sources are heterogeneous. Furthermore, it contains some diacritization errors and some useless entities. Therefore, we applied some operations to normalize this dataset and keep the necessary text:

1. Remove the lines which do not contain any useful data (empty lines or lines without diacritized Arabic text).
2. Split the sentences at XML tags and end of lines, then discard these symbols. After that, split the new sentences at some punctuation symbols: dots, commas, semicolons, double dots, interrogation, and exclamation marks without removing them.
3. Fix some diacritization errors, such as removing the extra *Sukoon* on the declarative ال , reversing the ا + *Tanween Fath* and diacritic + *Shadda* combinations, removing any diacritic preceded by anything other than Arabic letter or *Shadda*, and keeping the latest diacritic when having more than one (excluding *Shadda* + diacritic combinations).
4. Any sentence containing undiacritized words or having less than 2 Arabic words is discarded.

After this process, the resulted dataset will be a raw text file with one sentence per line and a single space between every two tokens. This file is further shuffled then divided into a training set containing 90% of the sentences, and the rest is distributed equally between the validation and the test sets[2]. After the division, we calculated some statistics and presented them in Table 3.

Table 3. Statistics about the processed Tashkeela dataset

	Train	Val	Test
All tokens	31774001	1760397	1766844
Numbers only	80417	4462	4396
Arabic words only	27657285	1532625	1537878
Unique undiacritized Arabic words	351816	100799	101263
Unique diacritized Arabic words	626693	152752	153311

We note that the train-test Out-of-Vocabulary ratio for the unique Arabic words is 9.53% when considering the diacritics and 6.83% when ignoring them.

4 Proposed Method

Our approach is a pipeline of different components, where each one does a part of the process of the diacritization of the undiacritized Arabic text of the input sentence. Only a human-readable, fully diacritized Arabic text is needed to train this architecture, without any additional morphological or syntactic information.

4.1 Preprocessing

At first, only the necessary characters of the sentence which affect the diacritization are kept. These are Arabic characters, numbers, and spaces. The numbers are replaced by 0 since their values will most likely not affect the diacritization of the surrounding words. The other characters are removed before the diacritization process and restored at the end.

Every filtered sentence is then separated into an input and an output. The input is the bare characters of the text, and the output is the corresponding diacritics for every character. Considering that an Arabic letter can have up to two diacritics where one of them is *Shadda*, the output is represented by two vectors; one indicates the primary diacritic corresponding to every letter, and the other indicates the presence or the absence of the *Shadda*. Figure 1 illustrates this process.

The input is mapped to a set of 38 numeric labels representing all the Arabic characters in addition to 0 and the white space. It is transformed into a 2D one-hot encoded array, where the size of the first dimension equals the length of the sentence, and the size of the second equals the number of the labels.

[2] Dataset available at https://sourceforge.net/projects/tashkeela-processed/.

Fig. 1. Transformation of the diacritized text to the input and output labels

After that, this array is extended to 3 dimensions by inserting the time steps dimension as the second dimension and moving the dimension of the label into the third position. The time steps are generated by a sliding window of size 1 on the first dimension. The number of time steps is fixed to 10 because this number is large enough to cover most of the Arabic words, along with a part of their previous words. The output of the primary diacritics is also transformed from a vector of labels to a 2D one-hot array. The output of *Shadda* marks is left as a binary vector. Figure 2 shows a representation of the input array and the two output arrays after the preprocessing of the previous example. The ∅ represents a padding vector (all zeros), and the numbers in the input and the second output indicate the indexes of the cells of the one-hot vectors set to 1.

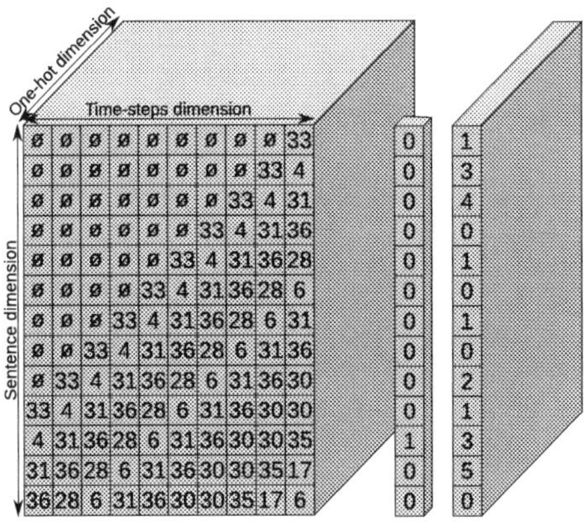

Fig. 2. Input and output arrays after the transformations

4.2 Deep Learning Model

The following component in this system is an RNN model, composed from a stack of two bidirectional LSTM [22,28] layers of 64 cells in each direction, and parallel dense layers of sizes 8 and 64. All of the previous layers use *hyperbolic tangent* (Tanh) as an activation function. The first parallel layer is connected to a single perceptron having the *sigmoid* activation function, while the second is connected to 7 perceptrons having *softmax* as an activation function. The first estimates the probability that the current character has a *Shadda*, and the second generates the probabilities of the primary diacritics for that character. A schema of this network is displayed in Fig. 3. The size, type, and number of layers were determined empirically and according to the previous researches that used deep learning approaches [2,13,14,27,32].

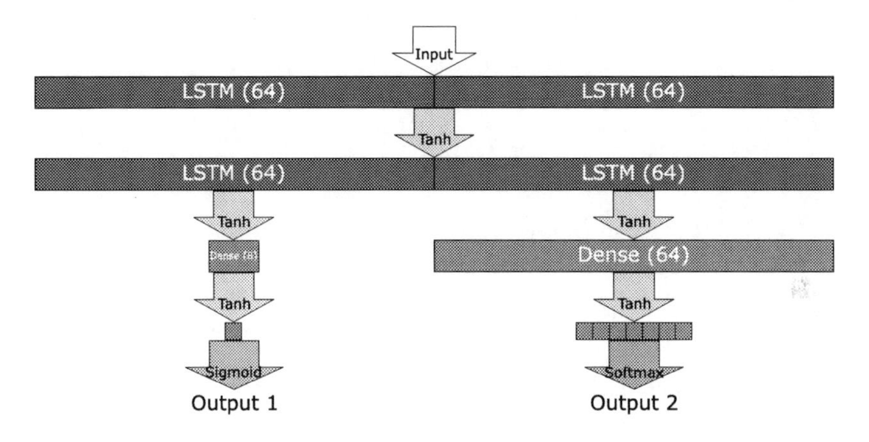

Fig. 3. Architecture of the Deep Learning model

4.3 Rule-Based Corrections

Rule-based corrections are linked to the input and output of the RNN to apply some changes to the output. These rules can select the appropriate diacritic for some characters in some contexts, or exclude the wrong choices in other contexts by nullifying their probabilities. Different sets of rules are applied to the outputs to eliminate some impossible diacritizations according to Arabic rules.

Shadda Corrections. The first output of the DL model representing the probability of the *Shadda* diacritic is ignored by nullifying its value if any of these conditions are met for the current character:

- It is a space, 0 or one of the following: ة, ا, أ, إ, آ, ء, ئ, ى.
- It is the first letter of the Arabic word.
- It has *Sukoon* as a predicted primary diacritic.

Primary Diacritics Corrections. The probabilities of the second output of the DL model are also altered by these rules when their respective conditions are met for the current character:

- If it is إ, set the current diacritic to *Kasra*, by setting the probability of its class to 1 and the others to 0.

- If it is ى or ة, set the diacritic of the previous character to *Fatha*.
- If it is ا and the last letter of the word, allow only *Fatha*, *Fathatan*, or no-diacritic choices by zeroing the probabilities of the other classes.
- If it is ا and not the last letter of the word, set *Fatha* on the previous character.
- If it is the first letter in the word, forbid *Sukoon*.
- If it is not the last character of the word, prohibit any *Tanween* diacritic from appearing on it.
- If it is the last letter, prohibit *Fathatan* unless this character is ء or ة.
- If it is a space, 0 or any of the following characters: آ, ى, ا, set the choice to no-diacritic.

4.4 Statistical Corrections

The output and the input of the previous phase are transformed and merged to generate a standard diacritized sentence. The sentence is segmented into space-delimited words and augmented by unique starting and ending entities. Every word in the sentence is checked up to 4 times in the levels of correction using the saved training data. If any acceptable correction is found at any level, the word will be corrected. Otherwise, it will be forwarded to the next level. In the case where many corrections get the same score in a single level, the first correction is chosen. If no correction is found, the predicted diacritization of the previous component is not changed.

Word Trigram Correction. In the first stage, trigrams are extracted from the undiacritized sentence and checked whether a known diacritization for its core word is available. The core word is the second one in the trigram, while the first and the third are considered previous and following contexts, respectively. If such a trigram is found, the most frequent diacritization for the core word in that context is selected. Despite its high accuracy, especially for the syntactic part, this correction rarely works since having the exact surrounding words in the test data is not common. This correction is entirely independent of the output of the DL model and the rule-based corrections. An example is shown in Fig. 4.

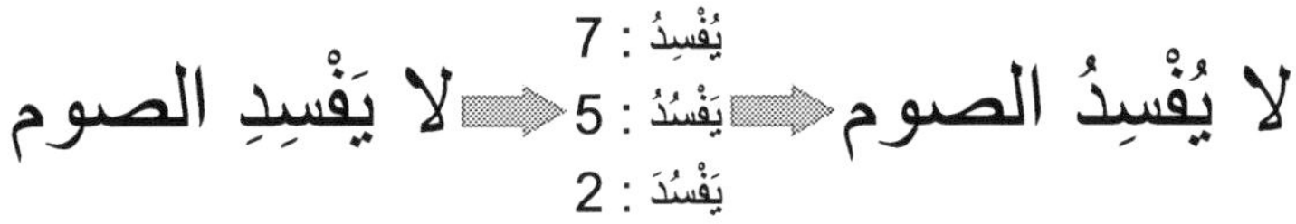

Fig. 4. Selecting the diacritization using the trigrams

Word Bigram Correction. In the second stage, the same processing as the previous one is applied for the remaining words but considering bigrams where the core word is the second one, and the first one represents the previous context. This correction works more often than the trigram-based one since it depends only on the previous word. Similarly, it does not depend on the output of the previous components.

Word Minimum Edit Distance Correction. In the third stage, when the undiacritized word has known compatible diacritizations, the Levenshtein distance [29] is calculated between the predicted diacritization and every saved diacritization for that word. The saved diacritization corresponding to the minimal edit distance is chosen, as shown in Fig. 5. Most predictions are corrected at this stage when the vocabulary of the test set is relatively similar to the training set.

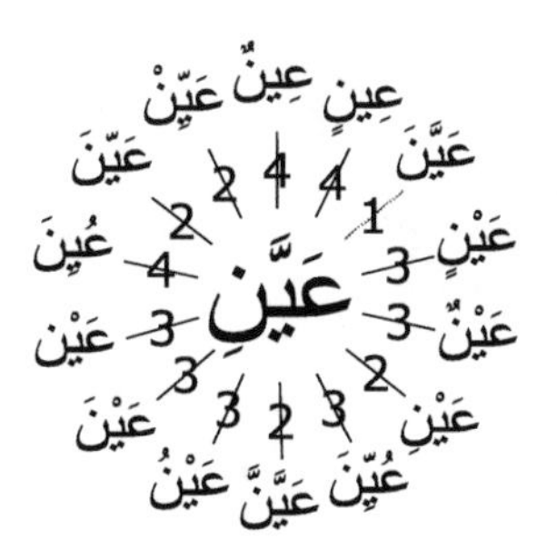

Fig. 5. Selecting the diacritization according to the minimal edit distance

Pattern Minimum Edit Distance Correction. Finally, if the word was never seen, the pattern of the predicted word is extracted and compared against the saved diacritized forms of that pattern. To generate the word pattern, the following substitutions are applied: أ, إ, ؤ, ئ, آ are all replaced by ء. ى is replaced by ا. The rest of the Arabic characters except ة and the long vowels (ا, و, ي) are substituted by the character ح. The diacritics and the other characters are not affected. The predicted diacritized pattern is compared to the saved diacritization forms of this pattern when available, and the closest one, according to the Levenshtein distance, is used as a correction, following the same idea of the previous stage. This correction is effective when the test data contains many words not seen in the training data.

5 Experiments

5.1 Implementation Details

The described architecture was developed using Python [41] 3.6 with NumPy [40] 1.16.5 and TensorFlow [1] 1.14.

The training data was transformed into NumPy arrays of input and output. The DL model was implemented using Keras, and each processed sentence of text is considered a single batch of data when fed into the DL model. The optimizer used for adjusting the model weights is ADADELTA [43] with an initial learning rate of 0.001 and ρ of 0.95.

The rule-based corrections are implemented as algebraic operations working on the arrays of the input and the output.

The statistical corrections use dictionaries as data structures, where the keys are the undiacritized n-grams/patterns, and the values are lists of the possible tuples of the diacritized form along with their frequencies in the training set.

5.2 System Evaluation

The DL model is trained for a few iterations to adjust its weights, while the dictionaries of the statistical corrections are populated while reading the training data in the first pass.

We report the accuracy of our system using the variants of the metrics DER and WER as explained in the introduction. These metrics do not have an agreed exact definition, but most of the previous works followed the definition of Zitouni et al. [46] which takes non-Arabic characters into account, while some of the new ones tend to follow the definition of Alansary et al. [7] and Fadel et al. [19] which excludes these characters. In our work, we chose the latter definition since the former can be significantly biased, as demonstrated in [19]. The calculation of these metrics should include the letters without diacritics, but they can be excluded as well, especially when the text is partially diacritized.

First, we used our testing set to measure the performances of our system. We got DER1 = 4.00%, WER1 = 12.08%, DER2 = 2.80%, and WER2 = 6.22%.

Table 4. Comparison of the performances of our system to the available baselines

System	Include no-diacritic letters				Exclude no-diacritic letters			
	DER1	WER1	DER2	WER2	DER1	WER1	DER2	WER2
Farasa	21.43%	58.88%	23.93%	53.13%	24.90%	57.28%	27.55%	51.84%
MADAMIRA	34.38%	76.58%	29.94%	59.07%	40.03%	75.39%	33.87%	57.22%
Mishkal	16.09%	39.78%	13.78%	26.42%	17.59%	35.63%	14.22%	21.92%
Tashkeela-Mode	49.96%	96.80%	52.96%	94.16%	58.50%	96.03%	60.92%	92.45%
Shakkala	3.73%	11.19%	2.88%	6.53%	4.36%	10.89%	3.33%	6.37%
Ours	**3.39%**	**9.94%**	**2.61%**	**5.83%**	**3.34%**	**7.98%**	**2.43%**	**3.98%**

The same testing data and testing method of Fadel et al. [19] were used as well in order to compare our system to the others evaluated in that work. The results are summarized in Table 4.

Results show that our system outperforms the best-reported system (Shakkala). These results can be justified as Shakkala does not perform any corrections on the output of the deep learning model, while ours includes a cascade of corrections that fix many of its errors.

When training and testing our system on the text extracted from the LDC's ATB part 3 [30], it archives DER1 = 9.32%, WER1 = 28.51%, DER2 = 6.37% and WER2 = 12.85%. Its incomplete diacritization mainly causes the higher error rates for this dataset in a lot of words, in addition to its comparatively small size, which prevents our system from generalizing well.

5.3 Error Analysis

To get a deeper understanding of the system performances, we study the effect of its different components and record the errors committed at each level. We performed the tests taking all and only Arabic characters into account on our test part of the *Tashkeela* dataset.

Contribution of the Components. In order to show the contribution in error reduction of every component, two evaluation setups were used.

Firstly, only the DL model and the static rules are enabled at first, then the following component is enabled at every step, and the values of the metrics are recalculated. Table 5a shows the obtained results.

Secondly, all the components are enabled except one at a time. The same calculations are done and displayed in Table 5b.

Table 5. Reduction of the error rates according to the enabled components

(a) Incremental enabling

Components	Metrics			
	DER1	WER1	DER2	WER2
DL + rules	23.20%	59.32%	23.75%	51.52%
+Trigrams	12.99%	32.25%	13.19%	27.46%
+Bigrams	7.09%	17.54%	6.69%	13.20%
+Unigrams	4.06%	12.18%	2.87%	6.35%
All enabled	4.00%	12.08%	2.80%	6.22%

(b) One disabled at a time

Components	Metrics			
	DER1	WER1	DER2	WER2
No trigrams	4.38%	13.39%	2.97%	6.65%
No bigrams	6.44%	19.53%	4.88%	11.46%
No unigrams	6.51%	16.57%	5.88%	11.87%
No patterns	4.06%	12.18%	2.87%	6.35%
All enabled	4.00%	12.08%	2.80%	6.22%

The contributions of the unigram and bigram corrections are the most important considering their effect on the error rates in both setups. The effect of the trigram correction is more visible on the syntactic diacritization rather than the morphological diacritization since the former is more dependant on the context.

The contribution of the pattern corrections is not very noticeable due to the position of this component in the pipeline, limiting its effect only to the OoV words.

Error Types. We use our system to generate diacritization for a subset of the sentences of our testing set. We limit our selection to 200 sentences where there is at least one word wrongly diacritized. We counted and classified a total of 426 errors manually. We present the results in Table 6.

Table 6. Diacritization errors count from 200 wrong test sentences

Error	Syntactic	Replacement	Non-existence	Prediction missing	Label missing	Total
Count	224	103	48	26	25	426

We found that 52.58% of the mistakes committed by our diacritization system are caused by the syntactic diacritization, which specify the role of the word in the sentence. The syntactic diacritization is so hard that even the Arabic native speakers often commit mistakes of this type when speaking. Since this is manual verification, we do not just suppose that the diacritic of the last character of the Arabic word is the syntactic one as what is done in the calculations of DER2 and WER2, but we select the diacritics which have a syntactic role according to Arabic rules, no matter where they appear.

A replacement error is when the system generates a diacritization that makes a valid Arabic word, but it is wrong according to the test data. 24.18% of the errors of our system are considered in this type.

Non-existence error happens when the diacritization system generates a diacritization making a word that does not exist in the standard Arabic. 11.27% of our system's errors are in this type.

The remaining error types are prediction missing and label missing, which indicate that the system has not predicted any diacritic where it should do, and the testing set has missing/wrong diacritics, respectively. These types are generally caused by the mistakes of diacritization in training and testing sets.

6 Conclusion

In this work, we developed and presented our automatic Arabic diacritization system, which follows a hybrid approach combining a deep learning model, rule-based corrections, and two types of statistical corrections. The system was trained and tested on a large part of the Tashkeela corpus after being cleaned and normalized. On our test set, the system scored DER1 = 4.00%, WER1 = 12.08%, DER2 = 2.80% and WER2 = 6.22%. These values were calculated when taking all and only Arabic words into account.

Our method establishes new state-of-the-art results in the diacritization of raw Arabic texts, mainly when the classical style is used. It performs well even on the documents that contain unseen words or non-Arabic words and symbols. We made our code publicly available as well[3].

In the next work, we will focus on improving the generalization of the system to better handle the out-of-vocabulary words, while reducing the time and memory requirements.

Acknowledgments. To Dr. Yasser Hifny for his help concerning the train and the test of our system on the diacritized text of the ATB part 3 dataset.

References

1. Abadi, M., et al.: TensorFlow: a system for large-scale machine learning. In: 12th {USENIX} Symposium on Operating Systems Design and Implementation ({OSDI} 2016), pp. 265–283 (2016)
2. Abandah, G., Arabiyat, A., et al.: Investigating hybrid approaches for Arabic text diacritization with recurrent neural networks. In: 2017 IEEE Jordan Conference on Applied Electrical Engineering and Computing Technologies (AEECT), pp. 1–6. IEEE (2017)
3. Abandah, G.A., Graves, A., Al-Shagoor, B., Arabiyat, A., Jamour, F., Al-Taee, M.: Automatic diacritization of Arabic text using recurrent neural networks. Int. J. Doc. Anal. Recogn. (IJDAR) **18**(2), 183–197 (2015)
4. Abdelali, A., Darwish, K., Durrani, N., Mubarak, H.: Farasa: a fast and furious segmenter for Arabic. In: Proceedings of the 2016 Conference of the North American Chapter of the Association for Computational Linguistics: Demonstrations, pp. 11–16 (2016)
5. Al-Badrashiny, M., Hawwari, A., Diab, M.: A layered language model based hybrid approach to automatic full diacritization of Arabic. In: Proceedings of the Third Arabic Natural Language Processing Workshop, pp. 177–184 (2017)
6. Al Sallab, A., Rashwan, M., Raafat, H.M., Rafea, A.: Automatic Arabic diacritics restoration based on deep nets. In: Proceedings of the EMNLP 2014 Workshop on Arabic Natural Language Processing (ANLP), pp. 65–72 (2014)
7. Alansary, S.: Alserag: an automatic diacritization system for Arabic. In: Hassanien, A.E., Shaalan, K., Gaber, T., Azar, A.T., Tolba, M.F. (eds.) AISI 2016. AISC, vol. 533, pp. 182–192. Springer, Cham (2017). https://doi.org/10.1007/978-3-319-48308-5_18
8. Alghamdi, M., Muzaffar, Z., Alhakami, H.: Automatic restoration of Arabic diacritics: a simple, purely statistical approach. Arab. J. Sci. Eng. **35**(2), 125 (2010)
9. Alosaimy, A., Atwell, E.: Diacritization of a highly cited text: a classical Arabic book as a case. In: 2018 IEEE 2nd International Workshop on Arabic and Derived Script Analysis and Recognition (ASAR), pp. 72–77. IEEE (2018)
10. Ananthakrishnan, S., Narayanan, S., Bangalore, S.: Automatic diacritization of Arabic transcripts for automatic speech recognition. In: Proceedings of the 4th International Conference on Natural Language Processing, pp. 47–54 (2005)
11. Anwar, M.: Tashkeela-model (2018). https://github.com/Anwarvic/Tashkeela-Model

[3] Available at https://github.com/Hamza5/Pipeline-diacritizer.

12. Azmi, A.M., Almajed, R.S.: A survey of automatic Arabic diacritization techniques. Nat. Lang. Eng. **21**(3), 477–495 (2015)
13. Barqawi, A., Zerrouki, T.: Shakkala, Arabic text vocalization (2017). https://github.com/Barqawiz/Shakkala
14. Belinkov, Y., Glass, J.: Arabic diacritization with recurrent neural networks. In: Proceedings of the 2015 Conference on Empirical Methods in Natural Language Processing, pp. 2281–2285 (2015)
15. Chennoufi, A., Mazroui, A.: Morphological, syntactic and diacritics rules for automatic diacritization of Arabic sentences. J. King Saud Univ. Comput. Inf. Sci. **29**(2), 156–163 (2017)
16. Darwish, K., Magdy, W., et al.: Arabic information retrieval. Found. Trends® Inf. Retrieval **7**(4), 239–342 (2014)
17. Darwish, K., Mubarak, H., Abdelali, A.: Arabic diacritization: stats, rules, and hacks. In: Proceedings of the Third Arabic Natural Language Processing Workshop, pp. 9–17 (2017)
18. Elshafei, M., Al-Muhtaseb, H., Alghamdi, M.: Statistical methods for automatic diacritization of Arabic text. In: The Saudi 18th National Computer Conference. Riyadh, vol. 18, pp. 301–306 (2006)
19. Fadel, A., Tuffaha, I., Al-Jawarneh, B., Al-Ayyoub, M.: Arabic text diacritization using deep neural networks. arXiv preprint arXiv:1905.01965 (2019)
20. Fashwan, A., Alansary, S.: Shakkil: an automatic diacritization system for modern standard Arabic texts. In: Proceedings of the Third Arabic Natural Language Processing Workshop, pp. 84–93 (2017)
21. Gal, Y.: An hmm approach to vowel restoration in Arabic and Hebrew. In: Proceedings of the ACL-02 Workshop on Computational Approaches to Semitic Languages, pp. 1–7. Association for Computational Linguistics (2002)
22. Graves, A., Schmidhuber, J.: Framewise phoneme classification with bidirectional LSTM and other neural network architectures. Neural Networks **18**(5–6), 602–610 (2005)
23. Habash, N., Rambow, O.: Arabic diacritization through full morphological tagging. In: Human Language Technologies 2007: The Conference of the North American Chapter of the Association for Computational Linguistics; Companion Volume, Short Papers, pp. 53–56 (2007)
24. Hadj Ameur, M.S., Moulahoum, Y., Guessoum, A.: Restoration of Arabic diacritics using a multilevel statistical model. In: Amine, A., Bellatreche, L., Elberrichi, Z., Neuhold, E.J., Wrembel, R. (eds.) Computer Science and Its Applications, pp. 181–192. Springer, Cham (2015). https://doi.org/10.1007/978-3-319-19578-0_15
25. Hifny, Y.: Open vocabulary arabic diacritics restoration. IEEE Signal Process. Lett. **26**(10), 1421–1425 (2019). https://doi.org/10.1109/LSP.2019.2933721
26. Hifny, Y.: Higher order n-gram language models for Arabic diacritics restoration. In: The Twelfth Conference on Language Engineering (2012)
27. Hifny, Y.: Hybrid LSTM/MaxEnt networks for Arabic syntactic diacritics restoration. IEEE Signal Process. Lett. **25**(10), 1515–1519 (2018)
28. Hochreiter, S., Schmidhuber, J.: Long short-term memory. Neural Comput. **9**(8), 1735–1780 (1997)
29. Levenshtein, V.I.: Binary codes capable of correcting deletions, insertions and reversals. Soviet Physics Doklady **10**, 707 (1966)
30. Maamouri, M., Bies, A., Buckwalter, T., Jin, H., Mekki, W.: Arabic treebank: Part 3 (full corpus) v 2.0 (mpg+ syntactic analysis). Linguistic Data Consortium, Philadelphia (2005)

31. Metwally, A.S., Rashwan, M.A., Atiya, A.F.: A multi-layered approach for Arabic text diacritization. In: 2016 IEEE International Conference on Cloud Computing and Big Data Analysis (ICCCBDA), pp. 389–393. IEEE (2016)
32. Moumen, R., Chiheb, R., Faizi, R., El Afia, A.: Arabic diacritization with gated recurrent unit. In: Proceedings of the International Conference on Learning and Optimization Algorithms: Theory and Applications, p. 37. ACM (2018)
33. Nelken, R., Shieber, S.M.: Arabic diacritization using weighted finite-state transducers. In: Proceedings of the ACL Workshop on Computational Approaches to Semitic Languages, pp. 79–86. Association for Computational Linguistics (2005)
34. Pasha, A., et al.: MADAMIRA: a fast, comprehensive tool for morphological analysis and disambiguation of Arabic. LREC **14**, 1094–1101 (2014)
35. Rashwan, M.A., Al-Badrashiny, M.A., Attia, M., Abdou, S.M., Rafea, A.: A stochastic Arabic diacritizer based on a hybrid of factorized and unfactorized textual features. IEEE Trans. Audio Speech Lang. Process. **19**(1), 166–175 (2011)
36. Rashwan, M.A., Al Sallab, A.A., Raafat, H.M., Rafea, A.: Deep learning framework with confused sub-set resolution architecture for automatic Arabic diacritization. IEEE/ACM Trans. Audio Speech Lang. Process. (TASLP) **23**(3), 505–516 (2015)
37. Said, A., El-Sharqwi, M., Chalabi, A., Kamal, E.: A hybrid approach for Arabic diacritization. In: Métais, E., Meziane, F., Saraee, M., Sugumaran, V., Vadera, S. (eds.) NLDB 2013. LNCS, vol. 7934, pp. 53–64. Springer, Heidelberg (2013). https://doi.org/10.1007/978-3-642-38824-8_5
38. Shaalan, K., Abo Bakr, H.M., Ziedan, I.: A hybrid approach for building Arabic diacritizer. In: Proceedings of the EACL 2009 Workshop on Computational Approaches to Semitic Languages, pp. 27–35. Association for Computational Linguistics (2009)
39. Shahrour, A., Khalifa, S., Habash, N.: Improving Arabic diacritization through syntactic analysis. In: Proceedings of the 2015 Conference on Empirical Methods in Natural Language Processing, pp. 1309–1315 (2015)
40. Van Der Walt, S., Colbert, S.C., Varoquaux, G.: The numpy array: a structure for efficient numerical computation. Comput. Sci. Eng. **13**(2), 22 (2011)
41. Van Rossum, G., Drake, F.L.: The Python Language Reference Manual. Network Theory Ltd., Network (2011)
42. Zayyan, A.A., Elmahdy, M., binti Husni, H., Al Ja'am, J.M.: Automatic diacritics restoration for modern standard Arabic text. In: 2016 IEEE Symposium on Computer Applications & Industrial Electronics (ISCAIE), pp. 221–225. IEEE (2016)
43. Zeiler, M.D.: Adadelta: an adaptive learning rate method. arXiv preprint arXiv:1212.5701 (2012)
44. Zerrouki, T.: Mishkal, Arabic text vocalization software (2014). https://github.com/linuxscout/mishkal
45. Zerrouki, T., Balla, A.: Tashkeela: novel corpus of Arabic vocalized texts, data for auto-diacritization systems. Data Brief **11**, 147 (2017)
46. Zitouni, I., Sorensen, J.S., Sarikaya, R.: Maximum entropy based restoration of Arabic diacritics. In: Proceedings of the 21st International Conference on Computational Linguistics and the 44th Annual Meeting of the Association for Computational Linguistics, pp. 577–584. Association for Computational Linguistics (2006)

A Mixed Semantic Features Model for Chinese NER with Characters and Words

Ning Chang[1], Jiang Zhong[1,2](✉), Qing Li[1], and Jiang Zhu[3]

[1] Chongqing University, Chongqing 400044, People's Republic of China
zhongjiang@cqu.edu.cn

[2] Key Laboratory of Dependable Service Computing in Cyber Physical Society, Chongqing University, Chongqing 400044, People's Republic of China

[3] Chengdu Library and Information Center, Chinese Academy of Sciences, Chengdu 610041, People's Republic of China

Abstract. Named Entity Recognition (NER) is an essential part of many natural language processing (NLP) tasks. The existing Chinese NER methods are mostly based on word segmentation, or use the character sequences as input. However, using a single granularity representation would suffer from the problems of out-of-vocabulary and word segmentation errors, and the semantic content is relatively simple. In this paper, we introduce the self-attention mechanism into the BiLSTM-CRF neural network structure for Chinese named entity recognition with two embedding. Different from other models, our method combines character and word features at the sequence level, and the attention mechanism computes similarity on the total sequence consisted of characters and words. The character semantic information and the structure of words work together to improve the accuracy of word boundary segmentation and solve the problem of long-phrase combination. We validate our model on MSRA and Weibo corpora, and experiments demonstrate that our model can significantly improve the performance of the Chinese NER task.

Keywords: Chinese named entity recognition · Self-attention · Mixed semantic feature · Entity boundary segmentation

1 Introduction

In recent years, named entity recognition (NER) has received a lot of attention in the field of natural language processing (NLP), and it is the basis of many

Supported by National Key Research and Development Program of China Grant 2017YFB1402400, in part by the Graduate Research and Innovation Foundation of Chongqing under Grant CYB18058, in part by the Key Research Program of Chongqing Science and Technology Bureau No. cstc2019jscx-fxyd0142, in part by the Fundamental Research Funds for the Central Universities under Grant 2018CDYJSY0055.

J. M. Jose et al. (Eds.): ECIR 2020, LNCS 12035, pp. 356–368, 2020.
https://doi.org/10.1007/978-3-030-45439-5_24

downstream NLP tasks. NER refers to the identification of entities with specific meaning in the text, usually including names of people, places, institutions, proper nouns, and so on. For English text, this problem has been studied extensively [13,20,23]. However, Chinese NER still faces challenges such as Chinese word segmentation, and it is often difficult to define what constitutes a word in Chinese.

Most methods of existing state-of-the-art models for Chinese NER are usually based on word segmentation, and train neural network and Conditional Random Field (CRF) to perform sequence labeling on word-level [17]. However the effect of the word segmentation depends heavily on the quality of the dictionaries and segmentation tools, and it's possible to lead to error propagation if the boundaries are partitioned improperly at the very start. Moreover, it can not deal with unseen words. There are also some models which recognize entities in character-level, which solve the problem of out-of-vocabulary (OOV) [19]. However, fully character-based models cannot express enough semantic information and word structure, and could lead to wrong word boundaries.

In order to take advantage of both character-level semantic information and word structure content, some models mix word embedding and its corresponding character vectors, and then feed mixed representation into neural network for NER [2,22,26]. The generic model mentioned above is shown in Fig. 1(a), these methods divide each sequence into several characters, and then represent these character vectors as a comprehensive representation through LSTM networks or other models. About word vectors, they concatenate each word vector with the representation of its corresponding characters, and then form a new multi-granularity representation of the word. In the process of generating the final word representation, the intermediate dimensional transformation may lead to original information loss. Moreover, for each word, the concatenation of the two granularity representations at the word-level does not express well the relationship between characters and words. These drawbacks affect the accuracy of Chinese word boundary segmentation and entity recognition.

In this paper, we incorporate the self-attention mechanism into the Chinese named entity recognition model to compute the weighted sum of character and word vectors, and integrate the semantic features of the two representations. Different from the previously mentioned model, our model captures character content features and the information of token structure in word level (as shown in Fig. 1(b)). Our model uses two sequences of character and word segmentation as input, and outputs the final character-based recognition tags through the attention mechanism. The model preserves the character-level semantic representation and the word tokens structure completely, and uses self-attention to assign the weight of both. Multi-granularity semantic and structural features are combined with word representation to enrich character representation and reduce the loss of original information. Moreover, the character level and the word segmentation structure are complementary to each other, and a single character can correct the word boundary error caused by the word segmentation level.

Moreover, phrases that do not appear in the prior dictionary can also be identified. As the example in Fig. 1 shows, given the sequence of "Beijing/People/Park" that is segmented using the dictionary, our model could add the segmentation structural information into character-based semantic information. When predicting tags of characters, we can determine the phrase boundary based on the comprehensive context and correctly identify "Beijing people's park" as a phrase to be marked.

We experiment with our model on MSRA and Weibo data sets, and the results show that using the self-attention mechanism to fuse two granularity semantic and structure representations in sequence context can significantly improve performance.

The contributions of our paper are as follows:

- We improve the accuracy of word boundary segmentation by combining two granularity features. Our model retains the primitiveness of character semantics and participle structures completely, and the two embedding information assist each other. Character semantics combined with word tokens structure could modify word boundary segmentation.
- We investigate a method to enhance the recognition of Chinese long phrases that do not appear in prior dictionaries. Our model uses a self-attention mechanism to integrate features of word segmentation into a character-level sequence, and predicts it in conjunction with the context of the sentence to merge the short tags into long phrases.

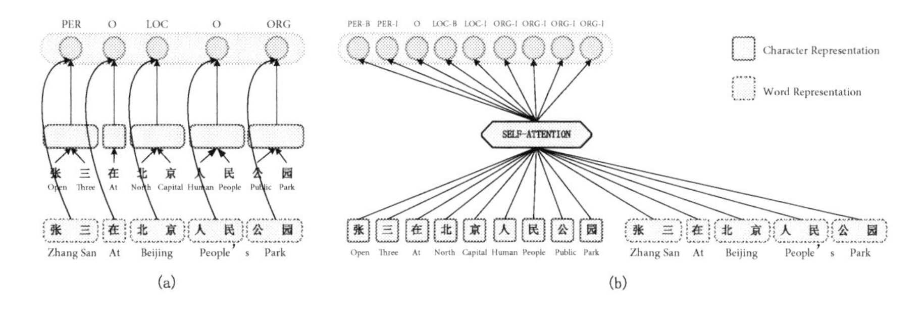

Fig. 1. Example of how previous models (a) and our model (b) combine two granularity representations of characters and words.

2 Related Work

NER. Early named entity recognition methods are based on rules and statistical machine learning such as Hidden Markov Model (HMM) [1], Conditional Random Fields (CRF) [12], and Support Vector Machines (SVM) [11]. In recent years, with the development of machine learning, more and more neural network models are used for the NER task. Collobert et al. [4] propose a unified neural network architecture that can be used in various NLP tasks. Zhou et al. [27] formulate Chinese NER as a joint identification and categorization task. Huang et al. [10] first apply BiLSTM-CRF model to NER, and achieve the advance results at that time. The BiLSTM-CRF model is now also the benchmark model for many pieces of research. Lample et al. [13] use BiLSTM-CRF as the basic model, rely on character-based word representations learned from the supervised corpus and unsupervised word representations learned from unannotated corpora. For Chinese NER, Zhang et al. [26] investigate a lattice-structured LSTM model, utilize information on words and character sequences, and solve the problem of Chinese words boundaries. Dong et al. [6] utilize both character-level and radical-level representations based on bidirectional LSTM-CRF. Besides, incorporating the five-stroke information into the network also achieves outstanding performance [24]. [14] add gazetteer-enhanced sub-tagger on hybrid semi-Markov CRF architecture and observe some promising results. And [5] also propose a neural multi-digraph model with the information of gazetteers.

Self-attention. Vaswani et al. [21] first proposed a self-attention mechanism for machine translation to connect all positions with a constant number of sequentially executed operations, and attract great attention. Subsequently, a large number of studies begin to use the attention mechanism. Zukov et al. [29] use no language-specific features, and the model they proposed is based on RNN structure, coupled with a self-attention mechanism for NER. Yang et al. [25] propose a novel adversarial transfer learning framework and first introduce a self-attention mechanism to the Chinese NER task. And then Zhu et al. [28] propose a convolutional attention network for Chinese named entity recognition. They use a character-based CNN with local-attention and GRU with self-attention to get information from characters of the sentence.

Joint Character and Word Embedding. Some models join characters with words for sequence tagging. Lample et al. [13] feed the characters of a word into the bidirectional LSTM, and connect the final output of the forward and backward network as character representation. This character-level representation is then concatenated with its corresponding word representation. Rei et al. [18] use the same structure [13] to represent character-level representation. Instead of connecting two-level representations directly, an attention mechanism is used to calculate the weighted sum of character embedding and word embedding. Ma et al. [15] utilize CNN to compute character representation for each word, and concatenate it with word embedding before feeding into the BiLSTM network.

The main benefit of Chinese characters is they can solve the problem of phrases that is not in the dictionary, and can flexibly determine the phrase boundary. Besides, word-level modeling can provide information about the structure of common words. We propose a model based on self-attention which use the sequence-level joint representation of characters and words to take advantage of two granularity embedding.

3 Methodology

In this chapter, we will introduce our methodology in detail. Our model utilizes BiLSTM-CRF as our basic structure, and extends a self-attention mechanism to obtain the long distance dependencies of the character encoder and word encoder sequence. As illustrated in Fig. 2, the architecture of our model mainly consists of character and word embedding, Bi-LSTM network with self-attention and CRF for tagging. We will describe our method in the following sections.

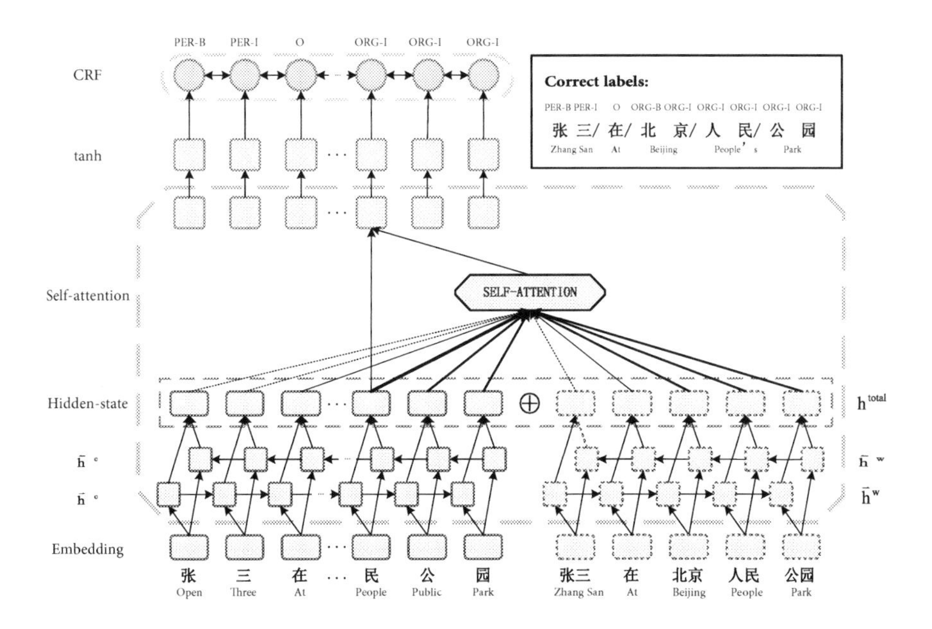

Fig. 2. The general architecture of our proposed model.

3.1 Characters and Words Representation in Sequence

Word embedding, also known as distributed word representation, can capture both the semantic and syntactic information of the words from a large unlabeled corpus. We use open source Chinese word vector corpus of Tencent AI Lab[1], which includes more than 8 million Chinese words, and each word corresponds to a 200-dimensional vector. For a sentence, we utilize *jieba*[2] to perform word segmentation. And every word is disintegrated into individual characters. Furthermore, characters in a sentence also contain the rich context of the entities, and Chinese character-based embedding could alleviate problems of long

[1] https://ai.tencent.com/ailab/nlp/embedding.html.
[2] https://github.com/fxsjy/jieba.

phrases that are not in dictionaries. Our model uses both granularity levels of embedding information to learn the mixture semantic of characters and words.

In this paper, we use Bi-LSTM [9] as our basic structure to use forward and backward information of character and word embedding. We denote the two embedding sequences separately as $[E_1^c, E_2^c, ..., E_n^c]$ and $[E_1^w, E_2^w, ..., E_m^w]$. And they are generated by a look-up layer, and are fed into two parallel Bi-LSTM structures respectively, which have the same structure, but with different parameters. The output character and the word level hidden state are represented as h^c and h^w. Join the two hidden layers to form a total hidden state (represented as $h^{total} = [h^c, h^w]$), where the front part is the characters representation, and the latter part is the words semantic feature. Then a self-attention mechanism operation is performed on the total hidden state sequence.

3.2 Multi-granularity Representation Fusion by Using Self-attention

Solve the Problem of Boundary Segmentation. In the process of word segmentation, there may be problems with word boundary errors. As shown in the example in Fig. 1, the first three characters may be incorrectly split into a person's name in the sentence, which would lead to error propagation, and cause severe bias effects on subsequent predictions. The previous general model can not solve the problem of word segmentation very well, and cause some content loss when combining the embedded information of characters and words. Our model combines two granular hidden states at the sequence level to preserve the original features intact. Also, the self-attention trains the weight information of the total sequence, and preserves the semantic information of the context characters to perform a calculation with the information of the word sequence structure. The character information will correct the error problem of word boundary segmentation, and correctly identify "Zhang San" as a person name, while the third character as a preposition.

Solve the Problem of Phrase Combination. In Chinese, long phrases are usually composed of short phrase sequences in order. For example, Beijing People's Park is composed of Beijing/ people/ park. Compared with English, there is usually no need for prepositional connections in phrases, which leads to the poor distinction of the boundaries in long phrases. It is also a difficult point in the recognition of Chinese named entities. Dictionary-based word segmentation usually divides sentences into short words. For long phrases that do not appear in the dictionary, there is currently no good solution. This paper proposes a method to improve the above problem by using two granularity semantic representations to assist each other. The model uses the self-attention mechanism to calculate the similarity on different levels of representation subspace, sequentially calculates each character with all tokens in the total sequence. This method captures the structural information of the word sequence, in order to compute similarity and correlation with character context information to further identify the combined boundaries of the long phrases.

In addition, the attention mechanism uses the weighted sum calculation to generate the output, which effectively solves the problem of the gradient disappearing. And the self-attention mechanism can be calculated in parallel, which greatly improves efficiency.

3.3 CRF for Tag Prediction

We quote a standard Conditional Random Field (CRF) layer on top of the attention layer. The CRF can use the state feature function and the state transfer function to maximize the characteristics of the text. Besides, it can consider the context information and the annotation information of adjacent words. The feature functions are defined as follows:

$$f_j\left(s, j, l_i, l_{i-1}\right) = \begin{cases} t_j\left(l_{i-1}, l_i, s, i\right) & State\ transfer\ function \\ s_j\left(l_i, s, i\right) & \text{State feature function} \end{cases} \tag{1}$$

Where s indicates the sentence we want to predict. l is the label sequence of the sentence, and l_i represents the label of i-th token. i is the current location. The state transition function defines the probability of the $(i-1)$-th token label l_{i-1} move to the label l_i of the next i-th token in the sentence s. And the state feature function indicates the probability that the current i-th token is marked as l_i.

Then we normalize the score to get the probability that the label sequence is l given the sentence s. Given all predicted tag sequences l, the probability of label sequence s is calculated as follows:

$$p(l|s) = \frac{\exp[s\,\mathrm{core}(l|s)]}{\sum_{l'} \exp\left[\mathrm{score}\left(l'|s\right)\right]} \tag{2}$$

Where l represents all possible tag sequences.

The output of the self-attention mechanism is independent of each other. Although the context information is taken into account when performing the matrix transformation, the outputs do not affect each other. Our model uses CRF for label prediction. By considering the transition characteristics between output labels, we constrain the final label and improve the accuracy of entity label prediction.

4 Experiments

4.1 Datasets

We use corpora provided by Microsoft Research Asia (MSRA) and Weibo corpus [17] extracted from Sina Weibo to experiment with the model presented in this paper. MSRA contains three entity types: Person (PER), Location (LOC) and Organization (ORG). And Weibo dataset is annotated with four types of entities (in addition to the above three entities, there is also a Geo-Political entity type, GPE). We train on both name mentions and nominal mentions in the Weibo data set. The detailed statistics of the corpora are summarized in Table 1.

We preprocess the datasets and annotate the entity type using BIO rules, which indicates Begin, Inside and Outside of a named entity.

Table 1. The statistics of datasets.

Corpus		Train set	Test set	Dev set
MSRA		46317	4376	—
Weibo	Named mention entity	957	153	211
	Nominal mention entity	898	226	198

4.2 Experimental Settings

Our experiments employ character-level precision (P), recall (R), and F1-score (F) as the evaluation criteria. We use *Jieba* for segmentation, and utilize word embedding dataset published by Tencent AI Lab to perform embedding, and the dimension of word embedding is 200, the same as character embedding.

Pytorch library is used to build our model. We train the model using an *Adam* optimizer with an initial learning rate of 0.001, and the network is fine-tuned by back-propagating. For the over-fitting and vanishing gradient problems, we employ the dropout method with a probability of 0.5. We control the length of the sentence to be 80, and the number of words after sentence segmentation to be 40. Otherwise, we would pad the shorter sequences, truncate the longer parts. Detailed hyper-parameters are listed in Table 2.

Table 2. Hyper-parameter settings.

Parameters	Values
Character embedding dim	200
Word embedding dim	200
Hidden dim	50
Optimizer	Adam
Initial learning rate	0.001
Dropout rate	0.5
Batch size	64
Epoch	40

4.3 Evaluation of Components

Considering that the character-based models are not dependent on the quality of dictionaries and are more flexible, our model would use character-based output in the subsequent experiments.

Ablation experiments are designed to verify the necessity of each part in our model and its impact on the experimental results. We gradually add each component to the baseline architecture BiLSTM-CRF. The results are shown in Table 3.

To evaluate the effects of two embedding approaches, we perform comparison experiments on character embedding and word embedding respectively. In the third comparative experiment, the embedding for each word and its corresponding characters compose a concatenation to be the input of the Bi-LSTM layer (as [15] did).

Experimental results show that character-based model performance is better than word-based models on the two data sets. At the same time, the model using two embedding methods for prediction has a slight improvement compared with the original two models, but the effect is not obvious. By contrast, adding a self-attention mechanism can significantly improve the performance of NER.

Attention can obtain sentence context information from the long-distance relationship between tokens, overcoming the limitations of recurrent neural networks. In this model, self-attention can capture the dependence of characters and words at the same time over a long distance.

Table 3. Experiments of each component on MSRA and Weibo datasets.

Models	MSRA			Weibo
	Precision	Recall	F1-score	F1-score (overall)
Baseline (Char.Emb)	90.74	89.85	90.29	57.51
Baseline (Word.Emb)	91.21	88.60	89.89	57.02
Baseline (Char+Word.Emb)	92.79	89.93	91.34	57.63
Our model	**95.92**	**94.80**	**95.36**	**61.46**

4.4 Comparison with Previous Work

In this section, we compare our BiLSTM+Self-Attention+CRF model based on a mixture of characters and words with the previous proposed advanced models on the MSRA and Weibo data sets. The comparison results are listed in the Tables 4 and 5.

MSRA Dataset. Chen et al. [3] first apply Conditional Random Fields (CRF) for sequence tagging, and achieve 86.20% F1-score in MSRA corpus. Zhou et al. [27] formulate NER as a joint identification to recognize entity-level features, which effectively improves performance. And Cao et al. [2] also use the information of CWS for NER. Zhang et al. [24] and Ding et al. [5] add additional features, and the latter achieve 94.4% F1-score. Zhu et al. [28] investigate a Convolution Attention Network to capture the information from adjacent characters and sentence contexts, which achieves F1-score of 92.97%. Our model utilizes self-attention on character+word hidden state and gets effective performance improvement with 95.36 F1-score.

Table 4. Results on MSRA dataset.

Models	Precision	Recall	F1-score
Conditional Probabilistic Models [3]	91.22	81.71	86.20
Joint Identification and Categorization [27]	91.86	88.75	90.28
Adversarial Transfer Learning with Self-Attention [2]	91.30	89.58	90.64
Five-Stroke based Cnn-birnn-crf [24]	92.04	91.31	91.67
Convolutional Attention Network [28]	93.53	92.42	92.97
Multi-digraph Model with Gazetteers [5]	94.6	94.2	94.4
Our model	**95.92**	**94.80**	**95.36**

Table 5. Results on Weibo dataset.

Models	Named entities	Nominal mentions	Overall
Joint Trained Embedding [16]	51.96	61.05	56.05
Word Segmentation Representation Learning [17]	55.28	62.97	58.99
BiLSTM with F-score driven [7]	50.60	59.32	54.82
Unified Model for Cross-domain and Semi-supervised [8]	54.50	62.17	58.23
Lattice Network [26]	53.04	62.25	58.79
Multi-digraph Model with Gazetteers [5]	63.1	56.3	59.5
Our model	**59.79**	**63.22**	**61.46**

Weibo Dataset. We compare our model with the latest models on Weibo corpus. Weibo-NER is in the domain of social media. Results of named mentions, nominal mentions, and the total are demonstrated in Table 5 respectively. As there are many non-standard data in social media data, such as spelling errors, and informal words, the overall result of social media corpus is lower than that of MSRA data set. We can see that the model we proposed has achieved state-of-the-art performance.

Peng et al. [16] propose joint training for embedding and achieve 56.05 F1-score. Peng et al. [17] utilize word boundary tags as features to provide richer information and improve the F1-score to 58.99%. He et al. [8] propose a unified model for cross domain and improve F1-score to 58.23% from 54.82% [7]. Zhang et al. [26] investigate a lattice network which explicitly leverages word and word sequence information, and achieve F1-score of 58.79%. Our proposed model has a significant improvement in the named entities, which improves 1.96% compared with Ding et al. [5]. And overall performance is significantly better than other models.

From the experimental results, we can see that our model has improved on both datasets compared with previous models. On the MSRA dataset, our model has improved 0.96, and 1.96% on the Weibo dataset. Because MASA data is standard, previous studies have achieved valid results on this data set. While there are many unregistered words in the Weibo dataset, and the recognition model based on two granularity representations with self-attention can effectively improve the recognition results.

The model improves on existing approaches to reduce out-of-vocabulary and word segmentation issues by using self-attention to fuse the information of the two granularity. The word-level structure make judgment on segmentation of the common words, and character-based semantic information can make more flexible combination of phrase.

5 Conclusion

This paper incorporates self-attention mechanism into BiLSTM-CRF neural network for Chinese named entity recognition. Our model uses self-attention to capture multi-granularity information through the total sequence, which combines the semantic and structural features of characters and words to predict entity tags. We solve the problems of word boundary segmentation and long-phrase combination, and the experimental results show that our method has improved the accuracy of Chinese named entity recognition.

Future work will focus on more granular information representations, such as sentence and paragraph levels, and apply this work to specialized entity identification in a variety of areas.

References

1. Bikel, D.M., Miller, S., Schwartz, R., Weischedel, R.: Nymble: a high-performance learning name-finder. In: Conference on Applied Natural Language Processing (1997)
2. Cao, P., Chen, Y., Liu, K., Zhao, J., Liu, S.: Adversarial transfer learning for Chinese named entity recognition with self-attention mechanism. In: Proceedings of the 2018 Conference on Empirical Methods in Natural Language Processing, pp. 182–192 (2018)
3. Chen, A., Peng, F., Shan, R., Sun, G.: Chinese named entity recognition with conditional probabilistic models. In: Proceedings of the Fifth SIGHAN Workshop on Chinese Language Processing, pp. 173–176 (2006)
4. Collobert, R., Weston, J., Bottou, L., Karlen, M., Kavukcuoglu, K., Kuksa, P.: Natural language processing (almost) from scratch. J. Mach. Learn. Res. **12**(Aug), 2493–2537 (2011)

5. Ding, R., Xie, P., Zhang, X., Lu, W., Li, L., Si, L.: A neural multi-digraph model for Chinese NER with gazetteers. In: Proceedings of the 57th Annual Meeting of the Association for Computational Linguistics, Florence, Italy, pp. 1462–1467. Association for Computational Linguistics, July 2019. https://doi.org/10.18653/v1/P19-1141
6. Dong, C., Zhang, J., Zong, C., Hattori, M., Di, H.: Character-based LSTM-CRF with radical-level features for Chinese named entity recognition. In: Lin, C.-Y., Xue, N., Zhao, D., Huang, X., Feng, Y. (eds.) ICCPOL/NLPCC 2016. LNCS (LNAI), vol. 10102, pp. 239–250. Springer, Cham (2016). https://doi.org/10.1007/978-3-319-50496-4_20
7. He, H., Sun, X.: F-score driven max margin neural network for named entity recognition in Chinese social media. arXiv preprint arXiv:1611.04234 (2016)
8. He, H., Sun, X.: A unified model for cross-domain and semi-supervised named entity recognition in Chinese social media. In: Thirty-First AAAI Conference on Artificial Intelligence (2017)
9. Hochreiter, S., Schmidhuber, J.: Long short-term memory. Neural Comput. **9**(8), 1735–1780 (1997)
10. Huang, Z., Xu, W., Yu, K.: Bidirectional LSTM-CRF Models for Sequence Tagging. arXiv:1508.01991 [cs], August 2015
11. Isozaki, H., Kazawa, H.: Efficient support vector classifiers for named entity recognition. In: International Conference on Computational Linguistics, pp. 1–7. Association for Computational Linguistics (2002)
12. Lafferty, J., McCallum, A., Pereira, F.C.: Conditional random fields: probabilistic models for segmenting and labeling sequence data (2001)
13. Lample, G., Ballesteros, M., Subramanian, S., Kawakami, K., Dyer, C.: Neural architectures for named entity recognition. CoRR abs/1603.01360 (2016). http://arxiv.org/abs/1603.01360
14. Liu, T., Yao, J.Q., Lin, C.Y.: Towards improving neural named entity recognition with gazetteers. In: Proceedings of the 57th Annual Meeting of the Association for Computational Linguistics, pp. 5301–5307 (2019)
15. Ma, X.: End-to-end sequence labeling via bi-directional LSTM-CNNs-CRF. arXiv preprint arXiv:1603.01354 (2016)
16. Peng, N., Dredze, M.: Named entity recognition for Chinese social media with jointly trained embeddings. In: Proceedings of the 2015 Conference on Empirical Methods in Natural Language Processing, pp. 548–554 (2015)
17. Peng, N., Dredze, M.: Improving named entity recognition for Chinese social media with word segmentation representation learning. arXiv preprint arXiv:1603.00786, pp. 149–155 (2016). http://aclweb.org/anthology/P16-2025
18. Rei, M., Crichton, G.K., Pyysalo, S.: Attending to characters in neural sequence labeling models. arXiv preprint arXiv:1611.04361 (2016)
19. Shao, Y., Hardmeier, C., Tiedemann, J., Nivre, J.: Character-based joint segmentation and POS tagging for Chinese using bidirectional RNN-CRF. arXiv preprint arXiv:1704.01314 (2017)
20. Shen, Y., Yun, H., Lipton, Z.C., Kronrod, Y., Anandkumar, A.: Deep active learning for named entity recognition. arXiv preprint arXiv:1707.05928 (2017)
21. Vaswani, A., et al.: Attention Is All You Need. arXiv:1706.03762 [cs], June 2017
22. Xiang, Y., et al.: Chinese named entity recognition with character-word mixed embedding. In: Proceedings of the 2017 ACM on Conference on Information and Knowledge Management, pp. 2055–2058. ACM (2017)

23. Xu, M., Jiang, H., Watcharawittayakul, S.: A local detection approach for named entity recognition and mention detection. In: Proceedings of the 55th Annual Meeting of the Association for Computational Linguistics (Volume 1: Long Papers), vol. 1, pp. 1237–1247 (2017)
24. Yang, F., Zhang, J., Liu, G., Zhou, J., Zhou, C., Sun, H.: Five-stroke based CNN-BiRNN-CRF network for Chinese named entity recognition. In: Zhang, M., Ng, V., Zhao, D., Li, S., Zan, H. (eds.) NLPCC 2018. LNCS (LNAI), vol. 11108, pp. 184–195. Springer, Cham (2018). https://doi.org/10.1007/978-3-319-99495-6_16
25. Yang, Y., Zhang, M., Chen, W., Zhang, W., Wang, H., Zhang, M.: Adversarial Learning for Chinese NER from Crowd Annotations. arXiv:1801.05147 [cs], January 2018
26. Zhang, Y., Yang, J.: Chinese NER using lattice LSTM. arXiv preprint arXiv:1805.02023 (2018)
27. Zhou, J., Qu, W., Zhang, F.: Chinese named entity recognition via joint identification and categorization. Chin. J. Electron. **22**(2), 225–230 (2013)
28. Zhu, Y., Wang, G., Karlsson, B.F.: CAN-NER: Convolutional Attention Network for Chinese Named Entity Recognition. arXiv:1904.02141 [cs], April 2019
29. Zukov-Gregoric, A., Bachrach, Y., Minkovsky, P., Coope, S., Maksak, B.: Neural named entity recognition using a self-attention mechanism. In: 2017 IEEE 29th International Conference on Tools with Artificial Intelligence (ICTAI), pp. 652–656. IEEE (2017)

VGCN-BERT: Augmenting BERT with Graph Embedding for Text Classification

Zhibin Lu(✉), Pan Du, and Jian-Yun Nie

University of Montreal, Montreal, Canada
{zhibin.lu,pan.du}@umontreal.ca, nie@iro.umontreal.ca

Abstract. Much progress has been made recently on text classification with methods based on neural networks. In particular, models using attention mechanism such as BERT have shown to have the capability of capturing the contextual information within a sentence or document. However, their ability of capturing the global information about the vocabulary of a language is more limited. This latter is the strength of Graph Convolutional Networks (GCN). In this paper, we propose VGCN-BERT model which combines the capability of BERT with a Vocabulary Graph Convolutional Network (VGCN). Local information and global information interact through different layers of BERT, allowing them to influence mutually and to build together a final representation for classification. In our experiments on several text classification datasets, our approach outperforms BERT and GCN alone, and achieve higher effectiveness than that reported in previous studies.

Keywords: Text classification · BERT · Graph Convolutional Networks

1 Introduction

Text classification is a fundamental problem in natural language processing (NLP) and has been extensively studied in many real applications. In recent years, we witnessed the emergence of text classification models based on neural networks such as convolutional neural networks (CNN) [15], recurrent neural networks (RNN) [13], and various models based on attention [27]. BERT [8] is one of the self-attention models that uses multi-task pre-training technique based on large corpora. It often achieves excellent performance, compared to CNN/RNN models and traditional models, in many tasks [8] such as Named-entity Recognition (NER), text classification and reading comprehension.

The deep learning models excel by embedding both semantic and syntactic information in a learned representation. However, most of them are known to be limited in encoding long-range dependency information of the text [2]. The utilization of self-attention helps alleviate this problem, but the problem still remains. The problem stems from the fact that the representation is generated

J. M. Jose et al. (Eds.): ECIR 2020, LNCS 12035, pp. 369–382, 2020.
https://doi.org/10.1007/978-3-030-45439-5_25

from a sentence or a document only, without taking into account explicitly the knowledge about the language (vocabulary). For example, in the movie review below:

> *"Although it's a bit smug and repetitive, this documentary engages your brain in a way few current films do."*

both negative and positive opinions appear in the sentence. Yet the positive attitude *"a way few current films do"* expresses a very strong opinion of the innovative nature of the movie in an implicit way. Without connecting this expression more explicitly to the meaning of *"innovation"* in the context of movie review comments, the classifier may underweight this strong opinion and the sentence may be wrongly classified to be negative. On this example, self-attention that connects the expression to other tokens in the sentence may not help.

In recent studies, approaches have also been developed to take into account the global information between words and concepts. The most representative work is Graph Convolutional Networks (GCN) [16] and its variant Text GCN [32], in which words in a language are connected in a graph. By performing convolution operations on neighbor nodes in the graph, the representation of a word will incorporate those of the neighbors, allowing to integrate the global context of a domain-specific language to some extent. For example, the meaning of *"new"* can be related to that of *"innovation"* and *"surprised"* through the connections between them. However, GCNs that only take into account the global vocabulary information may fail to capture local information (such as word order), which is very important in understanding the meaning of a sentence. This is shown in the following examples, where the position of *"work"* in the sentence will change the meaning depending on its context:

- *"skip work to see it at the first opportunity."*
- *"skip to see it, work at the first opportunity."*

In this paper, inspired by GCN [16,32] and self-attention mechanism in BERT, we propose to combine the strengths of both mechanisms in the same model. We first construct a graph convolutional network on the vocabulary graph based on the word co-occurrence information, which aims at encoding the global information of the language, then feed the graph embedding and word embedding together to a self-attention encoder in BERT. The word embedding and graph embedding then interact with each other through the self-attention mechanism while learning the classifier. This way, the classifier can not only make use of both local information and global information, but also allow them to guide each other via the attention mechanism so that the final representation built up for classification will integrate gradually both local and global information. We also expect that the connections between words in the initial vocabulary graph can be spread to more complex expressions in the sentence through the layers of self-attention.

We call the proposed model VGCN-BERT. Our source code is available at https://github.com/Louis-udm/VGCN-BERT.

We carry out experiments on 5 datasets of different text classification tasks (sentiment analysis, grammaticality detection, and hate speech detection). On all these datasets, our approach is shown to outperform BERT and GCN alone.

The contribution of this work is twofold:

- *Combining global and local information*: There has not been much work trying to combine local information captured by BERT and global information of a language. We demonstrate that their combination is beneficial.
- *Interaction between local and global information through attention mechanism*: We propose a tight integration of local information and global information, allowing them to interact through different layers of networks.

2 Related Work

2.1 Self-attention and BERT

As aforementioned, attention mechanisms [28,31] based on various deep neural networks, in particular the self-attention mechanism proposed by Vaswan et al. [27], have greatly improved the performance in text classification tasks. The representation of a word acquired through self-attention can incorporate the relationship between the word and all other words in a sentence by fusing the representations of the latter.

BERT (Bidirectional Encoder Representations from Transformers) [8], which leverages a multi-layer multi-head self-attention (called transformer) together with a positional word embedding, is one of the most successful deep neural network model for text classification in the past years. The attention mechanism in each layer of the encoder enhances the new representation of the input data with contextual information by paying multi-head attentions to different parts of the text. A pre-trained BERT model based on 800M words from BooksCorpus and 2,500M words from English Wikipedia is made available. It has also been widely used in many NLP tasks, and has proven effective. However, as most of other attention-based deep neural networks, BERT mainly focuses on local consecutive word sequences, which provides local context information. That is, a word is placed in its context, and this generates a contextualized representation. However, it may be difficult for BERT to account for the global information of a language.

2.2 Graph Convolutional Networks (GCN)

Global relations between words in a language can be represented as a graph, in which words are nodes and edges are relations. Graph Neural Network (GNN) [2,5] based on such a graph is good at capturing the general knowledge about the words in a language. A number of variants of GNN have been proposed and applied to text classification tasks [7,12,16,21,33], of which Kipf et al. [16] creatively presented Graph Convolutional networks (GCN) based on spectral graph theory. GCN first builds a symmetric adjacency matrix based on a given relationship graph (such as a paper citation relationship), and then the representation of each node is fused according to the neighbors and corresponding relationships in the graph during the convolution operation.

Text GCN is a special case of GCN for text classification proposed by Yao et al. [32] recently. Different from general GCN, it is based on a heterogeneous graph where both words and documents are nodes. The relationships among nodes, however, are measured in three different ways, which are co-occurrence relations among words, tf-idf measure between documents and words, and self similarity among documents. In terms of convolution, Text GCN uses the same algorithm as GCN. GCN and its variants are good at convolving the global information in the graph into a sentence, but they do not take into account local information such as the order between words. When word order and other local information are important, GCN may be insufficient. Therefore, it is natural to combine GCN with a model capturing local information such as BERT.

2.3 Existing Combinations of GCN and BERT

Some recent studies have combined GCN with BERT. Shang et al. [23] applied a combination to the medication recommendation task, which predict a medical code given the electronic health records (EHR), i.e., a sequence of historical medical codes, of a patient. They first embed the medical codes from a medical ontology using Graph Attention Networks (GAT), then feed the embedding sequence of the medical code in an EHR into BERT for code prediction. Nevertheless, the order in the code sequence is discarded in the transformer since it is not applicable in their scenario, making it incapable of capturing all the local information as in our text classification tasks.

Jong et al. [14] proposed another combination to the citation recommendation task using paper citation graphs. This model simply concatenates the output of GCN and the output of BERT for downstream predictive tasks. We believe that interactions between the local and global information are important and can benefit the downstream prediction tasks. In fact, through layers of interactions, one could allow the information captured in GCN be applied to the input text, and the representation of the input text be spread over GCN. This will produce the effect we illustrated in the earlier example of movie review (*a way few current films do* vs. *innovation*). This is the approach we propose in this paper.

One may question about the necessity to explicitly use graph embedding to cope with global dependency information, as some studies [18,25] have shown that word embedding trained on a corpus, such as Word2Vec [19], GloVe [22], FastText [10], can capture some global connections between words in a language. We believe that a vocabulary graph can still provide additional information given the fact that the connections between words observed in word embeddings are limited within a small text window (usually 5 words). Long-range connections are missing. In addition, by building a vocabulary graph on an application-specific document collection, one can capture application-dependent dependencies, in addition to the general dependencies in the pre-trained models.

3 Proposed Method

The global language information can take multiple forms. In this paper, we consider lexical relations in a language, i.e. a vocabulary graph. Any vocabulary graph can be used to complement BERT (e.g. Wordnet). In this paper, we consider a graph constructed using word co-occurrences with documents. Local information from a text is captured by BERT. The interaction between them is achieved by first selecting the relevant part of the global vocabulary graph according to the input sentence and transforming it into an embedding representation. We use multiple layers of attention mechanism on concatenated representation of input text and the graph. These processes are illustrated in Fig. 1. We will provide more details in the following subsections.

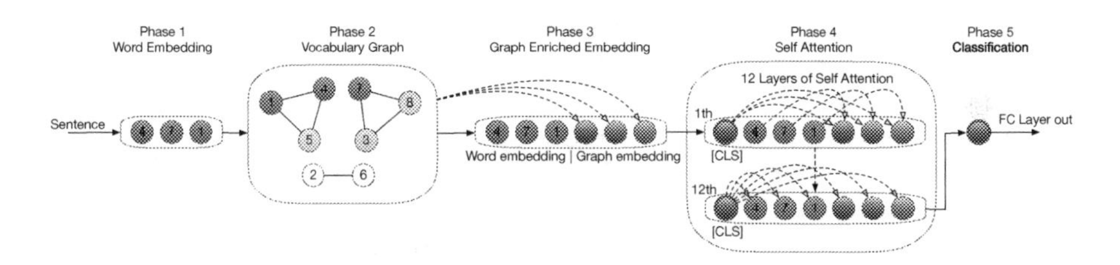

Fig. 1. Illustration of VGCN-BERT. The embeddings of input sentence (Phase 1) are combined with the vocabulary graph (Phase 2) to produce a graph embedding, which is concatenated to the input sentence (Phase 3). Note that from the vocabulary graph, only the part relevant to the input is extracted and embedded. In Phase 4, several layers of self-attention are applied to the concatenated representation, allowing interactions between word embeddings and graph embedding. The final embedding at the last layer is fed in a fully connected layer (Phase 5) for classification.

3.1 Vocabulary Graph

Our vocabulary graph is constructed using normalized point-wise mutual information (NPMI) [3], as shown in Eq. 1:

$$\mathrm{NPMI}(i,j) = -\frac{1}{\log p(i,j)} \log \frac{p(i,j)}{p(i)p(j)} \tag{1}$$

where i and j are words, $p(i,j) = \frac{\#W(i,j)}{\#W}$, $p(i) = \frac{\#W(i)}{\#W}$, $\#W(*)$ is the number of sliding windows containing a word or a pair of words, and $\#W$ is the total number of sliding windows. To obtain long-range dependency, we set the window to the whole sentence. The range of value of NPMI is [-1,1]. A positive NPMI value implies a high semantic correlation between words, while a negative NPMI value indicates little or no semantic correlation. In our approach, we create an edge between two words if their NPMI is larger than a threshold. Our experiments show that the performance is better when the threshold is between 0.0 and 0.3.

3.2 Vocabulary GCN

A general GCN [16] is a multi-layer (usually 2 layers) neural network that convolves directly on a graph and induces embedding vectors of nodes based on properties of their neighborhoods. Formally, consider a graph $G = (P, E)^1$, where P (with $|P| = n$) and E are sets of nodes and edges, respectively. For a single convolutional layer of GCN, the new representation is calculated as follows:

$$H = \tilde{A}XW, \tag{2}$$

where $X \in \mathbb{R}^{n \times m}$ is the input matrix with n nodes and m dimensions of the feature, $W \in \mathbb{R}^{m \times h}$ is a weight matrix, $\tilde{A} = D^{-\frac{1}{2}}AD^{-\frac{1}{2}}$ is the normalized symmetric adjacency matrix, where $D_{ii} = \sum_j A_{ij}$. The normalization operation for A is to avoid numerical instabilities and exploding/vanishing gradients when used in a deep neural network model [16].

The graph nodes of GCN are "task entities" such as documents that need to be classified. It requires all entities, including those from training set, validation set, and test set, to be presented in the graph, so that no node representation is missing in downstream tasks. This limits the application of GCN in many predictive tasks, where the test data is unseen during the training process.

In our case, we aim to convolve the related words instead of the documents in the corpus for classification. Therefore, the graph of our proposed GCN is constructed on the vocabulary instead of the documents. Thus, for a single document, assuming the document is a row vector $\boldsymbol{x}$ consisting of words in the vocabulary, a layer of convolution is defined in Eq. 3:

$$\boldsymbol{h} = (\tilde{A}\boldsymbol{x}^T)^T W = \boldsymbol{x}\tilde{A}W, \tag{3}$$

where $\tilde{A}^T = \tilde{A}$ represent the vocabulary graph. $\boldsymbol{x}\tilde{A}$ extracts the part of vocabulary graph relevant to the input sentence $\boldsymbol{x}$. W holds the weights of the hidden state vector for the single document, with dimension $|V| \times h$. Then for m documents in a mini-batch, the one-layer graph convolution in Eq. 3 becomes:

$$H = X\tilde{A}W, \tag{4}$$

and the corresponding 2-layer Vocabulary GCN with ReLU function is as follows:

$$\mathbf{VGCN} = \mathrm{ReLU}(X_{mv}\tilde{A}_{vv}W_{vh})W_{hc}, \tag{5}$$

where m is the mini-batch size, v is the vocabulary size, h is the hidden layer size, c the class size or sentence embedding size. Every row of X_{mv} is a vector containing document features, which can be a bag-of-words vector, or word embedding of BERT. The above equation aims to produce a layer of convolution of the graph, which captures the part of the graph relevant to the input (through $X_{mv}\tilde{A}_{vv}$), then performs 2 layers of convolution, combining words from input sentence with their related words in vocabulary graph.

[1] In order to distinguish from notations $(v, V, |V|)$ of vocabulary, this paper uses notations $(p, P, |P|)$ to represent the point(vertex) of the graph.

3.3 Integrating VGCN into BERT

When BERT is applied to text classification, a typical solution contains three parts. The first part is the word embedding module with the position information of the word; the second part is the transformer module using multi-layer multi-head self-attention stacking; and the third part is the fully connected layer using the output sentence embedding for classification.

Self-attention operates with a query Q against a key K and value V pair. The attention score is calculated as follows:

$$\text{Attention}(Q, K, V) = \text{Softmax}\left(\frac{QK^T}{\sqrt{d_k}}\right)V, \tag{6}$$

where the denominator is a scaling factor used to control the scale of the attention score, d_k is the dimension of the query and key vectors. Using these attention scores, every word can get a weighted vector representation encoding the contextual information.

Instead of using only word embeddings of the input sentence in BERT, we feed both the vocabulary graph embedding obtained by Eq. 5 and the sequence of word embeddings to BERT transformer. This way, not only the order of the words in the sentence is retained, but also the background information obtained by VGCN is utilized. The overall VGCN-BERT model is schematically illustrated in Fig. 1. Through the attention score calculated by Eq. 6, local embedding and global embedding are fully integrated after layer-by-layer interaction in 12-layer and 12-heads self-attention encoder. The corresponding VGCN can then be formulated as:

$$\mathbf{G_{embedding}} = \text{ReLU}(X_{mev}\tilde{A}_{vv}W_{vh})W_{hg}, \tag{7}$$

where W_{hg}, which was originally used for classification, becomes the output of size g of graph embedding (hyperparameter) whose dimension is the same as every word embedding; m is the size of the mini-batch; e is the dimension of word embedding, and v is the vocabulary size.

4 Experiment

We evaluate VGCN-BERT and compare it with baseline models on 5 datasets to verify whether our model can leverage both local and global information.

4.1 Baselines

In addition to the original BERT model, we also use several other neural network models as baselines.

- **MLP:** Multilayer perceptron with 2 hidden layers (512 and 100 nodes), and bag-of-words model with TF weighting.

- **Bi-LSTM** [11]**:** The BERT's pre-trained word embeddings are used as input to the Bi-LSTM model.
- **Text GCN:** The original Text GCN model uses the same input feature as MLP model, and we use the same training parameters as in [32].
- **VGCN:** This model only uses VGCN, corresponding to Eq. 7, but the output dimension becomes the class size. BERT's pre-trained word embeddings are used as input. The output of VGCN is relayed to a fully connected layer with Softmax function to produce the classification score. This model only uses the global information from vocabulary graph.
- **BERT:** We use the small version (Bert-base-uncased) pre-trained BERT [8].
- **Vanilla-VGCN-BERT:** Vanilla combination of BERT and VGCN is similar to [14], which produces two separate representations through BERT and GCN, and then concatenates them. ReLU and a fully connected layer are applied to the combined representation for classification. The main difference of this model with ours is that it does not allow interactions between the input text and the graph.

4.2 Datasets

We ran our experiments on the following five datasets:

- **SST-2.** The Stanford Sentiment Treebank is a binary single-sentence classification task consisting of sentences extracted from movie reviews with human annotations of their sentiment [24]. We use the public version[2] which contains 6,920 examples in training set, 872 in validation set and 1,821 in test set, for a total of 4,963 positive reviews and 4,650 negative reviews. The average length of reviews is 19.3 words.
- **MR** is also a movie review dataset for binary sentiment classification, in which each review only contains one sentence [20][3]. We used the public version in [26][4]. It contains 5,331 positive and 5,331 negative reviews. The average length is 21.0 words.
- **CoLA.** The Corpus of Linguistic Acceptability is a binary single-sentence classification task. CoLA is manually annotated for acceptability (grammaticality) [29]. We use the public version which contains 8,551 training data and 1,043 ation data[5], for a total of 6,744 positive and 2,850 negative cases. The average length is 7.7 words. Since we do not have the label for the test set, we split 5% of the training set as validation set and use the original validation set as the test set.
- **ArangoHate** [1] is a resampled dataset merging the datasets from [30] and [6]. It contains 2,920 hateful documents and 4,086 normal documents. The average length is 13.3 words. Since the dataset is not pre-divided into training, validation and test sets, we randomly split it into three sets at the ratio of 85:5:10.

[2] https://github.com/kodenii/BERT-SST2.
[3] http://www.cs.cornell.edu/people/pabo/movie-review-data/.
[4] https://github.com/mnqu/PTE/tree/master/data/mr.
[5] https://github.com/nyu-mll/GLUE-baselines.

- **FountaHate** is a large four-label dataset for hate speech and offensive language detection [9]. It contains 99,996[6] tweets with cross-validated labels and is classified into 4 labels: normal (53,851), spam (14,030), hateful (27,150) and abusive (4,965). The average length is 15.7 words. Since the dataset is not pre-divided into training, validation and test sets, we split it into three sets at the ratio of 85:5:10 after shuffle.

4.3 Preprocessing and Setting

We removed URL strings and @-mentions to retain the text content, then the text was lower-cased and tokenized using NLTK's *TweetTokenizer*[7]. We use BERTTokenizer function to split text, so that the vocabulary for GCN is always a subset of pre-trained BERT's vocabulary. When computing NPMI on a dataset, the whole sentence is used as the text window to build the vocabulary graph. The threshold of NPMI is set as 0.2 for all datasets to filter out non-meaningful relationships between words.

In the VGCN-BERT model, the graph embedding output size is set as 16, and the hidden dimension of graph embedding as 128. We use the *Bert-base-uncased* version of pre-trained BERT, and set the max sequence length as 200. The model is then trained in 9 epochs with a dropout rate of 0.2. The following are other parameter settings for different datasets.

- SST-2: mini-batch = 16, learning rate = 1e−5, L_2 loss weight decay = 0.01.
- CoLA and MR: mini-batch = 16, learn. rate = 8e−6, L_2 loss decay = 0.01.
- ArangoHate: mini-batch = 16, learn. rate = 1e−5, and L_2 loss decay = 1e−3.
- FountaHate: mini-batch = 12, learn. rate = 4e−6, and L_2 loss decay = 2e−4.

These parameters are set based on our preliminary tests. We also use the default fine-tuning learning rate and L_2 loss weight decay as in [8]. The baseline methods are set with the same parameters as in the original papers.

4.4 Loss Function

We use the cross-entropy as the loss function for all models, except for FountaHate dataset where we use the mean squared error as the loss function in order to leverage the annotators' voting information.

We use Adam as training optimizer for all models. For cases where the label distributions are uneven (CoLA (2.4:1), ArangoHate (1.4:1) and FountaHate (10.9:5.5:2.8:1)), *comput_class_weight* function[8] from scikit-learn [4] is used as the weighted loss function. The weight of each the classes (W_c) is calculated by

$$W_{classes} = \frac{\#dataset}{\#classes \cdot \#every\_class}, \tag{8}$$

[6] The final version provided by the author is more than the one described in the paper.

[7] http://www.nltk.org/api/nltk.tokenize.html.

[8] https://scikit-learn.org/stable/modules/generated/sklearn.utils.class_weight.compute_class_weight.html.

where $\#dataset$ is the total size of dataset and $\#classes$ is the number of classes and $\#every\_class$ is the count of every class.

4.5 Evaluation Metrics

We adopt the two most widely used metrics to evaluate the performance of the classifiers - the weighted average F1-score, and the macro F1-score [17].

$$\textbf{Weighted avg F1} = \sum_{i=1}^{C} F1_{c_i} * W_{c_i}, \qquad \textbf{Macro F1} = \frac{1}{C}\sum_{i=1}^{C} F1_{c_i} \tag{9}$$

4.6 Experimental Result

The main results on weighted average F1-Score and macro F1-Score on test sets are presented in Table 1. The main observation is that VGCN-BERT outperforms all the baseline models (except against Vanilla-VGCN-BERT on MR dataset). In particular, it outperforms both VGCN and BERT alone, confirming the advantage to combine them.

Among the models that only use local information, we see that BERT outperforms MLP, Bi-LSTM. Between the models that exploit a vocabulary graph - VGCN and Text-GCN, the performance is similar.

Vanilla-VGCN-BERT and VGCN-BERT are two models that combine local and global information. In general, these models perform better than the other baseline models. This result confirms the benefit of combining local information and global information.

Comparing VGCN-BERT with Vanilla-VGCN-BERT, we see that the former generally performs better. The difference is due to the interactions between local and global information. The superior performance of VGCN-BERT clearly shows the benefit of allowing interactions between the two types of information.

4.7 Visualization

To better understand the behaviors of BERT, and its combination with VGCN, we visualize the attention distribution of the [CLS] token in the self-attention module of BERT, VGCN-BERT and Vanilla-VGCN-BERT models. As the vocabulary graph is embedded into vectors of 16 dimensions, it is not obvious to show what meaning corresponds to each dimension. To facilitate our understanding, we show the top two words from the sub-graph related to the input sentence, which are strongly connected to each of the 16 dimensions of graph embedding. More specifically, w each word embedding of a document is input to Eq. 7, we only need to broadcast the result of XA and element-multiply it by W to obtain the representation value of the words involved. The equation for obtain the involved words' id is as follow:

$$Z = (\boldsymbol{x}A)^T \odot W, \tag{10}$$

$$\text{IDs involved} = \arg\text{sort}(Z[:, g]), \tag{11}$$

Table 1. Weighted average F1-Score and (Macro F1-score) on the test sets. We run 5 times under the same preprocessing and random seed. Macro F1-score and Weighted F1-Score are the same on SST-2 and MR. Bold indicates the highest score and underline indicates the second highest score.

Model	SST-2	MR	CoLA	ArangoHate	FountaHate
MLP	80.78	75.55	61.39 (53.20)	84.71 (84.42)	79.22 (65.33)
Text-GCN	80.45	75.67	56.18 (52.30)	84.77 (84.43)	78.74 (64.54)
Bi-LSTM	81.32	76.39	62.88 (55.25)	84.92 (84.58)	79.04 (65.13)
VGCN	81.64	76.42	63.59 (54.82)	85.97 (85.69)	79.00 (64.04)
BERT	<u>91.49</u>	86.24	<u>81.22</u> (<u>77.02</u>)	87.99 (87.75)	80.59 (66.61)
Vanilla-VGCN-BERT	91.38	**86.49**	80.70 (76.30)	<u>88.01</u> (<u>87.79</u>)	<u>81.11</u> (<u>67.86</u>)
VGCN-BERT	**91.93**	<u>86.35</u>	**83.68** (**80.46**)	**88.43** (**88.22**)	**81.26** (**68.45**)

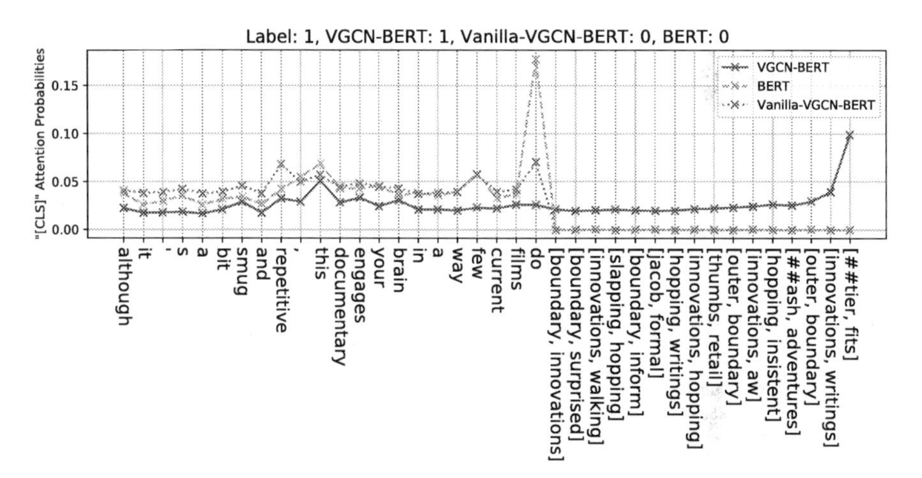

Fig. 2. Visualization of the attention that the token [CLS] (used as sentence embedding) pays to other tokens. The first part corresponds to word embeddings of the sentence. The second part is the graph embedding. [word1, word2] indicates the approximate meaning of a dimension in graph embedding.

where x is a document in row vector, $g \in [1, G]$, $G = 16$ is the size of graph embedding. For example, the first dimension of graph embedding shown in Fig. 2 corresponds roughly to the meaning of *"[boundary, innovations]"*.

In Fig. 2, we show the attention paid to each word (embedding) and each dimension of graph embedding (second part). As BERT does not use graph embedding, the attention paid to graph embedding is 0. In VGCN-BERT, we see that graph embedding draws an important part of attention.

For the movie review *"Although it's a bit smug and repetitive, this documentary engages your brain in a way few current films do."*, the first half of the sentence is explicitly negative, while the remaining part expresses a positive attitude in an implicit way, which makes the sentence difficult to judge. For this example, BERT pays a very high attention to *"do"*, and a quite high attention to *"this"*. These words do not bear much meaning in sentiments. The final classification results by BERT is 0 (negative) while the true label is 1 (positive).

Vanilla-VGCN-BERT concatenates graph embedding with BERT without interaction between them. We can see that still no attention is paid to graph embedding, showing that such a simplistic combination cannot effectively leverage vocabulary information.

Finally, for VGCN-BERT, we see that a considerable part of attention is paid to graph embedding. The graph embedding is produced by integrating gradually the local information in the sentence with the global information in the graph. At the end, several dimensions of the graph embedding imply the meaning of *"innovation"*, to which quite high attentions are paid. This results in classifying the sentence to the correct class (positive).

The meaning of *"innovation"* is not produced immediately, but after a certain number of layers in BERT. In fact, through the layers of BERT, local information in the input sentence is combined to generate a higher level representation. In this example, at a certain layer, the expression *"a way few current films do"* is grouped and represented as an embedding similar to the meaning of *"innovation"*. From then, the meaning related to *"innovation"* in the graph embedding is capture through self-attention, and reinforced later on through interactions between the local and global information.

5 Conclusion and Future Work

In this study, we propose a new VGCN-BERT model to integrate a vocabulary graph embedding module with BERT. The goal is to complement the local information captured by BERT with the global information on the vocabulary, and allow both types of information to interact through the layers of attention mechanism. Our experiments on classification on 5 datasets show that the graph embedding does bring useful global information to BERT and this improves the performance. In comparison with BERT and VGCN alone, our model can clearly lead to better results, showing that VGCN-BERT can indeed take advantage of both mechanisms.

As future work, we will consider using other types of vocabulary graph such as Wordnet, in addition to a graph created by co-occurrences. We believe that Wordnet contains useful connections between words that NPMI cannot cover. It is thus possible to combine several lexical resources into the vocabulary graph.

References

1. Arango, A., Perez, J., Poblete, B.: Hate Speech Detection is Not as Easy as You May Think: A Closer Look at Model Validation. Paris (2019)
2. Battaglia, P.W., et al.: Relational inductive biases, deep learning, and graph networks. arXiv preprint arXiv:1806.01261 (2018)

3. Bouma, G.: Normalized (pointwise) mutual information in collocation extraction. In: Proceedings of the Biennial GSCL Conference 2009, University of Potsdam (2009). https://pdfs.semanticscholar.org/1521/8d9c029cbb903ae7c729b2c644c24994c201.pdf
4. Buitinck, L., et al.: API design for machine learning software: experiences from the scikit-learn project. In: ECML PKDD Workshop: Languages for Data Mining and Machine Learning, pp. 108–122 (2013)
5. Cai, H., Zheng, V.W., Chang, K.: A comprehensive survey of graph embedding: problems, techniques and applications. IEEE Trans. Knowl. Data Eng. **30**(9), 1616–1637 (2018)
6. Davidson, T., Warmsley, D., Macy, M., Weber, I.: Automated hate speech detection and the problem of offensive language. In: Proceedings of the 11th International AAAI Conference on Web and Social Media, ICWSM 2017, pp. 512–515 (2017)
7. Defferrard, M., Bresson, X., Vandergheynst, P.: Convolutional neural networks on graphs with fast localized spectral filtering. In: NIPS, pp. 3844–3852 (2016)
8. Devlin, J., Chang, M.W., Lee, K., Toutanova, K.: Bert pre-training of deep bidirectional transformers for language understanding. arXiv preprint arXiv:1810.04805 (2018)
9. Founta, A.M., et al.: Large scale crowdsourcing and characterization of twitter abusive behavior. In: 11th International Conference on Web and Social Media, ICWSM 2018. AAAI Press (2018)
10. Grave, E., Mikolov, T., Joulin, A., Bojanowski, P.: Bag of tricks for efficient text classification. In: Proceedings of the 15th Conference of the European Chapter of the Association for Computational Linguistics, EACL 2017, Valencia, Spain, 3–7 April 2017, Volume 2: Short Papers, pp. 427–431 (2017). https://www.aclweb.org/anthology/E17-2068/
11. Graves, A., Mohamed, A., Hinton, G.: Speech recognition with deep recurrent neural networks. In: Acoustics, Speech and Signal Processing (ICASSP), pp. 6645–6649 (2013)
12. Henaff, M., Bruna, J., LeCun, Y.: Deep convolutional networks on graph-structured data. arXiv preprint arXiv:1506.05163 (2015)
13. Hochreiter, S., Schmidhuber, J.: Long short-term memory. Neural Comput. **9**(8), 1735–1780 (1997)
14. Jeong, C., Jang, S., Shin, H., Park, E., Choi, S.: A context-aware citation recommendation model with BERT and graph convolutional networks. arXiv:1903.06464 (2019)
15. Kim, Y.: Convolutional neural networks for sentence classification. In: EMNLP, pp. 1746–1751 (2014)
16. Kipf, T.N., Welling, M.: Semi-supervised classification with graph convolutional networks. In: ICLR (2017)
17. Lever, J., Krzywinski, M., Altman, N.: Classification evaluation. Nat. Methods **13**(8), 603–604 (2016). https://doi.org/10.1038/nmeth.3945
18. Levy, O., Goldberg, Y.: Neural word embedding as implicit matrix factorization. In: Advances in Neural Information Processing Systems 27: Annual Conference on Neural Information Processing Systems 2014, Montreal, Quebec, Canada, 8–13 December 2014, pp. 2177–2185 (2014). http://papers.nips.cc/paper/5477-neural-word-embedding-as-implicit-matrix-factorization
19. Mikolov, T., Chen, K., Corrado, G., Dean, J.: Efficient estimation of word representations in vector space. In: 1st International Conference on Learning Representations, ICLR 2013, Scottsdale, Arizona, USA, 2–4 May 2013, Workshop Track Proceedings (2013). http://arxiv.org/abs/1301.3781

20. Pang, B., Lee, L.: Seeing stars: exploiting class relationships for sentiment categorization with respect to rating scales. In: ACL, pp. 115–124 (2005)
21. Peng, H., et al.: Large-scale hierarchical text classification with recursively regularized deep graph-CNN. In: WWW, pp. 1063–1072 (2018)
22. Pennington, J., Socher, R., Manning, C.D.: Glove: global vectors for word representation. In: Proceedings of the 2014 Conference on Empirical Methods in Natural Language Processing (EMNLP), vol. 14, pp. 1532–1543 (2014)
23. Shang, J., Ma, T., Xiao, C., Sun, J.: Pre-training of graph augmented transformers for medication recommendation. In: Proceedings of the Twenty-Eighth International Joint Conference on Artificial Intelligence, IJCAI 2019, Macao, China, 10–16 August 2019, pp. 5953–5959 (2019). https://doi.org/10.24963/ijcai.2019/825
24. Socher, R., et al.: Recursive deep models for semantic compositionality over a sentiment treebank. In: Proceedings of the 2013 Conference on Empirical Methods in Natural Language Processing (EMNLP), vol. 14, pp. 1631–1642 (2013)
25. Srinivasan, B., Ribeiro, B.: On the equivalence between node embeddings and structural graph representations. In: Proceedings of International Conference on Learning Representations 2020 (2020). https://openreview.net/forum?id=SJxzFySKwH
26. Tang, J., Qu, M., Mei, Q.: PTE: predictive text embedding through large-scale heterogeneous text networks. In: KDD, pp. 1165–1174. ACM (2015)
27. Vaswani, A., et al.: Attention is all you need. Long Beach (2017)
28. Wang, Y., Huang, M., Zhao, L., et al.: Attention-based LSTM for aspect-level sentiment classification. In: EMNLP, pp. 606–615 (2016)
29. Warstadt, A., Singh, A., Bowman, S.R.: Neural network acceptability judgments. arXiv preprint arXiv:1805.12471 (2018)
30. Waseem, Z.: Are You a Racist or Am I Seeing Things? Annotator Influence on Hate Speech Detection on Twitter (2016)
31. Yang, Z., Yang, D., Dyer, C., He, X., Smola, A., Hovy, E.: Hierarchical attention networks for document classification. In: NAACL, pp. 1480–1489 (2016)
32. Yao, L., Mao, C., Luo, Y.: Graph convolutional networks for text classification. In: AAAI (2019)
33. Zhang, Y., Liu, Q., Song, L.: Sentence-state LSTM for text representation. In: ACL, pp. 317–327 (2018). https://aclanthology.info/papers/P18-1030/p18-1030

Retrieval

A Computational Approach for Objectively Derived Systematic Review Search Strategies

Harrisen Scells[1](✉), Guido Zuccon[1], Bevan Koopman[2], and Justin Clark[3]

[1] The University of Queensland, St Lucia, Australia
h.scells@uq.net.au
[2] CSIRO, Brisbane, Australia
[3] Institute for Evidence-Based Healthcare, Bond University, Gold Coast, Australia

Abstract. Searching literature for a systematic review begins with a manually constructed search strategy by an expert information specialist. The typical process of constructing search strategies is often undocumented, ad-hoc, and subject to individual expertise, which may introduce bias in the systematic review. A new method for *objectively* deriving search strategies has arisen from information specialists attempting to address these shortcomings. However, this proposed method still presents a number of manual, ad-hoc interventions, and trial-and-error processes, potentially still introducing bias into systematic reviews. Moreover, this method has not been rigorously evaluated on a large set of systematic review cases, thus its generalisability is unknown. In this work, we present a computational adaptation of this proposed objective method. Our adaptation removes the human-in-the-loop processes involved in the initial steps of creating a search strategy for a systematic review; reducing bias due to human factors and increasing the objectivity of the originally proposed method. Our proposed computational adaptation further enables a formal and rigorous evaluation over a large set of systematic reviews. We find that our computational adaptation of the original objective method provides an effective starting point for information specialists to continue refining. We also identify a number of avenues for extending and improving our adaptation to further promote supporting information specialists.

Keywords: Systematic reviews · Boolean queries · Query formulation

1 Introduction

The goal of a systematic review is to synthesise *all* relevant literature for a highly focused research question. Systematic reviews are used extensively in evidence based medicine (this is the domain we consider in the rest of the paper), both for healthcare decision making and institutional policy mandates concerning health topics. While systematic reviews strive to be methodical and comprehensive, there are still a number of processes associated with them which introduce bias

J. M. Jose et al. (Eds.): ECIR 2020, LNCS 12035, pp. 385–398, 2020.
https://doi.org/10.1007/978-3-030-45439-5_26

and subjectivity. Arguably, the process which contributes the most bias is the construction of *search strategies*. The main element of a search strategy is a complex Boolean query. This is issued to one or more publication databases (e.g., PubMed, EMBASE). Retrieved studies are first screened (i.e., the title and abstracts are assessed) for potential inclusion in the review. Then, the full-text of screened studies deemed potentially relevant to the review are assessed to determine if they should be synthesised in the final review [5].

The most common method for developing a search strategy is the *conceptual* method [2,12] (although other methods have been investigated that do not produce a Boolean query [6,7]). Here, the query is formulated by dividing the research question of a systematic review into multiple high-level concepts, and then choosing suitable synonyms for each concept. Query formulation is typically performed by trained information specialists (e.g., librarians), who use domain expertise and intuition to decide, for example, what keywords to add to a query and where they should be added, what kind of field restrictions should be applied, and when to stop formulating. Often, information specialists also have access to a handful of studies (seeds) that the researchers are certain will be included in the synthesis of the review. In the conceptual approach, information specialists use the few seed studies to repeatedly gauge the effectiveness of the queries they formulate in an ad-hoc manner.

An *objective* [4,20] method for deriving systematic review search strategies has recently been proposed which aims to avoid the unrigorous, subjective aspects of the conceptual approach. In this method, a small set of 'gold standard' studies are first identified—these serve to (semi-)automatically identify keywords to add to the query, and to validate its effectiveness. The gold standard set is akin to the seed studies considered in the conceptual approach, but generally much larger (conceptual: a handful; objective: in the order of 10s-100s). Despite the name, this method is still ad-hoc and involves manual trial-and-error with respect to choosing a subset of the identified keywords to add to the query, and where to place keywords in the query. In addition, this method has only been evaluated on a handful of use-case systematic reviews, thus its effectiveness and generalisability beyond these cases is as yet unknown.

In this paper, we propose a computational adaptation of the objective methodology proposed by Hausner et al. [4] for objectively deriving medical systematic review search strategies. Our approach does not require manual human involvement, nor trial-and-error procedures, and, in fact, is capable of generating a query automatically, given a set of relevant studies as input. Furthermore, we evaluate this method on a large set of 40 systematic reviews from a collection used for the evaluation of automation methods in this context [6] and further replicate a small study by Hausner et al. [4]. We also consider the cost factors of systematic review development in our evaluation. The primary goal of this research is to develop a more transparent and less subjective method to search strategy development by computationally adapting and extending the current objective approach. Achieving total recall of the relevant literature for a study is important. However, the effectiveness of systematic reviews is often hampered by the fact that they are resource-intensive and often become out-of-date at the time of publication [21]: it takes on average 2 years and AUD$350K to create a

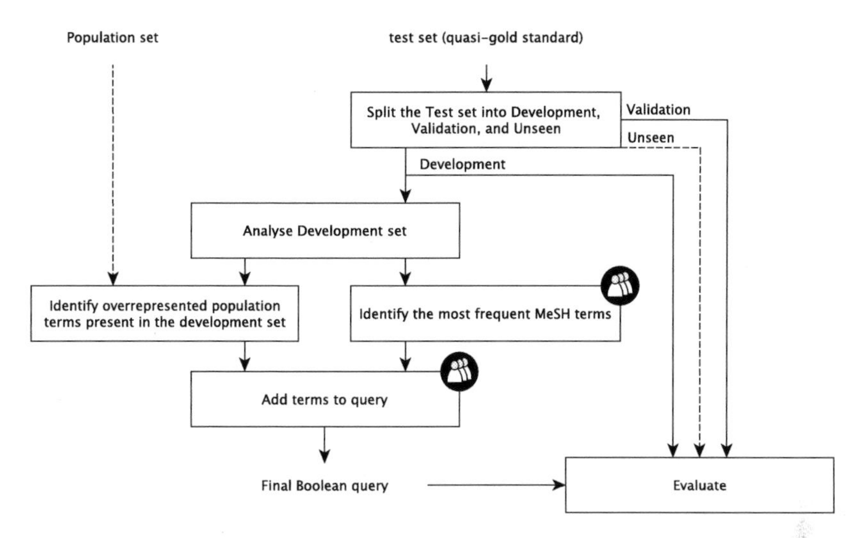

Fig. 1. The process used for deriving queries using the objective method. The dashed line signifies an extension of the objective method not in the originally proposed method. The symbol refers to the processes in the objective method which currently require manual intervention. Automating these manual processes is the main focus of this work.

systematic review [9,13] and currently only 36% of Cochrane SRs are deemed up-to-date. The largest time and monetary cost involved in systematic review creation is the *screening* phase, which is directly influenced by the number of studies retrieved by the search strategy. Furthermore, the exponentially growing body of published research casts doubts on how effective the reviews are in identifying all relevant research; potentially introducing bias.

With the method presented in this research, our overarching goal is to automatically and objectively derive suitable queries which can be used as a starting point for query formulation, to derive more effective search strategies (higher precision while maintaining recall) than manually derived queries. The method of this study is expected not to replace information specialists, but to provide support by reducing bias and subjectivity in the search development process.

2 Computational, Objective Method

Our method for automatically and objectively deriving search strategies is an adaptation of the method originally proposed by Hausner et al. [4]. A high-level process overview of the objective method is shown in Fig. 1. This figure highlights manual aspects of the original method which we computationalise, and extensions to the original method our method makes, which seek to further reduce bias. The following two sections first describe the original method, and then the specific computational adaptations and extensions we make.

2.1 The Objective Method

The objective method [4] starts by identifying studies either by scanning the references of similar, already published systematic reviews, or by issuing broad queries to medical databases (e.g., PubMed, MEDLINE) and screening a subset for identifying gold-standard references. These references form the 'test' set, which is divided into a development set ($\frac{2}{3}$) and a validation set ($\frac{1}{3}$). The title and abstract of each of the references in the test set is then analysed by ranking each term by the frequency it appears in each of the references (i.e., document frequency – DF). Next, these terms are filtered to include the top 20% of terms according to DF. At the same time, a population set (i.e., a background collection) of 10,000 studies is identified by issuing an empty search to PubMed and restricting results to the last 12 months (the default ranking of PubMed is by descending publication date). The previously filtered terms are filtered yet again to include the bottom 2% of terms according to DF. Finally, the 20 most frequent MeSH terms are identified from studies in the development set. A Boolean query is then developed by dividing terms into three categories (which form three clauses, linked with the Boolean `AND` operator): (1) terms relating to health conditions, (2) terms relating to a treatment, and (3) terms relating to the types of study design to be included. Through a time consuming process of trial-and-error, filtered terms and MeSH terms are then added to one of the three clauses of the query depending on the category of the term. Terms inside a category are combined using an `OR` operator, and the three categories are then combined with an `AND` operator. The effectiveness of the query is then compared to the validation set.

2.2 Automating the Objective Method

We propose a number of computational modifications to this method which seek to further remove human subjectivity from the process. We also improve the process by which evaluation of the resulting queries is undertaken in a number of ways. In our modified methodology, we begin with the same test set, however we split into ($\frac{2}{4}$) development, ($\frac{1}{4}$) validation, and ($\frac{1}{4}$) unseen. The unseen set is used to approximate how the query will perform on studies which are not assessed (i.e., a study which is relevant, but which may never be retrieved by the query, and therefore never screened for potential relevance). It also allows us to develop the query on the development set, tune parameters values on the validation set, and study their effectiveness on the unseen set. We then follow the same method of filtering terms using the development set and the population set, as well as identifying MeSH terms to use. To automatically assign a category for a term, the semantic type of a term is used. The semantic type is obtained automatically by mapping terms to UMLS concepts via MetaMap [1] (version 2018 with options set to their default values and using UMLS2018AA). If a term does not map to a concept in MetaMap, it is discarded. Once a semantic type is obtained for a term, it is automatically categorised in a two fold process: (i) if the semantic type is present in Table 1a, then the term is mapped accordingly [22],

Table 1. Processes for mapping terms to a Hausner et al. [4] category in a query.

Uzuner et al. [22] Relationship	Hausner et al. [4] Category
Test	→ Treatment
Treatment	→ Treatment
Diagnosis	→ Condition

(a) How a relationship as identified by Uzuner et al. [22] maps to a category.

Semantic Group	Hausner et al. [4] Category
ACTI	→ Treatment
ANAT	→ Condition
CHEM	→ Treatment
CONC	→ None
DEVI	→ Treatment
DISO	→ Condition
GENE	→ Condition
GEOG	→ Study Type
LIVB	→ Condition
OBJC	→ Treatment
OCCU	→ Condition
ORGA	→ Study Type
PHEN	→ Condition
PHYS	→ Condition
PROC	→ Treatment

(b) How a semantic group maps to a category.

MeSH top-level heading	Hausner et al. [4] Category
Anatomy	→ Condition
Organisms	→ Condition
Diseases	→ Condition
Chemicals and Drugs	→ Treatment
Analytical, Diagnostic and Therapeutic Techniques, and Equipment	→ Treatment
Psychiatry and Psychology	→ Condition
Phenomena and Processes	→ Condition
Disciplines and Occupations	→ Condition
Anthropology, Education, Sociology, and Social Phenomena	→ None
Technology, Industry, and Agriculture	→ None
Humanities	→ None
Information Science	→ Study Type
Named Groups	→ Condition
Health Care	→ None
Publication Characteristics	→ Study Type
Geographicals	→ Study Type

(c) How a top-level MeSH heading maps to a category.

(ii) if the semantic type is not present, then the semantic group of the semantic type is used to broadly categorise the term according to Table 1b. Note that in step (ii), some terms may be discarded due to the semantic group they belong to, denoted by 'None' in the table. Following this process, the identified MeSH terms are then added to one or more of the three categories according to the top-level MeSH parent in Table 1c. Once all of the identified terms are categorised, the computational assembly of the Boolean query takes place.

We also take a computational approach to the assembly of the Boolean query. A naïve approach could involve trying all combinations of terms in a category with all combinations of all other terms in all other categories (similar to the manual trial-and-error employed in the original method). The complexity of this approach, however, presents itself as infeasible: $O(n!^3)$ (where n is the number of terms for a given category, assuming all categories have the same number of terms, in the worst case). Instead, we compute the maximum number of studies in the development set retrievable using the filtered terms and MeSH terms by first representing the set of relevant studies retrieved for each category as a binary array (e.g., health conditions $\boldsymbol{c} = [1,1,1,1,1,1,1,1]$, treatments $\boldsymbol{t} = [0,1,1,1,1,1,1,1]$ and study type $\boldsymbol{s} = [1,1,0,1,1,1,1,1]$), where 1 indicates that the relevant study referred by that position in the array is retrieved. Then we perform conjunction (bitwise `AND`) on the three binary arrays to obtain a new binary array (i.e., $\boldsymbol{c} \wedge \boldsymbol{t} \wedge \boldsymbol{s} = \boldsymbol{q} = [0, 1, 0, 1, 1, 1, 1, 1]$) which represents the set of relevant studies in the development set that can be retrieved by the query that includes all terms (i.e. the maximal query). Note that when there are no terms present for a category[1] then the category is removed from the conjunction which forms $\boldsymbol{q}$. The logical conjunction of the three vectors has the same effect as executing the query, thus greatly increasing the number of comparisons that can be made (i.e, it reduces computation time). Further note that it is not guaranteed that the set of categories which contain terms from the development set, when combined using a Boolean `AND` operator, will retrieve all the relevant studies in the validation set—this is true regardless of using our technique for speeding up query assembly, or trying all combinations. Next, in an iterative manner, each term from each category is temporarily removed and a new binary vector ($\boldsymbol{v}_i$) is computed, containing the set of relevant studies in the development set retrieved without that term. If $\boldsymbol{q} \wedge \boldsymbol{v}_i = \boldsymbol{q}$, that is, if the removal of the term has no effect on the number of relevant studies in the development set retrieved by the rest of the terms, then the removal of that term from the category is made permanent. In other words, that term contributes nothing overall to the query (or its contribution is redundant as its contributing studies are also retrieved by other query terms) and is removed from its respective clause. Note that this technique could also be used in an interactive system to highlight to a user those terms that do not contribute to the set of retrieved documents, or alternatively for evaluating existing search strategies. The iteration proceeds by considering one candidate term for removal at a time; terms are ordered descending by the sum of the components of their document vectors, i.e., their total document frequency, thus the order of terms removed is deterministic. The complexity of this approach is $O(3n)$: each term in the query is required to be only tested once for inclusion in the final query, rather than for all possible combinations. The resulting query is guaranteed to retrieve the maximum number of relevant studies possible in the development set (based on the terms which have been identified in the previous process).

We further propose to tune the term cut-off thresholds parameters for the filtering steps. Rather than fixing the thresholds at 20% for development and 2% for population (as done by Hausner et al. [4]), we perform a grid search (independently for each query) over combinations of thresholds to find the parameters best suited (optimising for F_1, F_3, recall) for a particular query. We also apply the same strategy to identify the number of MeSH terms to add to a query. The development set is used to identify terms; then we evaluate queries on the validation set to select the best combination of parameters. The query can then be evaluated fairly on the unseen set.

[1] E.g., there are no terms that can be categorised into Study Type ($\mathbf{s}$), but there are terms categorised into Conditions ($\mathbf{c}$) and Treatments ($\mathbf{t}$).

Table 2. Evaluation results on unseen documents, with and without MeSH terms applied to queries. Relaxed indicates original queries where MeSH explosion is removed and phrases converted to Boolean `OR` clauses. Significant differences (paired two-tailed t-test $p < 0.05$) between original queries indicated by $\star$. Significant differences between relaxed queries indicated by †. Highest values are **bolded**. Original queries do not achieve 100% recall because (i) errors in reporting of queries [3, 14], and (ii) all queries are issued to PubMed even if the original query was reported as a MEDLINE query (i.e., translated automatically [17]).

		$F_{0.5}$	F_1	F_3	NNR	Precision	Recall
	Original	0.0078†	0.0123†	0.0558†	1040.03	0.0062†	0.9384
	Original (Relaxed)	0.0015*	0.0024*	0.0115*	230824.58	0.0012*	0.9078
Automated objective tuned for:	F_1+MeSH	0.0056†	0.0086†	0.0340†	**895.38**	0.0046†	0.5329*†
	F_1	0.0148†	0.0194†	0.0442†	2186.58	0.0129†	0.2418*†
	F_3+MeSH	0.0060†	0.0094†	0.0384†	921.29	0.0049†	0.5095*†
	F_3	**0.0166**†	**0.0219**†	**0.0510**†	1217.09	**0.0146***†	0.2672*†
	Recall+MeSH	0.0005*†	0.0007*†	0.0035*†	84809.31*	0.0004*†	**0.9523**
	Recall	0.0002*†	0.0004*†	0.0017*†	102020.38*	0.0002*†	0.8561*

3 Empirical Evaluation

We evaluate the computational method for objectively deriving systematic review search strategies on the CLEF 2018 Technology Assisted Reviews (TAR) collection [6]. This collection of diagnostic test accuracy systematic review protocols contains 75 topics (i.e., systematic reviews use-cases).[2] Diagnostic test accuracy reviews are highly specific and are considered one of the most difficult types of systematic reviews to search literature for [11]. Each topic comprises the title of the review, the Boolean query used to retrieve studies, and relevance assessments for the studies retrieved by the query. To determine the effectiveness of queries, we execute them in PubMed through the entrez API [16]. In our experiments, the test set for each topic is derived from the studies labelled relevant at an abstract level: these are studies that were retrieved by the Boolean query of the original systematic review and were screened for inclusion. We set the minimum size of the development set to 25, therefore excluding topics from the collection where the number of studies labelled relevant was less than 50 (development = 2/4 of total size). This number was chosen as the size of the development set in the study by Hausner et al. [4] was 25 (for a single topic). For comparison, the development set in a study by Simon et al. [20] was 78 (single topic). After removing topics in this way, 40 topics remained (still considerably larger than the previous studies), and the average number of relevant

[2] The CLEF 2018 TAR collection is a superset of queries from the 2017 TAR collection. The CLEF 2017 TAR collection was not used as the overlap of queries in the 2017 and 2018 collections for our purposes was the same, once we removed topics that had less than 50 relevant studies.

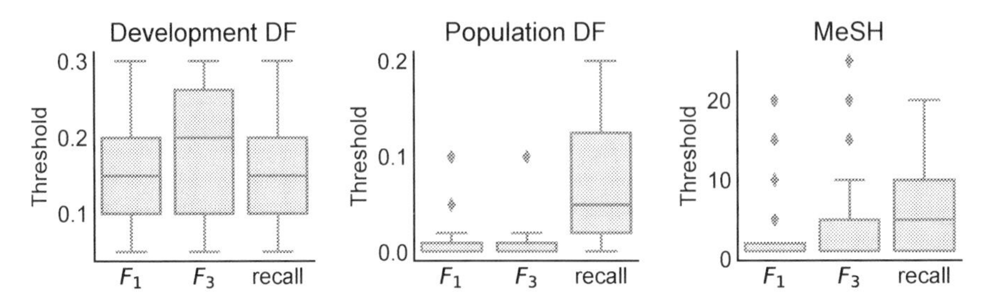

Fig. 2. Variance across topics for the best selected values for different parameters, via optimisation of the considered evaluation measures on validation set.

studies per topic was 180.65 ± 157.8 (min: 52, max: 604). When optimising the threshold parameters, we performed a grid search over the values [0.05, 0.30] with step 0.05 for the development and 0.001, 0.01, 0.02, 0.05, 0.10, 0.20 for the population sets. The number of MeSH terms to add to a query were parametrised to 1, 5, 10, 15, 20, 25. Evaluation on the final query is performed on the validation (as it was for Hausner et al. [4]) and unseen sets. As in the work by Simon et al. [20], queries are evaluated using precision, recall (sensitivity), and number needed to read (NNR). Additionally, we compute F_β for $\beta = 0.5, 1, 3$ (standard values used to evaluate automatic systematic review methods [15]).

4 Results

We first compare the results obtained using the computational method for objectively deriving search strategies against those the queries originally used to retrieve studies. We then study the effect selection of terms has on queries. Next, we contrast the differences between adding versus not adding MeSH terms and report the differences between the most effective and the least effective query when MeSH terms are not added and when they are added. And finally, we compare our adaptation to the original method.

Empirical results obtained by applying our method to queries is reported in Table 2. Our approach produces queries that are tuned for different evaluation measures. We show that for the F_β variations, NNR, precision, and recall, there are queries which outperform the original queries for each of these measures. While tuning parameters produces gains over the original queries, it introduces a trade-off. Generally, after tuning, queries with gains for precision measures obtain significant losses for recall measures compared to the original queries (e.g., queries tuned for F_1 without adding MeSH terms obtain the highest precision, but suffer a significant loss in recall). Likewise, where there are gains in recall, there are significant losses in precision compared to the original queries (e.g., queries tuned for recall with the addition of MeSH terms obtained the highest recall, however suffer a significant loss in precision). Figure 2 highlights the differences in parameter choices tuned for each evaluation measure. Higher

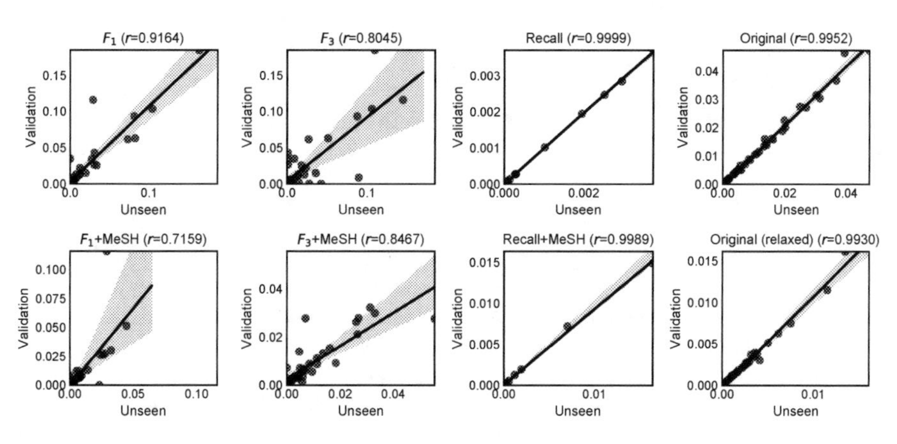

Fig. 3. Correlation of F_1 between validation and unseen for the results presented in Table 2. Pearson's r correlation is signified in the title of each subplot.

DF thresholds for the development and lower DF thresholds for the population lead to queries with higher precision and lower recall. Lower DF thresholds for the development and higher thresholds for the population lead to queries with higher recall and lower precision. Furthermore, adding more MeSH terms increased recall but at the expense of precision, as expected. Finally, while using the validation set appears to be a good indication of how the query will perform on unseen data, over-fitting leads to the trade-off in precision and recall. The correlation between the performance on the validation data and the unseen data is presented in Fig. 3. The figure suggests that performance obtained when tuning parameters on the validation set are strongly correlated with those obtained on unseen data (for the same parameters values). However, we find that as more weight is placed on precision, the correlation between performance on the unseen data and validation data becomes weaker.

We further analyse the queries by studying the terms that were considered for inclusion by threshoding: we focus on queries that did not have MeSH terms added. Table 3 provides a comparison between the highest performing topic in terms of F_1 on the unseen set (Table 3a, topic `CD009135`: *Rapid tests for the diagnosis of visceral leishmaniasis in patients with suspected disease*) and the lowest performing topic (Table 3b, topic `CD010276`: *Diagnostic tests for oral cancer and potentially malignant disorders in patients presenting with clinically evident lesions*). Both topics contain high prevalence terms from the title of the systematic review – indicating they are likely relevant to the topic. However, this is the case for terms identified in the best performing query, as well as for those in the worst. This suggests that the identification of terms likely plays only a partial role in the effectiveness of the query – and that selection and location within the Boolean syntax of the query may also be conducive of effectiveness.

We also study the interplay between the number of studies provided in the development set and query effectiveness. One may hypothesise that a higher number of studies in the development set is associated to higher effectiveness

Table 3. Prevalence of the top 10 terms in development and Population sets for the most effective (Sub-table 3a), and least effective (Sub-table 3a) topics in F_1.

(a) Prevalence (p) of terms for CD009135.

	development n=39		Population n= 30m	
	p	n	p	n
visceral	0.9231	36	0.0026	76907
in	0.8974	35	0.0235	707649
test	0.8974	35	0.0847	2553063
leishmaniasis	0.8462	33	0.0009	27225
to	0.7436	29	0.0237	715449
patients	0.6154	24	0.1826	5503941
positive	0.5641	22	0.0513	1546457
specificity	0.5385	21	0.0401	1209795
is	0.5385	21	0.0223	671619
sensitivity	0.4872	19	0.0423	1275222

(b) Prevalence (p) of terms for CD010276.

	development n=27		Population n= 30m	
	p	n	p	n
oral	0.9630	26	0.0356	1072216
in	0.9630	26	0.0235	707649
to	0.8519	23	0.0237	715449
patients	0.6667	18	0.1826	5503941
specificity	0.6667	18	0.0401	1209795
lesions	0.6296	17	0.0202	610428
sensitivity	0.5926	16	0.0423	1275222
detection	0.5556	15	0.0289	872069
is	0.5556	15	0.0223	671619
malignant	0.5185	14	0.0122	369252

(given that topic CD009135 contains 39 references in the development set, while topic CD010276 contains 27). Indeed there is a moderate positive correlation (Pearson's $r = +0.51$) between the size of the development set (used to derive terms) and F_1 on unseen data. Similarly, we study the interplay between the number of terms in the final query and effectiveness. We found a weak negative correlation (Pearson's $r = -0.2$) between the number of terms in queries and the actual effectiveness on the unseen set (F_1). This suggests that including few representative terms is more conducive of effectiveness than many broad terms.

Differences between queries with MeSH terms and those without are also analysed. The results in Table 2 suggest a trade-off in precision and recall when MeSH terms are added. When MeSH terms are added, we observe a higher recall, as expected, but lower precision than when MeSH terms were not added. Queries with MeSH terms did not retrieve a significantly higher number of studies than those without. We now study the effect of adding MeSH terms in more detail, specifically on queries where parameters were tuned for F_3, where the highest gains were observed overall. The most effective and least effective queries in terms of F_1 were for topics CD009579, *Circulating antigen tests and urine reagent strips for diagnosis of active schistosomiasis in endemic areas* (precision: 0.0330, recall: 0.1765, F_1: 0.0556), and CD009647, *Clinical symptoms, signs and tests for identification of impending and current water-loss dehydration in older people* (precision: 0.0002, recall: 0.4286, F_1: 0.0003). The MeSH terms identified for addition to these queries are listed in Table 4. Figure 5 presents the two queries for comparison. Sub-figs. 5a and b contain the aforementioned best and worst queries derived by optimising F_1. The identified MeSH terms lead to small improvements in recall when they are added to the query, at the expenses of a substantial drop in precision. When observing the performance on

Table 4. Top 10 MeSH terms identified in development set for CD009579 (left, highest F_1) and CD009647 (right, lowest F_1).

CD009579	CD009647
Parasite Egg Count	Aged, 80 and over
Schistosomiasis haematobia	Water-Electrolyte Balance
Antigens, Helminth	Body Water
Sensitivity and Specificity	Osmolar Concentration
Schistosoma haematobium	Electric Impedance
Schistosomiasis mansoni	Dehydration
Hematuria	Sodium
Schistosoma mansoni	Reproducibility of Results
Feces	Prospective Studies
Prevalence	Urine

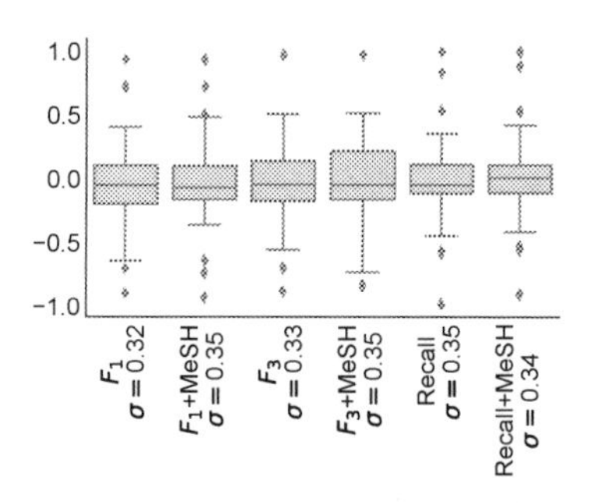

Fig. 4. Normalised differences in no. of terms in queries between our method and relaxed original.

```
((schistosomiasis OR cca OR used OR found OR hematuria OR either) AND (strips OR urinalysis
OR dipstick OR dipsticks OR mg OR Parasite Egg Count) AND (village OR villages OR kg))
```

(a) Highest F_1; query for topic CD009579.

```
((found OR urine OR balance OR Aged, 80 and over) AND fluid)
```

(b) Lowest F_1; query for topic CD009647.

Fig. 5. Computationally derived queries. Queries refer to the most effective (Sub-fig. 5a) and the least effective (Sub-fig. 5b) in terms of F_1 (by optimising for F_1) among those in the collection. The queries do not contain field restrictions for space reasons. MeSH terms indicated by *italics* (`[Mesh Terms:noexp]`). In all other cases the `[Title/Abstract]` field restriction was applied.

the validation data, topic CD009579 still performs better than topic CD009647. This suggests that this particular topic was more difficult to search for. This is reflected in the queries originally formulated for these topics. The original query for topic CD009579 (precision: 0.0019, recall: 1, F_1: 0.0038) performs better than the original query for topic CD009647 (precision: 2.6^{10e-5}, recall: 1, F_1: 5.2^{10e-5}). Note that although the performance of the queries are similar, the manually formulated queries have many more terms (CD009579 – derived: 15, original (relaxed): 40; CD009647 – derived: 6, original (relaxed): 199). However, looking at the distribution of query lengths for each method in Fig. 4, not only is there little variation in the total number of terms in queries, but there is little variance in amount of terms added in the automatic method and in the relaxed versions of the original queries.

Finally we study a query derived manually (by Hausner et al. [4], Fig. 7) and the same query derived computationally (Fig. 6). Table 5 presents the differences in effectiveness given the number of documents retrieved, NNR, precision, and recall on the same set of validation documents (note that only the development and validation sets are used to make a fair comparison to the manual method; thus no tuning was used). The query derived objectively using our computational

```
1. prostate.ti,ab.
2. psa.ti,ab.
3. used.ti,ab.
4. either.ti,ab.
5. seed.ti,ab.
6. symptom.ti,ab.
7. ml.ti,ab.
8. toxicities.ti,ab.
9. prostatic.ti,ab.
10. Prostatic Neoplasms/
11. or/1-10
12. beam.ti,ab.
13. brachytherapy.ti,ab.
14. radical.ti,ab.
15. prostatectomy.ti,ab.
16. ebrt.ti,ab.
17. cox.ti,ab.
18. androgen.ti,ab.
19. implantation.ti,ab.
20. consensus.ti,ab.
21. pretreatment.ti,ab.
22. sexual.ti,ab.
23. neoadjuvant.ti,ab.
24. mailed.ti,ab.
25. implant.ti,ab.
26. curative.ti,ab.
27. or/12-26
29. 11 and 27
```

Fig. 6. Computationally derived objective query.

```
1. cancer.ti,ab,sh.
2. adenocarcinoma.ti,ab,sh.
3. 1 or 2
4. prostat*.ti,ab,sh.
5. 3 and 4
6. Prostatic Neoplasms/
7. 5 or 6
8. seed*.rs.
9. permanent*.ti,ab,sh.
10. 8 or 9
11. implant*.ti,ab,sh.
12. 10 and 11
13. Brachytherapy/
14. Brachytherapy.ti,ab,sh.
15. or/12-14
16. 7 and 15
```

Fig. 7. Manually derived objective query (commonalities in *italics*).

Table 5. Difference in effectiveness between the computationally derived query and the manually derived query.

	# Ret	NNR	Precision	Recall
Manual	78913	6070.23	0.0002	1.0000
Computational	48945	3496.14	0.0003	1.0000

method retrieves less documents, but maintains recall: this results in a saving of approximately USD$90,000 (considering double screening and the costs/times per study reported by McGowan et al. [13]).

5 Conclusion

We presented a computational approach to objectively deriving search strategies for systematic reviews. This approach adapts and extends the proposal of Hausner et al. [4], to further reduce human subjectivity in an otherwise objective methodology. The computational method can be used as a starting point for query formulation, as demonstrated by our results. The manual objective method included human intervention; our computational adaptations and extensions approximated the steps a human would take. To better approximate these steps, we will set up an interactive query formulation study with information specialists. The feedback and results from this can be used to improve computational methods and would provide us with the means to fairly compare our computational approach with the ad-hoc method.

We have identified a number of avenues for further extending the fully automatic approach and its empirical evaluation. Firstly, randomness is introduced in this method when the test set is split into development, validation, and unseen. The use of 3-fold cross validation would reduce experimental bias in the subsequent phases. Next, we have observed that in the current approach only unigrams are used as candidate terms for possible inclusion in queries (this is also the case in the original method and has already been identified as an issue [4]). This is a limitation because the semantic context that may have been encoded as a phrase (e.g., in an n-gram such as "myocardial infarction") is lost. We suggest that by automatically identifying medical phrases using automatic tools such as MetaMap, this shortcoming can be overcome.

A limitation of this work is that a prospective study was not undertaken. New queries formulated using our method may have retrieved unjudged but relevant studies. A future extension of this work could be to use our proposed method to identify the proportion of new relevant studies retrieved (if any). Another task to be considered is to automatically further refine derived queries. Scells and colleagues [18,19] have found that automatic generation and refinement techniques can improve the effectiveness of existing Boolean queries.

The computational method presented is envisioned to be integrated into tools for assisting researchers conducting systematic reviews (for example, query suggestion [8]). The aim is not to replace humans constructing search strategies—at the very least, a number of gold standard studies are still needed to seed this approach, (as is typically the case in this context [10]). Query formulation is currently a highly subjective and error-prone process, and reducing subjectivity and mistakes in search strategy construction can only lead to less biased, reproducible, and timely systematic reviews.

Acknowledgements. Harrisen is the recipient of a CSIRO PhD Top Up Scholarship. Dr Guido Zuccon is the recipient of an Australian Research Council DECRA Research Fellowship (DE180101579) and a Google Faculty Award. This research is supported by the National Health and Medical Research Council Centre of Research Excellence in Informatics and E-Health (1032664).

References

1. Aronson, A.R.: Effective mapping of biomedical text to the UMLS Metathesaurus: the MetaMap program. In: Proceedings of the AMIA Symposium, p. 17. American Medical Informatics Association (2001)
2. Clark, J.: Systematic reviewing. In: Suhail, A.R., Doi, G.M.W. (eds.) Methods of Clinical Epidemiology, pp. 187–211. Springer, Heidelberg (2013). https://doi.org/10.1007/978-3-642-37131-8_12
3. Golder, S., Loke, Y., McIntosh, H.M.: Poor reporting and inadequate searches were apparent in systematic reviews of adverse effects. J. Clin. Epidemiol. **61**, 440–448 (2008)
4. Hausner, E., Waffenschmidt, S., Kaiser, T., Simon, M.: Routine development of objectively derived search strategies. Syst. Rev. **1**(1), 19 (2012)
5. Higgins, J.P.T., Green, S.: Cochrane handbook for systematic reviews of interventions version 5.1.0 [updated March 2011]. The Cochrane Collaboration (2011)
6. Kanoulas, E., Spijker, R., Li, D., Azzopardi, L.: CLEF 2018 technology assisted reviews in empirical medicine overview. In: CLEF 2018 Evaluation Labs and Workshop: Online Working Notes (2018)
7. Karimi, S., Pohl, S., Scholer, F., Cavedon, L., Zobel, J.: Boolean versus ranked querying for biomedical systematic reviews. BMC MIDM **10**(1), 1 (2010)

8. Kim, Y., Seo, J., Croft, W.B.: Automatic Boolean query suggestion for professional search. In: Proceedings of the 34th International ACM SIGIR Conference on Research and Development in Information Retrieval (2011)
9. Lau, J.: Systematic review automation thematic series (2019)
10. Lee, G.E., Sun, A.: Seed-driven document ranking for systematic reviews in evidence-based medicine. In: The 41st International ACM SIGIR Conference on Research & Development in Information Retrieval, pp. 455–464 (2018)
11. Leeflang, M., Deeks, J., Takwoingi, Y., Macaskill, P.: Cochrane diagnostic test accuracy reviews. Syst. Rev. **2**, 82 (2013). pubmed pmid: 24099098. pubmed central pmcid: Pmc3851548. Technical report, Epub 2013/10/09. Eng
12. Lefebvre, C., Manheimer, E., Glanville, J.: Searching for studies. Cochrane Handbook for Systematic Reviews of Interventions: Cochrane Book Series, pp. 95–150 (2008)
13. McGowan, J., Sampson, M.: Systematic reviews need systematic searchers (IRP). J. Med. Libr. Assoc. **93**(1), 74 (2005)
14. McGraw, K.A., Anderson, M.J., et al.: Analysis of the reporting of search strategies in cochrane systematic reviews. J. Med. Libr. Assoc. **97**(1), 21 (2009)
15. O'Mara-Eves, A., Thomas, J., McNaught, J., Miwa, M., Ananiadou, S.: Using text mining for study identification in systematic reviews: a systematic review of current approaches. Syst. Rev. **4**(1), 5 (2015)
16. Sayers, E.: A general introduction to the e-utilities. Entrez Programming Utilities Help [Internet]. National Center for Biotechnology Information, Bethesda (2010)
17. Scells, H., Locke, D., Zuccon, G.: An information retrieval experiment framework for domain specific applications. In: The 41st International ACM SIGIR Conference on Research & Development in Information Retrieval (2018)
18. Scells, H., Zuccon, G.: Generating better queries for systematic reviews. In: The 41st International ACM SIGIR Conference on Research & Development in Information Retrieval, SIGIR 2018 (2018)
19. Scells, H., Zuccon, G., Koopman, B.: Automatic Boolean query refinement for systematic review literature search. In: Proceedings of the 2019 World Wide Web Conference (2019)
20. Simon, M., Hausner, E., Klaus, S.F., Dunton, N.E.: Identifying nurse staffing research in medline: development and testing of empirically derived search strategies with the pubmed interface. BMC Med. Res. Methodol. **10**(1), 76 (2010)
21. Tsafnat, G., Glasziou, P., Choong, M.K., Dunn, A., Galgani, F., Coiera, E.: Systematic review automation technologies. SR **3**(1), 74 (2014)
22. Uzuner, Ö., South, B.R., Shen, S., DuVall, S.L.: 2010 i2b2/VA challenge on concepts, assertions, and relations in clinical text. J. Am. Med. Inform. Assoc. **18**(5), 552–556 (2011)

You *Can* Teach an Old Dog New Tricks: Rank Fusion applied to Coordination Level Matching for Ranking in Systematic Reviews

Harrisen Scells[1(✉)], Guido Zuccon[1], and Bevan Koopman[2]

[1] The University of Queensland, St Lucia, Australia
h.scells@uq.net.au
[2] CSIRO, Brisbane, Australia

Abstract. Coordination level matching is a ranking method originally proposed to rank documents given Boolean queries that is now several decades old. Rank fusion is a relatively recent method for combining runs from multiple systems into a single ranking, and has been shown to significantly improve the ranking. This paper presents a novel extension to coordination level matching, by applying rank fusion to each sub-clause of a Boolean query. We show that, for the tasks of systematic review screening prioritisation and stopping estimation, our method significantly outperforms the state-of-the-art learning to rank and bag-of-words-based systems for this domain. Our fully automatic, unsupervised method has (i) the potential for significant real-world cost savings (ii) does not rely on any intervention from the user, and (iii) is significantly better at ranking documents given only a Boolean query in the context of systematic reviews when compared to other approaches.

Keywords: Coordination level matching · Rank fusion · Systematic reviews · Boolean queries · Information retrieval

1 Introduction

The goal of medical systematic review literature search is to retrieve all research publications relevant to a highly focused research question that satisfies an inclusion criteria. This is so that *all* literature relevant to the review's research question can be synthesised in the systematic review [23]. Search takes place using a Boolean query that is formulated by highly trained information specialists using their own intuition and domain knowledge, in order to capture the information need of the systematic review [8]. Afterwards, every study retrieved by the Boolean query is *screened* (assessed) for inclusion using the titles and abstracts of studies (abstract level assessment). Identified relevant abstracts are further processed by acquiring the full-text for additional assessment, information extraction and synthesis [23].

J. M. Jose et al. (Eds.): ECIR 2020, LNCS 12035, pp. 399–414, 2020.
https://doi.org/10.1007/978-3-030-45439-5_27

The process of creating a medical systematic review typically involves large monetary and temporal costs; the average Cochrane review costs $350K to create [35] and it takes up to two years to publish – thus often rendering the result of the systematic review already out-of-date at the time of publication. The process that incurs the most cost when creating a systematic review is the screening of studies retrieved by the Boolean query; often a large set of studies is retrieved, but only a handful are relevant.

A number of solutions have arisen to address the amount of time spent screening documents, including: *screening prioritisation* (which seeks to re-rank the set of retrieved documents to show more relevant documents first, thus starting the full-text screening earlier), and *stopping estimation* (which seeks to predict at what point continuing to screen will no longer contribute gain) [25,26,40]. In this paper, we propose and evaluate a Boolean query ranking function aimed at tackling these two tasks. The proposed method incorporates intuitions from both coordination level matching of Boolean queries and search engine rank fusion.

This paper proposes an extension to coordination level matching (CLM) by exploiting the query-document relationship with rank fusion. CLM is a ranking function originally proposed for Boolean queries that scores documents using the occurrences of documents retrieved by different clauses of the query. The proposed extension, *coordination level fusion* (CLF), has many advantages over CLM that enable it to use multiple weighting schemes (rankers) and different fusion methods dependent on the Boolean clauses. We use CLF to rank studies in the screening prioritisation task of systematic reviews. We further plan to study the use of a cut-off threshold tuned on training data to control when the screening of studies should be stopped based on the CLF retrieval score. The empirical results obtained on the CLEF Technology Assisted Review datasets [25,26] show that CLF significantly outperforms existing state-of-the-art methods that consider similar settings, including the ranking method currently used in PubMed (a popular database to search for literature for systematic reviews).

2 Related Work

Systematic reviews are costly and often out-of-date by the time they are published due to the amount of time involved in their creation. A wide range of systematic review creation processes have been considered for automation or improvement using semi-automatic techniques [40], including: query formulation [27,46,48], screening prioritisation [1–3,7,28,29,37,47,54,56], stopping prediction [7,16,24], assessment of bias [33,43], among others. This paper proposes a technique for screening prioritisation, thus the remainder of this section focuses on this specific task.

Active learning has been explored extensively for screening prioritisation and automatic assessment [1,10,37,56]. However, the main drawbacks of active learning are that a poor initial ranking will slow down the rate of learning, and that explicit human effort is required to update the ranking. While current practice prescribes all documents must be screened (therefore explicit assessments could

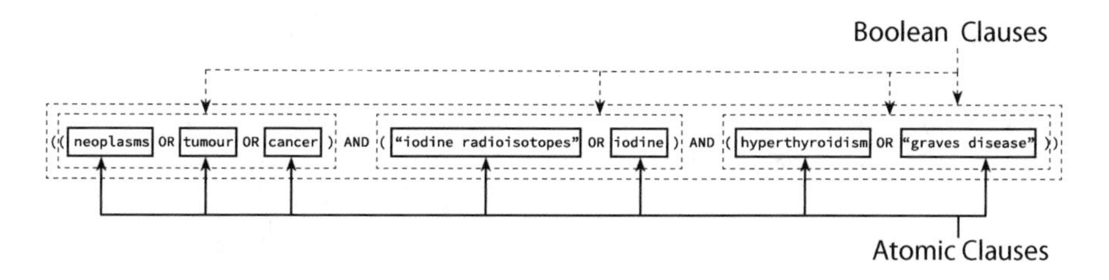

Fig. 1. Types of clauses in a Boolean query. Dashed lines surround Boolean clauses, dotted lines surround atomic clauses.

be used for active learning), an initially poor ranking would require many assessments before the system is able to identify relevant documents. Thus the analysis of the full-text of eligible documents may be delayed. Automatic assessment has been suggested to be used in place of a second researcher performing screening [40]. Fully automatic methods of screening prioritisation allow for other processes of systematic reviews to begin earlier *and* do not require the effort of humans, saving more time (and costs). In this paper, we do not consider screening prioritisation methods based on active learning. However, we note that CLF could be used as the first pass ranking in the context of an active learning method. Then, active learning could be used to augment CLF to performing re-ranking in the presence of continuous, iterative relevance feedback. We leave the study of CLF in an active learning setting for future work.

The CLEF Technology Assisted Reviews (TAR) track [25,26] considers both screening prioritisation and stopping prediction tasks. The screening prioritisation task has gained substantial interest from CLEF participants, with submitted methods including active learning [12,13], relevance feedback [4,18,21,36,38,39,52,55], automatic supervised [9,17,30,47,51], and automatic unsupervised methods (which do not rely on any relevance feedback or human intervention) [2,3,7,54]. Meanwhile, the stopping prediction task has seen little participation and naïve techniques like static score-based cut-offs [24], as well as techniques based on continuous relevance feedback [16] are used. Many of the participants also do not use the Boolean queries directly, instead resorting only to the title of the review (a sentence), which is contrived and unrealistic in the context of systematic review literature search. This work overcomes these shortcomings by only using the Boolean query to rank documents, with no additional effort required by the information specialist.

Several approaches to ranking documents retrieved by Boolean queries were proposed in the '80s and '90s outside of the context of systematic review creation. Most of these approaches rely on users explicitly weighting terms in the query [42], probabilistic retrieval using fuzzy set theory [6,41] and term dependencies [15]. A drawback of these methods is their heavy reliance on the users to impose a ranking over retrieved documents (e.g., the requirement that users must specify individual term weightings). Users often are unable to provide such weights, or it creates an additional hindrance in using the retrieval system.

A ranking function for Boolean queries which relies solely on the structure of the Boolean query, without further user intervention, is Coordination Level Matching (CLM) [31]. The intuition behind CLM is that nested sub-clauses of a Boolean query could be considered as separate but related queries, and therefore documents that appear in multiple clauses should be ranked higher. For example, a very common way information specialists formulate Boolean queries for systematic review literature search is to break a search down into three of four categories based on the Population, Intervention, Controls, Outcomes (PICO) framework [8]. Query terms from each category become a clause in the Boolean query, grouped together by a single `AND` operator [8]. Formally, in CLM the score of a document d is the number of Boolean clauses of the query Q that are satisfied by it. A clause can be considered as both a single atomic keyword, and the grouping of several keywords or other nested groupings by a single Boolean operator (Boolean clause). Figure 1 visualises the differences between atomic clauses and Boolean clauses.

Rankings produced by CLM typically perform poorly (as supported by our empirical findings in Sect. 5.1). This is because the amount of information about the query being exploited to produce a document ranking is low. CLM has been noted to be more effective when weighting occurrences of documents by, for example, IDF or TF-IDF [14]. Which weighting scheme to use for CLM is then unclear, and some documents may be ranked higher than others using different weighting schemes. Moreover, when computing scores, CLM does not account for the different Boolean operators present in the query, i.e., scores are summed in the same manner irrespective of the operator used, e.g., `AND`, `OR`.

The CLF method proposed in this paper exploits rank fusion [49], i.e., the combination of multiple document rankings, typically returned by different systems or weighting schemes for the same query (although recent work has applied fusion to different query variations [5]). There are many methods for fusion of rankings, and they can be classified into two main categories [22]: score-based [49] and rank-based [32]. Score-based methods fuse rankings using the original scores of documents in different rankings to infer the new fused ranking. As systems and weighting schemes will typically assign wildly different scores to documents, scores are often normalised before fusion (e.g., using min-max normalisation). Rank-based methods fuse rankings using only the rank positions of documents (similarly to electoral vote fusion [32]).

The novelty of our contribution is that by combining insights from decades-old research about ranking documents directly with Boolean queries with relatively more recent research about the fusion of ranked lists, significant gains in effectiveness can be obtained.

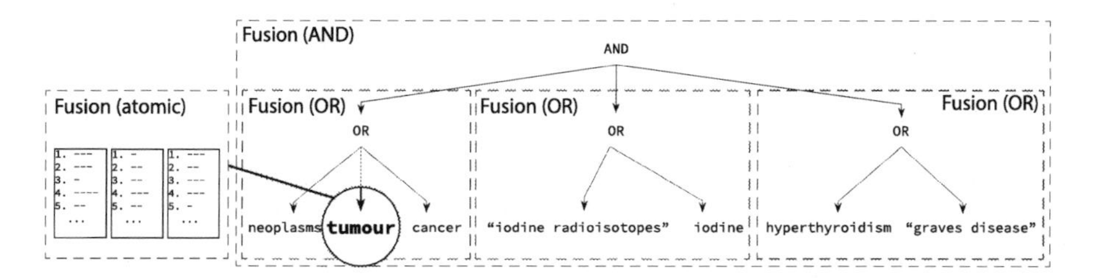

Fig. 2. Bottom-up visualisation of the fusion of ranked lists using the CLF method. First one or more ranked lists of an atomic clause are fused, then the results of each Boolean clause are fused. Each clause that has fusion applied to is encapsulated in a dashed box. The nested clauses which it encapsulates are included inside it. Each applicable fusion method is labelled within each respective box. Note that all atomic clauses use the same range of weighting schemes: in this figure only one is shown for space reasons.

3 Coordination Level Fusion

In this paper, we propose Coordination Level Fusion (CLF), a novel method that extends the traditional Coordination Level Matching (CLM) [31] by integrating rank fusion into the Boolean retrieval model by exploiting the semantic and syntactic aspects of the Boolean query.

CLM's intuition is that documents retrieved by many clauses should be considered more likely to be relevant. We note that this intuition is supported by axioms put forward in axiomatic analyses of ranking functions [19], and, more importantly for our work, it is similar to the intuition of rank fusion, namely, the *chorus effect*: the fact that "several retrieval approaches suggest that an item is relevant to a query" [53]. CLF leverages this intuition to further boost relevant documents higher up the ranking, using the agreement from multiple weighting schemes (rankers) *and* the agreement afforded by the structure of Boolean queries. Next, we describe the CLF method for ranking documents.

3.1 Producing a Ranking

We assume that a set R of rankings $r_1, r_2, \ldots, r_k$ is available for each atomic Boolean clause (i.e., a term in the Boolean query, see Fig. 1) . These rankings could be produced by any weighting scheme available, e.g., IDF, BM25, etc. A ranking is an ordered list of documents: $r = <d_0, d_1, ..., d_k>$ with $s(d_i, r_j)$ representing the score of document d_i within ranking r_j. In CLF, these rankings are recursively fused, first at an atomic clause level, then at the level of (often nested) Boolean operators, until the highest level of the Boolean query is considered (typically represented by an AND operator): at this level, rankings are again fused together to produce a single, final ranking. This is achieved by applying the CLF fusion function to each document d as:

```
(((oesophag*[All Fields] OR endocapsule[All Fields] OR microcam[All Fields] OR esophag*[All
Fields] OR enteroscop*[All Fields] OR pillcam[All Fields] OR videocapsule*[All Fields]) AND
("Esophageal and Gastric Varices"[Mesh Terms:noexp] OR (gastroesophag*[All Fields] OR
oesophag*[All Fields] OR oesophago gastric varix[All Fields] OR paraoesophag*[All Fields] OR
oesophago gastric varic*[All Fields] OR periesophag*[All Fields] OR perioesophag*[All Fields]
 OR esophag*[All Fields]))) AND (23593613[pmid] OR 23029720[pmid] OR 22379346[pmid] OR
22346246[pmid] OR 22155754[pmid] OR 21814064[pmid] OR 21624583[pmid] OR 21429016[pmid] OR
21372764[pmid] OR 21274889[pmid] OR 20490679[pmid] OR 20684186[pmid] OR 20682230[pmid] OR
20363433[pmid] OR 20135731[pmid] OR 20054320[pmid] OR 19809355[pmid] OR 19743993[pmid]))
```

Fig. 3. Example query formatted to be issued to PubMed for re-ranking. Constructing the query like above ensures only the documents specified (e.g., document number 23593613) are retrieved, and therefore re-ranked.

$$f_{CLF}(R,T,d) = \begin{cases} \sum_{r_j \in R} s(d,r_j) & \text{if } T = \texttt{AND} \\ |d \in R| \cdot \sum_{r_j \in R} s(d,r_j) & \text{if } T = \texttt{OR}/\texttt{Atomic} \end{cases} \tag{1}$$

where R is the set of rankings associated with the clauses of the Boolean query considered at the current level, and T is the type of Boolean operator applied. In this work, we consider T as being either identifying an atomic clause, or the AND and OR operators. The queries we consider do not have NOT clauses (therefore we do not have a fusion method for this operator). According to Eq. 1, CLF performs CombSUM fusion [49] if the Boolean clause is AND ($T =$ AND). Likewise, CombMNZ fusion [49] is used when dealing with atomic clauses or the OR operator. Figure 2 visualises how fusion is performed for different Boolean clauses. When scoring exploded MeSH terms, the score provided by a weighting scheme is the summed score of each child in the subsumption (similar for phrases). Both CombSUM and CombMNZ boost the documents which multiple rankers estimate to be highly relevant (i.e., the chorus effect), however CombMNZ at the OR and atomic levels is used to combat less accurate estimates of relevance (i.e., the dark horse effect). That is, documents where only a single ranker estimates them as highly relevant are not boosted.

3.2 Stopping Prediction

The task of stopping prediction in systematic review literature search is that: given a ranking of the set of documents retrieved by the Boolean query, at what position should screening stop? We model this task with an equivalent description: given a set of documents retrieved by a Boolean query, what is the subset of documents which does not need to be screened? In this work, stopping prediction is performed by exploiting the scores of documents for each atomic term after fusion. Rather than setting a fixed cut-off on scores similar to participants in the CLEF TAR task [24], here a gain-based approach is used. Our approach is as follows: Given that researchers will screen documents starting at the first document and continuing to the next document for the entire list, they are accumulating gain from documents (equal to the document score) as they continue down the list of documents. Once enough gain from documents has been accumulated, they can stop screening. To model this, we use a κ parameter to control what percentage of the total gain a researcher can accumulate before stopping. The stopping point therefore becomes the position of the document in the ranked list where the cumulative gain exceeds the total allowable gain. When κ is set to 1, no documents are discarded. In the task of screening prioritisation, where documents are assessed, κ is set to 1.

4 Experimental Setup

Empirical evaluation is conducted on the CLEF TAR 2017 and 2018 collections [25,26]. For the 2018 collection, evaluation is performed on topics from Task 2. Experiments are compared with respect to two baselines: a ranking obtained by submitting queries directly to PubMed (explained in detail below), and a ranking obtained by using CLM. The results of the CLF rankings are also compared to the rankings produced by the participants of the CLEF TAR task. Note that many of these participants do not rank directly according to the terms and structure of the Boolean query (while we do), and often consider the query as a bag-of-words, and incorporate terms from the title for re-ranking. Also note that many of the participants used feedback from the relevance assessments and created active learning solutions. The comparisons between participants and our results only consider those which reported to not use relevance assessments and do not use human intervention to rank (fully automatic, thus excluding active learning settings). In other words, we experiment considering the first round of retrieval.

All experiments are run using the QueryLab domain-specific Information Retrieval framework [45]. To obtain statistics for ranking documents, the documents retrieved by each query are fetched from PubMed and indexed by QueryLab. No stopwording or stemming is applied. The particular queries in this collection contain terms which are explicitly stemmed. Therefore, we use the PubMed Entrez API [44] to identify the original terms in documents from the explicitly stemmed term (this backward approach to stemming is to allow information specialists fine-grained control over their search). The title, abstract, MeSH headings, and publication date of each PubMed document is stored in four separate fields. When a title was not available for a document, the book title field was used instead; if no book title was available, the field was left empty (this replicates how searching on the title field works in PubMed). All of the experimental code to reproduce the experiments is made available at https://github.com/ielab/clf.

The following weighting schemes are used in our experiments to produce document rankings for an atomic clause: IDF, TF-IDF, BM25, InL2 of Divergence from Randomness, PubMed, term position, text score, publication date, and document length. The PubMed weighting scheme uses the state-of-the-art learning to rank system of Pubmed [20]. The best match ranking system of PubMed uses a three-stage ranking system: first, documents are retrieved using the Boolean query; then, documents are ranked using BM25; finally, top-ranked documents are re-ranked using LambdaMART trained on click data, using document features such as document length, publication date, and past usage. Note that the PubMed best match ranker can *only* rank documents given a term or phrase, *not* a Boolean query. After the first stage, the Boolean query is translated into a bag-of-words type of query, similar to those seen in web search (it is often the case that the query translation results in fewer documents retrieved). Therefore, by embedding the PubMed ranker into CLF, the query translation step may be skipped entirely. The term position weighting scheme is defined as the relative position of a term in a document (0 if the term does not appear in the document). Publication date scores documents higher if the document is newer (accounting for recency, linearly). Document length scores documents higher the longer the document is. Text score weights documents by the fields a term appears in: for example, a document is scored higher if a term appears in the title and the body than if the term appears only in the body. When queries are submitted to PubMed, they are modified to restrict them to only the PMIDs reported in the CLEF topic file (in order to account for minor discrepancies in retrieval after different time periods, see Fig. 3 for an example), and set the retrieval mode in PubMed to 'relevance' in order to obtain a ranked list of documents by relevance (instead of the default ranking by publication date). Prior to fusion for any clause, ranked lists are normalised using min-max normalisation. Z-score and softmax normalisation were also considered, however through early empirical testing, min-max normalisation provided the consistently higher effectiveness compared to z-score and softmax. When there are ties in the ranking, the document which has a more recent publication date is ranked higher. The different modifications made to CLF used in this paper are taxonimised below:

CLM – The basic form of coordination level matching using the approach described in Sect. 1.

CLF+PubMed – CLF, using the PubMed ranker via the PubMed Entrez API.

CLF+weighting – CLF, using the weighting schemes described in the paragraph above (excluding the PubMed weighting scheme).

CLF+weighting+PubMed – CLF, using all of the weighting schemes from **CLF+weighting** in addition to the PubMed ranker from **CLF+PubMed**.

CLF+weighting+qe – CLF, using all the weighting schemes from **CLF+weighting**, but with a naíve query expansion method using terms from the topic titles and terms specific to DTA systematic reviews (obtained from an information specialist). Here, two additional Boolean `OR` clauses are constructed, each containing terms from the title and DTA specific terms respectively. Terms from the title have stopwording and Porter stemming applied.

CLF+weighting+PubMed+qe – CLF, using all of the weighting schemes from **CLF+weighting**, in addition to the PubMed ranker from **CLF+PubMed**, and the approach to query expansion from **CLF+weighting+qe**.

4.1 Evaluation

Evaluation is performed differently depending on the task. For the screening prioritisation task, rank-based measures are used. For comparison between the CLEF TAR participants (of which we acquired the runs), the MAP measure is included. The nDCG measure is included as a more realistic model of user behaviour. Reciprocal rank (RR) is used to demonstrate the effectiveness of systems in an active learning scenario (to show how soon the first relevant document would be shown and an update to the ranking potentially triggered). Precision after R documents (Rprec) is used to show the theoretical best possible precision obtainable in the stopping task, along with last relevant (Last Rel) that reports at what rank position the very last relevant document was shown. Participant runs are chosen for comparison if they are a fully automatic, unsupervised method, which does not use the training data or explicit relevance feedback, and do not set a threshold (as categorised in the TAR overview papers [25,26]). Note that the tables in the CLEF TAR overview papers contain errors regarding these aspects, instead each of the participant's papers were considered to individually determine which runs to directly compare our methods to. For the stopping prediction task, several standard set-based measures are used: precision, recall, $F_{\beta=\{0.5,1,3\}}$, total cost, and reliability [11]. Reliability is a loss measure (i.e., where smaller values are better) specifically designed for the TAR task. It has two components: $loss_r = 1-(\text{recall})^2$ and $loss_e = (n/(R+100)*100/N)^2$, where n is the number of documents retrieved, N is the size of the collection, and R is the total number of relevant documents. Therefore, Reliability $= loss_r + loss_e$. Participants runs are chosen if they are fully automatic, supervised or unsupervised (thus we consider approaches that used training data), do not use explicit relevance feedback, and do set a threshold. Runs are evaluated using `trec_eval` or the evaluation scripts that are provided by the CLEF TAR organisers, where applicable.

When used for predicting when to stop screening, κ is tuned on training queries using a grid search to determine the best value. The parameter space searched in these experiments is $\{0.05, 0.075, 0.1, 0.2, 0.3, 0.4, 0.5, 0.75, 0.9, 0.95\}$. Note that κ can be set at a clause-level, therefore it is possible for it to be adaptive based on the clause. We leave learning an adaptive κ for future work, and here we fix κ to a set value across all clauses.

5 Results

5.1 Screening Prioritisation

Tables 1 and 2 present the results of the screening prioritisation task for the 2017 and 2018 CLEF TAR collections. Comparing CLM to CLF (without query expansion), CLF is statistically significantly better than CLM in all of the evaluation measures presented in both 2017 and 2018 tables (using a two-tails t-test where $p < 0.05$). Comparing the CLM and CLF methods to the state-of-the-art PubMed ranking, CLM is often statistically significantly worse than the

PubMed ranker, whereas some CLF-based methods are able to perform statistically significantly better than the PubMed method. Next, the best performing CLF method (**CLF+weighting+PubMed+qe**) and the best performing CLEF participant method for each year is compared. For 2017 topics, the best performing methods are Sheffield-run-2 (documents ranked with TF-IDF vector space model using terms from topic title and terms extracted from the Boolean query) and Sheffield-run-4 (same as Sheffield-run-2 except a PubMed stopword list is used) [3]. The CLF method does not perform statistically significantly better than these two methods in any evaluation measure considered (however in all measures apart from MAP and last relevant, CLF is better). For 2018 topics, the best performing method is Sheffield-general-terms (same as Sheffield-run-4 from 2017, however terms specifically designed to identify systematic reviews are added to the query) [2]. Comparing this method to CLF, the CLF method performs statistically significantly better in RR (and has gains in all evaluation measures apart from last relevant). Overall, CLF is able to obtain the highest MAP overall for 2018 topics, and the highest overall nDCG, RR, and Rprec for both 2017 topics and 2018 topics, performing statically significantly better than the state-of-the-art PubMed ranker.

Table 1. Results for CLEF TAR 2017. The first row of results is obtained by issuing queries to PubMed, the next set of rows is are results of the various configurations of CLF, and the last set of rows are the relevant runs from participants for that year. Two-tailed t-test between the PubMed ranker and the other methods with $p < 0.05$ is indicated by $*$ and $p < 0.01$ by $\dagger$.

	MAP	nDCG	RR	Rprec	Last Rel
PubMed	0.1597	0.5378	0.4292	0.1786	2974.00
CLM	$0.0483^{*\dagger}$	$0.3941^{*\dagger}$	$0.1344^{*\dagger}$	$0.0415^{*\dagger}$	3763.76^{*}
CLF+PubMed	0.1313	0.5129	0.3722	0.1387	3119.06
CLF+weighting	0.1494	0.5247	0.4213	0.1696	3307.76
CLF+weighting+PubMed	0.1643	0.5422	0.4028	0.1754	3048.10
CLF+weighting+qe	0.1960	0.5735	0.5326	0.2239	3301.73
CLF+weighting+PubMed+qe	0.2165^{*}	**0.5939***	**0.6037***	**0.2302**	3028.03
Sheffield-run-1	0.1700	0.5404	0.3644	0.1788	2678.33
Sheffield-run-2	**0.2183**	0.5930	0.5085	0.2190	2441.70
Sheffield-run-3	0.1986	0.5770	0.4700	0.2115	2404.96
Sheffield-run-4	0.2179	0.5937	0.5099	0.2185	2382.46^{*}
ECNU-run1	$0.0905^{*\dagger}$	$0.4517^{*\dagger}$	$0.1849^{*\dagger}$	$0.0907^{*\dagger}$	3633.16^{*}
QUT-bool	0.1293	0.4221^{*}	0.3465	0.1535	$1972.20^{*\dagger}$
QUT-pico	0.1197	0.4067^{*}	0.3088	0.1565	**1873.53***†

Table 2. Results for CLEF TAR 2018. Presentation of results and statistical significance is indicated the same was as in Table 1.

	MAP	nDCG	RR	Rprec	Last Rel
PubMed	0.1918	0.5971	0.5085	0.2131	**3479.40**
CLM	0.0483*†	0.4413*†	0.1338*	0.0316*†	7194.76
CLF+PubMed	0.1734*†	0.5938	0.4942	0.2002	6363.13
CLF+weighting	0.2012	0.6186	0.5331	0.2139	6061.06
CLF+weighting+PubMed	0.2363*	0.6390	0.5289	0.2435	5937.93
CLF+weighting+qe	0.2397*	0.6501	0.5969	0.2662*†	5931.13
CLF+weighting+PubMed+qe	**0.2722***†	**0.6767***†	**0.6649**	**0.2882***†	5743.26
ECNU-TASK2-RUN1-TFIDF	0.1415*	0.5682	0.4212	0.1862	7173.00
sheffield-general-terms	0.2584*	0.6495*	0.4723	0.2779*	5519.20
sheffield-query-terms	0.2243	0.6184	0.4012	0.2425	5736.70

Table 3. Results of CLF for stopping prediction for CLEF TAR 2017. The first row are the results from the original queries, the second row is when CLF with $\kappa = 0.4$. Two-tailed t-test between the original results and the other methods with $p < 0.05$ is indicated by $*$ and $p < 0.01$ by †.

	Precision	Recall	F_1	$F_{0.5}$	F_3	Total Cost	Reliability
No stopping	0.0415	**1.0000**	0.0752	0.0505	0.2345	3918.70	0.5441
CLF/0.4	**0.1040***†	0.7836*†	**0.1545***†	**0.1186***†	**0.3286***†	1324.63*†	**0.1259***†
ecnu-run2	0.0397	0.7075*†	0.0696	0.0478	0.2085	**1000.00***†	0.4445
ecnu-run3	0.0399	0.7164*†	0.0700	0.0480	0.2102	**1000.00***†	0.4433
sis.t1	0.0461*†	0.9868	0.0834*†	0.0561*†	0.2544*†	3435.03*†	0.4453*†
sis.t1.5	0.0482*†	0.9727*	0.0865*†	0.0585*†	0.2596*†	3165.56*†	0.3843*†
sis.2	0.0517*†	0.9531*†	0.0919*†	0.0626*†	0.2684*†	2824.6667*†	0.3309*†
sis.t2.5	0.0577*†	0.9382*†	0.1007*†	0.0695*†	0.2815*†	2536.80*†	0.2724*†

Table 4. Results of CLF for stopping prediction for CLEF TAR 2018. The first row are the results from the original queries, the second row is when CLF with $\kappa = 0.4$. Significance is indicated the same as in Table 3.

	Precision	Recall	F_1	$F_{0.5}$	F_3	Total Cost	Reliability
No stopping	0.0471	**1.0000**	0.0851	0.0573	0.2622	4640.23	0.3981
CLF/0.4	**0.1225***†	0.8582*†	**0.1827***†	**0.1400***†	**0.3794***†	**1140.06***†	0.4330*†

5.2 Stopping Prediction

Tables 3 and 4 present the results of the stopping prediction task using the cut-off parameter κ. A κ value of 0.4 through parameter tuning on training data was found to provide the least loss in Reliability, and was therefore chosen for the test queries for both 2017 and 2018. Results of the parameter tuning process on the training portion of the CLEF 2017 and 2018 topics are presented in Fig. 4. The CLF method used in this task was **CLF+weighting+PubMed+qe** as it obtained the highest performance on the screening task.

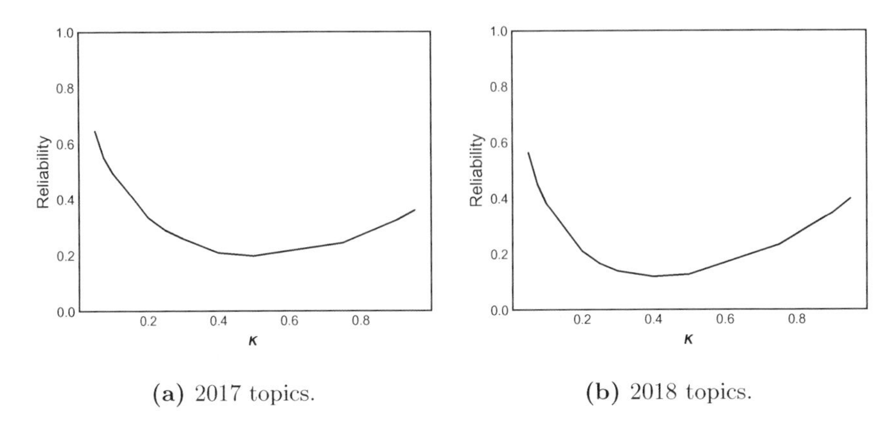

(a) 2017 topics. **(b)** 2018 topics.

Fig. 4. Tuning the κ parameter on the training portions of the 2017 (left) and 2018 (right) CLEF TAR topics. Lowest value for both plots is 0.4.

Examining first Table 3, CLF obtains the highest precision, F_1, $F_0.5$, F_3, and lowest loss in reliability. CLF also obtains the second-lowest total cost, and maintains both a low total cost and reliability for this set of queries. Losses in recall are within a tolerable threshold [11].Table 4, reveals similar results to the 2017 topics. Significant improvements over the original queries in terms of precision, F_1, $F_{0.5}$, F_3, and total cost, with a tolerable reduction in recall can be observed. However, the Reliability on this set of queries is higher (thus worse). Given that the total cost is low, this indicates that the $loss_r$ component of Reliability does not decrease at the same rate as $loss_e$ increases for these topics. There were no participants which contributed a comparable run to the 2018 TAR task, therefore no comparisons to other systems can be made for this collection.

While there is a drop in recall, there are real monetary savings associated with the increase in precision. Across the 2017 and 2018 topics, the CLF method provides savings between approximately USD$5000 and USD$12,000, according to estimates reported by McGowan et al. [35] when considering double screening.

6 Conclusion and Future Work

In this paper, a novel approach to ranking documents for systematic review literature search using rank fusion applied to coordination level matching was presented. The method, dubbed Coordination Level Fusion (CLF), outperformed the current state of the art for two different tasks. For the screening prioritisation task, CLF significantly outperformed the existing PubMed ranking system, as well as participants that submitted comparable runs to the CLEF TAR tasks. The results of the screening prioritisation task demonstrate the applicability of CLF to systematic review literature search when prioritisation is considered, and suggest it may also be applied to obtain an effective early ranking in settings that consider active learning. For the stopping prediction task, CLF could significantly reduce the cost of screening with tolerable losses in recall. The results of the stopping prediction task demonstrate the applicability of CLF to specific systematic reviews where total recall is not essential, such as in rapid reviews [34].

There are many aspects about CLF that require further investigation. First, we propose to study the effectiveness of CLF within an active learning setting. In this context, CLF can be used as the first ranker, before relevance feedback is collected. Then, feedback could be further weaved into CLF by devising and integrating weighting schemes that account for this. We also plan to investigate the use of CLF as a method for query performance prediction (e.g., as a post-retrieval predictor using reference lists [50], or as a candidate selection function in query transformation chain frameworks [48]). In terms of extending CLF, the weighting schemes themselves can be weighted (i.e., one weighting scheme may have more importance over others); e.g., using the linear combination fusion method [53] which assigns weights to each ranker being fused. The problem then is learning the weight to assign to each weighting scheme (ranker) used for rank fusion. Rather than using fusion methods like CombMNZ, it is foreseeable to use a different combination of weights for each Boolean clause considered.

Acknowledgements. Harrisen is the recipient of a CSIRO PhD Top Up Scholarship. Dr Guido Zuccon is the recipient of an Australian Research Council DECRA Research Fellowship (DE180101579) and a Google Faculty Award. This research is supported by the National Health and Medical Research Council Centre of Research Excellence in Informatics and E-Health (1032664).

References

1. Abualsaud, M., Ghelani, N., Zhang, H., Smucker, M.D., Cormack, G.V., Grossman, M.R.: A system for efficient high-recall retrieval. In: The 41st International ACM SIGIR Conference on Research & Development in Information Retrieval, pp. 1317–1320 (2018)
2. Alharbi, A., Briggs, W., Stevenson, M.: Retrieving and ranking studies for systematic reviews: University of Sheffield's approach to CLEF eHealth 2018 task 2. In: CEUR Workshop Proceedings, vol. 2125. CEUR Workshop Proceedings (2018)
3. Alharbi, A., Stevenson, M.: Ranking abstracts to identify relevant evidence for systematic reviews: The University of Sheffield's approach to CLEF eHealth 2017 task 2. In: CLEF (Working Notes) (2017)
4. Anagnostou, A., Lagopoulos, A., Tsoumakas, G., Vlahavas, I.P.: Combining inter-review learning-to-rank and intra-review incremental training for title and abstract screening in systematic reviews. In: CLEF (Working Notes) (2017)
5. Benham, R., Culpepper, J.S., Gallagher, L., Lu, X., Mackenzie, J.: Towards efficient and effective query variant generation. In: DESIRES, pp. 62–67 (2018)
6. Buell, D.A.: A general model of query processing in information retrieval systems. Inf. Process. Manag. **17**(5), 249–262 (1981)
7. Chen, J., et al.: ECNU at 2017 eHealth task 2: Technologically assisted reviews in empirical medicine. In: CLEF (Working Notes) (2017)
8. Clark, J.: Systematic reviewing. In: Doi, S.A.R., Williams, G.M. (eds.) Methods of Clinical Epidemiology. Springer, Heidelberg (2013). https://doi.org/10.1007/978-3-642-37131-8_12

9. Cohen, A.M., Smalheiser, N.R.: UIC/OHSU CLEF 2018 task 2 diagnostic test accuracy ranking using publication type cluster similarity measures. In: CEUR Workshop Proceedings, vol. 2125 (2018)
10. Cohen, A., Hersh, W., Peterson, K., Yen, P.: Reducing workload in systematic review preparation using automated citation classification. JAMIA **13**(2), 206–219 (2006)
11. Cormack, G.V., Grossman, M.R.: Engineering quality and reliability in technology-assisted review. In: Proceedings of the 39th International ACM SIGIR Conference on Research and Development in Information Retrieval, pp. 75–84 (2016)
12. Cormack, G.V., Grossman, M.R.: Technology-assisted review in empirical medicine: Waterloo participation in CLEF eHealth 2017. In: CLEF (Working Notes) (2017)
13. Cormack, G.V., Grossman, M.R.: Technology-assisted review in empirical medicine: Waterloo participation in clef eHealth 2018. In: CLEF (Working Notes) (2018)
14. Crestani, F.: Exploiting the similarity of non-matching terms at retrieval time. Inf. Retrieval **2**(1), 27–47 (2000). https://doi.org/10.1023/A:1009973415168
15. Croft, W.B.: Boolean queries and term dependencies in probabilistic retrieval models. J. Am. Soc. Inf. Sci. **37**(2), 71–77 (1986)
16. Di Nunzio, G.M.: A study of an automatic stopping strategy for technologically assisted medical reviews. In: Pasi, G., Piwowarski, B., Azzopardi, L., Hanbury, A. (eds.) ECIR 2018. LNCS, vol. 10772, pp. 672–677. Springer, Cham (2018). https://doi.org/10.1007/978-3-319-76941-7_61
17. Di Nunzio, G.M., Beghini, F., Vezzani, F., Henrot, G.: An interactive two-dimensional approach to query aspects rewriting in systematic reviews. IMS unipd at CLEF eHealth task 2. In: CLEF (Working Notes) (2017)
18. Di Nunzio, G.M., Ciuffreda, G., Vezzani, F.: Interactive sampling for systematic reviews. IMS unipd at CLEF 2018 eHealth task 2. In: CLEF (Working Notes) (2018)
19. Fang, H., Zhai, C.: An exploration of axiomatic approaches to information retrieval. In: Proceedings of the 28th Annual International ACM SIGIR Conference on Research and Development in Information Retrieval, pp. 480–487. ACM (2005)
20. Fiorini, N., et al.: Best match: new relevance search for pubmed. PLoS Biol. **16**(8), e2005343 (2018)
21. Hollmann, N., Eickhoff, C.: Ranking and feedback-based stopping for recall-centric document retrieval. In: CLEF (Working Notes) (2017)
22. Hsu, D.F., Taksa, I.: Comparing rank and score combination methods for data fusion in information retrieval. Inf. Retrieval **8**(3), 449–480 (2005). https://doi.org/10.1007/s10791-005-6994-4
23. Higgins, J.P.T., Green, S.: Cochrane handbook for systematic reviews of interventions version 5.1.0 [updated March 2011]. The Cochrane Collaboration (2011)
24. Kalphov, V., Georgiadis, G., Azzopardi, L.: SiS at CLEF 2017 eHealth TAR task. In: CEUR Workshop Proceedings, vol. 1866, pp. 1–5 (2017)
25. Kanoulas, E., Li, D., Azzopardi, L., Spijker, R.: CLEF 2017 technologically assisted reviews in empirical medicine overview. In: CLEF 2017 (2017)
26. Kanoulas, E., Spijker, R., Li, D., Azzopardi, L.: CLEF 2018 technology assisted reviews in empirical medicine overview. In: CLEF 2018 Evaluation Labs and Workshop: Online Working Notes (2018)
27. Karimi, S., Pohl, S., Scholer, F., Cavedon, L., Zobel, J.: Boolean versus ranked querying for biomedical systematic reviews. BMC MIDM **10**(1), 1 (2010)

28. Lagopoulos, A., Anagnostou, A., Minas, A., Tsoumakas, G.: Learning-to-rank and relevance feedback for literature appraisal in empirical medicine. In: Bellot, P., et al. (eds.) CLEF 2018. LNCS, vol. 11018, pp. 52–63. Springer, Cham (2018). https://doi.org/10.1007/978-3-319-98932-7_5
29. Lee, G.E., Sun, A.: Seed-driven document ranking for systematic reviews in evidence-based medicine. In: Proceedings of the 41st Annual International ACM SIGIR Conference on Research & Development in Information Retrieval, SIGIR 2018, pp. 455–464 (2018)
30. Lee, G.E.: A study of convolutional neural networks for clinical document classification in systematic reviews: SysReview at CLEF eHealth 2017 (2017)
31. Losee, R.: Probabilistic retrieval and coordination level matching. J. Am. Soc. Inf. Sci. **38**(4), 239–244 (1987)
32. Macdonald, C., Ounis, I.: Voting for candidates: adapting data fusion techniques for an expert search task. In: Proceedings of the 15th ACM International Conference on Information and Knowledge management, pp. 387–396. ACM (2006)
33. Marshall, I.J., Kuiper, J., Wallace, B.C.: RobotReviewer: evaluation of a system for automatically assessing bias in clinical trials. J. Am. Med. Inform. Assoc. **23**, 193–201 (2015)
34. Marshall, I.J., Marshall, R., Wallace, B.C., Brassey, J., Thomas, J.: Rapid reviews may produce different results to systematic reviews: a meta-epidemiological study. J. Clin. Epidemiol. **109**, 30–41 (2019)
35. McGowan, J., Sampson, M.: Systematic reviews need systematic searchers (IRP). J. Med. Libr. Assoc. **93**(1), 74 (2005)
36. Minas, A., Lagopoulos, A., Tsoumakas, G.: Aristotle university's approach to the technologically assisted reviews in empirical medicine task of the 2018 CLEF eHealth lab. In: CLEF (Working Notes) (2018)
37. Miwa, M., Thomas, J., O'Mara-Eves, A., Ananiadou, S.: Reducing systematic review workload through certainty-based screening. JBI **51**, 242–253 (2014)
38. Norman, C., Leeflang, M., Névéol, A.: LIMSI@CLEF eHealth 2018 task 2: technology assisted reviews by stacking active and static learning. In: CLEF (Working Notes) (2018)
39. Norman12, C., Leeflang, M., Névéol, A.: LIMSI@CLEF eHealth 2017 task 2: logistic regression for automatic article ranking (2017)
40. O'Mara-Eves, A., Thomas, J., McNaught, J., Miwa, M., Ananiadou, S.: Using text mining for study identification in systematic reviews: a systematic review of current approaches. Syst. Rev. **4**(1), 5 (2015)
41. Radecki, T.: A probabilistic approach to information retrieval in systems with boolean search request formulations. J. Am. Soc. Inf. Sci. **33**(6), 365–370 (1982)
42. Salton, G., Fox, E.A., Wu, H.: Extended Boolean information retrieval. Technical report. Cornell University (1982)
43. Savoie, I., Helmer, D., Green, C.J., Kazanjian, A.: Beyond medline: reducing bias through extended systematic review search. Int. J. Technol. Assess. Health Care **19**(1), 168–178 (2003)
44. Sayers, E.: A general introduction to the e-utilities. Entrez Programming Utilities Help [Internet]. National Center for Biotechnology Information, Bethesda (2010)
45. Scells, H., Locke, D., Zuccon, G.: An information retrieval experiment framework for domain specific applications. In: The 41st International ACM SIGIR Conference on Research & Development in Information Retrieval (2018)
46. Scells, H., Zuccon, G.: Generating better queries for systematic reviews. In: Proceedings of the 41st Annual International ACM SIGIR Conference on Research & Development in Information Retrieval, SIGIR 2018 (2018)

47. Scells, H., Zuccon, G., Deacon, A., Koopman, B.: QUT ielab at CLEF eHealth 2017 technology assisted reviews track: initial experiments with learning to rank. In: CEUR Workshop Proceedings: Working Notes of CLEF 2017: Conference and Labs of the Evaluation Forum, vol. 1866, pp. Paper-98. CEUR Workshop Proceedings (2017)
48. Scells, H., Zuccon, G., Koopman, B.: Automatic Boolean query refinement for systematic review literature search. In: The World Wide Web Conference, pp. 1646–1656 (2019)
49. Shaw, J.A., Fox, E.A.: Combination of multiple searches. NIST Special Publication SP, pp. 105–105 (1995)
50. Shtok, A., Kurland, O., Carmel, D.: Query performance prediction using reference lists. ACM Trans. Inf. Syst. (TOIS) **34**(4), 19 (2016)
51. Singh, G., Marshall, I., Thomas, J., Wallace, B.: Identifying diagnostic test accuracy publications using a deep model. In: CEUR Workshop Proceedings, vol. 1866. CEUR Workshop Proceedings (2017)
52. Singh, J., Thomas, L.: IIIT-H at CLEF eHealth 2017 task 2: technologically assisted reviews in empirical medicine. In: CLEF (Working Notes) (2017)
53. Vogt, C.C., Cottrell, G.W.: Fusion via a linear combination of scores. Inf. Retrieval **1**(3), 151–173 (1999)
54. Wu, H., Wang, T., Chen, J., Chen, S., Hu, Q., He, L.: ECNU at 2018 eHealth task 2: technologically assisted reviews in empirical medicine. Methods **4**(5), 7 (2018)
55. Yu, Z., Menzies, T.: Data balancing for technologically assisted reviews: undersampling or reweighting. In: CLEF (Working Notes) (2017)
56. Zou, J., Li, D., Kanoulas, E.: Technology assisted reviews: finding the last few relevant documents by asking yes/no questions to reviewers. In: The 41st International ACM SIGIR Conference on Research & Development in Information Retrieval, pp. 949–952 (2018)

Counterfactual Online Learning to Rank

Shengyao Zhuang(✉) and Guido Zuccon

The University of Queensland, St Lucia, Australia
{s.zhuang,g.zuccon}@uq.edu.au

Abstract. Exploiting users' implicit feedback, such as clicks, to learn rankers is attractive as it does not require editorial labelling effort, and adapts to users' changing preferences, among other benefits. However, directly learning a ranker from implicit data is challenging, as users' implicit feedback usually contains bias (e.g., position bias, selection bias) and noise (e.g., clicking on irrelevant but attractive snippets, adversarial clicks). Two main methods have arisen for optimizing rankers based on implicit feedback: counterfactual learning to rank (CLTR), which learns a ranker from the historical click-through data collected from a deployed, logging ranker; and online learning to rank (OLTR), where a ranker is updated by recording user interaction with a result list produced by multiple rankers (usually via interleaving).

In this paper, we propose a counterfactual online learning to rank algorithm (COLTR) that combines the key components of both CLTR and OLTR. It does so by replacing the online evaluation required by traditional OLTR methods with the counterfactual evaluation common in CLTR. Compared to traditional OLTR approaches based on interleaving, COLTR can evaluate a large number of candidate rankers in a more efficient manner. Our empirical results show that COLTR significantly outperforms traditional OLTR methods. Furthermore, COLTR can reach the same effectiveness of the current state-of-the-art, under noisy click settings, and has room for future extensions.

1 Introduction

Traditional learning to rank (LTR) requires labelled data to permit the learning of a ranker: that is, a training dataset with relevance assessments for every query-document pair is required. The acquisition of such labelled datasets presents a number of drawbacks: they are expensive to construct [5,25], there may be ethical issues in privacy-sensitive tasks like email search [37], and they cannot capture changes in user's preferences [19].

The reliance on users implicit feedbacks such as clicks is an attractive alternative to the construction of editorially labelled datasets, as this data does not present the aforementioned limitations [15]. However, this does not come without its own drawbacks and challenges. User implicit feedback cannot be directly treated as (pure) relevance labels because it presents a number of biases, and part of this implicit user signal may actually be noise. For example, in web

J. M. Jose et al. (Eds.): ECIR 2020, LNCS 12035, pp. 415–430, 2020.
https://doi.org/10.1007/978-3-030-45439-5_28

search, users often examine the search engine result page (SERP) from top to bottom. Thus, higher ranked documents have a higher probability to be examined, attracting more clicks (position bias), which in turn may infer these results as relevant even when they are not [7,18,24]. Other types of biases may affect this implicit feedback including selection and presentation bias [2,16,40]. In addition, clicks on SERP items may be due to noise, e.g., sometimes users may click for unexpected reasons (e.g., clickbaits and serendipity), and these noisy clicks may hurt the learnt ranker. Hence, in order to leverage the benefits of implicit feedback, LTR algorithms have to be robust to these biases and noises. There are two main categories of approaches to learning a ranker from implicit feedback [14]:

(1) **Offline LTR:** Methods in this category learn a ranker using historical click-through log data collected from a production system (logging ranker). A representative method in this category is Counterfactual Learning to Rank (CLTR) [18], where a user's observation probability (known as propensity) is adopted to construct an unbiased estimator which is used as the objective function to train the ranker.
(2) **Online LTR (OLTR):** Methods in this category interactively optimize a ranker given the current user's interactions. A representative method in this category is Dueling Bandit Gradient Descent (DBGD) [39], where multiple rankers are used to produce an interleaved[1] results list to display to the user and collect clicks. This signal is used to unbiasedly indicate which rankers that participated in the interleaving process are better (Online Evaluation) and to trigger an update of the ranker in production.

The aim of the counterfactual and the online evaluations is similar: they both attempt to unbiasedly evaluate the effectiveness of a ranker and thus can provide LTR algorithms with reliable updating information.

In this paper, we introduce Counterfactual Online Learning to Rank (COLTR), the first online LTR algorithm that combines the key aspects of both CLTR and OLTR approaches to obtain an effective ranker that can learn online from user feedback. COLTR uses the DBGD framework from OLTR to interactively update the ranker used in production, but it uses the counterfactual evaluation mechanism of CLTR in place of online evaluation. The main challenge we address is that counterfactual evaluation cannot be directly used in online learning settings because the propensity model is unknown. This is resolved by mirroring solutions developed for learning in the bandit feedback problem (and specifically the Self-Normalized Estimator [34]) within the considered ranking task – this provides a position-unbiased evaluation of rankers. Our empirical results show that COTLR significantly improves the traditional DBGD baseline algorithm. In addition, because COTLR does not require interleaving or multileaving, which is the most computationally expensive part in online evaluation [28], COLTR is more efficient than DBGD. We also find that COLTR performance is at par with the current state-of-the-art OLTR method

[1] Two rankers: interleaving [12,26]; more than two rankers: multileaving[28,30].

[22] under noisy click settings, while presenting a number of avenues for further improvement.

2 Related Work

The goal of counterfactual learning to rank (CLTR) is to learn a ranker from historical user interaction logs obtained with the ranker used in production. An advantage of this approach is that candidate rankers are trained and evaluated offline, i.e., before being deployed in production, thus avoiding exposing users to rankers of lesser quality compared to that currently in production. However, unlike traditional supervised LTR methods [20], users interaction data provides only partial feedback which cannot be directly treated as absolute relevance labels [14,16]. This is because clicks may have not been observed on some results because of position or selection bias, and clicks may have instead been observed because of noise or errors. As a result, much of the prior work has focused on removing these biases and noise.

According to position bias, users are more likely to click on top-ranked search results than those at the bottom of the SERP [2,16,18]: in CLTR this probability is referred to as *propensity*. Joachims et al. [18] developed an unbiased (with respect to position) LTR that relies on clicks using a modified SVMRank approach that optimizes the empirical risk computed using the Inverse Propensity Scoring (IPS) estimator. The IPS is an unbiased estimator which can indicate the effectiveness of a ranker given propensity (the probability that the user will examine a document) and click data [18]. However, this approach requires a propensity model to compute the IPS score. To estimate this, randomization experiments are usually required when collecting the interaction data and the propensity model is estimated under offline setting [37,38].

Aside from position bias, selection bias is also important, and it dominates problems in other ranking tasks such as recommendation and ad placement. Selection bias refers to the fact that users can only interact with items presented to them. Typically, in ad placement systems, the assumption is made that users examine the displayed ads with certainty if only one item is shown: thus no position bias. However, users are given the chance to click on the displayed item only, so clicks are heavily biased due to selection. User interactions with this kind of systems are referred to as bandit feedback [17,33,34]. The Counterfactual Risk Minimization (CRM) learning principle [33] is used to remove the bias from bandit feedback. Instead of a deterministic ranker, this group of methods assume the system relies on the hypothesis that a probability distribution is available over the candidate items, which is used to sample items to show to users. Importance sampling [3] is commonly used to remove selection bias.

Online Learning to Rank aims to optimize the production ranker interactively by exploiting user clicks [10,22,23,29]. Unlike CLTR, OLTR algorithms do not require a propensity model to handle position or selection bias. Instead, they assume that relevant documents are more likely to receive more clicks than non-relevant documents and exploits clicks to identify the gradient's direction.

Dueling Bandit Gradient Descent (DBGD) based algorithms [39] are commonly used in OLTR. The traditional DBGD uses online evaluation to unbiasedly compare two or more rankers given a user interaction [12,29]. Subsequent methods developed more reliable or more efficient online evaluation methods, including Probabilistic Interleaving (PIGD) which has been proven to be unbiased [12]. The Probabilistic Multileaving extension (PMGD) [28], compares multiple rankers at each interaction, resulting in the best DBGD-based algorithm, which reaches a better convergence given less training impressions [23]. However, this method suffers from a high computational cost because it requires sampling ranking assignments to infer outcomes. Further variations that reuse historical interaction data to accelerate the learning in DBDG have also been investigated [10].

The current state-of-the-art OLTR algorithm is Pairwise Differentiable Gradient Descent (PDGD) [22], which does not require sampling candidate rankers to create interleaved results lists for online evaluation. Instead, PDGD creates a probability distribution over the document set and constructs the result list by sampling documents from this distribution. Then the gradients are estimated from pairwise documents preferences based on user clicks. This algorithm provides much better performance than traditional DBGD-based methods in terms of final convergence and user online experience.

3 Counterfactual Online Learning to Rank

3.1 Counterfactual Evaluation for Online Learning to Rank

The proposed COLTR method uses counterfactual evaluation to estimate the effectiveness of candidate rankers based on the click data collected by the logging ranker. This is unlike DBGD and other OLTR methods that use interleaving. In the counterfactual learning to rank setting, the IPS estimator is used to eliminate position bias [18], providing an unbiased estimation. However, the IPS estimator requires that the propensities of result documents are known. The propensity of a document is the probability that the user will examine the document. In offline LTR settings, propensities are estimated using offline click-through data, via a randomization experiment [38]. Offline click-through data is not available in the online setting we consider, and thus the use of IPS in such an online setting becomes a challenge. To overcome this, we adapt the counterfactual estimator used in batch learning from logged bandit feedback [32, 34]. This type of counterfactual learning treats rankers as policies and samples documents from a probability distribution to create the result list. This allows us to use importance sampling to fix the distribution mismatch between policies and to use Monte Carlo approximation to estimate the risk function $\mathcal{R}(f_{\theta'})$:

$$\mathcal{R}(f_{\theta'}) = \frac{1}{k}\sum_{i=1}^{k} \delta_i \frac{p(d_i|f_{\theta'}, D)}{p(d_i|f_{\theta}, D)} \tag{1}$$

Where k is the number of documents in the result list, θ is the feature weights of the logging ranker, $\theta^{'}$ is the new ranker's feature weights which need to be estimated, and δ is the reward function. Following the Counterfactual Risk Minimization (CRM) learning principle [33], we set:

$$\delta_i = \begin{cases} 0, & \text{if the user clicked or did not examine } d_i \\ 1, & \text{if the user examined but did not click } d_i \end{cases} \tag{2}$$

In counterfactual learning to rank, the user examination is modelled as propensity. In learning from logged bandit feedback, only the examined documents are considered. In the online setting, however, it is unclear how to determine which documents the user has examined (e.g. a user may have considered a snippet, but did not click on it). We make the assumption that users always examine documents from top to bottom, and thus consider the documents ranked above the one that was clicked last as having been examined. With this in place, the reward function described in Eq. 2 can be used to assign rewards to documents in the result list.

Unlike traditional DBGD-based OLTR which ranks documents according to the scores assigned by the ranking function (i.e., deterministically), COLTR creates the result list to be provided to the user for gathering feedback by sampling documents from a known probability distribution. That is, document d_i is drawn from a distribution $p(d_i|f_\theta, D)$ computed by the logging ranker θ. We use softmax to convert document scores into a probability distribution:

$$p(d_i|f_\theta, D) = \frac{e^{\frac{f_\theta(d_i)}{\tau}}}{\sum_{d \in D} e^{\frac{f_\theta(d)}{\tau}}} \tag{3}$$

where τ is the temperature parameter, which is commonly used in the field of reinforcement learning to control the sharpness of the probability distribution [31]. For high values of τ ($\tau \rightarrow \infty$), the distribution becomes uniform. For low values ($\tau \rightarrow 0$), the probability of the document with the highest score tends to 1. After a document has been picked, the probability distribution will be renormalized to avoid sampling duplicates. This kind of probabilistic ranker has been used in previous works [4,14,22].

While it has been proved that the risk estimator in Eq. 1 is an unbiased estimator, it does suffer from the propensity overfitting problem [34], i.e., the learning algorithm may learn a ranker that assigns small probability values over all the documents d_i in the result list, as this can minimize the risk function. To address this problem, we use the self-normalized risk estimator $\mathcal{R}^{SN}(f_{\theta'})$ (similar to [34]):

$$\mathcal{R}^{SN}(f_{\theta'}) = \frac{\mathcal{R}(f_{\theta'})}{\mathcal{S}(f_{\theta'})} \tag{4}$$

where:

$$\mathcal{S}(f_{\theta'}) = \frac{1}{k} \sum_{i=1}^{k} \frac{p(d_i|f_{\theta'}, D)}{p(d_i|f_\theta, D)} \tag{5}$$

Algorithm 1. Counterfactual Online Learning to Rank (COLTR).

1: **Input**: Initial weights θ_1, ranking function f, reward function δ, number of candidate ranker n, learning rate α, step size η, variance control λ;
2: **for** $t \leftarrow 1....\infty$ **do**
3: $q_t \leftarrow recive\_query(t)$
4: $D_t \leftarrow get\_canditate\_set(q_t)$
5: $L_t \leftarrow sample\_list(f_{\theta_t}, D_t)$ // Eq.3
6: $\delta_t \leftarrow recive\_clicks(L_t)$ // Eq.2
7: $C \leftarrow [\ ]$ // create an empty candidate ranker pool
8: **for** $i \leftarrow 1....n$ **do**
9: $u_i \leftarrow sample\_unit\_vector()$
10: $\theta_i \leftarrow \theta_t + \eta u_i$ // create a candidate ranker
11: $append(C, \theta_i)$ // add the new ranker to the candidate pool
12: **end for**
13: $W \leftarrow infer\_winners(\delta_t, \theta_t, C, L_t, D_t, \lambda)$ //counterfactual evaluation, see Alg.2
14: $\theta_{t+1} \leftarrow \theta_t + \alpha \frac{1}{|W|} \sum_{j \in W} u_j$ //update θ_t to the mean of winners' unit vector
15: **end for**

Intuitively, if propensity overfitting does occur, $\mathcal{S}(f_{\theta'})$ will be small, giving a penalty to $\mathcal{R}^{SN}(f_{\theta'})$.

Following the CRM principle, the aim of the learning algorithm is to find a ranker with feature weights θ that can optimize the self-normalized risk estimator, as well as its empirical standard deviation; formally:

$$\theta^{CRM} = argmin \left(\mathcal{R}^{SN}(f_{\theta}^{'}) + \lambda \sqrt{\frac{Var(\mathcal{R}^{SN}(f_{\theta}^{'}))}{k}} \right) \tag{6}$$

The $Var(\mathcal{R}^{SN}(f_{\theta}^{'}))$ is the empirical variance of $\mathcal{R}^{SN}(f_{\theta'})$, to compute which we use an approximate variance estimation [27], where $\lambda = 1$ controls the impact of empirical variance:

$$Var(\mathcal{R}^{SN}(f_{\theta}^{'})) = \frac{\sum_{i=1}^{k} \left(\delta_i - \mathcal{R}^{SN}(f_{\theta'})\right)^2 \left(\frac{p(d_i|f_{\theta'},D)}{p(d_i|f_{\theta},D}\right)^2}{\left(\sum_{i=1}^{k} \frac{p(d_i|f_{\theta'},D)}{p(d_i|f_{\theta},D)}\right)^2} \tag{7}$$

3.2 Learning a Ranker with COLTR

The previous section described the counterfactual evaluation that can be used in an online learning to rank setting. Next, we introduce the COLTR algorithm that can leverage the counterfactual evaluation to update the current ranker weights θ_t. COLTR uses the DBGD framework to optimize the current production ranker, but it does not rely on interleaving or multileaving comparisons.

Algorithm 2. Counterfactual Evaluation ($infer\_winners(\delta_t, \theta_t, C, L_t, D_t, \lambda)$).

1: **Input**: rewards δ_t, logging ranker θ_t candidate ranker set C, result list L_t, candidate document set D_t, variance control λ;
2: $\mathcal{R} \leftarrow [\theta_t]$, $k \leftarrow length(L_t)$
3: **for** θ_i in C **do**
4: $r_i \leftarrow 0$, $s_i \leftarrow 0$
5: **for** d_j in L_t **do**
6: $p \leftarrow p(d_j|f_{\theta_t}, D_t)$ // compute the logging probability using Eq.3
7: $p' \leftarrow p(d_j|f_{\theta_i}, D_t)$ // compute the new ranker probability using Eq.3
8: $r_i \leftarrow r_i + \delta_i \frac{p'}{p}$ //compute the $R(\theta_i)$ using Eq.1
9: $s_i \leftarrow s_i + \frac{p'}{p}$ //compute the $S(\theta_i)$ using Eq.5
10: **end for**
11: $r_i^{SN} \leftarrow \frac{r_i}{s_i}$, $v_i \leftarrow Var(r_i^{SN})$ //Eq.4 and Eq.7
12: $append(\mathcal{R}, r_i^{SN} + \lambda\sqrt{\frac{v_i}{k}})$ //Eq.6
13: **end for**
14: **return** $where(\mathcal{R} < \mathcal{R}[0]) - 1$ // indexes of candidate ranker that has lower risk

Algorithm 1 describes the COLTR updating process: similar to DBGD, it requires the initial ranker weights θ_1, the learning rate α which is used to control the update speed, and the step size η which controls the gradient size. At each timestamp t, i.e., at each round of user interactions (line 2), the search engine receives a query q_t issued by a user (line 3). Then the candidate document set D_t is generated given q_t (line 4), and the results list L_t is created by sampling documents d_i without replacement from the probability distribution computed by Eq. 3 (line 5). The results list is then presented to the user and clicks observed. Then the reward label vector δ_t is generated according to Eq. 2 (line 6)[2]. Next, an empty candidate ranker pool C is created (line 7) and candidate rankers are generated and added to the pool (lines 8–12). Counterfactual evaluation is used to compute the risk associated to each ranker, as described in Algorithm 2. The rankers with a risk lower than the logging ranker are said to win and are placed in the set W (line 13). Finally, the current ranker weights are updated by adding the mean of the winners' unit vector (line 14) modulated by the learning rate α.

The method COLTR uses for computing gradients is similar to that of DBGD with Multileaving (PMGD) [29]. However, COLTR is more efficient. In fact, it does not need to generate an interleaved or multileaved result list for exploring user preferences. When the length of the result list is large, the computational cost for multileaving becomes considerable. In addition, using online evaluation to infer outcomes is very expensive, especially for probabilistic multileaving evaluation [28]: this type of evaluation requires sampling a large number of ranking assignments to decide which ones are the winner rankers – a computationally expensive operation. In contrast, the time complexity for counterfactual evaluation increases linearly with the number of candidate rankers (the for loop in

[2] Note that the length of δ_t is equal to the length of L_t.

Algorithm 2, line 3[3]). To compute the probabilities of sampling documents for the logging and new rankers (Algorithm 2, line 6 and 7), the document scores in Eq. 3 need to be renormalized after each rank: this attracts additional computational cost. For efficiency reasons, we approximate these probabilities by assuming independence, so that we can compute the probabilities only once[4]. As a result, COLTR can efficiently compare a large number of candidate rankers at each interaction.

4 Empirical Evaluation

Datasets. We used four publicly available web search LTR datasets to evaluate COLTR. Each dataset contains query-document pair features and (graded) relevance labels. All feature values are normalised using MinMax at the query level. The datasets are split into training, validation and test sets using the splits according to the datasets. The smallest datasets in our experiments are MQ2007 (1,700 queries) and MQ2008 (800 queries) [25], which are a subset of LETOR 4.0. They rely on the Gov2 collection and the query set from the TREC Million Query Track [1]. Query-document pairs are represented with respect to 46 features and 3-graded relevance (from 0, not relevant, to 2, very relevant). In addition to these datasets, we use the larger MLSR-WEB10K [25] and Yahoo! Learning to Rank Challenge datasets [5]. Data for these datasets comes from commercial search engines (Bing and Yahoo, respectively), and relevance labels are assigned on a five-point scale (0 to 4). MLSR-WEB10K contains 10,000 queries and 125 retrieved documents on average, which are represented with respect to 136 features; while, Yahoo! is the largest dataset we consider, with 29,921 queries and 709,877 documents, represented using 700 features.

Simulating User Behaviour. Following previous OLTR work [9,11,22,23,29,41], we use the cascade click model (CCM) [6,8] to generate user clicks. This click model assumes users examine documents in the result list from top to bottom and decide to click with a probability $p(click = 1|R)$, where R is the relevance grade of the examined document. After a document is clicked, the user may stop examining the remainder of the list with probability $p(stop = 1|R)$. In line with previous work, we study three different user behaviours and the corresponding click models. The *perfect* model simulates the user who clicks on every relevant document in the result list and never clicks on non-relevant documents. The *navigational* model simulates the user looking for a single highly relevant document and thus is unlikely to continue after finding the first relevant one. The *informational* model represents the user that searches for topical information and that exhibits a much nosier click behaviour. We use the settings used by previous work for instantiating these click behaviours, e.g., see Table 1 in [23]. In our experiments, the issuing of queries is simulated by uniformly sampling

[3] The length of the results list L_t is fixed.

[4] We empirically observed that this assumption does not deteriorate the effectiveness of the method.

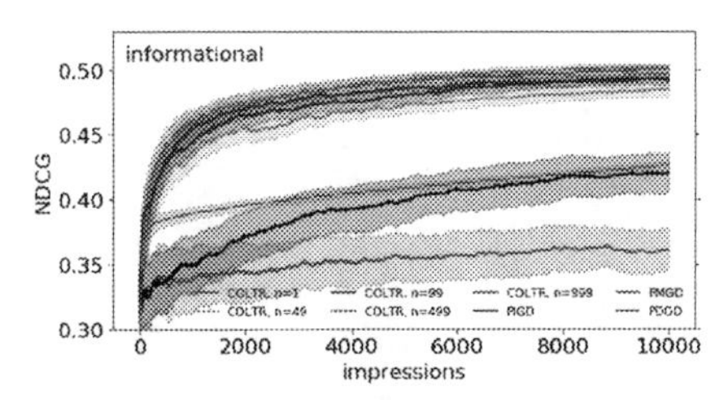

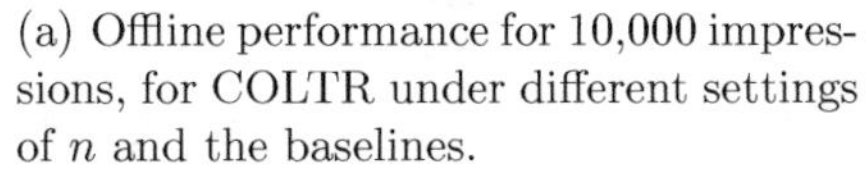
(a) Offline performance for 10,000 impressions, for COLTR under different settings of n and the baselines.

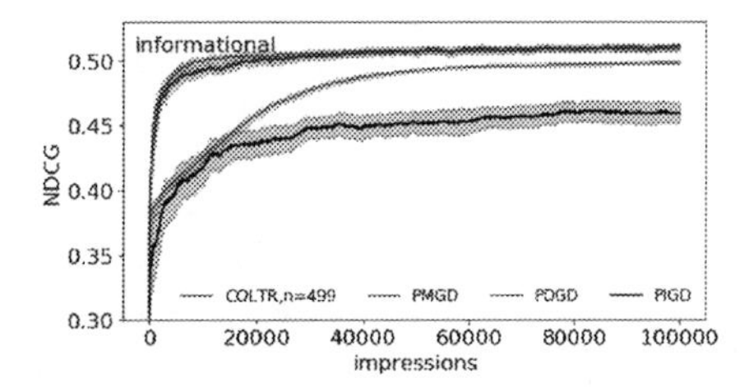

(b) Long term offline performance (100,000 impressions). For clarity, only $n = 499$ is reported for COLTR.

Fig. 1. Offline performance on the MQ2007 with the informational click model.

from the training dataset (line 3 in Algorithm 1). Then a result list is generated in answer to the query and the list is then displayed to the user. Finally, user's clicks on displayed results are simulated using CCM.

Baselines. Three baselines are considered for comparison with COLTR. The traditional DBGD with probabilistic interleaving (PIGD) [12] is used as a representative OLTR method – note that COLTR also uses DBGD, but with counterfactual evaluation in place of the interleaving method. For PIGD, only one candidate ranker is sampled at each interaction; sampling occurs by randomly varying feature weights on the unit sphere with step size $\eta = 1$, and updating the current ranker with learning rate $\alpha = 0.01$. The Probabilistic Multileaving Gradient Descent method (PMGD) [23] is also used in our experiments, as it is the DBGD-based method that has been reported to achieve the highest performance so far for this class of approaches [21]. For this baseline, we use the same parameters settings reported in previous work [22], where the number of candidates was set to $n = 49$, step size to $\eta = 1$ and learning rate to $\alpha = 0.01$. The third baseline we consider is the Pairwise Differentiable Gradient Descent (PDGD) [22], which is the current state-of-the-art OLTR method. We set PDGD's parameters according to Oosterhuis et al. [22], and specify learning rate $\alpha = 0.1$ and use zero initialization. For COLTR, we use $\eta = 1$. We use a learning rate decay for α: in this case, α starts at 0.1 and decreases according to $\alpha = \alpha * 0.99966$ after each update. We set the temperature parameter $\tau = 0.1$ when sampling documents and test different numbers of candidate rankers from $n = 1$ to $n = 999$. For all experiments, we only display $k = 10$ documents to the user, and all methods are used to optimize a linear ranker. Note, we do not directly compared with Counterfactual LTR approaches like that of Joachims et al. [18] because we consider an online setup (while counterfactual LTR requires a large datasets of previous interactions, and the estimation of propensity, which is unfeasible to be performed in an online setting).

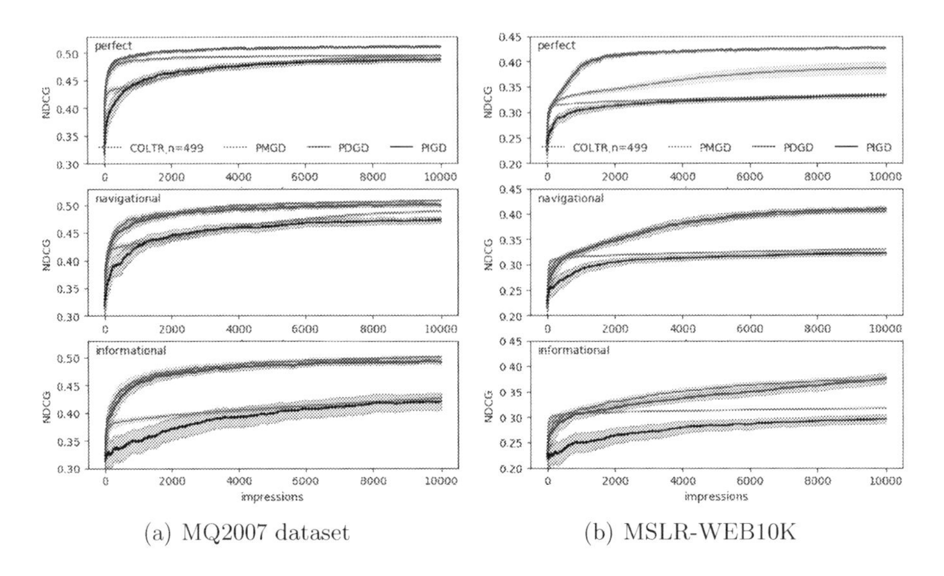

(a) MQ2007 dataset (b) MSLR-WEB10K

Fig. 2. Offline performance under three different click models

Evaluation Measures. The Effectiveness of the considered OLTR methods is measured with respect to both *offline* and *online* performance. For offline performance, we average the nDCG@10 scores of the production ranker over the queries in the held-out test set. This measure indicates the effectiveness of the learned ranker. The offline performance of each method is measured for 10,000 impressions, and the final offline performance is also recorded. Online performance is computed as the nDCG@10 score produced by the result list displayed to the user during the training phase [13]. This measure indicates the quality of the user experience during training. A discount factor γ is used to ensure that long-term impressions have less impact, i.e. $\sum_{t=1} nDCG(L_t) \cdot \gamma^{t-1}$. Following previous work [21–23], we choose $\gamma = 0.9995$ so that impressions after the horizon of 10,000 have less than a 1% impact. We repeated each experiment 125 times, spread over different training folds. The evaluation results are averaged and statistically significant differences between system pairs are computed using a two-tailed t-test.

5 Results Analysis

5.1 Offline Performance: Final Ranker Convergence

We first investigate how the number of candidate rankers impacts offline performance. Figure 1(a) displays the offline nDCG of COLTR and the baselines under the informational click setting when a different number of candidate rankers is used by COLTR (recall that PIGD uses two rankers and PMGD uses 49 rankers). Consider COLTR with one candidate ranker in addition to the production ranker ($n = 1$) and PIGD: both are considering a single alternative ranker to that in production. From the figure, it is clear that PIGD achieves a better offline performance than COLTR. However, when more candidate rankers are considered, e.g., n is increased to 49, the offline performance of COLTR becomes significantly higher than that of PIGD. Furthermore, COLTR is also better than PMGD when the same number of candidate rankers are considered. Moreover, COLTR allows to efficiently compare a large number of candidate rankers at each interaction (impression), and thus can test with a larger set of candidate rankers. We find that increasing the number of candidate rankers can help boosting the offline performance of COLTR and achieve a higher final converge. When $n = 499$, COLTR can reach significantly better ($p < 0.01$) offline performance than PDGD, the current state-of-the-art OLTR method. However, beyond $n = 499$ there are only minor improvements in offline performance, achieved at a higher computational cost – thus, in the remaining experiments, we consider only $n = 499$.

We also consider long-term convergence. Figure 1(b) displays the results for COLTR (with $n = 499$) and the baselines after 100,000 impressions. Because a learning rate decay is used in COLTR, the learning rate becomes insignificant after 30,000 impressions. In order to prevent this to happen, we stop the learning rate decay when $\alpha < 0.01$, and we leave $\alpha = 0.01$ constant for the remaining impressions. The figure shows that, contrary to the results in Fig. 1(a), PMGD can reach much higher performance than PIGD when enough impressions are considered – this finding is consistent with previously reported observations [22]. Nevertheless, both COLTR and PDGD are still significantly better than PIGD and PMGD, and have similar convergence: their offline performance is less affected by the long term impressions.

Figure 2 displays the offline performance across datasets of varying dimensions (small: MQ2007, and large: MSLR-WEB10K) under three different click models and for 10,000 impressions. The results show that PDGD and COLTR outperform PIGD and PMGD for all click models. We also find that, overall, COLTR and the current state-of-the-art online LTR approach, PDGD have very similar learning curves across all click models and datasets, apart for the perfect click model on the MSLR-WEB10K dataset, for which COLTR is severely outperformed by PDGD. Note, the trends observed in Fig. 2 found also for the majority of the remaining datasets. For space reasons, we omit these results from the paper, but we make them available as an online appendix at http://ielab.io/COLTR. Table 1 reports the final convergence performance for all datasets and click models (including statistical significance analysis), displaying similar trends across the considered datasets.

Table 1. Offline nDCG performance obtained under different click models. Significant gains and losses of COLTR over PIGD, PMGD and PDGD are marked by △, ▽ ($p < 0.05$) and ▲, ▼ ($p < 0.01$) respectively.

		MQ2007	MQ2008	MSLR10K	Yahoo!
Perfect	PIGD	0.488	0.684	0.333	0.677
	PMGD	0.495	0.689	0.336	0.716
	PDGD	0.511	0.699	0.427	0.734
	COLTR, n = 499	0.495▲ ▼	0.682 ▽ ▼	0.388▲ ▲ ▼	0.718 ▲ ▲ ▼
Navig.	PIGD	0.473	0.670	0.322	0.642
	PMGD	0.489	0.681	0.330	0.709
	PDGD	0.500	0.696	0.410	0.718
	COLTR, n = 499	0.508 ▲ ▲ ▲	0.689▲ △ ▽	0.405▲ ▲ ▽	0.718 ▲ ▲
Inform.	PIGD	0.421	0.641	0.296	0.605
	PMGD	0.426	0.687	0.317	0.677
	PDGD	0.492	0.693	0.375	0.709
	COLTR, n = 499	0.500 ▲ ▲ ▲	0.686▲ ▼	0.374▲ ▲	0.706 ▲ ▲ ▽

5.2 Online Performance: User Experience

Along with the performance obtained by rankers once training is over, the user experience obtained during training should also be considered. Table 2 reports the online performance of all methods, for all datasets and click models. The state-of-the-art PDGD has the best online performance across all conditions. COLTR outperforms PIGD and PMGD when considering the perfect click model. For other click models, COLTR is better than PIGD but it does provide less cumulative online performance than PMGD, even if it achieves a better offline performance. We posit that this is because PMGD uses a deterministic ranking function to create the result list the user observes, and via multileaving it guarantees that the interleaved result list is not worse than that of the worst candidate ranker. COLTR instead uses a probabilistic ranking function, and if the document sampling distribution is too similar to a uniform distribution, the result list may incorrectly contain many non-relevant documents: this results in a bad online performance. A uniform sampling distribution is obtained because noisy clicks result in some candidate rankers randomly winning the counterfactual evaluation and thus slowing down the gradient convergence and achieving an "elastic effect", where the weight vectors go forward in one interaction, and backwards in the next. This will cause the margins between the documents' scores assigned by the ranking function to become too small and thus the softmax function will not generate a "deterministic" distribution. This also explains why the online performance is much better when clicks are perfect: the gradient directions corresponding to the winning candidates are likely similar, leading the current ranker moving fast through large gradient updates (no elastic effect).

Table 2. Online cumulative nDCG performance under different click models. Significant gains and losses of COLTR over PIGD, PMGD and PDGD are marked by $^{\triangle}$, $^{\triangledown}$ ($p < 0.05$) and $^{\blacktriangle}$, $^{\blacktriangledown}$ ($p < 0.01$) respectively.

		MQ2007	MQ2008	MSLR10K	Yahoo!
Perfect	PIGD	795.6	1184.8	549.8	1202.0
	PMGD	824.8	1225.6	587.6	1284.7
	PDGD	936.1	1345.5	718.5	1407.8
	COLTR, n = 499	933.0$^{\blacktriangle}$ $^{\blacktriangle}$	1344.2 $^{\blacktriangle}$ $^{\blacktriangle}$	641.7$^{\blacktriangle}$ $^{\blacktriangle}$ $^{\blacktriangledown}$	1370.0 $^{\blacktriangle}$ $^{\blacktriangle}$ $^{\blacktriangledown}$
Navig.	PIGD	766.3	1152.1	533.6	1174.1
	PMGD	796.4	1195.9	581.3	1258.4
	PDGD	883.0	1309.0	642.8	1358.9
	COLTR, n = 499	790.7 $^{\blacktriangle}$ $^{\triangledown}$ $^{\blacktriangledown}$	1112.0 $^{\blacktriangledown}$ $^{\blacktriangledown}$ $^{\blacktriangledown}$	542.9$^{\blacktriangle}$ $^{\blacktriangledown}$ $^{\blacktriangledown}$	1194.8$^{\blacktriangle}$ $^{\blacktriangledown}$ $^{\blacktriangledown}$
Inform.	PIGD	681.8	1068.3	483.8	1149.6
	PMGD	745.7	1188.3	575.8	1237.9
	PDGD	859.5	1297.5	600.6	1325.4
	COLTR, n = 499	780.9 $^{\blacktriangle}$ $^{\blacktriangle}$ $^{\blacktriangledown}$	1138.7 $^{\blacktriangle}$ $^{\blacktriangledown}$ $^{\blacktriangledown}$	522.1 $^{\blacktriangle}$ $^{\blacktriangledown}$ $^{\blacktriangledown}$	1186.5 $^{\blacktriangle}$ $^{\blacktriangledown}$ $^{\blacktriangledown}$

6 Conclusion

In this paper, we have presented a novel online learning to rank algorithm that combines the key aspects of counterfactual learning and OLTR. Our method, *counterfactual online learning to rank* (COLTR), replaces online evaluation, which is the most computational expensive step in the traditional DBGD-style OLTR methods, with counterfactual evaluation. COLTR does not derive a gradient function and use it to optimise an objective, but still samples different rankers, akin to the online evaluation practice. As a result, COLTR can evaluate a large number of candidate rankers at a much lower computational expense.

Our empirical results, based on publicly available web search LTR datasets, also show that the COLTR can significantly outperform DBGD-style OLTR methods across different datasets and click models for offline performance. We also find that COLTR achieves the same offline performance as the state-of-the-art OLTR model, the PDGD, across all datasets under noisy click settings. This means COLTR can provide a robust and effective ranker to be deployed into production, once trained online. However, due to the uniform sampling distribution employed by COLTR to select among candidate documents, COLTR has worse online performance than PMGD and PDGD.

Future work will investigate the difference between gradients provided by PDGD and COLTR, as they both use a probabilistic ranker to create the result list. This analysis could provide further indications about the reasons why the online performance of COLTR is limited. Other improvements could be implemented for COLTR. First, instead of stochastically learning at each interaction, historical user interaction data could be used to perform batch learning, which may provide even more reliable gradients under noisy clicks. Note that this extension is possible, and methodologically simple for COLTR, but not for PDGD. Second, the use of the exploration variance reduction method [35,36] could be investigated to reduce the gradient exploration space: this may solve the uniform sampling distribution problem.

Acknowledgements. Dr Guido Zuccon is the recipient of an Australian Research Council DECRA Research Fellowship (DE180101579) and a Google Faculty Award.

References

1. Allan, J., Carterette, B., Aslam, J.A., Pavlu, V., Dachev, B., Kanoulas, E.: Million query track 2007 overview. In: TREC Proceedings (2007)
2. Baeza-Yates, R.: Bias on the web. Commun. ACM **61**(6), 54–61 (2018)
3. Bottou, L., et al.: Counterfactual reasoning and learning systems: the example of computational advertising. J. Mach. Learn. Res. **14**(1), 3207–3260 (2013)
4. Cao, Z., Qin, T., Liu, T.Y., Tsai, M.F., Li, H.: Learning to rank: from pairwise approach to listwise approach. In: Proceedings of the 24th International Conference on Machine Learning, pp. 129–136. ACM (2007)
5. Chapelle, O., Chang, Y.: Yahoo! learning to rank challenge overview. In: Proceedings of the Learning to Rank Challenge, pp. 1–24 (2011)
6. Chuklin, A., Markov, I., Rijke, M.D.: Click models for web search. Synth. Lect. Inf. Concepts Retrieval Serv. **7**(3), 1–115 (2015)
7. Guan, Z., Cutrell, E.: An eye tracking study of the effect of target rank on web search. In: Proceedings of the SIGCHI Conference on Human Factors in Computing Systems, CHI 2007, pp. 417–420. ACM, New York (2007)
8. Guo, F., Liu, C., Wang, Y.M.: Efficient multiple-click models in web search. In: Proceedings of the Second ACM International Conference on Web Search and Data Mining, pp. 124–131. ACM (2009)
9. He, J., Zhai, C., Li, X.: Evaluation of methods for relative comparison of retrieval systems based on clickthroughs. In: Proceedings of the 18th ACM Conference on Information and Knowledge Management, pp. 2029–2032. ACM (2009)
10. Hofmann, K., Schuth, A., Whiteson, S., de Rijke, M.: Reusing historical interaction data for faster online learning to rank for IR. In: Proceedings of the Sixth ACM International Conference on Web Search and Data Mining, pp. 183–192. ACM (2013)
11. Hofmann, K., Whiteson, S., de Rijke, M.: Balancing exploration and exploitation in learning to rank online. In: Clough, P., et al. (eds.) ECIR 2011. LNCS, vol. 6611, pp. 251–263. Springer, Heidelberg (2011). https://doi.org/10.1007/978-3-642-20161-5_25
12. Hofmann, K., Whiteson, S., De Rijke, M.: A probabilistic method for inferring preferences from clicks. In: Proceedings of the 20th ACM International Conference on Information and Knowledge Management, pp. 249–258. ACM (2011)
13. Hofmann, K., et al.: Fast and reliable online learning to rank for information retrieval. In: SIGIR Forum, vol. 47, p. 140 (2013)
14. Jagerman, R., Oosterhuis, H., de Rijke, M.: To model or to intervene: a comparison of counterfactual and online learning to rank from user interactions. In: Proceedings of the 42nd International ACM SIGIR Conference on Research and Development in Information Retrieval, SIGIR 2019, pp. 15–24. Association for Computing Machinery (2019)

15. Joachims, T.: Optimizing search engines using clickthrough data. In: Proceedings of the Eighth ACM SIGKDD International Conference on Knowledge Discovery and Data Mining, pp. 133–142. ACM (2002)
16. Joachims, T., Granka, L.A., Pan, B., Hembrooke, H., Gay, G.: Accurately interpreting clickthrough data as implicit feedback. SIGIR **5**, 154–161 (2005)
17. Joachims, T., Swaminathan, A., de Rijke, M.: Deep learning with logged bandit feedback. In: The Sixth International Conference on Learning Representations (ICLR) (2018)
18. Joachims, T., Swaminathan, A., Schnabel, T.: Unbiased learning-to-rank with biased feedback. In: Proceedings of the Tenth ACM International Conference on Web Search and Data Mining, pp. 781–789. ACM (2017)
19. Lefortier, D., Serdyukov, P., De Rijke, M.: Online exploration for detecting shifts in fresh intent. In: Proceedings of the 23rd ACM International Conference on Conference on Information and Knowledge Management, pp. 589–598. ACM (2014)
20. Liu, T.Y., et al.: Learning to rank for information retrieval. Found. Trends Inf. Retrieval **3**(3), 225–331 (2009)
21. Oosterhuis, H., de Rijke, M.: Balancing speed and quality in online learning to rank for information retrieval. In: Proceedings of the 2017 ACM on Conference on Information and Knowledge Management, pp. 277–286. ACM (2017)
22. Oosterhuis, H., de Rijke, M.: Differentiable unbiased online learning to rank. In: Proceedings of the 27th ACM International Conference on Information and Knowledge Management, pp. 1293–1302. ACM (2018)
23. Oosterhuis, H., Schuth, A., de Rijke, M.: Probabilistic multileave gradient descent. In: Ferro, N., et al. (eds.) ECIR 2016. LNCS, vol. 9626, pp. 661–668. Springer, Cham (2016). https://doi.org/10.1007/978-3-319-30671-1_50
24. Pan, B., Hembrooke, H., Joachims, T., Lorigo, L., Gay, G., Granka, L.: In google we trust: users' decisions on rank, position, and relevance. J. Comput.-Mediat. Commun. **12**(3), 801–823 (2007)
25. Qin, T., Liu, T.Y.: Introducing LETOR 4.0 datasets. arXiv preprint arXiv:1306.2597 (2013)
26. Radlinski, F., Kurup, M., Joachims, T.: How does clickthrough data reflect retrieval quality? In: Proceedings of the 17th ACM Conference on Information and Knowledge Management, pp. 43–52. ACM (2008)
27. Rubinstein, R.Y., Kroese, D.P.: Simulation and the Monte Carlo Method, vol. 10. Wiley, Hoboken (2016)
28. Schuth, A., et al.: Probabilistic multileave for online retrieval evaluation. In: Proceedings of the 38th International ACM SIGIR Conference on Research and Development in Information Retrieval, pp. 955–958. ACM (2015)
29. Schuth, A., Oosterhuis, H., Whiteson, S., de Rijke, M.: Multileave gradient descent for fast online learning to rank. In: Proceedings of the Ninth ACM International Conference on Web Search and Data Mining, pp. 457–466. ACM (2016)
30. Schuth, A., Sietsma, F., Whiteson, S., Lefortier, D., de Rijke, M.: Multileaved comparisons for fast online evaluation. In: Proceedings of the 23rd ACM International Conference on Conference on Information and Knowledge Management, pp. 71–80. ACM (2014)
31. Sutton, R.S., Barto, A.G.: Reinforcement Learning: An Introduction. MIT Press, Cambridge (2011)
32. Swaminathan, A., Joachims, T.: Batch learning from logged bandit feedback through counterfactual risk minimization. J. Mach. Learn. Res. **16**(1), 1731–1755 (2015)

33. Swaminathan, A., Joachims, T.: Counterfactual risk minimization: learning from logged bandit feedback. In: International Conference on Machine Learning, pp. 814–823 (2015)
34. Swaminathan, A., Joachims, T.: The self-normalized estimator for counterfactual learning. In: Advances in Neural Information Processing Systems, pp. 3231–3239 (2015)
35. Wang, H., Kim, S., McCord-Snook, E., Wu, Q., Wang, H.: Variance reduction in gradient exploration for online learning to rank. In: Proceedings of the 42nd International ACM SIGIR Conference on Research and Development in Information Retrieval, SIGIR 2019 (2019)
36. Wang, H., Langley, R., Kim, S., McCord-Snook, E., Wang, H.: Efficient exploration of gradient space for online learning to rank. In: The 41st International ACM SIGIR Conference on Research & Development in Information Retrieval, pp. 145–154. ACM (2018)
37. Wang, X., Bendersky, M., Metzler, D., Najork, M.: Learning to rank with selection bias in personal search. In: Proceedings of the 39th International ACM SIGIR conference on Research and Development in Information Retrieval, pp. 115–124. ACM (2016)
38. Wang, X., Golbandi, N., Bendersky, M., Metzler, D., Najork, M.: Position bias estimation for unbiased learning to rank in personal search. In: Proceedings of the Eleventh ACM International Conference on Web Search and Data Mining, pp. 610–618. ACM (2018)
39. Yue, Y., Joachims, T.: Interactively optimizing information retrieval systems as a dueling bandits problem. In: Proceedings of the 26th Annual International Conference on Machine Learning, pp. 1201–1208. ACM (2009)
40. Yue, Y., Patel, R., Roehrig, H.: Beyond position bias: examining result attractiveness as a source of presentation bias in clickthrough data. In: Proceedings of the 19th International Conference on World Wide Web, pp. 1011–1018. ACM (2010)
41. Zoghi, M., Whiteson, S.A., De Rijke, M., Munos, R.: Relative confidence sampling for efficient on-line ranker evaluation. In: Proceedings of the 7th ACM International Conference on Web Search and Data Mining, pp. 73–82. ACM (2014)

A Framework for Argument Retrieval

Ranking Argument Clusters by Frequency and Specificity

Lorik Dumani(✉), Patrick J. Neumann, and Ralf Schenkel(✉)

Trier University, 54286 Trier, Germany
{dumani,s4paneum,schenkel}@uni-trier.de

Abstract. Computational argumentation has recently become a fast growing field of research. An argument consists of a claim, such as *"We should abandon fossil fuels"*, which is supported or attacked by at least one *premise*, for example *"Burning fossil fuels is one cause for global warming"*. From an information retrieval perspective, an interesting task within this setting is finding the best supporting and attacking premises for a given query claim from a large corpus of arguments. Since the same logical premise can be formulated differently, the system needs to avoid retrieving duplicate results and thus needs to use some form of clustering. In this paper we propose a principled probabilistic ranking framework for premises based on the idea of TF-IDF that, given a query claim, first identifies highly similar claims in the corpus, and then clusters and ranks their premises, taking clusters of claims as well as the stances of query and premises into account. We compare our approach to a baseline system that uses BM25F which we outperform even with a primitive implementation of our framework utilising BERT.

Keywords: Argumentation retrieval · Argument clustering · Argument ranking · Argument search

1 Introduction

Computational argumentation is an emerging research area that has recently received increasing interest. It deals with representing and analysing arguments for controversial topics, which includes mining argument structures from large text corpora [8]. A widely accepted definition for an argument is that it consists of a *claim* or a standpoint, for instance *"We should abandon fossil fuels"*, which is supported or attacked by at least one *premise*, for example *"Burning fossil fuels is one cause for global warming"* or *"Poor people cannot afford alternative fuels"* [21]. The claim is the central and usually also a controversial component, which should not be accepted by the reader without further support (by premises) [28].

This work has been funded by the Deutsche Forschungsgemeinschaft (DFG) within the project ReCAP, Grant Number 375342983 - 2018–2020, as part of the Priority Program "Robust Argumentation Machines (RATIO)" (SPP-1999).

J. M. Jose et al. (Eds.): ECIR 2020, LNCS 12035, pp. 431–445, 2020.
https://doi.org/10.1007/978-3-030-45439-5_29

From an information retrieval perspective, an interesting task within this setting is finding the best supporting (pro) and attacking (con) premises for a given query claim [31]. This has applications in many domains, including journalism and politics, and in general is relevant for making informed decisions. By now, existing (Web) search engines like GOOGLE only provide the most relevant documents to the user, but cannot structure their results in terms of claims and premises. There is a relatively large body of work on how arguments can be mined from text (see [8] for a recent survey). In this paper, we build upon established research on argument search engines and focus on effectively retrieving premises for a query claim from a large corpus of already mined arguments. Here, a query can be either a controversial topic (e.g. *"fossil fuels"*) or statement (e.g. *"we should abandon fossil fuels"*), and the task of the system is to retrieve a ranked list of pro and con premises for the query. Since the same logical premise can be formulated semantically similar, an argument retrieval system has to avoid retrieving duplicate results and thus needs to use some form of clustering.

Previous approaches in this area have focused on estimating the relevance of premises in combination with the corresponding claims, using BM25F [30] for example. The novel contribution of this paper is a principled probabilistic ranking framework for premises that, given a query claim, first determines highly similar claims in the corpus, and then clusters and ranks their premises, taking clusters of claims as well as the stances of query and premises into account.

The remainder of this paper is structured as follows: Sect. 2 discusses related work. Section 3 introduces necessary notation and Sect. 4 presents our probabilistic ranking framework. Section 5 describes details of the implementation of our framework in which we use BERT [11] to capture the vectors of premises and applied hierarchical clustering. In Sect. 6 we evaluate our approach with a large corpus [12] consisting of 63,250 claims and about 695,000 premises and compare it to a baseline system that uses BM25F. Section 7 concludes the paper and discusses ideas for future work.

2 Related Work

Stab et al. [27] present ARGUMENTEXT [4]. Their argument retrieval system first retrieves relevant documents, then it identifies relevant arguments. We do not address the argument mining task. Our work is more similar to the work of Wachsmuth et al. [30] who present ARGS [3], one of the first prototypes of an argument search engine. ARGS operates on arguments crawled from five debate portals (such as debate.org and idebate.org). Given a user's keyword query, the system retrieves, ranks, and presents premises supporting and attacking the query, taking similarity of the query with the premise, its corresponding claim, and other contextual information into account. They apply a standard BM25F ranking model implemented on top of Lucene. In our prior work [12], we build on the work of Wachsmuth et al. and systematically compared 196 methods for identification of similar claims by textual similarity, using a comparable large corpus of (claim, premise) pairs crawled from several debate portals. The results imply that matching similar claims to a query claim with Divergence from Randomness (DFR) [2] yields slightly better results than BM25 [24]. Thus, we will make use of DFR to find the most similar claims to a query claim.

The work on argument quality and ranking is also a subarea addressed in the community. Habernal and Gurevych address the relevance of premises [15]. They confronted users in a crowdsourced task with pairs of premises to decide which premise is more convincing. Then, they used a bidirectional LSTM to predict which of two given arguments is better. In a follow-up work [14], they also investigate in the constitution of convincing arguments. Wachsmuth et al. [32] consider the problem of judging the relevance of arguments. An overview of the work on computational argumentation quality in natural language, including theories and approaches is provided by them. Their work can be used to determine the quality of arguments and thus also for the ranking.

Reimers et al. [23] deal with clustering premises. ELMo [22] and BERT [11] were used to classify and cluster topic-dependent arguments. They improve the baseline for both tasks but also recognise that arguments can address multiple aspects and therefore belong to multiple clusters. We build upon this work by using BERT to cluster claims as well as premises. As they do, we use a hard clustering algorithm and leave soft clustering algorithms for future work since this paper intends to set up the foundation and show the potential of the framework.

3 Problem Definition and Notations

We assume that we work with a large corpus of argumentative text, for example collections of political speeches or forum discussions, that has already been mined and transferred into claims with the corresponding premises and stances.

We consider the following problem: Given a controversial claim or topic, for example "*We should abandon fossil fuels*", a user searches for the most important premises from the corpus supporting or attacking it. It is important to take into account that even if different claims or premises are semantically equivalent, they will usually be formulated in different ways, so we will consider clusters of claims (and clusters of premises) with the same meaning instead of isolated claims and premises. Finding this clustering of premises and claims as well as choosing a good representative of each result cluster to show to the user are additional tasks of the system.

We will now introduce some notations used in the remainder of the paper. Let $\mathcal{C}$ be the set of all claims in our corpus. A *claim cluster* $\gamma_j \subseteq \mathcal{C}$ is a subset of claims with the same meaning, and a *claim clustering* $\Gamma = \{\gamma_1, \gamma_2, \ldots\}$ is a disjoint partitioning of $\mathcal{C}$ into claim clusters. The function $\gamma : \mathcal{C} \rightarrow \Gamma$ assigns to a claim $c_i \in \mathcal{C}$ its corresponding cluster γ_j (which exists and is unique).

Let $\mathcal{P}$ be the set of all premises in the corpus. We write $p \rightarrow c$ if $p \in \mathcal{P}$ appears as a premise for $c \in \mathcal{C}$ in the corpus, and $p^+ \rightarrow c$ if p supports c. Similar to claim clusters, we consider premise clusters $\pi_j \subseteq \mathcal{P}$ of premises with the same meaning and the corresponding *premise clustering* $\Pi = \{\pi_1, \pi_2, \ldots\}$ as a disjoint partitioning of $\mathcal{P}$ into premise clusters. The function $\pi : \mathcal{P} \rightarrow \Pi$ assigns to a premise $p_i \in \mathcal{P}$ its corresponding premise cluster π_j.

For a premise cluster π_j, $C(\pi_j) \subseteq \mathcal{C}$ denotes the set of claims attacked or supported by premises in π_j. Note that two subsets $C(\pi_j), C(\pi_l)$ with $j \neq l$ may overlap for different premise clusters because the same premise or premises from the same cluster (e.g. '*it is very expensive*') can support or attack very different claims (e.g. '*nuclear energy*' and '*health care*'). Figure 1 gives an example of a corpus with similar claims and premises.

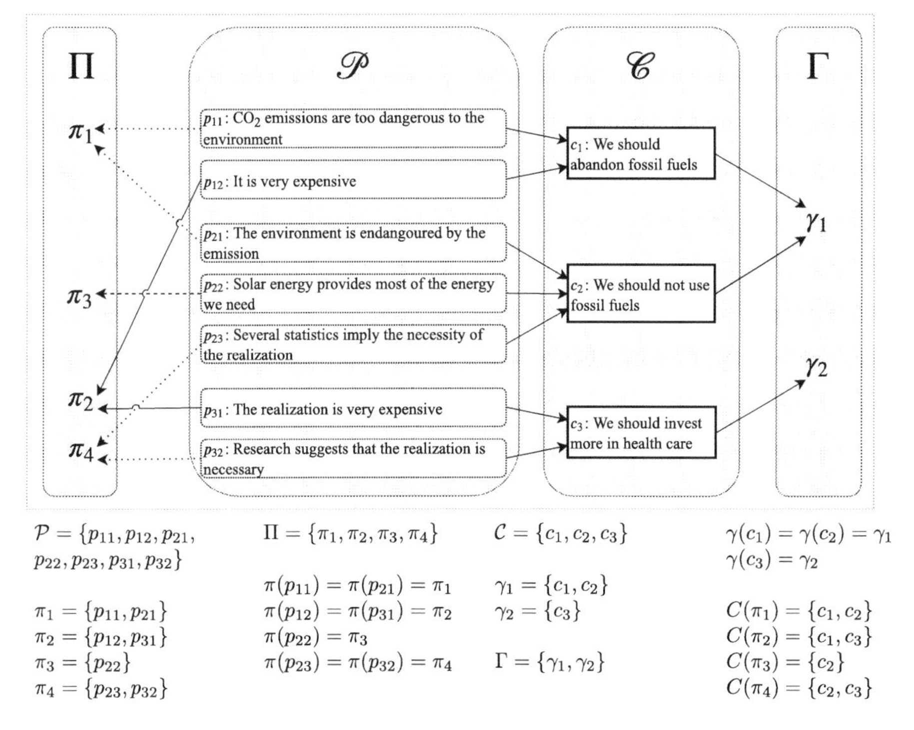

Fig. 1. Example for a corpus with clusters of similar claims $\Gamma = \{\gamma_1, \gamma_2, \ldots\}$ and clusters of similar premises $\Pi = \{\pi_1, \pi_2, \ldots\}$.

A claim may come with a stance, and different claims may have different stances, even though they deal with the same topic. To see why this is important, consider the following example claims and their stances: c_1 = *"We should use fossil fuels"* (positive stance), c_2 = *"We should abandon fossil fuels" (negative stance)*, c_3 = *"Fossil fuels"* (neutral stance), and c_4 = *"Should fossil fuels be used?"* (neutral stance). We treat claims with neutral stances as if they had a positive stance. For a query asking for *"increase usage of fossil fuels"*, supporting premises would be premises that support c_1, c_3, c_4, but also premises that attack c_2. Similarly, attacking premises would be those attacking c_1, c_3, c_4 or supporting c_2. Let q and c be query and claim on the same topic, then if q and c have the same stance, a premise supporting c will also support q. Also, if q and c have opposite stance, a premise supporting c will attack q. We write $q \uparrow\uparrow c$ if the stances of q and c are aligned and $q \uparrow\downarrow c$ otherwise. We further assume that all claims within the same cluster have the same stance.

4 Probabilistic Ranking Framework

4.1 Probability of Premise Clusters

Given a query claim q, the goal is to find the best clusters of supporting and attacking premises π^+, π^- for q in the corpus. Here, $P(\pi^+|q)$ defines the probability that a user would pick π as the supporting cluster of premises for q amongst all premise clusters in the corpus. Furthermore, $P(\pi^-|q)$ is defined analogously for attacking clusters.

To compute these probabilities, we first consider single premises and claims and extend this to clusters afterwards; we then will discuss how stances can be taken into account. We will restrict the examination to supporting premises, attacking premises are computed analogously.

First we estimate the probability $P(p^+|q)$ that the user picks the supporting premise p for query claim q. We assume the following user model: To pick a supporting premise, the user initially selects a matching claim c for q amongst all claims in the corpus with probability $P(c|q)$, and then picks a premise p with probability $P(p^+|c, q)$ amongst all supporting premises of this claim. Considering that p may support multiple claims, $P(p^+|q)$ can thus be written as

$$P(p^+|q) = \sum_{c \in C} P(c|q) \cdot P(p^+|c, q) \tag{1}$$

where $\sum_{c \in C} P(c|q) = 1$. Since $P(p^+|c, q) = 0$ if p is not a premise of c as the user picks only premises of c, we can restrict the summation to claims for which p appears as premise. In addition, we assume that $P(p^+|c, q) = P(p^+|c)$, i.e. p is picked as support for c independently from q.

To include the stances of query and claims, we must consider that an attacking premise of a claim with opposite stance to the query can also be picked as a supporting premise of the query. This results in the following updated expression:

$$P(p^+|q) = \sum_{c:p \rightarrow c} P(c|q) \cdot \left(P(q \uparrow\uparrow c) \cdot P(p^+|c) + P(q \uparrow\downarrow c) \cdot P(p^-|c)\right) \tag{2}$$

with $P(p^-|c)$ describing the probability that p is picked as an attacking premise of claim c, $P(q \uparrow\uparrow c)$ being the probability that q and c have the same stance, and $P(q \uparrow\downarrow c)$ being the probability that q and c have opposite stance.

Finally, to compute the probability of picking a premise cluster instead of a single premise, we additionally need to aggregate over all premises in the cluster; this works since premise clusters are disjoint by construction:

$$P(\pi_j^+|q) = \sum_{p \in \pi_j} P(p^+|q) \tag{3}$$

Note that if the user does not make a distinction between supporting and attacking clusters, but instead just wants good premise clusters, we can extend the experiment such that the user first throws a fair coin to decide if he will pick a supporting or attacking premise cluster. This leads to the following probability for picking premise cluster π_j:

$$P(\pi_j|q) = \frac{P(\pi_j^+|q) + P(\pi_j^-|q)}{2} \tag{4}$$

4.2 Estimating the Probabilities

We now present possible estimators for each of the probabilities used in our ranking framework. While we think that these estimators are reasonable, there are clearly many other ways for their estimation, for example taking argument quality [32] into account; this is left for future work.

$P(c|q)$ denotes the probability that c is "relevant" for query q, which can be estimated using standard text retrieval approaches; in our experiments, we will use Divergence from Randomness [2]. Since most retrieval approaches are not probabilistic in nature, we need to recalibrate the computed scores such that their values correspond to probabilities.

$P(p^+|c)$ is the probability that p is chosen amongst all supporting premises of c. Here, we will not use textual similarity of p and c since good premises supporting or attacking a claim often have only small textual overlap with the claim. As an example, consider a user searching for premises supporting the claim "*we should abandon fossil fuels*". A good premise could be "*wind and solar energy can already provide most of the needed energy*", which does not overlap at all with the claim. Instead, we will estimate this based on two different frequency statistics: the *premise frequency* $pf(p^+, c)$, which describes the frequency with which premise p is used as support for claims within c's claim cluster, i.e. with the same meaning as c, and the *claim frequency* $cf(p^+)$, which is the number of claim clusters for which premise p is used as support. Intuitively, we prefer premises that appear frequently within a claim cluster, and we may want to give lower weight to premises that appear within most or even all claim clusters. This is exactly the same principle used in the TF-IDF term weight [25]. We therefore use the *inverse claim frequency* $icf(p^+)$ in a form similar to standard IDF. Since the same "semantic" premise can appear in different textual formulations, we will consider its premise cluster instead of the actual premise when computing $pf(p^+)$ and $icf(p^+)$. We can formalise this as follows:

(i) $pf(p^+, c) = |\{p'^+ \rightarrow c' : p' \in \pi(p+), c' \in \gamma(c)\}|$
(ii) $icf(p^+) = \log\left(\frac{|\Gamma|}{|\{\gamma \in \Gamma:\ \exists p'^+ \in \pi(p), \exists c' \in \gamma \text{ such that } p'^+ \rightarrow c'\}|}\right)$

We then estimate $P(p^+|c)$ as

$$P(p^+|c) = \frac{pf(p^+, c) \cdot icf(p^+)}{Z} \tag{5}$$

where Z is a normalisation term computed as the sum of the unnormalised *pf·icf* products over all candidate premises; this is not needed for ranking the premises.

Estimating the probability that two claims (or, more generally, two statements) have the same stance is a surprisingly hard problem that has not yet been solved, especially if two statements have different stances [16]. We therefore omit this part of the framework in this paper and instead focus on the evaluation of the other parts, which form the core of the framework.

5 Implementation

Now we describe the concrete implementation of the framework, i.e. the clustering of claims as well as the clustering of premises.

Clustering the Claims. We cluster the claims in an offline operation with hierarchical clustering. For each claim, we calculate its embeddings using BERT [11][1]. This allowed us to create an agglomerative clustering [17], i.e. a bottom-up approach[2]. Compared to k-means [20], hierarchical clustering has the advantage of not needing to provide the number of resulting clusters beforehand. In general, only few parameters are expected here, which leads to less overfitting. For example, it expects only a method to determine the distance between two vectors and a method to link clusters. For the former we have taken the often used Euclidean distance function, and for the latter the widely used average linkage method [26], which calculates the mean of two clusters for connecting both. In order to determine a cutoff value for the clustering, we took the implementation of Langfelder et al. [19], which produces a dynamic tree cut. Contrasting constant height cut, amongst others it is capable of identifying nested clusters.

Clustering the Premises and Computing Results. Since there are usually many more premises than claims, precomputing their clustering is not viable. Instead, we use an approximation that clusters relevant premises at query time. After a query claim q arrives in the system, the top K most similar claims $R = \{r_i | 1 \leq i \leq K\}$ are retrieved from the corpus using Divergence from Randomness [2]. At the same time, we obtain $P(c|q)$ (after normalisation). Then the corresponding claim clusters are determined and all their premises $M = \{p | \exists c \in R, \exists c' \in \gamma(c) \text{ such that } p \rightarrow c'\}$ are retrieved from the corpus. From the set M, an expanded set M' is then constructed by adding, for each premise in M, its N most similar premises from the corpus, according to the state-of-the-art standard retrieval method BM25[3]. This ensures that our premise set is large enough

[1] We use the python framework FLAIR which supports document embeddings [1] and choose the pretrained model "large-uncased" where the output vectors have 4,096 dimensions.

[2] We perform the clustering with the scripting language R and the packages STATS and FASTCLUSTER.

[3] Note that we use DFR to find similar claims, but not to find similar premises, because we only have a study supporting the former [12]. Also, claims and premises differ in length as well as details and information [13].

to compute claim frequencies. Using BERT embeddings again, this expanded premise set is first hierarchically clustered and then a dynamic tree cut is made.

Unfortunately, BERT does not support more than 512 tokens, but some premises are longer. We have thus implemented the three variants BERT_{512}, BERT_{sw}, and BERT_{sent}. With BERT_{512} we simply truncate a premise after 512 tokens, i.e. the embeddings only refer to the first 512 tokens of a text. With BERT_{sw} we utilised a sliding window, i.e. for premises with more than 512 tokens we always considered only text spans with a maximum length of 512, but always shifted the window to the right by 256 until the end of the premise in order to keep as much context information as possible. Hence, for a text s that has more than 512 tokens, we get $\left\lceil \frac{|s|}{256} \right\rceil$ embeddings, of which the average is calculated pointwise at the end. With BERT_{sent} we determine embeddings for each sentence of a premise and finally form the average of all embeddings for a premise pointwise.

After the clustering, premise frequency and claim frequency are computed for each premise in the original set M as well as the final probabilities for each premise cluster. Lastly, the clusters have to be presented to the user in an adequate format. Therefore, a premise is chosen from each cluster as a representative. In our implementation, this is the premise p with the longest text.

6 Evaluation

Now we describe the evaluation of our approach which clusters and ranks premises with respect to given queries. First we explain the dataset and the baseline we used, then we describe the setup of the ground-truth of premise clusters and the evaluation metrics. Finally, we present the evaluation results.

6.1 Dataset and Baseline

We used the dataset of our prior work [12] which consists of 63,250 claims and about 695,000 premises extracted from four debate portals. After clustering, the 63,250 claims were distributed over a total of 10,611 clusters. The average cluster size is about 6.1, the median is 5.

The final evaluation corpus in this prior work consists of triples of the form (query claim, result claim, result premise) for a total of 232 query claims which are all related to the topic "energy". Result claims are these which were identified by pooling the top five similar claims for a query claim using standard IR methods. The result premises are associated with the corresponding result claims. Using this final evaluation corpus, we randomly selected 30 query claims and extracted 1,221 individual triples. As the premises later had to be clustered manually, we made sure that the union of the result premises of all result claims for each of the 30 query claims did not exceed the number 50.

The relevance of each premise for the corresponding query claim was assessed by two annotators on a three-fold relevance scale as "very relevant", "relevant", and "not relevant". Note that the actual result claims were not shown to the assessors. The inter-annotator agreement, measured with Krippendorff's α [18], was 0.480 on a nominal scale and 0.689 on an interval scale, indicating that the annotation is robust. Disagreements between the annotations were discussed in order to achieve an agreement. After removing 26 triples because their premises were annotated as "spam" or "other", we obtained a final corpus $corp_{eval}$ of 1,195 triples consisting of 389 very relevant, 139 relevant, and 667 not relevant premises for the 30 queries.

As a baseline system, we implemented the approach proposed by Wachsmuth et al. [30] that indexes premises together with their claims and uses a BM25F scoring model [24], giving more importance to the claim than to the premise[4]. Since they gave no parameter settings, we use the default values 1.2 and 0.75 for k_1 and b, respectively [7]. As Wachsmuth et al. describe, the three fields *conclusion*, *full arguments*, and *discussion* were added to the BM25F method. In the field 'conclusion' we store the result claim, in the field 'full argument' the premise together with the result claim. The field 'discussion' reflects the context and contains the whole debate, i.e. the result claim and all its premises.

6.2 Ground-Truth and Evaluation Metrics

In order to setup a ground-truth for our experiments, we derive a ground-truth corpus $corp_{gt}$ by including only the 528 triples from $corp_{eval}$ where the premises were assessed either as relevant or as very relevant to the query claim.

For each of the 30 query claims, the premises of $corp_{gt}$ were clustered by two annotators. They were shown all result premises for a query claim, then they clustered them based on their subjectively perceived semantic similarity. One annotated, the other checked. Again, discordances were discussed in order to achieve an agreement. Please note, that the annotators were instructed to assign only premises with the same relevance level to the same (ground-truth) cluster, which also served as a pre-filter to reduce complexity.

Since we are searching for similar claims to a query claim in the first step, it is essential to know their stances in order to identify the stances of the premises to a query, so that the clustering of the premises can be divided into pros and cons. However, as it is (still) an unsolved problem to match the stance with a good probability [16], we will ignore the stance in this experiment and tackle this task in future work.

For each query, the ground-truth G then consists of clusters $G_1, \ldots, G_t$ such that each G_i contains premises with the same meaning and with the same relevance level assigned by the assessors. The relevance level assigned to premises in cluster G_i is denoted by $rel(G_i)$. We assume that the clusters are numbered such that $i \leq j$ implies $rel(G_i) \geq rel(G_j)$. Note that premises assessed as irrelevant are not included in any ground-truth cluster.

The user now asks for a summary of premises supporting and attacking the query claim. A good system will now retrieve, for a given query, a list of premises that (1) covers the different premises clusters in the ground-truth, (2) retrieves

[4] To implement a BM25F scoring model, we used the code described in [5,6].

premises from highly relevant clusters before premises from "only" relevant clusters, and (3) does not retrieve multiple premises from the same cluster. Note that this setup is different from standard adhoc retrieval since the system must identify the various aspects of the results. It also differs from diversity-aware and novelty-aware approaches [9] since the user is interested in all aspects of the query, but asks for a single representative result per aspect only.

To evaluate the quality of the retrieved results, we use a simplified variant of α-nDCG [10], which we will later extend to work with clusters as results. We consider two sub-tasks here. In Task A, the system retrieves a list of premise clusters, whereas in Task B, the system needs to additionally decide for one representative premise from each cluster to show to the user. A system that would not at all consider premise clusters, for example by indexing and searching directly at the level of premises, can solve Task B only.

We will now first explain how to evaluate Task B with a simplified variant of α-nDCG [10] where we set $\alpha = 1.0$ and consider each ground-truth cluster as an information nugget. The system returns a sorted list of premises $R = (r_1, r_2, ..., r_k)$ where r_1 is the topmost result; we assume that there are no ties in the ranking (otherwise, ties will be broken arbitrarily). To compute the gain of the result at rank i, we first check if it appears in any ground-truth cluster; if not, its gain is 0. Otherwise, let g_j be the ground-truth cluster of r_i. If no result of this cluster has appeared up to rank $i-1$, the gain of r_i is $rel(g_j)$; otherwise, its gain is 0 since it does not contribute a novel aspect. As in standard nDCG, the discount for rank i is computed as $\frac{1}{\log_2 i}$ if $i \geq 2$ and 1 otherwise. In the ideal gain vector needed for computing nDCG, the component at position i is the relevance level $rel(G_i)$ of G_i, which is ideal since ground-truth clusters are ordered by descending relevance level.

To illustrate the principles of our metrics for Task B, consider the ground-truth shown in Fig. 2. The left visualises the ground-truth for a query with three clusters: G_1 which is highly relevant (score 2), and G_2 and G_3 which are relevant (score 1). On the right are the premises that the system has returned, sorted by their estimated relevance. The ideal gain vector for this ground-truth is $2, 1, 1$, corresponding to an ideal discounted cumulative gain of $2 + 1 + \frac{1}{\log_2(3)} \approx 3.63$. The gains for the result list retrieved by the system are $2, 1, 0, 0, 0, 0, 0, 1$ (since duplicate results from the same cluster are assigned a gain of 0), corresponding to a discounted cumulative gain of $2 + 1 + \frac{1}{\log_2(8)} = \frac{10}{3}$. The nDCG of this result list is thus (approximately) $\frac{10/3}{3.63} = 0.92$.

Task A is more difficult to evaluate since we do not have a list of premises, but of premise clusters (i.e. sets of premises); existing nDCG variants cannot be applied here since they operate on lists of documents, not clusters. To be able to apply the evaluation machinery introduced for Task B, we generate all possible result lists from the list of clusters, compute nDCG for each list, and aggregate the per-list values using either average, max, or min. If, e.g. our system returns two clusters $\pi_1 = \{p_1, p_2\}, \pi_2 = \{p_3, p_4\}$, then the result lists $(p_1, p_3), (p_1, p_4), (p_2, p_3), (p_2, p_4)$ are generated.

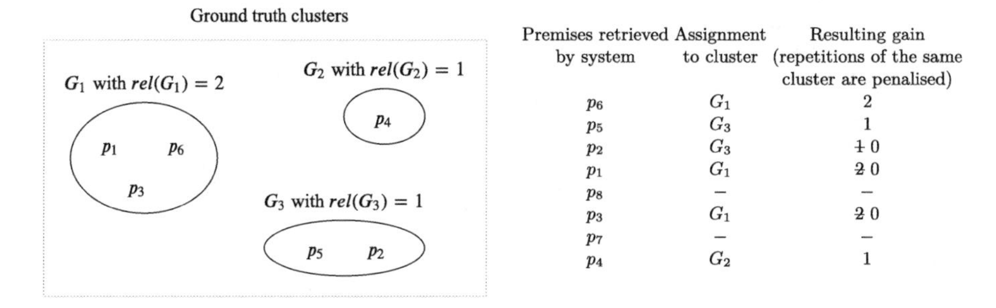

Fig. 2. Example of premise clusters with graded relevance assessments.

6.3 Evaluation Results

The results of our evaluation can be found in Tables 1 and 2. Table 1 shows the evaluation of Task B, i.e. the mean nDCG@{5,10} values for all queries. Since this process requires the selection of a representative and is difficult to decide even for humans, we have simply taken the longest premise. The table reveals that the implementation BERT_{sw}, which calculates the premises' embeddings using the 'sliding window' method, performs best. For BERT_{sw} and BERT_{512}, the observed improvements over the baseline BM25F are statistically significant for nDCG@5 (tested with Welch's t-test [33] with $p = 0.05$).

Table 1. The evaluation results for Task B showing the mean nDCG values for the baseline and clustering methods BERT_m with premise preprocessing method m for the 30 queries. The p-values are related to the baseline.

Method	Representative	mean nDCG@5	p-value (nDCG@5)	mean nDCG@10	p-value (nDCG@10)
BM25F	–	.6383	–	.6087	–
BERT_{512}	longest premise	.6458	.008	.6097	.085
BERT_{sw}	longest premise	**.6782**	.002	**.6467**	.084
BERT_{sent}	longest premise	.5943	–	.5615	–

Since the baseline only returns a ranked list and not a ranked list of clusters, we interpret this list as clusters each with one entry in Table 2. We can infer from Table 2 that BERT_{sw} performs best. Using Welch's t-test with $p = 0.05$ once more, the observed improvement over the baseline is statistically significant for the mean average nDCG@5 but not for nDCG@10. Still, the results imply that BERT_{sw} is at least as good as the baseline for nDCG@10. Note that BERT_{sw} has not even been fine-tuned. Moreover, the results in Table 2 unambiguously underline the importance of clustering and even more the choice of the correct representative. If we always chose the best representative, then we always have

Table 2. The evaluation results for Task A showing the nDCG values for baseline BM25F as well as the mean average, minimum, and maximum nDCG values for clustering methods BERT_m with premise preprocessing method m. The p-values are related to the baseline.

Method	mean average nDCG	mean minimum DCG@5	mean maximum nDCG@5	p-value (mean average nDCG@5)	mean average nDCG@10	mean minimum nDCG@10	mean maximum nDCG@10	p-value (mean average nDCG@10)
BM25F	.6383	–	–	–	.6087	–	–	–
BERT_{512}	.6309	.4409	.7744	.439	.5987	.4292	.7381	.586
BERT_{sw}	**.6523**	.4561	.8053	.012	**.6187**	.4383	.7638	.203
BERT_{sent}	.5994	.4274	.7449	–	.5665	.4135	.7067	–

the maximum value and vice versa. Note that the premises used in our experiment are extracted from debate portals and thus are not always premises in the sense of argumentation theory, as they often consider more than one aspect.

7 Conclusion and Future Work

Clustering and ranking premises is a very difficult, but important task, since a user searching for premises wants them to be presented in a compact and complete format. In this paper, we made use of the idea of TF-IDF and presented a framework for clustering and ranking premises. We used premises from debate portals, which are partially from moderated websites, and of high quality but usually very long. We showed that ranking premises by their frequency and specificity has great potential since our implementation using BERT and a hard clustering algorithm outperforms the baseline BM25F although the model was not fine-tuned and the premises actually cover many aspects, so a premise could be assigned to several clusters.

In future work we will integrate soft clustering algorithms, for which we first have to break down the premises into their individual parts (e.g. *Argumentative Discourse Units* and *Elementary Discourse Units*) [29]. In addition, we will train different fine-tunings for different sentence embedding models in order to achieve better results. In our implementation, the clustering of the 695,000 premises was not precalculated, instead it was determined dynamically for a smaller subset, since this is a very computationally intensive task. Therefore, we will also precalculate the clusters of premises. To stay within the scope of this paper, we have assumed a flat hierarchy for argument graphs, where an argument consists of a claim and many premises, as they occur e.g. in debate portals. In the future we will extend our framework with more complex structures with more layers.

Acknowledgement. We would like to thank Manuel Biertz, Christin Katharina Kreutz, Alex Witry, and Tobias Zeimetz for their invaluable help in the annotations.

References

1. Akbik, A., Blythe, D., Vollgraf, R.: Contextual string embeddings for sequence labeling. In: Proceedings of the 27th International Conference on Computational Linguistics, COLING 2018, Santa Fe, New Mexico, USA, 20–26 August 2018, pp. 1638–1649 (2018). https://aclweb.org/anthology/C18-1139/
2. Amati, G., van Rijsbergen, C.J.: Probabilistic models of information retrieval based on measuring the divergence from randomness. ACM Trans. Inf. Syst. **20**(4), 357–389 (2002). https://doi.org/10.1145/582415.582416
3. Args. https://www.args.me/index.html. Accessed 08 Jan 2020
4. ArgumenText. http://www.argumentsearch.com/. Accessed 08 Jan 2020
5. BM25F in lucene. github. https://github.com/o19s/lucene-bm25f/. Accessed 23 Sept 2019
6. Open source connections. BM25F in lucene. https://opensourceconnections.com/blog/2016/10/19/bm25f-in-lucene/. Accessed 23 Sept 2019
7. Standard values for k1 and b for BM25. https://www.elastic.co/guide/en/elasticsearch/guide/current/pluggable-similarites.html/. Accessed 23 Sept 2019
8. Cabrio, E., Villata, S.: Five years of argument mining: a data-driven analysis. In: Proceedings of the Twenty-Seventh International Joint Conference on Artificial Intelligence, IJCAI 2018, Stockholm, Sweden, 13–19 July 2018, pp. 5427–5433 (2018). https://doi.org/10.24963/ijcai.2018/766
9. Clarke, C.L.A., Craswell, N., Soboroff, I., Ashkan, A.: A comparative analysis of cascade measures for novelty and diversity. In: King, I., Nejdl, W., Li, H. (eds.) Proceedings of the Forth International Conference on Web Search and Web Data Mining, WSDM 2011, Hong Kong, China, 9–12 February 2011, pp. 75–84. ACM (2011). https://doi.org/10.1145/1935826.1935847
10. Clarke, C.L.A., et al.: Novelty and diversity in information retrieval evaluation. In: Proceedings of the 31st Annual International ACM SIGIR Conference on Research and Development in Information Retrieval, SIGIR 2008, Singapore, 20–24 July 2008, pp. 659–666. ACM (2008). https://doi.org/10.1145/1390334.1390446
11. Devlin, J., Chang, M., Lee, K., Toutanova, K.: BERT: pre-training of deep bidirectional transformers for language understanding. In: Proceedings of the 2019 Conference of the North American Chapter of the Association for Computational Linguistics: Human Language Technologies, NAACL-HLT 2019, Minneapolis, MN, USA, 2–7 June 2019, Volume 1 (Long and Short Papers), pp. 4171–4186 (2019). https://aclweb.org/anthology/papers/N/N19/N19-1423/
12. Dumani, L., Schenkel, R.: A systematic comparison of methods for finding good premises for claims. In: Proceedings of the 42nd International ACM SIGIR Conference on Research and Development in Information Retrieval, SIGIR 2019, Paris, France, 21–25 July 2019, pp. 957–960 (2019). https://doi.org/10.1145/3331184.3331282
13. Gleize, M., et al.: Are you convinced? Choosing the more convincing evidence with a Siamese network. In: Proceedings of the 57th Conference of the Association for Computational Linguistics, ACL 2019, Florence, Italy, 28 July–2 August 2019, Volume 1: Long Papers, pp. 967–976 (2019). https://www.aclweb.org/anthology/P19-1093/
14. Habernal, I., Gurevych, I.: What makes a convincing argument? Empirical analysis and detecting attributes of convincingness in web argumentation. In: Proceedings of the 2016 Conference on Empirical Methods in Natural Language Processing, EMNLP 2016, Austin, Texas, USA, 1–4 November 2016, pp. 1214–1223 (2016). http://aclweb.org/anthology/D/D16/D16-1129.pdf

15. Habernal, I., Gurevych, I.: Which argument is more convincing? Analyzing and predicting convincingness of web arguments using bidirectional LSTM. In: Proceedings of the 54th Annual Meeting of the Association for Computational Linguistics, ACL 2016, Berlin, Germany, 7–12 August 2016, Volume 1: Long Papers (2016). https://www.aclweb.org/anthology/P16-1150/
16. Hanselowski, A., et al.: A retrospective analysis of the fake news challenge stance-detection task. In: Proceedings of the 27th International Conference on Computational Linguistics, COLING 2018, Santa Fe, New Mexico, USA, 20–26 August 2018, pp. 1859–1874 (2018). https://www.aclweb.org/anthology/C18-1158/
17. Jain, A.K., Dubes, R.C.: Algorithms for Clustering Data. Prentice-Hall, Upper Saddle River (1988)
18. Krippendorff, K.: Estimating the reliability, systematic error and random error of interval data (1970)
19. Langfelder, P., Zhang, B., Horvath, S.: Dynamic tree cut: in-depth description, tests and applications (2009). https://horvath.genetics.ucla.edu/html/CoexpressionNetwork/BranchCutting/Supplement.pdf
20. Lloyd, S.P.: Least squares quantization in PCM. IEEE Trans. Inf. Theory **28**(2), 129–136 (1982). https://doi.org/10.1109/TIT.1982.1056489
21. Peldszus, A., Stede, M.: From argument diagrams to argumentation mining in texts: a survey. Int. J. Cogn. Inform. Nat. Intell. **7**(1), 1–31 (2013). https://doi.org/10.4018/jcini.2013010101
22. Peters, M.E., et al.: Deep contextualized word representations. In: Proceedings of the 2018 Conference of the North American Chapter of the Association for Computational Linguistics: Human Language Technologies, NAACL-HLT 2018, New Orleans, Louisiana, USA, 1–6 June 2018, Volume 1 (Long Papers), pp. 2227–2237 (2018). https://www.aclweb.org/anthology/N18-1202/
23. Reimers, N., Schiller, B., Beck, T., Daxenberger, J., Stab, C., Gurevych, I.: Classification and clustering of arguments with contextualized word embeddings. In: Proceedings of the 57th Conference of the Association for Computational Linguistics, ACL 2019, Florence, Italy, 28 July–2 August 2019, Volume 1: Long Papers, pp. 567–578 (2019). https://www.aclweb.org/anthology/P19-1054/
24. Robertson, S.E., Zaragoza, H.: The probabilistic relevance framework: BM25 and beyond. Found. Trends Inf. Retrieval **3**(4), 333–389 (2009). https://doi.org/10.1561/1500000019
25. Salton, G., Wong, A., Yang, C.: A vector space model for automatic indexing. Commun. ACM **18**(11), 613–620 (1975). https://doi.org/10.1145/361219.361220
26. Sokal, R.R., Michener, C.D.: A statistical method for evaluating systematic relationships. Univ. Kansas Sci. Bull. **38**, 1409–1438 (1958)
27. Stab, C., et al.: ArgumenText: searching for arguments in heterogeneous sources. In: Proceedings of the 2018 Conference of the North American Chapter of the Association for Computational Linguistics, NAACL-HLT 2018, New Orleans, Louisiana, USA, 2–4 June 2018, Demonstrations, pp. 21–25 (2018). https://www.aclweb.org/anthology/N18-5005/
28. Stab, C., Gurevych, I.: Identifying argumentative discourse structures in persuasive essays. In: Proceedings of the 2014 Conference on Empirical Methods in Natural Language Processing, EMNLP 2014, 25–29 October 2014, Doha, Qatar, A Meeting of SIGDAT, A Special Interest Group of the ACL, pp. 46–56 (2014). https://www.aclweb.org/anthology/D14-1006/

29. Stede, M., Afantenos, S.D., Peldszus, A., Asher, N., Perret, J.: Parallel discourse annotations on a corpus of short texts. In: Proceedings of the Tenth International Conference on Language Resources and Evaluation LREC 2016, Portorož, Slovenia, 23–28 May 2016 (2016). http://www.lrec-conf.org/proceedings/lrec2016/summaries/477.html
30. Wachsmuth, H., et al.: Building an argument search engine for the web. In: Proceedings of the 4th Workshop on Argument Mining (ArgMining@EMNLP), pp. 49–59 (2017). https://doi.org/10.18653/v1/W17-5106. https://www.aclweb.org/anthology/W17-5106/
31. Wachsmuth, H., Stein, B., Ajjour, Y.: "PageRank" for argument relevance. In: Proceedings of the 15th Conference of the European Chapter of the Association for Computational Linguistics, EACL 2017, Valencia, Spain, 3–7 April 2017, Volume 1: Long Papers, pp. 1117–1127 (2017). https://aclweb.org/anthology/E17-1105/
32. Wachsmuth, H., et al.: Computational argumentation quality assessment in natural language. In: Proceedings of the 15th Conference of the European Chapter of the Association for Computational Linguistics, EACL 2017, Valencia, Spain, 3–7 April 2017, Volume 1: Long Papers. pp. 176–187 (2017), https://aclweb.org/anthology/E17-1017/
33. Welch, B.L.: The generalization of 'student's' problem when several different population variances are involved. Biometrika **34**(1–2), 28–35 (1947). https://doi.org/10.1093/biomet/34.1-2.28

Relevance Ranking Based on Query-Aware Context Analysis

Ali Montazeralghaem(✉), Razieh Rahimi, and James Allan

Center for Intelligent Information Retrieval,
University of Massachusetts Amherst, Amherst, MA, USA
{montazer,rahimi,allan}@cs.umass.edu

Abstract. Word mismatch between queries and documents is a long-standing challenge in information retrieval. Recent advances in distributed word representations address the word mismatch problem by enabling semantic matching. However, most existing models rank documents based on semantic matching between query and document terms without an explicit understanding of the relationship of the match to relevance. To consider semantic matching between query and document, we propose an unsupervised semantic matching model by simulating a user who makes relevance decisions. The primary goal of the proposed model is to combine the exact and semantic matching between query and document terms, which has been shown to produce effective performance in information retrieval. As semantic matching between queries and entire documents is computationally expensive, we propose to use local contexts of query terms in documents for semantic matching. Matching with smaller query-related contexts of documents stems from the relevance judgment process recorded by human observers. The most relevant part of a document is then recognized and used to rank documents with respect to the query. Experimental results on several representative retrieval models and standard datasets show that our proposed semantic matching model significantly outperforms competitive baselines in all measures.

Keywords: Semantic matching · Local context · Retrieval model

1 Introduction

In basic retrieval models such as BM25 [30] and the language modeling framework [29], the relevance score of a document is estimated based on explicit matching of query and document terms. These retrieval models have been improved in several directions; in this study, we focus on two of them: (1) semantic matching, and (2) simulating human relevance decision making.

First, different choices of words between the authors of documents and users interested in those documents impose the long-standing challenge of term mismatch between query and documents. Basic retrieval models suffer from the

J. M. Jose et al. (Eds.): ECIR 2020, LNCS 12035, pp. 446–460, 2020.
https://doi.org/10.1007/978-3-030-45439-5_30

term mismatch problem, since semantically related terms do not contribute to the relevance scores of documents.

Several techniques have been developed to address the term mismatch problem, including query expansion [10,24,31,37,39], latent models [4,8,14], and retrieval using distributed representations of words [13]. Query expansion techniques using global or local analysis of documents have shown improvements in the performance of retrieval models; however, these techniques suffer from query-independent analysis of documents in a large corpus or query drift [7], respectively. Latent models have been used for matching queries and documents represented in latent semantic space. Although semantic matching is required for information retrieval, exact matching, especially when query terms are new or rare, still provides strong evidence of relevance [23]. Thus, these latent models alone do not perform well for information retrieval [3]. Translation models were initially proposed to address the term mismatch problem in the language modeling framework for information retrieval. These models estimate the likelihood that a query can be generated as a translation of a given document. Ganguly et al. [13] used word embedding [22,28] to estimate document language models through a noisy channel to address the term mismatch problem. Although there is a large body of research on semantic matching of terms in queries and documents, many studies fail to capture important IR heuristics such as proximity and term dependencies [21].

The second direction considers how people actually make relevance decisions. Relevance in almost all retrieval models is measured by comparing query terms with terms in the entire text of a document. This fundamental choice of input to scoring functions is not compatible with how a person perceives a document as relevant or non-relevant to his/her information need. This mismatch can lead to non-optimal performance of retrieval systems [18,19]. Kong et al. [38] described that a person first tries to locate pieces of a document that are likely to be related to the query. For each piece, the person then makes relevance decision based on the local context of the piece. If the piece is found to be relevant to the query, the document is judged as relevant, otherwise other pieces are considered for evaluation. Surprisingly little attention has been given to relevance ranking based on simulating how a person makes relevance decisions. Wu et al. [36] proposed a retrieval model simulating human relevance decision making and using context of query terms, however their model is not based on semantic similarity between query and the context of query terms.

Simulating human relevance decision making, we propose a novel model for semantic matching in information retrieval. The document's relevance to a query is thus estimated based on local contexts in the document. Local contexts for determining relevance consist of query terms' window-based pieces of text. These local contexts reduce the amount of texts considered for estimation of relevance to a query, while no information related to the query will be missed. Having local contexts, we compare each piece of text with the query based on both exact and semantic term matching. The proposed semantic matching model relaxes the assumption of independence between query terms to some extent, in that semantic similarity of terms in the local context of a query term is weighted by similarity of the query term with other query terms.

Our model can thus produce document ranking effectively and efficiently. Finally, our proposed model for relevance ranking provides the basis for natural integration of semantic term matching and local document context analysis into any retrieval model.

2 Related Work

2.1 Semantic Matching

In this section, we review the existing models for semantic term matching in information retrieval.

Query Expansion. Query expansion has a long history in information retrieval, where a query is expanded with terms relevant to query terms. Query expansion can be done using global or/and local analysis [37]. Global expansion methods use word associations obtained independent of queries, such as corpus-wide information [16] or external resources such as WordNet [33]. Local query expansion methods mainly analyze the top retrieved documents for a query. Pseudo-relevance feedback is a well-known model of automatic query expansion, and has shown improvements in the performance of retrieval. Several models for pseudo-relevance feedback has been developed [2,20,24–26,31].

In addition, some models for query expansion use word embeddings [1,15]. Kuzi et al. find terms that are similar to the entire query or its terms using word embeddings and use them to expand query in the relevance model [17]. Diaz et al. [10] propose to use locally-trained embeddings for query expansion, where documents sampled from the top retrieved documents for queries are used to train word embeddings.

Latent Models. Latent models represent queries and documents in a latent space of reduced dimensionality using the term-document matrix, such as LSA [9], PLSA [14], and LDA [4]. However, these models do not perform well for information retrieval [3]. Wei and Croft [35] use LDA topics to estimate document language models in the language modeling framework, and shown improvements in the performance of retrieval. Therefore, we compare our model with this LDA-based language model as a representative of this group.

Embedding-Based Retrieval Models. Vulić and Moens [34] use (bilingual) word embeddings for monolingual and bilingual information retrieval, where queries and documents are represented as an average of the embeddings of their terms. However, the proposed model did not improve the performance of retrieval, unless it is combined with a basic retrieval model. Zheng and Callan [41] proposed a supervised model for re-weighting query terms in traditional retrieval models, BM25 and language modeling framework, based on embeddings of query terms. Although this model has shown to improve the performance of traditional retrieval models, the retrieval is still based on only exact matching of query and document terms.

Ganguly et al. [13] proposed a generalized estimate of document language models using a noisy channel, which captures semantic term similarities computed using word embeddings. In this model, words that are semantically related to any word of a document contribute to the new estimate of document language models, while we only consider words that are semantically related to terms in local contexts of documents. Zamani and Croft [39] show that their query expansion model outperforms the generalized language model [13], therefore we only report the result of embedding-based estimation of query language models.

2.2 Local Context Analysis

Wu et al. [36] proposed a model for information retrieval by simulating how human makes relevance decisions. They consider context of query terms in documents to propose a novel retrieval method. They used related terms in context of query terms in document as expansion terms.

Kong et al. [38] introduced three principles for combining relevance evidence from different pieces of a document to make the final relevance decision: (1) Disjunctive Relevance Decision (DRD) principle, (2) Aggregate Relevance (AR) principle, and (3) Conjunctive Relevance Decision (CRD) principle. Following TREC guideline for relevance judgment stating that a document is relevant if any piece of it is relevant to the query.

2.3 Log-Logistic Retrieval Model

We briefly introduce the information-based retrieval model based on the Log-logistic distribution, which is used in our proposed ranking model. Document scores in this model is computed as follow:

$$\mathrm{RSV}(Q,D) = \sum_{w \in Q} \mathrm{count}(w,Q) \log(\frac{\mathrm{tf}(w,D) + \lambda_w}{\lambda_w}), \tag{1}$$

where $\mathrm{tf}(w,D) = \mathrm{count}(w,D) \times \log(1 + c\frac{\mathrm{avdl}}{|D|})$, c is a free parameter, avdl is average document length in the collection and $\lambda_w = \frac{N_w}{N}$ where N_w is the number of documents containing w and N is the number of documents in the collection.

3 The Proposed Semantic Model

The relevance ranking problem is to score a document $D = \{d_1, d_2, \ldots, d_n\}$ with respect to a query $Q = \{q_1, q_2, \ldots, q_m\}$ based on a combination of exact and semantic matching between query and document terms.

According to the TREC relevance judgment process, a document is relevant to a query if any piece of the document is relevant to the query[1]. Taking this

[1] TREC: Text Retrieval Conference (TREC) data - English relevance judgements (2000) https://trec.nist.gov/data/reljudge_eng.html.

definition, we first need to determine locations in the document that are likely related to the query.

Determining Query-Related Locations. Wu et al. [36] proposed the query-centric assumption, which states that relevant information only occurs in the contexts around query terms in documents. The validity of this assumption is confirmed with a user study [18,19]. Following this assumption, a local context of query term q_i inside document D is denoted by $C(q_i, D)$ and is determined by a window around one occurrence of q_i in D. A symmetric window of size h centred at the occurrence of query term q_i in the document, gives a local context of following document terms.

$$C(q_i, D) = [d_{j-h}, \ldots, d_j = q_i, \ldots, d_{j+h}]. \tag{2}$$

Thus, each local context of a query term has a length of $2h + 1$. For simplicity, we refer to the local context of a query term in a document as the document context.

The next step is to estimate the relevance score of each query-centric context $c(q_i, D)$ with respect to query Q. For this purpose, a scoring function based on exact and semantic matching is desired. In addition, the scoring function should satisfy the constraints defined on information retrieval models so that ranking based on these scores provides reasonable rankings [12]. The Log-logistic model [5] has shown to be effective for information retrieval [6]. Therefore, we derive our local context scoring function based on the Log-logistic model as follows:

$$S(Q, C(q_i, D)) = \sum_{q_j \in Q} \log\left(\frac{\text{sim}(q_j, C(q_i, D)) + \lambda_{q_j}}{\lambda_{q_j}}\right), \tag{3}$$

where $\lambda_{q_j} = N_{q_j}/N$ is computed based on N_{q_j} representing the number of documents in the collection containing q_j and the total number of documents in the collection, N. This scoring function is obtained by replacing the normalized frequency of a query term in a document used in the Log-logistic scoring function of Eq. (1) by semantic similarity of the query term with terms in the document context. The similarity of term q_j with respect to the local context $C(q_i, D)$ which is a short text, is then estimated based on

$$\text{sim}(q_j, C(q_i, D)) = \sum_{w \in C(q_i, D)} \text{sim}(q_j, w) \times \mathbb{1}_{(\text{sim}(q_j, w) > \theta)}, \tag{4}$$

where $\mathbb{1}_{(.)}$ represents the indicator function taking on a value of 1 if the similarity between two terms is above the threshold parameter θ, and 0 otherwise. This indicator function is added to filter out the impacts of words unrelated to the query. Using this estimation of similarity, exact occurrences of query terms as well as words semantically related to query terms contribute to the relevance scores of local contexts in Eq. (3). And when parameter θ is set to the similarity value of a term to itself, the scoring function reduces to exact term matching.

The underlying assumption in the scoring function of Eq. (3) as well as many well-established retrieval models is independence between query terms. However, the similarity between query-centric (q_i) and current query term (q_i)

does not consider in this function. As an example, consider the query 303 of TREC Robust dataset, "Hubble telescope achievements". And assume that we want to score a local context of query term "Hubble" which is a space telescope in a document. Thus, term "Hubble" is in the center of this local context. Although terms "Hubble" and "telescope" should logically have a higher similarity degree than terms "Hubble" and "achievements" in any term association resource, we believe occurrences of terms related to "achievements" in the local context makes it more likely to be relevant to the query than those of "telescope", because we already know that "Hubble" exists in the context and occurring "telescope" can not have much information of relevance compared to "achievements". Therefore, to account for this observation, we proposed the following function to score a document context with respect to a query.

$$S(Q, C(q_i, D)) = \sum_{q_j \in Q} \log(\frac{\text{sim}(q_j, C(q_i, D)) + \lambda_{q_j}}{\lambda_{q_j}}) \times \text{dis}(q_i, q_j), \tag{5}$$

where $\text{dis}(q_i, q_j)$ denotes the semantic difference between the query term q_i in the center of the local context and the current query term q_j. We add $\text{dis}(q_i, q_j)$ to this function because we want to promote occurring query terms that semantically are far from query-centric. We compute $\text{dis}(q_i, q_j) = a - \text{sim}(q_i, q_j)$, where a is a constant that its value is obtained as $a = 1 + \text{sim}(t, t)$.

Herein, we use word embeddings to compute term similarities. Therefore, the similarity of a query term with a document context in Eq. (4) is computed as follows:

$$\text{sim}(q_j, C(q_i, D)) = \sum_{w \in C(q_i, D)} \cos(\boldsymbol{q_j}, \boldsymbol{w}) \times \mathbb{1}_{(\cos(\boldsymbol{q_j}, \boldsymbol{w}) > \theta)}, \tag{6}$$

where term vectors denote their embeddings in a continues space, and cos(.) function computes the cosine similarity between two vectors. Accordingly, the dissimilarity between query terms is computed as:

$$\text{dis}(q_i, q_j) = 2 - \cos(\boldsymbol{q_i}, \boldsymbol{q_j}). \tag{7}$$

the cosine similarity gives a value in the range of $[-1, 1]$ meaning that in case of perfect similarity ($\cos(\boldsymbol{q_i}, \boldsymbol{q_j}) = 1$) the dissimilarity is minimum $\text{dis}(q_i, q_j) = 1$ (note that because we use Eq. 7 in Eq. 5, the minimum value of dissimilarity should not be 0) and when $\cos(\boldsymbol{q_i}, \boldsymbol{q_j}) = -1$ dissimilarity is maximum i.e., $\text{dis}(q_i, q_j) = 3$.

Document Relevance Score. Obtaining the relevance score of each local context of query terms in a document, we then need to score the document with respect to the query based on the scores of its local contexts. We start by estimating the relevance score of a document with respect to each query term. Let $\zeta(q_i, D)$ denote the set of all local contexts of query term q_i in document D,

$$\zeta(q_i, D) = \{C_1(q_i, D), C_2(q_i, D), \ldots, C_k(q_i, D)\}, \tag{8}$$

where k equals to frequency of term q_i in the document, $\text{TF}(q_i, D)$.

The relevance scores of local contexts in the set $\zeta(q_i, D)$ should be aggregated to estimate the relevance score of document D with respect to q_i, denoted by $S_L(q_i, D)$. Following the aggregation principles introduced by Kong et al. [38], we consider two different aggregation function. The first *disjunctive relevance decision* principle indicates that a document is relevant if any one of its local contexts is relevant to the query. Accordingly, we estimate $S_L(q_i, D)$ using

$$S_L(q_i, D) = \max_{C \in \zeta(q_i, D)} S(Q, C(q_i, D)). \tag{9}$$

Therefore, a document is scored according to its most relevant part to the query.

The second *aggregate relevance* principle states that a document with more pieces relevant to the query should get a higher relevance score. Following this relevance principle, we compute the relevance score of a document as

$$S'_L(q_i, D) = \sum_{C \in \zeta(q_i, D)} S(Q, C(q_i, D)). \tag{10}$$

The max aggregation function conforms to the TREC definition of document relevance[2]; a document is relevant if any part of the document is relevant, regardless of how small that part is compared to the entire document. We also observed better retrieval performance using this aggregation. Therefore, we adopt the max aggregation function for our proposed model and show the effectiveness of using the other function in one experiment.

Normalizing Local Relevance Scores. Relevance scores of documents with respect to query terms in Eq. (9) are not theoretically bonded above, because we do not consider any normalization over semantic similarity of query term q_j with respect to the query-centric context in Eq. 4.

A transformation function $f(.)$ to normalize local relevance scores of documents needs to satisfy three constraints: (1) vanishes at 0, (2) upper bounded to 1, (3) $f'(x) > 0$ to make sure that as the value of x increases, the output of function also increases. The simple yet effective function $f(x) = \frac{x}{x+\sigma}$ used in multiple information retrieval models [27] satisfies the three mentioned constraints. Therefore, to normalize the local relevance scores, we use this function as follows:

$$S_N(q_i, D) = \frac{S_L(q_i, D)}{S_L(q_i, D) + \sigma}, \tag{11}$$

where $\sigma > 0$ is a free parameter in this function.

Final Document Scores. Having the relevance score of a document with respect to each query term, the final score of the document can be calculated. For this purpose, we use weighted sum of scores of each query term to consider the importance of each query term in ranking. Therefore, the final score of a document is estimated as follow:

$$score(Q, D) = \sum_{q_i \in Q} S_N(q_i, D) \times \mathcal{W}(q_i, D), \tag{12}$$

where $\mathcal{W}(qi, D)$ is the importance of query term q_i given document D.

[2] https://trec.nist.gov/data/reljudge_eng.html.

Query Term Importance. Importance of each query term has two sides: (1) global importance which is mainly computed by the inverse document frequency (idf) of the term in the collection. (2) local importance which can be the weight of the query term in the document. Existing retrieval models provide document ranking based on these two factors. Therefore, we can use any basic retrieval function to weigh query terms. In the experiments section, we show the results of using BM25, language modeling framework, and Log-logistic models for weighting query terms.

4 Discussion

Computational Time: To compute the semantic similarity between query terms and document, we use the query-centric contexts. To do that, we first extract the positions of query terms in the document by Indri. Then, we find the neighbors of the query terms based on their positions in the document. By finding the neighbors of the query terms, we compute the semantic similarity between query terms and the query-centric contexts (i.e., the neighbors of the query terms). In other words, we do not consider all terms in a document to compute the semantic similarity. In contrast, the generalized language model [13] take into account all terms in the document to find the semantic similarity between the query and document which makes their approach so expensive. Note that our approach still is more effective since is more compatible with the human judgment process. Our approach is also more efficient compared to the topic modeling language model (LDA) [35], since they also use all terms in a document to model term associations. Also, the running time for each of the Gibbs sampling in LDA increases linearly with the number of documents N and the number of topics K i.e., $O(NK)$ which makes their approach even more expensive compared to the generalized language model.

Zamani and Croft [39] proposed an embedding query expansion named EQE1 to estimate query language model. They used word embedding similarities to find terms in the entire vocabulary that are semantically related to the query terms. This method is much faster than the previous ones (i.e., LDA model [35] and generalized language model [13]) but is not optimal since it needs to compute similarity scores between query terms and all terms in the vocabulary.

Properties of the Proposed Method: Fang et al. [11], proposed seven constraints for IR models and showed that it is necessary to satisfy them to get good performance. To compute the semantic similarity score between a query-centric context and query terms, it is also necessary to satisfy these constraints. For example, if we have a query with two terms, and two query-centric contexts, the context that can interpret more distinct query words should be assigned a higher score. Clinchant et al. [6] showed that Log-Logistic model satisfies all constraints in the PRF framework. Therefore, we modify this model in our approach to compute semantic similarity between query-centric context and query terms. In other words, by modifying Log-Logistic model and considering query-centric context as a small document, we make sure that our model satisfies all constraints proposed by Fang et al. [11].

Tao and Zhai [32], proposed proximity based constraints for IR models. There are similar constraints for pseudo relevance feedback [25]. They showed that by comparing two documents that match the same number of query words, it is more desirable to rank the document in which all query terms are close to each other above another one. We argue that this assumption also is valid in semantic space. By using the query-centric window, we implicitly promote documents that have query terms close in semantic or exact matching. Therefore, our method can capture important IR characteristics i.e. exact/semantic matching signals, proximity heuristics, and query term importance.

5 Experiments

In this section, we aim to address the following research questions.

RQ1: How does our model perform with different retrieval functions to weight query terms? (see Eq. 12)
RQ2: How does our model perform compared to existing retrieval models?
RQ3: What is the effect of different principle functions for combining relevance evidence, including Aggregate Relevance (AR) principle and Disjunctive Relevance Decision (DRD), in our model?

Experimental Setup. We use three standard TREC collections in our experiments: Robust, Gov2, and WT10g. We use the title of topics as queries. We use standard INQUERY stopword list to remove stopwords and no stemming is performed. The experiments are carried out using the Indri and the Lemur toolkits[3]. In our experiments, we use pre-trained word embeddings of Glove, trained on Wikipedia 2014 and Gigaword 5 corpus [28]. Statistically significant differences of performances are determined using the two-tailed paired t-test computed at a 95% confidence level over average precision per query.

Baseline Methods. We compare our proposed model with three categories of existing models for information retrieval: basic retrieval models, retrieval models that consider semantic similarity or topic modeling, and proximity-based retrieval models. Since our proposed model is an unsupervised semantic matching, we do not compare our model with supervised approaches and neural network based models that require labeled data. Three basic retrieval models are used as baseline: **BM25:** An effective and widely-used retrieval method [30]. **LM:** Standard language modeling approach for information retrieval where document language models are smoothed using the Jelinek-Mercer smoothing method [40].

Logistic: An information-based retrieval method using the Log-logistic distribution [6]. For proximity-based retrieval models, we choose the sequential dependence model (**SDM**) as a baseline, which considers term dependencies in the

[3] http://lemurproject.org/.

language modeling framework using Markov random fields [21]. Baseline models that take advantages of semantic similarity or topic modeling are: **EQE1:** An embedding-based model for query expansion proposed by Zamani and Croft [39]. EQE1 is shown to outperform other embedding-based expansion models, e.g., embedding-based expansion of document language models [13] and heuristic-based query expansion using word embeddings [1]. Therefore, we only compare our model with the EQE1 model in the experiments. Note that in this experiment, we only consider methods that select expansion terms based on word embeddings and not other information sources such as the top retrieved documents for each query (PRF). **LDA:** An LDA-based estimation of document language models [35].

We compare baseline models with some variants of our proposed model. Variants of our model starting with prefix **LCA** or **LCD** indicate that documents are scored using aggregate relevance or disjunctive relevance, respectively.

Parameter Setting. In all experiments, parameter c in Eq. 1, σ in Eq. 11, θ in Eq. 6, the parameter λ of the smoothing method of the LM baseline, and parameters b and k_1 of the BM25 baseline are set using 2-fold cross validation to optimize MAP performance over the queries of each collection. The value of parameters c, σ, and θ are selected from $\{1, 3, 6, \cdots, 12\}$, $\{1, 5, 10, \cdots, 20\}$, and $\{0.3, 0.4, 0.5, \cdots, 1.0\}$, respectively. The parameter λ of the LM baseline is swept between $\{0, 0.1, \cdots, 1\}$ and the value of b and k_1 of BM25 baseline are chosen from $\{0.75, 1.0\}$ and $\{1.0, 2.0\}$, respectively. In all experiments, the dimensions of embedding vectors is 200. We set the LDA hyper-parameters α and β to $50/K$ and 0.001, respectively, where K is the number of topics in LDA. K is set to 800 as suggested in [35]. For the SDM model, the weight of the unigram component, the ordered and unordered window are selected from $\{0, 0.1, \cdots, 1\}$. We also made sure that they sum to 1.

5.1 Effectiveness of Different Weighting Functions

In this section, without loss of generality, we use three simple but effective retrieval functions including LM, BM25, and Log-Logistic in our model to weight query terms in Eq. (12), which aims to answer **RQ1**. The results of this experiment are reported in Table 1. According to this table, semantic matching in the local contexts of documents improves retrieval effectiveness in all cases. The improvements are statistically significant in most cases. This shows the effectiveness of our approach in integration of semantic matching into retrieval models. One can also observe that using Log-Logistic function for weighting the importance of query terms in our model outperforms using BM25 or the language model framework in most cases. This observation demonstrates that relevance scores of local contexts are more compatible with the scores of the Log-logistic model. Therefore, in the next experiments, we use this retrieval model as the weighing function in our model.

Table 1. Performance of the proposed method with different retrieval models as the weighing function in Eq. (12). ▲ indicates that the improvements over corresponding retrieval model are statistically significant.

Dataset	Metric	LM	LCD-LM	BM25	LCD-BM25	Logistic	LCD-Logistic
Robust	*MAP*	0.2048	0.2351$^{\blacktriangle}$	0.2264	0.2458$^{\blacktriangle}$	0.2258	0.2481$^{\blacktriangle}$
	P@10	0.3566	0.3916$^{\blacktriangle}$	0.4088	0.4249$^{\blacktriangle}$	0.4004	0.4329$^{\blacktriangle}$
	nDCG@10	0.3580	0.3948$^{\blacktriangle}$	0.4171	0.4327$^{\blacktriangle}$	0.4132	0.4391$^{\blacktriangle}$
	Recall	0.5093	0.5501$^{\blacktriangle}$	0.5156	0.5507$^{\blacktriangle}$	0.5195	0.5539$^{\blacktriangle}$
	GMAP	0.1043	0.1324$^{\blacktriangle}$	0.1235	0.1432$^{\blacktriangle}$	0.1257	0.1462$^{\blacktriangle}$
WT10g	*MAP*	0.1119	0.1469$^{\blacktriangle}$	0.1773	0.1899$^{\blacktriangle}$	0.1744	0.1945$^{\blacktriangle}$
	P@10	0.1806	0.2224$^{\blacktriangle}$	0.2622	0.2867$^{\blacktriangle}$	0.2704	0.2908$^{\blacktriangle}$
	nDCG@10	0.1695	0.2239$^{\blacktriangle}$	0.2904	0.3132$^{\blacktriangle}$	0.2842	0.3110$^{\blacktriangle}$
	Recall	0.5430	0.5896$^{\blacktriangle}$	0.5883	0.5989	0.5792	0.6058$^{\blacktriangle}$
	GMAP	0.0400	0.0646$^{\blacktriangle}$	0.0680	0.0847$^{\blacktriangle}$	0.0656	0.0871$^{\blacktriangle}$
Gov2	*MAP*	0.2521	0.2850$^{\blacktriangle}$	0.2671	0.2937$^{\blacktriangle}$	0.2679	0.2926$^{\blacktriangle}$
	P@10	0.5332	0.5672$^{\blacktriangle}$	0.5669	0.5858$^{\blacktriangle}$	0.5479	0.5655$^{\blacktriangle}$
	nDCG@10	0.4343	0.4578	0.4681	0.4907$^{\blacktriangle}$	0.4535	0.4739$^{\blacktriangle}$
	Recall	0.6028	0.6336$^{\blacktriangle}$	0.6188	0.6487$^{\blacktriangle}$	0.6166	0.6534$^{\blacktriangle}$
	GMAP	0.1701	0.1946$^{\blacktriangle}$	0.1897	0.2096$^{\blacktriangle}$	0.1863	0.2126$^{\blacktriangle}$

Table 2. Performance of proposed method and baselines. The superscript ▲ indicates that the improvements over all other baselines are statistically significant.

Dataset	Metric	LM	BM25	Logistic	SDM	LDA	EQE1	LCD-Logistic
Robust	*MAP*	0.2048	0.2264	0.2258	0.2273	0.2299	0.2278	**0.2481**$^{\blacktriangle}$
	P@10	0.3566	0.4088	0.4004	0.3936	0.4123	0.4040	**0.4329**$^{\blacktriangle}$
	nDCG@10	0.3580	0.4171	0.4132	0.4069	0.4192	0.4094	**0.4391**$^{\blacktriangle}$
	Recall	0.5093	0.5156	0.5192	0.5188	0.5209	0.5244	**0.5539**$^{\blacktriangle}$
	GMAP	0.1043	0.1235	0.1257	0.1212	0.1187	0.1186	**0.1462**$^{\blacktriangle}$
WT10g	*MAP*	0.1119	0.1773	0.1744	0.1845	–	0.1867	**0.1945**$^{\blacktriangle}$
	P@10	0.1806	0.2622	0.2704	0.2776	–	0.2750	**0.2908**$^{\blacktriangle}$
	nDCG@10	0.1695	0.2904	0.2842	0.2916	–	**0.3110**	**0.3110**
	Recall	0.5430	0.5883	0.5792	0.5930	–	0.6046	**0.6058**
	GMAP	0.0400	0.0680	0.0656	0.0783	–	0.0792	**0.0871**$^{\blacktriangle}$
Gov2	*MAP*	0.2521	0.2671	0.2679	0.2695	–	0.2731	**0.2926**$^{\blacktriangle}$
	P@10	0.5332	0.5669	0.5479	0.5608	–	**0.5682**	0.5655
	nDCG@10	0.4343	0.4681	0.4535	0.4615	–	0.4671	**0.4739**$^{\blacktriangle}$
	Recall	0.6028	0.6188	0.6166	0.6250	–	0.6232	**0.6534**$^{\blacktriangle}$
	GMAP	0.1701	0.1897	0.1863	0.1859	–	0.1901	**0.2126**$^{\blacktriangle}$

5.2 Performance of the Proposed Model

In this section, we compare our model with the baselines, which aims to address **RQ2**. The results of this experiment are reported in Table 2. According to this table, the proposed method (i.e., LCD-Logistic) outperforms SDM. This shows that our model in addition to using the proximity of query terms in a document improves the retrieval performance by exploiting the semantic similarity of

Table 3. Comparing different principle function for combining local context relevance scores on Gov2 only.

Metric	LCA-Logistic	LCD-Logistic
MAP	0.2730	**0.2926**
P@10	0.5628	**0.5655**
nDCG@10	0.4585	**0.4739**
Recall	0.6297	**0.6534**
GMAP	0.1897	**0.2126**

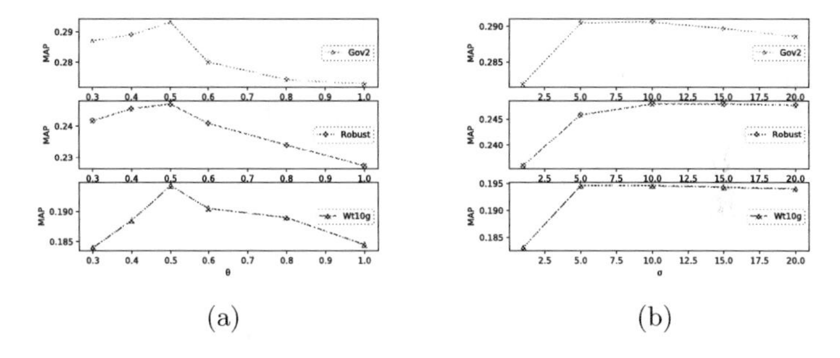

Fig. 1. Sensitivity of the proposed method (LCD-Logistic) to the θ and σ

terms. Unlike SDM, the EQE1 and LDA baseline models consider semantic similarity and topic modeling in document ranking, respectively. According to the results in this table, LCD-Logistic outperforms EQE1 and LDA-based retrieval models. We only report the results of the LDA model on Robust due to the prohibit training time of LDA on the other two collections. These comparisons show the importance of capturing semantic similarity in the local context for information retrieval. We also report GMAP to evaluate our method in confrontation with hard queries. We also report the recall metric in this table. According to the results, LCD-Logistic improves recall in Robust and Gov2 collections substantially.

5.3 Different Principle Functions For combining Relevance Evidences

This section aims to answer **RQ3**. The results of this experiment are shown in Table 3[4]. Comparing LCD and LCA variants of our model together shows that the LCD variant outperforms LCA in all cases. This means that using the most relevant local context of a document to score it, following the disjunctive relevance principle, has better performance.

[4] Note that for the sake of space, in this experiment, we just consider the Gov2 collection.

Parameter Sensitivity Analysis: Fig. 1a shows the sensitivity of the proposed method to the similarity threshold in Eq. 4. According to this figure, the best value for the similarity threshold is 0.5 in all datasets. It is worth noting that by setting the value of θ to 1, we just consider exact matching in the local context of query terms. Figure 1b shows sensitivity of LCD-Logistic to the normalization parameter σ in Eq. 11. According to this figure, the best value for this parameter in all collections is 10.

6 Conclusion

We propose a new model for semantic matching in information retrieval. Our model is designed based on simulating human judgment process to find high-quality similarity scores between query and document. The proposed method is designed to be able to capture important IR heuristics, e.g, proximity of query terms in documents, semantic matching between query and document terms, and importance of query terms. We showed that our model can be integrated into any retrieval models and improve their performance significantly.

Acknowledgements. This work was supported in part by the Center for Intelligent Information Retrieval and in part by NSF grant #IIS-1617408. Any opinions, findings and conclusions or recommendations expressed in this material are those of the authors and do not necessarily reflect those of the sponsor.

References

1. ALMasri, M., Berrut, C., Chevallet, J.P.: A comparison of deep learning based query expansion with pseudo-relevance feedback and mutual information. In: Advances in Information Retrieval (2016)
2. Ariannezhad, M., Montazeralghaem, A., Zamani, H., Shakery, A.: Iterative estimation of document relevance score for pseudo-relevance feedback. In: Jose, J., et al. (eds.) ECIR 2017. LNCS, vol. 10193, pp. 676–683. Springer, Cham (2017). https://doi.org/10.1007/978-3-319-56608-5_65
3. Atreya, A., Elkan, C.: Latent semantic indexing (LSI) fails for TREC collections. SIGKDD Explor. Newsl. **12**(2), 5–10 (2011)
4. Blei, D.M., Ng, A.Y., Jordan, M.I.: Latent Dirichlet allocation. J. Mach. Learn. Res. **3**, 993–1022 (2003)
5. Clinchant, S., Gaussier, E.: Information-based models for ad hoc IR. In: Proceedings of the 33rd International ACM SIGIR Conference on Research and Development in Information Retrieval, pp. 234–241. ACM (2010)
6. Clinchant, S., Gaussier, E.: A theoretical analysis of pseudo-relevance feedback models. In: ICTIR 2013, pp. 6:6–6:13 (2013)
7. Collins-Thompson, K.: Reducing the risk of query expansion via robust constrained optimization. In: Proceedings of the 18th ACM Conference on Information and Knowledge Management, pp. 837–846. ACM (2009)
8. Deerwester, S., Dumais, S.T., Furnas, G.W., Landauer, T.K., Harshman, R.: Indexing by latent semantic analysis. J. Am. Soc. Inf. Sci. **41**(6), 391–407 (1990)

9. Deerwester, S.C., Dumais, S.T., Landauer, T.K., Furnas, G.W., Harshman, R.A.: Indexing by latent semantic analysis. J. Am. Soc. Inf. Sci. Technol. **41**, 391–407 (1990)
10. Diaz, F., Mitra, B., Craswell, N.: Query expansion with locally-trained word embeddings. arXiv preprint arXiv:1605.07891 (2016)
11. Fang, H., Tao, T., Zhai, C.: A formal study of information retrieval heuristics. In: Proceedings of the 27th Annual International ACM SIGIR Conference on Research and Development in Information Retrieval, pp. 49–56. ACM (2004)
12. Fang, H., Tao, T., Zhai, C.: Diagnostic evaluation of information retrieval models. ACM Trans. Inf. Syst. **29**(2), 7:1–7:42 (2011)
13. Ganguly, D., Roy, D., Mitra, M., Jones, G.J.: Word embedding based generalized language model for information retrieval. In: SIGIR 2015, pp. 795–798 (2015)
14. Hofmann, T.: Probabilistic latent semantic indexing. In: SIGIR 1999, pp. 50–57 (1999)
15. Imani, A., Vakili, A., Montazer, A., Shakery, A.: Deep neural networks for query expansion using word embeddings. In: Azzopardi, L., Stein, B., Fuhr, N., Mayr, P., Hauff, C., Hiemstra, D. (eds.) ECIR 2019. LNCS, vol. 11438, pp. 203–210. Springer, Cham (2019). https://doi.org/10.1007/978-3-030-15719-7_26
16. Jones, K.S.: Automatic keyword classification for information retrieval. Libr. Q. **41**(4), 338–340 (1971)
17. Kuzi, S., Shtok, A., Kurland, O.: Query expansion using word embeddings. In: Proceedings of the 25th ACM International on Conference on Information and Knowledge Management, pp. 1929–1932. ACM (2016)
18. Li, X., Liu, Y., Mao, J., He, Z., Zhang, M., Ma, S.: Understanding reading attention distribution during relevance judgement. In: CIKM 2018, pp. 733–742 (2018)
19. Li, X., Mao, J., Wang, C., Liu, Y., Zhang, M., Ma, S.: Teach machine how to read: reading behavior inspired relevance estimation. In: SIGIR (2019)
20. Lv, Y., Zhai, C.: A comparative study of methods for estimating query language models with pseudo feedback. In: Proceedings of the 18th ACM Conference on Information and Knowledge Management, pp. 1895–1898. ACM (2009)
21. Metzler, D., Croft, W.B.: A Markov random field model for term dependencies. In: SIGIR 2005, pp. 472–479 (2005)
22. Mikolov, T., Sutskever, I., Chen, K., Corrado, G.S., Dean, J.: Distributed representations of words and phrases and their compositionality. In: Advances in Neural Information Processing Systems, pp. 3111–3119 (2013)
23. Mitra, B., Diaz, F., Craswell, N.: Learning to match using local and distributed representations of text for web search. In: Proceedings of the 26th International Conference on World Wide Web, pp. 1291–1299. International World Wide Web Conferences Steering Committee (2017)
24. Montazeralghaem, A., Zamani, H., Shakery, A.: Axiomatic analysis for improving the log-logistic feedback model. In: Proceedings of the 39th International ACM SIGIR Conference on Research and Development in Information Retrieval, pp. 765–768 (2016)
25. Montazeralghaem, A., Zamani, H., Shakery, A.: Term proximity constraints for pseudo-relevance feedback. In: Proceedings of the 40th International ACM SIGIR Conference on Research and Development in Information Retrieval, pp. 1085–1088 (2017)
26. Montazeralghaem, A., Zamani, H., Shakery, A.: Theoretical analysis of interdependent constraints in pseudo-relevance feedback. In: The 41st International ACM SIGIR Conference on Research & Development in Information Retrieval, pp. 1249–1252 (2018)

27. Paik, J.H.: A novel TF-IDF weighting scheme for effective ranking. In: SIGIR 2013, pp. 343–352 (2013)
28. Pennington, J., Socher, R., Manning, C.: Glove: global vectors for word representation. In: Proceedings of the 2014 Conference on Empirical Methods in Natural Language Processing (EMNLP), pp. 1532–1543, October 2014
29. Ponte, J.M., Croft, W.B.: A language modeling approach to information retrieval. In: Proceedings of the 21st Annual International ACM SIGIR Conference on Research and Development in Information Retrieval, pp. 275–281 (1998)
30. Robertson, S.E., Walker, S.: Some simple effective approximations to the 2-Poisson model for probabilistic weighted retrieval. In: Croft, B.W., van Rijsbergen, C.J. (eds.) SIGIR 1994, pp. 232–241. Springer, London (1994). https://doi.org/10.1007/978-1-4471-2099-5_24
31. Salton, G.: The SMART Retrieval System-Experiments in Automatic Document Processing. Prentice-Hall Inc., Upper Saddle River (1971)
32. Tao, T., Zhai, C.: An exploration of proximity measures in information retrieval. In: Proceedings of the 30th Annual International ACM SIGIR Conference on Research and Development in Information Retrieval, pp. 295–302. ACM (2007)
33. Voorhees, E.M.: Query expansion using lexical-semantic relations. In: Croft, B.W., van Rijsbergen, C.J. (eds.) SIGIR 1994, pp. 61–69. Springer, London (1994). https://doi.org/10.1007/978-1-4471-2099-5_7
34. Vulić, I., Moens, M.F.: Monolingual and cross-lingual information retrieval models based on (bilingual) word embeddings. In: SIGIR 2015, pp. 363–372 (2015)
35. Wei, X., Croft, W.B.: LDA-based document models for ad-hoc retrieval. In: SIGIR 2006, pp. 178–185 (2006)
36. Wu, H.C., Luk, R.W., Wong, K.F., Kwok, K.: A retrospective study of a hybrid document-context based retrieval model. Inf. Process. Manag. **43**(5), 1308–1331 (2007)
37. Xu, J., Croft, W.B.: Query expansion using local and global document analysis. In: SIGIR 1996, pp. 4–11 (1996)
38. Kong, Y.K., Luk, R., Lam, W., Ho, K.S., Chung, F.L.: Passage-based retrieval based on parameterized fuzzy operators. In: The SIGIR 2004 Workshop on Mathematical/Formal Methods for Information Retrieval (2004)
39. Zamani, H., Croft, W.B.: Embedding-based query language models. In: ICTIR 2016, pp. 147–156 (2016)
40. Zhai, C., Lafferty, J.: A study of smoothing methods for language models applied to ad hoc information retrieval. In: ACM SIGIR Forum, vol. 51, pp. 268–276. ACM (2017)
41. Zheng, G., Callan, J.: Learning to reweight terms with distributed representations. In: SIGIR 2015, pp. 575–584 (2015)

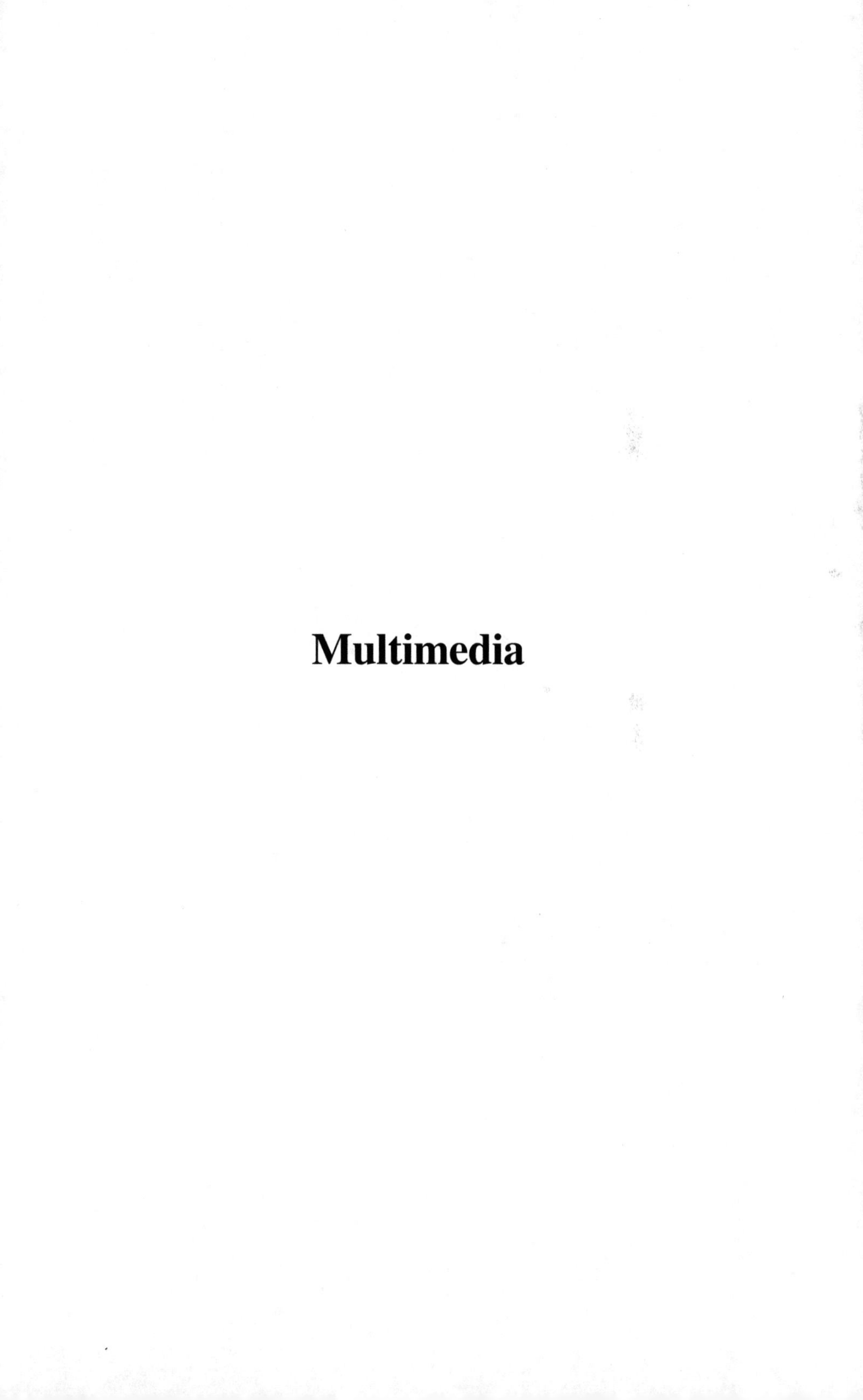

Multimedia

Multimodal Entity Linking for Tweets

Omar Adjali[1(✉)], Romaric Besançon[1], Olivier Ferret[1], Hervé Le Borgne[1], and Brigitte Grau[2]

[1] CEA, LIST, Laboratoire Analyse Sémantique Texte et Image, 91191 Gif-sur-Yvette, France
{omar.adjali,romaric.besancon,olivier.ferret,herve.le-borgne}@cea.fr
[2] Université Paris-Saclay, CNRS, LIMSI, 91400 Orsay, France
brigitte.grau@limsi.fr

Abstract. In many information extraction applications, entity linking (EL) has emerged as a crucial task that allows leveraging information about named entities from a knowledge base. In this paper, we address the task of multimodal entity linking (MEL), an emerging research field in which textual and visual information is used to map an ambiguous mention to an entity in a knowledge base (KB). First, we propose a method for building a fully annotated Twitter dataset for MEL, where entities are defined in a Twitter KB. Then, we propose a model for jointly learning a representation of both mentions and entities from their textual and visual contexts. We demonstrate the effectiveness of the proposed model by evaluating it on the proposed dataset and highlight the importance of leveraging visual information when it is available.

Keywords: Information extraction · Entity linking · Multimodality

1 Introduction

Entity linking (EL) is a crucial task for natural language processing applications that require disambiguating named mentions within textual documents. It consists in mapping ambiguous named mentions to entities defined in a knowledge base (KB). To address the EL problem, most of the state-of-the-art approaches use some form of textual representations associated with the target mention and its corresponding entity. [3] first proposed to link a named mention to an entity in a knowledge base, followed by several other local approaches [9,10,14,15,35] where mentions are individually disambiguated using lexical mention-entity measures and context features extracted from their surrounding text. In contrast, global approaches [19,21,39,42] use global features at the document-level to disambiguate all the mentions within the document and can also exploit semantic relationships between entities in the KB. These approaches have shown their

This research was partially supported by Labex DigiCosme (project ANR11LABEX0045DIGICOSME) operated by ANR as part of the program "Investissements d'Avenir" Idex Paris Saclay (ANR11IDEX000302).

J. M. Jose et al. (Eds.): ECIR 2020, LNCS 12035, pp. 463–478, 2020.
https://doi.org/10.1007/978-3-030-45439-5_31

effectiveness on standard EL datasets such as TAC KBP [27], CoNLL-YAGO [23], and ACE [2] which contain structured documents that provide a rich context for disambiguation. However, EL becomes more challenging when a limited, informal, textual context is available, such as in social media posts. On the other hand, social media posts often include images to illustrate the text that could be exploited to improve the disambiguation by adding a visual context.

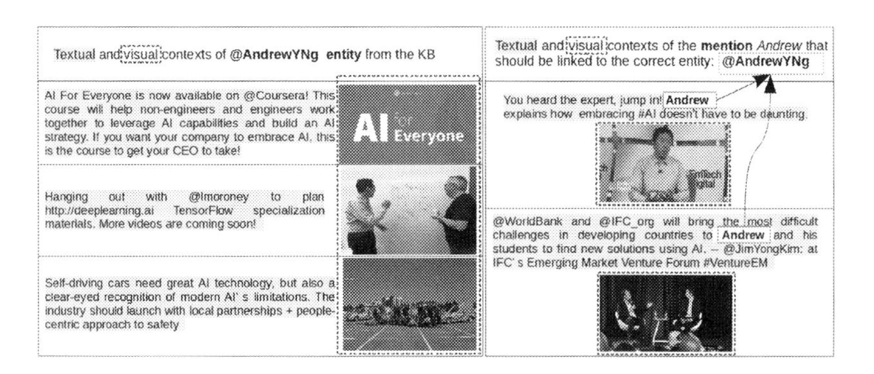

Fig. 1. An illustrative example of multimodal entity linking on our Twitter dataset. The example depicts a subset of (text, image) pairs representing the entity @AndrewNg (left), and some example of (text, image) pairs related to the mention *Andrew* (right).

In this paper, we address the problem of multimodal entity linking (MEL), by leveraging both semantic textual and visual information extracted from mention and entity contexts. We propose to apply MEL on Twitter posts as they form a prototypical framework where the textual context is generally poor and visual information is available. Indeed, an important mechanism in Twitter communication is the usage of a user's screen name (@UserScreenName) in a tweet text. This helps to explicitly mention that the corresponding user is somehow related to the posted tweet. One observation we made is that some Twitter users tend to mention other users without resorting to screen names but rather use their first name, last name, or acronyms (when the Twitter user is an organization). Consider the following example tweet (see Fig. 1):

Andrew *explains how embracing AI doesn't have to be daunting*

In this example, the mention *Andrew* refers to the user's screen name @AndrewYNg. Obviously, the mention *Andrew* could have referred to any Twitter user whose name includes Andrew. Such a practice, when it occurs, may lead to ambiguities that standard EL systems may not be able to resolve, as most of Twitter users do not have entries in standard knowledge bases. We thus propose to link ambiguous mentions to a specific Twitter KB, composed of Twitter users.

Our contributions are: **(i)** we investigate the multimodal entity linking task. To the best of our knowledge, this is the first EL work that combines textual and visual contexts on a Twitter dataset; **(ii)** we construct a new large-scale dataset for multimodal EL and above all, define a method for building such datasets at convenience; **(iii)** we present a new EL model-based for learning a multimodal joint representation of tweets and show, on the proposed dataset, the interest of adding visual features for the EL task. Code and data are available at https://github.com/OA256864/MEL_Tweets.

2 Related Work

Entity Linking on Twitter. Recently, several research efforts proposed to meet the challenges posed by the EL task on Twitter media posts. Collective approaches are preferred and leverage global information about tweets in relation with the target mention. [33] collectively resolves a set of mentions by aggregating all their related tweets to compute mention-mention and mention-entity similarities while [25] takes advantage of both semantic relatedness and mention co-referencing. In a more original way, [44] determines the user's topics of interest from all its posted tweets in order to collectively link all its named entity mentions. Similarly [24] considers social (user's interest + popularity) and temporal contexts. Other collective approaches include in their EL model additional non-textual features. For example, [16] and [6] use global information of tweets that are close in space and time to the tweet of the target mention. Finally, [13] proposes a joint cross-document co-reference resolution and disambiguation approach including temporal information associated with their corpus to improve EL performance. While these works yield interesting results using non-textual features, they often depend on the availability of social data and do not exploit visual information.

Multimodal Representation Learning. Joint multimodal representations are used in several multimodal applications such as visual question answering [1,26], text-image retrieval [4,7,48] and image captioning [28,30] and exploits different approaches e.g., Canonical Correlation Analysis [46], linear ranking-based models and non-linear deep learning models to learn projecting image features and text features into a joint space [17]. In our work, we take inspiration from [47] for our joint representation learning model as it showed satisfying results for multimodal learning.

To our knowledge, [36] is the only other work that leverages multimodal information for an entity disambiguation task in a social media context. However, they use a dataset of 12K annotated image-caption pairs from Snapchat (which is not made available), whereas our work relies on a much larger base (85K samples for the benchmark and 2M to build the KB) and is more in line with the state-of-the-art which mostly uses Twitter as case study for EL on social media. Furthermore, they use Freebase as KB, which does not contain image information and does not allow a multimodal representation at the KB level, leading to a very different disambiguation model.

3 Problem Formulation

In Twitter terms, each user account has a unique screen name (@AndrewYNg), a user name (Andrew Ng), a user description (Stanford CS adjunct faculty) and a timeline containing all the tweets (text+image) posted by the user (see Fig. 1). On this basis, we consider the following definitions:

- an entity e corresponds to a Twitter user account u (generally associated with a person or an organization);
- a mention m corresponds to an ambiguous textual occurrence of an entity e_j mentioned in a tweet t that does not belong to the timeline of e_j;
- the knowledge base is the set of entities (i.e. Twitter accounts).

Formally, we denote the knowledge base $KB = \{e_j\}$ as a set of entities, each entity being defined as a tuple $e_j = (s_j, u_j, TL_j)$ including their screen name s_j, user name u_j and timeline TL_j (we did not use the user description in the representation of the entity). The timeline contains both texts and images. A mention m_j is defined as a pair (w_i, t_i) composed of the word (or set of words) w_i characterizing the mention and the tweet t_i in which it occurs: t_i contains both text and images.

The objective of the task consists in finding the most similar entity $e^*(m_i)$ to the mention m_i in the KB according to a given similarity. From a practical point of view, we do not compute the similarities between the mention and all entities in the KB. We first select a subset of the entities that are good candidates to disambiguate m_i. It is defined as $Cand(m_i) = \{e_j \in KB | w_i \sim u_j\}$ where $w_i \sim u_j$ indicates that the mention words w_i are close to the entity name u_j. In our case, due to the nature of the dataset, we only use a simple inclusion (i.e. the mention words are present in the entity name) but a more complex lexical distance that takes into account more variations could be considered [36]. The disambiguation of m_i is then formalized as finding the best multimodal similarity measure between the tweet containing the mention and the timeline of the correct entity, both containing text and images:

$$e^*(m_i) = \underset{e_j \in Cand(m_i)}{\text{argmax}} \; sim(t_i, TL_j) \tag{1}$$

4 Twitter-MEL Dataset

One important contribution of our work is a novel dataset for multimodal entity linking on short texts accompanied by images derived from Twitter posts. The process to build this dataset is mostly automatic and can, therefore, be applied to generate a new dataset at convenience.

All the tweets and user's metadata were collected using the Twitter official API[1]. The dataset creation process comprises two phases: the construction of the knowledge base and the generation of ambiguous mentions. As mentioned

[1] https://dev.twitter.com.

in Sect. 1, our multimodal EL task aims at mapping mentions to entities (i.e. twitter users) from our Twitter knowledge base, which are characterized by a timeline, namely the collection of the most recent tweets posted by a given user.

As a first step, we established a non-exhaustive initial list of Twitter user's screen names using Twitter lists in order to have users that are likely to produce a sufficient number of tweets and be referred by a sufficient number of other users. A Twitter list is a curated set of Twitter accounts[2] generally grouped by topic. From this initial list of users, we started building the KB by collecting the tweets of each user's timeline along with its meta-information ensuring that both re-tweets and tweets without images were discarded. Moreover, as explained previously, users tend to create ambiguous mentions in tweets when they employ any expression (for example first or last name) other than Twitter screen names to mention other users in their post (see Sect. 1). Consequently, we have drawn inspiration from this usage to elaborate a simple process for both candidate entity and ambiguous mention generation.

4.1 Selection of Possibly Ambiguous Entities

To ensure that the KB contains sufficiently ambiguous entities, i.e. entities with possible ambiguous mentions, to make the EL task challenging, we expanded the KB from the initial lists with ambiguous entities. More precisely, we first extracted the last name from each Twitter account name of the initial list of users. Then, we used these names as search queries in the Twitter API user search engine to collect data about similar users[3]. Users that have been inactive for a long period, non-English users and non-verified user accounts were filtered. Furthermore, to ensure more diversity in the entities and not only use person names, we manually collected data about organization accounts. We relied on Wikipedia acronym disambiguation pages to form groups of ambiguous (organization) entities that share the same acronym.

4.2 Generation of Ambiguous Mentions

After building the KB, we used the collected entities to search for tweets that mention them. The Twitter Search API[4] returns a collection of relevant tweets matching the specified query. Thus, for each entity in the KB, **(i)** we set its screen name (@user) as the query search; **(ii)** we collect all the retrieved tweets; **(iii)** we filtered out tweets without images. Given the resulting collections of tweets mentioning the different entities of the KB, we systematically replaced the screen name mentioned in the tweet with its corresponding ambiguous mention: last names for *person* entities and acronyms for *organization* entities. Finally, we kept track of the ground truths of each tweet in the dataset reducing the cost of a manual annotation task and resulting in a dataset composed of annotated pairs of

[2] https://help.twitter.com/en/using-twitter/twitter-lists.

[3] Only the first 1,000 matching results are available with the Twitter API.

[4] Twitter API searches within a sampling of tweets published in the past 7 days.

text and image. Although ambiguous mentions are synthetically generated, they are comparable to some extent with real-world ambiguous mentions in tweets. We applied a named entity recognition (NER) system [12] on the tweets of the dataset which achieved a 77% accuracy score on the generated mentions. This suggests that these latter are somehow close to real-world named mentions.

4.3 Dataset Statistics and Analysis

Altogether, we collected and processed 14M tweets, 10M timeline tweets for entity characterization and 4M tweets with ambiguous mentions (mention tweets) covering 20k entities in the KB. Filtering these tweets drastically reduced the size of our data set. Regarding mention tweets, a key factor in the reduction is the elimination of noisy tweets where mentions have no syntactic role in the tweet text (e.g. where the mention is included in a list of recipients of the message). Discarding these irrelevant tweets as well as tweets without image left a dataset of 2M timeline tweets and 85k mention tweets. After 3 months of data collection, we found that only 10% of tweet posts are accompanied by images. In the end, we randomly split the set of mention tweets into training (40%), validation (20%) and test (40%) while ensuring that 50% of mention tweets in the test set correspond to entities unseen during the training.

Table 1. Statistics on timeline and entity distributions in MEL dataset.

	Mean	Median	Max	Min	StdDev
nb Tweets/timeline (text+image)	127.9	52	3,117	1	222.2
nb ambiguous entities/mention	16.5	16	67	2	12

Table 1 shows the timeline tweet distribution of all entities in our KB. As noted by [24], this distribution reveals that most Twitter users are information seekers, i.e. they rarely tweet, in contrast to users that are content generators who tweet frequently. Along with user's popularity, this has an influence on the number of mention tweets we can collect. We necessarily gathered more mention tweets from content generator entities, as they are more likely to be mentioned by others than information seeker entities.

5 Proposed MEL Approach

Visual and textual representations of mentions and entities are extracted with pre-trained networks. We then learn a two-branch feed-forward neural network to minimize a triplet-loss defining an implicit joint feature space. This network provides a similarity score between a given mention and an entity that is combined with other external features (such as popularity or other similarity measures) as the input of a multi-layer perceptron (MLP) which performs a binary classification on a (mention, entity) pair.

5.1 Features

Textual Context Representation. We used the unsupervised Sent2Vec [37] model to learn tweet representations, pre-trained on large Twitter corpus. We adopted this model as training on the same type of data (short noisy texts) turns out to be essential for performing well on the MEL task (see Sect. 6.2). The Sent2Vec model extends the CBOW model proposed by [34] to learn a vector representation of words. More precisely, it learns a vector representation of a sentence S by calculating the average of the embeddings of the words (unigrams) making up the sentence S and the n-grams present in S:

$$V_S = \frac{1}{|R(S)|} \sum_{w \in R(S)} v_w \tag{2}$$

where R(S) is the set of n-grams (including unigrams) in the sentence S, and v_w is the embedding vector of the word w. Therefore, the textual context of a mention m_i within the tweet t_i is represented by the sentence embedding vector of t_i. We produce then for each mention two continuous vector representations (D=700), a sentence embedding $U_m^{(i)}$ inferred using only tweet unigrams and a sentence embedding $B_m^{(i)}$ inferred using tweet unigrams and bigrams. Combining their vectors is generally beneficial [37]. An entity context being represented by a set of tweets (see Sect. 3), given an entity e_i, we average the unigram and bigram embeddings of all e_i's timeline tweets yielding two average embedding vectors $U_e^{(i)}$, $B_e^{(i)}$ representing the entity textual context used as features.

BM25 Features

Given that the disambiguation task aims at finding the correct entity for a tweet, it can be viewed as an IR problem, where we try to associate a given tweet with the most relevant timeline in the KB. We, therefore, consider, as a baseline, a *tf-idf*-like model to match the mention with the entity: in our case, both the tweet and the timeline are represented as bag-of-words vectors and we used the standard BM25 weighting scheme to perform the comparison.

Popularity Features. Given an entity e representing a Twitter user u, we consider 3 popularity features represented by: N_{fo} the number of followers, N_{fr} the number of friends and N_t the number of tweets posted by u.

Visual Context Features. The visual features are extracted with the Inception_v3 model [45], pre-learned on the 1.2M images of the ILSVRC challenge [43]. We use its last layer (D = 1,000), which encodes high-level information that may help discriminating between entities. For an entity e_i, we retain a unique feature vector that is the average of the feature vectors of all the images within its timeline, similarly to the process for the textual context. The visual feature vector of a mention is extracted from the image of the tweet that contains the mention.

5.2 Joint Multimodal Representation Learning

The proposed model measures the multimodal context similarity between a mention and its candidate entities. Figure 2 shows the architecture of the proposed joint representation learning model. It follows a triplet loss structure with two sub-models as in [22,41], one processing the mention contexts and the other processing the entity contexts. Each sub-model has a structure resembling the similarity model proposed by [47], i.e., it comprises three branches (unigram embedding, bigram embedding, image feature), each in turn including 2 fully connected (FC) layers with a Rectified Linear Unit (ReLU) activation followed by a normalization layer [32]. Then, the output vectors of the three branches are merged using concatenation, followed by a final FC layer. Other merging approaches exist such as element-wise product and compact bilinear pooling [17,47]. However, in our work, simple concatenation showed satisfying results as a first step. Moreover, the mention and entity inputs differ in our task, a mention being characterized by one (text, image) pair and an entity by a large set of (text, image) pairs. Thus, we investigated the performance of our model when the parameters of the two sub-models are partially or fully shared. We found that shared parameters yielded better accuracy results on the validation set, thus the weight of the FC layers of both branches are shared.

In summary, using the extracted visual and textual features $\{U_m, B_m, I_m\}$, $\{U_e, B_e, I_e\}$ respectively of a mention m_i and an entity e_i, our model is trained to project each modality through the branches into an implicit joint space. The resulting representations are concatenated into C_m and C_e and passed through the final FC layer yielding two multimodal vectors J_m and J_e. The objective is to minimize the following triplet loss function:

$$\min_W \sum_{e^- \neq e^+} \max(0, 1 - \|f_m(m), f_e(e^+)\| - \|f_m(m), f_m(e^-)\|)$$

where m, e^+, e^-, are the mention, positive and negative entities respectively. $f_m(\cdot)$ is the mention sub-network output function, $f_e(\cdot)$ the entity sub-network function and $\|\cdot\|$ is the L_2-norm. The objective aims at minimizing the L_2 distance between the multimodal representation of m and the representation of a positive entity p^+, and maximizing the distance to the negative entity e^- multimodal representation. For MEL, we calculate the cosine similarity between the two vectors of a pair (m_i, e_i) given by $f_m(\cdot)$ and $f_e(\cdot)$ to represent their multimodal context similarity: $sim(m_i, e_i) = cosine(J_m^{(i)}, J_e^{(i)})$.

Finally, we propose to integrate the popularity of the entity to the estimation of the similarity by combining the multimodal context similarity score and popularity features through a MLP. Other external features may also be combined at the MLP level, as in our case the BM25 similarity. This MLP is trained to minimize a binary cross-entropy, with label 0 (resp. 1) for negative (resp. positive) entities w.r.t the mention.

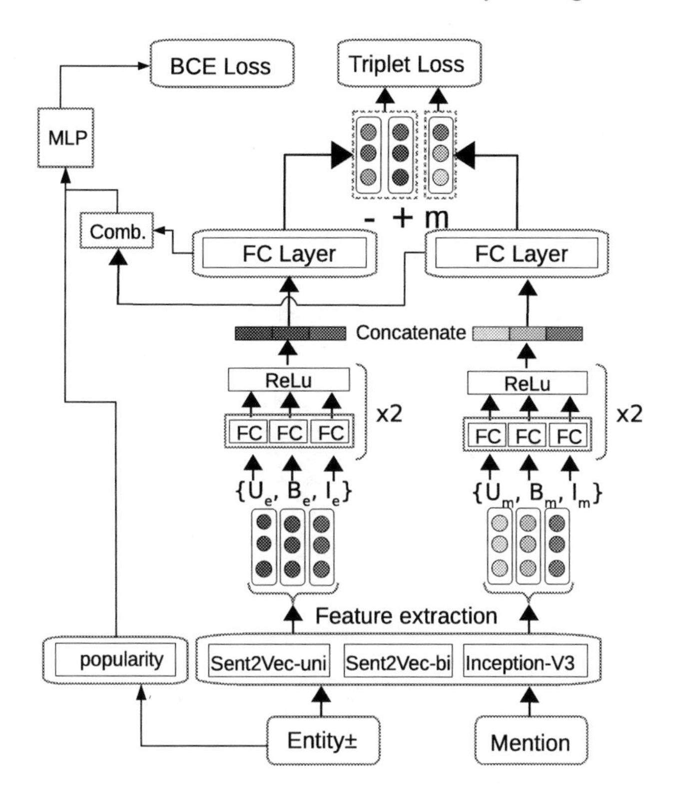

Fig. 2. Triplet ranking model for learning entity and mention joint representations.

6 Experiments Analysis and Evaluation Setup

6.1 Parameter Settings

We initialized the weights of each FC layer using the Xavier initialization [20]. Training is performed over 100 epochs, each covering 600k (mention, entity) pairs of which 35k are positive samples. The triplet loss is minimized by stochastic gradient descent, with a momentum of 0.9, an initial learning rate of 0.1 and a batch size of 256. The learning rate is divided by 10 when the loss does not change by more than $1e^{-4}$ during 6 epochs. After 50 epochs, if the accuracy on the validation set does not increase during 5 successive evaluations, the training is early stopped. Regarding the BCE branch, the MLP has two hidden layers with one more neuron than the number of inputs and *tanh* non-linearities. The BCE loss is minimized using L-BGFS with a learning rate of 10^{-5}. The network is implemented in PyTorch [38].

Since our approach is not directly comparable with previous works [36], we compare the results with different configurations and baselines. As 50% of mentions are unique in each split of our dataset, we do not take the standard most frequent entity prediction as a baseline, which links the entity that was seen most in

Table 2. Features and models used in our experiments.

Features	Description (see Sect. 5.1)
Popularity (Pop)	Baseline feature where the most popular entity is selected
BM25	Standard textual context similarity with BM25 weighting
S2V-uni	Similarity measured between the unigram embeddings extracted using the Sent2Vec language model
S2V-bi	Similarity measured between the bigram embeddings extracted using the Sent2Vec language model
S2V	For easy readability, we use the S2V notation to represent the combination of S2V-uni and S2V-bi
Img	Similarity measured between the image features extracted using the pretrained Inception-V3 model
ET(X)	Combination of features X using an Extra-Trees classifier
JMEL(X)	Combination of features X with our joint multimodal representation model (see Sect. 5.2)

training to its corresponding mention. We rather consider the popularity features and standard textual similarity measures (BM25), presented in Sect. 5.1. Moreover, for comparison sake, we also report in Table 3 the performance of a baseline combination of the features using an Extra-Trees classifier [18], compared with our Joint Multimodal Entity Linking approach (JMEL), which combines the textual and visual features at the representation level. Table 2 summarizes the features and the combination models used in these experiments.

In order to more thoroughly assess the contribution of textual and visual information in the MEL task, we compare the performance of the proposed model when the visual features are combined with sentence embeddings that are derived from different models the following notable models: Skip-Thought [31] (D = 4,800) trained on a large corpus of novels, Sent2Vec [37] unigram and bigram vectors (D = 700) trained respectively on English Wikipedia and on a Twitter corpus, InferSent [8] (D = 2,048) Bi-LSTM+max pooling model trained on the Standford SNLI corpus, BERT [12] trained on BooksCorpus [49] and on English Wikipedia and ELMo [40] trained on the One Billion Word Benchmark [5]. For BERT and ELMo, sentences are represented by the average word embedding.

Table 3. Multimedia entity linking results (accuracy).

	Valid	Test
Single features		
Popularity	0.369	0.590
BM25	0.415	0.433
S2V-uni	0.482	0.513
S2V-bi	0.487	0.523
Img	0.290	0.299
Combination of features with an ExtraTrees Classifier		
ET(S2V)	0.495	0.529
ET(S2V + Img)	0.507	0.542
ET(S2V + Img + Pop)	0.585	0.627
ET(S2V + Img + Pop + BM25)	0.654	0.671
Combination of features with our JMEL model		
JMEL(S2V)	0.628	0.724
JMEL(S2V + Img)	0.639	0.731
JMEL(S2V + Img + Pop)	**0.767**	**0.776**
JMEL(S2V + Img + Pop + BM25)	**0.795**	**0.803**

6.2 Results

Table 3 reports the accuracy results on the validation and test sets for the binary classification task about the correctness of the first entity selected from the KB for a given mention. First, we note that the baseline points out an imbalance in our dataset, as 59% of the mentions in the test set correspond to the most popular entity among the candidate entities, compared to 36.9% in the validation set. Note that for some Entity linking datasets, popularity can achieve up to 82% accuracy score [11]. We also observe that our popularity baseline outperforms the combination of textual and visual features with the Extra-Trees classifier: this indicates that the features extracted from the textual and visual contexts, when naively used and combined, produce poor results. In contrast, our model achieves significant improvements on both the validation and test sets compared with the popularity baseline and the Extra-Trees combination. We see that combining additional features in the JMEL model (popularity and BM25) also provides significant performance gain. Regarding the visual modality, although considering it alone leads to poor results, its integration in a global model always improve the performance, compared to text-only features.

Image and Sentence Representation Impact Analysis. Table 4 reports the MEL accuracy on the validation and test sets with various sentence representation models. It shows that the integration of visual features always improves the performance of our approach, whatever the sentence embedding model we use, even though the level of improvement varies depending on the sentence

Table 4. Impact of sentence embeddings on EL results.

Sent. Embedding	Valid		Test	
	Txt	Txt+Img	Txt	Txt+Img
S2V-uni(Twitter)	0.592	**0.611**	0.698	**0.708**
S2V-uni(Wiki)	0.499	**0.538**	0.625	**0.654**
S2V-bi(Twitter)	0.616	**0.637**	0.709	**0.716**
S2V-bi(Wiki)	0.511	**0.547**	0.639	**0.663**
InferSent(GloVe)	0.559	**0.579**	0.666	**0.683**
InferSent(fastText [29])	0.551	**0.570**	0.671	**0.689**
Avg. BERT	0.580	**0.594**	0.641	**0.687**
Avg. ELMo	0.524	**0.563**	0.605	**0.655**
Skip-Thought	0.464	**0.511**	0.575	**0.605**
S2V(Twitter)	0.628	**0.639**	0.724	**0.731**
S2V(Wiki)	0.524	**0.551**	0.652	**0.666**

model. For example, while the results of averaging BERT word embeddings and InferSent are comparable using images on the test set, InferSent performs significantly better than BERT using text only. This emphasizes the role of the visual context representation of mentions and entities to help in the EL task. If we look at the performance of the various sentence embeddings models, we can see that the sent2vec model trained on a Twitter corpus outperforms all other embeddings: this reveals the importance of training data in a transfer learning setting. Indeed, we observe that the sent2vec model trained on English Wikipedia produces worse results than the model trained on the target task data (Twitter). Hence, we can assume that other models that achieve good results when trained on a general corpus (such as InferSent or BERT) would get better results if trained on a Twitter collection.

Error Analysis. We identified several potential sources of errors that may be addressed in future work. First, in our approach we characterize entity contexts with a collection of text/image pairs and by taking their mean, we consider all these pairs equally important to represent an entity. However, by manually checking some entities, we note that each timeline may contain a subset of outlier pairs and more specifically, images that are not representative of an entity. Sampling strategies may be employed to select the most relevant images and to discard the misleading outliers. Moreover, our model fails on some difficult cases where the visual and textual contexts of entity candidates are indistinguishable. For example, it fails on the mention "*post*", by linking it to the entity *@nationalpost* instead of *@nypost*, two entities representing news organizations whose posts cover various topics. One additional bias is that we restricted our dataset and KB to have only tweets with images. Meanwhile, we observed that tweets without images tend to have more textual context. Thus, it would be interesting to include tweets without images for further experiments.

7 Conclusion

We explore a novel approach that makes use of text and image information for the entity linking task applied to tweets. Specifically, we emphasize the benefit of leveraging visual features to help in the disambiguation. We propose a model that first extracts textual and visual contexts for both mentions and entities and learns a joint representation combining textual and visual features. This representation is used to compute a multimodal context similarity between a mention and an entity. Preliminary experiments on a dedicated dataset demonstrated the effectiveness of the proposed model and revealed the importance of leveraging visual contexts in the EL task. Furthermore, our work is all the more relevant with the emergence of social media, which offer abundant textual and visual information. In that perspective, we propose a new multimodal EL dataset based on Twitter posts and a process for collecting and constructing a fully annotated multimodal EL dataset, where entities are defined in a Twitter KB.

Further exploration is still needed concerning certain points in our model: in particular, our future work includes exploring attention mechanisms for text and images and experimenting different sampling strategies for the triplet loss.

References

1. Agrawal, A., et al.: VQA: visual question answering. Int. J. Comput. Vis. **123**(1), 4–31 (2017)
2. Bentivogli, L., Forner, P., Giuliano, C., Marchetti, A., Pianta, E., Tymoshenko, K.: Extending English ACE 2005 corpus annotation with ground-truth links to Wikipedia. In: Proceedings of the 2nd Workshop on The Peoples Web Meets NLP: Collaboratively Constructed Semantic Resources, pp. 19–27 (2010)
3. Bunescu, R., Paşca, M.: Using encyclopedic knowledge for named entity disambiguation. In: 11th Conference of the European Chapter of the Association for Computational Linguistics (2006)
4. Chami, I., Tamaazousti, Y., Le Borgne, H.: AMECON: abstract meta-concept features for text-illustration. In: International Conference on Multimedia Retrieval (ICMR), Bucharest, Romania (2017)
5. Chelba, C., et al.: One billion word benchmark for measuring progress in statistical language modeling. In: Fifteenth Annual Conference of the International Speech Communication Association (2014)
6. Chong, W.-H., Lim, E.-P., Cohen, W.: Collective entity linking in tweets over space and time. In: Jose, J.M., et al. (eds.) ECIR 2017. LNCS, vol. 10193, pp. 82–94. Springer, Cham (2017). https://doi.org/10.1007/978-3-319-56608-5_7
7. Chowdhury, M., Rameswar, P., Papalexakis, E., Roy-Chowdhury, A.: Webly supervised joint embedding for cross-modal image-text retrieval. In: ACM International Conference on Multimedia (2018)

8. Conneau, A., Kiela, D., Schwenk, H., Barrault, L., Bordes, A.: Supervised learning of universal sentence representations from natural language inference data. In: Proceedings of the 2017 Conference on Empirical Methods in Natural Language Processing, pp. 670–680. Association for Computational Linguistics, Copenhagen, Denmark (2017)
9. Cucerzan, S.: Large-scale named entity disambiguation based on Wikipedia data. In: Proceedings of the 2007 Joint Conference on Empirical Methods in Natural Language Processing and Computational Natural Language Learning (EMNLP-CoNLL), pp. 708–716 (2007)
10. Daher, H., Besançon, R., Ferret, O., Borgne, H.L., Daquo, A.-L., Tamaazousti, Y.: Supervised learning of entity disambiguation models by negative sample selection. In: Gelbukh, A. (ed.) CICLing 2017. LNCS, vol. 10761, pp. 329–341. Springer, Cham (2018). https://doi.org/10.1007/978-3-319-77113-7_26
11. Dai, H., Song, Y., Qiu, L., Liu, R.: Entity linking within a social media platform: a case study on Yelp. In: Proceedings of the 2018 Conference on Empirical Methods in Natural Language Processing, pp. 2023–2032 (2018)
12. Devlin, J., Chang, M.W., Lee, K., Toutanova, K.: BERT: pre-training of deep bidirectional transformers for language understanding. In: Proceedings of the 2019 Conference of the North American Chapter of the Association for Computational Linguistics: Human Language Technologies, pp. 4171–4186 (2019)
13. Dredze, M., Andrews, N., DeYoung, J.: Twitter at the grammys: a social media corpus for entity linking and disambiguation. In: Proceedings of The Fourth International Workshop on Natural Language Processing for Social Media, pp. 20–25 (2016)
14. Dredze, M., McNamee, P., Rao, D., Gerber, A., Finin, T.: Entity disambiguation for knowledge base population. In: Proceedings of the 23rd International Conference on Computational Linguistics, pp. 277–285. Association for Computational Linguistics (2010)
15. Eshel, Y., Cohen, N., Radinsky, K., Markovitch, S., Yamada, I., Levy, O.: Named entity disambiguation for noisy text. In: Proceedings of the 21st Conference on Computational Natural Language Learning, CoNLL 2017, pp. 58–68. Association for Computational Linguistics, Vancouver, Canada (2017)
16. Fang, Y., Chang, M.W.: Entity linking on microblogs with spatial and temporal signals. Trans. Assoc. Comput. Linguist. **2**, 259–272 (2014)
17. Fukui, A., Park, D.H., Yang, D., Rohrbach, A., Darrell, T., Rohrbach, M.: Multimodal compact bilinear pooling for visual question answering and visual grounding. In: Proceedings of the 2016 Conference on Empirical Methods in Natural Language Processing, pp. 457–468 (2016)
18. Geurts, P., Ernst, D., Wehenkel, L.: Extremely randomized trees. Mach. Learn. **63**(1), 3–42 (2006). https://doi.org/10.1007/s10994-006-6226-1
19. Globerson, A., Lazic, N., Chakrabarti, S., Subramanya, A., Ringaard, M., Pereira, F.: Collective entity resolution with multi-focal attention. In: Proceedings of the 54th Annual Meeting of the Association for Computational Linguistics, pp. 621–631 (2016)
20. Glorot, X., Bengio, Y.: Understanding the difficulty of training deep feedforward neural networks. In: Proceedings of the Thirteenth International Conference on Artificial Intelligence and Statistics, pp. 249–256 (2010)
21. Guo, Z., Barbosa, D.: Entity linking with a unified semantic representation. In: Proceedings of the 23rd International Conference on World Wide Web, pp. 1305–1310. ACM (2014)

22. He, H., Gimpel, K., Lin, J.: Multi-perspective sentence similarity modeling with convolutional neural networks. In: Proceedings of the 2015 Conference on Empirical Methods in Natural Language Processing, pp. 1576–1586 (2015)
23. Hoffart, J., et al.: Robust disambiguation of named entities in text. In: Proceedings of the Conference on Empirical Methods in Natural Language Processing, pp. 782–792. Association for Computational Linguistics (2011)
24. Hua, W., Zheng, K., Zhou, X.: Microblog entity linking with social temporal context. In: Proceedings of the 2015 ACM SIGMOD International Conference on Management of Data, pp. 1761–1775. ACM (2015)
25. Huang, H., Cao, Y., Huang, X., Ji, H., Lin, C.Y.: Collective tweet Wikification based on semi-supervised graph regularization. In: Proceedings of the 52nd Annual Meeting of the Association for Computational Linguistics, pp. 380–390 (2014)
26. Jabri, A., Joulin, A., van der Maaten, L.: Revisiting visual question answering baselines. In: Leibe, B., Matas, J., Sebe, N., Welling, M. (eds.) ECCV 2016. LNCS, vol. 9912, pp. 727–739. Springer, Cham (2016). https://doi.org/10.1007/978-3-319-46484-8_44
27. Ji, H., Grishman, R., Dang, H.T., Griffitt, K., Ellis, J.: Overview of the TAC 2010 knowledge base population track. In: Third Text Analysis Conference, TAC 2010 (2010)
28. Johnson, J., Karpathy, A., Fei-Fei, L.: DenseCap: fully convolutional localization networks for dense captioning. In: Proceedings of the IEEE Conference on Computer Vision and Pattern Recognition, pp. 4565–4574 (2016)
29. Joulin, A., Grave, E., Bojanowski, P., Douze, M., Jégou, H., Mikolov, T.: FastText.zip: compressing text classification models. arXiv preprint arXiv:1612.03651 (2016)
30. Karpathy, A., Fei-Fei, L.: Deep visual-semantic alignments for generating image descriptions. In: Proceedings of the IEEE Conference On Computer Vision and Pattern Recognition, pp. 3128–3137 (2015)
31. Kiros, R., et al.: Skip-thought vectors. In: Advances in Neural Information Processing Systems, pp. 3294–3302 (2015)
32. Lei Ba, J., Kiros, J.R., Hinton, G.E.: Layer normalization. arXiv preprint arXiv:1607.06450 (2016)
33. Liu, X., Li, Y., Wu, H., Zhou, M., Wei, F., Lu, Y.: Entity linking for tweets. In: Proceedings of the 51st Annual Meeting of the Association for Computational Linguistics, pp. 1304–1311 (2013)
34. Mikolov, T., Chen, K., Corrado, G., Dean, J.: Efficient estimation of word representations in vector space. arXiv preprint arXiv:1301.3781 (2013)
35. Milne, D., Witten, I.H.: Learning to link with Wikipedia. In: Proceedings of the 17th ACM Conference on Information and Knowledge Management, pp. 509–518. ACM (2008)
36. Moon, S., Neves, L., Carvalho, V.: Multimodal named entity disambiguation for noisy social media posts. In: Proceedings of the 56th Annual Meeting of the Association for Computational Linguistics, pp. 2000–2008 (2018)
37. Pagliardini, M., Gupta, P., Jaggi, M.: Unsupervised learning of sentence embeddings using compositional n-gram features. In: 2018 Conference of the North American Chapter of the Association for Computational Linguistics, NAACL 2018 (2018)
38. Paszke, A., et al.: Automatic differentiation in PyTorch. In: NIPS 2017 Autodiff Workshop (2017)
39. Pershina, M., He, Y., Grishman, R.: Personalized page rank for named entity disambiguation. In: Proceedings of the 2015 Conference of the North American

Chapter of the Association for Computational Linguistics: Human Language Technologies, pp. 238–243 (2015)
40. Peters, M.E., et al.: Deep contextualized word representations. In: Proceedings of NAACL-HLT, pp. 2227–2237 (2018)
41. Rao, J., He, H., Lin, J.: Noise-contrastive estimation for answer selection with deep neural networks. In: Proceedings of the 25th ACM International on Conference on Information and Knowledge Management, pp. 1913–1916. ACM (2016)
42. Ratinov, L., Roth, D., Downey, D., Anderson, M.: Local and global algorithms for disambiguation to Wikipedia. In: Proceedings of the 49th Annual Meeting of the Association for Computational Linguistics: Human Language Technologies, pp. 1375–1384. Association for Computational Linguistics (2011)
43. Russakovsky, O., et al.: ImageNet large scale visual recognition challenge. Int. J. Comput. Vis. **115**(3), 211–252 (2015). https://doi.org/10.1007/s11263-015-0816-y
44. Shen, W., Wang, J., Luo, P., Wang, M.: Linking named entities in tweets with knowledge base via user interest modeling. In: Proceedings of the 19th ACM SIGKDD International Conference on Knowledge Discovery and Data Mining, pp. 68–76. ACM (2013)
45. Szegedy, C., Vanhoucke, V., Ioffe, S., Shlens, J., Wojna, Z.: Rethinking the inception architecture for computer vision. In: Proceedings of the IEEE Conference on Computer Vision and Pattern Recognition, pp. 2818–2826 (2016)
46. Tran, T.Q.N., Le Borgne, H., Crucianu, M.: Aggregating image and text quantized correlated components. In: IEEE Conference on Computer Vision and Pattern Recognition (CVPR), Las Vegas, USA (2016)
47. Wang, L., Li, Y., Huang, J., Lazebnik, S.: Learning two-branch neural networks for image-text matching tasks. IEEE Trans. Pattern Anal. Mach. Intell. **41**(2), 394–407 (2019)
48. Wang, L., Li, Y., Lazebnik, S.: Learning deep structure-preserving image-text embeddings. In: Proceedings of the IEEE Conference on Computer Vision and Pattern Recognition, pp. 5005–5013 (2016)
49. Zhu, Y., et al.: Aligning books and movies: towards story-like visual explanations by watching movies and reading books. In: Proceedings of the IEEE International Conference on Computer Vision, pp. 19–27 (2015)

MEMIS: Multimodal Emergency Management Information System

Mansi Agarwal[1(✉)], Maitree Leekha[1(✉)], Ramit Sawhney[2], Rajiv Ratn Shah[3], Rajesh Kumar Yadav[1], and Dinesh Kumar Vishwakarma[1]

[1] Delhi Technological University, New Delhi, India
r18522mansi@dpsrkp.net, maitreeleekha@yahoo.in
[2] Netaji Subhas Institute of Technology, New Delhi, India
[3] Indraprastha Institute of Information Technology, New Delhi, India

Abstract. The recent upsurge in the usage of social media and the multimedia data generated therein has attracted many researchers for analyzing and decoding the information to automate decision-making in several fields. This work focuses on one such application: disaster management in times of crises and calamities. The existing research on disaster damage analysis has primarily taken only unimodal information in the form of text or image into account. These unimodal systems, although useful, fail to model the relationship between the various modalities. Different modalities often present supporting facts about the task, and therefore, learning them together can enhance performance. We present **MEMIS**, a system that can be used in emergencies like disasters to identify and analyze the damage indicated by user-generated multimodal social media posts, thereby helping the disaster management groups in making informed decisions. Our leave-one-disaster-out experiments on a multimodal dataset suggest that not only does fusing information in different media forms improves performance, but that our system can also generalize well to new disaster categories. Further qualitative analysis reveals that the system is responsive and computationally efficient.

Keywords: Disaster management · Multimodal systems · Social media

1 Introduction

The amount of data generated every day is colossal [10]. It is produced in many different ways and many different media forms. Analyzing and utilizing this data to drive the decision-making process in various fields intelligently has been the primary focus of the research community [22]. Disaster Response Management is one such area. Natural calamities occur frequently, and in times of such crisis, if the large amount of data being generated across different platforms is harnessed

M. Agarwal and M. Leekha—The authors contributed equally, and wish that they be regarded as joint First Authors.

Rajiv Ratn Shah is partly supported by the Infosys Center for AI, IIIT Delhi.

J. M. Jose et al. (Eds.): ECIR 2020, LNCS 12035, pp. 479–494, 2020.
https://doi.org/10.1007/978-3-030-45439-5_32

well, the relief groups will be able to make effective decisions that have the potential to enhance the response outcomes in the affected areas.

To design an executable plan, disaster management and relief groups should combine information from different sources and in different forms. However, at present, the only primary source of information is the textual reports which describe the disaster's location, severity, etc. and may contain statistics of the number of victims, infrastructural loss, etc. Motivated by the cause of humanitarian aid in times of crises and disasters, we propose a novel system that leverages both textual and visual cues from the mass of user-uploaded information on social media to identify damage and assess the level of damage incurred.

In essence, we propose MEMIS, a system that aims to pave the way to automate a vast multitude of problems ranging from automated emergency management, community rehabilitation via better planning from the cues and patterns observed in such data and improve the quality of such social media data to further the cause of immediate response, improving situational awareness and propagating actionable information.

Using a real-world dataset, CrisisMMD, created by Alam *et al.* [1], which is the first publicly available dataset of its kind, we present the case for a novel multimodal system, and through our results report its efficiency, effectiveness, and generalizability.

2 Literature Review

In this section, we briefly discuss the disaster detection techniques of the current literature, along with their strengths and weaknesses. We also highlight how our approach overcomes the issues present in the existing ones, thereby emphasizing the effectiveness of our system for disaster management.

Chaudhuri *et al.* [7] examined the images from earthquake-hit urban environments by employing a simple CNN architecture. However, recent research has revealed that often fine-tuning pre-trained architectures for downstream tasks outperform simpler models trained from scratch [18]. We build on this by employing transfer learning with several successful models from the ImageNet [9], and observed significant improvements in the performance of our disaster detection and analysis models, in comparison to a simple CNN model.

Sreenivasulu *et al.* [24] investigated microblog text messages for identifying those which were informative, and therefore, could be used for further damage assessment. They employed a Convolutional Neural Network (CNN) for modeling the text classification problem, using the dataset curated by Alam *et al.* [1]. Extending on their work on CrisisMMD, we experimented with several other state-of-the-art architectures and observed that adding recurrent layers improved the text modeling.

Although researchers in the past have designed and experimented with unimodal disaster assessment systems [2,3], realizing that multimodal systems may outperform unimodal frameworks [16], the focus has now shifted to leveraging information in different media forms for disaster management [20]. In addition to using several different media forms and feature extraction techniques, several researchers have also employed various methods to combine the information obtained from these modalities, to make a final decision [19]. Yang *et al.* [28] developed a multimodal system- MADIS which leverages both text and image modalities, using hand-crafted features such as TF-IDF vectors, and low-level color features. Although their contribution was a step towards advancing damage assessment systems, the features used were relatively simple and weak, as opposed to the deep neural network models, where each layer captures complex information about the modality [17]. Therefore, we utilize the latent representation of text and image modalities, extracted from their respective deep learning models, as features to our system. Another characteristic that is essential for a damage assessment system is generalizability. However, most of the work carried out so far did not discuss this practical perspective. Furthermore, to the best of our knowledge, so far no work has been done on developing an end-to-end multimodal damage identification and assessment system.

To this end, we propose MEMIS, a multimodal system capable of extracting information from social media, and employs both images and text for identifying damage and its severity in real-time (refer Sect. 3). Through extensive quantitative experimentation in the leave-one-disaster-out training setting and qualitative analysis, we report the system's efficiency, effectiveness, and generalizability. Our results show how combining features from different modalities improves the system's performance over unimodal frameworks.

3 A Real-Time Tweet Processing Pipeline

In this section, we describe the different modules of our proposed system in greater detail. The architecture for the system is shown in Fig. 1. The internal methodological details of the individual modules are in the next section.

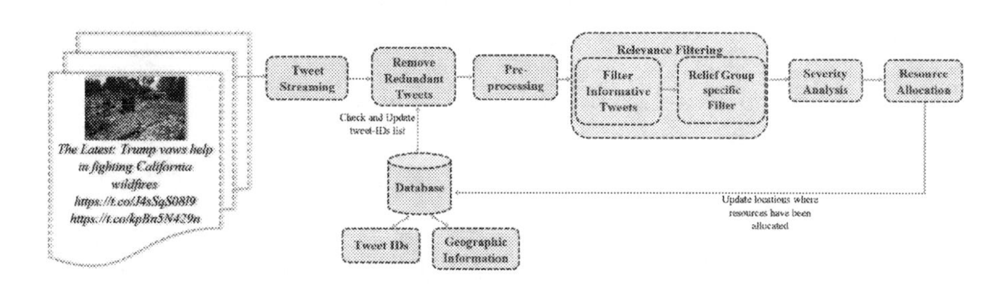

Fig. 1. System architecture of MEMIS

3.1 Tweet Streaming

The Tweet Streaming module uses the Twitter Streaming API[1] to scrap real-time tweets. As input to the API, the user can enter filtering rules based on the

[1] https://developer.twitter.com/en/docs/tutorials/consuming-streaming-data.

available information like hashtags, keywords, phrases, and location. The module outputs all the tweets that match these defined cases as soon as they are live on social media. Multiple rules can be defined to extract tweets for several disasters at the same time. Data from any social media platform can be used as input to the proposed framework. However, in this work, we consume disaster-related posts on Twitter. Furthermore, although the proposed system is explicitly for multimodal tweets having both images and text, we let the streaming module filter both unimodal and multimodal disaster tweets. We discuss in Sect. 5.5 how our pipeline can be generalized to process unimodal tweets as well, making it more robust.

3.2 Remove Redundant Tweets

A large proportion of the tweets obtained using the streaming module may be retweets that have already been processed by the system. Therefore, to avoid overheads, we maintain a list of identifiers (IDs) of all tweets that have been processed by the system. In case an incoming tweet is a retweet that has already been processed by the system before, we discard it. Furthermore, some tweets may also have location or geographic information. This information is also stored to maintain a list of places where relief groups are already providing services currently. If a streamed geo-tagged tweet is from a location where the relief groups are already providing aid, the tweet is not processed further.

3.3 Relevance Filtering

A substantial number of tweets streamed from the social media platforms are likely to be irrelevant for disaster response and management. Furthermore, different relief groups have varying criteria for what is relevant to them for responding to the situation. For instance, a particular relief group could be interested only in reaching out to the injured victims, while another provides resources for infrastructural damages. Therefore, for them to make proper use of information from social media platforms, the relevant information must be filtered.

We propose two sub-modules for filtering: (*i*) the first filters the informative tweets, *i.e.*, the tweets that provide information relevant to a disaster, which could be useful to a relief group, (*ii*) the second filter is specific to the relief group, based on the type of damage response they provide. To demonstrate the system, in this work, we filter tweets that indicate infrastructural damage or physical damage in buildings and other structures.

3.4 Severity Analysis and Resource Allocation

Finally, once the relevant tweets have been filtered, we analyze them for the severity of the damage indicated. The system categorizes the severity of infrastructural damage into three levels: high, medium and low. Based on the damage severity assessment by the system, the relief group can provide resources and services to a particular location. This information must further be updated in the database storing the information about all the places where the group is providing aid currently. Furthermore, although not shown in the system diagram, we must also remove a location from the database once the relief group's activity is over, and it is no longer actively providing service there. This ensures that if there is an incoming request from that location after it was removed from the database, it can be entertained.

4 Methodology

In this section, we discuss the implementation details of the two main modules of the system for Relevance Filtering and Severity Analysis. We begin by describing the data pre-processing required for the multimodal tweets, followed by the deep learning-based models that we use for the modules.

4.1 Pre-processing

Image Pre-processing: The images are resized to 299×299 for the transfer learning model [29] and then normalized in the range $[0, 1]$ across all channels (RGB).

Text Pre-processing: All *http* URLs, retweet headers of the form *RT*, punctuation marks, and twitter user handles specified as *@username* are removed. The tweets are then lemmatized and transformed into a stream of tokens that can be fed as input to the models used in the downstream modules. These tokens act as indices to an embedding matrix, which stores the vector representation for tokens corresponding to all the words maintained in the vocabulary. In this work, we use 100 dimensional FastText word-embeddings [6], trained on the CrisisMMD dataset [1] that has been used in this work. The system as a whole, however, is independent of the choice of vector representation.

4.2 Unimodal Models

For the proposed pipeline, we use Recurrent Convolutional Neural Network (RCNN) [14] as the text classification model. It adds a recurrent structure to the convolutional block, thereby capturing contextual information with long term dependencies and the phrases which play a vital role at the same time. Furthermore, we use the Inception-v3 model [25], pre-trained on the ImageNet Dataset [9] for modelling the image modality. The same underlying architectures, for both text and image respectively, are used to filter the tweets that convey useful information regarding the presence of infrastructural damage in the Relevance Filtering modules, and the analysis of damage in the Severity Analysis module. Therefore, we effectively have three models for each modality: first for filtering the informative tweets, then for those pertaining to the infrastructural damage (or any other category related to the relief group), and finally for assessing the severity of damage present.

4.3 Combining Modalities

In this subsection, we describe how we combine the unimodal predictions from the text and image models for different modules. We also discuss in each case about how the system would treat a unimodal text or image only input tweet.

Gated Approach for Relevance Filtering. For the two modules within Relevance Filtering, we use a simplistic approach of combining the outputs from the text and image models by using the OR function ($\oplus$). Technically speaking, we conclude that the combined output is positive if at least one of the unimodal models predicts so. Therefore, if a tweet is predicted as informative by either the text, or the image, or both the models, the system predicts the tweet as informative, and it is considered for further processing in the pipeline. Similarly, if at least one of the text and the image modality predicts an informative tweet as containing infrastructural damage, the tweet undergoes severity analysis. This simple technique helps avoid missing any tweet that might have even the slightest hint of damage, in either or both the modalities. Any false positive can also be easily handled in this approach. If, say, a non-informative tweet is predicted as informative in the first step at Relevance Filtering, it might still be the case that in the second step, the tweet is predicted as not containing any infrastructural damage. Furthermore, in case a tweet is unimodal and has just the text or the image, then the system can take the default prediction of the missing modality as negative (or `False` for a boolean OR function), which is the identity for the OR operation. In that case, the prediction based on the available modality will guide the analysis (Fig. 2).

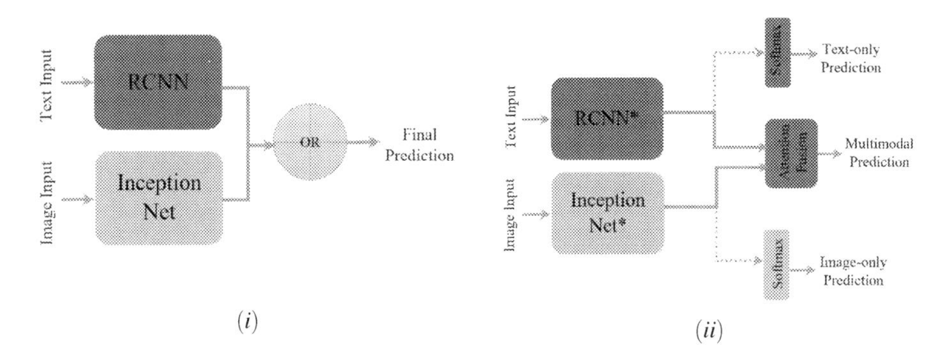

Fig. 2. Internal architecture of the (*i*) Relevance Filtering modules using an OR function to combine the predictions of the unimodal text and image models. (*ii*) Severity Analysis module that uses attention fusion to combine the text and image modalities, when both available, and switching to the unimodal models when either is missing. * indicates the model architecture till the penultimate layer, excluding the softmax.

Attention Fusion for Severity Analysis. The availability of data from different media sources has encouraged researchers to explore and leverage the potential boost in performance by combining unimodal classifiers trained on individual modalities [5,27]. Here, we use attention fusion to combine the feature interpretations from the text and image modalities for the severity analysis module [12,26]. The idea of attention fusion is to attend particular input features as compared to others while predicting the output class. The features, *i.e.*, the outputs of the penultimate layer or the layer before the softmax, of the text and image models are concatenated. This is followed by a softmax layer to learn the attention weights for each feature dimension, i.e., the attention weight α_i for a feature x_i is given by:

$$\alpha_i = \text{softmax}(\sum_{j=1}^{p} W_{ji} \cdot x_j) = \frac{exp(\sum_{j=1}^{p} W_{ji} \cdot x_j)}{\sum_{i=1}^{p} exp(\sum_{j=1}^{p} W_{ji} \cdot x_j)} \tag{1}$$

Therefore, the input feature after applying the attention weights is,

$$\beta_i = \alpha_i \cdot x_i \tag{2}$$

where, $i, j \in 1, 2, .., p$, and p is the total number of dimensions in the multimodal concatenated feature vector. W is the weight matrix learned by the model. This vector of attended features is then used to classify the given multimodal input. With this type of fusion, we can also analyze how the different modalities are interacting with each other employing their attention weights. Moving from the Relevance Filtering to the Severity Analysis module, we strengthen our fusion technique by using attention mechanism. This is required since human resources are almost always scarce, and it is necessary to correctly assess the requirements at different locations based on the severity of the damage. As opposed to an OR function, using attention, we are able to combine the most important information as seen by the different modalities to together analyze the damage severity.

In this case, the treatment of unimodal tweets is not that straightforward, since the final prediction using attention fusion occurs after concatenation of the latent feature vectors of the individual modalities. Therefore, in case the text or image is missing, we use the unimodal model for the available modality. In other words, we use attention mechanism only when both the modalities are present to analyze damage severity, else we use the unimodal models.

5 Evaluation

5.1 Dataset

Recently, several datasets on crisis damage analysis have been released to foster research in the area [21]. In this work, we have used the first multimodal, labeled, publicly available damage related to the Twitter dataset, CrisisMMD, created by Alam *et al.* [1]. It was collected by crawling the blogs posted by users during seven natural disasters, which can be grouped into 4 disaster categories, namely-Floods, Hurricanes, Wildfires and Earthquakes. CrisisMMD introduces three hierarchical tasks:

1. **Informativeness.** This initial task classifies each multimodal post as informative or non-informative. Alam *et al.* define a multimodal post as informative if it serves to be useful in identifying areas where damage has occurred due to disaster. It is therefore a binary classification problem, with the two classes being informative and non-informative.
2. **Infrastructural Damage.** The damage in an informative tweet may be of many different kinds [1,4]. CrisisMMD identifies several categories for the type of damage, namely- Infrastructure and utility damage, Vehicle damage, Affected individuals, Missing or found people, Other relevant information, None. Alam *et al.* [1] also noted that the tweets which signify physical damage in structures, where people could be stuck, are especially beneficial for the rescue operation groups to provide aid. Out of the above-listed categories, the tweets having Infrastructure and utility damage are therefore identified in this task. This again is modelled as a classification problem with two classes- infrastructural and non-infrastructural damage.
3. **Damage Severity Analysis.** This final task uses the text and image modalities together to analyze the severity of infrastructural damage in a tweet as- high, medium, or low. We add another label, no-damage, to support the pipeline framework that can handle false positives as well. Specifically, if a tweet having no infrastructural damage is predicted as positive, it can be detected here as having no damage. This is modelled as a multi-class classification problem.

The individual modules of the proposed pipeline essentially model the above three tasks of CrisisMMD. Specifically, the two Relevance Filtering modules model the first and the second tasks, respectively, whereas the Severity Analysis module models the third task (Table 1).

Table 1. CrisisMMD Class Distribution: For Tasks 1 and 2, the text and images have been labeled separately, and therefore, the number of samples in the respective classes are separated by a / *i.e.*, `text/image`. Task 3 has a single tweets level label signifying the severity of damage.

Task 1	**Informative**			**Non informative**
	12877/9375			5249/8751
Task 2	**Infrastructural**			**Non infrastructural**
	1428/3624			16698/14502
Task 3	**Low**	**Mild**	**Severe**	**No-damage**
	566	842	2216	14502

5.2 Experimental Settings

To evaluate how well our system can generalize to new disaster categories, we train our models for all the three tasks in a leave-one-disaster-out (LODO) training paradigm. Therefore, we train on 3 disaster categories and evaluate the performance on the left-out disaster. To handle the class imbalance, we also used SMOTE [8] with the word embeddings of the training fold samples for linguistic baselines. We used Adam Optimizer with an initial learning rate of 0.001, the values of $\beta 1$ and $\beta 2$ as 0.9 and 0.999, respectively, and a batch size of 64 to train our models. We use F1-Score as the metric to compare the model performance. All the models were trained on a GeForce GTX 1080 Ti GPU with a memory speed of 11 Gbps.

5.3 Results

To demonstrate the effectiveness of the proposed system for multimodal damage assessment on social media, we perform an ablation study, the results for which have been described below.

Design Choices. We tried different statistical and deep learning techniques for modelling text- TF-IDF features with **SVM**, Naive Bayes (**NB**) and Logistic Regression (**LR**); and in the latter category, **CNN** [13], Hierarchical Attention model (**HAttn**), bidirectional LSTM (**BiLSTM**) and **RCNN** [14]. As input to the deep learning models, we use 100-dimensional Fasttext word embeddings [6] trained on the dataset. By operating at the character n-gram level, Fasttext tends to capture the morphological structure well. Thus, helping the otherwise out of vocabulary words (such as hash-tags) to share semantically similar embeddings with its component words. As shown in Table 2, the RCNN model performed the best on all three tasks of the Relevance Filtering and Severity Analysis modules. Specifically, the average LODO F1-Scores of RCNN on the three tasks are 0.82, 0.76, and 0.79, respectively. Furthermore, the architecture considerably reduces the effect of noise in social media posts [14].

Table 2. Text and Image Baselines: F1-Scores for Leave one disaster out evaluation

		Unimodal Text Baselines							Unimodal Image Baselines			
Module	**Disaster Category**	**SVM**	**NB**	**LR**	**CNN**	**HAttn**	**BiLSTM**	**RCNN**	**CNN**	**VGG-16**	**Res-50**	**IncV3**
Relevance Filtering-1	**Floods**	0.47	0.43	0.48	0.64	0.69	0.62	**0.70**	0.31	0.63	0.75	**0.77**
	Hurricanes	0.42	0.51	0.49	0.71	**0.79**	0.73	**0.78**	0.29	0.61	**0.71**	**0.71**
	Wildfires	0.52	0.47	0.51	0.76	0.80	0.71	**0.81**	0.33	0.56	0.65	**0.75**
	Earthquakes	0.43	0.54	0.59	0.69	0.77	0.70	**0.78**	0.35	0.65	0.67	**0.73**
Relevance Filtering-2	**Floods**	0.51	0.46	0.53	**0.70**	0.65	0.68	**0.70**	0.67	0.66	0.69	**0.70**
	Hurricanes	0.47	0.54	0.50	0.69	0.72	0.65	**0.75**	0.61	0.61	0.72	**0.74**
	Wildfires	0.49	0.42	0.47	0.74	0.60	0.75	**0.79**	0.77	0.79	**0.81**	**0.81**
	Earthquakes	0.41	0.40	0.44	0.76	0.81	0.71	**0.83**	0.78	0.81	0.80	**0.82**
Severity Analysis	**Floods**	0.54	0.48	0.50	0.68	**0.80**	0.73	**0.79**	0.71	0.73	0.77	**0.81**
	Hurricanes	0.51	0.45	0.48	0.71	0.76	0.70	**0.82**	0.72	0.72	0.73	**0.80**
	Wildfires	0.43	0.49	0.48	0.75	**0.81**	0.74	**0.80**	0.69	0.72	0.79	**0.79**
	Earthquakes	0.41	0.47	0.42	0.66	0.71	0.74	**0.76**	0.65	0.71	0.74	**0.76**

For images, we fine-tuned the **VGG-16** [23], **ResNet-50** [11] and **InceptionV3** [25] models, pre-trained on the ImageNet Dataset [9]. We also trained a CNN model from scratch. Experimental results in Table 2 reveal that InceptionV3 performed the best, and the average F1-Score with LODO training for the three tasks are 0.74, 0.77, and 0.79, respectively. The architecture employs multiple sized filters to get a thick rather than a deep architecture, as very deep networks are prone to over-fitting. Such a design makes the network computationally less expensive, which is a prime concern for our system as we want to minimize latency to give quick service to the disaster relief groups.

Ablation Study. Table 3 highlights the results of an ablation study over the best linguistic and vision models, along with the results obtained when the predictions by these individual models are combined as discussed in Sect. 4.3. The results for all the modules demonstrate the effectiveness of multimodal damage assessment models. Specifically, we observe that for each disaster category in the LODO training paradigm, the F1-Score for the multimodal model is always better than or compares with those of the text and image unimodal models.

5.4 Qualitative Analysis

In this section, we analyze some specific samples to understand the shortcomings of using unimodal systems, and to demonstrate the effectiveness of our proposed multimodal system. Table 4 records these sample tweets along with their predictions as given by the different modules. In **green** are the correct predictions, whereas the incorrect ones are shown in **red** They have been discussed below in order:

1. The image in the first sample portrays the city landscape from the top, damaged by the calamity. Due to the visual noise, the image does not give much information about the intensity of damage present, and therefore, the image model incorrectly predicts the tweet as mildly damaged. On the other hand, the text model can identify the severe damage indicated by phrases like 'hit hard'. Combining the two predictions by using attention fusion, therefore, helps in overcoming the unimodal misclassifications.

Table 3. Ablation Study: Leveraging both textual and visual cues helps improve the leave one disaster out model performance in terms of F1-Score.

Module	Technique	Floods	Hurricanes	Wildfires	Earthquakes
Relevance Filtering-1	Unimodal Text RCNN	0.70	**0.78**	0.81	0.78
	Unimodal Image IncV3	0.77	0.71	0.75	0.73
	Text ⊕ Image	**0.79**	**0.78**	**0.83**	**0.79**
Relevance Filtering-2	Unimodal Text RCNN	0.70	0.75	0.79	**0.83**
	Unimodal Image IncV3	0.70	0.74	0.81	0.82
	Text ⊕ Image	**0.73**	**0.77**	**0.83**	**0.83**
Severity Analysis	Unimodal Text RCNN	0.79	**0.82**	0.80	0.76
	Unimodal Image IncV3	0.81	0.80	0.79	0.76
	Attention Fusion	**0.84**	**0.82**	**0.85**	**0.86**

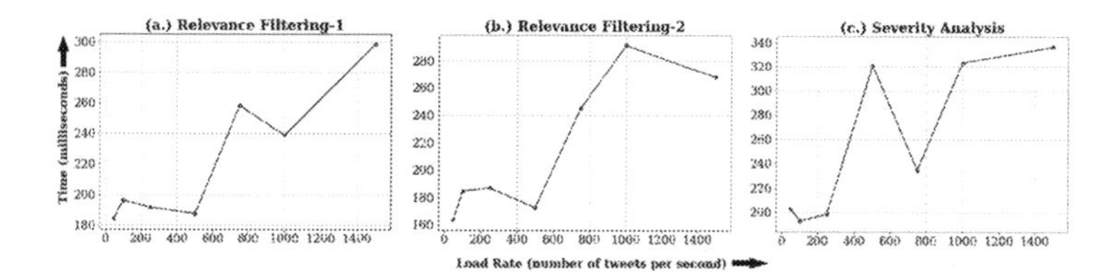

Fig. 3. Latency analysis

2. In this tweet, the text uses several keywords, such as ‘damaged’ and ‘earthquake’, which misleads the text model in predicting it as severely damaged. However, the image does not hold the same perspective. By combining the feature representations, attention fusion can correctly predict the tweet as having mild damage.

3. The given tweet is informative and therefore, it is considered for damage analysis. However, the text classifier, despite the presence of words like ‘killed’ and ‘destroyed’, incorrectly classifies it to the non-infrastructural damage class. The image classifier correctly identifies the presence of damage, and therefore, the overall prediction for the tweet is infrastructural damage, which is correct. Furthermore, both the text and image models are unable to identify the severity of damage present, but the proposed system can detect the presence of severe damage using attention fusion.

4. The sample shows how the Severity Analysis module combines the text and visual cues by identifying and attending to more important features than others. This helps in modelling the dependency between the two modalities, even when both, individually give incorrect predictions. The image in the tweet shows some hurricane destroyed structures, depicting severe damage. However, the text talks about ‘raising funds and rebuilding’, which does not indicate severe damage. The multimodal system learns to attend the text features more and correctly classifies the sample as having no damage, even though both the individual models predicted incorrectly. Furthermore, in this particular example, even by using the OR function, the system could not correctly classify it as not having infrastructural damage. Yet, the damage Severity Analysis module identifies this false positive and correctly classifies it.

Table 4. Qualitative Analysis: **T** and **I** indicate predictions by the text and image unimodal models, respectively.

Tweet	Relevance Filtering-1	Relevance Filtering-2	Severity Analysis
 **1.** *RT* ***@UMDCSA****: Dominica was hit hard from Hurricane Maria. Please keep them in your thoughts and prayers https://t.co/SpKYztnltV*	**I:** Informative **T:** Informative **T ⊕ I:** Informative	**I:** Infrastructural **T:** Infrastructural **T ⊕ I:** Infrastructural	**I:** Mild **T:** Severe **Attn Fusion:** Severe
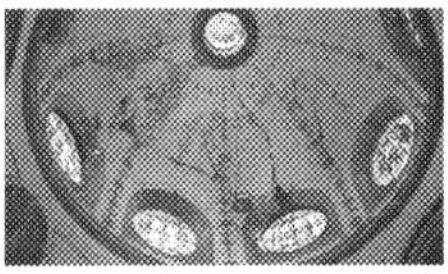 **2.** *RT @latimes: 2,000 historic buildings in Mexico have been damaged by the earthquake https://t.co/57pb1Pse1o https://t.co/7LjHlp3ljB*	**I:** Informative **T:** Informative **T ⊕ I:** Informative	**I:** Infrastructural **T:** Infrastructural **T ⊕ I:** Infrastructural	**I:** Mild **T:** Severe **Attn Fusion:** Mild
 **3.** *Cyclone Mora : 4 killed in Manipur, 140 houses destroyed in Mizoram :: https://t.co/dMlEngezZ4*	**I:** Informative **T:** Informative **T ⊕ I:** Informative	**I:** Infrastructural **T:** Non-infrastructural **T ⊕ I:** Infrastructural	**I:** Mild **T:** No-damage **Attn Fusion:** Severe
 **4.** *https://t.co/lLQwAQ4zhk Help us raise funds to rebuild our playgrounds after Hurricane Maria https://t.co/chCGXCltp2*	**I:** Informative **T:** Informative **T ⊕ I:** Informative	**I:** Infrastructural **T:** Infrastructural **T ⊕ I:** Infrastructural	**I:** Severe **T:** Severe **Attn Fusion:** No-damage

5.5 Discussion

In this section, we discuss some of the practical and deployment aspects of our system, as well as some of its limitations.

Latency Analysis. We simulate an experiment to analyze the computational efficiency of the individual modules in terms of the time they take to process a tweet, *i.e.,* the latency. We are particularly interested in analyzing the Relevance Filtering and Severity Analysis modules. We developed a simulator program to act as the Tweet Streaming module that publishes tweets at different load rates (number of tweets in 1 second) to be processed by the downstream modules. The modules also process the incoming tweets at the same rate. We calculate the average time for processing a tweet by a particular module as the total processing time divided by the total number of tweets used in the experiment. We used 15, 000 multimodal tweets from CrisisMMD, streamed at varying rates. The performance of the two Relevance Filtering modules and the Severity Analysis module as we gradually increase the load rate is shown in the Fig. 3.

As a whole, including all the modules, we observed that on an average, the system can process 80 tweets in 1 minute. This experiment was done using an Intel i7-8550U CPU having 16 GB RAM. One can expect to see an improvement if a GPU is used over a CPU.

Generalization. The proposed system is also general and robust, especially in three aspects. Firstly, the results of our LODO experiments indicate that the system can perform well in case it is used for analyzing new disasters, which were not used for training the system. This makes it suitable for real-world deployment where circumstance with new disaster categories cannot be foreseen. Furthermore, we also saw how the two main modules of the system work seamlessly, even when one of the modalities is missing. This ensures that the system can utilize all the information that is available on the media platforms to analyze the disaster. Finally, the second module in Relevance Filtering can be trained to suit the needs of several relief groups that target different types of damage, and therefore, the system is capable of being utilized for many different response activities.

Limitations. Although the proposed system is robust and efficient, some limitations must be considered before it can be used in real-time. Firstly, the system is contingent on the credibility *i.e.,* the veracity of the content shared by users on social media platforms. It may so happen that false information is spread by some users to create panic amongst others [15]. In this work, we have not evaluated the content for veracity, and therefore, it will not be able to differentiate such false news media. Another aspect that is also critical to all systems that utilize data generated on social media is the socio-economic and geographic bias. Specifically, the system will only be able to get information about the areas where people have access to social media, mostly the urban cities, whereas damage in the rural locations may go unnoticed since it did not appear on Twitter or any other platform. One way to overcome this is to make use of aerial images, that can provide a top view of such locations as the rural lands. However, this again has a drawback as to utilize aerial images effectively, a bulk load of data would have to be gathered and processed.

6 Conclusion

Identifying damage and human casualties in real-time from social media posts is critical to providing prompt and suitable resources and medical attention, to save as many lives as possible. With millions of social media users continuously posting content, an opportunity is present to utilize this data and learn a damage recognition system. In this work, we propose MEMIS, a novel Multimodal Emergency Management Information System for identifying and analyzing the level of damage severity in social media posts with the scope for betterment in disaster management and planning. The system leverages both textual and visual cues to automate the process of damage identification and assessment from social media data. Our results show how the proposed multimodal system outperforms the state-of-the-art unimodal frameworks. We also report the system's responsiveness through extensive system analysis. The leave-one-disaster-out training setting proves the system is generic and can be deployed for any new unseen disaster.

References

1. Alam, F., Ofli, F., Imran, M.: CrisisMMD: multimodal twitter datasets from natural disasters. CoRR abs/1805.00713 (2018). http://arxiv.org/abs/1805.00713
2. Alam, F., Ofli, F., Imran, M.: Processing social media images by combining human and machine computing during crises. Int. J. Hum. Comput. Interact. **34**, 311–327 (2018)
3. Alam, F., Ofli, F., Imran, M.: CrisisDPS: crisis data processing services. In: Proceedings of the 16th ISCRAM Conference (2019)
4. Alam, F., Ofli, F., Imran, M., Aupetit, M.: A Twitter tale of three hurricanes: Harvey, Irma, and Maria. ArXiv abs/1805.05144 (2018)
5. Asvadi, A., Garrote, L., Premebida, C., Peixoto, P., Nunes, U.J.C.: Multimodal vehicle detection: fusing 3D-lidar and color camera data. Pattern Recognit. Lett. **115**, 20–29 (2017)
6. Bojanowski, P., Grave, E., Joulin, A., Mikolov, T.: Enriching word vectors with subword information. CoRR abs/1607.04606 (2016). http://arxiv.org/abs/1607.04606
7. Chaudhuri, N., Bose, I.: Application of image analytics for disaster response in smart cities, pp. 3036–3045 (2019). http://hdl.handle.net/10125/59740
8. Chawla, N.V., Bowyer, K.W., Hall, L.O., Kegelmeyer, W.P.: SMOTE: synthetic minority over-sampling technique. J. Artif. Intell. Res. **16**(1), 321–357 (2002). http://dl.acm.org/citation.cfm?id=1622407.1622416

9. Deng, J., Dong, W., Socher, R., Li, L.J., Li, K., Fei-Fei, L.: ImageNet: a large-scale hierarchical image database. In: 2009 IEEE Conference on Computer Vision and Pattern Recognition, pp. 248–255. IEEE (2009). https://doi.org/10.1109/CVPR.2009.5206848
10. Forbes: How much data do we create every day? The mind-blowing stats everyone should read (2018). Accessed 26 Apr 2019
11. He, K., Zhang, X., Ren, S., Sun, J.: Deep residual learning for image recognition. In: Proceedings of the IEEE Conference on Computer Vision and Pattern Recognition, pp. 770–778 (2016). https://doi.org/10.1109/CVPR.2016.90
12. Hua, X.-S., Zhang, H.-J.: An attention-based decision fusion scheme for multimedia information retrieval. In: Aizawa, K., Nakamura, Y., Satoh, S. (eds.) PCM 2004. LNCS, vol. 3332, pp. 1001–1010. Springer, Heidelberg (2004). https://doi.org/10.1007/978-3-540-30542-2_123
13. Kim, Y.: Convolutional neural networks for sentence classification. CoRR abs/1408.5882 (2014). http://arxiv.org/abs/1408.5882
14. Lai, S., Xu, L., Liu, K., Zhao, J.: Recurrent convolutional neural networks for text classification. In: Proceedings of the Twenty-Ninth AAAI Conference on Artificial Intelligence, AAAI 2015, pp. 2267–2273. AAAI Press (2015). http://dl.acm.org/citation.cfm?id=2886521.2886636
15. Mahata, D., Talburt, J.R., Singh, V.K.: From chirps to whistles: discovering event-specific informative content from Twitter. In: Proceedings of the ACM Web Science Conference, p. 17. ACM (2015)
16. Mouzannar, H., Rizk, Y., Awad, M.: Damage identification in social media posts using multimodal deep learning. In: Proceedings of the 15th ISCRAM Conference (2018)
17. Nanni, L., Ghidoni, S., Brahnam, S.: Handcrafted vs. non-handcrafted features for computer vision classification. Pattern Recognit. **71**, 158–172 (2017)
18. Pan, S.J., Yang, Q.: A survey on transfer learning. IEEE Trans. Knowl. Data Eng. **22**(10), 1345–1359 (2009)
19. Pouyanfar, S., Tao, Y., Tian, H., Chen, S.C., Shyu, M.L.: Multimodal deep learning based on multiple correspondence analysis for disaster management. World Wide Web **22**(5), 1893–1911 (2019). https://doi.org/10.1007/s11280-018-0636-4
20. Rizk, Y., Jomaa, H.S., Awad, M., Castillo, C.: A computationally efficient multi-modal classification approach of disaster-related Twitter images. In: SAC (2019)
21. Said, N., et al.: Natural disasters detection in social media and satellite imagery: a survey. Multimed. Tools Appl. **78**(22), 31267–31302 (2019). https://doi.org/10.1007/s11042-019-07942-1
22. Shah, R., Zimmermann, R.: Multimodal Analysis of User-generated Multimedia Content. Springer, Cham (2017). https://doi.org/10.1007/978-3-319-61807-4
23. Simonyan, K., Zisserman, A.: Very deep convolutional networks for large-scale image recognition. arXiv preprint arXiv:1409.1556 (2014). https://arxiv.org/abs/1409.1556
24. Sreenivasulu, M., Sridevi, M.: Detecting informative Tweets during disaster using deep neural networks. In: 2019 11th International Conference on Communication Systems & Networks (COMSNETS), pp. 709–713 (2019)
25. Szegedy, C., Vanhoucke, V., Ioffe, S., Shlens, J., Wojna, Z.: Rethinking the inception architecture for computer vision. In: Proceedings of the IEEE Conference on Computer Vision and Pattern Recognition, pp. 2818–2826 (2016)
26. Vaswani, A., et al.: Attention is all you need. In: Advances in Neural Information Processing Systems, pp. 5998–6008 (2017)

27. Wang, X., Gong, G., Li, N.: Multimodal fusion of EEG and fMRI for epilepsy detection. IJMSSC **9**, 1850010 (2017)
28. Yang, Y., et al.: MADIS: a multimedia-aided disaster information integration system for emergency management. In: 8th International Conference on Collaborative Computing: Networking, Applications and Worksharing (CollaborateCom), pp. 233–241 (2012). https://doi.org/10.4108/icst.collaboratecom.2012.250525
29. Yosinski, J., Clune, J., Bengio, Y., Lipson, H.: How transferable are features in deep neural networks? In: Advances in Neural Information Processing Systems, pp. 3320–3328 (2014)

Interactive Learning for Multimedia at Large

Omar Shahbaz Khan[1(✉)], Björn Þór Jónsson[1,4], Stevan Rudinac[2], Jan Zahálka[3], Hanna Ragnarsdóttir[4], Þórhildur Þorleiksdóttir[4], Gylfi Þór Guðmundsson[4], Laurent Amsaleg[5], and Marcel Worring[2]

[1] IT University of Copenhagen, Copenhagen, Denmark
omsh@itu.dk
[2] University of Amsterdam, Amsterdam, Netherlands
[3] Czech Technical University in Prague, Prague, Czech Republic
[4] Reykjavik University, Reykjavík, Iceland
[5] CNRS–IRISA, Rennes, France

Abstract. Interactive learning has been suggested as a key method for addressing analytic multimedia tasks arising in several domains. Until recently, however, methods to maintain interactive performance at the scale of today's media collections have not been addressed. We propose an interactive learning approach that builds on and extends the state of the art in user relevance feedback systems and high-dimensional indexing for multimedia. We report on a detailed experimental study using the ImageNet and YFCC100M collections, containing 14 million and 100 million images respectively. The proposed approach outperforms the relevant state-of-the-art approaches in terms of interactive performance, while improving suggestion relevance in some cases. In particular, even on YFCC100M, our approach requires less than 0.3 s per interaction round to generate suggestions, using a single computing core and less than 7 GB of main memory.

Keywords: Large multimedia collections · Interactive multimodal learning · High-dimensional indexing · YFCC100M

1 Introduction

A dominant trend in multimedia applications for industry and society today is the ever-growing scale of media collections. As the general public has been given tools for unprecedented media production, storage and sharing, media generation and consumption have increased drastically in recent years. Furthermore, upcoming multimedia applications in countless domains—from smart urban spaces and business intelligence to health and wellness, lifelogging, and entertainment—increasingly require joint modelling of multiple modalities [20,47]. Finally, users expect to be able to work very efficiently with large-scale collections, even with the limited computing resources they have at their immediate disposal. All these trends contribute to making scalability a greater concern than ever before.

J. M. Jose et al. (Eds.): ECIR 2020, LNCS 12035, pp. 495–510, 2020.
https://doi.org/10.1007/978-3-030-45439-5_33

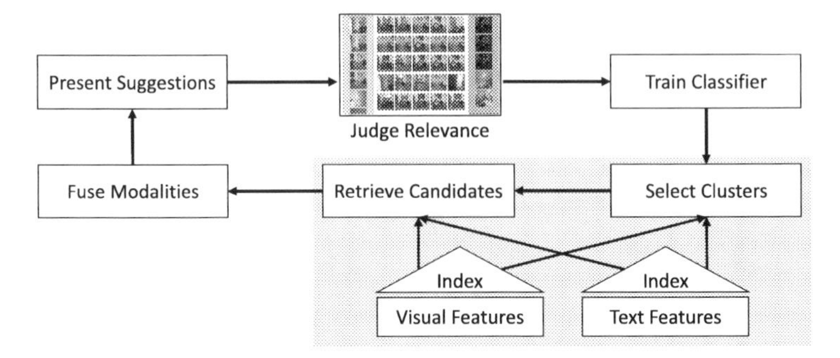

Fig. 1. An outline of the user relevance feedback approach proposed in this paper. The shaded area indicates that the traditional relevance feedback pipeline is enhanced with a novel query mechanism to a state-of-the-art cluster-based high-dimensional index.

User relevance feedback, a form of interactive learning, provides an effective mechanism for addressing various analytic tasks that require alternating between search and exploration. Figure 1 shows an example of such a relevance feedback process, where positive and negative relevance judgments from the user are used to train a classifier, which in turn is used to provide new suggestions to the user, with the process continuing until the user completes the interaction. There has been relatively little work on user relevance feedback and truly scalable and interactive multimedia systems in general in the last decade, however, which recently raised serious concerns in the multimedia community [39]. Clearly, the time has come to re-visit interactive learning with an aim towards scalability.

We propose Exquisitor, a highly scalable and interactive approach for user relevance feedback on large media collections. As illustrated in Fig. 1, the proposed approach tightly integrates high-dimensional indexing with the interactive learning process. To the best of our knowledge, our approach is the first scalable interactive learning method to go beyond utilizing clustering in the preprocessing phase only. To evaluate the approach, we propose a new zero-shot inspired evaluation protocol over the ImageNet collection, and use an existing protocol for the large-scale YFCC100M collection. We show that our approach outperforms state-of-the-art approaches in terms of both suggestion relevance and interactive performance. In particular, our approach requires less than 0.3 s per interaction round to generate suggestions from the YFCC100M collection, using a single CPU core and less than 7 GB of main memory.

The remainder of this paper is organized as follows. In Sect. 2, we discuss interactive learning from a scalability perspective, setting the stage for the novel approach. In Sect. 3, we then present the proposed approach in detail, and compare its performance to the state of the art in Sect. 4, before concluding.

2 Related Work

As outlined in the introduction, combining interactive learning with high dimensional indexing is a step towards unlocking the true potential of multimedia collections and providing added value for users. In this section we first describe the state of the art in interactive learning. Then, based on the identified advantages and limitations of interactive learning algorithms, we provide a set of requirements that high-dimensional indexing should satisfy for facilitating interactivity on extremely large collections. Finally, we use those requirements for reflecting on the state of the art in high-dimensional indexing.

Interactive Learning: Interactive learning has long been a cornerstone of facilitating access to document collections [1,16,18,27] and it became an essential tool of multimedia researchers from the early days of content-based image and video retrieval [15,36]. The most popular flavour of interactive learning is user relevance feedback that presents the user, in each interaction round, with the items for which the classification model is most confident [36]. User relevance feedback has frequently been used in the best performing entries of benchmarks focusing on interactive video search and exploration [28,41]. However, those solutions were designed for collections far smaller than YFCC100M, which is the challenge we take in this paper. Linear models for classification, such as Linear SVM are still amongst the most frequent choices in relevance feedback applications [22,31,48] due to their simplicity, interpretability and explainability as well as the ability to produce accurate results with few annotated samples and scale to very large collections.

To the best of our knowledge, Blackthorn [48] is the most efficient interactive multimodal learning approach in the literature. Its efficiency is achieved through adaptive data compression and feature selection, multi-core processing, and a classification model capable of scoring items directly in the compressed domain. Compared to product quantization [17], a popular alternative optimized for k-NN search, Blackthorn was found to yield significantly more accurate results over YFCC100M with similar latency (1.2 s), while consuming modest computational resources (16 CPU cores with 5 GB of main memory).

Indexing Requirements: We have identified the following requirements for high-dimensional indexing to enhance the performance of interactive learning:

- **R1** *Short and Stable Response Time:* A successful indexing approach in interactive learning combines good result quality with response time guarantees [44].
- **R2** *Preservation of Feature Space Similarity Structure:* Since interactive classifiers compute relevance based on a similarity structure on the feature space, the space partitioning of the high-dimensional indexing algorithm must preserve this similarity structure.
- **R3** *k Farthest Neighbours:* Relevance feedback approaches typically try to inform the user by presenting the most confidently relevant items based on the judgments observed so far, which are the items farthest from the classification boundary. As results are intended for display on screen, the index should thus return k farthest neighbours (k-FN).

We are not aware of any work in the high-dimensional literature targeting approximate k-FN where the query is a classification boundary. We therefore next review the related work and discuss how well different classes of high-dimensional indexing methods can potentially satisfy these three requirements.

High-Dimensional Indexing. Scalable high-dimensional indexing methods generally rely on approximation through some form of quantization. One class of methods uses scalar quantization. The NV-tree, for example, is a large-scale index that uses random projections at its core [25,26], recursively projecting points onto segmented random lines. LSH is another indexing method that uses random projections acting as locality preserving hashing functions [2,8]. Recently, multimedia researchers have considered hashing for multimedia applications, but typically at a much smaller scale than considered here [13,29,42]. LSH has been considered in the context of hyperplane-based nearest-neighbour queries [5,45] and point-based farthest-neighbour queries [7,32,46], but not in the context of *hyperplane-based farthest-neighbour* queries. We argue that LSH and related methods fail to satisfy the three requirements above: they focus on quality guarantees rather than performance guarantees (**R1**); hashing creates "slices" in high-dimensional space, making ranking based on distance to a decision boundary difficult (**R2**); and they typically focus on ϵ-range queries, giving no guarantees on the number of results returned (**R3**).

A second class of methods is based on vector quantization, typically using clustering approaches, such as k-means, to determine a set of representative feature vectors to use for the quantization. These methods create Voronoï cells in the high-dimensional space, which satisfy **R2** well. Some methods, such as BoW-based methods, only store image identifiers in the clusters, thus failing to support **R3**, while others store the entire features, allowing to rank the results from the farthest clusters. Finally, many clustering methods seek to match well the distribution of data in the high-dimensional space. Typically, these methods end with a large portion of the collection in a single cluster, which in turn takes very long to read and score, thus failing to satisfy **R1** [12].

Product quantization (PQ) [17] and its variants [4,10,14] cluster the high-dimensional vectors into low-dimensional subspaces that are indexed independently. PQ better captures the location of points in the high-dimensional space, which in turn improves the quality of the approximate results that are returned. One of the main aims of PQ is data compression, however, and PQ-based methods essentially transform the Euclidean space, complicating the identification of furthest neighbours (**R2**). PQ-compression was compared directly with the Blackthorn compression method designed for interactive learning [48] and was shown as having inferior performance. The extended Cluster Pruning (eCP) algorithm [11,12], however, is an example of a vector quantifier which attempts to balance cluster sizes for improved performance, thus aiming to satisfy all three requirements; we conclude that eCP is our prime candidate.

3 The Exquisitor Approach

In this section, we describe Exquisitor, a novel interactive learning approach that tightly integrates high-dimensional indexing with the interactive learning process, facilitating interactive learning at the scale of the YFCC100M image collection using very moderate hardware resources. Figure 1 shows an outline of the Exquisitor approach. We start by considering the multimodal data representation and classifier, before describing the indexing and retrieval algorithms in separate sub-sections. To facilitate the exposition in this section, we occasionally use actual examples from the YFCC100M collection.

3.1 Media Representation and Classification Model

Similar to [48], we choose to represent each image with two semantic feature vectors, one for visual content using deep-learning-based feature vectors and the second for textual content by extracting LDA topics from any textual metadata associated with the images. Although more descriptive approaches for extracting text features exist, in this case the LDA is effective in yielding discriminative representation for different items.

Directly working with these representations, however, is infeasible. In our case, using 1,000 and 100 dimensions for the visual and text domains, respectively, the feature vectors would require 8.8 KB of main memory per image, or around 880 GB for the YFCC100M collection, which is far beyond the storage capacity of typical hardware. We use the data compression method presented in [48] that preserves semantic information with over 99% compression rate.

Consistent with the state of the art in user relevance feedback, the classifier used in Exquisitor is Linear SVM. The choice is further motivated by the algorithm's speed, reasonable performance and compatibility with the sparse compressed representation. Note that the choice of interactive classifier and features in each respective modality made in this paper is not an inherent setting of Exquisitor; they can be replaced as deemed fit. The choices made in this paper are in line with the choices made in the state of the art Exquisitor competes against (most notably [48]), providing a level field for experimental evaluation.

3.2 Data Indexing

The data indexing algorithm used in Exquisitor is based on the extended Cluster Pruning (eCP) algorithm [12]. As motivated in Sect. 2, the goal is to individually cluster each of the two feature representations with a vector quantizer, using a hierarchical index structure to facilitate efficient selection of clusters to process for suggestions. For each collection, cluster representatives are selected randomly and clusters are formed by assigning images to the nearest cluster based on Euclidean distance, computed efficiently directly in compressed space. The indexing algorithm recursively selects 1% of the images at each level as representatives for the level above, until fewer than 100 representatives remain to form the root of the index. As an example, the bottom level of the index for each modality in the YFCC100M collection consists of 992,066 clusters, organized in a 3 level deep index hierarchy, which gives on average 100 images per cluster and per internal node. When building the indices, the average cluster size was chosen to be small, as previous studies show that searching more small clusters yields better results than searching fewer large clusters [11,40].

3.3 Suggestion Retrieval

The retrieval of suggestions has the following three phases: identify b most relevant clusters, select r most relevant candidates per modality, and fuse modalities to retrieve k most relevant suggestions.[1]

Identify b Most Relevant Clusters: In each interaction round, the index of representatives is used to identify, for each modality, the b clusters most likely to contain useful candidates for suggestions. This search expansion parameter, b, affects the size of the subset that will be scored and can be used to balance between search quality and latency at run-time. All cluster representatives are scored by the interactive classifier and the b clusters farthest from the separating plane in the positive direction are selected as the most relevant clusters. In Sect. 4.3 we evaluate the effects of b on the YFCC100M collection.

We observe that with the YFCC100M collection, both modalities have 1–2 clusters that are very large, with more than 1M items. These clusters require a significant effort to process, without improving suggestion quality. In the experiments reported here, we have therefore omitted clusters larger than 1M.

Select r Most Relevant Candidates per Modality: Once the most relevant b clusters have been identified, the compressed feature vectors within these clusters are scored to suggest the r most relevant media items for each modality. The method of scoring individual feature vectors is the same as when selecting the most relevant clusters.

Some notes are in order here. First, in this scoring phase, media items seen in previous rounds are not considered candidates for suggestions. Second, an item already seen in the first modality is not considered as a suggestion in the second modality. Third, if all b clusters are small, the system may not be able to identify r candidates, in which case it simply returns all the candidates found. Finally, we observe that treating all b clusters equally results in an over-emphasis on items that score very highly in only one modality, but have a low score in the other modality. This can be troublesome if the relevant items have a decent score in both modalities. By segmenting the b clusters into S_c segments of size b/S_c this dominance can be avoided; we explore the impact of S_c in Sect. 4.3.

Modality Fusion for k Most Relevant Suggestions: Once the r most relevant candidates from each modality have been identified, the modalities must be fused by aggregating the candidate lists to produce the final list of k suggestions. First, for each candidate in one modality, the score in the other modality

[1] In the case of unimodal retrieval, the latter two phases can be merged.

is computed if necessary, by directly accessing the compressed feature vector, resulting in $2r$ candidates with scores in both modalities.[2] Second, the rank of each item in each modality is computed by sorting the $2r$ candidates. Finally, the average rank is used to produce the final list of suggestions.

Multi-core Processing: If desired, Exquisitor can take advantage of multiple CPU cores. With w cores available, the system creates w worker processes and assigns b/w clusters to each worker. Each worker produces r suggestions in each modality and fuses the two modalities into k candidates, as described above. The top k candidates overall are then selected by repeating the modality fusion process for the suggestions produced by the workers.

4 Experimental Evaluation

In this section, we experimentally analyse the interactive performance of Exquisitor. We first outline the baseline comparison architectures from the literature. We then describe two detailed experiments. In the first experiment, we propose a new experimental protocol for interactive learning based on the popular ImageNet benchmark dataset, and show that (a) the Linear SVM model is capable of discovering new classes in the data, and (b) with high-dimensional indexing, performance is significantly improved. In the second experiment, we then use a benchmark experimental protocol from the literature defined over the YFCC100M collection, and show that at this scale the Exquisitor approach outperforms the baseline architectures significantly, both in terms of retrieval quality and interactive performance.

4.1 Baseline Approaches

In the experiments we compare Exquisitor with the following state-of-the-art approaches from the literature.

Blackthorn: To the best of our knowledge, Blackthorn [48] is the only direct competitor in the literature for interactive learning at the YFCC100M scale. Unlike Exquisitor, Blackthorn uses no indexing or prior knowledge about the structure of the collection, instead using data compression and multi-core processing for scalability.

kNN+eCP: This baseline is representative of pure query-based approaches using a k-NN query vector based on relevance weights [23, 34], an approach that was initially introduced for text retrieval [35] but has been adapted for CBIR with relevance feedback [37].

SVM+LSH, kNN+LSH: These baselines represent SVM-based and k-NN-based approaches using LSH indexing. We replace the eCP index with a multi-probing LSH index [30] using the FALCONN library [3].

[2] To facilitate late modality fusion, the location of each feature vector in each cluster index is stored; each vector requires ~800 KB of RAM for the YFCC100M collection.

All comparison architectures are compiled with g++. Experiments are performed using dual 8-core 2.4 GHz CPUs, with 64 GB RAM and 4 TB local SSD storage. Note, however, that even the YFCC100M collection requires less than 7 GB of SSD storage and RAM, and most experiments use only a single CPU core.

While tuning LSH performance is difficult, due to the many parameters that interact in complex ways (L is the number of tables, B is the number of buckets in each table, and p is the number of buckets to read from each table at query time), we have strived to find parameter settings that (a) lead to a similar cell size distribution as eCP and (b) yield the best performance.

4.2 Experiment 1: Discovering ImageNet Concepts

Zero-shot learning is a method which trains a classifier to find target classes without including the target classes when training the model. Taking inspiration from zero-shot learning, the objective of this experiment is to simulate a user that is looking for a concept that is on their mind, but is not directly represented in the data; a successful interactive learning approach should be able to do this.

Image Collection: ImageNet is an image database based on the WordNet hierarchy. It is a well-curated collection targeting object recognition research as the images in the collection are categorized into approximately 21,000 WordNet synsets (synonym sets) [9]. The collection contains 14,198,361 images, each of which is represented with the 1,000 ILSVRC concepts [38]. Due to images being categorized into multiple WordNet synsets, the ImageNet collection contains duplicate images, each labelled differently, which can lead to false negatives.

Experimental Protocol: The protocol for the experiment is constructed by randomly selecting 50 concepts from the 1,000 ILSVRC concepts. For each concept a simulated user (henceforth called actor) is created, which knows which images belong to its concept and is charged with the task of finding items belonging to that concept. We have then created and indexed 5 different collections of visual features, where the feature value of the concepts belonging to 10 different actors have been set to 0 to introduce the zero-shot setting.

The workload for each actor proceeds as follows. Initially, 10 images from the concept and 100 random images are used as positive and negative examples, respectively, to create the first round of suggestions, simulating a situation where the exploration process has already started. In each round of the interactive learning process, the actor considers the suggested images from the system and designates images from its concept as positive examples, while 100 additional negative examples are drawn randomly from the entire collection. This is repeated for 10 interaction rounds, with performance statistics collected in each round. To combat the duplicate images problem, we first run the workload using the original data where the concepts are known in order to establish an upper bound baseline for each approach.

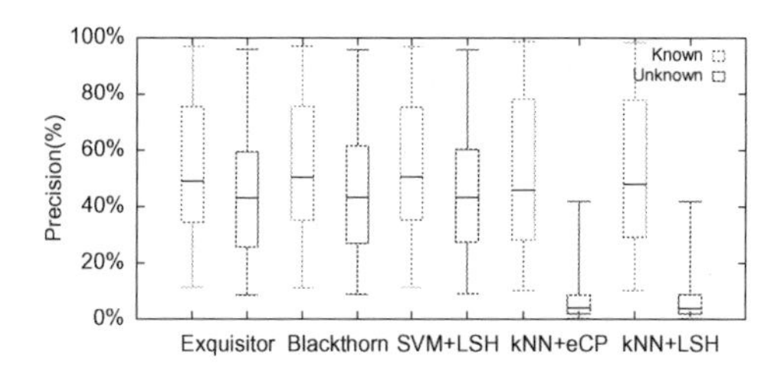

Fig. 2. Average precision per round across all ImageNet actors for each interactive learning approach. The blue columns depict the known case, while the red depict the unknown case. (Color figure online)

Table 1. Average latency per interaction round across all ImageNet actors.

Approach	Latency
Exquisitor	0.008 s
Blackthorn (1w)	0.130 s
Blackthorn (16w)	0.017 s
SVM+LSH	0.008 s
kNN+eCP	0.008 s
kNN+LSH	0.004 s

Results: Figure 2 compares the average precision across the 10 rounds for each of the approaches under study, for both the case when the concept is *known* (blue columns) and *unknown* (red columns). For Exquisitor and eCP+kNN, the search expansion parameter b is set to 256, while SVM+LSH and kNN+LSH have the following settings for the LSH index: $L = 10$, $B = 2^{14}$, and $p = 20$.

Overall, the figure shows that precision for the known case is nearly 50% on average for the SVM-based approaches, and only slightly lower for the k-NN-based approaches. When the feature value for the actor's concept is not known, however, the average precision drops only slightly for the SVM-based approaches, while the k-NN-based approaches perform very poorly. These results indicate that the Linear-SVM model is clearly superior to the k-NN approach.

Turning to the average time required for each iteration of the learning process, Table 1 compares the approaches under study. Overall, we note that the four approaches relying on high-dimensional indexing perform very well using a single computing core, requiring less than 10 ms to return suggestions. At the moderate scale of the ImageNet collection, eCP and LSH perform similarly. Running Blackthorn with 16 cores is 2x slower, however, while running Blackthorn using a single core is about 16x slower.

As mentioned above, precision is impacted by the ImageNet collection itself containing duplicates. A visual inspection of the results of some of the worst-performing actors suggest that with known data, the majority of the non-relevant images are such duplicates. For the unknown case, a similar trend is seen for the SVM-based approaches, but not for the k-NN-based approaches, which clearly are unable to steer the query vectors for suggestions to a more relevant part of the collection. Figure 3 shows some examples of this, for the actor for concept "knee pad". As the figure shows, with any SVM-based approach the irrelevant images are also knee pads, but tagged to another related concept, while for the k-NN-based approach, no relevant images were found and the irrelevant images bear no relationship to knee pads.

4.3 Experiment 2: Performance at YFCC100M Scale

The goal of this experiment is to study the scalability of the Exquisitor approach, in comparison to the baseline approaches from the literature. To that end, we apply the only interactivee learning evaluation protocol from the literature that we are aware of at YFCC100M scale [48].

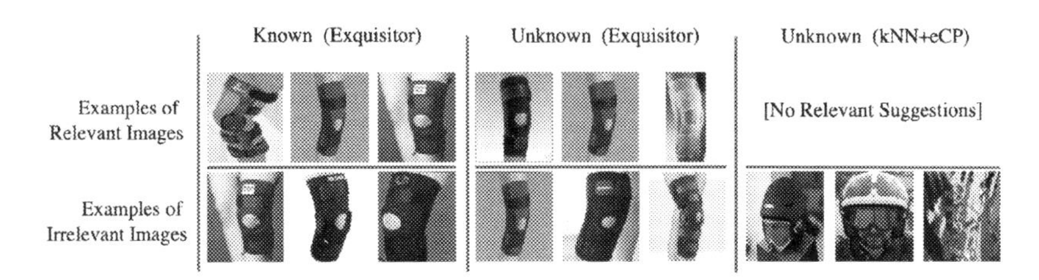

Fig. 3. Examples of relevant and irrelevant suggestions for different approaches for the ImageNet actor for the concept "knee pad".

Collection: The YFCC100M collection contains 99,206,564 Flickr images, their associated annotations (i.e. title, tags and description), a range of metadata produced by the capturing device, the online platform, and the user (e.g., geolocation and time stamps). The visual content is represented using the 1,000 ILSVRC concepts [38] extracted using the GoogLeNet convolutional neural network [43]. The textual content is encoded by (a) treating the title, tags, and description as a single text document, and (b) extracting 100 LDA topics for each image using the gensim toolkit [33].

The YFCC100M collection, being large and uncurated, displays some interesting phenomena worth mentioning. First, a non-trivial proportion of images are a standard Flickr "not found" image.[3] A similar situation arises in the text modality, with many images lacking text information altogether, resulting in zero-valued vectors. Such images are essentially noise, potentially crowding out more suitable candidates. Second, with the collection being massive and the data being compressed and clustered, discriminativeness of feature vectors becomes a problem: non-identical images may be mapped to identical feature vectors.

Experimental Protocol: For this experiment we follow the experimental interactive learning protocol in [48]. This evaluation protocol is inspired by the MediaEval Placing Task [6,24], in which actors simulating user behaviour look for images from 50 world cities.

To illustrate the tradeoffs between the interactive performance and result quality, we focus our analysis on precision and latency (response time) per interaction round. It is worth noting that due to both the scale of YFCC100M and

[3] The image collection was actually downloaded very shortly after release, but already then this had become a significant issue.

its unstructured nature, precision is lower than in experiments involving small and well-curated collections.

Impact of Search Expansion Parameter: We start by exploring the impact of the search expansion parameter b for the eCP index. Figure 4 analyses the impact of b, the number of clusters read and scored, on the precision (fraction of relevant items seen) in each round of the interactive exploration. The x-axis shows how many clusters are read for scoring at each round, ranging from $b = 1$ to $b = 512$ (note the logarithmic scale of the axis), while the y-axis shows the average precision across the first 10 rounds of analysis. The figure shows precision for two Exquisitor variants, with $S_c = 1$ and $S_c = 16$. In both cases, only one worker is used, $w = 1$. For comparison, the figure also shows the average precision for Blackthorn, the state-of-the-art SVM-based alternative.

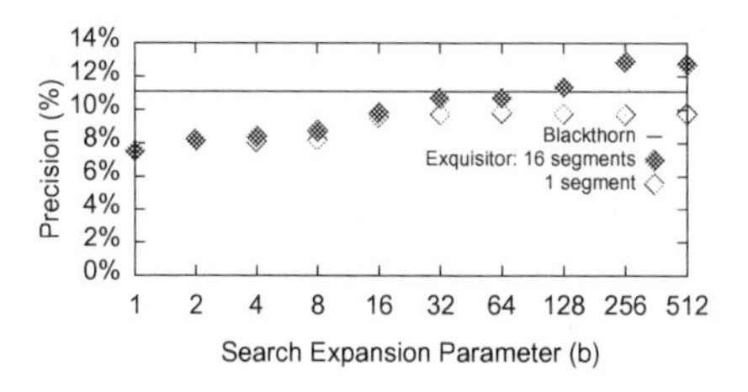

Fig. 4. Average precision over 10 rounds of analysis across all YFCC100M actors. Exquisitor: varying b; $w = 1$; $S_c = 1, 16$.

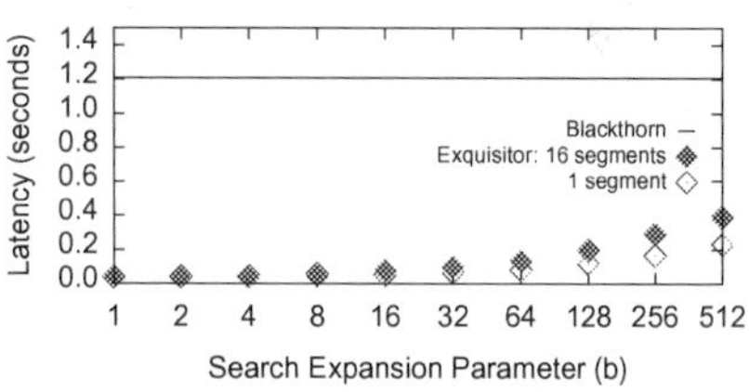

Fig. 5. Average latency over 10 rounds of analysis across all YFCC100M actors. Exquisitor: varying b; $w = 1$; $S_c = 1, 16$.

As Fig. 4 shows, result quality is surprisingly good when scoring only a single cluster in each interaction round, returning about two-thirds of the precision of the state-of-the-art algorithm. As more clusters are considered, quality then improves further. As expected, dividing the b clusters into $S_c = 16$ chunks results in better quality, an effect that becomes more pronounced as b grows. In particular, with $b = 256$, Exquisitor returns significantly better results than Blackthorn. The reason is that by assigning the b relevant clusters to $S_c = 16$ segments, Exquisitor is able to emphasize the bi-modal media items as explained in Sect. 3.3. Note that as further clusters are added with Exquisitor ($b = 512$ and beyond), the results become more and more similar to the Blackthorn results.

Figure 5, on the other hand, shows the latency per interaction round. The figure again shows the two Exquisitor variants, with $S_c = 1$ and $S_c = 16$; in both cases, one worker is used, $w = 1$. For comparison, as before, it also shows the average latency for Blackthorn (with 16 CPU cores). Unsurprisingly, Fig. 5 shows linear growth in latency with respect to b (recall the logarithmic x-axis). With $b = 256$, each interaction round takes less than 0.3 s with $S_c = 16$, and about 0.17 s with $S_c = 1$. Both clearly allow for interactive performance; the remainder of our experiments focus on $b = 256$. If even shorter latency is desired, however, fewer clusters can be read: $b = 32$, for example, also gives a good tradeoff between latency and result quality. This latency is produced using only a single CPU core, meaning that the latency is ~4x better than Blackthorn, with 16x fewer computing cores, for an improvement of ~64x, or nearly two orders of magnitude. With this knowledge we see b as a parameter that is determined by collection size and the task a user is dealing with, but, as a general starting point we recommend $b = 256$ for large collections.

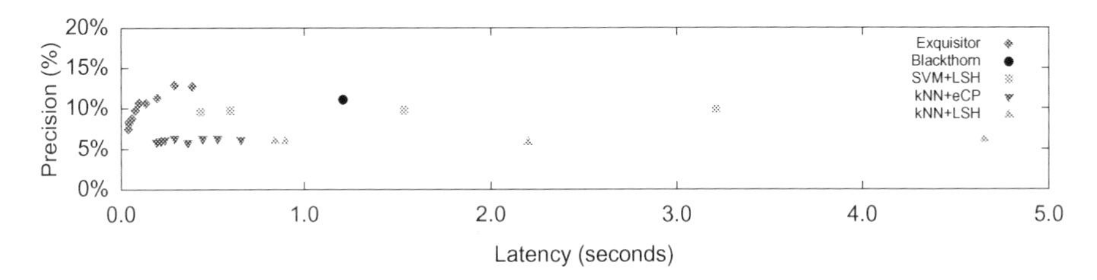

Fig. 6. Average precision vs. latency over 10 rounds of analysis across all YFCC100M actors. Exquisitor, kNN+eCP: $b = 1 - 512$. LSH: $L = 10$, $B = [2^{10}, 2^{18}]$, $p = [15, 40]$.

Comparison: Figure 6 shows the tradeoff between result quality, measured by average precision across 10 rounds of interaction, and the average latency required to produce the suggestions in each round. For Exquisitor, the figure essentially summarizes Figs. 4 and 5. For kNN+eCP, the dots represent the same b parameter values, while for the LSH-based approaches a variety of parameter values are represented. The figure clearly demonstrates that Exquisitor is the best approach in both precision and response time compared to all the baseline approaches, achieving better precision than Blackthorn, requiring less than 0.3 s compared to Blackthorn's 1.2 s. Both k-NN-bases approaches get stuck at 6% which is to be expected since the k-NN query narrows down the scope of the search making it impossible to get out of local optima. SVM+LSH performs better, with precision nearly as good as Blackthorn and response time close to Exquisitor. Overall, however, Exquisitor performs better partly due to being able to utilize the SVM during cluster selection with k-FN queries, and partly due to the cluster segments allowing better multi-modal results.

5 Conclusions

In this paper, we presented Exquisitor, a new approach for exploratory analysis of very large image collections with modest computational requirements. Exquisitor combines state-of-the-art large-scale interactive learning with a new cluster-based retrieval mechanism, enhancing the relevance capabilities of interactive learning by exploiting the inherent structure of the data. Through experiments conducted on YFCC100M, the largest publicly available multimedia collection,

Exquisitor achieves higher precision and lower latency, with less computational resources. Additionally, through a modified zero-shot learning experiment on ImageNet, we determine the Exquisitor approach to be excellent at solving cumbersome classification tasks. Exquisitor also introduces customizability that is, to the best of our knowledge, previously unseen in large-scale interactive learning by: (i) allowing a tradeoff between low latency (few clusters) and high quality (many clusters); and (ii) combatting data skew by omitting huge (and thus likely nondescript) clusters from consideration. Exquisitor has recently been used successfully in interactive media retrieval competitions such as the Lifelog Search Challenge [21] and Video Browser Showdown [19]. In conclusion, Exquisitor provides excellent performance on very large collections while being efficient enough to bring large-scale multimedia analytics to standard desktops and laptops, and even high-end mobile devices.

Acknowledgments. This work was supported by a PhD grant from the IT University of Copenhagen and by the European Regional Development Fund (project Robotics for Industry 4.0, CZ.02.1.01/0.0/0.0/15 003/0000470).

References

1. Allan, J.: Incremental relevance feedback for information filtering. In: Proceedings of the 19th Annual International ACM SIGIR Conference on Research and Development in Information Retrieval, pp. 270–278. ACM, New York (1996)
2. Andoni, A., Indyk, P.: Near-optimal hashing algorithms for approximate nearest neighbor in high dimensions. In: Proceedings of the IEEE Symposium on the Foundations of Computer Science, pp. 459–468. IEEE Computer Society, Berkeley (2006)
3. Andoni, A., Indyk, P., Laarhoven, T., Razenshteyn, I., Schmidt, L.: Practical and optimal LSH for angular distance. In: Cortes, C., Lawrence, N.D., Lee, D.D., Sugiyama, M., Garnett, R. (eds.) Advances in Neural Information Processing Systems, vol. 28, pp. 1225–1233. Curran Associates, Inc., Red Hook (2015)
4. Babenko, A., Lempitsky, V.S.: The inverted multi-index. IEEE Trans. Pattern Anal. Mach. Intell. **37**(6), 1247–1260 (2015)
5. Basri, R., Hassner, T., Zelnik-Manor, L.: Approximate nearest subspace search. IEEE Trans. Pattern Anal. Mach. Intell. **33**(2), 266–278 (2011)
6. Choi, J., Hauff, C., Laere, O.V., Thomee, B.: The placing task at mediaeval 2015. In: Proceedings of the MediaEval 2015 Workshop. CEUR, Wurzen (2015)
7. Curtin, R.R., Gardner, A.B.: Fast approximate furthest neighbors with data-dependent candidate selection. In: Amsaleg, L., Houle, M.E., Schubert, E. (eds.) SISAP 2016. LNCS, vol. 9939, pp. 221–235. Springer, Cham (2016). https://doi.org/10.1007/978-3-319-46759-7_17
8. Datar, M., Immorlica, N., Indyk, P., Mirrokni, V.S.: Locality-sensitive hashing scheme based on p-stable distributions. In: Proceedings of ACM Symposium on Computational Geometry, pp. 253–262. ACM, Brooklyn (2004)
9. Deng, J., Dong, W., Socher, R., Li, L.J., Li, K., Fei-Fei, L.: ImageNet: a large-scale hierarchical image database. In: 2009 IEEE Conference on Computer Vision and Pattern Recognition, pp. 248–255 (2009)

10. Ge, T., He, K., Ke, Q., Sun, J.: Optimized product quantization. IEEE Trans. Pattern Anal. Mach. Intell. **36**(4), 744–755 (2014)
11. Gudmundsson, G.Þ., Amsaleg, L., Jónsson, B.Þ.: Impact of storage technology on the efficiency of cluster-based high-dimensional index creation. In: Yu, H., Yu, G., Hsu, W., Moon, Y.-S., Unland, R., Yoo, J. (eds.) DASFAA 2012. LNCS, vol. 7240, pp. 53–64. Springer, Heidelberg (2012). https://doi.org/10.1007/978-3-642-29023-7_6
12. Gudmundsson, G.Þ., Jónsson, B.Þ., Amsaleg, L.: A large-scale performance study of cluster-based high-dimensional indexing. In: Proceedings of International Workshop on Very-large-scale Multimedia Corpus, Mining and Retrieval (VLS-MCMR), pp. 31–36. ACM, Firenze (2010)
13. Hansen, C., Hansen, C., Simonsen, J.G., Alstrup, S., Lioma, C.: Unsupervised neural generative semantic hashing. In: Proceedings of the 42nd International ACM SIGIR Conference on Research and Development in Information Retrieval, SIGIR 2019, pp. 735–744. ACM, New York (2019)
14. Heo, J., Lin, Z., Yoon, S.: Distance encoded product quantization. In: Proceedings of the IEEE International Conference on Computer Vision and Pattern Recognition, pp. 2139–2146. IEEE Computer Society, Columbus (2014)
15. Huang, T., et al.: Active learning for interactive multimedia retrieval. Proc. IEEE **96**(4), 648–667 (2008)
16. Iwayama, M.: Relevance feedback with a small number of relevance judgements: incremental relevance feedback vs. document clustering. In: Proceedings of the 23rd Annual International ACM SIGIR Conference on Research and Development in Information Retrieval, pp. 10–16. ACM, New York (2000)
17. Jégou, H., Douze, M., Schmid, C.: Product quantization for nearest neighbor search. IEEE Trans. Pattern Anal. Mach. Intell. **33**(1), 117–128 (2011)
18. Joachims, T.: A probabilistic analysis of the Rocchio algorithm with TFIDF for text categorization. In: Proceedings of the Fourteenth International Conference on Machine Learning, ICML 1997, pp. 143–151. Morgan Kaufmann Publishers Inc., San Francisco (1997)
19. Jónsson, B.Þ., Khan, O.S., Koelma, D.C., Rudinac, S., Worring, M., Zahálka, J.: Exquisitor at the video browser showdown 2020. In: Ro, Y.M., et al. (eds.) MMM 2020. LNCS, vol. 11962, pp. 796–802. Springer, Cham (2020). https://doi.org/10.1007/978-3-030-37734-2_72
20. Jónsson, B.Þ., Worring, M., Zahálka, J., Rudinac, S., Amsaleg, L.: Ten research questions for scalable multimedia analytics. In: Tian, Q., Sebe, N., Qi, G.-J., Huet, B., Hong, R., Liu, X. (eds.) MMM 2016. LNCS, vol. 9517, pp. 290–302. Springer, Cham (2016). https://doi.org/10.1007/978-3-319-27674-8_26
21. Khan, O.S., Jónsson, B.Þ., Zahálka, J., Rudinac, S., Worring, M.: Exquisitor at the lifelog search challenge 2019. In: Proceedings of the ACM Workshop on Lifelog Search Challenge, pp. 7–11. ACM (2019)
22. Kovashka, A., Parikh, D., Grauman, K.: WhittleSearch: interactive image search with relative attribute feedback. Int. J. Comput. Vis. **115**(2), 185–210 (2015)
23. Lan, M., Tan, C.L., Su, J., Lu, Y.: Supervised and traditional term weighting methods for automatic text categorization. IEEE Trans. Pattern Anal. Mach. Intell. **31**(4), 721–735 (2008)
24. Larson, M., et al.: Automatic tagging and geotagging in video collections and communities. In: Proceedings of the 1st ACM International Conference on Multimedia Retrieval, pp. 51:1–51:8. ACM, New York (2011)

25. Lejsek, H., Ásmunðsson, F.H., Jónsson, B.Þ., Amsaleg, L.: NV-tree: an efficient disk-based index for approximate search in very large high-dimensional collections. IEEE Trans. Pattern Anal. Mach. Intell. **31**(5), 869–883 (2009)
26. Lejsek, H., Jónsson, B.Þ., Amsaleg, L.: NV-tree: nearest neighbors at the billion scale. In: Proceedings of the ACM International Conference on Multimedia Retrieval. ACM, Trento (2011)
27. Lewis, D.D., Schapire, R.E., Callan, J.P., Papka, R.: Training algorithms for linear text classifiers. In: Proceedings of the 19th Annual International ACM SIGIR Conference on Research and Development in Information Retrieval, SIGIR 1996, pp. 298–306. ACM, New York (1996)
28. Lokoč, J., Bailer, W., Schoeffmann, K., Muenzer, B., Awad, G.: On influential trends in interactive video retrieval: video browser showdown 2015–2017. IEEE Trans. Multimed. **20**(12), 3361–3376 (2018)
29. Lu, X., Zhu, L., Cheng, Z., Nie, L., Zhang, H.: Online multi-modal hashing with dynamic query-adaption. In: Proceedings of the 42nd International ACM SIGIR Conference on Research and Development in Information Retrieval, pp. 715–724. ACM, New York (2019)
30. Lv, Q., Josephson, W., Wang, Z., Charikar, M., Li, K.: Multi-probe LSH: efficient indexing for high-dimensional similarity search. In: Proceedings of the 33rd international conference on Very large data bases, pp. 950–961. VLDB Endowment (2007)
31. Mironică, I., Ionescu, B., Uijlings, J., Sebe, N.: Fisher kernel temporal variation-based relevance feedback for video retrieval. Comput. Vis. Image Underst. **143**, 38–51 (2016)
32. Pagh, R., Silvestri, F., Sivertsen, J., Skala, M.: Approximate furthest neighbor with application to annulus query. Inf. Syst. **64**, 152–162 (2017)
33. Řehůřek, R., Sojka, P.: Software framework for topic modelling with large corpora. In: Proceedings of the LREC 2010 Workshop on New Challenges for NLP Frameworks, pp. 45–50. ELRA, Valletta, May 2010
34. Robertson, S.E., Spärck Jones, K.: Simple, proven approaches to text retrieval. Technical report, University of Cambridge, Computer Laboratory (1994)
35. Rocchio, J.J.: Relevance feedback in information retrieval. Technical report, University of Harvard, Computer Laboratory (1965)
36. Rui, Y., Huang, T.S., Mehrotra, S.: Content-based image retrieval with relevance feedback in MARS. In: Proceedings of International Conference on Image Processing (ICIP), pp. 815–818. IEEE Computer Society, Santa Barbara (1997)
37. Rui, Y., Huang, T.S., Mehrotra, S.: Content-based image retrieval with relevance feedback in mars. In: Proceedings of International Conference on Image Processing, vol. 2, pp. 815–818. IEEE (1997)
38. Russakovsky, O., et al.: Imagenet large scale visual recognition challenge. Int. J. Comput. Vis. **115**(3), 211–252 (2015)
39. Schoeffmann, K., Bailer, W., Gurrin, C., Awad, G., Lokoč, J.: Interactive video search: where is the user in the age of deep learning? In: Proceedings of ACM Multimedia, pp. 2101–2103. ACM, Seoul (2018)
40. Sigurðardóttir, R., Hauksson, H., Jónsson, B.Þ., Amsaleg, L.: The quality vs. time tradeoff for approximate image descriptor search. In: Proceedings of IEEE EMMA workshop. IEEE, Tokyo (2005)
41. Snoek, C., Worring, M., de Rooij, O., van de Sande, K., Yan, R., Hauptmann, A.: VideOlympics: real-time evaluation of multimedia retrieval systems. IEEE MM **15**(1), 86–91 (2008)

42. Sun, C., Song, X., Feng, F., Zhao, W.X., Zhang, H., Nie, L.: Supervised hierarchical cross-modal hashing. In: Proceedings of the 42nd International ACM SIGIR Conference on Research and Development in Information Retrieval, pp. 725–734. ACM, New York (2019)
43. Szegedy, C., et al.: Going deeper with convolutions. In: Proceedings of IEEE CVPR, pp. 1–9. IEEE Computer Society, Boston (2015)
44. Tavenard, R., Jégou, H., Amsaleg, L.: Balancing clusters to reduce response time variability in large scale image search. In: International Workshop on Content-Based Multimedia Indexing. IEEE, Madrid (2011)
45. Vijayanarasimhan, S., Jain, P., Grauman, K.: Hashing hyperplane queries to near points with applications to large-scale active learning. IEEE Trans. Pattern Anal. Mach. Intell. **36**(2), 276–288 (2014)
46. Xu, X., et al.: Reverse furthest neighbors query in road networks. J. Comput. Sci. Technol. **32**(1), 155–167 (2017)
47. Zahálka, J., Worring, M.: Towards interactive, intelligent, and integrated multimedia analytics. In: Proceedings of the IEEE Conference on Visual Analytics Science and Technology (VAST), Paris, France, pp. 3–12 (2014)
48. Zahálka, J., Rudinac, S., Jónsson, B.T., Koelma, D.C., Worring, M.: Blackthorn: large-scale interactive multimodal learning. IEEE Trans. Multimed. **20**(3), 687–698 (2018)

Visual Re-Ranking via Adaptive Collaborative Hypergraph Learning for Image Retrieval

Noura Bouhlel(✉), Ghada Feki, and Chokri Ben Amar

REGIM: Research Groups in Intelligent Machines, University of Sfax, National Engineering School of Sfax (ENIS), BP 1173, 3038 Sfax, Tunisia
noura.bouhlel.tn@ieee.org

Abstract. Visual re-ranking has received considerable attention in recent years. It aims to enhance the performance of text-based image retrieval by boosting the rank of relevant images using visual information. Hypergraph has been widely used for relevance estimation, where textual results are taken as vertices and the re-ranking problem is formulated as a transductive learning on the hypergraph. The potential of the hypergraph learning is essentially determined by the hypergraph construction scheme. To this end, in this paper, we introduce a novel data representation technique named adaptive collaborative representation for hypergraph learning. Compared to the conventional collaborative representation, we consider the data locality to adaptively select relevant and close samples for a test sample and discard irrelevant and faraway ones. Moreover, at the feature level, we impose a weight matrix on the representation errors to adaptively highlight the important features and reduce the effect of redundant/noisy ones. Finally, we also add a nonnegativity constraint on the representation coefficients to enhance the hypergraph interpretability. These attractive properties allow constructing a more informative and quality hypergraph, thereby achieving better retrieval performance than other hypergraph models. Extensive experiments on the public MediaEval benchmarks demonstrate that our re-ranking method achieves consistently superior results, compared to state-of-the-art methods.

Keywords: Image retrieval · Visual re-ranking · Hypergraph learning · Collaborative representation · Ridge regression

1 Introduction

Empowered by the ubiquitous access to computer devices and the Internet, an ever-growing amount of digital images has been emerged [25]. In light of this, image retrieval is considered as an active research topic that aims at retrieving relevant images to a user query from a large database of digital images [11,14,21,26]. Until recently, most of the popular search engines (e.g., Flickr) are built upon the textual information associated with images [4,7,24]. Nevertheless, they cannot comprehensively describe the rich content of images since

J. M. Jose et al. (Eds.): ECIR 2020, LNCS 12035, pp. 511–526, 2020.
https://doi.org/10.1007/978-3-030-45439-5_34

they totally ignore the visual information [10]. Besides, they suffer from the fact that the textual information is often noisy, ambiguous and language-dependent [8,12]. As a consequence, the retrieved results may be noisy and irrelevant which may affect the retrieval performance [17,24]. To tackle those issues, visual re-ranking has been introduced to refine the text-based retrieval results using the visual information [4,19,32,35]. Namely, it attempts to boost the rank of relevant images with respect to the textual query [24]. Recently, the hypergraph learning has been widely used in many applications for its capability in capturing complex relationships among samples [4,15,23]. In case of visual re-ranking, the textual results are taken as vertices and the re-ranking problem is formulated as a transductive learning on the hypergraph [2,9]. The potential of the hypergraph learning is essentially determined by the hypergraph construction scheme [22]. Most of previous hypergraph learning methods adopt a neighborhood-based strategy to build the hypergraph, in which textual results are taken as vertices and each vertex is linked to its k nearest neighbors by an hyperedge. While obvious, this method suffers from the following drawbacks: (1) it is sensitive to noise (2) lacks the ability to discover the real neighborhood structure (3) the parameter k is fixed as global parameter for all samples regardless their local data distribution. To tackle those issues, recent works have proposed to leverage the regularized regression models, namely the sparse representation and the ridge regression for hypergraph construction [22]. Compared to the neighborhood-based hypergraph, the sparse hypergraph achieves superior performance in revealing the local data structure and handling the noisy data. However, it cannot discover related samples to one hyperedge centroid as thoroughly as possible. Moreover, the sparse constraint makes the hypergraph construction very expensive [41]. Recently, the ridge regression has gained considerable attention not only for its effectiveness in data representation but also for its computational efficiency [41]. In contrast to sparse representation which aims at encouraging the competition between samples to represent a datum, the ridge regression attempt to include all samples in the representation process. That's why this framework is often called the collaborative representation. Owing to these desirable properties, in this paper, we put a particular emphasis on the collaborative representation and we propose an adaptive collaborative hypergraph learning for visual re-ranking. The proposed data representation technique adaptively preserve the locality structure and discard irrelevant/outlier samples with respect to a test sample by integrating a distance-regularizer on the representation coefficients. At the feature level, we impose a weight matrix on the representation errors to adaptively highlight the important features and reduce the effect of redundant/noisy ones. Moreover, to enhance the representation interpretability, a nonnegativity constraint is added in such a way that the representation coefficients can directly reveal the similarity among samples. This way, we obtain a more informative and quality hypergraph which not only captures the grouping information but also reveal the local neighborhood structure and exhibit more discriminative power and robustness to noisy data. Extensive experiments on the public MediaEval benchmarks demonstrate that our re-ranking method achieves consistently superior results, compared to state of-the-art methods.

2 Related Works

In recent years, many visual re-ranking methods have been proposed in the literature. According to the statistical analysis model used, they could be classified as supervised and unsupervised methods. The former cast the re-ranking to a classification problem that aims at separating relevant from irrelevant images using data from the initial results as training samples. For instance, authors in [30] built a supervised classification model using expert annotations to assign a relevance score to each image. The latter assumes that relevant samples are probably to be close to each other than to irrelevant ones. It aims at discovering and mining patterns using pair-wise similarities. Clearly, there are two paramount ways. The first is to leverage clustering to group images with respect to their visual closeness. For instance, a Hierarchical Clustering is applied in [1] and [29] to cluster samples by relevance. Authors in [28] apply a graph-based clustering method where a similarity graph is initially built to represent relationships among images. The second way is to adopt the graph-based learning for its effectiveness in modeling the intrinsic structure within data. VisualRank proposed by Jing and Baluja [20] is the most popular graph-based re-ranking method. It applies a random walk on an affinity graph where images are taken as nodes and their visual similarities as probabilistic hyper-links. In [39], a manifold ranking process is applied over the data manifold, with the aim of naturally finding the most relevant images. Although promising results are achieved, how to represent complex and high-order relationships hidden in data still the performance bottleneck for graph-based re-ranking. As a generalization of the graph learning, the hypergraph learning is receiving increasing attention in recent years owing to its ability in modeling complex data structure in a more flexible and elegant way [3,23]. Considering the visual re-ranking, the hypergraph learning is widely used for relevance estimation. For instance, in [2], authors construct a k-nearest neighbor graphs based on the visual similarity between images. Then, a hypergraph ranking is performed to learn the images' relevance scores. Although efficient, this method suffers from some drawbacks. First, the neighborhood strategy cannot capture the local data distribution of each datum since it uses a fixed number of neighbors k for all samples [35]. Second, the neighborhood strategy is very sensitive to noisy data due to the use of the Euclidean distance as similarity measure [22,37]. To address those limitations, some researchers have proposed to exploit the regression models for data representation. The most widely used model is the sparse representation (SR) in which each sample is represented as a linear combination of the remaining samples [15,36]. Compared to the neighborhood-based hypergraph, the sparse hypergraph achieves superior performance in revealing the local data structure and handling the noisy data. However, it cannot discover related samples to one hyperedge centroid as thoroughly as possible. Moreover, the sparse constraint makes the hypergraph construction very expensive.

Recently, the collaborative representation has gained considerable attention not only for its effectiveness in data representation but also for its computational efficiency [41]. Therefore, in this paper, we put a particular emphasis on the collaborative representation and we propose an adaptive collaborative hypergraph learning for visual re-ranking.

3 The Proposed Hypergraph Model for Visual Re-Ranking

3.1 Adaptive Collaborative Representation Representation

For clarity, we first introduce some important notations used throughout this paper. The matrix $X = [x_1, ..., x_N] \in \mathbb{R}^{d \times N}$ is a collection of N data samples where $x_i \in \mathbb{R}^d$ denotes the i-th data sample. $||Z||_F$ is the Frobenius norm of matrix Z. $\mathbf{1}$ and $\boldsymbol{1}$ are a matrix and a vector whose elements are equal to 1, $\odot$ denotes te element-wise multiplication. For a scalar v, we define $(v)_+$ as $(v)_+ = max(v, 0)$ [27].

Problem Formulation. Conventionally, the collaborative representation aims to solve the following least square problem:

$$\underset{Z}{argmin} \left\|X - XZ\right\|_2^2 + \lambda \left\|Z\right\|_2^2 \tag{1}$$

In this paper, we propose an adaptive collaborative representation formulated as follows:

$$\underset{Z,W}{argmin} \left\|W^{1/2} \odot (X - XZ)\right\|_F^2 + \frac{\beta}{2} \left\|W\right\|_F^2 + \lambda \left\|Z\right\|_F^2 + \gamma tr(D^T Z)$$
$$\text{s.t } W \geq 0, W^T\mathbf{1} = \mathbf{1}, Z \geq 0, diag(Z) = 0, Z\mathbf{1} = \mathbf{1} \tag{2}$$

Specifically, the objective function contains the following terms:

1. The self-representation term: It represents the reconstruction error between the estimated and the real data. Many references have pointed out that redundant/noisy features are likely to have large reconstruction errors [23, 40]. Based on this assumption, we regularize the reconstruction errors by a non-negative weight matrix W. Hence, we adaptively highlight the important features while reducing the effect of redundant/noisy ones.
2. The ℓ_2-regularizer on the weight matrix: This term as well as the constraint $W^T\mathbf{1} = \mathbf{1}$ are imposed to avoid the trivial solution of W as in [42].
3. The regularization term on the representation matrix: It shrinks the representation coefficients towards zero by imposing an ℓ_2--regularizer on their sizes. Indeed, all samples will collaborate during the representation process of a test sample since their coefficients will never become exactly zero.

4. The locality-preserving term: The collaborative representation does not consider the data locality which has been observed to be critical for many learning tasks [34]. For this purpose, we incorporate a locality-preserving term in our model so that (1) the local structure is preserved (i.e, close samples will have close representation) and (2) irrelevant/outliers samples are discarded. Mathematically, each element of the distance matrix D is defined as: $d_{ij} = ||x_i - x_j||_2^2$.
5. Finally, we add the following constraints on the representation matrix Z:
   - $Z \geq 0$: A non-negative representation coefficient z_{ij} can directly reveal the similarity between the samples x_i and x_j [45].
   - $diag(Z) = 0$: this constraint is used to avoid that a sample is represented as a linear combination of itself.
   - $Z\mathbf{1} = \mathbf{1}$: the sum of each row of Z is set to be equal to 1 which ensure that all samples are selected in the joint representation.

The ADMM-Based Optimization. There are two unknown variables in the problem (2), e.g., Z and W. To make the problem (2) separable, some auxiliary variables are added as follows:

$$\underset{Z,W}{argmin} \left\| W^{1/2} \odot E \right\|_F^2 + \frac{\beta}{2} \|W\|_F^2 + \lambda \|J\|_F^2 + \gamma tr(D^T Z)$$
$$\text{s.t } W \geq 0, W^T\mathbf{1} = \mathbf{1}, Z \geq 0, diag(Z) = 0, Z\mathbf{1} = \mathbf{1}, E = X - XZ, J = Z \quad (3)$$

Considering the problem (3) as a two-block optimization problem, we adopt the alternating direction method (ADM) to solve it [38]. Thus, we define the augmented Lagrangian function as:

$$\begin{aligned}\mathfrak{L}(Z, W, E, J, C_1, C_2) = & \left\| W^{1/2} \odot E \right\|_F^2 + \frac{\beta}{2} \|W\|_F^2 + \lambda \|J\|_F^2 + \gamma tr\left(D^T Z\right) \\ & + \frac{\mu}{2} \left(\left\| X - XZ - E + \frac{C_1}{\mu} \right\|_F^2 + \left\| Z - J + \frac{C_2}{\mu} \right\|_F^2 \right) \end{aligned} \quad (4)$$

where C_1, C_2 are the Lagrangian multipliers and μ is a penalty parameter.

Then, we solve each unknown variable while fixing the other variables in an alternate way.

Step 1: The variable W is obtained by minimizing the following problem while fixing the other variables:

$$\min_W \left\| W^{1/2} \odot E \right\|_F^2 + \frac{\beta}{2} ||W||_F^2 \ \text{ s.t } W \geq 0, W^T\mathbf{1} = 1 \quad (5)$$

Solving the problem (5) is equivalent to solve:

$$\min_{w_{ij} \geq 0, \sum_j w_{ij} = 1} \sum_{i,j} \left(w_{ij} + \frac{e_{ij}^2}{\beta} \right)^2 \quad (6)$$

The problem (6) can be written in the vector form since it is independent for different i [27].

$$\min_{w_i \geq 0, w_i^T \mathbf{1}=1} \left\| w_i + \frac{h_i}{\beta} \right\|_2^2 \tag{7}$$

where $H = E \odot E$

The associated Lagrangian function is:

$$\mathfrak{L}(w_i, c, m_i) = \frac{1}{2} \left\| w_i + \frac{h_i}{\beta} \right\|_2^2 - c(w_i^T \mathbf{1} - 1) - m_i^T w_i \tag{8}$$

where c and m_i are the Lagrangian multipliers associated to the boundary constraints on w_i.

Given the fact that $m_{ij} w_{ij} = 0$ according to the KKT condition [42], we have:

$$w_{ij} = \left(c - \frac{h_{ij}}{\beta} \right)_+ \tag{9}$$

Finally, we update the Lagrangian multiplier c according to the constraint $w_i^T \mathbf{1} = 1$ as follows:

$$\sum_{i=1}^{N} (c - \frac{h_{ij}}{\beta}) = 1 \Rightarrow c = \frac{1}{N} + \frac{1}{N\beta} \sum_{j=1}^{N} h_{ij} \tag{10}$$

Step 2: We can obtain the error matrix E by solving the following problem:

$$\min_{E} ||W^{1/2} \odot E||_F^2 + \frac{\mu}{2} ||E - G||_F^2 \; where \; G = X - XZ + \frac{C_1}{\mu} \tag{11}$$

The problem (11) is equivalent to :

$$\sum_{i,j} \min_{e_{ij}} \left(e_{ij} - \frac{\mu g_{ij}}{\mu + 2w_{ij}} \right)^2 \tag{12}$$

Then, the optimal solution of each element e_{ij} is

$$e_{ij} = \frac{\mu g_{ij}}{\mu + 2w_{ij}} \tag{13}$$

Step 3: We can obtain the matrix J by solving the following problem:

$$\min_{J} \lambda \, ||J||_F^2 + \frac{\mu}{2} ||Z - J + \frac{C_2}{\mu}||_F^2 \tag{14}$$

The close-form of J can be obtained by setting the derivative of (14) w.r.t J to zero:

$$J^* = \frac{\mu G}{\mu + 2\lambda} \; where \; G = Z + \frac{C_2}{\mu} \tag{15}$$

Step 4: The variable Z can be obtained by solving the following problem:

$$\begin{aligned} \underset{Z}{min} \; \gamma tr(D^T Z) + \frac{\mu}{2}\left(||M_1 - XZ||_F^2 + ||Z - M_2||_F^2\right) \\ \text{s.t } Z \geq 0, diag(Z) = 0, Z\mathbf{1} = \mathbf{1} \end{aligned} \tag{16}$$

where $M_1 = X - E + \frac{C_1}{\mu}$ and $M_2 = J - \frac{C_2}{\mu}$

Considering the following unconstrained problem:

$$\underset{Z}{argmin} \; \gamma tr(D^T Z) + \frac{\mu}{2}\left(||M_1 - XZ||_F^2 + ||Z - M_2||_F^2\right) \tag{17}$$

The problem (17) has a closed-form solution obtained by setting its derivative equal to zero:

$$\widehat{Z} = \left(X^T X + I\right)^{-1} \left(X^T M_1 + M_2 - \frac{\gamma}{\mu} D\right) \tag{18}$$

Then, the optimal solution Z of the problem (16) can be obtained more efficiently by solving the following problem:

$$\min_{Z \geq 0, diag(Z)=0, Z\mathbf{1}=\mathbf{1}} ||Z - \widehat{Z}||_F^2 \Leftrightarrow \min_{z_{ij} \geq 0, z_{ii}=0, \sum_i z_{ij}=1} \left(z_{ij} - \widehat{z_{ij}}\right)^2 \tag{19}$$

We obtain the optimal solution for each row z_i as in problem (6):

$$z_i = \left(\eta_i I_f^T + \bar{z}_i\right)_+ \tag{20}$$

where I_f is a column vector whose elements are equal to one expect the $i-$th is set equal to zero. $\bar{z_i}$ is defined as:

$$\bar{z_i} = \begin{cases} \widehat{z_{ij}} & i \neq j \\ 0 & otherwise \end{cases} \tag{21}$$

η_i is the Lagrangian multiplier which is calculated as:

$$\eta_i = \frac{1 + \bar{z}_i \boldsymbol{1}}{N - 1} \tag{22}$$

Step 5: We update the Lagrangian multipliers and the penalty parameter as follows, respectively:

$$C_1 = C_1 + \mu\,(X - XZ - E) \tag{23}$$

$$C_2 = C_2 \mu\,(Z - J) \tag{24}$$

$$\mu = min(\mu_{max}, \mu\rho) \tag{25}$$

Convergence and Computational Complexity. In this section, we first analyze the computational complexity of the proposed representation model. Clearly, the most computationally-demanding step in the ADMM-based Optimization is the step 4 which includes matrix multiplication and matrix inverse operations. It costs $O(N^3)$ for $N \times N$ matrix. Fortunately, the term $\left(X^T X + I\right)^{-1}$ can be pre-calculated before the iteration loop since it is independent from all variables and. The first two steps are efficiently calculated since they can be considered as element-wise operations. The third step mainly involves matrix addition operation. Hence, their computational complexities can be ignored compared to the fourth step.

3.2 The Proposed Hypergraph Construction Scheme

In this work, we assume that the representation vectors corresponding to two similar samples should be close since they can be similarly represented using remaining ones. More formally, we measure the similarity between two data samples as follows:

$$A(i,j) = z_i \cdot z_j \tag{26}$$

In terms of hypergraph, such information is very useful to characterize the incidence relations between hyperedges and their vertices:

$$h(v_i, e_j) = \begin{cases} A\left(i,j\right), & \text{if } z_{ij} \geq \theta \\ 0, & otherwise \end{cases} \tag{27}$$

Here, we set θ as the mean values of $\{z_{ik}\}_{k=1}^{N}$. According to this formulation, each vertex v_i is associated to hyperedge e_j based on whether it has prominently contributed in the representation of its centroid v_j. Moreover, for each centroid, the number of neighbors is adaptively selected. Hence, its distinctive neighborhood structure is well preserved.

3.3 The Hypergraph-Based Re-Ranking

In this work, we formulate the visual re-ranking problem as a transductive learning framework on the adaptive collaborative hypergraph model $G = (V, E, \omega)$:

$$\arg\underset{f}{min}\left\{\Omega(f) + \mu R_{emp}(f)\right\} \tag{28}$$

where the vector f is constituted of the relevance scores to be learned.

Following the Zhou' works [44], the regularization term can be written as follows:

$$\Omega(f) = f^T(I - \Theta)f = f^T\left(I - D_v^{-1/2} H W D_e^{-1} H^T D_v^{-1/2}\right) f \tag{29}$$

The empirical loss $R_{emp}(f)$ guarantees that final ranking scores are close to the initial ones. It is defined as:

$$R_{emp}(f) = \|f - y\|^2 = \sum_{v_i \in V} (f(v_i) - f(v_i))^2 \tag{30}$$

Where the initial ranking vector y is uniformly defined as:

$$y_i = 1 - \frac{i}{N} \tag{31}$$

By substituting (29) and (30) into (28) and setting the derivative of (28) with respect to f to 0, we have

$$f(I - \Theta) + \mu(f - y) = 0 \Rightarrow f = \frac{\mu}{1+\mu}(I - \frac{\Theta}{1+\mu})^{-1} y \tag{32}$$

Table 1. Description of databases

Database	Description	No. of images
Landmark-30 [16]	30 one-concept locations queries	8923
Landmark-123 [16]	123 one-concept locations queries	36452
General-65 [18]	65 complex and multi-concept queries	20000
General-70 [18]	70 complex and multi-concept queries	30000

4 Experiments

4.1 Experimental Settings

In this section, we have conducted visual re-ranking experiments on four public databases designed within the MediaEval 2014 [16] and MediaEval 2016 [18] competitions and listed in Table 1. In particular, the MediaEval 2014 benchmark consists of information for 153 one-concept location queries (e.g., buildings, museums, roads,bridges, sites, monuments, etc) with about 300 photos per location [16]. The MediaEval 2016 benchmarks consists of 135 complex and general-purpose multi-concept queries (e.g., animals at zoo, sunset in the city, accordion player, etc)[18]. We choose those databases for the following reasons: (1) they are consisted of real-world images (i.e. images are initially retrieved from Flickr in response to a textual query) (2) they are publicly available and (3) annotations are carried out by experts [17].

We use the convolutional neural networks based descriptors to represent images of all databases for its impressive performance in image retrieval [43]. In all experiments, we followed the rules of the MediaEval competitions. Indeed, in evaluation, a photo is considered to be relevant if it is a common photo representation of the query [16,18]. Experiments were carried out for different cut-off points, $X \in \{5, 10, 20, 30, 40, 50\}$. For performance evaluation, we adopt the precision $P@20$ as the official ranking for both MediaEval 2014 and MediaEval 2016 benchmarks was set to a cut-off of 20 images [16,18]. For fair comparison, we conducted all experiments on the same platform, i.e., Matlab platform running on Windows7, with an Intel (R)-Core(TM) i7-4500U 3.40 GHz processor and 8 GB memory. Moreover, we manually tuned the parameters of all other methods to obtain their optimal results.

4.2 Performance Comparison with State-of-the-art Methods

This experiment is conducted in order to compare our method with other methods that achieved best performance during the MediaEval competitions. In this experiment, we select only those visual-based methods. Comparison results are reported in Table 2. First, it can be observed that our method achieves a consistent improvement over the Flickr baseline on all databases. For examples, at a cut-off point $X = 20$, the precision gains of ACR-HG over Flickr are 6.67%, 8.29%, 10.07% and 6.49% on Landmark-30, Landmark-123, General-65 and General-70 respectively. Second, our method almost always outperforms other methods on all databases. For example, on Landmark-123, the precision of our method is $P@20 = 0.8894$ while other methods achieve 0.769 (TUW)[28], 0.7561 (SocSens) [31] and 0.748 (PeRCeiVe)[29]. On the General-70 database, which is a complex and general-purpose multi-concept database, we achieve a $P@20 = 0.7921$ compared to $P@20 = 0.5437$ achieved by the best team (LAPI) [6]. Our method, which not only models the complex and high-order relationships among visual samples via hypergraph but also capture the overall contextual information by the means of collaborative representation, achieves the best performance among the compared methods. This clearly demonstrates the validity of our method for visual re-ranking not only on for landmark image retrieval but also for multi-topic image retrieval.

Table 2. Performance comparison to state-of-the-art re-ranking methods.

Methods	P@20	
	Landmark-30	Landmark-123
Flickr	0.8333	0.8065
PeRCeiVe [29]	0.866	0.748
SocSens [31]	0.815	0.7561
TUW [28]	0.805	0.769
ACR-HG (ours)	**0.9**	**0.8894**

Methods	P@20	
	General-70	General-65
Flickr	0.6914	0.5531
UPMC [33]	0.631	0.5200
LAPI [6]	0.6514	0.5437
RECOD [13]	0.6821	0.5180
ACR-HG (ours)	**0.7921**	**0.618**

Table 3. Performance comparison to graph/hypergraph-based methods

Methods	P@20			
	Landmark30	Landmark-123	General-70	General-65
Flickr	0.8333	0.8065	0.6914	0.5531
VR [20]	0.8517	0.8314	0.74	0.5492
MR [5]	0.8251	0.8045	0.7293	0.5383
Knn-HG [2]	0.865	0.8537	0.7364	0.5461
SR-HG [36]	0.88	0.8541	0.6971	0.5531
CR-HG [41]	0.8883	0.8728	0.7564	0.5758
ACR-HG (ours)	**0.9**	**0.8894**	**0.7921**	**0.618**

4.3 Performance Comparison for Hypergraph Learning

In this experiment, we aim to validate the superiority of our hypergraph model over the conventional graph/hypergraph models. Results are showed in Table 3. From the results, the following observations can be drawn:

- Despite their ability in refining the initial retrieval results, graph-based re-ranking methods are almost outperformed by the hypergraph-based ones. This demonstrates that, in contrast to graph model, hypergraph model has and inherent ability to capture the local group information and latent high-order relationships among samples.
- The experimental results reveal also the good robustness and discriminative power of representation based hypergraph learning compared to neighborhood based hypergraph learning. On different databases, the representation based hypergraph ranking achieves the highest precision compared to hypergraph ranking based on neighborhood relationships. In particular, our method consistently and significantly achieves the best relevance improvement among other representation based hypergraph ranking.
- The adaptive collaborative representation has bring more robustness and discriminative power to the hypergraph than the collaborative representation. For instance, the precision gains of ACR-HG over the CR-HG are 1.17%, 1.66%, 3.57% and 4.22% on Landmark-30, Landmark-123, General-70 and General-65 respectively. One explanation is that the adaptive collaborative representation impose a locality-preserving regularizer on the representation coefficients which enable to capture the global and local structures of data during the hypergraph learning.

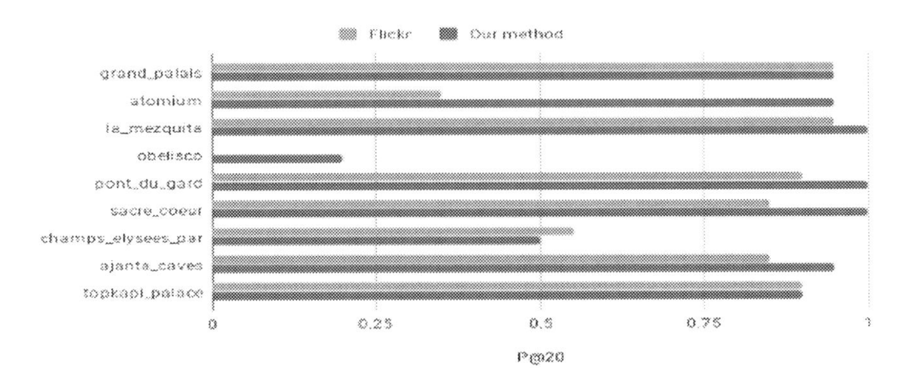

Fig. 1. Evolution curve of relevance for different landmark query topics

4.4 Performance Evaluation per Topic Class

The aim of this experiment is the investigate the performance stability of our method for different query topics. Comparison results are presented in Figs. 1 and 2. We find that our method outperforms Flickr for almost all query topics. The experimental results also reveal that, the relevance of retrieval results is higher for landmarks queries compared to complex queries. One explanation is that, non -relevant images were likely to be arisen when the query is ambiguous or involve multiple topics. For example, the query 'baby in stroller' may give rise to images that contain an empty stroller. Another interesting observations, is that the retrieval performance is degraded for some queries (e.g. 'baby in stroller'). This can be attributed to the fact that a high relevance score for a non-relevant image will be propagated to its visually similar neighbors since only the visual information is used for building the hypergraph.

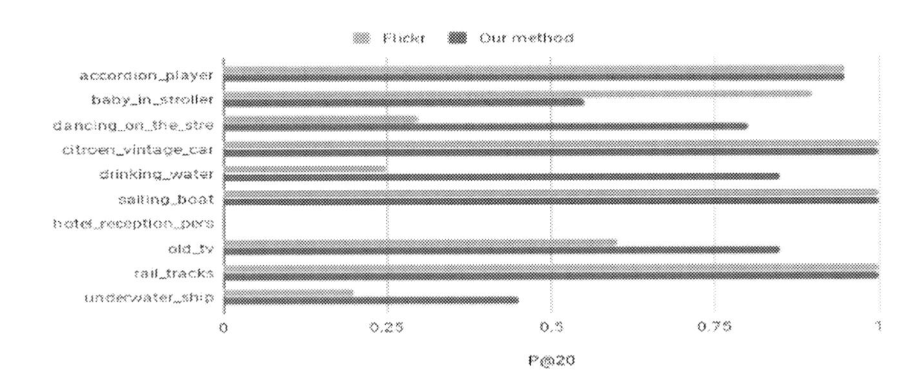

Fig. 2. Evolution curve of relevance for different general multi-concept query topics

5 Conclusion

In this paper, we proposed a novel hypergraph-based visual re-ranking method to enhance the performance of text-based image retrieval. At the core of our method is the data representation. Particularly, we proposed a novel representation technique called adaptive collaborative representation to build a more informative hypergraph. By constraining the self-representation term with an weighted matrix, the effect of those redundant and useless features can be adaptively minimized so that a more robust hypergraph can be constructed. In addition, our data representation technique has the advantage of simultaneously capturing both global and local structures of data during hypergraph learning by introducing a locality-preserving term. Based on the obtained representation matrix, we showed how to generate consistent hyperedge connections and hyperedge weights. Finally, a transductive learning is successfully performed upon the constructed hypergraph to learn the images' relevance scores. Experimental results performed on public MediaEval benchmarks demonstrate that our method achieves consistently superior results compared to state-of-the art re-ranking methods.

Acknowledgements. The research leading to these results has received funding from the Ministry of Higher Education and Scientific Research of Tunisia under the grant agreement number LR11ES48.

References

1. Boteanu, B., Mironică, I., Ionescu, B.: Hierarchical clustering pseudo-relevance feedback for social image search result diversification. In: Proceedings - International Workshop on Content-Based Multimedia Indexing (2015)
2. Bouhlel, N., Feki, G., Ben Ammar, A., Ben Amar, C.: A hypergraph-based reranking model for retrieving diverse social images. In: Felsberg, M., Heyden, A., Krüger, N. (eds.) CAIP 2017. LNCS, vol. 10424, pp. 279–291. Springer, Cham (2017). https://doi.org/10.1007/978-3-319-64689-3_23
3. Bouhlel, N., Ksibi, A., Ben Ammar, A., Ben Amar, C.: Semantic-aware framework for mobile image search. In: International Conference on Intelligent Systems Design and Applications, ISDA, vol. 2016-June, pp. 479–484. IEEE (2016)
4. Cai, J., Zha, Z.J., Wang, M., Zhang, S., Tian, Q.: An attribute-assisted reranking model for web image search. IEEE Trans. Image Process. **24**(1), 261–272 (2015)
5. Cheng, X.Q., Du, P., Guo, J., Zhu, X., Chen, Y.: Ranking on data manifold with sink points. IEEE Trans. Knowl. Data Eng. **25**(1), 177–191 (2013)
6. Constantin, M.G., Boteanu, B., Ionescu, B.: LAPI at mediaeval 2016 predicting media interestingness task, October 2016
7. Feki, G., Fakhfakh, R., Ben Ammar, A., Ben Amar, C.: Knowledge structures: which one to use for the query disambiguation? In: 2015 15th International Conference on Intelligent Systems Design and Applications (ISDA), pp. 499–504, December 2015
8. Feki, G., Fakhfakh, R., Ben Ammar, A., Ben Amar, C.: Query disambiguation: user-centric approach. J. Inform. Assur. Secur. **11**, 144–156 (2016)

9. Feki, G., Fakhfakh, R., Bouhlel, N., Ben Ammar, A., Ben Amar, C.: REGIM @ 2016 retrieving diverse social images task. In: Working Notes Proceedings of the MediaEval 2016 Workshop, 20–21 October 2016, Hilversum, The Netherlands (2016)
10. Feki, G., Ksibi, A., Ben Ammar, A., Ben Amar, C.: Improving image search effectiveness by integrating contextual information. In: 2013 11th International Workshop on Content-Based Multimedia Indexing (CBMI), pp. 149–154 (2013)
11. Feki, G., Ammar, A.B., Amar, C.B.: Adaptive semantic construction for diversity-based image retrieval. In: KDIR 2014 - Proceedings of the International Conference on Knowledge Discovery and Information Retrieval, Rome, Italy, 21–24 October 2014, pp. 444–449 (2014)
12. Feki, G., Ammar, A.B., Amar, C.B.: Towards diverse visual suggestions on Flickr. In: Ninth International Conference on Machine Vision, ICMV 2016, Nice, France, 18–20 November 2016, p. 103411Z (2016)
13. Ferreira, C., et al.: Recod @ mediaeval 2016: Diverse social images retrieval, October 2016
14. Guedri, B., Zaied, M., Ben Amar, C.: Indexing and images retrieval by content. In: 2011 International Conference on High Performance Computing Simulation, pp. 369–375 (2011)
15. Hong, C., Zhu, J.: Hypergraph-based multi-example ranking with sparse representation for transductive learning image retrieval. Neurocomputing **101**, 94–103 (2013)
16. Ionescu, B., Popescu, A., Lupu, M., GÎnscă, A.L., Boteanu, B., Müller, H.: Div150Cred: a social image retrieval result diversification with user tagging credibility dataset. In: Proceedings of the 6th ACM Multimedia Systems Conference, MMSys 2015, pp. 207–212. ACM, New York (2015)
17. Ionescu, B., Popescu, A., Radu, A.-L., Müller, H.: Result diversification in social image retrieval: a benchmarking framework. Multimed. Tools Appl. **75**(2), 1301–1331 (2014). https://doi.org/10.1007/s11042-014-2369-4
18. Ionescu, B., Zaharieva, M.: Retrieving diverse social images at MediaEval 2016: challenge, dataset and evaluation. In: Gravier, G., et al. (eds.) Working Notes Proceedings of the MediaEval 2016 Workshop, pp. 20–22 (2016)
19. Jing, P., Su, Y., Xu, C., Zhang, L.: HyperSSR: a hypergraph based semi-supervised ranking method for visual search reranking. Neurocomputing **274**, 50–57 (2018)
20. Jing, Y., Baluja, S.: VisualRank: applying PageRank to large-scale image search. IEEE Trans. Pattern Anal. Mach. Intell. **30**(11), 1877–1890 (2008)
21. Ksibi, A., Feki, G., Ben Ammar, A., Ben Amar, C.: Effective diversification for ambiguous queries in social image retrieval. In: Wilson, R., Hancock, E., Bors, A., Smith, W. (eds.) CAIP 2013. LNCS, vol. 8048, pp. 571–578. Springer, Heidelberg (2013). https://doi.org/10.1007/978-3-642-40246-3_71
22. Liu, Q., Sun, Y., Wang, C., Liu, T., Tao, D.: Elastic net hypergraph learning for image clustering and semi-supervised classification. IEEE Trans. Image Process. **26**(1), 452–463 (2017)
23. Liu, Y., Shao, J., Xiao, J., Wu, F., Zhuang, Y.: Hypergraph spectral hashing for image retrieval with heterogeneous social contexts. Neurocomputing **119**, 49–58 (2013)
24. Mei, T., Rui, Y., Li, S., Tian, Q.: Multimedia search reranking: a literature survey. ACM Comput. Surv. **46**(3), 1–38 (2014)
25. Mejdoub, M., Fonteles, L., BenAmar, C., Antonini, M.: Fast indexing method for image retrieval using tree-structured lattices. In: 2008 International Workshop on Content-Based Multimedia Indexing, pp. 365–372, June 2008

26. Mejdoub, M., Fonteles, L., Ben Amar, C., Antonini, M.: Embedded lattices tree: an efficient indexing scheme for content based retrieval on image databases. J. Vis. Commun. Image Represent. **20**(2), 145–156 (2009)
27. Nie, F., Wang, X., Jordan, M.I., Huang, H.: The constrained Laplacian rank algorithm for graph-based clustering. In: 30th AAAI Conference on Artificial Intelligence, AAAI 2016, no. 1, pp. 1969–1976 (2016)
28. Sabetghadam, S., Palotti, J.R.M., Rekabsaz, N., Lupu, M., Hanbury, A.: TUW @ mediaeval 2015 retrieving diverse social images task. In: Working Notes Proceedings of the MediaEval 2015 Workshop, 14–15 September 2015, Wurzen, Germany (2015)
29. Spampinato, C., Palazzo, S.: PeRCeiVe lab@UNICT at MediaEval 2014 diverse images: random forests for diversity-based clustering. In: MediaEval (2014)
30. Spyromitros-Xioufis, E., Papadopoulos, S., Ginsca, A.L., Popescu, A., Kompatsiaris, Y., Vlahavas, I.: Improving diversity in image search via supervised relevance scoring. In: ICMR 2015 - Proceedings of the 2015 ACM International Conference on Multimedia Retrieval, ICMR 2015, pp. 323–330. ACM, New York (2015)
31. Spyromitros-Xioufis, E., Papadopoulos, S., Kompatsiaris, I., Vlahavas, I.: SocialSensor: finding diverse images at mediaeval 2014, vol. 1263, October 2014
32. Tian, X., Yang, L., Wang, J., Wu, X., Hua, X.S.: Bayesian visual reranking. IEEE Trans. Multimed. **13**(4), 639–652 (2011)
33. Tollari, S.: UPMC at MediaEval 2016 retrieving diverse social images task. In: CEUR Workshop Proceedings, vol. 1739 (2016)
34. Wang, J., Yang, J., Yu, K., Lv, F., Huang, T., Gong, Y.: Locality-constrained linear coding for image classification. In: 2010 IEEE Computer Society Conference on Computer Vision and Pattern Recognition, pp. 3360–3367, June 2010
35. Wang, M., Li, H., Tao, D., Lu, K., Wu, X.: Multimodal graph-based reranking for web image search. IEEE Trans. Image Process. **21**(11), 4649–4661 (2012)
36. Wang, M., Liu, X., Wu, X.: Visual classification by ℓ_1-hypergraph modeling. IEEE Trans. Knowl. Data Eng. **27**(9), 2564–2574 (2015)
37. Wang, Y., Lin, X., Wu, L., Zhang, W.: Effective multi-query expansions: robust landmark retrieval. In: MM 2015 - Proceedings of the 2015 ACM Multimedia Conference, MM 2015, pp. 79–88. ACM, New York (2015)
38. Wen, J., Fang, X., Xu, Y., Tian, C., Fei, L.: Low-rank representation with adaptive graph regularization. Neural Netw. **108**, 83–96 (2018)
39. Xu, B., Bu, J., Chen, C., Wang, C., Cai, D., He, X.: EMR: a scalable graph-based ranking model for content-based image retrieval. IEEE Trans. Knowl. Data Eng. **27**(1), 102–114 (2015)
40. Yang, J., Luo, L., Qian, J., Tai, Y., Zhang, F., Xu, Y.: Nuclear norm based matrix regression with applications to face recognition with occlusion and illumination changes. IEEE Trans. Pattern Anal. Mach. Intell. **39**(1), 156–171 (2017)
41. Zhang, L., Yang, M., Feng, X.: Sparse representation or collaborative representation: which helps face recognition? In: Proceedings of the IEEE International Conference on Computer Vision, pp. 471–478 (2011)
42. Zheng, J., Yang, P., Chen, S., Shen, G., Wang, W.: Iterative re-constrained group sparse face recognition with adaptive weights learning. Trans. Image Process. **26**(5), 2408–2423 (2017)
43. Zheng, L., Yang, Y., Tian, Q.: SIFT meets CNN: a decade survey of instance retrieval. IEEE Trans. Pattern Anal. Mach. Intell. **40**(5), 1224–1244 (2018). https://doi.org/10.1109/TPAMI.2017.2709749

44. Zhou, D., Huang, J., Schölkopf, B.: Learning with hypergraphs: clustering, classification, and embedding. In: Advances in Neural Information Processing Systems 19, vol. 19, no. Figure 1, pp. 1601–1608 (2007)
45. Zhuang, L., Gao, H., Lin, Z., Ma, Y., Zhang, X., Yu, N.: Non-negative low rank and sparse graph for semi-supervised learning. In: Proceedings of the IEEE Computer Society Conference on Computer Vision and Pattern Recognition, pp. 2328–2335 (2012)

Motion Words: A Text-Like Representation of 3D Skeleton Sequences

Jan Sedmidubsky(✉), Petra Budikova, Vlastislav Dohnal, and Pavel Zezula

Masaryk University, Brno, Czechia
{xsedmid,budikova,dohnal,zezula}@fi.muni.cz

Abstract. There is a growing amount of human motion data captured as a continuous 3D skeleton sequence without any information about its semantic partitioning. To make such unsegmented and unlabeled data efficiently accessible, we propose to transform them into a text-like representation and employ well-known text retrieval models. Specifically, we partition each motion synthetically into a sequence of short segments and quantize the segments into motion words, i.e. compact features with similar characteristics as words in text documents. We introduce several quantization techniques for building motion-word vocabularies and propose application-independent criteria for assessing the vocabulary quality. We verify these criteria on two real-life application scenarios.

Keywords: 3D skeleton sequence · Motion word · Motion vocabulary · Quantization · Border problem · Text-based processing

1 Introduction

In recent years, we have witnessed a rapid development of motion capture devices and 3D pose-estimation methods [2] that enable recording human movements as a sequence of *poses*. Each pose keeps the 3D coordinates of important *skeleton joints* in a specific time moment. Effective and efficient processing of such spatio-temporal data is very desirable in many application domains, ranging from computer animation, through sports and medicine, to security [5,7,9].

To illustrate the range of possible tasks over motion data, let us assume that we have the 3D skeleton data from a figure skating competition. Existing research mainly focuses on *action recognition* [23], i.e. categorizing the figure performed in a given, manually selected motion segment. This is typically solved using convolutional [1,17] or recurrent [10,20,22] neural-network classifiers. However, this approach is not applicable to other situations where motion data are captured as long continuous sequences without explicit knowledge of semantic partitioning. In such cases, other techniques need to be applied, e.g., *subsequence search* to find all competitors who performed the triple Axel jump, or *similarity joins* to identify different performances of the same choreography, similar choreographies, or the most common figures. These techniques require identifying query-relevant

J. M. Jose et al. (Eds.): ECIR 2020, LNCS 12035, pp. 527–541, 2020.
https://doi.org/10.1007/978-3-030-45439-5_35

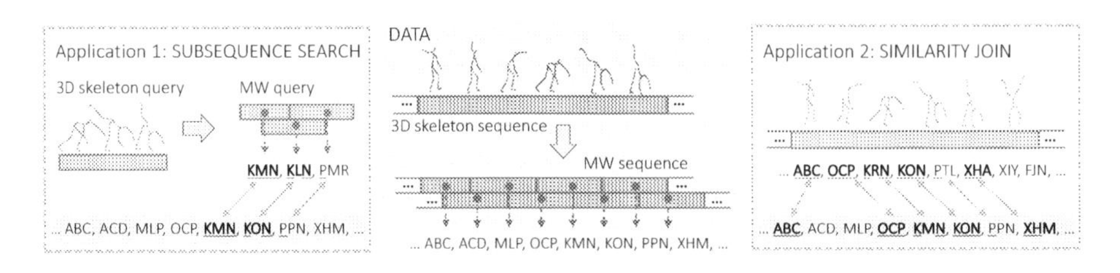

Fig. 1. Representing motions by motion words: both data and queries are transformed into MW sequences and efficiently organized and processed by text-based approaches.

subsequences within the continuous motion data. To allow efficient evaluation of such queries, the data need to be automatically segmented and indexed.

Since a universal semantic segmentation is hardly achievable, we suggest to partition each motion sequence synthetically into short fixed-size *segments* whose length is smaller than the expected size of future queries. In this way, we transform the input motion into an ordered sequence of segments, structurally similar to a text document. To complete the analogy, we quantize the segments into compact representations, denoted as *motion words* (MWs), having similar properties as words in text documents. Individual MWs deal with the spatial variability of the short segments, whereas the temporal variability of longer motions is captured by the MW order and quantified by mature text-retrieval models [12]. We believe that such universal text-based representation is applicable for a wide range of applications that need to process continuous motion data efficiently, as illustrated in Fig. 1.

In this paper, we mainly focus on effective quantization of the motion segments to build a *vocabulary* of motion words. The most desirable MW property is that two MWs *match* each other if their corresponding segments exhibit similar movement characteristics, and do *not match* if the segments are dissimilar. This is challenging with the quantization approach, since it is in general not possible to divide a given space in such way that all pairs of similar objects are in the same partition. Some pairs of similar segments thus get separated by partition borders and become non-matching, which we denote as the *border problem*. We answer this challenge by designing two MW construction techniques that reduce the border problem but still enable efficient organization using text retrieval techniques. Furthermore, we recommend generic (application-independent) criteria for selection of a suitable vocabulary for specific application needs, and verify the usability of such criteria on two real-life applications.

2 Related Work and Our Contributions

Most existing works that process *continuous* 3D skeleton sequences in an *unsupervised* way focus on subsequence search [18], unsupervised segmentation [8], or anticipating future actions based on the past-to-current data [4]. In [18], the continuous sequences are synthetically partitioned into a lot of overlapping and variable-size segments that are represented by 4,096D deep features. However, indexing a large number of such very high-dimensional features is costly. To move towards more efficient processing, the approaches in [3,11] quantize the segment features using a single k-means clustering. However, with such simple quantization the border problem appears frequently, which decreases the effectiveness of applications with an increasing number of clusters (i.e. the vocabulary size).

In our research, we also take inspiration from image processing where high-dimensional image features are quantized into visual words. There are two lines of research that are important to us: fundamental quantization techniques, and reducing the border problem. The image quantization strategies have evolved from basic k-means clustering used in [21], through cluster hierarchies [14], approximate k-means [16], to recent deep neural-network approaches [24]. The influence of the border problem can be reduced using a weighted combination of the nearest visual words for each feature [16], or by a consensus voting of multiple independent vocabularies [6].

Contributions of This Paper

We propose an effective quantization of unlabeled 3D skeleton data into sequences of motion words that can be efficiently managed by text-retrieval techniques. In contrast to previous works, we give a particular attention to the border problem. Specifically,

- we systematically analyze the process of MW vocabulary construction and discuss possible solutions of the border problem (Sect. 3);
- we propose application-independent criteria that do not require labeled data for selecting a suitable MW vocabulary for a given task (Sect. 3.3);
- we implement three vocabulary construction techniques that differ in dealing with the border problem, and evaluate their quality (Sect. 4);
- we verify the suitability of the proposed criteria by evaluating the best-ranked vocabularies in the context of two real-world applications (Sect. 5).

3 MW Vocabulary Construction

The motion-word (MW) approach assumes that the continuous 3D skeleton data are cut into short, possibly overlapping segments which are consequently transformed into the motion words. The segment and overlap lengths are important parameters of the whole system and have also been studied in our experiments, however their thorough analysis is out of the scope of this paper. Therefore, we assume that a suitable segmentation is available, and focus solely on transforming the segment space into the space of motion words, denoted as the *MW vocabulary*.

The MW vocabulary consists of a finite set of motion words and a Boolean-valued *MW matching function* that determines whether two MWs are considered equal: $match^{MW} : MW \times MW \rightarrow \{0,1\}$. The Boolean matching of words

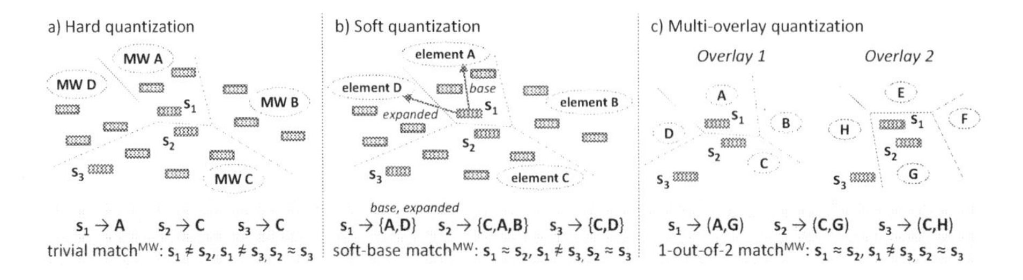

Fig. 2. Comparison of the hard, soft, and multi-overlay quantization of segments.

is a standard text-processing primitive required by most text retrieval techniques [12]. The transformation from segments to MWs has to be *similarity-preserving*: with a high probability, similar segment pairs need to be mapped to matching MWs and dissimilar segment pairs to non-matching MWs. Noticeably, the vocabulary construction can be investigated independently of a particular application, since it only considers the distribution of segments in the segment space. We propose to build the MW vocabulary using *quantization* of the segment space, which can be seen as analogous to the word stemming in text processing.

In the following, we first review the standard quantization approach that leads to a *basic MW model* and discuss its limitations, namely the border problem. Next, we introduce a *generalized MW model* with two techniques for reducing the border problem. Lastly, we present the evaluation methodology that we propose for comparing the quality of different vocabularies.

3.1 Basic MW Model

Basic data quantization is usually performed by the k-means algorithm that divides the segment space into non-overlapping partitions [3,11,21]. Each partition can be assigned a one-dimensional identifier, which constitutes the motion word. Each motion segment is associated with exactly one MW, which we denote as the *hard quantization* (Fig. 2a). To compare two *hard MWs*, a trivial MW matching function is defined: it returns 1 for identical words and 0 otherwise.

Using this approach, the 3D skeleton data are transformed into a sequence of scalar MWs to be readily processed by the standard text-retrieval tools. However, the hard quantization makes it difficult to preserve the similarity between the segments. Unless the input data are inherently well-clustered, which is not likely in the high-dimensional segment space, it is not possible to avoid the border problem, i.e. the situations when two similar segments get assigned to different MWs (s_1 and s_2 in Fig. 2a). Moreover, finding a good clustering is computationally expensive. Therefore, approximate or sampling methods are often used for large data, which makes the border problem even more pronounced.

3.2 Generalized MW Model

We believe that the border problem can be reduced significantly if we allow a given segment to be associated with several partitions of the input space. Therefore, we define the *generalized MW* as a collection of *MW elements*, where each element corresponds to a single partition of the input space. In contrast to the basic model where individual MWs are atomic and mutually exclusive, the generalized MWs may share some MW elements. This allows us to define a more fine-grained MW matching function that better approximates the original similarity between the motion segments.

As illustrated in Fig. 2b and c, we adopt the following two orthogonal principles of selecting the MW elements for a given segment.

- *Soft quantization.* Recall again that the border problem occurs when two similar segments are separated into different partitions. Intuitively, at least one of these segments has to lie near the partition border. Segment s_1 in Fig. 2a lies outside the partition D but is close to its borders, so there is a good chance that some segments similar to s_1 are in D. Therefore, it could be helpful to associate s_1 also with D. Following this idea, we define the *soft MW* for s_1 as an *ordered set* of one or more MW elements, where the first *base element* identifies the partition containing s_1 and the remaining *expanded elements* refer to the partitions that are sufficiently close to s_1 (see Fig. 2b for illustration). A naive MW matching function could return 1 whenever the intersection of two soft MWs is non-empty, however this tends to match even segments that are not so close in the segment space (s_1 and s_3 in Fig. 2b). Therefore, our *soft-base* matching function returns 1 only if the intersection contains at least one base element.
- *Multi-overlay quantization*: So far, we have assumed that the MW elements are taken from a single partitioning of the segment space. However, it is also possible to employ several independent *partitioning overlays* obtained by different methods. A single overlay may incorrectly separate a pair of similar segments, but it is less probable that the same pair will be separated by the other independent overlays. We define the *multi-overlay MW* as an *n-tuple* of MW elements that are assigned to a given segment in the individual overlays. To decide whether two MWs match, the consensus of m out of n MW elements is used. The matching function returns 1 if the multi-overlay MWs agree on at least m positions of the respective n-tuples (see Fig. 2c).

By allowing the MWs to be compound, we improve the quantization quality but create new challenges regarding indexability. The generalized MWs are no longer scalar and cannot be simply treated the same way as text words. However, existing text retrieval tools can be adjusted to index both the soft and multi-overlay MWs, as briefly discussed in Sect. 4.4.

3.3 Evaluation Methodology

For evaluating MW vocabularies, we need to consider two different aspects: (i) *vocabulary quality* – measured by the application-independent ability to perform a similarity-preserving transformation from the segment space to the MW space, and (ii) *vocabulary usefulness* – measured by effectiveness of the application employing the specific vocabulary. Our objective is to show that both vocabulary quality and vocabulary usefulness are related, so we can choose a suitable vocabulary without evaluating it within the real application, i.e. not needing the application ground truth (GT).

In the following, we introduce the dataset used for both types of evaluation, and describe the application-independent vocabulary quality measures that are examined in Sect. 4. The vocabulary usefulness is discussed in Sect. 5.

Dataset. We adopt the HDM05 dataset [13] of 3D skeleton sequences, which consists of 2,345 labeled actions categorized in 130 classes. The actions capture exercising and daily movement activities with the sampling frequency of 120 Hz and track 31 skeleton joints. The action length ranges from 13 frames (108 ms) to 900 frames (7.5 s). We use this dataset to evaluate the MW usefulness in two applications: a *kNN classification of actions*, and a *similar action search*. These applications do not require complex retrieval algorithms and allow us to clearly show the effect of MWs on application effectiveness.

Both the vocabulary construction and the application-independent quality assessment are designed for completely *unlabeled* segment data, which we extract from the HDM05 dataset as follows. We divide each action synthetically into a sequence of overlapping segments. As recommended in [3], we fix the segment length to 80 frames and the segment overlap to 64 frames, so the segments are shifted by 16 frames. This generates **28 k** segments in total, with 12 segments per action on average. We also down-sample the segments to 12 frames per second. The *similarity* of any two segments is determined by the Dynamic Time Warping (DTW), where the pose distance inside DTW is computed as the sum of Euclidean distances between the 3D coordinates of the corresponding joints.

Estimating GT for Unlabeled Segments. The similarity-preserving property states that similar segments should be mapped to matching MWs, whereas dissimilar segments to non-matching MWs. To be able to check this property for a given vocabulary, we need a ground truth (GT) of similar and dissimilar segment pairs. Since the segments have no semantic labels, we can only use pairwise distances to estimate the GT. Using the distance distribution of all segments from our dataset, we determine two threshold distances that divide the segment pairs into similar pairs, dissimilar pairs, and a grey zone. In particular, the 0.5^{th} percentile distance becomes the similarity threshold T_{sim} and all segment pairs with the mutual distance lower than T_{sim} are the GT's similar pairs. The 40th percentile becomes the dissimilarity threshold T_{dissim} which defines the dissimilar pairs. Both the thresholds are set tightly to eliminate the chance that semantically unrelated segments are considered similar and vice versa. The segment pairs with mutual distance between T_{sim} and T_{dissim} form the grey zone and are ignored in the vocabulary quality evaluations.

Vocabulary Quality Measures. To assess how well a given MW vocabulary manages to match a given segment with similar segments, we use standard IR measures of *precision* (P) and *recall* (R) computed over the above-described GT of similar and dissimilar segment pairs: $P = \frac{tp}{tp+fp}$ and $R = \frac{tp}{tp+fn}$, where the *true positives (tp)* are pairs of similar segments mapped to matching MWs, *false positives (fp)* are dissimilar segments with matching MWs, etc. To quantify the trade-off between P and R, we employ the F_β *score* $= (1+\beta^2) \cdot \frac{P \cdot R}{(\beta^2 \cdot R)+P}$, where the positive real β is used to adjust the importance of the precision and recall according to the target application preferences.

As already mentioned, we test our vocabularies in context of two applications with different needs. The kNN classification requires high precision of retrieved actions for correct decision, but some positives can be missed. On the other hand, action search typically requires high recall. With these two applications in mind, we select the following two F scores for our experiments: $F_{0.25}$ score that emphasizes precision, as required by the classification task, and F_1 score that is the harmonic mean of both precision and recall and complies to the needs of a search-oriented application.

4 Implementation and Evaluation

To create a vocabulary, we use a Voronoi partitioning of the segment space. It assumes a set of sites (*pivots*) is selected beforehand by a particular selection algorithm. The Voronoi cell of pivot p is formed by all segments closer to p than to the other pivots. The pivots' IDs become the motion words or MW elements. Regarding the pivot selection, we must keep in mind that the segment space may not be the Euclidean space, which is our case with DTW that brakes the triangle inequality. So a particular pivot selection algorithm must respect that an artificial data item (e.g., a mean vector) cannot be computed.

In the following, we introduce algorithms implementing the aforementioned MW vocabulary construction principles, and show how the quality measures introduced in Sect. 3.3 can be used to tune the parameters of the algorithms.

4.1 Hard Quantization

Firstly, we analyze the viability of three pivot selection techniques: the *k-medoids*, the *hierarchical k-medoids*, and a *random selection*. We also study the influence of the number of pivots, which determines the cardinality of the vocabulary.

Implementation. The *k-medoids* algorithm is a variation of the k-means clustering that is mostly used for quantization. It works in iterations, gradually moving from a random set of pivots to more optimal ones. With the k-medoids, the pivots must be selected from existing motion segments. The optimization criterion is to minimize the sum of distances to other segments within the cluster. The algorithm does not guarantee to find the global optimum and is very

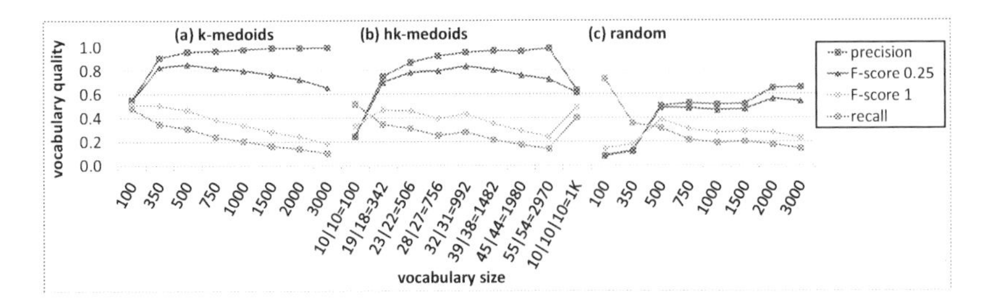

Fig. 3. Vocabulary quality in relation to vocabulary method and varying vocabulary size: (a) k-medoids, (b) hk-medoids and (c) random pivot selection.

costly since the distances of all pivot-object pairs need to be computed in each iteration. The *hierarchical k-medoids* (*hk-medoids*) seeks the pivots by recursive application of k-medoids, which allows using much smaller values of k in each iteration to create a vocabulary of the same size. The pivots for the next level are always selected from the parental cell, so the data locality is preserved. We use a constant number of pivots per level and similar pivot numbers across levels. For example, the set-up 39|38 denotes 39 pivots in the root level and 38 pivots in the second level, which creates 1,482 cells. Finally, we also try a *random* selection of pivots where a pivot too close to another one is omitted. This is the most efficient approach which is known to perform well in permutation-based indexes [15].

Experimental Evaluation. Using the three algorithms, we create vocabularies of sizes ranging from 100 up to 3,000 MWs, and compare their quality. The results presented in Fig. 3 are averages over five runs. In general, the higher the precision is the more pivots are used, and vice-versa for the recall. A good vocabulary is prepared by techniques choosing the pivots in correspondence to the distribution of segments, thus the random selection should be rejected, since its precision is low. Focusing on the vocabulary size, the $F_{0.25}$ score that favors precision guides us to pick the k-medoids with 350 or 500 pivots and the hk-medoids of the 32|31 breakdown. In the F_1 score, the optimum is 100 or 350 pivots by k-medoids, and 19|18 or 10|10|10 by hk-medoids.

The k-medoids with 350 pivots has been identified as the most promising hard quantization method, therefore we use it in the following trials exclusively. We also experimented with the best settings of the other algorithms and obtained analogous trends, so we do not include them.

4.2 Soft Quantization

Secondly, it is vital for the soft quantization to assign additional MW elements of neighboring cells only to the segments that are close to the cell borders. We limit such closeness by the distance D and bound the number of MW elements to the maximum number K. We study the influence of D and K on the quality measures, which should show that the border problem is reduced.

Implementation. The distance of a segment s located in the Voronoi cell of pivot p_1 to the borderline of the cell of p_2 is estimated as $\frac{|DTW(p_1,s)-DTW(p_2,s)|}{2}$. We gradually check all pivots p_i and expand the segment's MW with the MW element p_i until the estimated distance exceeds D. The value of D must be smaller than the similarity threshold T_{sim} discussed in Sect. 3.3, since it identifies objects that should be assigned the same MW. There can be many neighboring cells, so we constrain the MW elements to the K closest ones.

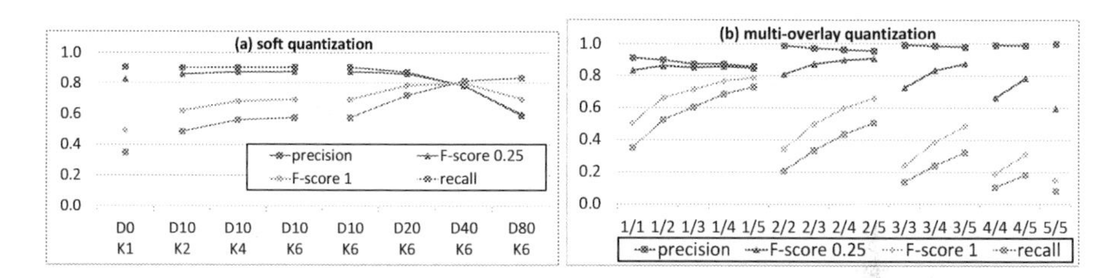

Fig. 4. Quality of vocabulary in relation to vocabulary construction method: (a) soft quantization for 350 pivots: varying K for $D10$ and varying D for $K6$; (b) multi-overlay quantization: 1 to 5 overlays with 350 pivots each, varying number of matching overlays.

Experimental Evaluation. We vary the values of D from 10 ($1/8 \cdot T_{sim}$) to 80 (T_{sim}), and K from 2 to 6. The relevant results are shown in Fig. 4a. Increasing K for a small D ($D10$, $K2$–6) leads to improved recall and nearly constant precision. On the other hand, multiplying D ($D10$–80, $K6$) produces extensive MWs, which reduces the border problem (recall is boosted), but it negatively effects precision. For the classification task ($F_{0.25}$ score), $D10$, $K6$ and $D20$, $K6$ are the best, while $D40$, $K6$ is the optimum for the search (F_1 score).

4.3 Multi-overlay Quantization

Thirdly, independent sets of pivots are likely to provide different Voronoi partitionings, thus increasing a chance of similar segments to share the same cell. We create up to 5 overlays and vary the number of overlays required to agree.

Implementation. Since the k-medoids algorithm provides a locally optimal solution, we ran it five times to obtain different sets of 350 pivots for the multi-overlay quantization. Noticeably, the quality of hard vocabularies created from individual sets of pivots differs up to 5% in both the F scores.

Experimental Evaluation. In Fig. 4b, we present the results for all combinations of the five overlays, where the notation m/n refers to the m-out-of-n matching function. The combination 1/1 corresponds to hard quantization. When we fix m to 1 and add more overlays, the border problem is reduced, as witnessed by a major improvement of recall and only a marginal drop in precision. Similar trends can be observed also for higher values of m, but the actual values of recall get lower when we require more overlays to agree. The most restrictive combination 5/5 requires all overlays to meet and performs similarly to the hard quantization with more than 3,000 pivots (see Fig. 3a). The best $F_{0.25}$ score is for the 2/5 setup and the best F_1 score is for the 1/5 setting.

4.4 Discussion

By thorough experimentation, we have observed that the k-medoids clustering is the best hard quantization method but its quality can still be significantly improved by the soft and multi-overlay principles. The suppression of the border problem is mainly attributed to the increased number of correctly matched segment pairs (true positives) by both these principles. Although some new false positives are introduced, they decrease the overall precision only marginally.

Since the k-medoids clustering has high computation complexity, we have also considered cheaper techniques, i.e. the random clustering, with the soft and multi-overlay approach. However, the experimental results were not much competitive, so the k-medoids still remains a reasonable choice for quantization.

Our vocabulary construction techniques are universal, but the created vocabulary is clearly data-dependent. Since our evaluation data are relatively small (28,104 segments), the optimal vocabulary size is 350 MWs for the hard quantization. For larger and more diverse data, we expect the quality measures to recommend a larger vocabulary.

Finally, a successful application also requires fast access to the data, which calls for indexes. The hard vocabulary can be directly organized in an inverted file. The soft-assigned vocabulary just expands the query, so the inverted file is sought multiple times (proportional to the number of MW elements in the query). The multi-overlay vocabulary can be managed in separate search indexes (one per overlay) and the query results merged to compute the m-out-of-n matching.

5 Motion Words in Applications

In this section, we experimentally verify that: (i) the MW representation preserves important characteristics of complex 3D skeleton data and causes no drop in application effectiveness (Sect. 5.2), and (ii) the vocabulary quality measures well approximate the usefulness of different vocabularies in applications. Both these aspects are evaluated in context of two applications: the *action classification* that aims at recognizing the correct class of a given action using a kNN classifier, and the *action search* where the goal is to retrieve all actions relevant to a query, i.e. the actions belonging to the same class as the query.

5.1 Evaluation Methodology of Classification/Search Applications

The input for both classification and search applications is the dataset of $2,345$ synthetically-segmented actions discussed in Sect. 3.3. On average, each action is transformed into a sequence of 12 MWs. To compare two MW sequences, we again adopt the DTW sequence alignment function. Realize that the MW matching function inside DTW deals with the spatial variability of short segments, whereas DTW considers the temporal dimension of the whole actions.

Both applications are evaluated on the basis of k-nearest neighbor (kNN) queries. We use the standard leave-one-out approach to evaluate 2, 345 kNN queries in a sequential way by computing the distance between the specific query action and each of the remaining dataset actions. For the classification task, we fix k to 4 and apply a 4NN classifier (similar to the classifier proposed in [19]). We measure the application *effectiveness* as the average classification accuracy over all 2, 345 queries. For the search task, the value of k is adjusted for each query individually based on the number of available actions belonging to the same class as the query action. Such adaptive value of k allows us to focus on recall as well as precision. The effectiveness of the search application is then determined as the average recall over all the queries. Note that the recall is always the same as the precision in the search task with the adaptive value of k.

5.2 Usefulness and Efficiency of MWs

We quantify the usefulness of the MW concept by evaluating application effectiveness with different vocabularies and comparing it to the baseline case that uses no quantization (i.e. the action segments are represented by original 3D skeleton data). The most interesting results are summarized in Table 1.

For classification, we observe that the baseline case achieves the effectiveness of 77.70%. Worse results have been expected for any MW quantization due to the dimensionality reduction of the original segment data. A standard hard quantization – the single-level k-medoids – indeed achieves the worst result (74.97%). Surprisingly, the soft-assignment $D10$, $K6$ vocabulary reaches basically the same effectiveness (77.61%) as the baseline, and the 2/5 multi-overlay quantization is actually better (80.30%). Thus, the best MW vocabulary not only preserves important motion characteristics but also aggregates many tiny variations in joint positions that confuse DTW on raw 3D skeleton data (the baseline case).

A similar trend can be observed on the search application where the hard quantization has the worst result too. As the recall is very important for the search task, the 1/5 multi-overlay vocabulary is now the best candidate that also outperforms the baseline case. Compared to the state-of-the art approaches [3,11] that employ the hard quantization, the proposed generalized MWs reach much better effectiveness, e.g., about 25% higher recall in the search application (increase from 44.21% to 55.62%).

From the performance point of view, it takes almost 1.5 h to evaluate all the 2, 345 kNN queries with the baseline segment representation. Using any of the MW representations, the evaluation finishes in 30 s, which is an improvement by two orders of magnitude.

5.3 Concordance of Vocabulary Quality and Usefulness

Remember that in Sect. 3.3 we proposed to quantify the vocabulary quality by the F_β score, where the parameter β is set according to the precision/recall preference of the target application. For classification and search, we proposed to use $F_{0.25}$ and F_1, respectively. Here, we verify whether such F_β scores correspond to the actual usefulness of individual vocabularies. To do so, we apply the vocabularies discussed in Sects. 4.1, 4.2 and 4.3 to our real-life applications and measure the application effectiveness.

The results in Fig. 5 confirm that the estimated quality of vocabularies (red line in Fig. 5a and yellow line in Fig. 5b) shares the same *trend* with the actual vocabulary usefulness measured by the real classification (grey dashed line) and search (grey solid line) effectiveness. Therefore, the F_β score can be used for selecting the most suitable vocabulary for a given application, instead of a tedious and costly experimenting with all candidate vocabularies within the application.

Table 1. Effectiveness of classification and search applications with different segment representations (MW representations use the best-ranked vocabularies with 350 pivots).

Application	Segments as raw 3D skel. data	MW segment representations					
		Hard quant.	Soft assignment		Multi overlays		
			$D10$, $K6$	$D20$, $K6$	1/4	1/5	2/5
Classification	77.70%	74.97%	77.61%	76.42%	76.33%	75.69%	**80.30%**
Search	53.84%	44.21%	47.92%	50.26%	54.97%	**55.62%**	50.29%

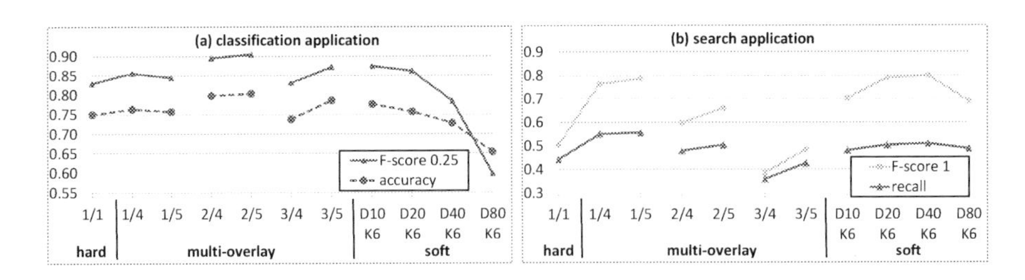

Fig. 5. Comparison of F_β score and actual effectiveness (accuracy, recall) for selected vocabularies in the (a) classification and (b) search applications. (Color figure online)

6 Conclusions

This paper studies the possibility of transforming unlabeled 3D skeleton data into text-like representations that allow efficient processing. In particular, we focused on quantizing short synthetic motion segments into compact, similarity-preserving motion words (MWs). In contrast to existing works on motion quantization, we recognize the border problem and try to minimize it using the soft-assignment and multi-overlay partitioning principles. We also proposed a methodology for application-independent evaluation of the MW vocabulary quality. The experimental results on two real-world motion processing tasks confirm that we are able to construct MW vocabularies which preserve or even slightly increase application effectiveness and significantly improve processing efficiency.

We believe that these achievements open new possibilities for efficient analysis of 3D motion data. In the future, we will study more thoroughly the preparation and preprocessing of the short segments, and develop scalable indexing and search algorithms for the MW data. In particular, we plan to enrich the segmentation process to include several segment sizes, which should help us deal with possible speed variability of semantically related motions. Before the actual quantization, the segments can be replaced by characteristic features extracted, e.g., by state-of-the-art neural networks. To index and search MW sequences, we intend to employ the shingling technique and adapted inverted files.

Acknowledgements. This research has been supported by the GACR project No. GA19-02033S.

References

1. Ahmad, Z., Khan, N.M.: Towards improved human action recognition using convolutional neural networks and multimodal fusion of depth and inertial sensor data. In: 20th International Symposium on Multimedia (ISM), pp. 223–230. IEEE (2018)
2. Alldieck, T., Magnor, M.A., Bhatnagar, B.L., Theobalt, C., Pons-Moll, G.: Learning to reconstruct people in clothing from a single RGB camera. In: IEEE Conference on Computer Vision and Pattern Recognition (CVPR). IEEE (2019)
3. Aristidou, A., Cohen-Or, D., Hodgins, J.K., Chrysanthou, Y., Shamir, A.: Deep motifs and motion signatures. ACM Trans. Graph. **37**(6), 187:1–187:13 (2018). https://doi.org/10.1145/3272127.3275038
4. Butepage, J., Black, M.J., Kragic, D., Kjellstrom, H.: Deep representation learning for human motion prediction and classification. In: IEEE Conference on Computer Vision and Pattern Recognition (CVPR), pp. 6158–6166. IEEE (2017)
5. Demuth, B., Röder, T., Müller, M., Eberhardt, B.: An information retrieval system for motion capture data. In: Lalmas, M., MacFarlane, A., Rüger, S., Tombros, A., Tsikrika, T., Yavlinsky, A. (eds.) ECIR 2006. LNCS, vol. 3936, pp. 373–384. Springer, Heidelberg (2006). https://doi.org/10.1007/11735106_33
6. Dohnal, V., Homola, T., Zezula, P.: MDPV: metric distance permutation vocabulary. Inf. Retr. J. **18**(1), 51–72 (2015)
7. Kabary, I.A., Schuldt, H.: Using hand gestures for specifying motion queries in sketch-based video retrieval. In: de Rijke, M., et al. (eds.) ECIR 2014. LNCS, vol. 8416, pp. 733–736. Springer, Cham (2014). https://doi.org/10.1007/978-3-319-06028-6_84
8. Krüger, B., Vögele, A., Willig, T., Yao, A., Klein, R., Weber, A.: Efficient unsupervised temporal segmentation of motion data. IEEE Trans. Multimed. **19**(4), 797–812 (2017)

9. Liu, B., Cai, H., Ju, Z., Liu, H.: RGB-D sensing based human action and interaction analysis: a survey. Pattern Recogn. **94**, 1–12 (2019)
10. Liu, J., Wang, G., Duan, L., Hu, P., Kot, A.C.: Skeleton based human action recognition with global context-aware attention LSTM networks. IEEE Trans. Image Process. **27**(4), 1586–1599 (2018)
11. Liu, X., He, G., Peng, S., Cheung, Y., Tang, Y.Y.: Efficient human motion retrieval via temporal adjacent bag of words and discriminative neighborhood preserving dictionary learning. IEEE Trans. Hum.-Mach. Syst. **47**(6), 763–776 (2017). https://doi.org/10.1109/THMS.2017.2675959
12. Manning, C.D., Raghavan, P., Schütze, H.: Introduction to Information Retrieval. Cambridge University Press, Cambridge (2008). https://doi.org/10.1017/CBO9780511809071
13. Müller, M., Röder, T., Clausen, M., Eberhardt, B., Krüger, B., Weber, A.: Documentation Mocap Database HDM05. Technical Report CG-2007-2, Universität Bonn (2007)
14. Nistér, D., Stewénius, H.: Scalable recognition with a vocabulary tree. In: International Conference on Computer Vision and Pattern Recognition (CVPR), pp. 2161–2168 (2006)
15. Novak, D., Zezula, P.: PPP-codes for large-scale similarity searching. In: Hameurlain, A., Küng, J., Wagner, R., Decker, H., Lhotska, L., Link, S. (eds.) Transactions on Large-Scale Data- and Knowledge-Centered Systems XXIV. LNCS, vol. 9510, pp. 61–87. Springer, Heidelberg (2016). https://doi.org/10.1007/978-3-662-49214-7_2
16. Philbin, J., Chum, O., Isard, M., Sivic, J., Zisserman, A.: Object retrieval with large vocabularies and fast spatial matching. In: International Conference on Computer Vision and Pattern Recognition (CVPR) (2007)
17. Sedmidubsky, J., Elias, P., Zezula, P.: Effective and efficient similarity searching in motion capture data. Multimed. Tools Appl. **77**(10), 12073–12094 (2017). https://doi.org/10.1007/s11042-017-4859-7
18. Sedmidubsky, J., Elias, P., Zezula, P.: Searching for variable-speed motions in long sequences of motion capture data. Inf. Syst. **80**, 148–158 (2019). https://doi.org/10.1016/j.is.2018.04.002
19. Sedmidubsky, J., Zezula, P.: Probabilistic classification of skeleton sequences. In: Hartmann, S., Ma, H., Hameurlain, A., Pernul, G., Wagner, R.R. (eds.) DEXA 2018. LNCS, vol. 11030, pp. 50–65. Springer, Cham (2018). https://doi.org/10.1007/978-3-319-98812-2_4
20. Sedmidubsky, J., Zezula, P.: Augmenting spatio-temporal human motion data for effective 3D action recognition. In: 21st IEEE International Symposium on Multimedia (ISM), pp. 204–207. IEEE Computer Society (2019). https://doi.org/10.1109/ISM.2019.00044
21. Sivic, J., Zisserman, A.: Video Google: a text retrieval approach to object matching in videos. In: 9th International Conference on Computer Vision (ICCV), pp. 1470–1477. IEEE (2003)
22. Zhao, R., Wang, K., Su, H., Ji, Q.: Bayesian graph convolution LSTM for skeleton based action recognition. In: IEEE International Conference on Computer Vision (ICCV), pp. 6882–6892. IEEE (2019)

23. Zheng, W., Li, L., Zhang, Z., Huang, Y., Wang, L.: Relational network for skeleton-based action recognition. In: International Conference on Multimedia and Expo (ICME), pp. 826–831. IEEE (2019)
24. Zhu, H., Long, M., Wang, J., Cao, Y.: Deep hashing network for efficient similarity retrieval. In: 30th Conference on Artificial Intelligence (AAAI), pp. 2415–2421. AAAI Press (2016)

Deep Learning III

Reinforced Rewards Framework for Text Style Transfer

Abhilasha Sancheti[1(✉)], Kundan Krishna[2], Balaji Vasan Srinivasan[3], and Anandhavelu Natarajan[3]

[1] University of Maryland, College Park, USA
sancheti@cs.umd.edu
[2] Language Technologies Institute, Carnegie Mellon University, Pittsburgh, USA
kundank@andrew.cmu.edu
[3] Adobe Research, Bangalore, India
{balsrini,anandvn}@adobe.com

Abstract. Style transfer deals with the algorithms to transfer the stylistic properties of a piece of text into that of another while ensuring that the core content is preserved. There has been a lot of interest in the field of text style transfer due to its wide application to tailored text generation. Existing works evaluate the style transfer models based on content preservation and transfer strength. In this work, we propose a reinforcement learning based framework that directly rewards the framework on these target metrics yielding a better transfer of the target style. We show the improved performance of our proposed framework based on automatic and human evaluation on three independent tasks: wherein we transfer the style of text from formal to informal, high excitement to low excitement, modern English to Shakespearean English, and vice-versa in all the three cases. Improved performance of the proposed framework over existing state-of-the-art frameworks indicates the viability of the approach.

Keywords: Style transfer · Rewards · Content preservation · Transfer strength

1 Introduction

Text style transfer deals with transforming a given piece of text in such a way that the stylistic properties change to that of the target text while preserving the core content of the given text. This is an active area of research because of its wide applicability in the field of content creation including news rewriting, generating messages with a particular style to maintain the personality of a brand, etc. The stylistic properties may denote various linguistic phenomenon,

A. Sancheti and K. Krishna—This work was done while the authors were working at Adobe Research, Bangalore, India.

J. M. Jose et al. (Eds.): ECIR 2020, LNCS 12035, pp. 545–560, 2020.
https://doi.org/10.1007/978-3-030-45439-5_36

from syntactic changes [7,23] to sentiment modifications [4,10,18] or extent of formality in a sentence [16].

Most of the existing works in this area either use copy-enriched sequence-to-sequence models [7] or employ an adversarial [4,15,18] or much simpler generative approaches [10] based on the disentanglement of style and content in text. On the other hand, more recent works like [19] and [3] perform the task of style transfer without disentangling style and content, as practically this condition cannot always be met. However, all of these works use word-level objective function (eg. cross-entropy) while training which is inconsistent with the desired metrics (content preservation and transfer strength) to be optimized in style transfer tasks. These metrics are generally calculated at a sentence-level and use of word level objective functions is not sufficient. Moreover, discreteness of these metrics makes it even harder to directly optimize the model over these metrics.

Recent advancements in Reinforcement Learning and its effectiveness in various NLP tasks like sequence modelling [8], abstractive summarization [14], and a related one machine translation [21] have motivated us to leverage reinforcement learning approaches in style transfer tasks.

In this paper, we propose a reinforcement learning (RL) based framework which adopts to optimize sequence-level objectives to perform text style transfer. Our reinforced rewards framework is based on a sequence-to-sequence model with attention [1,12] and copy-mechanism [7] to perform the task of text style transfer. The sentence generated by this model along with the ground truth sentence is passed to a content module and a style classifier which calculates the metric scores to finally obtain the reward values. These rewards are then propagated back to the sequence-to-sequence model in the form of loss terms.

The rest of our paper is organized as follows: we discuss related work on text style transfer in Sect. 2. The proposed reinforced rewards framework is introduced in Sect. 3. We evaluate our framework and report the results on formality transfer task in Sect. 4, on affective dimension like excitement in Sect. 5 and on Shakespearean-Modern English corpus in Sect. 6. In Sect. 7, we discuss few qualitative sample outputs. Finally, we conclude the paper in Sect. 8.

2 Related Work

Style transfer approaches can be broadly categorized as style transfer with parallel corpus and style transfer with non-parallel corpus.

Parallel corpus consists of input-output sentence pairs with mapping. Since such corpora are not readily available and difficult to curate, efforts here are limited. [23] introduced a parallel corpus of 30K sentence pairs to transfer Shakespearean English to modern English and benchmark various phrase-based machine translation methods for this task. [7] use a copy-enriched sequence-to-sequence approach for Shakespearizing modern English and show that it outperforms the previous benchmarks by [23]. Recently, [16] introduced a parallel corpus of formal and informal sentences and benchmark various neural frameworks to transfer sentences across different formality levels. Our approach contributes in this field of parallel style transfer and extends the work by [7] by directly optimizing the metrics used for evaluating the style transfer tasks.

Another class of explorations are in the area of non-parallel text style transfer [4,10,15,18] which does not require mapping between the input and output sentences. [4] compose a non-parallel dataset for paper-news titles and propose models to learn separate representations for style and content using adversarial frameworks. [18] assume a shared latent content distribution across a given corpora and propose a method that leverages refined alignment of latent representations to perform style transfer. [10] define style in terms of attributes (such as, sentiment) localized to parts of the sentence and learn to disentangle style from content in an unsupervised setting. Although these approaches perform well on the transfer task, content preservation is generally observed to be low due to the non-parallel nature of the data. Along this line, parallel style transfer approaches have shown better performance in benchmarks despite the data curation challenges [16].

Style transfer models are primarily evaluated on **content preservation** and **transfer strength**. But the existing approaches do not optimize on these metrics and rather teach the model to generate sentences to match the ground truth. This is partly because of the reliance on a differentiable training objective and discreteness of these metrics makes it challenging to differentiate the objective. Leveraging recent advancements in reinforcement learning approaches, we propose a reinforcement learning based text style transfer framework which directly optimizes the model on the desired evaluation metrics. Though there exists some prior work on reinforcement learning for machine translation [21], sequence modelling [8] and abstractive summarization [14] dealing model optimization for qualitative metrics like Rouge [11], they do not consider style aspects which is one of the main requirements of style transfer tasks. More recently, efforts [5,22] have been made to incorporate RL in style transfer tasks in a non-parallel setup. However, our work is in the field of parallel text style transfer which is not much explored.

Our work is different from these related works in the sense that we take care of content preservation and transfer strength with the use of a content module (to ensure content preservation) and cooperative style discriminator (style classifier) without explicitly separating content and style. We illustrate the improvement in the performance of the framework on the task of transferring text between different levels of formality [16]. Furthermore, we present the generalizability of the proposed approach by evaluating it on a self-curated excitement corpus as well as modern English to Shakespearean corpus [7].

3 Reinforced Rewards Framework

The proposed approach takes an input sentence $x = x_1 \ldots x_l$ from source style s_1 and translates it to sentence $y = y_1 \ldots y_m$ with style s_2, where x and y are represented as a sequence of words. If x is given by (c_1, s_1) where c_1 represents the content and s_1 the style of the source, our objective is to generate $y = (c_1, s_2)$ which has same content as the source but with the target style.

Our approach is based on a copy-enriched sequence-to-sequence framework [7] which allows the model to retain factual parts of the text while changing the style specific text using an attention mechanism. At the time of training, the framework takes in the source style and the target style sentence as input to the attention based sequence-to-sequence encoder-decoder model. The words in the input sentence are mapped into an embedding space and the sentence is encoded into a latent space by the LSTM encoder. The network learns to pay attention to the words in the source sentence and creates a context vector based on the attention. The decoder model is a mixture of RNN and pointer (PTR) network where the RNN predicts the probability distribution over the vocabulary and the pointer network predicts the probability over the words in the input sentence based on the context vector. A weighted average of the two probabilities yields the final probability distribution at time step t given by,

$$P_t(w) = \delta P_t^{RNN}(w) + (1 - \delta) P_t^{PTR}(w),$$

where δ is computed based on encoder outputs and previous decoder hidden states. The decoder generates the transferred sentence by selecting the most probable word at each time step. This model is trained to minimize cross entropy loss given by

$$L_{ml} = -\sum_{t=1}^{m} \log(p(P_t(y_t^*))),$$

where m is the maximum length of the output sentence and y_t^* is the ground truth word at time t in the transferred sentence. While this framework optimizes for generating sentences close to the ground truth, it does not explicitly teach the network to preserve the content and generate sentences in target style. To achieve this, we introduce a style classifier and a content module which takes in the generated sentence from the sequence-to-sequence model along with the ground truth target sentence to provide reward to the sentence, as shown in Fig. 1. We leverage BLEU [13] score to measure the reward for preserving content and because of the lack of any formal score for transfer strength, we use a cooperative discriminator to provide score to the generated sentence. This score from the discriminator is used as a measure to reward for transfer strength. These rewards are then back propagated as explicit loss terms to penalize the network for incorrect generation.

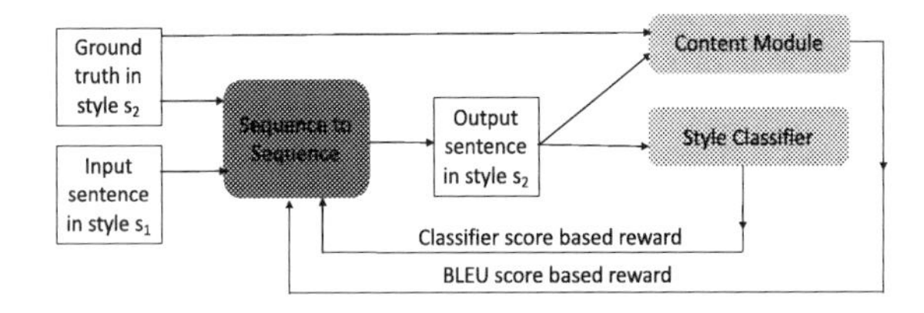

Fig. 1. Model overview

3.1 Content Module: Rewarding Content Preservation

To preserve the content while transferring the style, we leverage Self-Critic Sequence Training (SCST) [17] approach and optimize the framework with BLEU scores as the reward. SCST is a policy gradient method for reinforcement learning and is used to train end-to-end models directly on non-differentiable metrics. We use BLEU score as reward for content preservation because it measures the overlap between the ground truth and the generated sentences. Teaching the network to favor this would result in high overlap with the ground truth and subsequently preserve the content of the source sentence since ground truth ensures this preservation.

We produce two output sentences y^s and y', where y^s is sampled from the distribution $p(y^s_t|y^s_{1:t-1}, x)$ at each decoding time step and y' (baseline output) is obtained by greedily maximizing the output distribution at each time step. The BLEU score between the sampled and greedy sequences is computed as the reward and the corresponding content-preservation loss is given by,

$$L_{cp} = (r(y') - r(y^s)) \sum_{t=1}^{m} \log(p(y^s_t|y^s_{1:t-1}, x)),$$

where the log term is the log likelihood on sampled sequence and the difference term is the difference between the reward (BLEU score) for the greedily sampled y' and multinomially sampled y^s sentences. Note that our formulation is flexible and does not require the metric to be differentiable because rewards are used as weights to the log-likelihood loss. Minimizing L_{cp} is equivalent to encouraging the model to generate sentences which have higher reward as compared to the baseline y' and thus increasing the reward expectation of the model. The framework can now be trained end to end by using this loss function along with the cross entropy loss to preserve the content of the source sentence in the transferred sentence.

3.2 Style Classifier: Rewarding Transfer Strength

To optimize the model to generate sentences which belong to the target style, it is possible to use a similar loss function as above and use it with the SCST framework [17]. However, that will require a formal measure for the target style aspect. Here, we present an alternate framework where such a formal measure is not readily available. We train a convolutional neural network based style classifier as proposed by [9] on the training dataset. This style classifier predicts the likelihood that an input sentence is in the target style, and the likelihood is taken as a proxy to the reward for style of a sentence and appended to a discriminator-based loss function extended from [6]. Based on the transfer direction, we add the following term to the cross-entropy loss,

$$L_{ts} = \begin{cases} -\log(1 - s(y')), & \text{high to low level} \\ -\log(s(y')), & \text{low to high level} \end{cases}$$

In this formulation, y' is the greedily generated output from the decoder and s(y') is the likelihood score predicted by the classifier for y'. When transfer is done from high to low level of style, minimization of L_{ts} will encourage generation of sentences such that the classifier score is as low as possible. When the sentences are transferred from low to high level of style then the formulation ensures that the generated sentences have a score as high as possible. The framework is trained end-to-end using this loss function to generate the sentences which belong to the target style.

3.3 Training and Inference

The overall loss function thus can be written as a combination of the 3 loss functions,

$$Loss = \alpha L_{ml} + \beta L_{cp} + \gamma L_{ts}$$

We train various models using this loss function and different training methodologies (setting $\alpha = 1.0$, $\beta = 0.125$, $\gamma = 1.0$ after hyper-parameter tuning) as described in the next section. During the inference phase, the model predicts a probability distribution over the vocabulary based on the sentence generated so far and the word having the highest probability is chosen as the next word till the maximum length of the output sentence is reached. Note that unlike training phase in which case both the input and ground truth transferred sentences are available to the model, only the input sentence is made available to the model.

4 Experiments: Reinforcing Formality (GYAFC Dataset)

We evaluate the proposed approach on the GYAFC [16] dataset which is a parallel corpus for formal-informal text. We present the transfer task results in both the directions - formal to informal and vice-versa. This dataset (from Entertainment and Music domain) consists of ~56K informal-formal sentence pairs: ~52K in train, ~1.5K in test and ~2.5K in validation split.

We use both human and automatic evaluation measures for content preservation and transfer strength to illustrate the performance of the proposed approach.

Content preservation measures the degree to which the target style model outputs have the same meaning as the input style sentence. Following [16], we measure preservation of content using BLEU [13] score between the ground truth and the generated sentence since the ground truth ensures that content of the source style sentence is preserved in it. For human evaluation, we presented 50 randomly selected model outputs to the Mechanical turk annotators and requested them to rate the outputs on a Likert [2] scale of 6 as described in [16].

Transfer strength measures the degree to which style transfer was carried out. We reuse the classifiers that we built to provide rewards to the generated sentences (Sect. 3.2). A score above 0.5 from the classifier represents that the generated sentence belongs to the target style and to the source style otherwise. We define accuracy as the fraction of generated sentences which are classified to be in the target style. The higher the accuracy, higher is the transfer strength. For human evaluation, we ask the Mechanical turk annotators to rate the generated sentence on a Likert scale of 5 as described in [16].

Following [4] who illustrate the trade-off between the two metrics - content preservation and transfer strength, we combine the two evaluation measures and present an **overall score** for the transfer task since both the measures are central to different aspects of text style transfer task. The trade-off arises because the best content preservation can be achieved by simply copying the source sentence. However, the transfer strength in such scenario will be the worst. We compute overall score in the following way

$$\text{Overall} = \frac{\text{BLEU} \times \text{Accuracy}}{\text{BLEU} + \text{Accuracy}}$$

which is similar to F1-score since content preservation can be considered as measuring recall of the amount of source content retained in the target style sentence and transfer strength acts as a measure of precision with which the transfer task is carried out. In the above formulation, both BLEU and accuracy scores are normalized to be between 0 and 1.

We first ran an **ablation study** to demonstrate the improvement in performance of the model with introduction of the two loss terms in the various settings differing in the way training is being carried out. Below we provide details about each of the settings.

CopyNMT: Trained with L_{ml}
TS: Trained with L_{ml} followed by $\alpha L_{ml} + \gamma L_{ts}$
CP: Trained with L_{ml} followed by $\alpha L_{ml} + \beta L_{cp}$
TS+CP: Trained with L_{ml} followed by $\alpha L_{ml} + \beta L_{cp} + \gamma L_{ts}$
TS→CP: Trained with L_{ml} followed by $\alpha L_{ml} + \gamma L_{ts}$ and finally with $\alpha L_{ml} + \beta L_{cp}$
CP→TS: Trained with L_{ml} followed by $\alpha L_{ml} + \beta L_{cp}$ and finally with $\alpha L_{ml} + \gamma L_{ts}$

Table 1. Ablation study to demonstrate the improvement of the addition of the loss terms on formality transfer task.

Models	**Informal to Formal**			**Formal to Informal**		
	BLEU↑	Accuracy↑	Overall↑	BLEU↑	Accuracy↑	Overall↑
CopyNMT	0.263	0.774	0.196	0.280	0.503	0.180
TS	0.240	0.801	0.184	0.271	0.527	0.179
CP	0.272	0.749	0.199	0.281	0.487	0.178
TS+CP	0.259	0.772	0.194	0.271	0.527	0.179
CP→TS	0.227	**0.817**	0.178	0.259	**0.5441**	0.175
TS→CP	**0.286**	0.723	**0.205**	**0.298**	0.516	**0.189**

Training with L_{ml} alone in all the above settings is done for 10 epochs with all the hyper-parameters set as default in the off-the-shelf implementation of [7]. Each of the iterative model training is done using the model with the best performance on validation set for 5 more epochs. We can observe from Table 1 that L_{ts} and L_{cp} helps in improving the accuracy which measures transfer strength (TS) and BLEU score which measures content preservation (CP) respectively as compared to CopyNMT. When all the three loss terms are used simultaneously (TS+CP) the resulting performance lies between TS and CP, indicating that there is a trade-off between the two metrics and improvement in one metric is at the cost of another as observed by [4]. This phenomenon is evident from the results of TS→CP and CP→TS where the network gets a bit biased towards the latter optimization. Moreover, improvement in CP→TS and TS→CP as compared to TS and CP respectively suggests that incremental training better helps in teaching the framework. Since the performance on both transfer strength and content preservation metrics plays an important role in text style transfer task, we chose TS→CP, which has the maximum overall score, over the other models for further analysis.

Baselines: We compare the proposed approach TS→CP against the state-of-the-art cross-aligned autoencoder style transfer approach (Cross-Aligned) by

Table 2. Comparison of TS→CP with the baselines on the three transfer tasks in both the directions. All the scores are normalized to be between 0 and 1.

Models	Informal to Formal			Formal to Informal		
	BLEU↑	**Accuracy↑**	**Overall↑**	**BLEU↑**	**Accuracy↑**	**Overall↑**
Transformer [20]	0.125	**0.933**	0.110	0.099	**0.894**	0.089
Cross-Aligned [18]	0.116	0.670	0.098	0.117	0.766	0.101
CopyNMT [7]	0.263	0.774	0.196	0.280	0.503	0.180
TS→CP (Proposed)	**0.286**	0.723	**0.205**	**0.298**	0.516	**0.189**
	Exciting to Non-exciting			**Non-exciting to Exciting**		
Transformer [20]	0.077	**0.922**	0.071	0.069	0.605	0.062
Cross-Aligned [18]	0.059	0.818	0.055	0.061	0.547	0.054
CopyNMT [7]	0.143	0.919	0.124	0.071	**0.813**	0.065
TS→CP (Proposed)	**0.153**	**0.922**	**0.131**	**0.088**	0.744	**0.078**
	Modern to Shakespearean			**Shakespearean to Modern**		
Transformer [20]	0.027	**0.736**	0.026	0.046	**0.915**	0.043
Cross-Aligned [18]	0.044	0.614	0.041	0.049	0.537	0.044
CopyNMT [7]	0.104	0.495	0.085	0.111	0.596	0.093
TS→CP (Proposed)	**0.127**	0.489	**0.100**	**0.137**	0.567	**0.110**

[18][1], parallel style transfer approach (CopyNMT) by [7][2] and neural encoder-decoder based transformer model [20][3].

Results: It can be seen from Table 2 that even though the transformer model has the best accuracy, it fails in preserving the content. Closer look at the outputs (formal to informal transfer task in Table 4) reveal that it generates sentences in target style but the sentences do not preserve the meaning of the input and sometimes are out of context (discussed in the Sect. 7). Cross-Aligned performs the worst in informal to formal transfer task among all the other approaches because it is generating a lot of unknowns and is not able to preserve content. TS→CP, on the other hand, has the highest overall score and performs the best in preserving the content. We also observed that the dataset had many sentences containing proper nouns like name of the songs, person or artists. In such cases, copy mechanism helps in retaining the proper nouns whereas other models are not able to do so. This is evident from the higher BLEU scores for our proposed model. Table 3 presents the human evaluation results aggregated over three annotators per sample. It can be seen that in at least 70% of the cases, annotators rated model outputs from TS→CP as better than the three baselines on both the evaluated metrics except for the content preservation as compared to CopyNMT in formal to informal task wherein, both the models perform equally good. One reason behind this is that both the models use copy-mechanism.

Table 3. Human evaluation results of 50 randomly selected model outputs. The values represent the % of times annotators rated model outputs from TS→CP (R) as better than the baseline CopyNMT (C), Transformer (T) and Cross-Aligned (S) over the metrics. I-F (E-NE) refers to informal to formal (exciting to non-exciting) task.

Task	**Transfer strength**			**Content preservation**		
	R > C	R > T	R > S	R > C	R > T	R > S
I-F	88.67	81.34	70.00	70.00	72.67	83.67
F-I	73.34	88.67	61.22	59.34	79.34	91.80
E-NE	64.00	79.34	68.00	60.67	71.34	71.73
NE-E	76.67	70.67	68.00	69.34	74.00	70.00

[1] We use the off-the-shelf implementation provided by the authors at https://github.com/shentianxiao/language-style-transfer.

[2] https://github.com/harsh19/Shakespearizing-Modern-English.

[3] https://github.com/pytorch/fairseq We also tried using the model proposed by [5] to compare against out proposed approach but we couldn't get stable performance on our datasets.

5 Experiments: Beyond Formality (Excitement Dataset)

In order to demonstrate the generalizability of our approach on an affective style dimension like excitement (the feeling of enthusiasm and eagerness), we curated our own dataset using reviews from Yelp dataset[4] which is a subset of Yelp's businesses, reviews, and user data. We request human annotators to provide rewrites for given exciting sentences such that they sound as non-exciting/boring as possible. Reviews with rating greater than or equal to 3 were filtered out and considered as exciting to get the non-exciting/boring rewrites. We also asked the annotators to rate the given and transferred sentences on a Likert scale of 1 (No Excitement at all) to 5 (Very high Excitement). The dataset thus curated was split into train (~36K), test (1K) and validation (2K) sets. We evaluate the transfer quality on content preservation and transfer strength metrics as defined in Sect. 4.

For measuring the transfer strength we train a classifier as described in Sect. 3.2. We use the annotations provided by the human annotators on these sentences to get the labels for the two styles. Sentences with a rating greater than or equal to 3 were considered as exciting and non-exciting otherwise.

Results: The transfer task in this case is to convert the input sentence with high excitement (exciting) to a sentence with low excitement (non-exciting) and vice-versa. We can observe from Table 2 that model performance in the case of excitement transfer task is similar to what we observed in the formality transfer task. However, CopyNMT performs the best in transferring style in case of non-exciting to exciting transfer task because the model has picked up on expressive words ('awesome', 'great', and 'amazing') which helps in boosting the transfer strength. TS→CP (with highest overall score) consistently outperforms Cross-Aligned in all the metrics and both the directions. Table 3 presents the human evaluation results on this transfer task. We notice that humans preferred outputs from our proposed model at least 60% of the times on both the measures as compared to the other three baselines. This provides an evidence that the proposed RL-based framework indeed helps in improving generation of more content preserving sentences which align with the target style.

6 Experiments: Beyond Affective Elements (English Dataset)

Besides affective style dimensions, our approach can also be extended to other style transfer tasks like converting modern English to Shakespearean English. To illustrate the performance of our model on this task we experimented with the corpus used in [7]. The dataset consists of ~21K modern-Shakespearean English sentence pairs with ~18K in train, ~1.5K in test and ~1.2K in validation split.

[4] https://www.yelp.com/dataset.

We use the same evaluation measures as in the previous two tasks for illustrating the model performance and generalizability of the approach. For this task we present only the automatic evaluation results because manual evaluation of this task is not easy since it requires an understanding of Shakespearean english and finding such population is a difficult task due to limited availability.

Results: We can observe from Table 2 that model performance in the case of this transfer task is also similar to what we have observed in the earlier two transfer tasks. Although Cross-Aligned has better accuracy than TS→CP, it fails to preserve the content (sample 3 of Table 6). Similar is the case with transformer which outperforms others in accuracy but is not able to retain the content (sample 1 of Table 6). TS→CP outperforms the three baselines in preserving the content with the highest overall score. This establishes the viability of our approach to various types of text style transfer tasks.

These experiments further indicate that our proposed reinforcement learning framework improves the transfer strength and content preservation of parallel style transfer frameworks and is also generalizable across various stylistic expression.

7 Discussion

In this section, we provide few qualitative samples from the baselines and the proposed reinforcement learning based model. We can observe from the transformer model output for Input 1 and 2 in formal to informal column of Table 4 that it generates sentences with correct target style but does not preserve the content. It either adds random content or deletes the required content ('band' instead of 'better' in 1 and 'hot' instead of 'talented' in 2). As mentioned earlier, in sample output 3 of Table 4, Cross-Aligned is unable to retain the content and tend to generate unknown tokens. CopyNMT, even though is able to preserve content, tend to generate repeated token like 'please' in sample input 2 (informal to formal task) which results in lower BLEU score than our proposed approach. Transformer model outputs for exciting to non-exciting task in samples 1 and 2 of Table 5, miss specific content words like 'environment' and 'alisha' respectively. However, it is able to generate the sentences in target style. Similary, Cross-Aligned and CopyNMT are also not able to retain the name of the server in sample 2 of Table 5. Sample 2 of Shakespearean to Modern English and 1 of Modern to Shakespearean English task in Table 6 provide evidence for high accuracy and lower BLEU scores for transformer model. From sample 2 of Shakespearean to modern English transfer task, we can observe that Cross-Aligned although can generate the sentence in the target style is not able to preserve the entities like 'father' and 'child'. On the other hand, TS→CP can not only generate the sentences in the target style but is also able to retain the entities. There are few cases when CopyNMT is better in preserving the content

Table 4. Sample model outputs and target style reference for Informal to Formal and Formal to Informal style transfer task. The first line is the source style sentence (input), second line is the reference output and the following lines correspond to the outputs from the baselines and the RL-based model.

	Model	Informal to Formal	Formal to Informal
1	Input	I want to be on TV!	I do not understand what that has to do with who's better looking?
	Reference	I would like to be on television	I don't know what the hell that has to do with who's better looking but OKAY!
	Transformer	I want to be on TV	I don't know what that's better looking with the band that do u?
	Cross-Aligned	I want to be on TV!	I do n't know that that do to have to talk of more better?
	CopyNMT	I would like to be on TV	I don't understand what that has to do with who's better looking for?
	TS→CP	I would like to be on TV	I don't understand what that has to do with who better?
2	Input	When you find out please let me know	I think that she is so talented, if she does not win, I am going to be really disappointed
	Reference	Please let me know when you find out	He is so talented, if she didn't win, I'd be really disappointed!
	Transformer	Keep me informed as soon as you know anything	I don't think she's hot, but i'm going to win so she'll win
	Cross-Aligned	If you find out please let me know	I think she is so funny, she doesn't win, I'm not sure to be gonna be cute
	CopyNMT	When you find out please please please me know?	I think she's so talented, she's not that i'm going to be really disappointed
	TS→CP	Please inform me if you find out	I think she is so talented, if she doesn't win, I'm gonna be really disappointed
3	Input	I dono I think that is the DUMBEST show EVER!!!!!!	Our mother is so unintelligent that she was hit & by a cop and told the police that she was mugged
	Reference	I don't think it's a very intelligent show	Your mama is so stupid, she got hit by a cop and told the police that she got mugged
	Transformer	I do not think that the show is appropriate	Your mama is so stupid that she sat on the ocean and said she was a bus
	Cross-Aligned	I think that I am $\langle unk \rangle$ the show $\langle unk \rangle$ $\langle unk \rangle$!	Yo mama is so fat that she had a $\langle unk \rangle$ and got a bunch of that's and she was $\langle unk \rangle$
	CopyNMT	I am not sure that is the DUMBEST show EVER!	Your mama is so unintelligent she she hit hit cop and told the police that she was
	TS→CP	I think that is the DUMBEST show EVER!	Your mama is so unintelligent she got hit by a cop and told that she was so

as compared to other models, for instance, sample 1 of formal to informal transfer task and sample 3 of non-exciting to exciting transfer task since it leverages copy-mechanism.

Another point to notice is the lexical level changes made to reflect the target style. For example, the use of 'would', 'don't' and 'inform' instead of 'want', 'dono' and 'let me know' respectively for transforming informal sentences into formal ones. Use of colloquial words like 'u', 'gonna' and 'mama' for converting the formal sentences to informal can be observed from the sample outputs. Not

Table 5. Sample model outputs and target style reference for Exciting to Non-exciting and Non-exciting to Exciting style transfer task. The first line is the source style sentence (input), second line is the reference output and the following lines correspond to the outputs from the baselines and RL-based model.

	Model	Exciting to Non-exciting	Non-exciting to Exciting
1	Input	Delicious food and good environment	A good choice if you are in the phoenix area
	Reference	Good food and environment	A must visit if in the phoenix area
	Transformer	I recommend this food	If you're in the phoenix area, this is the place to go
	Cross-Aligned	Good food and good drinks	A great spot if you're in the area area
	CopyNMT	The food was good	This is a great choice of if you are in the phoenix area
	TS→CP	Good food and atmosphere	If you're in the phoenix area, this is a great choice if you're in the phoenix area
2	Input	Our server alisha was amazing	The food menu is reasonable and happy hour specials are good
	Reference	Our server alisha did a good job	Reasonable food menu and great happy hour specials
	Transformer	Our server was good	They have a great happy hour menu and the food is very good
	Cross-Aligned	Our server server was good	The food is great and happy hour prices are awesome
	CopyNMT	Our server was good	The food menu is great and the food is amazing
	TS→CP	Our server alisha was very good	The food menu is reasonable and happy hour specials are great
3	Input	The patio is amazing too	Acceptable food and beers with live music sometimes
	Reference	I like the patio also	Good food and great beers with occasional live music
	Transformer	The patio ... . great	Live bands, good food and great beer
	Cross-Aligned	The patio is pretty good	Awesome food and great selection of music and music
	CopyNMT	The patio is good	Great food and great drinks and live music
	TS→CP	The patio is good	Great food, great beers, and great music

only lexical level changes but structural transformations can also be observed as in 'Please inform me if you find out'. In case of excitement transfer task, use of strong expressive words like 'amazing' and 'great' makes the sentence sound more exciting while less expressive words such as 'okay' and 'good' makes the sentence less exciting. Use of 'thou' for you and 'hither' for here are more frequently used in Shakespearean English than in modern English. These sample outputs indeed provide an evidence that our model is able to learn these lexical or structural level differences in various transfer tasks, be it formality, beyond formality or beyond affective dimensions.

Table 6. Sample model outputs and target style reference for Modern to Shakespearean English and Shakespearean to Modern English transfer task. The first line is the source style sentence (input), second line is the reference output and the following lines correspond to the outputs from the baselines and the RL-based model.

	Model	Modern to Shakespearean	Shakespearean to Modern
1	Input	Don't you see that I'm out of breath?	Good morrow to you both
	Reference	Do you not see that I am out of breath?	Good morning to you both
	Transformer	Do you not hear me?	Good morning to you
	Cross-Aligned	Do you not think I had out of breath?	Good morrow to you
	CopyNMT	Do not see see I breath of breath?	Good morning, you both
	TS→CP	Do you not see that I am out of breath?	Good morning to you both
2	Input	Do you love me?	Well, well, thou hast a careful father, child
	Reference	Dost thou love me?	Well, well, you have a careful father, child
	Transformer	Do you love me?	Well, good luck
	Cross-Aligned	Dost thou love me?	Well, sir, be a man, Give it this
	CopyNMT	Do you love?	Well, well, you hast a father father, child
	TS→CP	Dost thou love me?	Well, well, you have a careful father, child
3	Input	Come here, man	Thou know'st my daughter's of a pretty age
	Reference	Come hither, man	You know how young my daughter is
	Transformer	Come, man	You are my daughter
	Cross-Aligned	Come hither, Iago	You know how noble my name is
	CopyNMT	Come hither, man	You know'st my daughter's age
	TS→CP	Come hither, man	You're know'st my daughter's of a pretty age

8 Conclusion and Future Work

The primary contribution of this work is a reinforce rewards based sequence-to-sequence model which explicitly optimizes over content preservation and transfer strength metrics for style transfer with parallel corpus. Initial results are promising and generalize to other stylistic characteristics as illustrated in our experimental sections. Leveraging this approach for simultaneously changing multiple stylistic properties (for e.g. high excitement and low formality) is a subject of further research.

References

1. Bahdanau, D., Cho, K., Bengio, Y.: Neural machine translation by jointly learning to align and translate. arXiv preprint arXiv:1409.0473 (2014)
2. Bertram, D.: Likert scales (2007). Accessed 2 Nov 2013
3. Dai, N., Liang, J., Qiu, X., Huang, X.: Style transformer: unpaired text style transfer without disentangled latent representation. arXiv preprint arXiv:1905.05621 (2019)
4. Fu, Z., Tan, X., Peng, N., Zhao, D., Yan, R.: Style transfer in text: exploration and evaluation. arXiv preprint arXiv:1711.06861 (2017)

5. Gong, H., Bhat, S., Wu, L., Xiong, J., Hwu, W.-H.: Reinforcement learning based text style transfer without parallel training corpus. arXiv preprint arXiv:1903.10671 (2019)
6. Holtzman, A., Buys, J., Forbes, M., Bosselut, A., Golub, D., Choi, Y.: Learning to write with cooperative discriminators. In: Proceedings of the 56th Annual Meeting of the Association for Computational Linguistics (Volume 1: Long Papers), pp. 1638–1649. Association for Computational Linguistics (2018). http://aclweb.org/anthology/P18-1152
7. Jhamtani, H., Gangal, V., Hovy, E., Nyberg, E.: Shakespearizing modern language using copy-enriched sequence to sequence models. In: Proceedings of the Workshop on Stylistic Variation, pp. 10–19. Association for Computational Linguistics (2017). https://doi.org/10.18653/v1/W17-4902. http://aclweb.org/anthology/W17-4902
8. Keneshloo, Y., Shi, T., Reddy, C.K., Ramakrishnan, N.: Deep reinforcement learning for sequence to sequence models. arXiv preprint arXiv:1805.09461 (2018)
9. Kim, Y.: Convolutional neural networks for sentence classification. arXiv preprint arXiv:1408.5882 (2014)
10. Li, J., Jia, R., He, H., Liang, P.: Delete, retrieve, generate: a simple approach to sentiment and style transfer. arXiv preprint arXiv:1804.06437 (2018)
11. Lin, C.Y.: ROUGE: a package for automatic evaluation of summaries. Text Summarization Branches Out (2004)
12. Luong, M.T., Pham, H., Manning, C.D.: Effective approaches to attention-based neural machine translation. arXiv preprint arXiv:1508.04025 (2015)
13. Papineni, K., Roukos, S., Ward, T., Zhu, W.J.: BLEU: a method for automatic evaluation of machine translation. In: Proceedings of the 40th Annual Meeting on Association for Computational Linguistics, pp. 311–318. Association for Computational Linguistics (2002)
14. Paulus, R., Xiong, C., Socher, R.: A deep reinforced model for abstractive summarization. arXiv preprint arXiv:1705.04304 (2017)
15. Prabhumoye, S., Tsvetkov, Y., Salakhutdinov, R., Black, A.W.: Style transfer through back-translation. In: Proceedings of the 56th Annual Meeting of the Association for Computational Linguistics (Volume 1: Long Papers), pp. 866–876. Association for Computational Linguistics (2018). http://aclweb.org/anthology/P18-1080
16. Rao, S., Tetreault, J.: Dear sir or madam, may i introduce the GYAFC dataset: corpus, benchmarks and metrics for formality style transfer. In: Proceedings of the 2018 Conference of the North American Chapter of the Association for Computational Linguistics: Human Language Technologies (Long Papers), vol. 1, pp. 129–140 (2018)
17. Rennie, S.J., Marcheret, E., Mroueh, Y., Ross, J., Goel, V.: Self-critical sequence training for image captioning. In: CVPR, vol. 1, p. 3 (2017)
18. Shen, T., Lei, T., Barzilay, R., Jaakkola, T.: Style transfer from non-parallel text by cross-alignment. In: Advances in Neural Information Processing Systems, pp. 6830–6841 (2017)
19. Subramanian, S., Lample, G., Smith, E.M., Denoyer, L., Ranzato, M., Boureau, Y.L.: Multiple-attribute text style transfer. arXiv preprint arXiv:1811.00552 (2018)
20. Vaswani, A., et al.: Attention is all you need. In: Advances in Neural Information Processing Systems, pp. 5998–6008 (2017)

21. Wu, L., Tian, F., Qin, T., Lai, J., Liu, T.Y.: A study of reinforcement learning for neural machine translation. In: Proceedings of the 2018 Conference on Empirical Methods in Natural Language Processing, pp. 3612–3621. Association for Computational Linguistics (2018). http://aclweb.org/anthology/D18-1397
22. Xu, J., et al.: Unpaired sentiment-to-sentiment translation: a cycled reinforcement learning approach. arXiv preprint arXiv:1805.05181 (2018)
23. Xu, W., Ritter, A., Dolan, B., Grishman, R., Cherry, C.: Paraphrasing for style. In: Proceedings of COLING 2012, pp. 2899–2914 (2012)

Recognizing Semantic Relations: Attention-Based Transformers vs. Recurrent Models

Dmitri Roussinov[1(✉)], Serge Sharoff[2], and Nadezhda Puchnina[3]

[1] University of Strathclyde, 16 Richmond Street, Glasgow G1 1XQ, UK
dmitri.roussinov@strath.ac.uk
[2] University of Leeds, Leeds LS2 9JT, UK
s.sharoff@leeds.ac.uk
[3] University of Tallinn, Narva maantee 25, 10120 Tallinn, Estonia
np486061@tlu.ee

Abstract. Automatically recognizing an existing semantic relation (such as "is a", "part of", "property of", "opposite of" etc.) between two arbitrary words (phrases, concepts, etc.) is an important task affecting many information retrieval and artificial intelligence tasks including query expansion, common-sense reasoning, question answering, and database federation. Currently, two classes of approaches exist to classify a relation between words (concepts) X and Y: (1) path-based and (2) distributional. While the path-based approaches look at word-paths connecting X and Y in text, the distributional approaches look at statistical properties of X and Y separately, not necessary in the proximity of each other. Here, we suggest how both types can be improved and empirically compare them using several standard benchmarking datasets. For our *distributional* approach, we are suggesting using an attention-based transformer. While they are known to be capable of supporting knowledge transfer between different tasks, and recently set a number of benchmarking records in various applications, we are the first to successfully apply them to the task of recognizing semantic relations. To improve a *path-based* approach, we are suggesting our original neural word path model that combines useful properties of convolutional and recurrent networks, and thus addressing several shortcomings from the prior path-based models. Both our models significantly outperforms the state-of-the-art within its type accordingly. Our transformer-based approach outperforms current state-of-the-art by 1–12% points on 4 out of 6 standard benchmarking datasets. This results in 15–40% error reduction and is closing the gap between the automated and human performance by up to 50%. It also needs much less training data than prior approaches. For the ease of re-producing our results, we make our source code and trained models publicly available.

1 Introduction

During the last few years, Recurrent Neural Networks (RNNs) and Convolutional Neural Networks (CNNs) have resulted in major breakthroughs and are behind

J. M. Jose et al. (Eds.): ECIR 2020, LNCS 12035, pp. 561–574, 2020.
https://doi.org/10.1007/978-3-030-45439-5_37

the current state-of-the-art algorithms in language processing, computer vision, and speech recognition [9]. Meanwhile, modeling higher level abstract knowledge still remains a challenging problem even for them. This includes classification of semantic relations: given a pair of concepts (words or word sequences) to identify the best semantic label to describe their relationship. The possible labels are typically "is a", "part-of", "property-of", "made-of", etc. This information is useful in many applications. For example, knowing that *London* is a *city* is needed for a Question Answering system to answer the question *What cities does the River Thames go through?* Information retrieval benefits from query expansion with more specific words, e.g. *transportation disasters* → *railroad disasters.* For the task of database federation, an attribute in one database (e.g. with values *France, Germany,* and *UK*) often needs to be automatically matched with an attribute called *country* in another database. Knowing the semantic relations allows large-scale knowledge base construction [11,23,33], automated inferencing [6,29], query understanding [31], post-search navigation [7], and personalized recommendation [34]. The connection between word meanings and their usage is prominent in the theories of human cognition [12] and human language acquisition [2]. While manually curated dictionaries exist, they are known to be out-of-date, not covering specialized domains, designed to be used by people, and exist for only a few well resourced languages (English, German, etc.). Therefore, here we are interested in methods for automated discovery (knowledge acquisition, taxonomy mining, etc.). As our Sect. 2 elaborates, this problem has been a subject of extensive exploration for more than three decades. Our results here suggest that *knowledge transfer*, that was recently demonstrated to be useful for the other tasks, can be also successfully applied to recognizing semantic relations leading to substantial performance improvements and needing much less training data.

The automated approaches to detecting semantic relations between concepts (words or phrases) can be divided into two major groups: (1) path-based and (2) distributional methods. *Path-based approaches* (e.g. [25]) look for certain patterns in the joint occurrences of words (phrases, concepts, etc.) in the corpus. Thus, every word pair of interest (x,y) is represented by the set of *word paths* that connect x and y in a raw text corpus (e.g. Wikipedia). *Distributional approaches* (e.g. [30]) are based on modeling the occurrences of each word, x or y, separately, not necessary in the proximity to each other. Our goal here is to *improve and compare both classes of approaches.*

Attention-based transformers (e.g. [28]) have been recently shown more effective than convolutional and recurrent neural models for several natural text applications, leading to new state-of-the-art results on several benchmarks including GLUE, MultiNLI, and SQuAD [4,8]. At the same time, we are not aware of any applications of attention-based transformers to the task of recognizing semantic relations, so we are the first to successfully apply them to this task. Thus, our contributions are as follows:

(1) We develop a novel neural path-based model that combines useful properties of convolutional and recurrent networks. Our approach resolves several

shortcomings of the prior models within that type. As a result, it outperforms the state-of-the art path-based approaches on 3 out of 6 well known benchmarking datasets, and on par on the other 3. (2) Our distributional approach worked better than our neural path-based model and outperformed current state-of-the-art by 1–12% points (15–40% error reduction) on 4 out of 6 (same) standard datasets, and on par on the remaining 2. (3) We show that the datasets that are not improved are those where we have already reached the human performance. (4) We illustrate that even our best model still has certain limitations which are not always revealed by the standard datasets.

We make our code and data publicly available.[1] The next section overviews the prior related work. It is followed by the description of the models, followed by our empirical results.

2 Related Work

The approaches to automatically classifying semantic relations between words can be divided into two major groups: (1) path-based and (2) distributional. *Path-based approaches* look for certain patterns in the joint occurrences of words (phrases, concepts, etc.) in some validation text corpus. Thus, every word pair of interest *(x,y)* is represented by the set of *word paths* that connect x and y in a raw text corpus (e.g. Wikipedia). The earliest *path-based* approach is typically attributed to "Hearst Patterns" [5] – a set of 6 regular expressions to detect "is-a" relations (e.g. *Y such as X*). Later works successfully involved trainable templates and larger texts (e.g. [18,27]). However, a major limitation in relying on patterns in the word paths is the sparsity of the feature space [14]. Distributed representation do not have such limitations, thus with deep neural representations ("embeddings", e.g. [13,17]) becoming popular, a number of successful models were developed that used word embeddings as features (concatenation, dot product or difference) and surpassed the path-based methods in performance [15,20], to the point that *the path-based approaches were perceived not be adding anything to the distributional ones.*

However, [10] noted that supervised distributional methods tend to perform "lexical memorization:" instead of learning a relation between the two terms, they learn an independent property of a single term in the pair. For example, if the training set contains pairs such as *(dog, animal)*, *(cat, animal)*, and *(cow, animal)*, the algorithm learns to classify any new *(x, animal)* pair as true, regardless of x. Shwartz et al. [24,25] (one of our baselines) successfully combined distributional and path-based approaches to improve the state-of-the performance, and thus proving that path-based information is also crucial for that. In their approach, each word path connecting a pair of concepts (X, Y) is mapped by an RNN into a *context vector*. Those vectors are averaged across all existing paths and fed to a two-layer fully connected network.

There have been several related studies following [24]: [26] did extensive comparison of supervised vs. unsupervised approaches to detecting "is-a" relation.

[1] https://github.com/dminus1/meta-cats.

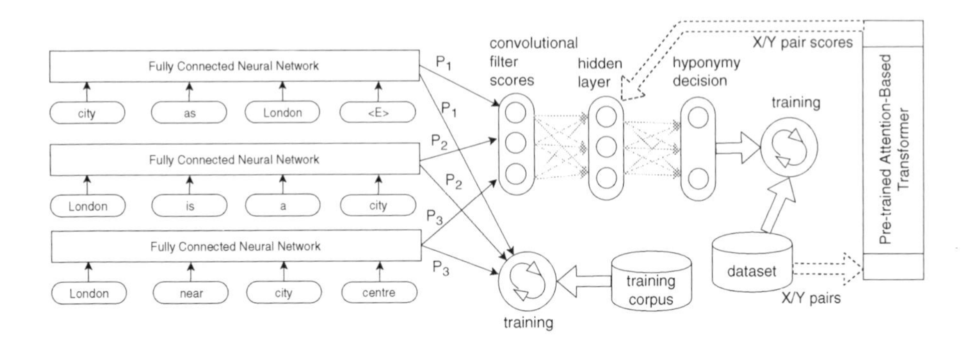

Fig. 1. Our path-based neural approach to semantic relationship classification.

[32] looked at how additional word paths can be predicted even if they are not in the corpus [19] also looked at "is-a" relation and confirmed the importance of modeling word paths in addition to purely distributional methods. Still, the results in [24,25] remain unsurpassed within the class of *word-path models.* Our baseline for distributional approaches is [30], who suggested using hyperspherical relation embeddings and improved the results of [24] on 3 out of 4 datasets.

3 Compared Models for Semantic Relations

3.1 Path-Based Neural Model

Intuitive Description. Our proposed path-based neural model combines useful properties of convolutional and recurrent networks, while resolving several shortcomings of the current state-of-the-art model [25] as we explain below. Figure 1 presents an informal intuitive illustration. We jointly train our semantic classification along with an unsupervised language modeling (LM) task.

The output of LM is the probability of occurrence of any input word sequence. *We use some of those probabilities as features for our relation classification model.* Inspired by the success of convolutional networks (CNNs), we use a fixed set of trainable filters (also called *kernels*), which learn to respond highly to certain patterns that are indicative of specific semantic relations. For example, a specific filter f_i can learn to respond highly to *is a* (and similar) patterns. At the same time, our recurrent LM may suggest that there is a high probability of occurrence of the sequence *green is a color* in raw text corpus. Combining those two facts suggests that *green* belongs to the category *color* (true *is-a* relation between them). Figure 1 shows only three convolutional filters (and the probabilities of the sequences P_1, P_2, P_3), while in our current study we used up to 16.

Thus, the LM probabilities act as approximate ("soft") pattern matching scores: (1) similar patterns receive similar scores with the same filter and (2) similar filters produce similar scores for the same pattern. LM also reduces the need for using many filters as explained by the following intuitive example: While training, LM can encounter many examples of sequences like *green is a popular color* and *green is a relaxing color.* By modeling the properties of a language, LM learns that removing an adjective in front of a noun does not normally result in a large drop of the probability of occurrence, so the sequence *green is a color* also scores highly even if it never occurs in the corpus.

Since the current state-of-the art path-based approach [25] aggregates the word paths connecting each target pair by averaging the context vectors representing all the paths, we believe their approach has two specific drawbacks that our approach does not: (1) when averaging is applied, the different occurrences of word patterns are forced to compete against each other, so the more rare occurrences can be dominated by more common ones and their impact on classification decision neglected as a result. By using LM we avoid facing the question how to aggregate the context vectors representing each path existing in the corpus. (2) The other relative strength of our approach over the baseline comes from the fact that our model does not "anonymize" the word paths unlike [25], which uniformly uses "x" and "y" for the path ends regardless of which words the target pair (x,y) actually represents. Without the use of LM, this anonymizing is unavoidable to generalize to the previously unseen (x,y) pairs, but it also misses the opportunity for the model to transfer knowledge from similar words.

Formal Definitions. Language Model (LM) is a probability distribution over sequences of words: $p(w_1, ..., w_m)$, where $w_1, ..., w_m$ is any arbitrary sequence of words in a language. We train LM jointly with our semantic relation classification task by minimizing cross-entropy costs, equally weighted for both tasks. As nowadays de-facto standard for a LM, we use a recurrent neural network (specifically a GRU variation [3], which works as well as LSTM while being faster to train). Thus, the probability of a word w_m in the language to follow a sequence of words $w_1, ..., w_{m-1}$ is determined by using the RNN to map the sequence $w_1, ..., w_{m-1}$ into its context vector:

$$\vec{v}_{w_1,...,w_{m-1}} = \text{RNN}(w_1, ..., w_{m-1}) \quad (1)$$

and then applying a linear mapping and the softmax function:

$$p(w_m|w_1, ..., w_{m-1}) = \text{softmax}\left(W \cdot \vec{v}_{w_1,...,w_{m-1}} + b\right) \quad (2)$$

where W is a trainable matrix, b is a trainable bias, and *softmax* is a standard function to scale any given vector of scores to probabilities.

As any typical neural LM, our LM also takes distributed representations of words as inputs: all the words are represented by their trainable *embedding* vectors $v_1, ..., v_m$.[2] This is important for our model and allows us to *treat LM as a function defined over arbitrary vectors* $p(v_m|v_1, ..., v_{m-1})$ *rather than over words.*

To classify semantic relations, we only look at the word paths that connect the target word pairs. Thus, we only make use of probabilities of the form

[2] We deliberately do not use the arrow over the word vectors to simplify the notation.

$p(v_y|v_x, v_1, ..., v_k)$, where (x, y) is one of the *target pairs* of words - those in the dataset that are used in training or testing the semantic relations, (v_x, v_y) are their embedding vectors. The sequence of vectors $v_1, ..., v_k$ defines a trainable *filter*, and k is its size. While vectors $v_1, ..., v_k$ have the same dimensions as the word embeddings, they are additional parameters in the model that we introduce. They are trained with the other ones (word embeddings + RNN matrices + the decision layer) by back propagation. It is possible since due to the smoothness of a neural LM, the entire model is differentiable.

Thus, we formally define the score of each of our convolutional filters (kernels) the following way:

$$f_i = \log p(v_y|v_x, v_1^i, ..., v_k^i) \tag{3}$$

where $p()$ is determined by our language model as the probability of the word with the embedding vector v_y to follow the sequence of words with the vectors $v_x, v_1^i, ..., v_k^i$. We apply *log* in order to deal with high variation in the orders of magnitude of $p()$. Finally, we define the vector of filter scores by concatenating the individual scores: $\overrightarrow{f} = [f_1, f_2, f_3, ..f_N]$, where N is the total number of filters (16 in our study here).

Filter scores $\overrightarrow{f}$ are mapped into a semantic relation classification decision by using a neural network with a single hidden layer. Thus, we define:

$$\overrightarrow{h_1} = \tanh(W_2 \cdot \overrightarrow{f} + b_2) \tag{4}$$

where W_2 is a trainable matrix and b_2 is a trainable "bias" vector. The classification decision is made based on the output activations:

$$c = \text{argmax}\,(W_3 \cdot \overrightarrow{h_1} + b_3) \tag{5}$$

where W_3 and b_3 are also trainable parameters. As traditional with neural networks, we train to minimize the cross-entropy cost:

$$cost = -\log((\text{softmax}\,(W_3 \cdot \overrightarrow{h_1} + b_3))[c_l]) \tag{6}$$

where c_l is the correct (expected) class label. We used stochastic gradient descent for cost minimization.

3.2 Distributional Model: Attention-Based Transformer

The diagram on Fig. 2 illustrates how attention-based transformer [28] operates. Instead of recurrent units with "memory gates" essential for RNN-s, attention-based transformers use additional word positional embeddings which allows them to be more flexible and parallelizable than recurrent mechanisms which have to process a sequence in a certain direction. The conversions from the inputs to the outputs are performed by several layers, which are identical in their architecture, varying only in their trained parameters. In order to obtain the vectors on the layer above, the vectors from layer immediately below are simply weighted and

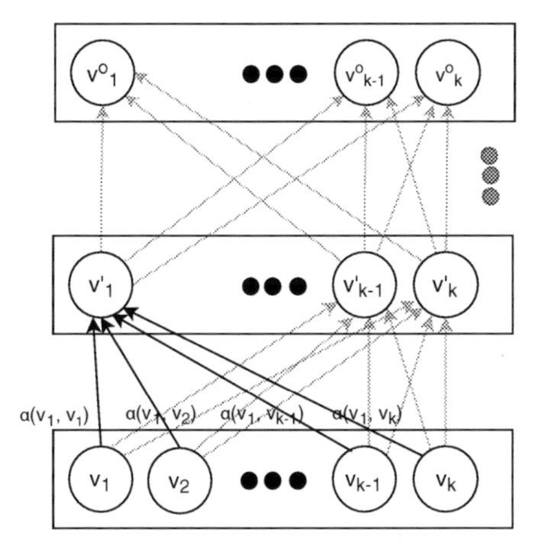

Fig. 2. Attention-based transformer used in our distributional approach to semantic relationship classification.

added together. After that, they are transformed by a standard nonlinearity function. We use *tanh*:

$$\vec{v_i}' = \tanh(W \cdot \sum_{t=1}^{k} \alpha_t \vec{v_t}) \tag{7}$$

here, $\vec{v_i}'$ is the vector in the i-th position on the upper layer, $\vec{v_t}$ is the vector in the t-th position on the lower layer, W is a trainable matrix (same regardless of i but different at different layers), and α_t is a trainable function of vectors $\vec{v_i}$ and $\vec{v_t}$, such as the weights for all $\vec{v_t}$ add up to 1. We use a scaled dot product of the vectors $\vec{v_i}$ and $\vec{v_t}$:

$$\alpha_t = \vec{v_i} \cdot W' \cdot \vec{v_t} \tag{8}$$

where W' is a trainable matrix (also same regardless of i and t at the same layer but different at different layers). The normalization to 1 is accomplished by using a *softmax* function.

This mechanism allows rich vector representations to be formed at the highest layers that can capture the entire content of a word sequence (e.g. a sentence or a word pair) so it can be effectively used for any AI applications such as text classification or generation. As it is commonly done with the transformers, we make our output classification decision based on the first vector on the top level. We do not use a hidden layer here, so we apply our formula 5 above to h_1 defined as the following:

$$\vec{h_1} = \vec{v_0^u} \tag{9}$$

where $\{\vec{v_t^u}\}$ is the vector sequence produced by the transformer for the top level.

Table 1. The relation types and statistics in each dataset.

Dataset	Dataset relations	#inst.	#uniq. X	#uniq. Y
Hypenet Lexical	is a	20335	16044	5148
Hypenet Random	is a	49475	38020	12600
K& H+N	is a, part of	57509	1551	16379
BLESS	is a, part of, event, attribute	26546	201	8089
ROOT09	is a	12762	1218	3436
EVALution	is a, part of, attribute, opposite, made of	7378	1631	1497

4 Empirical Evaluation

4.1 The Datasets

Table 1 summarizes general statistics of the datasets. We used the same datasets as our baselines: the first two are from [25] and were built using a similar methodology: the relations used in them have been primarily taken from various sources including WordNet, DBPedia, Wikidata and Yago. Thus, their x-s are primarily named entities (places, films, music albums and groups, people, companies, etc.). The important difference is that in order to create the split between training, testing and validation sets for HypeNet Lexical, the lexical separation procedure was followed [10], so that there is no overlap in words (neither x nor y) between them. This reduces "lexical memorization" effect mentioned above. The last four datasets are from [24], which originate from various preceding studies: K&H+N [15], BLESS [1], ROOT09 [20], EVALution [21]. Most of the relations for them were also taken WordNet. BLESS dataset also contains *event* and *attribute* relations, connecting a concept with a typical activity/property, e.g. *(alligator, swim)* and *(alligator, aquatic)*. EVALution dataset contains the largest number of semantic relations including antonyms, e.g. *(good, bad)*. To make our comparison more direct, we used exactly the same splits into training, development (validation) and testing subsets as in the baselines. We also used exactly the same word paths data, as it is made publicly available by the authors.

4.2 Experimental Setups

Since we sought to keep the number of hyper-parameters to the minimum, we set the word embedding size, the RNN context vector size, and the hidden layer size to be the same within all our path-based models. We tested their values in the {50,100,500,1000} set. This size is the only hyper-parameter that was varied in our experiments. We used the static learning rate of 0.01. As it is commonly done, we report the results computed on the test sets with the hyper-parameter and the number of training iterations that maximize the F_1 scores on the validation sets,

Table 2. F_1 scores of our tested models compared to the state-of-the-art baselines. Datasets: HyperNet Lexical (HL), Hypenet Random (HR), BLESS (B), ROOT09 (R), EVALution(E).

Model	HL	HR	K&H	B	R	E
Prior Shwartz et al. (2016)	0.660	0.890	0.983	0.889	0.788	0.595
Word path models:						
Shwartz et al. (2016)	0.700	0.901	0.985	0.893	0.814	0.600
Our neural model	0.740	0.899	**0.990**	0.927	0.832	0.602
Distributional:						
Wang et al. (2019)	N/A	N/A	**0.990**	0.938	0.861	0.620
Our transformer-based	**0.821**	**0.905**	0.987	**0.950**	**0.905**	**0.701**
Human	0.90	0.90	0.98	0.96	0.95	0.82

thus using exactly the same metrics and procedures as were used to obtained the baseline results: scikit-learn [16] with the "weighted" set-up, which computes the metrics for each relation, and reports their average, weighted by support (the number of true instances for each relation). For HypeNet datasets, that was accordingly set to "binary". We also verified through personal communications with the authors of [24] that our metrics are numerically identical for the same sets of predicted labels.

For our path-based models, all the trainable parameters were initialized by a normal distribution around 0 average and standard deviation of 1. We used the same transformer architecture and hyper-parameters as in [4] (BERT mono-lingual English uncased version) which has 12 layers and the output vector size of 768, resulting in the total number of trainable parameters of 110 million. As it is commonly done when using a pre-trained transformer, we initialize our weights to those that were already trained by [4] for a language model and next sentence prediction tasks on a copy of English Wikipedia text and the BookCorpus. For consistency with the data used during pre-training, we add the same special markers before, between and after our input word sequences x and y.

4.3 Comparing Against the Baselines

Table 2 presents our results. For additional comparison, we also include "Before Baseline" row, which lists the baselines used in [24,25]. For HypeNet Random and Evalution datasets, we put the larger values that we obtained in our re-implementation of the distributional methods that they used rather than their reported values. The following can be observed:

(1) Our **neural word path model** has been able to improve the state-of-the-art on three (3) out of six (6) datasets: Hypenet L, Bless and Root09. The differences are statistically significant at the level of .01. On the remaining three (3) datasets (HypeNet Random, K&H+N and Evalution), our results

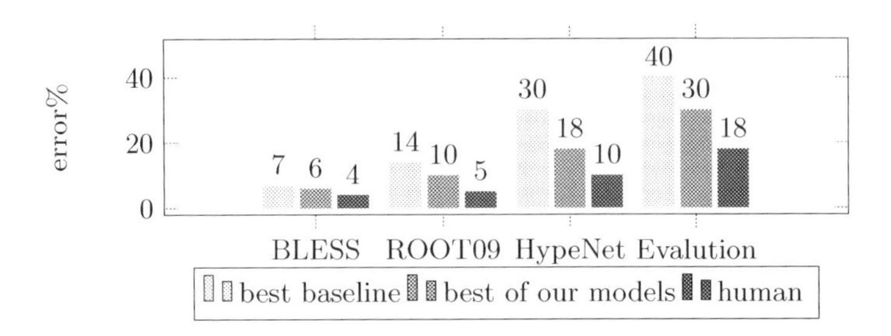

Fig. 3. Visual illustration of the error reduction relatively to the baseline and human performance on the tasks.

are the same as with the baseline performance (no statistically significant difference at the level .05). The baseline did not improve on those datasets over the prior work either. The scores for HypeNet Random and K&H+N are already high due to "lexical memorization" mentioned above. Since the compared models used exactly the same data, the obtained results clearly suggest that *our neural model is better* than the current state-of-the-art word-path model [24,25].

(2) Our **transformer-based** model has also demonstrated tangible gains over state-of-the-art baselines regardless of the class of the approach (both path-based and distributional) on four (4) out of six (6) datasets by 1–12% points (15–40% error reduction). Those differences are statistically significant at the level of .01. There are no statistically significant differences on HypeNet Random and K&H+N. This suggests that *an attention-based transformer is a very powerful mechanism for modeling semantic relations.* Although they have been shown to be very effective in many other applications where knowledge transfer between tasks is essential, this is the first study that has used them for semantic relations.

(3) On four (4) out of six (6) datasets, our **distributional model worked better** than our neural word path model. The differences are statistically significant at the level of .01. There are no statistically significant differences on the remaining two.

We estimated the human performance on our datasets by giving 100 randomly selected word pairs to 3 independent graders, who were allowed to look up the meanings online (last row). It can be seen that the state-of-the-art approaches have already achieved the human level on the datasets where no improvement was detected (HypeNet Random and K&H+N), so this may explain why our approaches did not substantially improve them any further. Fig. 3 illustrates the effect of error reduction on the four datasets on which our approaches improved the state-of-the-art. For comparison, we plot the semantic relation classification error calculated as $100 - F_1$ score rounded to the nearest integer. It can be seen that our approaches have approximately reduced the errors on those "unsolved" datasets half-way from the baseline to the human level. We believe that this result is truly remarkable!

Table 3. Average Precision (AP) scores of our tested models compared to other recent strong baselines on binary category verification ("is-a" relation only).

Model	BLESS	EVALution
Chang et al. 2018	0.186	0.364
Roller et al. 2018	0.760	0.480
Nguyen et al. 2017	0.454	0.538
Yin and Roth 2018	0.595	0.623
Our path-based	0.939	0.603
Our transformer	**0.986**	**0.739**

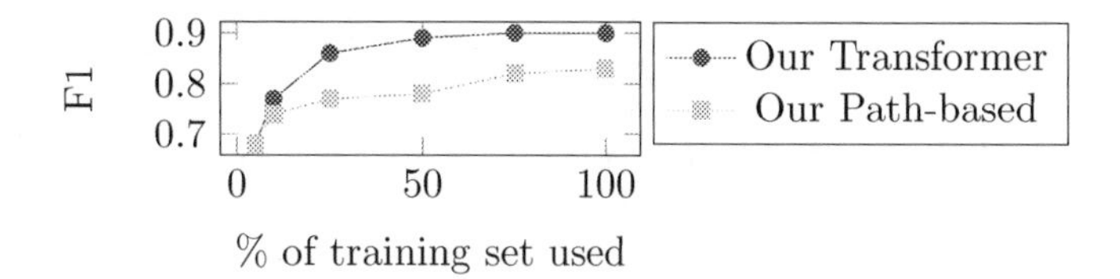

Fig. 4. Using only portion of Root dataset for training.

For additional comparison, we also include our results along with the results of other recent works that looked at semantic relations classification even though those works did not claim to exceed the state-of-the-art approaches presented in [24,25]. We report the metric of Average Precision (AP) used in those studies and the results on the two datasets (Bless and Evalution) also commonly used in them. We did not use the other datasets for comparison since they are much smaller and relying on manual part of speech tags (nouns, verbs, adjectives etc.). As Table 3 illustrates, our path-based model outperforms all but one, and our transformer-based model sizably exceeds all of those results reported.

We have also tested the influence of training size on the model by comparing its performance with 5%, 10%, 25%, 50% and 75% of randomly selected training subsets. Due to the size limit, we show only the results on Root09 (Fig. 4). The results suggest the importance of the dataset size and the possibility of further improvements when more training data is available for the path-based. At the same time, out transformer-based model needs much less training to reach its top possible performance. We also verified that all the components of our path-based model here are essential to exceed the baselines, specifically: using a hidden layer, using all the available word paths, using all 16 filters. Larger number of filters did not result in any gains, but increased the training time.

We also tried to play an adversarial role and fed more challenging pairs to the trained models to see when they are starting to fail. Our attention-based transformer model trained for HypeNet Lexical dataset (named entities mostly) erroneously classified all the 100 examples created by combining random general words and the word "air" (e.g. "car air", "circle air", "new air") as "airline."

It also erroneously classified all the 30 correct airline names that we tried as "airports" in addition to correctly classifying them as "airline." The proportion of correct airline names classified as "recording label" was 60%, which is lower than for the correct category, but still alarmingly high. Meanwhile, general words (like "car", "book", "new", etc.) are very rarely classified as members of any categories in this dataset since the model correctly sees that they are not named entities. Those observations suggest that what the transformer actually learns for this datasets is to use the combined properties of a word sequence (n-gram) to check if it can possibly be a named entity, and then if it *topically* fits the category (e.g. "aviation" in general). Those two conditions are sufficient to make a positive classification and to obtain high scores since very few test categories in the dataset are closely related (e.g. "airport" and "airline"). While our neural path models don't make such mistakes, their mistakes are primarily due to no word paths existing between the candidates in the training corpus, which was already noted in the related prior work. This suggests that a combination of those two approaches may provide additional gains over each. We have left more formal exploration of those observations for future studies.

5 Conclusions

We have considered the task of automatically recognizing semantic relations between words (phrases, concepts, etc.) such as "is a", "part of", "property of", "opposite of" etc., which is an important task affecting many applications including knowledge base construction, inference, query understanding, personalized recommendation and post-search navigation. Using six standard datasets, we have demonstrated that both distributional and word path state-of-the-art approaches can be tangibly improved. Out of those two approaches that we suggested, the transformer-based distributional approach worked significantly better. It has decreased the gap between the current strong baselines and human performance by roughly 50% for those datasets that still had room for improvement. We are not aware of any other work applying a pre-trained attention-based transformer (ABT) for this task. Since ABT-s are currently known to be the first practically useful mechanism for knowledge transfer between natural language tasks, our work paves the way to making knowledge transfer to be a default feature in any modern NLP tool.

It will also lead to integrating training of the transformer with the semantic classification task on a deeper level, which can be accomplished by customizing its pre-training (weight-initialization) algorithm to include word semantic information available from existing taxonomies, which we are planning to undertake in future, along with experimenting with cross-lingual knowledge transfer (e.g. [22]), when a model uses English data to predict semantic relations in other, less resourced, languages.

References

1. Baroni, M., Lenci, A.: How we BLESSed distributional semantic evaluation. In: Proceedings of 2011 Workshop on Geometrical Models of Natural Language Semantics (2011)
2. Bybee, J.L., Beckner, C.: Usage-based theory. In: The Oxford Handbook of Linguistic Analysis (2015)
3. Cho, K., van Merrienboer, B., Gulcehre, C., Bougares, F., Schwenk, H., Bengio, Y.: Learning phrase representations using RNN encoder-decoder for statistical machine translation. In: 55th Annual Meeting of the Association for Computational Linguistics (2014)
4. Devlin, J., Chang, M.W., Lee, K., Toutanova, K.: Bert: pre-training of deep bidirectional transformers for language understanding. arXiv preprint arXiv:1810.04805 (2018)
5. Hearst, M.A.: Automatic acquisition of hyponyms from large text corpora. In: ACL 1992 (1992)
6. Inkpen, D., Zhu, X., Ling, Z.H., Chen, Q., Wei, S.: Neural natural language inference models enhanced with external knowledge. In: ACL (2018)
7. Keller, M., Mühlschlegel, P., Hartenstein, H.: Search result presentation: supporting post-search navigation by integration of taxonomy data. In: WWW Conference (2013)
8. Lample, G., Conneau, A.: Cross-lingual language model pretraining. arXiv preprint arXiv:1901.07291 (2019)
9. LeCun, Y., Bengio, Y., Hinton, G.E.: Deep learning. Nature **521**, 7553 (2015)
10. Levy, O., Remus, S., Biemann, C., Dagan, I.: Do supervised distributional methods really learn lexical inference relations? In: Proceedings of NAACL-HLT (2015)
11. Mahdisoltani, F., Biega, J., Suchanek, F.M.: Yago3: a knowledge base from multilingual wikipedias. In: CIDR (2015)
12. Mikoajczak-Matyja, N.: The associative structure of the mental lexicon: hierarchical semantic relations in the minds of blind and sighted language users. In: Psychology of Language and Communication (19) (2015)
13. Mikolov, T., Sutskever, I., Chen, K., Corrado, G.S., Dean, J.: Distributed representations of words and phrases and their compositionality. In: NIPS 2013 (2013)
14. Nakashole, N., Weikum, G., Suchanek, F.: PATTY: a taxonomy of relational patterns with semantic types. In: 2012 Joint Conference EMNLP and CoNLL (2012)
15. Necsulescu, S., Mendes, S., Jurgens, D., Bel, N., Navigli, R.: Reading between the lines: overcoming data sparsity for accurate classification of lexical relationships. In: SEM 2015 (2015)
16. Pedregosa, F., et al.: Scikit-learn: machine learning in Python. J. Mach. Learn. Res. **12**, 2825–2830 (2011)
17. Pennington, J., Socher, R., Manning, C.: Glove: global vectors for word representation. In: EMNLP 2014 (2014)
18. Riedel, S., Yao, L., McCallum, A., Marlin, M.B.: Relation extraction with matrix factorization and universal schemas. In: NAACL-HLT 2013 (2013)
19. Roller, S., Kiela, D., Nickel, M.: Hearst patterns revisited: automatic hypernym detection from large text corpora. In: ACL 2018 (Short Paper) (2018)
20. Santus, E., Lenci, A., Chiu, T.S., Lu, Q., Huang, C.R.: Nine features in a random forest to learn taxonomical semantic relations. In: LREC 2016 (2016)

21. Santus, E., Yung, F., Lenci, A., Huang, C.R.: Evalution 1.0: an evolving semantic dataset for training and evaluation of distributional semantic models. In: Proceedings of the 4th Workshop on Linked Data in Linguistics: Resources and Applications (2015)
22. Sharoff, S.: Finding next of kin: cross-lingual embedding spaces for related languages. Nat. Lang. Eng. **26**(2), 163–182 (2019)
23. Shen, J., et al.: Yago3: a knowledge base from multilingual wikipedias. In: KDD (2018)
24. Shwartz, V., Dagan, I.: Path-based vs. distributional information in recognizing lexical semantic relations. In: COLING 2016 (2016)
25. Shwartz, V., Goldberg, Y., Dagan, I.: Improving hypernymy detection with an integrated path-based and distributional method. In: ACL 2016 (2016)
26. Shwartz, V., Santus, E., Schlechtweg, D.: Hypernyms under siege: linguistically-motivated artillery for hypernymy detection. In: EACL 2017 (2017)
27. Snow, R., Jurafsky, D., Ng, A.Y.: Learning syntactic patterns for automatic hypernym discovery. In: NIPS 2004 (2004)
28. Vaswani, A., et al.: Attention is all you need. In: NIPS 2017 (2017)
29. Vulic, I., Gerz, D., Kiela, D., Hill, F., Korhonen, A.: Hyperlex: a large-scale evaluation of graded lexical entailment. Comput. Linguist. **43**(4), 781–835 (2017)
30. Wang, C., He, X., Zhou, A.: Spherere: distinguishing lexical relations with hyperspherical relation embeddings. In: ACL 2019 (2019)
31. Wang, Z., Zhao, K., Wang, H., Meng, X., Wen, J.R.: Query understanding through knowledge-based conceptualization. In: IJCAI (2015)
32. Washio, K., Kato, T.: Filling missing paths: modeling co-occurrences of word pairs and dependency paths for recognizing lexical semantic relations. In: NAACL-HLT 2018 (2018)
33. Wu, W., Li, H., Wang, H., Zhu, K.Q.: Probase: a probabilistic taxonomy for text understanding. In: SIGMOD (2012)
34. Zhang, Y., Ahmed, A., Josifovski, V., Smola, A.J.: Taxonomy discovery for personalized recommendation. In: WSDM (2014)

Early Detection of Rumours on Twitter via Stance Transfer Learning

Lin Tian[1], Xiuzhen Zhang[1](✉), Yan Wang[2], and Huan Liu[3]

[1] RMIT University, Melbourne, Australia
{lin.tian,xiuzhen.zhang}@rmit.edu.au
[2] Macquarie University, Sydney, Australia
yan.wang@mq.edu.au
[3] Arizona State University, Tempe, USA
huan.liu@asu.edu

Abstract. Rumour detection on Twitter is an important problem. Existing studies mainly focus on high detection accuracy, which often requires large volumes of data on contents, source credibility or propagation. In this paper we focus on early detection of rumours when data for information sources or propagation is scarce. We observe that tweets attract immediate comments from the public who often express uncertain and questioning attitudes towards rumour tweets. We therefore propose to learn user attitude distribution for Twitter posts from their comments, and then combine it with content analysis for early detection of rumours. Specifically we propose convolutional neural network (CNN) CNN and BERT neural network language models to learn attitude representation for user comments without human annotation via transfer learning based on external data sources for stance classification. We further propose CNN-BiLSTM- and BERT-based deep neural models to combine attitude representation and content representation for early rumour detection. Experiments on real-world rumour datasets show that our BERT-based model can achieve effective early rumour detection and significantly outperform start-of-the-art rumour detection models.

Keywords: Twitter · Rumour detection · Stance detection · Transfer learning · CNN · BERT

1 Introduction

Nowadays, people tend to acquire more information from online social media platforms than traditional media channels. Especially Twitter allows users to freely publish short messages called "tweets" and has become a popular platform for spreading information. On the other hand, Twitter has also become an ideal place for rumor and misinformation propagation [25]. In 2013, the Associated Press (AP) Twitter account was hacked and published a tweet that two explosions rocked the White House and President was injured. The tweet led Dow Jones Industrial Average dropped 143.5 points and Standard & Poor's 500

J. M. Jose et al. (Eds.): ECIR 2020, LNCS 12035, pp. 575–588, 2020.
https://doi.org/10.1007/978-3-030-45439-5_38

Index lost more than $136 billion in a short time period after the event [6]. In this paper, rumours refer to any unconfirmed information, including misinformation, regardless of the intention of the information source.

To assess the truthfulness of rumours and combat misinformation, manual fact checking websites such as snopes.com and emergent.info heavily rely on human observers to report potential rumors and employ professional journalists to fact-check their truthfulness, which is costly and time consuming. Automatic rumour detection is thus desirable to reduce the time and human cost [11,28].

Automatic rumour detection has attracted significant research [28]. There are mainly three types of rumour detection approaches based on the type of data used. Content-based methods focus on rumour detection using the textual contents of tweets and their user comments [12,25,30]. Generally tweet contents have direct signals for misinformation and content analysis for rumour detection is desirable. Feature-based models exploit features other than tweet contents such as author profile information for rumour detection [3,9,10,13,23]. Propagation-based methods exploit patterns in tweet propagation for rumour detection [14,16,18,27]. Most existing approaches rely on large volumes of training data that are only possible when users have shown sufficient usage or tweets have been propagated for a while, and therefore are not designed for early detection.

Early detection of rumours is most desirable, as it can trigger efforts for effective mitigation of rumours and misinformation at an early stage. But early rumour detection is a challenging task due to the lack of prominent signals in propagation and user metadata within the short period after tweet publication. It is shown by previous research [30] that users post comments to tweets early and they contain questioning or enquiring phrases (e.g. "Is this true?" or "Really?") that can be exploited for early detection of rumours. But the reliance on fixed expressions implies low recall for the approach.

In this paper, we propose early rumour detection based on only tweet contents and their immediate user comments that are readily available at the early stage. Our main idea is to exploit the wisdom of the public crowd. As shown in previous studies [11,30], the crowd shows attitudes such as disagreeing and questioning toward rumours. We therefore hypothesize that attitudes of the crowd to a tweet contains signals for identifying rumour tweets. We propose to mine the user comments to predict crowd attitudes and detect rumours. But we face the challenge that there do not exist annotations of attitudes for tweet comments. We specifically address the following research questions:

- Can crowd attitudes be exploited for effective early rumour detection?
- How to learn attitude representation from tweet comments without costly human annotation?

Towards answering these research questions, we made several contributions. To address the issue of lack of attitude annotations for user comments, we propose CNN- and BERT-based deep neural models to learn attitude representation from user comments via transfer learning from resources for stance prediction [1,5,20,24,29]. We further propose CNN-BiLSTM and BERT neural models to integrate attitude representation and content representation for tweets and their comments for rumour detection. Experiments on real-world Twitter rumour datasets show that our proposed models, especially the BERT-based model, outperform state-of-the-art rumour detection models.

2 Related Work

Rumour classification and rumour verification attract significant attention from the research community in shared tasks like RumourEval [8]. According to the type of data used, rumour detection approaches can be divided into three major categories, content-based, feature-based and propagation-based.

Content-based methods focus on rumour detection based on the textual contents of posts, including the original tweets, user comments and retweets. Generally textual contents have direct signals for misinformation and deep analysis of the Twitter messages is desirable for rumour detection. Zhao et al. [30] used a set of expressions (such as "is this true?", "what?") from user comments that express questioning and enquiring as signals for rumours. Limitations from the signal expressions lead to low recall for rumour detection. In [12] a RNN model is trained to automatically learn representations from tweets for rumour detection. In [25], linguistic features of different writing styles and sensational headlines from tweets are exploited to detect misinformation.

Feature-based methods use non-textual features such as user profile data for rumour and misinformation detection [3,9,10,13,23]. In [3] user registration age and number of followers are used for credibility assessment. In [11], features such as belief identification are used for rumour detection. Other studies [10,13,23] build time series model for information propagation and integrate other social and contextual features to detect rumours. Generally the feature-based approaches can be applied only when the original tweets have attracted significant attention on the social network after some time and therefore are not adequate for early detection of rumours or misinformation.

Propagation-based methods exploit tweet propagation information [18] to build classification models such as kernel-based methods [14,27] for rumour classification. Recently a neural network model [16] is proposed, where an extended tree-structured recursive neural network (RvNN) is constructed to model information propagation. Propagation-based approaches require large amounts of metadata and intensive pre-processing to model the propagation process.

Research shows that the public respond differently to rumours than non-rumours [11,16,18,22]. However most existing research treats rumour detection and stance detection as separate tasks. In one exception [7], crowd stance is examined as a feature to classify true and false rumours. In another exception [15] a multi-task learning problem for rumour and stance detection is formulated. It is found that the proposed multi-task model is inferior to models designed specifically for rumour detection.

Stance detection [1,5,20,24,29] aims to automatically detect user attitudes towards given posts, whether the user is in favour of, against or neutral toward the target post. Some deep neural models are proposed for the task and achieve reasonable performance [1,5,29].

More generally transfer learning is widely applied to NLP tasks. As one transfer learning strategy, feature transfer can utilise the feature representation from the source to target domains in order to reduce the target task error rate. In [26], multiple shared layers are created to capture cross-domain features and domain-specific features. To minimise the feature differences between the source and target domains, Cao et al. [2] fine-tuned a shared embedding layer to automatically transfer features from the source to the target domain.

3 Problem Formulation

The task of rumour detection can be formulated as a supervised classification problem. Consider a set of n source tweets $S = \{s_1, s_2, s_3, ..., s_n\}$. Each source tweet s_i, $s_i \in S$, is associated with a label l indicating its rumour class label and a set of comments $C_i = \{c_{i1}, c_{i2}, c_{i3}, ..., c_{im}\}$. Based on the observation that users respond to rumours and non-rumours differently, comments C_i reflect the attitudes of users towards source tweet s_i; significant variation in user attitudes to s_i indicates the uncertainty from the public towards the truthfulness of s_i. Conversely unamimous attributes towards a source tweet likely indicates that truthfulness of the source tweet is clear. The problem of rumour detection for tweet s_i can thus be decomposed to two sub-problems, stance detection from user comments C_i and rumour detection for tweet s_i.

We propose to formulate the task of rumour detection as a transfer learning problem. To achieve rumour classification for a tweet message s_i where user comments C_i do not have attitude annotation, we propose to learn representation for attitudes for user comments via transfer learning based on the readily available annotated resources for stance prediction in the literature [1,5,20,24,29]. The idea is to pretrain a model on the stance data source to learn stance representation and then transfer and integrate this knowledge to the neural model for tweet and comment contents for rumour detection.

The SemEval [19] dataset with stance annotation is employed in our study but generally other stance resources can also be used. The SemEval dataset is a public Twitter dataset where each tweet is annotated with one of three stance labels "Favor", "Against" and "Neither". The tweets are about six target topics, including "Atheism","Climate Change is a Real Concern","Feminist Movement","Hillary Clinton","Donald Trump" and "Legalization of Abortion".

4 Methodology

We propose two approaches to learn vector representation for different stance classes based on the SemEval dataset and then transfer the knowledge to the model for rumour classification, as detailed next.

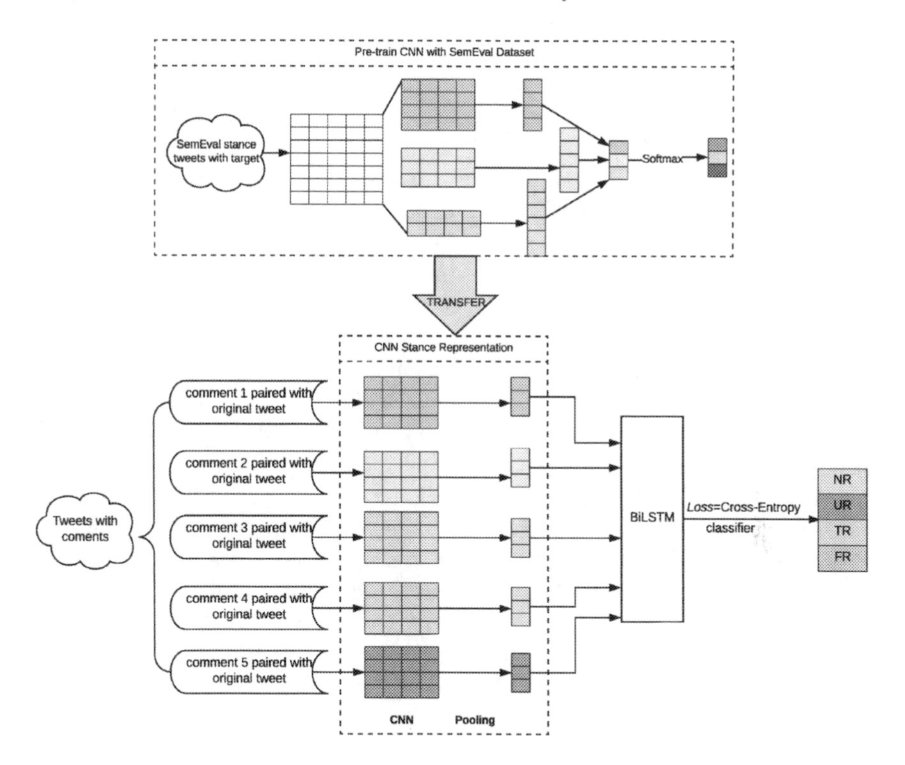

Fig. 1. Our Stance-CNN+BiLSTM model

4.1 Stance-CNN+BiLSTM

Our first model, namely Stance-CNN+BiLSTM, models crowd stances in each comment for a tweet. Specifically we pre-train a CNN model on the SemEval dataset based on the stance labels and then transfer the knowledge to learn attitude representation for each tweet comment. The CNN architecture has the ability to learn high-level feature representation for the interaction between low-level input based on annotated labels. The attitude representation for comments are then integrated into a CNN-biLSTM (bi-directional Long Short Term Memory) model for rumour prediction for tweets with comments.

The model architecture is shown in Fig. 1. The CNN model has convolutional layers and max pooling layers to capture high level features for each comment. Vectors generated from the CNN model become the input for BiLSTM for rumour detection for tweets, where the chronological order of comments and their stance variations from content representations are captured and employed to classify tweets into rumours and non-rumours. In addition, BiLSTM has the ability of ignoring unnecessary features using the delete gate. The entire model was trained to minimise the categorical cross-entropy error: $Loss = -\sum_{c=1}^{M} y_{o,c} \log(p_{o,c})$, where M is number of rumors labels, y is the binary indicator and p stands for the predicted probability.

4.2 Stance-BERT

Our second model, namely Stance-BERT, models the stance distribution for a tweet and its comments via transfer learning from tweet pairs generated from the SemEval dataset. BERT (Bidirectional Encoder Representations from Transformers) is a pre-trained transformer language model to generate deep bidirectional context representations by jointly conditioning on both left and right context in all layers [4]. The main idea of Stance-BERT is to leverage the structure of BERT to capture the complex stance distribution for tweets based on their comments, and to further integrate with a second BERT architecture modelling the language patterns for tweets and comments for rumour classification.

The architecture of our proposed stance-BERT model is shown in Fig. 2. As shown in Fig. 2, the first BERT model is to learn stance distribution for a tweet and its comments. Input are tweet pairs constructed from the SemEval dataset. If tweet A holds the "Favour" stance for topic A, and tweet B also holds the "Favour" stance for topic A, then it can be inferred that tweet A and tweet B has the same Agree stance for topic A; in other words, the new instance, the (tweet A, tweet B) pair, has the label "Favour-Favour". Similarly if tweet C has "Favour" stance for topic A and tweet D has "Against" stance for topic A, then we generate an instance (tweet C, tweet D) with the label "Favour-Against". As there are three stance labels in the original SemEval dataset, there are six combinations for labels, which are "Favour-Favour" (FF), "Against-Against" (AA) and Neither-Neither (NN), "Against-Favour" (AF), "Against-Neither" (AN), and "Favour-Neither" (FN). The six label combinations are used to label tweet pairs. Using the tweet pairs with combined stance labels as input the first BERT model is trained, which is then transferred to learn representation for (tweet, comment) pairs. This formulation of tweet pairs is aimed to capture the different language patterns of (source-tweet, comment) pair for different stance combinations.

To transfer the stance knowledge from the first BERT model for rumour prediction, the stance language patterns flow from the first BERT model to the second BERT model; the feature vector for stance representation is transferred to Twitter comments. Based on the degree of consistency among comments, the second BERT structure is trained for rumour classification.

At the first stage, the uncased BERT base model is fine-tuned with tweet pairs generated based on the SemEval dataset. The generated representation vector for [CLS] are then concatenated and input to the second BERT model to further fine-tune the BERT model for rumour classification based on the original tweet and comments. The second stage of the model has one additional output layer with softmax function for rumour classification, namely $\hat{y} = Softmax(Wh + b)$, where h is the linear vector, W and b are the weights and bias in the output layer.

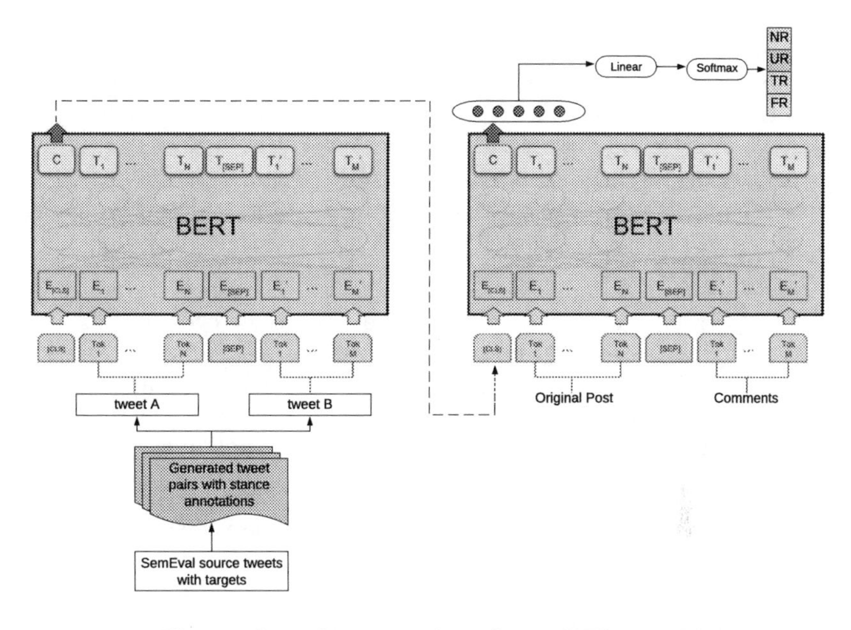

Fig. 2. The architecture of our Stance-BERT model

5 Experiments

We first describe the datasets and then the performance for early rumour detection by our models compared with other baseline models. We further evaluate our stance-transfer models against their counterparts without stance transfer.

Table 1. The Twitter15 and Twitter16 datasets

	Twitter15				Twitter16			
	NR	FR	TR	UR	NR	FR	TR	UR
#tweets	374	370	372	374	205	205	207	201
#comments	25867	21059	14948	15105	17006	7876	5397	9970
Min delay (mins)	1.08	1.50	1.48	1.96	1.02	3.45	1.76	2.25
Max delay (mins)	2714.26	1731.08	1248.85	1161.39	2690.72	2075.73	216.65	1748.02

5.1 Datasets and Experiment Setup

We use two public Twitter datasets [14], namely Twitter15 and Twitter16 (Table 1), for our experiments. In each dataset, tweets and their associated retweets and user response comments are included. Twitter15 and Twitter16 contain 1490 and 818 source tweet posts respectively. Four different rumour labels are applied with these two datasets, including True Rumour (TR), Non-Rumour

(NR), False Rumour (FR) and Unverified Rumour (UR). We removed retweets from the original datasets since retweets are not providing any new information in terms of contents. The comments and retweet contents are not included in the original dataset, only tweet ids are provided. We therefore crawled all the comments through Twitter API according to the tweets ids and user ids.

We compare our models against state-of-the-art rumour detection models:

- Stance-BERT: our BERT-based stance transfer learning models.
- Stance-CNN+LSTM: our CNN+LSTM-based stance transfer learning model.
- SVM [30]: SVM with linguistic features from tweets and comments.
- MT-ES [15]: Multi-task learning model for stance and rumour classification.
- GRU-RNN [12]: RNN model with GRU units for capturing rumour representations with sequential structure of relevant posts.
- TD-RvNN [16]: Propagation tree-based recursive neural network model.

We implemented the SVM model using scikit-learn package in Python and TD-RvNN model with Theano. The SVM model is implemented with radial basis function kernel where $C = 1.0$. All other neutral network models are based on Tensorflow v1.14. We use overall macro F_1 and F_1 scores for each class as model performance evaluation metrics. Five-fold cross-validation experiments are applied for evaluation of models.

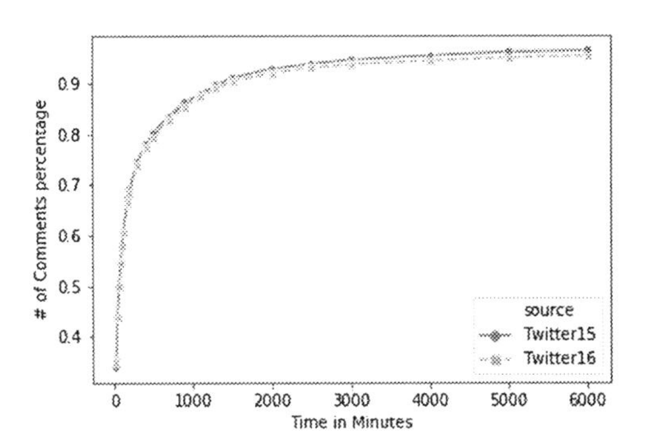

Fig. 3. Number of comments over time delay in minutes

5.2 Analysis of Early Comments for Tweets

We first evaluate the feasibility of using comments for early rumour detection. Figure 3 plots the number of comments with increasing time delay from when the original source tweet was published in the Twitter15 and Twitter16 datasets. It can be seen that over 50% comments appear within the first 60 min since the original tweet was published. Over 80% of comments appear within the first 100 min since publication of the original tweet. The number of comments plateaus at 1000 min since publication of the original source tweet. Our analysis confirms that it is feasible to use comments for early rumour detection [30]. Our default setting for early rumour detection is 60 min.

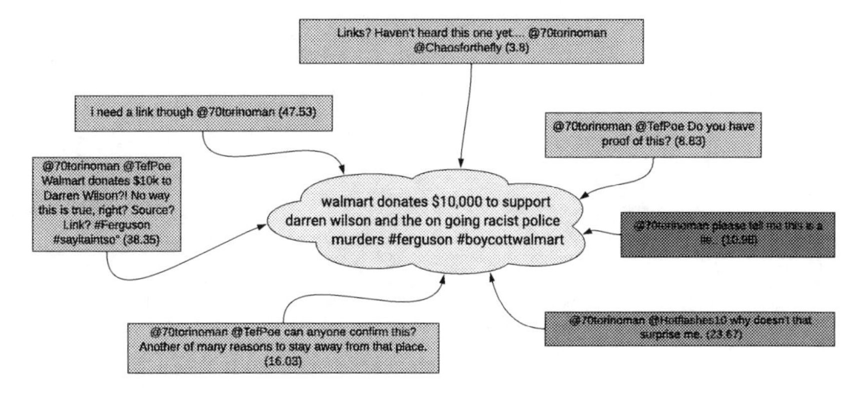

(a) An example false-rumour tweet with comments

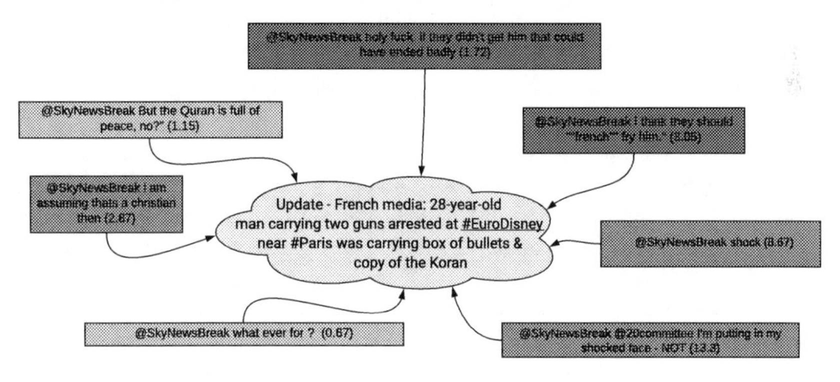

(b) An example non-rumour tweet with comments

Fig. 4. Different types of tweets and their comments. The green, blue and yellow boxes indicate the Favour, Against and Neutral user stances for comments, and numbers in brackets indicate time delay in minutes. (Color figure online)

We next analyse the user stances expressed in comments for different types of rumours in our datasets. Figure 4 shows examples of different types of tweets. Figure 4(a) shows an example false rumour (misinformation) tweet and its comments. It can be seen that most comments contain questioning phrases such as "No way this is tru, right?" and "Source?" [30]. On the other hand Fig. 4(b) shows an example non-rumour (truthful information) tweet and its comments. It can be seen that there are more presence of Favour stance in the comments. Note also that the first user comment appeared at only 0.67 min after publication of the original tweet.

5.3 Our Stance-Tranfer Models Versus Baseline Models

As shown in Table 2, our stance-based models Stance-CNN+BiLSTM and Stance-BERT yield significantly better performance than all other methods over-

Table 2. Rumour detection results (F1 score) based on the 60-min window. Bold indicates the best result for each column. Stars (*) indicate statistical significance against four baselines with Bonferroni correction under the corrected t-test [21] in 5-fold cross validation experiments.

	Twitter15					Twitter16				
	MacroF1	NR	FR	TR	UR	MacroF1	NR	FR	TR	UR
SVM [30]	0.345	0.380	0.330	0.320	0.350	0.338	0.420	0.190	0.330	0.410
MT-ES [15]	0.460	0.350	0.480	0.600	0.410	0.470	0.390	0.480	0.600	0.410
GRU-RNN [12]	0.644	0.684	0.634	0.688	0.571	0.609	0.617	0.715	0.577	0.527
TD-RvNN [16]	0.700	0.630	0.710	0.800	0.660	0.695	0.580	0.670	0.840	0.690
Stance-CNN+LSTM	0.735*	0.680	0.735	0.785	0.740	0.740*	0.690	0.680	0.780	0.810
Stance-BERT	**0.823**	**0.850**	**0.796**	**0.852**	**0.794**	**0.825***	**0.826**	**0.766**	**0.856**	**0.850**

all. Especially Stance-BERT performs consistently the best for each class. Only for the True Rumour class, it seems that stance-CNN+LSTM performs slightly worse than TD-RvNN, the propagation tree-based model. This can be explained by that for the true rumours, the stance information is harder to capture. It appears that the tree-structure neural network TD-RvNN model performs worse than our models in general. It confirms that the structural information can contribute the rumor detection to some extent, but for early detection, the average length of tree nodes can only get up to 5, and can not capture sufficient propagation signals for effective rumour detection.

It can be observed that the SVM and MT-ES models performance badly compared with other baselines. Even though the SVM model uses some expression to capture the stance information from user comments, but only 19.6% and 22.2% tweets contains these keywords. It fails due to very low recall across all classes and results in the low F_1 scores across each class. The unsatisfactory performance of MT-ES shows that the multi-task formulation of stance and rumour detection is far less effective than our transfer learning formulation for the rumour detection task.

Figure 5 plots the performance of different models in terms of the size of time windows, from 20 min to 100 min, after publication of the source tweet. It can be seen that our stance-BERT model can achieve better performance at the very early stage. The stable performance of Stance-BERT confirms the strong language signals for stance in the early user comments.

5.4 Stance-Based Models Versus Non-stance Models

To evaluate the utility of stance features for rumour detection, we compare our models Stance-CNN+BiLSTM and Stance-BERT against their non-stance variants. As shown in Table 3, Stance-CNN+biLSTM outperforms its non-stance counterpart CNN+biLSTM for the overall MacroF1, and generally outperforms CNN-LSTM for each class. Stance-BERT always outperforms its variants by big margins. Note that Stance-BERT based on tweet-comment pairs outperforms Stance-BERT (comment) based on comments. Moreover, Stance-BERT always outperforms the other non-stance models BERT(comment), BERT(tweet) and BERT(tweet-comment). These results confirm our hypothesis that the stance feature extracted from user comments data can effectively contribute to rumour detection at the early stage. Moreover our approach of modelling stance for tweet-comment pairs is especially effective.

By transfer learning using the language model BERT, it better captures the language features. In more specific terms, BERT can adjust the weights associated with the model to better represent text originating from comments. This means that during classifier fine-tuning, the starting points of the weights are closer to values that correctly model Twitter data. Closer values mean that the model has a better chance of finding good representations, even with very limited amount of training data.

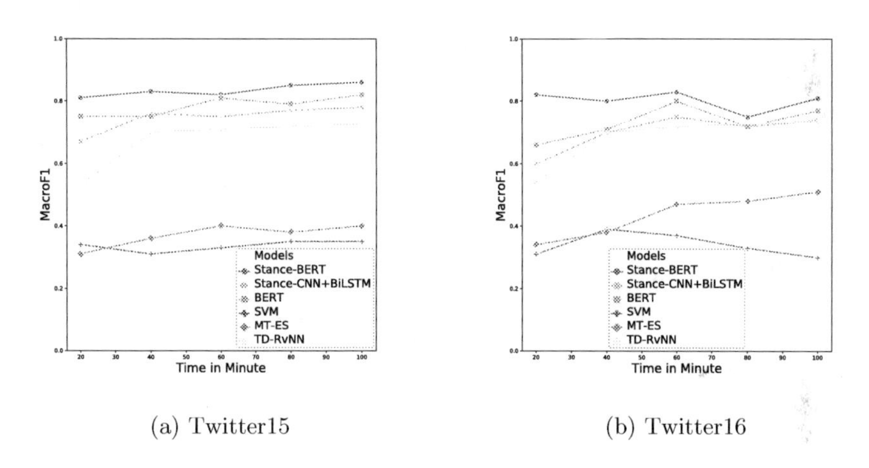

(a) Twitter15 (b) Twitter16

Fig. 5. Early rumour detection accuracy at different time windows

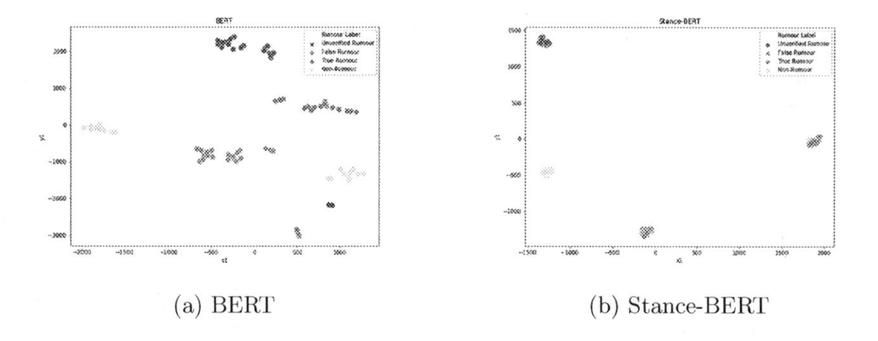

(a) BERT (b) Stance-BERT

Fig. 6. t-NSE of [CLS] hidden state

To evaluate the utility of stance transfer, we randomly selected 80 samples within the 40-min window from the Twitter15 dataset and use t-SNE [17] to visualize the embeddings of [CLS] for BERT (without stance transfer) and Stance-BERT, which shows the hidden state for sequence embedding. As shown in Fig. 6, Stance-BERT clearly performs better than BERT by grouping the same type of rumours into clusters. It confirms that transferred stance knowledge work effectively with rumour data. In addition, the clear boundaries among different types of rumours shows that strong stance signals exist in the user comments, which confirms our hypothesis that stance can help directly on rumour detection at the early stage.

Table 3. Results (F1 score) for comparing stance models against non-stance models. Best results for each column are in bold. Stars (*) indicate statistical significance with Bonferroni correction under corrected t-test [21] in five-fold cross validation experiments.

	Twitter15					Twitter16				
	MacroF1	NR	FR	TR	UR	MacroF1	NR	FR	TR	UR
Stance-CNN+BiLSTM	0.735*	0.735	0.680	0.735	0.785	0.740*	0.690	0.680	0.780	0.810
CNN-LSTM	0.682	0.590	0.794	0.644	0.700	0.664	0.560	0.602	0.708	0.784
Stance-BERT	**0.823***	**0.850**	**0.796**	0.852	0.794	**0.825***	0.826	**0.766**	0.856	**0.850**
Stance-BERT(comment)	0.747	0.712	0.747	0.810	0.717	0.677	0.683	0.580	0.767	0.677
BERT(comment)	0.708	0.744	0.670	0.676	0.740	0.660	0.728	0.456	0.740	0.722
BERT(tweet)	0.762	0.784	0.710	0.824	0.730	0.781	0.802	0.656	**0.862**	0.804
BERT(tweet-comment)	0.814	0.836	0.774	**0.858**	0.786	0.797	**0.828**	0.718	0.846	0.796

6 Conclusion

We proposed stance transfer learning models based on user comments for early detection of rumours on Twitter. To address the lack of stance annotation for user comments on Twitter, we proposed to design deep CNN model and fine-fune BERT model to learn stance representation for user comments via transfer learning from public resources. We further propose CNN-BiLSTM and BERT-based models to integrate stance representation into the representation for tweets for rumour detection. Experiments on two public Twitter datasets showed that user comments contain early signals for detection rumour tweets. Especially our model based on BERT achieves consistently good performance for early rumour detection and significantly outperforms state-of-the-art baselines. For future work, we will investigate making use of non-content information to further improve the performance of early rumour detection.

Acknowledgement. This research is supported in part by the Australian Research Council Discovery Project DP200101441.

References

1. Augenstein, I., Rocktäschel, T., Vlachos, A., Bontcheva, K.: Stance detection with bidirectional conditional encoding (2016). arXiv preprint arXiv:1606.05464
2. Cao, Z., Li, W., Li, S., Wei, F.: Improving multi-document summarization via text classification. In: Thirty-First AAAI Conference on Artificial Intelligence (2017)
3. Castillo, C., Mendoza, M., Poblete, B.: Information credibility on Twitter. In: Proceedings of the 20th International Conference on World Wide Web, pp. 675–684. ACM (2011)
4. Devlin, J., Chang, M.W., Lee, K., Toutanova, K.: Bert: Pre-training of deep bidirectional transformers for language understanding (2018). arXiv preprint arXiv:1810.04805
5. Dey, K., Shrivastava, R., Kaushik, S.: Topical stance detection for twitter: a two-phase LSTM model using attention. In: Pasi, G., Piwowarski, B., Azzopardi, L., Hanbury, A. (eds.) ECIR 2018. LNCS, vol. 10772, pp. 529–536. Springer, Cham (2018). https://doi.org/10.1007/978-3-319-76941-7_40
6. Domm, P.: False rumor of explosion at white house causes stocks to briefly plunge; Ap confirms its twitter feed was hacked. CNBC. COM, vol. 23 (2013)
7. Dungs, S., Aker, A., Fuhr, N., Bontcheva, K.: Can rumour stance alone predict veracity? In: Proceedings of the 27th International Conference on Computational Linguistics, pp. 3360–3370 (2018)
8. Gorrell, G., et al.: SemEval-2019 task 7: RumourEval, determining rumour veracity and support for rumours. In: Proceedings of the 13th International Workshop on Semantic Evaluation. pp. 845–854. Association for Computational Linguistics, Minneapolis, Minnesota, USA, June 2019. https://doi.org/10.18653/v1/S19-2147https://www.aclweb.org/anthology/S19-2147
9. Gupta, A., Kumaraguru, P., Castillo, C., Meier, P.: TweetCred: real-time credibility assessment of content on twitter. In: Aiello, L.M., McFarland, D. (eds.) SocInfo 2014. LNCS, vol. 8851, pp. 228–243. Springer, Cham (2014). https://doi.org/10.1007/978-3-319-13734-6_16
10. Kwon, S., Cha, M., Jung, K., Chen, W., Wang, Y.: Prominent features of rumor propagation in online social media. In: 2013 IEEE 13th International Conference on Data Mining, pp. 1103–1108. IEEE (2013)
11. Liu, X., Nourbakhsh, A., Li, Q., Fang, R., Shah, S.: Real-time rumor debunking on twitter. In: Proceedings of the 24th ACM International on Conference on Information and Knowledge Management, pp. 1867–1870. ACM (2015)
12. Ma, J., et al.: Detecting rumors from microblogs with recurrent neural networks. In: Ijcai, pp. 3818–3824 (2016)
13. Ma, J., Gao, W., Wei, Z., Lu, Y., Wong, K.F.: Detect rumors using time series of social context information on microblogging websites. In: Proceedings of the 24th ACM International on Conference on Information and Knowledge Management, pp. 1751–1754. ACM (2015)
14. Ma, J., Gao, W., Wong, K.F.: Detect rumors in microblog posts using propagation structure via kernel learning. In: Proceedings of the 55th Annual Meeting of the Association for Computational Linguistics (Volume 1: Long Papers), pp. 708–717 (2017)
15. Ma, J., Gao, W., Wong, K.F.: Detect rumor and stance jointly by neural multi-task learning. In: Companion of the the Web Conference 2018 on the Web Conference 2018, International World Wide Web Conferences Steering Committee, pp. 585–593 (2018)

16. Ma, J., Gao, W., Wong, K.F.: Rumor detection on Twitter with tree-structured recursive neural networks. In: Proceedings of the 56th Annual Meeting of the Association for Computational Linguistics (Volume 1: Long Papers), vol. 1, pp. 1980–1989 (2018)
17. Maaten, L.V.D., Hinton, G.: Visualizing data using t-SNE. J. Mach. Learn. Res. **9**(Nov), 2579–2605 (2008)
18. Mendoza, M., Poblete, B., Castillo, C.: Twitter under crisis: can we trust what we RT? In: Proceedings of the First Workshop on Social Media Analytics, pp. 71–79. ACM (2010)
19. Mohammad, S., Kiritchenko, S., Sobhani, P., Zhu, X., Cherry, C.: Semeval-2016 task 6: detecting stance in tweets. In: Proceedings of the 10th International Workshop on Semantic Evaluation (SemEval-2016), pp. 31–41 (2016)
20. Mohtarami, M., Baly, R., Glass, J., Nakov, P., Màrquez, L., Moschitti, A.: Automatic stance detection using end-to-end memory networks (2018). arXiv preprint arXiv:1804.07581
21. Nadeau, C., Bengio, Y.: Inference for the generalization error. In: Advances in Neural Information Processing Systems, pp. 307–313 (2000)
22. Qazvinian, V., Rosengren, E., Radev, D.R., Mei, Q.: Rumor has it: identifying misinformation in microblogs. In: Proceedings of the Conference on Empirical Methods in Natural Language Processing, pp. 1589–1599. Association for Computational Linguistics (2011)
23. Rath, B., Gao, W., Ma, J., Srivastava, J.: From retweet to believability: utilizing trust to identify rumor spreaders on twitter. In: Proceedings of the 2017 IEEE/ACM International Conference on Advances in Social Networks Analysis and Mining 2017, pp. 179–186. ACM (2017)
24. Riedel, B., Augenstein, I., Spithourakis, G.P., Riedel, S.: A simple but tough-to-beat baseline for the fake news challenge stance detection task (2017). arXiv preprint arXiv:1707.03264
25. Shu, K., Sliva, A., Wang, S., Tang, J., Liu, H.: Fake news detection on social media: a data mining perspective. ACM SIGKDD Explor. Newslett. **19**(1), 22–36 (2017)
26. Shu, X., Qi, G.J., Tang, J., Wang, J.: Weakly-shared deep transfer networks for heterogeneous-domain knowledge propagation. In: Proceedings of the 23rd ACM International Conference on Multimedia, pp. 35–44. ACM (2015)
27. Wu, K., Yang, S., Zhu, K.Q.: False rumors detection on Sina Weibo by propagation structures. In: 2015 IEEE 31st International Conference on Data Engineering, pp. 651–662. IEEE (2015)
28. Yang, F., Liu, Y., Yu, X., Yang, M.: Automatic detection of rumor on Sina Weibo. In: Proceedings of the ACM SIGKDD Workshop on Mining Data Semantics, p. 13. ACM (2012)
29. Zarrella, G., Marsh, A.: Mitre at semeval-2016 task 6: Transfer learning for stance detection (2016). arXiv preprint arXiv:1606.03784
30. Zhao, Z., Resnick, P., Mei, Q.: Enquiring minds: early detection of rumors in social media from enquiry posts. In: Proceedings of the 24th International Conference on World Wide Web, International World Wide Web Conferences Steering Committee, pp. 1395–1405 (2015)

Learning to Rank Images with Cross-Modal Graph Convolutions

Thibault Formal[1], Stéphane Clinchant[1(✉)], Jean-Michel Renders[1], Sooyeol Lee[2], and Geun Hee Cho[2]

[1] Naver Labs Europe, Meylan, France
{thibault.formal,stephane.clinchant,jean-michel.renders}@naverlabs.com
[2] Naver Corporation - Search Intelligence Team - Image & Video, Seongnam-si, South Korea
{sooyeol.lee,geunhee.cho}@navercorp.com

Abstract. We are interested in the problem of cross-modal retrieval for web image search, where the goal is to retrieve images relevant to a text query. While most of the current approaches for cross-modal retrieval revolve around learning how to represent text and images in a shared latent space, we take a different direction: we propose to generalize the cross-modal relevance feedback mechanism, a simple yet effective unsupervised method, that relies on standard information retrieval heuristics and the choice of a few hyper-parameters. We show that we can cast it as a supervised representation learning problem on graphs, using graph convolutions operating jointly over text and image features, namely cross-modal graph convolutions. The proposed architecture directly learns how to combine image and text features for the ranking task, while taking into account the context given by all the other elements in the set of images to be (re-)ranked. We validate our approach on two datasets: a public dataset from a `MediaEval` challenge, and a small sample of proprietary image search query logs, referred as `WebQ`. Our experiments demonstrate that our model improves over standard baselines.

Keywords: Cross-modal retrieval · Learning to rank · Graph convolutions

1 Introduction

This paper considers the typical image search scenario, where a user enters a text query, and the system returns a set of ranked images. More specifically, we are interested in re-ranking a subset of candidate images retrieved from the whole image collection by an efficient base ranker, following standard multi-stage ranking architectures in search engines [36]. Directly including visual features in the ranking process is actually not straightforward due to the semantic gap between text and images: this is why the problem has initially been addressed using standard text-based retrieval, relying for instance on text crawled from the image's

J. M. Jose et al. (Eds.): ECIR 2020, LNCS 12035, pp. 589–604, 2020.
https://doi.org/10.1007/978-3-030-45439-5_39

webpage (e.g. surrounding text, title of the page etc.). In order to exploit visual information, and therefore improve the quality of the results –especially because this text is generally noisy, and hardly describes the image semantic–, many techniques have been developed since. For instance, some works have focused on building similarity measures by fusing mono-modal similarities, using either simple combination rules, or more complex propagation mechanisms in similarity graphs. More recently, techniques have emerged from the computer vision community, where text and images are embedded in the same latent space (a.k.a. joint embedding), allowing to directly match text queries to images. The latter are currently considered as state-of-the-art techniques for the cross-modal retrieval task. However, they are generally evaluated on artificial retrieval scenarios (e.g. on MSCOCO dataset [34]), and rarely considered in a re-ranking scenario, where mechanisms like pseudo-relevance feedback (PRF) [31] are highly effective.

We propose to revisit the problem of cross-modal retrieval in the context of re-ranking. Our first contribution is to derive a general formulation of a differentiable architecture, drawing inspiration from cross-modal retrieval, learning to rank, neural information retrieval and graph neural networks. Compared to joint embedding approaches, we tackle the problem in a different view: instead of learning new (joint) embeddings, we focus on designing a model that *learns to combine* information from different modalities. Finally, we validate our approach on two datasets, using simple instances of our general formulation, and show that the approach is not only able to reproduce PRF, but actually outperform it.

2 Related Work

Cross-Modal Retrieval. In the literature, two main lines of work can be distinguished regarding cross-modal retrieval: the first one focuses on designing effective cross-modal similarity measures (e.g. [2,10]), while the second seeks to learn how to map images and text into a shared latent space (e.g. [15,18,19,54]).

The first set of approaches simply combines different mono-media similarity signals, relying either on simple aggregation rules, or on unsupervised cross-modal PRF mechanisms, that depend on the choice of a few but critical hyper-parameters [2,10,11,45]. As it will be discussed in the next section, the latter can be formulated as a two-step PRF propagation process in a graph, where nodes represent multi-modal objects and edges encode their visual similarities. It has been later extended to more general propagation processes based on random walks [28].

Alternatively, joint embedding techniques aim at learning a mapping between textual and visual representations [15,18,19,23,52–55,61]. Canonical Correlation Analysis (CCA) [17] and its deep variants [5,27,58], as well as bi-directional ranking losses [8,9,52,53,55,61] (or triplet losses) ensure that, in the new latent space, an image and its corresponding text are correlated or close enough w.r.t. to the other images and pieces of text in the training collection. Other objective functions utilize metric learning losses [35], machine translation-based measures [44] or even adversarial losses [51].

These approaches suffer from several limitations [61]: they are sensitive to the triplet sampling strategy as well as the choice of appropriate margins in the ranking losses. Moreover, constituting a training set that ensures good learning and generalization is not an easy task: the text associated to an image should describe its visual content (e.g. *"a man speaking in front of a camera in a park"*), and nothing else (e.g. *"the President of the US, the 10th of March"*, *"John Doe"*, *"joy and happiness"*).

Building a universal training collection of paired (image, text) instances, where text describes faithfully the content of the image in terms of elementary objects and their relationships, would be too expensive and time-consuming in practice. Consequently, image search engines rely on such pairs crawled from the Web, where the link between image and text (e.g. image caption, surrounding sentences etc.) is tenuous and noisy.

To circumvent this problem, query logs could be used but, unfortunately –and this is our second argument regarding the limitations–, real queries are never expressed in the same way as the ones considered when evaluating joint embedding methods (e.g. artificial retrieval setting on MSCOCO [34] or Flickr-30K [43] datasets, where the query is the full canonical textual description of the image). In practice, queries are characterised by very large intent gaps: they do not really describe the content of the image but, most of the time, contain only a few words, and are far from expressing the true visual needs. What does it mean to impose close representations for all images representing *"Paris"* (e.g. *"the Eiffel Tower"*, *"Louvre Museum"*), even if they can be associated to the same textual unit?

Neural Information Retrieval. Neural networks, such as RankNet and LambdaRank, have been intensively used in IR to address the learning to rank task [7]. More recently, there has been a growing interest in designing effective IR models with neural models [1,12,13,20,25,26,37,38,41,56], by learning the features useful for the ranking task directly from text.

While standard strategies focus on learning a global ranking function that considers each query-document pair in isolation, they tend to ignore the difference in distribution in the feature space for different queries [4]. Hence, some recent works have been focusing on designing models that exploit the context induced by the re-ranking paradigm, either by explicitly designing differentiable PRF models [32,40], or by encoding the ranking context –the set of elements to re-rank–, using either RNNs [4] or attention mechanisms [42,62]. Consequently, the score for a document takes into account all the other documents in the candidate list. Because of their resemblance with structured problems, this type of approaches could benefit from the recent body of work around graph neural networks, which operate on graphs by learning how to propagate information to neighboring nodes.

Graph Neural Networks. Graph Neural Networks (GNNs) are extensions of neural networks that deal with structured data encoded as a graph. Recently,

Graph Convolutional Networks (GCNs) [30] have been proposed for semi-supervised classification of nodes in a graph. Each layer of a GCN can generally be decomposed as: (i) node features are first transformed (e.g. linear mapping), (ii) node features are convolved, meaning that for each node, a differentiable, permutation-invariant operation (e.g. sum, mean, or max) of its neighbouring node features is computed, before applying some non-linearity, (iii) finally, we obtain a new representation for each node in the graph, which is then fed to the next layer. Many extensions of GCNs have been proposed (e.g. GraphSAGE [21], Graph Attention Network [50], Graph Isomorphism Network [57]), some of them directly tackling the recommendation task (e.g. PinSAGE [59]). But to the best of our knowledge, there is no prior work on using graph convolutions for the (re-)ranking task.

3 Learning to Rank Images

3.1 Cross-Modal Similarity Measure

Our goal is to extend and generalize simple yet effective unsupervised approaches which have been proposed for the task [2,3,10,11,45], that can be seen as an extension of pseudo-relevance feedback methods for multi-modal objects. Let $d \in D$ denote a document to re-rank, composed of text and image. We denote by $s_V(.,.)$ a normalized similarity measure between two images, and by $s_T(q,d)$ the textual relevance score of document d w.r.t. query q. The cross-modal similarity score is given by:

$$\forall d \in D, s_{CM}(q,d) = \sum_{d_i \in NN_T^K(q)} s_T(q,d_i) s_V(d_i,d) \tag{1}$$

where $NN_T^K(q)$ denotes the set of K most relevant documents w.r.t. q, based on text, i.e. on $s_T(q,.)$. The model can be understood very simply: similarly to PRF methods in standard information retrieval, the goal is to boost images that are visually similar to *top* images (from a text point of view), i.e. images that are likely to be relevant to the query but were initially badly ranked (which is likely to happen in the web scenario, where text is crawled from source page and can be very noisy).

Despite showing good empirical results, cross-modal similarities are fully unsupervised, and lack some dynamic behaviour, like being able to adapt to different queries. Moreover, they rely on a single relevance score $s_T(q,.)$, while it could actually be beneficial to learn how to use a larger set of features such as the ones employed in learning to rank models.

3.2 Cross-Modal Graph Convolution

In [3], the authors made a parallel between the cross-modal similarity from Eq. (1) and random walks in graphs: it can be seen as a kind of multimodal label propagation in a graph. This motivates us to tackle the task using graph convolutions. We therefore represent each query $q \in \mathcal{Q}$ as a graph $\mathcal{G}_q$, as follows:

- The set of nodes is the set of candidate documents d_i to be re-ranked for this query: typically from a few to hundreds of documents, depending on the query.
- Each node i is described by a set of n learning to rank features $x_{q,d_i} \in \mathbb{R}^n$.
- $v_i \in \mathbb{R}^d$ denotes the (normalized) visual embedding for document d_i.
- As we do not have an explicit graph structure, we consider edges given by a k–nearest neighbor graph, based on a similarity between the embeddings v_i[1].
- We denote by $\mathcal{N}_i$ the neighborhood of node i, i.e. the set of nodes j such that there exists an edge from j to i.
- We consider edge weights, given by a similarity function between the visual features of its two extremity nodes $f_{ij} = \boldsymbol{g}(v_i, v_j)$.

Our goal is to learn how to propagate features in the above graph. Generalizing convolution operations to graphs can generally be expressed as a message passing scheme [16]:

$$h_i^{(l+1)} = \gamma(h_i^{(l)}, \sum_{j \in \mathcal{N}_i} \boldsymbol{\phi}(h_i^{(l)}, h_j^{(l)}, f_{ij})) \tag{2}$$

where $\boldsymbol{\gamma}$ and $\boldsymbol{\phi}$ denote differentiable functions, e.g. MLPs (Multi Layer Perceptron). By choosing $\boldsymbol{\phi}(h_i^{(l)}, h_j^{(l)}, f_{ij}) = \boldsymbol{\tau}(h_i^{(l)})\boldsymbol{g}(v_i, v_j)$ and $\boldsymbol{\gamma}(x, y) = y$, Eq. (2) reduces to:

$$h_i^{(l+1)} = \sum_{j \in \mathcal{N}_i} \boldsymbol{\tau}(h_j^{(l)})\boldsymbol{g}(v_i, v_j) \tag{3}$$

This graph convolution can be reduced to the cross-modal similarity in Eq. (1). Indeed, assuming that $h_i := s_T(q, d_i)$, $\boldsymbol{\tau}(.)$ is a top-k filtering function, $\boldsymbol{g}(v_i, v_j) := s_V(d_i, d)$, and $\mathcal{N}_i := \mathcal{N}$ is the whole set of candidates to re-rank, then:

$$\sum_{j \in \mathcal{N}} \tau_k(s_T(q, d_i))\boldsymbol{g}(v_i, v_j) = \sum_{d_i \in NN_T^K(q)} s_T(q, d_i) s_V(d_i, d) \tag{4}$$

In other words, one layer defined with Eq. (3) includes the standard cross-modal relevance feedback as a special case. Equation (3) is more general, and can easily be used as a building block in a differentiable ranking architecture.

3.3 Learning to Rank with Cross-Modal Graph Convolutions

In the following, we derive a simple convolution layer from Eq. (3), and we introduce the complete architecture –called DCMM for Differentiable Cross-Modal

[1] With the special case of considering that all the nodes are connected to each other, i.e. $k = |\mathcal{G}_q| - 1$.

Model–, summarized in Fig. 1. Learning to rank features x_{q,d_i} are first encoded with an MLP(.;$\boldsymbol{\theta}$) with ReLU activations, in order to obtain node features h_i^0. Then, the network splits into two branches:

- The first branch simply projects linearly each $h_i^{(0)}$ to a real-valued score $s_T(q,d_i) = \boldsymbol{w_0}^T h_i^{(0)}$, that acts as a pure text-based score[2].
- The second branch is built upon one or several layer(s) of cross-modal convolution, simply defined as:

$$h_i^{(l+1)} = \text{ReLU}(\sum_{j \in \mathcal{N}_i} \boldsymbol{W}^{(l)} h_j^{(l)} \boldsymbol{g}(v_i, v_j)) \tag{5}$$

For the edge function $\boldsymbol{g}$, we consider two cases: the cosine similarity $g_{cos}(v_i,v_j) = \cos(v_i,v_j)$, defining the first model (referred as **DCMM-cos**), and a simple learned similarity measure parametrized by a vector $\boldsymbol{a}$ such that $\boldsymbol{g}_{edge}(v_i,v_j) = v_i^T diag(\boldsymbol{a}) v_j$, defining our second model (referred as **DCMM-edge**).

After the convolution(s), the final embedding for each node $h_i^{(L)}$ is projected to a real-valued score $s_{conv}(q,d_i)$, using either a linear layer ($s_{conv}(q,d_i) = \boldsymbol{w_L}^T h_i^{(L)}$) or an MLP ($s_{conv}(q,d_i) = MLP(h_i^{(L)}, \boldsymbol{\omega})$). Finally, the two scores are combined to obtain the final ranking score:

$$s(q,d_i) = \boldsymbol{w_0}^T h_i^{(0)} + s_{conv}(q,d_i)$$

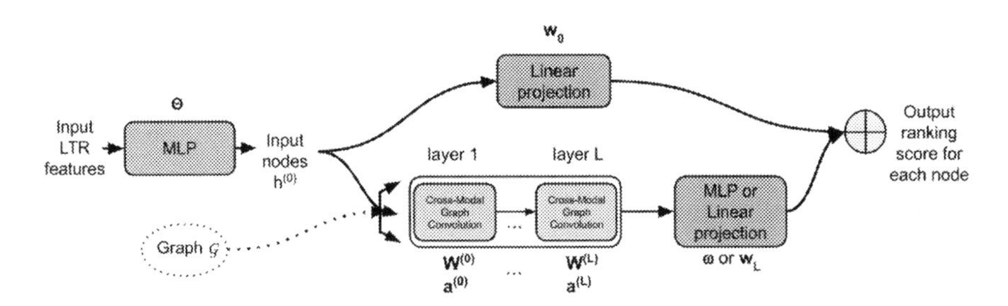

Fig. 1. Differentiable Cross-Modal Model architecture - schematic view. Note that the upper part is a standard MLP-based LTR model.

The model is trained using backpropagation and any standard learning to rank loss: pointwise, pairwise or listwise. It is worth to remark that, by extending PRF mechanisms for cross-modal re-ranking, our model is actually closer to listwise context-based models introduced in Sect. 2 than current state-of-the-art

[2] In addition to improve the results, keeping a separate branch focusing on learning to rank images solely from input nodes (i.e. learning to rank features) actually stabilizes the training, thanks to the shared input transformation.

cross-modal retrieval models. It is listwise by design[3]: an example in a batch is not a single image in isolation, but all the candidate images for a given query, encoded as a graph, that we aim to re-rank together in a one shot manner. In our experiments, we used the pairwise BPR loss [46], from which we obtained the best results[4]. Let's consider a graph (i.e. the set of candidate documents for query q) in the batch, and all the feasible pairs of documents $D_q^{+,-}$ for this query (by feasible, we mean all the pairs that can be made from positive and negative examples in the graph). Then the loss is defined:

$$\mathcal{L}(\boldsymbol{\theta}, \boldsymbol{w_0}, \gamma_{\text{conv}}, \boldsymbol{\omega}) = - \sum_{d^+, d^- \in D_q^{+,-}} \log \sigma(s(q, d^+) - s(q, d^-))$$

Note that contrary to previous works on listwise context modeling, we consider a *set* of objects to re-rank, and not a sequence (for instance in [4], a RNN encoder is learned for re-ranking). In other words, we discard the rank information of the first ranker into the re-ranking process: we claim that the role of the first retriever is to be recall-oriented, and not precision-oriented. Thus, using initial order might be a too strong prior, and add noise information. Moreover, in the case of implicit feedback (clicks used as weak relevance signals), using rank information raises the issue of biased learning to rank (sensitivity to position and trust biases). It is also worth to emphasize that, contrary to most of the works around graph convolution models, our graph structure is somehow implicit: while edges between nodes generally indicate a certain relationship between nodes (for instance, connection between two users in a social network), in our case a connection represents the visual similarity between two nodes.

4 Experiments

In the following, we introduce the two datasets we used to validate our approach –a public dataset from a `MediaEval`[5] challenge, and an annotated set of queries sampled from image search logs of Naver, the biggest commercial search engine in Korea–, as well as our experimental strategy. We emphasize on the fact that we restrict ourselves to two relatively small datasets and few features as input for the models. Even though the formulation from Eq. (3) is very general, our claim is that a simple model, i.e. containing few hundreds to thousands parameters, should be able to reproduce PRF mechanisms introduced in Sect. 3. When adapting the approach to larger datasets, the model capacity can be adjusted accordingly, in order to capture more complex relevance patterns. Note that

[3] Note the difference between a model that has been trained using a listwise loss function but uses a pointwise scoring function (i.e. the score depends only on the document itself), and a model that directly uses a listwise scoring function.

[4] While it seems appealing to directly use listwise losses, we actually obtained slightly worse performance when experimenting with the AttRank loss from [4] and the ApproxNDCG loss from [6].

[5] http://www.multimediaeval.org/mediaeval2019/.

we did not consider in our study standard datasets generally used to train joint embeddings such as MSCOCO [34] or Flickr30k [43], because the retrieval scenario is rather artificial, compared to web search: there are no explicit queries, and a text is only relevant to a single image. Furthermore, we have tried to obtain the Clickture [24] dataset without success[6], and therefore cannot report on it.

4.1 Datasets

MediaEval. We first conduct experiments on the dataset from the "MediaEval17, Retrieving Diverse Social Images Task" challenge[7]. While this challenge also had a focus on diversity aspects, we solely consider the standard relevance ranking task. The dataset is composed of a ranked list of images (up to 300) for each query, retrieved from Flickr using its default ranking algorithm. The queries are general-purpose queries (e.g. q = *autumn color*), and each image has been annotated by expert annotators (binary label, i.e. relevant or not). The goal is to refine the results from the base ranking. The training set contains 110 queries for 33340 images, while the test set contains 84 queries for 24986 images.

While we could consider any number of learning to rank features as input for our model, we choose to restrict ourselves to a very narrow set of weak relevance signals, in order to remain comparable to its unsupervised counterpart, and ensure that the gain does not come from the addition of richer features. Hence, we solely rely on four relevance scores, namely tf-idf, BM25, Dirichlet smoothed LM [60] and DESM score [39], between the query and each image's text component (the concatenation of the image title and tags). We use an Inception-ResNet model [48] pre-trained on ImageNet to get the image embeddings ($d = 1536$).

WebQ. In order to validate our approach on a real world dataset, we sample a set of 1000 queries[8] from the image search logs of Naver. All images appearing in the top-50 candidates for these queries within a period of time of two weeks have been labeled by three annotators in terms of relevance to the query (binary label). Because of different query characteristics (in terms of frequency, difficulty etc.), and given the fact that new images are continuously added to/removed from the index, the number of images per query in our sample is variable (from around ten to few hundreds). Note that, while we actually have access to a much larger amount of click logs, we choose to restrict the experiments to this small sample in order keep the evaluations simple. Our goal here is to show that we are able to learn and reproduce some PRF mechanisms, without relying on large amount of data. Moreover, in this setting, it is easier to understand model's behaviour, as we avoid to deal with click noise and position bias. After removing queries without relevant images (according to majority voting among the three annotators), our sample includes 952 queries, and 43064 images, indexed through

[6] As of today, the data link seems broken and we got no response from the person in charge of this dataset.

[7] http://www.multimediaeval.org/mediaeval2017/diverseimages/.

[8] Our sample includes head, torso and tail queries.

various text fields (title of the page, image caption etc.). We select seven of such fields, that might contain relevant pieces of information, and for which we compute two simple relevance features w.r.t. query q: BM25 and DESM [39] (using embeddings trained on a large query corpus from an anterior period). We also add an additional feature, which is a mixture of the two above, on the concatenation of all the fields. Image embeddings ($d = 2048$) are obtained using a ResNet-152 model [22] pre-trained on ImageNet.

4.2 Evaluation Methodology

Given the limited number of queries in both collections, we conducted 5-fold cross-validation, by randomly splitting the queries into five folds. The model is trained on 4 folds (with 1 fold kept for validation, as we use early stopping on nDCG), and evaluated on the remaining one; this procedure is repeated 5 times. Then, the average validation nDCG is used to select the best model configuration. Note that for the `MediaEval` dataset, we have access to a separate test set, so we modify slightly the evaluation methodology: we do the above 5-fold cross-validation on the training set, without using a validation fold (hence, we do not use early stopping, and the number of epochs is a hyperparameter to tune). Once the best model has been selected with the above strategy, we re-train it on the full training set, and give the final performance on the test set. We report the nDCG, MAP, P@20, and nDCG@20 for both datasets.

We train the models using stochastic gradient descent with the Adam optimizer [29]. We set the batch size (i.e. number of graphs per batch) to $bs = \{5\}, \{32\}$ for respectively `MediaEval` and `WebQ`, so that training fits on a single NVIDIA Tesla P100 GPU. The hyper-parameters we tune for each dataset are: (1) the learning rate $\in \{1e{-}3, 1e{-}4, 5e{-}5\}$, (2) the number of layers $\in \{2, 3\}$ for the input MLP, as well as the number of hidden units $\in \{4, 8, 16, 32\}, \{8, 16, 32, 64\}$, (2) the dropout rate [47] in the MLP layers $\in \{0, 0.2\}$, (4) the number of graph convolutions $\in \{1, 2, 3, 4\}$ as well as the number of hidden units $\in \{4, 8, 16\}, \{8, 16, 32\}$, (5) the dropout rate of the convolution layers $\in \{0, 0.2, 0.5\}$ and (6) the number of visual neighbors to consider when building the input graph, $\in \{1, 3, 5, 10, 20, 50, 80, 100, 120, |\mathcal{G}| - 1\}$, $\{1, 3, 5, 10, 15, 20, 30, |\mathcal{G}| - 1\}$ for respectively `MediaEval` and `WebQ`. For `MediaEval`, we also tune the number of epochs $\in \{50, 100, 200, 300, 500\}$, while for `WebQ`, we set it to 500, and use early stopping with patience set to 80. All node features are query-level normalized (mean-std normalization). The models are implemented using PyTorch and PyTorch geometric[9] [14] for the message passing components.

4.3 Baselines

In order to be fair, we want to compare methods with somewhat similar feature sets. Obviously, for the supervised methods, results can be improved by either

[9] https://github.com/rusty1s/pytorch_geometric.

adding richer/more features, or increasing models' capacity. For both datasets, we compare our **DCMM** model to the following baselines:

- A learning to rank model only based on textual features (**LTR**).
- The cross-modal similarity introduced in Sect. 3.1 [2,3,10,11,45] (**CM**).
- The above LTR model with the cross-modal similarity as additional input feature (**LTR+CM**), to verify that it is actually beneficial to learn the cross-modal propagation in DCMM in a end-to-end manner.

For the cross-modal similarity, we use as proxy for $s_T(q,.)$ a simple mixture of term-based relevance score (Dirichlet-smoothed LM and BM25 for respectively `MediaEval` and `WebQ`) and DESM score, on a concatenation of all text fields. From our experiments, we observe that it is actually beneficial to recombine the cross-modal similarity with the initial relevance $s_T(q,.)$, using a simple mixture. Hence, three parameters are tuned (the two mixture parameters, and the number of neighbors for the query), following the evaluation methodology introduced in Sect. 4.2[10]. The LTR models are standard MLPs: they correspond to the upper part of architecture Fig. 1 (text branch), and are tuned following the same strategy.

We do not compare our models with joint embedding approaches on those datasets for the reasons mentioned in Sect. 2, but also due to our initial experiments on `Medieval` which gave poor results. For the sake of illustration, on `MediaEval`, 64% of the queries have no lemmas in common with training queries (and 35% for `WebQ`): given the relatively small size of these datasets, the models cannot generalize to unseen queries. This illustrates an "extreme" example of the generalization issues –especially on tail queries– of joint embedding techniques. In the meantime, as our model is fed with learning to rank features, especially term-based relevance scores like BM25, it could be less sensitive to generalization issues, for instance on new named entities. However, we want to emphasize that both approaches are not antagonist, but can actually be complementary. As our model can be seen as an extension of listwise learning to rank for bi-modal objects (if edges are removed, the model reduces to a standard MLP-based learning to rank), it can take as input node features matching scores from joint embeddings models. The model being an extension of PRF, we actually see the approaches at different stages of ranking.

4.4 Results and Analysis

Table 1 gathers the main results of our study. Without too much surprise, going from pure text ranker to a model using both media types improves the results by a large margin (all the models are significantly better than the text-based LTR model, so we do not include these tests on Table 1 for clarity). Moreover, results

[10] Note that, when used as a LTR feature, we obviously do not recombine the CM score with initial relevance score, as it will be redundant with other text features. Hence, we directly use score from Eq. (1), and tune only two parameters.

Table 1. Comparison of the methods on both datasets (test metrics). Significant improvement w.r.t. the cross-modal similarity (CM sim) is indicated with $*$ (*p-value* < 0.05). The number of trained parameters are indicated for the convolution models: ranging from few hundreds to few thousands, i.e. orders of magnitude less than joint embeddings models.

		# params	P@20	nDCG@20	nDCG	MAP
`MediaEval'17`	LTR		0.758	0.767	0.912	0.707
	CM [11]		0.843	0.857	0.939	0.784
	LTR + CM		0.852	0.868	0.942	0.789
	DCMM-cos	268	**0.871**	**0.876**	**0.944**	0.803
	DCMM-edge	3314	0.861	0.871	**0.944**	**0.806***
`WebQ`	LTR		0.69	0.801	0.884	0.775
	CM [11]		0.724	0.84	0.901	0.815
	LTR + CM		0.724	0.839	0.901	0.813
	DCMM-cos	1868	0.729	0.847	0.905	0.821
	DCMM-edge	15522	**0.738***	**0.857***	**0.91***	**0.83***

indicate that combining initial features with the unsupervised cross-modal similarity in a LTR model allows to slightly improve results over the latter (not significantly though) for the `MediaEval` dataset, while it has no effect on `WebQ`: this is likely due to the fact that features are somehow redundant in our setting, because of how $s_T(q,.)$ is computed for the cross-modal similarity; the same would not hold if we would consider a richer set of features for the LTR models. Furthermore, the DCMM-cos model outperforms all the baselines, with larger margins for `MediaEval` than for `WebQ`; the only significant result (*p-value* < 0.05) is obtained for the MAP on `MediaEval`. Nevertheless, it shows that this simple architecture –the most straightforward extension of cross-modal similarity introduced in Sect. 3.1–, with a handful of parameters (see Table 1) and trained on small datasets, is able to reproduce PRF mechanisms. Interestingly, results tend to drop as we increase the number of layers (best results are obtained with a single convolution layer), no matter the number of neighbors chosen to define the visual graph. While it might be related to the relative simplicity of the model, it actually echoes common observations in PRF models (e.g. [3]): if we propagate too much, we also tend to diffuse information too much. Similarly, we can also make a parallel with over-smoothing in GNNs [33], which might be more critical for PRF, especially considering the simplicity of this model.

The DCMM-edge shows interesting results: on `WebQ`, we manage to improve results significantly w.r.t. to CM sim, while on `MediaEval`, results are slightly worse than DCMM-cos (except for the MAP). It might be due to the fact that images in the latter are more alike to the ones used to train image signatures, compared to the (noisy) web images in `WebQ`; hence, learning a new metric between images has less impact. Interestingly, for both datasets, best results are obtained with more than a single layer; we hypothesize that the edge function plays the role of a simple filter for edges, allowing to propagate information from useful nodes across more layers. Note that the number of layers needed for the task is tied with how we define our input graph: the less neighbors we consider for each node, the more layers might be needed, in order for each node to gather information from useful nodes. In Fig. 2, we observe that if the number of neighbors is too small (e.g. 3 or 5), then the model needs more layers to improve performance. On the other side, when considering too many neighbors (e.g. 20 or all), the nodes already have access to all the useful neighbors, hence adding layers only reduces performances. We need to find the right balance between the number of neighbors and the number of convolution layers, so that the model can learn to propagate relevant signals (e.g. 10 neighbors and 3 layers for `WebQ`).

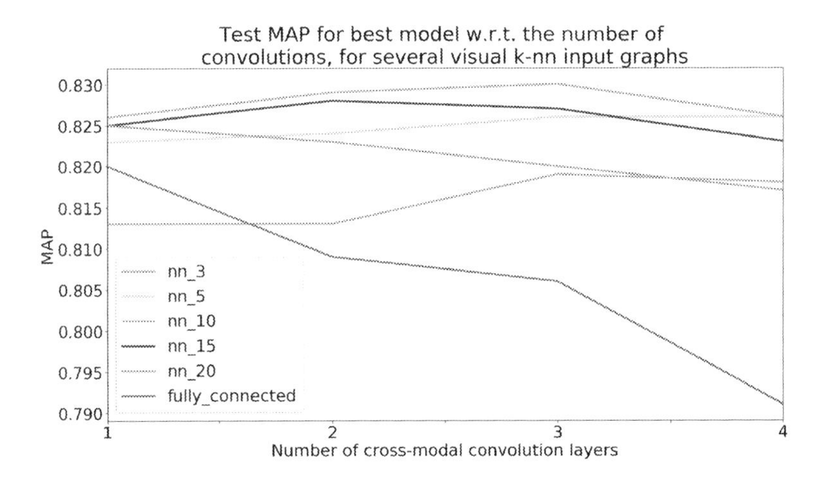

Fig. 2. Impact of the number of convolutions layers and top-k neighbors for `WebQ`.

5 Conclusion

In this paper, we have proposed a reformulation of unsupervised cross-modal PRF mechanisms for image search as a differentiable architecture relying on graph convolutions. Compared to its unsupervised counterpart, our novel approach can integrate any set of features, while providing a high flexibility in the design of the architecture. Experiments on two datasets showed that a simple model derived from our formulation achieved comparable –or better– performance compared to cross-modal PRF.

There are many extensions and possible directions stemming from the relatively simple model we have studied. Given enough training data (e.g. large amount of click logs), we could for instance learn to dynamically filter the visual similarity by using an attention mechanism to choose which nodes to attend, similarly to Graph Attention Networks [50] and Transformer model [49], discarding the need to set the number of neighbors in the input graph. Finally, our approach directly addressed the cross-modal retrieval task, but its application to the more general PRF problem in IR remains possible.

References

1. Learning Deep Structured Semantic Models for Web Search using Clickthrough Data. In: ACM International Conference on Information and Knowledge Management (CIKM), October 2013. https://www.microsoft.com/en-us/research/publication/learning-deep-structured-semantic-models-for-web-search-using-clickthrough-data/
2. Ah-Pine, J.M., Cifarelli, C.M., Clinchant, S.M., Csurka, G.M., Renders, J.M.: XRCE's Participation to ImageCLEF 2008. In: 9th Workshop of the Cross-Language Evaluation Forum (CLEF 2008), Aarhus, Denmark, September 2008. https://hal.archives-ouvertes.fr/hal-01504444
3. Ah-Pine, J., Csurka, G., Clinchant, S.: Unsupervised visual and textual information fusion in CBMIR using graph-based methods. ACM Trans. Inf. Syst. **33**(2), 9:1–9:31 (2015). https://doi.org/10.1145/2699668
4. Ai, Q., Bi, K., Guo, J., Croft, W.B.: Learning a deep listwise context model for ranking refinement. CoRR abs/1804.05936 (2018). http://arxiv.org/abs/1804.05936
5. Andrew, G., Arora, R., Bilmes, J., Livescu, K.: Deep canonical correlation analysis. In: International Conference on Machine Learning, pp. 1247–1255 (2013)
6. Bruch, S., Zoghi, M., Bendersky, M., Najork, M.: Revisiting approximate metric optimization in the age of deep neural networks. In: Proceedings of the 42nd International ACM SIGIR Conference on Research and Development in Information Retrieval (SIGIR 2019), pp. 1241–1244 (2019)
7. Burges, C.J.: From ranknet to lambdarank to lambdamart: an overview. Technical report, June 2010. https://www.microsoft.com/en-us/research/publication/from-ranknet-to-lambdarank-to-lambdamart-an-overview/
8. Carvalho, M., Cadène, R., Picard, D., Soulier, L., Thome, N., Cord, M.: Cross-modal retrieval in the cooking context: learning semantic text-image embeddings. CoRR abs/1804.11146 (2018). http://arxiv.org/abs/1804.11146
9. Chen, K., Bui, T., Chen, F., Wang, Z., Nevatia, R.: AMC: attention guided multi-modal correlation learning for image search. CoRR abs/1704.00763 (2017). http://arxiv.org/abs/1704.00763
10. Clinchant, S., Renders, J.-M., Csurka, G.: Trans-media pseudo-relevance feedback methods in multimedia retrieval. In: Peters, C., et al. (eds.) CLEF 2007. LNCS, vol. 5152, pp. 569–576. Springer, Heidelberg (2008). https://doi.org/10.1007/978-3-540-85760-0_71
11. Csurka, G., Ah-Pine, J., Clinchant, S.: Unsupervised visual and textual information fusion in multimedia retrieval - a graph-based point of view. CoRR abs/1401.6891 (2014). http://arxiv.org/abs/1401.6891
12. Dai, Z., Xiong, C., Callan, J., Liu, Z.: Convolutional neural networks for soft-matching n-grams in ad-hoc search. In: Proceedings of the Eleventh ACM International Conference on Web Search and Data Mining. WSDM 2018, pp. 126–134. ACM, New York (2018). https://doi.org/10.1145/3159652.3159659

13. Fan, Y., Guo, J., Lan, Y., Xu, J., Zhai, C., Cheng, X.: Modeling diverse relevance patterns in ad-hoc retrieval. CoRR abs/1805.05737 (2018). http://arxiv.org/abs/1805.05737
14. Fey, M., Lenssen, J.E.: Fast graph representation learning with PyTorch geometric. In: ICLR Workshop on Representation Learning on Graphs and Manifolds (2019)
15. Frome, A., et al.: DeViSE: a deep visual-semantic embedding model (2013)
16. Gilmer, J., Schoenholz, S.S., Riley, P.F., Vinyals, O., Dahl, G.E.: Neural message passing for quantum chemistry. CoRR abs/1704.01212 (2017). http://arxiv.org/abs/1704.01212
17. Gong, Y., Ke, Q., Isard, M., Lazebnik, S.: A Multi-View Embedding Space for Modeling Internet Images, Tags, and Their Semantics. Int. J. Comput. Vis. **106**(2), 210–233 (2013). https://doi.org/10.1007/s11263-013-0658-4
18. Gong, Y., Wang, L., Hodosh, M., Hockenmaier, J., Lazebnik, S.: Improving image-sentence embeddings using large weakly annotated photo collections. In: Fleet, D., Pajdla, T., Schiele, B., Tuytelaars, T. (eds.) ECCV 2014. LNCS, vol. 8692, pp. 529–545. Springer, Cham (2014). https://doi.org/10.1007/978-3-319-10593-2_35
19. Gordo, A., Larlus, D.: Beyond instance-level image retrieval: leveraging captions to learn a global visual representation for semantic retrieval (2017)
20. Guo, J., Fan, Y., Ai, Q., Croft, W.B.: A deep relevance matching model for ad-hoc retrieval. CoRR abs/1711.08611 (2017). http://arxiv.org/abs/1711.08611
21. Hamilton, W.L., Ying, R., Leskovec, J.: Inductive representation learning on large graphs. CoRR abs/1706.02216 (2017). http://arxiv.org/abs/1706.02216
22. He, K., Zhang, X., Ren, S., Sun, J.: Deep residual learning for image recognition. CoRR abs/1512.03385 (2015). http://arxiv.org/abs/1512.03385
23. Hu, P., Zhen, L., Peng, D., Liu, P.: Scalable deep multimodal learning for cross-modal retrieval. In: Proceedings of the 42nd International ACM SIGIR Conference on Research and Development in Information Retrieval, pp. 635–644 (2019)
24. Hua, X.S., et al.: Clickage: towards bridging semantic and intent gaps via mining click logs of search engines, pp. 243–252, October 2013. https://doi.org/10.1145/2502081.2502283
25. Hui, K., Yates, A., Berberich, K., de Melo, G.: A position-aware deep model for relevance matching in information retrieval. CoRR abs/1704.03940 (2017). http://arxiv.org/abs/1704.03940
26. Hui, K., Yates, A., Berberich, K., de Melo, G.: RE-PACRR: a context and density-aware neural information retrieval model. CoRR abs/1706.10192 (2017). http://arxiv.org/abs/1706.10192
27. Kan, M., Shan, S., Chen, X.: Multi-view deep network for cross-view classification. In: Proceedings of the IEEE Conference on Computer Vision and Pattern Recognition, pp. 4847–4855 (2016)
28. Khasanova, R., Dong, X., Frossard, P.: Multi-modal image retrieval with random walk on multi-layer graphs. In: 2016 IEEE International Symposium on Multimedia (ISM), pp. 1–6 (2016)
29. Kingma, D.P., Ba, J.: Adam: a method for stochastic optimization. CoRR abs/1412.6980 (2014). http://arxiv.org/abs/1412.6980
30. Kipf, T.N., Welling, M.: Semi-supervised classification with graph convolutional networks. CoRR abs/1609.02907 (2016). http://arxiv.org/abs/1609.02907
31. Lavrenko, V., Croft, W.B.: Relevance based language models. In: Proceedings of the 24th Annual International ACM SIGIR Conference on Research and Development in Information Retrieval. SIGIR 2001, pp. 120–127. ACM, New York (2001). https://doi.org/10.1145/383952.383972

32. Li, C., et al.: NPRF: a neural pseudo relevance feedback framework for ad-hoc information retrieval. CoRR abs/1810.12936 (2018). http://arxiv.org/abs/1810.12936
33. Li, Q., Han, Z., Wu, X.: Deeper insights into graph convolutional networks for semi-supervised learning. CoRR abs/1801.07606 (2018). http://arxiv.org/abs/1801.07606
34. Lin, T., et al.: Microsoft COCO: common objects in context. CoRR abs/1405.0312 (2014). http://arxiv.org/abs/1405.0312
35. Liong, V.E., Lu, J., Tan, Y.P., Zhou, J.: Deep coupled metric learning for cross-modal matching. IEEE Trans. Multimedia **19**(6), 1234–1244 (2016)
36. Liu, S., Xiao, F., Ou, W., Si, L.: Cascade ranking for operational e-commerce search. In: Proceedings of the 23rd ACM SIGKDD International Conference on Knowledge Discovery and Data Mining - KDD 2017 (2017). https://doi.org/10.1145/3097983.3098011
37. Mitra, B., Craswell, N.: An updated duet model for passage re-ranking. CoRR abs/1903.07666 (2019). http://arxiv.org/abs/1903.07666
38. Mitra, B., Diaz, F., Craswell, N.: Learning to match using local and distributed representations of text for web search. CoRR abs/1610.08136 (2016). http://arxiv.org/abs/1610.08136
39. Mitra, B., Nalisnick, E.T., Craswell, N., Caruana, R.: A dual embedding space model for document ranking. CoRR abs/1602.01137 (2016). http://arxiv.org/abs/1602.01137
40. Nogueira, R., Cho, K.: Task-oriented query reformulation with reinforcement learning. CoRR abs/1704.04572 (2017). http://arxiv.org/abs/1704.04572
41. Pang, L., Lan, Y., Guo, J., Xu, J., Xu, J., Cheng, X.: DeepRank: a new deep architecture for relevance ranking in information retrieval. CoRR abs/1710.05649 (2017). http://arxiv.org/abs/1710.05649
42. Pei, C., et al.: Personalized context-aware re-ranking for e-commerce recommender systems. CoRR abs/1904.06813 (2019). http://arxiv.org/abs/1904.06813
43. Plummer, B.A., Wang, L., Cervantes, C.M., Caicedo, J.C., Hockenmaier, J., Lazebnik, S.: Flickr30k entities: collecting region-to-phrase correspondences for richer image-to-sentence models. CoRR abs/1505.04870 (2015). http://arxiv.org/abs/1505.04870
44. Qi, J., Peng, Y.: Cross-modal bidirectional translation via reinforcement learning. In: IJCAI, pp. 2630–2636 (2018)
45. Renders, J.M., Csurka, G.: NLE@MediaEval'17: combining cross-media similarity and embeddings for retrieving diverse social images. In: MediaEval (2017)
46. Rendle, S., Freudenthaler, C., Gantner, Z., Schmidt-Thieme, L.: BPR: Bayesian personalized ranking from implicit feedback. In: Proceedings of the Twenty-Fifth Conference on Uncertainty in Artificial Intelligence. UAI 2009, pp. 452–461. AUAI Press, Arlington (2009). http://dl.acm.org/citation.cfm?id=1795114.1795167
47. Srivastava, N., Hinton, G., Krizhevsky, A., Sutskever, I., Salakhutdinov, R.: Dropout: a simple way to prevent neural networks from overfitting. J. Mach. Learn. Res. **15**, 1929–1958 (2014). http://jmlr.org/papers/v15/srivastava14a.html
48. Szegedy, C., Ioffe, S., Vanhoucke, V.: Inception-v4, inception-ResNet and the impact of residual connections on learning. CoRR abs/1602.07261 (2016). http://arxiv.org/abs/1602.07261
49. Vaswani, A., et al.: Attention is all you need. CoRR abs/1706.03762 (2017). http://arxiv.org/abs/1706.03762
50. Veličković, P., Cucurull, G., Casanova, A., Romero, A., Liò, P., Bengio, Y.: Graph attention networks. In: International Conference on Learning Representations (2018). https://openreview.net/forum?id=rJXMpikCZ

51. Wang, B., Yang, Y., Xu, X., Hanjalic, A., Shen, H.T.: Adversarial cross-modal retrieval. In: Proceedings of the 25th ACM International Conference on Multimedia. MM 2017, pp. 154–162 (2017)
52. Wang, L., Li, Y., Huang, J., Lazebnik, S.: Learning two-branch neural networks for image-text matching tasks. IEEE Trans. Pattern Anal. Mach. Intell. **41**(2), 394–407 (2018)
53. Wang, L., Li, Y., Lazebnik, S.: Learning deep structure-preserving image-text embeddings. In: Proceedings of the IEEE Conference on Computer Vision and Pattern Recognition, pp. 5005–5013 (2016)
54. Weston, J., Bengio, S., Usunier, N.: WSABIE: scaling up to large vocabulary image annotation (2011)
55. Wu, Y., Wang, S., Huang, Q.: Learning semantic structure-preserved embeddings for cross-modal retrieval. In: Proceedings of the 26th ACM International Conference on Multimedia. MM 2018, pp. 825–833. ACM (2018)
56. Xiong, C., Dai, Z., Callan, J., Liu, Z., Power, R.: End-to-end neural ad-hoc ranking with kernel pooling. CoRR abs/1706.06613 (2017). http://arxiv.org/abs/1706.06613
57. Xu, K., Hu, W., Leskovec, J., Jegelka, S.: How powerful are graph neural networks? CoRR abs/1810.00826 (2018). http://arxiv.org/abs/1810.00826
58. Yan, F., Mikolajczyk, K.: Deep correlation for matching images and text. In: Proceedings of the IEEE Conference on Computer Vision and Pattern Recognition, pp. 3441–3450 (2015)
59. Ying, R., He, R., Chen, K., Eksombatchai, P., Hamilton, W.L., Leskovec, J.: Graph convolutional neural networks for web-scale recommender systems. CoRR abs/1806.01973 (2018). http://arxiv.org/abs/1806.01973
60. Zhai, C., Lafferty, J.: A study of smoothing methods for language models applied to ad hoc information retrieval. In: Proceedings of the 24th Annual International ACM SIGIR Conference on Research and Development in Information Retrieval. SIGIR 2001, pp. 334–342. ACM, New York (2001). https://doi.org/10.1145/383952.384019
61. Zhang, Y., Lu, H.: Deep cross-modal projection learning for image-text matching. In: Ferrari, V., Hebert, M., Sminchisescu, C., Weiss, Y. (eds.) ECCV 2018. LNCS, vol. 11205, pp. 707–723. Springer, Cham (2018). https://doi.org/10.1007/978-3-030-01246-5_42
62. Zhu, L., Chen, Y., He, B.: A domain generalization perspective on listwise context modeling. CoRR abs/1902.04484 (2019). http://arxiv.org/abs/1902.04484

Diagnosing BERT with Retrieval Heuristics

Arthur Câmara(✉) and Claudia Hauff

Delft University of Technology, Delft, The Netherlands
{a.barbosacamara,c.hauff}@tudelft.nl

Abstract. Word embeddings, made widely popular in 2013 with the release of word2vec, have become a mainstay of NLP engineering pipelines. Recently, with the release of BERT, word embeddings have moved from the term-based embedding space to the contextual embedding space—each term is no longer represented by a single low-dimensional vector but instead each term and *its context* determine the vector weights. BERT's setup and architecture have been shown to be general enough to be applicable to many natural language tasks. Importantly for Information Retrieval (IR), in contrast to prior deep learning solutions to IR problems which required significant tuning of neural net architectures and training regimes, "vanilla BERT" has been shown to outperform existing retrieval algorithms by a wide margin, including on tasks and corpora that have long resisted retrieval effectiveness gains over traditional IR baselines (such as Robust04). In this paper, we employ the recently proposed axiomatic dataset analysis technique—that is, we create diagnostic datasets that each fulfil a retrieval heuristic (both term matching and semantic-based)—to explore what BERT is able to learn. In contrast to our expectations, we find BERT, when applied to a recently released large-scale web corpus with ad-hoc topics, to *not* adhere to any of the explored axioms. At the same time, BERT outperforms the traditional query likelihood retrieval model by 40%. This means that the axiomatic approach to IR (and its extension of diagnostic datasets created for retrieval heuristics) may in its current form not be applicable to large-scale corpora. Additional—different—axioms are needed.

1 Introduction

Over the course of the past few years, IR has seen the introduction of a large number of successful deep learning approaches for solving all kinds of tasks previously tackled with hand-crafted features (within the learning to rank framework) or traditional retrieval models such BM25.

In 2017, with the introduction of the transformer architecture [33], a second wave of neural architectures for NLP has emerged. Approaches (and respective models) like BERT [7], XLNet [41] and GPT-2 [24] have shown that it is indeed possible for one general architecture to achieve state-of-the-art performance across very different NLP tasks (some of which are also related to IR tasks, such as question answering, reading comprehension, etc.).

J. M. Jose et al. (Eds.): ECIR 2020, LNCS 12035, pp. 605–618, 2020.
https://doi.org/10.1007/978-3-030-45439-5_40

Ad-hoc retrieval, the task of ranking a set of documents given a single query, has long resisted the success of neural approaches, especially when employed across standard IR test collections such as Robust04[1], which come with hundreds of topics (and thus relatively little training data). Often, the proposed neural approaches require a very careful design of their architecture. Also, the training regime and the input data transformations have to be *just right* [17] in order to beat or come close to well-tuned traditional IR baselines such as RM3 [2,16][2]. With the introduction of BERT in late 2018, this has finally changed. Recently, a range of BERT-inspired approaches have been shown to clearly surpass all strong IR baselines on Robust04 [6,40] and other IR corpora.

It is still an open question though what exactly makes BERT and similar approaches perform so well on IR tasks. While recent works try to understand what BERT learns most often by analysing attention values, e.g., [4,12,21,32], analysing BERT under the IR light requires a different set of tools. While most NLP tasks optimise for precision, recall or other objective metrics, the goal of ad-hoc retrieval is to optimise for *relevance*, a complex multidimensional and somewhat subjective concept [3].

In this paper, we set out to explore BERT under the IR lens, employing the concept of *diagnostic datasets*, an IR model analysis approach (inspired by similar NLP and computer vision approaches) proposed last year by Rennings et al. [25]. The idea behind these datasets is simple: each dataset is designed to fulfil one *retrieval axiom* [9], i.e., a heuristic that a good retrieval function should fulfil[3]. Each dataset contains query-documents instances (most often, a query and two documents) that the investigated model should rank in the correct order as determined by the heuristic. The extent to which a model correctly ranks those instances is allowing us to gain insights into what type of information the retrieval model pays attention to (or not) when ranking documents. While traditional retrieval models such as BM25 [26] can be analysed analytically, neural nets with their millions or even billions of learnt weights can only be analysed in such an empirical manner.

More concretely, we attempt to analyse a version of BERT, DistilBERT (that was shown to attain 97% of "vanilla" BERT performance [28]), fine-tuned on the TREC 2019 Deep Learning track dataset[4]. We extend previous work [25] by incorporating additional axioms (moving from term matching to semantic axioms). We find that DistilBERT to outperform the traditional query likelihood (QL) model by 40%. In contrast to our expectations however, we find that BERT does not adhere to any of the axioms we incorporate in our work. This implies that the currently existing axioms are *not sufficient* and *not applicable* to capture

[1] Robust04 is a test collection employed at the TREC 2004 robust retrieval task [35], consisting of 528K newswire documents, 250 topics and 311K relevance judgements.

[2] We want to emphasise here that this observation is specific to IR corpora with few training topics; for the very few corpora with hundreds of thousands of released topics (such as MSMarco) this observation does not hold.

[3] As a concrete example, consider the TFC1 [9] heuristic: *The more occurrences of a query term a document has, the higher its retrieval score.*

[4] https://microsoft.github.io/TREC-2019-Deep-Learning/.

the heuristics that a strong supervised model learns (at least for the corpus and model we explore); it is not yet clear to what extent those results generalise beyond our model and corpus combination but it opens up a number of questions about the axiomatic approach to IR.

2 Related Work

Axiomatic Information Retrieval. The use of axioms (or "retrieval heuristics") as a means to improve and understand information retrieval techniques is well established. It is an analytic technique to explore retrieval models and how best to improve them. In their seminal work, Fang et al. [9,10] introduced a number of term-matching based *heuristics* that models should follow in order to be successful in retrieval tasks. Subsequently, Fang et al. [11] proposed a set of axioms based on semantic matching and thus allowing non-exact matches to be accounted for in axiomatic retrieval. We apply these axioms in our work—albeit in a slightly adapted manner. Other applications for axioms in IR include document re-ranking based on a Learning to Rank scenario [13] and query expansion [8] by exploring similar axioms. It should be noted, that—while sensible—it cannot be assumed that these axioms are a good fit for all kinds of corpora; they represent a general notion of how a good retrieval function should behave. Recently, Rennings et al. [25] introduced *diagnostic datasets* extracted from actual corpora that each fulfil one axiom. In contrast to the axiomatic approach, which requires an analytical evaluation of the retrieval functions under investigation, a diagnostic corpus enables us to analyse models' axiomatic performance that are too large to be analysed analytically (such as neural models with millions or even billions of parameters[5]). Our work continues in that direction with a larger number of axioms (9 vs. 4) and the analysis of the current neural state-of-the-art (i.e., BERT).

Neural IR Models. Neural IR models, i.e., deep learning based approaches that tackle IR problems, have seen a massive rise in popularity in the last few years, with considerable success [20]. Models like DRMM [19], ARCII [14] and aNMM [39] have been shown to be suitable for a range of IR tasks when sufficient training data is available; it remains at best unclear at smaller data scale whether the reported successes are not just an artefact of weak baselines [17].

Recently, a new wave of approaches, based on the transformer architecture [33] has shown that, finally, neural models can significantly outperform traditional and well-tuned retrieval methods such as RM3 [2]. Yang et al. [40] have shown that BERT, fine-tuned on the available TREC microblog datasets, and combined with a traditional retrieval approach such as query likelihood significantly outperforms well-tuned baselines, even on Robust04 which has shown to be a notoriously difficult dataset for neural models to do well on. With similar success, Dai and Callan [6] have recently employed another BERT variant on Robust04 and ClueWeb09. Lastly we point out, that works are now also

[5] As a concrete example, our BERT model contains 66 million parameters.

beginning to appear, e.g., [18], that use the contextual word embeddings produced by BERT in combination with another strong neural model, again with strong improvements over the existing baselines.

Analysing Neural IR Models. As we aim to analyse BERT, we also consider how others have tackled this problem. Analysing neural models—whether for IR, NLP or another research domain—is not a trivial task. By now a great number of works have tried to light up the black box of the neural learning models [1], with varying degrees of success. Within IR, Pang et al. [22] have aimed to paint a complete and high-level picture of the neural IR area, comparing the behaviour of different approaches, and showing that interaction and representation-based models focus on different characteristics of queries and documents. While insightful, such work does not enable us to gain deep insights into a single model. Closer to our work, Rosset et al. [27] employ axioms to generate artificial documents for the training of neural models and the regularization of the loss function. In contrast, we employ axioms to *analyze* retrieval models.

Another direction of research has been the development of interpretation tools such as DeepSHAP [12] and LIRME [34] that aim to generate *local* explanations for neural IR models. Recently, in particular BERT (due to its successes across a wide range of tasks and domains) has become the focus of analysis—not within IR though. While approaches like [4] explore the attention values generated by the model's attention layers, Tenney et al. [32] argue that BERT is re-discovering classical NLP pipeline approaches in its layers, *"in an interpretable and localizable way"*, essentially repeating traditional NLP steps in a similar order as an expert would do, with steps like POS tagging, parsing, NER and coreference resolution happening within its layers in the expected order. Finally, Niven et al. [21] raise some critical points about BERT, arguing that it only *"exploits spurious statistical cues in the dataset"*; they showcase this by creating adversarial datasets that can significantly harm its performance.

3 Diagnostic Datasets

The usage of diagnostic datasets as a means to analyse neural models is common in NLP, e.g. [15,36,37] as there are a large number of fine-grained linguistic tasks (anaphora resolution, entailment, negation, etc.) that datasets can be created for with relative ease. In contrast, in IR the central notion is relevance and although we know that it can be decomposed into various types (topical, situational, etc.) of relevance [29], we have no easy way of creating datasets for each of these—it remains a time-intensive and expensive task. This also explains why corpora such as Robust04 remain useful and in use for such a long time. Instead, like Rennings et al. [25] we turn to the axiomatic approach to IR and create diagnostic datasets—one for each of our chosen retrieval heuristics. It has been shown that, generally, retrieval functions that fulfil these heuristics achieve a greater effectiveness than those that do not. In contrast to [25] which restricted itself to four term matching axioms, we explore a wider range of axioms, covering term frequency, document length, lower-bounding term frequency, semantic term matching and term proximity constraints. In total, we explore 9 axioms, all of which are listed in Table 1 with a short informal description of their main assumption of what a sensible retrieval function should fulfil. We note that this covers most of the term-matching and semantic-matching axioms that have been proposed. We have eliminated a small number from our work as we do not consider them relevant to BERT (e.g., those designed for pseudo-relevance feedback [5]).

As the axiomatic approach to IR has been designed to *analytically* analyse retrieval functions, in their original version they assume very specific artificial query and document setups. As a concrete example, let us consider axiom `STMC1` [11]. It is defined as follows: *given a single-term query* $Q = \{q\}$ *and two single-term documents* $D_1 = \{d_1\}$, $D_2 = \{d_2\}$ *where* $d_1 \neq d_2 \neq q$, *the retrieval score of* D_1 *should be higher than* D_2 *if the semantic similarity between* q *and* d_1 *is higher than that between* q *and* d_2. This description is sufficient to mathematically analyse classic retrieval functions, but not suitable for models with more than a handful of parameters. We thus turn to the creation of datasets that *exclusively* contain instances of query/documents (for `STMC1` an instance is a triple, consisting of one query and two documents) that satisfy a particular axiom. As single-term queries and documents offer no realistic test bed, we *extend* (moving beyond single-term queries and documents) and *relax* (moving beyond strict requirements such as equal document length) the axioms in order to extract instances from existing datasets that fulfil the requirements of the extended and relaxed axiom. Importantly, this process requires no relevance judgements—we can simply scan all possible triples in the corpus (consisting of queries and documents) and add those to our diagnostic dataset that fulfil our requirements. We then score each query/document pair with our BERT model[6] and determine whether the score order of the documents is in line with the axiom. If it is, we consider our model to have classified this instance correctly.

While Table 1 provides an informal overview of each heuristic, we now formally describe each one in more detail. Due to the space limitations, we focus on a mathematical notation which is rather brief. For completeness, we first state the original axiom and then outline how we extend and relax it in order to create a diagnostic dataset from it. For axioms `TFC1`, `TFC2`, `LNC2` and `M-TDC` we follow the process described in [25]. We make use of the following notation: Q is a query and consists of terms $q_1, q_2, ...$; D_i is a document of length $|D_i|$ containing terms $d_{i_1}, d_{i_2}, ...$; the count of term w in document D is $c(w, D)$; lastly, $S(Q, D)$ is the retrieval score the model assigns to D for a given Q. Apart from the proximity heuristic `TP`, the remaining heuristics are based on the bag-of-word assumption, i.e., the order of terms in the query and documents do not matter.

TFC1—Original. Assume $Q = \{q\}$ and $|D_1| = |D_2|$. If $c(q, D_1) > c(q, D_2)$, then $S(Q, D_1) > S(Q, D_2)$.

[6] Note, that scoring each document independently for each query is an architectural choice, there are neural architectures that take a query/doc/doc triplet as input and output a preference score.

TFC1—Adapted. In order to extract query/document/document triples from actual corpora, we need to consider multi-term queries and document pairs of approximately the same length. Let $Q = \{q_1, q_2, .., q_{|Q|}\}$ and $|D_1| - |D_2| \leq abs(\delta)$. $S(Q, D_1) > S(Q, D_2)$ holds, when D_1 has at least the same query term count as D_2 for all but one query term (and for this term D_1's count is higher), i.e., $c(q_i, D_1) \geq c(q_i, D_2)\ \forall q_i \in Q$ and $\sum_{q_i \in Q} c(q_i, D_1) > \sum_{q_i \in Q} c(q_i, D_2)$.

TFC2—Original. Assume $Q = \{q\}$ and $|D_1| = |D_2| = |D_3|$. If $c(q, D_1) > 0$, $c(q, D_2) - c(q, D_1) = 1$ and $c(q, D_3) - c(q, D_2) = 1$, then $S(Q, D_2) - S(Q, D_1) > S(Q, D_3) - S(Q, D_2)$.

TFC2—Adapted. Analogous to TFC1, queries can contain multiple terms and documents only have to have approximately the same length. Let $Q = \{q_1, q_2, .., q_{|Q|}\}$ and $max_{D_i, D_j \in \{D_1, D_2, D_3\}}(|D_i| - |D_j| \leq abs(\delta))$. If every document contains at least one query term, and D_3 has more query terms than D_2 and D_2 has more query terms than D_1, and the difference of query terms count between D_2 and D_1 should be the same as between D_3 and D_2, for all query terms, i.e. $\sum_{q \in Q} c(q, D_3) > \sum_{q \in Q} c(q, D_2) > \sum_{q \in Q} c(q, D_1) > 0$ and $c(q, D_2) - c(q, D_1) = c(q, D_3) - c(q, D_2) \forall q \in Q$, then $S(Q, D_2) - S(Q, D_1) > S(Q, D_3) - S(Q, D_2)$.

M-TDC—Original. Let $Q = \{q_1, q_2\}$, $|D_1| = |D_2|$, $c(q_1, D_1) = c(q_2, D_2)$ and $c(q_2, D_1) = c(q_1, D_2)$. If $idf(q_1) \geq idf(q_2)$ and $c(q_1, D_1) > c(q_1, D_2)$, then $S(Q, D_1) \geq S(Q, D_2)$.

M-TDC—Adapted. Again, Let Q contain multiple terms and $|D_1| - |D_2| \leq abs(\delta)$. D_1, D_2 also differ in at least one query term count ($\exists q_i \in Q$, such that $c(q_i, D_1) \neq c(q_i, D_2)$). If, for all query term pairs the conditions hold that $c(q_i, D_1) \neq c(q_j, D_j)$, $idf(q_i) \geq idf(q_j)$, $c(q_i, D_1) = c(q_j, D_2)$, $c(q_j, D_1) = c(q_i, D_2)$, $c(q_i, D_1) > c(q_i, D_2)$ and $c(q_i, Q) \geq c(q_j, Q)$, then $S(Q, D_1) \geq S(Q, D_2)$.

LNC1—Original. Let Q be a query and D_1, D_2 be two documents. If for some $q' \notin Q$, $c(q', D_2) = c(q', D_1) + 1$ and for any $q \in Q$ $c(q, D_2) = c(q, D_1)$, then $S(Q, D_1) \geq S(Q, D_2)$.

LNC1—Adapted. The axiom can be used with no adaptation.

LNC2—Original. Let Q be a query. $\forall k > 1$, if D_1 and D_2 are two documents such that $|D_1| = k \cdot |D_2|$, and $\forall q \in Q, c(q, D_1) = k \cdot c(q, D_2)$, then $S(Q, D_1) \geq S(Q, D_2)$.

LNC2—Adapted. The axiom can be used with no adaptation.

TP—Original. Let $Q = \{q_1, q_2, ...q_{|Q|}\}$ be a query and D' a document generated by switching the position of query terms in D. Let $\sigma(Q, D)$ be a function that measures the distance of query terms $q_i \in Q$ inside a document D. If $\sigma(Q, D) > \sigma(Q, D')$, then $S(Q, D) < S(Q, D')$.

TP—Adapted. Let $\Gamma(D, Q) = min_{(q_1, q_2 \in Q \cap D, q_1 \neq q_2)} Dis(q_1, q_2; D)$ be a function that computes the minimum distance between every pair of query terms in D. If $\Gamma(D_1, Q) < \Gamma(D_2, Q)$, then $S(Q, D_1) > S(Q, D_2)$.

For the following semantic axioms, let us define the function $\sigma(t_1, t_2)$ as the cosine distance between the embeddings of terms t_1 and t_2. We also define $\sigma'(T_1, T_2)$, where T can be either a document D or a query Q, as an extension to σ, defined by $\sigma'(T_1, T_2) = cos(\frac{\sum_{i \in T_1} t_i}{|T_1|}, \frac{\sum_{i \in T_2} t_i}{|T_2|})$, the cosine distance between the average term embeddings for each document.

STMC1—Original. Let $Q = \{q\}$ be a one-term query, $D_1 = \{d_{1_1}\}$ and $D_2 = \{d_{2_1}\}$ be two single term documents, such that $d_{1_1} \neq d_{2_1}$, $q \neq d1_1$ and $q \neq d_{2_1}$. If $\sigma(q, d_{1_1}) > \sigma(q, d_{2_1})$, then $S(Q, D_1) > S(Q, D_2)$.

STMC1—Adapted. We allow D_1 and D_2 to be arbitrarily long, covering the same number of query terms (i.e. $|D_1 \bigcap Q| = |D_2 \bigcap Q|$). Assume $\{D_i\} - \{Q\}$ be the document D_i without query terms, If $\sigma'(\{D_1\} - \{Q\}, Q) > \sigma'(\{D_2\} - \{Q\}, Q)$, then $S(Q, D_1) > S(Q, D_2)$.

STMC2—Original. Let $Q = \{q\}$ be a one-term query and d a non-query term such that $\sigma(d, q) > 0$. If $D_1 = \{q\}$ and $|D_2| = k, (k \geq 1)$, composed entirely of d's (i.e.vc$(d, D_2) = k)$, then $S(Q, D_1) \geq S(Q, D_2)$.

STMC2—Adapted. We allow Q to be a multiple query term, D_1 to contain non-query terms and D_2 to contain query terms. If $\sum_{t_i, t_i \notin Q} c(t_i, D_2) > \sum_{q_i \in Q} c(q_1, D_1) > 0$, $\sigma'(\{D_1\} - \{Q\}, \{D_2\} - \{Q\}) > \delta$ then $S(Q, D_1) \geq S(Q, D_2)$.

STMC3—Original. Let $Q = \{q_1, q_2\}$ be a two-term query and d a non-query term such that $\sigma(d, q_2) > 0$. If $|D_1| = |D_2| > 1$, $c(q_1, D_1) = |D_1|$, $c(q1, D_2) = |D_2| - 1$ and $c(d, D_2) = 1$, then $S(Q, D_1) \leq S(Q, D_2)$.

STMC3—Adapted. Let D_1 and D_2 be two arbitrary long documents that covers the same number of query terms (i.e. $|D_1 \bigcap Q| = |D_2 \bigcap Q|$). If $|D_1| - |D_2| \leq abs(\delta)$, $\sum_{q_i \in Q} c(q_i, D_1) > \sum_{q_i \in Q} c(q_i, D_2)$ and $\sigma'(\{D_2\} - \{Q\}, Q) > \sigma'(\{D_1\} - \{Q\}, Q)$, then $S(Q, D_1) > S(Q, D_2)$.

4 Experiments

We create diagnostic datasets for each of these axioms by extracting instances of queries and documents that already exist in the dataset. In this section, we explain how these datasets were generated and how we employed them to evaluate BERT.

4.1 TREC 2019 Deep Learning Track

In order to extract diagnostic datasets, we used the corpus and queries for the Document Ranking Task from the TREC 2019 Deep Learning track[7]. This is the only publicly available ad-hoc retrieval dataset that was built specifically for the training of deep neural models, with 3,213,835 web documents and 372,206

[7] https://microsoft.github.io/TREC-2019-Deep-Learning/.

Table 1. Overview of retrieval heuristics employed in our work. The diagnostic datasets for heuristics marked with a blue background were first discussed in [25]. The naming of the heuristics is largely taken from the papers proposing them.

Heuristic Instance	Informal description
Term frequency constraints	
TFC1 [9]	The more occurrences of a query term a document has, the higher its retrieval score.
TFC2 [9]	The increase in retrieval score of a document gets smaller as the absolute query term frequency increases.
M-TDC [9,30]	The more discriminating query terms (i.e., those with high IDF value) a document contains, the higher its retrieval score.
Length normalization constraints	
LNC1 [9]	The retrieval score of a document decreases as terms not appearing in the query are added.
LNC2 [9]	A document that is duplicated does not have a lower retrieval score than the original document.
Semantic term matching constraints	
STMC1 [11]	A document's retrieval score increases as it contains terms that are more semantically related to the query terms.
STMC2 [11]	The document terms that are a syntactic match to the query terms contribute at least as much to the document's retrieval score as the semantically related terms.
STMC3 [11]	A document's retrieval score increases as it contains more terms that are semantically related to *different* query terms.
Term proximity constraint	
TP [31]	A document's retrieval score increases as the query terms appearing in it appear in closer proximity.

queries (367,013 queries in the training set and 5,193 in the development set). The queries and documents, while stripped of HTML elements, are not necessarily well-formed as seen in the following examples from the training set:

- `what is a flail chest`
- `a constitution is best described as a(n) _________.`
- `)what was the immediate impact of the success of the manhattan project?`

The queries consist on average of 5.89(±2.51) words while documents consist on average of 1084.88(±2324.22) words.

Most often, one relevant document exists per query (1.04 relevant documents on average). These relevance judgements were made by human judges on a passage-level: if a passage within a document is relevant, the document is considered relevant. Unlike other TREC datasets, like Robust04, there is no topic description or topic narrative.

At the time of this writing, the relevance judgements for the test queries were not available. Therefore, we split the development queries further, in a new *dev* and *test* dataset, in a 70%–30% fashion. In the rest of this paper, when we refer to the *test* or *dev* dataset, we are referring to this split. The *train* split remains the same as the original dataset.

4.2 Retrieval and Ranking

We begin by indexing the document collection using the Indri toolkit[8], and retrieve the top-100 results using a traditional retrieval model with just one hyperparameter, namely, QL, with Indri's default setting (Dirichlet smoothing [42] and $\mu = 2500$). Finally, for BERT, we employ Hugging Face's library[9] of DistilBERT [28], a distilled version of the original BERT model, with fewer parameters (66 million instead of 340 million), and thus more efficient to train, but with very similar results.

We fine-tuned our BERT[10] model with 10 negative samples for each positive sample from the training dataset, randomly picked from the top-100 retrieved from QL. We set the maximum input length to 512 tokens. For fine-tuning we used the sequence classification model. It is implemented by adding a fully-connected layer on top of the `[CLS]` token embedding, which is the specific output token of the BERT model that our fine-tuning is based upon.

Given the limitation of BERT regarding the maximum number of tokens, we limited the document length to its first 512 tokens, though we note that alternative approaches exist (e.g., in [40] the BERT scores across a document's passages/sentences are aggregated). In Table 3, we report the retrieval effectiveness in terms of nDCG and MRR for the documents limited to 512 tokens. We rerank the top-100 retrieved documents based on its first 512 tokens. It is clear that BERT is vastly superior to QL with a 40% improvement in $nDCG$ and 25% improvement in MRR.

4.3 Diagnostic Datasets

Given the adapted axioms defined in Sect. 3, we now proceed on describing how to extract actual datasets from our corpus.

TFC1*, *TFC2*, `M-TDC`, *LNC1 We add tuples of queries and documents $\{q, d_i, d_j\}$ (or $\{q, d_i, d_j, d_k\}$ for `TFC2`) for every possible pair of documents $\{d_i, d_j\}$ in the top 100 retrieved by QL that follow the assumptions from Sect. 3, with $\delta = 10$. We also compute `IDF` for `M-TDC` on the complete corpus of documents, tokenized by WordPiece [38].

---

[8] https://www.lemurproject.org/indri.php.

[9] https://github.com/huggingface/transformers.

[10] Code for fine-tuning DistilBERT and generating the diagnostic datasets is available at https://github.com/ArthurCamara/bert-axioms.

Table 2. Overview of the number of instances in each diagnostic dataset (row I), the number of instances within each diagnostic dataset that contain a relevant document (row II) and the fraction of instances among all of row II where the order of the documents according to the axiom is in line with the relevance judgments. LNC2 is based on new documents, thus, it does not have a fraction of agreement

	TFC1	TFC2	M-TDC	LNC1	LNC2	TP	STMC1	STMC2	STMC3
Diagnostic dataset size	119,690	10,682	13,871	14,481,949	7452	3,010,246	319,579	7,321,319	217,104
Instances with a *relevant* document	1,416	17	11	138,399	82	20,559	19,666	70,829	1,626
Fraction of instances agreeing with relevance	0.91	0.29	0.82	0.50	–	0.18	0.44	0.63	0.35

LNC2 We create a new dataset, appending the document to itself $k \in \mathbb{Z}$ times until we reach up to 512 tokens[11]. In contrast to Rennings et al. [25] we only perform this document duplication for our test set, i.e., BERT does not "see" this type of duplication during its training phase. On average, the documents were multiplied $k = 2.6408 \pm 1.603$ times, with a median of $k = 2$.
TP We simply add to our dataset every pair of documents $\{q, d_i, d_j\}$ in the top 100 retrieved documents by QL for a given topic that follow the stated TP assumptions.
STMC1, STMC2, STMC3 We define σ as the cosine distance between the embeddings of the terms and σ' as the cosine distance between the average embeddings. We trained the embeddings using GLoVe [23] on the entire corpus. For STMC3, we set $\delta = 0.2$.

4.4 Results

In Table 2 we list the number of diagnostic instances we created for each diagnostic dataset. In addition, we also performed a sanity check on the extent to which the document order determined by each axiom corresponds to the relevance judgements. Although only a small set of instances from each diagnostic dataset contains a document with a relevant document (row II in Table 2) we already see a trend: apart from TFC1 and M-TDC where 91% and 82% of the diagnostic instances have an agreement between axiomatic ordering and relevance ordering, the remaining axioms are actually not in line with the relevance ordering for most of the instances. This is a first indication that we have to consider an alternative set of axioms, better fit for such a corpus, in future work.

In Table 3 we report the fraction of instances both QL and BERT fulfil for each diagnostic dataset. As expected, QL *correctly* (as per the axiom) ranks the document pairs or triples most of the time, with the only outlier being TP, where QL performs essentially random—again, not a surprise given that QL is a bag-of-words model. In contrast—with the exception of LNC2, where BERT's ranking is

[11] Note that we only append the document to itself if the final size does not exceed 512.

essentially the opposite of what the LNC2 axiom considers correct (with only 6% of the instances ranked correctly)—BERT has not learnt anything that is related to the axioms as the fraction of correctly ranked instances hovers around 50% (which is essentially randomly picking a document order). Despite this lack of axiomatic fulfilment, BERT clearly outperforms QL, indicating that the existing axioms are not suitable to analyse BERT.

The reverse ranking BERT proposes for nearly all of the LNC2 instances can be explained by the way the axiom is phrased. It is designed to avoid over penalising documents, and thus a duplicated document should always have a retrieval score that is not lower than the original document. The opposite argument though could be made too (and BERT ranks accordingly), that a duplicated document should not yield a higher score than the original document as it does not contain novel/additional information. As we did not provide LNC2 instances in the training set, BERT is not able to rank according to the axiom, in line with the findings of other neural approaches as shown by Rennings et al. [25].

Finally, we observe that, counter-intuitively, BERT does not show a performance better than QL for semantic term matching constraints. For instance, one may expect that BERT would fare quite well on STMC1, given its semantic nature. However, our results indicate that BERT is actually considers term matching as one of its key features. In order to further explore this tension between semantic and syntactic term matching, we split the queries in our test set by their term overlap between the query and the relevant document (if several relevant documents exist for a query, we randomly picked one of them). If a query/document has no (or little) term overlap, we consider this as a semantic match.

Table 3. Overview of the retrieval effectiveness (nDCG columns) and the fraction of diagnostic dataset instances each model ranks correctly.

	nDCG	MRR	TFC1	TFC2	M-TDC	LNC1	LNC2	TP	STMC1	STMC2	STMC3
QL	0.2627	0.3633	**0.99**	**0.70**	**0.88**	**0.50**	**1.00**	0.39	0.49	**0.70**	**0.70**
DistilBERT	**0.3633**	**0.4537**	0.61	0.39	0.51	**0.50**	0.00	**0.41**	**0.50**	0.51	0.51

The results of this query split can be found in Fig. 1. We split the query set roughly into three equally sized parts based on the fraction of query terms appearing in the relevant document (as an example, if a query/document pair has a fraction of 0.5, half of all query terms appear in the document). We report results for all queries (Fig. 1 (left)), as well as only those where the relevant document appears in the top-100 QL ranking (Fig. 1 (right)). We find that BERT outperforms QL across all three splits, indicating that BERT is indeed able to pick up the importance of syntactic term matching. At the same time, as expected, BERT is performing significantly better than QL for queries that require a large amount of semantic matching. That brings into question on why, then, the axiomatic performance across our semantic axiomatic datasets does not reflect that. One hypothesis is that the semantic similarity we measure (based on context-free word embeddings) is different to the semantic similarity measured via contextual word embeddings. This in itself is an interesting avenue for future work, since it brings a new question on *what* that semantic relationship is, and how to accurately measure it.

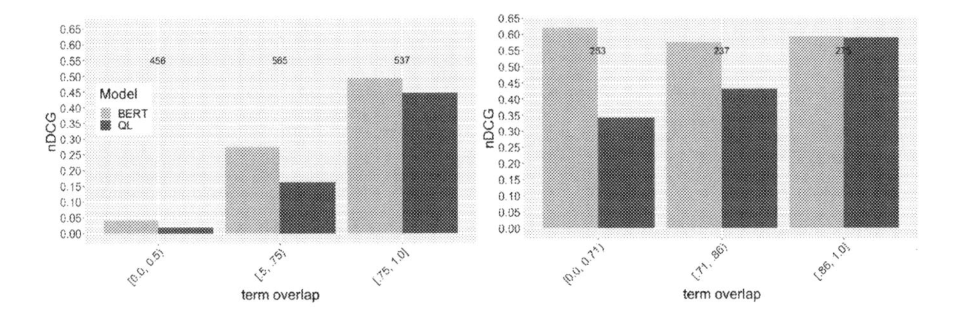

Fig. 1. The test queries are split into three sets, depending on the fraction of term overlap between the query and its corresponding relevant document. On the left, we plot all queries, on the right only those queries for which the relevant document appears in the top-100 ranked documents of the QL ranking.

5 Discussion and Conclusion

In this paper, we set out to analyze BERT with the help of the recently proposed *diagnostic datasets for IR based on retrieval heuristics* approach [25]. We expected BERT to perform better at fulfilling some of the proposed semantic axioms. Instead, we have shown that BERT, while significantly better than traditional models for ad-hoc retrieval, does not fulfil most retrieval heuristics, created by IR experts, that are supposed to produce better results for ad-hoc retrieval models. We argue that based on these results, the axioms are not suitable to analyse BERT and it is an open question what type of axioms would be able to capture some performance aspects of BERT and related models. In fact, how to arrive at those additional axioms, based on the knowledge we have now gained about BERT is in itself an open question.

Acknowledgement. This research has been supported by NWO project SearchX (639.022.722).

References

1. Linzen, T., Chrupala, G., Alishahi, A.: Proceedings of the 2018 EMNLP Workshop BlackboxNLP: Analyzing and Interpreting Neural Networks for NLP. ACL (2018)
2. Abdul-Jaleel, N., et al.: UMass at TREC 2004: Novelty and HARD. Computer Science Department Faculty Publication Series, p. 189 (2004)

3. Borlund, P.: The concept of relevance in IR. J. Am. Soc. Inf. Sci. Technol. **54**(10), 913–925 (2003)
4. Clark, K., Khandelwal, U., Levy, O., Manning, C.D.: What does BERT look at? An analysis of BERT's Attention. CoRR abs/1906.04341 (2019)
5. Clinchant, S., Gaussier, E.: Is document frequency important for PRF? In: Amati, G., Crestani, F. (eds.) ICTIR 2011. LNCS, vol. 6931, pp. 89–100. Springer, Heidelberg (2011). https://doi.org/10.1007/978-3-642-23318-0_10
6. Dai, Z., Callan, J.: Deeper text understanding for IR with contextual neural language modeling. In: SIGIR, pp. 985–988. ACM (2019)
7. Devlin, J., Chang, M., Lee, K., Toutanova, K.: BERT: pre-training of deep bidirectional transformers for language understanding. In: NAACL-HLT (1), pp. 4171–4186. ACL (2019)
8. Fang, H.: A re-examination of query expansion using lexical resources. In: ACL, pp. 139–147. ACL (2008)
9. Fang, H., Tao, T., Zhai, C.: A formal study of information retrieval heuristics. In: SIGIR, pp. 49–56. ACM (2004)
10. Fang, H., Zhai, C.: An exploration of axiomatic approaches to information retrieval. In: SIGIR, pp. 480–487. ACM (2005)
11. Fang, H., Zhai, C.: Semantic term matching in axiomatic approaches to information retrieval. In: SIGIR, pp. 115–122. ACM (2006)
12. Fernando, Z.T., Singh, J., Anand, A.: A study on the interpretability of neural retrieval models using DeepSHAP. In: SIGIR, pp. 1005–1008. ACM (2019)
13. Hagen, M., Völske, M., Göring, S., Stein, B.: Axiomatic result re-ranking. In: CIKM, pp. 721–730. ACM (2016)
14. Hu, B., Lu, Z., Li, H., Chen, Q.: Convolutional neural network architectures for matching natural language sentences. In: NIPS, pp. 2042–2050 (2014)
15. Jia, R., Liang, P.: Adversarial examples for evaluating reading comprehension systems. In: EMNLP, pp. 2021–2031. ACL (2017)
16. Lavrenko, V., Croft, W.B.: Relevance-based language models. In: SIGIR, pp. 120–127. ACM (2001)
17. Lin, J.: The neural hype and comparisons against weak baselines. SIGIR Forum **52**(2), 40–51 (2018)
18. MacAvaney, S., Yates, A., Cohan, A., Goharian, N.: CEDR: contextualized embeddings for document ranking. In: SIGIR, pp. 1101–1104. ACM (2019)
19. McDonald, R., Brokos, G., Androutsopoulos, I.: Deep relevance ranking using enhanced document-query interactions. In: EMNLP, pp. 1849–1860. ACL (2018)
20. Mitra, B., Craswell, N.: An introduction to neural information retrieval. Found. Trends Inf. Retrieval **13**(1), 1–126 (2018)
21. Niven, T., Kao, H.: Probing neural network comprehension of natural language arguments. In: ACL (1), pp. 4658–4664. ACL (2019)
22. Pang, L., Lan, Y., Guo, J., Xu, J., Cheng, X.: A deep investigation of deep IR models. CoRR abs/1707.07700 (2017)
23. Pennington, J., Socher, R., Manning, C.D.: GloVe: global vectors for word representation. In: EMNLP, pp. 1532–1543. ACL (2014)
24. Radford, A., Wu, J., Child, R., Luan, D., Amodei, D., Sutskever, I.: Language models are unsupervised multitask learners (2019)
25. Rennings, D., Moraes, F., Hauff, C.: An axiomatic approach to diagnosing neural IR models. In: Azzopardi, L., Stein, B., Fuhr, N., Mayr, P., Hauff, C., Hiemstra, D. (eds.) ECIR 2019. LNCS, vol. 11437, pp. 489–503. Springer, Cham (2019). https://doi.org/10.1007/978-3-030-15712-8_32

26. Robertson, S.E., Zaragoza, H.: The probabilistic relevance framework: BM25 and beyond. Found. Trends Inf. Retrieval **3**(4), 333–389 (2009)
27. Rosset, C., Mitra, B., Xiong, C., Craswell, N., Song, X., Tiwary, S.: An axiomatic approach to regularizing neural ranking models. In: SIGIR, pp. 981–984. ACM (2019)
28. Sanh, V., Debut, L., Chaumond, J., Wolf, T.: DistilBERT, a Distilled Version of BERT: Smaller. Faster, Cheaper and Lighter (2019)
29. Saracevic, T.: Relevance reconsidered. In: CoLIS 2, pp. 201–218. ACM (1996)
30. Shi, S., Wen, J.R., Yu, Q., Song, R., Ma, W.Y.: Gravitation-based model for information retrieval. In: SIGIR, pp. 488–495. ACM (2005)
31. Tao, T., Zhai, C.: An exploration of proximity measures in information retrieval. In: SIGIR, pp. 295–302. ACM (2007)
32. Tenney, I., Das, D., Pavlick, E.: BERT rediscovers the classical NLP pipeline. In: ACL (1), pp. 4593–4601. ACL (2019)
33. Vaswani, A., et al.: Attention is all you need. In: NIPS, pp. 5998–6008 (2017)
34. Verma, M., Ganguly, D.: LIRME: locally interpretable ranking model explanation. In: SIGIR, pp. 1281–1284. ACM (2019)
35. Voorhees, E.M.: Overview of the TREC 2004 robust retrieval track. In: TREC (2004)
36. Wang, A., Singh, A., Michael, J., Hill, F., Levy, O., Bowman, S.R.: GLUE: a multi-task benchmark and analysis platform for natural language understanding. In: BlackboxNLP@EMNLP, pp. 353–355. ACL (2018)
37. Weston, J., Bordes, A., Chopra, S., Mikolov, T.: Towards AI-complete question answering: a set of prerequisite toy tasks. In: ICLR (Poster) (2016)
38. Wu, Y., et al.: Google's neural machine translation system: bridging the gap between human and machine translation. CoRR abs/1609.08144 (2016)
39. Yang, L., Ai, Q., Guo, J., Croft, W.B.: aNMM: ranking short answer texts with attention-based neural matching model. In: CIKM, pp. 287–296. ACM (2016)
40. Yang, W., Zhang, H., Lin, J.: Simple applications of BERT for ad hoc document retrieval. CoRR abs/1903.10972 (2019)
41. Yang, Z., Dai, Z., Yang, Y., Carbonell, J.G., Salakhutdinov, R., Le, Q.V.: XLNet: generalized autoregressive pretraining for language understanding. In: NeurIPS, pp. 5754–5764 (2019)
42. Zhai, C., Lafferty, J.: A study of smoothing methods for language models applied to ad hoc information retrieval. In: ACM SIGIR Forum, vol. 51, pp. 268–276. ACM (2017)

Queries

Generation of Synthetic Query Auto Completion Logs

Unni Krishnan[1], Alistair Moffat[1(✉)], Justin Zobel[1], and Bodo Billerbeck[2]

[1] The University of Melbourne, Melbourne, Australia
ammoffat@unimelb.edu.au
[2] Microsoft Australia and RMIT University, Melbourne, Australia

Abstract. Privacy concerns can prohibit research access to large-scale commercial query logs. Here we focus on generation of a synthetic log from a publicly available dataset, suitable for evaluation of query auto completion (QAC) systems. The synthetic log contains plausible string sequences reflecting how users enter their queries in a QAC interface. Properties that would influence experimental outcomes are compared between a synthetic log and a real QAC log through a set of side-by-side experiments, and confirm the applicability of the generated log for benchmarking the performance of QAC methods.

1 Introduction

Query auto completion (QAC) systems offer a list of completions while users enter queries in a search interface. Users can either submit one of the completions as their *query*, or *advance* their *partial query* by selecting a completion and then continuing to type [33]. A detailed QAC log capturing the sequence of partial queries, along with the completions presented and the user interactions with them, is required in order to evaluate a QAC system [37,38]. However, concerns about the privacy of query logs and regulatory requirements such as GDPR mean that there is a need for alternative ways of obtaining logs for academic purposes.

Here we explore a framework for generating synthetic QAC logs, extending the work of Krishnan et al. [33], who suggest converting a QAC log to an abstracted format (an *abstract QAC log*) that records only the length of each partial query and the lengths of words used, minimizing privacy concerns but removing the possibility of performing evaluations on actual strings. Synthetic QAC log generation seeks to produce a list of *plausible* synthetic partial query sequences by mapping the word lengths from the abstract QAC log to strings from a publicly available dataset. An example of the proposed process is shown in Fig. 1. On the left are partial queries typed by a user. The abstracting process converts these to the digit sequences shown in the middle column, describing the strings but not their characters; and then the corresponding synthesized strings are shown at the right. Note that it is neither necessary nor sufficient for the synthetic log to contain semantically valid phrases. Comparison between the

J. M. Jose et al. (Eds.): ECIR 2020, LNCS 12035, pp. 621–635, 2020.
https://doi.org/10.1007/978-3-030-45439-5_41

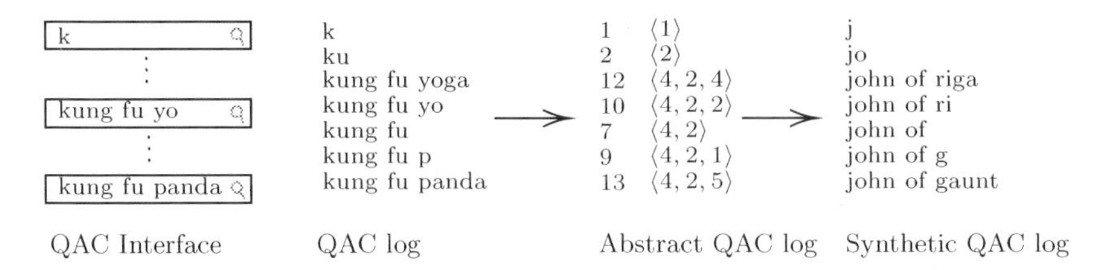

Fig. 1. A sample QAC log entry, the corresponding abstract QAC log entry, and a synthetic QAC log entry generated from the string "`john of gaunt`".

original and synthetic logs across a range of properties show that the synthetic log can be used to evaluate QAC system performance. Moreover, the synthetic log eliminates the privacy concerns associated with the original.

2 Background

A QAC system retrieves a *candidate set* matching the partial query P, drawing from a target string collection, with strings in the target collection having an associated *score*. Query Auto Completion systems typically match P against past queries from a log; or, in the absence of logs, they can also be synthesized [14,40]. Methods of ranking the candidates include static popularity [9], search context [9,29], forecast popularity [15], personalized ranking parameters [15,29,42], and diversity [16]. It is also possible to choose an initial candidate set based on popularity and then apply a second ranking criteria to obtain the final strings [15,16]. QAC implementation strategies vary based on how the partial query P is matched against the target strings [32]. A common approach is to use a trie [3,4,25,27] to retrieve candidates that have P as a prefix; or inverted index-based approaches [10,11,23] that offer completions independent of the ordering of the words in the partial query. The functionality of a QAC system can be extended beyond character level matches by including contextual cues [11] or synonyms [12,28]. Error-tolerant QAC approaches [17,28,36,47] allow up to a fixed number of character mismatches to account for possible typing errors.

User interactions are a key factor in implementation and evaluation of QAC systems [26,33,38] and have been captured using a wide range of models [31–33,37,38,43,44]. In particular, users are not limited to entering single characters, and can alter the partial query by selecting a completion or deleting characters already entered. Until now, the *test collections* used to evaluate traditional search systems have been anonymized commercial logs [1,19,34,48] or synthesized logs [5,30,45,49]. QAC system evaluations have typically been performed over large publicly available string collections [10,14,23], with strings taken sequentially from left-to-right to generate partial query sequences [10]. However this approach does not account for the full range of possible interactions [32,33]. In this work, we explore an approach to generation of synthetic partial query sequences that addresses this gap.

Abstract QAC log	seedsig list	targsig list	Surrogate log
...	...	...	...
$\langle 2\rangle, \langle 2,1\rangle, \langle 2,5\rangle$	$\langle 2,5\rangle$	$\langle 9\rangle$	`autopilot`
$\langle 1\rangle, \langle 2\rangle, \langle 5\rangle, \langle 5,2\rangle$	$\langle 5,2\rangle$	$\langle 5,8\rangle$	`stack overflow`
$\langle 6,3\rangle$	$\langle 6,3\rangle$	$\langle 9\rangle$	`christmas`
$\langle 6\rangle, \langle 2\rangle, \langle 1\rangle, \langle 7,9\rangle$	$\langle 7,9\rangle$	$\langle 6,3\rangle$	`coffee mug`
$\langle 1\rangle, \langle 2\rangle\ \langle 6\rangle, \langle 6,3\rangle$	$\langle 6,3\rangle$	$\langle 4,4\rangle$	`main page`
...	...	...	...

Fig. 2. The synthetic pattern generation process. Signatures of FinalP from the abstract QAC log are used to form the seedsig list, and the signatures from surrogate log form targsig list. A match between the signature $\langle 6,3\rangle$ in the seedsig list and in the targsig list might correspond to the target string "`coffee mug`".

Terminology. A partial query P is the string currently displayed in the search box. An *interaction* updates P and results in loggable changes. A new *conversation* starts when the user begins a query, and continues until either explicitly terminated by the user or as a result of a session timeout. The last partial query from a conversation is referred to as FinalP. A QAC log records a set of conversations as a sequence of partial queries. A *surrogate log* is a dictionary of strings that can be used as substitutes for final partial queries, with each such string having an associated score reflecting its popularity. For a string T, the ordered tuple $\langle |\mathsf{w}_1|, |\mathsf{w}_2| \ldots |\mathsf{w}_k|\rangle$ representing the lengths of its words $\mathsf{w}_1, \mathsf{w}_2, \ldots \mathsf{w}_k$ in T is referred to as its *signature*. For each partial query P, an abstract QAC log records only its signature and its length $|\mathsf{P}|$, including whitespace. The signature of each FinalP in the abstract QAC log is referred to as seedsig. The signature of a string in the surrogate log is its targsig.

Problem Definition. For each conversation in the abstract QAC log, find a *target string* in the surrogate log with the same signature as the final partial query FinalP. Then, starting from the first interaction in the conversation, apply the word lengths from the abstract QAC log to the target string in order to obtain a plausible partial query sequence. For example, consider the last conversation in Fig. 2, with its signature sequence $\langle 1\rangle, \langle 2\rangle, \langle 6\rangle, \langle 6,3\rangle$. The seedsig for this conversation is $\langle 6,3\rangle$. The string "`coffee mug`" in the surrogate log has the same signature and hence might be selected as a target string. Mapping the signature sequence from the conversation, we get the synthetic partial query sequence "`c`", "`co`", "`cof`", "`coffee`", and then "`coffee mug`".

3 Generation Process

Depending on the distribution of word lengths in the surrogate log, for every final partial query in the abstract QAC log, there may not be a string having the same signature. Moreover, strings are not entered by the users in the word order of the target collection. For instance, a user looking for the Wikipedia main page might enter the queries "`wikipedia`", "`main page wikipedia`", or "`wiki`". We narrow down the possible ways of matching a seedsig with the list of signatures in targsig list to the following hierarchical *modes*:

1. *Exact.* The targsig is equal to the seedsig. For example, the seedsig $\langle 3, 3, 2\rangle$ only matches with the target signature $\langle 3, 3, 2\rangle$. There might be zero, or multiple strings in the surrogate log that match.
2. *Prefix.* The seedsig is a prefix of the targsig. For example, seedsig $\langle 3, 3, 2\rangle$ matches $\langle 3, 3, 2, 4\rangle$ and $\langle 3, 3, 2, 4, 7\rangle$, but not $\langle 9, 3, 3, 2\rangle$.
3. *Match by Drop* (MbD). The seedsig is an ordered subset of the targsig, For example, $\langle 3, 3, 2\rangle$ matches $\langle 3, 4, 3, 2\rangle$ and $\langle 4, 3, 1, 3, 6, 2\rangle$, but not $\langle 3, 4, 2, 3\rangle$.
4. *Bag of Numbers* (BoN). Relaxing the ordering requirement, a BoN match occurs if the sets of word lengths in seedsig are a subset of those in targsig. For example, seedsig $\langle 3, 3, 2\rangle$ matches $\langle 2, 3, 3, 4\rangle$ and $\langle 5, 2, 3, 6, 3\rangle$, but not $\langle 3, 4, 2, 6\rangle$. In a BoN match the target string's words are reordered to match the seed signature ordering.

Locating Target Strings. The first step locates, for a given seedsig, a set of matching signatures in the targsig list. The set of target candidates is maintained in lexicographically sorted order, so that the exact-match signatures for a given seedsig can be found via two binary searches, establishing a range $[\mathsf{r}_{beg}, \mathsf{r}_{end})$. A similar process can be used to find the prefix-match range, which is a larger contiguous block in the lexicographically sorted array of target candidates.

If a target signature S_t matches with seedsig using MbD, any targsig having S_t as a prefix will also have an MbD match with seedsig. For example, under MbD the seedsig $\langle 3, 5\rangle$ matches $\langle 3, 2, 5\rangle$. Then signatures $\langle 3, 2, 5, 4\rangle$ and $\langle 3, 2, 5, 9, 16\rangle$ will also be a match because they have $\langle 3, 2, 5\rangle$ as a prefix. Using this property, once a matching signature S_t for MbD is found, we can add the prefix match range for S_t to the MbD range for the current seedsig. In contrast to exact and prefix match ranges, MbD ranges may not form a continuous range over SigList.

For BoN matching, the tokens in the signature are sorted to form a canonical representation. For example, $\langle 3, 2, 4\rangle$ becomes $\langle 2, 3, 4\rangle$. This list is sorted into lexicographical order. A BoN match between the canonical representations of a seedsig and a target signature S_{tc} can be verified by a linear scan over the tokens. The matching ranges for BoN in the modified SigList are calculated by finding prefix match ranges for each S_{tc} that matches with seedsig using BoN.

Handling Deletions. To include deletions, we assume that something different was initially typed (the *replacement* word) and was converted to the word from the target string after the deletions. In particular, suppose that the current conversation contains the deletion of a word w_k starting from the ith interaction and ending with the jth interaction. Then, for every interaction preceding the jth interaction, w_k is substituted by a *replacement* word constructed via a set of deletion heuristics. If character sequences are deleted and re-entered several times during the conversation, several replacement words will be required. The last replacement word should be close to the original word, and the penultimate

Table 1. Deriving synthetic partial query sequence for a conversation containing multiple word deletions. The fourth column shows three target strings used.

Sequence	|P|	Signature	Modified target string	Synthetic P
1	12	$\langle 5,6 \rangle$	"`black mirror`"	"`black mirror`"
2	11	$\langle 5,5 \rangle$	"`black mirror`"	"`black mirro`"
3	5	$\langle 5 \rangle$	"`black mirror`"	"`black`"
4	6	$\langle 5 \rangle$	"`black hole`"	"`black`"
5	7	$\langle 5,1 \rangle$	"`black hole`"	"`black h`"
6	8	$\langle 5,2 \rangle$	"`black hole`"	"`black ho`"
7	7	$\langle 5,1 \rangle$	"`black hole`"	"`black h`"
8	8	$\langle 5,2 \rangle$	"`black hawk down`"	"`black ha`"
9	15	$\langle 5,4,4 \rangle$	"`black hawk down`"	"`black hawk down`"

replacement prior to it should similarly be a modification of the last replacement word. Table 1 gives an example of the partial query sequence generated by rewriting target strings when the conversation contains multiple deletions.

Users delete characters either due to typing errors [6,20] made within that word, or to switch to an entirely different word. Among various typing errors that can occur in any character level entry systems [22], the following error categories that are predominantly discussed in past research [6,20,21,46] are included in the generation process.

Deletion. A deletion error occurs when the user initially missed out one of the characters in the word, for example, in the correction "`acount`" → "`account`". Deletion errors are more frequent at character repetitions, and occur more commonly at the beginning of a word [6]. To simulate a deletion error, if there is a deletion sequence from ith interaction to the jth interaction in which more than two characters of w_k are removed, then w_k from the $j+1$th interaction (after the deletions) is examined to see if the last two characters in w_k are repeating. The error is then simulated by deleting one of the repeating characters of w_k in the interactions prior to $j+1$.

Insertion. These arise when an extra letter is initially typed, for example, the correction "`sherriff`" → "`sheriff`". Our experimental results show that only a low fraction of insertion errors are observed in a QAC log and for that reason, they are not included in the synthetic QAC log generation process.

Substitution. These arise when the user enters one of the neighboring keys instead of the intended key, for example, the correction "`disturv`" → "`disturb`". Single character deletions are assumed to be substitution errors. The substituted character is found via a probability distribution for mistyped keys around the current key.

Transposition. This is when the user swaps two characters in a word either not knowing the correct spelling or because of mis-ordered keystrokes, for

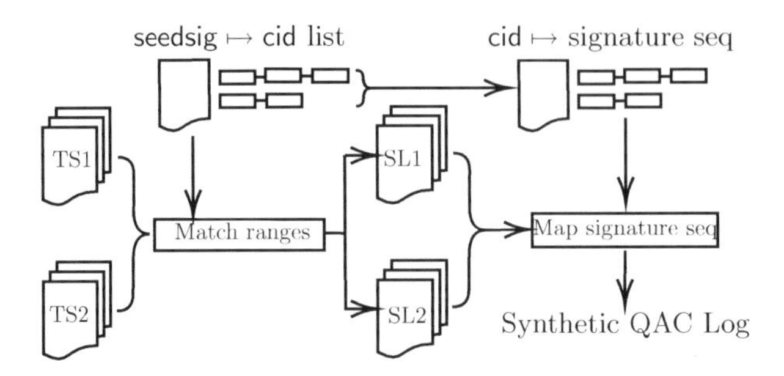

Fig. 3. Framework for generating a synthetic QAC log. See the text for details.

example, the correction "wierd" → "weird". Deletions of length two are considered to be transposition errors and the last two characters of the word are swapped to get the replacement word.

In a QAC interface, typing errors may not be the only reason why users delete characters. Sometimes users replace certain parts of a well-formed partial query, for instance to get a different set of completions. In such cases, referred as *multichar* deletions, the corrected word will differ from the original word in two or more character positions. To allow for such cases, the deleted part of the word is replaced with another string starting with the word's remaining prefix, but differing in the next character. For example, consider the word "hawk" in Table 1. A deletion chain removed all of the characters except the "h"; and hence (working backwards), "hawk" was replaced by an alternate word that starts with "h" but not "ha", such as "hole".

Generation Framework. The overall framework for generation of a synthetic QAC log is illustrated in Fig. 3. The seed signatures are stored with a mapping from signatures to the list of cids having the same signature. Signatures from the surrogate log are precomputed and stored in lexicographically sorted order (TS1). Additionally, the strings in the surrogate log are re-ordered based on TS1 ordering to obtain the permutation SL1. The permutations TS2 stores a sorted list of within sorted signatures to support BoN matching, and SL2 is the corresponding permutation of the surrogate log. To generate a synthetic QAC log entry, a signature is selected from seedsig ↦ cid list and a list of target signatures retrieved based on the four modes. Using the mappings SL1 and SL2, strings from the surrogate log corresponding to the target signatures are then obtained.

The strings are converted to target strings by aligning the word lengths with the seedsig. A target string can be generated from multiple originals. If, say, seedsig is $\langle 9, 4\rangle$, then "wikipedia main page" gives the target string "wikipedia page" using MbD; and another string "personal web page wikipedia" gives the same target using BoN mode. If the same target string is produced by more than one string from the surrogate log, then the one with lower score is discarded.

The signature sequences obtained from the file providing the cid ↦ signature sequence mapping are then applied to the list of target strings, to obtain a synthetic partial query sequence for each conversation. Partial query sequences are generated for each cid from the current seedsig ↦ cid list until either the cid list or the target strings are exhausted. Finally, after the generation process is completed, the synthetic QAC log is re-ordered so that the conversations in the synthetic QAC log have one-to-one correspondence with the abstract QAC log.

4 Experiments

Datasets Used. The Wikipedia clickstream dataset,[1] generated from Wikipedia request logs containing tuples of the form (*referrer*, *resource*, *frequency*), is used to generate the surrogate log. The number of requests for an article (*resource*) is the *frequency*. Data dumps from January to March 2019 were aggregated by updating the *frequency* of each *resource* by the mean *frequency* over the dumps. We refer to the resulting dataset as `Wiki-Clickstream` and the synthetic QAC log generated from `Wiki-Clickstream` as `Wiki-Synth`.

A QAC log was formed by randomly sampling the logs from Bing[2] QAC system over a period of one week from 13 August 2018. This `Bing-QAC-2018` records the partial queries entered by the users and a unique cid for each conversation. The abstracted version of `Bing-QAC-2018`, referred to as `Bing-Abs-QAC-2018`, is generated by recording the signature and total length of each partial query along with a unique cid for each conversation. Note that `Wiki-Synth` was generated from `Bing-Abs-QAC-2018` only (with no use of `Bing-QAC-2018`), and that the subsequent comparisons between `Wiki-Synth` and `Bing-QAC-2018` were performed on secure Microsoft servers and in accordance with Microsoft privacy requirements.

Preprocessing. Conversations in `Bing-Abs-QAC-2018` where an intermediate signature contained more words than the signature of the final partial query, or contained words that were longer than the corresponding words in the final partial query, were removed. The strings from `Bing-QAC-2018` were *transliterated* from UTF-8 to ASCII encoding using `Unidecode`[3] Python package. Conversations containing non-converting characters in any partial query were discarded. The strings from `Wiki-Clickstream` were transliterated in the same way and non-converting strings were removed. The conversion from UTF-8 to ASCII resulted in the loss of 0.26 million conversations from `Bing-QAC-2018` and 0.02 million strings from `Wiki-Clickstream`. After the preprocessing, there were 1.44 million conversations in `Bing-Abs-QAC-2018` and 5.11 million strings in `Wiki-Clickstream`. In `Bing-Abs-QAC-2018`, 7.53% of signatures were unique, of which 2.26% had no matches in `Wiki-Clickstream`. A total of 0.23 million conversations from

[1] https://meta.wikimedia.org/wiki/Research:Wikipedia_clickstream, accessed 29th October, 2019.
[2] https://www.bing.com, accessed 29th October, 2019.
[3] https://pypi.org/project/Unidecode, accessed 29th October, 2019.

`Bing-Abs-QAC-2018` were not included in `Wiki-Synth` because there were not enough matching strings in `Wiki-Clickstream` to map their signature sequence.

Sampling of Target Strings. The frequency of search queries has been found to exhibit a power law distribution [8] with the probability distribution $p(x) \propto x^{-\alpha}$. Therefore, while generating a synthlog, the target strings need to be sampled to obtain a power law distribution based on their frequencies. Strings in the `Wiki-Clickstream` log have a score that tends to follow a power law distribution with $\alpha = 3.09$. Using this observation we sample target strings based on a weighted probability distribution over their scores, so that the probability of a string T given a seed signature $\mathsf{seedsig}$ is

$$Prob(\mathsf{T} \mid \mathsf{seedsig}) = \frac{score(\mathsf{T})}{\sum_{i=1}^{n} score(\mathsf{T}_i)} \tag{1}$$

where $\mathsf{T}_1, \mathsf{T}_2, \ldots, \mathsf{T}_n$ are the set of target strings retrieved for $\mathsf{seedsig}$. The frequency distribution of the resulting target strings used as the FinalP in `Wiki-Synth` is analyzed below, where we compare partial query frequencies.

Language Model for Finding Replacement Words. For deletion types other than multichar, the replacement word will be a modification of the original word based on the error category. For multichar deletions, the heuristics discussed above are used to find replacement words. The *best* replacement word based on contextual information among the candidate words that passes the heuristics is selected using a 4-gram language model (LM) trained on the title strings from `Wiki-Clickstream`. The language model is generated using KenLM [24],[4] which is based on modified Kneser-Ney smoothing and provides fast model construction and querying. For example, an LM-based replacement for the target string "`live queen`", is "`live together`" while a random replacement gives "`live teufelshorner`"; for the target string "`web server`", a random replacement yields "`web castelvetere`" and an LM-based replacement gives "`web content`". A similar character bigram model trained on the Microsoft spelling-correction dataset[5] is used to find the most likely next mistyped character, given the previous character, to simulate substitution errors.

Comparison of Synthetic QAC Logs with QAC Logs. A synthetic QAC log should have some properties similar to the QAC log, and these can be used to validate the generation process. A comparison of these properties is given in Table 2. Other properties are desirable if a synthetic QAC log is to be a substitute in experiments for a QAC log. While comparing these properties, the partial queries from both the logs are treated as the *test queries* that will be queried against a collection of strings acting as the *test collection*. Extending our assumption that a user's goal was to enter FinalP, we claim that these strings could get

[4] https://github.com/kpu/kenlm, accessed 29th October, 2019.
[5] https://www.microsoft.com/en-us/download/details.aspx?id=52418, accessed 29th October, 2019.

Table 2. Basic statistics of `Bing-QAC-2018` and `Wiki-Synth`.

Attribute	`Bing-QAC-2018`	`Wiki-Synth`
Number of conversations (millions)	1.74	1.21
Number of interactions (millions)	11.75	7.90
Percentage of unique partial queries	38.99	39.41
Percentage of unique final partial queries	52.30	63.32
Lengths of partial queries – mean (characters)	10.37	9.67
Lengths of partial queries – std.dev. (characters)	9.68	8.53

Table 3. Goodness of power law fit against other distributions.

Frequency distribution	Exponential		Log-normal	
	R	p	R	p
`Bing-QAC-2018`, P	11885.4	<0.01	0.04	0.8
`Wiki-Synth`, P	10895.3	<0.01	0.16	0.15
`Bing-QAC-2018`, FinalP	5903.6	<0.01	−2.42	0.04
`Wiki-Synth`, FinalP	1211.9	0.01	0.12	0.04

eventually indexed by the QAC system and subsequently be queried. Thus, in our experiments, the list of FinalP strings are considered as a representative sample of a larger hypothetical test collection.

Heap's Law. Heap's law gives an estimate of the number of unique terms V in a collection as a function of the number of terms N. The relationship is given by $V = kN^{\beta}$ and the typical values of the parameters are $30 \leq k \leq 100$, with $\beta \approx 0.50$ for English text [39] and $\beta \approx 0.60$ for web documents [7]. The values for Heap's law parameters estimated from `Bing-QAC-2018` are $k = 14.50$, $\beta = 0.67$ and the parameters from `Wiki-Synth` are $k = 32.61$, $\beta = 0.57$. The difference in growth rate can be explained by the nature of the logs. The growth rate of V is expected to be higher in a collection containing numbers and spelling errors [39]. Each of the FinalP strings from `Wiki-Synth` is a Wikipedia title that comes from a curated collection whereas FinalP strings in `Bing-QAC-2018` are strings entered by users. Considering, all partial queries, the parameters estimated are $k = 5.54$, $\beta = 0.71$ for `Bing-QAC-2018` and $k = 20.94$, $\beta = 0.60$ for `Wiki-Synth`.

Frequency of Partial Query Strings. The likelihood of a query being repeated over time as a result of the power-law distribution affects the performance of various query processing strategies [13,35]. The frequency distributions of partial queries and final queries from `Wiki-Synth` and `Bing-QAC-2018` are given in Fig. 4 (left) with dashed lines showing the power-law fits for the corresponding distributions. The distributions give similar exponents except for the distribution of FinalP from `Wiki-Synth`, which has $\alpha = 3.02$, indicating a steeper decay in the

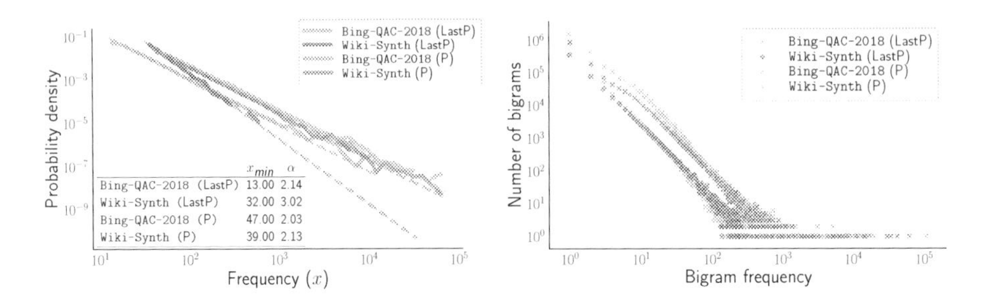

Fig. 4. Power law fit (left) and bigram frequency distribution (right).

probability of higher frequency partial query strings. The goodness of power-law fit against exponential and log-normal fits is compared using likelihood ratio R [2,18] along with the significance level p, as reported in Table 3. Except for the frequency distribution of FinalP from `Bing-QAC-2018`, which is better explained with a log-normal distribution ($R = -2.42, 0.04$), the other three distributions suit a power-law distribution, perhaps as a result of the sampling process followed. For some datasets, it is not surprising to get a better log-normal fit [18].

N-Gram Frequency. The distribution of terms in a collection can be modeled using a decay law which estimates the collection frequency of the ith most common term as $\mathsf{F}_i = Ci^k$, where $k = -1$ and C is constant. Similarity in term distribution with a real log is considered to be a desirable property of a synthetic log [45]. Figure 4 (right) shows the bigram frequency distribution from partial queries and final partial queries. We find close correspondence between the bigram frequency distributions (Kolmogorov-Smirnov statistic, $D = 0.05, 0.63$ for FinalP and $D = 0.06, 0.04$ for all partial queries). Unigram frequencies follow a similar trend. While the bulk of the distribution can be explained with the decay law, the tail of the distributions coming from rare n-grams show deviations.

Empirical Entropy. The empirical entropy of the FinalP strings gives a measure of their compressibility. The kth-order empirical entropy of a string $\mathsf{T}[1 \dots n]$ over an alphabet set Σ is given by

$$\mathsf{H}_k(\mathsf{T}) = \frac{1}{n} \sum_{w \in \Sigma^k} |\mathsf{T}_w| \cdot \mathsf{H}_0(\mathsf{T}_w), \text{ with } \mathsf{H}_0(\mathsf{T}) = \frac{1}{n} \sum_{c \in \Sigma} n_c \cdot \log \frac{n}{n_c}, \tag{2}$$

where T_w is formed by collecting the characters that immediately follow *context* w in T, and where n_c is the frequency of character c in T. A lower bound on the number of bits required to encode T is given by $n\mathsf{H}_k(\mathsf{T})$ [41]. For a list of strings $\mathsf{T}_1, \mathsf{T}_2, \dots, \mathsf{T}_l$, it can be observed that $\sum_i n_i \mathsf{H}_k(\mathsf{T}_i) \leq n_c \mathsf{H}_k(\mathsf{T}_c)$ where T_c is the concatenation of strings T_i. This can be shown by extending the proof given by Navarro [41, Chapter 2], applying Jensen's inequality to $\sum_i n_i \mathsf{H}_k(\mathsf{T}_i)$. Therefore, we consider $n_c \mathsf{H}_k(\mathsf{T}_c)$ as the worst case lower bound for the space required to encode the FinalP strings from `Bing-QAC-2018` and `Wiki-Synth`. The values of $\mathsf{H}_k(\mathsf{T}_c)$ for $k = 0 \dots 4$ computed from `Bing-QAC-2018` and `Wiki-Synth` are given

Table 4. Values of H_k and number of contexts C_t per length n ($\times 10^{-3}$) of the concatenated string formed by the FinalP strings of `Bing-QAC-2018` and `Wiki-Synth`.

Dataset	$k = 0$		$k = 1$		$k = 2$		$k = 3$		$k = 4$	
	H_0	C_t/n	H_1	C_t/n	H_2	C_t/n	H_3	C_t/n	H_4	C_t/n
`Bing-QAC-2018`	4.39	0.00	3.64	0.00	2.50	0.17	1.41	2.51	0.64	11.27
`Wiki-Synth`	4.28	0.00	3.43	0.00	2.32	0.08	1.27	0.97	0.54	4.95

Table 5. Example for pre-correction and post-correction strings. In the table "`pal`" is taken as the pre-correction string and "`pla`" is taken as the post-correction string. This correction pair will be classified as a transposition error.

Inter.	Partial query	Word len.	Edit dist.
2	`coffee pa`	2	
3	`coffee pal`	3	
4	`coffee pa`	2	
5	`coffee p`	1	
6	`cofee pl`	2	1
6	`coffee pla`	**3**	**1**
6	`cofee places`	6	4

in Table 4. The values for H_k tend to be similar between the two logs. Slightly higher values of H_k and increased C_t/n for `Bing-QAC-2018` can be explained by the higher degree of randomness expected from web search queries, compared to the results obtained for Heap's law coefficients.

Identification of Typing Errors. Similarities in typing patterns between the synthetic QAC log and the original QAC log are verified by analyzing the typing errors present in both logs. Typing errors are identified by extracting and comparing a set of *pre-correction* and *post-correction* strings from the conversations, extending the method proposed by Baba and Suzuki [6]. When the kth word w_k in P gets deleted from the ith interaction, w_k from the $i-1$th interaction is taken as the pre-correction string. To find the post-correction string, the edit distance between the pre-correction string and the words w_k from the interactions after the end of current deletion sequence are calculated until $|\mathsf{w}_k|$ from interaction j is less than that from interaction $j+1$. From this list, the word having minimum edit distance with the pre-correction string is taken as the post correction string. Table 5 gives an example for how pre-correction and post-correction strings are calculated from a deletion sequence. The edit distance between two strings is calculated as a Damerau-Levenshtein distance, so transposition of two characters is given a cost of 1. The pre-correction and post-correction pairs are then compared to classify the possible typing errors as one of the types discussed above. The results obtained by analyzing the typing errors from `Bing-QAC-2018`

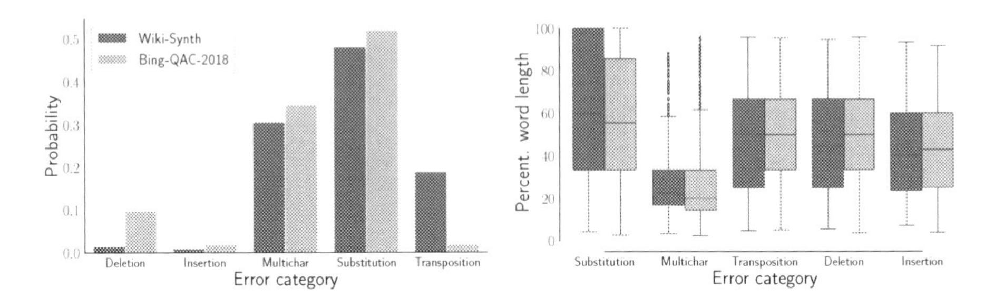

Fig. 5. Comparison between frequency (left) and position (right) of typing errors.

and the synthetic QAC log are in Fig. 5. The fraction of transposition errors was found to be higher in `Wiki-Synth` compared to `Bing-QAC-2018` while the latter contains more deletion errors than `Wiki-Synth`. As a result of the method of simulating substitution errors, they tend to occur more towards the end of words in `Wiki-Synth`. The remainder of the error categories are distributed across similar word positions in both the logs, suggesting similarities in typing patterns.

5 Conclusions

We have explored a method for generating a synthetic QAC log from an abstract QAC log, by mapping the word lengths of the abstract QAC log to those of a publicly available string collection, and applying a range of corrective techniques. Synthetic QAC formation can also be posed as a language generation problem relying on various models of QAC systems [31,33].

We have demonstrated that the synthetic log generated from a pre-existing string collection encompasses many of the properties found in the original QAC log from which the abstract QAC log was derived. In particular, an analysis of typing errors found in actual QAC logs is reported, along with a description of how they were introduced into the synthetic log. As a result, there is a close correspondence between the real QAC log and the synthetic QAC log across a range of properties, each of which might influence the computational cost of providing the completion strings. That is, experiments using the synthetic QAC log can be expected to provide close approximations to behaviors that would be observed on a real QAC log. A particular example is efficiency, which we plan to examine in future work, comparing the performance of a range of QAC implementations using both synthetic and real QAC logs.

Acknowledgments. This work was supported by the Microsoft Research Centre for Social Natural User Interfaces (SocialNUI) at The University of Melbourne. We are grateful to Peter Bailey (Microsoft Australia) for his support of that project, and for additionally facilitating the work that is reported here.

References

1. Adar, E.: User 4xxxxx9: anonymizing query logs. In: Proceedings of the WWW Query Log Analysis Workshop (2007). http://www.cond.org/anonlogs.pdf
2. Alstott, J., Bullmore, E., Plenz, D.: Powerlaw: a Python package for analysis of heavy-tailed distributions. PLoS One **9**(1), 1–11 (2014)
3. Askitis, N., Sinha, R.: HAT-trie: a cache-conscious trie-based data structure for strings. In: Proceedings of the Australasian Conference on Computer Science, pp. 97–105 (2007)
4. Askitis, N., Zobel, J.: Redesigning the string hash table, burst trie, and BST to exploit cache. ACM J. Exp. Algorithmics **15**, 1–7 (2010)
5. Azzopardi, L., de Rijke, M., Balog, K.: Building simulated queries for known-item topics: an analysis using six European languages. In: Proceedings of the SIGIR, pp. 455–462 (2007)
6. Baba, Y., Suzuki, H.: How are spelling errors generated and corrected? A study of corrected and uncorrected spelling errors using keystroke logs. In: Proceedings of the ACL, pp. 373–377 (2012)
7. Baeza-Yates, R., Saint-Jean, F.: A three level search engine index based in query log distribution. In: Nascimento, M.A., de Moura, E.S., Oliveira, A.L. (eds.) SPIRE 2003. LNCS, vol. 2857, pp. 56–65. Springer, Heidelberg (2003). https://doi.org/10.1007/978-3-540-39984-1_5
8. Baeza-Yates, R., Tiberi, A.: Extracting semantic relations from query logs. In: Proceedings of the KDD, pp. 76–85 (2007)
9. Bar-Yossef, Z., Kraus, N.: Context-sensitive query auto-completion. In: Proceedings of the WWW, pp. 107–116 (2011)
10. Bast, H., Weber, I.: Type less, find more: fast autocompletion search with a succinct index. In: Proceedings of the SIGIR, pp. 364–371 (2006)
11. Bast, H., Weber, I.: The CompleteSearch engine: interactive, efficient, and towards IR & DB integration. In: Proceedings of the CIDR, pp. 88–95 (2007)
12. Bast, H., Majumdar, D., Weber, I.: Efficient interactive query expansion with complete search. In: Proceedings of the CIKM, pp. 857–860 (2007)
13. Beitzel, S.M., Jensen, E.C., Chowdhury, A., Grossman, D., Frieder, O.: Hourly analysis of a very large topically categorized web query log. In: Proceedings of the SIGIR, pp. 321–328 (2004)
14. Bhatia, S., Majumdar, D., Mitra, P.: Query suggestions in the absence of query logs. In: Proceedings of the SIGIR, pp. 795–804 (2011)
15. Cai, F., Liang, S., de Rijke, M.: Time-sensitive personalized query auto-completion. In: Proceedings of the CIKM, pp. 1599–1608 (2014)
16. Cai, F., Reinanda, R., de Rijke, M.: Diversifying query auto-completion. ACM Trans. Inf. Syst. **34**(4), 25:1–25:33 (2016)
17. Chaudhuri, S., Kaushik, R.: Extending autocompletion to tolerate errors. In: Proceedings of the SIGMOD, pp. 707–718 (2009)
18. Clauset, A., Shalizi, C.R., Newman, M.E.J.: Power-law distributions in empirical data. SIAM Rev. **51**(4), 661–703 (2009)
19. Cooper, A.: A survey of query log privacy-enhancing techniques from a policy perspective. ACM Trans. Web **2**(4), 19:1–19:27 (2008)
20. Damerau, F.J.: A technique for computer detection and correction of spelling errors. Commun. ACM **7**(3), 171–176 (1964)
21. Dhakal, V., Feit, A.M., Kristensson, P.O., Oulasvirta, A.: Observations on typing from 136 million keystrokes. In: Proceedings of the CHI, pp. 646:1–646:12 (2018)

22. Gentner, D.R., Grudin, J.T., Larochelle, S., Norman, D.A., Rumelhart, D.E.: A glossary of terms including a classification of typing errors. In: Cooper, W.E. (ed.) Cognitive Aspects of Skilled Typewriting, pp. 39–43. Springer, New York (1983). https://doi.org/10.1007/978-1-4612-5470-6_2
23. Hawking, D., Billerbeck, B.: Efficient in-memory, list-based text inversion. In: Proceedings of the Australasian Document Computing Symposium, pp. 5.1–5.8 (2017)
24. Heafield, K.: KenLM: faster and smaller language model queries. In: Proceedings of the Workshop on Statistical Machine Translation, pp. 187–197 (2011)
25. Heinz, S., Zobel, J., Williams, H.: Burst tries: a fast, efficient data structure for string keys. ACM Trans. Inf. Syst. **20**(2), 192–223 (2002)
26. Hofmann, K., Mitra, B., Radlinski, F., Shokouhi, M.: An eye-tracking study of user interactions with query auto completion. In: Proceedings of the CIKM, pp. 549–558 (2014)
27. Hsu, B.-J.P., Ottaviano, G.: Space-efficient data structures for top-k completion. In: Proceedings of the WWW, pp. 583–594 (2013)
28. Ji, S., Li, G., Li, C., Feng, J.: Efficient interactive fuzzy keyword search. In: Proceedings of the WWW, pp. 371–380 (2009)
29. Jiang, J., Ke, Y., Chien, P., Cheng, P.: Learning user reformulation behavior for query auto-completion. In: Proceedings of the SIGIR, pp. 445–454 (2014)
30. Jordan, C., Watters, C., Gao, Q.: Using controlled query generation to evaluate blind relevance feedback algorithms. In: Proceedings of the JCDL, pp. 286–295 (2006)
31. Kharitonov, E., Macdonald, C., Serdyukov, P., Ounis, I.: User model-based metrics for offline query suggestion evaluation. In: Proceedings of the SIGIR, pp. 633–642 (2013)
32. Krishnan, U., Moffat, A., Zobel, J.: A taxonomy of query auto completion modes. In: Proceedings of the Australasian Document Computing Symposium, pp. 6:1–6:8 (2017)
33. Krishnan, U., Billerbeck, B., Moffat, A., Zobel, J.: Abstraction of query auto completion logs for anonymity-preserving analysis. Inf. Retrieval J. **22**(5), 499–524 (2019). https://doi.org/10.1007/s10791-019-09359-8
34. Kumar, R. Novak, J., Pang, B. Tomkins, A.: On anonymizing query logs via token-based hashing. In: Proceedings of the WWW, pp. 629–638 (2007)
35. Lempel, R., Moran, S.: Predictive caching and prefetching of query results in search engines. In: Proceedings of the WWW, pp. 19–28 (2003)
36. Li, G., Ji, S., Li, C., Feng, J.: Efficient fuzzy full-text type-ahead search. VLDB J. **20**(4), 617–640 (2011). https://doi.org/10.1007/s00778-011-0218-x
37. Li, L., Deng, H., Dong, A., Chang, Y., Zha, H., Baeza-Yates, R.: Analyzing user's sequential behavior in query auto-completion via Markov processes. In: Proceedings of the SIGIR, pp. 123–132 (2015)
38. Li, Y., Dong, A., Wang, H., Deng, H., Chang, Y., Zhai, C.: A two-dimensional click model for query auto-completion. In: Proceedings of the SIGIR, pp. 455–464 (2014)
39. Manning, C., Raghavan, P., Schütze, H.: Introduction to Information Retrieval. Cambridge University Press, Cambridge (2008)
40. Maxwell, D., Bailey, P., Hawking, D.: Large-scale generative query autocompletion. In: Proceedings of the Australasian Document Computing Symposium, pp. 9:1–9:8 (2017)
41. Navarro, G.: Compact Data Structures: A Practical Approach. Cambridge University Press, Cambridge (2016)

42. Shokouhi, M.: Learning to personalize query auto-completion. In: Proceedings of the SIGIR, pp. 103–112 (2013)
43. Smith, C.L., Gwizdka, J., Feild, H.: Exploring the use of query auto completion: search behavior and query entry profiles. In: Proceedings of the CHIIR, pp. 101–110 (2016)
44. Smith, C.L., Gwizdka, J., Feild, H.: The use of query auto-completion over the course of search sessions with multifaceted information needs. Inf. Process. Manag. **53**(5), 1139–1155 (2017)
45. Webber, W., Moffat, A.: In search of reliable retrieval experiments. In: Proceedings of the Australasian Document Computing Symposium, pp. 26–33 (2005)
46. Wobbrock, J.O., Myers, B.A.: Analyzing the input stream for character- level errors in unconstrained text entry evaluations. ACM Trans. Comput.-Hum. Interact. **13**(4), 458–489 (2006)
47. Xiao, C., Qin, J., Wang, W., Ishikawa, Y., Tsuda, K., Sadakane, K.: Efficient error-tolerant query autocompletion. Proc. VLDB **6**(6), 373–384 (2013)
48. Xiong, L., Agichtein, E.: Towards privacy-preserving query log publishing. In: Proceedings of the WWW Query Log Analysis Workshop (2007)
49. Zobel, J., Moffat, A., Ramamohanarao, K.: Inverted files versus signature files for text indexing. ACM Trans. Database Syst. **23**(4), 453–490 (1998)

What Can Task Teach Us About Query Reformulations?

Lynda Tamine[(✉)], Jesús Lovón Melgarejo, and Karen Pinel-Sauvagnat

Université Paul Sabatier, IRIT, Toulouse, France
{tamine,sauvagnat}@irit.fr, jesus.lovon-melgarejo@univ-tlse3.fr

Abstract. A significant amount of prior research has been devoted to understanding query reformulations. The majority of these works rely on time-based sessions which are sequences of contiguous queries segmented using time threshold on users' activities. However, queries are generally issued by users having in mind a particular task, and time-based sessions unfortunately fail in revealing such tasks. In this paper, we are interested in revealing in which extent time-based sessions vs. task-based sessions represent significantly different background contexts to be used in the perspective of better understanding users' query reformulations. Using insights from large-scale search logs, our findings clearly show that task is an additional relevant search unit that helps better understanding user's query reformulation patterns and predicting the next user's query. The findings from our analyses provide potential implications for model design of task-based search engines.

Keywords: Tasks · Query reformulation · Query suggestion

1 Introduction

Query reformulation is a critical user behaviour in modern search engines and it is still addressed by a significant amount of research studies [10–12,17,23,26,33]. A salient behavioural facet that has been widely captured and analysed by those studies is query history. The latter is generally structured into "query sessions" which are sequences of queries submitted by a user while completing a search activity with a search system. In the literature review, there are many definitions of query sessions. The widely used definitions are the following [19,25]: (1) a *Time-based session*, also called physical session in [6], is a set of consecutive queries automatically delimited using a time-out threshold on user's activities. Time-gap values of 30 min and 90 min have been the most commonly used in previous research [4,6,9,19]; (2) a *Task-based session*, also called mission in [6], is a set of queries that are possibly neither consecutive nor within the same time-based session. The queries belong to related information needs that are driven by a goal-oriented search activity, called search *task* (eg., *job search* task). The latter could be achieved by subsets of consecutive related queries called logical sessions in [6] or subtasks in [9].

J. M. Jose et al. (Eds.): ECIR 2020, LNCS 12035, pp. 636–650, 2020.
https://doi.org/10.1007/978-3-030-45439-5_42

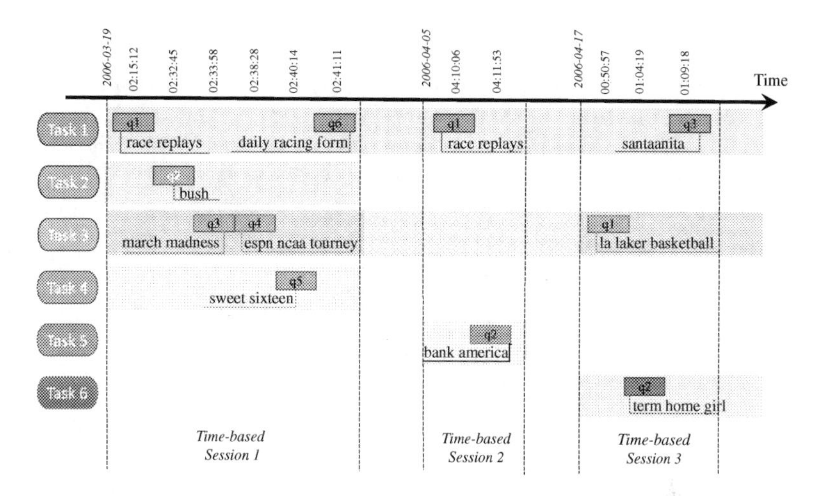

Fig. 1. Examples of time-based sessions and tasks, with the associated queries. Sample of the Webis-SMC-12 Search Corpus [6] for a given user.

Previous research [4,7,20,21] showed that: (1) users have a natural multi-tasking behaviour by intertwining different tasks during the same time-based session; and that (2) users possibly interleave the same task at different timestamps in the same time-based session or throughout multiple time-based sessions (ie., *multi-session tasks*). Such long-term tasks are acknowledged as being *complex tasks* [7,9]. Figure 1 shows a sample of 3 time-based search sessions extracted from the Webis-SMC-12 Search Corpus [6] for a single user. The sessions are manually annotated with tasks. As can be seen, 6 tasks (Task 1 - Task 6) are performed by the user during these 3 sessions. We can observe that all these sessions are *multi-tasking*, since they include queries that relate to multiple tasks (eg., Session 1 is multi-tasking since it includes queries that relate to Task 1, 2, 3 and 4). We can also see that Task 1 and Task 3 are *interleaved* within and across sessions (eg., Task 1 is interleaved within Session 1 and across Session 1, 2 and 3). Thus, Tasks 1 and 3 are *multi-session tasks.*

While it is well-known that time-based session detection methods fail in revealing tasks [6,19], most of previous research work has employed time-based sessions as the focal units of analysis for understanding query reformulations [10–12,26,33]. Other works rather studied users' query reformulations from the task perspective through user studies [15,17,29]. However, the authors analysed low-scale pre-designed search tasks conducted in controlled laboratory settings. In addition to their limited ability to observe natural search behaviour, there is a clear lack of comparability in search tasks across those studies.

To design support processes for task-based search systems, we argue that we need to: (1) fully understand how user's task performed in natural settings drives the query reformulations changes; and (2) gauge the level of similarity of these changes trends with those observed in time-based sessions. Our ultimate goal is to gain insights regarding the relevance of using user's tasks as the focal units of search to both understand and predict query reformulations. With this in mind, we perform large-scale log analyses of users naturally engaged in tasks to examine query reformulations from both the time-based session vs. task-based session perspectives. Moreover, we show the role of the task characteristics in predicting the next user's query. Our findings clearly show that task is an additional relevant search unit that helps to better understand user's query reformulation patterns and to predict the next user's query.

2 Related Work

2.1 Query Reformulation Understanding

Query reformulation has been the focus of a large body of work. A high number of related taxonomies have been proposed [5,11,16]. To identify query reformulation patterns, most of the previous works used large-scale log analyses segmented into time-based sessions. Different time gaps have been used including 10–15 min [8], 30 min [4,19] and 90 min [6,9]. In a significant body of work, authors categorised the transitions made from one query to the subsequent queries through syntactic changes [11,12,23,26] and query semantic changes [10,12,33]. Syntactic changes include word substitution, removing, adding and keeping. The results highlighted that the query and its key terms evolve throughout the session regardless of the query position in the session. Moreover, such strategies are more likely to cause clicks on highly ranked documents. Further experiments on semantic query changes through generalisation vs. specialisation [10,12] showed that a trend exists toward going from generalisation to specialisation. This behavioural pattern represents a standard building-box strategy while specialisation occurs early in the session.

Another category of work rather employed lab user studies to understand how different task characteristics impact users' query reformulations [15,17,18,28,31,32]. The results mainly revealed that: (1) the domain knowledge of the task doer significantly impacts query term changes. For instance, Wildemuth [31] found that search tactics changed while performing the task as users' domain knowledge evolved; (2) the cognitive complexity and structure of the task (eg., simple, hierarchical, parallel) has a significant effect on users' query reformulation behavior. For instance, Liu et al. [17] found that specialisation in parallel tasks was significantly less frequent than in simple and hierarchical tasks.

A few work [4,22] used large-scale web search logs annotated with tasks to understand query reformulations. The findings in [4] were consistent with log-based studies [26] showing that page visits have significant influence on the vocabulary of subsequent queries. Odijk et al. [22] studied the differences in users' reformulation strategies within successful vs. unsuccessful tasks. Using a crowd-sourcing methodology, the authors showed that query specialisation through term adding is substantially more common in successful tasks than in unsuccessful tasks. It also appeared that actions such as formulating the same query than the previous one and reformulating completely a new query are rather relevant signals of unsuccessful tasks.

2.2 Contributions over Previous Work

We make several contributions over prior work. First, to the best of our knowledge, no previous study examined the differences in query reformulation strategies from the two perspectives of time-based sessions and task-based sessions viewed as background contexts. Insights gleaned from our data analysis have implications for designing task-based search systems. Second, although there has been intensive research on query reformulation, we provide a new insight into the variation of query reformulation strategies. The latter are analysed in relation with search episode size (*Short*, *Medium* and *Long*) and search stage (*Start*, *Middle* and *End*) from two different viewpoints (stream of query history and the search task progress). Third, building on the characterisation of search tasks, we provide insights on how considering task features might improve a supervised predictive model of query reformulations.

3 Analytical Set up

3.1 Datasets

This analysis is carried out using the freely available Webis-SMC-12 Search Corpus[1] [1,6] extracted from the 2006 AOL query log which is a very large collection of web queries. The released corpus comprises 8800 queries. We remove the repeated successive queries that were automatically generated following a click instead of a user's reformulation. We also remove all non-alphanumeric characters from the queries and apply a lowercasing. The cleaned data finally include 4734 queries submitted by 127 unique users. The query log is automatically segmented into time-based sessions using a time-gap threshold on users' activities. Since there is so far no agreement about the most accurate time-out threshold for detecting session boundaries [9,19], we consider the two widely used time-gap values between successive queries: 30 min as done in [4,19] and 90 min as done in [6,9]. We also use the provided manual annotations to segment the query log into task-based sessions. For care of simplicity, we subsequently refer to time-based session as "*Session*" and we refer to task-based session as "*Task*".

Table 1 presents the data collection statistics. One immediate observation is that the average number of queries in tasks (3.45) is higher than that of the sessions (eg., 2.04 in the 30 min-sessions) as reported in [9,19]. The total percentage of multi-tasking sessions is roughly 13% (resp. 16%) of the 30 min-session (resp. 90 min-session). Higher statistics (50%) were reported in [19]. However, we found that there are only 30.28% (resp. 31.27%) of the 30-min sessions (resp. 90-min sessions) that include only 1 task that is non interleaved throughout the user's search history. Thus, the 70% remaining sessions are either multi-tasking or include interleaved tasks that reoccur in multiple sessions. Similar statistics were observed in previous work (eg., 68% in [9]). Another interesting observation is that a high percentage of tasks (23.23%) are interleaved, which is roughly comparable to that of previous studies (eg., 17% in [14]), or spanned over multiple sessions (e.g, 27.09% of tasks spanned over multiple 30-min sessions).

[1] http://www.webis.de/research/corpora.

Table 1. The Webis search corpus statistics based on automatic segmentation of sessions (30 min, 90 min) and manual annotation of tasks.

	Sessions		Tasks
	30 min	90 min	
# of sessions/tasks	2318	2024	1373
Avg number of queries	2.04	2.34	3.45
Avg query length (#terms)	2.51	2.47	2.41
Multi-tasking sessions	12.87%	15.82%	-
Multi-session tasks	27.09%	25.42%	-
Interleaving tasks	-	-	23.23%

Table 2. Overview of query reformulation features.

Notation	Description	Measurement
$Sim(q_i, q_{i+1})$	Jaccard query pair similarity	$\frac{\lvert s(q_i) \cap s(q_{i+1}) \rvert}{\lvert s(q_i) \cup s(q_{i+1}) \rvert}$
$Rr(q_i, q_{i+1})$	Ratio of term-retention	$\frac{\lvert s(q_i) \cap s(q_{i+1}) \rvert}{\lvert s(q_i) \rvert}$
$Rm(q_i, q_{i+1})$	Ratio of term-removal	$\frac{\lvert s(q_i) - s(q_{i+1}) \rvert}{\lvert s(q_i) \rvert}$
$Ra(q_i, q_{i+1})$	Ratio of term-adding	$\frac{\lvert s(q_{i+1}) - s(q_i) \rvert}{\lvert s(q_{i+1}) \rvert}$

3.2 Query Reformulation Features

To study query reformulations, we consider the three usual categories of syntactic changes [11,13,26] between successive query pairs (q_i, q_{i+1}) composed of $s(q_i)$ and $s(q_{i+1})$ term sets respectively: (1) query term-retention Rr; (2) query term-removal Rm acts as search generalisation [12,13]; and (3) query term-adding Ra acts as search specialisation [12,13]. For each query pair, we compute the similarity and the query reformulation features presented in Table 2, both at the sessions and tasks levels (Sect. 5).

4 Query Characteristics

4.1 Query Length

Here, our objective is twofold: (1) we investigate how query length (ie., # query terms) varies across the search stages within sessions and tasks of different sizes (ie., # queries); and (2) we examine in what extent the trends of query length changes observed within tasks are similar to those observed within sessions.

To make direct comparisons of trends between sessions and tasks with different sizes in a fair way, we first statistically partition the search sessions and tasks into three balanced categories (*Short*, *Medium* and *Long*). To do so, we compute the cumulative distribution function (CDF) of session size values for the 30-min and the 90-min sessions, as well as the CDF of task size values in

Table 3. Classification of sessions and tasks regarding the number of related queries. If applicable, query positions boundaries to delimit the search stages in sessions and tasks of different sizes.

	Short	Medium		Long		
Sessions (30 min, 90 min)	1	2		≥3		
		Query position boundary		*Query position boundary*		
		Start	Middle	Start	Middle	End
		1	2	1–2	3	≥4
Tasks	1	2		≥3		
		Query position boundary		*Query position boundary*		
		Start	Middle	Start	Middle	End
		1	2	1–3	4–8	≥9

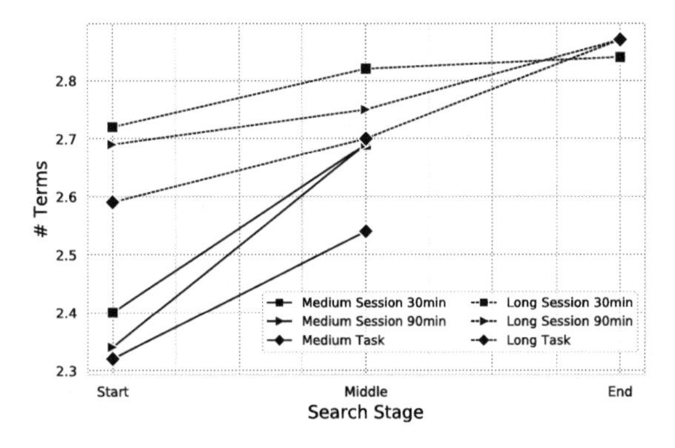

Fig. 2. Average query length variation along sessions vs. tasks of different sizes.

relation with the number of included queries. Then, we compute the CDF of the search stage values in relation with the query position boundary (*Start*, *Middle* and *End*) along each size-based category of sessions vs. tasks. Since short sessions and tasks only contain 1 query and consequently do not contain query reformulations, we do not distinguish between the search stages nor consider this category of sessions and tasks in the remainder of the paper. Table 3 shows the statistics of the search stages (*Start, Middle, End*) with respect to *Medium* and *Long* sessions and tasks.

Based on those categorisations, Fig. 2 shows the variation of the query length limit within each category of sessions and tasks and along the different search stages. We can see two clear trends. First, queries in both longer sessions and longer tasks generally tend to contain more terms (2.60–2.87 vs. 2.41–2.51 in average). This trend remains along all the different search stages. Regarding sessions, previous studies [2] have also shown similar trends in log-based data. Regarding tasks, our results suggest that long tasks require to issue more search terms. One could argue that long tasks, that more likely involve complex

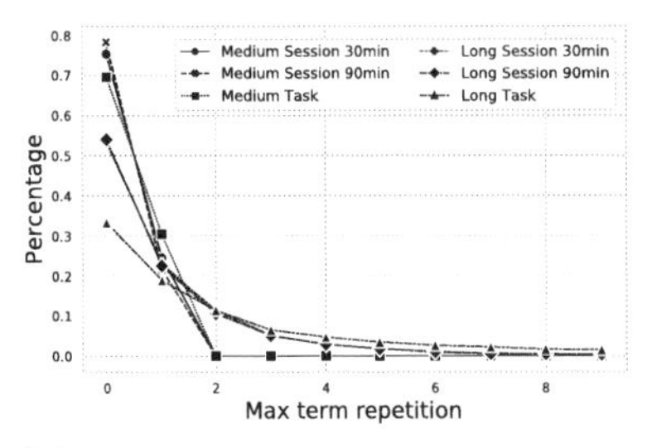

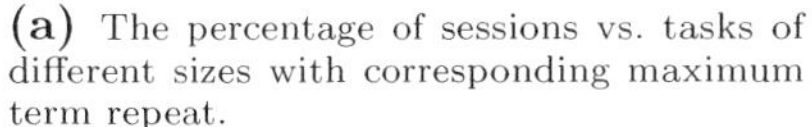
(a) The percentage of sessions vs. tasks of different sizes with corresponding maximum term repeat.

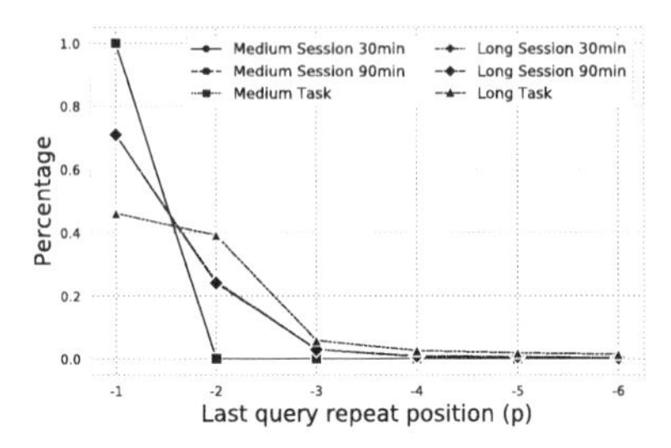

(b) The percentage of queries containing the same term to a previous query over different positions.

Fig. 3. Term repetition trends over sessions vs. tasks.

information needs, lead users to formulate more informative queries. We also relate this observation with previous findings [2] showing that increased success is associated with longer queries, particularly in complex search tasks. Second we can surprisingly see that in general, queries observed within sessions whatever their sizes, are slightly longer in average than queries issued within tasks of the same category except at the end of the search stage. By cross-linking with the CDF results presented in Table 3, we expect that this observation particularly relates to long sessions. One possible explanation is that since long sessions are more likely to be multi-tasking (eg., there are 1.57 task in average in the long 90-min sessions vs. 1.29 in the 30-min sessions), the average query length is particularly increased within sessions that include queries at late search stages of the associated tasks (*Middle*, *End*).

4.2 Query Term Repeat

Inspired by [13], we examine query term frequency along the search with respect to session vs. task search context. In contrast to [13], our underlying intent here is rather to learn more about the impact of search context (ie., session vs. task) on the level of query term reuse. For a query q_i belonging to session S and task T and not submitted at the beginning (ie., $i > 1$), we compute the frequency of each of its terms from the previous queries within the same session q_j^S (resp. same task q_j^T), $j = 1..i-1$. Then, we take the maximal value Tr as "maximum term repeat" for query q_i if the latter contains at least one term used Tr times in previous queries.

Figure 3a plots the average "maximum term repeat values" for all the queries within all the sessions and tasks ranged by size (*Short*, *Medium* and *Long*). We can see that the term repeat trend across sessions is similar to that reported in [13]. By comparing between the term repeat trends in sessions and tasks, we clearly observe that there are less reformulated queries that do not share any identical terms with the previous queries in tasks (eg., 70% of medium tasks) in comparison to sessions (eg., 75–78% of medium sessions). Interestingly, we can see that the difference is particularly higher in the case of long tasks and long sessions (33% vs. 53–54%). However, we can notice that even if the percentage of queries sharing an increased number of terms with previous queries decreases for both medium sessions and medium tasks, the difference is reversed between long sessions and long tasks. It is more likely that query terms are renewed during long tasks which could be explained by shifts in information needs related to the same driving long-term task.

Figure 3b shows the percentage of reformulated queries for which each reused term occurs at the first time at a given position within sequences from length 1 to 6. It appears that the sources of reused query terms in both tasks and sessions are limited to the two previous queries. More particularly, while we find terms used in the previous query in all (100%) of the reformulated queries in medium sessions and medium tasks, it is more likely to observe reformulated queries containing terms from the two previous queries in long sessions than in long tasks (71% of sessions vs. 46% of tasks). To sum up, the context used for driving query actions is limited to the two previous queries even for long sessions and tasks, with however, a lower level of term reuse in long tasks.

5 Query Reformulation

5.1 User Actions

Given each query q_i belonging to session S (resp. task T), Table 4 gives the query reformulation feature values (See Table 2) for both *Medium* (M) and *Long* (L) sessions and tasks and are computed over: (1) the short-term context (SC), by considering the query reformulation pair observed within the same session S (resp. task T) $(q_i, q_{i+1})^S$ (resp. $(q_i, q_{i+1})^T$), $i \geq 1$; and (2) the long-term context (LC), by considering the set of successive query reformulation pairs within the same session S (resp. task T), $(q_k, q_{k+1})^S$ (resp. $(q_k, q_{k+1})^T$), $1 \leq k \leq i$. Significance of the differences between the "Within Session" scenario and the "Within Task" scenario considering either the short-term context (SC) or the long-term context (LC) is computed using the non-paired student t-test. We can see from Table 4 that for the whole set of search actions (ie., term-retention *Rr*, term-removal *Rm* and term-adding *Ra*) and similarity values (ie., *Avg Sim*), most of the differences between task-based and session-based scenarios are highlighted as significant. More particularly, we can make two key observations: (1) successive queries in both medium and long tasks are significantly more similar (*Avg Sim* of 0.27 and 0.25 respectively) than they are in medium and long sessions for both time-out thresholds (*Avg Sim* of 0.20–0.23) with higher ratios of term-retention (34% vs. 25–29%); and (2) the query history along long tasks exhibits a higher topical cohesion (*Avg Sim* of 0.24) than it does in long sessions (*Avg Sim* of 0.18–0.20) with a higher ratio of term-retention (30% vs. 23–26%) and a lower ratio of term-adding (70% vs. 74–77%) for tasks. All these results are consistent with those obtained through the analysis of query term repeat (Sect. 4.2). They suggest that longer tasks more likely include topically and lexically closer information needs that might drive subtasks in comparison with long sessions.

Unlikely, the latter might include multiple and topically different information needs that belong to distinct tasks.

5.2 Similarity Analysis over the Search Progress

To better understand the changes trends along the search, we also examine (Fig. 4) the query reformulation similarities at different stages of the search sessions vs. tasks by considering both short-term context (SC) and long-term context (LC). We can make from Fig. 4 two important observations: (1) successive query reformulations within tasks are clearly more similar ($m = 0.25$, $sd = 0.27$, $avg = 0.27$) at the different search stages than they are within both the 30-min and 90-min sessions (eg., $m = 0.0$, $sd = 0.27$, $avg = 0.23$ for the 30-min sessions) regardless of their sizes; and (2) the overall similarity of query reformulations observed over the search history in both long sessions and long tasks tends to decrease along the search (eg., decrease from $m = 0.21$ to $m = 0.13$ for tasks). These results indicate that the queries tend to be lexically dissimilar while the search evolves. This observation might be explained by different reasons

Table 4. Reformulation and similarity feature values in sessions vs. tasks. Significant differences ($p < 0.01$) of the "Within Session" scenario in comparison to the "Within Task" scenario are highlighted using a star '*'.

Features	Within Task (Baseline)				Within Session							
	SC		LC		30 min				90 min			
					SC		LC		SC		LC	
	M	L	M	L	M	L	M	L	M	L	M	L
Avg Sim.	0.27	0.25	0.35	0.24	0.23	0.22*	0.23*	0.20*	0.20*	0.21*	0.20*	0.18*
Rr	0.34	0.34	0.41	0.30	0.29	0.29*	0.29*	0.26*	0.25*	0.28*	0.25*	0.23*
Rm	0.61	0.63	0.54	0.69	0.66	0.68*	0.66*	0.72*	0.71*	0.70*	0.71*	0.75*
Ra	0.66	0.66	0.59	0.70	0.71	0.71*	0.71*	0.74*	0.75*	0.72*	0.75*	0.77*

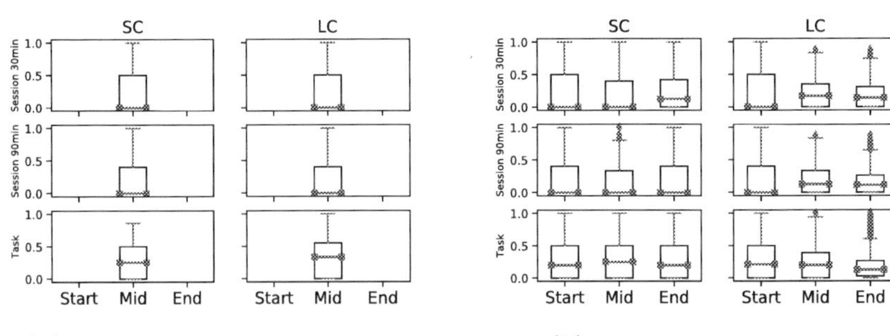

(a) Medium sessions and medium tasks. **(b)** Long sessions and long tasks

Fig. 4. Plots of query reformulation similarities along different search stages of sessions and tasks of different sizes.

depending on the context used (session vs. task) to make the observation. As outlined earlier through query length analysis (Sect. 4.1), sessions might include different ongoing tasks that lead to formulate lexically distinct queries. Unlikely, tasks might include different ongoing related subtasks. However, queries are still overall more similar ($m = 0.13$, $sd = 0.23$, $avg = 0.20$) across the search stages in long tasks than they are in long sessions ($m = 0.11$, $sd = 0.17$, $avg = 0.16$), particularly at the end of the search stage. This observation might be related to the better cohesiveness of tasks with increased number of queries since, unlike sessions, they are goal-oriented.

5.3 Summary

Through the analyses presented in the previous sections, we have shown that there are significant differences in query reformulation patterns depending potentially on the context used (session or task) to make the observations. The results also indicate that time threshold value used to segment the sessions has no impact on the differences trends. In general, the most significant differences are observed regarding long tasks. Informed by these findings, we show in the final contribution of this paper the potential of the task features studied in Sects. 4 and 5 for enhancing the performance of a query reformulation predictive model.

6 Predicting Query Reformulation Using Task Context

Given a session $S = \{q_1, q_2, \ldots, q_{M-1}, q_M\}$, we aim to predict for each query sequence $S_k \subset S, S_k = \{q_1, q_2 \ldots, q_{k-1}, q_k\}$, $1 < k < M$, the target query q_k given the context C_{q_k} defined by queries $\{q_1, q_2 \ldots, q_{k-1})$, where q_{k-1} is the anchor query.

6.1 Experimental Setting

Evaluation Protocol. As usually done in previous work for query auto-completion [13] and next query prediction [3,24,27], we adopt a train-test methodology. We first sort the 30 min-sessions time-wise and partition them into two parts. We use the first 60 day-data for training the predictive model and the remaining 30 days for testing. We use 718 sessions (including 2418 queries) which represent 70% of the dataset as our training set, and 300 sessions (including 998 queries) which represent 30% of the dataset as our testing set. To enable the evaluation of the learning approach, we first produce a set of ground truth suggestions for each test query. To do so, we follow a standard procedure [3,13,27]: for each session in the training-test sets, we select as the candidate set, the top-20 queries q_k that follows each anchor query q_{k-1}, ranked by query frequency. To assess the contributions of the task context features in predicting the next user's query, we use the *Baseline Ranker*, a competitive learning to rank query suggestion model that relies on contextual features [3,27].

Model Training. We design the task-aware *Baseline Ranker* which we refer to as *TaskRanker*. For training purpose, we first generate from the 718 training sessions, 1395 task-based query sequences that are built with respect to the task labels provided in the Webis-SMC-12 Search Corpus. We remove the task-based query sequences with only 1 query candidate. For instance, using task labels provided in Fig. 1, we built and then select from *Session 1* the task-based query sequences $\{q1, q6\}; \{q3, q4\}$ with respectively $q6$ and $q4$ as the ground truth queries. Besides, to guarantee the candidate set includes the target query, we remove the task-based query sequences whose ground truth is not included in the associated candidate sets. After filtering, we obtain 215 cleaned task-based query sequences used for training the *TaskRanker* model. Similarly to [3,27], we use the state-of-the-art boosted regression tree ranking algorithm LamdaMART as our supervised ranker. We tune the LamdaMART model with parameters of 500 decision trees across all experiments. We use 2 sets of features (30 in total): (1) 10 features related to the analyses conducted in previous sections of the paper (Sects. 4, 5). We use the *user-action related features* including ratios of term-retention (*Rr*), term-adding (*Ra*), term-removal (*Rm*), and term-repeat (*Tr*), that are measured using both the short-term (SC) and long-term (LC) contexts. We also use *query-similarity related features* (*Avg Sim*) based on the similarity of the target query q_k with short-term context SC (anchor query q_{k-1}) and long-term context LC (with the previous queries in C_{q_k}); (2) 20 features that are similar to those previously used for a learning to rank suggestion model, and described in detail in [3,27]. This set of features includes (a) pairwise and suggestion features based on target query characteristics and anchor query characteristics including length and frequency in the dataset; (b) contextual features that include n-gram similarity values between the suggestion and the 10 most recent queries. Note that we extended the *Baseline Ranker* released by Sordoni et al. [27][2].

Baselines and Evaluation Metric. We use the conventional models widely used in the literature [3,13,27] namely the *Most Popular Suggestion (MPS)*, and the traditional *Baseline Ranker* which we refer to as *SessionRanker*. The MPS relies on query frequency to rank candidates. Unlike the *TaskRanker*, the *SessionRanker* is trained on session-based query sequences that are built from the same subset of the 718 training sessions. For instance, we built from *Session 1* presented in Fig. 1, the session-based query sequences $\{q1, q2\}; \{q1, q2, q3\}; \{q1, q2, q3, q4\}; \{q1, q2, q3, q4, q5\}; \{q1, q2, q3, q4, q5, q6\}$ with respectively $q2$, $q3$, $q4$, $q5$ and $q6$ as the ground truth queries. We obtain 1700 session-based query sequences that are then cleaned, similarly to the *TaskRanker* by removing query sequences with only 1 query candidate and those with ground truth not included in the associated candidate sets. Finally, the *SessionRanker* has been trained on 302 cleaned session-based query sequences.

[2] https://github.com/sordonia/hred-qs.

Similarly to the *TaskRanker*, we use the same sets of features (30 in total) learned here at the session level, and we tune it using the LamdaMART model. We use the *Mean Reciprocal Rank* (*MRR*) which is the commonly used metric for evaluating next query prediction models [3,24,27]. The MRR performance of the *TaskRanker* and the baselines is measured using the same test subset that includes 150 cleaned session-based query sequences built up on the subset of 698 session-based query sequences generated from the 300 test sessions. The task annotations of the testing test are ignored.

6.2 Prediction Results

Table 5 shows the MRR performance for the *TaskRanker* and the baselines. The *TaskRanker* achieves an improvement of +152.8% with respect to the MPS model and an improvement of +10.2% with respect to the *SessionRanker* model. The differences in MRR are statistically significant by the t-test ($p < 0.01$). It has been shown in previous work [3,27] that session size has an impact on the performance of context-aware next query prediction models. Thus, we report in Fig. 5 separate MRR results for each of the *Medium* (2 queries) and the *Long* sessions (≥3 queries) studied in our analyses (Sects. 4 and 5). As can be seen, the task-based contextual features particularly help predicting the next query in long sessions (+14, 1% in comparison to the *SessionRanker*, $p = 7 \times 10^{-3}$). Prediction performance for *Medium* sessions is slightly but not significantly lower (−1, 3% in comparison to the *SessionRanker*, $p = 0.65$). This result can be expected from

Table 5. Next query prediction performance. All improvements are significant by the t-test ($p < 0.01$).

Model	MRR	Improvement
MPS	0.3677	+152.8%
SessionRanker	0.8433	+10.2%
TaskRanker	**0.9296**	–

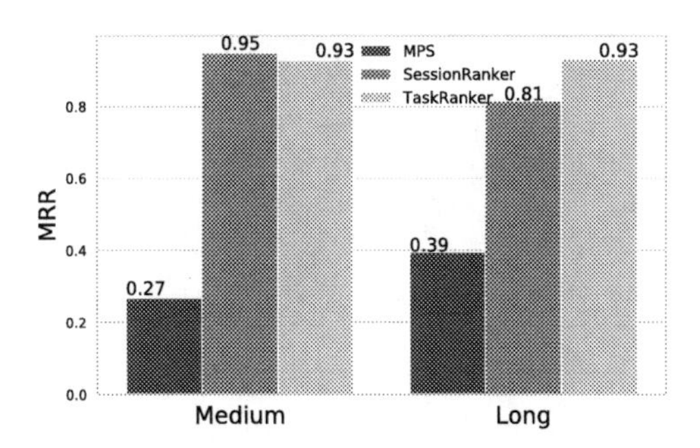

Fig. 5. Performance of *TaskRanker* compared to the baseline models on sessions with different sizes.

the findings risen from our analyses, since *Long* sessions include queries related to 89.9% of *Long* tasks whose cohesive contexts enable more accurate predictions of user's future search intent.

7 Conclusion and Implications

Better understanding user's query reformulations is important for designing task completion engines. Through the analysis of large-scale query logs annotated with task labels, we have revealed significant differences in the query changes trends along the search depending on the retrospective context used, either session or task. We found that queries are even longer in longer tasks with however a lower level of term reuse in tasks than in sessions. In addition, terms are particularly renewed in long tasks indicating clear shifts in information needs. Using lexical similarity measures, we have also shown that the query reformulations exhibit a clearer cohesiveness within tasks than within sessions along the different search stages, with however a decreasing level of similarity. Finally, we provided insights on the usefulness of task features to enhance the user's next query prediction accuracy. Given the crucial lack of query logs with annotated tasks, we acknowledge that the predictive model has been trained and tested with limited amount of data. However, the features used are based on the analysis performed on a large-scale data provided in the Webis corpus. Thus, we believe that the trend of our results would remain reliable.

There are several promising research directions for future work. Firstly, evidence related to the characterization of tasks through query length variation and query reformulation similarities along the search, presented in Sects. 4 and 5, may benefit research on automatic task boundary detection. In Sect. 6, we showed that learning from query streams annotated with tasks helps the query suggestion process particularly for long-term tasks. It will be interesting to design a predictive model of *query trails* associated with subtasks, by analogy to search trails [30]. This might help users in completing complex tasks by issuing fewer queries. This would decrease the likeliness of search struggling as shown in previous work [22].

References

1. Webis corpus archive. https://zenodo.org/record/3265962#.Xc8HoS2ZPOQ. https://doi.org/10.5281/zenodo.3265962
2. Agapie, E., Golovchinsky, G., Qvarfordt, P.: Leading people to longer queries. In: Proceedings of the SIGCHI Conference on Human Factors in Computing Systems, CHI 2013, pp. 3019–3022 (2013)
3. Dehghani, M., Rothe, S., Alfonseca, E., Fleury, P.: Learning to attend, copy, and generate for session-based query suggestion. In: Proceedings of the 2017 ACM on Conference on Information and Knowledge Management, CIKM 2017, pp. 1747–1756 (2017)

4. Eickhoff, C., Teevan, J., White, R., Dumais, S.: Lessons from the journey: a query log analysis of within-session learning. In: Proceedings of the 7th ACM International Conference on Web Search and Data Mining, WSDM 2014, pp. 223–232 (2014)
5. Guo, J., Xu, G., Li, H., Cheng, X.: A unified and discriminative model for query refinement. In: Proceedings of the 31st Annual International ACM SIGIR Conference on Research and Development in Information Retrieval, SIGIR 2008, pp. 379–386 (2008)
6. Hagen, M., Gomoll, J., Beyer, A., Stein, B.: From search session detection to search mission detection. In: Proceedings of the 10th Conference on Open Research Areas in Information Retrieval, OAIR 2013, pp. 85–92 (2013)
7. Hassan Awadallah, A., White, R.W., Pantel, P., Dumais, S.T., Wang, Y.M.: Supporting complex search tasks. In: Proceedings of the 23rd ACM International Conference on Conference on Information and Knowledge Management, CIKM 2014, pp. 829–838 (2014)
8. He, D., Göker, A.: Detecting session boundaries from web user logs. In: In Proceedings of of the BCS-IRSG 22nd Annual Colloquium on Information Retrieval Research, pp. 57–66 (2000)
9. He, J., Yilmaz, E.: User behaviour and task characteristics: a field study of daily information behaviour. In: Proceedings of the 2017 Conference on Conference Human Information Interaction and Retrieval, CHIIR 2017, pp. 67–76 (2017)
10. He, Y., Tang, J., Ouyang, H., Kang, C., Yin, D., Chang, Y.: Learning to rewrite queries. In: Proceedings of the 25th ACM International on Conference on Information and Knowledge Management, CIKM 2016, pp. 1443–1452 (2016)
11. Huang, J., Efthimiadis, E.N.: Analyzing and evaluating query reformulation strategies in web search logs. In: Proceedings of the 18th ACM Conference on Information and Knowledge Management, CIKM 2009, pp. 77–86 (2009)
12. Jansen, B.J., Booth, D.L., Spink, A.: Patterns of query reformulation during web searching. J. Am. Soc. Inf. Sci. Technol. **60**(7), 1358–1371 (2009)
13. Jiang, J.Y., Ke, Y.Y., Chien, P.Y., Cheng, P.J.: Learning user reformulation behavior for query auto-completion. In: Proceedings of the 37th International ACM SIGIR Conference on Research & #38; Development in Information Retrieval, SIGIR 2014, pp. 445–454 (2014)
14. Jones, R., Klinkner, K.L.: Beyond the session timeout: automatic hierarchical segmentation of search topics in query logs. In: Proceedings of the 17th ACM Conference on Information and Knowledge Management, CIKM 2008, pp. 699–708 (2008)
15. Kinley, K., Tjondronegoro, D.W., Partridge, H.L., Edwards, S.L.: Human-computer interaction: the impact of users' cognitive styles on query reformulation behaviour during web searching. In: Australasian Conference on Computer-Human Interaction (OZCHI 2012), Melbourne, Vic, August 2012. https://doi.org/10.1145/2414536.2414586
16. Lau, T., Horvitz, E.: Patterns of search: analyzing and modeling web query refinement. In: Kay, J. (ed.) UM99 User Modeling. CICMS, vol. 407, pp. 119–128. Springer, Vienna (1999). https://doi.org/10.1007/978-3-7091-2490-1_12
17. Liu, C., Gwizdka, J., Liu, J., Xu, T., Belkin, N.J.: Analysis and evaluation of query reformulations in different task types. In: Proceedings of the 73rd ASIS&T Annual Meeting on Navigating Streams in an Information Ecosystem, ASIS&T 2010, vol. 47, pp. 17:1–17:10 (2010)

18. Lu, K., Joo, S., Lee, T., Hu, R.: Factors that influence query reformulations and search performance in health information retrieval: a multilevel modeling approach. J. Assoc. Inf. Sci. Technol. **68**(8), 1886–1898 (2017)
19. Lucchese, C., Orlando, S., Perego, R., Silvestri, F., Tolomei, G.: Identifying task-based sessions in search engine query logs. In: Proceedings of the Fourth ACM International Conference on Web Search and Data Mining, WSDM 2011, pp. 277–286 (2011)
20. Mehrotra, R., Bhattacharya, P., Yilmaz, E.: Characterizing users' multi-tasking behavior in web search. In: Proceedings of the 2016 ACM on Conference on Human Information Interaction and Retrieval, CHIIR 2016, pp. 297–300 (2016)
21. Mehrotra, R., Bhattacharya, P., Yilmaz, E.: Uncovering task based behavioral heterogeneities in online search behavior. In: Proceedings of the 39th International ACM SIGIR Conference on Research and Development in Information Retrieval, SIGIR 2016, pp. 1049–1052. ACM, New York (2016). https://doi.org/10.1145/2911451.2914755
22. Odijk, D., White, R.W., Hassan Awadallah, A., Dumais, S.T.: Struggling and success in web search. In: Proceedings of the 24th ACM International on Conference on Information and Knowledge Management, CIKM 2015, pp. 1551–1560 (2015)
23. Rieh, S.Y., Xie, H.I.: Analysis of multiple query reformulations on the web: the interactive information retrieval context. Inf. Process. Manag. **42**(3), 751–768 (2006)
24. Santos, R.L.T., Macdonald, C., Ounis, I.: Learning to rank query suggestions for adhoc and diversity search. Inf. Retrieval **16**(4), 429–451 (2013). https://doi.org/10.1007/s10791-012-9211-2
25. Silverstein, C., Marais, H., Henzinger, M., Moricz, M.: Analysis of a very large web search engine query log. SIGIR Forum **33**(1), 6–12 (1999)
26. Sloan, M., Yang, H., Wang, J.: A term-based methodology for query reformulation understanding. Inf. Retrieval J. **18**(2), 145–165 (2015). https://doi.org/10.1007/s10791-015-9251-5
27. Sordoni, A., Bengio, Y., Vahabi, H., Lioma, C., Grue Simonsen, J., Nie, J.Y.: A hierarchical recurrent encoder-decoder for generative context-aware query suggestion. In: Proceedings of the 24th ACM International on Conference on Information and Knowledge Management, CIKM 2015, pp. 553–562 (2015)
28. Tamine, L., Chouquet, C.: On the impact of domain expertise on query formulation, relevance assessment and retrieval performance in clinical settings. Inf. Process. Manag. **53**(2), 332–350 (2017)
29. Vakkari, P.: A theory of the task-based information retrieval. J. Doc. **57**, 44–60 (2001)
30. White, R.W., Huang, J.: Assessing the scenic route: measuring the value of search trails in web logs. In: Proceedings of the 33rd International ACM SIGIR Conference on Research and Development in Information Retrieval, SIGIR 2010, pp. 587–594 (2010)
31. Wildemuth, B.M.: The effects of domain knowledge on search tactic formulation. JASIST **55**, 246–258 (2004)
32. Wildemuth, B.M., Kelly, D., Boettcher, E., Moore, E., Dimitrova, G.: Examining the impact of domain and cognitive complexity on query formulation and reformulation. Inf. Process. Manag. **54**(3), 433–450 (2018)
33. Ãzmutlu, H., Cavdur, F.: Application of automatic topic identification on excite web search engine data logs. Inf. Process. Manag. **41**, 1243–1262 (2005). https://doi.org/10.1016/j.ipm.2004.04.018

A Regularised Intent Model for Discovering Multiple Intents in E-Commerce Tail Queries

Subhadeep Maji[1(✉)], Priyank Patel[1], Bharat Thakarar[1], Mohit Kumar[2], and Krishna Azad Tripathi[1]

[1] Flipkart Internet Private Limited, Bangalore, India
{subhadeep.m,priyank.patel,bharat.thakarar,krishna.tripathi}@flipkart.com
[2] Udaan.com, Bangalore, India
mohitkum@udaan.com

Abstract. A substantial portion of the query volume for e-commerce search engines consists of infrequent queries and identifying user intent in such *tail* queries is critical in retrieving relevant products. The intent of a query is defined as a labelling of its tokens with the product attributes whose values are matched against the query tokens during retrieval. Tail queries in e-commerce search tend to have multiple correct attribute labels for their tokens due to multiple valid matches in the product catalog. In this paper, we propose a latent variable generative model along with a novel data dependent regularisation technique for identifying multiple intents in such queries. We demonstrate the superior performance of our proposed model against several strong baseline models on an editorially labelled data set as well as in a large scale online A/B experiment at Flipkart, a major Indian e-commerce company.

1 Introduction

E-commerce companies offer a wide selection of products from many categories and the number of unique queries submitted to their search engines can be of the order of millions per month. A substantial portion of these queries are infrequent; we observed that approximately 35% of the unique queries at Flipkart, a major Indian e-commerce company occur less than 50 times a month. Such *tail* queries [11,24] lack sufficient click-through data and tend to have poor retrieval performance [11,14,17]. Improving performance on these queries has a large business impact from the long term benefits of greater customer satisfaction [2,7].

E-commerce search is a faceted search on a structured catalog of products defined by a set of specifications represented as key-value pairs. Two products from the Jewellery and Home Furnishing categories at Flipkart are shown in Fig. 1 along with some of their specifications. Specifications like 'plating' and 'shape' are product attributes that take values 'silver' and 'rectangle' respectively. The intent of a search query is defined as a labelling of its tokens with the

M. Kumar—This work was done while the author was at Flipkart.

J. M. Jose et al. (Eds.): ECIR 2020, LNCS 12035, pp. 651–665, 2020.
https://doi.org/10.1007/978-3-030-45439-5_43

Fig. 1. Specifications of products from Jewellery and Home Furnishing.

product attributes whose values are matched against the query tokens during retrieval. The intent of two search queries is illustrated in Table 1.

Queries in e-commerce search can have multiple correct intents due to multiple valid matches between their tokens and the values of product attributes. An example of this is shown in Table 1 where the attributes 'color', 'plating', and 'base material' are all correct labels for the token 'silver' in the query 'silver oxidised earring'. This phenomenon is particularly prevalent in tail queries; an analysis of an editorially labelled sample of tail queries at Flipkart revealed that approximately 42% of tail queries had multiple correct intents. Existing techniques for identifying user intent in search queries are either supervised [17,19] or semi-supervised [11,19] and require labelled or partially labelled queries. Extending them to identify multiple intents in tail queries is difficult due to a lack of sufficient click-through data from which labels can be derived [14,17,25]. We address this shortcoming of existing techniques in our current work.

We start with an empirical study of the product catalog and search query logs at Flipkart and base our current work on its conclusions. We propose a latent variable generative model for the observed ordered pairs of query tokens that has the corresponding ordered pairs of attribute labels as the latent variables. This addresses the lack of labelled data for tail queries. We observed that tail queries tend to have multiple intents due to multiple attributes having similar high

Table 1. Labellings of multi-intent queries 'silver oxidised earring' and 'rectangle room mat' by the baselines and our proposed model (RIM) which identifies all correct intents.

	silver	oxidised	earring	rectangle	room	mat
Correct intent	color, plating, material	model	store	shape, pattern	place-of-use	store
LR	color, store	model	store	type	type	store
CRF	color, store, model	model	store	store	place-of-use	store
Bi-LSTM-1	model	model	store	key-features	model	model
Bi-LSTM-2	color	model	store	key-features	place-of-use	store
UMM	color, store	model	store	type	place-of-use	store
RIM	color, plating, material	model	store	shape, pattern	place-of-use	store

empirical probabilities of generating the same tokens. We propose a similarity measure between attribute pairs and use it to regularize our model in a way that the learnt posterior distributions have similar probabilities for similar attribute pairs. This addresses the problem of identifying multiple intents in tail queries. We finally demonstrate the superior performance of our proposed model against several strong baselines on an editorially labelled data set and in a large scale online A/B experiment at Flipkart where we achieved statistically significant improvements of 3.03% in click-through rate and 15.45% in add-to-cart ratio.

2 Definitions and Preliminaries

E-commerce product catalogs are typically divided into various categories where every product belongs to a single category. Examples of such categories are Jewellery, Furniture, and Home Furnishing (bed sheets, table covers, curtains, etc.). Sample products from Jewellery and Home Furnishing are shown in Fig. 1. We define *tail* queries as queries that occur less than 50 times a month.

The attributes that describe the products within a category are denoted by $\mathcal{A}$ and the values these attributes can take are denoted by $\mathcal{V}$. Every product can thus be represented by a set of attribute-value pairs (a, v) where v may consist of multiple tokens. For example, some of the values that the attribute 'material' can take in the Jewellery category are 'rose gold', 'silver', 'bronze', 'stainless steel', etc. The vocabulary of tokens that constitute all the attribute values is denoted by $\mathcal{W}$. A query is denoted by $\mathbf{x}$ and is defined as a sequence of n tokens $(x_1, x_2, \ldots, x_n)$. The intent of this query is denoted by $\mathbf{z}$ and is defined as a corresponding assignment of n attribute sets $(z_1, z_2, \ldots, z_n)$, where $z_i \subseteq \mathcal{A}$. We let z_i be a set so that a query can have multiple intents. In our current work, we focus on intent identification within a category and assume a query to category mapping is available; a fairly standard assumption in vertical search engines [3].

Constructing Intent Labels from Click Logs: Manual intent labelling of queries is a laborious task requiring significant domain expertise. However, for queries that occur sufficiently often in the click logs, matches between the query tokens and the attribute-values of the clicked products provide a natural means of obtaining the attribute labels. Following [19], for a particular query we find matches between its tokens and the tokens of the attribute-values of every product that is clicked for this query. We then aggregate these matches across attributes to construct intent labels for every token in the query. This process is applied to queries that occur at least 500 times in a month with a click-through rate of at least 40%. Using such frequent queries with high click-through rates lets us construct reliable and fairly noise-free attribute labels for them. Applying this process to tail queries will result in fairly noisy attribute labels [11,14,17]. The labelled data set thus constructed is denoted by $\mathcal{D}^L$ and is referred to as the *click-log labelled data* in this paper. The average number of such labelled queries $\mathcal{D}^L$ for the Jewellery, Furniture, and Home Furnishing categories is ≈5k while the average number of unique queries $\mathcal{D}$ that occur at least 10 times a month in these categories is ≈50k. The labelled queries are much fewer than the unique queries which shows the limitations of constructing intent labels from click logs.

3 Empirical Data Analysis

Query intent understanding on a large scale product catalog presents unique challenges and we discuss two distinct characteristics here. The fraction of unique queries with a particular attribute pattern in the click-log labelled data has a long-tailed distribution as shown in Fig. 2a. Two example attribute patterns for the query 'silver oxidised earring' are 'color, model name, store name' and 'plating, model name, store name' as shown in Fig. 1. From Fig. 2a, it is noteworthy that the most frequent attribute pattern represents on average only 5% of the unique queries in the three categories. This makes supervised learning difficult since most attribute patterns have very few example queries. Moreover, this analysis is for the relatively frequent queries $\mathcal{D}^L$ and we expect this distribution to have an even longer tail for tail queries.

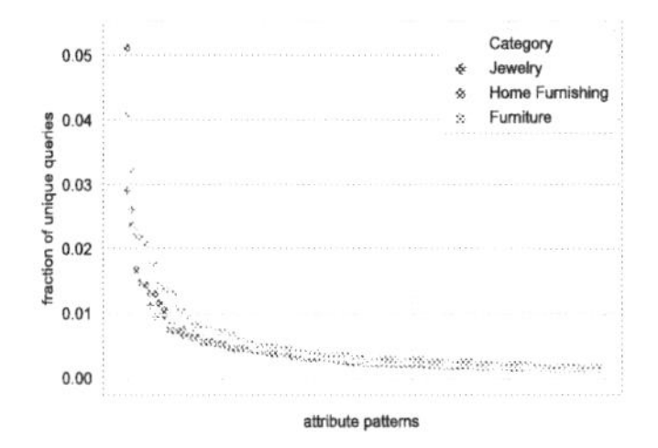

(a) The proportion of unique queries in 100 most frequent attribute patterns in the labelled data.

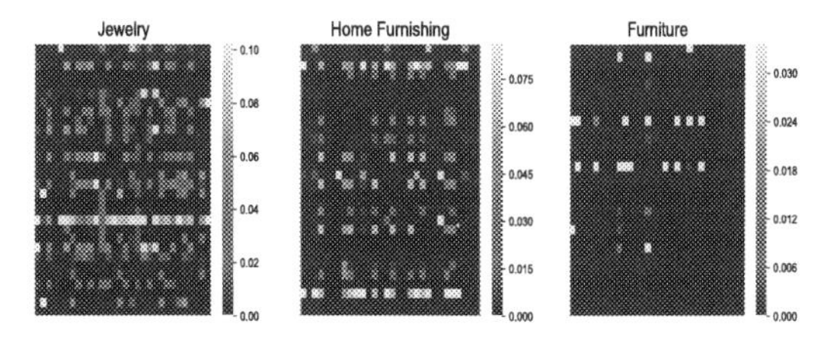

(b) Each point in the heat map is the normalized overlap between the vocabularies of a pair of attributes. Each graph visualises a random set of 30 attribute pairs.

Fig. 2. Empirical data analysis

The average number of attributes $\mathcal{A}$ for the three categories is ≈130 while the average size of the vocabulary $\mathcal{W}$ is ≈20k. Many pairs of attributes have a significant degree of overlap between their vocabularies. We illustrate this in Fig. 2b where the non-zero entries in the heat map indicate an overlap between the vocabularies of a particular pair of attributes. For example, the attributes 'plating' and 'base material' in the Jewellery category have an overlap of ≈30% in their vocabularies. This overlap indicates the possibility of multiple attributes being the correct labels for a token in a query and thus the query having multiple correct intents. We use this characteristic to develop a regularisation technique that improves our model's ability to capture multiple intents in queries.

4 The Latent Variable Generative Model

Tail queries have very few clicks and thus the click log mining technique of Sect. 2 can not be used to derive labels for them. Generative models are naturally suited to an unsupervised setting where labels are absent. The authors of [5] propose a simple generative process for queries which generates query tokens independently by first sampling an attribute and then sampling a token from that attribute's vocabulary. However, modelling dependence is important since the attribute label for a token depends on the other tokens in a query. For example, consider the queries 'cotton sofa cushion' and 'cotton bed sheets double bed'. The correct label for 'cotton' in the first query is 'filling material' while in the second query is 'fabric'. This highlights the need for a richer generative model that captures token interactions and attribute co-occurrences in a query.

We propose a latent variable generative model for queries where the observed variables are ordered pairs of tokens and the latent variables are the corresponding ordered pairs of attribute labels. The generative process is defined over all ordered pairs of tokens in a query and not just the adjacent ones. For example, there are 3 ordered pairs of tokens in the query 'silver oxidised earring': ('silver', 'oxidised'), ('oxidised', 'earring'), and ('silver', 'earring').

Let $c_{\mathbf{x}}$ be the set of all ordered pairs of tokens in a query $\mathbf{x}$. We define ψ as a $|\mathcal{A}| \times |\mathcal{A}|$ matrix of parameters specifying the attribute co-occurrence probabilities, i.e., $\sum_a \psi_{a,a'} = 1$ for each a'. We similarly define ϕ as a $|\mathcal{W}| \times |\mathcal{A}|$ matrix of parameters specifying the probability of generating a token from an attribute, i.e., $\sum_w \phi_{w,a} = 1$ for each a. We assume that the ith ordered token pair $x_i = (x_{i1}, x_{i2})$ is generated from a corresponding ordered attribute pair $z_i = (z_{i1}, z_{i2})$ as follows: Sample an attribute z_{i1} uniformly at random and then sample the attribute $z_{i2} \sim \text{Mult}(\psi_{\cdot,z_{i1}})$ conditioned on z_{i1}. The token pair x_i is then generated by sampling $x_{i1} \sim \text{Mult}(\phi_{\cdot,z_{i1}})$ and $x_{i2} \sim \text{Mult}(\phi_{\cdot,z_{i2}})$. The joint probability of x_i and z_i is thus given by

$$p(x_i, z_i) = p(x_{i1}|z_{i1})\, p(x_{i2}|z_{i2})\, p(z_{i2}|z_{i1})\, p(z_{i1}) = \phi_{x_{i1},z_{i1}}\, \phi_{x_{i2},z_{i2}}\, \psi_{z_{i2},z_{i1}}\, \frac{1}{|\mathcal{A}|}.$$

Therefore, our model represents queries as a set of all ordered pairs of its tokens and we assume all pairs to be independent to get $p(\mathbf{x}) \approx p(c_{\mathbf{x}}) = \prod_i \sum_{z_i} p(x_i, z_i)$. This assumption is critical for computational tractability while still capturing the interactions between the tokens as well as the co-occurrences between the attributes. The observed log-likelihood which we optimize using the standard Expectation Maximization (EM) algorithm is

$$l_o(q, \phi, \psi) = \sum_i \left[\mathbb{E}_{q_i} \left[\log\left(\frac{p(x_i, z_i)}{q_i(z_i)} \right) \right] + \text{KL}(q_i \,\|\, p(z_i|x_i)) \right]. \tag{1}$$

Via a standard derivation, the E-Step update for the token pair x_i is given by

$$q_i((z_{i1} = a, z_{i2} = a')) = \frac{\phi_{x_{i1},a}\phi_{x_{i2},a'}\psi_{a,a'}}{\sum_{(a,a')} \phi_{x_{i1},a}\phi_{x_{i2},a'}\psi_{a,a'}}, \tag{2}$$

where $q_i((z_{i1} = a, z_{i2} = a'))$ is the posterior probability of the attribute pair (z_{i1}, z_{i2}) being (a, a') given the token pair (x_{i1}, x_{i2}). Via a standard derivation, the M-Step updates for the parameters ϕ and ψ are given by

$$\phi_{w,a}^{(o)} = \frac{\sum_i \left[\mathbb{1}[x_{i1}=w]\, q_i((a,\cdot)) + \mathbb{1}[x_{i2}=w]\, q_i((\cdot,a))\right]}{\sum_i \left[q_i((a,\cdot)) + q_i((\cdot,a))\right]}, \quad \psi_{a,a'}^{(o)} = \frac{\sum_i q_i((a,a'))}{\sum_i q_i((\cdot,a'))}, \quad (3)$$

where $q_i((\cdot, a)) = \sum_{a'} q_i((z_{i1} = a', z_{i2} = a))$ and $q_i((a, \cdot))$ is defined similarly.

Since our model is defined over pairs of tokens, computing the attribute assignments for each token in a query during posterior inference requires an approximation. We follow [18] and approximate the posterior distribution of the attribute assignments by decomposing it over pairs of tokens as follows

$$p(\mathbf{z}|\mathbf{x}) \approx p((z_1, z_2)|(x_1, x_2)) \prod_{i=3}^{n-1} p((z_{i-1}, z_i)|(x_{i-1}, x_i)).$$

We compute multiple attribute assignments at each position in the query using a standard forward-backward algorithm to obtain multiple intents per token.

5 Regularisation for Learning Multiple Intents

Queries with multiple intents have multiple attribute labels for one or more of their tokens, for example, the token ‘silver’ in the query ‘silver oxidised earring’ shown in Table 1. As illustrated in Fig. 2b, certain attributes have a significant overlap between their vocabularies. We use this observation to define a similarity measure between attributes using background estimates of the generative model's parameters. We then use this similarity measure to devise a data dependent regularisation technique that distributes the generative model's posterior across attributes with significantly overlapping vocabularies which improves its ability to detect multiple intents.

5.1 Background Parameter Estimates

We use the product catalog and the click-log labelled data to derive background estimates for the generative model's parameters. To derive the estimates for ϕ, we first iterate over all products in a category and construct the set $\{(a, v, \kappa_{v,a})\}$, where a is an attribute, v is an attribute value and $\kappa_{v,a}$ is the number of products with v as the attribute-value for the attribute a. We then define the estimate

$$\widetilde{\phi}_{w,a} = \frac{C(w,a) + \varphi_a}{\sum_w C(w,a) + \varphi_a |\mathcal{W}|},$$

where $C(w,a) = \frac{C^U(w,a) + C^L(w,a)}{2}$, $C^U(w,a) = \sum_v \frac{\mathbb{1}_{[w \in v]} \log \kappa_{v,a}}{|v|}$, $C^L(w,a)$ is the number of times the token w is labelled with the attribute a in the click-log labelled data set $\mathcal{D}^L$, and $\varphi_a = \frac{\varphi}{\max_w C(w,a)}$ is a smoothing factor with $\varphi > 0$ being a hyper-parameter.

To derive the estimates for ψ, we first iterate over all products in a category and construct the set $\{(a, a', \kappa_{a,a'})\}$, where a and a' are attributes and $\kappa_{a,a'}$ is the number of products having both attributes a and a'. We then define the estimate

$$\widetilde{\psi}_{a,a'} = \frac{C(a,a') + \omega_{a'}}{\sum_a C(a,a') + \omega_{a'}|\mathcal{A}|},$$

where $C(a,a') = \frac{C^U(a,a') + C^L(a,a')}{2}$, $C^U(a,a') = \log \kappa_{a,a'}$, $C^L(a,a')$ is the number of times the attribute pair (a,a') co-occur in the click-log labelled data set $\mathcal{D}^L$, and $\omega_{a'} = \frac{\omega}{\max_a C(a,a')}$ is a smoothing factor with $\omega > 0$ being a hyper-parameter.

5.2 Attribute Similarity Regularisation

The probability of the model generating the token w from the attribute a is given by the model parameter $\phi_{w,a}$ and its background estimate is $\widetilde{\phi}_{w,a}$. Thus, if $\widetilde{\phi}_{w,a} \approx \widetilde{\phi}_{w,b}$ and both background estimates are high, then the model should pick both attributes a and b as relevant labels for the token w. Analogously, if two attribute pairs (a,a') and (b,b') have similar high background estimated probabilities of generating the token pair (w,w'), then the model should pick both attribute pairs (a,a') and (b,b') as relevant labels for the token pair (w,w'). We quantify this notion by defining

$$g_{(w,w')}((a,a'),(b,b')) = (\widetilde{\phi}_{w,a}\widetilde{\phi}_{w,b})^2(\widetilde{\phi}_{w',a'}\widetilde{\phi}_{w',b'})^2.$$

Note that g is high when $\widetilde{\phi}_{w,a} \approx \widetilde{\phi}_{w,b}$, $\widetilde{\phi}_{w',a'} \approx \widetilde{\phi}_{w',b'}$ and the individual $\widetilde{\phi}$'s are high. We use this notion of attribute similarity to define a regularisation term that distributes the generative model's posterior across attribute pairs with similar vocabularies. Let $g_{(w,w')}$ denote a square positive matrix of size $|\mathcal{A}|^2 \times |\mathcal{A}|^2$ over the attribute pairs. Alternating normalization of the rows and columns (the Sinkhorn-Knopp algorithm [23]) of $g_{(w,w')}$ will generate a doubly stochastic matrix $\bar{g}_{(w,w')}$ that we will use instead of $g_{(w,w')}$ as the measure of similarity. For a token pair x, the regularisation term penalizes large differences in the posterior probabilities $p((a,a')|x)$ and $p((b,b')|x)$ if $\bar{g}_x((a,a'),(b,b'))$ is high and is given in the following regularised log-likelihood

$$l_o(q,\phi,\psi) - \alpha \sum_i \left[\frac{1}{2} \sum_{z,\bar{z}} \bar{g}_{x_i}(z,\bar{z}) \big(p(z|x_i) - p(\bar{z}|x_i)\big)^2 \right], \tag{4}$$

where $x_i = (x_{i1}, x_{i2})$ is the ith token pair, z and $\bar{z}$ are attribute pairs, and $\alpha \in (0,1)$ is a hyper-parameter. Unfortunately, maximizing the above regularised log-likelihood becomes intractable due to a coupling of the model parameters in the M-step optimization. So we establish the following upper bound on the regularisation term in (4) that gives us a lower bound on the regularised log-likelihood that is tractable to maximize.

Theorem 1. *Let* $\mathbf{z}$ *and* $\mathbf{x}$ *be discrete random variables and* $\bar{g}_{\mathbf{x}}$ *be a* $|\mathbf{z}| \times |\mathbf{z}|$ *doubly stochastic matrix. Then, for any distribution* $q_{\mathbf{x}}$*, we have*

$$\tfrac{1}{2}\textstyle\sum_{z,\bar{z}} \bar{g}_{\mathbf{x}}(z,\bar{z})\big(p(z|\mathbf{x}) - p(\bar{z}|\mathbf{x})\big)^2 \leq \tfrac{1}{2}\sum_{z,\bar{z}} \bar{g}_{\mathbf{x}}(z,\bar{z})\big(q_{\mathbf{x}}(z) - q_{\mathbf{x}}(\bar{z})\big)^2 + \min\left[1, 5\sqrt{2\mathrm{KL}(q_{\mathbf{x}} \,\|\, p(\mathbf{z}|\mathbf{x}))}\right]$$

Applying this bound on the posterior distribution $p(z_i|x_i)$ and the approximate posterior distribution q_i gives the following lower bound on (4)

$$\mathbb{E}_{q_i}\left[\log\left(\frac{p(x_i,z_i)}{q_i(z_i)}\right)\right]-\alpha\left[\frac{1}{2}\sum_{z,\bar{z}}\bar{g}_{x_i}(z,\bar{z})\big(q_i(z)-q_i(\bar{z})\big)^2\right]-\frac{5\alpha}{4}. \quad (5)$$

Proof. See Online Supplementary Material.

Thus, the regularised E-step optimization is

$$\max_{q_i}\mathbb{E}_{q_i}\left[\log\left(\frac{p(x_i,z_i)}{q_i(z_i)}\right)\right]-\alpha\left[\frac{1}{2}\sum_{z,\bar{z}}\bar{g}_{x_i}(z,\bar{z})\big(q_i(z)-q_i(\bar{z})\big)^2\right], \quad (6)$$

subject to $\sum_{z_i} q_i(z_i) = 1$, where we have dropped the constant term involving α. The optimization in (6) can be done via projected gradient descent [21]. In our experiments, we observed that 3 to 4 iterations were usually sufficient for convergence and that our method results in the posterior distribution being distributed over similar attribute pairs instead of being concentrated on one of them. The M-step updates for this model are exactly the same as in (3).

6 Experiments and Analysis

We evaluated our proposed model against several strong baseline models on data sets proprietary to Flipkart. To the best of our knowledge, there are no publicly available data sets for evaluating query intent algorithms for e-commerce search or similar domains and all previous related work [11,16,18,19,24] has been evaluated on such proprietary data sets. We selected the Jewellery, Home Furnishing, and Furniture categories for experimental evaluation. These categories at Flipkart have a high business value in spite of low query volume and thus very sparse click data leading to more tail queries as compared to more popular categories like Electronics or Lifestyle. The click-log labelled data set $\mathcal{D}^L$ and the unlabelled data set $\mathcal{D}$ used to train all models were obtained from one month of query logs. We restricted $\mathcal{D}$ to queries with at least 10 occurrences over that month to filter out queries with misspellings.

6.1 Baseline Models

There is little prior work on understanding the intent of e-commerce search queries, especially in our setting where we have access to labelled as well as unlabelled query logs in addition to data from the product catalog. Prior work on intent understanding can be broadly classified into supervised and unsupervised methods. The unsupervised baseline model we compare against is UMM [5] described in Sect. 4. The supervised baseline models we compare against are Multinomial Logistic Regression (LR), the Linear Chain CRF from the query intent understanding work in [11,19], and the Bi-LSTM-CRF from [10]. The recent work [26] on understanding intent in Google shopping queries is not applicable in our setting since it focuses on a different problem of understanding overall query intent and not token level attribute labelling as ours. These supervised baseline models were trained on the click-log labelled data set $\mathcal{D}^L$ with elastic-net regularisation whose hyper-parameters were selected by 3-fold cross-validation with F_1 score as the performance metric.

Multinomial LR and Linear Chain CRF: Each training instance consisted of a query token x_i at position i and its attribute label z_i taken from the queries in $\mathcal{D}^L$. We extended the features from [19] by defining additional catalog features in terms of matches between the query tokens and the catalog attributes and additional syntactic features in terms of the surface form of the tokens. The catalog features were unigram and bigram TF-IDF matches with the vocabulary of each attribute in a category. The syntactic features were whether a unigram is a stopword, is a short word with less than 4 characters, or is alphanumeric.

Bi-LSTM-CRF: We implemented two variants of the Bi-LSTM-CRF from [10]. The first, Bi-LSTM-1, used 100-dimensional word embeddings trained on the product descriptions from the catalog (using fastText [13]) as its features. The second, Bi-LSTM-2, additionally used the catalog features described above. It is important to note that we have a much stronger set of features compared to the standard implementations of a Bi-LSTM-CRF since we incorporate where a unigram or a bigram matches in the attribute space.

We evaluated all baseline models against the all pairs mixture model (PMM) described in Sect. 4 and the all pairs mixture model with attribute similarity regularisation (RIM) described in Sect. 5.2.

6.2 Evaluation of Intent Labellings

A team of search quality experts at Flipkart labelled a random sample of tail queries from the query logs using their domain expertise. We randomly selected 900 queries with multiple intents (300 queries per category) from this labelled set on which to evaluate all models and refer to it as the *golden set*. We further created 5 randomized 80/20 splits of the golden set to get multiple test and validation sets. We computed marginal distributions at each token position in a query for all models and considered only those labellings that were above a threshold tuned on a validation set. We chose F_1 score as the performance metric and since we are interested in queries with multiple intents, we follow [8] and get the overall F_1 score per query by micro-averaging the F_1 score per query token. We used the same validation and test sets for all models in each run.

The performance of all models on the test sets is summarized in Fig. 3 and Table 2. RIM outperforms PMM as well as all baseline models with an average improvement of 12.5% in F_1 score over UMM, the best performing baseline model. RIM achieves an average improvement of 13.4%, 15.2%, and 8.6% in F_1 score over UMM for the Furniture, Home Furnishing, and Jewellery categories. Moreover, RIM and PMM together outperform all baseline models which demonstrates the effectiveness of modelling pairwise dependencies between the query tokens. All the supervised baseline models including Bi-LSTM-CRF, a

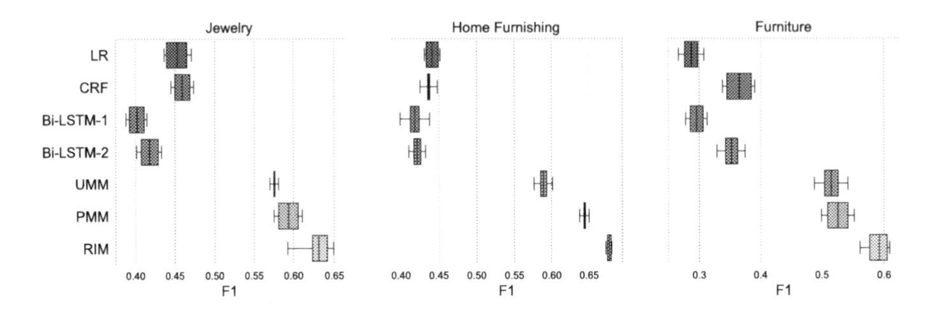

Fig. 3. Box plots of F_1 scores on the held-out test splits of the golden set for all models.

state-of-the-art model for slot-tagging problems, perform much worse than the unsupervised baseline model UMM due to a lack of sufficient labelled data. RIM's performance improvements over PMM demonstrate the effectiveness of our data dependent attribute similarity regularisation for queries with multiple intents. An example of this is illustrated in Table 3 where RIM's posterior is distributed over the correct attribute labels whereas that of PMM is distributed over the correct and incorrect attribute labels.

Table 2. Average F_1 scores on the held-out test splits of the golden set for all models. The results for RIM are statistically significant against all baselines with p-value < 0.01.

	LR	CRF [11,19]	Bi-LSTM-1 [10]	Bi-LSTM-2 [10]	UMM [5]	PMM	RIM
Jewellery	0.45	0.46	0.40	0.42	0.58	0.59	**0.63**
Home Furnishing	0.44	0.44	0.42	0.42	0.59	0.64	**0.68**
Furniture	0.29	0.36	0.29	0.35	0.52	0.53	**0.59**
Average	0.39	0.42	0.37	0.39	0.56	0.59	**0.63**

Table 3. The marginal posterior distributions for the token 'silver' in the query 'silver oxidised earring' returned by PMM and RIM. Here, $\delta < 10^{-4}$ and the correct attribute labels are color, plating, and base material.

	color	plating	base material	store	model	ideal for	body material
PMM	0.381	0.148	0.121	0.143	0.059	0.033	0.015
RIM	0.693	0.135	0.081	δ	δ	δ	δ

6.3 Performance in an Online A/B Experiment

The intent inferred for a search query plays a major role in determining and retrieving the most relevant products for that query at Flipkart as is standard in e-commerce search [15]. Thus, the quality of the inferred intent very strongly influences a user's propensity to click and add-to-cart the products retrieved for a search query. Hence, we measure the click-through rate (CTR) and the add-to-cart ratio as the relevant metrics in the online A/B experiment. The add-to-cart ratio (i.e., search conversion) is defined as the fraction of searches leading to a product being added to the shopping cart. We deployed RIM and UMM in the production search system at Flipkart and compared the performance of the models against each other in a standard A/B experiment configuration where we treated UMM as the control condition. More than 10 million users visit Flipkart daily and we randomly assigned 15% of the users to each condition and conducted the test over 10 days. Since the models were trained for the Jewellery, Home Furnishing, and Furniture categories, only those queries belonging to these categories were considered for comparison. The query to category mapping was obtained by a separate production system at Flipkart. We would have ideally liked to restrict the experiment to tail queries with multiple intents only in order to better demonstrate the capabilities of RIM. However, in practice it is difficult to determine on the fly if a query has multiple intents. Thus, we conducted the experiment on all tail queries. The query volume affected by the experiment was $\approx$75k tail queries (with $\approx$ 36k unique queries). The results of this online A/B experiment are summarized in Table 4.

Table 4. Results of the online A/B experiment comparing RIM against the best baseline model UMM. Statistical significance with p-value < 0.01 is denoted by $*$ and that with p-value < 0.001 is denoted by $\wedge$.

	Tail CTR (%)	Tail Add-to-Cart (%)
Jewellery	$+2.78^{*}$	$+10.22^{*}$
Home Furnishing	$+2.13^{\wedge}$	$+13.46^{\wedge}$
Furniture	$+4.19^{\wedge}$	$+22.67^{\wedge}$
Average	+3.03	+15.45

RIM significantly improves both the CTR and the add-to-cart ratio for tail queries across all categories. The average improvement in CTR is 3.03% while that in add-to-cart ratio is 15.45%. The results for all categories were statistically significant as measured by a paired sample t-test with p-value < 0.01. The much larger improvement in add-to-cart ratio as compared to CTR is noteworthy. On further analysis, we found that most tail queries express a very specific product need and when the search system is able to infer the correct query intent and retrieve the relevant products, the customers are satisfied, as indicated by add-to-cart, with fewer clicks. We are thus able to demonstrate the effectiveness of the proposed model in a large scale real world setting. We finally illustrate the retrieval quality with intents inferred by RIM compared to the existing production system for two queries in Figs. 4 and 5. Both queries are tail queries drawn from the online A/B experiment and RIM correctly identifies their intents.

Fig. 4. Top retrieved products for the query 'small pillow cover pack' by the existing production system (left) and with intents inferred by RIM (right). The production system retrieves irrelevant bed sheets. RIM correctly identifies 'pillow cover' as 'store/model' and 'small' as 'size/shape'.

Fig. 5. Top retrieved products for the query 'glass top wooden dining table 6 seater' by the existing production system (left) and with intents inferred by RIM (right). The products highlighted in red are wooden top tables and thus irrelevant for the query. RIM correctly identifies 'glass' as 'top material'. (Color figure online)

7 Related Work

The existing work on query understanding has mainly focused on learning query intent in a supervised manner by using click-through data [6,9,12,22] and this restricts their generalization to combinations of frequent attribute patterns only. However, tail queries exhibit tail attribute patterns and this is the focus of our current work. The existing methods for understanding intent of tail queries can be broadly divided into two major types: (a) Those that identify a mapping between tail queries and similar frequent queries [11,24], and (b) Those that learn query intent from partially labelled queries [16,17,19]. Fusing the results of a tail query with those of a similar frequent query as a way of improving retrieval metrics is suggested in [11]. However, the underlying assumption that a tail query is a frequent query that is expressed differently does not hold in our case. Transferring the intent of frequent queries to tail queries using an external knowledge base is studied in [24]. However, building domain specific knowledge bases is difficult. Learning query intent from partially labelled queries along with side-supervision in the form of derived attributes for some query tokens is studied in [19]. However, it is difficult to obtain partial labellings for tail queries because most tokens in tail queries will be marked as 'unknown' due to the sparsity of the click-through data as observed in [17]. A hidden-unit linear-chain CRF that allows for non-linearities is introduced in [16]. However, its formulation too requires partially labelled queries. The availability of derived labelled data by performing rule-based labelling of unlabelled sequences is assumed in [4]. However, the rule-based labelling is domain specific and is difficult to extend. The CRF auto-encoder [1] and its application to tasks like POS tagging [20] is promising especially since it does not require labelled data. However, the CRF auto-encoder has difficulty scaling to the label space for query intent understanding that is much larger than that for POS tagging. The most recent work on understanding intent of e-commerce search queries is described in [26] for Google shopping. However, it is not applicable in our setting since it focuses on a different problem of understanding overall query intent and not token level attribute labelling.

8 Conclusion and Future Work

In this paper, we investigated the problem of discovering multiple intents in tail queries for e-commerce search. We introduced a latent variable generative model for queries to overcome the lack of sufficient labelled data. To improve this model's ability to identify multiple intents, we then introduced a novel data dependent regularisation technique derived from empirical evidence of overlap in attribute vocabularies. We finally demonstrated the superior performance of our regularised intent model against several strong baseline models on an editorially labelled data set as well as in a large scale online A/B experiment at Flipkart, a major Indian e-commerce company. In the future, we plan to investigate deep generative intent models and knowledge graph representation of the product catalog to further improve intent understanding.

References

1. Ammar, W., Dyer, C., Smith, N.A.: Conditional random field autoencoders for unsupervised structured prediction. In: Proceedings of the 27th International Conference on Neural Information Processing Systems, vol. 2, pp. 3311–3319. MIT Press, Cambridge (2014)
2. Anderson, C.: The Long Tail: Why the Future of Business is Selling Less of More. Hyperion (2006)
3. Arguello, J., Diaz, F., Callan, J., Crespo, J.F.: Sources of evidence for vertical selection. In: Proceedings of the 32nd International ACM SIGIR Conference on Research and Development in Information Retrieval. SIGIR 2009, pp. 315–322. ACM, New York (2009). https://doi.org/10.1145/1571941.1571997
4. Dredze, M., Talukdar, P.P., Crammer, K.: Sequence learning from data with multiple labels. In: Learning from Multi-Label Data at ECML PKDD, 2009, vol. 39 (2009)
5. Duan, H., Zhai, C., Cheng, J., Gattani, A.: Supporting keyword search in product database: a probabilistic approach. In: Proceedings of the VLDB Endowment, vol. 6, no. 14, pp. 1786–1797 (2013)

6. Gao, J., He, X., Nie, J.Y.: Clickthrough-based translation models for web search: from word models to phrase models. In: Proceedings of the 19th ACM International Conference on Information and Knowledge Management, CIKM 2010, pp. 1139–1148. ACM, New York (2010)
7. Goel, S., Broder, A., Gabrilovich, E., Pang, B.: Anatomy of the long tail: ordinary people with extraordinary tastes. In: Proceedings of the Third ACM International Conference on Web Search and Data Mining, WSDM 2010, pp. 201–210. ACM, New York (2010)
8. Han, J., Fan, J., Zhou, L.: Crowdsourcing-assisted query structure interpretation. In: Proceedings of the Twenty-Third International Joint Conference on Artificial Intelligence, IJCAI 2013, pp. 2092–2098. AAAI Press (2013)
9. Hashemi, H.B., Asiaee, A., Kraft, R.: Query intent detection using convolutional neural networks. In: International Conference on Web Search and Data Mining, Workshop on Query Understanding (2016)
10. Huang, Z., Xu, W., Yu, K.: Bidirectional LSTM-CRF models for sequence tagging. CoRR abs/1508.01991 (2015). http://arxiv.org/abs/1508.01991
11. Huo, S., Zhang, M., Liu, Y., Ma, S.: Improving tail query performance by fusion model. In: Proceedings of the 23rd ACM International Conference on Conference on Information and Knowledge Management. CIKM 2014, pp. 559–568. ACM, New York (2014)
12. Joachims, T.: Optimizing search engines using clickthrough data. In: Proceedings of the Eighth ACM SIGKDD International Conference on Knowledge Discovery and Data Mining, KDD 2002, pp. 133–142. ACM, New York (2002)
13. Joulin, A., Grave, E., Bojanowski, P., Mikolov, T.: Bag of tricks for efficient text classification. In: Proceedings of the 15th Conference of the European Chapter of the Association for Computational Linguistics: Volume 2, Short Papers, pp. 427–431. Association for Computational Linguistics (2017)
14. Kang, C., Lin, X., Wang, X., Chang, Y., Tseng, B.L.: Modeling perceived relevance for tail queries without click-through data. CoRR abs/1110.1112 (2011)
15. Karmaker Santu, S.K., Sondhi, P., Zhai, C.: On application of learning to rank for e-commerce search. In: Proceedings of the 40th International ACM SIGIR Conference on Research and Development in Information Retrieval, SIGIR 2017, pp. 475–484. ACM, New York (2017)
16. Kim, Y.B., Jeong, M., Stratos, K., Sarikaya, R.: Weakly supervised slot tagging with partially labeled sequences from web search click logs. In: Proceedings of the 2015 Conference of the North American Chapter of the Association for Computational Linguistics: Human Language Technologies, pp. 84–92. Association for Computational Linguistics (2015)
17. Kiseleva, J., Agichtein, E., Billsus, D.: Mining query structure from click data: A case study of product queries. In: Proceedings of the 20th ACM International Conference on Information and Knowledge Management, CIKM 2011, pp. 2217–2220. ACM, New York (2011). https://doi.org/10.1145/2063576.2063930
18. Konishi, T., Ohwa, T., Fujita, S., Ikeda, K., Hayashi, K.: Extracting search query patterns via the pairwise coupled topic model. In: Proceedings of the Ninth ACM International Conference on Web Search and Data Mining, WSDM 2016, pp. 655–664. ACM, New York (2016)
19. Li, X., Wang, Y.Y., Acero, A.: Extracting structured information from user queries with semi-supervised conditional random fields. In: Proceedings of the 32nd International ACM SIGIR Conference on Research and Development in Information Retrieval, SIGIR 2009, pp. 572–579. ACM, New York (2009)

20. Lin, C.C., Ammar, W., Dyer, C., Levin, L.: Unsupervised POS induction with word embeddings. In: Proceedings of the 2015 Conference of the North American Chapter of the Association for Computational Linguistics: Human Language Technologies, pp. 1311–1316. Association for Computational Linguistics (2015)
21. Murphy, K.P.: Machine Learning: A Probabilistic Perspective. The MIT Press, Cambridge (2012)
22. Radlinski, F., Szummer, M., Craswell, N.: Inferring query intent from reformulations and clicks. In: Proceedings of the 19th International Conference on World Wide Web, WWW 2010, pp. 1171–1172. ACM, New York (2010). https://doi.org/10.1145/1772690.1772859
23. Sinkhorn, R.: A relationship between arbitrary positive matrices and doubly stochastic matrices. Ann. Math. Stat. **35**(2), 876–879 (1964)
24. Song, Y., Wang, H., Chen, W., Wang, S.: Transfer understanding from head queries to tail queries. In: Proceedings of the 23rd ACM International Conference on Conference on Information and Knowledge Management, CIKM 2014, pp. 1299–1308. ACM, New York (2014)
25. Szpektor, I., Gionis, A., Maarek, Y.: Improving recommendation for long-tail queries via templates. In: Proceedings of the 20th International Conference on World Wide Web. WWW 2011, pp. 47–56. ACM, New York (2011). https://doi.org/10.1145/1963405.1963416
26. Wu, C.Y., Ahmed, A., Kumar, G.R., Datta, R.: Predicting latent structured intents from shopping queries. In: Proceedings of the 26th International Conference on World Wide Web. WWW 2017, pp. 1133–1141, International World Wide Web Conferences Steering Committee, Republic and Canton of Geneva, Switzerland (2017)

Utilising Information Foraging Theory for User Interaction with Image Query Auto-Completion

Amit Kumar Jaiswal(✉), Haiming Liu, and Ingo Frommholz

Institute for Research in Applicable Computing, University of Bedfordshire, Luton, UK
{amitkumar.jaiswal,haiming.liu,ingo.frommholz}@beds.ac.uk

Abstract. Query Auto-completion (QAC) is a prominently used feature in search engines, where user interaction with such explicit feature is facilitated by the possible automatic suggestion of queries based on a prefix typed by the user. Existing QAC models have pursued a little on user interaction and cannot capture a user's information need (IN) context. In this work, we devise a new task of QAC applied on an image for estimating patch (one of the key components of Information Foraging Theory) probabilities for query suggestion. Our work supports query completion by extending a user query prefix (one or two characters) to a complete query utilising a foraging-based probabilistic patch selection model. We present iBERT, to fine-tune the BERT (Bidirectional Encoder Representations from Transformers) model, which leverages combined textual-image queries for a solution to image QAC by computing probabilities of a large set of image patches. The reflected patch probabilities are used for selection while being agnostic to changing information need or contextual mechanisms. Experimental results show that query auto-completion using both natural language queries and images is more effective than using only language-level queries. Also, our fine-tuned iBERT model allows to efficiently rank patches in the image.

Keywords: Query auto completion · Interactive information retrieval · Information Foraging Theory

1 Introduction

Query auto-completion (QAC) is an action of signalling full queries once the user starts typing a prefix of a few characters that eases user query compositions [4]. It is also termed as (dynamic) query suggestion [17], query completion [35] and real-time query expansion [37]. Popular features such as QAC make people more dependent on search engines to find any relevant information. However, such kind of factor lets users express their queries only ambiguously, which are then overly vague to be completely interpreted by search engines. This makes query auto-completion a bottleneck construct in the usability of search engines [5].

J. M. Jose et al. (Eds.): ECIR 2020, LNCS 12035, pp. 666–680, 2020.
https://doi.org/10.1007/978-3-030-45439-5_44

Also, users often apply several rounds of search to reformulate their queries further to adhere to their information needs given they find some relevant results. Past work [6,20] demonstrated the use of information scent to model users' information need during web search, and it has been used to understand the factors affecting search and what takes a user to stop the search. Despite the good observation, the exploitation of information scent (from Information Foraging Theory [27]) is under-explored in case of ambiguous queries and have not been extended to take into account an image in query expansion (or suggestion) tasks. For the users' convenience, current search engines generally endue query suggestions for them in order to describe their queries more explicitly. They have been explored extensively in query auto-completion tasks, especially the traditional approach known as Most Popular Completion (MPC) [3] which at the extreme is incapable of anticipating a query it has never seen before. Solutions further improved by recent semantically-driven models [23,24] and neural model [26] approaches which are the current state-of-the art in QAC. However, most of the language embedding models [13] have obtained strong results on multiple benchmarks for understanding the polarity of word compositions. Unsupervised pre-trained natural language embeddings [7,21] successfully model long term dependencies with the purpose of predicting masked terms and assessing if sentences ensue one another, which showed strong results on several natural language processing and information retrieval tasks. Empirically, recent advances in sequence models have been adapted to span a prefix to full text and index [12] but despite the attainment, it has not been generalised to take an image into account. Also, deep neural networks are mature enough and capable of segmenting regions within an image [9,10].

To address the above mentioned gaps, we move one step forward to present a method that extends and modifies the state-of-the-art approaches in query completion and text embedding. We apply our ideas to an image search scenario where we assume patches are regions of images that are relevant to the user's information need. Our work is concerned with providing users of image search engines with a useful query suggestion (via a visually-oriented patch form) during interaction, to further amplify their exploratory search experience. Hence, finding useful patches for query expansion in an image based on textual queries (or descriptions) is the primary focus of our work. Past work [11,30] used both the query and image for typical retrieval and segmentation tasks. In our task formulation, we rely only upon a given arbitrary text prefix rather than having the entire text query which is used to perform search based on the image and supported by a modified deep language model [12] to find the most relevant patch in the image. We break down the task into three sub-tasks: (a) completing the query from user query prefix and an image; (b) finding patch probabilities based on the complete user query, and (c) aligning and segmenting all patches in the image. We summarise our contributions of this paper as follows:

1. To the best of our knowledge, we are the first to present a method for image query auto-completion where a user query prefix is adapted upon an image.

2. We elaborate the analogy of query auto-completion based on Information Foraging Theory and propose an explainable strategy for the observed challenges of query formulation and the varying users' information need.
3. We propose iBERT inspired by [7] to compute probabilities of patches and rank them efficiently in the image.

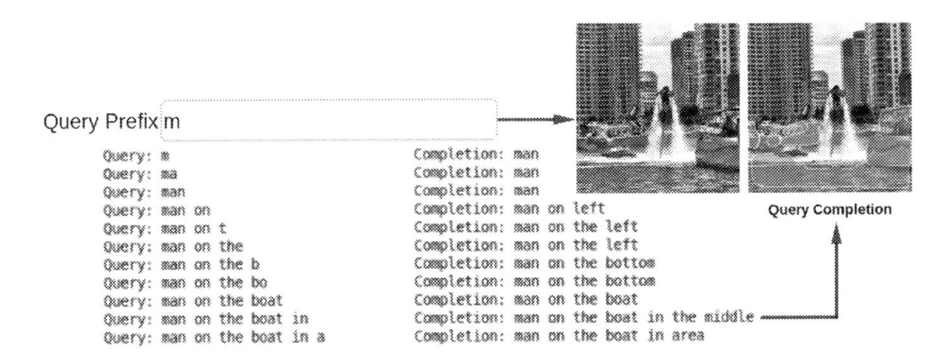

Fig. 1. Query auto-completion using our extended LSTM language model

2 Related Work

This section details a brief overview of query auto-completion, image search suggestion, Information Foraging Theory and BERT pre-trained language embedding model. We will investigate the latter approach experimentally in the following section.

Query Auto-Completion: Query auto-completion is an important aspect for information retrieval systems which allow it to predict what could be the next character (or query item) right after the first key was pressed by a user. The predictions in IR systems are generally driven by the query logs (or query history) which are the factual queries that users have previously entered as they were trying to satisfy their information need [14,37]. [3] introduced a method called NearestCompletion that addresses the situation of "context" which depicts the users' preceding queries in suggestion-based IR systems. The authors' proposed MPC mechanism relies on the entire popularity of the queries conforming to the provided prefix. Recent work reported in [15] studies user reformulation behaviour by leveraging textual features, whereas [31] introduced personalised query auto-completion and found that utilising a user's long-term search logs and locations as well as both context-based textual features and demographic features is more effective. More recent advances in QAC using neural language models are proposed in [26] using recurrent neural networks that effectuate the performance on immediately unseen queries. A generalised and adaptable language model for personalised QAC is introduced in [12]. We extend this adaptable language model to query completion in an image search scenario in the following section.

Query Suggestion in Image Search: Query suggestion and query completion differs in their end goal in which the former search aspect outputs a list of ranked queries against an input query, whereas the latter search aspect outputs queries with the first few characters (or text) similar to the user's input. Recent work [39] introduced a learning-based personalised suggestion framework for query suggestion which uses both visual and textual queries. Their work uses users' click-through data. A new paradigm of attention-based mechanisms for referring expressions in image segmentation [30] is proposed which contains a keyword-aware network and query attention model that demonstrates the relationships with various image regions for a given query. Inspired by the idea of attention models, we modify this mechanism for patch alignments within images via information scent in the following section.

Information Foraging Theory: Information Foraging Theory (IFT) [27] is a theoretical framework for understanding information access behaviour, derived from the ecological science concept of optimal foraging theory which applies to how humans access information. IFT stands on three different models, namely information scent model, information patch model and information diet model, which can illustrate users' search preferences and behaviours [19]: (1) The information within a certain environment scattered in form of *patches* (images, text snippets, documents) consisting of *information features* (colors, words) refers to the *information patch model*; (2) A user can go from one patch to another via a *cue* (e.g., typing a query by following perceptual or heuristic cues [32]), which meets the user's information need. The goal of such cues is to characterise the contents that will be envisaged by trailing the links, which refers to *the information scent model*; (3) Different types of information sources will vary in their information access costs. Users will assess the information sources based on information gain per unit cost or varied profitability, and then the users will narrow or expand diversities of information sources based on their profitability. This user behaviour refers to *the information diet model.*

One of the main IFT concepts are *information patches.* For instance, sections and their associated features in search engine results can be considered patches. From a foraging perspective in image search, the searcher is the predator (or forager [38]), the information patch is any segment or a region within an image (or image itself) in a given information environment. The piece of information a user is looking for is the prey, and the consumed (or gained) information is the information diet. Something on the user interface that informs users about a specific place they should look next is referred to as a *cue* of the information scent.

Language Embeddings: Nowadays, many information retrieval or natural language processing tasks rely on language embeddings, such as word2vec [22], Glove[1], and fastText[2]. They use vector word embeddings for word representation to transform a distinct space of human language into a continuous space,

[1] https://nlp.stanford.edu/projects/glove/.
[2] https://fasttext.cc.

which will be further processed usually through a neural network. In query auto-completion, embeddings have been employed for distributed representation of queries based on a convolutional latent semantic model [23]. Word embeddings have been used to compute query similarity for query auto-completion [29], incorporating the features with the Most Popular Completion model. Very recent work [7] introduced a pre-trained deep language model known as BERT which has shown promising results on several IR and natural language processing tasks. However, it is still not well-explored how to leverage such pre-trained language models for QAC, which poses certain challenges both regarding the task and training. Based on this work, we describe our proposed BERT-based model for computing patch probabilities in the following section.

3 Our Model

Let a set of patches $p_k \in P$, where P is the complete set of recognisable patch classes, be given. The user inputs a query prefix q_p, an incomplete query to retrieve an image I. With the given q_p, we auto-complete the expected query q. We formulate the auto-completion query task as the probability maximisation of a given query adapted on an image as shown in Eq. (1)

$$q_{a^*} = \operatorname*{argmax}_{q} P(q|q_p, I) = \operatorname*{argmax}_{\{t_1 t_2 ... t_n\}} P(t_1 t_2 ... t_n | q_p, I) \tag{1}$$

where q_{a^*} is the adapted query on an image, $t_i \in S$ is the term in position i in a sequence S.

We consider the task of estimating patch probabilities provided an auto-completion query q_{a^*} as a multi-label problem where each class of patches can independently exist. Let $P_{q_{a^*}}$ be the set of patches attributed to in q_{a^*}. As $\hat{q_{p_k}}$ is the estimate of $P(p_k \in P_{q_{a^*}})$ and $y_k = \mathbb{1}[p_k \in P_{q_{a^*}}]$, the sigmoid cross entropy loss function is minimized by the patch selection model:

$$\mathcal{L}_{f_{selection}} = -\sum_{k} y_k \log(\hat{q_{p_k}}) + (1 - y_k) \log(1 - \hat{q_{p_k}}) \tag{2}$$

An overview of the proposed end-to-end architecture shown in Fig. 2. The user types his/her query prefix for the given image to autocomplete and we perform image feature extraction using a pre-trained Convolutional Neural Network (CNN). Then, we feed the image features into the extended Long Short-Term Memory (LSTM) language model together with the query prefix which has a context-dependent weight matrix with an adaptation matrix constructed from a context-driven embedding model. These two constructs from image and text as visual features and textual queries are applied to complete a query. The completed query is then passed to iBERT (fine-tuned BERT language embedding model) to compute the patch probabilities, which in are utilised for patch selection. More details are provided in the next section.

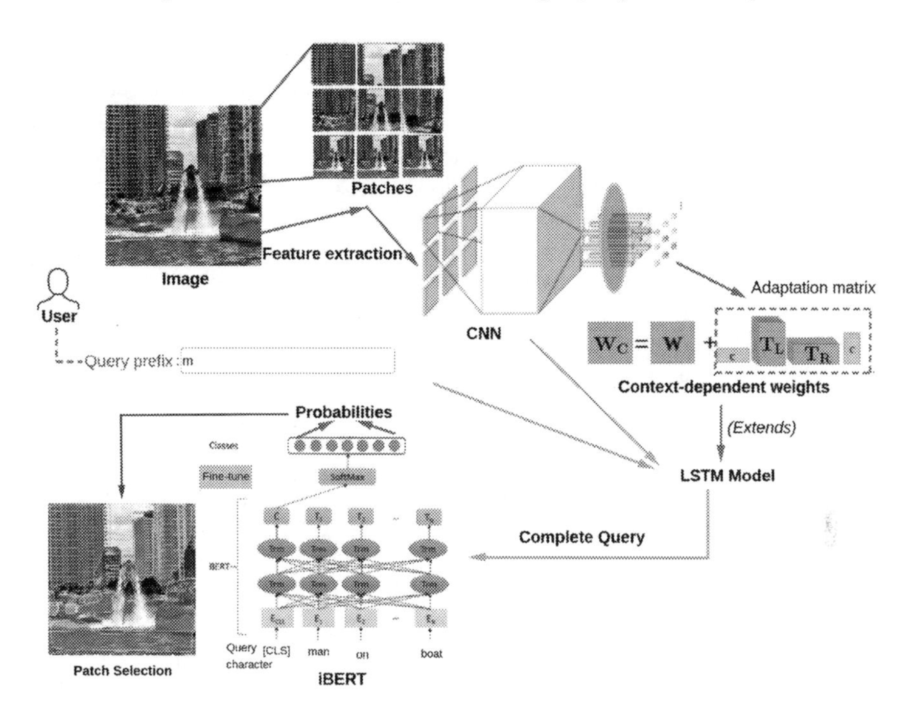

Fig. 2. The end-to-end architecture of Image Query Auto-Completion: User query prefix with the image features generated from a pre-trained CNN are input to an extended LSTM model (by incorporating a context-dependent weight matrix) which predicts a complete query. The resulting query is fed into a fine-tuned BERT pre-trained embedding model which outputs patch probabilities for patch selection.

3.1 Image Query Auto Completion

The challenge of query auto-completion is to predict and generate queries from prefixes that have never been seen in the training set. An initial attempt using neural language models has been introduced in [33]. The benefit of using character-level neural language models is providing more fine-grained predictions but they suffer from the semantic understanding that word-level models provide. For a prefix that has not been seen before (such as an incomplete word), their model enriches the shared information among comparable prefixes to create prediction nonetheless. In our scenario, we are given a prefix to complete a query conditioned on an image. To solve this new QAC problem, we exploit and extend the Long Short-Term Memory (LSTM) language model [12] with combined input and forget gates to auto complete queries. The language model is made up of a single-layer character-level LSTM with layer normalisation [2]. Our extension and modification to this language model is that we replace user embeddings with a low-dimensional representation of images. We adapt this LSTM language model alongside a context-dependent weight matrix $\mathbf{W}$ replaced by $\mathbf{W_C} = \mathbf{W} + \mathbf{M_A}$. We are providing a character embedding $w_c \in \mathbb{R}^e$, a preceding hidden state $h_{c-1} \in \mathbb{R}^h$, where $\mathbf{M_A}$ is the adaptation matrix constructed by the product ($\times_i$ denotes the i-th-order tensor product) of the context c with two basis tensors, $\mathbf{T_L} \in \mathbb{R}^{u \times (e+h) \times v}$ and $\mathbf{T_R} \in \mathbb{R}^{v \times h \times u}$. Alternatively, the two basis tensors i.e., $\mathbf{T_L}$ and $\mathbf{T_R}$ are re-shaped to $\mathbb{R}^{u \times (v(e+h))}$ and $\mathbb{R}^{vh \times u}$. So the next predicted hidden state and the adaptation matrix can be equated as follows:

$$\begin{aligned} h_c &= \sigma([w_c, h_{c-1}]\mathbf{W_C} + b) \\ \mathbf{M_A} &= (c \times_1 \mathbf{T_L})(\mathbf{T_R} \times_3 c) \end{aligned} \tag{3}$$

We combine the context-driven weight matrix and the immediate preceding hidden state followed by the generated adaptation matrix which able to alter each query completion to be personalised to a particular image representation. We perform feature extraction on an input image using a Convolutional Neural Network (CNN) trained on ImageNet (pre-trained CNN), where we retrain only the last two fully connected layers shown in Fig. 2. The generated image feature vector is then fed into the LSTM language model via the adaptation matrix. We apply beam search decoding [34] in the generated array of predicted characters to select the optimal completion for the user query prefix.

3.2 iBERT - BERT for Patch Probability

We describe our approach to compute the probability of image patches which addresses an important aspect of query auto-completion systems. We assume that during the search process, users are typically interested in some part of the image as well as the image itself if it matches the mental picture of their belief [36]. Our work focuses on a new perspective of query auto-completion on images and the proposed model finds image patches which match the user context based on the query prefix using Eq. (1). BERT (Bidirectional Encoder Representations from Transformers) [7] shows promising results in multiple tasks of natural language processing and information retrieval [25] and is presently the state-of-the-art embedding model. We propose to fine-tune the BERT model as a transfer learning task for patch selection, using images composed of several patches (regions of an image), hence the name iBERT[3]. To the best of our knowledge, BERT has not yet been retraced for the QAC task. We use the BERT embedding model, which has a twelve layer implementation, extending it by adding a dense layer with 10% dropout which then is mapped to the final pooled layer connected the object class, and which outputs patch probabilities as shown in Fig. 2.

3.3 Information Foraging Explanation

Our goal of using Information Foraging Theory [27] from a cognitive viewpoint is to find explanations for the observed behaviour in query auto-completion and to model the information need within query sessions. IFT postulates that

[3] The lowercase "i" represents image patch.

the human information seekers follow an information scent to navigate from one information region to another in an information environment that is instinctively patchy in nature, and from one information patch to another within a region. IFT implies that foragers adapt their behaviour to the structure of the information environment in which they prevail such that the entire system (encompassing the information seeker, the information environment, and the interactions among these two) tries to maximise the ratio of the expected value of the information gained to the total cost of the interaction. Following the IFT analogy, when users start typing a prefix to auto-complete, their perceptual cues (such as mental beliefs [36]) either allow them to type the next character or to access the provided suggestion (under the query field) which acts as a distal cue and visually inspires the user to acquire them instantly to forage or seek. Query auto-completion, from an IFT perspective as query-level user interaction, is initiated by the user typing as little as a single-character query prefix. The user then may follow suggestions in case a completion is generated (which again follows the earlier mentioned strategy). In case the query prefix is unknown to the system (e.g. by being entered for the first time) the information scent associated with a result might be too poor [6] to immediately infer information needs. In this case we are applying beam search to generate the query based on image features. Suggestions are based on information scent values as described in the following subsection. These query suggestions represent the diversity of information scent patterns which elicits a varied distribution of relevant queries in the search field.

Patch Selection. This section describes the foraging-based strategy for patch selection. The technicalities of ranking patches (after patch selection) in the image (from image search results) are illustrated in Sect. 3.2. We utilise IFT to infer the user's information need utilising the Inferring User Need by Information Scent (IUNIS) algorithm [6] which was proposed to weigh each page vector along with the two factors i.e., TF-IDF weight and time, that were used to quantify the associated information scent with the page. In our image search scenario, we have images as search results where an image is considered as a set of patches containing features such as color, shape, texture, etc. In our proposed iBERT model, we use information scent to inspect patches based on image features and select patches which have higher probability estimated by the iBERT model. *Probabilistic Patch Selection Model* (PPSM) is a first attempt to reflect users' information need coherently by means of information scent. PPSM is used for a task that extends finding patches and makes the quantification of semantic uncertainties an important choice in selection. The important requirement for PPSM is a model (iBERT) that identifies patches in an image which are relevant to the user's information need (query). Inspired by the concept of TF-IDF in IR, we represent the categorical distribution of frequency (f_{p_i}) of each patch in an image (from the search results) in a given query session Q_s and the ratio of total number of query session (Q_T) during the entire search process to the number of query sessions (N_q) that contain the given patches (p_i) found in Q_s. We also consider the time spent (T) on the resulting images in a given query session to estimate the information scent (IS) within a query session as:

$$IS(Q_s) = \sum_{i=1}^{n} f_{p_i} \log(\frac{Q_T}{N_q}) T(p_i). \tag{4}$$

The user effort in terms of time is a function of patches which can be diverse and of different image class category. To generalise this for finding the information scent of a patch which then is assessed to select patches with higher information scent and then compared against the patch probability obtained via iBERT to distinguish the result. If we assume that the generated auto-completions induce several suggested queries (representing different information needs) simultaneously, every suggestion is in a competition to be discriminated as evident to the user. In the same way, an image contains multiple related or unrelated patches within it, and users find it difficult to judge which patches are relevant among images, which is due to the high uncertainty of correlated features within an image spread via patches. This motivates us to estimate the information scent of an image patch. There are two ways to compute the information scent of an image; one is to hire individual judges to rate scent on a scale [27] and the second approach is an algorithmic approach [28]. To estimate the information scent of a patch, we consider that PPSM constitutes patches that are probability distributions over images as *observations*. We assume image features as activators to perceptual cues because the user interpretation to image features when matched gives rise to a selection of an item (i.e., patch). The distributions are independent Bernoulli distributions of the features. Each observation is allocated to a patch, but the number of patches is not necessarily fixed i.e., the model is a non-parametric mixture with a product of independent Bernoullis as observation model. Therefore, the log-probability of selecting an image I for patch p_i

$$p(I \mid p_i) = \prod_{q_p} {r_{pf}}^{i_f} (1 - r_{pf})^{1-i_f} \tag{5}$$

where $r_p f = f((\pi_i, s_i), (1-\pi_i)s_i)$ is the Bernoulli rate for patch p to emit feature f, i_f is the image containing feature f, and $r_p f$ is a function of prior parameters representing activators (perceptual cues) for the selected patch. There can be a situation when most patches have only one observation (image) and features are very sparse i.e., the possibility of multiple perceptual cues per patch (i.e., $\pi_s \ll 1$) is low. To interpret Bernoulli's prior parameters such as s_i, we find the probability to observe a feature ($f \in i$ meaning $i = 1$) provided that it has been observed for a patch p ($k = 1$) is:

$$p(i = 1 | k = 1, n = 1) = \frac{s_1 \pi_s + n}{s_1 + n} = \frac{s_1 \pi_s + 1}{s_1 + 1} \approx \frac{1}{s_1 + 1} \tag{6}$$

if $\pi_s \ll 1$. The probability of observing a feature in a new image, given that it has been observed before, is a measure of its reliability. We use this probabilistic model to compare the results based on the probabilities of patches obtained from iBERT.

4 Experiments

4.1 Dataset

We use two well-known and diverse datasets: a visual dataset with large-scale knowledge bases that provide a rich collection of language annotations for visual concepts known as *Visual Genome* [18] with over 100k images where most image categories fall within a long tail, and the *ReferIt* dataset [16] which contains ∼42k image regions with descriptions. These two datasets fit well for our tasks. The Visual Genome dataset includes images, region descriptions, question-answers, objects, relationships, and attributes. The region descriptions confer a substitution for queries as they refer to several objects in various regions of every image. Few region descriptions are referring phrases and few of them are quite alike to descriptions. For example, referring descriptions are “guy sitting on the couch”, “white keyboard on the desk” and non-referring descriptions are “couch is brown” and “mouse is in the charger”. The huge number of instances from the Visual Genome dataset makes it quite convenient for our task. The ReferIt dataset is a collection of referring expressions engaged to images which quite intently resemble probable user queries of images. We separately train models for query auto-completion and patch selection using both datasets.

4.2 Training

We combine query and image as pairs by utilising the region descriptions from the Visual Genome dataset and referring to expressions from the ReferIt dataset. During training, we taken 85% of the Visual Genome data as the training set consists of 16,000 images and 740,000 corresponding region descriptions in which there are approximately 40–45 text descriptions per image. The training data from the ReferIt dataset consists of 9,000 images and 54,000 referring expression with approximately 4–6 referring expression per image.

For the query auto-completion task, we train our extended LSTM language model where the dimension of image representation is 128, $r = 64$ is the rank of the matching personalised matrix (component from Fig. 2). We use character embeddings with dimension 24, the dimension of the LSTM hidden units is 512, and a maximum length of 50 characters per query with Adam optimizer at a learning rate of 5e-4 for 50,000 iterations as well as a batch size of 32. For the patch selection task, we train our proposed iBERT model using pairs of (region description, patch set) from the Visual Genome dataset, giving rise to a training set of approximately 1.73 million samples. The extra 0.3 million samples are split into test and validation set. We conduct training for the patch selection model that fine-tunes BERT having twelve layers with batch size of 32 for 250,000 iteration using Adam as optimizer at a learning rate of 5e-5 in which the performance increases steeply for the initial 10% of iterations. We use a NVIDIA Tesla T4 GPU which takes a day and half for the complete training activity.

4.3 Performance Measure

We evaluate the quality of our predictions and estimations using the following performance metrics:

Mean Reciprocal Rank: The most standard metrics for QAC tasks is the mean reciprocal rank (MRR), which is the average of the reciprocal ranks of the final queries in the QAC outcomes. The MRR for the query auto-completion system Q_A provided the test dataset D_T is as follows:

$$MRR(Q_A) = \frac{1}{\mid D_T \mid} \sum_{q \in D_T} RR(q, Q_A(q_p))$$

where q_p is a prefix of query q and $Q_A(q_p)$ is the list ranked for candidate completions of q_p from Q_A. RR denotes the reciprocal rank of q if q is present in $Q_A(q_p)$, in other cases reciprocal is 0.

Language Perplexity: Perplexity is a measure to encapsulate uncertainty of the model for a given query prefix. This metric has been explored earlier for an information retrieval task [8] and its correlation with the standard precision-recall measures has been investigated [1]. The average inverse probability is perplexity. A better model has lower perplexity.

$$Perplexity(q_p) = \sqrt[N]{\prod_{i=1}^{N} \frac{1}{P(q_i|q_{i-1})}}$$

where N is the normalised length of the query and $P(q_i|q_{i-1})$ is the probability of the complete query given the immediate preceding query prefix.

We evaluate the patch selection by F1 score.

4.4 Results and Discussion

We report the evaluation result in Table 1. We perform our evaluation in two parts. Firstly, we evaluate the quality of our query completion (query prefix of length one or more character) by mean reciprocal rank and perplexity. Secondly, we evaluate the patch selection task by F1 score. We evaluate the query completion task on Visual Genome and ReferIt datasets which have character vocabulary sizes of 89 and 77. We match index T_q of the true query prefix in the top 10 predicted completions where we estimate the MRR score as $\sum_n \frac{1}{T_q}$ and reinstate the reciprocal rank with 0 in case if query does not appear in the top 10 completions. The perplexity comparison on both collection of test queries utilising corresponding contexts i.e., images and indiscriminate noise. The perplexity on the Visual Genome and ReferIt test queries with both contexts is shown in Table 2. During the evaluation on the Visual Genome and ReferIt test sets (or queries), we analyse the query prefix with different length for the corresponding context (noise and image). We found that mean reciprocal rank is altered by the query prefix length, as long-tailed queries are comparatively more difficult than queries of average length to match. Hence, we examine quite better performance for all prefix lengths on the ReferIt dataset (from Table 2).

Table 1. Evaluation results of the query completion task. Our MRR score is in bold face.

Model	MRR (Seen+Unseen)
MPC [3]	0.171
Character n-gram (n = 7)	0.287
Mitra10K+MPC+λMART [24]	0.278
Mitra100K+MPC+λMART [24]	0.298
NQLM(S)+WE+MPC [26]	0.345
NQLM(L)+WE+MPC [26]	0.355
NQLM(L)+WE+MPC+λMART [26]	0.354
FactorCell [12]	0.309
E-LSTM LM(Ours)[a]	**0.764**

[a]E-LSTM LM: Extended LSTM Language Model

Table 2. Perplexity of image query auto-completion on both datasets utilising an image and indiscriminate noise. Inclusion of image results in a better (lower) perplexity

Dataset	Context	
	Image	Indiscriminate noise
Visual Genome	2.35	3.81
ReferIt	2.63	3.45

We evaluated our proposed iBERT model for finding patch probabilities which is used to select and rank patches in the image. We achieve an F1 score[4] of **0.7638** over 3,000 patch classes.

5 Conclusion and Future Work

In this work, we propose an extended LSTM language model for a new task of query auto-completion adapted upon an image. The language model enriches both image features and text information in which the surplus of beam search over our model is efficiently able to predict future queries at least on a single character prefix. The significant increase in MRR is due to the inclusion of

[4] F1 score for the baseline methods shown in Table 1 were not available.

visual information within textual queries as explained by IFT model. Also, we present iBERT for patch selection to efficiently rank them in the image and eventually predicts the most suitable image for the auto-completed query, and compare against the result from probabilistic patch selection model. This work is among the first attempt to apply foraging-based strategy to QAC. The self-explanatory power of IFT to understand user interaction at query level leads the foundation of probabilistic patch selection model to devise users' information need. Our future work is to generalise the referring expression with contextual model to distinguish referring and non-referring region descriptions. We intend to aggregate information from textual queries and visual descriptions to scale it for multimodal query auto-completion in a single model.

Acknowledgement. This work is supported by the Quantum Access and Retrieval Theory (QUARTZ) project, which has received funding from the European Union's Horizon 2020 research and innovation programme under the Marie Sklodowska-Curie grant agreement No. 721321, and partially supported for computing resources by Google Cloud grant.

References

1. Azzopardi, L., Girolami, M., Van Rijsbergen, K.: Investigating the relationship between language model perplexity and IR precision-recall measures (2003)
2. Ba, J.L., Kiros, J.R., Hinton, G.E.: Layer normalization. arXiv preprint arXiv:1607.06450 (2016)
3. Bar-Yossef, Z., Kraus, N.: Context-sensitive query auto-completion. In: Proceedings of the 20th International Conference on World Wide Web, pp. 107–116. ACM (2011)
4. Cai, F., De Rijke, M., et al.: A survey of query auto completion in information retrieval. Found. Trends® Inf. Retrieval **10**(4), 273–363 (2016)
5. Cao, H., et al.: Context-aware query suggestion by mining click-through and session data. In: Proceedings of the 14th ACM SIGKDD International Conference on Knowledge Discovery and Data Mining, pp. 875–883. ACM (2008)
6. Chi, E.H., Pirolli, P., Chen, K., Pitkow, J.: Using information scent to model user information needs and actions and the web. In: Proceedings of the SIGCHI Conference on Human Factors in Computing Systems, pp. 490–497. ACM (2001)
7. Devlin, J., Chang, M.W., Lee, K., Toutanova, K.: Bert: pre-training of deep bidirectional transformers for language understanding. arXiv preprint arXiv:1810.04805 (2018)
8. Hauff, C., Murdock, V., Baeza-Yates, R.: Improved query difficulty prediction for the web. In: Proceedings of the 17th ACM Conference on Information and Knowledge Management, pp. 439–448. ACM (2008)
9. He, K., Gkioxari, G., Dollár, P., Girshick, R.: Mask R-CNN. In: Proceedings of the IEEE International Conference on Computer Vision, pp. 2961–2969 (2017)
10. Hu, R., Dollár, P., He, K., Darrell, T., Girshick, R.: Learning to segment every thing. In: Proceedings of the IEEE Conference on Computer Vision and Pattern Recognition, pp. 4233–4241 (2018)
11. Hu, R., Xu, H., Rohrbach, M., Feng, J., Saenko, K., Darrell, T.: Natural language object retrieval. In: Proceedings of the IEEE Conference on Computer Vision and Pattern Recognition, pp. 4555–4564 (2016)

12. Jaech, A., Ostendorf, M.: Personalized language model for query auto-completion. arXiv preprint arXiv:1804.09661 (2018)
13. Jaiswal, A.K., Holdack, G., Frommholz, I., Liu, H.: Quantum-like generalization of complex word embedding: a lightweight approach for textual classification. In: Proceedings of the Conference "Lernen, Wissen, Daten, Analysen", LWDA 2018, Mannheim, Germany, 22–24 August 2018, pp. 159–168 (2018). http://ceur-ws.org/Vol-2191/paper19.pdf
14. Ji, S., Li, G., Li, C., Feng, J.: Efficient interactive fuzzy keyword search. In: Proceedings of the 18th International Conference on World Wide Web, pp. 371–380. ACM (2009)
15. Jiang, J.Y., Ke, Y.Y., Chien, P.Y., Cheng, P.J.: Learning user reformulation behavior for query auto-completion. In: Proceedings of the 37th International ACM SIGIR Conference on Research & Development in Information Retrieval, pp. 445–454. ACM (2014)
16. Kazemzadeh, S., Ordonez, V., Matten, M., Berg, T.: Referitgame: referring to objects in photographs of natural scenes. In: Proceedings of the 2014 Conference on Empirical Methods in Natural Language Processing (EMNLP), pp. 787–798 (2014)
17. Kharitonov, E., Macdonald, C., Serdyukov, P., Ounis, I.: User model-based metrics for offline query suggestion evaluation. In: Proceedings of the 36th International ACM SIGIR Conference on Research and Development in Information Retrieval, pp. 633–642. ACM (2013)
18. Krishna, R., et al.: Visual genome: connecting language and vision using crowdsourced dense image annotations. Int. J. Comput. Vis. **123**(1), 32–73 (2017)
19. Liu, H., Mulholland, P., Song, D., Uren, V., Rüger, S.: Applying information foraging theory to understand user interaction with content-based image retrieval. In: Proceedings of the Third Symposium on Information Interaction in Context, pp. 135–144. ACM (2010)
20. Maxwell, D., Azzopardi, L.: Information scent, searching and stopping. In: Pasi, G., Piwowarski, B., Azzopardi, L., Hanbury, A. (eds.) ECIR 2018. LNCS, vol. 10772, pp. 210–222. Springer, Cham (2018). https://doi.org/10.1007/978-3-319-76941-7_16
21. McCann, B., Bradbury, J., Xiong, C., Socher, R.: Learned in translation: contextualized word vectors. In: Advances in Neural Information Processing Systems, pp. 6294–6305 (2017)
22. Mikolov, T., Sutskever, I., Chen, K., Corrado, G.S., Dean, J.: Distributed representations of words and phrases and their compositionality. In: Advances in Neural Information Processing Systems, pp. 3111–3119 (2013)
23. Mitra, B.: Exploring session context using distributed representations of queries and reformulations. In: Proceedings of the 38th International ACM SIGIR Conference on Research and Development in Information Retrieval, pp. 3–12. ACM (2015)
24. Mitra, B., Craswell, N.: Query auto-completion for rare prefixes. In: Proceedings of the 24th ACM International on Conference on Information and Knowledge Management, pp. 1755–1758. ACM (2015)
25. Mitra, B., Rosset, C., Hawking, D., Craswell, N., Diaz, F., Yilmaz, E.: Incorporating query term independence assumption for efficient retrieval and ranking using deep neural networks. arXiv preprint arXiv:1907.03693 (2019)
26. Park, D.H., Chiba, R.: A neural language model for query auto-completion. In: Proceedings of the 40th International ACM SIGIR Conference on Research and Development in Information Retrieval, pp. 1189–1192. ACM (2017)

27. Pirolli, P., Card, S.: Information foraging. Psychol. Rev. **106**(4), 643 (1999)
28. Pirolli, P., Card, S.K., Van Der Wege, M.M.: Visual information foraging in a focus+ context visualization. In: Proceedings of the SIGCHI Conference on Human Factors in Computing Systems, pp. 506–513. ACM (2001)
29. Shao, T., Chen, H., Chen, W.: Query auto-completion based on word2vec semantic similarity. In: Journal of Physics: Conference Series, vol. 1004, p. 012018. IOP Publishing (2018)
30. Shi, H., Li, H., Meng, F., Wu, Q.: Key-Word-aware network for referring expression image segmentation. In: Ferrari, V., Hebert, M., Sminchisescu, C., Weiss, Y. (eds.) ECCV 2018. LNCS, vol. 11210, pp. 38–54. Springer, Cham (2018). https://doi.org/10.1007/978-3-030-01231-1_3
31. Shokouhi, M.: Learning to personalize query auto-completion. In: Proceedings of the 36th International ACM SIGIR Conference on Research and Development in Information Retrieval, pp. 103–112. ACM (2013)
32. Sundar, S.S., Knobloch-Westerwick, S., Hastall, M.R.: News cues: information scent and cognitive heuristics. J. Am. Soc. Inform. Sci. Technol. **58**(3), 366–378 (2007)
33. Sutskever, I., Martens, J., Hinton, G.E.: Generating text with recurrent neural networks. In: Proceedings of the 28th International Conference on Machine Learning (ICML 2011), pp. 1017–1024 (2011)
34. Vijayakumar, A.K., et al.: Diverse beam search: decoding diverse solutions from neural sequence models. arXiv preprint arXiv:1610.02424 (2016)
35. Weber, I., Castillo, C.: The demographics of web search. In: Proceedings of the 33rd International ACM SIGIR Conference on Research and Development in Information Retrieval, pp. 523–530. ACM (2010)
36. White, R.: Beliefs and biases in web search. In: Proceedings of the 36th International ACM SIGIR Conference on Research and Development in Information Retrieval, pp. 3–12. ACM (2013)
37. White, R.W., Marchionini, G.: Examining the effectiveness of real-time query expansion. Inf. Process. Manag. **43**(3), 685–704 (2007)
38. Wittek, P., Liu, Y.H., Darányi, S., Gedeon, T., Lim, I.S.: Risk and ambiguity in information seeking: eye gaze patterns reveal contextual behavior in dealing with uncertainty. Front. Psychol. **7**, 1790 (2016)
39. Wu, C.C., Mei, T., Hsu, W.H., Rui, Y.: Learning to personalize trending image search suggestion. In: Proceedings of the 37th International ACM SIGIR Conference on Research & Development in Information Retrieval, pp. 727–736. ACM (2014)

Using Image Captions and Multitask Learning for Recommending Query Reformulations

Gaurav Verma[1(✉)], Vishwa Vinay[1(✉)], Sahil Bansal[2], Shashank Oberoi[3], Makkunda Sharma[4], and Prakhar Gupta[5]

[1] Adobe Research, Bangalore, India
{gaverma,vinay}@adobe.com
[2] IBM Research, Bangalore, India
[3] Adobe Inc., Noida, India
[4] Indian Institute of Technology Delhi, Delhi, India
[5] Carnegie Mellon University, Pittsburgh, USA

Abstract. Interactive search sessions often contain multiple queries, where the user submits a reformulated version of the previous query in response to the original results. We aim to enhance the query recommendation experience for a commercial image search engine. Our proposed methodology incorporates current state-of-the-art practices from relevant literature – the use of generation-based sequence-to-sequence models that capture session context, and a multitask architecture that simultaneously optimizes the ranking of results. We extend this setup by driving the learning of such a model with captions of clicked images as the target, instead of using the subsequent query within the session. Since these captions tend to be linguistically richer, the reformulation mechanism can be seen as assistance to construct more *descriptive* queries. In addition, via the use of a pairwise loss for the secondary ranking task, we show that the generated reformulations are more *diverse*.

Keywords: Query reformulations · Seq-to-seq translation · Captions

1 Introduction

A successful search relies on the engine accurately interpreting the intent behind a user's query and returning likely relevant results ranked high. There has been much progress allowing search engines to respond effectively even to short keyword queries on rare intents [5,9,25]. Despite this, recommendation of queries is an integral part of all search experiences – either in the form of *query autocomplete* (queries that match the prefix the user has currently typed into the search box) or *query suggestions* (reformulation options once an initial query has been provided). In this work, we focus on the query suggestion task.

S. Bansal, S. Oberoi, and M. Sharma contributed equally to this work.

J. M. Jose et al. (Eds.): ECIR 2020, LNCS 12035, pp. 681–696, 2020.
https://doi.org/10.1007/978-3-030-45439-5_45

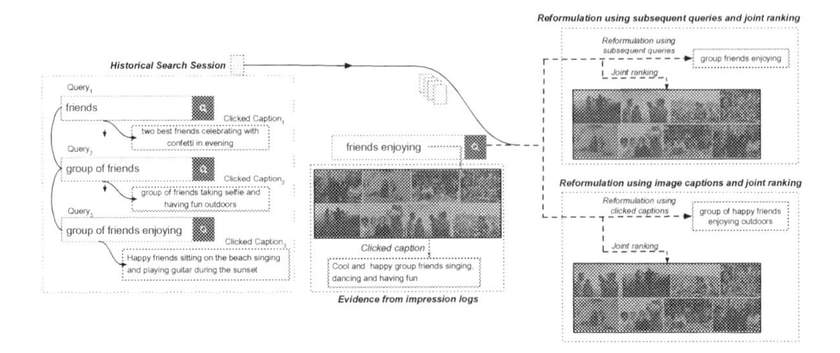

Fig. 1. The basic idea behind our work. We generate query reformulations using *(a)* subsequent queries within sessions, and *(b)* the captions of clicked images, as supervision signals. In both the cases, the task of generating reformulations is done while jointly optimizing the ranking of results.

Original algorithms for this scenario relied on extracting co-occurrence patterns between query pairs, and their constituent terms, within historical logs [3,12,16,18]. Such methods often work well for frequent queries. Recent work utilizing generative approaches common in natural language processing (NLP) scenarios offer generalization in terms of being able to provide suggestions even for rare queries [10,21]. More specifically, the work by Sordoni et al. [26] focuses on generating query suggestions that are aware of the context of the user's current session. The current paper is most similar to this work in terms of motivation and the core technical component.

The experiments described here are based on data from a commercial stock image search engine. In this setting, the items in the index are professionally taken high quality images to be used in commercial publishing material. The users of such a system exhibit similar properties to what might be expected on general purpose search engines - i.e., the use of relatively short queries often with multiple reformulations within a session. The logged data therefore contains not only the sequence of within-session queries, but also impression logs listing what images were shown in response to a query and which amongst those were clicked.

The availability of usage data, which provides implicit relevance signals, allows the building of a query reformulation model that includes aspects that have been shown to be useful in related literature: session context capturing information from previous queries in the session, as well as properties of relevant results via a multitask component. Building on state-of-the-art models in this manner, we specialize the solution to our setting by utilizing a novel supervision signal for the reformulation model in the form of linguistically rich captions available for the clicked results (in our case, images) across sessions (Fig. 1).

2 Related Work

A user of a search system provides an input query, typically a short list of keywords, into the search box and expects content relevant to their need ranked high in the result list. There are many reasons why a single iteration of search may not be successful – mis-specified queries (including spelling errors), imperfect ranking, ambiguous intent, and many more. As a result, it is useful to think of a search session as a series of interactions – where the user enters a query, examines and potentially interacts with the returned results, and constructs a refined query that is expected to more accurately represent their intent. Search engines therefore mine historical behavior of users on this query and similar ones in an attempt to optimize the entire search session [24].

Being able to effectively extract these signals from historical logs starts with understanding and interpreting user behavior appropriately. For example, Huang et al. [17] pointed out that successful reformulations, especially those involving changes to words and their order, can be identified as those that retrieve new items which are presented higher in the subsequent results. An automatic reformulation experience involves implementing lessons from such analyses. The first of these is the use of previous queries within the current search sessions to inform the subsequent suggestions – i.e., modeling the *session context*. Earlier papers (e.g. [7]) explicitly captured co-occurrence within sessions which, while being an intuitive and simple strategy, had the disadvantage of not being able to account for rarer queries. Newer efforts (e.g. [21]) therefore utilize distributed representations of terms and queries to help generalize to unseen queries.

Such efforts are part of a wider expansion of techniques originally common within NLP domains to Information Retrieval (IR) scenarios. Conceptually, a generation-based model for query reformulation is obtained by mapping a query to the subsequent one in the same session. Such a model incorporates two signals known to be useful from traditional IR: (1) sequence of terms within a query & (2) sequence of queries within a session. Recent papers have investigated models anchored in the original generic NLP settings but customized to the characteristics of search queries. For example, Dehghani et al. [11] suggest a 'copy' mechanism within the sequence-to-sequence (seq-to-seq) models [27] to allow for terms to be carried over across queries in the session. In the current paper, we consider the work of Sordoni et al. [26] as a reference for the core seq-to-seq model. The model, referred to here as *H*ierarchical *R*ecurrent *E*ncoder *D*ecoder (*HRED*), is a standard encoder-decoder setup, where word embeddings are aggregated into a query representation, a sequence of which in turn leads to a session representation. A decoder for the hierarchically organized query and session encoders is trained to predict the sequence of query words that compose the subsequent query in the session. Along with being a strong baseline, it serves to illustrate the core components of our work: (*a*) use of a novel supervision signal in the form of captions of clicked results, and (*b*) jointly optimizing ranking along with query reformulation. These extensions could similarly be done with other seq-to-seq models used for query suggestion.

Our motivation for using captions of clicked images as supervision signal stems from the fact that captions are often succinct summaries of the content of the actual images as the creators are incentivized to have their images found. In particular, captions indicate which objects are present in the image, their corresponding attributes, as well as relationships with other objects in the same image – for example, *"A beautiful girl* ***wearing*** *a yellow shirt* ***standing near*** *a red car"*. These properties make the captions a good target.

Multitask learning [8] has been shown to have success in scenarios where related tasks benefit from common signals. A recent paper [1] shows benefits of such a pairing in a search setting. Specifically, Ahmad et al. show that coupling with a classifier distinguishing clicked results from those skipped helps improve a query suggestion model. We extend this work by utilizing a pairwise loss function commonly used in learning-to-rank [6]. We show that not only does this provide the expected increase in the effectiveness of the ranker component, but also increases the diversity of suggested reformulations. Such diversity has been shown to be important for the query suggestion user experience [20].

We begin by providing details of the mathematical notation in the next section, before describing our models in detail. The subsequent experimental section provides empirical evidence of the benefits that our design choices bring.

3 Notation and Model Architectures

3.1 Notation

We define a session as a sequence of queries, $\mathcal{S} = \{q_1, \ldots, q_n\}$. Each query q_i in session $\mathcal{S}$ has a set of displayed images associated with it, $\mathcal{I}_i = \{I_i^1, \ldots, I_i^m\}$. A subset of images in $\mathcal{I}_i$ are clicked, we refer to the top-ranked clicked image as I_i^{clicked}. All the images in the set $\mathcal{I}_i$ have a caption describing them, the entire set of which is represented as $\mathcal{C}_i = \{C_i^1, \ldots, C_i^m\}$. It follows that every I_i^{clicked} will also have an associated caption with it, given as C_i^{clicked}. Given this, for every successful query q_i in session $\mathcal{S}$, we will have an associated clicked image I_i^{clicked} and a corresponding caption C_i^{clicked}. We consider the size of impression m (number of images) to be fixed for all q_i.

Our models treat each query q_i in any given session, as a sequence of words, $q_i = \{w_1, \ldots, w_{l_q}\}$. Captions are represented similarly - as sequences of words, $C_i^j = \{w_1, \ldots, w_{l_c}\}$. We use LSTMs [15] to model the sequences, owing to their demonstrated capabilities in modeling various natural language tasks, ranging from machine translation [27] to query suggestion [11].

The input to our models is a query q_i in the session $\mathcal{S}$, and the desired output is a target reformulation q_{reform}. This target reformulation q_{reform} can either be *(i)* the subsequent query q_{i+1} in the same session S, or *(ii)* the caption C_i^{clicked} corresponding to the clicked image I_i^{clicked}. Note that obtaining contextual query suggestions via a *translation model* that has learnt a mapping between successive queries within a session (i.e., *(i)*) has been previously proposed in our reference baseline papers [1,26]. In the current paper, we utilize a

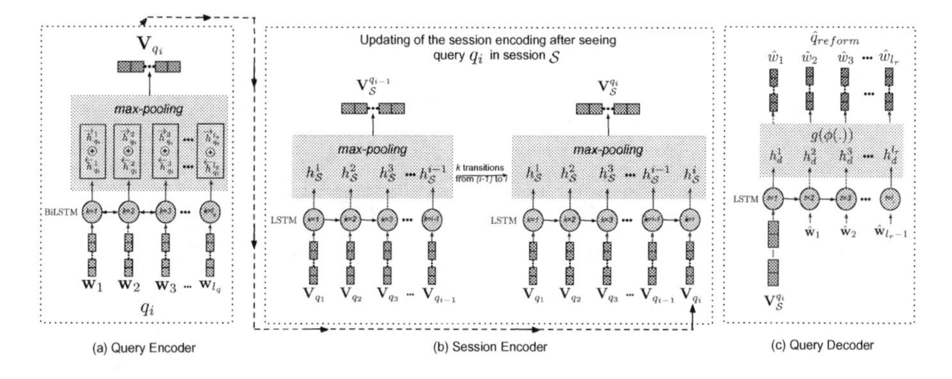

Fig. 2. An illustration of the *(a)* query encoder, *(b)* session encoder, and *(c)* query decoder

linguistically richer supervision signal, in the form of captions of clicked images (i.e., *(ii)*), and analyze the behavior of the different models across three high level axes - relevance, descriptiveness and diversity of generated reformulations.

3.2 Model Architectures

In this paper, we evaluate two base models – *HRED* and *HRED* with **Cap**tions (*HREDCap*), and to study the effect of multitask learning, we add a ranker component to each of these models; giving us two more multitask variants – *HRED + Ranker* and *HREDCap + Ranker*. The underlying architecture of *HRED* and *HREDCap* (and the corresponding variants) is essentially the same, but *HRED* has been trained by using q_{i+1} as target and *HREDCap* has been trained using $C_i^{clicked}$ as target. *HRED* comprises of a query encoder, a session encoder, and a query decoder; all of which are descried below.

Query Encoder: The query encoder generates a query level encoding $\mathbf{V}_{q_i}$ for every $q_i \in \mathcal{S}$. This is done by first representing the query q_i using vector embeddings of corresponding words $\{\mathbf{w}_1, \ldots, \mathbf{w}_{l_q}\}$, and then sequentially feeding them into a bidirectional LSTM (BiLSTM) [14]. As shown in Fig. 2(a), the query encoder takes each of these word representations as input to the BiLSTM at every encoding step and updates the hidden states based on the forward and backward pass over the input query. The forward and backward hidden states are concatenated, and after applying attention [2] over the concatenated hidden states, we obtain a fixed size vector representation $\mathbf{V}_{q_i}$ for the query $q_i \in \mathcal{S}$.

Session Encoder: The encoded representation $\mathbf{V}_{q_i}$ of query $q_i \in \mathcal{S}$ is used by the session encoder, along with encoded representations $\{\mathbf{V}_{q_1}, \ldots, \mathbf{V}_{q_{i-1}}\}$ of previous queries within the same session, to capture the context of the ongoing session thus far. The session encoder, which is modeled by a unidirectional LSTM [15], updates the session context $\mathbf{V}_{\mathcal{S}}^{q_i}$ after each new $\mathbf{V}_{q_i}$ is presented to it. Figure 2(b) illustrates one such update where the session encoding is updated

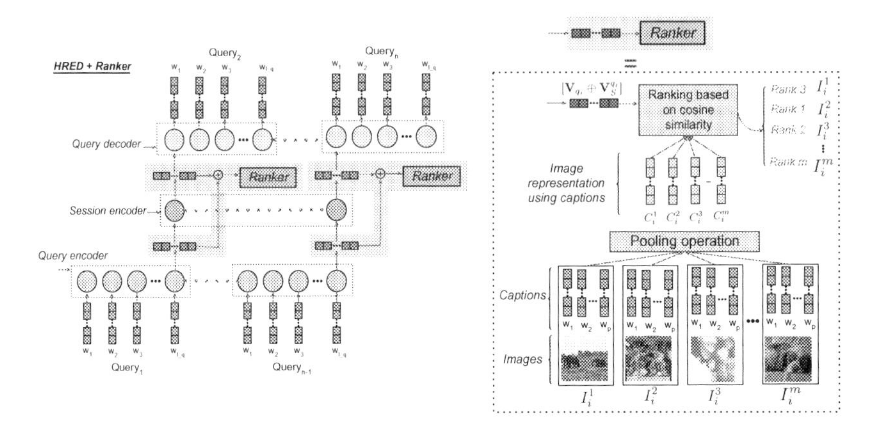

Fig. 3. The proposed architecture of our multitask model: *HRED + Ranker* (left). For the sake of brevity, we have shown the ranker component separately (right). For *HREDCap + Ranker*, the supervision signals are obtained from captions of clicked images and not subsequent queries.

from $\mathbf{V}_{\mathcal{S}}^{q_{i-1}}$ to $\mathbf{V}_{\mathcal{S}}^{q_i}$ after $\mathbf{V}_{q_i}$ is provided as input to the session encoder by the query encoder. Since it is unreasonable to assume access to future queries in the session while generating a reformulation for the current query, we use a unidirectional LSTM to model the forward sequence of queries within a session. Accordingly, the session encoder updates its hidden state based on the forward pass over the query sequence. As shown in Fig. 2(b), max-pooling is applied over each dimension of the hidden state to obtain the session encoding $\mathbf{V}_{\mathcal{S}}^{q_i}$.

Query Decoder: The generated session encoding $\mathbf{V}_{\mathcal{S}}^{q_i}$ is used as input by a query decoder to generate a reformulation $\hat{q}_{\text{reform}} = \{\hat{w}_1, \ldots, \hat{w}_{l_r}\}$ for the query $q_i \in \mathcal{S}$. As shown in Fig. 2(c), the reformulation is generated word by word using a single layer unidirectional LSTM. With each unfolding of the decoder LSTM at step $t \in \{1, \ldots, l_r\}$, a new word $\hat{w}_t$ is generated as per the following probability:[1]

$$\begin{aligned} \hat{w}_t &= \arg\max_{w^i \in \mathcal{V}} P(\hat{w}_t = w^i \mid \hat{w}_{1:t-1}, \mathbf{V}_{\mathcal{S}}^{q_i}) \\ P(\hat{w}_t = w^i \mid \hat{w}_{1:t-1}, \mathbf{V}_{\mathcal{S}}^{q_i}) &= g(\phi(h_d^t)) \end{aligned} \tag{1}$$

Here, h_d^t is the hidden state of the decoder at decoding step t, $\hat{w}_{1:t-1}$ denotes the previous words generated by the decoder, and $\phi(h_d^t)$ is a non-linear operation over h_d^t. The softmax function $g(.)$ provides a probability distribution over the entire vocabulary $\mathcal{V}$. w^i is used to denote the i-th word in $\mathcal{V}$. The joint probability of generating a reformulation $\hat{q}_{\text{reform}} = \{\hat{w}_1, \ldots, \hat{w}_{l_r}\}$ can be decomposed into

[1] For $t = 1$, $P(\hat{w}_t = w^i \mid \hat{w}_{1:t-1}, \mathbf{V}_{\mathcal{S}}^{q_i})$ reduces to $P(\hat{w}_t = w^i \mid \mathbf{V}_{\mathcal{S}}^{q_i})$. However, for the sake of readability, this special consideration for $t = 1$ has been skipped for the following equations.

the ordered conditionals as $P(\hat{q}_{\text{reform}} \mid q_i) = \prod_{t=1}^{l_r} P(\hat{w}_t \mid \hat{w}_{1:t-1}, \mathbf{V}_{\mathcal{S}}^{q_i})$. During training, the decoder compares each word $\hat{w}_t$ in the generated reformulation $\hat{q}_{\text{reform}}$ with the corresponding word w_t in the target reformulation q_{reform}, and aims to minimize the negative log-likelihood. For a given reformulation by the decoder, the loss is

$$\mathcal{L}_{\text{reform}} = -\sum_{t=1}^{l_r} \log P(\hat{w}_t = w_t \mid \hat{w}_{1:t-1}, \mathbf{V}_{\mathcal{S}}^{q_i}) + \mathcal{L}_{reg} \tag{2}$$

Here, $\mathcal{L}_{reg} = -\lambda \sum_{w^i \in \mathcal{V}} P(w^i \mid \hat{w}_{1:t-1}, \mathbf{V}_{\mathcal{S}}^{q_i}) \cdot \log P(w^i \mid \hat{w}_{1:t-1}, \mathbf{V}_{\mathcal{S}}^{q_i})$ is a regularization term added to prevent the predicted probability distribution over the words in the vocabulary from being highly skewed. λ is a regularization hyperparameter. The training loss is the sum of $\mathcal{L}_{\text{reform}}$ over all query reformulations generated by the decoder during training.

To summarize, the model encodes the queries, generates session context encodings, and generates the reformulated query using the decoder while updating the model parameters using the gradients of $\mathcal{L}_{\text{reform}}$.

Ranker Component: This additional component is responsible for ranking the m retrieved results for $q_i \in \mathcal{S}$. As shown in Fig. 3 (right), the ranker takes as input the concatenation of query and session encoding $[\mathbf{V}_{q_i} \oplus \mathbf{V}_{\mathcal{S}}^{q_i}]$, for every $q_i \in \mathcal{S}$. The concatenated vector representation $[\mathbf{V}_{q_i} \oplus \mathbf{V}_{\mathcal{S}}^{q_i}]$ is used to compute the similarity between the query q_i and its candidate results. The concatenation of these encodings is done to ensure that both current query information (as captured in $\mathbf{V}_{q_i}$) and ongoing session context (as captured in $\mathbf{V}_{\mathcal{S}}^{q_i}$) is used by the ranker. To obtain a representation of the images, we use their corresponding captions. Formally, for every query $q_i \in \mathcal{S}$ each image $I_i^j \in \mathcal{I}_i$ is represented using $\mathbf{C}_i^j$. The average of the vector embeddings of words $\{w_1, \ldots, w_{l_c}\}$ in $\mathbf{C}_i^j$ is computed for the image I_i^j. The cosine similarities between $[\mathbf{V}_{q_i} \oplus \mathbf{V}_{\mathcal{S}}^{q_i}]$ and the image representations $\mathbf{C}_i^j \in \mathcal{C}_i$ are used to rank order the retrieved results. The j-th element of the similarity vector $\mathbf{S}_i$ represents the similarity between $[\mathbf{V}_{q_i} \oplus \mathbf{V}_{\mathcal{S}}^{q_i}]$ and $\mathbf{C}_i^j$.

$$S_i^j = sim([\mathbf{V}_{q_i} \oplus \mathbf{V}_{\mathcal{S}}^{q_i}], \mathbf{C}_i^j) \tag{3}$$

During training, the ranker tries to learn model parameters based on one of the following two objectives:

(i) **Cross Entropy Loss:** As described in [1], we utilize the ‘clicked’ versus ‘not-clicked’ boolean event to train a classifier, where the ranker scores the m retrieved results based on the probability of being clicked by the user. In the following equation, $\mathbf{R}_i$ for query q_i is an m-dimensional vector, where each value in the vector indicates whether the corresponding image was clicked or not. I.e., $R_i^j = 0$ if I_i^j was not clicked, and $R_i^j = 1$ if I_i^j was clicked. A sigmoid of the scores from Eq. 3 is taken as the probability of click. Using the $\mathbf{R}_i$ as labels, the ranker can now be trained using a standard cross entropy loss function:

$$\mathcal{L}_{\text{rank}} = BCE(\sigma(\mathbf{S}_i), \mathbf{R}_i) \tag{4}$$

(ii) **Pairwise Ranking Loss:** As described in [6], the original boolean labels in $\mathbf{R}_i$ can be used to construct an alternate event space where labels $M_{jk} = 1$ when the image at rank j was clicked while the one at k was not. Pairwise ranking loss allows to better model the preferences of certain results over the others.

$$\mathcal{L}_{\text{rank}} = -\frac{1}{m^2} \sum_{j=1}^{m} \sum_{\substack{k=1 \\ k \neq j}}^{m} M_{jk} * \log \hat{M}_{jk} + (1 - M_{jk}) * \log(1 - \hat{M}_{jk}) \tag{5}$$

$$\text{where, } \hat{M}_{jk} = P(S_i^j > S_i^k \mid [\mathbf{V}_{q_i} \oplus \mathbf{V}_{\mathcal{S}}^{q_i}]) = \sigma(S_i^j - S_i^k)$$

Since *HRED + Ranker* and *HREDCap + Ranker* are multitask models, their training objective is a weighted combination of $\mathcal{L}_{\text{reform}}$ and $\mathcal{L}_{\text{rank}}$.

$$\mathcal{L}_{\text{multitask}} = \alpha \cdot \mathcal{L}_{\text{reform}} + (1 - \alpha) \cdot \mathcal{L}_{\text{rank}} \tag{6}$$

Here, α is a hyperparameter used for controlling the relative contribution of the two losses. As mentioned earlier, either the regular binary cross-entropy loss or the pairwise-ranking loss can be used for $\mathcal{L}_{\text{rank}}$. We experiment using both and report our results on the effect of using one over the other. The models that are trained using cross entropy loss are appended with *(CE)*, and the models that are trained using pairwise ranking objective are denoted as *(RO)*.

It is worth noting that since for a given query q_i there can be more than one clicked images, our ranker component allows $\mathbf{R}_i$ to take the value 1 at more than a single place. However, while training the reformulation model, we only consider the caption of the *highest ranked* clicked image.

4 Experiments

Dataset: We use logged impression data from Adobe Stock[2]. The query logs contain information about the queries that were issued by users, and the images that were presented in response to those queries. Additionally, they contain information about which of the displayed images were clicked by the user. We consider the top-10 ranked results, i.e., the number of results to be considered for each query is $m = 10$. The queries are segmented into sessions (multiple queries by the same user within a 30 min time window), while maintaining the sequence in which they were executed by a user. We retain both multi-query sessions as well as single-query sessions, leading to a dataset comprising $1,301,888$ sessions, $2,122,079$ queries, and $10,185,979$ unique images. We note that ~24.8% of the sessions are single-query sessions, while rest all are multi-query sessions; each of which, on average, comprise of 2.19 queries. Additionally, we remove all non-alphanumeric characters from the user-entered queries, while keeping spaces, and convert all characters to lowercase.

To obtain the train, test and validation set, we first shuffle the sessions and split them in a $80:10:10$ ratio, respectively. While it is possible for a query to

[2] https://stock.adobe.com/.

be issued by different users in distinct sessions, a given search session occurs in only one of these sets. These sets are kept the same for all experiments, to ensure consistency while comparing the performance of trained models. The validation set is used for hyperparameter tuning.

Experimental Setup: We construct a global vocabulary $\mathcal{V}$ of size 37, 648 comprising of words that make up the queries and captions for images. Each word in the vocabulary is represented using a 300-dimensional vector $\mathbf{w}_i$. Each $\mathbf{w}_i \in \mathcal{V}$ is initialized using pre-trained GloVe vectors [23]. Words in our vocabulary $\mathcal{V}$ that do not have a pre-trained embedding available in GloVe (1, 941 in number), are initialized using samples from a standard normal distribution. Since the average number of words in a query, average number of words in a caption, and average number of queries within a session are 2.31, 5.22, and 1.63, we limit their maximum sizes to 5, 10, and 5, respectively. For queries and captions that contain less than 5 and 10 words respectively, we pad them using '$< p >$' tokens. The number of generated words in $\hat{q}_{\text{reform}}$ was limited to 10, i.e., $l_r = 10$.

During training, we use Adam optimizer [19] with a learning rate initialized to 10^{-3}. Across all the models, the regularization coefficient λ is set to be 0.1. For multitask models, the loss trade-off hyperparameter α is set to 0.45. The sizes of the hidden states of query level encoder $\overrightarrow{h}_q$ and $\overleftarrow{h}_q$ are set to 256, and that of session level encoder $h_{\mathcal{S}}$ is set to 512. The size of the decoder's hidden state is kept to be 256. We train all the models for a maximum of 30 epochs, using batches of size 512, with early stopping based on the loss over the validation set. The best trained models are quantitatively and qualitatively evaluated and we discuss the results in the upcoming section.

At test time, we use a beam search-based decoding approach to generate multiple reformulations [2]. For our experiments, we set the beam width $K = 3$. The choice of K was governed by observations that will be discussed later, while analyzing the diversity and relevance of generated reformulations. These three reformulations are rank ordered using their generation probability.

We experiment with a range of hyperparameters and find that the evaluation results are stable with respect to our hyperparameter choices. However, our motivation is less about training the most accurate models, as we wish to measure the effect of the supervision signal and training objective when used alongside the baseline models. While presenting the results in Tables 1 and 2, we report the average of values over 10 different runs, as well the standard deviations.

5 Evaluation and Results

In this section, we evaluate the performance of the aforementioned models using multiple metrics for each of the two tasks: query reformulation and ranking. The metrics used here are largely inspired from [11], and we discuss these below briefly. Towards the end of the section we also provide some qualitative results.

Table 1. Performance of models based on reformulation and ranking metrics

Model	Query reformulation			Ranking
	BLEU (%) ($\uparrow$)	sim_{emb} (%) ($\uparrow$)	Diversity *Top K* = 3 ($\uparrow$)	MRR *Baseline*: 0.31 ($\uparrow$)
HRED	6.92 ± 0.06	40.7 ± 1.3	0.37 ± 0.01	-
HRED + Ranker (CE)	7.63 ± 0.07	43.5 ± 1.2	0.42 ± 0.02	0.35 ± 0.02
HRED + Ranker (RO)	7.51 ± 0.07	40.8 ± 1.4	0.43 ± 0.02	0.39 ± 0.01
HREDCap	7.13 ± 0.09	37.8 ± 1.4	0.39 ± 0.04	-
HREDCap + Ranker (CE)	7.95 ± 0.11	39.4 ± 1.2	0.44 ± 0.06	0.38 ± 0.02
HREDCap + Ranker (RO)	7.68 ± 0.10	37.6 ± 1.4	0.45 ± 0.05	0.41 ± 0.02

5.1 Evaluation Metrics

Evaluation for query reformulation involves comparing the generated reformulation $\hat{q}_{\text{reform}}$ with the target reformulation q_{reform}. For all the models, irrespective of whether they utilize the next query within the session q_{i+1} as the target reformulation, or the caption C_i^{clicked} corresponding to the clicked image, the ground truth reformulation q_{reform} is always taken to be q_{i+1}[3]. This consistency has been maintained across all models to ensure that their performance is comparable, no matter what signal was used to train the reformulation model. The metrics used here cover three aspects: 'Relevance' (BLEU & sim_{emb}), 'Ranking' (MRR) and 'Diversity' (analyzed later).

BLEU Score: This metric [22], commonly used in machine translation scenarios, quantifies the similarity between a predicted sequence of words and the target sequence of words using n-gram precision. A higher BLEU score corresponds to a higher similarity between the predicted and target reformulations.

Embedding Based Query Similarity: This metric takes semantic similarity of words into account, instead of their exact overlap. A phrase-level embedding is calculated using vector extrema [13], for which pretrained GLoVe embeddings were used. The cosine similarity between the phrase-level vectors for the two queries is given by sim_{emb}. A higher value of sim_{emb} is taken to signify a greater semantic similarity between the prediction and the ground truth. Unlike BLEU, we expect sim_{emb} to provide a notion of similarity of the generated query to the target that allows for replacement words that are similar to the observed ones.

Mean Reciprocal Rank (MRR): The ranker's effectiveness is evaluated using MRR [28], which is given as the reciprocal rank of the first relevant (i.e., clicked) result averaged over all queries, across all sessions. A higher value of MRR will signify a better ranker in the proposed multitask models. To have a standard point of reference to compare against, we computed the observed MRR for the

[3] For sessions with less than 5 queries in a session, if q_i is the last query of the session, the model is trained to predict the 'end of session' token as the first token of q_{i+1}. The subsequent predicted tokens are encouraged to be the padding token '$< p >$'.

queries in the test set and found it to be 0.31. This means that on average, for queries in our test set, the first image clicked by the users was at rank ~ 3.1.

5.2 Main Results

Having discussed the metrics, we will now present the performance of our models on the two tasks under consideration, namely query reformulation and ranking. Table 1 provides these results as well as the effect of different ranking losses – denoted by (RO) and (CE) respectively.

Evaluation Based on Reformulation: For the purpose of this evaluation, we fix the beam width $K = 3$ and report the average of maximum values among all the candidate reformulations, across all queries in our test set.

While comparing *HRED* and *HRED + Ranker* (both *CE* and *RO*), we observe that the multitask version performs better across *all* metrics. A similar trend can be observed when comparing *HREDCap* with its multitask variants. For all the three metrics for query reformulations, the best performing model is a multitask model – this validates the observations from [1] in our context.

When comparing the two core reformulation models – *HRED* & *HREDCap*, we find that the richer captions data that *HREDCap* sees is aiding the model – while *HRED* scores better sim_{emb}, *HREDCap* wins out on BLEU & Diversity. The drop in sim_{emb} values can be explained by noting that on average captions contain more words than queries (5.22 in comparison to 2.31), and hence similarity-based measures, due to additional words in the captions, will not be as high as overlap-based measures (i.e., BLEU). **Evaluation based on Ranking:** To evaluate the performance of the ranker component in our proposed multitask models, we use MRR. We use the observed MRR of clicked results in the test set (0.31) as the baseline. We also analyze the effect of using the pairwise objective as opposed to the binary cross entropy loss.

Looking at the results presented in Table 1, three trends emerge. Firstly, all the proposed multitask models perform better than the baseline. The best performing model, i.e., *HREDCap + Ranker* with pairwise loss (RO), outperforms the baseline by about 32%. Secondly, we observe that using pairwise loss leads to an increase in MRR, for both of the cases under consideration, with only marginal drop in reformulation metrics – we revisit this observation in the next section. Lastly, the multitask models that use captions perform better than multitask models that use subsequent queries.

5.3 Analysis

In this section, we concentrate on the following two aspects of the generated query reformulations: (a) diversity, and (b) descriptiveness.

Diverse Query Reformulations due to Multitasking: The importance of suggesting diverse queries to enhance user search experience is well established within the IR community. The mechanism to obtain a diverse set of reformulation alternatives is via the use of beam search based decoding. In scenarios where a set of top-K candidates are required, we take inspiration from Ma et al. [20] to evaluate the predictions of our models for their diversity. For a beam width of K, a reformulation model will generate $\mathcal{R}_{gen} = \{r_1, r_2, \ldots, r_K\}$ candidate reformulations for a given original query. We quantify the diversity in the candidate reformulations by comparing each candidate reformulation r_i with other reformulations $r_j \in \mathcal{R}_{gen} : i \neq j$. The diversity of a set of K queries is evaluated as

$$D(\mathcal{R}_{\text{gen}}) = 1 - \frac{1}{K(K-1)} * \left(\sum_{r_i \in \mathcal{R}_{\text{gen}}} \sum_{r_j \in \mathcal{R}_{\text{gen}}:\ j \neq i} sim_{emb}(r_i, r_j) \right)$$

In Table 1, it can be observed that multitask models generate more diverse reformulations than models trained just for the task of query reformulation. This is particularly evident when comparing the effect of the ranking loss.

From Fig. 4, it can be noted that as more candidate reformulations are taken into consideration, i.e., as the beam width K is increased, the average relevance of the reformulations decreases across all the models. However, the diverseness of $\mathcal{R}_{gen}$ flattens after $K = 3$. This was the reason for setting the beam width to 3 while presenting results in Table 1.

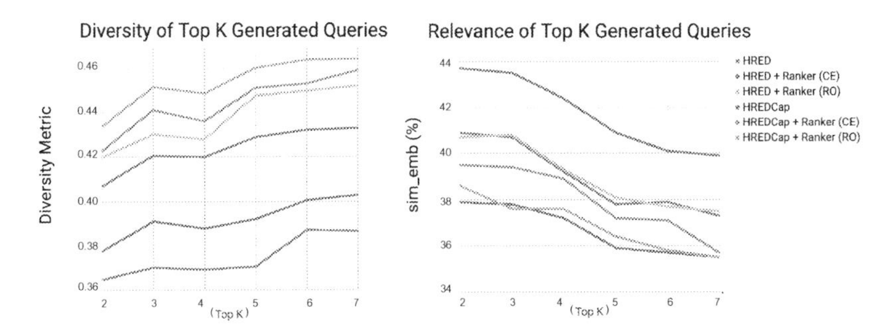

Fig. 4. The trade-off between relevance (as quantified by sim_{emb}) and diversity. As K is increased, the relevance of generated predictions drops across all models.

Descriptive Reformulations using Captions: The motivation for generating more descriptive reformulations is of central importance to our idea of using image captions. To this end, we analyze the generated reformulations to assess if this is indeed the case. We start by noting (see Table 2) that captions corresponding to clicked images for queries in our test set contain, on average, more words than the queries. Following this, we analyze the generated reformulations by two of our multitask models – (i) *HRED + Ranker (RO)*, which guides the process of query reformulation using subsequent queries within a session, and (ii) *HREDCap + Ranker (RO)*, which guides the process of query reformulation using captions corresponding to clicked images. For this entire analysis, we removed stop words [4] from all the queries and captions under consideration.

As can be noted from Table 2, reformulations using captions tend to contain more words than reformulations without them. However, number of words in a query is only a facile proxy for its descriptiveness. Acknowledging this, we perform a secondary aggregate analysis on the number of novel words inserted into the reformulation and number of words dropped from the original query. We identify novel words as words that were not present in the original query q_i but have been generated in the reformulation $\hat{q}_{\text{reform}}$, and dropped words as the words that were present in the original query but are absent from the generated reformulation. Table 2 indicates that, on average, the model trained using captions tends to insert more novel words while reformulating the query, and at the same time drops fewer words from the query. Interestingly, models trained using subsequent queries inserts almost as many words into the reformulation as it drops from the original query.

To analyze this further, we compute the average similarity between the novel words that were inserted and the words that were dropped, by averaging the GloVe vector based similarity between words, across all queries in our test set. For *HRED + Ranking (RO)* this average similarity is **0.64**, while for *HREDCap + Ranker (RO)* it is **0.41**. A higher similarity value for the former suggests that the model largely *substitutes* the existing words with words having similar semantic meaning. Using captions, on the other hand, is more likely to generate novel words which bring in additional meaning.

Table 2. Analyzing the effect of using captions on length of generated query reformulations, along with influence on generating novel words while dropping the existing ones.

Avg. # of words in queries	2.31 ± 0.92 word(s)	
Avg. # of words in captions	5.22 ± 2.37 word(s)	
Models →	HRED + Ranker (RO)	HREDCap + Ranker (RO)
Avg. # generated words	2.18 ± 0.61 word(s)	4.91 ± 1.16 word(s)
Avg. # novel words	1.04 ± 0.13 word(s)	2.56 ± 0.47 word(s)
Avg. # dropped words	1.14 ± 0.15 word(s)	0.89 ± 0.17 word(s)
Avg. similarity b/w insertions and drops	0.64 ± 0.03	0.41 ± 0.04

5.4 Qualitative Results

In Table 3, we present a few examples depicting the descriptive nature of generated reformulations. The generated reformulations by *HRED + Ranker* are compared against those by *HREDCap + Ranker*. We only present the top ranked reformulation among top-K reformulations. We note that using captions as target generates reformulations that are more descriptive and the process of generation results in more insertions of novel words, in comparison to using subsequent queries as targets. These qualitative observations, along with quantitative observations discussed earlier, reinforce the efficacy of using captions of clicked images for the task of query reformulation.

Table 3. Qualitative results comparing the generated reformulation by *HRED + Ranker* and *HREDCap + Ranker*. The words in **bold** are novel insertions.

		Queries	Clicked caption	*HRED + Ranker (RO)*	*HREDCap + Ranker (RO)*
Session$_1$	q_1	traffic	rush hour traffic	traffic jam	traffic **jam during rush hour**
	q_2	traffic **jam**	traffic jams in the city, road, rush hour	**city** traffic jam	traffic **during rush hour** in **city**
	q_3	traffic jam pollution	blurred silhouettes of cars by steam of exhaust	traffic jam **cars**	**dirt** and **smoke** from **cars** in traffic jam
Session$_2$	q_1	sleeping baby	sleeping one year old baby girl	**cute** sleeping baby	**little** baby sleeping **peacefully**
	q_2	sleeping baby cute	baby boy in white sunny bedroom	sleeping baby	baby sleeping in **bed peacefully**
	q_3	white bed sleeping baby	carefree little baby sleeping with white soft toy	baby sleeping in bed	**little** baby sleeping in white bed **peacefully**
Session$_3$	q_1	chemistry	three dimensional illustration of molecule model	chemical **reaction**	**molecules** and **structures** in chemistry
	q_3	molecule reaction	chemical reaction between molecules	reaction molecules	molecules reacting in **chemistry**
	q_3	molecule collision	frozen moment of two particle collision	collision molecules	molecules colliding **chemistry reaction**

6 Conclusion

In this paper, we build upon recent advances in sequence-to-sequence models based approaches for recommending queries. The core technical component of our paper is the use of a novel supervision signal for training seq-to-seq models for query reformulation – i.e., captions of clicked images instead of subsequent queries within a session, as well as the use of a pairwise preference based objective for the secondary ranking task. The effect of these are evaluated alongside baseline model architectures for this setting. Our extensive analysis evaluated the model and training method combinations towards being able to generate a set of descriptive, relevant and diverse reformulations.

Although the experiments were done on data from an image search engine, we believe that similar improvements can be observed if content properties from textual documents can be integrated into the seq-to-seq models. Future work will look into the influence of richer representations on the behavior of the ranker, and in turn on the characteristics of the reformulations.

References

1. Ahmad, W.U., Chang, K.W., Wang, H.: Multi-task learning for document ranking and query suggestion. In: International Conference on Learning Representations (2018)
2. Bahdanau, D., Cho, K., Bengio, Y.: Neural machine translation by jointly learning to align and translate. arXiv preprint arXiv:1409.0473 (2014)

3. Beeferman, D., Berger, A.: Agglomerative clustering of a search engine query log. In: Proceedings of the Sixth ACM SIGKDD International Conference on Knowledge Discovery and Data Mining, pp. 407–416. ACM (2000)
4. Bird, S., Loper, E.: NLTK: the natural language toolkit. In: Proceedings of the ACL 2004 on Interactive Poster and Demonstration Sessions, p. 31. Association for Computational Linguistics (2004)
5. Broder, A.Z., Fontoura, M., Gabrilovich, E., Joshi, A., Josifovski, V., Zhang, T.: Robust classification of rare queries using web knowledge. In: Proceedings of the 30th Annual International ACM SIGIR Conference on Research and Development in Information Retrieval, pp. 231–238. ACM (2007)
6. Burges, C., et al.: Learning to rank using gradient descent. In: Proceedings of the 22nd International Conference on Machine Learning, ICML 2005, pp. 89–96. ACM (2005)
7. Cao, H., et al.: Context-aware query suggestion by mining click-through and session data. In: Proceedings of the 14th ACM SIGKDD International Conference on Knowledge Discovery and Data Mining, pp. 875–883. ACM (2008)
8. Caruana, R.: Multitask learning. Mach. Learn. **28**(1), 41–75 (1997)
9. Chirita, P.A., Firan, C.S., Nejdl, W.: Personalized query expansion for the web. In: Proceedings of the 30th Annual International ACM SIGIR Conference on Research and Development in Information Retrieval, pp. 7–14. ACM (2007)
10. Cho, K., et al.: Learning phrase representations using RNN encoder-decoder for statistical machine translation. arXiv preprint arXiv:1406.1078 (2014)
11. Dehghani, M., Rothe, S., Alfonseca, E., Fleury, P.: Learning to attend, copy, and generate for session-based query suggestion. In: Proceedings of the 2017 ACM on Conference on Information and Knowledge Management, CIKM 2017, pp. 1747–1756 (2017)
12. Fonseca, B.M., Golgher, P., Pôssas, B., Ribeiro-Neto, B., Ziviani, N.: Concept-based interactive query expansion. In: Proceedings of the 14th ACM International Conference on Information and Knowledge Management, pp. 696–703. ACM (2005)
13. Forgues, G., Pineau, J., Larchevêque, J.M., Tremblay, R.: Bootstrapping dialog systems with word embeddings. In: Nips, Modern Machine Learning and Natural Language Processing Workshop, vol. 2 (2014)
14. Graves, A., Schmidhuber, J.: Framewise phoneme classification with bidirectional LSTM and other neural network architectures. Neural Netw. **18**(5–6), 602–610 (2005)
15. Hochreiter, S., Schmidhuber, J.: Long short-term memory. Neural Comput. **9**(8), 1735–1780 (1997)
16. Huang, C.K., Chien, L.F., Oyang, Y.J.: Relevant term suggestion in interactive web search based on contextual information in query session logs. J. Am. Soc. Inf. Sci. Technol. **54**(7), 638–649 (2003)
17. Huang, J., Efthimiadis, E.N.: Analyzing and evaluating query reformulation strategies in web search logs. In: Proceedings of the 18th ACM Conference on Information and Knowledge Management, pp. 77–86. ACM (2009)
18. Jones, R., Rey, B., Madani, O., Greiner, W.: Generating query substitutions. In: Proceedings of the 15th International Conference on World Wide Web, pp. 387–396. ACM (2006)
19. Kingma, D.P., Ba, J.: Adam: a method for stochastic optimization. CoRR arXiv:1412.6980 (2014)
20. Ma, H., Lyu, M.R., King, I.: Diversifying query suggestion results. In: AAAI, vol. 10 (2010)

21. Mitra, B.: Exploring session context using distributed representations of queries and reformulations. In: Proceedings of the 38th International ACM SIGIR Conference on Research and Development in Information Retrieval, pp. 3–12. ACM (2015)
22. Papineni, K., Roukos, S., Ward, T., Zhu, W.J.: BLEU: a method for automatic evaluation of machine translation. In: Proceedings of the 40th Annual Meeting on Association for Computational Linguistics, pp. 311–318. Association for Computational Linguistics (2002)
23. Pennington, J., Socher, R., Manning, C.: GloVe: global vectors for word representation. In: Proceedings of the 2014 Conference on Empirical Methods in Natural Language Processing (EMNLP), pp. 1532–1543 (2014)
24. Silvestri, F.: Mining query logs: turning search usage data into knowledge. Found. Trends® Inf. Retr. **4**(1), 171–174 (2009)
25. Song, Y., He, L.W.: Optimal rare query suggestion with implicit user feedback. In: Proceedings of the 19th International Conference on World Wide Web, pp. 901–910. ACM (2010)
26. Sordoni, A., Bengio, Y., Vahabi, H., Lioma, C., Simonsen, J.G., Nie, J.: A hierarchical recurrent encoder-decoder for generative context-aware query suggestion. CoRR arXiv:1507.02221 (2015)
27. Sutskever, I., Vinyals, O., Le, Q.V.: Sequence to sequence learning with neural networks. In: Advances in Neural Information Processing Systems, pp. 3104–3112 (2014)
28. Voorhees, E.M., Dang, H.T.: Overview of the TREC 2003 question answering track. In: TREC, vol. 2003, pp. 54–68 (2003)

IR - General

Curriculum Learning Strategies for IR

An Empirical Study on Conversation Response Ranking

Gustavo Penha(✉) and Claudia Hauff

TU Delft, Delft, The Netherlands
{g.penha-1,c.hauff}@tudelft.nl

Abstract. Neural ranking models are traditionally trained on a series of random batches, sampled uniformly from the entire training set. Curriculum learning has recently been shown to improve neural models' effectiveness by sampling batches non-uniformly, going from easy to difficult instances during training. In the context of neural Information Retrieval (IR) curriculum learning has not been explored yet, and so it remains unclear (1) how to *measure the difficulty* of training instances and (2) *how to transition* from easy to difficult instances during training. To address both challenges and determine whether curriculum learning is beneficial for neural ranking models, we need large-scale datasets and a retrieval task that allows us to conduct a wide range of experiments. For this purpose, we resort to the task of *conversation response ranking*: ranking responses given the conversation history. In order to deal with challenge (1), we explore *scoring functions* to measure the difficulty of conversations based on different input spaces. To address challenge (2) we evaluate different *pacing functions*, which determine the velocity in which we go from easy to difficult instances. We find that, overall, by just intelligently sorting the training data (i.e., by performing curriculum learning) we can improve the retrieval effectiveness by up to 2% (The source code is available at https://github.com/Guzpenha/transformers_cl.).

Keywords: Curriculum learning · Conversation response ranking

1 Introduction

Curriculum Learning (CL) is motivated by the way humans teach complex concepts: teachers impose a certain order of the material during students' education. Following this guidance, students can exploit previously learned concepts to more easily learn new ones. This idea was initially applied to machine learning over two decades ago [8] as an attempt to use a similar strategy in the training of a recurrent network by *starting small* and gradually learning more difficult examples. More recently, Bengio et al. [1] provided additional evidence that curriculum strategies can benefit neural network training with experimental results on different tasks such as shape recognition and language modelling. Since then, empirical successes were observed for several computer vision [14,49] and natural language processing (NLP) tasks [36,42,60].

J. M. Jose et al. (Eds.): ECIR 2020, LNCS 12035, pp. 699–713, 2020.
https://doi.org/10.1007/978-3-030-45439-5_46

In supervised machine learning, a function is learnt by the learning algorithm (the *student*) based on inputs and labels provided by the *teacher*. The teacher typically samples randomly from the entire training set. In contrast, CL imposes a structure on the training set based on a notion of difficulty of instances, presenting to the student easy instances before difficult ones. When defining a CL strategy we face two challenges that are specific to the domain and task at hand [14]: (1) arranging the training instances by a sensible measure of *difficulty*, and, (2) determining the *pace* in which to present instances—going over easy instances too fast or too slow might lead to ineffective learning.

We conduct here an empirical investigation into those two challenges in the context of IR. Estimating relevance—a notion based on human cognitive processes—is a complex and difficult task at the core of IR, and it is still unknown *to what extent CL strategies are beneficial for neural ranking models*. This is the question we aim to answer in our work.

Given a set of queries—for instance user utterances, search queries or questions in natural language—and a set of documents—for instance responses, web documents or passages—neural ranking models learn to distinguish relevant from non-relevant query-document pairs by training on a large number of labeled training pairs. Neural models have for some time struggled to display significant and additive gains in IR [53]. In a short time though, BERT [7] (released in late 2018) and its derivatives (e.g. XLNet [56], RoBERTa [25]) have proven to be remarkably effective for a range of NLP tasks. The recent breakthroughs of these large and heavily pre-trained language models have also benefited IR [54,55,57].

In our work we focus on the challenging IR task of conversation response ranking [50], where the query is the dialogue history and the documents are the candidate responses of the agent. The set of responses are not generated on the go, they must be retrieved from a comprehensive dialogue corpus. A number of deep neural ranking models have recently been proposed for this task [43,50,52,61,62], which is more complex than retrieval for single-turn interactions, as the ranking model has to determine where the important information is in the previous user utterances (dialogue history) and how it is relevant to the current information need of the user. Due to the complexity of the relevance estimation problem displayed in this task, we argue it to be a good test case for curriculum learning in IR.

In order to tackle the first challenge of CL (determine what makes an instance difficult) we study different *scoring functions* that determine the difficulty of query-document pairs based on four different input spaces: conversation history $\{\mathcal{U}\}$, candidate responses $\{\mathcal{R}\}$, both $\{\mathcal{U},\mathcal{R}\}$, and $\{\mathcal{U}, \mathcal{R}, \mathcal{Y}\}$, where $\mathcal{Y}$ are relevance labels for the responses. To address the second challenge (determine the pace to move from easy to difficult instances) we explore different *pacing functions* that serve easy instances to the learner for more or less time during the training procedure. We empirically explore how the curriculum strategies perform for two different response ranking datasets when compared against vanilla (no curriculum) fine-tuning of BERT for the task. Our main findings are that (i) CL improves retrieval effectiveness when we use a difficulty criteria based on a supervised model that uses all the available information $\{\mathcal{U}, \mathcal{R}, \mathcal{Y}\}$, (ii) it is best to give the model more time to assimilate harder instances during training by introducing difficult instances in earlier iterations, and, (iii) the CL gains over the no curriculum baseline are spread over different conversation domains, lengths of conversations and measures of conversation difficulty.

2 Related Work

Neural Ranking Models. Over the past few years, the IR community has seen a great uptake of the many flavours of deep learning for all kinds of IR tasks such as ad-hoc retrieval, question answering and conversation response ranking. Unlike traditional learning to rank (LTR) [24] approaches in which we manually define features for queries, documents and their interaction, neural ranking models learn features directly from the raw textual data. Neural ranking approaches can be roughly categorized into representation-focused [17,38,47] and interaction-focused [13,48]. The former learns query and document representations separately and then computes the similarity between the representations. In the latter approach, first a query-document interaction matrix is built, which is then fed to neural net layers. Estimating relevance directly based on interactions, i.e. interaction-focused models, has shown to outperform representation-based approaches on several tasks [16,27].

Transfer learning via large pre-trained Transformers [46]—the prominent case being BERT [7]—has lead to remarkable empirical successes on a range of NLP problems. The BERT approach to learn textual representations has also significantly improved the performance of neural models for several IR tasks [33,37,54,55,57], that for a long time struggled to outperform classic IR models [53]. In this work we use the no-CL BERT as a strong baseline for the conversation response ranking task.

Curriculum Learning. Following a curriculum that dictates the ordering and content of the education material is prevalent in the context of human learning. With such guidance, students can exploit previously learned concepts to ease the learning of new and more complex ones. Inspired by cognitive science research [35], researchers posed the question of whether a machine learning algorithm could benefit, in terms of learning speed and effectiveness, from a similar curriculum strategy [1,8]. Since then, positive evidence for the benefits of curriculum training, i.e. training the model using easy instances first and increasing the difficulty during the training procedure, has been empirically demonstrated in different machine learning problems, e.g. image classification [11,14], machine translation [21,30,60] and answer generation [23].

Processing training instances in a meaningful order is not unique to CL. Another related branch of research focuses on *dynamic* sampling strategies [2,4,22,39], which unlike CL that requires a definition of what is easy and difficult before training starts, estimates the importance of instances during the training procedure. Self-paced learning [22] simultaneously selects easy instances to focus on and updates the model parameters by solving a biconvex optimization problem. A seemingly contradictory set of approaches give more focus to difficult or more uncertain instances. In active learning [4,6,44], the most uncertain instances with respect to the current classifier are employed for training. Similarly, hard example mining [39] focuses on difficult instances, measured by the model loss or magnitude of gradients for instance. Boosting [2,59] techniques give more weight to difficult instances as training progresses. In this work we focus on CL, which has been more successful in neural models, and leave the study of dynamic sampling strategies in neural IR as future work.

The most critical part of using a CL strategy is defining the difficulty metric to sort instances by. The estimation of instance difficulty is often based on our prior knowledge on what makes each instance difficult for a certain task and thus is domain dependent (cf. Table 1 for curriculum examples). CL strategies have not been studied yet in neural ranking models. To our knowledge, CL has only recently been employed in IR within the LTR framework, using LambdaMart [3], for ad-hoc retrieval by Ferro et al. [9]. However, no effectiveness improvements over randomly sampling training data were observed. The representation of the query, document and their interactions in the traditional LTR framework is dictated by the manually engineered input features. We argue that neural ranking models, which learn how to represent the input, are better suited for applying CL in order to learn increasingly more complex concepts.

Table 1. Difficulty measures used in the curriculum learning literature.

Difficulty criteria	Tasks
Sentence length	Machine translation [30], language generation [42], reading comprehension [58]
Word rarity	Machine translation [30,60], language modeling [1]
External model confidence	Machine translation [60], image classification [14,49], ad-hoc retrieval [9]
Supervision signal intensity	Facial expression recognition [12], ad-hoc retrieval [9]
Noise estimate	Speaker identification [34], image classification [5]
Human annotation	Image classification [45] (through weak supervision)

3 Curriculum Learning

Before introducing our experimental framework (i.e., the scoring functions and the pacing functions we investigate), let us first formally introduce the specific IR task we explore—a choice dictated by the complex nature of the task (compared to e.g. ad-hoc retrieval) as well as the availability of large-scale training resources such as `MSDialog` [32] and UDC [26].

Conversation Response Ranking. Given a historical dialogue corpus and a conversation, (i.e., the user's current utterance and the conversation history) the task of conversation response ranking [43,50,52] is defined as the ranking of the most relevant response available in the corpus. This setup relies on the fact that a large corpus of historical conversation data exists and adequate replies (that are coherent, well-formulated and informative) to user utterances can be found in it [51]. Formally, let $\mathcal{D} = \{(\mathcal{U}_i, \mathcal{R}_i, \mathcal{Y}_i)\}_{i=1}^{N}$ be an information-seeking conversations data set consisting of N triplets: dialogue context, response candidates and response labels. The dialogue context $\mathcal{U}_i$ is composed of the previous utterances $\{u^1, u^2, ..., u^\tau\}$ at the turn τ of the dialogue. The candidate responses $\mathcal{R}_i = \{r^1, r^2, ..., r^k\}$ are either the true response ($u^{\tau+1}$) or negative sampled candidates[1]. The relevance labels $\mathcal{Y}_i = \{y^1, y^2, ..., y^k\}$ indicate the responses' binary relevance scores, 1 if $r = u^{\tau+1}$ and 0 otherwise. The task is then to learn a ranking function $f(.)$ that is able to generate a ranked list for the set of candidate responses $\mathcal{R}_i$ based on their predicted relevance scores $f(\mathcal{U}_i, r)$.

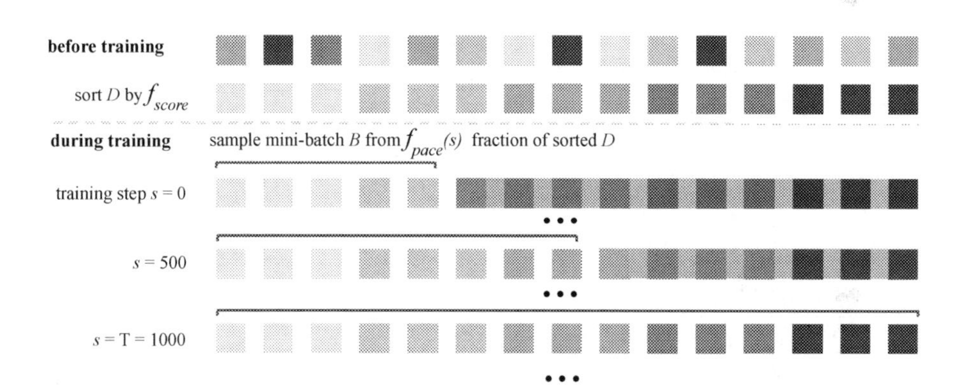

Fig. 1. Our curriculum learning framework is defined by two functions. The scoring function $f_{score}(instance)$ defines the instances' difficulty (darker/lighter blue indicate higher/lower difficulty). The pacing function $f_{pace}(s)$ indicates the percentage of the dataset available for sampling according to the training step s. (Color figure online)

Curriculum Framework. When training neural networks, the common training procedure is to divide the dataset $\mathcal{D}$ into $\mathcal{D}_{train}, \mathcal{D}_{dev}, \mathcal{D}_{test}$ and randomly (i.e., uniformly—every sample has the same likelihood of being sampled) sample mini-batches $\mathcal{B} = \{(\mathcal{U}_i, \mathcal{R}_i, \mathcal{Y}_i)\}_{i=1}^{k}$ of k instances from $\mathcal{D}_{train}$ where $k \ll N$, and perform an optimization procedure sequentially in $\{\mathcal{B}_1, ..., \mathcal{B}_M\}$. The CL framework employed here is inspired by previous works [30,49]. It is defined by two functions: the *scoring function* which determines the difficulty of instances and the *pacing function* which controls the pace with which to transition from easy to hard instances during training. More specifically, the scoring function

[1] In a production setup the ranker would either retrieve responses from the entire corpus or re-rank the responses retrieved by a recall-oriented retrieval method.

$f_{score}(\mathcal{U}_i, \mathcal{R}_i, \mathcal{Y}_i)$, is used to sort the training dataset. The pacing function $f_{pace}(s)$ determines the percentage of the sorted dataset available for sampling according to the current training step s (one forward pass plus one backward pass of a batch is considered to be one step). The neural ranking model samples uniformly from the initial $f_{pace}(s) * |D_{train}|$ instances sorted by f_{score}, while the rest of the dataset is not available for sampling. During training $f_{pace}(s)$ goes from δ (percentage of initial training data) to 1 when $s = T$. Both δ and T are hyperparameters. We provide an illustration of the training process in Fig. 1.

Table 2. Overview of our curriculum learning scoring functions.

Input space	Name	Definition	Difficulty notion
baseline	*random*	$f_{score} = Uniform(0, 1)$	
$(\mathcal{U})$	$\#_{turns}$	$f_{score}(\mathcal{U}) = \|\mathcal{U}\|$	Information spread
	$\overline{\#_{\mathcal{U}words}}$	$f_{score}(\mathcal{U}) = \frac{\sum_{i=0}^{\|\mathcal{U}\|} word\_count(u_i)}{\|\mathcal{U}\|}$	
$(\mathcal{R})$	$\overline{\#_{\mathcal{R}words}}$	$f_{score}(\mathcal{R}) = \frac{\sum_{i=0}^{\|\mathcal{R}\|} word\_count(r_i)}{\|\mathcal{R}\|}$	Distraction in responses
$(\mathcal{U}, \mathcal{R})$	σ_{SM}	$f_{score}(\mathcal{U}, \mathcal{R}) = \sqrt{\frac{\sum_{i=0}^{\|\mathcal{R}\|}(SM(\mathcal{U},r_i) - \overline{SM(\mathcal{U},\mathcal{R})})^2}{\|\mathcal{R}\|-1}}$	Responses heterogeneity
	σ_{BM25}	$f_{score}(\mathcal{U}, \mathcal{R}) = \sqrt{\frac{\sum_{i=0}^{\|\mathcal{R}\|}(BM25(\mathcal{U},r_i) - \overline{BM25(\mathcal{U},\mathcal{R})})^2}{\|\mathcal{R}\|-1}}$	
$(\mathcal{U}, \mathcal{R}, \mathcal{Y})$	$BERT_{pred}$	$f_{score}(\mathcal{U}, \mathcal{R}, \mathcal{Y}) = -(BERT\_pred(\mathcal{U}, r_i^+) - BERT\_pred(\mathcal{U}, r_i^-))$	Model confidence
	$\overline{BERT_{loss}}$	$f_{score}(\mathcal{U}, \mathcal{R}, \mathcal{Y}) = \frac{\sum_{i=0}^{\|\mathcal{R}\|} BERT\_loss(\mathcal{U},r_i)}{\|\mathcal{R}\|}$	

Scoring Functions. In order to measure the difficulty of a training triplet composed of $(\mathcal{U}_i, \mathcal{R}_i, \mathcal{Y}_i)$, we define pacing functions that use different parts of the input space: functions that leverage (i) the text in the dialogue history $\{\mathcal{U}\}$ (ii) the text in the response candidates $\{\mathcal{R}\}$ (iii) interactions between them, i.e., $\{\mathcal{U}, \mathcal{R}\}$, and, (iv) all available information including the labels for the training set, i.e., $\{\mathcal{U}, \mathcal{R}, \mathcal{Y}\}$. The seven[2] scoring functions we propose are defined in Table 2; we now provide intuitions of why we believe each function to capture some notion of instance difficulty.

- $\#_{turns}(\mathcal{U})$ and $\overline{\#_{\mathcal{U}words}}(\mathcal{U})$: The important information in the context can be spread over different utterances and words. Bigger dialogue contexts means there are more places where the important part of the user information need can be spread over. $\overline{\#_{\mathcal{R}words}}(\mathcal{R})$: Longer responses can distract the model as to which set of words or sentences are more important for matching. Previous

[2] The function *random* is the baseline—instances are sampled uniformly (no CL).

work shows that it is possible to fool machine reading models by creating longer documents with additional distracting sentences [18].

- $\sigma_{SM}(\mathcal{U}, \mathcal{R})$ and $\sigma_{BM25}(\mathcal{U}, \mathcal{R})$: Inspired by query performance prediction literature [40], we use the variance of retrieval scores to estimate the amount of heterogeneity of information, i.e. diversity, in the response candidate. Homogeneous ranked lists are considered to be easy. We deploy a semantic matching model (SM) and BM25 to capture both semantic correspondences and keyword matching [19]. SM is the average cosine similarity between the first k words from $\mathcal{U}$ (concatenated utterances) with the first k words from r using pre-trained word embeddings.
- $BERT_{pred}(\mathcal{U}, \mathcal{R}, \mathcal{Y})$ and $\overline{BERT_{loss}}(\mathcal{U}, \mathcal{R}, \mathcal{Y})$: Inspired by CL literature [14], we use external model prediction confidence scores as a measure of difficulty[3]. We fine-tune BERT [7] on $\mathcal{D}_{train}$ for the conversation response ranking task. For $BERT_{pred}$ easy dialogue contexts are the ones that the BERT confidence score for the positive response r^+ candidate is higher than the confidence for the negative response candidate r^-. The higher the difference the easier the instance is. For $\overline{BERT_{loss}}$we consider the loss of the model to be an indicator of the difficulty of an instance.

Table 3. Overview of our curriculum learning pacing functions. δ and T are hyperparameters.

Pacing function	Definition
baseline_training	$f_{pace}(s) = 1$
step	$f_{pace}(s) = \begin{cases} \delta, & \text{if } s \leq T * 0.33 \\ 0.66, & \text{if } s > T * 0.33, s \leq T * 0.66 \\ 1, & \text{if } s > T * 0.66 \end{cases}$
root	$f_{pace}(s, n) = min\left(1, \left(s\frac{1-\delta^n}{T} + \delta^n\right)^{\frac{1}{n}}\right)$
linear	$f_{pace}(s, n) = root(s, 1)$
root_n	$f_{pace}(s, n) = root(s, n)$
geom_progression	$f_{pace}(s) = min\left(1, 2^{\left(s\frac{log_2 1 - log_2 \delta}{T} + log_2 \delta\right)}\right)$

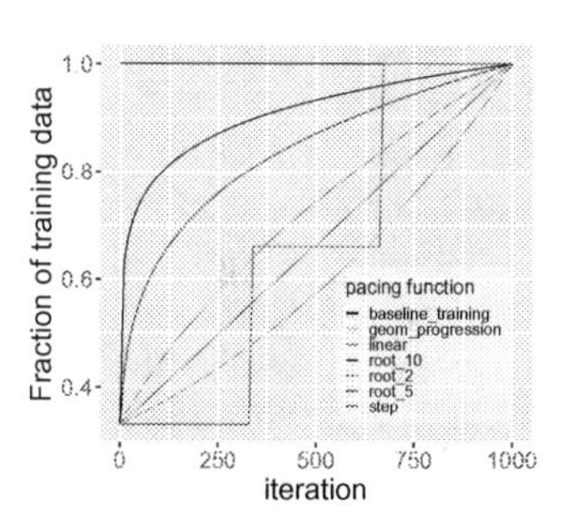

Fig. 2. Example with $\delta = 0.33$ and $T = 1000$.

Pacing Functions. Assuming that we know the difficulty of each instance in our training set, we still need to define how are we going to transition from easy to hard instances. We use the concept of pacing functions $f_{pace}(s)$; they should each have the following properties [30,49]: (i) start at an initial value of training instances $f_{pace}(0) = \delta$ with $\delta > 0$, so that the model has a number of instances to train in the first iteration, (ii) be non-decreasing, so that harder instances are added to the training set, and, (iii) eventually all instances are available for sampling when it reaches T iterations, $f_{pace}(T) = 1$.

[3] We note, that using BM25 average precision as a scoring function failed to outperform the baseline.

As intuitively visible in the example in Fig. 2, we opted for pacing functions that introduce more difficult instances at different paces—while *root_10* introduces difficult instances very early (after 125 iterations, 80% of all training data is available), *geom_progression* introduces them very late (80% is available after $\sim$ 800 iterations). We consider four different types of pacing functions, formally defined in Table 3. The *step* function [1,14,41] divides the data into S fixed sized groups, and after $\frac{T}{S}$ iterations a new group of instances is added, where S is a hyperparameter. A more gradual transition was proposed by Platanios et al. [30], by adding a percentage of the training dataset linearly with respect to the total of CL iterations T, and thus the slope of the function is $\frac{1-\delta}{T}$ (*linear* function). They also proposed *root_n* functions motivated by the fact that difficult instances will be sampled less as the training data grows in size during training. By making the slope inversely proportional to the current training data size, the model has more time to assimilate difficult instances. Finally, we propose the use of a geometric progression that instead of quickly adding difficult examples, it gives easier instances more training time.

4 Experimental Setup

Datasets. We consider two large-scale information-seeking conversation datasets (cf. Table 4) that allow the training of neural ranking models for conversation response ranking. `MSDialog`[4] [32] contain 246 K context-response pairs, built from 35.5 K information seeking conversations from the Microsoft Answer community, a question-answer forum for several Microsoft products. `MANtIS`[5] [29] was created by us and contains 1.3 million context-response pairs built from conversations of 14 different sites of Stack Exchange. Each `MANtIS` conversation fulfills the following conditions: (i) it takes place between exactly two users (the information *seeker* who starts the conversation and the information *provider*); (ii) it consists of at least 2 utterances per user; (iii) one of the provider's utterances contains a hyperlink, providing grounding; (iv) if the final utterance belongs to the seeker, it contains positive feedback. We created `MANtIS` to consider *diverse* conversations from different domains besides technical ones. We include `MSDialog` [31,32,52] here as a widely used benchmark.

Implementation Details. As strong neural ranking model for our experiments, we employ BERT [7] for the conversational response ranking task. We follow recent research in IR that employed fine-tuned BERT for retrieval tasks [28,55] and obtain strong baseline (i.e., no CL) results for our task. The best model by Yang et al. [52], which relies on external knowledge sources for `MSDialog`, achieves a MAP of 0.68 whereas our BERT baselines reaches a MAP of 0.71 (cf. Table 5). We fine-tune BERT[6] for sentence classification, using the

[4] `MSDialog` is available at https://ciir.cs.umass.edu/downloads/msdialog/.

[5] `MANtIS` is available at https://guzpenha.github.io/MANtIS/.

[6] We use the PyTorch-Transformers implementation https://github.com/huggingface/pytorch-transformers and resort to *bert-base-uncased* with default settings.

Table 4. Dataset used. $\mathcal{U}$ is the dialogue context, r a response and u an utterance.

	MSDialog			MANtIS		
Number of domains	75			14		
	Train	Valid	Test	Train	Valid	Test
Number of $(\mathcal{U}, r)$ pairs	173k	37k	35k	904k	199k	197k
Number of candidates per $\mathcal{U}$	10	10	10	11	11	11
Average number of turns	5.0	4.8	4.4	4.0	4.1	4.1
Average number of words per u	55.8	55.8	52.7	98.2	107.2	110.4
Average number of words per r	67.3	68.8	67.7	91.0	100.1	94.6

CLS token[7]; the input is the concatenation of the dialogue context and the candidate response separated by SEP tokens. When training BERT we employ a balanced number of relevant and non-relevant context and response pairs[8]. We use cross entropy loss and the Adam optimizer [20] with learning rate of $5e-5$ and $\epsilon = 1e-8$.

For σ_{SM}, as word embeddings we use pre-trained fastText[9] embeddings with 300 dimensions and a maximum length of $k = 20$ words of dialogue contexts and responses. For σ_{BM25}, we use default values[10] of $k_1 = 1.5$, $b = 0.75$ and $\epsilon = 0.25$. For CL, we fix T as 90% percent of the total training iterations—this means that we continue training for the final 10% of iterations after introducing all samples—and the initial number of instances δ as 33% of the data to avoid sampling the same instances several times.

Evaluation. To compare our strategies with the baseline where no CL is employed, for each approach we fine-tune BERT five times with different random seeds—to rule out that the results are observed only for certain random weight initialization values—and for each run we select the model with best observed effectiveness on the development set. The best model of each run is then applied to the test set. We report the effectiveness with respect to Mean Average Precision (MAP) like prior works [50,52]. We perform paired Student's t-tests between each scoring/pacing-function variant and the baseline run without CL.

5 Results

We first report the results for the pacing functions (Fig. 3) followed by the main results (Table 5) comparing different scoring functions. We finish with an error analysis to understand when CL outperforms our no-curriculum baseline.

[7] The BERT authors suggest CLS as a starting point for sentence classification tasks [7].

[8] We observed similar results to training with 1 to 10 ratio in initial experiments.

[9] https://fasttext.cc/docs/en/crawl-vectors.html.

[10] https://radimrehurek.com/gensim/summarization/bm25.html.

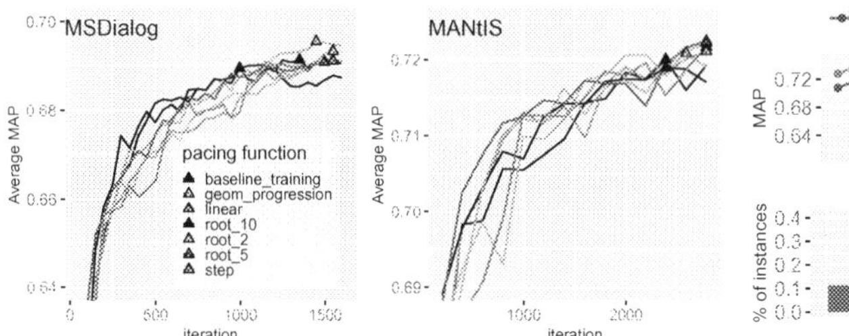

Fig. 3. Average development MAP for 5 different runs, using different curriculum learning pacing functions. △ is the maximum observed MAP.

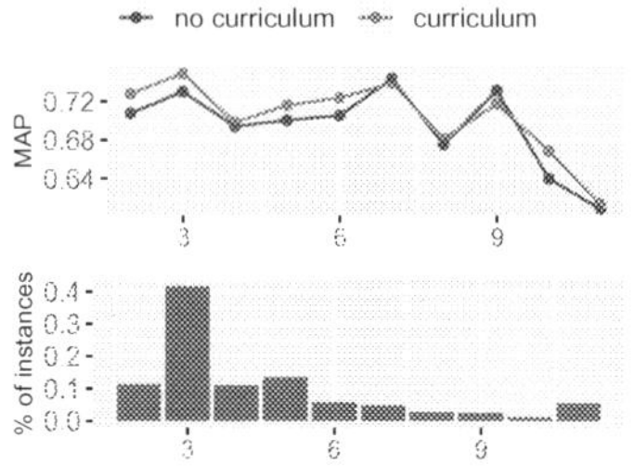

Fig. 4. `MSDialog` test set MAP of curriculum learning and baseline by number of turns.

Pacing Functions. In order to understand how CL results are impacted by the pace we go from easy to hard instances, we evaluate the different proposed *pacing functions*. We display the evolution of the development set MAP (average of 5 runs) during training on Fig. 3 (we use development MAP to track effectiveness during training). We fix the scoring function as $BERT_{pred}$; this is the best performing scoring function, more details in the next section. We see that the pacing functions with the maximum observed average MAP are *root_2* and *root_5* for `MSDialog` and `MANtIS` respectively[11]. The other pacing functions, *linear*, *geom_progression* and *step*, also outperform the standard training baseline with statistical significance on the test set and yield similar results to the *root_2* and *root_5* functions.

Our results are aligned with previous research on CL [30], that giving more time for the model to assimilate harder instances (by using a root pacing function) is beneficial to the curriculum strategy and is better than no CL with statistical significance on both development and test sets. For the rest of our experiments we fix the pacing function as *root_2*, the best pacing function for `MSDialog`. Let's now turn to the impact of the scoring functions.

Scoring Functions. The most critical challenge of CL is defining a measure of difficulty of instances. In order to evaluate the effectiveness of our scoring functions we report the test set results across both datasets in Table 5. We observe that the scoring functions which do not use the relevance labels $\mathcal{Y}$ are not able to outperform the no CL baseline (*random* scoring function). They are based on features of the dialogue context $\mathcal{U}$ and responses $\mathcal{R}$ that we hypothesized make them difficult for a model to learn. Differently, for $\overline{BERT_{loss}}$ and $BERT_{pred}$ we observe statistically significant results on both datasets across different runs. They differ in two ways from the unsuccessful scoring functions: they have access

[11] If we increase the n of the root function to bigger values, e.g. *root_10*, the results drop and get closer to not using CL. This is due to the fact that higher n generate root functions with a similar shape to standard training, giving the same amount of time to easy and hard instances (cf. Fig. 2).

Table 5. Test set MAP results of 5 runs using different curriculum learning scoring functions. Superscripts †/‡ denote statistically significant improvements over the baseline where no curriculum learning is applied ($f_{score} = random$) at 95%/99% confidence intervals. Bold indicates the highest MAP for each line.

run	$random$	$\#turns$	$\#\mathcal{U}words$	$\#\mathcal{R}words$	σ_{SM}	σ_{BM25}	$BERT_{pred}$	$BERT_{loss}$
MSDialog								
1	0.7142	0.7220 †	0.7229 †	0.7182	0.7239 †‡	0.7175	**0.7272** †‡	0.7244 †‡
2	0.7044	0.7060	0.7053	0.6968	0.7032	0.7003	**0.7159** †‡	0.7194 †‡
3	0.7126	0.7215 †	0.7163	0.7171	0.7174	0.7159	**0.7296** †‡	0.7225 †‡
4	0.7031	0.7065	0.7043	0.6993	0.7026	0.6949	0.7154 †‡	**0.7204** †‡
5	0.7148	0.7225 †	0.7203	0.7169	0.7171	0.7134	0.7322 †‡	**0.7331** †‡
AVG	0.7098	0.7157	0.7138	0.7097	0.7128	0.7084	**0.7241**	0.7240
SD	0.0056	0.0086	0.0086	0.0106	0.0095	0.0101	0.0079	0.0055
MANtIS								
1	0.7203	0.7192	0.7198	0.7194	0.7166	0.7200	0.7257 †‡	**0.7268** †‡
2	0.6984	0.6993	0.6989	0.6996	0.6964	0.7009	**0.7067** †‡	0.7051 †‡
3	0.7200	0.7197	0.7134	0.7206	0.7153	0.7153	**0.7282** †‡	0.7221
4	0.7114	0.7117	0.7002	0.6978	0.7140	0.7084	**0.7240** †‡	0.7184 †‡
5	0.7156	0.7174	0.7193 †	0.7162	0.7147	0.7185	**0.7264** †‡	0.7258 †‡
AVG	0.7131	0.7135	0.7103	0.7107	0.7114	0.7126	**0.7222**	0.7196
SD	0.0090	0.0085	0.0102	0.0111	0.0084	0.0079	0.0088	0.0088

to the training labels $\mathcal{Y}$ and the difficulty of an instance is based on what a previously trained model determines to be hard, and thus not our intuition.

Our results bear resemblance to Born Again Networks [10], where a student model which is identical in parameters and architecture to the teacher model outperforms the teacher when trained with knowledge distillation [15], i.e., using the predictions of the teacher model as labels for the student model. The difference here is that instead of transferring the knowledge from the teacher to the student through the labels, we transfer the knowledge by imposing a structure/order on the training set, i.e. curriculum learning.

Error Analysis. In order to understand when CL performs better than random training samples, we fix the scoring ($BERT_{pred}$) ad pacing function (*root_2*) and explore the test set effectiveness along several dimensions (cf. Figs. 4 and 5). We report the results only for `MSDialog`, but the trends hold for `MANtIS` as well.

We first consider the number of turns in the conversation in Fig. 4. CL outperforms the baseline approach for the types of conversations appearing most frequently (2–5 turns in `MSDialog`). The CL-based and baseline effectiveness drops for conversations with a large number of turns. This can be attributed to two factors: (1) employing pre-trained BERT in practice allows only a certain maximum number of tokens as input, so longer conversations can lose important information due to truncating; (2) for longer conversations it is harder to identify the important information to match in the history, i.e information spread.

Next, we look at different conversation domains in Fig. 5 (left), such as *physics* and *askubuntu*—are the gains in effectiveness limited to particular domains? The

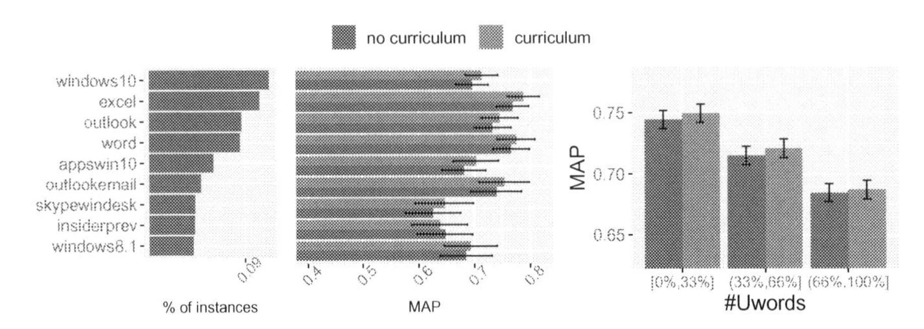

Fig. 5. Test set MAP for `MSDialog` across different domains (left) and instances' difficulty (right) according to $\overline{\#_{\mathcal{R}words}}$ for curriculum learning and the baseline.

error bars indicate the confidence intervals with confidence level of 95%. We list only the most common domains in the test set. The gains of CL are spread over different domains as opposed to concentrated on a single domain.

Lastly, using our scoring functions we sort the test instances and divide them into three buckets: first 33% instances, 33%–66% and 66%–100%. In Fig. 5 (right), we see the effectiveness of CL against the baseline for each bucket using $\overline{\#_{\mathcal{U}words}}$ (the same trend holds for the other scoring functions). As we expect, the bucket with the most difficult instances according to the scoring function is the one with lowest MAP values. Finally, the improvements of CL over the baseline are again spread across the buckets, showing that CL is able to improve over the baseline for different levels of difficulty.

6 Conclusions

In this work we studied whether CL strategies are beneficial for neural ranking models. We find supporting evidence for curriculum learning in IR. Simply reordering the instances in the training set using a difficulty criteria leads to effectiveness improvements, requiring no changes to the model architecture—a similar relative improvement in MAP has justified novel neural architectures in the past [43,50,61,62]. Our experimental results on two conversation response ranking datasets reveal (as one might expect) that it is best to use all available information $(\mathcal{U}, \mathcal{R}, \mathcal{Y})$ as evidence for instance difficulty. Future work directions include considering other retrieval tasks, different neural architectures and an investigation of the underlying reasons for CL's workings.

Acknowledgements. This research has been supported by NWO projects SearchX (639.022.722) and NWO Aspasia (015.013.027).

References

1. Bengio, Y., Louradour, J., Collobert, R., Weston, J.: Curriculum learning. In: ICML, pp. 41–48 (2009)
2. Breiman, L.: Arcing classifier. Ann. Stat. **26**(3), 801–849 (1998)
3. Burges, C.J.: From ranknet to lambdarank to lambdamart: an overview. Learning **11**(23–581), 81 (2010)
4. Chang, H.S., Learned-Miller, E., McCallum, A.: Active bias: training more accurate neural networks by emphasizing high variance samples. In: NeurIPS, pp. 1002–1012 (2017)
5. Chen, X., Gupta, A.: Webly supervised learning of convolutional networks. In: ICCV, pp. 1431–1439 (2015)
6. Cohn, D.A., Ghahramani, Z., Jordan, M.I.: Active learning with statistical models. J. Artif. Intell. Res. **4**, 129–145 (1996)
7. Devlin, J., Chang, M.W., Lee, K., Toutanova, K.: BERT: pre-training of deep bidirectional transformers for language understanding. In: NAACL, pp. 4171–4186 (2019)
8. Elman, J.L.: Learning and development in neural networks: the importance of starting small. Cognition **48**(1), 71–99 (1993)
9. Ferro, N., Lucchese, C., Maistro, M., Perego, R.: Continuation methods and curriculum learning for learning to rank. In: CIKM, pp. 1523–1526 (2018)
10. Furlanello, T., Lipton, Z., Tschannen, M., Itti, L., Anandkumar, A.: Born-again neural networks. In: ICML, pp. 1602–1611 (2018)
11. Gong, C., Tao, D., Maybank, S.J., Liu, W., Kang, G., Yang, J.: Multi-modal curriculum learning for semi-supervised image classification. IEEE Trans. Image Process. **25**(7), 3249–3260 (2016)
12. Gui, L., Baltrušaitis, T., Morency, L.P.: Curriculum learning for facial expression recognition. In: FG, pp. 505–511 (2017)
13. Guo, J., Fan, Y., Ai, Q., Croft, W.B.: A deep relevance matching model for ad-hoc retrieval. In: CIKM, pp. 55–64 (2016)
14. Hacohen, G., Weinshall, D.: On the power of curriculum learning in training deep networks. arXiv preprint arXiv:1904.03626 (2019)
15. Hinton, G., Vinyals, O., Dean, J.: Distilling the knowledge in a neural network. arXiv preprint arXiv:1503.02531 (2015)
16. Hu, B., Lu, Z., Li, H., Chen, Q.: Convolutional neural network architectures for matching natural language sentences. In: NeurIPS, pp. 2042–2050 (2014)
17. Huang, P.S., He, X., Gao, J., Deng, L., Acero, A., Heck, L.: Learning deep structured semantic models for web search using clickthrough data. In: CIKM, pp. 2333–2338 (2013)
18. Jia, R., Liang, P.: Adversarial examples for evaluating reading comprehension systems. In: EMNLP, pp. 2021–2031 (2017)
19. Rao, J., Liu, L., Tay, Y., Yang, W., Shi, P., Lin, J.: Bridging the gap between relevance matching and semantic matching for short text similarity modeling. In: EMNLP (2019)
20. Kingma, D.P., Ba, J.: Adam: A method for stochastic optimization. arXiv preprint arXiv:1412.6980 (2014)
21. Kocmi, T., Bojar, O.: Curriculum learning and minibatch bucketing in neural machine translation. In: RANLP, pp. 379–386 (2017)
22. Kumar, M.P., Packer, B., Koller, D.: Self-paced learning for latent variable models. In: NeurIPS, pp. 1189–1197 (2010)

23. Liu, C., He, S., Liu, K., Zhao, J.: Curriculum learning for natural answer generation. In: IJCAI, pp. 4223–4229 (2018)
24. Liu, T.Y., et al.: Learning to rank for information retrieval. Found. Trends® Inf. Retr. **3**(3), 225–331 (2009)
25. Liu, Y., et al.: RoBERTa: a robustly optimized BERT pretraining approach. arXiv preprint arXiv:1907.11692 (2019)
26. Lowe, R., Pow, N., Serban, I., Pineau, J.: The ubuntu dialogue corpus: a large dataset for research in unstructured multi-turn dialogue systems. In: SIGDIAL, pp. 285–294 (2015)
27. Nie, Y., Li, Y., Nie, J.Y.: Empirical study of multi-level convolution models for IR based on representations and interactions. In: SIGIR, pp. 59–66 (2018)
28. Nogueira, R., Cho, K.: Passage re-ranking with BERT. arXiv preprint arXiv:1901.04085 (2019)
29. Penha, G., Balan, A., Hauff, C.: Introducing MANtIS: a novel multi-domain information seeking dialogues dataset. arXiv preprint arXiv:1912.04639 (2019)
30. Platanios, E.A., Stretcu, O., Neubig, G., Poczos, B., Mitchell, T.: Competence-based curriculum learning for neural machine translation. In: NAACL, pp. 1162–1172 (2019)
31. Qu, C., Yang, L., Croft, W.B., Zhang, Y., Trippas, J., Qiu, M.: user intent prediction in information-seeking conversations. In: CHIIR (2019)
32. Qu, C., Yang, L., Croft, W.B., Trippas, J.R., Zhang, Y., Qiu, M.: Analyzing and characterizing user intent in information-seeking conversations. In: SIGIR, pp. 989–992 (2018)
33. Qu, C., Yang, L., Qiu, M., Croft, W.B., Zhang, Y., Iyyer, M.: BERT with history answer embedding for conversational question answering. In: SIGIR, pp. 1133–1136 (2019)
34. Ranjan, S., Hansen, J.H., Ranjan, S., Hansen, J.H.: Curriculum learning based approaches for noise robust speaker recognition. TASLP **26**(1), 197–210 (2018)
35. Rohde, D.L., Plaut, D.C.: Language acquisition in the absence of explicit negative evidence: how important is starting small? Cognition **72**(1), 67–109 (1999)
36. Sachan, M., Xing, E.: Easy questions first? a case study on curriculum learning for question answering. In: ACL, vol. 1, pp. 453–463 (2016)
37. Sakata, W., Shibata, T., Tanaka, R., Kurohashi, S.: FAQ retrieval using query-question similarity and BERT-based query-answer relevance. arXiv preprint arXiv:1905.02851 (2019)
38. Shen, Y., He, X., Gao, J., Deng, L., Mesnil, G.: A latent semantic model with convolutional-pooling structure for information retrieval. In: CIKM, pp. 101–110 (2014)
39. Shrivastava, A., Gupta, A., Girshick, R.: Training region-based object detectors with online hard example mining. In: CVPR, pp. 761–769 (2016)
40. Shtok, A., Kurland, O., Carmel, D.: Predicting query performance by query-drift estimation. In: ICTIR, pp. 305–312 (2009)
41. Soviany, P., Ardei, C., Ionescu, R.T., Leordeanu, M.: Image difficulty curriculum for generative adversarial networks (CuGAN). arXiv preprint arXiv:1910.08967 (2019)
42. Subramanian, S., Rajeswar, S., Dutil, F., Pal, C., Courville, A.: Adversarial generation of natural language. In: Rep4NLP, pp. 241–251 (2017)
43. Tao, C., Wu, W., Xu, C., Hu, W., Zhao, D., Yan, R.: One time of interaction may not be enough: go deep with an interaction-over-interaction network for response selection in dialogues. In: ACL, pp. 1–11 (2019)

44. Tong, S., Koller, D.: Support vector machine active learning with applications to text classification. J. Mach. Learn. Res. **2**(Nov), 45–66 (2001)
45. Tudor Ionescu, R., Alexe, B., Leordeanu, M., Popescu, M., Papadopoulos, D.P., Ferrari, V.: How hard can it be? Estimating the difficulty of visual search in an image. In: CVPR, pp. 2157–2166 (2016)
46. Vaswani, A., et al.: Attention is all you need. In: NeurIPS, pp. 5998–6008 (2017)
47. Wan, S., Lan, Y., Guo, J., Xu, J., Pang, L., Cheng, X.: A deep architecture for semantic matching with multiple positional sentence representations. In: AAAI, pp. 2835–2841 (2016)
48. Wan, S., Lan, Y., Xu, J., Guo, J., Pang, L., Cheng, X.: Match-SRNN: modeling the recursive matching structure with spatial RNN. In: IJCAI, pp. 2922–2928. AAAI Press (2016)
49. Weinshall, D., Cohen, G., Amir, D.: Curriculum learning by transfer learning: theory and experiments with deep networks. In: ICML, pp. 5235–5243 (2018)
50. Wu, Y., Wu, W., Xing, C., Zhou, M., Li, Z.: Sequential matching network: a new architecture for multi-turn response selection in retrieval-based chatbots. In: ACL, vol. 1, pp. 496–505 (2017)
51. Yang, L., et al.: A hybrid retrieval-generation neural conversation model. arXiv preprint arXiv:1904.09068 (2019)
52. Yang, L., et al.: Response ranking with deep matching networks and external knowledge in information-seeking conversation systems. In: SIGIR, pp. 245–254 (2018)
53. Yang, W., Lu, K., Yang, P., Lin, J.: Critically examining the neural hype: weak baselines and the additivity of effectiveness gains from neural ranking models. In: SIGIR, pp. 1129–1132, New York, NY, USA (2019)
54. Yang, W., et al.: End-to-end open-domain question answering with BERTserini. In: NAACL, pp. 72–77 (2019)
55. Yang, W., Zhang, H., Lin, J.: Simple applications of BERT for ad hoc document retrieval. arXiv preprint arXiv:1903.10972 (2019)
56. Yang, Z., Dai, Z., Yang, Y., Carbonell, J., Salakhutdinov, R., Le, Q.V.: XLNet: generalized autoregressive pretraining for language understanding. arXiv preprint arXiv:1906.08237 (2019)
57. Yilmaz, Z.A., Yang, W., Zhang, H., Lin, J.: Cross-domain modeling of sentence-level evidence for document retrieval. In: EMNLP, pp. 3481–3487 (2019)
58. Yu, Y., Zhang, W., Hasan, K., Yu, M., Xiang, B., Zhou, B.: End-to-end answer chunk extraction and ranking for reading comprehension. arXiv preprint arXiv:1610.09996 (2016)
59. Zhang, D., Kim, J., Crego, J., Senellart, J.: Boosting neural machine translation. In: IJCNLP, pp. 271–276 (2017)
60. Zhang, X., et al.: An empirical exploration of curriculum learning for neural machine translation. arXiv preprint arXiv:1811.00739 (2018)
61. Zhang, Z., Li, J., Zhu, P., Zhao, H., Liu, G.: Modeling multi-turn conversation with deep utterance aggregation. In: ACL, pp. 3740–3752 (2018)
62. Zhou, X., et al.: Multi-turn response selection for chatbots with deep attention matching network. In: ACL, pp. 1118–1127 (2018)

Accelerating Substructure Similarity Search for Formula Retrieval

Wei Zhong[1(✉)], Shaurya Rohatgi[2], Jian Wu[3], C. Lee Giles[2], and Richard Zanibbi[1(✉)]

[1] Rochester Institute of Technology, Rochester, USA
{wxz8033,rxzvcs}@rit.edu
[2] Pennsylvania State University, State College, USA
[3] Old Dominion University, Norfolk, USA

Abstract. Formula retrieval systems using substructure matching are effective, but suffer from slow retrieval times caused by the complexity of structure matching. We present a specialized inverted index and rank-safe dynamic pruning algorithm for faster substructure retrieval. Formulas are indexed from their Operator Tree (OPT) representations. Our model is evaluated using the NTCIR-12 Wikipedia Formula Browsing Task and a new formula corpus produced from Math StackExchange posts. Our approach preserves the effectiveness of structure matching while allowing queries to be executed in real-time.

Keywords: Math information retrieval · Query processing optimization · Dynamic pruning

1 Introduction

In information retrieval, a great deal of research has gone into creating efficient search engines for large corpora. However, few have addressed substructure search in structural content, e.g., in Mathematical Information Retrieval (MIR) [21] where efficient substructure similarity search is needed to identify shared subexpressions effectively. For example, in math formula search, to discern that $a + b$ and $b + a$ are equivalent (by commutativity), but that $ab + cd$ and $a + bcd$ are different, applying tokenization and counting common token frequencies is insufficient. Instead, a hierarchical representation of mathematical operations is needed and we may want to identify shared substructures.

In the most recent math similarity search competition,[1] effective systems all take a tree-based approach by extracting query terms from tree representations. For example, an Operator Tree (OPT) is used in Fig. 1 to represent math formulas where operands are represented by leaves and operators are located at internal nodes. This facilitates searching substructures shared by two math expressions. For example, we can extract paths from their tree representations

[1] The NTCIR-12 Wikipedia Formula Browsing Task.

J. M. Jose et al. (Eds.): ECIR 2020, LNCS 12035, pp. 714–727, 2020.
https://doi.org/10.1007/978-3-030-45439-5_47

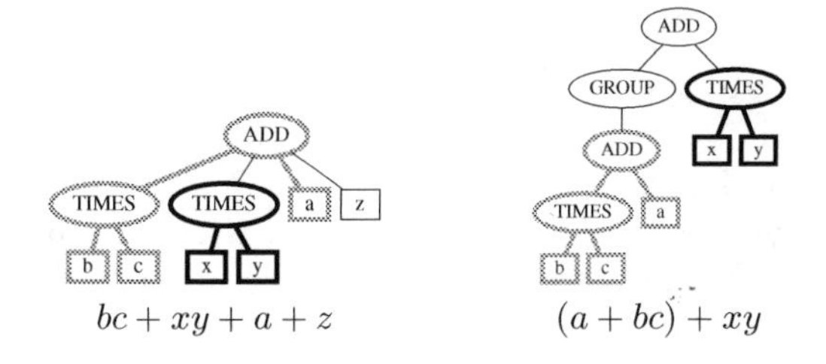

Fig. 1. Operator trees (OPTs) for two similar formulas. OPTs represent the application of operations (at internal nodes in circles) to operands (at the leaves in squares). Two common substructures are highlighted in black and gray.

and find their shared subtrees by matching their common paths grouped by subtree root nodes. However, in order to carry structure information, it is common to see structural queries with over tens or even hundreds of path tokens which is unusual for normal fulltext search. This makes query processing costly for realistic math search tasks.

In text similarity search, query processing can be accelerated through dynamic pruning [18], which typically estimates score upperbounds to prune documents unlikely to be in the top K results. However, effective substructure search requires additional matching or alignment among query terms, and this makes it hard to get a good score estimation and it prevents us applying traditional dynamically pruning effectively. In fact, reportedly few state-of-the-art MIR systems have achieved practical query run times even when given a large amount of computing resources [11,20]. In this paper we try to address this problem by introducing a specialized inverted index and we propose a dynamic pruning method based on this inverted index to boost formula retrieval efficiency.

2 Related Work

Recently there has been an increasing amount of research on similarity search for math formulas, with most focusing on search effectiveness [5,7,11,23]. There are many emerging issues regarding effectiveness, including handling mathematical semantics, and identifying interchangeable symbols and common subexpressions. However, the efficiency of math formula search systems is often not addressed.

A number of MIR systems apply text search models to math retrieval, extracting sequential features from formulas and use variants of TF-IDF scoring [12,14,16]. These approaches incorporate a bag-of-words model, and use frequency to measure formula similarity. Inevitably, they need to index different combinations of sequences or substrings to handle operator commutativity and subexpression identification. This index augmentation results in a non-linearly increasing index size in the number of indexed "words" [12] and thus hurts efficiency for large corpora. On the other hand, recent results [10,20,23] reveal that effective systems for formula retrieval use tree-based approaches distinct from text-based methods. However, tree-based systems usually need to calculate costly graph matching or edit distance metrics [9,22], which generally have non-linear time complexity. Recently, a path-based approach [23] was developed to search substructures in formula OPTs approximately by assuming that identical formulas have the same leaf-root path set. Although at the time of writing, it obtains the best effectiveness for the NTCIR-12 dataset, the typically large number of query paths means that query run times are not ideal - maximum run times can be a couple of seconds.

Dynamic pruning has been recognized as an effective way to reduce query processing times [2,8,13,18]. Dynamic pruning speeds up query processing by skipping scoring calculations or avoiding unnecessary reads for documents which are unlikely to be ranked in the top K results. Pruning methods can be based on different query processing schemes: Document-at-a-time (DAAT) requires all relevant posting lists be merged simultaneously. Term-at-a-time (TAAT) or score-at-a-time (SAAT) processes one posting list at a time for each term, requiring additional memory to store partial scores, and posting lists in this case are usually sorted by document importance (e.g, impact score [1]), with promising documents placed at the front of inverted lists. Pruning strategies are *rank-safe* (or *safe up to rank K*) [19] if they guarantee that the top K documents are ranked in the same order before and after pruning. The most well-known rank-safe pruning strategies for DAAT are MaxScore [8,17,19] and WAND variants [3,6]. Shan et al. [15] show that MaxScore variants (e.g. BMM, LBMM) outperform other dynamic pruning strategies for long queries, and recently Mallia et al. [2] report a similar finding over a range of popular index encodings.

3 Preliminaries

Baseline Model. This work is based on our previous work [23] which extracts prefixes from OPT leaf-root paths as index or query terms. The OPT is parsed from a formula in LaTeX. For indexed paths, they are mapped to corresponding posting lists in an inverted index where the IDs of expressions containing the path are appended. For query paths, the corresponding posting lists are merged and approximate matching is performed on candidates one expression at a time. The similarity score is measured from matched common subtree(s).

Because math symbols are interchangeable, paths are tokenized for better recall, e.g., variables such as a, b, c are tokenized into VAR. In our tokenized path representation uppercase words denote token types, which may be for operators as well as operands (e.g., TIMES for symbols representing multiplication). In Fig. 1, when indexing "bc + xy + a + z," its expression ID (or ExpID) will be appended to posting lists associated with tokenized prefix paths from its OPT representation, i.e., VAR/TIMES, VAR/ADD and VAR/TIMES/ADD. At query processing, the shared structures highlighted in black and gray are found by matching these tokenized paths (two paths match if and only if they have the same tokenized paths, for example, "a/+" and "z/+" can be matched) and common subtree roots are identified by grouping paths by their root nodes. As a result, the posting list entry also stores the root node ID for indexed paths, in order to reconstruct matches substructures at merge time.

At query time, the similarity score is given by the size of matched common subtrees. Specifically, the model chooses a number of "widest" matched subtree(s) (e.g., $a + bc$ is the widest matched in Fig. 1 because it has 3 common leaves and is "wider" than the other choices) and measure formula similarity based on the size of these common subtrees.

The original Approach0 model [23] matches up to three widest common subtrees and scores similarity by a weighted sum of the number of matched leaves (operands) and operators from different common subtrees $\hat{T}_q^i, \hat{T}_d^i$ of a common forest π. Operators and operand (leaf) nodes weights are controlled by parameter α, while the weight of rooted substructures from largest to smallest are given by β_i. In the following, $|\cdot|$ indicates the size of a set:

$$\sum_{i=1}^{3} \beta_i \left(\alpha \cdot \left| \text{operators}(\hat{T}_d^i) \right| + (1-\alpha) \cdot \left| \text{leaves}(\hat{T}_d^i) \right| \right) , \qquad (\hat{T}_q^i, \hat{T}_d^i) \in \pi \tag{1}$$

Interestingly, while multiple subtree matching boosts effectiveness, using just the widest match still outperforms other systems in terms of highly relevant results [23]. The simplified similarity score based on widest common subtree between query and document OPTs T_q, T_d is the widest match $w^*_{Q,D}$, formally

$$w^*_{Q,D} = \max_{\hat{T}_q, \hat{T}_d \in \text{CFS}(T_q, T_d)} |\text{leaves}(\hat{T}_d)| \tag{2}$$

where $\text{CFS}(T_q, T_d)$ are all the *common formula subtrees* between T_q and T_d. In addition to subtree isomorphism, a *formula subtree* requires leaves in a subtree to match leaves in the counterpart, in other words, subtrees are matched bottom-up from operands in OPTs. In Fig. 1, the value of $w^*_{Q,D}$ is 3, produced by the widest common subtrees shown in gray.

Dynamic Pruning. In dynamic pruning, the top K scored hits are kept throughout the querying process, with the lowest score in the top K at a given point defining the threshold θ. Since at most K candidates will be returned, dynamic pruning strategies work by estimating score upperbounds before knowing the precise score of a hit so that candidate hits with a score upperbound less or equal to θ can be pruned safely, because they will not appear in the final top K results. Moreover, if a subset of posting lists alone cannot produce a top K result from their upperbounds, they are called a *non-requirement set*, the opposite being the *requirement set*. Posting lists in the non-requirement with IDs less than the currently evaluating IDs in the requirement set can be skipped safely, because posting lists in the non-requirement set alone will not produce a top K candidate.

4 Methodology

In this paper, we apply dynamic pruning to structural search. As structure search has more query terms in general, we focus on a MaxScore-like strategy suggested

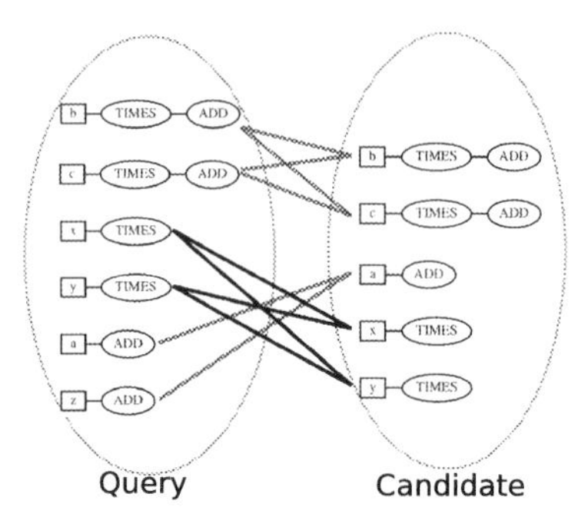

Fig. 2. Bipartite graph of hit path set for formulas in Fig. 1 (original leaf symbol is used here to help identify paths). Edges are established if paths from the two sides are the same after tokenization. Edges with shared end points (i.e., same root-end nodes) in original OPTs have the same color (black or gray).

by [2,15], since they do not need to sort query terms at merge iterations (which is expensive for long queries). Our approach is different from the original MaxScore, as upperbound scores are also calculated from the query tree representation. We also use the simplified scoring Eq. (2) where a subset of query terms in the widest matched common subtrees $\hat{T}_q^*, \hat{T}_d^*$ contribute to the score. In contrast, typical TF-IDF scoring has all hit terms contribute to the rank score.

When we merge posting lists, a set of query paths match paths from a document expression one at a time, each time a *hit path set* for matched query and candidate paths are examined. Define $\mathcal{P}(T)$ to be all paths extracted from OPT T, i.e., $\mathcal{P}(T) = \{p : p \in \text{leafroot\_paths}(T^n), n \in T\}$ where T^n is the entire subtree of T rooted at n with all its descendants. We model the hit path set by a bipartite graph $G(Q, D, E)$ where $Q = \{q : q \in \mathcal{P}(T_q)\}, D = \{d : d \in \mathcal{P}(T_d)\}$ are query and document path sets, and edges are ordered pairs $E = \{(q, d) : \text{tokenized}(q) = \text{tokenized}(d), q \in Q, d \in D\}$ representing a potential match between a query path to a document path. Since an edge is established only for paths with the same token sequence, we can partition the graph into disconnected smaller bipartite graphs $G_t = G(Q_t, D_t, E_t)$, each identified by tokenized query path t:

$$
\begin{aligned}
Q_t &= \{q : q \in Q, \text{tokenized}(q) = t\} \\
D_t &= \{d : d \in D, \text{tokenized}(d) = t\} \\
E_t &= \{(q, d) : (q, d) \in E, \text{tokenized}(q) = \text{tokenized}(d)\}
\end{aligned}
$$

Figure 2 shows the hit path set of the example in Fig. 1, this example can be partitioned into independent subgraphs associated with tokenized paths VAR/TIMES/ADD, VAR/TIMES and VAR/ADD. Each partition is actually a complete bipartite graph (fully connected) because for any edge between Q_t and D_t, it is in edge set E_t. And for each complete bipartite graph $G(Q_t, D_t, E_t)$, we can obtain their maximum matching sizes from $\min(|Q_t|, |D_t|)$ easily.

On the other hand, to calculate score $w^*_{Q,D}$, we need to find a pair of query and document nodes at which the widest common subtree $\hat{T}^*_q, \hat{T}^*_d$ are rooted (see Eq. 2), so we also define the matching candidate relations filtered by nodes. Let $G^{(m,n)} = G(Q^{(m)}, D^{(n)}, E^{(m,n)})$ be the subgraph matching between query subtree rooted at m and document subtree rooted at n where

$$\begin{aligned} Q^{(m)} &= \{q : q \in Q, \text{root\_end}(q) = m\} \\ D^{(n)} &= \{d : d \in D, \text{root\_end}(d) = n\} \\ E^{(m,n)} &= \{(q,d) : (q,d) \in E, \text{root\_end}(q) = m, \text{root\_end}(d) = n\} \end{aligned}$$

Then, similarity score $w^*_{Q,D}$ can be calculated from selecting the best matched node pairs and summing their partition matches. Specifically, define *token paths* of tree T rooted at n as set $\mathfrak{T}(n) = \{t : t = \text{tokenized}(p), p \in \text{leafroot\_paths}(T^n)\}$,

$$\begin{aligned} w^*_{Q,D} &= \max_{m \in T_q, n \in T_d} \nu(G^{(m,n)}) &(3)\\ &= \max_{m \in T_q, n \in T_d} \sum_{t \in \mathfrak{T}(m)} \nu(G^{(m,n)}_t) &(4)\\ &= \max_{m \in T_q, n \in T_d} \sum_{t \in \mathfrak{T}(m)} \min(|Q^{(m)}_t|, |D^{(n)}_t|) &(5) \end{aligned}$$

where $\nu(G)$ is the maximum matching size of bipartite graph G.

Denote $w_{m,t} = |Q^{(m)}_t|$, we call $w_{m,t} \geq \min(|Q^{(m)}_t|, |D^{(n)}_t|)$ as our (precomputed) partial score upperbound. It is analogous to text search where each posting list has a partial score upperbound, the TF-IDF score upperbound is merely their sum. In our case, the sum for partial score upperbounds is only for one node or a subtree.

In the following we propose three strategies to compute $w^*_{Q,D}$ upperbound from partial score upperbounds and assign non-requirement set.

Max Reference (MaxRef) Strategy. In MaxScore [17,19], each posting list has a partial score upperbound, however, our scoring function implies each posting list can be involved with multiple partial score upperbounds. One way to select the non-requirement set in our case is using an upperbound score $MaxRef_t$ (for each posting list t) which is the maximum partial score from the query nodes by which this posting list gets "referenced", and if a set of posting lists alone has a sum of MaxRef scores less or equal to θ, they can be safely put into the non-requirement set.

The rank safety can be justified, since each posting list corresponds to a unique tokenized path t, and $\text{MaxRef}_t = \max_m w_{m,t}$. Then for $m \in T_q, n \in T_d$,

$$\sum_t \min(|Q^{(m)}_t|, |D^{(n)}_t|) \leq \sum_t w_{m,t} \leq \sum_t \text{MaxRef}_t \qquad (6)$$

then the selection of non-requirement set (named **Skip** set for short) such that $\sum_{t \in \textbf{Skip}} \text{MaxRef}_t \leq \theta$ follows $w^*_{Q,D} \leq \theta$ for all non-requirement set posting lists.

Greedy Binary Programming (GBP) Strategies. Inequality (6) is relaxed twice, so it spurs the motivation to get tighter upperbound value by maximizing the number of posting lists in the non-requirement set, so that more posting lists are likely to be skipped. Define partial upperbound matrix $\mathbf{W} = \{w_{i,j}\}_{|T_q| \times |\mathfrak{T}|}$ where $\mathfrak{T} = \{\mathfrak{T}(m), m \in T_q\}$ are all the token paths from query OPT ($\mathfrak{T}$ is essentially the same as tokenized $\mathcal{P}(T_q)$), and a binary variable $\boldsymbol{x}_{|\mathfrak{T}| \times 1}$ indicating which corresponding posting lists are placed in the non-requirement set. One heuristic objective is to maximize the number of posting lists in the non-requirement set (GBP-NUM):

$$\text{maximize} \quad \mathbf{1} \cdot \boldsymbol{x} \tag{7}$$

$$s.t. \quad \mathbf{W}\boldsymbol{x} \leq \theta \tag{8}$$

However, maximizing the number of posting lists in the non-requirement set does not necessarily cause more items to be skipped, because the posting lists can be very short. Instead, we can maximize the total length of posting lists in the non-requirement set. In this case, the vector of ones in objective function (7) is replaced with posting list length vector $\mathbf{L} = \left[L_1, L_2, \ldots L_{|\mathfrak{T}|}\right]$, where L_i is the length of posting list i. We call this strategy GBP-LEN. The two GBP strategies are rank-safe since constraints in inequality (8) implies $\sum_{t \in \mathbf{Skip}} w_{m,t} \leq \theta$.

Both strategies require solving binary programming problems, which are known to be NP-complete and thus too intensive for long queries. Instead, we greedily follow one branch of the binary programming sub-problems to obtain a feasible (but not optimal) solution in $O(|T_q||\mathfrak{T}|^2)$.

5 Implementation

Figure 3 illustrates formula query processing using a modified inverted index for dynamic pruning. For each internal node m of the query OPT, we store the number of leaves of m as $w_m = |Q^{(m)}|$. Each query node points to tokenized path entries in a dictionary, where each reference is associated with $w_{m,t} = |Q_t^{(m)}|$ identified by tokenized path t (denoted as m/w_m of t). In Fig. 3, node $q1$ from the query has 6 leaves, which is also the upperbound number of path matches for $q1$, i.e, $|Q^{(1)}|$. Since $q1$ consists of 2 tokenized leaf-root paths VAR/TIMES/ADD and VAR/ADD, $q1$ is linked to two posting lists, each associated with a partial score upperbound (5 and 1).

Each posting list maps to a token path $t \in \mathfrak{T}$ with a dynamic counter for the number of query nodes referring to it (initially $|Q_t|$). Query nodes are pruned by our algorithm when its subtree width is no longer greater than the current threshold, because the corresponding subexpression cannot be in the top-K results. In this case the reference counter decreases. A posting list is removed if its reference counter is less than one.

Each posting list entry identified by an ExpID stores n and $w_{n,t} = |D_t^{(n)}|$ values of subtree token path t rooted at n (denoted as n/w_n of t). As an example, in Fig. 3, the hit OPT (of ExpID 12) has 5 paths tokenized as

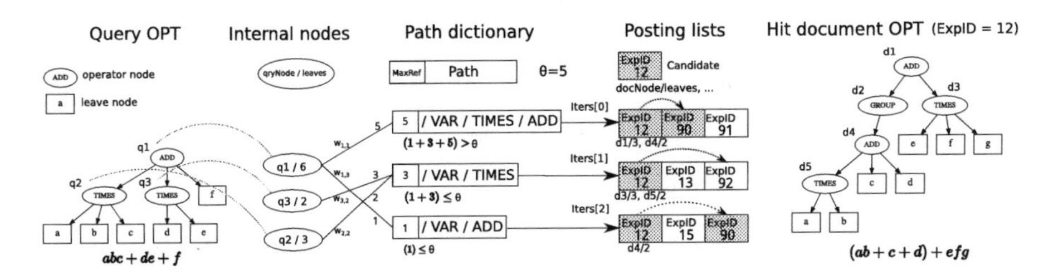

Fig. 3. Indices for formula search with dynamic pruning. For MaxRef strategy, the top posting list is the only one in the requirement set. The bottom two posting lists are advanced by skipping to next candidate ExpID.

$t =$ VAR/TIMES/ADD, 2 rooted at $d4$ and 3 rooted at $d1$. The information $(d1/3, d4/2)$ is stored with corresponding posing list t. In our implementation, each posting list is traversed by an iterator ($iters[t]$), and its entries are read by $iters[t].read()$ from the current position accessed by iterator.

Query processing is described in Algorithm 1. RequirementSet returns selected iterators of the requirement set. Assignment according to different pruning strategies is described in Sect. 4. In the MaxRef strategy, we sort posting lists by descending MaxRef values, and take as many posting lists as possible into non-requirement set from the lowest MaxRef value. At merging, a *candidate* ID is assigned by the minimal ExpID of current posting list iterators in the requirement set. Requirement set iterators are advanced by one using the *next()* function, while iterators in the non-requirement set are advanced directly to the ID equal to or greater than the current candidate by the *skipTo()* function. In Fig. 3 for example, the posting list corresponding to VAR/TIMES/ADD is in the requirement set under the MaxRef strategy, while the other two are not: Document expression 13 and 15 will be skipped if the next candidate is 90. For ease of testing termination, we append a special ExpID *MaxID* at the end of each posting list, which is larger than any ExpID in the collection.

At each iteration, a set of *hitNodes* is inferred containing query nodes associated with posting lists whose current ExpIDs are candidate ID. qryNodeMatch calculates matches for hit nodes according to Eq. 5, pruning nodes whose maximum matching size is smaller than previously examined nodes. Given query hit node $q1$ in Fig. 3, function qryNodeMatch returns

$$\max_{n \in T_d} \nu(G^{(1,n)}) = \max(\min(5,2) + \min(1,2),\ \min(5,3)) = 3$$

Then the algorithm selects the best matched query node and its matched width (i.e., *widest* in Algorithm 1) is our structural similarity $w^*_{Q,D}$.

After obtaining $w^*_{Q,D}$, we compute a metric for the similarity of symbols (e.g., to differentiate $E = mc^2$ and $y = ax^2$) and penalize larger formulas, to produce a final *overall similarity score* [23] for ranking. Because of this additional layer, we need to relax our upperbound further. According to the overall scoring function in [23], our relaxing function u can be defined by assuming perfect symbol similarity score in overall scoring function, specifically

$$u(w) = \frac{w}{|\text{leaves}(T_q)| + w} \left[(1-\eta) + \eta \frac{1}{\log(1+n_d)} \right] \tag{9}$$

where in our setting, parameters $\eta = 0.05, n_d = 1$. Whenever threshold θ is updated, we will examine all the query nodes, if a query node m has an upperbound less or equal to the threshold, i.e., $u(w_m) \leq \theta$, then the corresponding subtree of this node is too "small" to make it into top K results. As a result, some of the posting lists (or iterators) may also be dropped due to zero reference.

Algorithm 1. Formula searching algorithm with pruning

```
1: function QRYNODEMATCH(iters, m, candidate, widest, θ)
2:     nodeMatch[ ] := 0; ℓ := | leaves(m)|          ▷ ℓ is the leftover estimated upperbound.
3:     for each m/w_m of tokenized path t rooted at m do
4:         Let i be the iterator index associated with t
5:         if iters[i].expID < candidate then
6:             iters[i].skipTo(candidate)
7:         if iters[i].expID = candidate then
8:             for each n/w_n of t from iters[i].read() do
9:                 nodeMatch[n] := nodeMatch[n] + min(w_m, w_n)
10:        ℓ := ℓ − w_m;
11:        estimate := max(nodeMatch) + ℓ                              ▷ Update estimation.
12:        if estimate ≤ widest or u(estimate) ≤ θ then
13:            return 0
14:    return max(nodeMatch)
15:
16: function FORMULASEARCH(iters, strategy)
17:    θ := 0; reqs := REQUIREMENTSET(θ, strategy)
18:    heap := data structure to hold top K results
19:    while true do
20:        candidate := minimal ID in current expIDs of reqs
21:        if candidate equals MaxID then              ▷ Search terminated, return results.
22:            return top K results
23:        Let G(Q, D, E) be the hit path set bipartite graph.
24:        widest := 0; hitNodes := {root_end(q) : (q, d) ∈ E}
25:        for m in hitNodes do            ▷ Calculate maximum match for each hit query node.
26:            if | leaves(m)| ≤ widest then
27:                continue
28:            maxMatch := QRYNODEMATCH(iters, m, candidate, widest, θ)
29:            if maxMatch > widest then widest := maxMatch        ▷ Find the widest width.
30:        if widest > 0 then
31:            score := calculate final score (including symbol similarity).        ▷ See [23].
32:            if heap is not full or score > θ then
33:                Push candidate or replace the lowest scored hit in heap.
34:                if heap is full then                          ▷ Update current threshold.
35:                    θ := minimal score in current top K results
36:                    Drop small query nodes and unreferenced iterators.
37:                    reqs := REQUIREMENTSET(θ, strategy)          ▷ Update requirement set.
38:        for iters[i] in reqs do                          ▷ Advance posting list iterators.
39:            if iters[i].expID = candidate then iters[i].next()
```

6 Evaluation

We first evaluate our system[2] on the NTCIR-12 Wikipedia Formula Browsing Task [20] (NTCIR-12 for short), which is the most current benchmark for formula-only retrieval. The dataset contains over 590,000 math expressions taken from English Wikipedia. Since work in formula retrieval is relatively new, there are only 40 queries in NTCIR-12 that can be compared with other published systems. However, these queries are well designed to cover a variety of math expressions in different complexity. There are 20 queries containing wildcards in this task (using wildcard specifier **\qvar** to match arbitrary subexpression or symbols, e.g., query "\qvar{a}^{2} + \qvar{b}^{3}" can match "$x^2 + (y+1)^3$"). We add support for wildcards by simply treating internal nodes (representing a rooted subexpression) of formulas as additional "leaves" (by ignoring their descendants), and the wildcard specifiers in a query are treated as normal leaves to match those indexed wildcard paths.

Since the corpus of NTCIR-12 is not large enough to show the full impact of pruning, we also evaluate query run times on a corpus containing over 1 million math related documents/threads from Math StackExchange (MSE) Q&A website[3] and we run the same query set from NTCIR-12. Run times are shown for the posting list merging stage (e.g., time for parsing the query into OPT is excluded) and unless specified, posting lists are compressed and cached into memory. Each system had five independent runs, and we report results from overall distribution. The resulting uncompressed index size for NTCIR-12 and MSE corpora are around 2 GB and 16 GB in size, with 961,604 and 5,764,326 posting lists respectively. The (min, max, mean, standard deviation) for posting list lengths are (1, 262309, 16.95, 737.84) and (1, 7916296, 73.74, 9736.72).

Table 1 reports run time statistics. Non-pruning (exhaustive search) baselines with K = 100 are also compared here. Almost consistently, GBP-LEN strategy achieves the best efficiency with smaller variance. This is expected since GBP-LEN models the skipping possibility better than GBP-NUM. Although GBP-NUM gives a tighter theoretic upperbound than MaxRef, it only maximizes the number of posting lists in the non-requirement set and may lead to bad performance when these posting lists are short.

There are a few times the best minimal run times are from other strategies, for those with meaningful gaps, i.e., in Wiki dataset of non-wildcard queries when K = 1000, MaxRef outperforms in standard deviation and maximum run time to a notable margin; however, it likely results from a small threshold due to large K, so that the efficiency on the small sized NTCIR dataset is less affected by pruning (small θ means less pruning potential) compared to the time complexity added from assigning to the requirement set. The latter is more dominant in GBP runs. In wildcard queries, however, many expressions can match the query thus the threshold value is expected to be larger than that in the non-wildcard case.

[2] Source code: https://github.com/approach0/search-engine/tree/ecir2020.

[3] MSE corpus: https://www.cs.rit.edu/~dprl/data/mse-corpus.tar.gz.

Table 1. Query merge time performance (in milliseconds) for different strategies.

Dataset	Runs K	Strategy	Non-wildcards μ	σ	median	min	max	Wildcards μ	σ	median	min	max
Wiki Dataset	100	Baseline	540.12	569.44	360.50	7.00	2238.00	426.73	383.47	225.50	8.00	1338.00
	100	MaxRef	90.29	74.14	79.00	3.00	312.00	145.50	121.19	136.00	**7.00**	573.00
		GBP-NUM	84.90	80.44	52.50	3.00	321.00	138.82	102.55	135.00	9.00	428.00
		GBP-LEN	**67.49**	**61.40**	**45.00**	**2.00**	**218.00**	**125.27**	**97.28**	**103.50**	9.00	**404.00**
	200	MaxRef	107.71	**82.64**	102.00	**5.00**	**322.00**	160.10	121.40	149.00	9.00	583.00
		GBP-NUM	105.34	99.51	71.50	**5.00**	357.00	155.52	110.61	153.00	**8.00**	479.00
		GBP-LEN	**89.63**	83.20	**62.00**	**5.00**	330.00	**142.78**	**103.11**	**143.50**	9.00	**446.00**
	1000	MaxRef	154.51	**93.75**	157.50	**6.00**	**361.00**	211.86	140.01	186.00	10.00	662.00
		GBP-NUM	159.80	143.70	120.50	**6.00**	626.00	208.91	136.42	178.50	10.00	591.00
		GBP-LEN	**144.25**	126.95	**105.00**	**6.00**	622.00	**195.70**	**122.25**	**176.00**	**9.00**	**536.00**
MSE Dataset	100	Baseline	15134.10	15186.78	11161.00	157.00	55499.00	13450.57	12554.19	7075.50	304.00	47513.00
	100	MaxRef	1083.23	1274.23	745.50	28.00	5922.00	3188.66	2458.91	2925.00	85.00	10412.00
		GBP-NUM	1202.24	1240.21	815.00	37.00	4987.00	2943.79	2025.96	2987.00	**84.00**	8775.00
		GBP-LEN	**562.83**	**635.26**	**382.50**	**24.00**	**2313.00**	**2257.95**	**1491.59**	**2346.50**	86.00	**4494.00**
	200	MaxRef	1261.21	1368.93	1012.50	30.00	6439.00	3416.77	2753.09	3032.50	160.00	12412.00
		GBP-NUM	1378.19	1398.08	998.50	39.00	5863.00	3174.93	2283.05	3125.00	**159.00**	10099.00
		GBP-LEN	**697.32**	**739.11**	**478.00**	**27.00**	**2925.00**	**2504.90**	**1683.16**	**2382.50**	**159.00**	**6049.00**
	1000	MaxRef	2030.05	1746.17	1796.50	53.00	7816.00	4123.26	3510.01	3473.00	287.00	16981.00
		GBP-NUM	1952.52	1746.05	1530.50	60.00	7197.00	3786.89	2744.99	3493.50	**281.00**	11323.00
		GBP-LEN	**1217.16**	**1083.53**	**764.50**	**47.00**	**3756.00**	**3304.69**	**2403.09**	**2812.00**	285.00	**9895.00**

System	Non-Wildcard Full	Non-Wildcard Partial	Wildcard Full	Wildcard Partial	All queries Full	All queries Partial
MCAT	.5678	.5698	**.4725**	.5015	.5202	.5356
Tangent-S	.6361	.5872	.4699	**.5368**	**.5530**	**.5620**
base-best	**.6726**	**.5950**	-	-	-	-
base-opd-only	.6586	.5153	-	-	-	-
Ours (pruning)	.6586	.5173	.3678	.3973	.5132	.4573
Ours (exhaustive)	.6586	.5173	.3678	.3973	.5132	.4573

Fig. 4. Bpref [4] scores. Bpref chosen because we did not participate in NTCIR-12 and did not contribute to the pooling.

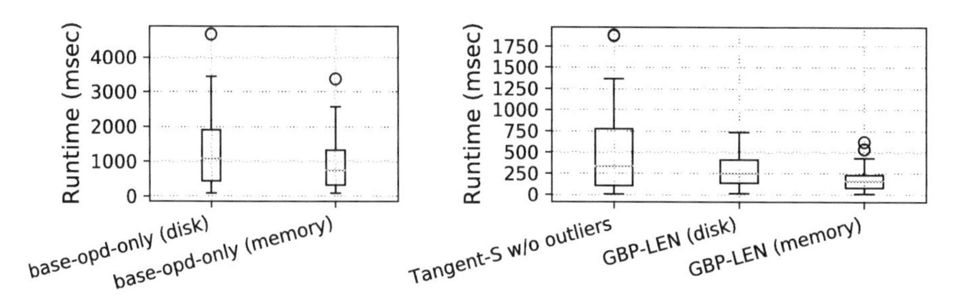

Fig. 5. Average run times on the same machine (Environment: Intel Core i5 @ 3.60 GHz per core, 16 GB memory and SSD drive) for NTCIR-12 Wiki Formula Browsing Task.

Secondly, we have compared our system effectiveness (Fig. 4) and efficiency (Fig. 5) with Tangent-S [5], MCAT [11] and our baseline system without pruning [23], which are all structure-based formula search engines that have obtained the best published Bpref scores on NTCIR-12 dataset. In addition, ICST system [7] also obtains effective results for math and text mixed task, but they do training on previous Wiki dataset and their system is currently not available.

All systems are evaluated in a single thread for top-1000 results. We use our best performance strategy, i.e., GBP-LEN, having an on-disk version with posting lists uncompressed and always read from disk, and an in-memory version with compression. For the baseline system, only 20 non-wildcard queries are reported because it does not support wildcards. We compare the baseline best performed run (base-best) which uses costly multiple tree matching as well as its specialized version (base-opd-only) which considers only the largest matched tree width (see Eq. 2). Tangent-S has a few outliers as a result of its costly alignment algorithm to rerank structure and find the Maximum Subtree Similarity [22], its non-linear complexity makes it expensive for some long queries (especially in wildcard case). And MCAT reportedly has a median query execution time around 25 s, using a server machine and multi-threading [11]. So we remove Tangent-S outliers and MCAT from runtime boxplot. For space, we only include the faster base-opd-only baseline in Fig. 5.

We outperform Tangent-S in efficiency even if we exclude their outlier queries, with higher Bpref in non-wildcard fully relevant results. Our efficiency is also better than the baseline system, even if the latter only considers less complex non-wildcard queries. However, our overall effectiveness is skewed by bad performance of wildcard queries because a much more expensive phase is introduced to boost accuracy by other systems to handle inherently difficult "structural wildcards."

Our pruning strategies are rank-safe (pruning and exhaustive version shows the same Bpref scores) but there is a minor Bpref difference between ours and baseline (base-opd-only) due to parser changes we have applied to support wildcards (e.g., handle single left brace array as seen in a wildcard query) and they happen to slightly improve accuracy in partially relevant cases.

7 Conclusion

We have presented rank-safe dynamic pruning strategies that produce an upper-bound estimation of structural similarity in order to speedup formula search using subtree matching. Our dynamic pruning strategies and specialized inverted index are different from traditional linear text search pruning methods and they further associate query structure representation with posting lists. Our results show we can obtain substantial improvement in efficiency over the baseline model, while still generating highly relevant non-wildcard search results. Our approach can process a diverse set of structural queries in real time.

References

1. Anh, V.N., Moffat, A.: Pruned query evaluation using pre-computed impacts. In: Proceedings of the 29th Annual International ACM SIGIR Conference on Research and Development in Information Retrieval, pp. 372–379. ACM (2006)

2. Mallia, A., Siedlaczek, M., Suel, T.: An experimental study of index compression and DAAT query processing methods. In: Azzopardi, L., Stein, B., Fuhr, N., Mayr, P., Hauff, C., Hiemstra, D. (eds.) ECIR 2019. LNCS, vol. 11437, pp. 353–368. Springer, Cham (2019). https://doi.org/10.1007/978-3-030-15712-8_23
3. Broder, A.Z., Carmel, D., Herscovici, M., Soffer, A., Zien, J.: Efficient query evaluation using a two-level retrieval process. In: Proceedings of the Twelfth International Conference on Information and Knowledge Management, pp. 426–434. ACM (2003)
4. Buckley, C., Voorhees, E.M.: Retrieval evaluation with incomplete information. In: Proceedings of the 27th Annual International ACM SIGIR Conference on Research and Development in Information Retrieval, pp. 25–32. ACM (2004)
5. Davila, K., Zanibbi, R.: Layout and semantics: combining representations for mathematical formula search. In: Proceedings of the 40th International ACM SIGIR Conference on Research and Development in Information Retrieval, pp. 1165–1168. ACM (2017)
6. Ding, S., Suel, T.: Faster top-k document retrieval using block-max indexes. In: Proceedings of the 34th International ACM SIGIR Conference on Research and Development in Information Retrieval, pp. 993–1002. ACM (2011)
7. Gao, L., Yuan, K., Wang, Y., Jiang, Z., Tang, Z.: The math retrieval system of ICST for NTCIR-12 MathIR task. In: NTCIR (2016)
8. Jonassen, S., Bratsberg, S.E.: Efficient compressed inverted index skipping for disjunctive text-queries. In: Clough, P., et al. (eds.) ECIR 2011. LNCS, vol. 6611, pp. 530–542. Springer, Heidelberg (2011). https://doi.org/10.1007/978-3-642-20161-5_53
9. Kamali, S., Tompa, F.W.: Structural similarity search for mathematics retrieval. In: Carette, J., Aspinall, D., Lange, C., Sojka, P., Windsteiger, W. (eds.) CICM 2013. LNCS (LNAI), vol. 7961, pp. 246–262. Springer, Heidelberg (2013). https://doi.org/10.1007/978-3-642-39320-4_16
10. Davila, K., Joshi, R., Setlur, S., Govindaraju, V., Zanibbi, R.: Tangent-V: math formula image search using line-of-sight graphs. In: Azzopardi, L., Stein, B., Fuhr, N., Mayr, P., Hauff, C., Hiemstra, D. (eds.) ECIR 2019. LNCS, vol. 11437, pp. 681–695. Springer, Cham (2019). https://doi.org/10.1007/978-3-030-15712-8_44
11. Kristianto, G.Y., Topic, G., Aizawa, A.: MCAT math retrieval system for NTCIR-12 MathIR task. In: NTCIR (2016)
12. Lin, X., Gao, L., Hu, X., Tang, Z., Xiao, Y., Liu, X.: A mathematics retrieval system for formulae in layout presentations. In: Proceedings of the 37th International ACM SIGIR Conference on Research and Development in Information Retrieval, SIGIR 2014. ACM, New York (2014)
13. Macdonald, C., Ounis, I., Tonellotto, N.: Upper-bound approximations for dynamic pruning. ACM Trans. Inf. Syst. (TOIS) **29**(4), 17 (2011)
14. Miller, B.R., Youssef, A.: Technical aspects of the digital library of mathematical functions. Ann. Math. Artif. Intell. **38**(1–3), 121–136 (2003). https://doi.org/10.1023/A:1022967814992
15. Shan, D., Ding, S., He, J., Yan, H., Li, X.: Optimized top-k processing with global page scores on block-max indexes. In: Proceedings of the Fifth ACM International Conference on Web Search and Data Mining, WSDM 2012, pp. 423–432. ACM, New York (2012)
16. Sojka, P., Líška, M.: Indexing and searching mathematics in digital libraries. In: Davenport, J.H., Farmer, W.M., Urban, J., Rabe, F. (eds.) CICM 2011. LNCS (LNAI), vol. 6824, pp. 228–243. Springer, Heidelberg (2011). https://doi.org/10.1007/978-3-642-22673-1_16

17. Strohman, T., Turtle, H., Croft, W.B.: Optimization strategies for complex queries. In: Proceedings of the 28th Annual International ACM SIGIR Conference on Research and Development in Information Retrieval, pp. 219–225. ACM (2005)
18. Tonellotto, N., Macdonald, C., Ounis, I., et al.: Efficient query processing for scalable web search. Found. Trends Inf. Retr. **12**(4–5), 319–500 (2018)
19. Turtle, H., Flood, J.: Query evaluation: strategies and optimizations. Inf. Process. Manag. **31**(6), 831–850 (1995)
20. Zanibbi, R., Aizawa, A., Kohlhase, M., Ounis, I., Topic, G., Davila, K.: NTCIR-12 MathIR task overview. In: NTCIR (2016)
21. Zanibbi, R., Blostein, D.: Recognition and retrieval of mathematical expressions. Int. J. Doc. Anal. Recogn. **15**(4), 331–357 (2012)
22. Zanibbi, R., Davila, K., Kane, A., Tompa, F.W.: Multi-stage math formula search: using appearance-based similarity metrics at scale. In: Proceedings of the 39th International ACM SIGIR Conference on Research and Development in Information Retrieval, SIGIR 2016. ACM, New York (2016)
23. Zhong, W., Zanibbi, R.: Structural similarity search for formulas using leaf-root paths in operator subtrees. In: Azzopardi, L., Stein, B., Fuhr, N., Mayr, P., Hauff, C., Hiemstra, D. (eds.) ECIR 2019. LNCS, vol. 11437, pp. 116–129. Springer, Cham (2019). https://doi.org/10.1007/978-3-030-15712-8_8

Quantum-Like Structure in Multidimensional Relevance Judgements

Sagar Uprety[1(✉)], Prayag Tiwari[2], Shahram Dehdashti[3], Lauren Fell[3], Dawei Song[1,4], Peter Bruza[3], and Massimo Melucci[2]

[1] The Open University, Milton Keynes, UK
sagar.uprety@open.ac.uk
[2] University of Padova, Padua, Italy
[3] Queensland University of Technology, Brisbane, Australia
[4] Beijing Institute of Technology, Beijing, China

Abstract. A large number of studies in cognitive science have revealed that probabilistic outcomes of certain human decisions do not agree with the axioms of classical probability theory. The field of Quantum Cognition provides an alternative probabilistic model to explain such paradoxical findings. It posits that cognitive systems have an underlying quantum-like structure, especially in decision-making under uncertainty. In this paper, we hypothesise that relevance judgement, being a multi-dimensional, cognitive concept, can be used to probe the quantum-like structure for modelling users' cognitive states in information seeking. Extending from an experiment protocol inspired by the Stern-Gerlach experiment in Quantum Physics, we design a crowd-sourced user study to show violation of the Kolmogorovian probability axioms as a proof of the quantum-like structure, and provide a comparison between a quantum probabilistic model and a Bayesian model for predictions of relevance.

Keywords: Multidimensional relevance · User behaviour · Quantum Cognition

1 Introduction

Relevance in Information Retrieval (IR) is widely accepted to be a cognitive feature, driving all our information interactions. All areas of research within IR thus strive to improve relevance of documents to a user's information need (IN). These research areas of IR can be broadly divided into two: system-oriented and user-oriented IR. Whereas the system-oriented viewpoint ties relevance to be an objective property of the document and query content, the user-oriented approach to IR views relevance as a cognitive property. Although IR fundamentally involves user interaction and decision-making, the user-oriented approach has been found harder to implement, especially in evaluating performance of IR systems. This is because of the variability in user judgements of relevance [5].

J. M. Jose et al. (Eds.): ECIR 2020, LNCS 12035, pp. 728–742, 2020.
https://doi.org/10.1007/978-3-030-45439-5_48

System-oriented IR thus sought to standardise IR evaluation, in which the user-cognitive notion of relevance was replaced by an objective, topical relevance. This led to evaluation methodologies based on the Cranfield and TREC type test collections. The user and all of his/her contexts were removed from the evaluation process.

Recent surge in availability of online user data has led to incorporation of more user context in the computation of relevance, e.g. in learning based ranking algorithms. This context is based on the user's past interactions with the system, in addition to user attributes like age, interests, etc. and current attributes like location, type of device, etc. The common feature in these various contexts is that they are static. They are determined before the point of user's interaction with the IR system. However, the process of IR is interactive and dynamic. In this paper, we focus on another type of context driving user interactions - dynamic context. Dynamic context is one which changes user's cognitive state *during* information interaction.

One well-known example of when a dynamic context affects relevance is the phenomenon of Order Effect [8]. Order effects have been investigated and found to exist in IR in the presentation order of documents [4,6,9,24]. For example, in a recent study reported in [22], two groups of participants were presented with a pair of documents D_1 and D_2 in two different orders. For some of such pairs, it was found that the relevance of a document judged by users is different depending on the order it was presented. Although the phenomenon may appear to have an intuitive explanation, it violates one of the fundamental assumptions of classical probability theory - joint distributions, where, for two random variables representing relevance of the documents - R_1, R_2, $P(R_1, R_2) = P(R_2, R_1)$, i.e., the order of judging the documents does not matter. Order effects violate this fundamental assumption. Such order effects have also been investigated and reported in between the different dimensions of relevance, like Topicality, Understandability, Reliability, etc. [1,19,20], where different orders of dimensions considered to judge a document lead to different relevance judgements.

The field of Quantum Cognition [2] offers a generalised framework to model probabilistic outcomes of human decision-making. It has been successful in modelling and predicting order effects [16,23] and other paradoxical findings where axioms of classical probability theory are violated [3,14]. Conceptually, it challenges the notion that cognitive states have pre-defined values and that a measurement merely records them. Instead, the act of measurement creates a definite state out of an indefinite state and in doing so, changes the initial state of the cognitive system. In terms of relevance, we cannot pre-assign relevance of a document for a user. Instead, relevance is defined only at the point of interaction of the user's cognitive state with the document. Therefore, judgement of document D_2 first, changes user's initial state and the subsequent judgement of relevance of D_1 is different than when D_1 is judged before D_2. Should relevance of the documents for a user be a pre-defined entity, it would not be influenced by judgement of other documents and a joint distribution over relevance of the two documents would exist. We also say that these two measurements of relevance are incompatible with each other. That is, it is not possible to jointly consider the relevance of the two documents, at the same time. At the mathematical level, measurements in quantum theory are represented by operators, which in general, do not commute with each other.

In a classical system, all measurements will commute with each other. However, conversely, commutativity of measurements does not necessarily imply that the system is classical. Therefore, the type of measurements becomes imperative in identifying a quantum system. Even then, not all measurements on quantum systems generate data violating the classical probability theory. The system needs to be probed in a way which exploits the underlying quantum structure. In physics, this was done by experiments such as Stern-Gerlach and double-slit experiments [15] which showed the violation of classical probability principles for microscopic particles like electrons and photons. In cognitive science too, several experiments performed by Tversky, Kahneman and colleagues showed such violations in human decision-making under uncertainty [17].

Recently, an experiment protocol inspired by the Stern-Gerlach experiment in Physics has provided a new way to probe cognitive systems such that they exhibit a quantum-like structure [7]. By quantum-like structure we mean the representation of a system using the mathematical framework of quantum theory in order to model and predict the experimental data. In [19], this experiment was performed in an IR scenario involving judgement of relevance with respect to different dimensions. Extending from the Stern-Gerlach protocol, in this paper we design a new experiment to show the violation of classical probability theory in multidimensional relevance judgements. We hypothesise that multidimensional relevance judgement has an underlying quantum-like structure, which when subject to appropriate measurement design can exhibit violations of classical probability theory. Specifically, we investigate the violation of a particular axiom of Kolmogorovian probability theory [11]. Our results show that the experimental data indeed violates classical probability theory, and a quantum framework provides more accurate predictions to describe the data. This experiment not only shows the necessity of the quantum framework as an alternative for constructing probabilistic models, but also gives novel insights into user behaviour in IR. This understanding can contribute to improvement of interactive IR systems and we also discuss such implications in this paper.

2 Stern-Gerlach Inspired Protocol for Multidimensional Relevance

The basis of the research reported in this paper is the cognitive analogue of the Stern-Gerlach (S-G) experiment, originally conducted in [19]. The S-G experiment [15] was an important milestone in quantum physics as it showed the non-classical behaviour of microscopic systems. The key was a particular design of the experiment which exploited the incompatibility between measurement of electron spin states along different axes. An electron has a particular property

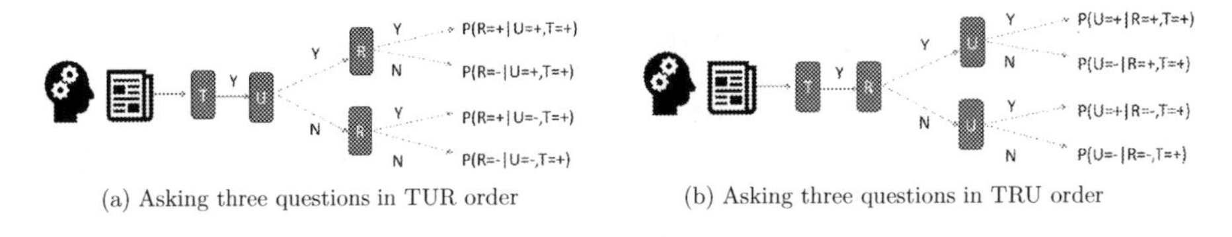

(a) Asking three questions in TUR order

(b) Asking three questions in TRU order

Fig. 1. S-G type experiment to construct a complex-valued Hilbert space

called spin, having two possible values - up (+), down (-), which can be measured along different axes. An electron may have spin state + along the x-axis but state − along y-axis. So the outcome of measurement of the spin property of the electron depends upon the axis of measurement. Also, any measurement of spin disturbs the system. If a measurement of spin is made along X axis and Z axis, then a third measurement along X axis may give a different answer than the first one. This phenomena is called measurement incompatibility, where two measurements cannot be jointly conducted on a system - one measurement disturbs the system and the other would then measure the changed system.

The S-G experiment also describes the minimum number of measurements required from a system to construct a complex-valued Hilbert Space structure. In particular, we need three incompatible measurements each with two mutually exclusive outcomes. We can use this arrangement of measuring properties of a quantum system to measure relevance of a document in IR. For this, we consider three dimensions of relevance: Topicality (T) - whether a document is topically relevant to a query, Understandability (U) - how easy it is to understand the content of the document, and Reliability (R) - how much can the document be relied upon. Each of these three dimensions can be posed as questions requiring a Yes/No type answer (denoted as + and − respectively) for a document. These three dimensions are important factors considered by users for deciding relevance. Besides, they are tied to a single document, unlike diversity or novelty, which is always considered in comparison with other documents. Certain dimensions like Interest, Habit, etc. are difficult to ascertain via crowdsourcing. As reported in [1], the different relevance dimensions can exhibit incompatibility for certain query-document pairs.

In [19], three query-document pairs were designed in such a way as to potentially exhibit incompatibility between judgement of relevance with respect to different dimensions. The content of the documents was altered to introduce uncertainty in judging each of the three dimensions. The participants were presented with three questions related to three relevance dimensions, for each query-document pair, in line with the S-G design. Figure 1 shows the three questions asked to two different groups in different orders. More details about this design can be found in [19] and [7]. This setup enables one to construct a complex-valued Hilbert space, which models the quantum-like structure of the user's cognitive state during information interaction.

2.1 Constructing Complex-Valued Hilbert Space

The first step in building a quantum probabilistic model is to construct a representation for the user's cognitive state. In the quantum framework, a complex-valued Hilbert space is used to represent a quantum system, and the state of the system is represented as a vector in this Hilbert space.

Following the convention used in Quantum Physics, we represent any complex-valued vector A in a finite dimensional Hilbert space as a ket vector $|A\rangle$ and its complex conjugate as a bra vector $\langle A|$. The norm of this vector is the square root of its inner product with its conjugate - $|\langle A|A\rangle|^{1/2}$. For two such vectors, their projection onto each other is given as the square of their inner product - $|\langle A|B\rangle|^2$. Each vector is written as a linear combination of the vectors of the basis in which it is represented. For the purpose of representing the cognitive state of a person judging a document as topically relevant or topically irrelevant, we consider a basis formed by two orthogonal vectors $|T+\rangle$ and $|T-\rangle$ respectively. Before a user considers a judgement of topicality, the cognitive state is indefinite with respect to considering the document as topically relevant or irrelevant. Both potentialities exist. We say that the cognitive state collapses to either $|T+\rangle$ or $|T-\rangle$ after the judgement. Before the judgement, we can represent the indefinite cognitive state in terms of probabilities of its potential responses. This is represented as a linear combination of the two basis states, weighted each by real or complex coefficients (called probability amplitudes), such that the square of the probability amplitude gives the probability of collapsing to the respective state. The initial state S is thus written as:

$$|S\rangle = t\,|T+\rangle + \sqrt{1-t^2}\,|T-\rangle \tag{1}$$

Query	P(T)	P(U|T)	P(R|U,T)	P(R|T)	P(U|R,T)
Query 1	P(T=+) = 0.762	P(U = +|T+) = 0.578	P(R = +|U=+,T=+) = 0.587	P(R = +|T+) = 0.546	P(U = +|R=+,T=+) = 0.600
			P(R = -|U=+,T=+) = 0.413		P(U = -|R=+,T=+) = 0.400
		P(U = -|T+) = 0.422	P(R = +|U=-,T=+) = 0.370	P(R = -|T+) = 0.454	P(U = +|R=-,T=+) = 0.407
			P(R = -|U=-,T=+) = 0.630		P(U = -|R=-,T=+) = 0.593
Query 2	P(T=+) = 0.671	P(U = +|T+) = 0.802	P(R = +|U=+,T=+) = 0.844	P(R = +|T+) = 0.731	P(U = +|R=+,T=+) = 0.882
			P(R = -|U=+,T=+) = 0.156		P(U = -|R=+,T=+) = 0.118
		P(U = -|T+) = 0.198	P(R = +|U=-,T=+) = 0.526	P(R = -|T+) = 0.269	P(U = +|R=-,T=+) = 0.480
			P(R = -|U=-,T=+) = 0.474		P(U = -|R=-,T=+) = 0.520
Query 3	P(T=+) = 0.899	P(U = +|T+) = 0.977	P(R = +|U=+,T=+) = 0.738	P(R = +|T+) = 0.646	P(U = +|R=+,T=+) = 0.963
			P(R = -|U=+,T=+) = 0.262		P(U = -|R=+,T=+) = 0.037
		P(U = -|T+) = 0.023	P(R = +|U=-,T=+) = 0.000	P(R = -|T+) = 0.354	P(U = +|R=-,T=+) = 0.889
			P(R = -|U=-,T=+) = 1.000		P(U = -|R=-,T=+) = 0.111

Fig. 2. Probabilities for the questions of TUR and TRU for the three queries

In the S-G inspired experiment design, we ask the user sequential questions about judgement of Topicality (T), Understandability (U) and Reliability (R) in the order TUR or TRU, as shown in Fig. 1. Therefore we represent the cognitive state w.r.t Understandability and Reliability in term of Topicality:

$$|U+\rangle = u\,|T+\rangle + \sqrt{1-u^2}\,|T-\rangle\,, |U-\rangle = \sqrt{1-u^2}\,|T+\rangle - u\,|T-\rangle \tag{2}$$

$|U-\rangle$ is constructed using the fact that $|U+\rangle$ and $|U-\rangle$ are orthogonal. u^2 is the probability that users judge a document Understandable, given that they have judged it as Topically relevant.

Refer to [19] (Section 3) or [15] (Chapter 1) for the necessity of using a complex-valued probability amplitude in the representation of Reliability in term of Topicality:

$$|R+\rangle = r\,|T+\rangle + \sqrt{1-r^2}e^{i\theta_r}\,|T-\rangle\,, |R-\rangle = \sqrt{1-r^2}e^{-i\theta_r}\,|T+\rangle - r\,|T-\rangle \tag{3}$$

The parameters (u, r and θ_r) comprise the construction of the Hilbert space for user's cognitive state w.r.t the interaction between the three dimensions. The parameter t defines the initial state. The experiment design of Fig. 1 was carried out in [19] for three queries. The results are listed in Fig. 2.

3 Formulation of Research Hypotheses

Using the complex-valued Hilbert Space of multidimensional relevance, this paper aims to design an extended experiment to test the following research hypotheses: (1) Fundamental axioms of classical Kolmogorov probability are violated in a multidimensional relevance judgement scenario; (2) Probabilities obtained from the experiment can be better predicted with quantum than classical (Bayesian) probabilistic models. In the following two subsections, we mathematically formulate these hypotheses.

3.1 Violation of Kolmogorov Probability and Quantum Correction

Quantum probabilities are generalisation of Kolmogorov probabilities. In fact, Kolmogorov probabilities are related to set theory which formalises Boolean logic. The following proposition gives one of their fundamental properties [11]:

$$0 = \delta = P(A \vee B) - P(A) - P(B) + P(A \wedge B) \tag{4}$$

where A, B are subsets of the set of all alternatives Ω, and $P(A)$, $P(B)$ are the corresponding probabilities. The axiom will be violated if the value of δ is different from zero.

In the quantum probability theory, the computation of probabilities are represented by projection operators for the events $U\pm$ and $R\pm$ corresponding to relevance or non-relevance with respect to Understandability and Reliability.

The analogue of relation (4) in quantum mechanics is given by the following definition [21]:

$$\mathcal{D}(U\pm, R\pm) = \Pi(U \pm \vee R\pm) - \Pi(U\pm) - \Pi(R\pm) + \Pi(U \pm \wedge R\pm) \quad (5)$$

where projection operators $\Pi(U\pm)$ and $\Pi(R\pm)$ are given by:

$$\Pi(U\pm) = |U\pm\rangle \langle U\pm| , \qquad \Pi(R\pm) = |R\pm\rangle \langle R\pm| \quad (6)$$

It is possible to prove that this quantum correction term $\mathcal{D}(U\pm, R\pm)$ is proportional to the commutator of the projection operators of $U\pm$ and $R\pm$ [21] and can be thus obtained as:

$$\mathcal{D}(U\pm, R\pm) = \left[\Pi(U\pm), \Pi(R\pm)\right] \left(\Pi(U\pm) - \Pi(R\pm)\right)^{-1} \quad (7)$$

where $[A, B]$ stands for the commutator for two operators A and B. The projection operator $\Pi(U+)$ is equal to the outer product of the state $|U+\rangle$ with itself, where the vector $|U+\rangle$ is computed using Eq. 2. In order to construct the vector, first the Topicality basis is represented as the standard basis and hence the orthogonal vectors $|T+\rangle$ and $|T-\rangle$ are given as:

$$|T+\rangle = \begin{pmatrix} 1 \\ 0 \end{pmatrix}, \quad |T-\rangle = \begin{pmatrix} 0 \\ 1 \end{pmatrix} \quad (8)$$

Thus, vectors $|U+\rangle$ and $|U-\rangle$ are given as:

$$|U+\rangle = \begin{pmatrix} u \\ \sqrt{1-u^2} \end{pmatrix}, \quad |U-\rangle = \begin{pmatrix} \sqrt{1-u^2} \\ -u \end{pmatrix} \quad (9)$$

Then the projector $\Pi(U+)$ is given as:

$$\Pi(U+) = |U+\rangle \langle U+| = \begin{pmatrix} u \\ \sqrt{1-u^2} \end{pmatrix} \begin{pmatrix} u & \sqrt{1-u^2} \end{pmatrix} = \begin{pmatrix} u^2 & u\sqrt{1-u^2} \\ u\sqrt{1-u^2} & 1-u^2 \end{pmatrix}$$

Similarly, $\Pi(R+)$ is:

$$\Pi(R+) = |R+\rangle \langle R+| = \begin{pmatrix} r \\ \sqrt{1-r^2}e^{i\theta_r} \end{pmatrix} \begin{pmatrix} r & \sqrt{1-r^2}e^{-i\theta_r} \end{pmatrix}$$

$$= \begin{pmatrix} r^2 & r\sqrt{1-r^2}e^{-i\theta_r} \\ r\sqrt{1-r^2}e^{i\theta_r} & 1-r^2 \end{pmatrix}$$

From the values of u, r and θ_r obtained in [19], these projection operators can be constructed. The quantum analogue of δ, can then be calculated from Eq. (7). Value of δ obtained from our experiment is compared to that predicted by the classical (always zero) and quantum probability frameworks.

3.2 Quantum Probabilities vs Classical Probabilities

The violation of Kolmogorovian probability axiom by a given system would likely lead to inaccurate predictions on the system using Kolmogorovian probability. This subsection formulates computation of conditional probabilities of relevance judgement along one dimension given another, using classical vs. quantum frameworks. They will be compared for our experimental data in Sect. 5.

For an initial state of the system $|S\rangle$, the probability of event $|T+\rangle$ in the quantum framework is given by $P(T+) = |\langle T+|S\rangle|^2 = t^2$, i.e., square of projection of vector $|S\rangle$ onto vector $|T+\rangle$. The probability for sequence $U+$ following $T+$ is given as [2]:

$$P(U+,T+) = |\langle U+|T+\rangle|^2 |\langle T+|S\rangle|^2 \tag{10}$$

The quantum framework does not define joint probability of events T and U, as in general $P(T+,U+) \neq P(U+,T+)$. As we can see $P(T+,U+) = |\langle T+|U+\rangle|^2 |\langle U+|S\rangle|^2$, which for $\langle U+|S\rangle \neq \langle T+|S\rangle$ is not equal to $P(U+,T+)$ in Eq. 10. The conditional probabilities are given according to Luder's rule [2,10] as:

$$\begin{aligned} P_q(U+|T+) &= P(U+,T+)/P(T+|S) \\ &= \frac{|\langle U+|T+\rangle|^2 |\langle T+|S\rangle|^2}{|\langle T+|S\rangle|^2} \\ &= |\langle U+|T+\rangle|^2 = u^2 \end{aligned} \tag{11}$$

Note that subscript q is added to distinguish from classical conditional probability. Then $P_q(R+|U+,T+)$ is given as (see [19] Sect. 4.2 for derivation):

$$\begin{aligned} P_q(R+|U+,T+) &= |\langle R+|U+\rangle|^2 \\ &= (ur)^2 + (1-u^2)(1-r^2) + 2ur\sqrt{(1-u^2)(1-r^2)}\cos\theta_r \end{aligned} \tag{12}$$

In contrast, classical probability theory has the basic assumption of commutativity of two events. Therefore the joint probability distribution always exists, which is the basis of calculating conditional probabilities in Bayes' rule. Consequently, for events T, U and R we have:

$$P(U+,R+,T+) = P(R+,U+,T+) \tag{13}$$

which can be written in terms of conditional probabilities as:

$$P(T+)P(R+|T+)P(U+|R+,T+) = P(T+)P(U+|T+)P(R+|U+,T+) \tag{14}$$

This enables calculation of conditional probabilities using the Bayes rule:

$$P(R+|U+,T+) = \frac{P(U+|R+,T+)P(R+|T+)}{P(U+|T+)} \tag{15}$$

Similarly, the other conditional probabilities can be obtained. Again, note that the probabilities in Eqs. (15) and (12) are different because of the difference in the underlying assumption of commutativity or joint probability.

4 Experiment

4.1 Methodology

The main aim of this experiment is to investigate the violation of Eq. 4. We already have the single question probabilities from the experiment in [19] and we need to obtain the probabilities of conjunction and disjunction. We do so by posing questions about Understandability and Reliability at the same time, as a pair, rather than sequentially. Each of the dimensions have two outcomes (e.g. Reliable or Not Reliable) and therefore we construct four pairs of statements, as listed in Fig. 4. For the disjunction measurement, we ask the participants to select whether they agree with at least one of the two statements or none of them (corresponding to a Boolean Or condition). For a conjunction measurement on each of the four statement pairs, we ask the participants whether they agree with both of the questions or not. Figure 5a and b show the designs for the disjunction and conjunction questions for a query-document pair. We now have a total of eight such questions and we follow a between-subjects design such that a participant is shown only one of these eight questions randomly. Note that we are able to use the probabilities from the experiment in [19] because our experiment is a between-subjects design. The same participant is not asked all the questions - to avoid memory bias. The design is summarised in the following steps for each of the three query-document pairs:

1. The participants are shown information need, query and document snippet.
2. Next, they are asked a Yes/No question about the Topicality of the document. This is to prepare the cognitive state of all participants by projecting their initial/background state onto the Topicality subspace of the underlying Hilbert space constructed in the previous experiment in [19].
3. Lastly, they are randomly shown one of the eight possible conjunction or disjunction questions and asked to choose the appropriate answer (Fig. 3).

4.2 Participants and Material

We recruited 335 participants for the experiment using the online crowd-sourcing platform Prolific (prolific.ac). The study was designed using the survey platform Qualtrics (qualtrics.com/uk). The participants were paid at a rate of £6.30/h. We sought the participants' consent and complied with the local data protection guidelines. The study was approved by The Open University UK's Human Research Ethics Committee with reference number HREC/3063/Uprety.

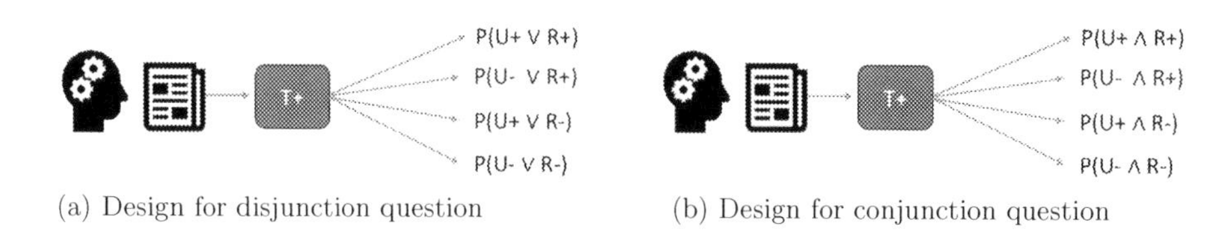

(a) Design for disjunction question (b) Design for conjunction question

Fig. 3. Experiment design

Statement Pairs
A. It is Easy to Understand the information presented in the document snippet. B. The information presented in the document snippet is Reliable.
A. It is Not Easy to Understand the information presented in the document snippet. B. The information presented in the document snippet is Reliable.
A. It is Easy to Understand the information presented in the document snippet. B. The information presented in the document snippet is Not Reliable.
A. It is Not Easy to Understand the information presented in the document snippet. B. The information presented in the document snippet is Not Reliable.

Fig. 4. Four pairs of statements for conjunction and disjunction questions

We use the same set of three query-document pairs for our experiment as used in [19], as we have reused some of their data. Each participant was shown the three queries (and the documents) and were asked to judge the topicality of the document and one of the eight questions (so we obtain probabilities like $P(U+ \vee R+ | T+)$, etc.) Thus the participants can be said to be divided into eight groups for a between-subjects design.

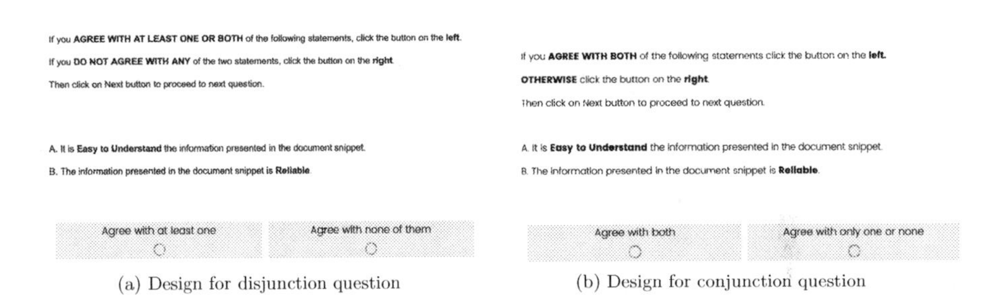

(a) Design for disjunction question

(b) Design for conjunction question

Fig. 5. Conjunction/disjunction question design

5 Results and Discussion

5.1 Violation of Kolmogorov Probability Axiom

The probabilities of conjunction and disjunction of the Understandability and Reliability questions are reported in Fig. 6. In order to compute the δ reported in Eq. 4, we also need the two probabilities related to single questions $U+$ and $R+$, apart from the conjunction and disjunction probabilities. These single question probabilities are obtained from the results in [19] (listed in Fig. 2). Then, we calculate $\delta = P(U \pm \vee R \pm |T+) + P(U \pm \wedge R \pm |T+) - P(R + |T+) - P(U + |T+)$. In Fig. 6 we see that δ is different from zero for all the three queries, although according to classical probability we expect that δ would be zero in all cases. Eq. (7), based on the projection operators in quantum probability, gives predictions of δ, as are shown in the last column of the table.

The violation of classical probability is a result of non-commutative structure of operators for U and R. As we can see, if operators of U and R commute with each other, the quantum correction term in the Eq. (7) approaches zero (the commutator is zero). In fact, the probability values obtained may violate some of the other basic axioms of classical/Kolmogorovian probability. For example, for Query 2, we can see that $P(U- \wedge R+|T+) = 0.414$ and $P(U-|T+) = 0.198$ which clearly violates $P(A,B) < P(A)$. Also, for this query, $P(U- \wedge R-|T+)$ is greater than both $P(U-|T+)$ and $P(R-|T+)$. This type of violation has been termed as conjunction fallacy in the cognitive science literature [18]. Quantum models have been previously used to explain such violation [3] where the fundamental notion of incompatibility in judgements is identified as the potential cause.

Query	P(U ∨ R | T =+)	P(U ∧ R | T = +)	δ	D(U,R)
Query 1	P(U+ ∨ R+|T=+) = 0.641	P(U+ ∧ R+|T=+) = 0.308	-0.175	-0.124
	P(U- ∨ R+|T=+) = 0.826	P(U- ∧ R+|T=+) = 0.393	0.251	0.032
	P(U+ ∨ R-|T=+) = 0.774	P(U+ ∧ R-|T=+) = 0.241	-0.017	-0.032
	P(U- ∨ R-|T=+) = 0.656	P(U- ∧ R-|T=+) = 0.435	0.215	0.124
Query 2	P(U+ ∨ R+|T=+) = 0.792	P(U+ ∧ R+|T=+) = 0.692	-0.049	-0.533
	P(U- ∨ R+|T=+) = 0.714	P(U- ∧ R+|T=+) = 0.414	0.199	0.071
	P(U+ ∨ R-|T=+) = 0.692	P(U+ ∧ R-|T=+) = 0.321	-0.058	-0.071
	P(U- ∨ R-|T=+) = 0.536	P(U- ∧ R-|T=+) = 0.368	0.437	0.533
Query 3	P(U+ ∨ R+|T=+)= 0.943	P(U+ ∧ R+|T=+) = 0.700	0.02	-0.623
	P(U- ∨ R+|T=+) = 0.562	P(U- ∧ R+|T=+) = 0.234	0.127	0.331
	P(U+ ∨ R-|T=+) = 0.907	P(U+ ∧ R-|T=+) = 0.378	-0.046	-0.623
	P(U- ∨ R-|T=+) = 0.535	P(U- ∧ R-|T=+) = 0.283	0.354	0.331

Fig. 6. Probabilities for conjunction and disjunction questions and associated violation from Kolmogorovian probability

5.2 Comparison of Quantum and Classical Probability Predictions

Figure 7 shows a comparison between quantum and classical probabilities with the experimental data for first two queries. The data for Query 3 had many probabilities close to 0 (see Fig. 2) and hence the sample became too small for a meaningful comparison. The probabilities are calculated for prediction of judgement of Reliability given the participant has judged Understandability and Topicality (positively), using equations derived in Sect. 2.1. Bayesian probabilities, in some cases, are significantly different from experimental data ($P(R+|U-,T+)$ for query 1 and $P(R-|U-,T+)$ for query 2). Quantum probabilities are consistently closer to the experimental data.

The Bayesian probabilities, as mentioned earlier, are based on the chain rule $P(R+, U+, T+) = P(R+|U+, T+)P(U+|T+)P(T+)$. The fundamental assumption here is that the variables corresponding to R, U and T can be jointly measured. In terms of the judgement process, this implies that a user can jointly consider information regarding the Reliability, Understandability and Topicality of a document with respect to the query. The incompatibility revealed in [19] and the order effects shown in [1] suggest that this is not always the case in general. Therefore we see Bayesian predictions deviate from the experimental data. As the quantum probability theory based on the Hilbert space model is free from this assumption of compatibility, it provides a promising alternative model that gives predictions closer to the experimental data. In fact, the modelling of incompatibility of different judgement perspectives forms one of the pillars of the Quantum Cognition research framework.

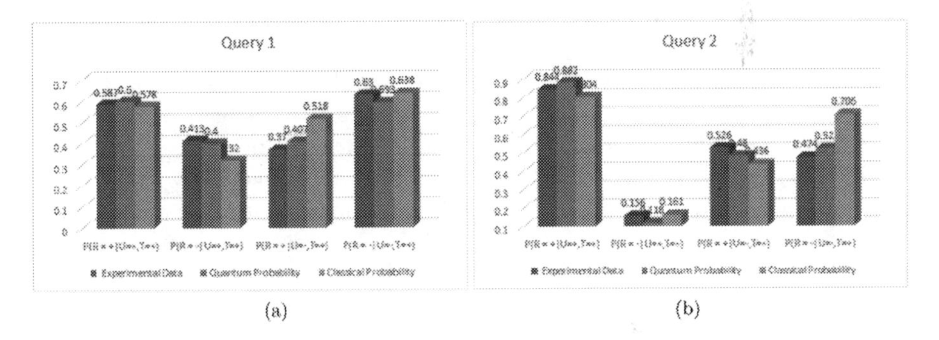

Fig. 7. How quantum and classical probabilities compare with the experimental data for query 1 and query 2

6 Implications for IR

Quantum models can capture richer cognitive interactions, by way of generalising some of the constraints of classical models like commutativity. Here we discuss a few cases where our findings can inform the design of IR systems and algorithms.

The impossibility of jointly modelling Reliability and Understandability (which leads to the Kolmogorovian axiom violations) can be attributed to the fact that humans make decisions in a sequential manner and consideration of one dimension affects the judgement of the next dimension. Therefore, different orders of consideration of dimensions would lead to different final relevance judgements, *making the order a factor in the variability of relevance judgements by users*. When using an IR system to perform a task or make an important decision, there might be a particular order of dimensions which can lead the user to make an optimal decision. For example, for a health related query, a user might find a document difficult to understand, which may affect his or her judgement of Reliability and hence the overall relevance. However, if another user first judges reliability and finds it highly reliable, the judgement of understandability might be different. The IR system can help users to consider the optimum sequence of dimensions and thus maximise the utility, by providing extra information. For example, if the system can also provide information about the Reliability of the document in terms of a Reliability score or ratings by other users, it can reduce uncertainty in judgement and thus minimise the influence of judgement of other dimensions. Thus, for the given medical document, the low understandability might not affect the perception of Reliability.

Secondly, quantum probabilistic models can replace Bayesian models used in IR algorithms for ranking and evaluation. For example, in [13], a multidimensional evaluation metric is proposed where the gain provided by a document is written as a function of the joint probability of relevance with respect to different dimensions, e.g. $P(T, U, R, ...)$. Similar assumptions have also been made in [12,25]. For documents exhibiting incompatibility between different dimensions, predictions from such a model will be inaccurate. A probabilistic model based on non-commutative operator algebra, accounting for the incompatibility between different dimensions, needs to be considered.

Finally, these results of violation of classical probability theory calls for further user behaviour experiments to be conducted in IR that further exploit the Quantum-like Structure in human judgements. It would require novel experimental protocols like that of Stern-Gerlach, Double-slit experiment, etc., to generate data beyond the modelling capacity of classical probability theory. Such experiments in themselves might lead us to new insights into user behaviour in IR and information based decision-making in general.

7 Conclusion

Extending a quantum-inspired experiment protocol, in this work, we begin with the hypothesis that the multidimensional property of relevance has an underlying quantum cognitive structure which can be shown as violation of certain classical (Kolmogorovian) probability axioms. A particular experimental design is reported which can exploit the quantum cognitive structure. The data shows violation of one of Kolmogorovian probability axioms. We further show that quantum probability theory is a better alternative to model multidimensional relevance judgements than its classical counterpart, i.e. Bayesian model. Finally, we highlight important implications of our research findings to the design of IR algorithms system and user experiments.

Acknowledgements. Authors affiliated to the universities in UK, Italy and China are funded by the European Union's Horizon 2020 research and innovation programme under the Marie Sklodowska-Curie grant agreement No. 721321, National Key Research and Development Program of China (grant No. 2018YFC0831704) and Natural Science Foundation of China (grant No. U1636203). Authors affiliated to QUT, Australia are supported by the Asian Office of Aerospace Research and Development (AOARD) grant: FA2386-17-1-4016.

References

1. Bruza, P., Chang, V.: Perceptions of document relevance. Front. Psychol. **5**, 612 (2014). https://doi.org/10.3389/fpsyg.2014.00612
2. Busemeyer, J.R., Bruza, P.D.: Quantum Models of Cognition and Decision, 1st edn. Cambridge University Press, New York (2012)
3. Busemeyer, J.R., Pothos, E.M., Franco, R., Trueblood, J.S.: A quantum theoretical explanation for probability judgment errors. Psychol. Rev. **118**(2), 193–218 (2011). https://doi.org/10.1037/a0022542
4. Clemmensen, M.L., Borlund, P.: Order effect in interactive information retrieval evaluation: an empirical study. J. Doc. **72**(2), 194–213 (2016). https://doi.org/10.1108/JD-04-2015-0051
5. Cool, C., Belkin, N.J.: Interactive information retrieval: history and background. Facet, 1–14 (2011). https://doi.org/10.29085/9781856049740.003
6. Eisenberg, M., Barry, C.: Order effects: a study of the possible influence of presentation order on user judgments of document relevance. J. Am. Soc. Inf. Sci. **39**(5), 293–300 (1988)
7. Fell, L., Dehdashti, S., Bruza, P., Moreira, C.: An experimental protocol to derive and validate a quantum model of decision-making. In: Proceedings of the 41st Annual Meeting of the Cognitive Science Society (COGSCI 2019) (2019)
8. Hogarth, R.M., Einhorn, H.J.: Order effects in belief updating: the belief-adjustment model. Cognitive Psychol. **24**(1), 1–55 (1992). https://doi.org/10.1016/0010-0285(92)90002-j
9. Huang, M.H., Wang, H.Y.: The influence of document presentation order and number of documents judged on user's judgments of relevance. J. Am. Soc. Inf. Sci. Technol. **55**(11), 970–979 (2004). https://doi.org/10.1002/asi.20047
10. Khrennikov, A.: Basics of quantum theory for quantum-like modeling information retrieval. In: Aerts, D., Khrennikov, A., Melucci, M., Toni, B. (eds.) Quantum-Like Models for Information Retrieval and Decision-Making. SSTEAMH, pp. 51–82. Springer, Cham (2019). https://doi.org/10.1007/978-3-030-25913-6_4
11. Kolmogorov, A.N.: Foundations of the Theory of Probability. Martino Fine Books, Eastford (2013)
12. Palotti, J., Goeuriot, L., Zuccon, G., Hanbury, A.: Ranking health web pages with relevance and understandability. In: Proceedings of the 39th International ACM SIGIR Conference on Research and Development in Information Retrieval SIGIR 2016, pp. 965–968. ACM, New York (2016). https://doi.org/10.1145/2911451.2914741
13. Palotti, J., Zuccon, G., Hanbury, A.: MM: a new framework for multidimensional evaluation of search engines. In: Proceedings of the 27th ACM International Conference on Information and Knowledge Management CIKM 2018, pp. 1699–1702. ACM, New York (2018). https://doi.org/10.1145/3269206.3269261
14. Pothos, E.M., Busemeyer, J.R.: A quantum probability explanation for violations of 'rational' decision theory. Proc. R. Soc. B: Biol. Sci. **276**(1665), 2171–2178 (2009). https://doi.org/10.1098/rspb.2009.0121
15. Sakurai, J.J., Napolitano, J.: Modern Quantum Mechanics. Cambridge University Press, Cambridge (2017)
16. Trueblood, J.S., Busemeyer, J.R.: A quantum probability account of order effects in inference. Cognitive Sci. **35**(8), 1518–1552 (2011). https://doi.org/10.1111/j.1551-6709.2011.01197.x

17. Tversky, A., Kahneman, D.: Judgment under uncertainty: heuristics and biases. Science **185**(4157), 1124–1131 (1974). https://doi.org/10.1126/science.185.4157.1124
18. Tversky, A., Kahneman, D.: Extensional versus intuitive reasoning: the conjunction fallacy in probability judgment. Psychol. Rev. **90**(4), 293–315 (1983). https://doi.org/10.1037/0033-295x.90.4.293
19. Uprety, S., Dehdashti, S., Fell, L., Bruza, P., Song, D.: Modelling dynamic interactions between relevance dimensions. In: Proceedings of the 2019 ACM SIGIR International Conference on Theory of Information Retrieval - ICTIR 2019. ACM Press (2019). https://doi.org/10.1145/3341981.3344233
20. Uprety, S., Song, D.: Investigating order effects in multidimensional relevance judgment using query logs. In: Proceedings of the 2018 ACM SIGIR International Conference on Theory of Information Retrieval ICTIR 2018, pp. 191–194. ACM, New York (2018). https://doi.org/10.1145/3234944.3234972
21. Vourdas, A.: Probabilistic inequalities and measurements in bipartite systems. J. Phys. A: Math. Theor. **52**(8), 085301 (2019). https://doi.org/10.1088/1751-8121/aafe97
22. Wang, B., Zhang, P., Li, J., Song, D., Hou, Y., Shang, Z.: Exploration of quantum interference in document relevance judgement discrepancy. Entropy **18**(12), 144 (2016). https://doi.org/10.3390/e18040144
23. Wang, Z., Busemeyer, J.R.: A quantum question order model supported by empirical tests of an a priori and precise prediction. Topics Cognitive Sci. **5**(4), 689–710 (2013)
24. Xu, Y., Wang, D.: Order effect in relevance judgment. J. Am. Soc. Inf. Sci. Technol. **59**(8), 1264–1275 (2008). https://doi.org/10.1002/asi.20826
25. Zuccon, G.: Understandability biased evaluation for information retrieval. In: Ferro, N., et al. (eds.) ECIR 2016. LNCS, vol. 9626, pp. 280–292. Springer, Cham (2016). https://doi.org/10.1007/978-3-319-30671-1_21

Question Answering, Prediction, and Bias

Temporal Latent Space Modeling for Community Prediction

Hossein Fani[1,2](✉), Ebrahim Bagheri[2], and Weichang Du[1]

[1] Faculty of Computer Science, University of New Brunswick, Fredericton, Canada
{hfani,wdu}@unb.ca

[2] Laboratory for Systems, Software and Semantics (LS3), Ryerson University, Toronto, Canada
{hossein.fani,bagheri}@ryerson.ca

Abstract. We propose a temporal latent space model for user community *prediction* in social networks, whose goal is to predict *future* emerging user communities based on past history of users' topics of interest. Our model assumes that each user lies within an unobserved latent space, and similar users in the latent space representation are more likely to be members of the same user community. The model allows each user to adjust its location in the latent space as her topics of interest evolve over time. Empirically, we demonstrate that our model, when evaluated on a Twitter dataset, outperforms existing approaches under two application scenarios, namely news recommendation and user prediction on a host of metrics such as mrr, ndcg as well as precision and f-measure.

1 Introduction

Social networks have been an effective medium for communication and social interaction. Predicting users' behaviour, interactions, and influence are of interest due to their wide range of applications such as personalized recommendations and marketing campaigns. Community-level analytics provide the means to understand social network dynamics at a higher collective level. In order to support community-level models, various community detection methods have been proposed, which employ information such as users' social connections and content engagement to identify communities. The objective of these models is to identify past and current user communities; however, little work has been done on *community prediction*, that is, to determine how the community structure of a social network will look like in a future yet-to-be-observed time interval.

In this paper, we focus on an instance of this problem, namely *content-based* (*topical*) *future community prediction*. Specifically, given a sequence of users' contributions towards a set of topics from time interval 1 to T, the goal is to predict topical user communities in a future interval $\mathrm{T}+1$. To perform topical future community prediction, we construct graph snapshots G_t for each time interval t in which users are linked based on pairwise topical similarity

J. M. Jose et al. (Eds.): ECIR 2020, LNCS 12035, pp. 745–759, 2020.
https://doi.org/10.1007/978-3-030-45439-5_49

at time interval t. Given the sequence of graph snapshots $[G_{1:T}]$ from time 1 to T, we propose temporal latent space modeling to predict inter-user topical similarities in G_{T+1} after which a community detection method yields future user communities.

Latent space modeling [26] has been successfully employed for link prediction in graphs where given the observed links in the graph, the location of each node in a latent space is learned such that the closer two nodes are in that space, the higher the probability of a link between them would be. In other words, similarity in latent space translates into links in graph space. Latent space modeling preserves homophily [23] where links between nodes are considered clues for similarity, and so, densely connected groups of nodes imply communities.

Different approaches based on matrix factorization and deep neural networks [1,31–33] have been proposed to learn the latent space model of users in the social network. For instance, Akbari et al. [1] propose to learn a single multi-modal latent space representation from users' social views including network structure as well as contents, inter-user interactions (e.g., reply or retweet), and prior knowledge (if any). However, such studies are concerned with static graphs, where the latent representation of the users are fixed. They overlook the fact that latent space representations need to evolve over time and, hence, fall short when identifying user communities of the *future*.

Temporal tensor-factorization approaches [11], or temporal latent space models [34] go beyond static networks and assume that the network is dynamic and changes over time. Such models endeavor to learn low-rank latent space representations based on the intuition that nodes can move in latent space over time. While suitable for predicting links in a social network, dynamic link prediction models are inherently deficient when the communities need to take users' content similarity into account, i.e., identify *content-based* user communities in the future. Temporal content-based user community prediction is of interest due to the following reasons: (*i*) there are many users on a social network that have similar interests but are not explicitly linked to each other, (*ii*) an explicit social link does not necessarily indicate user interest similarity but could be owing to sociological processes such as conformity, sociability or other factors such as friendship and kinship [10,28], (*iii*) there are some cases where the network structure is not accessible [4] or misleading, e.g., when links are fraudulent due to link-farming [21], and (*iv*) empirical research has shown that link evolution happens at a much lower pace compared to content changes [24], particularly because links are often not removed when they become effectively 'dead'.

Temporal content-based user community detection methods currently exist that incorporate temporal aspects of users' content and stress that users of the same community would ideally show similar interest patterns for similar topics over time [12,17,18]. However, users' temporal content is only used for pairwise user similarity calculation to build content-based user communities as opposed to user community *prediction*. As a result, they have limited applicability for identifying user communities of the future. Regression techniques such as autoregressive integrated moving average (arima) and support vector regression (svr) that leverage temporal information to predict users' future interests have shown promising results and can be utilized to identify user communities in the future based on users' pairwise content similarity [3]. However, they require building predictive models on a per user basis and, hence, are practically prohibitive.

In this work, we propose temporal latent space modeling to predict content-based user communities in the future. First, contrary to *non*-temporal methods [31–33], our approach incorporates temporality. Second, in contrast to Zhu et al.'s method [34] and the likes [11,14,27,35] that focus on dynamics of social network structure, our approach employs dynamics of social content. Third, although we use temporal information to predict future users' topical interests similar to regression methods, we train only one model for all users and, thus, significantly reduce computational cost compared to regression techniques. Last, unlike Fani et al. [12] and Hu et al. [18] who employ users' temporal and topical interests for performing pairwise user similarities in order to identify user communities up until 'now', our work in this paper employs such information for predicting user communities 'in the future', which is a step forward compared to the state of the art. We perform experiments on a Twitter corpus and compare our work for user community prediction with several state-of-the-art baselines in the context of news recommendation and user prediction. The results show that our method, which incorporates the temporal evolution of users' topics of interest within latent space, exhibits a stronger predictive power compared to the baselines.

The key contributions of this paper are as follows:

1. We propose a temporal latent space model for user community prediction in social networks that (*i*) allows users to change their latent representations as their topics of interest evolve over time, and (*ii*) users who are similar not only in their contribution towards the same set of topics of interest, but also have similar temporal behaviour, remain close in d-dimensional latent space.
2. We illustrate how temporal latent space modeling can be effectively employed to predict future emerging content-based user communities given the past history of users' topics of interest.
3. We perform experimentation on a Twitter dataset to demonstrate the superiority of our proposed model compared to the state of the art methods under two application scenarios, namely, news recommendation and user prediction on a host of metrics such as mrr, ndcg as well as precision and f-measure.

The rest of the paper is organized as follows: In Sect. 2 we describe the related work. Sections 3 and 4 are dedicated to the problem definition and the details of our proposed approach. Section 5 presents our experimental work after which Sect. 6 concludes the paper.

2 Related Work

In this paper, we assume that an existing state of the art technique such as those proposed in [5,30,36] can be employed for extracting and modeling users' topics of interest. Therefore, we will not be engaged with the process of identifying topics and will only focus on determining content-based user communities in a future time interval based on the temporal interest of users from the past up until now towards those topics. Given this focus, the related works to this paper are largely centered around two areas: (1) temporal user community detection; and, (2) temporal latent space modeling.

2.1 Temporal User Community Detection

There is a rich line of research on user community detection; ranging from link-based community detection methods [16,19,31], which rely only on network structure, to content (topic)-based approaches, which mainly focus on information content generated by the users [4,22]. More recently, several effective approaches have been proposed which integrate both the network structure (links) and content to improve community detection performance [1,7]. All these works assume that the user's topics of interest remain stable across time. However, very few consider the notion of temporality in users' topics of interest [12,17,18], particularly in online social networks such as Twitter.

From among the work that consider temporality, Hu et al. [18] have proposed a probabilistic generative model jointly over text, time and links, namely community level diffusion (cold), to simultaneously identify both user communities and topics in order to uncover inter-community influence dynamics. The generative process can be summarized in three steps. First, per community topic distribution is sampled according to Dirichlet. Then for each community-topic pair, temporal distribution (timestamp) of a topic for a community is also sampled according to Dirichlet. Finally, a user chooses a community based on her community membership distribution and selects a topic from the community according to the community's topic distribution to generate a post (e.g., a tweet). The time of the post is from the temporal distribution of that topic for that community. Contrary to the unified generative model, Fani et al. [12] have proposed a neural embedding approach to identify temporally like-minded user communities given the user generated textual content. They model the users' temporal contribution towards topics of interest by introducing the notion of regions of like-mindedness between users. These regions cover users who share not only similar topical interests but also similar temporal behaviour. By considering the identified set of regions of like-mindedness as a context, a neural network is trained such that the probability of a user in a region is maximized given other users in the same region. The final weights of the neural networks form the low-dimensional vector representation of each user that incorporates both topics of interest and their temporal nature. Finally, a graph partitioning technique is applied on a weighted user graph in which the similarity of two users is based on the cosine similarity of their respective vectors to identify like-minded user communities. While both of these works sketch different architectures to incorporate temporality, they have shown performance improvement in modelling content-based user communities, particularly in time-sensitive applications such as item recommendation. However, in both work, users' temporal content is only used to build user communities up until now and have limited applicability for identifying user communities of the future.

2.2 Temporal Latent Space Modeling

Temporal latent space modeling aims at learning the evolution of a temporal graph node and/or an edge over time into a *dynamic* low dense d-dimensional vector in latent space which is allowed to modify its position in the latent space over time. This can be employed for different underlying applications such as link prediction and node classification [13]. Various works such as Temporal tensor-factorization [11,34], and neural embeddings [14,27,35] have been proposed in this respect. For instance, Singer et al. [27] extend the prior neural-based embedding approaches on static graphs, e.g., node2vec [15], to temporal graphs. They propose a semi-supervised algorithm, namely tNodeEmbed, that learns to combine a node's historical temporal embeddings into a final embedding such that it can optimize for a given underlying task, e.g., link prediction. tNodeEmbed initializes node embeddings for time interval t using a static node embedding method. Since initialization for each time interval happens in isolation and independently, the coordinates of the embeddeding space are not guaranteed to align. tNodeEmbed learns a rotation matrix to align coordinates of node embeddings within the time intervals given the fact that a node's temporal behaviour between two consecutive time intervals is gradual and does not follow a bursty pattern (temporal smoothness). Each node embedding at time t, after alignment, is fed to a recurrent neural network (rrn) with long short term memory (lstm) to output the final temporal embedding of the node by optimizing for the specific task. Zhu et al. [34], however, propose temporal latent space modeling for the task of dynamic link prediction via non-negative matrix factorization followed by a scalable inference algorithm based on block coordinate gradient descent (bcgd) to obtain embeddings in linear time.

While suitable for predicting links in a social network structure, proposed temporal latent space models are inherently deficient when the communities need to take users' content similarity into account, i.e., identify *content-based* user communities in the future. To the best of our knowledge, no approach investigates the application of temporal latent space modeling for temporal content-based community detection, which is the main objective of this paper.

3 Problem Definition

Given a set of topics $\mathcal{Z}$ from a social network, such as Twitter, within T time intervals extracted by a topic detection method (e.g., lda) and a set of users $\mathcal{U}$, we represent the topic preference of user $u \in \mathcal{U}$ towards topic set $\mathcal{Z}$ at time interval $t : 1 \leq t \leq \mathrm{T}$ as a vector $\mathbf{x}_{ut} = [x_{ut,1:|\mathcal{Z}|}]$, namely *topic preference vector*, where $x_{ut,z} \in \mathcal{R}^{[0,1]}$ indicates the preference by user u for topic z at time interval t. We let temporal graph $G_t = (\mathcal{U}, \mathcal{E}_t, s)$ represent the content-based similarity between the users of the social network whose nodes are users in $\mathcal{U}$ and $\mathcal{E}_t$ is the set of weighted undirected edges whose weights are based on a similarity function s, which is defined as the cosine similarity of topic preference vectors of the users at time interval t, i.e., $\forall u, v \in \mathcal{U} : s(u, v : t) = \frac{\mathbf{x}_{ut} \cdot \mathbf{x}_{vt}}{|\mathbf{x}_{ut}||\mathbf{x}_{vt}|}$. Given $[G_{1:\mathrm{T}}]$, we aim to accurately predict a set of induced subgraphs in $G_{\mathrm{T}+1}$ to form content-based user communities at time interval $\mathrm{T}+1$.

4 Proposed Approach

Our approach consists of three subsequent phases: users' topic preference detection, temporal latent space inference, and community prediction. In the following, we lay out the details of each step.

4.1 Topic Preference Detection

To instantiate the topic preference vector, we find (i) a set of topics $\mathcal{Z}$ that have been observed within T time intervals, and (ii) u's degree of interest at time interval t towards each topic $z \in \mathcal{Z}$, i.e., $x_{ut,z}$ in topic preference vector $\mathbf{x}_{ut} = [x_{ut,1} \,..\, x_{ut,z} \,..\, x_{ut,|\mathcal{Z}|}]$. We derive the set of topics from the collection of users' posts using lda [5]. To this end, we view all tweets authored by each user u at time interval t as a single document d_{ut}. Given the document corpus $\mathcal{D} = \{d_{ut} | \forall u \in \mathcal{U}, 1 \leq t \leq \mathrm{T}\}$ and the number of topics $|\mathcal{Z}|$, lda distills $\mathcal{D}$ into two probability distributions: (i) distribution of words in each topic (ϕ_z); and (ii) distribution of each topic z in each document $(\theta_{d_{ut,z}})$ showing u's degree of interest toward z at time t. Formally, $x_{ut,z} = \theta_{d_{ut,z}}$. Once users' topic preference vectors have been identified at time interval t, we are able to calculate the similarity function s for all pairs of users and build the temporal graph G_t.

4.2 Temporal Latent Space Inference

Within time period T, the stream of graphs $[G_1 \,..G_t..\, G_\mathrm{T}]$ could be considered as a dynamic graph $\mathscr{G}$ which is evolving over time. We map each user u up until time interval $t \leq \mathrm{T}$ to a low-rank d-dimensional latent space, denoted by $\mathbf{y}_{ut}$, while imposing the following assumptions: (i) users change their latent representations over time, (ii) two users that are close to each other in $\mathscr{G}$ remain close in latent space, (iii) two users who are close in latent space share similar topics of interest with each other.

Formally, given a dynamic network $\mathscr{G}$, we find a d-dimensional latent space representation for $\forall u \in \mathcal{U}$ for time interval $1 \leq t \leq \mathrm{T}$ that minimizes the *quadratic loss* with temporal regularization:

$$\arg\min \Big[\sum_{t=1}^{\mathrm{T}} \sum_{u,v \in \mathcal{U}} \overbrace{|s(u, v : t) - \mathbf{y}_{ut}\mathbf{y}_{vt}^\top|_F^2}^{\text{quadratic loss}} + \lambda \sum_{t=1}^{\mathrm{T}} \sum_{u \in \mathcal{U}} \overbrace{(1 - \mathbf{y}_{ut}\mathbf{y}_{u(t-1)}^\top)}^{\text{temporal smoothness}} \Big] \tag{1}$$
$$\forall u \in \mathcal{U}; \mathbf{y}_{ut} \geq 0, \mathbf{y}_{ut}\mathbf{y}_{ut}^\top = 1$$

where in the quadratic loss component, $s(u, v : t)$ is the similarity score for a pair of users u and v in G_t, $\mathbf{y}_{ut}$ is the d-dimensional latent representation for u up until time interval t, λ is a regularization parameter, and the temporal smoothness component $(1 - \mathbf{y}_{ut}\mathbf{y}_{u(t-1)}^\top)$ penalizes user u for a sudden change in its location in latent space. Our model maps each user to a point in a unit hypersphere rather than simplex, because sphere modeling gives a clearer boundary between similar users and dissimilar users when mapping all user pairs into latent space. It is worth noting that in our proposed model, a user's position in latent space up until time interval t depends on preceding movement of the user in the latent space since the first time interval $1 \leq t' < t$ via observation of $[G_1 .. G_t]$. This is contrary to static models that obtain latent representation based solely on G_t.

Optimizing Eq. 1 is expensive in terms of space and time complexity as it requires all graphs in $\mathscr{G}$ to jointly update all temporal latent representations for users in all time intervals. To optimize Eq. 1, we use the *local* block coordinate gradient descent (bcgd) algorithm [34], in which inference happens sequentially. Specifically, we optimize users' latent representation locally by minimizing the following objective function at each time interval t:

$$\arg\min \sum_{u,v \in \mathcal{U}} (s(u, v : t) - \mathbf{y}_{ut}\mathbf{y}_{vt}^\top)^2 + \sum_{u \in \mathcal{U}} (1 - \mathbf{y}_{ut}\mathbf{y}_{u(t-1)}^\top) \tag{2}$$

The local bcgd algorithm infers users' latent representation from a single graph snapshot G_t and prior initialization from $\mathbf{y}_{u(t-1)}$. The algorithm iteratively updates $\mathbf{y}_{ut}$ until it converges and then moves to the computation of temporal latent space in the next time interval $t + 1$. This local sequential update schema greatly reduces the computational cost in practice. We refer readers to [34] for in-depth analysis of the local bcgd algorithm.

4.3 User Community Detection in the Future

Our goal is to predict those user communities whose members share similar temporal expositions toward similar topics of interest in the future graph $G_{\mathrm{T}+1}$. To do so, we first estimate the future graph $G_{\mathrm{T}+1}$. Based on our model, the topical similarity between two users depends only on their latent representations. In other words, the more two latent representations for a pair of users are close, the more similar the users are in terms of topics of interest. As a result, given $\forall u, v \in \mathcal{U} : \mathbf{y}_{u(\mathrm{T}+1)}$ and $\mathbf{y}_{v(\mathrm{T}+1)}$, we are able to predict future graph $G_{\mathrm{T}+1} = (\mathcal{U}, \mathcal{E}_{\mathrm{T}+1}, s)$ assuming $s(u, v : \mathrm{T}+1) = \mathbf{y}_{u(\mathrm{T}+1)}\mathbf{y}_{v(\mathrm{T}+1)}^\top$. However, $\mathbf{y}_{u(\mathrm{T}+1)}$ and $\mathbf{y}_{v(\mathrm{T}+1)}$ are not available and have to be approximated based on temporal latent representations up until time interval T, i.e., $\mathbf{y}_{u(\mathrm{T}+1)} = \eta(f(\mathbf{y}_{u1}, .. \mathbf{y}_{ut}, .. \mathbf{y}_{u\mathrm{T}}))$ where η is a link function and f is a temporal function.

In order to ensure that our proposed model considers all user information from time interval 1 to T when learning the representation for $\mathbf{y}_{u(\mathrm{T}+1)}$, we define the user representation at each time interval to be a summarized representation of that user's activities in all previous time intervals. In other words, $\mathbf{y}_{u(\mathrm{T}+1)}$ will encode information for user u in time intervals 1 to T. Similarly, the user representation for u for the time period 1 to $\mathrm{T}-1$ is captured in $\mathbf{y}_{u\mathrm{T}}$. As such, a user's latent position at time interval T depends on her latent representation at time interval $\mathrm{T}-1$ (which already captures user information up to $\mathrm{T}-2$), notationally, $\mathbf{y}_{u(\mathrm{T}+1)} = \eta(f(\mathbf{y}_{u\mathrm{T}}))$. Without loss of generality, we choose f as the identity function and η as the identity link function. Hence, user's latent representation at time T becomes the proxy for her latent representation at time $\mathrm{T}+1$, as suggested by Zhu et al. [34], i.e.,

$$\mathbf{y}_{u(\mathrm{T}+1)} \simeq \mathbf{y}_{u\mathrm{T}} \tag{3}$$

$$s(u, v : \mathrm{T}+1) = \mathbf{y}_{u\mathrm{T}} \mathbf{y}_{v\mathrm{T}}^{\top} \tag{4}$$

where $\mathbf{y}_{u\mathrm{T}}$ encapsulates the latent representations of all snapshots from 1 to $\mathrm{T}-1$.

Now, given $G_{\mathrm{T}+1}$, we employ a graph partitioning heuristic to extract clusters of users that form our final user communities in the future. We leverage the Louvain method [6] as it is a linear heuristic for the problem of graph partitioning based on modularity optimization. Louvain can be applied to weighted graphs, does not require *a priori* knowledge about the number of communities, and is computationally efficient on large graphs [25]. The application of Louvain on $G_{\mathrm{T}+1}$ produces a set of induced subgraphs such as $G_{\mathrm{T}+1}[\mathcal{C}]$ whose vertex set $\mathcal{C} \subset \mathcal{U}$ and edge set consists of all of the edges in $\mathcal{E}_{\mathrm{T}+1}$ that have both endpoints in $\mathcal{C}$. Subgraphs with $|\mathcal{C}| \geq 2$ form instances of user communities.

5 Evaluation

5.1 Dataset and Experimental Setup

We adopted a Twitter dataset consisting of 2,948,742 tweets authored by 135,731 users in Nov. and Dec. 2010. The two month time period is sampled on a daily basis, i.e., $\mathrm{T}+1=61$. The settings in each step of our method are as follows:

Topic Preference Detection. We applied lda using Mallet api after removing stopwords. The number of topics used for reporting results in this paper has been set to $|\mathcal{Z}| = 50$ noting that other topic sizes did not change the findings of this paper. We created $\forall u \in \mathcal{U} : \mathbf{x}_{ut} = [x_{ut,1} \; .. \; x_{ut,z} \; .. \; x_{ut,50}]$ for $t = 1$ up to day 60 as our observation to build $\mathscr{G} = [G_1 \; .. \; G_t \; .. \; G_{\mathrm{T}=60}]$ in order to predict $G_{\mathrm{T}+1=61}$ at the future day 61.

Model Training. We adopt sequential (local) version of block coordinate gradient descent proposed by Zhu et al. [34]. By setting the temporal smoothness (regularization) parameter $\lambda = 0.01$, we performed experiments on increasing number of dimensions $d \in \{10, 20, ..., 100\}$ for learning temporal latent representation of users in 1,000 iterations.

User Community Detection in Future. We apply Louvain with resolution parameter 1.0 using Pajek[1] to identify subgraphs.

[1] mrvar.fdv.uni-lj.si/pajek/.

5.2 Baselines

We compare our work against four categories of baselines:

Community Prediction Baseline. To the best of our knowledge, the most related baseline to our work is a temporal content-based latent space model proposed by **Appel et al.** [2] where *shared* matrix factorization has been used to embed social network dynamics and temporal content in a shared feature space followed by a traditional clustering technique, such as k-means, to identify user communities.

Temporal Community Detection Baselines. Hu et al. [18] and **Fani et al.** [12] are two temporal content-based community detection baselines. The former is a generative process for predefined number of topics and communities. This method is a mixture model in which all users are members of all communities with a probability distribution. We only consider the community with the highest probability as the user's community. The latter is based on temporal user embeddings. This method learns a mapping from the user space to a low-rank latent space that incorporates both topics of interest and their temporality.

Non-temporal Community Detection Baselines. Ye et al. [33] and **Louvain** [6] are *non*-temporal link-based community detection methods from two extremes of neural-based non-negative matrix factorization (nmf) and modularity optimization, respectively. To select the best setting for each method, we performed experiments on increasing number of communities $\mathrm{C} = \{5, 10, 20, 30\}$ for Appel et al., Hu et al., and Ye et al., and varying embedding dimensions $d = \{100, 200 .. 500\}$ and $d = \{5, 10, 20, 30\}$ for Fani et al. and Appel et al., respectively. The final communities which are based on the users' temporal content until day 60 are used to predict communities in day 61.

Collaborative Filtering Baselines. Temporal collaborative filtering methods are able to predict users' topics of interest in future and, hence, can be used for the task of content-based community prediction, among which we choose the strongest methods, namely, **timesvd++** [20] and **rrn** [29] as our baselines. We performed grid search over the bin size in $\{1, 2, 4 .. 64\}$ and factor size in $\{10, 20, 40, 80\}$ to select the best settings.

5.3 Evaluation Methodology

Contrary to small real social networks or synthetic ones, gold standard communities are often not available for real world applications [8]. As such, well-defined quality measures such as rand index or normalized mutual information (nmi) that require comparison to the gold standard cannot be used. On the other hand and in the absence of a golden standard, quality functions such as *modularity* are not helpful either since they are based on the explicit links between users. In our approach and the baselines, the links between the users are inferred through a learning process and are not explicit. For instance, a near perfect method may result in low modularity because graph edges are sparse and do not form densely

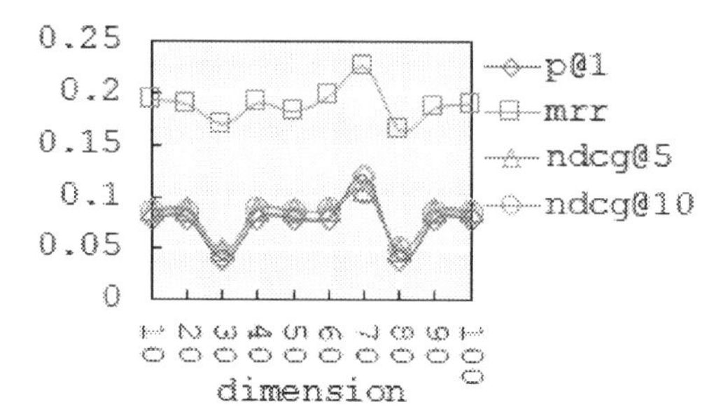

Fig. 1. The impact of dimension size on our method.

connected user sets. Conversely, a weak method may connect topically dissimilar users together forming communities of users that do not share similar interests but have a high modularity. So, the communities that achieve high structural quality in an inferred similarity graph are not necessarily optimal.

Alternatively, the performance of community detection methods can be measured through observations made at the application level. In these *extrinsic* evaluation strategies, a user community detection method is considered better *iff* its output communities improve an underlying application. We deploy two applications, namely news recommendation, and user prediction. By using these applications, we explore whether our proposed method is able to provide stronger performance compared to the state of the art.

To this end, we first build a gold standard dataset for these applications by collecting news articles to which a user has explicitly linked in her tweets (or retweets). We postulate that users post news articles since they are interested in the topics of the news articles. We build the gold standard from a set of news articles whose urls have been posted by user u at time $\mathrm{T}+1$. We see each entry as a triple $(u,\ a,\ \mathrm{T}+1)$ consisting of the news article a, user u, and the time interval $\mathrm{T}+1$ to form our gold standard.

5.4 Results

We compare the quality of the communities predicted by our method against the baselines in the context of news recommendation and user prediction.

News Recommendation. To evaluate communities of the future in the context of the news recommendation, we recommend news articles in two steps:

1. For each community $\mathcal{C}$, we recommend news articles in a ranked list based on the similarity of the article a and the community's overall topic preference vector at time $\mathrm{T}+1$. The overall topic preference vector for a community is the sum over all users' topic preference vector belonging to the community, i.e., $\sum_{u\in\mathcal{C}} \mathbf{x}_{u(\mathrm{T}+1)}$.
2. We recommend news article a to user $u \in \mathcal{C}$ based on the same ranked list as her community's list. A true community is one whose members are interested

in the same topics of interest in the future. As a result, at time $T + 1$, a news article is about the same topics of interest as the community's overall interests *iff* all the members post about the same or similar news articles.

Table 1. Comparison with baselines. Asterisk (*) indicates statistically significant improvement over other baselines using paired t-test at $p < 0.05$.

Method	News recommendation			User prediction		
	mrr	ndcg5	ndcg10	precision	recall	f-measure
Community prediction						
Our approach	**0.225***	**0.108***	**0.105***	**0.012***	0.035	**0.015***
Appel et al. [PKDD'18]	0.176	0.056	0.055	0.007	0.094	0.0105
Temporal community detection						
Hu et al. [SIGMOD'15]	0.173	0.056	0.049	0.007	0.136	0.013
Fani et al. [CIKM'17]	0.065	0.040	0.040	0.007	0.136	0.013
Non-temporal link-based community detection						
Ye et al. [CIKM'18]	0.139	0.056	0.055	0.008	0.208	0.014
Louvain [JSTAT'08]	0.108	0.048	0.055	0.004	0.129	0.007
Collaborative filtering						
rrn [WSDM'17]	0.173	0.073	0.08	0.004	**0.740***	0.008
timesvd++ [KDD'08]	0.141	0.058	0.064	0.003	0.657	0.005

We evaluate the recommended list of news articles using standard retrieval metrics such as mrr, ndcg5, and ndcg10. Foremost, we analyze the effect of dimension d on our inference algorithm. We vary d from 10 to 100 and report the performance in Fig. 1. As seen, the overall trend indicates that the recommendation performance in terms of all ranking metrics increases with the number of dimensions up to an extremum at $d = 70$. Next, we compare our proposed method at its best setting ($d = 70$) against the baselines at their best settings in Table 1. As shown, our proposed method outperforms other baselines in terms of all ranking metrics in the context of news recommendation. We attribute the accuracy of our proposed approach to the fact that it directly models and leverages the impact of users' pairwise similarity over their topics of interest within the time dimension, i.e., sequence of similarity graphs, which has been overlooked in all of the other baselines. For instance, Hu et al. is neither a predictive model nor aware of temporal similarity among users, and Ye et al. and Louvain do not take temporal information into account at all. However, it is worth noting that due to capturing sequences of inter-user similarities indirectly through collaborative filtering, rrn was able to become the runner-up in terms of ndcg5 and ndcg10.

User Prediction. The other application with which we evaluate our approach is the user prediction application. Here, given the user communities of the future,

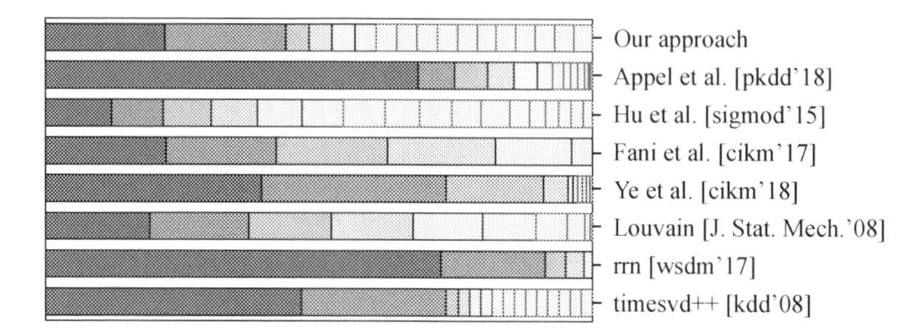

Fig. 2. User distribution in communities. Our method leads to a higher number of communities with a proportional distribution of users in the communities while the baseline methods have a higher skewness. Disproportionate distribution of users in communities can lead to poor application-level performance.

the goal is to predict which users will post news article a at time $\mathrm{T}+1$. To do so, we consider members of the *closest* community to a news article in terms of topics of interest at time $\mathrm{T}+1$ to be the potential posters. We use precision, recall, and f-measure to report user prediction performance. We further compare our method at its best which happens to be at $d=80$, against the baselines at their best setting in Table 1. In terms of precision, our proposed method was able to outperform other baselines. In terms of *recall*; however, some of the baselines could achieve higher performance and our method is not as strong. The reason for such high recall for some baselines is the fact that the they cluster users into very few, yet large user communities, as seen in Fig. 2. For instance, rrn was able to excel in recall due to its low number of communities. In an extreme, if a method only identifies one community that includes all of the users, recall would be 1. As such the lower the number of the communities is, the higher the recall would be. However, this comes at the cost of precision. Overall, the f-measure metric points to higher quality communities identified based on our proposed work. This reinforces the fact that when users' pairwise similarity with respect to the topics of interest over time are explicitly embedded in a sequence of graphs, it will lead to higher quality user communities in the future. Further, Fig. 2 shows that unlike some of the baselines where the majority of the users are placed in only a few communities and the other communities only have a few members (leading to higher recall but poor performance on precision), our approach could proportionally distribute users across different communities and hence show superior performance over precision and f-measure.

6 Conclusion and Future Work

Our work is among the first to explore the idea of *predicting* topical user communities on social networks. We learn to represent users within a latent space that preserves users' topical similarities over time. Our experiments show that our approach is able to predict communities of like-minded users with respect to topics of interest in future yet-to-be-observed time interval and outperform the state of the art. The area that we would like to work on in our future work pertains to the fact that our approach penalizes significant and sudden changes in the position of a user's representations in latent space. In other words, our approach favors smooth transition of user representations across different time intervals. However, there may be cases where sudden change in the position of the user representation in latent space may be warranted such as in reaction to bursty topics. As future work, we plan to generalize our approach to support for such cases based on intuitions from Deng et al. [9].

References

1. Akbari, M., Chua, T.: Leveraging behavioral factorization and prior knowledge for community discovery and profiling. In: Proceedings of the Tenth ACM International Conference on Web Search and Data Mining, WSDM 2017, Cambridge, United Kingdom, 6–10 February 2017, pp. 71–79 (2017)
2. Appel, A.P., Cunha, R.L.F., Aggarwal, C.C., Terakado, M.M.: Temporally evolving community detection and prediction in content-centric networks. In: Berlingerio, M., Bonchi, F., Gärtner, T., Hurley, N., Ifrim, G. (eds.) ECML PKDD 2018. LNCS (LNAI), vol. 11052, pp. 3–18. Springer, Cham (2019). https://doi.org/10.1007/978-3-030-10928-8_1
3. Arabzadeh, N., Fani, H., Zarrinkalam, F., Navivala, A., Bagheri, E.: Causal dependencies for future interest prediction on Twitter. In: Proceedings of the 27th ACM International Conference on Information and Knowledge Management, CIKM 2018, Torino, Italy, 22–26 October 2018, pp. 1511–1514 (2018)
4. Barbieri, N., Bonchi, F., Manco, G.: Efficient methods for influence-based network-oblivious community detection. ACM TIST **8**(2), 32:1–32:31 (2017)
5. Blei, D.M., Ng, A.Y., Jordan, M.I.: Latent Dirichlet allocation. J. Mach. Learn. Res. **3**, 993–1022 (2003)
6. Blondel, V.D., Guillaume, J.-L., Lambiotte, R., Lefebvre, E.: Fast unfolding of communities in large networks. J. Stat. Mech: Theory Exp. **2008**(10), P10008 (2008)
7. Cao, J., Wang, H., Jin, D., Dang, J.: Combination of links and node contents for community discovery using a graph regularization approach. Future Gener. Comput. Syst. **91**, 361–370 (2019)
8. Chakraborty, T., Cui, Z., Park, N.: Metadata vs. ground-truth: a myth behind the evolution of community detection methods. In: Companion of the The Web Conference 2018 on The Web Conference 2018, WWW 2018, Lyon, France, 23–27 April 2018, pp. 45–46 (2018)
9. Deng, D., Shahabi, C., Demiryurek, U., Zhu, L., Yu, R., Liu, Y.: Latent space model for road networks to predict time-varying traffic. In: Proceedings of the 22nd ACM SIGKDD International Conference on Knowledge Discovery and Data Mining, San Francisco, CA, USA, 13–17 August 2016, pp. 1525–1534 (2016)
10. Diehl, C.P., Namata, G., Getoor, L.: Relationship identification for social network discovery. In: Proceedings of the Twenty-Second AAAI Conference on Artificial Intelligence, Vancouver, British Columbia, Canada, 22–26 July 2007, pp. 546–552 (2007)
11. Dunlavy, D.M., Kolda, T.G., Acar, E.: Temporal link prediction using matrix and tensor factorizations. TKDD **5**(2), 10:1–10:27 (2011)

12. Fani, H., Bagheri, E., Du, W.: Temporally like-minded user community identification through neural embeddings. In: Proceedings of the 2017 ACM on Conference on Information and Knowledge Management, CIKM 2017, Singapore, 06–10 November 2017, pp. 577–586 (2017)
13. Milani Fard, A., Bagheri, E., Wang, K.: Relationship prediction in dynamic heterogeneous information networks. In: Azzopardi, L., Stein, B., Fuhr, N., Mayr, P., Hauff, C., Hiemstra, D. (eds.) ECIR 2019. LNCS, vol. 11437, pp. 19–34. Springer, Cham (2019). https://doi.org/10.1007/978-3-030-15712-8_2
14. Goyal, P., Chhetri, S.R., Canedo, A.: dyngraph2vec: capturing network dynamics using dynamic graph representation learning. Knowl.-Based Syst. (2019)
15. Grover, A., Leskovec, J.: node2vec: scalable feature learning for networks. In: Proceedings of the 22nd ACM SIGKDD International Conference on Knowledge Discovery and Data Mining, San Francisco, CA, USA, 13–17 August 2016, pp. 855–864 (2016)
16. He, D., Liu, D., Jin, D., Zhang, W.: A stochastic model for detecting heterogeneous link communities in complex networks. In: Proceedings of the Twenty-Ninth AAAI Conference on Artificial Intelligence, Austin, Texas, USA, 25–30 January 2015, pp. 130–136 (2015)
17. Hu, Z., Yao, J., Cui, B.: User group oriented temporal dynamics exploration. In: Proceedings of the Twenty-Eighth AAAI Conference on Artificial Intelligence, Québec City, Québec, Canada, 27–31 July 2014, pp. 66–72 (2014)
18. Hu, Z., Yao, J., Cui, B., Xing, E.P.: Community level diffusion extraction. In: Proceedings of the 2015 ACM SIGMOD International Conference on Management of Data, Melbourne, Victoria, Australia, 31 May–4 June 2015, pp. 1555–1569 (2015)
19. Jin, D., Chen, Z., He, D., Zhang, W.: Modeling with node degree preservation can accurately find communities. In: Proceedings of the Twenty-Ninth AAAI Conference on Artificial Intelligence, Austin, Texas, USA, 25–30 January 2015, pp. 160–167 (2015)
20. Koren, Y.: Collaborative filtering with temporal dynamics. Commun. ACM **53**(4), 89–97 (2010)
21. Labatut, V., Dugué, N., Perez, A.: Identifying the community roles of social capitalists in the Twitter network. In: 2014 IEEE/ACM International Conference on Advances in Social Networks Analysis and Mining, ASONAM 2014, Beijing, China, 17–20 August 2014, pp. 371–374 (2014)
22. Li, C., Cheung, W.K., Ye, Y., Zhang, X., Chu, D., Li, X.: The author-topic-community model for author interest profiling and community discovery. Knowl. Inf. Syst. **44**(2), 359–383 (2015)
23. McPherson, M., Smith-Lovin, L., Cook, J.M.: Birds of a feather: homophily in social networks. Ann. Rev. Sociol. **27**(1), 415–444 (2001)
24. Myers, S.A., Leskovec, J.: The bursty dynamics of the Twitter information network. In: Proceedings of the 23rd International Conference on World Wide Web, WWW 2014, pp. 913–924. ACM, New York (2014)
25. Rotta, R., Noack, A.: Multilevel local search algorithms for modularity clustering. ACM J. Exp. Algorithmics **16** (2011)
26. Sarkar, P., Moore, A.W.: Dynamic social network analysis using latent space models. In: Advances in Neural Information Processing Systems 18 [Neural Information Processing Systems, NIPS 2005, Vancouver, British Columbia, Canada, 5–8 December 2005], pp. 1145–1152 (2005)
27. Singer, U., Guy, I., Radinsky, K.: Node embedding over temporal graphs. In: Proceedings of the Twenty-Eighth International Joint Conference on Artificial Intelligence, IJCAI 2019, Macao, China, 10–16 August 2019, pp. 4605–4612 (2019)

28. Snijders, T.A., Lomi, A.: Beyond homophily: Incorporating actor variables in statistical network models. Netw. Sci. **7**(1), 1–19 (2019)
29. Wu, C., Ahmed, A., Beutel, A., Smola, A.J., Jing, H.: Recurrent recommender networks. In: Proceedings of the Tenth ACM International Conference on Web Search and Data Mining, WSDM 2017, Cambridge, United Kingdom, 6–10 February 2017, pp. 495–503 (2017)
30. Yan, X., Guo, J., Lan, Y., Cheng, X.: A biterm topic model for short texts. In: 22nd International World Wide Web Conference, WWW 2013, Rio de Janeiro, Brazil, 13–17 May 2013, pp. 1445–1456 (2013)
31. Yang, L., Cao, X., He, D., Wang, C., Wang, X., Zhang, W.: Modularity based community detection with deep learning. In: Proceedings of the Twenty-Fifth International Joint Conference on Artificial Intelligence, IJCAI 2016, New York, NY, USA, 9–15 July 2016, pp. 2252–2258 (2016)
32. Yang, L., Cao, X., Jin, D., Wang, X., Meng, D.: A unified semi-supervised community detection framework using latent space graph regularization. IEEE Trans. Cybern. **45**(11), 2585–2598 (2015)
33. Ye, F., Chen, C., Zheng, Z.: Deep autoencoder-like nonnegative matrix factorization for community detection. In: Proceedings of the 27th ACM International Conference on Information and Knowledge Management, CIKM 2018, Torino, Italy, 22–26 October 2018, pp. 1393–1402 (2018)
34. Zhu, L., Guo, D., Yin, J., Steeg, G.V., Galstyan, A.: Scalable temporal latent space inference for link prediction in dynamic social networks. IEEE Trans. Knowl. Data Eng. **28**(10), 2765–2777 (2016)
35. Zuo, Y., Liu, G., Lin, H., Guo, J., Hu, X., Wu, J.: Embedding temporal network via neighborhood formation. In: Proceedings of the 24th ACM SIGKDD International Conference on Knowledge Discovery & Data Mining, KDD 2018, London, UK, 19–23 August 2018, pp. 2857–2866 (2018)
36. Zuo, Y., Zhao, J., Xu, K.: Word network topic model: a simple but general solution for short and imbalanced texts. Knowl. Inf. Syst. **48**(2), 379–398 (2015). https://doi.org/10.1007/s10115-015-0882-z

KvGR: A Graph-Based Interface for Explorative Sequential Question Answering on Heterogeneous Information Sources

Hans Friedrich Witschel(✉), Kaspar Riesen, and Loris Grether

FHNW University of Applied Sciences and Arts Northwestern Switzerland, 4600 Olten, Switzerland
{hansfriedrich.witschel,kaspar.riesen,loris.grether}@fhnw.ch

Abstract. Exploring a knowledge base is often an iterative process: initially vague information needs are refined by interaction. We propose a novel approach for such interaction that supports sequential question answering (SQA) on knowledge graphs. As opposed to previous work, we focus on exploratory settings, which we support with a visual representation of graph structures, helping users to better understand relationships. In addition, our approach keeps track of context – an important challenge in SQA – by allowing users to make their focus explicit via subgraph selection. Our results show that the interaction principle is either understood immediately or picked up very quickly – and that the possibility of exploring the information space iteratively is appreciated.

Keywords: Sequential question answering · Graph databases · Natural language interfaces

1 Introduction

Today's information repositories are numerous, diverse and often very large. There is an increasing demand for accessing and querying these repositories using questions posed in natural language. While there is a long history of research in the fields of Question Answering (over both structured and unstructured content) and Natural Language Interfaces to Databases (NLIDB), as further elaborated in Sect. 2, the field of *(Complex) Sequential Question Answering* [5,14] is still rather new.

Possibly fuelled by the rise of chatbot technology and the resulting expectations of users, it claims that a more interactive approach to both fields will better meet user needs. Its main assumption is that users do not simply ask a question to a knowledge base and then quit. Instead, users tend to break down complex questions into a series of simple questions [5]. In addition, as known from exploratory search [12], users who do not have a very clearly articulated information need and/or who aim at getting familiar with a new field of knowledge tend to ask series of questions where one answer triggers the next question. That is, a user might ask a rather "fuzzy" first question (such as "what are

J. M. Jose et al. (Eds.): ECIR 2020, LNCS 12035, pp. 760–773, 2020.
https://doi.org/10.1007/978-3-030-45439-5_50

important topics in the field of 'Information Retrieval'?") and then – when studying the answer – start to think of new questions, concerning some of the new concepts found in that answer. Although the concept of exploratory search is well known from the field of information retrieval, this exploratory motivation for performing sequential question answering (over structured knowledge bases) has not been studied so far. In any case, sequential question answering raises the major challenge of keeping track of context: since they assume the context to be known from the prior questions and answers, users tend to leave away sentence elements [14].

Especially in exploratory search settings, answers to fuzzy questions can be very complex, involving a large number of concepts and relations. Hence, researchers have proposed various kinds of visualisations in order to aid users in grasping such complexity and studying relationships between concepts [1,20].

In our work, we aim at building a context-aware sequential question answering system, especially suited for exploratory search. To this end, the solution is based on a knowledge graph – which integrates information from various structured and unstructured data sources, see Sect. 3.1. Since the visualization of graphs provides an intuitive overview of complex structures and relationships [2], our system allows users to ask questions in natural language, but provides answers via a visual representation of subgraphs of the underlying knowledge graph. It supports both the user and the system in keeping track of the context/current focus of the search via a novel interaction concept that combines pointing/clicking and asking questions in natural language, described in Sect. 3.2.

We will show empirically that users appreciate the new interaction concept and its ability to define context and focus graphically, see Sect. 4.

2 Related Work

Both question answering and natural language interfaces to databases (NLIDB, see [6] for a survey) have a long history. They share many characteristics since both support querying of knowledge bases using natural language. Many question answering systems retrieve answers from textual (i.e. unstructured) resources, but there are also many approaches based on structured content, often in the form of ontologies [11].

In NLIDB, many challenges have been addressed, e.g. making systems domain-independent [10] or overcoming specific difficulties with certain query languages, above all SQL [22]. Recent advances in this area are relying on sequence-to-sequence models [7,17], based on encoding and decoding of sequences via deep (reinforcement) learning. An obvious drawback of these supervised learning approaches – as opposed to earlier hand-crafted rule-based grammars – is the amount of training data required. Although large hand-annotated datasets have been published [24,25], trained models cannot be expected to be fully domain-independent.

While the fields of Question Answering (over structured data), Semantic Parsing and NLIDB are obviously quite advanced, researchers have only recently begun to study the domain of "Sequential Question Answering" (SQA). This new focus on interactive, dialog-driven access to knowledge bases is based on the insight that users rarely pose a question to such a knowledge base and then quit [3,14]. Instead, a more common and natural access pattern consists in posing a series of questions. Most researchers in SQA assume that the motivation for dialogs comes from the need to decompose complex questions into simple ones [5,14]. Some researchers propose to perform such decomposition algorithmically [15], while others provide evidence that it is more natural and realistic to assume that humans will like to perform this decomposition themselves, resulting in a series of simple, but inter-related questions [5]. A key challenge in any form of sequential or conversational question answering is the resolution of ellipses (e.g. omissions of arguments in relations) or anaphora which are very frequent in a dialogue where the user expects the system to keep track of the context [5,9,14].

These approaches all assume that a searcher always accesses a knowledge base with a clear question in mind. As outlined above, we advocate a wider perspective on SQA, including scenarios of an exploratory nature. In information retrieval, it has been thoroughly accepted that there exist situations in which users are unable to clearly articulate information needs, e.g. when trying to get acquainted with a new field where terminology is still unknown [12]. Thus, users would like to *explore*, and often their questions become better articulated as they learn more about the new field.

In order to support them in grasping relationships between new concepts in the – often very complex – answers to their fuzzy questions, IR researchers have proposed result set visualisations that provide a better overview than the typical ranked lists of document references [1,20].

Using visualisations, especially of graphs/ontologies as an output of retrieval systems has also been proposed, mainly in QA and NLIDB that are based on knowledge graphs [2,13,23].

Visualising graph query results is different from visualising graphs in general; the former resembles generation of results snippets in text retrieval [16]. However, we can learn and employ mechanisms from general approaches to analysing large graphs, e.g. by applying global ranking mechanisms (such as PageRank) or by summarizing properties of selected nodes [8]. As pointed out in [19], visual graph analysis requires, besides the visual representation of graph structures, to have good interaction mechanisms and algorithmic analysis, such as aggregation/merging of nodes, identification of certain graph structures (such as cliques) or node ranking mechanisms such as PageRank.

Additional challenges originate in the fuzziness of natural language and the potential resulting number of (partially) matching result graphs. Graph summarization approaches have been proposed as a solution [21,23] – where summarized/aggregated graph structures play the role of snippets. Another approach [4] uses result previews to narrow down result sets via "early" user interaction.

2.1 Contribution

While approaches to semantic parsing, NLIDB and question answering over structured data are well studied, there is a recent rise in interest in better studying and supporting the interaction in sequential question answering (SQA) scenarios.

However, the emerging field of SQA lacks – in our opinion – a clear idea of why users want to engage in a conversation. We claim that one important motivation can be found in exploratory settings where users need to first gain insights by interacting with a knowledge base, before being able to ask the "right" questions. Another challenge in SQA is keeping track of context: in their survey on semantic parsing, Kamath & Das [6] mention "adding human in the loop for query refinement" as a promising future research direction in cases where the system is uncertain in its predictions.

Our contribution consists mainly in proposing a new interaction paradigm which allows users to ask questions in natural language and to receive answers in the form of visualised subgraphs of a knowledge graph. Users can then interact with that subgraph to define the focus of their further research, before asking the next question. With this human involvement, we can show empirically both how the human benefits from clarifying the search direction while exploring the knowledge graph and how the machine is supported in understanding incomplete questions better because their context is made explicit.

We further use a robust query relaxation approach to trade precision for recall when recall is low. Our approach is domain-independent and does not require training data – it only requires a specification of node type names and their possible synonyms. It can be seen as a "traditional" and simple grammar-based approach – the focus is not on sophisticated semantic parsing (we might add e.g. sequence-to-sequence models later), but on the interactive process of graph exploration via natural language.

3 The Retrieval System

3.1 Graph-Based Integration of Heterogeneous Information Sources

The knowledge graph underlying our experiments was constructed out of a collection of heterogeneous sources and stored in a Neo4j graph database[1]. For our experiments, we chose books as a domain and aimed at retrieving all information – from various sources – which users (leisure-time readers, students, ...) might find relevant, ranging from core bibliographic information, over author-related information (affiliation/prizes won) to reviews and social media coverage of books.

To populate it, we implemented a collection of parsers for a variety of data sources.[2]:

[1] https://neo4j.com/.

[2] Note: these are meant to illustrate different *types* of data, the concrete sources do not matter in general.

- For *structured data*, we built an XML parser (which can be applied to structured XML databases, but also for semi-structured XML files) and an RDF parser. The XML parser was used to integrate a sample of data from the bibliographic platform iPEGMA[3], while the RDF parser was applied to the DBPedia SPARQL endpoint[4] to retrieve data about books, persons, their institutes and awards. The iPEGMA data covers mostly German books while DBPedia data is focused on English books.
- In terms of *semi-structured data*, our HTML parser can process web content and a special Twitter parser deals with Tweets (and uses the HTML parser to process web pages linked from tweets). We applied the HTML parser to the websites literaturkritik.de and www.complete-review.com to retrieve book reviews and related book metadata in German and English. The Twitter parser was applied to a collection of Twitter accounts of major publishers whose timelines were analysed for tweets referring to books.
- We also integrated a sentiment analysis service (Aylien Text API[5]) as a typical example of analysis of the *unstructured* part of webpages, i.e. the plain text. In our case, we applied the service to the book reviews from literaturkritik.de to find out whether reviews were positive or negative. For www.complete-review.com, this information could be parsed directly from the web page.

In Neo4j, it is not required to define a schema (i.e. node or relation types) before inserting nodes or relationships. We used this property heavily: each parser has a configuration file in which one can define node and relation types to be extracted. We have developed a special syntax with which one can define the patterns to be searched within the various data sources to retrieve the corresponding data. This means that parsers can be extended to find new types of nodes and relationships and/or cover new data sources of known type, without the need to modify the program code of the parser. Typically, the specifications for various data sources have overlapping node types, thus resulting in a data integration task. In order to match identical nodes (e.g. the same book) found in different data sources, the definitions also specify a "uniquneness attribute" (similar to a primary key in relational databases). As a result, the knowledge base consists of a single integrated graph.

We have chosen a graph database because graphs are a very natural way of modeling relationships and are easy to visualise and interact with [2].

3.2 The Interaction Concept

As laid out in Sect. 2, most previous work sees sequential question answering as a conversation in which complex questions are broken down into simpler ones. For instance, Iyyer et al. [5] assume that users have already at the initial state

[3] https://ipegma.com/.
[4] http://dbpedia.org/sparql.
[5] https://aylien.com/text-api/.

of a conversation a complex question in mind – which they then decompose into simpler ones.

In contrast, our new interaction concept aims at supporting scenarios that are more *exploratory* in nature (cf. exploratory search in text retrieval [12]). In such settings, users often ask series of questions that emerge one from another – i.e. the answer to a first question triggers the next one etc. – without the final goal of such a conversation being clear initially.

We propose a novel interaction mechanism for such an exploratory "conversation", where questions are posed in natural language, but answers are given in the form of subgraph visualisations, with a possibility to interact and select parts of subgraphs for further exploration (again via asking questions). Note that it does not play a role whether a user starts from general concepts to "zoom in" to more specific ones or vice versa.

In exploratory search, it is typical that – since the nature of the problem is unclear to the user – queries are imprecise or "tentative" [20]. This implies very often that the *answers* – much more than the questions or queries – can be quite complex. As pointed out in [1], systems that support exploration hence often offer visualisation of search results as well as interaction mechanisms for further exploration.

In our case, results are (possibly large) subgraphs of a given knowledge graph. By studying such a subgraph and interacting with it, a user may *learn* about important concepts and relations in a domain – and this leads to asking the next question(s). A next question may aim at either *filtering* the current subgraph or further broadening the scope by *expanding* a subgraph region with further related nodes.

The design of our interaction concept was informed by a questionnaire which was filled out by a sample of 16 students. Participants received a description of a situation (e.g. having read a good book) and were asked to formulate some questions that they would have in such a situation. We analysed their answers, looking for common patterns of questions and expected result sets.

Our resulting interaction concept is very simple: based on an initial keyword search or question in step 0, a user finds an entry point into the graph, i.e. an initial subgraph G_0.

From this point on – provided that the user would like to continue the current session – there are two main possibilities for exploration in each step i:

1. Use the graphical user interface, e.g. expand the subgraph G_i by unhiding all nodes related to a chosen node.
2. Select a node or a set of nodes N_{G_i} as a "context" and ask a question about it. Selection can be done
   (a) directly via one or more clicks on nodes or
   (b) by selecting all nodes of a certain type via a button.

Each interaction leads to a new graph G_{i+1}.

While option 1 is not new, option 2 can lead to a new form of sequential question answering, with questions being asked in natural language and answers given as visualisations of subgraphs. This combination is user-friendly since on the one hand – as a basis of all NLIDB research and conversational interfaces – natural language is the most natural form of expressing information needs. On the other hand, researchers in both information retrieval [1] and graph querying [23] communities use visualisations for improving the user-friendliness of exploratory search.

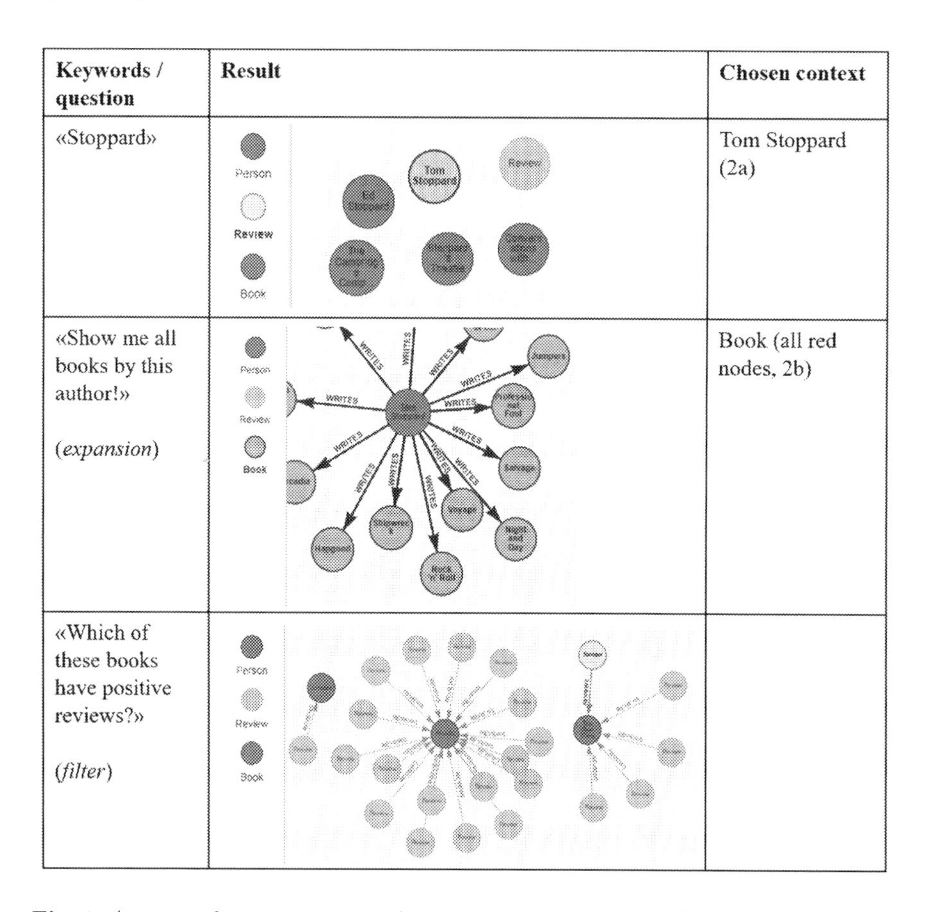

Keywords / question	Result	Chosen context
«Stoppard»		Tom Stoppard (2a)
«Show me all books by this author!» (*expansion*)		Book (all red nodes, 2b)
«Which of these books have positive reviews?» (*filter*)		

Fig. 1. An exemplary conversation between a user and KvGR (Color figure online)

In addition, we claim (and will later show empirically) that, while it is not natural for users to repeat entity names from an earlier question, it is rather natural for them to select preliminary results and thus make context explicit. We will show that such selection is even often helpful for their own understanding of how a question-answer-sequence develops and what they have learned so far/what they want to learn next.

Since the user specifies the context explicitly when using option 2, it is easy for our system to fill in missing parts of questions by simply assuming that they originate from that context.

Figure 1 illustrates the interaction concept with a small "exploration session" from the book domain (see Sect. 3.1). In short, the session consists in a user searching for an author, then demanding to see all books from that author and finally asking which of these books have positive reviews. Note how the visualisation of the result graph helps her to get a quick overview of complex structures – for instance to see at a glance which books have many vs. few positive reviews (yellow nodes) in the last result.

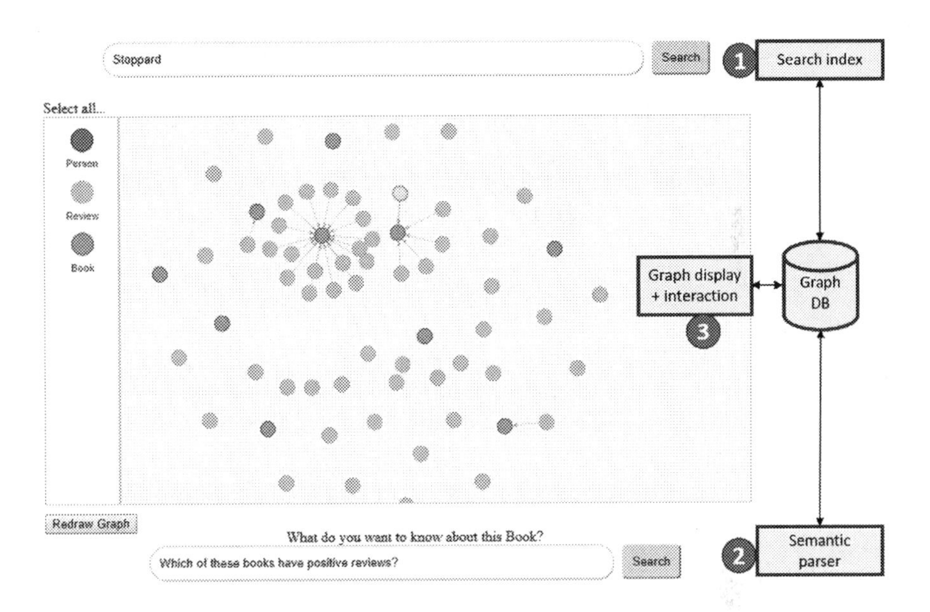

Fig. 2. A screenshot of the KvGR core UI components

3.3 The KvGR Architecture

In order to realise the interaction described in the previous section, KvGR builds several components on top of the knowledge graph (see Sect. 3.1). All of these components are visible on the user interface, the numbers in Fig. 2 refer to the corresponding (backend) components in the following enumeration:

1. **Fielded keyword search**: each node in the knowledge graph is treated as a document and its (textual) attributes as fields. Field weights are domain-specific – in the book domain the "title" field of books will have a higher weight than e.g. the "genre" field. The number of shown nodes is limited by applying a cut-off to node scores.
2. **Semantic parser**, see Sect. 3.4
3. **Graph visualisation and interaction**, allowing common basic graph interactions, plus selecting a context, see Sect. 3.2.

3.4 Semantic Parser

Since semantic parsing is not the core contribution of our work, we have built a simple, but robust grammar for parsing. It takes advantage of the interaction concept and the basic principles of graphs, but makes no further assumptions about the graph schema – it can be adapted easily to new domains simply by providing a lexicon of node types (see below).

The grammar consists of JAPE rules in GATE[6], which annotate occurrences of graph nodes in user utterances, based on a simple lookup mechanism using a lexicon with manually maintained synonyms. Each annotation is associated with a number of features, see Fig. 3.

The annnotated questions are then passed to a Cypher generator, which simply takes all nodes found in an utterance and generates a relationship pattern that is matched against the graph.

We illustrate our parser with the example shown in Fig. 3.

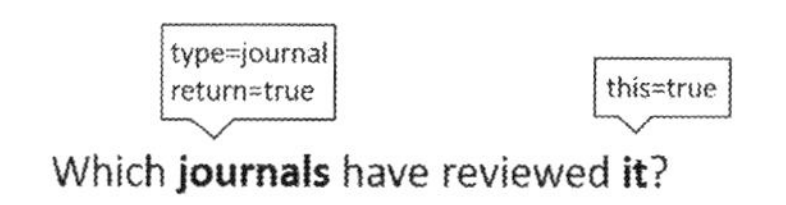

Fig. 3. A user utterance, with annotated nodes

The parts of the question recognised as nodes are put in bold font, their extracted features are presented in the box above. The grammar has marked "journal" as a return node type and "it" as referring to a current user selection ("this=true"). Here, the interaction concept is exploited: because the user has selected a book (let us assume, the book with id 629025), the system can assume that the pronoun "it" refers to that current selection (the same would apply to a phrase like "this book").

This information is enough for the Cypher generator to generate a Cypher query as follows:

```
match (x:Journal)-[]-(y:Book) where ID(y) IN [629025] return x,y
```

This query, however, will not retrieve anything since the question contains an ellipsis: it should actually be formulated as "Which **journals** have published a **review** about **it**?". That is, the system needs to extend the pattern to allow an intermediate node type related to both the current selection and the return type nodes.

To this end, we have implemented a *query relaxation mechanism* which will first try out the above Cypher query and then – if nothing is returned – will relax the query by allowing an ellipsis like this:

```
match (x:Journal)-[]-(z)-[]-(y:Book) where ID(y) IN [629025] return x,y,z
```

[6] https://gate.ac.uk/sale/thakker-jape-tutorial/GATE%20JAPE%20manual.pdf.

The system does not know/specify that the intermediate node z is of type Review – thus a negative impact on retrieval precision might result, which we trade for recall here.

4 Evaluation and Discussion

In order to evaluate our main hypothesis – namely that our new interaction mechanism effectively supports users in iteratively refining an exploratory search – we performed user tests in an exploratory search scenario.

To make the sessions more comparable, we pre-defined the information needs: the "story" started with a keyword search for the topic "criminal law" and was continued with some typical questions about e.g. prominent authors in that field, authors who had won prizes, their institutes, as well as books with positive reviews in that field. Before each session, participants were instructed about the features of the system via a short demo. Within the session, the predefined information needs were explained and users were asked to interact with the system to satisfy them. When users got stuck with interaction or query formulation, help was offered.

Following the popular "five-user assumption" of usability testing [18], we recruited 5 participants, 2 colleagues from our School of Business and 3 of our students. All subjects were not previously aware of our project. This selection was made for practical feasibility reasons – we are aware of the bias, in terms of user characteristics, that it introduces.

4.1 Observations

Participants received overall 5 different information needs (q_1 to q_5). The first one (q_1) started from a single node (the topic "criminal law"), i.e. a context selection was not required. All subsequent ones required participants to select a subset of the nodes that were currently displayed (e.g. all books or all persons). The last information need (q_5) was formulated in a complex way ("which authors that have written a book about criminal law have also written a review?") and required participants to recognise that a partial result to the question was already available from a previous step.

We observed the participants' difficulties to (a) formulate queries that the semantic parser would understand correctly, (b) grasp the principle of breaking down complex information needs into simpler ones (here, participants would typically try to extend the previous query by adding more constraints) and to (c) remember to select a subset of nodes as a context for their next query.

Table 1 shows the number of participants facing these problems for each of the test queries. In terms of query reformulation, there is no clear pattern – we observed a number of ways in which our grammar can be improved.

Grasping the process of iterative refinement shows a clear learning curve: while two participants had understood the principle immediately from the introductory demo, the other three needed only one experience with q_2 to grasp it.

Table 1. Number of test persons encountering problems for each test query

Type of problem	q_1	q_2	q_3	q_4	q_5
Reformulation required	1	5	3	0	2
Struggle with breakdown	–	3	0	0	2
No context selection	–	3	2	1	1

We observed that the problems with q5 resulted merely from participants not accurately understanding the complex question – they both said that it would have been different if it had been their own information need.

Remembering to select a subset of nodes as a context was harder: while two participants never forgot to do it, one needed q_2, another one q_2 and q_3 to remember it; one participant could not get used to it until the end of the test. The persons who struggled expressed their expectation that – if they did not select any nodes, but asked a question like "which of these persons..." – the system should automatically assume that it referred to all currently visible persons. Since this is easy to build into our system, we can conclude that context selection will not be an issue once the principle of iterative refinement has been grasped.

4.2 Feedback

Besides observing the query formulation and interaction strategies of the users – including their need for help – we asked the users to give us feedback on the following points:

- **Intuitiveness of context selection:** three participants stated that they found it intuitive and natural to select a context for their query and to break down complex questions. The other two expressed their expectation for the system to identify context automatically (see above).
- **Results of elliptic queries**: queries containing "intermediate nodes", e.g. a query "show me all authors who have written about criminal law" would show not only authors, but also their books, although the question did not ask for books. Only one participant had difficulties in understanding what was shown (because the legend was not clear to him). When judging the result, 4 participants said that seeing the books was interesting, especially for someone wishing to explore criminal law as a new area, while 3 participants remarked that the result was not strictly what they had asked for. Two participants stated that they would appreciate to see a list of persons – in addition to the graph visualisation.
- **General feedback** on the interaction was very positive. Despite the observed difficulties that did occur with query formulation, all participants said that they were impressed with the ability of the system to understand queries in natural language. Four participants mentioned explicitly that the visual representation helped them to better understand relationships and to see "how

things belong together". One participant said that it sparked his curiosity to explore further. All participants stated that the interaction mechanism was either "intuitive" or at least "easy to learn" (because, as they stated, "the effect of what you do is always visible") and three of them mentioned expressly that they liked the refinement process of breaking down complex queries.
Participants also came forth with a number of suggestions for improvement: two participants stated that they would appreciate if the system could understand – besides fully formulated questions – keyword-based inputs. The same participants and a third one expressed their wish to have result lists, in addition to a graph. The main reason mentioned for this was the lack of a ranking provided in the graph. The participants said that they would not know where to start looking if a result graph grew too large.
- **Comparison to traditional interfaces**, especially ones with list-based result presentation: participants said that our system would be more effective in supporting "detailed investigation" that required to "understand relationships", whereas traditional list-based systems would be better suited to get an overview of e.g. the most important books on criminal law because of their clear ranking.

5 Conclusions

In this work, we have proposed a novel context-aware sequential question answering system, especially suited for exploratory search, based on graph visualisation for result presentation and iterative refinement of information needs. This refinement in turn is based on the selection of subsets of nodes for context definition and natural language questions towards this context.

Our results are somewhat limited by the specific scenario and use case that we explored and the small user group involved. However, they do show quite clearly that users either understand the principle immediately or pick it up very quickly – and that they appreciate the possibility of exploring the information space iteratively. Having to explicitly select context is hard to get used to for some, and should be automated. The visual representation of results was well received for its support of understanding relationships. On the other hand, it became clear that ranking or highlighting the more "relevant" nodes will be needed to help users focus, especially when results get larger.

Thus, our main goal for future work will be to investigate the best way to incorporate node scoring into the system – either visually (e.g. via node sizes) or by providing ranked result lists in addition to and linked to the graph. Because of the limitations of our participant selection strategy, further test with a more varied user group will also be required. Finally, it might be interesting to explore the possibility for users to combine search results (sub-graphs) of queries before exploring the combined results further.

References

1. Ahn, J.W., Brusilovsky, P.: Adaptive visualization for exploratory information retrieval. Inf. Process. Manag. **49**(5), 1139–1164 (2013)
2. Bhowmick, S.S., Choi, B., Li, C.: Graph querying meets hci: state of the art and future directions. In: Proceedings of the 2017 ACM International Conference on Management of Data, pp. 1731–1736. ACM (2017)
3. Guo, D., Tang, D., Duan, N., Zhou, M., Yin, J.: Dialog-to-action: conversational question answering over a large-scale knowledge base. In: Advances in Neural Information Processing Systems, pp. 2942–2951 (2018)
4. Hung, H.H., Bhowmick, S.S., Truong, B.Q., Choi, B., Zhou, S.: QUBLE: towards blending interactive visual subgraph search queries on large networks. VLDB J. Int. J. Very Large Data Bases **23**(3), 401–426 (2014). https://doi.org/10.1007/s00778-013-0322-1
5. Iyyer, M., Yih, W.T., Chang, M.W.: Search-based neural structured learning for sequential question answering. In: Proceedings of the 55th Annual Meeting of the Association for Computational Linguistics (Volume 1: Long Papers), pp. 1821–1831 (2017)
6. Kamath, A., Das, R.: A survey on semantic parsing. arXiv preprint arXiv:1812.00978 (2018)
7. Keneshloo, Y., Shi, T., Ramakrishnan, N., Reddy, C.K.: Deep reinforcement learning for sequence to sequence models. arXiv preprint arXiv:1805.09461 (2018)
8. Koutra, D., Jin, D., Ning, Y., Faloutsos, C.: Perseus: an interactive large-scale graph mining and visualization tool. Proc. VLDB Endowment **8**(12), 1924–1927 (2015)
9. Kumar, V., Joshi, S.: Incomplete follow-up question resolution using retrieval based sequence to sequence learning. In: Proceedings of the 40th International ACM SIGIR Conference on Research and Development in Information Retrieval, pp. 705–714. ACM (2017)
10. Llopis, M., Ferrández, A.: How to make a natural language interface to query databases accessible to everyone: an example. Comput. Stand. Interfaces **35**(5), 470–481 (2013)
11. Lopez, V., Uren, V., Sabou, M., Motta, E.: Is question answering fit for the semantic web?: a survey. Seman. Web **2**(2), 125–155 (2011)
12. Marchionini, G.: Exploratory search: from finding to understanding. Commun. ACM **49**(4), 41–46 (2006)
13. Park, C.S., Lim, S.: Efficient processing of keyword queries over graph databases for finding effective answers. Inf. Process. Manag. **51**(1), 42–57 (2015)
14. Saha, A., Pahuja, V., Khapra, M.M., Sankaranarayanan, K., Chandar, S.: Complex sequential question answering: towards learning to converse over linked question answer pairs with a knowledge graph. In: Thirty-Second AAAI Conference on Artificial Intelligence (2018)
15. Talmor, A., Berant, J.: The web as a knowledge-base for answering complex questions. arXiv preprint arXiv:1803.06643 (2018)
16. Turpin, A., Tsegay, Y., Hawking, D., Williams, H.E.: Fast generation of result snippets in web search. In: Proceedings of the 30th Annual International ACM SIGIR Conference on Research and Development in Information Retrieval, pp. 127–134. ACM (2007)
17. Utama, P., et al.: An end-to-end neural natural language interface for databases. arXiv preprint arXiv:1804.00401 (2018)

18. Virzi, R.: Refining the test phase of usability evaluation: how many subjects is enough. Hum. Factors **34**(4), 457–468 (1992)
19. Von Landesberger, T., et al.: Visual analysis of large graphs: state-of-the-art and future research challenges. Comput. Graph. Forum **30**, 1719–1749 (2011). Wiley Online Library
20. White, R.W., Kules, B., Bederson, B.: Exploratory search interfaces: categorization, clustering and beyond: report on the XSI 2005 workshop at the Human-Computer Interaction Laboratory, University of Maryland. In: ACM SIGIR Forum, vol. 39, pp. 52–56. ACM (2005)
21. Wu, Y., Yang, S., Srivatsa, M., Iyengar, A., Yan, X.: Summarizing answer graphs induced by keyword queries. Proc. VLDB Endowment **6**(14), 1774–1785 (2013)
22. Xu, X., Liu, C., Song, D.: SQLNet: generating structured queries from natural language without reinforcement learning. arXiv preprint arXiv:1711.04436 (2017)
23. Yang, S., et al.: SLQ: a user-friendly graph querying system. In: Proceedings of the 2014 ACM SIGMOD International Conference on Management of Data, pp. 893–896. ACM (2014)
24. Yu, T., et al.: Spider: a large-scale human-labeled dataset for complex and cross-domain semantic parsing and text-to-SQL task. arXiv preprint arXiv:1809.08887 (2018)
25. Zhong, V., Xiong, C., Socher, R.: Seq2sql: generating structured queries from natural language using reinforcement learning. arXiv preprint arXiv:1709.00103 (2017)

Answering Event-Related Questions over Long-Term News Article Archives

Jiexin Wang[1](✉), Adam Jatowt[1], Michael Färber[2], and Masatoshi Yoshikawa[1]

[1] Department of Social Informatics, Kyoto University, Kyoto, Japan
wang.jiexin.83m@st.kyoto-u.ac.jp
[2] Karlsruhe Institute of Technology, Karlsruhe, Germany

Abstract. Long-term news article archives are valuable resources about our past, allowing people to know detailed information of events that occurred at specific time points. To make better use of such heritage collections, this work considers the task of large scale question answering on long-term news article archives. Questions on such archives are often event-related. In addition, they usually exhibit strong temporal aspects and can be roughly categorized into two types: (1) ones containing explicit temporal expressions, and (2) ones only implicitly associated with particular time periods. We focus on the latter type as such questions are more difficult to be answered, and we propose a retriever-reader model with an additional module for reranking articles by exploiting temporal information from different angles. Experimental results on carefully constructed test set show that our model outperforms the existing question answering systems, thanks to an additional module that finds more relevant documents.

Keywords: News archives · Question answering · Information Retrieval

1 Introduction

With the application of digital preservation techniques, more and more old news articles are being digitized and made accessible online. News article archives help users to learn detailed information on events that occurred at specific time points in the past and constitute part of our heritage [1]. Some professionals, like historians, sociologists, or journalists need to deal with these time-aligned document collections for a variety of purposes [2]. Moreover, average users can verify information about the past using original, primary resources. However, it is difficult for users to efficiently make use of news archives due to their large sizes and complexities. Large scale question answering systems (QA systems) can solve the problem, with the aim to identify the most correct answer from relevant documents for a particular information need, expressed as a natural language question. User questions on these archives are often event-related and include temporal aspects. They can be divided into two types: (1) those with explicit temporal expressions (e.g., "NATO signed a peace treaty with which country in 1999?"), and (2) those only implicitly associated to time periods, hence not

J. M. Jose et al. (Eds.): ECIR 2020, LNCS 12035, pp. 774–789, 2020.
https://doi.org/10.1007/978-3-030-45439-5_51

Table 1. Examples of questions in our test set, their answers, and dates of their events

Questions	Answers	Event dates
Which party, led by Buthelezi, threatened to boycott the South African elections?	Inkatha Freedom Party	1993.08
What bill was signed by Clinton for firearms purchases?	Brady Bill	1993.11
Which federal prosecutor that led the investigation for the leak of identity of Valerie Plame?	Patrick J. Fitzgerald	2003.11
Riot in Los Angeles occurred because of the acquittal of how many officers in police department?	Four	1992.04
Which American professional pitcher died because his small airplane crashed in New York?	Cory Lidle	2006.10

containing any temporal expression (e.g., "How many members of International Olympic Committee were expelled or resigned because of the bribery scandal?"). We focus on the latter type, which is more challenging, as the temporal information cannot be obtained directly. Table 1 shows some examples of the questions that we use.

This paper presents a large scale question answering system called QANA (Question Answering in News Archives) designed specifically for answering event-related questions on news article archives. It exploits the temporal information of a question, of a document content and of its timestamp for reranking candidate documents. In the experiments, we use New York Times (NYT) archive as the underlying knowledge source and a carefully constructed test set of questions which are associated with past events. The questions are selected from existing datasets and history quiz websites, and they lack any temporal expressions which makes them particularly difficult to be answered. Experimental results show that our proposed system improves retrieval effectiveness and outperforms the existing QA systems commonly used for large scale question answering.

We make the following contributions: (a) we propose a new subtask of QA, which uses long-term news archives as the data source, (b) we build effective models for solving this task by exploiting temporal characteristics of both questions and documents, (c) we perform experiments to prove their effectiveness and construct a novel dedicated test set for evaluating QA on news archives.

The remainder of this paper is structured as follows. The next section overviews the related work. In Sect. 3, we introduce our model. Section 4 describes experimental settings and results. Finally, we conclude the paper in Sect. 5.

2 Related Work

Question Answering System. Current large scale question answering systems usually consist of two modules: (1) IR module (called also a document retriever module) responsible for selecting relevant articles from an underlying corpus and

(2) Machine Reading Comprehension (MRC) module (called also a document reader module) used to extract answer spans from relevant articles, typically, by using neural network models.

Latest MRC models, especially those that use Bert [3] can even surpass human-level performance (based on EM (Exact Match) and F1 scores) on both SQuAD 1.1 [4] and SQuAD 2.0 [5], the two most widely-used MRC datasets, where each question is connected with a given reading passage. However, recent studies [6–8] indicate that IR module is a bottleneck having a significant impact on the performance of the whole system (degraded performance of MRC component due to noisy input). Hence, few works tried to improve the IR task. Chen et al. [9] propose one of the most well-known large scale question answering system, DrQA whose IR component is based on a TF-IDF retriever that uses bigrams with TF-IDF matching. Wang et al. [7] introduce R^3 model, where IR component and MRC component are trained jointly by reinforcement learning. Ni et al. [10] propose ET-RR model, which improves IR part by identifying essential terms of a question and reformulating the query.

Nonetheless, as the existing question answering systems are essentially designed for synchronic document collections (e.g., Wikipedia), they are incapable of utilizing temporal information like document timestamp when answering questions on long-term news article archives, despite temporal information constituting an important feature of events reported by news articles. The questions and documents are then processed in the same way as on synchronic collections. Even though some temporal question answering systems that can exploit temporal information of question and document content have been proposed in the past [11,12], they are still designed for synchronic document collections (e.g., Wikipedia or Web) and they do not use document timestamps. Besides, they are based on traditional rule-based methods and their performance is rather poor.

In addition, there are very few resources available for temporal question answering. Jia et al. [13] propose a dataset with 1,271 temporal question-answer pairs where 209 pairs do is without any explicit temporal expression. However, only few pairs can be used in our case, as most are about events which happened long time ago (e.g., Viking Invasion of England) or are not event-related.

Our approach contains an additional module that is used for reranking documents which improves the retrieval of correct documents by exploiting temporal information from different angles. We not only utilize the inferred time scope information from the questions themselves, but also combine it with the document timestamp information and with temporal information embedded inside document content. To the best of our knowledge, no studies, as well as no available datasets that can help to design a question answering system on news article archives have been proposed so far. Building a system that makes full use of the past news articles and satisfies different user information needs is however of great importance due to the continuously growing document archives.

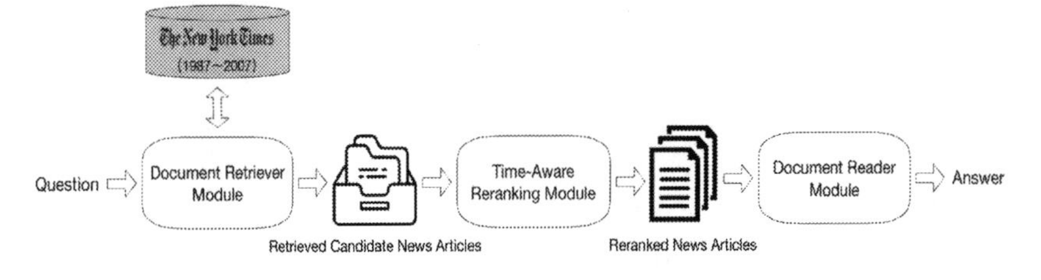

Fig. 1. The architecture of QANA system

Temporal Information Retrieval. In Information Retrieval (IR) domain, several research studies have already been proposed for temporal ranking of documents [14–16]. Li and Croft [17] introduce a time-based language model that takes into account timestamp information of documents to favor recent documents. Metzler et al. [18] propose a method that analyzes query frequencies over time to infer the implicit temporal information of queries and exploit this information for ranking results. Arikan et al. [19] design a temporal retrieval model that integrates temporal expressions of document content into query-likelihood language modeling. Berberich et al. [20] propose a similar model but also consider uncertainty in temporal expressions. However, in [19] and [20], the temporal scopes of queries are explicitly given in their setting and the proposed methods do not utilize timestamp information. Kanhabua and Nørvåg [21] propose three different methods to determine the implicit temporal scope of queries and exploit this temporal information to improve the retrieval effectiveness by reranking documents. [21] is probably the most related work to ours as it also linearly combines both textual and temporal similarity to rerank documents, however, that work does not use any temporal information embedded in document content and the linear combination is done in a static way. In our experiments, for comparison with [21] we will replace our ranking method in the reranking module with the best one proposed in [21].

All the above-mentioned temporal information retrieval methods are designed for short queries instead of questions, and none of them exploits both timestamps and content temporal information. We are the first to adapt and improve concepts from temporal information retrieval to the QA research domain, showing significant improvement in answering questions on long-term news archives.

3 Methodology

In this section, we present the proposed system that is designed specifically for answering questions over news archives. We focus on questions for which the time periods are not given explicitly, and so further knowledge is required for obtaining or inferring their time periods (e.g. "Who replaced Goss as the director of the Central Intelligence Agency?"). Figure 1 shows the architecture of QANA system which is composed of three modules: *Document Retriever Module*, *Time-Aware Reranking Module* and *Document Reader Module*. Compared with the architectures of other common large scale question answering systems, we add an additional component: Time-Aware Reranking Module which exploits temporal information from different angles for selecting the best documents.

3.1 Document Retriever Module

This module firstly performs keyword extraction and expansion, then retrieves candidate documents from the underlying document archive. First, single-token nouns, compound nouns, and verbs from each question are extracted based on analyzing part of speech (POS) and dependency information using spaCy[1]. After removing common stop words, the module expands keywords with their synonyms taken from WordNet [22]. The synonyms are further filtered by keeping those whose POS types match the original term in the question, and whose word embeddings[2] similarity to question terms is over 0.5. Finally, a query is issued to Solr [24] search engine which returns the top 300 documents ranked by BM25.

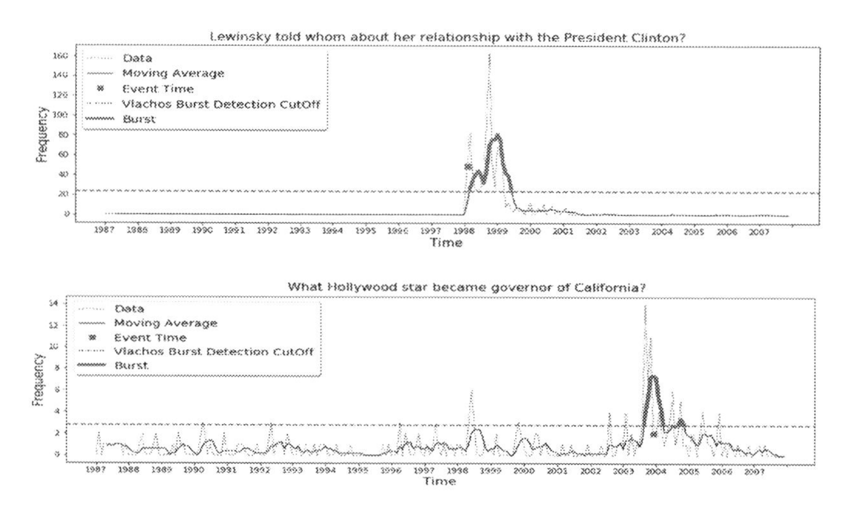

Fig. 2. Burst detection results of two questions (Color figure online)

3.2 Time-Aware Reranking Module

In this module, temporal information is exploited from different angles to rerank retrieved candidate documents. Since the time scope information of questions is not provided explicitly, the module firstly determines candidate periods of the time scope $T(Q)$ of a question Q. These are supposed to represent when an event mentioned in the question could occur. Each inferred candidate period is assigned a weight to indicate its importance. Then, the module contrasts the query time

[1] https://spacy.io/.

[2] We use Glove [23] embeddings trained on the Common Crawl dataset with 300 dimensions.

scope against the information derived from the document timestamp $t_{pub}(d)$ and the temporal information embedded inside document content $T_{text}(d)$, in order to compute two temporal scores $S_{pub}^{temp}(d)$ and $S_{text}^{temp}(d)$ for each candidate document d. Finally, both the textual relevance score $S^{rel}(d)$ and the final temporal score $S^{temp}(d)$ are used for document reranking.

Query Time Scope Estimation. Although the time scope information of the questions is not given explicitly, the distributions of relevant documents over time should provide information regarding temporal characteristics of the questions. Examining the timeline of a query's result set should allow us to characterize how temporally dependent the topic is. For example, in Fig. 2, the dashed lines of the data show the distribution of relevant documents obtained from the NYT archive per month for two example questions: "Lewinsky told whom about her relationship with the President Clinton?", and "Which Hollywood star became governor of California?". We use a cross mark to indicate the time of each corresponding event, which is also the true time scope of the question.

We can see that the actual time scope (January, 1988) of the first question is reflected relatively well by its distribution of relevant documents as generally these documents are located between 1998 and 1999. However, still most of the relevant documents are published in October rather than January, because another event - the impeachment of Bill Clinton - occurred at that time. On the other hand, the distribution of relevant documents corresponding to the second question is more complex as it contains many peaks, and documents are not located in a specific short time period, and the number of relevant documents published around the actual event time is relatively small when compared to the total number of relevant documents. However, the distribution line near the actual time of the event (November, 2003) still reveals useful features, i.e, the highest peak (maxima) of the dashed line of the data is near the event time. Therefore, the characteristics of the distribution of relevant documents over time can be used for inferring hidden time scopes of questions.

We perform burst detection on the retrieved relevant time-aligned documents, as the time and the duration of bursts are likely to signify the start point and the end point of events underlying the questions. More specifically, we apply burst detection method used by Vlachos et al. [25], which is a simple yet effective approach[3]. Bursts are detected as points with values higher than β standard deviations above the mean value of the moving average (MA). The procedure of calculating the candidate periods of time scope $T(Q)$ of question Q is as follows:

[3] Note that other techniques could be used to perform burst detection (e.g., [26–28]).

Algorithm 1: Query Time Scope Estimation

```
INPUT: Timestamp sequence T_pub(Q), window size w, cutoff parameter β
OUTPUT: Candidate periods of question time scope T(Q)
1  T(Q) ← ∅;
2  calculate moving average MA_w of w for sequence T_pub(Q);
3  cutoff ← mean(MA_w) + β · std(MA_w);
4  T(Bursts) ← {t_i | MA_w(t_i) > cutoff}, and further represented by
   (t(Burst_1), t(Burst_2), ...), t_i is a time point;
5  C ← {t(Burst_0)};
6  foreach t(Burst_j) ∈ T(Bursts) do
7      if t(Burst_j) == t(Burst_{j+1}) − 1          // test if two bursts are adjacent //
8      then
9          C ← C ∪ {t(Burst_{j+1})}                  // add t(Burst_{j+1}) to C if true //
10     else
11         t_i^s ← C.selectFirstElement();
12         t_i^e ← C.selectLastElement();
13         T(Q) ← T(Q) ∪ {(t_i^s, t_i^e)};
14     end
15 end
```

$T_{pub}(Q)$ can be easily obtained by collecting timestamp information of each retrieved candidate document, $T(Q)$ is a list of tuples of t_i^s and t_i^e, which are two border time points of the ith estimated time period. There are two parameters in our burst detection: w and β. For simplicity, moving Average MA_w of $T_{pub}(Q)$ of each question is calculated using w equal to 4, corresponding to four months. Following [25] that uses typical values of β equal to 1.5–2.0, we use 2.0 in the experiments. In Fig. 2, the red solid lines show the bursts of previously mentioned two example questions. The inferred time scope of the first question is [('1998-03', '1999-05')], while the time scope of the second question contains three periods: [('2003-08', '2004-02'), ('2004-06', '2004-06'), ('2004-09', '2004-10')]. Note that the second time period of the second time scope is actually a single time point (shown as a single small red point in the graph).

After calculating $T(Q)$, each candidate period is assigned a weight indicating its importance, which is obtained by dividing the number of documents published within the period over the total number of documents published in all the candidate periods of time scope $T(Q)$. For example, for the second example question, the number of documents published within the period ('2003-08', '2004-02') is 43, while the total number of documents published within all the periods $T(Q)$ is 55, so the weight assigned to this period is $\frac{43}{55}$. We use $W(T(Q))$ to represent the weight list, such that $W(T(Q)) = [w(t_1^s, t_1^e), ..., w(t_m^s, t_m^e)]$, where m is the number of candidate periods of time scope $T(Q)$.

Timestamp-Based Temporal Score Calculation. After obtaining candidate periods of time scope $T(Q)$, the module computes the timestamp-based temporal score $S_{pub}^{temp}(d)$ of each candidate document d as shown in Eq. 1. We calculate $S_{pub}^{temp}(d)$ based on the intuition that articles published within or soon after time period of the question have high probability of containing detailed information of the event mentioned in the question. The calculation way is as follows:

$$\begin{aligned} S_{pub}^{temp}(d) &= P(T(Q)|t_{pub}(d)) = P(\{(t_1^s,t_1^e),...(t_m^s,t_m^e)\}|t_{pub}(d)) \\ &= \frac{1}{m} \sum_{(t_i^s,t_i^e)\in T(Q)} P((t_i^s,t_i^e)|t_{pub}(d)) \end{aligned} \tag{1}$$

$S_{pub}^{temp}(d)$ is estimated as $P(T(Q)|t_{pub}(d))$, which is the average probability of generating m candidate periods of time scope $T(Q)$. Then, the probability of generating a period (t_i^s,t_i^e) given document timestamp $t_{pub}(d)$ is defined as:

$$P((t_i^s,t_i^e)|t_{pub}(d)) = \begin{cases} 0.0 & when\ t_i^s > t_{pub}(d) \\ w(t_i^s,t_i^e)\cdot(1.0 - \frac{|t_i^s - t_{pub}(d) + t_i^e - t_{pub}(d)|}{2\cdot TimeSpan(D)}) & elsewhere \end{cases} \tag{2}$$

$TimeSpan(D)$ is the length of time span of news archive D. In the experiments, we use NYT archive with monthly granularity, so $TimeSpan(D)$ equals to 246 units, corresponding to the number of all months in the archive. $w(t_i^s,t_i^e)$ is the weight indicating the importance of (t_i^s,t_i^e) over candidate periods of time scope $T(Q)$ (as explained before). $P((t_i^s,t_i^e)|t_{pub}(d))$ equals to 0.0 when document d is published before t_i^s, as such document usually cannot provide much information on the events that occurred after its publication. Otherwise, $P((t_i^s,t_i^e)|t_{pub}(d))$ can be larger when the timestamp is closer to the time period (t_i^s,t_i^e), and when the importance weight $w(t_i^s,t_i^e)$ of this period is large.

Content-Based Temporal Score Calculation. Next, we compute another temporal score, $S_{text}^{temp}(d)$, of a candidate document d based on the relation between temporal information embedded in d's content and the candidate periods of time scope $T(Q)$. We compute $S_{text}^{temp}(d)$ because some news articles, even the ones published long time ago after the events mentioned in questions, may retrospectively refer to these events, providing salient information on them, and can thus help to distinguish between similar events. For example, articles published near a certain US presidential election may also discuss previous elections for comparison or for other purposes. Such references are often in the form of temporal expressions that refer to particular points in the past.

Temporal expressions are detected and normalized by the combination of temporal tagger (we use SUTime [29]) and temporal signals[4] (words that help to identify temporal relations, e.g. "before","after","during"). The normalized result of each temporal expression is mapped to the time interval with the "start" and "end" information. For example, temporal expression "between 1999 and 2002" is normalized to [('1999-01', '2002-12')]. Special cases like "until January 1992" are normalized as [('', '1992-01')], since the "start" temporal information can not be determined. Finally, we can obtain a list of time scopes of temporal

[4] We use the list of temporal signals taken from [13].

expressions contained in a document d, denoted as $T_{text}(d) = \{\tau_1, \tau_2, ..., \tau_{m(d)}\}$ where $m(d)$ is the total number of temporal expressions found in d.

As each time scope τ_i has its "start" information, denoted as τ_i^s, and "end" information, τ_i^e, we create two lists $T_{text}^s(d)$, $T_{text}^e(d)$ containing all τ_i^s and all τ_i^e, respectively. Next, we construct two probability density functions by using kernel density estimation (KDE) based on these two lists. KDE is a technique closely related to histograms, which has characteristics that allow it to asymptotically converge to any density function. The probabilities of $t^s(Q)$ and $t^e(Q)$ denoted as $S_{text}^{temp\_b}(d)$, $S_{text}^{temp\_e}(d)$, respectively, can be then estimated using the probability density functions.

$$S_{text}^{temp\_b}(d) = \hat{f}\left(t^s(Q); h\right) = \frac{1}{m(d)} \sum_{i=1}^{m(d)} K_h\left(t^s(Q) - \tau_i^s\right) \tag{3}$$

where h is a bandwidth (equal to 4) and K is a Guassian Kernel defined by:

$$K_h(x) = \frac{1}{\sqrt{2\pi} \cdot h} exp\left(-\frac{x^2}{2 \cdot h}\right) \tag{4}$$

$S_{text}^{temp\_e}(d)$ is calculated in the same way but using τ_i^e and $t^e(Q)$, and $S_{text}^{temp}(d)$ is:

$$S_{text}^{temp}(d) = \frac{1}{2} \cdot \left(S_{text}^{temp\_b}(d) + S_{text}^{temp\_e}(d)\right) \tag{5}$$

Final Temporal Score Calculation & Document Ranking. After computing the two temporal scores, the final temporal score of d is given by:

$$S^{temp}(d) = \frac{1}{2} \cdot \left(S_{pub}^{temp'}(d) + S_{text}^{temp'}(d)\right) \tag{6}$$

where $S_{pub}^{temp'}(d)$ and $S_{text}^{temp'}(d)$ are the normalized values computed by dividing by the corresponding maximum scores among all candidate documents.

Additionally, document relevance score $S^{rel}(d)$ is used after normalization:

$$S^{rel}(d) = \frac{BM25(d)}{MAX\_BM25} \tag{7}$$

Finally, we rerank documents by a linear combination of their relevance scores and temporal scores:

$$S(d) = (1 - \alpha(Q)) \cdot S^{rel}(d) + \alpha(Q) \cdot S^{temp}(d) \tag{8}$$

$\alpha(Q)$ is an important parameter, which determines the proportion between document temporal score and its relevance score. For example, when $\alpha(Q)$ equals to 0.0, the relevance of the temporal information is completely ignored. As different questions have different shapes of the distributions of their relevant documents, we propose to dynamically determine $\alpha(Q)$ per each question. The idea is that when a question has many bursts, meaning that the event of the question is frequently mentioned at different times or many similar or related events occurred over time, then time should play lesser role. In this case we want to decrease $\alpha(Q)$ value to pay more attention to document relevance. In contrast, when only few bursts are found, which means that the question has obvious temporal character, time should be considered more. $\alpha(Q)$ is computed as follows:

$$\alpha(Q) = \begin{cases} 0.0 & when\ burst\_num = 0 \\ ce^{-(1-\frac{1}{burst\_num})} & elsewhere \end{cases} \tag{9}$$

c is a constant set to 0.25. $\alpha(Q)$ assumes small values when the number of bursts is high, while it is the highest for the case of a single burst. When the relevant document distribution of the question does not exhibit any bursts, which also means that the list of candidate periods of the question time scope ($T(Q)$) is empty, $\alpha(Q)$ is set to 0 and the reranking is based on document relevance.

3.3 Document Reader Module

For this module, we utilize a commonly used MRC model called BiDAF [30] which achieves Exact Match score 68.0 and F1 score 77.5 on the SQuAD 1.1 dev set. We use BiDAF model to extract answers of the top N reranked documents and we select the most common answer as the final answer. Note that BiDAF could be replaced by other MRC models, for example, the models that combine with Bert [3]. We use BiDAF for easy comparison with DrQA, whose reader component's performance is similar although a little better than the one of BiDAF.

4 Experiments

4.1 Experimental Setting

Document Archive and Test Set. As we mentioned before, NYT archive [31] is used as the underlying document collection, and is indexed using Solr. The archive contains over 1.8 million articles published from January 1987 to June 2007 and is often used for Temporal Information Retrieval researches [15,16].

To evaluate the performance of our approach, we first need a set of answerable questions. To the best of our knowledge, there was no previous proposal for answering questions on news archives or available question answering test sets designed for news archives. Thus we have to manually construct the test set making sure that the questions can be answered in NYT archive. We finally construct a test set of 200 questions[5] for NYT archive, that are carefully selected from other existing datasets and history quiz websites, and that (a) fall into the time frame of NYT archive, (b) their answers could be found in NYT archive and

[5] The test set is available at https://www.dropbox.com/s/ygy7xy4k80wmcfl/TestQuestion.csv?dl=0.

Table 2. Resources used for constructing the test set

Resources	Number
TempQuestions [13]	15
SQuAD 1.1 [4]	15
History quizzes from funtrivia[a]	50
Quizwise[b]	70
Wikipedia pages	50
Total	200

[a] http://www.funtrivia.com/quizzes/history/index.html
[b] https://www.quizwise.com/history-quiz

Table 3. Performance of different models using EM and F1

Model	Top 1		Top 5		Top 10		Top 15	
	EM	F1	EM	F1	EM	F1	EM	F1
DrQA-NYT [9]	22.50	27.58	28.00	32.78	29.50	34.11	32.00	36.87
DrQA-Wiki [9]	21.00	26.17	22.50	27.92	26.00	31.49	29.00	34.37
QA-NLM-U [21]	23.50	30.54	33.00	39.71	41.00	48.02	43.00	50.71
QA-Not-Rerank [30]	25.50	32.45	30.00	37.84	40.50	47.32	42.00	48.95
QANA-TempPub	26.00	33.69	36.00	42.75	39.50	47.19	44.00	50.71
QANA-TempCont	22.50	29.70	32.50	40.67	41.50	49.05	44.50	51.09
QANA	**26.50**	**34.27**	**37.00**	**43.76**	**42.00**	**49.20**	**45.50**	**52.71**

(c) they do not contain any temporal expressions[6]. The second condition was verified by manually selecting correct keywords from the questions and checking whether at least one retrieved document can infer the correct answer. Table 2 shows the distribution of resources used for creating the test set while Table 1 gives few examples.

Baselines and Methods. We test the following models:

1. DrQA-NYT [9]: DrQA system which uses NYT archive.
2. DrQA-Wiki [9]: DrQA system which uses Wikipedia as its unique knowledge source. We would like to test if Wikipedia could be sufficient for answering questions on events distant in the past.
3. QA-NLM-U [21]: QA system that uses the best reranking method in [21], while the Document Retriever Module and Document Reader Module are the same as the modules of QANA.

[6] We note that we have also tested QANA on 200 separate questions containing explicit temporal expressions, hence with time scopes directly given, and found that it outperforms the same baselines with even better results.

4. QA-Not-Rerank [30]: QANA system without reranking module, same as other large scale question answering systems. The Document Retriever Module and Document Reader Module are the same as the modules of QANA.
5. QANA-TempPub: QANA version that uses only temporal information of timestamp for reranking in Time-Aware Reranking Module.
6. QANA-TempCont: QANA version that only uses temporal information embedded in document content for Time-Aware Reranking Module.
7. QANA: QANA with complete Time-Aware Reranking Module.

4.2 Experimental Results

We measure the performance of the models under comparison using exact match (EM) and F1 score - the two standard measures commonly used in QA research. As shown in Table 3, QANA with full components outperforms other systems for all different N, which represent the numbers of reranked documents used in the Document Reader Module. The performance improvement is due to the use of temporal information for locating more correct documents which is derived from the question itself, document timestamp and document content. We then compare our model with others by considering the top 1 and top 5 documents. When comparing with the DrQA system, which is often used as QA baseline, the improvement is in the range of 17.77% to 32.14%, and from 24.25% to 33.49% on EM and F1 metrics, respectively.

Table 4. Performance of the models when answering questions having few relevant documents vs. when answering questions with many relevant documents

		Top 1		Top 5		Top 10		Top 15	
		EM	F1	EM	F1	EM	F1	EM	F1
Questions with few relevant documents	QA-Not-Rerank	31.00	40.48	35.00	43.93	46.00	55.79	48.00	55.12
	QANA	31.00	40.52	45.00	54.18	48.00	57.28	52.00	59.22
Questions with many relevant documents	QA-Not-Rerank	20.00	24.41	25.00	31.75	35.00	42.86	36.00	42.84
	QANA	22.00	28.02	29.00	33.33	36.00	41.11	39.00	46.21

We have also examined the performance of DrQA when using Wikipedia articles as its knowledge source. In this case, the results are worse than the ones of any other compared method that uses NYT (including DrQA), which implies that Wikipedia cannot successfully answer questions on distant past events, and they need to be answered using primary sources, i.e., news articles from the past.

When comparing with QA-NLM-U [21], the improvement ranges from 12.76% to 12.12% on EM score, and 12.21% to 10.19% on F1 score. In addition, when comparing with QA-Not-Rerank [30] that does not include reranking module, we can also observe an obvious improvement, when considering the top 5 and top 15 documents, ranging from 23.33% to 8.33%, and from 15.64% to 7.11% on EM and F1 metrics, respectively. Moreover, QANA-TempPub performs better than QANA-TempCont when using the top 1 and top 5 documents, but worse when using top 10 and top 15. In addition, we can observe that just using only timestamp information still allows achieving relatively good performance. Nevertheless, QANA with all the proposed components, which make use of the inferred time scope of the questions and the temporal information from both document timestamps and document content, achieves the best results.

We next evaluate the performance of QANA based on the number of relevant documents, and compare it with QA-Not-Rerank. We first rank questions by the number of documents they return, and then group them into two equal parts. As shown in Table 4, we can see that both the tested models achieve better results on questions with few relevant documents, as it is likely easier to locate more relevant documents from small number of documents. We also observe an improvement when comparing our model with QA-Not-Rerank, especially, for the top 5 and top 15 documents, which proves the effectiveness of the reranking method by utilizing temporal information.

Table 5. Performance of the models when answering questions with few bursts vs. when answering questions with many bursts

		Top 1		Top 5		Top 10		Top 15	
		EM	F1	EM	F1	EM	F1	EM	F1
Questions with few bursts	QA-Not-Rerank	30.20	37.24	38.54	44.32	45.83	52.55	50.00	56.79
	QANA	30.20	38.11	42.70	48.55	46.87	54.98	52.08	58.96
Questions with many bursts	QA-Not-Rerank	21.15	28.10	22.11	31.87	35.57	40.16	34.61	41.74
	QANA	23.07	30.72	31.73	39.33	37.50	43.86	39.42	46.95

Moreover, we also analyze the impact of the number of bursts on performance. About half of the questions (96 questions) have few bursts (less than or equal to 4). Table 5 shows that both QANA and QA-Not-Rerank perform much better when answering such questions. The events in questions with many bursts are likely to be similar to other events that occurred at different times, which causes the difficulty to distinguish between the events. As our system considers the importance of bursts by assigning weights to them, it significantly outperforms QA-Not-Rerank. Although $\alpha(Q)$ is smaller in this case (according to Eq. 9), it still plays an important part in selecting relevant documents. For example, if the number of bursts of a question is 10, $\alpha(Q)$ approximately equals to 0.1, which means that the final reranking is driven by about 10% of the temporal score.

Finally, we examine the effect of $\alpha(Q)$, which determines the proportion between temporal relevance score and query relevance score. As shown in Fig. 3, the model using dynamic alpha (depicted by dashed lines) performs always better than the model with static alpha, since the dynamic value is dependent on the distributions of relevant documents over time for each question. The dynamic approach flexibly captures the changes in importance of temporal information and relevance information, resulting in better overall performance.

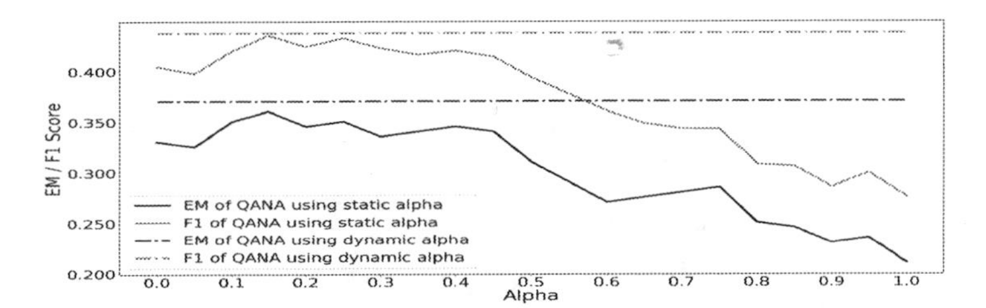

Fig. 3. QANA performance with different static alpha values vs. one with dynamic alpha for the top 5 reranked documents

5 Conclusions

In this work we propose a new research task of answering event-related questions on long-term news archives and we show effective solution for it. Unlike other common QA systems designed for synchronic document collections, questions on long-term news archives are usually influenced by temporal aspects, resulting from the interplay between the document timestamps, temporal information embedded in document content and query time scope. Therefore, exploiting temporal information is crucial for this type of QA, as also demonstrated in our experiments. We are also the first to incorporate and adapt temporal information retrieval approaches to QA systems.

Finally, our work makes few general observations. First, to answer event-related questions on long-span news archives one needs to (a) *infer the time scope embedded within a question*, and then (b) *rerank documents based on their closeness and order relation to this time scope*. Moreover, (c) *using temporal expressions in documents* further helps to select best candidates. Lastly, (d) *applying dynamic way to determine the importance between query relevance and temporal relevance is quite helpful.*

References

1. Korkeamäki, L., Kumpulainen, S.: Interacting with digital documents: a real life study of historians' task processes, actions and goals. In: Proceedings of the 2019 Conference on Human Information Interaction and Retrieval, CHIIR '19, pp. 35–43. ACM, New York (2019). https://doi.org/10.1145/3295750.3298931. ISBN 978-1-4503-6025-8
2. Bogaard, T., Hollink, L., Wielemaker, J., Hardman, L., Van Ossenbruggen, J.: Searching for old news: user interests and behavior within a national collection. In: Proceedings of the 2019 Conference on Human Information Interaction and Retrieval, pp. 113–121. ACM (2019)
3. Devlin, J., Chang, M.-W., Lee, K., Toutanova, K.: Bert: pre-training of deep bidirectional transformers for language understanding. arXiv preprint arXiv:1810.04805 (2018)
4. Rajpurkar, P., Zhang, J., Lopyrev, K., Liang, P.: Squad: 100,000+ questions for machine comprehension of text. arXiv preprint arXiv:1606.05250 (2016)
5. Rajpurkar, P., Jia, R., Liang, P.: Know what you don't know: Unanswerable questions for squad. arXiv preprint arXiv:1806.03822, 2018

6. Buck, C., et al.: Ask the right questions: Active question reformulation with reinforcement learning. arXiv preprint arXiv:1705.07830 (2017)
7. Wang, S., et al.: R3: reinforced ranker-reader for open-domain question answering. In: AAAI (2018)
8. Yang, W., et al.: End-to-end open-domain question answering with bertserini. arXiv preprint arXiv:1902.01718 (2019)
9. Chen, D., Fisch, A., Weston, J., Bordes, A.: Reading wikipedia to answer open-domain questions. arXiv preprint arXiv:1704.00051 (2017)
10. Ni, J., Zhu, C., Chen, W., McAuley, J.: Learning to attend on essential terms: An enhanced retriever-reader model for scientific question answering. arXiv preprint arXiv:1808.09492 (2018)
11. Pasca, M.: Towards temporal web search. In: Proceedings of the 2008 ACM symposium on Applied computing, pp. 1117–1121. ACM (2008)
12. Harabagiu, S., Bejan, C.A.: Question answering based on temporal inference. In: Proceedings of the AAAI-2005 workshop on inference for textual question answering, pp. 27–34 (2005)
13. Jia, Z., Abujabal, A., Saha Roy, R., Strötgen, J., Weikum, G.: TempQuestions: a benchmark for temporal question answering. In: Companion of the Web Conference 2018, pp. 1057–1062. International World Wide Web Conferences Steering Committee (2018)
14. Alonso, O., Gertz, M., Baeza-Yates, R.: On the value of temporal information in information retrieval. In: ACM SIGIR Forum, vol. 41, pp. 35–41. ACM (2007)
15. Campos, R., Dias, G., Jorge, A.M., Jatowt, A.: Survey of temporal information retrieval and related applications. ACM Comput. Surv. (CSUR) **47**(2), 15 (2015)
16. Kanhabua, N., Blanco, R., Nørvåg, K.: Temporal information retrieval. Found. Trends Inf. Retrieval **9**(2), 91–208 (2015). https://doi.org/10.1561/1500000043
17. Li, X., Croft, W.B.: Time-based language models. In: Proceedings of the Twelfth International Conference on Information and Knowledge Management, pp. 469–475. ACM (2003)
18. Metzler, D., Jones, R., Peng, F., Zhang, R.: Improving search relevance for implicitly temporal queries. In: Proceedings of the 32nd International ACM SIGIR Conference on Research and Development in Information Retrieval, pp. 700–701. Citeseer (2009)
19. Arikan, I., Bedathur, S., Berberich, K.: Time will tell: leveraging temporal expressions in IR. In: WSDM. Citeseer (2009)
20. Berberich, K., Bedathur, S., Alonso, O., Weikum, G.: A language modeling approach for temporal information needs. In: Gurrin, C., et al. (eds.) ECIR 2010. LNCS, vol. 5993, pp. 13–25. Springer, Heidelberg (2010). https://doi.org/10.1007/978-3-642-12275-0_5
21. Kanhabua, N., Nørvåg, K.: Determining time of queries for re-ranking search results. In: Lalmas, M., Jose, J., Rauber, A., Sebastiani, F., Frommholz, I. (eds.) ECDL 2010. LNCS, vol. 6273, pp. 261–272. Springer, Heidelberg (2010). https://doi.org/10.1007/978-3-642-15464-5_27
22. Miller, G.A.: WordNET: a lexical database for english. Commun. ACM **38**(11), 39–41 (1995)
23. Pennington, J., Socher, R., Manning, C.: GloVe: global vectors for word representation. In: Proceedings of the 2014 Conference on Empirical Methods in Natural Language Processing (EMNLP), pp. 1532–1543 (2014)
24. Grainger, T., Potter, T., Seeley, Y.: Solr in Action. Manning, Cherry Hill (2014)

25. Vlachos, M., Meek, C., Vagena, Z., Gunopulos, D.: Identifying similarities, periodicities and bursts for online search queries. In: Proceedings of the 2004 ACM SIGMOD International Conference on Management of Data, pp. 131–142. ACM (2004)
26. Fung, G.P.C., Yu, J.X., Yu, P.S., Lu, H.: Parameter free bursty events detection in text streams. In: Proceedings of the 31st International Conference on Very Large Data Bases, pp. 181–192. VLDB Endowment (2005)
27. Snowsill, T., Nicart, F., Stefani, M., De Bie, T., Cristianini, N.: Finding surprising patterns in textual data streams. In: 2010 2nd International Workshop on Cognitive Information Processing, pp. 405–410. IEEE (2010)
28. Kleinberg, J.: Bursty and hierarchical structure in streams. Data Min. Knowl. Discov. **7**(4), 373–397 (2003)
29. Chang, A.X., Manning, C.D.: SUTime: a library for recognizing and normalizing time expressions. In: LREC, vol. 2012, pp. 3735–3740 (2012)
30. Seo, M., Kembhavi, A., Farhadi, A., Hajishirzi, H.: Bidirectional attention flow for machine comprehension. arXiv preprint arXiv:1611.01603, 2016
31. Sandhaus, E.: The new york times annotated corpus. Linguist. Data Consortium, Philadelphia **6**(12), e26752 (2008)

bias goggles: Graph-Based Computation of the Bias of Web Domains Through the Eyes of Users

Panagiotis Papadakos[1,2(✉)] and Giannis Konstantakis[2]

[1] Institute of Computer Science, FORTH-ICS, Heraklion, Greece
papadako@ics.forth.gr
[2] Computer Science Department, University of Crete, Heraklion, Greece
jkonstan@csd.uoc.gr

Abstract. Ethical issues, along with transparency, disinformation, and bias, are in the focus of our information society. In this work, we propose the *bias goggles* model, for computing the bias characteristics of web domains to user-defined concepts based on the structure of the web graph. For supporting the model, we exploit well-known propagation models and the newly introduced `Biased-PR` PageRank algorithm, that models various behaviours of biased surfers. An implementation discussion, along with a preliminary evaluation over a subset of the greek web graph, shows the applicability of the model even in real-time for small graphs, and showcases rather promising and interesting results. Finally, we pinpoint important directions for future work. A constantly evolving prototype of the *bias goggles* system is readily available.

Keywords: Bias · Web graph · Propagation models · Biased PageRank

1 Introduction

There is an increasing concern about the potential risks in the consumption of abundant biased information in online platforms like Web Search Engines (WSEs) and social networks. Terms like echo chambers and filter-bubbles [26] depict the isolation of groups of people and its aftereffects, that result from the selective and restrictive exposure to information. This restriction can be the result of helpful personalized algorithms, that suggest user connections or rank highly information relevant to the users' profile. Yet, this isolation might inhibit the growth of informed and responsible humans/citizens/consumers, and can also be the result of malicious algorithms that promote and resurrect social, religious, ethnic, and other kinds of discriminations and stereotypes.

Currently, the community focus is towards the transparency, fairness, and accountability of mostly machine learning algorithms for decision-making, classification, and recommendation in social platforms like twitter. However, social

J. M. Jose et al. (Eds.): ECIR 2020, LNCS 12035, pp. 790–804, 2020.
https://doi.org/10.1007/978-3-030-45439-5_52

platforms and WSEs mainly act as gateways to information published on the web as common web pages (e.g., blogs and news). Unfortunately, users are unaware of the bias characteristics of these pages, except for obvious facts (e.g., a page in a political party's web site will be biased towards this party).

In this work, we propose the `bias goggles` model, where users are able to explore the biased characteristics of web domains for a specific biased concept (i.e., a bias goggle). Since there is no objective definition of what bias and biased concepts are [27], we let users define them. For these concepts, the model computes the *support* and the *bias score* of a web domain, by considering the *support* of this domain for each aspect (i.e., dimension) of the biased concept. These *support* scores are calculated by graph-based algorithms that exploit the structure of the web graph and a set of user-defined seeds representing each aspect of bias. As a running example we will use the biased concept of greek politics, that consists of nine aspects of bias, each one representing a popular greek party, and identified by a single seed; the domain of its homepage.

In a nutshell, the main contributions of this work are:

- the `bias goggles` model for computing the bias characteristics of web domains for a user-defined concept, based on the notions of `Biased Concepts` (`BCs`), `Aspects of Bias` (`ABs`), and the metrics of the *support* of the domain for a specific `AB` and `BC`, and its *bias score* for this `BC`,
- the introduction of the `Support Flow Graph (SFG)`, along with graph-based algorithms for computing the `AB` *support* score of domains, that include adaptations of the `Independence Cascade` (`IC`) and `Linear Threshold` (`LT`) propagation models, and the new `Biased-PageRank` (`Biased-PR`) variation that models different behaviours of a biased surfer,
- an initial discussion about performance and implementation issues,
- some promising evaluation results that showcase the effectiveness and efficiency of the approach on a relatively small dataset of crawled pages, using the new `AGBR` and `AGS` metrics,
- a publicly accessible prototype of `bias goggles`.

The rest of the paper is organized as follows: the background and the related work is discussed in Sect. 2, while the proposed model, and its notions and metrics are described in Sect. 3. The graph-based algorithms for computing the *support* score of a domain for a specific `AB` are introduced in Sect. 4. The developed prototype and related performance issues are discussed in Sect. 5, while some preliminary evaluation results over a relatively small dataset of web pages are reported in Sect. 6. Finally, Sect. 7 concludes the paper and outlines future work.

2 Background and Related Work

Social platforms have been found to strengthen users' existing biases [21] since most users try to access information that they agree with [18]. This behaviour leads to rating bubbles when positive social influence accumulates [24] and minimizes the exposure to different opinions [31]. This is also evident in WSEs, where the personalization and filtering algorithms lead to echo chambers and filter bubbles that reinforce bias [4,12]. Remarkably, users of search engines trust more the top-ranked search results [25] and biased search algorithms can shift the voting preferences of undecided voters by as much as 20% [8].

There is an increasingly growing number of discrimination reports regarding various protected attributes (e.g., race, gender, etc.) in various domains, like in ads [7,29] and recommendation systems [13], leading to efforts for defining principles of accountable[1], auditing [28] and de-bias algorithms [1], along with fair classifiers [6,14,34]. Tools that remove discriminating information[2], flag fake news[3], make personalization algorithms more transparent[4], or show political biases in social networks[5] also exist. Finally, a call for equal opportunities by design [16] has been raised regarding the risks of bias in the stages of the design, implementation, training and deployment of data-driven decision-making algorithms [3,11,20].

There are various efforts for measuring bias in online platforms [27]. Bias in WSEs has been measured as the deviation from the distribution of the results of a pool of search engines [23] and the coverage of SRPs towards US sites [30]. Furthermore, the presence of bias in media sources has been explored through human annotations [5], by exploiting affiliations [32], the impartiality of messages [33], the content and linked-based tracking of topic bias [22], and the quantification of data and algorithmic bias [19]. However, this is the first work that provides a model that allows users to explore the available web sources based on their own definitions of biased concepts. The approach exploits the web graph structure and can annotate web sources with bias metrics on any online platform.

3 The *bias goggles* Model

Below we describe the notions of `Biased Concepts` (BCs) and `Aspects of Bias` (`ABs`), along with the *support* of a domain for an `AB` and `BC`, and its *bias score* for a `BC`. Table 1 describes the used notation.

3.1 Biased Concepts (BCs) and Aspects of Bias (ABs)

The interaction with a user begins with the definition of a `Biased Concept` (`BC`), which is considered the goggles through which the user wants to explore the web domains. `BCs` are given by users and correspond to a concept that can range from a very abstract one (e.g., god) to a very specific one (e.g., political parties). For each `BC`, it is required that the users can identify at least two `Aspects of Bias` (`ABs`), representing its bias dimensions. `ABs` are given by the users and correspond to a non-empty set of seeds (i.e., domains) $\mathcal{S}$, that the user considers to fully

[1] http://www.fatml.org/resources/principles-for-accountable-algorithms.
[2] http://www.debiasyourself.org/.
[3] https://www.facebook.com/help/572838089565953.
[4] https://facebook.tracking.exposed/.
[5] http://politecho.org/.

Table 1. Description of the used notation. The first part describes the notation used for the Web Graph, while the second the notation for the proposed model.

Symbol	Description
$\mathcal{W}$	the set of crawled Web pages
$\mathtt{p}$	a page in $\mathcal{W}$
$\mathtt{dom(p)}$	the normalized SLD of page $\mathtt{p}$
$\mathtt{doms}(\mathcal{W})$	the set of normalized SLDs in $\mathcal{W}$
$\mathtt{dom}$	an SLD in $\mathtt{doms}(\mathcal{W})$
$\mathtt{link}_{\mathtt{p},\mathtt{p}'}$	a link from page $\mathtt{p}$ to $\mathtt{p}'$ | $\mathtt{p},\ \mathtt{p}' \in \mathcal{W}$
$\mathtt{link}_{\mathtt{dom},\mathtt{dom}'}$	a link from domain $\mathtt{dom}$ to $\mathtt{dom}'$ | $\mathtt{dom},\ \mathtt{dom}' \in \mathtt{doms}(\mathcal{W})$
$\mathtt{links}(\mathcal{W})$	the set of crawled links between pages in $\mathcal{W}$
$\mathtt{dom}(\mathtt{links}(\mathcal{W}))$	the set of crawled links between the domains in $\mathtt{doms}(\mathcal{W})$
$\mathtt{inv}(\mathtt{link}_{\mathtt{p},\mathtt{p}'})$	the inverse link of $\mathtt{link}_{\mathtt{p},\mathtt{p}'}$, i.e., $\mathtt{link}_{\mathtt{p}',\mathtt{p}}$
$\mathtt{inv}(\mathtt{link}_{\mathtt{dom},\mathtt{dom}'})$	the inverse link of $\mathtt{link}_{\mathtt{dom},\mathtt{dom}'}$, i.e., $\mathtt{link}_{\mathtt{dom}',\mathtt{dom}}$
$\mathtt{inv}(\mathtt{links}(\mathcal{W}))$	the set of inverse links between the pages in $\mathcal{W}$
$\mathtt{inv}(\mathtt{dom}(\mathtt{links}(\mathcal{W})))$	the set of inverse links between the domains in $\mathtt{doms}(\mathcal{W})$
$\mathtt{G}(\mathcal{W})$	the graph with $\mathtt{doms}(\mathcal{W})$ as nodes and $\mathtt{dom}(\mathtt{links}(\mathcal{W}))$ as edges
$\mathtt{outInvLinks}(\mathtt{dom})$	the set of $\mathtt{link}_{\mathtt{p},*} \in \mathtt{inv}(\mathtt{links}(\mathcal{W}))$ | $\mathtt{p} \in \mathcal{W},\ \mathtt{dom(p)} = \mathtt{dom}$
$\mathtt{outInvLinks}(\mathtt{dom},\mathtt{dom}')$	the set of $\mathtt{link}_{\mathtt{p},\mathtt{p}'} \in \mathtt{inv}(\mathtt{links}(\mathcal{W}))$ | $\mathtt{p},\mathtt{p}' \in \mathcal{W}$, $\mathtt{dom(p)} = \mathtt{dom}, \mathtt{dom(p')} = \mathtt{dom}'$
$\mathtt{neigh}(\mathtt{dom})$	the set of all $\mathtt{dom}' \in \mathtt{doms}(\mathcal{W})$ | $\mathtt{link}_{\mathtt{dom},\mathtt{dom}'} \in \mathtt{dom}(\mathtt{links}(\mathcal{W}))$
$\mathtt{invNeigh}(\mathtt{dom})$	the set of all $\mathtt{dom}' \in \mathtt{doms}(\mathcal{W})$ | $\mathtt{link}_{\mathtt{dom},\mathtt{dom}'} \in \mathtt{inv}(\mathtt{dom}(\mathtt{links}(\mathcal{W})))$
$\mathtt{w}_{\mathtt{dom},\mathtt{dom}'}$	the weight of the $\mathtt{link}_{\mathtt{dom},\mathtt{dom}'}$
$\mathtt{SFG}(\mathcal{W})$	the weighted graph with $\mathtt{doms}(\mathcal{W})$ as nodes and $\mathtt{inv}(\mathtt{dom}(\mathtt{links}(\mathcal{W})))$ as edges where $\mathtt{w}_{\mathtt{dom},\mathtt{dom}'} = \frac{\mathtt{outInvLinks}(\mathtt{dom},\mathtt{dom}')}{\mathtt{outInvLinks}(\mathtt{dom})}$
$\mathcal{S}$	a non-empty set of normalized domain urls (i.e., seeds)
$\mathtt{sign}(\mathcal{S})$	the signature of a set of seeds
$\mathtt{AB}_{\mathtt{sign}(\mathcal{S})}$	an Aspect of Bias (AB) as identified by $\mathtt{sign}(\mathcal{S})$
$\mathtt{seeds}(\mathtt{AB}_{\mathtt{sign}(\mathcal{S})})$	the set of seeds that define $\mathtt{AB}_{\mathtt{sign}(\mathcal{S})}$
$\mathcal{A}^u$	the universe of all available ABs
$\mathcal{A}$	a non-empty set of ABs | $\mathcal{A} \subseteq \mathcal{A}^u, \lvert\mathcal{A}\rvert \geq 2$
$\mathtt{BC}_{\mathcal{A}}$	a Biased Concept (BC) as defined by $\mathcal{A}$
$\boldsymbol{d}_{\mathcal{A}}$	an $\lvert\mathcal{A}\rvert$-dimensional vector holding the ABs of $\mathtt{BC}_{\mathcal{A}}$
$\boldsymbol{d}_{\mathcal{A}}[i]$	the AB stored in dimension i of $\boldsymbol{d}_{\mathcal{A}}$
$\mathtt{sup}(\boldsymbol{d}_{\mathcal{A}}[i], \mathtt{dom})$ $\mathtt{sup}(\mathtt{AB}_{\mathtt{sign}(\mathcal{S})}, \mathtt{dom})$	*support* score of domain dom regarding AB $\boldsymbol{d}_{\mathcal{A}}[i]$
$\boldsymbol{s}^{\mathtt{dom}}_{\boldsymbol{d}_{\mathcal{A}}}$	vector holding *support* scores $\forall \boldsymbol{d}_{\mathcal{A}}[i] \in \boldsymbol{d}_{\mathcal{A}}$ for domain dom
$\boldsymbol{s}^{\mathtt{dom}}_{\boldsymbol{d}_{\mathcal{A}}}[i]$	*support* score of dimension i of $\boldsymbol{s}^{\mathtt{dom}}_{\boldsymbol{d}_{\mathcal{A}}}$
$\mathtt{sup}(\boldsymbol{s}^{\mathtt{dom}}_{\boldsymbol{d}_{\mathcal{A}}})$ $\mathtt{sup}(\mathtt{BC}_{\mathcal{A}}, \mathtt{dom})$	*support* score of domain dom regarding BC $\mathtt{BC}_{\mathcal{A}}$
$\mathtt{bias}(\mathtt{BC}_{\mathcal{A}}, \mathtt{dom})$ $\mathtt{bias}(\boldsymbol{s}^{\mathtt{dom}}_{\boldsymbol{d}_{\mathcal{A}}})$	*bias score* of dom for BC $\mathtt{BC}_{\mathcal{A}}$
$\mathbf{1}_{\lvert\mathcal{A}\rvert}$	An $\lvert\mathcal{A}\rvert$-dimensional vector with *support* 1 in all dimensions

support this bias aspect. For example, consider the homepage of a greek political party as an aspect of bias in the biased concept of the politics in Greece. Notice, that an AB can be part of more than one BCs. Typically, an AB is denoted by $\mathtt{AB}_{\mathtt{sign}(\mathcal{S})}$, where $\mathtt{sign}(\mathcal{S})$ is the signature of the non-empty set of seeds $\mathcal{S}$. The $\mathtt{sign}(\mathcal{S})$ is the SHA1 hash of the lexicographic concatenation of the normalized Second Level Domains (SLDs)[6] of the urls in $\mathcal{S}$. We assume that all seeds in $\mathcal{S}$ are incomparable and support with the same strength this AB.

Assumption 1. ***Incomparable Seeds Support.*** *The domains in the set of seeds* $\mathcal{S}$ *are incomparable and equally supportive of the* $\mathtt{AB}_{\mathtt{sign}_{(\mathtt{S})}}$.

The user-defined BC of the set of ABs $\mathcal{A} \subseteq \mathcal{A}^{\mathcal{U}}$, where $|\mathcal{A}| \geq 2$ and $\mathcal{A}^{\mathcal{U}}$ the universe of all possible ABs in the set of domains $\mathtt{doms}(\mathcal{W})$ of the crawled pages $\mathcal{W}$, is denoted by $\mathtt{BC}_{\mathcal{A}}$ and is represented by the pair $< \boldsymbol{d}_{\mathcal{A}}, \mathtt{desc}_{\mathcal{A}} >$. The $\boldsymbol{d}_{\mathcal{A}}$ is an $|\mathcal{A}|$-dimensional vector with $|\mathcal{A}| \geq 2$, holding all $\mathtt{AB}_{\mathtt{sign}(\mathcal{S})} \in \mathcal{A}$ of this BC in lexicographic order. $\mathtt{desc}_{\mathcal{A}}$ is a user-defined textual description of this BC. In this work, we assume that all ABs of any BC are orthogonal and unrelated.

Assumption 2. ***Orthogonality of Aspects of Bias.*** *All* AB*s in a user-defined* BC *are considered orthogonal.*

Using the notation, our running example is denoted as $\mathtt{BC}_{\mathcal{R}} = < \mathtt{d}_{\mathcal{R}}, \mathtt{desc}_{\mathcal{R}} >$, where $\mathtt{d}_{\mathcal{R}}$ is a vector that holds lexicographically the SHA1 signatures of the nine ABs singleton seeds of greek political parties $\mathcal{R} = \{$ {"anexartitoiellines.gr"}, {"antidiaploki.gr"}, {"elliniki − lisi.gr"}, {"kke.gr"}, {"mera25.gr"}, {"nd.gr"}, {"syriza.gr"}, {"topotami.gr"}, {"xryshaygh.com"}}, and $\mathtt{desc}_{\mathcal{R}}$ = "politics in Greece" is its description.

3.2 Aspects of Bias Support and Biased Concepts Support

A core metric in the proposed model is the *support* score of a domain dom to an aspect of bias $\mathtt{AB}_{\mathtt{sign}(\mathcal{S})}$, denoted as $\mathtt{sup}(\mathtt{AB}_{\mathtt{sign}(\mathcal{S})}, \mathtt{dom})$. The *support* score ranges in $[0, 1]$, where 0 denotes an unsupportive domain for the corresponding AB, and 1 a fully supportive one. We can identify three approaches for computing this *support* for a dataset of web pages: (a) the graph-based ones that exploit the web graph structure and the relationship of a domain with the domains in $\mathtt{seeds}(\mathtt{AB}_{\mathtt{sign}(\mathcal{S})})$, (b) the content-based ones that consider the textual information of the respective web pages, and (c) the hybrid ones that take advantage of both the graph and the content information. In this work, we focus only on graph-based approaches and study two frequently used propagation models, the Independence Cascade (IC) and Linear Threshold (LT) models, along with the newly introduced Biased-PageRank (Biased-PR), that models various behaviours of biased surfers. The details about these algorithms are given in Sect. 4.

[6] We follow the standard URL normalization method (see https://en.wikipedia.org/wiki/URI_normalization) and get the SLD of an url.

In the same spirit, we are interested about the *support* of a specific domain $\mathtt{dom}$ to a biased concept $\mathtt{BC}_{\mathcal{A}}$, denoted by $\mathtt{sup}(\mathtt{BC}_{\mathcal{A}}, \mathtt{dom})$. The basic intuition is that we need a metric that shows the relatedness and *support* to all or any of the aspects in $\mathcal{A}$, which can be interpreted as the relevance of this domain with any of the aspects of the biased concept $\mathtt{BC}_{\mathcal{A}}$. A straightforward way to measure it, is the norm of the $\boldsymbol{s}^{\mathtt{dom}}_{\boldsymbol{d}_{\mathcal{A}}}$ vector that holds the *support* scores of $\mathtt{dom}$ for each $\mathtt{AB}$ in $\mathcal{A}$, normalized by the norm of the $\mathbf{1}_{|\mathcal{A}|}$ vector. This vector holds the *support* scores of a 'virtual' domain that fully supports all bias aspects in $\mathtt{BC}_{\mathcal{A}}$. Specifically,

$$\mathtt{sup}(\mathtt{BC}_{\mathcal{A}}, \mathtt{dom}) = \frac{\left\|\boldsymbol{s}^{\mathtt{dom}}_{\boldsymbol{d}_{\mathcal{A}}}\right\|}{\left\|\mathbf{1}_{|\mathcal{A}|}\right\|} = \frac{\sqrt{\sum_{i=1}^{|\mathcal{A}|} \boldsymbol{s}^{\mathtt{dom}}_{\boldsymbol{d}_{\mathcal{A}}}[i]^2}}{\sqrt{|\mathcal{A}|}} = \frac{\sqrt{\sum_{\mathtt{AB}_{\mathtt{sign}(\mathcal{S})} \in \mathcal{A}} \mathtt{sup}(\mathtt{AB}_{\mathtt{sign}(\mathcal{S})}, \mathtt{dom})^2}}{\sqrt{|\mathcal{A}|}} \tag{1}$$

The $\mathtt{sup}(\mathtt{BC}_{\mathcal{A}}, \mathtt{dom})$ value ranges in $[0, 1]$. By using the above formula two domains might have similar *support* scores for a specific $\mathtt{BC}$, while the *support* scores for the respective aspects might differ greatly. For example, consider two domains $\mathtt{dom}$ and $\mathtt{dom}'$, with $\mathtt{dom}$ fully supporting only one aspect in $\mathcal{A}$ and $\mathtt{dom}'$ fully supporting another aspect in $\mathcal{A}$. Then $\mathtt{sup}(\mathtt{BC}_{\mathcal{A}}, \mathtt{dom}) \sim \mathtt{sup}(\mathtt{BC}_{\mathcal{A}}, \mathtt{dom}')$. Below we introduce the *bias score* of a domain regarding a specific $\mathtt{BC}$, as a way to capture the leaning of a domain to specific $\mathtt{AB}$s of a $\mathtt{BC}$.

3.3 Bias Score of Domain Regarding a Biased Concept

The *bias score* of a domain regarding a $\mathtt{BC}$ tries to capture how biased the domain is over any of its $\mathtt{AB}$s, and results from the *support* scores that the domain has for each aspect of the $\mathtt{BC}$. For example, consider a domain $\mathtt{dom}$ that has a rather high *support* for a specific $\mathtt{AB}$, but rather weak ones for the rest $\mathtt{AB}$s of a specific $\mathtt{BC}$. This domain is expected to have a high *bias score*. On the other hand, the domain $\mathtt{dom}'$ that has similar *support* for all the available $\mathtt{AB}$s of a $\mathtt{BC}$ can be considered to be unbiased regarding this specific $\mathtt{BC}$.

We define the *bias score* of a domain $\mathtt{dom}$ for $\mathtt{BC}_{\mathcal{A}}$ as the distance of the $\boldsymbol{s}^{\mathtt{dom}}_{\boldsymbol{d}_{\mathcal{A}}}$ vector from the $\mathbf{1}_{|\mathcal{A}|}$ vector, multiplied by its support $\mathtt{sup}(\mathtt{BC}_{\mathcal{A}}, \mathtt{dom})$. The *bias score* takes values in $[0, 1]$. Specifically,

$$\mathtt{bias}(\boldsymbol{s}^{\mathtt{dom}}_{\boldsymbol{d}_{\mathcal{A}}}) = \mathtt{dist}(\boldsymbol{s}^{\mathtt{dom}}_{\boldsymbol{d}_{\mathcal{A}}}, \mathbf{1}_{|\mathcal{A}|}) * \mathtt{sup}(\mathtt{BC}_{\mathcal{A}}, \mathtt{dom}) \tag{2}$$

We use the cosine similarity to define the distance metric, as shown below:

$$\mathtt{dist}(\boldsymbol{s}^{\mathtt{dom}}_{\boldsymbol{d}_{\mathcal{A}}}, \mathbf{1}_{|\mathcal{A}|}) = 1 - \mathtt{cosSim}(\boldsymbol{s}^{\mathtt{dom}}_{\boldsymbol{d}_{\mathcal{A}}}, \mathbf{1}_{|\mathcal{A}|}) = 1 - \frac{\boldsymbol{s}^{\mathtt{dom}}_{\boldsymbol{d}_{\mathcal{A}}} \cdot \mathbf{1}_{|\mathcal{A}|}}{\left\|\boldsymbol{s}^{\mathtt{dom}}_{\boldsymbol{d}_{\mathcal{A}}}\right\| \left\|\mathbf{1}_{|\mathcal{A}|}\right\|} \tag{3}$$

4 Graph-Based Computation of Aspects of Bias Support

In this section, we discuss the graph-based algorithms that we use for computing the *support* score of a domain regarding a specific AB. We focus on the popular Independence Cascade (IC) and Linear Threshold (LT) propagation models, along with the newly introduced Biased-PageRank (Biased-PR) algorithm.

Let $\mathfrak{W}$ be the set of crawled web pages, $\texttt{doms}(\mathfrak{W})$ the set of normalized SLDs in $\mathfrak{W}$, $\texttt{links}(\mathfrak{W})$ the set of crawled links between the domains in $\texttt{doms}(\mathfrak{W})$, and $\texttt{G}(\mathfrak{W})$ the corresponding graph with $\texttt{doms}(\mathfrak{W})$ as nodes and $\texttt{links}(\mathfrak{W})$ as edges. With $\texttt{link}_{\texttt{dom},\texttt{dom}'}$ we denote a link from domain dom to $\texttt{dom}' \mid \texttt{dom}, \texttt{dom}' \in \texttt{doms}(\mathfrak{W})$, while $\texttt{inv}(\texttt{link}_{\texttt{dom},\texttt{dom}'})$ inverses the direction of a link and $\texttt{inv}(\texttt{links}(\mathfrak{W}))$ is the set of inverse links in $\mathfrak{W}$. Furthermore, for the links we assume that:

Assumption 3. ***Equally Supportive Links.*** *Any link* $\texttt{link}_{\texttt{dom},\texttt{dom}'}$ *from the domain* dom *to the domain* dom′ *in the set of crawled domains* $\mathfrak{W}$*, is considered to be of supportive nature (i.e.,* dom *has the same support stance as* dom′ *for any* AB*). All links in a domain are equally supportive and independent of the importance of the page they appear in.*

Although the above assumption might not be precise, since links from a web page to another are not always of supportive nature (e.g., a web page critizing another linked one), or of the same importance (e.g., links in the homepage versus links deeply nested in a site), it suffices for the purposes of this first study of the model. Identification of the nature of links and the importance of the pages they appear is left as future work. Given that the assumption holds, part or whole of the *support* of dom′ regarding any AB can flow to dom through $\texttt{inv}(\texttt{link}_{\texttt{dom},\texttt{dom}'})$. Specifically, we define the **Support Flow Graph** as:

Support Flow Graph (SFG) Definition. *The* SFG *of a set of web pages* $\mathfrak{W}$ *is the weighted graph that is created by inversing the links in* $\texttt{G}(\mathfrak{W})$ *(i.e., the graph with* $\texttt{doms}(\mathfrak{W})$ *as nodes and* $\texttt{inv}(\texttt{links}(\mathfrak{W}))$ *as edges). The weight of each edge is* $\texttt{w}_{\texttt{dom},\texttt{dom}'} = \frac{\texttt{outInvLinks(dom,dom}')}{\texttt{outInvLinks(dom)}}$ *(i.e., the number of outgoing inverse links of pages in the domain* dom *that link to pages in the domain* dom′*, divided by the total outgoing inverse links of pages in the domain* dom*), and takes a value in* $[0, 1]$.

So, given an $\texttt{SFG}(\mathfrak{W})$ and the $\texttt{seeds}(\texttt{AB}_{\texttt{sign}(\mathcal{S})})$ of an AB we can now describe how the *support* flows in the nodes of the $\texttt{SFG}(\mathfrak{W})$ graph. All algorithms described below return a map M holding $\texttt{sup}(\texttt{AB}_{\texttt{sign}(\mathcal{S})}, \texttt{dom})\ \forall\ \texttt{dom} \in \texttt{doms}(\mathfrak{W})$.

4.1 Independence Cascade (IC) Model

The IC propagation model was introduced by Kempe et al. [17], and a number of variations have been proposed in the bibliography. Below, we describe the basic form of the model as adapted to our needs. In the IC propagation model, we run n experiments. Each run starts with a set of activated nodes, in our case the $\texttt{seeds}(\texttt{AB}_{\texttt{sign}(\mathcal{S})})$, that fully *support* the $\texttt{AB}_{\texttt{sign}(\mathcal{S})}$. In each iteration there is a history independent and non-symmetric probability of activating the neighbors of the activated nodes associated with each edge, flowing the *support* to the neighbors of the activated nodes in the $\texttt{SFG}(\mathcal{W})$. This probability is represented by the weights of the links of an activated node to its neighbors, and each node, once activated, can then activate its neighbors. The nodes and their neighbors are selected in arbitrary order. Each experiment stops when there are no new activated nodes. After n runs we compute the average *support* score of nodes, i.e., $\texttt{sup}(\texttt{AB}_{\texttt{sign}(\mathcal{S})}, \texttt{dom})\ \forall\ \texttt{dom} \in \texttt{doms}(\mathcal{W})$. The algorithm is given in Algorithm 1.

Algorithm 1: IC *Support* Computation

input : $\texttt{SFG}(\mathcal{W})$: the Support Flow Graph of $\mathcal{W}$
$\texttt{seeds}(\texttt{AB}_{\texttt{sign}(\mathcal{S})})$: the set of seeds of $\texttt{AB}_{\texttt{sign}(\mathcal{S})}$
n : the number of experiments
output: a map M holding $\texttt{sup}(\texttt{AB}_{\texttt{sign}(\mathcal{S})}, \texttt{dom})\ \forall\ \texttt{dom} \in \texttt{doms}(\mathcal{W})$

```
1  L ← ∅                                         // list holding the support maps of each experiment
2  for i ← 1 to n do                                                        // for each experiment
3      A ← seeds(AB_sign(S))                                // set of active nodes for next iteration
4      I ← doms(W) \ A                                                   // set of inactive nodes
5      M ← mapWithZeros(doms(W))                                // map with 0 support for domains
6      while A ≠ ∅ do
7          C ← A                                                   // active nodes in this iteration
8          A ← ∅                                                   // active nodes in next iteration
9          foreach dom ∈ C do                                      // for each current active domain
10             N ← invNeigh(dom) ∩ I                              // get all inactive inverse neighbors
11             foreach dom' ∈ N do                                              // for each neighbor
12                 r ← random(0, 1)                                     // get a random value in [0,1]
13                 if r ≤ w_dom,dom' then                                     // successful experiment
14                     A ← A ∪ {dom'}                              // activate node for next iteration
15                     M ← setOne(M, dom')                                    // set dom' support to 1
16                     I ← I \ {dom'}                                   // remove dom' from inactive
17     L ← L ∪ {M}                                           // hold support values map to list
18 return average(L)                                            // map with average support values
```

4.2 Linear Threshold (LT) Model

The LT model is another widely used propagation model. The basic difference from the IC model is that for a node to become active we have to consider the *support* of all neighbors, which must be greater than a threshold $\theta \in [0, 1]$, serving as the resistance of a node to its neighbors joint *support*. Again, we use the *support* probabilities represented by the weights of the SFG links. The full algorithm, which is based on the static model introduced by Goyal et al. [10], is given in Algorithm 2. In each experiment the thresholds θ get a random value.

Algorithm 2: LT *Support* Computation

```
input : SFG(W) : the Support Flow Graph of W
        seeds(AB_sign(S)) : the set of seeds of AB_sign(S)
        n : the number of experiments
output: a map M holding sup(AB_sign(S), dom) ∀ dom ∈ doms(W)
1  L ← ∅                                      // list holding the support maps of each experiment
2  for i ← 1 to n do                                                   // for each experiment
3  |  N ← seeds(AB_sign(S))                        // set of active nodes for next iteration
4  |  A ← ∅                                                       // set of all active nodes
5  |  I ← doms(W) \ N                                              // set of inactive nodes
6  |  M ← mapWithZeros(doms(W))                           // map with 0 support for domains
7  |  T ← mapWithRandom(doms(W)                      // random value θ in [0,1] for each node
8  |  while N ≠ ∅ do
9  |  |  C ← N                                               // active nodes in this iteration
10 |  |  N ← ∅                                               // active nodes in next iteration
11 |  |  A ← A ∪ C                                                        // add to all nodes
12 |  |  foreach dom ∈ (⋃_{c∈C} invNeigh(c) ∩ I) do         // for inactive invNeigh of active
13 |  |  |  N ← neigh(dom) ∩ N                                      // get all active neighbors
14 |  |  |  jointSup ← 1 − ∏_{dom'∈N} (1 − w_{dom,dom'})  // compute joint support value in [0,1]
15 |  |  |  if jointSup ≥ getValue(T, dom) then         // joint support bigger than threshold
16 |  |  |  |  N ← N ∪ {dom}                               // activate node for next iteration
17 |  |  |  |  M ← setOne(M, dom)                                  // set dom support to 1
18 |  |  |  |  I ← I \ {dom}                                     // remove node from inactive
19 |  L ← L ∪ {M}                                        // hold support values map to list
20 return average(L)                                    // map with average support values
```

4.3 Biased-PageRank (Biased-PR) Model

We introduce the Biased-PR variation of PageRank [9] that models a biased surfer. The biased surfer always starts from the biased domains (i.e., the seeds of an AB), and either visits a domain linked by the selected seeds or one of the biased domains again, with some probability that depends on the modeled behaviour. The same process is followed in the next iterations. The Biased-PR differs to the original PageRank in two ways. The first one is how the score (*support* in our case) of the seeds is computed at any step. The *support* of all domains is initially 0, except from the *support* of the seeds that have the value $\mathtt{init}_{\mathtt{seeds}} = 1$. At any step, the *support* of each seed is the original PageRank value, increased by a number that depends on the behaviour of the biased surfer. We have considered three behaviours: (a) the Strongly Supportive (SS) one, where the support is increased by $\mathtt{init}_{\mathtt{seeds}}$ and models a constantly strongly biased surfer, (b) the Decreasingly Supportive (DS) one, where the support is increased by $\mathtt{init}_{\mathtt{seeds}}/\mathtt{iter}$, modeling a surfer that becomes less biased the more pages he/she visits, and (c) the Non-Supportive (NS) one, with no increment, modeling a surfer that is biased only on the initial visiting pages, and afterwards the *support* score is computed as in the original PageRank. Biased-PR differs also on how the biased surfer is teleported to another domain when he/she reaches a sink (i.e., a domain that has no outgoing links). The surfer randomly teleports with the same probability to a domain in any distance from the seeds. If a path from a node to any of the seeds does not exist, the distance of the node is the maximum distance of a connected node increased by one. Since the number of nodes at a certain distance from the seeds increase as we move away from the seeds, the teleporting probability for a node is greater the closer the node is to the seeds. We expect slower convergence for Biased-PR than the original PageRank, due to the initial zero scores of non-seed nodes. The algorithm is given in Algorithm 3.

Algorithm 3: Biased-PR *Support* Computation

```
input  : SFG(𝒲) : the Support Flow Graph of 𝒲
         seeds(AB_sign(𝒮)) : the set of seeds of AB_sign(𝒮)
         behaviour : bias user behaviour. One of SS, DS, NS
         θ_conv : converge threshold
         d : damping factor
output: a map M holding sup(AB_sign(𝒮), dom) ∀ dom ∈ doms(𝒲)

   // ----- INIT PART -----
1  init_s = 1                                               // initial support of seeds
2  iter ← 0                                                 // counts iterations
3  conv ← false                                            // holds if the algorithm has converged
4  M ← mapWithZeros(doms(𝒲))                              // map with 0 support for domains
5  foreach dom ∈ seeds(AB_sign(𝒮)) do                      // initialize support for each seed
6  |  M ← addSupport(M, dom, init_s)        // add init_s support value for domain dom in map M
   // D is a map with keys the distinct minimum distances of nodes from seeds in SFG,
   // and values the number of nodes with this minimum distance
7  D ← distinctMinDistancesAndCounts(doms(𝒲), seeds(AB_sign(𝒮)))
8  E ← mapWithZeros(doms(𝒲))                    // map that holds the teleportation probabilities
9  foreach dom ∈ doms(𝒲)) do                   // find teleportation probability for each node
10 |  minDist = findMinDistanceFromSeeds(dom, seeds(AB_sign(𝒮)))
11 |  E ← addProbability(E, dom, 1/(minDist * getValue(D, minDist)))     // compute probability
   // ----- MAIN PART -----
12 while !conv do                                               // alg has not finished
13 |  M' ← mapWithZeros(doms(𝒲))              // new map with ranks (0 support for domains)
14 |  foreach dom ∈ doms(𝒲) do                                       // for each node
15 |  |  tele ← getValue(D, dom)                       // find teleport probability of node
16 |  |  sup ← Σ_{dom' ∈ neigh(dom)} (getValue(M, dom')/w_{dom,dom'})   // compute joint support
17 |  |  if dom ∈ seeds(AB_sign(𝒮)) && behaviour == SS then        // support to seeds - SS
18 |  |  |  sup ← sup + init_s
19 |  |  if dom ∈ seeds(AB_sign(𝒮)) && behaviour == DS then        // support to seeds - DS
20 |  |  |  sup ← sup + init_s/(iter + 1)
21 |  |  final = (1 − d) * tele + d * sup                             // final support score
22 |  |  M' ← addSupport(M, dom, final)                              // add support to map
23 |  M' ← normalize(M')                                               // normalize values
24 |  conv ← checkConvergence(M, M', θ_conv)         // all supports changed less than θ_conv?
25 |  M ← M'                                              // prepare map for next iteration
26 |  iter ← iter + 1                                                // increase counter
27 return M                                                   // map with support values
```

5 Perfomance and Implementation Discussion

Due to size restrictions we provide a rather limited discussion about the complexities and the cost of tuning the parameters of each algorithm. The huge scale of the web graph has the biggest performance implication to the the graph-based computation of the ABs *support*. What is encouraging though, is that the algorithms are applied over the compact SFG graph, that contains the SLDs of the pages and their corresponding links. The complexity of IC is in $\Theta(n * |\mathtt{doms}\mathcal{W}| * |\mathtt{dom}(\mathtt{links}(\mathcal{W}))|)$, where n is the number of experiments. LT is much slower though since we have to additionally consider the joint *support* of the neighbors of a node. Finally, the Biased-PR converges slower than the original PageRank, since the algorithm begins only with the seeds, spreading the support to the rest nodes. Also, we must consider the added cost of computing the shortest paths of the nodes from the seeds. For the relatively small SFG used in our study (see Sect. 6), the SS converges much faster than the DS and NS, which need ten times more iterations.

For newly introduced ABs though, the computation of the *support* scores of the domains can be considered an offline process. Users can submit ABs and BCs into the *bias goggles* system and get notified when they are ready for use. However, what is important is to let users explore in real-time the domains space for any precomputed and commonly used BCs. This can be easily supported by providing efficient ways to store and retrieve the signatures of already known BCs, along with the computed *support* scores of domains of available ABs. Inverted files and trie-based data structures (e.g., the space efficient burst-tries [15] and the cache-consious hybrid or pure HAT-tries [2]) over the SLDs and the signatures of the ABs and BCs, can allow the fast retrieval of offsets in files where the *support* scores and the related metadata are stored. Given the above, the computation of the *bias score* and the *support* of a BC for a domain is lightning fast. We have implemented a prototype[7] that allows the exploration of predefined BCs over a set of mainly greek domains. The prototype offers a REST API for retrieving the bias scores of the domains, and exploits the open-source project crawler4j[8]. We plan to improve the prototype, by allowing users to search and ingest BCs, ABs and domains of interest, and develop a user-friendly browser plugin on top of it.

6 Experimental Evaluation Discussion

Evaluating such a system is a rather difficult task, since there are no formal definitions of what bias in the web is, and there are no available datasets for evaluation. As a result, we based our evaluation over BCs for which it is easy to find biased sites. We used two BCs for our experiments, the *greek politics* (BC1) with 9 ABs, and the *greek football* (BC2) with 6 ABs. For these BCs, we gathered well known domains, generally considered as fully supportive of only one of the ABs, without inspecting though their link coverage to the respective seeds, to avoid any bias towards our graph based approach. Furthermore, we did not include the original seeds to this collection. In total, we collected 50 domains for BC1 and 65 domains for BC2, including newspapers, radio and television channels, blogs, pages of politicians, etc. This collection of domains is our gold standard.

We crawled a subset of the greek web by running four instances of the crawler: one with 383 sites related to the greek political life, one with 89 sport related greek sites, one with the top-300 popular greek sites according to Alexa, and a final one containing 127 seeds related to big greek industries. We black-listed

[7] http://pangaia.ics.forth.gr/bias-goggles.
[8] https://github.com/yasserg/crawler4j.

Table 2. Experimental results over two BCs.

Alg.	n		t (s) BC1 - Political Parties (9 ABs)	bias AGBR	bias AGS		t (s) BC2 - Sports Teams (6 ABs)	bias AGBR	bias AGS
IC	1000		**0.399**	217.571	0.2619		**0.287**	320.408	0.1362
IC	$k/2$		0.622	230.798	0.3024		0.497	326.005	0.1365
IC	k		0.945	**234.535**	0.361		0.681	328.667	0.1675
LT	1000		129.0	230.611	0.2354		87.5	322.585	0.1528
LT	$k/2$		512.9	230.064	0.3621		322.5	**328.698**	0.1364
LT	k		966.1	231.087	0.3626		663.1	327.786	0.1659
Biased-PR	SS (40, 31)		34.8	227.999	**0.5569**		17.9	261.788	**0.4745**
Biased-PR	DSS (319, 391)		260.5	129.829	0.3730		219.4	163.165	0.4344
Biased-PR	NSS (306, 458)		231.4	32.602	0.3041		207.6	34.905	0.4052

popular sites like facebook and twitter to control the size of our data and avoid crawling non-greek domains. The crawlers were restricted to depth seven for each domain, and free to follow any link to external domains. In total we downloaded 893,095 pages including 531,296,739 links, which lead to the non-connected SFG graph with 90,419 domains, 288,740 links (on average 3.1 links per domain) and a diameter $k = 7,944$. More data about the crawled pages, the gold standard, and the SFG graph itself are available in the prototype's site.

Below we report the results of our experiments over an i7-5820K 3.3GHz system, with 6 cores, 15MB cache and 16GB RAM memory, and a 6TB disk. For each of the two BCs and for each algorithm, we run experiments for various iterations n and Biased-PR variations, for the singleton ABs of the 9 political parties and 6 sports teams. For Biased-PR we evaluate all possible behaviours of the surfer using the parameters $\theta_{conv} = 0.001$ and $d = 0.85$. We also provide the average number of iterations for convergence over all ABs for Biased-PR. We report the run times in seconds, along with the metrics Average Golden Bias Ratio (AGBR) and Average Golden Similarity (AGS), that we introduce in this work. The AGBR is the ratio of the average bias score of the golden domains, as computed by the algorithms for a specific BC, divided by the average bias score of all domains for this BC. The higher the value, the more easily we can discriminate the golden domains from the rest. On the other hand, the AGS is the average similarity of the golden domains to their corresponding ABs. The higher the similarity value, the more biased the golden domains are found to be by our algorithms towards their aspects. A high similarity score though, does not imply high support for the golden domains or high disimilarity for the rest. The perfect algorithm will have high values for all metrics. The results are shown in Table 2.

The difference in BC1 and BC2 results implies a less connected graph for BC2 (higher AGBR values for BC2), where the support flows to less domains, but with a greater interaction between domains supporting different aspects (smaller AGS values). What is remarkable is the striking time performance of IC, suggesting that it can be used in real-time and with excellent results (at least for AGBR). On the other hand, the LT is a poor choice, being the slowest of all and dominated in any aspect by IC. Regarding the Biased-PR only the SS variation offers exceptional performance, especially for AGS. The DS and NS variations are more expensive and have the worst results regarding AGBR, especially the NSS that avoids bias. In most cases, algorithms benefit from more iterations. The SS variation of Biased-PR needs only 40 iterations for BC1 and 31 for BC2 to converge, proving that less nodes are affected by the seeds in BC2. Generally, the IC and the SS variation of Biased-PR are the best options, with the IC allowing the real-time ingestion of ABs. But, we need to evaluate the algorithms in larger graphs and for more BCs.

We also manually inspected the top domains according to the *bias* and *support* scores for each algorithm and each BC. Generally the support scores of the domains were rather low, showcasing the value of other support cues, like the content and the importance of pages that links appear in. In the case of BC1, except from the political parties, we found various blogs, politicians homepages, news sites, and also the national greek tv channel, being biased to a specific political party. In the case of BC2 we found the sport teams, sport related blogs, news sites, and a political party being highly biased towards a specific team, which is an interesting observation. In both cases we also found various domains with high support to all ABs, suggesting that these domains are good unbiased candidates. Currently, the *bias goggles* system is not able to pinpoint false positives (i.e pages with non supportive links) and false negatives (i.e., pages with content that supports a seed without linking to it), since there is no content analysis. We are certain that such results can exist, although we were not able to find such an example in the top results of our study. Furthermore, we are not able to distinguish links that can frequently appear in users' content, like in the signatures of forum members.

7 Conclusion and Future Work

In this work, we introduce the *bias goggles* model that facilitates the important task of exploring the bias characteristics of web domains to user-defined biased concepts. We focus only on graph-based approaches, using popular propagation models and the new Biased-PR PageRank variation that models biased surfers behaviours. We propose ways for the fast retrieval and ingestion of aspects of bias, and offer access to a developed prototype. The results show the efficiency of the approach, even in real-time. A preliminary evaluation over a subset of the greek web and a manually constructed gold standard of biased concepts and domains, shows promising results and interesting insights that need futher research.

In the future, we plan to explore variations of the proposed approach where our assumptions do not hold. For example, we plan to exploit the supportive, neutral or oppositive nature of the available links, as identified by sentiment analysis methods, along with the importance of the web pages they appear in. Content-based and hybrid approaches for computing the *support* scores of domains are also in our focus, as well as the exploitation of other available graphs, like the graph of friends, retweets, etc. In addition interesting aspects include how the *support* and *bias scores* of multiple BCs can be composed, providing interesting insights about possible correlations of different BCs, as well as how the *bias* scores of domains change over time. Finally, our vision is to integrate the approach in a large scale WSE/social platform/browser, in order to study how users define bias, create a globally accepted gold standard of BCs, and explore how such tools can affect the consumption of biased information. In this way, we will be able to evaluate and tune our approach in real-life scenarios, and mitigate any performance issues.

References

1. Adomavicius, G., Bockstedt, J., Shawn, C., Zhang, J.: De-biasing user preference ratings in recommender systems. In: CEUR-WS, vol. 1253, pp. 2–9 (2014)
2. Askitis, N., Sinha, R.: Hat-trie: a cache-conscious trie-based data structure for strings. In: Proceedings of the Thirtieth Australasian Conference on Computer Science, vol. 62, pp. 97–105. Australian Computer Society Inc. (2007)
3. Bolukbasi, T., Chang, K., Zou, J.Y., Saligrama, V., Kalai, A.: Man is to computer programmer as woman is to homemaker? debiasing word embeddings. CoRR, abs/1607.06520 (2016)
4. Bozdag, E.: Bias in algorithmic filtering and personalization. Ethics Inf. Technol. **15**(3), 209–227 (2013)
5. Budak, C., Goel, S., Rao, J.M.: Fair and balanced? quantifying media bias through crowdsourced content analysis. Publ. Opin. Q. **80**(S1), 250–271 (2016)
6. Corbett-Davies, S., Pierson, E., Feller, A., Goel, S., Huq, A.: Algorithmic decision making and the cost of fairness. In: Proceedings of the 23rd ACM SIGKDD International Conference on Knowledge Discovery and Data Mining, pp. 797–806. ACM (2017)
7. Dwork, C., Hardt, M., Pitassi, T., Reingold, O., Zemel, R.S.: Fairness through awareness. In: ITCS, pp. 214–226 (2012)
8. Epstein, R., Robertson, R.E.: The search engine manipulation effect (seme) and its possible impact on the outcomes of elections. PNAS **112**(20), E4512–E4521 (2015)
9. Gleich, D.F.: Pagerank beyond the web. SIAM Rev. **57**(3), 321–363 (2015)
10. Goyal, A., Bonchi, F., Lakshmanan, L.V.: Learning influence probabilities in social networks. In: Proceedings of the Third ACM International Conference on Web Search and Data Mining, pp. 241–250. ACM (2010)
11. Hajian, S., Bonchi, F., Castillo, C.: Algorithmic bias: from discrimination discovery to fairness-aware data mining. In: KDD, pp. 2125–2126. ACM (2016)
12. Hannak, A., et al.: Measuring personalization of web search. In: WWW, pp. 527–538. ACM (2013)
13. Hannak, A., Soeller, G., Lazer, D., Mislove, A., Wilson, C.: Measuring price discrimination and steering on e-commerce web sites. In: Internet Measurement Conference, pp. 305–318 (2014)
14. Hardt, M., Price, E., Srebro, N.: Equality of opportunity in supervised learning. In: NIPS, pp. 3315–3323 (2016)
15. Heinz, S., Zobel, J., Williams, H.E.: Burst tries: a fast, efficient data structure for string keys. ACM Trans. Inf. Syst. (TOIS) **20**(2), 192–223 (2002)

16. House, W.: Big data: A report on algorithmic systems, opportunity, and civil rights. Executive Office of the President, White House, Washington, DC (2016)
17. Kempe, D., Kleinberg, J., Tardos, É.: Maximizing the spread of influence through a social network. In: Proceedings of the Ninth ACM SIGKDD International Conference on Knowledge Discovery and Data Mining, pp. 137–146. ACM (2003)
18. Koutra, D., Bennett, P. N., Horvitz, E.: Events and controversies: influences of a shocking news event on information seeking. In: WWW, pp. 614–624 (2015)
19. Kulshrestha, J., et al.: Quantifying search bias: investigating sources of bias for political searches in social media. In: CSCW (2017)
20. Lepri, B., Staiano, J., Sangokoya, D., Letouzé, E., Oliver, N.: The Tyranny of data? the bright and dark sides of data-driven decision-making for social good. In: Cerquitelli, T., Quercia, D., Pasquale, F. (eds.) Transparent Data Mining for Big and Small Data. SBD, vol. 11, pp. 3–24. Springer, Cham (2017). https://doi.org/10.1007/978-3-319-54024-5_1
21. Liu, Z., Weber, I.: Is twitter a public sphere for online conflicts? a cross-ideological and cross-hierarchical look. In: SocInfo, pp. 336–347 (2014)
22. Lu, H., Caverlee, J., Niu, W.: Biaswatch: a lightweight system for discovering and tracking topic-sensitive opinion bias in social media. In: Proceedings of the 24th ACM International on Conference on Information and Knowledge Management, pp. 213–222. ACM (2015)
23. Mowshowitz, A., Kawaguchi, A.: Measuring search engine bias. Inf. Process. Manag. **41**(5), 1193–1205 (2005)
24. Muchnik, L., Aral, S., Taylor, S.J.: Social influence bias: a randomized experiment. Science **341**(6146), 647–651 (2013)
25. Pan, B., Hembrooke, H., Joachims, T., Lorigo, L., Gay, G., Granka, L.: In google we trust: users' decisions on rank, position, and relevance. J. Comput.-Mediated Commun. **12**(3), 801–823 (2007)
26. Pariser, E.: The filter bubble: what the Internet is hiding from you. Penguin, UK (2011)
27. E. Pitoura, P. Tsaparas, G. Flouris, I. Fundulaki, P. Papadakos, S. Abiteboul, G. Weikum: On measuring bias in online information. ACM SIGMOD Record **46**(4), 16–21 (2018)
28. Sandvig, C., Hamilton, K., Karahalios, K., Langbort, C.: Auditing algorithms: research methods for detecting discrimination on internet platforms. Data and discrimination: converting critical concerns into productive inquiry (2014)
29. Skeem, J.L., Lowenkamp, C.T.: Risk, race, and recidivism: predictive bias and disparate impact. Criminology **54**(4), 680–712 (2016)
30. Vaughan, L., Thelwall, M.: Search engine coverage bias: evidence and possible causes. Inf. Process. Manag. **40**(4), 693–707 (2004)
31. Weber, I., Garimella, V.R.K., Batayneh, A.: Secular vs. islamist polarization in Egypt on twitter. In: ASONAM, pp. 290–297 (2013)
32. Wong, F.M.F., Tan, C.W., Sen, S., Chiang, M.: Quantifying political leaning from tweets and retweets. ICWSM **13**, 640–649 (2013)
33. Zafar, M.B., Gummadi, K.P., Danescu-Niculescu-Mizil, C.: Message impartiality in social media discussions. In: ICWSM, pp. 466–475 (2016)
34. Zafar, M.B., Valera, I., Rodriguez, M.G., Gummadi, K.P.: Fairness beyond disparate treatment & disparate impact: learning classification without disparate mistreatment. In: WWW (2017)

Deep Learning IV

Biconditional Generative Adversarial Networks for Multiview Learning with Missing Views

Anastasiia Doinychko[1,2(✉)] and Massih-Reza Amini[1]

[1] Laboratoire D'Informatique de Grenoble, Université Grenoble Alpes, Saint-Martin-d'hàres, France
Anastasiia.Doinychko@univ-grenoble-alpes.fr,
Massih-Reza.Amini@univ-grenoble-alpes.fr
[2] Mentor Graphics, Wilsonville, USA

Abstract. In this paper, we present a conditional GAN with two generators and a common discriminator for multiview learning problems where observations have two views, but one of them may be missing for some of the training samples. This is for example the case for multilingual collections where documents are not available in all languages. Some studies tackled this problem by assuming the existence of view generation functions to approximately complete the missing views; for example Machine Translation to translate documents into the missing languages. These functions generally require an external resource to be set and their quality has a direct impact on the performance of the learned multiview classifier over the completed training set. Our proposed approach addresses this problem by jointly learning the missing views and the multiview classifier using a tripartite game with two generators and a discriminator. Each of the generators is associated to one of the views and tries to fool the discriminator by generating the other missing view conditionally on the corresponding observed view. The discriminator then tries to identify if for an observation, one of its views is completed by one of the generators or if both views are completed along with its class. Our results on a subset of Reuters RCV1/RCV2 collections show that the discriminator achieves significant classification performance; and that the generators learn the missing views with high quality without the need of any consequent external resource.

Keywords: Bilingual text classification · Multiview learning · Conditional GAN

1 Introduction

We address the problem of multiview learning with Generative Adversarial Networks (GANs) in the case where some observations may have missing views without there being an external resource to complete them. This is a typical

J. M. Jose et al. (Eds.): ECIR 2020, LNCS 12035, pp. 807–820, 2020.
https://doi.org/10.1007/978-3-030-45439-5_53

situation in many applications where different sources generate different views of samples unevenly; like text information present in all Wikipedia pages while images are more scarce. Another example is multilingual text classification where documents are available in two languages and share the same set of classes while some are just written in one language. Previous works supposed the existence of view generating functions to complete the missing views before deploying a learning strategy [2]. However, the performance of the global multiview approach is biased by the quality of the generating functions which generally require external resources to be set. The challenge is hence to learn an efficient model from the multiple views of training data without relying on an extrinsic approach to generate altered views for samples that have missing ones.

In this direction, GANs provide a propitious and broad approach with a high ability to seize the underlying distribution of the data and create new samples [11]. These models have been mostly applied to image analysis and major advances have been made on generating realistic images with low variability [7,15,16]. GANs take their origin from the game theory and are formulated as a two players game formed by a generator G and a discriminator D. The generator takes a noise z and produces a sample $G(z)$ in the input space, on the other hand the discriminator determines whenever a sample comes from the true distribution of the data or if it is generated by G. Other works included an inverse mapping from the input to the latent representation, mostly referred to as BiGANs, and showed the usefulness of the learned feature representation for auxiliary discriminant problems [8,9]. This idea paved the way for the design of efficient approaches for generating coherent synthetic views of an input image [6,14,21].

In this work, we propose a GAN based model for bilingual text classification, called `Cond`^{2}`GANs`, where some training documents are just written in one language. The model learns the representation of missing versions of bilingual documents jointly with the association to their respective classes, and is composed of a discriminator D and two generators G_1 and G_2 formulated as a tripartite game. For a given document with a missing version in one language, the corresponding generator induces the latter conditionally on the observed one. The training of the generators is carried out by minimizing a regularized version of the cross-entropy measure proposed for multi-class classification with GANs [19] in a way to force the models to generate views such that the completed bilingual documents will have high class assignments. At the same time, the discriminator learns the association between documents and their classes and distinguishes between observations that have their both views and those that got a completed view by one of the generators. This is achieved by minimizing an aggregated cross-entropy measure in a way to force the discriminator to be certain of the class of observations with their complete views and uncertain of the class of documents for which one of the versions was completed. The regularization term in the objectives of generators is derived from an adapted feature matching technique [17] which is an effective way for preventing from situations where the models become unstable; and which leads to fast convergence.

We demonstrate that generated views allow to achieve state-of-the-art results on a subset of Reuters RCV1/RCV2 collections compared to multiview approaches that rely on Machine Translation (MT) for translating documents into languages in which their versions do not exist; before training the models. Importantly, we exhibit qualitatively that generated documents have meaningful translated words bearing similar ideas compared to the original ones; and that, without employing any large external parallel corpora to learn the translations as it would be the case if MT were used. More precisely, this work is the first to:

- Propose a new tripartite GAN model that makes class prediction along with the generation of high quality document representations in different input spaces in the case where the corresponding versions are not observed (Sect. 3.2);
- Achieve state-of-the art performance compared to multiview approaches that rely on external view generating functions on multilingual document classification; and which is another challenging application than image analysis which is the domain of choice for the design of new GAN models (Sect. 4.2);
- Demonstrate the value of the generated views within our approach compared to when they are generated using MT (Sect. 4.2).

2 Related Work

Multiview learning has been an active domain of research these past few years. Many advances have been made on both theoretic and algorithmic sides [5,12]. The three main families of techniques for (semi-)supervised learning are (kernel) Canonical Correlation Analysis (CCA), Multiple kernel learning (MKL) and co-regularization. CCA finds pairs of highly correlated subspaces between the views that is used for mapping the data before training, or integrated in the learning objective [3,10]. MKL considers one kernel per view and different approaches have been proposed for their learning. In one of the earliest work, [4] proposed an efficient algorithm based on sequential minimization techniques for learning a corresponding support vector machine defined over a convex nonsmooth optimization problem. Co-regularization techniques tend to minimize the disagreement between the single-view classifiers over their outputs on unlabeled examples by adding a regularization term to the objective function [18]. Some approaches have also tackled the tedious question of combining the predictions of the view specific classifiers [20]. However all these techniques assume that views of a sample are complete and available during training and testing.

Recently, many other studies have considered the generation of multiple views from a single input image using GANs [14,21,23] and have demonstrated the intriguing capacity of these models to generate coherent unseen views. The former approaches rely mostly on an encoder-encoder network to first map images into a latent space and then generate their views using an inverse mapping. This is a very exciting problem, however, our learning objective differs from these approaches as we are mostly interested in the classification of muti-view samples with missing views. The most similar work to ours that uses GANs for multiview classification is probably [6]. This approach generates missing views of images in the same latent space than the input image, while $\texttt{Cond}^2\texttt{GANs}$ learns the representations of the missing views in their respective input spaces conditionally on the observed ones which in general are from other feature spaces. Furthermore, $\texttt{Cond}^2\texttt{GANs}$ benefits from low complexity and stable convergence which has been shown to be lacking in the previous approach.

Another work which has considered multiview learning with incomplete views, also for document classification, is [2]. The authors proposed a Rademacher complexity bounds for a multiview Gibbs classifier trained on multilingual collections where the missing versions of documents have been generated by Machine Translation systems. Their bounds exhibit a term corresponding to the quality of the MT system generating the views. The bottleneck is that MT systems depend on external resources, and they require a huge amount of parallel collections containing documents and their translations in all languages of interest for their tuning. For rare languages, this can ultimately affect the performance of the learning models. Regarding these aspects our proposed approach differs from all the previous studies, as we do not suppose the existence of parallel training sets nor MT systems to generate the missing versions of the training observations.

3 Cond²GANs

In the following sections, we first present the basic definitions which will serve to our problem setting, and then the proposed model for multiview classification with missing views.

3.1 Framework and Problem Setting

We consider multiclass classification problems, where a bilingual document is defined as a sequence $\mathbf{x} = (x^1, x^2) \in \mathcal{X}$ that belongs to one and only one class $\mathbf{y} \in \mathcal{Y} = \{0,1\}^K$. The class membership indicator vector $\mathbf{y} = (y_k)_{1 \leq k \leq K}$, of each bilingual document, has all its components equal to 0 except the one that indicates the class associated with the example which is equal to one. Here we suppose that $\mathcal{X} = (\mathcal{X}_1 \cup \{\perp\}) \times (\mathcal{X}_2 \cup \{\perp\})$, where $x^v = \perp$ means that the v-th view is not observed. Hence, each observed view $x^v \in \mathbf{x}$ is such that $x^v \neq \perp$ and it provides a representation of $\mathbf{x}$ in a corresponding input space $\mathcal{X}_v \subseteq \mathbb{R}^{d_v}$. Following the conclusions of the co-training study [5], our framework is based on the following main assumption:

Assumption 1 ([5]). Observed views are not completely correlated, and are equally informative.

Furthermore, we assume that each example $(\mathbf{x}, \mathbf{y})$ is identically and independently distributed (i.i.d.) according to a fixed yet unknown distribution $\mathcal{D}$ over $\mathcal{X} \times \mathcal{Y}$, and that at least one of its views is observed. Additionally, we suppose to have access to a training set $\mathcal{S} = \{(\mathbf{x}_i, \mathbf{y}_i); i \in \{1, \ldots, m\}\} = \mathcal{S}_F \sqcup \mathcal{S}_1 \sqcup \mathcal{S}_2$ of size m drawn i.i.d. according to $\mathcal{D}$, where $\mathcal{S}_F = \{\left(\left(x_i^1, x_i^2\right), \mathbf{y}_i\right) \mid i \in \{1, \ldots, m_F\}\}$

denotes the subset of training samples with their both complete views and $\mathcal{S}_1 = \{((x_i^1, \perp), \mathbf{y}_i) \mid i \in \{1, \ldots, m_1\}\}$ (respectively $\mathcal{S}_2 = \{((\perp, x_i^2), \mathbf{y}_i) \mid i \in \{1, \ldots, m_2\}\}$) is the subset of training samples with their second (respectively first) view that is not observed (i.e. $m = m_F + m_1 + m_2$).

It is possible to address this problem using existing techniques; for example, by learning singleview classifiers independently on the examples of $\mathcal{S} \sqcup \mathcal{S}_1$ (respectively $\mathcal{S} \sqcup \mathcal{S}_2$) for the first (respectively second) view. To make predictions, one can then combine the outputs of the classifiers [20] if both views of a test example are observed; or otherwise, use one of the outputs corresponding to the observed view. Another solution is to apply multiview approaches over the training samples of $\mathcal{S}_F$; or over the whole training set $\mathcal{S}$ by completing the views of examples in $\mathcal{S}_1$ and $\mathcal{S}_2$ before using external view generation functions.

3.2 The Tripartite Game

As an alternative, the learning objective of our proposed approach is to generate the missing views of examples in $\mathcal{S}_1$ and $\mathcal{S}_2$, jointly with the learning of the association between the multiview samples (with all their views complete or completed) and their classes. The proposed model consists of three neural networks that are trained using an objective implementing a three players game between a discriminator, D, and two generators, G_1 and G_2. The game that these models play is depicted in Fig. 1 and it can be summarized as follows. At each step, if an observation is chosen with a missing view, the corresponding generator – G_1 (respectively G_2) if the first (respectively second) view is missing – produces the view from random noise conditionally on the observed view in a way to fool the discriminator. On the other hand, the discriminator takes as input an observation with both of its views complete or completed and, classifies it if the views are initially observed or tells if a view was produced by one of the generators. Formally, both generators G_1 and G_2 take as input; samples from the training subsets $\mathcal{S}_2$ and $\mathcal{S}_1$ respectively; as well as random noise drawn from a uniform distribution defined over the input space of the missing view and produce the corresponding pseudo-view, which is missing; i.e.

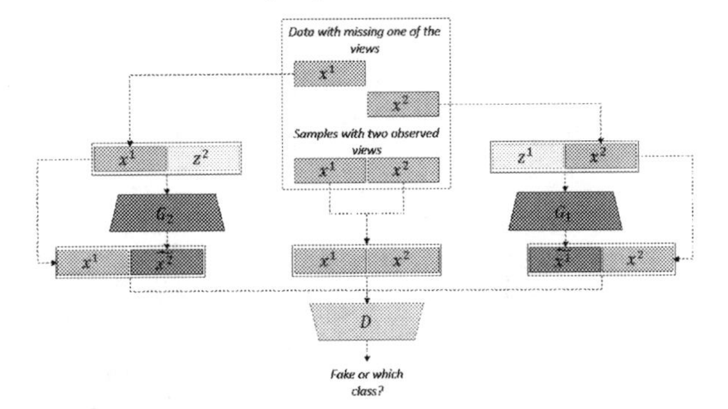

Fig. 1. A visual representation of the proposed GAN model composed of three neural networks; a discriminator D and two generators G_1 and G_2. The missing view of an observation is completed by the corresponding generator conditionally on its observed view. The discriminator is trained to recognize between observations having their views completed and those with complete initial views as well as their classes.

$G_1(z^1, x^2) = \tilde{x}^1$ and $G_2(x^1, z^2) = \tilde{x}^2$. These models are learned in a way to replicate the conditional distributions $p(x^1|x^2, z^1)$ and $p(x^2|x^1, z^2)$; and inherently define two probability distributions, denoted respectively by p_{G_1} and p_{G_2}, as the distribution of samples if both views where observed i.e. $(\tilde{x}^1, x^2) \sim p_{G_1}(x^1, x^2)$, $(x^1, \tilde{x}^2) \sim p_{G_2}(x^1, x^2)$. On the other hand, the discriminator takes as input a training sample; either from the set $\mathcal{S}_F$, or from one of the training subsets $\mathcal{S}_1$ or $\mathcal{S}_2$ where the missing view of the example is generated by one of the generators accordingly. The task of D is then to recognize observations from $\mathcal{S}_1$ and $\mathcal{S}_2$ that have completed views by G_1 and G_2 and to classify examples from $\mathcal{S}$ to their true classes. To achieve this goal we add a fake class, $K+1$, to the set of classes, $\mathcal{Y}$, corresponding to samples that have one of their views generated by G_1 or G_2. The dimension of the discriminator's output is hence set to $K+1$ which by applying softmax is supposed to estimate the posterior probability of classes for each multiview observation (with complete or completed views) given in input. For an observation $\mathbf{x} \in \mathcal{X}$, we use $D_{K+1}(\mathbf{x}) = p_D(y = K+1|\mathbf{x})$ to estimate the probability that one of its views is generated by G_1 or G_2. As the task of the generators is to produce good quality views such that the observation with the completed view will be assigned to its true class with high probability, we follow [17] by supplying the discriminator to not get fooled easily as stated in the following assumption:

Assumption 2 **([17]).** An observation $\mathbf{x}$ has one of its views generated by G_1 or G_2; if and only if $D_{K+1}(\mathbf{x}) > \sum_{k=1}^{K} D_k(\mathbf{x})$.

In the case where; $D_{K+1}(\mathbf{x}) \leq \sum_{k=1}^{K} D_k(\mathbf{x})$ the observation $\mathbf{x}$ is supposed to have its both views observed and it is affected to one of the classes following the rule; $\max_{k=\{1,\ldots,K\}} D_k(\mathbf{x})$. The overall learning objective of `Cond²GANs` is to train the generators to produce realistic views indistinguishable with the real ones, while the discriminator is trained to classify multiview observations having their complete views and to identify view generated samples. If we denote by p_{real} the marginal distribution of multiview observations with their both views observed (i.e. $(x^1, x^2) = p_{real}(x^1, x^2)$); the above procedure resumes to the following discriminator objective function $V_D(D, G_1, G_2)$:

$$\begin{aligned} \max_D V_D(D, G_1, G_2) = \; & \mathbb{E}_{(x^1,x^2,y)\sim\mathcal{S}_F}\left[\log p_D(y|x^1, x^2, y < K+1)\right] \\ & + \frac{1}{2}\mathbb{E}_{(\tilde{x}^1,x^2)\sim p_{G_1}}\left[\log p_D(y = K+1|\tilde{x}^1, x^2)\right] \\ & + \frac{1}{2}\mathbb{E}_{(x^1,\tilde{x}^2)\sim p_{G_2}}\left[\log p_D(y = K+1|x^1, \tilde{x}^2)\right]. \end{aligned} \tag{1}$$

In this way, we stated minmax game over $K+1$ component of discriminator. In addition to this objective, we made generators also learn from labels of completed samples. Therefore, the following equation defines objective for the generators

$V_{G_{1,2}}(D, G_1, G_2)$:

$$\max_{G_1,G_2} V_{G_{1,2}}(D, G_1, G_2) = \frac{1}{2}\mathbb{E}_{(x^2,y)\sim\mathcal{S}_2,z}\left[\log p_D(y|G_1(x^2,z),x^2)\right] + \frac{1}{2}\mathbb{E}_{(x^1,y)\sim\mathcal{S}_1,z}\left[\log p_D(y|x^1,G_2(x^1,z))\right]. \tag{2}$$

Note that, following Assumption 1, we impose the generators to produce equally informative views by assigning the same weight to their corresponding terms in the objective functions (Eqs. 1, 2).

3.3 Analyses and Convergence

From the outputs of the discriminator for all $\mathbf{x} \in \mathcal{X}$ we build an auxiliary function $\mathbf{D}(\mathbf{x}) = \sum_{k=1}^{K} p_D(y = k \mid \mathbf{x})$ equal to the sum of the first K outputs associated to the true classes. In this following, we provide a theoretical analysis of Cond^{2}GANs involving the auxiliary function $\mathbf{D}$ under nonparametric hypotheses.

Proposition 1. *For fixed generators G_1 and G_2, the objective defined in (Eq. 1) leads to the following optimal discriminator $\mathbf{D}^*_{G_1,G_2}$:*

$$\mathbf{D}^*_{G_1,G_2}(x^1,x^2) = \frac{p_{real}(x^1,x^2)}{p_{real}(x^1,x^2) + p_{G_{1,2}}(x^1,x^2)}, \tag{3}$$

where $p_{G_{1,2}}(x^1,x^2) = \frac{1}{2}(p_{G_1}(x^1,x^2) + p_{G_2}(x^1,x^2))$.

Proof. The proof follows from [11]. Let

$$\forall \mathbf{x} = (x^1,x^2), \mathbf{D}(\mathbf{x}) = \sum_{k=1}^{K} D_k(\mathbf{x})$$

From Assumption 2, and the fact that for any observation $\mathbf{x}$ the outputs of the discriminator sum to one i.e. $\sum_{k=1}^{K+1} D_k(\mathbf{x}) = 1$, the value function V_D writes:

$$V_D(\mathbf{D}, G_1, G_2) = \iint \log(\mathbf{D}(x^1,x^2))p_{real}(x^1,x^2)dx^1dx^2 +$$
$$\frac{1}{2}\iint \log(1-\mathbf{D}(x^1,x^2))p_{G_1}(x^1,x^2)dx^1dx^2 + \frac{1}{2}\iint \log(1-\mathbf{D}(x^1,x^2))p_{G_2}(x^1,x^2)dx^1dx^2$$

For any $(\alpha,\beta,\gamma) \in \mathbb{R}^3\backslash\{0,0,0\}$; the function $z \mapsto \alpha\log z + \frac{\beta}{2}\log(1-z) + \frac{\gamma}{2}\log(1-z)$ reaches its maximum at $z = \frac{\alpha}{\alpha+\frac{1}{2}(\beta+\gamma)}$, which ends the proof as the discrimintaor does not need to be defined outside the supports of p_{data}, p_{G_1} and p_{G_2}. □

By plugging back $\mathbf{D}^*_{G_1,G_2}$ (Eq. 3) into the value function V_D we have the following necessary and sufficient condition for attaining the global minimum of this function:

Theorem 1. *The global minimum of the function $V_D(G_1, G_2)$ is attained if and only if*

$$p_{real}(x^1, x^2) = \frac{1}{2}(p_{G_1}(x^1, x^2) + p_{G_2}(x^1, x^2)). \quad (4)$$

At this point, the minimum is equal to $-\log 4$.

Proof. By plugging back the expression of $\mathbf{D}^*$ (Eq. 3), into the value function V_D, it comes

$$V(\mathbf{D}^*, G_1, G_2) = \iint \log\left(\frac{p_{real}(x^1, x^2)}{p_{real}(x^1, x^2) + p_{G_{1,2}}(x^1, x^2)}\right) p_{real}(x^1, x^2) dx^1 dx^2 + \iint \log\left(\frac{p_{G_{1,2}}(x^1, x^2)}{p_{real}(x^1, x^2) + p_{G_{1,2}}(x^1, x^2)}\right) p_{G_{1,2}}(x^1, x^2) dx^1 dx^2$$

Which from the definition of the Kullback Leibler (KL) and the Jensen Shannon divergence (JSD) can be rewritten as

$$\begin{aligned} V_D(\mathbf{D}^*, G_1, G_2) &= -\log 4 + KL\left(p_{real} \,\Big\|\, \frac{p_{real} + p_{G_{1,2}}}{2}\right) + KL\left(p_{G_{1,2}} \,\Big\|\, \frac{p_{real} + p_{G_{1,2}}}{2}\right) \\ &= -\log 4 + 2JSD\big(p_{real} \,\|\, p_{G_{1,2}}\big) \end{aligned}$$

The JSD is always positive and $JSD\big(p_{real} \,\|\, p_{G_{1,2}}\big) = 0$ if and only if $p_{real} = p_{G_{1,2}}$ which ends the proof. □

From Eq. 4, it is straightforward to verify that $p_{real}(x^1, x^2) = p_{G_1}(x^1, x^2) = p_{G_2}(x^1, x^2)$ is a global Nash equilibrium but it may not be unique. In order to ensure the uniqueness, we add the Jensen-Shannon divergence between the distribution p_{G_1} and p_{real} and p_{G_2} and p_{real} the value function V_D (Eq. 1) as stated in the corollary below.

Corollary 1. *The unique global Nash equilibrium of the augmented value function:*

$$\bar{V_D}(\mathbf{D}, G_1, G_2) = V(\mathbf{D}, G_1, G_2) + JSD(p_{G_1}||p_{real}) + JSD(p_{G_2}||p_{real}), \quad (5)$$

is reached if and only if

$$p_{real}(x^1, x^2) = p_{G_1}(x^1, x^2) = p_{G_2}(x^1, x^2), \quad (6)$$

where $V_D(\mathbf{D}, G_1, G_2)$ is the value function defined in Eq. (1) and $JSD(p_{G_1}||p_{real})$ is the Jensen-Shannon divergence between the distribution p_{G_1} and p_{real}.

Proof. The proof follows from the positivness of JSD and the necessary and sufficient condition for it to be equal to 0. Hence, $\bar{V_D}(\mathbf{D}, G_1, G_2)$ reaches it minimum $-\log 4$, iff $p_{G_1} = p_{real} = p_{G_2}$. □

This result suggests that at equilibrium, both generators produce views such that observations with their completed view follow the same real distribution than those which have their both views observed.

3.4 Algorithm and Implementation

In order to avoid the collapse of the generators [17], we perform minibatch discrimination by allowing the discriminator to have access to multiple samples in combination. From this perspective, the minmax game (Eqs. 1, 2) is equivalent to the maximization of a cross-entropy loss, and we use minibatch training to learn the parameters of the three models. The corresponding empirical errors estimated over a minibatch $\mathcal{B}$ that contains m_b samples from each of the sets $\mathcal{S}_F$, $\mathcal{S}_1$ and $\mathcal{S}_2$ are:

$$\mathcal{L}_D(\mathcal{B}) = -\frac{1}{m_b}\sum_{\mathbf{x}\in\mathcal{B}\cap\mathcal{S}_F}\frac{1}{K+1}\sum_{k=1}^{K} y_k \log\left[D_k(x^1,x^2)\right] \\ -\frac{1}{2m_b}\sum_{\mathbf{x}\in\mathcal{B}\cap\mathcal{S}_1}\log\left[D_{K+1}(G_1(z^1,x^2),x^2))\right] - \frac{1}{2m_b}\sum_{\mathbf{x}\in\mathcal{B}\cap\mathcal{S}_2}\log\left[D_{K+1}(x^1,G_2(x^1,z^2))\right] \quad (7)$$

$$\mathcal{L}_{G_v}(\mathcal{B}) = -\frac{1}{m_b}\sum_{\mathbf{x}\in\mathcal{B}\cap\mathcal{S}_v}\frac{1}{K+1}\sum_{k=1}^{K} y_k \log\left[D_k(G_v(z^v,x^{3-v}),x^{3-v})\right] + \mathcal{L}_{FM}^v; v\in\{1,2\} \quad (8)$$

Minibatch stochastic training of Cond²GANs

Input: A training set $\mathcal{S} = \mathcal{S}_F \sqcup \mathcal{S}_1 \sqcup \mathcal{S}_2$
Initialization: Size of minibatches, m_b
Use *Xavier* initializer to initialize discriminator and generators parameters, respectively $\theta_d^{(0)}, \theta_{g1}^{(0)}, \theta_{g2}^{(0)}$
for $i = 0 \ldots T-1$ **do**
  Sample randomly a minibatch $\mathcal{B}_i$ of size $3m_b$ from $\mathcal{S}_1$, $\mathcal{S}_2$ and $\mathcal{S}_F$; create minibatches of noise vector z^1, z^2 from $\mathcal{U}(-1,1)$
  $\theta_d^{(i+1)} \leftarrow Adam(\mathcal{L}_D(\mathcal{B}_i), \theta_d^{(i)}, \alpha, \beta)$ # Update of D
  $\theta_{g1}^{(i+1)} \leftarrow Adam(\mathcal{L}_{G_1}(\mathcal{B}_i), \theta_{g1}^{(i)}, \alpha, \beta)$ # Update of G_1
  $\theta_{g2}^{(i+1)} \leftarrow Adam(\mathcal{L}_{G_2}(\mathcal{B}_i), \theta_{g2}^{(i)}, \alpha, \beta)$ # Update of G_2
end

In order to be inline with the premises of Corollary 1; we empirically tested different solutions and the most effective one that we found was the feature matching technique proposed in [17], which addressed the problem of instability for the learning of generators by adding a penalty term $\mathcal{L}_{FM}^v = \|\mathbb{E}_{p_{real}} f(x^1,x^2) - \mathbb{E}_{p_{G_v}} f(x^{3-v}, G_v(x^v))\|, v\in\{1,2\}$ to their corresponding objectives (Eq. (8)). Where, $\|.\|$ is the ℓ_2 norm and f is the sigmoid activation function on an intermediate layer of the discriminator. The overall algorithm of Cond²GANs is shown above. The parameters of the three neural networks are first initialized using *Xavier*. For a given number of iterations T, minibatches of size $3m_b$ are randomly sampled from the sets $\mathcal{S}_F$, $\mathcal{S}_1$ and $\mathcal{S}_2$. Minibatches of noise vectors are randomly drawn from the uniform distribution. Models parameters of the discriminator and both generators are then sequentially updated using *Adam* optimization algorithm [13]. We implemented our method by having two layers neural networks for each of the components of Cond²GANs. These neural nets are composed of 200 nodes in hidden layers with a sigmoid activation function. Since the values of the generated samples are supposed to approximate any possible real value, we do not use the activation function in the outputs of both generators.[1]

[1] We will release the code for reproducibility and research purpose.

4 Experiments

In this Section, we present experimental results aimed at evaluating how the generation of views by $\texttt{Cond}^2\texttt{GANs}$ can help to take advantage of existing training examples, with many having an incomplete view, in order to learn an efficient classification function. We perform experiments on a publicly available collection, extracted from Reuters RCV1/RCV2, that is proposed for multilingual multiclass text categorization[2] (Table 1). The dataset contains numerical feature vectors of documents originally presented in five languages (EN, FR, GR, IT, SP). In our experiments, we consider four pairs of languages with always English as one of the views ((EN,FR),(EN,SP),(EN,IT),(EN,GR)). Documents in different languages belong to one and only one class within the same set of classes ($K = 6$); and they also have translations into all the other languages. These translations are obtained from a state-of-the-art Statistical Machine Translation system [22] trained over the Europal parallel collection using about 8.10^6 sentences for the 4 considered pairs of languages.[3]

Table 1. The statistics of RCV1/RCV2 Reuters data collection used in our experiments.

Language	# docs	(%)	dim
EN	18,758	16.78	21,531
FR	26,648	23.45	24,893
GR	29,953	26.80	34,279
IT	24,039	21.51	15,506
SP	12,342	11.46	11,547
Total	111,740		

Class	Size (all lang.)	(%)
C15	18,816	16.84
CCAT	21,426	19.17
E21	13,701	12.26
ECAT	19,198	17.18
GCAT	19,178	17.16
M11	19,421	17.39

4.1 Experimental Setup

In our experiments, we consider the case where the number of training documents having their two versions is much smaller than those with only one of their available versions (i.e. $m_F \ll m_1 + m_2$). This corresponds to the case where the effort of gathering documents in different languages is much less than translating them from one language to another. To this end, we randomly select $m_F = 300$ samples having their both views, $m_1 = m_2 = 6000$ samples with one of their views missing and the remaining samples without their translations for test. In order to evaluate the quality of generated views by $\texttt{Cond}^2\texttt{GANs}$ we considered two scenarios. In the first one (denoted by $\mathrm{T}_{\mathrm{EN}\tilde{v}}$), we test on English documents by considering the generation of these documents with respect to the other view

2 https://archive.ics.uci.edu/ml/datasets/Reuters+RCV1+RCV2+Multilingual,+Multiview+Text+Categorization+Test+collection.
3 http://www.statmt.org/europarl/.

($v \in \{\mathtt{FR}, \mathtt{GR}, \mathtt{IT}, \mathtt{SP}\}$) using the corresponding generator. In the second scenario (denoted by $\mathtt{T}_{\widetilde{\mathtt{EN}}_v}$), we test on documents that are written in another language than English by considering their generation on English provided by the other generator. For evaluation, we test the following four classification approaches along with $\mathtt{Cond^2GANs}$; one singleview approach and four multiview approaches. In the singleview approach (denoted by $\mathbf{c}_v$) classifiers are the same as the discriminator and they are trained on the part of the training set with examples having their corresponding view observed. The multiview approaches are `MKL` [4], co-classification (`co-classif`) [1], unanimous vote ($\mathtt{mv}_b$) [2]. Results are evaluated over the test set using the accuracy and the F_1 measure which is the harmonic average of precision and recall. The reported performance are averaged over 20 random (train/test) sets, and the parameters of Adam optimization algorithm are set to $\alpha = 10^{-4}$, $\beta = 0.5$.

4.2 Experimental Results

On the value of the generated views. We start our evaluation by comparing the F_1 scores over the test set, obtained with $\mathtt{Cond^2GANs}$ and a neural network having the same architecture than the discriminator D of $\mathtt{Cond^2GANs}$ trained over the concatenated views of documents in the training set where the missing views are generated by Machine Translation. Figure 2 shows these results. Each point represents a class, where its abscissa (resp. ordinate) represents the test F_1 score of the Neural Network trained using MT (resp. one of the generators of $\mathtt{Cond^2GANs}$) to complete the missing views. All of the classes, in the different language pair scenarios, are above the line of equality, suggesting that the generated views by $\mathtt{Cond^2GANs}$ provide higher value information than translations provided by MT for learning the Neural Network. This is an impressive finding, as the resources necessary for the training of MT is large (8.10^6 pairs of sentences and their translations); while $\mathtt{Cond^2GANs}$ does both view completion and discrimination using only the available training data. This is mainly because both generators induce missing views with the same distribution than real pairs of views as stated in Corollary 1.

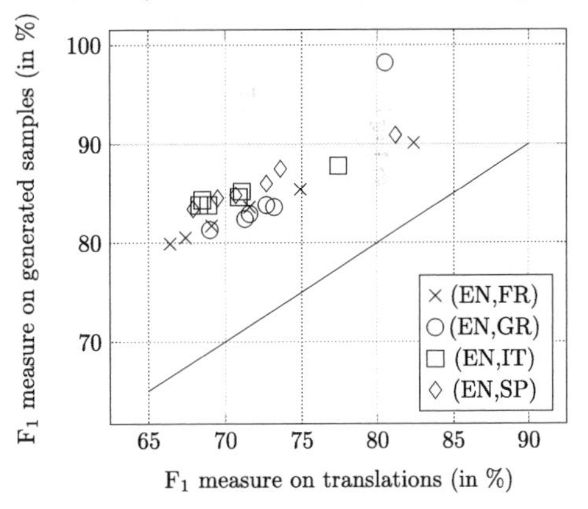

Fig. 2. F_1-score per class measured for test predictions made by a Neural-Network, with the same architecture than the discriminator of $\mathtt{Cond^2GANs}$, and trained over documents where their missing views are generated by MT, or by G_1 or G_2.

Comparison between multiview approaches. We now examine the gains, in terms of accuracy, of learning the different multiview approaches on a collection where for other approaches than Cond^{2}GANs the missing views are completed by one of the generators of our model. Table 2 summarizes these results obtained by Cond^{2}GANs, MKL, co-classif, and mv$_b$ for both test scenarios. In all cases Cond^{2}GANs, provides significantly better results than other approaches. This provides empirical evidence of the effectiveness of the joint view generation and class prediction of Cond^{2}GANs. Furthermore, MKL, co-classif and Cond^{2}GANs are binary classification models and tackle the multiclass classification case with one vs all strategy making them to suffer from class imbalance problem. Results obtained using the F_1 measure are in line with those of Table 2 and they are not reported for the sake of space.

Table 2. Test classification accuracy averaged over 20 random training/test sets. For each of the pairs of languages, the best result is in bold, and a $^{\downarrow}$ indicates a result that is statistically significantly worse than the best, according to a Wilcoxon rank sum test with $p < .01$.

Approaches	(EN,v = FR)		(EN,v = GR)		(EN,v = IT)		(EN,v = SP)	
	$\mathbf{T}_{\text{EN}\bar{v}}$	$\mathbf{T}_{\overline{\text{EN}}v}$	$\mathbf{T}_{\text{EN}\bar{v}}$	$\mathbf{T}_{\overline{\text{EN}}v}$	$\mathbf{T}_{\text{EN}\bar{v}}$	$\mathbf{T}_{\overline{\text{EN}}v}$	$\mathbf{T}_{\text{EN}\bar{v}}$	$\mathbf{T}_{\overline{\text{EN}}v}$
MKL	75.6$^{\downarrow}$	77.3$^{\downarrow}$	79.4$^{\downarrow}$	79.6$^{\downarrow}$	78.4$^{\downarrow}$	79.8$^{\downarrow}$	81.2$^{\downarrow}$	83.5$^{\downarrow}$
co-classif	81.4$^{\downarrow}$	83.2$^{\downarrow}$	84.3$^{\downarrow}$	81.6$^{\downarrow}$	82.7$^{\downarrow}$	82.5$^{\downarrow}$	85.1$^{\downarrow}$	86.2$^{\downarrow}$
mv$_b$	83.1$^{\downarrow}$	84.5$^{\downarrow}$	85.2$^{\downarrow}$	79.9$^{\downarrow}$	84.3$^{\downarrow}$	82.1$^{\downarrow}$	84.4$^{\downarrow}$	86.2$^{\downarrow}$
Cond^{2}GANs	**85.3**	**85.1**	**86.6**	**82.9**	**85.3**	**84.5**	**86.5**	**88.3**

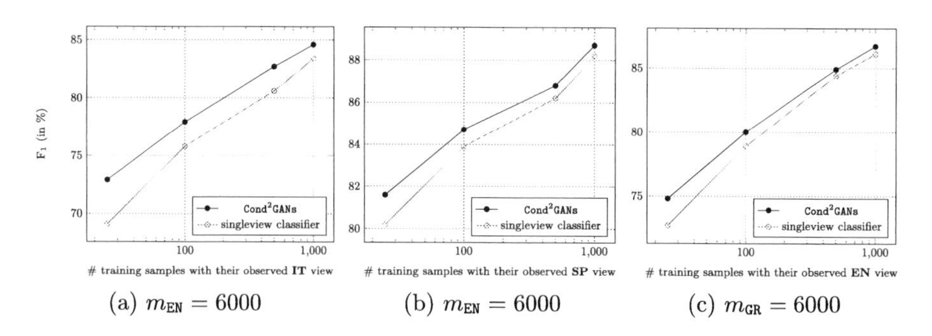

(a) $m_{\text{EN}} = 6000$ (b) $m_{\text{EN}} = 6000$ (c) $m_{\text{GR}} = 6000$

Fig. 3. F_1 measure of Cond^{2}GANs and a singleview classifier ($\mathbf{c}_v$) for an increasing number of training samples with the corresponding view that is observed. The number of training examples corresponding to the other view ($m_{\bar{v}} = 6000$); and the number of training examples with their both views observed is $m_F = 300$.

Impact of the Increasing Number of Observed Views. In Fig. 3, we compare F_1 measures between Cond^{2}GANs and one of the single-view classifiers with an increasing number of training samples, having the view corresponding to the singleview classifier observed; while the number of training examples with the other observed view is fixed. With an increasing number of training samples, the corresponding singleview classifier gains in performance. On the other hand, Cond^{2}GANs can leverage the lack of information from training examples having their other view observed, making that the difference of performance between these models for small number of training samples is higher.

5 Conclusion

In this paper we presented Cond^{2}GANs for multiview multiclass classification where observations may have missing views. The model consists of three neural-networks implementing a three players game between a discriminator and two generators. For an observation with a missing view, the corresponding generator produces the view conditionally on the other observed one. The discriminator is trained to recognize observations with a generated view from others having their views complete and to classify the latter into one of the existing classes. We evaluate the effectiveness of our approach on another challenging application than image analysis which is the domain of choice for the design of new GAN models. Our experiments on a subset of Reuters RCV1/RCV2 show the effectiveness of Cond^{2}GANs to generate high quality views allowing to achieve significantly better results, compared to the case where the missing views are generated by Machine Translation which requires a large collection of sentences and their translations to be tuned. As future study, we will be working on the generalization of the proposed model to more than 2 views. One possible direction is the use of an aggregation function of available views as a condition to the generators.

References

1. Amini, M.R., Goutte, C.: A co-classification approach to learning from multilingual corpora. Mach. Learn. J. **79**(1–2), 105–121 (2010)
2. Amini, M.R., Usunier, N., Goutte, C.: Learning from multiple partially observed views - an application to multilingual text categorization. In: Advances in Neural Information Processing Systems, vol. 22, 28–36 (2009)
3. Bach, F.R., Jordan, M.I.: Kernel independent component analysis. J. Mach. Learn. Res. **3**(Jul), 1–48 (2003)
4. Bach, F.R., Lanckriet, G.R.G., Jordan, M.I.: Multiple kernel learning, conic duality, and the SMO algorithm. In: Proceedings of the 21st International Conference on Machine Learning (ICML) (2004)
5. Blum, A., Mitchell, T.: Combining labeled and unlabeled data with co-training. In: Proceedings of the Eleventh Annual Conference on Computational Learning Theory (COLT), pp. 92–100 (1998)

6. Chen, M., Denoyer, L.: Multi-view generative adversarial networks. In: Ceci, M., Hollmén, J., Todorovski, L., Vens, C., Džeroski, S. (eds.) ECML PKDD 2017. LNCS (LNAI), vol. 10535, pp. 175–188. Springer, Cham (2017). https://doi.org/10.1007/978-3-319-71246-8_11
7. Denton, E.L., Chintala, S., Szlam, A., Fergus, R.: Deep generative image models using a laplacian pyramid of adversarial networks. In: Advances in Neural Information Processing Systems, vol. 28, 1486–1494 (2015)
8. Donahue, J., Krähenbühl, P., Darrell, T.: Adversarial feature learning. In: International Conference on Representation Learning (ICLR) (2017)
9. Dumoulin, V., et al.: Adversarially learned inference. In: International Conference on Representation Learning (ICLR) (2017)
10. Farquhar, J., Hardoon, D., Meng, H., Shawe-taylor, J.S., Szedmák, S.: Two view learning: SVM-2K, theory and practice. In: Advances in Neural Information Processing Systems vol. 18, pp. 355–362 (2006)
11. Goodfellow, I., et al.: Generative adversarial nets. In: Advances in Neural Information Processing Systems, vol. 27, pp. 2672–2680 (2014)
12. Goyal, A., Morvant, E., Germain, P., Amini, M.R.: PAC-Bayesian Analysis for a two-step Hierarchical Multiview Learning Approach. In: Joint European Conference on Machine Learning and Knowledge Discovery in Databases, pp. 205–221 (2017)
13. Kingma, D.P., Ba, J.: Adam: a method for stochastic optimization. In: International Conference on Representation Learning (ICLR) (2015)
14. Ma, L., Jia, X., Sun, Q., Schiele, B., Tuytelaars, T., Van Gool, L.: Pose guided person image generation. In: Advances in Neural Information Processing Systems, vol. 30, pp. 406–416 (2017)
15. Odena, A., Olah, C., Shlens, J.: Conditional image synthesis with auxiliary classifier GANs. In: Proceedings of the 34th International Conference on Machine Learning (ICML), pp. 2642–2651 (2017)
16. Radford, A., Metz, L., Chintala, S.: Unsupervised representation learning with deep convolutional generative adversarial networks. In: International Conference on Representation Learning (ICLR) (2016)
17. Salimans, T., et al.: Improved techniques for training GANs. In: Advances in Neural Information Processing Systems, vol. 29, pp. 2234–2242 (2016)
18. Sindhwani, V., Rosenberg, D.S.: An RKHS for multi-view learning and manifold co-regularization. In: Proceedings of the 25th International Conference on Machine Learning (ICML) (2008)
19. Springenberg, J.T.: Unsupervised and semi-supervised learning with categorical generative adversarial networks. In: International Conference on Representation Learning (ICLR) (2016)
20. Tian, L., Nie, F., Li, X.: A unified weight learning paradigm for multi-view learning. In: Proceedings of Machine Learning Research, pp. 2790–2800 (2019)
21. Tian, Y., Peng, X., Zhao, L., Zhang, S., Metaxas, D.N.: CR-GAN: learning complete representations for multi-view generation. In: Proceedings of the Twenty-Seventh International Joint Conference on Artificial Intelligence (IJCAI), pp. 942–948 (2018)
22. Ueffing, N., Simard, M., Larkin, S., Johnson, H.: NRC's portage system for WMT 2007. In: Proceedings of the Second Workshop on Statistical Machine Translation, pp. 185–188 (2007)
23. Zhao, B., Wu, X., Cheng, Z., Liu, H., Feng, J.: Multi-view image generation from a single-view. In: Proceedings of the 26th ACM International Conference on Multimedia (MM), pp. 383–391 (2018)

Semantic Path-Based Learning for Review Volume Prediction

Ujjwal Sharma[1(✉)], Stevan Rudinac[1], Marcel Worring[1], Joris Demmers[2], and Willemijn van Dolen[1]

[1] University of Amsterdam, Amsterdam, The Netherlands
{u.sharma,s.rudinac,m.worring,w.m.vandolen}@uva.nl
[2] Monash University, Melbourne, Australia
joris.demmers@monash.edu

Abstract. Graphs offer a natural abstraction for modeling complex real-world systems where entities are represented as nodes and edges encode relations between them. In such networks, entities may share common or similar attributes and may be connected by paths through multiple attribute modalities. In this work, we present an approach that uses semantically meaningful, bimodal random walks on real-world heterogeneous networks to extract correlations between nodes and bring together nodes with shared or similar attributes. An attention-based mechanism is used to combine multiple attribute-specific representations in a late fusion setup. We focus on a real-world network formed by restaurants and their shared attributes and evaluate performance on predicting the number of reviews a restaurant receives, a strong proxy for popularity. Our results demonstrate the rich expressiveness of such representations in predicting review volume and the ability of an attention-based model to selectively combine individual representations for maximum predictive power on the chosen downstream task.

Keywords: Heterogeneous information networks · Metapaths · Deep learning on graphs · Venue popularity prediction

1 Introduction

Multimodal graphs have been extensively used in modeling real-world networks where entities interact and communicate with each other through multiple information pathways or *modalities* [1,23,31]. Each modality encodes a distinct *view* of the relation between nodes. For example, within a social network, users can be connected by their shared preference for a similar product or by their presence in the same geographic locale. Each of these semantic contexts links the same user set with a distinct edge set. Such networks have been extensively used for applications like semantic proximity search in existing interaction networks [7], augmenting semantic relations between entities [36], learning interactions in an unsupervised fashion [3] and augmenting traditional matrix factorization-based collaborative filtering models for recommendation [27].

J. M. Jose et al. (Eds.): ECIR 2020, LNCS 12035, pp. 821–835, 2020.
https://doi.org/10.1007/978-3-030-45439-5_54

Each modality within a multimodal network encodes a different semantic relation and exhibits a distinct *view* of the network. While such views contain relations between nodes based on interactions within a single modality, observed outcomes in the real-world are often a complex combination of these interactions. Therefore, it is essential to compose these complementary interactions meaningfully to build a better representation of the real world. In this work, we examine a multimodal approach that attempts to model the review-generation process as the end-product of complex interactions within a restaurant network.

Restaurants share a host of attributes with each other, each of which may be treated as a modality. For example, they may share the same neighborhood, the same operating hours, similar kind of cuisine, or the same 'look and feel'. Furthermore, each of these attributes only uncovers a specific type of relation. For example, a view that only uses the location-modality will contain venues only connected by their colocation in a common geographical unit and will prioritize physical proximity over any other attribute. Broadly, each of these views is characterized by a semantic context and encodes modality-specific relations between restaurants. These views, although informative, are *complementary* and only record associations within the same modality. While each of these views encodes a part of the interactions within the network, performance on a downstream task relies on a suitable combination of views pertinent to the task [5].

In this work, we use metapaths as a semantic interface to specify which relations within a network may be relevant or meaningful and worth investigating. We generate bimodal low-dimensional embeddings for each of these metapaths. Furthermore, we conjecture that their relevance on a downstream task varies with the nature of the task and that this task-specific modality relevance should be learned from data. In this work,

- We propose a novel method that incorporates restaurants and their attributes into a multimodal graph and extracts multiple, bimodal low dimensional representations for restaurants based on available paths through shared visual, textual, geographical and categorical features.
- We use an attention-based fusion mechanism for selectively combining representations extracted from multiple modalities.
- We evaluate and contrast the performance of modality-specific representations and joint representations for predicting review volume.

2 Related Work

The principle challenge in working with multimodal data revolves around the task of extracting and assimilating information from multiple modalities to learn informative joint representations. In this section, we discuss prior work that leverages graph-based structures for extracting information from multiple modalities, focussing on the auto-captioning task that introduced such methods. We then examine prior work on network embeddings that aim to learn discriminative representations for nodes in a graph.

2.1 Graphs for Modelling Semantic Relationships

Graph-based learning techniques provide an elegant means for incorporating semantic similarities between multimedia documents. As such, they have been used for inference in large multimodal collections where a single modality may not carry sufficient information [2]. Initial work in this domain was structured around the task of captioning unseen images using correlations learned over multiple modalities (tag-propagation or auto-tagging). Pan *et al.* use a graph-based model to discover correlations between image features and text for automatic image-captioning [21]. Urban *et al.* use an *Image-Context Graph* consisting of captions, image features and images to retrieve relevant images for a textual query [32]. Stathopoulos *et al.* [28] build upon [32] to learn a similarity measure over words based on their co-occurrence on the web and use these similarities to introduce links between similar caption words. Rudinac *et al.* augment the *Image-Context Graph* with users as an additional modality and deploy it for generating visual-summaries of geographical regions [25]. Since we are interested in discovering multimodal similarities between restaurants, we use a graph layout similar to the one proposed by Pan *et al.* [21] for the image auto-captioning task but replace images with restaurants as central nodes. Other nodes containing textual features, visual features and users are retained. We also add categorical information like cuisines as a separate modality, allowing them to serve as semantic anchors within the representation.

2.2 Graph Representation Learning

Graph representation learning aims to learn mappings that embed graph nodes in a low-dimensional compressed representation. The objective is to learn embeddings where geometric relationships in the compressed embedding space reflect structural relationships in the graph. Traditional approaches generate these embeddings by finding the leading eigenvectors from the affinity matrix for representing nodes [16,24]. With the advent of deep learning, neural networks have become increasingly popular for learning such representations, jointly, from multiple modalities in an end-to-end pipeline [4,11,14,30,34].

Existing random walk-based embedding methods are extensions of the Random Walks with Restarts (RWR) paradigm. Traditional RWR-based techniques compute an affinity between two nodes in a graph by ascertaining the steady-state transition probability between them. They have been extensively used for the aforementioned auto-captioning tasks [21,25,28,32], tourism recommendation [15] and web search as an integral part of the PageRank algorithm [20]. Deep Learning-based approaches build upon the traditional paradigm by optimizing the co-occurrence statistics of nodes sampled from these walks. *DeepWalk* [22] uses nodes sampled from short truncated random walks as phrases to optimize a skip-gram objective similar to *word2vec* [17]. Similarly, *node2vec* augments this learning paradigm with second-order random walks parameterized by exploration parameters p and q which control between the importance of homophily and structural equivalence in the learnt representations [8]. For a homogeneous network, random walk based methods like *DeepWalk* and *node2vec* assume that while the probabilities of transitioning from one node to another can be different, every transition still occurs between nodes of the same type. For heterogeneous graphs, this assumption may be fallacious as all transitions do not occur between nodes of the same type and consequently, do not carry the same semantic context. Indeed, our initial experiments with *node2vec* model suggest that it is not designed to handle highly multimodal graphs. Clements *et al.* [5] demonstrated that in the context of content recommendation, the importance of modalities is strongly task-dependent and treating all edges in heterogeneous graphs as equivalent can discard this information. *Metapath2Vec* [6] remedies this by introducing unbiased walks over the network schema specified by a *metapath* [29], allowing the network to learn the semantics specified by the metapath rather than those imposed purely by the topology of the graph. Metapath-based approaches have been extended to a variety of other problems. Hu *et al.* use an exhaustive list of semantically-meaningful metapaths for extracting Top-N recommendations with a neural co-attention network [10]. Shi *et al.* use metapath-specific representations in a traditional matrix factorization-based collaborative filtering mechanism [27]. In this work, we perform random walks on sub-networks of a restaurant-attribute network containing restaurants and attribute modalities. These attribute modalities may contain images, text or categorical features. For each of these sub-networks, we perform random walks and use a variant of the heterogeneous skip-gram objective introduced in [6] to generate low-dimensional bimodal embeddings. Bimodal embeddings have several interesting properties. Training relations between two modalities provide us with a degree of modularity where modalities can be included or held-out from the prediction model without affecting others. It also makes training inexpensive as the number of nodes when only considering two modalities is far lower than in the entire graph.

3 Proposed Method

In this section, we begin by providing a formal introduction to graph terminology that is frequently referenced in this paper. We then move on to detail our proposed method illustrated in Fig. 1.

3.1 Definitions

Formally, a heterogeneous graph is denoted by $G = (V, E, \phi, \sigma)$ where V and E denote the node and edge sets respectively. For every node and edge, there exists mapping functions $\phi(v) \rightarrow \mathcal{A}$ and $\sigma(e) \rightarrow \mathcal{R}$ where $\mathcal{A}$ and $\mathcal{R}$ are sets of node types and edge types respectively such that $|\mathcal{A} + \mathcal{R}| > 2$.

For a heterogeneous graph $G = (V, E, \phi, \sigma)$, a network schema is a *metagraph* $M_G = (\mathcal{A}, \mathcal{R})$ where $\mathcal{A}$ is the set of node types in V and $\mathcal{R}$ is the set of edge types in E. A network schema enumerates the possible node types and edge types that can occur within a network.

A metapath $\mathcal{M}(\mathcal{A}_1, \mathcal{A}_n)$ is a path on the network schema M_G consisting of a sequence of ordered edge transitions: $\mathcal{M}(\mathcal{A}_1, \mathcal{A}_n) : [\mathcal{A}_1 \rightarrow \mathcal{A}_2 \rightarrow \ldots \rightarrow \mathcal{A}_n]$.

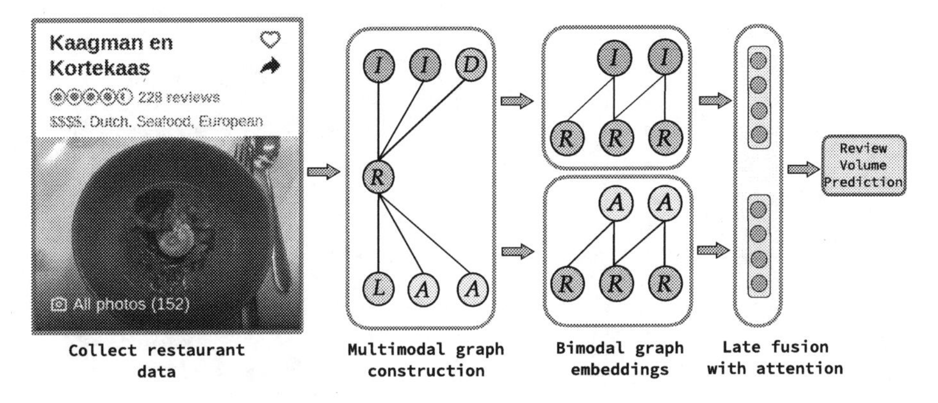

Fig. 1. Model pipeline: We use TripAdvisor to collect information for restaurants in Amsterdam. Each venue characteristic is then embedded as a separate node within a multimodal graph. In the figure above R nodes denote restaurants, I nodes denote images for a restaurant, D nodes are review documents, A nodes are categorical attributes for restaurants and L nodes are locations. Bimodal random walks are used to extract pairwise correlations between nodes in separate modalities which are embedded using a heterogeneous skip-gram objective. Finally, an attention-based fusion model is used to combine multiple embeddings together to regress the review volume for restaurants.

3.2 Graph Nodes

Let $G = (V, E)$ be the heterogeneous graph with a set of nodes V and edges E. We assume the graph to be undirected as linkages between venues and their attributes are inherently symmetric. Below, we describe the node types used to construct the graph (cf. Figs. 1 and 2).

1. **Restaurant Nodes:** For each of the N restaurants in a city, we introduce a node.
2. **Categorical Attribute Nodes:** A categorical feature node is added for each of the categories below.
   - **Cuisines:** Product/Cuisine type served in the restaurant.
   - **Meals:** Meals (Breakfast, Lunch, Dinner, etc.) which the restaurant serves.
   - **Features:** Additional services/attributes for the restaurants.
   - **Symbolic Price:** A discretized price bracket for venues.
3. **Location:** We split the region under consideration into rectangular geographical cells and bin restaurant locations into these cells. We add a node for every geographical cell that contains at least one restaurant within it.
4. **Image Features:** Images of a restaurant reflect its visual look-and-feel. These images may depict e.g. the interior, food and drinks, or people enjoying their meal. To extract a high-level semantic representation from these images, we deploy the ResNet-18 network [9] trained on the ImageNet Large Scale Visual Recognition Challenge (ILSVRC) dataset [26] with 1000 semantic concepts

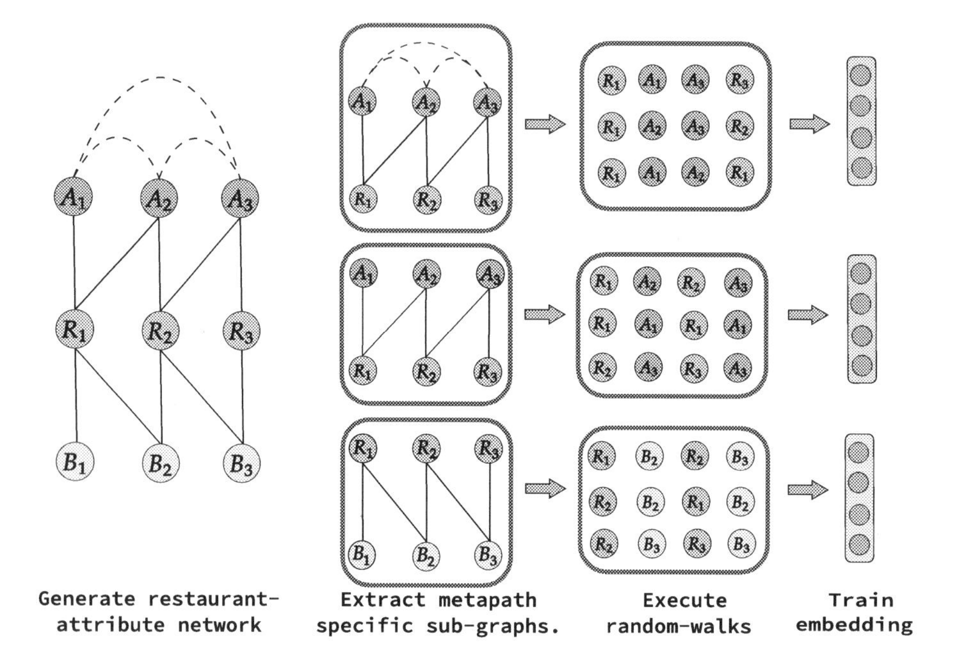

Fig. 2. Pipeline for embedding modality-specific sub-networks: In this illustrative example, we use a graph with 2 modalities. $\{R_1, \ldots, R_3\}$ are restaurant nodes, $\{A_1, \ldots, A_3\}$ are nodes from a modality with categorical information and $\{B_1, \ldots, B_3\}$ are nodes from a modality with similarity linkages. Random walks are performed over modality-specific sub-networks for each metapath and each model is trained separately to create metapath-specific embeddings.

and use the penultimate layer output as a compressed low-dimensional representation for the image. Since the number of available images for each venue may vary dramatically depending on its popularity, adding a node for every image can lead to an unreasonably large graph. To mitigate this issue, we cluster image features for each restaurant using the K-Means algorithm and use the cluster centers as representative image features for a restaurant, similar to Zahálka *et al.* [35]. We chose K = 5 as a reasonable trade-off between the granularity of our representations and tractability of generating embeddings for this modality.

5. **User Review Features:** The way patrons write about a restaurant and the usage of specialized terms can contain important information about a restaurant that may be missing from its categorical attributes. For example, usage of the Indian cottage cheese 'Paneer' can be found in similar cuisine types like Nepali, Surinamese, etc. and user reviews talking about dishes containing 'Paneer' can be leveraged to infer that Indian and Nepali cuisines share some degree of similarity. To model such effects, we collect reviews for every restaurant. Since individual reviews may not provide a comprehensive unbiased picture of the restaurant, we chose not to treat them individually,

but to consider them as a single document. We then use a distributed bag-of-words model from [13] to generate low-dimensional representations of these documents for each restaurant. Since the reviews of a restaurant can widely vary based on its popularity, we only consider the 10 most recent reviews for each restaurant to prevent biases from document length getting into the model.

6. **Users:** Since TripAdvisor does not record check-ins, we can only leverage explicit feedback from users who chose to leave a review. We add a node for each of the users who visited at least two restaurants in Amsterdam and left a review.

3.3 Graph Edges

Similar to [25,28,32], we introduce two kinds of edges in our graph:

1. **Attribute edges:** These are *heterogeneous* edges that connect a restaurant node to the nodes of its categorical attributes, image features, review features and users. In our graph, we instantiate them as undirected, unweighted edges.
2. **Similarity edges:** These are *homogeneous* edges between the feature nodes within a single modality. For image features, we use a radial basis function as a non-linear transformation of the euclidean distances between image feature vectors.

$$S_I(I_j, I_k) = \exp\left(-\frac{||I_j - I_k||^2}{2\sigma^2}\right) \tag{1}$$

   For document vectors, we use cosine similarity to find restaurants with similar reviews.

$$S_T(T_j, T_k) = \frac{T_j \cdot T_k}{||T_j||||T_k||} \tag{2}$$

   Adding a weighted similarity edge between every node in the same modality would yield an extremely dense adjacency matrix. To avoid this, we only add similarity links between a node and its K nearest neighbors in each modality. By choosing the nearest K neighbors, we make our similarity threshold adaptive allowing it to adjust to varying scales of distance in multiple modalities.

3.4 Bimodal Graph Embeddings

Metapaths can provide a modular and simple interface for injecting semantics into the network. Since metapaths, in our case, are essentially paths over the modality set, they can be used to encode inter-modality correlations. In this work, we generate embeddings with two specific properties:

1. All metapaths are binary and only include transitions over 2 modalities. Since venues/restaurants are always a part of the metapath, we only include one other modality.

2. During optimization, we only track the short-range context by choosing a small window size. Window size is the maximum distance between the input node and a predicted node in a walk. In our model, walks over the metapath only capture short-range semantic contexts and the choice of a larger window can be detrimental to generalization. For example, consider a random walk over the **Restaurant - Cuisine - Restaurant** metapath. In the sampled nodes below, restaurants are in red while cuisines are in blue.

   `[Kediri, Indonesian Cuisine, Indonesian Palace, Chinese Cuisine, China Town, Fast Food Cuisine, McDonalds]`

   Optimizing over a large context window can lead to McDonald's (fast-food cuisine) and Kediri (Indonesian cuisine) being placed close in the embedding space. This is erroneous and does not capture the intended semantics which should bring restaurants closer only if they share the exact attribute.

We use the metapaths in Table 1 to perform unbiased random walks on the graph detailed in Sect. 3.2. Each of these metapaths enforces similarity based on certain semantics.

Table 1. Metapaths used for the Restaurant-Attributes multimodal network.

No.	Metapath
1	Venues - Cuisines - Venues
2	Venues - Facilities - Venues
3	Venues - Meals - Venues
4	Venues - Price - Venues
5	Venues - Location cell - Venues
6	Venues - Users - Venues
7	Venues - Location cell - Location cell - Venues
8	Venues - Img. Feats. - Img. Feats - Venues
9	Venues - Review Feats. - Review Feats - Venues

We train separate embeddings using the heterogeneous skip-gram objective similar to [6]. For every metapath, we maximize the probability of observing the heterogeneous context $N_a(v)$ given the node v. In Eq. (3), $\mathcal{A}_m$ is the node type-set and V_m is the node-set for metapath m.

$$\arg\max_{\theta} \sum_{v \in V_m} \sum_{a \in \mathcal{A}_m} \sum_{c_a \in N_a(v)} \log p(c_a|v;\theta) \tag{3}$$

3.5 Modality Fusion

The original Metapath2vec model [6] uses multiple metapaths [29] to learn separate embeddings, some of which perform better than the others. On the DBLP

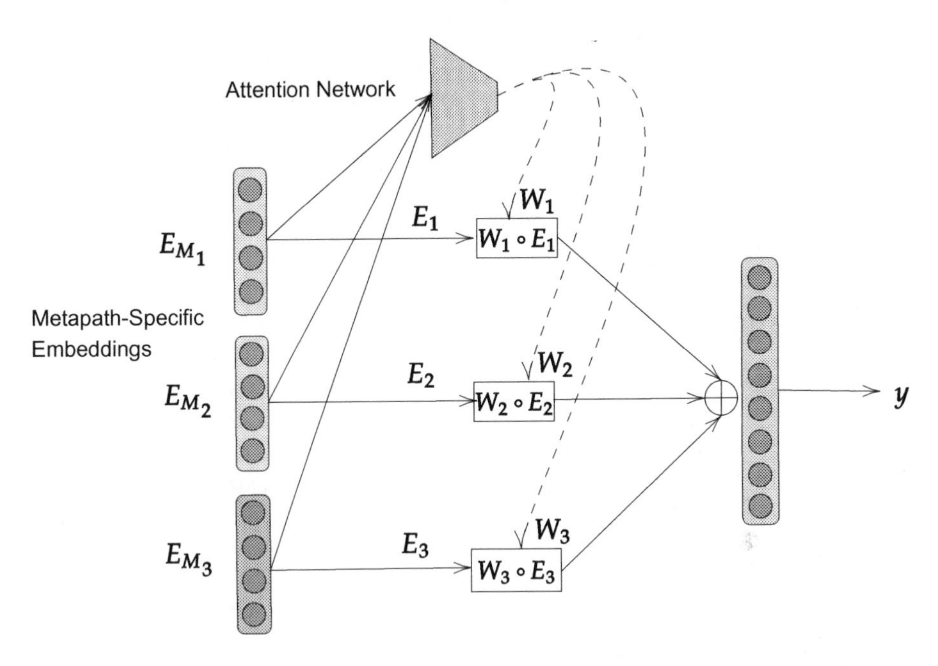

Fig. 3. Attention-weighted modality fusion: metapath-specific embeddings are fed into a common attention mechanism that generates an attention vector. Each modality is then reweighted with the attention vector and concatenated. This joint representation is then fed into a ridge regressor to predict the volume of ratings for each restaurant.

bibliographic graph that consists of Authors (A), Papers (P) and Venues (V), the performance of their recommended metapath 'A-P-V-P-A' was empirically better than the alternative metapath 'A-P-A' on the node classification task. At this point, it is important to recall that in our model, each metapath extracts a separate *view* of the same graph. These views may contain *complementary* information and it may be disadvantageous to only retain the best performing view. For an optimal representation, these complementary views should be fused.

In this work, we employ an embedding-level attention mechanism similar to the attention mechanism introduced in [33] that selectively combines embeddings based on their performance on a downstream task. Assuming S to be the set of metapath-specific embeddings for metapaths $m_1, m_2, \ldots, m_N$, following the approach outlined in Fig. 3, we can denote it as:

$$S = \{E_{m_1}, E_{m_2}, \cdots, E_{m_N}\} \tag{4}$$

We then use a two-layer neural network to learn an embedding-specific attention A_{m_n} for metapath m_n:

$$A_{m_n} = h \cdot ReLU(W \cdot E_{m_n} + b) \tag{5}$$

Further, we perform a softmax transformation of the attention network outputs to an embedding-specific weight

$$w_{m_n} = \frac{\exp(A_{m_n})}{\sum_{k=1}^{N} \exp(A_{m_k})} \tag{6}$$

Finally, we concatenate the attention-weighted metapath-specific embeddings to generate a fused embedding

$$O = [w_{m_1}.E_{m_1} | w_{m_2}.E_{m_2} | \ldots | w_{m_N}.E_{m_N}] \tag{7}$$

4 Experiments

We evaluate the performance of the embedding fusion model on the task of predicting the volume (total count) of reviews received by a restaurant. We conjecture that the volume of reviews is an unbiased proxy for the general popularity and footfall for a restaurant and is more reliable than indicators like ranking or ratings which may be biased by TripAdvisor's promotion algorithms. We use the review volume collected from TripAdvisor as the target variable and model this task as a regression problem.

4.1 Experiment Setup

Data Collection. We use publicly-available data from TripAdvisor for our experiments. To build the graph detailed in Sect. 3.2, we collect data for 3,538 restaurants in Amsterdam, The Netherlands that are listed on TripAdvisor. We additionally collect 168,483 user-contributed restaurant reviews made by 105,480 *unique* users, of which only 27,318 users visit more than 2 restaurants in the city. We only retain these 27,318 users in our graph and drop others. We also collect 215,544 user-contributed images for these restaurants. We construct the restaurant network by embedding venues and their attributes listed in Table 1 as nodes.

Bimodal Embeddings. We train separate bimodal embeddings by optimizing the heterogeneous skip-gram objective from Eq. (3) using stochastic gradient descent and train embeddings for all metapaths enumerated in Table 1. We use restaurant nodes as root nodes for the unbiased random walks and perform 80 walks per root node, each with a walk length of 80. Each embedding has a dimensionality of 48, uses a window-size of 5 and is trained for 200 epochs.

Embedding Fusion Models. We chose two fusion models in our experiments to analyze the efficacy of our embeddings:

1. Simple Concatenation Model: We use a model that performs a simple concatenation of the individual metapath-specific embeddings detailed in Sect. 3.4 to exhibit the baseline performance on the tasks detailed in Sect. 4. Simple concatenation is a well-established additive fusion technique in multimodal deep learning [18,19].

Table 2. Performance of individual metapath-specific embeddings on the review volume prediction task

Metapath	Mean squared error	Mean absolute error	Coefficient of determination
Venues - Cuisines - Venues	2.56	1.24	0.22
Venues - Facilities - Venues	1.93	1.09	0.40
Venues - Meals - Venues	2.41	1.23	0.28
Venues - Price - Venues	2.00	1.09	0.36
Venues - Location cell - Venues	3.13	1.45	0.09
Venues - Users - Venues	3.38	1.54	0.01
Venues - Location cell - Location cell - Venues	2.75	1.36	0.08
Venues - Img. Feats. - Img. Feats - Venues	2.73	1.33	0.12
Venues - Review Feats. - Review Feats - Venues	2.68	1.32	0.224
Modality concatenation + Ridge regression	1.29	0.89	0.62
Attention weighted modality concatenation + Ridge regression	0.99	0.77	0.70

2. Modality Fusion Model: We use our attention-based modality fusion model detailed in Sect. 3.5 to exhibit the effects of a learnable weighted-fusion of modalities on the evaluation tasks.

Each of the models uses a ridge regression algorithm to estimate the predictive power of each metapath-specific embedding on the volume regression task. This regressor is jointly trained with the attention model in the Attention-Weighted Model. All models are optimized using stochastic gradient descent with the Adam optimizer [12] with a learning rate of 0.1.

4.2 Results and Findings

In Table 2, we report the results from our experiments on the review-volume prediction task. We observe that metapaths with nodes containing categorical attributes perform significantly better than vector-based features. In particular, categorical attributes like Cuisines, Facilities, and Price have a significantly higher coefficient of determination (R^2) as compared to visual feature nodes. It is interesting to observe here that nodes like locations, images, and textual reviews are far more numerous than categorical nodes and part of their decreased performance may be explained by the fact that our method of short walks may not be sufficiently expressive when the number of feature nodes is large. In addition, as mentioned in related work, we performed these experiments with the *node2vec* model, but since it is not designed for heterogeneous multimodal graphs, it yielded performance scores far below the weakest single modality.

A review of the fusion models indicates that taking all the metapaths together can improve performance significantly. The baseline simple concatenation fusion model, commonly used in literature, is considerably better than the best-performing metapath (Venues - Facilities - Venues). The attention based-model builds significantly over the baseline performance and while it employs a similar concatenation scheme as the baseline concatenation model, the introduction of the attention module allows it to handle noisy and unreliable modalities. The significant increase in the predictive ability of the attention-based model can be attributed to the fact that while all modalities encode information, some of them may be less informative or reliable than others, and therefore contribute less to the performance of the model. Our proposed fusion approach is, therefore, capable of handling weak or noisy modalities appropriately.

5 Conclusion

In this work, we propose an alternative, modular framework for learning from multimodal graphs. We use metapaths as a means to specify semantic relations between nodes and each of our bimodal embeddings captures similarities between restaurant nodes on a single attribute. Our attention-based model combines separately learned bimodal embeddings using a late-fusion setup for predicting the review volume of the restaurants. While each of the modalities can predict the volume of reviews to a certain extent, a more comprehensive picture is only built by combining complementary information from multiple modalities. We demonstrate the benefits of our fusion approach on the review volume prediction task and demonstrate that a fusion of complementary views provides the best way to learn from such networks. In future work, we will investigate how the technique generalises to other tasks and domains.

References

1. Abrach, H., et al.: MANTIS: system support for MultimodAl NeTworks of in-situ sensors. In: Proceedings of the Second ACM International Workshop on Wireless Sensor Networks and Applications, WSNA 2003 (2003)
2. Arya, D., Rudinac, S., Worring, M.: HyperLearn: a distributed approach for representation learning in datasets with many modalities. In: MM 2019 - Proceedings of the 27th ACM International Conference on Multimedia (2019). https://doi.org/10.1145/3343031.3350572
3. Battaglia, P., Pascanu, R., Lai, M., Rezende, D.J., Kavukcuoglu, K.: Interaction networks for learning about objects, relations and physics. In: Proceedings of the 30th International Conference on Neural Information Processing Systems, NIPS 2016, pp. 4509–4517. Curran Associates Inc., USA (2016). http://dl.acm.org/citation.cfm?id=3157382.3157601

4. Chang, S., Han, W., Tang, J., Qi, G.J., Aggarwal, C.C., Huang, T.S.: Heterogeneous network embedding via deep architectures. In: Proceedings of the ACM SIGKDD International Conference on Knowledge Discovery and Data Mining (2015). https://doi.org/10.1145/2783258.2783296
5. Clements, M., De Vries, A.P., Reinders, M.J.T.: The task-dependent effect of tags and ratings on social media access. ACM Trans. Inf. Syst. (2010). https://doi.org/10.1145/1852102.1852107
6. Dong, Y., Chawla, N.V., Swami, A.: Metapath2vec: scalable representation learning for heterogeneous networks. In: Proceedings of the ACM SIGKDD International Conference on Knowledge Discovery and Data Mining (2017). https://doi.org/10.1145/3097983.3098036
7. Fang, Y., Zhao, X., Huang, P., Xiao, W., de Rijke, M.: M-HIN: complex embeddings for heterogeneous information networks via metagraphs. In: Proceedings of the 42Nd International ACM SIGIR Conference on Research and Development in Information Retrieval, SIGIR 2019, pp. 913–916. ACM, New York, NY, USA (2019). https://doi.org/10.1145/3331184.3331281, http://doi.acm.org/10.1145/3331184.3331281
8. Grover, A., Leskovec, J.: Node2vec: scalable feature learning for networks. In: Proceedings of the ACM SIGKDD International Conference on Knowledge Discovery and Data Mining (2016). https://doi.org/10.1145/2939672.2939754
9. He, K., Zhang, X., Ren, S., Sun, J.: Deep residual learning for image recognition. In: Proceedings of the IEEE Computer Society Conference on Computer Vision and Pattern Recognition (2016). https://doi.org/10.1109/CVPR.2016.90
10. Hu, B., Shi, C., Zhao, W.X., Yu, P.S.: Leveraging meta-path based context for top-n recommendation with a neural co-attention model. In: Proceedings of the ACM SIGKDD International Conference on Knowledge Discovery and Data Mining (2018). https://doi.org/10.1145/3219819.3219965
11. Huang, F., Zhang, X., Li, C., Li, Z., He, Y., Zhao, Z.: Multimodal network embedding via attention based multi-view variational autoencoder. In: ICMR 2018 - Proceedings of the 2018 ACM International Conference on Multimedia Retrieval (2018). https://doi.org/10.1145/3206025.3206035
12. Kingma, D.P., Ba, J.L.: Adam: a method for stochastic gradient descent. In: ICLR: International Conference on Learning Representations (2015)
13. Le, Q., Mikolov, T.: Distributed representations of sentences and documents. In: 31st International Conference on Machine Learning, ICML 2014 (2014)
14. Li, Z., Tang, J., Mei, T.: Deep collaborative embedding for social image understanding. IEEE Trans. Pattern Anal. Mach. Intell. (2018). https://doi.org/10.1109/TPAMI.2018.2852750
15. Lucchese, C., Perego, R., Silvestri, F., Vahabi, H., Venturini, R.: How random walks can help tourism. In: Baeza-Yates, R., et al. (eds.) ECIR 2012. LNCS, vol. 7224, pp. 195–206. Springer, Heidelberg (2012). https://doi.org/10.1007/978-3-642-28997-2_17
16. McAuley, J., Leskovec, J.: Image labeling on a network: using social-network metadata for image classification. In: Fitzgibbon, A., Lazebnik, S., Perona, P., Sato, Y., Schmid, C. (eds.) ECCV 2012. LNCS, vol. 7575, pp. 828–841. Springer, Heidelberg (2012). https://doi.org/10.1007/978-3-642-33765-9_59
17. Mikolov, T., Sutskever, I., Chen, K., Corrado, G., Dean, J.: Distributed representations of words and phrases and their compositionality. In: Advances in Neural Information Processing Systems (2013)

18. Ngiam, J., Khosla, A., Kim, M., Nam, J., Lee, H., Ng, A.Y.: Multimodal deep learning. In: Proceedings of the 28th International Conference on Machine Learning, ICML 2011 (2011)
19. Ouyang, W., Chu, X., Wang, X.: Multi-source deep learning for human pose estimation. In: Proceedings of the IEEE Computer Society Conference on Computer Vision and Pattern Recognition (2014). https://doi.org/10.1109/CVPR.2014.299
20. Page, L., Brin, S., Motwani, R., Winograd, T.: The PageRank citation ranking: bringing order to the web. Technical report, 1999-66. Stanford InfoLab, November 1999. Previous number: SIDL-WP-1999-0120. http://ilpubs.stanford.edu:8090/422/
21. Pan, J.Y., Yang, H.J., Faloutsos, C., Duygulu, P.: GCap: graph-based automatic image captioning. In: IEEE Computer Society Conference on Computer Vision and Pattern Recognition Workshops (2004). https://doi.org/10.1109/CVPR.2004.353
22. Perozzi, B., Al-Rfou, R., Skiena, S.: DeepWalk: online learning of social representations. In: Proceedings of the ACM SIGKDD International Conference on Knowledge Discovery and Data Mining (2014). https://doi.org/10.1145/2623330.2623732
23. Pirson, I., Fortemaison, N., Jacobs, C., Dremier, S., Dumont, J.E., Maenhaut, C.: The visual display of regulatory information and networks. Trends Cell Biol. (2000). https://doi.org/10.1016/S0962-8924(00)01817-1
24. Roweis, S.T., Saul, L.K.: Nonlinear dimensionality reduction by locally linear embedding. Science (2000). https://doi.org/10.1126/science.290.5500.2323
25. Rudinac, S., Hanjalic, A., Larson, M.: Generating visual summaries of geographic areas using community-contributed images. IEEE Trans. Multimed. (2013). https://doi.org/10.1109/TMM.2013.2237896
26. Russakovsky, O., et al.: ImageNet large scale visual recognition challenge. Int. J. Comput. Vis. **115**(3), 211–252 (2015). https://doi.org/10.1007/s11263-015-0816-y
27. Shi, C., Hu, B., Zhao, W.X., Yu, P.S.: Heterogeneous information network embedding for recommendation. IEEE Trans. Knowl. Data Eng. (2019). https://doi.org/10.1109/TKDE.2018.2833443
28. Stathopoulos, V., Urban, J., Jose, J.: Semantic relationships in multi-modal graphs for automatic image annotation. In: Macdonald, C., Ounis, I., Plachouras, V., Ruthven, I., White, R.W. (eds.) ECIR 2008. LNCS, vol. 4956, pp. 490–497. Springer, Heidelberg (2008). https://doi.org/10.1007/978-3-540-78646-7_47
29. Sun, Y., Han, J., Yan, X., Yu, P.S., Wu, T.: PathSim: meta path-based top-k similarity search in heterogeneous information networks. In: Proceedings of the VLDB Endowment (2011)
30. Tang, J., Qu, M., Wang, M., Zhang, M., Yan, J., Mei, Q.: LINE: large-scale information network embedding. In: WWW 2015 - Proceedings of the 24th International Conference on World Wide Web (2015). https://doi.org/10.1145/2736277.2741093
31. Uchida, K., Sumalee, A., Watling, D., Connors, R.: Study on optimal frequency design problem for multimodal network using probit-based user equilibrium assignment. Transp. Res. Rec. (2005). https://doi.org/10.3141/1923-25
32. Urban, J., Jose, J.M.: Adaptive image retrieval using a graph model for semantic feature integration. In: Proceedings of the ACM International Multimedia Conference and Exhibition (2006). https://doi.org/10.1145/1178677.1178696
33. Wang, X., et al.: Heterogeneous graph attention network. In: The Web Conference 2019 - Proceedings of the World Wide Web Conference, WWW 2019 (2019). https://doi.org/10.1145/3308558.3313562
34. Yang, C., Liu, Z., Zhao, D., Sun, M., Chang, E.Y.: Network representation learning with rich text information. In: IJCAI International Joint Conference on Artificial Intelligence (2015)

35. Zahalka, J., Rudinac, S., Worring, M.: Interactive multimodal learning for venue recommendation. IEEE Trans. Multimed. (2015). https://doi.org/10.1109/TMM.2015.2480007
36. Zhang, D., Yin, J., Zhu, X., Zhang, C.: MetaGraph2Vec: complex semantic path augmented heterogeneous network embedding. In: Phung, D., Tseng, V.S., Webb, G.I., Ho, B., Ganji, M., Rashidi, L. (eds.) PAKDD 2018. LNCS (LNAI), vol. 10938, pp. 196–208. Springer, Cham (2018). https://doi.org/10.1007/978-3-319-93037-4_16

An Attention Model of Customer Expectation to Improve Review Helpfulness Prediction

Xianshan Qu(✉), Xiaopeng Li, Csilla Farkas, and John Rose

CSE Department, University of South Carolina, Columbia, SC 29201, USA
xqu@email.sc.edu, xl4@email.sc.edu {farkas,rose}@cse.sc.edu

Abstract. Many people browse reviews online before making purchasing decisions. It is essential to identify the subset of helpful reviews from the large number of reviews of varying quality. This paper aims to build a model to predict review helpfulness automatically. Our work is inspired by the observation that a customer's expectation of a review can be greatly affected by review sentiment and the degree to which the customer is aware of pertinent product information. Consequently, a customer may pay more attention to that specific content of a review which contributes more to its helpfulness from their perspective. To model such customer expectations and capture important information from a review text, we propose a novel neural network which leverages review sentiment and product information. Specifically, we encode the sentiment of a review through an attention module, to get sentiment-driven information from review text. We also introduce a product attention layer that fuses information from both the target product and related products, in order to capture the product related information from review text. Our experimental results show an AUC improvement of 5.4% and 1.5% over the previous state of the art model on Amazon and Yelp data sets, respectively.

Keywords: Review helpfulness · Customer expectation · Attention mechanism

1 Introduction

E-commerce has become an important part of our daily life. Increasingly, people choose to purchase products online. According to a recent study [9], most online shoppers browse reviews before making decisions. It is essential for users to be able to find reliable reviews of high quality. Therefore, an automatic helpfulness evaluation mechanism is in high demand to help user evaluate these reviews.

Previous work typically derived useful information from different sources, such as a review content [6,15,28], metadata [4,15,17], and context [14,20,25]. However, such features are extracted from each source independently, without considering interactions. In particular, previous approaches do not take into

J. M. Jose et al. (Eds.): ECIR 2020, LNCS 12035, pp. 836–851, 2020.
https://doi.org/10.1007/978-3-030-45439-5_55

account the customer review evaluating process. A customer's perception of helpful information of a review is affected by the sentiment of the review and what the customer already knows about the product. Before reading a review text for a product, the customer is very likely to be aware of background information such as star rating, product attributes, etc. When a customer reads a review with a lower star rating, he may hold a negative opinion towards the item at first and mainly look into those aspects of the review supporting the lower star rating. Although review sentiment has been previously explored [7,15,17], previous work has not used review sentiment to identify useful information from review text. Moreover, the customer likely has some preconceptions of the product features they are most interested in. With these expectations in mind, the customer pays special attention to those aspects of the review text that they find most salient. There have been earlier efforts [4,6,13] at capturing useful information from a review by considering product information. However, the unique aspects of each product (different levels of importance of attributes, evaluation standard, etc.) were not fully identified in those efforts.

In order to address the above issues, we propose a novel neural network architecture to introduce sentiment and product information when identifying helpful content from a review text. The main contributions are summarized as follows:

- To our knowledge, we are the first to propose that customers may have different expectations for reviews that express different sentiments. We design a sentiment attention layer to model sentiment-driven changes in user focus on a review.
- We propose a novel product attention layer. The purpose of this layer is to automatically identify the important product-related attributes from reviews. This layer fuses information not only from related products, but also from the specific product.
- We evaluate the performance of our model on two real-world data sets: the Amazon data set and the Yelp data set. We consider two application scenarios: cold start and warm start. In the cold start scenario, our proposed model demonstrates an AUC improvement of 5.4% and 1.5% on Amazon and Yelp data sets, respectively, when compared to the state of the art model. We also validate the effectiveness of each of the attention layers of our proposed model in cold start and warm start scenarios.

2 Related Work

Previous studies have concentrated on mining useful features from the content (i.e., the review itself) and/or the context (other sources such as reviewer or user information) of the reviews [6,10,13,15,18–20,25,27,28].

Content features have been extracted and widely utilized. They can be roughly broken down into the following categories: *structural features* [6,10,13,15,27,28], *lexical features* [10,15,27,28], *syntactic features* [10,15,27], *emotional features* [15,28], *semantic features* [6,10,13,15,18,27,28], and *argument* *features* [12]. For instance, Kim et al. [10] investigated a variety of content features from Amazon product reviews, and found that features such as review length, unigrams and product ratings are most useful in measuring review helpfulness.

Context features have also been studied to improve helpfulness prediction [14, 20, 25]. For example, Lu et al. [14] examined social context that may reveal the quality of reviewers to enhance the prediction of the quality of reviews. While context information shows promise for improving helpfulness prediction, it may not be available across different platforms and is not appropriate for designing a universal model.

Deep Neural Networks have recently been proposed for helpfulness prediction of online reviews [1–4, 23]. Chen et al. [1] designed a word-level gating mechanism to represent the relative importance of each word. Fan et al. [3] proposed a multi-task paradigm to predict the star ratings of reviews and to identify the helpful reviews more accurately. They also utilized the metadata of the target product in addition to the textual content of a review to better represent a review [4].

The methods summarized above are representative of the research progress in review helpfulness prediction. Sentiment and product information have been explored previously [4, 7, 10, 15]. With respect to sentiment, Martin and Pu [15] extracted emotional words from review text to serve as important parameters for helpfulness prediction. However, previous research has not taken into account differences in customer expectations that can result from review sentiment perception. With respect to product information, Fan et al. [4] tried to better

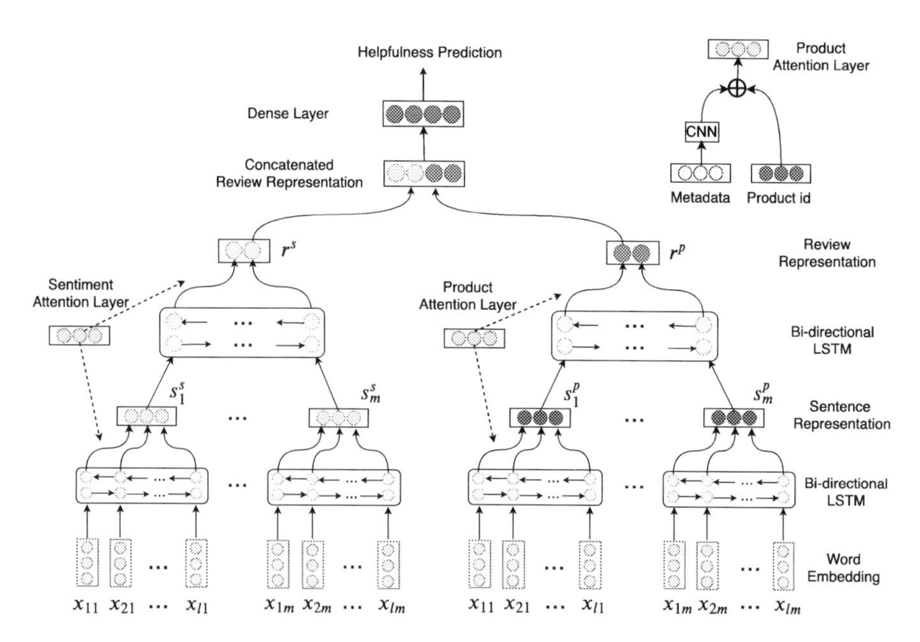

Fig. 1. The architecture of HSAPA.

represent the salient information in reviews by considering the metadata information (title, categories) of the target product. However, this information can be quite similar for products of the same type, so the unique aspects of each product (different degrees of importance of attributes, evaluation standard, etc.) can not be fully captured from reviews. Wu et al. [26] presented an architecture similar to the one we propose here. However, the design of the sentiment attention layer and product attention layer in our architecture is different from their attention layers. Moreover, their architecture is intended for classifying review sentiment.

3 Our Proposed Model

Our model, shown in Fig. 1, is built upon a hierarchical bi-directional LSTM. We incorporate sentiment and product information to improve review representations through two attention layers. As the main components of our model are the Hierarchical bi-directional LSTM, the Sentiment Attention layer, and the Product Attention layer, we refer to our model as HSAPA.

3.1 Hierarchical Bi-directional LSTM

A bi-directional LSTM model is able to learn past and future dependencies. This provides a better understanding of context [16]. The hierarchical architecture include two levels: the word level and the sentence level. These levels learn dependencies between words and sentences, respectively.

Word Encoder. A bi-directional LSTM consists of two LSTM networks that process data in opposite directions. At the word level, we feed the embedding of each word into a unit of both LSTMs, and get two hidden states. We then concatenate these two hidden states as a representation of a word. The process is defined as: $\overleftarrow{h}_{ij} = \overleftarrow{LSTM}(x_{ij})$, $\overrightarrow{h}_{ij} = \overrightarrow{LSTM}(x_{ij})$, $h_{ij} = [\overleftarrow{h}_{ij}, \overrightarrow{h}_{ij}]$, where x_{ij} is the embedding vector of the ith word of the jth sentence. $\overrightarrow{h}_{ij}$ and $\overleftarrow{h}_{ij}$ are hidden states learned from bi-directional LSTM. The state h_{ij} is the concatenation of these hidden states for the word x_{ij}.

Sentence Encoder. At the sentence level, a sentence representation is learned through an architecture similar to that used for the word level: $\overleftarrow{h}_j = \overleftarrow{LSTM}(s_j)$, $\overrightarrow{h}_j = \overrightarrow{LSTM}(s_j)$, $h_j = [\overleftarrow{h}_j, \overrightarrow{h}_j]$, where s_j refers to a weighted representation of the jth sentence after applying the attention layer. The state h_j is the final representation for the sentence s_j by concatenating the hidden states $\overrightarrow{h}_j$ and $\overleftarrow{h}_j$.

3.2 Sentiment Attention Layer

For reviews that express different types of sentiment (positive, negative, etc.), customers may have different expectations, and attend to different words or sentences of a review. Consider the following example:

I loved the simplicity of the mouse, ...and it was very comfortable ...*About 4 months of owning the mouse the scroll wheel seemed to be in always clicked in position, and would only stop after clicking it down hard for a couple seconds ...*

The above review has a star rating of 2 out of 5. For a review with an overall negative sentiment like this, we may pay more attention to its descriptions of bad aspects of the product than we do to the good aspects. Therefore, each word/sentence may contribute unequally to the helpfulness of a review, with regard to its sentiment.

In order to learn the sentiment-influenced importance of each word/sentence, we propose a custom attention layer. In this layer, we use an embedded vector to represent each type of sentiment. We use the star rating (ranging from 1 to 5) of each review to indicate its sentiment, and map each discrete star rating into a real-valued and continuous vector $Sent$. This vector is initialized randomly, and updated gradually through the training process by reviews with the corresponding star rating. $Sent$ can be interpreted as a high level representation of the sentiment-specific information. We measure the similarity between the sentiment and each word/sentence using a score function. The score function is defined as:

$$f(Sent, h_{ij}^s) = (v_w^s)^T \tanh(W_{wh}^s h_{ij}^s + W_{ws}^s Sent + b_w^s), \tag{1}$$

where v_w^s is a weight vector, and $(v_w^s)^T$ indicates its transpose, W_{wh}^s and W_{ws}^s are weight matrices, and b_w^s is the bias vector. At the word level, the input to the score function is the abstract sentiment representation $Sent$ and the hidden state of the ith word in the jth sentence h_{ij}^s. Next, we use the softmax function to normalize the scores to get the attention weights:

$$\alpha_{ij}^s = \frac{\exp(f(Sent, h_{ij}^s))}{\sum_{k=1}^{l} \exp(f(Sent, h_{kj}^s))}, \tag{2}$$

α_{ij}^s is the attention weight for the word representation h_{ij}^s.

The sentence representation is a weighted aggregation of word representations, the jth sentence is represented as Eq. 3. The number of words in the jth sentence is denoted by l. The representation of a review is also a weighted combination of sentence representations defined as Eq. 4, where h_j^s is the hidden state of the jth sentence s_j^s, which is learned through the bi-directional LSTM. The value m refers the number of sentences in a review.

$$s_j^s = \sum_{i=1}^{l} \alpha_{ij}^s h_{ij}^s \ . \tag{3}$$

$$r^s = \sum_{j=1}^{m} \beta_j^s h_j^s \ . \tag{4}$$

The value β_j^s indicates the corresponding attention score for h_j^s. The weight score β_j^s is calculated based on the score function $f(.)$ defined as:

$$f(Sent, h_j^s) = (v_s^s)^T \tanh(W_{sh}^s h_j^s + W_{ss}^s Sent + b_s^s) \ , \tag{5}$$

$$\beta_j^s = \frac{\exp(f(Sent, h_j^s))}{\sum_{k=1}^{m} \exp(f(Sent, h_k^s))} . \tag{6}$$

3.3 Product Attention Layer

As shown in the top right corner of Fig. 1, the Product Attention Layer consists of two components: related product information and unique product information. Metadata information is embedded and fed into a CNN model [11] to capture the related product information, and the product identifier is encoded to represent the unique product information.

3.3.1 Related Product Information

When reading a review, customers may refer to different attributes depending on the product the review references. For example, for a review of a mouse, we may expect to see the comments related to attributes such as scroll wheel, hand feel, etc. Such attributes are considered helpful and garner more attention.

We take advantage of the metadata information (such as title, product description, product category, etc.) of each product to learn common attributes shared by related products. We use the pretrained GloVe embedding [21] to initialize each token in the metadata into a 100-dimensional embedding. We extract important attributes from the metadata through a CNN model [11], which is widely used for different NLP tasks [8,24,30,31] such as text understanding, document classification, etc.

The CNN model consists of a convolution layer, a max-pooling layer, and a fully connected layer. In the convolution layer, each filter is applied to a window of words to generate the feature map. For example, we apply a filter $w \in \mathbb{R}^{hk}$ to a window of words $x_{i:i+h-1}$. Here k indicates the dimension of the word vector, and $x_{i:i+h-1}$ refers to the concatenation of h words from x_i to x_{i+h-1}. The context feature c_{ih} is generated as:

$$c_{ih} = \text{ReLU}(wx_{i:i+h-1} + b) \ , \tag{7}$$

where b is the bias item. A feature map of the text is then generated through $c_h = [c_{1h}, c_{2h}, ..., c_{nh}]$, where $c_{1h}, c_{2h}, ...c_{nh}$ refer to context features extracted from different sliding windows of the text, and c_h indicates the concatenation of these features. The feature map c_h is then fed into a max-pooling layer, and the maximum value is extracted as $c = max\{c_h\}$ as the important information extracted by a particular filter. A number of filters are used, and the extracted features are concatenated and fed into a fully connected layer to generate a vector $Prod_1$. $Prod_1$ is a representation of the important related product attributes in the metadata.

3.3.2 Unique Product Information

Although reviews for the same type of product may share the same important attributes, the degree of importance of these attributes may vary from product to product. Consider pet food for example. Some pet food may be of good quality and fair price, but the flavor may not appeal to a picky eater. Conversely, price may be the most salient feature for some brands.

In order to represent the unique characteristics of each product, the unique product identifier for each product is mapped into a continuous vector $Prod_2$. At the outset, $Prod_2$ is randomly initialized. During the training process, this vector is only updated when reviews specific to the product are used for training. Thus $Prod_2$ can be interpreted as a high level representation of product-specific information. The final product representation $Prod$ is generated by combining the two vectors: $Prod_1$ and $Prod_2$ as:

$$Prod = \tanh(W_1 Prod_1 + W_2 Prod_2 + b^p), \quad (8)$$

where W_1 and W_2 are weight matrices for $Prod_1$ and $Prod_2$ respectively, and b^p is the bias vector. We calculate the product attention weights based on the score function $f(.)$, and the input to the score function is the product representation $Prod$ and hidden state of a word h_{ij}^p:

$$f(Prod, h_{ij}^p) = (v_w^p)^T \tanh(W_{wh}^p h_{ij}^p + W_{wp}^p Prod + b_w^p), \quad (9)$$

where $(v_w^p)^T$ denotes the transpose of weight vector v_w^p, W_{wh}^p and W_{wp}^p are weight matrices, and b_w^p is the bias vector. Then we apply softmax function to get a normalized attention score α_{ij}^p. At the word level, the sentence representation is defined in Eq. 10, where α_{ij}^p indicates the product attention score of the word representation h_{ij}^p. The representation of a review can be obtained formally through Eq. 11, where β_j^p indicates the attention weight for hidden state of the jth sentence h_j^p.

$$s_j^p = \sum_{i=1}^{l} \alpha_{ij}^p h_{ij}^p, \quad (10)$$

$$r^p = \sum_{j=1}^{m} \beta_j^p h_j^p, \quad (11)$$

After applying the sentiment attention layer and the product attention layer separately, we obtain two different review representations r^s and r^p. These two representations are concatenated as the final representation of a review $r = [r^s, r^p]$. Then, we apply a fully connected layer on top of r, to classify the helpfulness of a review.

3.4 Loss Function

To minimize the difference between the predicted helpfulness value and the actual helpfulness label, we utilize cross entropy loss as the objective function. It is a commonly used loss function for binary classification, and is defined as:

$$Loss_{task} = -\sum_{i=1}^{N}(y_i log(p(y_i)) + (1 - y_i) log(1 - p(y_i))), \quad (12)$$

where y_i indicates the actual helpfulness label, $p(y_i)$ indicates the probability of helpfulness. N is the number of training observations. We present details on how these y_i are assigned in the following section.

4 Evaluation

In this section, we evaluate performance of our architecture in two application scenarios: cold start scenario and warm start scenario. Correspondingly, we split the data into training and test data differently for the two scenarios.

Data Sets. We evaluate our model on two publicly available data sets. One data set originates from Amazon reviews and was released by McAuley [5]. The other data set is from the Yelp Dataset Challenge 2018 [29]. We pre-process the data the same way as Fan et al. [4]: First, we join the product review with corresponding metadata information. Second, we filter out the reviews that have no votes. Last, we label reviews that receive more than 75% helpful votes out of total votes as helpful, and label the remaining reviews as unhelpful.

Table 1. Statistics of **Amazon** data set.

Category	# Products	# training data	# test data
Books	1,442,166	8,821,657	2,202,121
Clothing ...	477,958	1,478,488	372,662
Electronics	278,556	2,575,592	642,424
Grocery & ...	96,320	437,253	109,019
Health & ...	147,610	1,082,862	277,408
Home & Kitchen	220,687	1,480,520	360,402
Movies & TV	157,669	2,031,602	525,598
Pet Supplies	56,157	360,381	88,094
Tools & ...	134,446	637,594	162,161
Total	3,011,569	18,905,949	4,769,889

Table 2. Statistics of **Yelp** data set.

Category	# Products	# training data	# test data
Beauty & Spas	17,298	162,111	41,458
Health & Medical	15,458	102,592	25,687
Home Services	16,888	116,310	27,222
Restaurants	55,332	1,479,587	346,440
Shopping	29,489	220,431	55,743
Total	134,465	2,081,031	496,550

Evaluation Metric. In this study we use the Receiver Operating Characteristic Area Under the Curve (ROC AUC) statistic to evaluate the performance of our proposed model. This is a standard statistic used in the machine learning community to compare models. It is a robust statistic where imbalanced data sets are involved, and is a good metric for our problem where there are nearly four times as many helpful reviews as unhelpful reviews.

4.1 Application Scenario 1: Cold Start

In practice, a new product may have not yet received any helpful votes. Therefore assessment standards can't be captured from past voting information and can lead to the cold start problem. To evaluate model performance in this scenario, we randomly select 80% of the *products* and their corresponding reviews as the training data set. The remaining products and their reviews are employed as the test data set. Therefore, all of the reviews for a given product appear only in the training data set or test data set. Consequently, all products in this test data set face the cold start problem. The statistics of the two data sets are summarized in Tables 1 and 2. Even though the partitioning approach is the same as that reported by Fan et al. [4], a consequence of the random selection of products into test and training data sets is that the actual number of reviews differs from that of Fan et al. [4]. However, the difference is less than 1%, which is not statistically significant.

4.1.1 Model Comparison

We compare our proposed model with several baseline models. Below is a list of the hand-crafted features.

- Structural features (STR) as introduced by Xiong et al. [27] and Yang et al. [28], include the number of tokens, number of question sentences, the star rating. They are used to reveal a user's attitude towards a product.
- Emotional features (GALC) as introduced by Martin et al. [15] use the Geneva Affect Label Coder dictionary to define 36 emotion states.
- Lexical features (LEX) including unigrams and bigrams are commonly used and weighted by tf-idf to represent a text. They are usually used as baseline models [1,28].
- Semantic features (INQUIRER) employed by Yang et al. [28] leverage the existing linguistic dictionary INQUIRER to represent a review in semantic dimensions.

The baseline models that we use to compare our model are:

- Fusion (SVM) uses a Support Vector Machine to fuse features from the preceding feature list.
- Fusion (R.F.) uses a Random Forest to fuse features from the preceding feature list.
- Embedding-Gated CNN (EG-CNN) [1] introduces a word-level gating mechanism that weights word embeddings to represent the relative importance of each word.
- Multi-task Neural Learning (MTNL) [3] is based on a multi-task neural learning architecture with a secondary task that tries to predict the star ratings of reviews.
- Product-aware Review Helpfulness Net (PRH-Net) [4] is a neural network-based model that introduces target product information to enhance the representation of a review. Fan et al. evaluate this model on the two data sets we are using and claim that PRH-Net is the state of the art.

Table 3. Review helpfulness prediction of **Amazon** data set. The best performances are in bold.

Category (AC)	LEX	INQUIRER	FUSION (SVM)	FUSION (R.F.)	EG-CNN	MTNL	PRH-Net	HSAPA
AC1: Books	0.572	0.620	0.594	0.601	0.625	0.629	0.652	**0.712**
AC2: Clothing, Shoes & Jewelry	0.538	0.608	0.587	0.557	0.590	0.592	0.614	**0.679**
AC3: Electronics	0.555	0.627	0.584	0.588	0.615	0.618	0.644	**0.723**
AC4: Grocery & Gourmet Food	0.526	0.618	0.537	0.556	0.613	0.638	0.715	**0.718**
AC5: Health & Personal Care	0.533	0.617	0.599	0.565	0.617	0.624	0.672	**0.723**
AC6: Home & Kitchen	0.545	0.609	0.579	0.573	0.605	0.611	0.630	**0.697**
AC7: Movies & TV	0.562	0.637	0.605	0.617	0.648	0.652	0.675	**0.753**
AC8: Pet Supplies	0.542	0.603	0.548	0.558	0.580	0.619	0.679	**0.701**
AC9: Tools & Home Improv.	0.548	0.592	0.565	0.586	0.607	0.621	0.644	**0.699**
Average	0.547	0.615	0.578	0.578	0.611	0.623	0.658	**0.712**

Table 4. Review helpfulness prediction of **Yelp** data set. The best performances are in bold.

Category (YC)	LEX	INQUIRER	FUSION (SVM)	FUSION (R.F.)	EG-CNN	MTNL	PRH-Net	HSAPA
YC1: Beauty & Spas	0.500	0.570	0.521	0.541	0.571	0.581	0.642	**0.669**
YC2: Health & Medical	0.517	0.584	0.535	0.538	0.580	0.603	0.665	**0.683**
YC3: Home Services	0.528	0.627	0.584	0.588	0.563	0.618	0.732	**0.736**
YC4: Restaurants	0.516	0.582	0.569	0.554	0.581	0.605	0.658	**0.664**
YC5: Shopping	0.518	0.609	0.542	0.555	0.572	0.619	0.674	**0.695**
Average	0.516	0.584	0.541	0.544	0.573	0.601	0.674	**0.689**

4.1.2 Results and Findings

We use the same data sets as Fan et al. [4] in the cold start scenario. This allows us to directly compare the performance of our model with the results reported by Fan et al. [4]. We also randomly select 10% of the products and their corresponding reviews from the training set as a validation data set. We then performed a grid search of hyper-parameter space on the validation data set to determine the best choice of hyper-parameters. The models were then trained based on the entire training data set with these fixed hyper-parameters.

Tables 3 and 4 show the results on the Amazon data set and Yelp data set, respectively. In Table 3 we see that our model outperforms previous models on all categories of the Amazon data set. The average improvement in AUC is 5.4% over the next best model. We observe that the degree of improvement varies from category to category. In the categories AC3 (Electronics), our model achieves improvement of 7.9%. In contrast, for the category AC4 (Grocery & Gourmet Food), the improvement is only 0.3%. We note that the category AC4, has less data than most of other categories (Table 1). Only the category AC8 (Pet Supplies) contains fewer products and reviews. However, there are proportionally more reviews per product for the category AC8 than for the category AC4. We suspect that sentiment embedding and product embedding may not be learned well with such limited and divergent data. Therefore the improvement is not as high as that for the other categories. The results for the yelp data set are presented in Table 4. We find that our model also outperforms the previous models in all categories. The average improvement in AUC is 1.5% over the next best model. We note that the overall improvement is not as high as that demonstrated in the Amazon data set. This may be due to the relatively small number of products and reviews in the yelp data set. With the exception of the category YC4 (Restaurants), the other categories have fewer products and reviews than all of categories of Amazon data set.

The comparison results presented in Tables 3 and 4 show that our model outperforms the baseline models in the cold start scenario. In order to examine the significance of the improvement of our proposed model, we conducted a one-tailed t-test. As we are not certain as to whether the values reported in Fan et al. [4] refer to variance or standard deviation, we performed three tests: the one-sample t-test, and the two sample t-test where we evaluated both interpretations (variance and standard deviation) of the values reported in Fan et al. [4]. In all cases, the statistical results validate that our method is significantly better than the other baselines ($p < 0.001$).

In order to tease out the performance contribution of each of the components of our model, we evaluated different combinations of the components. The results are show in Table 5. Here HBiLSTM refers to the hierarchical bi-directional LSTM model without either of the attention layers. We use it as the baseline model for comparison. HSA refers to the combination of the HBiLSTM with the sentiment attention layer. HPA refers to the combination of the HBiLSTM with the product attention layer. Finally, HSAPA refers to the complete model which implements both attention layers. From Table 5, we see that adding a sentiment attention layer (HSA) to the base model (HBiLSTM) results in an average improvement in the AUC score of 2.0% and 2.6%, respectively on the Amazon and Yelp data sets. By adding a product attention layer (HPA) to the base model (HBiLSTM), the improvement is 0.7% and 1.3% on the Amazon and Yelp data sets respectively. Combining all three components results in an even larger increase in AUC score, 3.4% and 4.8%, respectively on the Amazon and Yelp data sets. We observe a synergistic effect resulting from the addition of the two attention layers. We also note that in both data sets, the improvement from the product attention layer is lower than that from the sentiment attention layer. This may be due to the fact that in the cold start scenario we have no information about the target product. Possibly the helpful attributes shared by related products may not be sufficiently accurate.

In order to verify that the gain in AUC is a consequence of the additional attention layers and not simply a result of adding more parameters, we conducted additional experiments. We adjusted the hyper-parameters of the HBiLSTM, HSA and HPA models to ensure they have approximately the same number of parameters as the complete model HSAPA. For example, for the category Grocery in the Amazon data set, the number of parameters of the complete model HSAPA is 30,194,490. We increased the number of hidden units in the

Table 5. Performance of each model in the Cold Start Scenario.

Data set	HBiLSTM	HSA	HPA	HSAPA
Amazon	0.678	**0.698**	**0.685**	**0.712**
Yelp	0.641	**0.667**	**0.654**	**0.689**

Table 6. Performance of each model on **Yelp** data set in the Warm Start Scenario. YC1-YC5 are described in Table 4.

Category	HBiLSTM	HSA	HPA	HSAPA
YC1	0.626	0.642	0.627	0.694
YC2	0.638	0.651	0.701	0.728
YC3	0.642	0.666	0.718	0.742
YC4	0.622	0.645	0.632	0.666
YC5	0.680	0.696	0.722	0.728
AVG.	0.642	**0.660**	**0.680**	**0.712**

Table 7. Performance of each model on **Amazon** data set in the Warm Start Scenario. AC1-AC10 are described in Table 3.

Category	HBiLSTM	HSA	HPA	HSAPA
AC1	0.682	0.712	0.704	0.775
AC2	0.657	0.667	0.717	0.723
AC3	0.693	0.708	0.705	0.725
AC4	0.682	0.699	0.758	0.779
AC5	0.689	0.718	0.767	0.782
AC6	0.665	0.696	0.703	0.744
AC7	0.739	0.764	0.818	0.826
AC8	0.666	0.709	0.727	0.746
AC9	0.664	0.702	0.727	0.745
AVG.	0.681	**0.708**	**0.736**	**0.760**

other three models to create new models with approximately the same number of parameters HBiLSTM: 30,420,604, HSA: 30,424,204, HPA: 30,412,858. Recall that the selection of hyper-parameters was determined by using a grid-search of the hyper-parameter space. Not surprisingly, the new models with more parameters do not demonstrate an improvement in performance in comparison to the models with hyper-parameters determined by grid-search. Our proposed model demonstrates improved performance, not simply because of greater modelling power due to more parameters, but because of the leveraging of sentiment and product related information by the sentiment and product attention layers.

4.2 Application Scenario 2: Warm Start

The warm start scenario is another commonly seen scenario in which some reviews for products have user votes, while other reviews haven't yet received user votes. For this scenario, we randomly select 80% of the reviews as the training data, and use the remaining reviews as the test data. The data statistics are essentially the same in scenario 2 as that in scenario 1 (Tables 1 and 2). As 80% of the reviews for products are in training data set, this partitioning produces a warm start scenario.

We also evaluated the contribution of each attention layer in the warm start scenario. Tables 6 and 7 show that, on average, the addition of the sentiment layer (HSA) to the base model increases the AUC by 1.8% and 2.7% on Yelp and Amazon data sets, respectively. We also find that the addition of the product attention layer (HPA) to the base model increases the AUC by 3.8% and 5.5% on Yelp and Amazon data sets, respectively. Comparing the results from warm start scenario to cold start Scenario, we make the following observations. First, the average performance of the base model (HBiLSTM) is very similar in both scenarios. Second, the AUC improvement from the product attention layer (HPA) is higher than that from the sentiment attention layer (HSA) on both data sets in the warm start scenario. This improvement is not seen in the case of the cold start scenario. In the cold start scenario the product embedding can only be learned from reviews of related products. In contrast, in the warm start scenario product information can be learned from both the target product and related products. This explains why the product attention layer can achieve better performance in warm start compared to cold start scenario. From the performance results described in the cold start scenario section (Tables 3 and 4), we see that HSAPA out-performs the other models. We also see that HSAPA has even better performance in the warm start scenario (Tables 6 and 7). In practice, one can expect a mix of cold and warm start scenarios where HSAPA can be expected to demonstrate superior performance than in cold start scenario.

Table 8. Highlighted words by sentiment and product attention scores in two review examples.

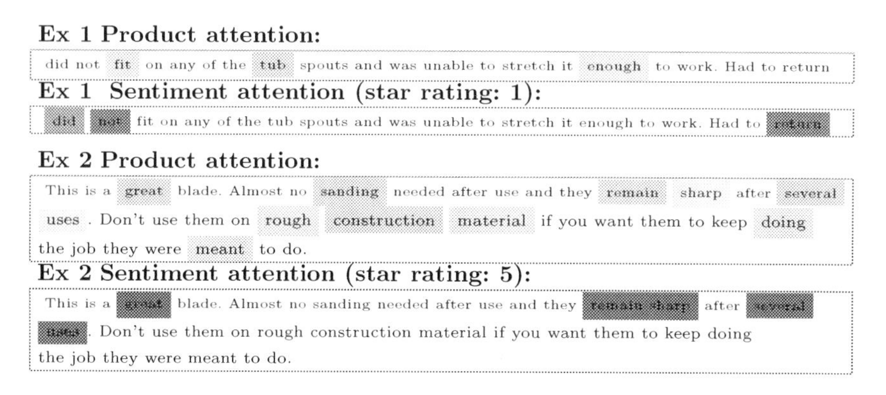

Ex 1 Product attention:

did not fit on any of the tub spouts and was unable to stretch it enough to work. Had to return

Ex 1 Sentiment attention (star rating: 1):

did not fit on any of the tub spouts and was unable to stretch it enough to work. Had to return

Ex 2 Product attention:

This is a great blade. Almost no sanding needed after use and they remain sharp after several uses . Don't use them on rough construction material if you want them to keep doing the job they were meant to do.

Ex 2 Sentiment attention (star rating: 5):

This is a great blade. Almost no sanding needed after use and they remain sharp after several uses. Don't use them on rough construction material if you want them to keep doing the job they were meant to do.

4.3 Visualization of Attention Layers

We demonstrate the visual examination of attention scores applied at the word level by randomly sampling two identical review examples (shown in Table 8). We use two colors: red and green to represent the sentiment attention scores and product attention scores respectively. The lightness/darkness of the color is proportional to the magnitude of the attention score. There are a few interesting patterns to note. First, for the sentiment attention layer, the words that are assigned large weights have sentiment that is close to the overall sentiment of the review. For instance, in the example 2 the overall sentiment of the review is positive (5 out of 5). Although there are several negative words such as "no" and "don't", the positive words/phrases like "great", "remain sharp" are still assigned higher attention weights. This observation is consistent with our previous hypothesis that the word importance in a review can be affected by review sentiment. Second, the attributes or descriptive words of an attribute of the product in a review text gain higher weights from the product attention layer. For instance, in the first example the descriptive words "fit", "enough" and the noun "tub" are assigned relatively high attention scores. Third, the combination of the important words captured by two attention layers can give us a brief and thorough summary of a review. It may also visually explain why the combination of these two can achieve a better result compared to a single attention layer.

5 Conclusion

In this paper, we describe our analysis of review helpfulness prediction and propose a novel neural network model with attention modules to incorporate sentiment and product information. We also describe the results of our experiments in two application scenarios: cold start and warm start. In the cold start scenario, our results show that the proposed model outperforms PRH-Net, the previous state of the art model. The increase in performance, measured by AUC, as compared with PRH-Net is 5.4% and 1.5% on Amazon and Yelp data sets, respectively. Furthermore, we evaluate the effect of each attention layer of proposed model in both scenarios. Both attention layers contribute to the improvement in performance. In the warm start scenario, the product attention layer is able to attain better performance than in cold scenario since it has access to reviews for targeted products. In this work, we evaluate review helpfulness from the perspective of review quality. For future work, we may rank the helpfulness of reviews by incorporating a user's own preferences [22] in order to make personalized recommendations.

Acknowledgments. This work was partially supported by 2018 IBM Faculty Award to the University of South Carolina.

References

1. Chen, C., et al.: Multi-domain gated CNN for review helpfulness prediction. In: Proceedings of the 2019 World Wide Web Conference (2019)
2. Chen, C., Yang, Y., Zhou, J., Li, X., Bao, F.S.: Cross-domain review helpfulness prediction based on convolutional neural networks with auxiliary domain discriminators. In: Proceedings of the 2018 Conference of the North American Chapter of the Association for Computational Linguistics: Human Language Technologies, (Short Papers), vol. 2 (2018)
3. Fan, M., Feng, Y., Sun, M., Li, P., Wang, H., Wang, J.: Multi-task neural learning architecture for end-to-end identification of helpful reviews. In: 2018 IEEE/ACM International Conference on Advances in Social Networks Analysis and Mining (ASONAM) (2018)
4. Fan, M., Feng, C., Guo, L., Sun, M., Li, P.: Product-aware helpfulness prediction of online reviews. In: Proceedings of the 2019 World Wide Web Conference (2019)
5. He, R., McAuley, J.: Ups and downs: modeling the visual evolution of fashion trends with one-class collaborative filtering. In: Proceedings of the 25th International Conference on World Wide Web (2016)
6. Hong, Y., Lu, J., Yao, J., Zhu, Q., Zhou, G.: What reviews are satisfactory: novel features for automatic helpfulness voting. In: Proceedings of the 35th International ACM SIGIR Conference on Research and Development in Information Retrieval (2012)

7. Huang, A.H., Chen, K., Yen, D.C., Tran, T.P.: A study of factors that contribute to online review helpfulness. Comput. Hum. Behav. **48**(C), 17–27 (2015)
8. Johnson, R., Zhang, T.: Effective use of word order for text categorization with convolutional neural networks. In: NAACL HLT 2015, the 2015 Conference of the North American Chapter of the Association for Computational Linguistics: Human Language Technologies (2015)
9. Kats, R.: Surprise! most consumers look at reviews before a purchase (2018). https://www.emarketer.com/content/surprise-most-consumers-look-at-reviews-before-a-purchase. Accessed 20 Aug 2019
10. Kim, S.M., Pantel, P., Chklovski, T., Pennacchiotti, M.: Automatically assessing review helpfulness. In: Proceedings of the 2006 Conference on Empirical Methods in Natural Language Processing (2006)
11. Kim, Y.: Convolutional neural networks for sentence classification. In: Proceedings of the 2014 Conference on Empirical Methods in Natural Language Processing (EMNLP) (2014)
12. Liu, H., et al.: Using argument-based features to predict and analyse review helpfulness. In: Proceedings of the 2017 Conference on Empirical Methods in Natural Language Processing (2017)
13. Liu, J., Cao, Y., Lin, C.Y., Huang, Y., Zhou, M.: Low-quality product review detection in opinion summarization. In: Proceedings of the 2007 Joint Conference on Empirical Methods in Natural Language Processing and Computational Natural Language Learning (EMNLP-CoNLL) (2007)
14. Lu, Y., Tsaparas, P., Ntoulas, A., Polanyi, L.: Exploiting social context for review quality prediction. In: Proceedings of the 19th International Conference on World Wide Web (2010)
15. Martin, L., Pu, P.: Prediction of helpful reviews using emotions extraction. In: Proceedings of the Twenty-Eighth AAAI Conference on Artificial Intelligence (2014)
16. Melamud, O., Goldberger, J., Dagan, I.: context2vec: learning generic context embedding with bidirectional LSTM. In: CoNLL (2016)
17. Mudambi, S.M., Schuff, D.: What makes a helpful online review? A study of customer reviews on amazon.com. MIS Q. **34**(1), 185–200 (2010)
18. Mukherjee, S., Popat, K., Weikum, G.: Exploring latent semantic factors to find useful product reviews. In: Proceedings of the 2017 SIAM International Conference on Data Mining (2017)
19. Ocampo Diaz, G., Ng, V.: Modeling and prediction of online product review helpfulness: a survey. In: Proceedings of the 56th Annual Meeting of the Association for Computational Linguistics (Volume 1: Long Papers) (2018)
20. O'Mahony, M.P., Smyth, B.: Learning to recommend helpful hotel reviews. In: Proceedings of the Third ACM Conference on Recommender Systems (2009)
21. Pennington, J., Socher, R., Manning, C.D.: Glove: global vectors for word representation. In: Empirical Methods in Natural Language Processing (EMNLP) (2014)
22. Qu, X., Li, L., Liu, X., Chen, R., Ge, Y., Choi, S.H.: A dynamic neural network model for CTR prediction in real-time bidding. In: 2019 IEEE International Conference on Big Data (Big Data) (2019)
23. Qu, X., Li, X., Rose, J.R.: Review helpfulness assessment based on convolutional neural network. arXiv abs/1808.09016 (2018). http://arxiv.org/abs/1808.09016
24. Severyn, A., Moschitti, A.: Twitter sentiment analysis with deep convolutional neural networks. In: Proceedings of the 38th International ACM SIGIR Conference on Research and Development in Information Retrieval (2015)

25. Tang, J., Gao, H., Hu, X., Liu, H.: Context-aware review helpfulness rating prediction. In: Proceedings of the 7th ACM Conference on Recommender Systems (2013)
26. Wu, Z., Dai, X., Yin, C., Huang, S., Chen, J.: Improving review representations with user attention and product attention for sentiment classification. In: Proceedings of the Thirty-Second AAAI Conference on Artificial Intelligence (2018)
27. Xiong, W., Litman, D.: Automatically predicting peer-review helpfulness. In: Proceedings of the 49th Annual Meeting of the Association for Computational Linguistics: Human Language Technologies: Short Papers, vol. 2 (2011)
28. Yang, Y., Yan, Y., Qiu, M., Bao, F.: Semantic analysis and helpfulness prediction of text for online product reviews. In: Proceedings of the 53rd Annual Meeting of the Association for Computational Linguistics and the 7th International Joint Conference on Natural Language Processing (Volume 2: Short Papers) (2015)
29. Yelp: Yelp dataset challenge (2018). https://www.yelp.com/dataset/challenge. Accessed 14 Sept 2018
30. Zhang, X., LeCun, Y.: Text understanding from scratch. arXiv abs/1502.01710 (2015)
31. Zhang, X., Zhao, J., LeCun, Y.: Character-level convolutional networks for text classification. In: Advances in Neural Information Processing Systems 28 (2015)

Abstracts of the IR Journal Papers

A Comparison of Filtering Evaluation Metrics Based on Formal Constraints

Enrique Amigó[1], Julio Gonzalo[1](✉), Felisa Verdejo[1], and Damiano Spina[2]

[1] UNED, Madrid, Spain
{enrique,julio,felisa}@lsi.uned.es
[2] RMIT University, Melbourne, Australia
damiano.spina@rmit.edu.au

Abstract. Although document filtering is simple to define, there is a wide range of different evaluation measures that have been proposed in the literature, all of which have been subject to criticism. Our goal is to compare metrics from a formal point of view, in order to understand whether each metric is appropriate, why and when, in order to achieve a better understanding of the similarities and differences between metrics. Our formal study leads to a typology of measures for document filtering which is based on (1) a formal constraint that must be satisfied by any suitable evaluation measure, and (2) a set of three (mutually exclusive) formal properties which help to understand the fundamental differences between measures and determining which ones are more appropriate depending on the application scenario. As far as we know, this is the first in-depth study on how filtering metrics can be categorized according to their appropriateness for different scenarios. Two main findings derive from our study. First, not every measure satisfies the basic constraint; but problematic measures can be adapted using smoothing techniques that and makes them compliant with the basic constraint while preserving their original properties. Our second finding is that all metrics (except one) can be grouped in three families, each satisfying one out of three formal properties which are mutually exclusive. In cases where the application scenario is clearly defined, this classification of metrics should help choosing an adequate evaluation measure. The exception is the Reliability/Sensitivity metric pair, which does not fit into any of the three families, but has two valuable empirical properties: it is strict (i.e. a good result according to reliability/sensitivity ensures a good result according to all other metrics) and has more robustness that all other measures considered in our study.

Keywords: Filtering · Evaluation metrics

Information Retrieval Journal (2019) 22:581–619
https://doi.org/10.1007/s10791-019-09355-y

How Do Interval Scales Help Us with Better Understanding IR Evaluation Measures?

Marco Ferrante[1], Nicola Ferro[2](✉), and Eleonora Losiouk[1]

[1] Department of Mathematics, University of Padua, Padua, Italy
{ferrante,elosiouk}@math.unipd.it
[2] Department of Information Engineering, University of Padua, Padua, Italy
ferro@dei.unipd.it

Abstract. Evaluation measures are the basis for quantifying the performance of IR systems and the way in which their values can be processed to perform statistical analyses depends on the scales on which these measures are defined. For example, mean and variance should be computed only when relying on interval scales.

In our previous work we defined a theory of IR evaluation measures, based on the representational theory of measurement, which allowed us to determine whether and when IR measures are interval scales. We found that common set-based retrieval measures – namely Precision, Recall, and F-measure – always are interval scales in the case of binary relevance while this does not happen in the multi-graded relevance case. In the case of rank-based retrieval measures – namely AP, gRBP, DCG, and ERR – only gRBP is an interval scale when we choose a specific value of the parameter p and define a specific total order among systems while all the other IR measures are not interval scales.

In this work, we build on our previous findings and we carry out an extensive evaluation, based on standard TREC collections, to study how our theoretical findings impact on the experimental ones. In particular, we conduct a correlation analysis to study the relationship among the above-mentioned state-of-the-art evaluation measures and their scales. We study how the scales of evaluation measures impact on non parametric and parametric statistical tests for multiple comparisons of IR system performance. Finally, we analyse how incomplete information and pool downsampling affect different scales and evaluation measures.

Keywords: Experimentation · Representational theory of measurement · Interval scale · IR evaluation measure · Formal framework

Reference

1. Ferrante, M., Ferro, N., Losiouk, E.: How do interval scales help us with better understanding IR evaluation measures? Inf. Retr. J. (2019). https://doi.org/10.1007/s10791-019-09362-z

Abstract of the paper originally published in [1].

Information Retrieval Journal (2020) 23:289–317
https://doi.org/10.1007/s10791-019-09362-z

The Impact of Result Diversification on Search Behaviour and Performance

David Maxwell[1](✉), Leif Azzopardi[2](✉), and Yashar Moshfeghi[2]

[1] TU Delft, Delft, The Netherlands
maxwelld90@acm.org
[2] University of Strathclyde, Glasgow, Scotland
{leif.azzopardi,yashar.moshfeghi}@strath.ac.uk

Abstract. *Result diversification* aims to provide searchers with a broader view of a given topic while attempting to maximise the chances of retrieving relevant material. Diversifying results also aims to reduce search bias by increasing the coverage over different aspects of the topic. As such, searchers should learn more about the given topic in general. Despite diversification algorithms being introduced over two decades ago, little research has explicitly examined their impact on search behaviour and performance in the context of *Interactive Information Retrieval (IIR)*. In this paper, we explore the impact of diversification when searchers undertake complex search tasks that require learning about different aspects of a topic *(aspectual retrieval)*. We hypothesise that by diversifying search results, searchers will be exposed to a greater number of aspects. In turn, this will maximise their coverage of the topic (and thus reduce possible search bias). As a consequence, diversification *should* lead to performance benefits, regardless of the task, but how does diversification affect search behaviours and search satisfaction?

Based on *Information Foraging Theory (IFT)*, we infer two hypotheses regarding search behaviours due to diversification, namely that *(i)* it will lead to searchers examining fewer documents per query, and *(ii)* it will also mean searchers will issue more queries overall. To this end, we performed a within-subjects user study using the *TREC AQUAINT* collection with 51 participants, examining the differences in search performance and behaviour when using *(i)* a non-diversified system (*BM25*) versus *(ii)* a diversified system (BM25+*xQuAD*) when the search task is either *(a)* ad-hoc or *(b)* aspectual. Our results show a number of notable findings in terms of search behaviour: participants on the diversified system issued more queries and examined fewer documents per query when performing the aspectual search task. Furthermore, we showed that when using the diversified system, participants were: more successful in marking relevant documents, and obtained a greater awareness of the topics (i.e. identified relevant documents containing more novel aspects). These findings show that search behaviour is influenced by diversification and task complexity. They also motivate further research into complex search

Work undertaken when a PhD student at the University of Glasgow, Scotland.

Information Retrieval Journal (2019) 22:422–446
https://doi.org/10.1007/s10791-019-09353-0

tasks such as aspectual retrieval – and how diversity can play an important role in improving the search experience, by providing greater coverage of a topic and mitigating potential bias in search results.

On the Impact of Group Size on Collaborative Search Effectiveness

Felipe Moraes, Kilian Grashoff, and Claudia Hauff(✉)

Delft University of Technology, Delft, The Netherlands
{f.moraes,c.hauff}@tudelft.nl

Abstract. While today's web search engines are designed for single-user search, over the years research efforts have shown that complex information needs—which are explorative, open-ended and multi-faceted—can be answered more efficiently and effectively when searching *in collaboration*. Collaborative search (and sensemaking) research has investigated techniques, algorithms and interface affordances to gain insights and improve the collaborative search process. It is not hard to imagine that the *size* of the group collaborating on a search task significantly influences the group's behaviour and search effectiveness. However, a common denominator across almost all existing studies is a fixed group size—usually either pairs, triads or in a few cases four users collaborating. Investigations into larger group sizes and the impact of group size dynamics on users' behaviour and search metrics have so far rarely been considered—and when, then only in a simulation setup. In this work, we investigate in a large-scale user experiment to what extent those simulation results carry over to the real world. To this end, we extended our collaborative search framework `SearchX` with algorithmic mediation features and ran a large-scale experiment with more than 300 crowd-workers. We consider the collaboration group size as a dependent variable, and investigate collaborations between groups of up to six people. We find that most prior simulation-based results on the impact of collaboration group size on behaviour and search effectiveness *cannot* be reproduced in our user experiment.

Information Retrieval Journal (2019) 22:476–498
https://doi.org/10.1007/s10791-018-09350-9

Fewer Topics? A Million Topics? Both?! On Topics Subsets in Test Collections

Kevin Roitero[1], J. Shane Culpepper[2], Mark Sanderson[2], Falk Scholer[2], and Stefano Mizzaro[1](✉)

[1] University of Udine, Udine, Italy
roitero.kevin@spes.uniud.it,mizzaro@uniud.it
[2] RMIT University, Melbourne, Australia
{shane.culpepper,mark.sanderson,falk.scholer}@rmit.edu.au

Abstract. When evaluating IR run effectiveness using a test collection, a key question is: What search topics should be used? We explore what happens to measurement accuracy when the number of topics in a test collection is reduced, using the Million Query 2007, TeraByte 2006, and Robust 2004 TREC collections, which all feature more than 50 topics, something that has not been examined in past work. Our analysis finds that a subset of topics can be found that is as accurate as the full topic set at ranking runs. Further, we show that the size of the subset, relative to the full topic set, can be substantially smaller than was shown in past work. We also study the topic subsets in the context of the power of statistical significance tests. We find that there is a trade off with using such sets in that significant results may be missed, but the loss of statistical significance is much smaller than when selecting random subsets. We also find topic subsets that can result in a low accuracy test collection, even when the number of queries in the subset is quite large. These negatively correlated subsets suggest we still lack good methodologies which provide stability guarantees on topic selection in new collections. Finally, we examine whether clustering of topics is an appropriate strategy to find and characterize good topic subsets. Our results contribute to the understanding of information retrieval effectiveness evaluation, and offer insights for the construction of test collections.

Information Retrieval Journal (2020) 23:49–85
https://doi.org/10.1007/s10791-019-09357-w

Neural Architecture for Question Answering Using a Knowledge Graph and Web Corpus

Uma Sawant[1], Saurabh Garg[2], Soumen Chakrabarti[3(✉)], and Ganesh Ramakrishnan[3]

[1] LinkedIn, Bangalore, India
[2] Carnegie-Mellon University, Pittsburgh, USA
[3] IIT Bombay, Mumbai, India
soumen@cse.iitb.ac.in

Abstract. In Web search, entity-seeking queries often trigger a special Question Answering (QA) system. It may use a parser to interpret the question to a structured query, execute that on a knowledge graph (KG), and return direct entity responses. QA systems based on precise parsing tend to be brittle: minor syntax variations may dramatically change the response. Moreover, KG coverage is patchy. At the other extreme, a large corpus may provide broader coverage, but in an unstructured, unreliable form. We present AQQUCN, a QA system that gracefully combines KG and corpus evidence. AQQUCN accepts a broad spectrum of query syntax, between well-formed questions to short "telegraphic" keyword sequences. In the face of inherent query ambiguities, AQQUCN aggregates signals from KGs and large corpora to directly rank KG entities, rather than commit to one semantic interpretation of the query. AQQUCN models the ideal interpretation as an unobservable or latent variable. Interpretations and candidate entity responses are scored as pairs, by combining signals from multiple convolutional networks that operate collectively on the query, KG and corpus. On four public query workloads, amounting to over 8,000 queries with diverse query syntax, we see 5–16% *absolute* improvement in mean average precision (MAP), compared to the entity ranking performance of recent systems. Our system is also competitive at entity set retrieval, almost doubling F1 scores for challenging short queries.

U. Sawant and S. Garg—Work done at IIT Bombay.

Information Retrieval Journal (2019) 22:324–349
https://doi.org/10.1007/s10791-018-9348-8

Evaluation Measures for Quantification: An Axiomatic Approach

Fabrizio Sebastiani[(✉)]

Istituto di Scienza e Tecnologie dell'Informazione,
Consiglio Nazionale delle Ricerche, 56124 Pisa, Italy
fabrizio.sebastiani@isti.cnr.it

Abstract. Quantification is the task of estimating, given a set σ of unlabelled items and a set of classes $\mathcal{C} = \{c_1, \ldots, c_{|\mathcal{C}|}\}$, the prevalence (or "relative frequency") in σ of each class $c_i \in \mathcal{C}$. While quantification may in principle be solved by classifying each item in σ and counting how many such items have been labelled with c_i, it has long been shown that this "classify and count" (CC) method yields suboptimal quantification accuracy. As a result, quantification is no longer considered a mere byproduct of classification, and has evolved as a task of its own. While the scientific community has devoted a lot of attention to devising more accurate quantification methods, it has not devoted much to discussing what properties an *evaluation measure for quantification* (EMQ) should enjoy, and which EMQs should be adopted as a result. This paper lays down a number of interesting properties that an EMQ may or may not enjoy, discusses if (and when) each of these properties is desirable, surveys the EMQs that have been used so far, and discusses whether they enjoy or not the above properties. As a result of this investigation, some of the EMQs that have been used in the literature turn out to be severely unfit, while others emerge as closer to what the quantification community actually needs. However, a significant result is that no existing EMQ satisfies all the properties identified as desirable, thus indicating that more research is needed in order to identify (or synthesize) a truly adequate EMQ.

Keywords: Quantification · Supervised prevelance estimation · Divergences · Evaluation measures

This is the abstract of a paper with the same title which is going to appear in the Special Issue on "Axiomatic Thinking for IR" of the *Information Retrieval Journal*, 2020.

Information Retrieval Journal (2020) 23:255–288
https://doi.org/10.1007/s10791-019-09363-y

Overcoming Low-Utility Facets for Complex Answer Retrieval

Sean MacAvaney[1(✉)], Andrew Yates[2], Arman Cohan[1,3], Luca Soldaini[1], Kai Hui[4], Nazli Goharian[1], and Ophir Frieder[1]

[1] Information Retrieval Laboratory, Computer Science Department, Georgetown University, 3700 O St NW, Washington, D.C. 20057, USA
sean@ir.cs.georgetown.edu

[2] Max Planck Institute for Informatics, Saarland Informatics Campus, Saarbrücken, Germany

[3] Allen Institute for Artificial Intelligence, Seattle, WA, USA

[4] SAP SE Machine Learning R&D, Berlin, Germany

Abstract. Many questions cannot be answered simply; their answers must include numerous nuanced details and context. Complex Answer Retrieval (CAR) is the retrieval of answers to such questions. These questions can be constructed from a topic entity (e.g., 'cheese') and a facet (e.g., 'health effects'). While topic matching has been thoroughly explored, we observe that some facets use general language that is unlikely to appear verbatim in answers, exhibiting low utility. In this work, we present an approach to CAR that identifies and addresses low-utility facets. First, we propose two estimators of facet utility: the hierarchical structure of CAR queries, and facet frequency information from training data. Then, to improve the retrieval performance on low-utility headings, we include entity similarity scores using embeddings trained from a CAR knowledge graph, which captures the context of facets. We show that our methods are effective by applying them to two leading neural ranking techniques, and evaluating them on the TREC CAR dataset. We find that our approach perform significantly better than the unmodified neural ranker and other leading CAR techniques, yielding state-of-the-art results. We also provide a detailed analysis of our results, verify that low-utility facets are indeed difficult to match, and that our approach improves the performance for these difficult queries.

Keywords: Complex answer retrieval · Knowledge graphs · Neural information retrieval · Reranking

Information Retrieval Journal (2019) 22:395–418
https://doi.org/10.1007/s10791-018-9343-0

Payoffs and Pitfalls in Using Knowledge-Bases for Consumer Health Search

Jimmy[1,2], Guido Zuccon[1(✉)], and Bevan Koopman[3]

[1] Queensland University of Technology (QUT), Brisbane, QLD, Australia
jimmy@hdr.qut.edu.au, g.zuccon@uq.edu.au
[2] University of Surabaya (UBAYA), Surabaya, Indonesia
[3] Australian e-Health Research Centre, CSIRO, Canberra, Australia
bevan.koopman@csiro.au

Abstract. Consumer health search (CHS) is a challenging domain with vocabulary mismatch and considerable domain expertise hampering people's ability to formulate effective queries. We posit that using knowledge bases for query reformulation may help alleviate this problem. How to exploit knowledge bases for effective CHS is nontrivial, involving a swathe of key choices and design decisions (many of which are not explored in the literature). Here we rigorously empirically evaluate the impact these different choices have on retrieval effectiveness. A state-of-the-art knowledge-base retrieval model–the Entity Query Feature Expansion model–was used to evaluate these choices, which include: which knowledge base to use (specialised vs. general purpose), how to construct the knowledge base, how to extract entities from queries and map them to entities in the knowledge base, what part of the knowledge base to use for query expansion, and if to augment the knowledge base search process with relevance feedback. While knowledge base retrieval has been proposed as a solution for CHS, this paper delves into the finer details of doing this effectively, highlighting both payoffs and pitfalls. It aims to provide some lessons to others in advancing the state-of-the-art in CHS.

Keywords: Knowledge base · Knowledge graph · Query expansion · Consumer health search

Information Retrieval Journal (2019) 22:350–394
https://doi.org/10.1007/s10791-018-9344-z

Author Index

Printed in the United States
By Bookmasters